PHYSICAL AND BIOLOGICAL HAZARDS OF THE WORKPLACE

PHYSICAL AND BIOLOGICAL HAZARDS OF THE WORKPLACE

Third Edition

Edited by

GREGG M. STAVE, MD, JD, MPH
Consultant, Occupational Medicine and Corporate Health, Chapel Hill, North Carolina and Assistant Consulting Professor, Division of Occupational and Environmental Medicine, Department of Community and Family Medicine, Duke University Medical Center, Durham, North Carolina

PETER H. WALD, MD, MPH
Enterprise Medical Director, USAA, San Antonio, Texas and Adjunct Professor of Public Health, San Antonio Regional Campus, University of Texas School of Public Health, Houston, Texas

WILEY

The authors and the publishers have exerted every effort to ensure that treatment recommendations including drug selection and dosage set forth in this text are in accord with current recommendations and practice at the time of publication. However, in view of ongoing research, changes in government regulations, and the constant flow of information relating to treatment, drug therapy, and drug reactions, the reader is urged to check the package insert for each drug for any change in indications and dosage and for added warnings and precautions. This is particularly important when the recommended agent is a new or infrequently employed drug.

Published by John Wiley & Sons, Inc., Hoboken, New Jersey
Published simultaneously in Canada

Library of Congress Cataloging-in-Publication Data:

Names: Stave, Gregg M., editor. | Wald, Peter H., 1955– editor.
Title: Physical and biological hazards of the workplace / edited by Gregg M. Stave, Peter H. Wald.
Description: Third edition. | Hoboken, NJ : John Wiley & Sons Inc., [2017] | Includes bibliographical references and index.
Identifiers: LCCN 2016019942 | ISBN 9781118928608 (hardback) | ISBN 9781119276524 (Adobe PDF) | ISBN 9781119276517 (epub)
Subjects: LCSH: Medicine, Industrial. | Industrial hygiene. | Occupational diseases. | BISAC: TECHNOLOGY & ENGINEERING / Industrial Health & Safety.
Classification: LCC RC963 .P48 2017 | DDC 616.9/803–dc23
LC record available at https://lccn.loc.gov/2016019942

Printed in the United States of America

Set in 10/12pt Times by SPi Global, Pondicherry, India

10 9 8 7 6 5 4 3 2 1

To our wives—Chris and Isabel

And to our children—Elise, Sam, and Ben

CONTENTS

ABOUT THE EDITORS

Gregg M. Stave, MD, JD, MPH, FACP, FACOEM, FACPM, is a consultant in occupational medicine and corporate health and Assistant Consulting Professor in the Division of Occupational and Environmental Medicine at Duke University Medical Center. Previously, he worked as a corporate medical director for Glaxo.

Dr. Stave received his undergraduate education at the Massachusetts Institute of Technology before attending the dual-degree MD/JD program at Duke University. He received his MPH degree in epidemiology from the University of North Carolina, Chapel Hill. Dr. Stave is board certified in internal medicine and preventive medicine (occupational medicine).

A Fellow of the American College of Physicians, American College of Occupational and Environmental Medicine, and the American College of Preventive Medicine, Dr. Stave is also a member of the Bar in North Carolina and the District of Columbia. He lives in Chapel Hill, North Carolina, with his wife and daughter.

Peter H. Wald, MD, MPH, FACP, FACOEM, FACMT, is the Enterprise Medical Director at USAA in San Antonio, Texas, and Adjunct Professor of Public Health, San Antonio Regional Campus, University of Texas School of Public Health, Houston, Texas.

Dr. Wald received his undergraduate education at Harvard before attending Tufts Medical School. He also holds an M.P.H. degree from the University of California, Berkeley. Dr. Wald is board certified in internal medicine, occupational medicine, and medical toxicology. Before joining USAA, he was a principal at WorkCare and the Corporate Medical Director for ARCO. Prior to that, he worked in the medical departments at Mobil Oil and the Lawrence Livermore National Laboratory.

Dr. Wald is a Fellow of the American College of Physicians, the American College of Occupational and Environmental Medicine, and the American College of Medical Toxicology. He lives in San Antonio with his wife.

LIST OF CONTRIBUTORS

Tony L. Alleman, MD, MPH, Medical Director, Occupational Medicine Clinics of South Louisiana, Lafayette, Louisiana

Manijeh Berenji, MD, MPH, Assistant Clinical Professor, Division of Occupational and Environmental Medicine, Department of Community and Family Medicine, Duke University Medical Center, Durham, North Carolina

David C. Caretto MD, MPH, Occupational and Environmental Medicine Physician, Mercy Medical Group, Sacramento, California

Richard Cohen, MD, MPH, Clinical Professor, Division of Occupational and Environmental Medicine, University of California School of Medicine, San Francisco, California

Dennis J. Darcey, MD, MPH, MSPH, Division Chief and Assistant Clinical Professor, Division of Occupational and Environmental Medicine, Department of Community and Family Medicine, Duke University Medical Center, Durham, North Carolina

David W. DeGroot, PhD, Research Physiologist, Epidemiology and Disease Surveillance Portfolio, Army Public Health Center (Provisional), Aberdeen Proving Ground, Maryland

Robert B. Dick, PhD, Captain, U.S. Public Health Service (retired), Visiting Scientist, National Institute for Occupational Safety and Health, Cincinnati, Ohio

Robert A. Dobie, MD, Clinical Professor, Department of Otolaryngology – Head and Neck Surgery, University of Texas Health Science Center at San Antonio, San Antonio, Texas

Aletheia S. Donahue, MD, Resident, Selikoff Centers for Occupational Medicine Icahn-Mount Sinai School of Medicine, New York, New York

Carol A. Epling, MD, MSPH, Assistant Professor, Division of Occupational and Environmental Medicine; and Director, Duke University Employee Occupational Health and Wellness, Department of Community and Family Medicine, Duke University Medical Center, Durham, North Carolina

Craig S. Glazer, MD, MSPH, Associate Professor of Medicine, Division of Pulmonary and Critical Care Medicine University of Texas Southwestern, Dallas, Texas

Gary N. Greenberg, MD, MPH, Medical Director, Urban Ministries of Wake County, Raleigh, North Carolina; Adjunct Associate Professor, Department of Public Health Leadership, University of North Carolina School of Public Health, Chapel Hill, North Carolina

Thomas R. Hales, MD, MPH, National Institute for Occupational Safety and Health (retired), Cincinnati, Ohio

James A. Hathaway, MD, MPH, Occupational Medical Consultant, Solvay USA Inc., Princeton, New Jersey

Jessica Herzstein, MD, MPH, Former Global Medical Director, Air Products, Adjunct Faculty, Milken Institute School of Public Health, George Washington University

Worthe S. Holt Jr., MD, MMM, Vice President, Office of the Chief Medical Officer, Humana, Inc., Louisville, Kentucky

Stephen D. Hudock, PhD, CSP, Manager, Musculoskeletal Health Cross-Sector and Lead Research Safety Engineer, Human Factors and Ergonomics Research Team, National Institute for Occupational Safety and Health, Cincinnati, Ohio

Harold R. Imbus, MD, MScD., Health and Hygiene, Inc. (retired), Greensboro, North Carolina

Robert R. Jacobs, PhD, Professor, Environmental and Occupational Health Sciences, Director Masters in Public Health Program, School of Public Health and Information Sciences, University of Louisville, Louisville, Kentucky

Jeffrey R. Jones, MPH, MS, CIH, Supervisor of Environmental Compliance and Safety, Environmental Programs and Planning, Port of Oakland (retired), Oakland, California

James Kubalik, MS, CSP, Manager Occupational Health, Safety, and Risk Management, B. Braun, Inc., Irvine, California

Ming-Lun Lu, PhD, Research Ergonomist, National Institute for Occupational Safety and Health, Cincinnati, Ohio

Christopher J. Martin, MD, MSc, Professor of Medicine and Emergency Medicine and Director, Institute for Occupational and Environmental, Health Robert C. Byrd Health Sciences Center, West Virginia University, Morgantown, West Virginia

John D. Meyer, MD, MPH, Assistant Professor, Selikoff Centers for Occupational Medicine, Icahn-Mount Sinai School of Medicine, New York, NY

Aubrey K. Miller, MD, MPH, Senior Medical Advisor, National Institute of Environmental Health Sciences, Bethesda, Maryland

Laura A. Pacha, MD, MPH, Director, Epidemiology and Disease Surveillance Portfolio, Army Public Health Center (Provisional), Aberdeen Proving Ground, Maryland

James A. Palmier, MD, MPH, MBA, Vice President and Medical Director, ExamOne/Quest Diagnostics, Lenexa, Kansas

Vern Putz-Anderson, PhD, Research Psychologist, National Institute for Occupational Safety and Health, Cincinnati, Ohio

Cecile S. Rose, MD, MPH, Professor of Medicine, Division of Environmental and Occupational Health Sciences, National Jewish Health Division of Pulmonary Medicine and Critical Care Sciences, University of Colorado, Denver, Colorado

Allene J. Scott, MD, MPH, MT (ASCP), Medical Consultant, UNUM Life Insurance Company of America, Portland, Maine.

Joseph R. Serio, MD, Staff Physician, Occupational Medicine Clinics of South Louisiana, Lafayette, Louisiana

James P. Seward, MD, MPP, MMM, Clinical Professor of Medicine, University of California, San Francisco, California; and Clinical Professor of Public Health, University of California, Berkeley, California

David H. Sliney, PhD, Associate, Department of Environmental Health Sciences, Bloomberg School of Public Health, The Johns Hopkins University, Baltimore, Maryland; Program Manager (retired), Laser/Optical Radiation Program, US Army Center for Health Promotion and Preventive Medicine, Aberdeen Proving Grounds, Maryland

Gregg M. Stave, MD, JD, MPH, Consultant, Occupational Medicine and Corporate Health, Chapel Hill, North Carolina; and Assistant Consulting Professor, Division of Occupational and Environmental Medicine, Department of Community and Family Medicine, Duke University Medical Center, Durham, North Carolina

Woodhall Stopford, MD, MSPH, Assistant Clinical Professor, Division of Occupational and Environmental Medicine, Department of Community and Family Medicine, Duke University Medical Center, Durham, North Carolina

Peter H. Wald, MD, MPH, Enterprise Medical Director, USAA, San Antonio, Texas; and Adjunct Professor of Public Health, San Antonio Regional Campus, University of Texas School of Public Health, Houston, Texas

Donald E. Wasserman, MSEE, MBA, Human Vibration Consultant, Frederick, Maryland

Thomas R. Waters[†]**, PhD,** Chief, Human Factors and Ergonomics Research Section, National Institute for Occupational Safety and Health, Cincinnati, Ohio

David G. Wilder, PhD, PE, CPE, Jolt/Vibration/Seating Lab, Director, Iowa Spine Research Center; Researcher, Center for Computer Aided Design; Professor, Biomedical Engineering Department, College of Engineering; Professor, Occupational and Environmental Health Department, College of Public Health; Orthopedic Surgery Department, College of Medicine, The University of Iowa, Iowa City, Iowa

William N. Yang, MD, MPH, Occupational Medicine Physician (retired), The Emory Clinic, Atlanta, Georgia

[†]Deceased October 29, 2014.

FOREWORD TO THE FIRST EDITION

It has been 15 years since the original publication of *Chemical Hazards of the Workplace*. That work was conceived as a handbook—which would serve as an authoritative guide to current concepts and practices aimed at protecting workers from chemical hazards. Over the intervening years, *Chemical Hazards of the Workplace* has been updated twice, but there has been a need for a similar work on the prevention and management of hazardous exposures from physical and biological agents.

I am happy to report that this void has now been filled expertly in the present volume prepared by Drs. Peter Wald and Gregg Stave. This exciting companion piece to *Chemical Hazards of the Workplace* is an important contribution to the practice of occupational and environmental health. It is arranged to function both as an introduction to and a review of physical and biological hazards. It provides practical information not previously available in a single source on emerging, reemerging, and classical hazards due to these agents. Topics range from electromagnetic fields, ionizing radiation, and ergonomics, to occupational exposures to tuberculosis, HIV, and hantavirus. The reader will find helpful current information on a broad array of hazardous agents with a selection of timely literature citations for follow-up review.

All health professionals involved in protecting worker health will find this work a valuable addition to their basic reference library.

James P. Hughes, M.D.

PREFACE

Fifteen years have passed since the second edition of this book appeared. During this time, physical and biological hazards have been increasingly recognized as important hazards of the workplace. We have seen new research along with revisions of government standards and guidelines for physical agents such as manual materials handling, shift work and high-pressure environments, and biological agents including tuberculosis and tick-borne diseases. In addition, we have seen the emergence or spread of biological hazards, including Ebola and Zika virus.

The reception of the first two editions has been very gratifying to us. Many of our colleagues have written to us with suggestions for new topics and agents. We have tried to preserve the unique style and format of the original edition, while updating and expanding existing content, and adding new agents that we felt have become important over the intervening years. We are deeply indebted to our contributors. Many of them have returned to update their original chapters, and many colleagues are new contributors to the third edition.

The primary focus of the book continues to be as a practical "how to" reference containing basic information about the physical and biological hazards for occupational health and safety professionals from an occupational health perspective. We are pleased that readers have told us that this is the book that they pull off the shelf when they need a quick introduction or refresher to a topic in physical and biological hazards, just before they go to talk to employees or patients. This is not meant to be a definitive reference book, but rather a first reference that provides a practical overview for the primary health practitioner. It is also intended to be useful for health professionals who have no formal occupational medicine training.

Our goal continues to be to bring you an introduction to the fascinating world of physical and biological hazards. We hope that the third edition will continue to assist all health professionals who are responsible for protecting the health and safety of workers.

Gregg M. Stave, MD, JD, MPH, FACP, FACOEM, FACPM
Peter H. Wald, MD, MPH, FACP, FACOEM, FACMT

ACGIH POLICY STATEMENT

1330 Kemper Meadow Drive • Cincinnati, OH 45240-4148, USA
Phone: 513-742-2020 • Fax: 513-742-3355
E-Mail: mail@acgih.org • http://www.acgih.org

Defining the Science of Occupational and Environmental Health®

POLICY STATEMENT ON THE USES OF TLVs® AND BEIs®

The Threshold Limit Values (TLVs®) and Biological Exposure Indices (BEIs®) are developed as guidelines to assist in the control of health hazards. These recommendations or guidelines are intended for use in the practice of industrial hygiene, to be interpreted and applied only by a person trained in this discipline. They are not developed for use as legal standards and ACGIH® does not advocate their use as such. However, it is recognized that in certain circumstances individuals or organizations may wish to make use of these recommendations or guidelines as a supplement to their occupational safety and health program. ACGIH® will not oppose their use in this manner, if the use of TLVs® and BEIs® in these instances will contribute to the overall improvement in worker protection. However, the user must recognize the constraints and limitations subject to their proper use and bear the responsibility for such use.

The Introductions to the TLV®/BEI® Book and the TLV®/BEI® *Documentation* provide the philosophical and practical bases for the uses and limitations of the TLVs® and BEIs®. To extend those uses of the TLVs® and BEIs® to include other applications, such as use without the judgment of an industrial hygienist, application to a different population, development of new exposure/recovery time models, or new effect endpoints, stretches the reliability and even viability of the database for the TLV® or BEI® as evidenced by the individual *Documentation*.

It is not appropriate for individuals or organizations to impose on the TLVs® or the BEIs® their concepts of what the TLVs® or BEIs® should be or how they should be applied or to transfer regulatory standards requirements to the TLVs® or BEIs®.

Approved by the ACGIH® Board of Directors on March 1, 1988.

Special Note to User

The values listed in this book are intended for use in the practice of industrial hygiene as guidelines or recommendations to assist in the control of potential workplace health hazards and for no other use. These values are *not* fine lines between safe and dangerous concentrations and *should not* be used by anyone untrained in the discipline of industrial hygiene. **It is imperative that the user of this book read the Introduction to each section and be familiar with the *Documentation* of the TLVs® and BEIs® before applying the recommendations contained herein.** ACGIH® disclaims liability with respect to the use of the TLVs® and BEIs®.

ACGIH STATEMENT OF POSITION

1330 Kemper Meadow Drive • Cincinnati, OH 45240-4148, USA
Phone: 513-742-2020 • Fax: 513-742-3355
E-Mail: mail@acgih.org • http://www.acgih.org

Defining the Science of Occupational and Environmental Health®

ACGIH® Statement of Position Regarding the TLVs® and BEIs®

The American Conference of Governmental Industrial Hygienists (ACGIH®) is a private not-for-profit, nongovernmental corporation whose members are industrial hygienists or other occupational health and safety professionals dedicated to promoting health and safety within the workplace. ACGIH® is a scientific association. ACGIH® is not a standards setting body. As a scientific organization, it has established committees that review the existing published, peer-reviewed scientific literature. ACGIH® publishes guidelines known as Threshold Limit Values (TLVs®) and Biological Exposure Indices (BEIs®) for use by industrial hygienists in making decisions regarding safe levels of exposure to various chemical and physical agents found in the workplace. In using these guidelines, industrial hygienists are cautioned that the TLVs® and BEIs® are only one of multiple factors to be considered in evaluating specific workplace situations and conditions.

Each year ACGIH® publishes its TLVs® and BEIs® in a book. In the introduction to the book, ACGIH® states that the TLVs® and BEIs® are guidelines to be used by professionals trained in the practice of industrial hygiene. The TLVs® and BEIs® are not designed to be used as standards. Nevertheless, ACGIH® is aware that in certain instances the TLVs® and the BEIs® are used as standards by national, state, or local governments.

Governmental bodies establish public health standards based on statutory and legal frameworks that include definitions and criteria concerning the approach to be used in assessing and managing risk. In most instances, governmental bodies that set workplace health and safety standards are required to evaluate health effects, economic and technical feasibility, and the availability of acceptable methods to determine compliance.

ACGIH® TLVs® and BEIs® are not consensus standards. Voluntary consensus standards are developed or adopted by voluntary consensus standards bodies. The consensus standards process involves canvassing the opinions, views and positions of all interested parties and then developing a consensus position that is acceptable to these parties. While the process used to develop a TLV® or BEI® includes public notice and requests for all available and relevant scientific data, the TLV® or BEI® does not represent a consensus position that addresses all issues raised by all interested parties (e.g., issues of technical or economic feasibility). The TLVs® and BEIs® represent a scientific opinion based on a review of existing peer-reviewed scientific literature by committees of experts in public health and related sciences.

ACGIH® TLVs® and BEIs® are health-based values. ACGIH® TLVs® and BEIs® are established by committees that review existing published and peer-reviewed literature in various scientific disciplines (e.g., industrial hygiene, toxicology, occupational medicine, and epidemiology). Based on the available information, ACGIH® formulates a conclusion on the level of exposure that the typical worker can experience without adverse health effects. The TLVs® and BEIs® represent conditions under which ACGIH® believes that nearly all workers may be repeatedly exposed without adverse health effects. They are not fine lines between safe and dangerous exposures, nor are they a relative index of toxicology. The TLVs® and BEIs® are not quantitative estimates of risk at different exposure levels or by different routes of exposure.

Since ACGIH® TLVs® and BEIs® are based solely on health factors, there is no consideration given to economic or technical feasibility. Regulatory agencies should not assume that it is economically or technically feasible for an industry or employer to meet TLVs® or BEIs®. Similarly, although there are usually valid methods to measure workplace exposures at TLVs® and BEIs®, there can be instances where such reliable test methods have not yet been validated. Obviously, such a situation can create major enforcement difficulties if a TLV® or BEI® was adopted as a standard.

ACGIH® does not believe that TLVs® and BEIs® should be adopted as standards without full compliance with applicable regulatory procedures including an analysis of other factors necessary to make appropriate risk management decisions. However, ACGIH® does believe that regulatory bodies should consider TLVs® or BEIs® as valuable input into the risk characterization process (hazard identification, dose-response relationships, and exposure assessment). Regulatory bodies should view TLVs® and BEIs® as an expression of scientific opinion.

ACGIH® is proud of the scientists and the many members who volunteer their time to work on the TLV® and BEI® Committees. These experts develop written *Documentation* that include an expression of scientific opinion and a description of the basis, rationale, and limitations of the conclusions reached by ACGIH®. The *Documentation* provides a comprehensive list and analysis of all the major published peer-reviewed studies that ACGIH® relied upon in formulating its scientific opinion. Regulatory agencies dealing with hazards addressed by a TLV® or BEI® should obtain a copy of the full written *Documentation* for the TLV® or BEI®. Any use of a TLV® or BEI® in a regulatory context should include a careful evaluation of the information in the written *Documentation* and consideration of all other factors as required by the statutes which govern the regulatory process of the governmental body involved.

- *ACGIH® is a not-for-profit scientific association.*
- *ACGIH® proposes guidelines known as TLVs® and BEIs® for use by industrial hygienists in making decisions regarding safe levels of exposure to various hazards found in the workplace.*
- *ACGIH® is not a standards setting body.*
- *Regulatory bodies should view TLVs® and BEIs® as an expression of scientific opinion.*
- *TLVs® and BEIs® are not consensus standards.*
- *ACGIH® TLVs® and BEIs® are based solely on health factors; there is no consideration given to economic or technical feasibility. Regulatory agencies should not assume that it is economically or technically feasible to meet established TLVs® or BEIs®.*
- *ACGIH® believes that TLVs® and BEIs® should NOT be adopted as standards without an analysis of other factors necessary to make appropriate risk management decisions.*
- *TLVs® and BEIs® can provide valuable input into the risk characterization process. Regulatory agencies dealing with hazards addressed by a TLV® or BEI® should review the full written Documentation for the numerical TLV® or BEI®.*

ACGIH® is publishing this Statement in order to assist ACGIH® members, government regulators, and industry groups in understanding the basis and limitations of the TLVs® and BEIs® when used in a regulatory context. This Statement was adopted by the ACGIH® Board of Directors on March 1, 2002.

Part I

PHYSICAL HAZARDS

Chapter 1

INTRODUCTION to PHYSICAL HAZARDS

PETER H. WALD

Physical hazards are hazards that result from energy and matter and the interrelationships between the two. Conceptually, physical hazards in the workplace can be subdivided into worker–material interfaces, the physical work environment, and energy and electromagnetic radiation. The consequences of exposure to these hazards can be modified by worker protection and a variety of human factors. This chapter will review the general principles of basic physics and worker protection.

Physics is the science of energy and matter and of the interrelationships between the two, grouped in traditional fields such as acoustics, optics, mechanics, thermodynamics, and electromagnetism. Quantum physics deals with very small energy forces; relativity deals with objects traveling at very high speeds (which causes time effects). Thus, physical hazards can be thought of as primarily hazards of energy, temperature, pressure, or time. This broad definition allows for the investigation of many hazards that are otherwise hard to classify but nevertheless represent important issues in the workplace. An understanding of these physical hazards requires familiarity with the two basic concepts of physics: classical mechanics, with its derivatives of thermodynamics and fluid dynamics, and electromagnetic radiation. For measurements, we have used Standard International (SI) units throughout this book, but we have included conversions to other units where they are in common usage. Table 1.1 reviews the standard unit prefixes for mathematics that are used in the physical hazards section. The mathematical equations for principles discussed in this section are included in tables that accompany the text. Although they are not necessary to understand the material, they are presented for those readers who wish to review them.

MECHANICS

Mechanics deals with the effects of forces on bodies or fluids at rest or in motion (Table 1.2). From mechanics, we can get to the study of sound, which is a result of the mechanical vibration of air molecules. The behavior of heat arises from the vibration of molecules. Temperature is proportional to the average random vibrational (in solids) or translational (in liquids and gases) kinetic energy. The physics of pressure arises from the laws of motion and temperature. The laws that govern electricity can be derived from special cases of mechanics (see below), and electromagnetic energy and waves are a direct result of the laws that govern electricity.

Classical mechanics is the foundation of all physics. Galileo (1564–1642) first described the study of kinematics. Kinematics is primarily concerned with uniform straight-line motion and motion where there is uniform acceleration. As a practical example, Galileo used these insights to predict the flight of projectiles. In uniform straight-line motion, velocity (v) is equal to the change in displacement (Δs) divided by the change in time (Δt). Acceleration (a) is the instantaneous change of velocity with respect to time, which is calculated by taking the derivative of velocity with respect to time. Where there is uniform acceleration, the new velocity is equal to the original velocity (v_0) plus acceleration times time. The distance traveled under acceleration is described by a combination of the component traveled at the original velocity plus the component traveled under acceleration. The mathematical equations for these forces are summarized in Table 1.3.

Sir Isaac Newton (1642–1727) originally described the study of mechanics in his 1687 *Philosophiæ Naturalis*

Physical and Biological Hazards of the Workplace, Third Edition. Edited by Gregg M. Stave and Peter H. Wald.

TABLE 1.1 Mathematical unit prefixes.

Prefix	Symbol	Multiplier
Tetra-	T	10^{12}
Giga-	G	10^{9}
Mega-	M	10^{6}
Kilo-	k	10^{3}
Deci-	d	10^{-1}
Centi-	c	10^{-2}
Milli-	m	10^{-3}
Micro-	μ	10^{-6}
Nano-	n	10^{-9}
Pico-	p	10^{-12}

TABLE 1.2 The disciplines of mechanics.

Solid mechanics	
Statics	The study of bodies at rest or equilibrium
Dynamics or kinetics	The study of forces or the change of motion that forces cause
Kinematics	The study of pure motion without reference to forces
Fluid dynamics	
Hydrostatics	The study of still liquids
Hydraulics	The study of the mechanics of moving liquids
Aerodynamics	A special subset of hydraulics that deals with the movement of air around objects

TABLE 1.3 Mathematical expressions of Galileo's description of kinematics.

$v=\Delta s/\Delta t$	Average straight-line velocity
$s=vt$	Distance traveled at constant velocity
$a=dv/dt$	Acceleration (derivative of velocity with respect to time)
$v=v_0+at$	Velocity at straight-line acceleration
$s=v_0t+\frac{1}{2}at^2$	Distance traveled at uniform acceleration

Variables: Δs = change in distance placement, Δt = change in time, v_0 = original velocity, v = velocity, a = acceleration, s = distance, t = time.

Principia Mathematica. He formulated three laws that serve as the foundation of classical mechanics (Table 1.4).

The first law is known as the law of inertia. It states that all matter resists being accelerated and will continue to resist until it is acted upon by an outside force. The second law states that the acceleration of this outside force will be related to the size of that net force (F) but inversely related to the mass (m) of the object. This relationship is described mathematically by the following expression:

$$a \propto \frac{F\text{net}}{m}$$

The third law states that when two bodies exert a force on each other, they do so with an action and reaction pair. The force between two bodies is always an interaction.

TABLE 1.4 Newton's laws of motion.

Newton's first law of motion: A body remains at rest, or if in motion it remains in uniform motion with a constant speed in a straight line, unless it is acted on by an unbalanced external force

Newton's second law of motion: The acceleration produced by an unbalanced force acting on a body is proportional to the magnitude of the net force, in the same direction as the force, and inversely proportional to the mass of the body

Newton's third law of motion: Whenever one body exerts a force upon a second body, the second body exerts a force upon the first body; these forces are equal in magnitude and oppositely directed

A good example of how all three laws operate can be seen at the bowling alley. When a bowling ball is sitting on the rack, the force of the ball pressing down on the rack (gravity) is equal and opposite to the force of the rack pressing up on the ball to resist gravity (the third law). The speed of the bowling ball at the end of the alley is dependent on the amount of acceleration imparted to it. An adult can apply more force to the ball than a child, so the adult's ball will go faster. However, if smaller balls (i.e., of less mass) are used, less force is required; therefore, a child can accelerate the ball to the same speed (the second law). Once the ball leaves your hand, no more net force is applied to the ball (if we ignore friction), and it travels down the alley at a constant speed (the first law).

Mechanics has been central to the advancement of physics. Two mechanical concepts are central to understanding what strategies to adopt in order to prevent injury and illness from physical hazards: kinetic energy and potential energy. In order for physical hazards to affect humans, they must possess energy to impart to the biological system. Energy is commonly described in terms of either force (F) or work (W). Force equals mass times the acceleration, and the result is a vector. $F=ma$ is the mathematical representation of Newton's second law. The work done on an object equals the amount of displacement times the force component acting along that displacement. In the special case of the force acting parallel to the displacement, work equals force times displacement. These two relationships are described mathematically in Table 1.5, equations 1 and 2.

Kinetic energy (KE) is the energy of a mass that is in motion relative to some fixed (inertial) frame. KE is related to the mass of the object, and the speed at which it is traveling (Table 1.5, equation 3). Potential energy (PE) is stored energy that can do work when it is released as kinetic energy.

Since mass and energy are conserved in all interactions, the sums of potential and kinetic energy from before and after an encounter are equal. The equation for kinetic energy is also important for electromagnetic radiation. An electric system can store electric energy in a magnetic field in an induction coil. The kinetic energy of the electric charges equals the amount of work done to set up the field in the coil, which is stored as potential energy.

TABLE 1.5 Mathematical expressions of force, work, and energy.

1. $F = kma$ $= ma$ (if units = kg m/s^2 = newtons)	Force
2. $W = (F \cos\theta)s = F \cdot s$ $= Fs$ (if F and s are parallel)	Work
3. $KE = \frac{1}{2}mv^2$	Kinetic energy—mechanical system
4. $PE = \frac{1}{2}LI^2$	Potential energy—electrical system

Variables: v = velocity, a = acceleration, I = the current (in amperes), k = a constant, L = the inductance of the coil (in henries), m = mass, s = displacement.

Work, kinetic energy and potential energy in this system are related to the inductance of the coil and the current (Table 1.5, equation 4). Potential energy is the potential to do work, and theoretically all this work can be turned into kinetic energy. The expressions for kinetic energy in the mechanical system and potential energy in the electric system have an identical form. This form shows the similarity between kinetic and potential energy in mechanics and electromagnetic radiation and lays the groundwork for examining the electromagnetic wave.

ELECTROMAGNETIC RADIATION

By far the most complicated concept related to the understanding of physical hazards is that of electromagnetic radiation (EMR). Energy can be transmitted directly by collision between two objects, or it can be transmitted by EMR. We see direct examples of energy transfer by EMR when we are warmed by the infrared rays of the sun or burned by its ultraviolet rays. EMR is a continuum of energies with different wavelengths and frequencies. Two similarities of all types of EMR are that they all move at the same speed and they are all produced by the acceleration or deceleration of electric charge. EMR has a dual, particle wave nature: its energy transfer is best described by a particle, but the behavior of the radiation is best described as a wave. All EMR travels at a constant speed, $c = 3 \times 10^8$ m/s (the speed of light). Each particle of energy, called a photon, is accompanied by an electric field (E-field) and a magnetic field (H-field); these fields are perpendicular to each other and perpendicular to the direction of travel of the wave (Figure 1.1).

It is important to remember that EMR is only produced when an electric charge is moving. Coulomb forces are forces between stationary charges, whereas magnetic forces are due to the motion of charges relative to each other. A moving electric charge (or electric field) induces a magnetic field, and a moving or changing magnetic field induces an electric field. In 1873, James Maxwell linked together these electric and magnetic phenomena into a unified field theory of EMR. As an electric charge moves, it induces a magnetic field, which in turn induces an electric

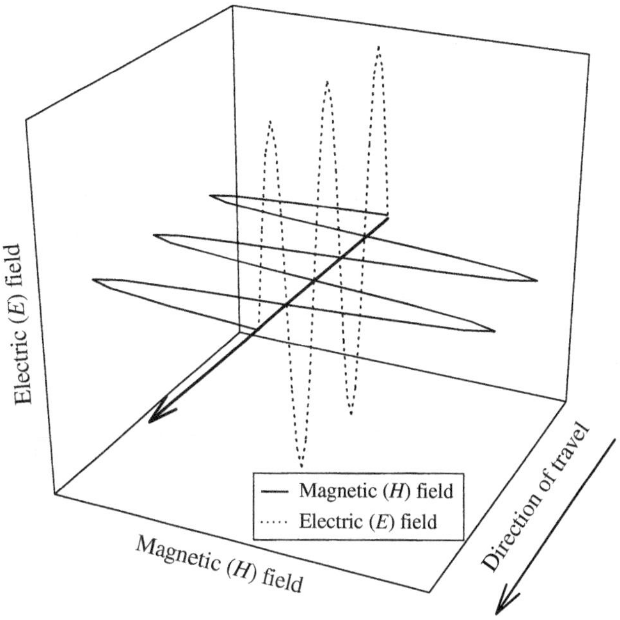

FIGURE 1.1 Stylized representation of an electromagnetic wave.

TABLE 1.6 Mathematical equations for electromagnetic radiation.

1. $E = hv$	Energy in joules
2. $c = v\lambda$	Wavelength and frequency related to the speed of light
3. $\lambda = c/v$, $v = c/\lambda$	Rearrangements of equation 2
4. $E = hc/\lambda$	Energy in joules, by substituting equation 3 into equation 1
5. $E = 12400/\lambda$	Energy in electron volts, where λ is in angstroms (1 Å = 10^{-10} m)
6. $E = 1:24 \times 10^{-6}/\lambda$	Energy in electron volts, where λ is in meters

Variables: λ = wavelength, v = frequency, c = speed of light (3×10^8 m/s), h = Planck's constant (6.626×10^{-34} Js).

field. The mutual interaction of these two fields is what allows the electromagnetic wave to propagate and what dictates its physical form in Figure 1.1.

The energy (E) in each photon in the wave can be calculated in joules (J) and is related to the frequency of the radiation in hertz (Hz). Energy is calculated by multiplying the frequency by Planck's constant (6.626×10^{-34} Js). The mathematical representation of this is shown in Table 1.6, equation 1.

Since the velocity at which the wave travels equals the frequency times the wavelength (Table 1.6, equation 2), we can discover the wavelength (λ) for each frequency by dividing 3×10^8 m/s (the speed of light or c) by the frequency (Table 1.6, equation 3). The energy of the wave can also be calculated in terms of the wavelength by substituting the speed of light divided by the wavelength for frequency (Table 1.6, equation 4). In biological systems, it is useful to determine photon energy in electron volts from the wavelength.

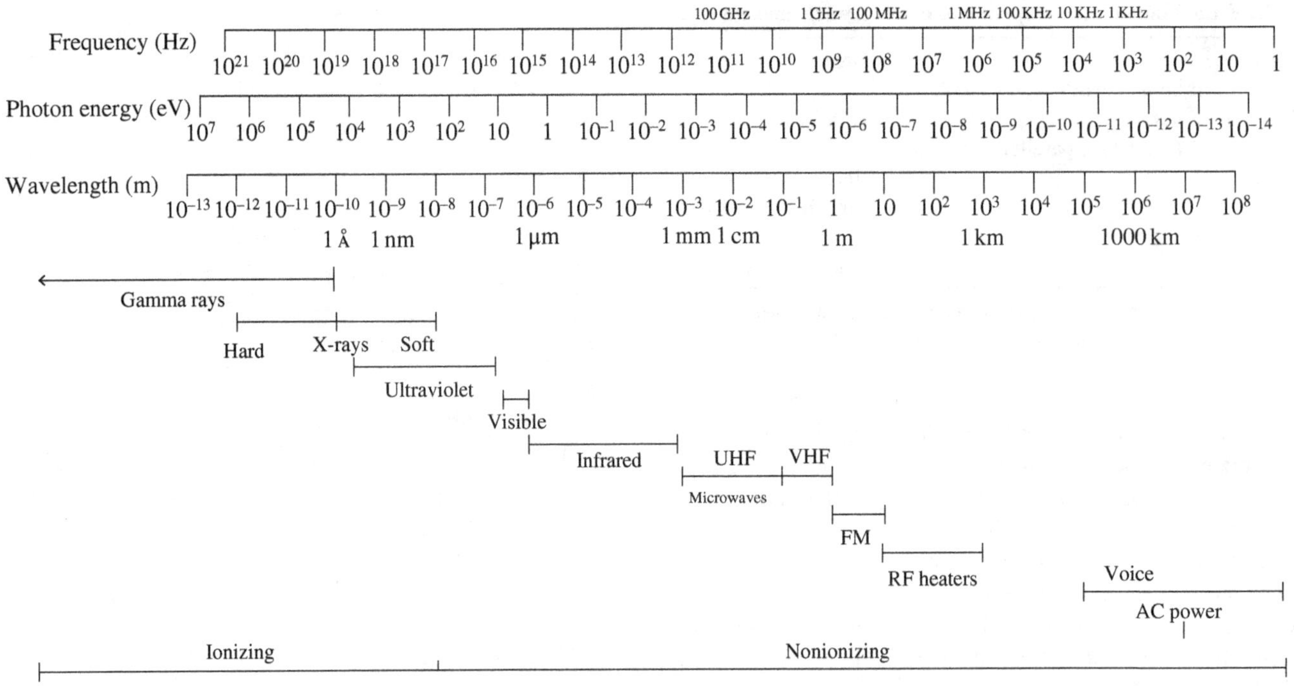

FIGURE 1.2 The electromagnetic spectrum.

This can be calculated from the wavelength in angstroms (Å) according to Table 1.6, equation 5. The electron volt is a convenient unit to use with biological systems, because it takes greater than roughly 10 electron volts (eV) to cause ionization in tissue.

We can also see from equations 1 and 2 in Table 1.6 that the energy of a given type of EMR varies directly with its frequency and inversely with its wavelength. Figure 1.2 shows a representative cross section of the electromagnetic spectrum, with the major classes noted. Notice that there are not strict divisions between the different classes of EMR. An important division in the EMR spectrum relates to the ability to ionize chemical bonds in biological tissue. As frequency increases from the radio bands, so does energy, until ionization potential is reached in the "hard" ultraviolet or "soft" X-ray bands.

A final important point about EMR involves the ways in which it can interact with objects. EMR interacts with biological tissues in one of the following three ways: (i) transmission, where the radiation passes through the tissue without any interaction; (ii) reflection, where the radiation is unable to pass through the air–tissue interface (also called the boundary layer) and is reflected back into space; and (iii) absorption, where the radiation is able to pass through the boundary layer and deposit its energy in the tissue. The frequency of the EMR determines what energy is released in the tissues (heat, electric potential, bond breaking, etc.). These interactions are summarized in Figure 1.3.

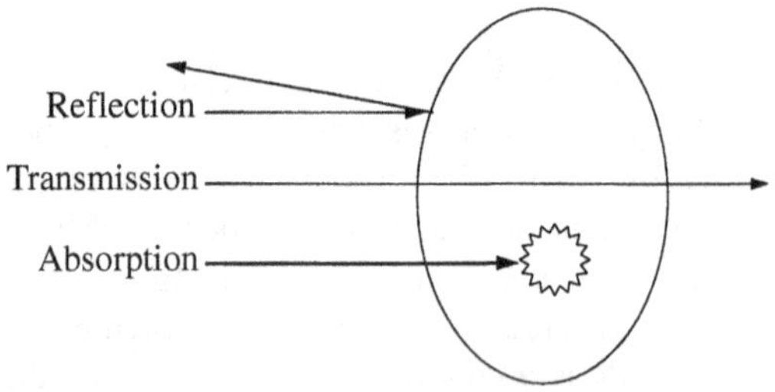

FIGURE 1.3 Interactions of electromagnetic radiation and biological tissue.

WORKER PROTECTION

Potential energy can also be called a potential hazard. The key to avoiding injuries and illnesses is to prevent the individuals in the workplace from being overexposed to the kinetic energy in the hazards. The major characteristics of the physical hazards covered in this text are reviewed in Table 1.7. Each of the following chapters will deal with the most appropriate method to prevent overexposure. However, there are certain recurring themes.

Since we are trying to prevent exposure, the first step is to educate the workforce. A good training program includes education about the potential hazards, the safest procedures to follow for each manufacturing or maintenance operation, correct tool selection and use for each job, use and care of personal protective equipment, and procedures to follow

TABLE 1.7 Major characteristics of the physical hazards.

Hazards	Occupational Settings	Measurement	Exposure Guidelines	Effects of Exposure	Surveillance
Worker–material interfaces					
Repetitive ergonomic hazards—extremities	Service and industrial operations	Repetition, force, posture	OSHA (pending)	Musculoskeletal strain, tunnel syndromes	Survey workers, observe tasks, measure physical parameters of the job
Manual materials handling—backs	Service and industrial operations	Repetition, force, posture	The National Institute for Occupational Safety and Health (NIOSH) lifting guide	Musculoskeletal strain, disk herniation	Survey workers, observe tasks, measure physical parameters of the job
Vibration	Whole body—vehicle/heavy equipment/industrial equipment operator, hand–arm-powered hand tool users	Frequency, motion	ISO 2631, ANSI S3 NIOSH criteria document, ACGIH	Whole body—low back pain. Hand–arm—hand–arm vibration syndrome (HAVS)	Survey workers, observe tasks, measure physical parameters of the job
Mechanical energy—direct injuries	Service and industrial operations	Velocity, distance, acceleration, force, weight, pressure, friction	None	Direct injury	Epidemiology of workplace injuries
The physical work environment					
Hot environments	Hot indoor or outdoor environments	Wet globe bulb temperature, core temperature	ISO 7243, NIOSH criteria document	Heat strain, heat stroke	Heart rate, core temperature, worker selection
Cold environments	Cold indoor or outdoor environments	Wind chill, core temperature	ACGIH-TLV®	Frostbite, trench/immersion foot, hypothermia	Worker selection
High-pressure environments	Divers, caisson workers	Pressure in atmosphere absolute (ATA), changes in pressure	OSHA, marine occupational health safety standards	Barotrauma, decompression sickness, indirect effects secondary to pressure acting on other gases (O_2, N_2)	Worker selection
Low-pressure environments	Aircraft crews, private pilots, astronauts	Pressure in atmosphere absolute (ATA), changes in pressure	None	Barotrauma, decompression sickness, hypoxia	Worker selection
Shift work	Service and industrial operations	Rotation, duration and hour changes of shift schedule	None	Sleep disturbance, gastrointestinal upset	Worker selection

(*Continued*)

TABLE 1.7 (Continued)

Hazards	Occupational Settings	Measurement	Exposure Guidelines	Effects of Exposure	Surveillance
Energy and electromagnetic radiation					
Ionizing radiation	Pilots, underground miners, radiographers, medical and dental X-ray personnel, operators of high-voltage equipment, nuclear power and fuel cycle workers, medical and scientific researchers, some commercial products manufacturing	Personal dosimetry of radiation exposures	NCRP, ICRP, NRC, UNSCEAR	Acute radiation injury, carcinogenesis	Personal dosimetry. Lung and whole-body scanning, biological monitoring as appropriate
Ultraviolet radiation	Outdoor workers, welders, printers	Wavelength, intensity	ACGIH-TLV®	Corneal photokeratitis, skin erythema, cataracts	None
Visible light and infrared radiation	Outdoor workers, welders, printers, glass blowers	Visible duration, wavelength, intensity. Infrared wavelength, intensity	ACGIH-TLV®	Visible scotoma, thermal burn, photosensitivity, urticaria. Infrared-thermal burns, cataracts	None
Laser radiation	Service and industrial operations, researchers, medical personnel, maintenance personnel	Wavelength, power, energy, duration	ACGIH-TLV®, (ANSI Z136.1)	Retinal and skin burns	None
Microwave and radiofrequency (MW/RF) and extremely low-frequency (ELF) radiation	MW/RF—communication workers, industrial heating and RF welding operations. ELF—electricians and electrical workers, telephone and cable workers, electric arc welders, movie projectionists	Frequency, electric field, magnetic field, power density, operating mode	ACGIH-TLV®, OSH	MW/RF—thermal effects. ELF—no proven effects	None
Noise	Service and industrial operations	Time-weighted dBA	NIOSH, OSHA	Noise-induced hearing loss	Hearing conservation program for all exposed workers
Electric power and electrocution injuries	Service and industrial operations	Current, voltage	OSHA	Electrical burns, electrocution	None

TABLE 1.8 Engineering and administrative controls for physical hazards.

Hazards	Engineering Controls	Administrative Controls
Worker–material interfaces		
Repetitive ergonomic hazards—extremities	Repetition—mechanical aids, automation, distribution of tasks across the shift and the workforce	More frequent or longer rest breaks, limit overtime, varying work tasks, rotation of workers between less and more ergonomically stressful jobs
	Force—decrease weight of tools/ containers, optimize handles, torque control devices	
	Postures—locate work for mechanical advantage	
Manual materials handling—backs	Same as above	Same as above
Vibration	Whole body—relocate worker away from vibration, mechanically isolate vibration, use vibration-isolating seats in vehicles	Hand–arm—removal from work for significantly affected workers
	Hand–arm—use antivibration tools	
Mechanical energy—direct injuries	Guards, interlocks, proper lighting, nonskid floors	None
The physical work environment		
Hot environments	Air conditioning, increase air movement, insulate and shield hot surfaces, decrease air humidity, shade work area, mechanize heavy work	Use recommended work/rest cycles, work during cool hours of the day, provide cool rest areas, use more workers for a given job, rotate workers between less and more physically stressful jobs, provide fluids for cooling and hydration
Cold environments	Enclose and heat work area	Use recommended work/rest cycles, provide appropriate clothing, provide shelter for break, provide fluids for warming and hydration
High-pressure environments	Engineer a “shirtsleeve” environment which avoids high-pressure work	Work under no decompression guidelines/tables. Adhere to recommended decompression guidelines
Low-pressure environments	Work remotely at low altitude	Wait 12–48 hours after diving to fly, schedule time for acclimation when working at altitude
Shift work	Automate processes to reduce the number of workers/shift	Rotate shifts forward, get worker input for desires of time off and shift design
Energy and electromagnetic radiation		
Ionizing radiation	Shielding, interlocks, increase worker distance to source, warning signs, enclose radionuclides	Worker removal if dose limit reached, minimize exposure times, use radionuclides only in designated areas using safe handling techniques, limited personnel access
Ultraviolet radiation	Enclosure, opaque shielding and/or tinted viewing windows, interlocks, increase worker distance to source, nonreflective surfaces, warning signs	Minimize exposure times, limited personnel access
Visible light and infrared radiation	Enclosure, shielding, interlocks, increase worker distance to source, nonreflective surfaces, warning signs	Limited personnel access
Laser radiation	Enclosure, interlocks, nonreflective surfaces, warning signs	Limited personnel access
Microwave and radiofrequency (MW/RF) and extremely low-frequency (ELF) radiation	MW/RF—wire mesh enclosure, interlocks, increase worker distance to source, warning signs	MW/RF—limited personnel access
	ELF—increase worker distance to source	
Noise	Enclose sources, warning signs	Limited personnel access
Electric power and electrocution injuries	Interlocks, warning signs	Limited personnel access

TABLE 1.9 Commonly used personal protective equipment for physical hazards.

Equipment Type	Hazard Category	Specific Hazard
Helmet	Direct injuries	1. Falling objects 2. Low clearances/"bump hazards"
Safety glasses	1. Direct injuries	1. Flying objects 2. Sparks
	2. Lasers	Retinal burns
Face shield	Direct injury	1. Flying objects 2. Molten metal, sparks
Welding helmet/goggles	1. Direct injury	1. Flying objects 2. Molten metal, sparks
	2. Ultraviolet radiation	Skin/conjunctival burns
Earplugs/earmuffs	Noise	Noise
Fall protection systems—safety belt, body harness, lines, and/or other hardware	Direct injury	Falls
Respirators	Ionizing radiation	α-Emitters: internal contamination
Clothing		
Leather	Heat	Burns
Aluminized	Heat	Heat stroke, burns
Lead	Ionizing radiation	γ-Emitter, X-rays
Fire resistant	Direct injury	Burns
Insulating	Cold	Hypothermia
Disposable	Ionizing radiation	α-Emitter: external contamination
Gloves		
Leather	Direct injury	Abrasions, lacerations
Rubber	Electric injury	Electrocution
Metal mesh	Direct injury	Lacerations
Antivibration	Vibration	Vibration
Footwear		
Steel toe	Direct injury	Falling objects
"Traction sole"	Direct injury	Slips, trips, falls
Rubber	Electric energy	Electrocution

in emergency situations, including fire and loss control, shutdown, rescue, and evacuation.

Substitution of less dangerous equipment or agents is the best protection from hazards, because it totally removes any chance of exposure. However, substitution is often not possible; therefore, worker protection from physical hazards generally focuses on engineering controls. Engineering and administrative controls for physical hazards are summarized in Table 1.8. Often, these controls involve isolation or shielding from the hazard. The most effective isolation involves physically restricting an individual from a hazard area by fencing off the area whenever the hazard is present. Interlocks that inactivate the equipment when the exclusion area is entered are often used to further enhance physical barriers. Alternatively, the hazard can be "locked out" when a worker is present in an area that would become hazardous if the equipment were energized (Chapter 5). This process of excluding maintenance workers from hazardous areas has been institutionalized in the Occupational Safety and Health Administration (OSHA) lockout/tagout (LOTO) standard (Code of Federal Regulations [CFR] 1910.147).

Another way to protect workers is to specifically shield them from the hazard. In some cases, an individual piece of equipment can be shielded to prevent exposure. With some higher-energy hazards, such as ionizing radiation, shielding may be needed in addition to isolation of the equipment. In special cases where it is not practical to shield the hazard (e.g., cold, low pressure), individual workers can be shielded with personal protective equipment, such as jackets or environment suits. In addition, it is sometimes possible to alter the process so as to decrease exposure. This is often the case with hazards affecting the worker–material interface, where engineering design is often inadequate. Personal protective equipment can also be used as an adjunct to engineering controls. Table 1.9 contains a summary of the most common personal protective equipment used for physical hazards.

The final strategy for hazard control is the use of administrative controls. These controls are implemented when exposures cannot be controlled to acceptable levels with substitution, engineering controls, or personal protective equipment. Administrative measures can be instituted to either rotate workers through different jobs to prevent repetitive motion injuries or to remove workers from ionizing radiation exposure once a predetermined exposure level is reached. Although this is not the preferred method of hazard control, it can be effective in some circumstances. Administrative controls are also reviewed in Table 1.8.

The best way to determine what hazards are present in a specific workplace is to go to the site and walk through the manufacturing or service process. There are a number of excellent texts available on evaluating workplaces from both an industrial hygiene and a safety perspective; they are included in the list of further reading at the end of this chapter. An additional point that will become obvious as you read through the text is that there are some significant measurement issues that need to be addressed by an appropriate health professional. Although larger employers will undoubtedly have such a person on staff, at the majority of smaller work sites, no such person will be available.

If you are unfamiliar with the measurement technology, make sure that you (or the employer) retain someone who knows how to perform an exposure assessment. Inaccurate measurements will invalidate the entire process of a prevention program. There are, of course, a number of physical hazards that do not require special measurements and can be handled quite nicely with relatively low-cost safety programs. Several excellent texts describing how to set up general safety programs are included in the further reading list at the end of this chapter.

Finally, remember that the human being is a biological system. For a given exposure, different people will respond differently because of interindividual variation. Most workplace standards are designed with a safety factor to protect against overexposure related to this variation (and to account for any knowledge gaps). In addition, a worker's perception of the hazard must also be taken into account. Some workers may have an exaggerated response to a nonexistent or low-threat hazard, whereas others may not respond appropriately to a series hazard with which they have "grown comfortable." The challenge in assessing and communicating the relative danger entailed by the hazard is to strike the right balance between these two competing tendencies.

The goal of the first section of this volume is to acquaint the reader with the types of physical hazards that may be present in the workplace. Once these hazards are identified at the site, he or she can refer to the specific chapter that addresses the salient measurement issues or offers general strategies to control exposures and monitor effects.

Further Reading

Balge MZ, Krieger GR. *Occupational health and safety*, 3rd edn. Chicago: National Safety Council Press, 2000.

Burgess WA. *Recognition of health hazards in industry: a review of materials processes*, 2nd edn. New York: John Wiley & Sons, Inc., 1995.

Hagan PE, Montgomery JF, O'Reilly JT. *Accident prevention manual for business and industry: administration and programs*, 14th edn. Chicago: National Safety Council Press, 2015.

Plog B, Quinlan P (eds.). *Fundamentals of industrial hygiene*, 6th edn. Chicago: National Safety Council Press, 2012.

Serway RA, Vuille C. *College physics*, 9th edn. Boston: Cengage Learning, 2011.

Spitz H, Albert RE. Ionizing Radiation. In Bingham E, Cohrssen B (eds.) *Patty's toxicology*, 6th edn. New York: John Wiley & Sons, Inc., 2012, pp. 1–23.

I Worker–Material Interfaces

Chapter 2

ERGONOMICS and UPPER EXTREMITY MUSCULOSKELETAL DISORDERS

Thomas R. Hales

Ergonomics has been defined as the science of fitting the job to the worker[1] or the art of matching job demands with worker capabilities. Upper extremity (UE) musculoskeletal disorders (MSDs) are soft tissue disorders of the muscles, tendons, ligaments, peripheral nerves, joints, cartilage, or supporting blood vessels in the neck, shoulder, arm, elbow, forearm, hand, or wrist. Examples of specific disorders include tension neck syndrome, rotator cuff tendinitis, epicondylitis, peritendinitis, and carpal tunnel syndrome (CTS).[2] When job demands overwhelm an employee's mental and/or physical capacity, employee health, comfort, and productivity can be adversely affected.[3] While comfort and productivity levels are important outcomes to consider, this chapter will focus upon the effect of workplace physical stressors (repetition, force, posture, and vibration) on the musculoskeletal system of the upper extremities. This chapter reviews the epidemiologic association between UE MSDs and work, and provides practical tools for healthcare providers to (i) assess physical stressors in the workplace and (ii) recognize, treat, and prevent UE MSDs.

OCCUPATIONAL SETTING

Magnitude of the problem

Bureau of Labor Statistics data

An injury or illness is work related if an event or exposure in the work environment either caused or contributed to the resulting condition or significantly aggravated a preexisting condition.[4] The Bureau of Labor Statistics (BLS) annually reports on the number of workplace injuries, illnesses, and fatal injuries in the United States. In addition to collecting private sector data, since 2008 the BLS began reporting injury and illness data on public sector workers in state and local governments (e.g., police and fire fighters).

MSDs are the most common type of occupational condition reported on the BLS survey, typically representing almost a third of all BLS-reported injuries and illnesses.[5] In 2014, the BLS estimated that 365 580 cases of MSDs occurred for an incidence rate of 33.8 cases per 10 000 full-time workers, a rate that is trending downward since 2011 (Figure 2.1). In 2014, workers who sustained an MSD required a median of 13 days to recuperate before returning to work, compared to 10 days in previous years.[5] This finding suggests that while MSDs are trending down, the cases may be becoming more severe. Carpal tunnel syndrome is probably the most well-known MSD, but sprains, strains, and tears are the most common diagnosis.[6] The BLS reports the MSD rate is higher among males. In 2014 the MSD incident rate was 37.5 per 10 000 full-time workers, compared to 29.7 per 10 000 among female workers, a trend that has persisted over the past decade.[5] The 45–54-year-old age group has the highest reported rate (40.4 per 10 000), followed by the 35–44-year-old age group (36.2 per 10 000) in 2014, again a trend that has persisted over the past decade[5] (Figure 2.2). It should be noted that the BLS data significantly underestimates the true number of these conditions.[7,8]

Workers' compensation data

A number of studies have described the magnitude of the problem of MSDs in terms of workers' compensation costs. In 1989, workers' compensation claims for policy holders

Physical and Biological Hazards of the Workplace, Third Edition. Edited by Gregg M. Stave and Peter H. Wald.

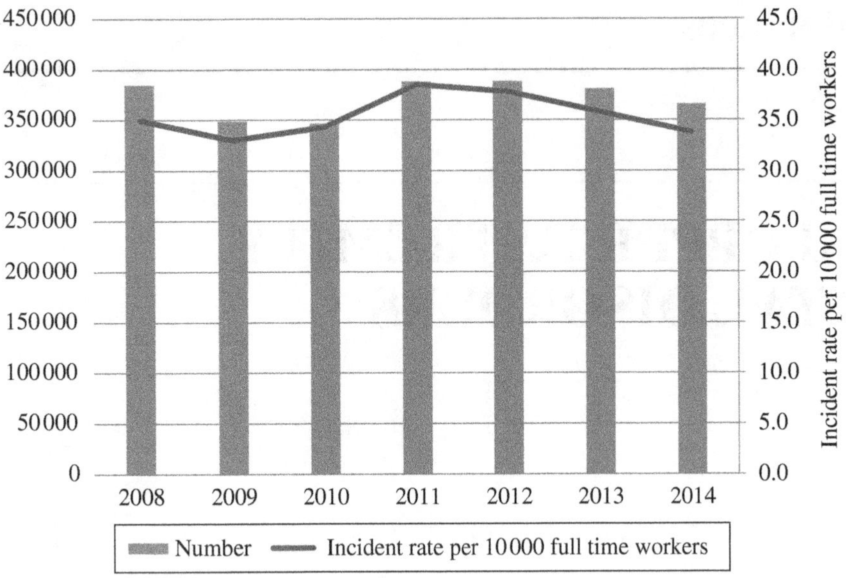

FIGURE 2.1 Number and incident rate of musculoskeletal disorders involving days away from work, 2008–2014.[5]

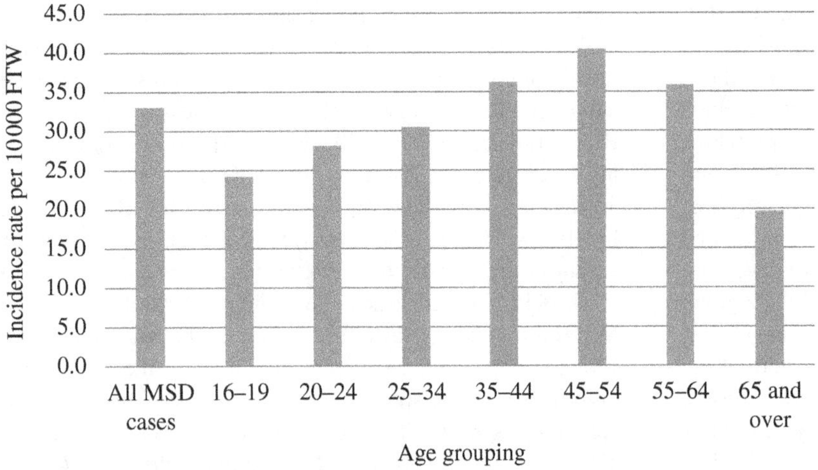

FIGURE 2.2 Incidence rate of musculoskeletal disorders involving days away from work by age group, 2014.[5] FTW, full-time workers.

in 45 states reported a mean cost of $8070 per UE cumulative trauma disorder claim.[9] They estimated the total direct US workers' compensation costs for UE cumulative trauma disorders to be $563 million in 1989.[9] For 1987–1995, the state of Washington adjudicated over 160000 UE MSD claims with a mean direct (medical and indemnity) cost of a claim ranging from $6593 to 15790.[10] For all MSD claims (neck and back in addition to UE), Washington state accepted 392925 claims resulting in $2.6 billion in direct costs and 20.5 million lost workdays.[11] These estimates do not include the indirect costs, such as administrative costs for claims processing, lost productivity, and the cost of recruiting and training replacements. It has been suggested that indirect costs are two to three times the direct compensation costs.[12]

While these numbers highlight the costs of UE MSD to society, they do not take into account those workers who suffer a UE MSD but are never recorded onto the Occupational Safety and Health Administration (OSHA) 300 Log or workers who choose not to file a claim.[13–15] For the state of Michigan in 1996, only 25% of workers with work-related MSD filed for workers' compensation.[13,14] Factors associated with filing a claim included increased length of employment, lower annual income, dissatisfaction with coworkers, physician restriction on activities, type of physician providing treatment, being off work for at least 7 days, decreased current health status, and increased severity of illness.[15] Other factors workers consider when deciding whether to file a claim include: Will the claim be contested, will there be employer retribution, and are there alternatives available for payment of medical costs?[16,17]

Occupations at risk

Case reports have given rise to a number of disorders named for the patient's occupation (Table 2.1).

These disorders are not unique to their occupations. In 2014, the BLS reported nursing assistants had the highest MSD rates, followed by emergency medical technicians/paramedics, fire fighters, and refuse/recyclable material collectors.[5] The National Health Interview Survey described cases of self-reported carpal tunnel syndrome to be highest among mail/message distributors (prevalence 3.2%), health assessment and treatment occupations (2.7%), and construction trades (2.5%).[18] The Wisconsin workers' compensation program reported wrist injury to be highest among dental hygienists, data entry keyers, and hand-grinding and polishing occupations.[19] Although the various occupations have different rates of MSD, the BLS and workers' compensation data point to almost all occupations reporting at least one case of work-related MSD.

TABLE 2.1 Work-related MSD named by occupation.

Bricklayer's shoulder
Carpenter's elbow
Golfer's elbow
Tennis elbow
Janitor's elbow
Stitcher's wrist
Cotton twister's hand
Telegraphist's cramp
Writer's cramp
Bowler's thumb
Jeweler's thumb
Cherry pitter's thumb
Gamekeeper's thumb
Carpetlayer's knee

Industries at risk

According to the BLS data, the transportation and warehousing industry had the highest number and rate of MSDs followed by the healthcare and social assistance industry (Figure 2.3). Work-place evaluations have also identified a high prevalence of UE MSD in the animal-slaughtering and processing industries (beef, pork, poultry, and fish).[20–23] Workers in the poultry industry with an astonishingly high prevalence (34%) were found to have carpal tunnel syndrome using self-reported hand and wrist symptoms, hand diagrams, and nerve conduction studies (median mononeuropathy) to define carpal tunnel syndrome.[20]

In summary, despite their numerous limitations, BLS and workers' compensation data are sufficient to confirm that the UE MSD problem is large and that rates significantly differ between industries and occupations signifying that workplace factors are important risk factors.

Epidemiology

One of the main purposes of epidemiologic studies is to identify factors that are associated (positively or negatively) with the development or recurrence of adverse medical conditions. No single epidemiologic study determines causality. Rather, results from epidemiologic studies can contribute to the evidence of causality. Over the past decade, several publications have reviewed the medical and ergonomic literature to determine whether scientific evidence supports a relationship between workplace physical factors and MSDs. The most comprehensive review was completed by Bernard et al. at the National Institute for Occupational Safety and Health (NIOSH).[24] This review focused on disorders that affected the neck (tension neck syndrome), upper extremities (shoulder tendinitis, epicondylitis, CTS, hand–wrist tendinitis, and

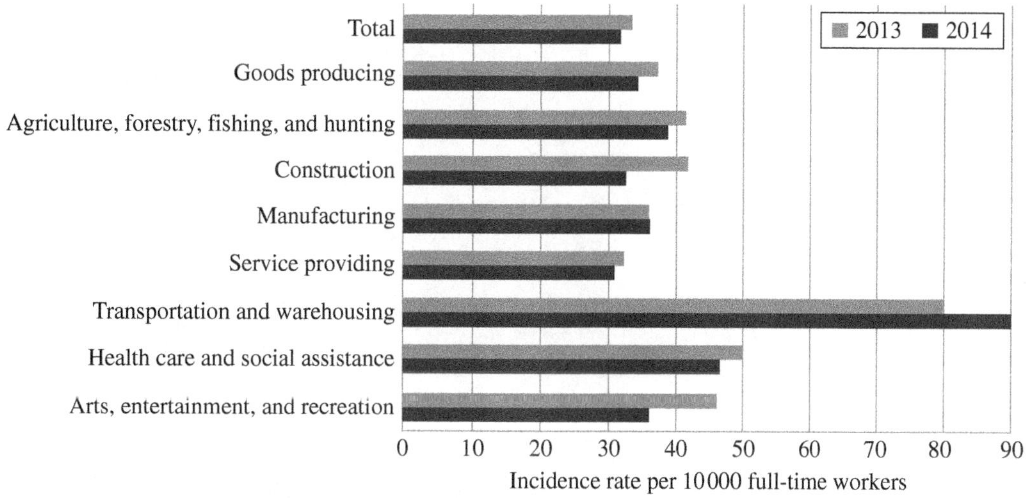

FIGURE 2.3 Musculoskeletal disorder incidence rates for selected private sector industries, 2013–2014.[6]

hand–arm vibration syndrome), and the lower back (work-related low back pain). A database search strategy initially identified 2000 studies, but laboratory, biomechanical, clinical treatment, and other nonepidemiologic studies were excluded, leaving 600 for systematic review. The review process consisted of three steps.

The first step gave the increased emphasis, or weight, to studies that had high participation rates (>70%), physical examinations, blinded assessment of health and exposure, and objective exposure assessment. The second step assessed for any other selection bias and any uncontrolled potential confounders. The final step summarized studies with regard to strength of the associations, consistency in the associations, temporal associations, and exposure–response (dose–response) relationships.

Bernard et al. concluded that a substantial body of credible epidemiologic research provides strong evidence of an association between MSDs and work-related physical factors. This is particularly true when there are high levels of exposure or exposure to more than one physical factor (e.g., repetition and forceful exertions). The strength of the associations for specific physical stressors varies from insufficient to strong (Table 2.2).

TABLE 2.2 Evidence for causal relationship between physical work factors and upper extremity musculoskeletal disorders.[24]

Body part and *risk factor*	Strong evidence (+++)	Evidence (++)	Insufficient evidence (+/0)	Evidence of no effect (–)
Neck and neck/shoulder				
Repetition		✓		
Force		✓		
Posture	✓			
Vibration			✓	
Shoulder				
Posture		✓		
Force			✓	
Repetition		✓		
Vibration			✓	
Elbow				
Repetition			✓	
Force		✓		
Posture			✓	
Combination	✓			
Hand/wrist				
Carpal tunnel syndrome				
Repetition		✓		
Force		✓		
Posture			✓	
Vibration		✓		
Combination	✓			
Tendinitis				
Repetition		✓		
Force		✓		
Posture		✓		
Combination	✓			
Hand–arm vibration syndrome				
Vibration	✓			

Source: Adapted from Bernard.[24]

The consistently positive findings from a large number of cross-sectional studies, strengthened by the available prospective studies, provide strong evidence for an increased risk of work-related MSDs for the neck, elbow, and hand–wrist for jobs that require high repetition, high force, awkward postures, and vibration. This conclusion was supported by subsequent prospective studies and reviews by the National Academy of Sciences and the Institute of Medicine.[16,25–27]

MEASUREMENT—ASSESSMENT

Physical stressors can be grouped into the following categories: repetition, force, posture, and vibration. They arise from excessive job demands, improperly designed workstations, tools, equipment, or inappropriate work techniques. A number of methods are available to measure/estimate these stressors. The method selected should be based on the purpose of the evaluation. The following grouping provides several options.

Survey methods

Employee or supervisor interviews, employee diaries, and employee-completed questionnaires are useful because of their low cost, rapid availability, and, for some, the ability to obtain historical data about previous exposures.[28–32] One of the most commonly used survey tools is the so-called Borg, or rating of perceived exertion, scale.[33,34] A 15-point scale (6–20) was created to reflect the linear relationship between physical workload and heart rate divided by 10 (e.g., a heart rate of 60 beats/minute corresponds to 6 on the scale). The scale is presented to the subject before the start of a job or job task with anchors of "no exertion at all = (6)" to "maximal exertion = (20)."[34] The subject is then asked to rate his or her exertion level after completing the job and/or job task. A 10-point Borg scale was also created to account for large muscle group exertion, rather than heart rate or total body exertion.[34,35]

The accuracy of self-assessment surveys has been questioned because of the potential for the worker to either underreport or overreport exposures. For example, highly motivated subjects might underestimate their exertion, while unmotivated subjects might overestimate their exertion.[36] This potential problem has led many to utilize observational checklists (described below).

Observational methods

Observational methods, such as observational checklists, are commonly employed to objectively assess the workplace for physical stressors. Some checklists can be used by healthcare providers with limited expertise,[37–41] others require some training,[29,42,43] while others require a considerable amount of experience and training.[44–47] Table 2.3 provides the reader with a simple checklist for healthcare providers with limited expertise. For healthcare providers or others with some training, the hand activity level (HAL) could be utilized. It is described in more detail in the Exposure Guidelines section"

Measuring workers

If the above checklist suggests that physical stressors exist in the workplace, quantitative measurement of those risk factors should be considered. However, quantitative measurement of ergonomic hazards can require the use of specialized equipment and training and expertise in its use and interpretation of the results.[29]

Methods used to generate quantitative information on physical stressors include electrogoniometers (dynamic measurements of posture), accelerometers, and imaging techniques (electronic and laser optical recordings). Two devices that may be useful outside of research settings are spring scales or gauges to estimate force requirements and simple goniometers to measure static postures. Both of these tools have been used successfully in workplaces due to their simplicity.

Internal forces can be measured using surface electromyography (EMG), but currently available equipment is expensive, and its use requires training and expertise to perform and interpret. Video and imaging systems as a means to measure posture have been used primarily in the laboratory setting where the camera's line of sight is perpendicular to the planes of the measured body segments. But given the dynamic nature of most job activities, their use in the workplace seems limited unless multiple cameras can be used from a variety of viewing angles. Goniometer use for measuring static postures is well established, but few jobs require continuous static postures. Electrogoniometers can measure dynamic postures, but their accuracy and associated analytic methods are not well established.

EXPOSURE GUIDELINES

American Conference of Governmental Industrial Hygienists (ACGIH)

The ACGIH provides guidelines for industrial hygienists to use while making decisions regarding safe levels of exposure to various hazards in the workplace. The organization issued a guideline known as the "hand activity level" (HAL) based on the hand, wrist, and forearm exposure to repetition and peak normal force for mono-task jobs.[48] Mono-task jobs are defined as jobs that repeatedly perform a similar set of motions or exertions for ≥4 hours per day.

TABLE 2.3 Physical stressors checklist.

Risk Factors	Yes	No
Hand force		
1.1 Gripping with the fingertips with something that weighs ≥2 pounds?		
1.2 Gripping with the whole hand with something that weighs ≥10 pounds?		
1.3 Lifting >10 pounds, more than twice a minute for >2 hours per day?		
1.4 Lifting >25 pounds above the shoulders more than 25 times per day?		
Posture		
2.1 Does the worker have to stand while performing the job?		
2.2 Are the hands working above the head (or the elbows above the shoulders) >2 hours per day?		
2.3 Are the wrists bent >30° in any direction for >2 hours per day?		
2.4 Does the worker use a twisting ("clothes-wringing") hand motion or use a pinch grip?		
Repetitiveness		
3.1 Is the cycle time longer than 30 seconds?		
3.2 Is the same motion with little change occurring every few seconds for >2 hours per day?		
Hand–arm vibration		
4.1 Does the worker use tools with high vibration levels (e.g., chain saws, chipping hammers, etc.) for >30 minutes per day?		
4.2 Does the worker use tools with moderate vibration levels (e.g., grinders, sanders, jigsaws, etc.) for >2 hours per day?		
Workstation hardware		
5.1 The workstation's orientation cannot be adjusted?		
5.2 The height of the work surface cannot be adjusted?		
Physical stress		
6.1 The worker's arms/hands/wrists come into contact with sharp edges?		
6.2 Is the worker using heavy gloves?		
Tool design		
7.1 Is the span of the tool's handle between 5 and 7 cm?		
7.2 Is the handle of the tool made from metal?		
7.3 The tool is not suspended.		
7.4 Is the tool handle slippery?		

The first step is selecting a job period that represents an average activity. Then observe (or videotape) the activity for several job cycles. The second step rates the HAL. This can be accomplished by two methods: (i) a trained observer using a validated rating scale based on exertion frequency, rest pauses, and speed of motion (Figure 2.4)[42,48] or (ii) calculated using information on the frequency of exertion and the work/recovery ratio (Table 2.4).[48] The third step identifies forceful exertions and postures by (i) observer ratings, (ii) worker ratings, (iii) biomechanical analysis, or (iv) instrumentation. Since the latter two methods (biomechanical analysis or instrumentation) require considerable expertise and equipment, the following discussion will focus on observer and worker ratings. Observer ratings of force utilize the same Latko et al. scale described earlier (Figure 2.4).[42] Factors that the observer should consider include the weight, shape, and friction of the work object, posture, glove fit and friction, mechanical assists, torque specification of power tools, quality control, and equipment maintenance. Worker ratings utilize the same Borg scale (1–10) described earlier.[34,35] Suppose, for example, a male worker rates his job's grip strength requirements as four (somewhat strong). To normalize this force, we measure the worker's grip strength (300 newtons (N)) and compare this to the average male strength (500 N). Therefore, the normalized peak force $= 4 \times 300\,\text{N}/500\,\text{N} = 2.4$. The precision of both the observer and worker ratings is improved by having multiple observers/workers rate the same job.

The HAL and the normalized force estimates can now be plotted and compared to the TLV® (Figure 2.5). Employees performing job tasks above the solid top line will be at significant risk of acquiring a UE MSD, and specific control measures should be utilized so that the force/repetition for a given level of hand activity is below this line. The dotted lower line represents an "action limit," the point at which general controls, including surveillance (discussed below), are recommended. The TLV® does not specifically account for awkward or extreme postures, contact stresses, low temperatures, and vibration; therefore, professional judgment is needed to account for these additional stressors. If any of these stressors are present on jobs, the TLV® and the action limit will be lower.

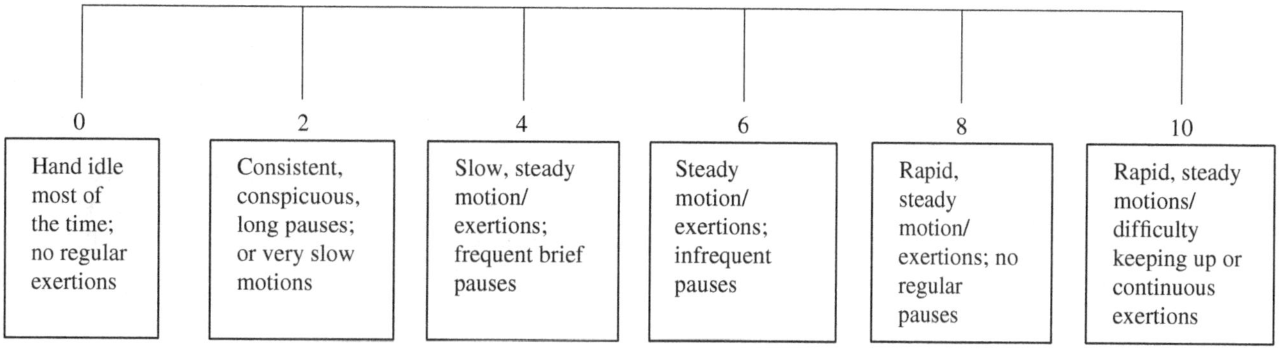

FIGURE 2.4 Visual analog scale for rating hand activity level (0–10) with verbal anchors.[42]

TABLE 2.4 Hand activity level (0–10) is related to exertion, frequency, and duty cycle (% of work cycle where force is greater than 5% of maximum).

Frequency (exertion/s)	Period (s/exertion)	Duty cycle (%)				
		0–20	20–40	40–60	60–80	80–100
0.125	8.0	1	1	–	–	–
0.25	4.0	2	2	3	–	–
0.5	2.0	3	4	5	5	6
1	1.0	4	5	5	6	7
2	0.5	–	5	6	7	8

Source: From American Conference of Governmental Industrial Hygienists (ACGIH®) TLV® Hand Activity Level Document.[48] From ACGIH®, *2015 TLVs® and BEIs®* Book. Copyright 2015. Reprinted with permission.

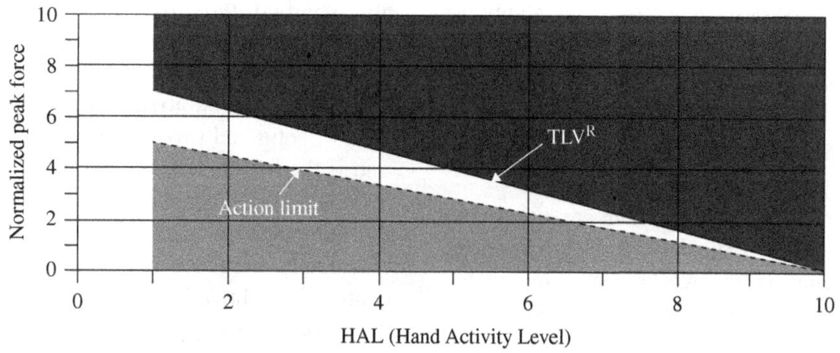

FIGURE 2.5 The TLV® for reduction of work related musculo-skeletal disorders based on "hand activity" or "HAL" and peak hand force. The top line depicts the TLV®. The bottom line is an Action Limit for which general controls are recommended. *Source*: Adapted from American Conference of Governmental Industrial Hygienist (ACGIH®) TLV® Hand Activity Draft Document. From ACGIH®, *2015 TLVs® and BEIs®* Book. Copyright 2015. Reprinted with permission.

Others

Since ACGIH TLV® does not account for all potential physical stressors, the reader is encouraged to review other proposed UE exposure guidelines, such as the strain index proposed by Moore and Garg[45,49] and the Rapid Upper Limb Assessment (RULA) proposed by McAtamney and Corlett.[46]

Posture is an important potential physical stressor.[29] Posture can be defined as the position of a part of the body relative to an adjacent part, as measured by the angle of the connecting joint. Standard posture definitions (neutral and nonneutral) and normal ranges of motion have been developed by the American Academy of Orthopaedic Surgeons.[50] Postural stress develops as a joint reaches its maximal deviation; therefore, postures should be maintained as close to neutral as possible. In addition to postures at the extreme end of a joint's range, tasks that require finger-pinching postures have been associated with UE musculoskeletal disorders. Kodak has proposed the following posture guidelines[51]:

- Keep the work surface height low enough to permit employees to work with their elbows at their sides and wrists near their neutral position.
- Keep reaches within 20 inches in front of the work surface so that the elbow is not fully extended when forces are applied.
- Keep motions within 20–30° of the wrist's neutral point.
- Avoid operations that require more than 90° of rotation around the wrist.
- Avoid gripping requirements in repetitive operations that spread the fingers and thumb apart more than 2.5 inches. Cylindrical grips should not exceed 2 inches in diameter, with 1.5 inches being the preferable size.

Federal Ergonomic Standard

In 2000, OSHA developed and issued an ergonomic standard. In 2001, Congress, under the Congressional Review Act, passed a resolution of disapproval, thereby eliminating the standard. In its place, OSHA has issued industry-specific guidelines.[52] OSHA has developed ergonomic guidelines to prevent MSDs for meatpacking plants (1993), retail grocery stores (2004), shipyards (2008), nursing homes (2009), foundries (2012), and poultry processing (2013).[52] In addition, OSHA continues to inspect and, if appropriate, cite companies for ergonomic hazards under its general duty clause.

California Ergonomic Standard

In 1993, the California State Legislature required its Occupational Safety and Health Standards Board to develop "standards for ergonomics in the workplace designed to minimize instances of injury from repetitive motion."[53] Subsequent legal challenges shaped its content, coverage, and start date. The standard, adopted in 1999, applies to a job, process, or operation where a repetitive motion injury (RMI) has occurred to more than one employee under the following conditions:

1. A licensed physician objectively identified and diagnosed the RMI.
2. The RMI was work related (≥50% caused by a repetitive job, process, or operation).
3. The employees with RMIs were performing a job process or operation of identical work activity (performing same repetitive motion task).
4. The employee reported the RMI to the employer in the last 12 months.

If the above conditions are met, the employer is required to develop an ergonomic program with the following three components: worksite evaluation, control of workplace exposures, and employee training. The worksite evaluation requires that each job, process, or operation of identical work activities be evaluated for exposures causing RMIs. If these exposures are found, they must be corrected in a timely manner or, if not capable of being corrected, have the exposures minimized to the extent feasible. In addition, employees must receive training on the following:

- The employer's ergonomic program
- The exposures which have been associated with RMIs
- The symptoms and consequences of injuries caused by repetitive motion
- The importance of reporting symptoms and injuries to the employer
- Methods used by the employer to minimize RMIs

NORMAL PHYSIOLOGY AND ANATOMY

Muscles

Muscle consists of muscle fibers (muscle cells), nerve elements (motor neurons, afferent neurons, receptors of different types), connective tissue, and blood vessels. Muscle fibers are classified into two types: Type I fibers, also known as slow-twitch or red muscle fibers, and Type II fibers, also known as fast-twitch or white muscle fibers. In muscle fibers, the smallest morphological contractile unit is the sarcomere, built of actin and myosin filaments. The smallest functional unit is the motor unit, which consists of a motor neuron cell and the muscle fibers that its branches supply. The muscles of the body are the generators of internal force that convert chemically stored energy into mechanical work. A muscle contracts its threadlike fibers, which shortens the length of the muscle, thereby generating a contractile force.

Myalgia is the medical term for the symptom of muscle pain. The most common type of myalgia, delayed-onset muscle soreness (DOMS), is a contraction-induced injury after vigorous or unaccustomed exercise. DOMS is a self-limiting condition that typically appears within the first 24 hours after exercise, peaks at 48–72 hours, and resolves within 1 week. Histologic and chemical changes are found in affected muscles, but these changes are not permanent and lead to a conditioning effect when occurring in a graduated manner.[54–56] Armstrong proposed the following theory for the pathogenesis of DOMS[54–56]:

- High mechanical forces, particularly those associated with eccentric exertions, cause structural damage of the muscle fibers and associated connective tissue structures.

- This structural damage alters the sarcolemma's permeability, producing a net influx of calcium into the cell. This calcium inhibits mitochondrial production of ATP and activates proteolytic enzymes that degrade Z-disks, troponin, and tropomyosin.
- The progressive degeneration of the sarcolemma is accompanied by diffusion of intracellular products into the interstitium and plasma; this attracts inflammatory cells that release lysosomal proteases, which further degrade the muscle proteins.
- Active phagocytosis and cellular necrosis lead to accumulation of histamine, kinin, and potassium, which stimulate regional nociceptors, resulting in the sensation of DOMS.

Eccentric contractions (muscle activation while the muscle is stretched), rather than isometric contractions, are felt to lead to DOMS.[57] Eccentric contractions can occur when muscles are exposed to either a single rapid stretch or a series of repetitive contractions.[54] Both models are consistent with DOMS requiring a temporary reduction in physical loading because of pain or discomfort. This is followed by a gradual increase in physical loading to stimulate healing and subsequent tissue-remodeling processes.

Muscle also undergoes a number of age-related changes, such as a 20% decrease in muscle mass, a 20% reduction in maximal isometric force, and a 35% decrease in the maximal rate of developing force and power.[58] This latter reduction is not due to differences in muscle recruitment strategies, but rather due to a change in the contractility of the muscle itself.[59] This translates into a marked decrease in the ability to sustain power over repeated contractions in older individuals. In addition, animal experiments have also demonstrated that older muscle damages more easily and heals more slowly.[60,61] These effects may help explain why older athletes seem to require greater rest intervals between training sessions and why workers in physically demanding jobs tend to change to less demanding jobs with age.[62]

Tendons

As a general rule, tendons transmit the contractile force generated by muscles to bone. Tendons are composed of collagen fibrils grouped into fibers that are collected together into fiber bundles that are united into fascicles.[63] A large number of fascicles form the tendon. The fiber bundles and fascicles are enclosed in thin films of loose connective tissue called the *endotenon*. This connective tissue contains blood vessels, lymphatic vessels, nerves, and elastic fibers and allows the fascicles to slide relative to one another. The whole tendon is wrapped in connective tissue called the *epitenon*. In some tendons, a further sheath, the *paratenon*, surrounds the tendon. The paratenon is merely a specialization of the areolar connective tissue through which many tendons run. A number of structures associated with tendons control and facilitate their movement. Where tendons wrap around bony pulleys or pass over joints, they are held in place by retaining ligaments (retinacula or fibrous sheaths that prevent bowstringing). Tendons glide beneath these retaining structures due to the lubrication provided by the synovial sheath.[64] In some regions, tendons are prevented from rubbing against adjacent structures by bursae. Although tendons generally have a good blood and nerve supply, regions of tendon subjected to friction, compression, or torsion are hypovascular or avascular. The general structure of tendons is modified in two regions: the sites where they attach to bone (enthesis) and the region where they are compressed against neighboring structures (around bony pulleys).[65] Fibrocartilage formation at the site of this compression loading is considered a normal/adaptive response. In summary, tendons have the capacity to change their structure and composition in response to mechanical stimulation. In most cases, this mechanical stress is beneficial and adaptive for maintaining cell activity and tissue function.

Peripheral nerves

Peripheral nerves carry signals to and from the central nervous system. A nerve fiber (neuron) consists of the nerve body, which is located in the anterior horn of the spinal cord (motor neuron) or in the dorsal root ganglia (sensory neuron), and a process extending into the periphery—the axon.[66] The axon is surrounded by Schwann cells. In myelinated fibers, a Schwann cell is wrapped around only one axon, in contrast to nonmyelinated fibers, where the Schwann cell wraps around several axons. Myelinated and nonmyelinated nerve fibers are organized in bundles, called *fascicles*, which are bound by supportive connective tissue, the *perineurium*. The bundles are usually organized in groups, held together by loose connective tissue called the *epineurium*. In between the nerve fibers and their basal membrane is the intrafascicular connective tissue—the *endoneurium*. The amount of connective tissue components varies between nerves and between various levels along the same nerve. The myelin insulation divides the axon into short, uninsulated regions (nodes of Ranvier) and longer, insulated regions (internodes). Conduction of nerve impulses proceeds by sequential activation of successive nodes without depolarization of the intervening internode (saltatory conduction).

PATHOPHYSIOLOGY AND PATHOGENESIS

Muscles

Myopathy is the medical term for measurable pathologic changes in a muscle, with or without symptoms. Myopathies can be due to a variety of congenital (e.g., muscular dystrophy) or acquired (e.g., inflammatory, metabolic, endocrine, or toxic) disorders. These diseases are not typically work related and will not be discussed further.

Muscle pain syndromes of unknown etiology can be classified into two categories: general and regional. General muscle pain involving all four quadrants of the body is called primary fibromyalgia. Primary fibromyalgia is not work related because, by definition, trauma-induced myalgia is excluded by the specific diagnostic criteria set by the American College of Rheumatology.[67] Regional muscle pain syndromes, not involving the whole body, often fall under the term myofascial syndrome. This has been defined as a painful condition of skeletal muscle characterized by the presence of one or more discrete areas (trigger points) that are tender when pressure is applied.[68] These muscle-related syndromes, a common example of which is tension neck syndrome, could be associated with work exposures. A variety of mechanisms have been proposed to account for this syndrome. A few are listed below.

Work and eccentric contractions

DOMS is a result of eccentric contractions that could occur on or off the job. DOMS has objective histologic and chemical changes, but these changes are part of the normal physiologic response. The pain or discomfort associated with DOMS, however, typically results in a temporary reduction in physical loading due to pain or discomfort. This is followed by a gradual increase in physical loading to stimulate healing and subsequent tissue remodeling. But if workers with physically demanding jobs have little control over the magnitude and duration of loading, the work can aggravate and hinder the healing process, thereby increasing the risk of developing a more chronic condition. Work-hardening programs are specifically designed to minimize this risk by prescribing graduated physical training regimens.

Work and gamma motor neurons

This theory starts with evidence that muscle pain, inflammation, ischemia, or sustained static muscle contractions are known to lead to the release of potassium chloride, lactic acid, arachidonic acid, bradykinin, serotonin, and histamine in the affected muscle.[69] These substances, in turn, are known to excite chemosensitive group II and IV afferents, which have a potent effect on gamma-muscle spindle systems and heighten the response of those spindles to stretch. Increased activity in the primary muscle spindle afferents may cause muscle stiffness, leading to further production of metabolites, more stiffness, and repetition of the cycle.

Work and the overload of Type I fibers

Another hypothesis for the pathogenesis of tension neck syndrome is that prolonged static contractions of the trapezius muscle result in an overload of Type I muscle fibers. Type I muscle fibers are used for low static contractions. Support for this hypothesis comes from findings on biopsy. When compared to healthy controls, Type I fibers in patients with chronic trapezius muscle pain (i) were larger, (ii) had a lower capillary-to-fiber ratio, (iii) had a more "ragged" appearance, and (iv) had reduced ATD and ADP levels.[70–72] Whether these findings are due to inadequate muscle recruitment[73] or inadequate tissue oxygenation is unknown.[74]

Work and muscle fatigue

Finally, much work has been done on the mechanisms of fatigue relating to muscle disorders. A complete review of these mechanisms can be found in Gandevai et al.[75]

Tendons

Physicians in sports medicine have suggested that tendon disorders fall into four main categories: paratendonitis, paratendonitis with tendinosis, tendinosis, and tendinitis.[76] These categories are based on clinical and histologic findings. It can be difficult to distinguish between these specific conditions on clinical evaluation alone. Because most conditions can be treated conservatively, histologic changes have been documented in only a subset of patients whose cases proceeded to surgery. Thus, many cases are defined simply as "tendinitis" based on history, examination, and impaired function.

Proposed mechanisms for work-related tendon disorders include[77] (i) ischemia in hypovascular tissues, (ii) microinjuries incurred at a rate that exceeds repair potential, (iii) thermal denaturation, (iv) dysregulation of paratendon–tendon function, and (v) inflammatory processes secondary to some, or all, of these other factors.

Shoulder disorders provide evidence for the ischemic theory. Work above one's head can have two effects: compression (impingement) and reduced local blood flow. Impingement comes from the narrow space between the humeral head and the tight coracoacromial arch. As the arm is raised in abduction, the rotator cuff tendons and the insertions on the greater tuberosity are forced under the coracoacromial arch.[78] Reduced local blood flow occurs when the supraspinatus muscle is statically contracted, increasing the intramuscular pressure higher than the arterial pressure of the vessel traversing the supraspinatus muscle belly and supplying it with oxygen.[79] These work factors, in combination with the fact that the entheses of the three tendons comprising the rotator cuff are hypovascular, lead to tendon degeneration manifested by microruptures and calcium deposits. Once the tendons are degenerated, exertion may trigger an inflammatory response resulting in active tendinitis.[80]

The pathogenesis and pathophysiology of lateral epicondylitis are less clear. Histologic evaluation of tennis elbow tendinosis identifies a noninflammatory response in the tendon. This histopathology reveals disorganized immature collagen formation in association with immature fibroblastic and vascular elements. This has been named angiofibroblastic tendinosis and is thought to be the result of an avascular degenerative process.[81] This pathology is located at the enthesis of the extensor carpi radialis brevis (ECRB) tendon. Factors associated with its development include age, systemic factors, direct trauma, and repetitive overuse from sports, occupation, and performing arts. It is theorized that multiple repetitive eccentric loading of the ECRB results in tension loading, microruptures of the peritendon, and secondary anoxia and degenerative consequences.

Tendon sheaths

Tendons can become trapped in their synovial sheaths due to a narrowing of their fibro-osseous canal. Tendons passing through stenotic canals frequently have a nodular or fusiform swelling and can be covered with granulation tissue.[82] Whether these tendon changes are a cause or an effect of the narrowing is unclear. If the narrowing occurs in the first dorsal compartment, the disorder is known as De Quervain's tenosynovitis.[82] If it occurs in the flexor digits or thumb (A1 pulley), it is known as trigger finger or trigger thumb.[83]

While the term tenosynovitis implies inflammation of the tendon sheath, inflammatory cells are rarely found on histology. The lack of inflammatory cells could represent a sampling bias, since typically only chronic severe cases are biopsied, when the inflammatory process could have already run its course.[84] The fundamental pathologic change is hypertrophy and/or fibrocartilaginous metaplasia; however, recent studies suggest that the fibrocartilaginous metaplasia represents an adaptive response to compressive forces.[65]

The pathogenesis of tendon entrapment disorders involves static compression, repeated compression, and acute trauma. The static compression model is based on clinicians' observations that De Quervain's tenosynovitis is related to repeated, prolonged, or unaccustomed exertions that involve the thumb in combination with nonneutral wrist or thumb postures. Tensile loading of the abductor pollicis longus or extensor pollicis brevis, in combination with their turning a corner at the extensor retinaculum, creates a compressive force. The retinaculum responds with functional hypertrophy or fibrocartilaginous metaplasia. The duration of compression is more important than the number (repetition) of compressions. The repeated compression theory relies on the same biomechanical argument, except that the number of episodes of loading (repetition) is more critical than the accumulated duration of loading.

Peripheral nerves

Although a number of peripheral nerve entrapment disorders exist, CTS is the most common and most studied and will be the only nerve entrapment disorder discussed in this chapter.

CTS is the entrapment of the median nerve within the carpal canal at the wrist. Three mechanisms have been suggested[85]: (i) friction associated with repetitive tendon motions, leading to flexor tendon sheath irritation and swelling, (ii) repeated direct mechanical trauma to the median nerve by structures within the carpal tunnel, and (iii) prolonged elevated pressure within the carpal tunnel, leading to ischemia, tissue swelling, and epineural fibrosis. The last mechanism is supported by the following: (i) carpal tunnel pressure (CTP) is almost always higher in patients with CTS than in normal subjects[86]; (ii) surgical decompression (carpal tunnel release surgery) seems to be effective at reducing the elevated CTP and improving symptoms[87]; (iii) histologic studies of the flexor tendon sheaths biopsied during carpal tunnel release show edema and vascular changes consistent with long-standing ischemia[88]; (iv) animal models of acute and chronic nerve compression show physiologic and histologic findings consistent with nerve ischemia[86]; (v) in human studies, acute elevation of CTP results in acute nerve dysfunction, with the critical threshold varying according to the subject's diastolic blood pressure[89]; and (vi) human studies have found a dose–response relationship between CTP and wrist posture,[90] fingertip loading,[91] and repetitive hand activity.[25–27,92] These findings are consistent with the static and dynamic biomechanical models of CTS.[93]

Psychosocial

Numerous studies have reported an association between psychosocial factors and UE MSD. Several models have been developed to explain these associations including the following:

- The balance theory of job design and stress[94]
- The biopsychosocial model[95]
- The ecological model[96]
- The workstyle model[97]

Figure 2.6 summarizes the salient features of these models.[98]

The biologic mechanism by which psychosocial stress and psychological strain lead to UE MSD has not been fully elucidated, but research suggests that it may involve neuroendocrine, vascular, and immunological pathways. For example, perceived stress can trigger the "fight-or-flight" response stimulating the endocrine system (e.g., cortisol, adrenalin, and noradrenalin). Stress can also lead to increased muscle tension,[99] decreased blood supply in the extremities,[100] breakdown of muscle protein and its repair,[101] and alteration of the inflammatory or immune response.[102] All these short-term responses might increase the risk of UE MSDs.

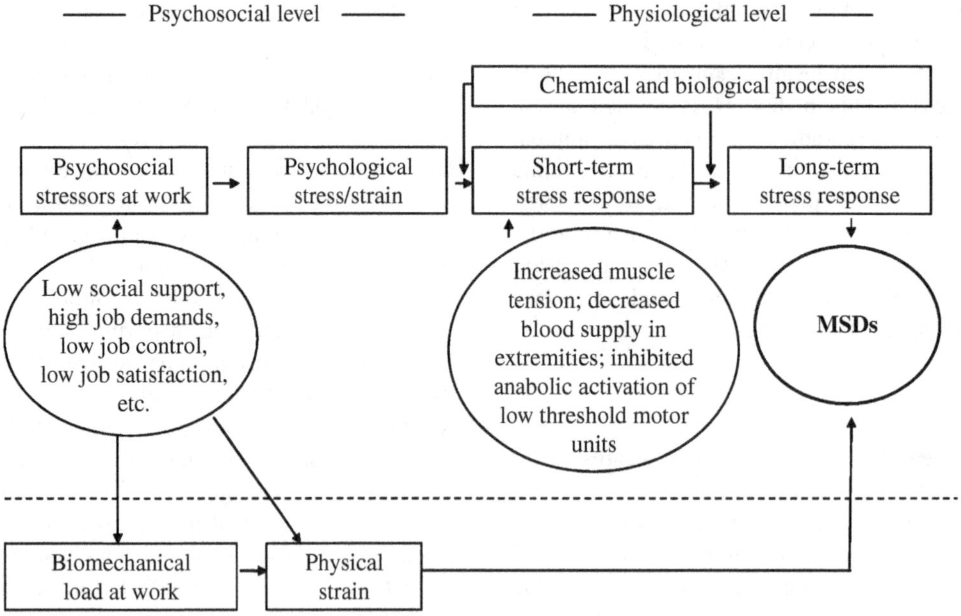

FIGURE 2.6 Explanatory model of the impact of psychosocial work environmental factors on the onset of musculoskeletal disorders. *Source*: Hauke A, Flintrop J, Brun E, et al. The impact of work-related psychosocial stressors on the onset of musculoskeletal disorders in specific body regions: a review and meta-analysis of 54 longitudinal studies. *Work & Stress: An International Journal of Work, Health & Organizations* 2011; 25(3):243–56. Reprinted with permission from Taylor and Francis, Copyright 2011.

Skeptics of the possible role between psychosocial factors/stress and UE MSD point out that support for the association comes from a plethora of cross-sectional studies. Due to inherent properties of their study design, cross-sectional studies have difficulty establishing the direction of the association (i.e., cause vs. effect). Longitudinal studies, on the other hand, can overcome this limitation. A review and meta-analysis of 54 longitudinal studies concluded that psychosocial factors are independent predictors for the onset of MSDs[98] and should play an important role in prevention and intervention programs.[103,104]

DIAGNOSIS AND TREATMENT

The successful treatment of UE MSDs relies on early reporting of symptoms, prompt evaluation, early diagnosis, and appropriate interventions. The interventions should include both the medical management of the disorder and the workplace (discussed in the Prevention section).

Clinical evaluation

Like all conditions with a potential occupational etiology, the health history is a fundamental component of the evaluation.[105] The history should (i) characterize the symptoms, (ii) provide an employee description of work activities, and (iii) identify predisposing conditions or factors. In addition, the history should inquire about previous occupations and job tasks, symptoms during previous jobs, home responsibilities, home use of power tools, hobbies, similar symptoms among coworkers, and broader changes in the workplace (e.g., changes in equipment, bonus or incentive programs, layoffs, and training). This information should be documented in the employee's medical record. If the employee description of work activities is unclear, or if further information is needed to understand employee job tasks and workplace conditions, this can be ascertained by visiting the workplace or viewing job tasks recorded on videotape. Simply reviewing a written description of job tasks may not provide an adequate understanding of the ergonomic stresses involved in the job.

After the history is taken and information obtained about workplace conditions and job tasks, the neck and UE should be examined.[106] A comprehensive examination would include inspection, palpation, assessment of the ranges of motion, evaluation of sensory and motor function, and applicable provocative maneuvers. Employees with underlying systemic disease (e.g., diabetes mellitus) may require a more complete examination involving other organ systems. Results of the examination findings, both positive and negative, should be documented in the employee's medical record.

Clinical diagnosis

Using the information from the clinical evaluation, an assessment or diagnosis should be made. Diagnoses should be consistent with the International Classification of Diseases, Tenth Revision (ICD-10) (Table 2.5). Terms such

TABLE 2.5 Common upper extremity musculoskeletal disorders by ICD-10 codes.

ICD-10 Code	Diagnosis
Nerve and Nerve root disorders	
G540	Brachial plexus lesions
G549	Nerve root and plexus disorder, unspecified
G5600	Carpal tunnel syndrome
G5610	Other lesions of median nerve
G5620	Lesion of ulnar nerve
G5630	Lesion of radial nerve
G5640	Causalgia of the upper limb
G5690	Unspecified mononeuropathy of the upper limb
Tendon or Tendon Sheath Disorders	
M66239	Spontaneous rupture of extensor tendons, forearm
M66249	Spontaneous rupture of extensor tendons, hand
M66339	Spontaneous rupture of flexor tendons, forearm
M66349	Spontaneous rupture of flexor tendons, hand
M6688	Spontaneous rupture of other tendons
M7520	Bicipital tendinitis of the shoulder
M7530	Calcific tendinitis of the shoulder
M7540	Impingement syndrome of the shoulder
M7580	Other shoulder lesions, unspecified
M7700	Medial epicondylitis
M7710	Lateral epicondylitis
M6530	Trigger Finger
M654	Radial styloid tenosynovitis [de Quervain]
M65849	Other synovitis and tenosynovitis, hand
M6790	Unspecified disorder of synovium and tendon
M70039	Crepitant synovitis (acute or chronic), wrist
M6740	Ganglion cysts
M67419	Ganglion, shoulder
M67429	Ganglion, elbow
M67439	Ganglion, wrist
M67449	Ganglion, hand
Joint Disorders	
M7010	Bursitis, hand
M7020	Bursitis, elbow (olecranon)
M7550	Bursitis, shoulder
M7720	Periarthritis, unspecified wrist
Muscle Disorders	
M609	Myositis, unspecified
M629	Disorder of muscle, unspecified
M791	Myalgia
M79609	Pain in limb
Soft Tissue Disorders	
R252	Cramp and spasm
R29898	Other symptoms and signs involving the musculoskeletal system
M7098	Unspecified soft tissue disorder related to use, overuse and pressure other
M7981	Nontraumatic hematoma of soft tissue
M7989	Other specified soft tissue disorders
Vascular	
I742	Embolism and thrombosis of arteries of the upper extremities

Source: Keyserling et al.[37] Reproduced with permission of Taylor and Francis.

as repetitive motion disorder (RMD), repetitive strain injury (RSI), overuse syndrome, and cumulative trauma disorder (CTD) may be useful for surveillance purposes or epidemiologic investigations, but they are not ICD-10 diagnoses and should not be used as individual medical diagnoses.

Once a diagnosis is made, an opinion is usually rendered regarding whether occupational factors caused, aggravated, or contributed to the condition. This tends to be the most difficult portion of the assessment. Unlike the classic occupational diseases, such as asbestosis or silicosis, most occupational MSDs do not have a pathognomonic finding or test specific to the exposure. Therefore, the importance of a thorough exposure assessment cannot be overemphasized.[107] In 1979, NIOSH published a guide for state agencies and physicians on the process of determining work relatedness of disease.[108] The process outlined in this guide, modified for MSD, is still relevant today:

- Has a disease condition been established by accepted clinical criteria?
- Does the literature support that the disease can result from the suspected agent?
- Has exposure to the agent been demonstrated?
- Has the exposure been of sufficient degree and duration to result in the diseased condition?
- Have nonoccupational factors been considered?
- Have other special circumstances been considered?

Clinical interventions

The goals for treatment are the elimination or reduction of symptoms and impairment and return to work under conditions that will not exacerbate the disorder. The expected duration of treatment, dates for follow-up evaluations, and time frames for improvement or resolution of symptoms should be specified at the initial evaluation and updated on subsequent evaluations.

Resting the symptomatic area

Resting the symptomatic area is a mainstay of conservative treatment. Reducing or eliminating employee exposure to musculoskeletal risk factors by changing the job conditions (forceful exertions, repetitive activities, extreme or prolonged static postures, vibration, direct trauma) is the most effective way to rest the symptomatic area. This allows employees to remain productive members of the workforce and is best accomplished by engineering and work practice controls in the workplace (see Prevention section).

Until effective controls are installed, employee exposure to biomechanical stressors can be reduced through restricted duty and/or temporary job transfer. The principle of restricted duty and temporary job transfer is to reduce or eliminate the total amount of time spent exposed to the same or similar musculoskeletal risk factors.[109,110] A variety of factors (e.g., symptom type, duration, and severity, response to treatment, and biomechanical stressors associated with work) must be considered when determining the length of time for which an employee is assigned to restricted duty. When trying to determine the length of time assigned to restricted work, the following principle applies: the degree of restriction should be proportional to symptom severity and intensity of the job's biomechanical stressors. In addition, caution must be used in deciding which jobs are suitable for job transfer, because different jobs may pose similar biomechanical demands on the same muscles and tendons.

Complete removal from the work environment should be reserved for severe disorders, workplaces where the only available jobs involve significant biomechanical stressors for the symptomatic area, or workplaces where significant modifications to the current or available jobs are not feasible. For purposes of removal from the work environment, severe disorders can be defined as those that negatively affect that employee's activities of daily living (e.g., difficulty in buttoning clothes, opening jars, and brushing hair). In addition, employees with MSDs should be advised about the potential risk posed by hobbies, recreational activities, and other personal habits that involve certain biomechanical stressors. Employees should modify their behaviors to reduce such stress.

Treatment

Clinical interventions should be tailored to the specific diagnosis. In general, involving a physical or occupational therapist has been shown to improve outcomes.[111–115] However, the effectiveness of various treatment modalities (traditional, complementary, and alternative) used by physiotherapists is less clear.[116–120] Table 2.6 lists a number of these treatment modalities. According to Crawford et al., there is evidence of the efficacy of conservative treatments for CTS, epicondylitis, rotator cuff tendonitis, bicipital tendonitis, and tension neck syndrome, but the evidence is less clear for tenosynovitis, tendonitis, de Quervain's disease, or diffuse nonspecific UE MSDs.[115]

Most (approximately 80%) UE MSDs improve with conservative measures, but about 20% of patients continue to have symptoms for a year or more.[121] Symptomatic employees should be followed until their symptoms improve. Employees who do not improve within the expected time frames should be reevaluated, or a second opinion should be obtained.[122] Surgical options should be reserved for severe, chronic cases that, after an adequate trial of conservative therapy (described above), prevent return to work or show objective signs of disease progression.[123] The length of time needed for an "adequate" trial of conservative therapy depends on many variables, including

TABLE 2.6 Interventions used by physiotherapists to treat upper extremity musculoskeletal disorders.

Stretching and strengthening exercises (passive and active)
Nonsteroidal anti-inflammatory drugs (oral or topical)
Orthotic devices (e.g., brace and splint)
Therapeutic ultrasound (thermal and mechanical)
Heat and cold therapy
General exercise
Massage
Various injections
Glucocorticoid (e.g., steroids)
Glycosaminoglycan polysulfate
Sodium hyaluronate
Autologous blood
Botulinum toxin
Platelet-rich plasma
Prolotherapy (proliferative injection therapy)
Acupuncture
Iontophoresis
Low-level lasers
Pulsed electromagnetic field
Extracorporeal shock wave therapy

an employee's ability to remain productive without jeopardizing his or her long-term health.

SURVEILLANCE

Surveillance is "the ongoing systematic collection, analysis, and interpretation of health and exposure data in the process of describing and monitoring a health event."[124] The goal of both hazard and health surveillance systems is the identification of hazardous exposures. Once hazardous exposures are recognized, intervention efforts can be targeted at those exposures with the purpose of preventing future health problems in (as yet) unaffected individuals.

Surveillance should not be confused with screening. Screening is the application of a clinical test to asymptomatic individuals at increased risk for a particular disease.[125] The goal of a screening test is the identification of individuals who need further medical evaluation or other interventions. Although clinical tests can identify individuals with MSDs early in their development, there are currently no known screening tests to predict which asymptomatic individuals will develop symptoms and disease. Exposure and health surveillance can be divided into two types: passive and active.

Passive surveillance

Passive health surveillance systems utilize existing databases to identify high-risk industries, occupations, or tasks that are associated with disease or injury. Examples of these databases in occupational medicine include the OSHA 300 Logs, workers' compensation records, medical department logs, clinical laboratory data, hospital discharge records, and accident reports. Jobs or tasks with an increased UE MSD injury rate can be targeted for further ergonomic and/or medical evaluation. In addition, jobs or departments can be followed over time, to monitor trends in these rates.

Although attractive because of their low cost, passive surveillance databases are sometimes developed for purposes other than surveillance and may have significant limitations.[126] These limitations include underreporting and exposure misclassification. For example, underreporting in the OSHA 300 Logs can occur for any of the following reasons: symptomatic employees not seeking medical care ("macho" attitude, ignorance that the condition could be work related, or fear of employer retaliation), restricted or no access to employee health facilities, or misunderstanding about when a case is to be recorded on the OSHA 300 Log.[7] Exposure misclassification can occur when employees use a general term to describe their job title; for example, an employee in the poultry industry may report his or her job title as "cutter" in a plant with five distinct cutting positions. Each one of these cutting jobs may be associated with a different ergonomic hazard, and the identification of high-risk jobs requires specific knowledge of the employees' cutting position.

Active surveillance

Active surveillance systems generate more accurate databases to identify high-risk positions. Direct symptom surveys are good examples of active disease surveillance tools in occupational medicine[127] and have been developed for musculoskeletal disorders.[128] Symptom surveys collect more accurate information and can serve a triage function if performed in a confidential manner. Active surveillance systems can also collect information on exposures. This information is typically established by plant personnel or nonmedical consultants; however, healthcare providers may be called upon to participate in a comprehensive ergonomic program. If exposure surveillance is a component of that program, the survey instrument in Tables 2.4 and Figures 2.4 and 2.5 could be used to establish an exposure surveillance database.

PREVENTION

The control of identified ergonomic hazards is the most effective means of preventing work-related UE MSDs and is the primary focus of any ergonomic program. Intervention strategies should follow a three-tiered approach: engineering controls, administrative controls, and medical treatment.[16,129]

Engineering controls

Studies have shown that engineering interventions can reduce ergonomic hazards and can lead to a reduction in MSDs.[16,130,131] In some situations, the interventions are obvious and represent common sense solutions. On the other hand, worksites frequently require a more comprehensive approach to the control of ergonomic hazards. This comprehensive approach should address the following risk factors: repetition, force, posture, and vibration.

To reduce repetitiveness, the following interventions could be used: (i) enlarged work content, (ii) automation of some job tasks, (iii) uniform spreading of work across the work shift, and (iv) job restructuring. To reduce force or mechanical stressors, the following interventions should be considered: (i) decreasing the weight of tools, containers, and parts; (ii) optimizing the size, shape, and friction of handles; and (iii) using torque control devices. Reduction of awkward or extreme postures could be achieved by (i) locating the work more appropriately and (ii) selecting tool design and location based upon workstation characteristics. Engineering controls for the reduction of vibration are reviewed in Chapter 4.

In some instances, however, engineering controls are not currently available. Until engineering controls become available, other aspects of an ergonomic program—administrative and medical treatment controls—can be implemented.

Administrative controls

Administrative controls can be defined as work practices or training used to reduce employee exposure to ergonomic stressors. Examples of work practice controls include (i) more frequent and longer rest breaks,[132] (ii) limiting overtime, (iii) varying work tasks or broadening job responsibilities, and (iv) periodic rotation of workers between stressful and less stressful jobs. Since job rotation exposes more workers to the more stressful job, it is suitable only where short-term performance of the stressful job poses no appreciable ergonomic hazard. Otherwise, other control methods must be utilized. Training programs range from fundamental instruction on the proper use of tools and materials to instruction on the use of protective devices. Improper work technique has been associated with the development of UE MSDs.[133,134] Finally, efforts to improve the workplace climate and culture with programs like participatory ergonomics are also considered a type of administrative control.[135,136]

Medical treatment

The goal of medical treatment as a component of an ergonomic program is to provide prompt evaluation and treatment to limit the severity, disability, and costs associated with these disorders. It should also serve to initiate a reevaluation of the ergonomic stresses associated with the affected worker's job and institution of appropriate control measures. Medical treatment in this sense is a secondary and tertiary prevention mechanism. Medical management programs are an important component of any successful ergonomic program.[137–139] However, they should always be used with engineering and administrative controls during the implementation of a complete ergonomic program.[129]

References

1. Chaffin DB, Andersson GBJ. *Occupational biomechanics*. New York: John Wiley & Sons, Inc., 1984.
2. Boocock MG, Collier JMK, McNair PJ, et al. A framework for the classification and diagnosis of work-related upper extremity conditions: systematic review. *Semin Arthritis Rheum* 2009; 38(4):296–311.
3. Eastman Kodak Company. *Ergonomic design for people at work*, Vol. 1. New York: Van Nostrand Reinhold Company, 1983:3.
4. BLS 2012. Occupational Safety and Health Definitions. U.S. Department of Labor, Bureau of Labor Statistics. Available at: http://www.bls.gov/iif/oshdef.htm. Accessed on October 18, 2015.
5. BLS 2015. Nonfatal Occupational Injuries and Illnesses Requiring Days Away from Work, 2014. U.S. Department of Labor, Bureau of Labor Statistics, BSDL 15-2205. Available at: http://www.bls.gov/news.release/archives/osh2_11192015.pdf. Accessed on December 22, 2015.
6. NIOSH 2009. Worker Health eChartbook. U.S. Department of Health and Human Services, Centers for Disease Control and Prevention, National Institute for Occupational Safety and Health. Available at: http://wwwn.cdc.gov/niosh-survapps/echartbook/Category.aspx?id=563. Accessed on October 20, 2015.
7. Rosenman KD, Kalush A, Reilly MJ, et al. How much work-related injury and illness is missed by the current national surveillance system? *J Occup Environ Med* 2006; 48(4):357–65.
8. Leigh JP, Marcin JP, Miller TR. An estimate of the U.S. Government's undercount of nonfatal occupational injuries. *J Occup Environ Med* 2004; 46(1):10–8.
9. Webster BS, Snook SH. The cost of compensable upper extremity cumulative trauma disorders. *J Occup Med* 1994; 36:713–27.
10. Silverstein B, Welp E, Nelson N, et al. Claims incidence of work-related disorders of the upper extremities: Washington State, 1987 through 1995. *Am J Public Health* 1998; 88:1827–33.
11. Silverstein B, Viikari-Juntura E, Kalat J. Use of a prevention index to identify industries at high risk for work-related musculoskeletal disorders of the neck, back, and upper extremity in Washington State, 1990–1998. *Am J Ind Med* 2002; 41:149–69.
12. Hagberg M, Silverstein B, Wells R, et al. Introduction. In: Kuorinka I, Forcier L, eds. *Work related musculoskeletal disorders (WMSDs): a reference book for prevention*. London: Taylor & Francis, 1995:1.
13. Pansky G, Synder T, Dembe A, et al. Under-reporting of work-related disorders in the workplace: a case study and review of the literature. *Ergonomics* 1999; 42:171–82.
14. Biddle J, Roberts K, Rosenman KD, et al. What percentage of workers with work-related illnesses receive workers' compensation benefits? *J Occup Environ Med* 1998; 40:325–31.

15. Rosenman KD, Gardiner JC, Wang J, et al. Why most workers with occupational repetitive trauma do not file workers' compensation. *J Occup Environ Med* 2000; 42:25–34.
16. National Research Council, Panel on Musculoskeletal Disorders and the Workplace. Commission on Behavioral and Social Sciences and Education, and Institute of Medicine. *Musculoskeletal disorders and the workplace: low back and upper extremities*. Washington, DC: National Academy Press, 2001. Available at: http://www.nap.edu/read/10032/chapter/1. Accessed on June 15, 2016.
17. Fan ZJ, Bonauto DK, Foley MP, et al. Underreporting of work-related injury or illness to workers' compensation: individual and industry factors. *J Occup Environ Med* 2006; 48(9):914–22.
18. Tanaka S, Wild DK, Seligman P, et al. Prevalence and work-relatedness of self-reported carpal tunnel syndrome among US workers: analysis of the Occupational Health Supplement Data of the 1988 National Health Interview Survey. *Am J Ind Med* 1995; 27:451–70.
19. Hanrahan LP, Moll MB. Injury surveillance. *Am J Public Health* 1989; 9(Suppl):38–45.
20. Ramsey J, Musolin K, Mueller C, and NIOSH 2015. Health Hazard Evaluation Report: Evaluation of Carpal Tunnel Syndrome and Other Musculoskeletal Disorders among Employees at a Poultry Processing Plant. NIOSH HHE Report No. 2014-0040-3232. Cincinnati, OH: U.S. Department of Health and Human Services, Centers for Disease Control and Prevention, National Institute for Occupational Safety and Health. Available at: https://www.cdc.gov/niosh/hhe/reports/pdfs/2014-0040-3232.pdf. Accessed on June 15, 2016.
21. Hales T, Habes D, Fine L, et al. 1989. Health Hazard Evaluation Report: John Morrell & Co. Sioux Falls, South Dakota, NIOSH HHE Report No. 88-180-1958. Cincinnati, OH: U.S. Department of Health and Human Services, Centers for Disease Control and Prevention, National Institute for Occupational Safety and Health.
22. Kim JY, Kim JI, Son JE, et al. Prevalence of carpal tunnel syndrome in meat and fish processing plants. *J Occup Health* 2004; 46(3):230–4.
23. Lipscomb H, Kucera K, Epling C, et al. Upper extremity musculoskeletal symptoms and disorders among a cohort of women employed in poultry processing. *Am J Ind Med* 2008; 51(1):24–36.
24. Bernard B, ed. *Musculoskeletal disorders and workplace factors: a critical review of epidemiologic evidence for work-related musculoskeletal disorders of the neck, upper extremity, and low back*. DHHS (NIOSH) Publication No. 97-141. Cincinnati, OH: National Institute for Occupational Safety and Health, 1997.
25. Fan ZJ, Harris-Adamson C, Gerr F, et al. Associations between workplace factors and carpal tunnel syndrome: a multi-site cross sectional study. *Am J Ind Med* 2015; 58(5):509–18.
26. Rempel D, Gerr F, Harris-Adamson C, et al. Personal and workplace factors and median nerve function in a pooled study of 2396 US workers. *J Occup Environ Med* 2015; 57(1):98–104.
27. Bonfiglioli R, Mattioli S, Armstrong T, et al. Validation of the ACGIH TLV for hand activity level in the OCTOPUS cohort: a two-year longitudinal study of carpal tunnel syndrome. *Scand J Work Environ Health* 2013; 39(2):155–63.
28. Radwin RG, Lavender SA. Work factors, personal factors, and internal loads: biomechanics of work stressors. In: National Resource Council, eds. *Work-related musculoskeletal disorders*. Washington, DC: National Academy Press, 1999: 116–51.
29. Lowe BD, Weir PL, Andrews DM, NIOSH 2014. Observation-Based Posture Assessment: Review of Current Practice and Recommendations for Improvement, DHHS (NIOSH) Publication No. 2014-131. Cincinnati, OH: U.S. Department of Health and Human Services, Centers for Disease Control and Prevention, National Institute for Occupational Safety and Health.
30. Wiktorin C, Karlqvist PT, Winkel J, et al. Validity of self-reported exposures to work postures and manual materials handling. *Scand J Work Environ Health* 1993; 19:208–14.
31. Viikari-Juntura E, Rauas S, Martikainen R, et al. Validity of self-report physical work load in epidemiologic studies on musculoskeletal disorders. *Scand J Work Environ Health* 1996; 22:251–9.
32. Buchholz B, Park J-S, Gold JE, et al. Subjective ratings of upper extremity exposures: inter-method agreement with direct measurement of exposures. *Ergonomics* 2008; 51(7):1064–77.
33. Borg GAV. *An introduction to Borg's RPE-Scale*. Ithaca, NY: Movement Publications, 1985.
34. Borg GAV. Psychological bases of perceived exertion. *Med Sci Sports Exerc* 1982; 4:377–81.
35. Borg GAV. *Borg's perceived exertion and pain scale*. Champaign, IL: Human Kinetics, 1998.
36. Spielholz P, Silverstein B, Morgan M, et al. Comparison of self-report, video observation and direct measurement methods for upper extremity musculoskeletal disorder physical risk factors. *Ergonomics* 2001; 44(6):588–613.
37. Keyserling WM, Stetson DS, Silverstein BA, et al. A checklist for evaluating ergonomic risk factors associated with upper extremity cumulative trauma disorders. *Ergonomics* 1993; 36:807–31.
38. Lifshitz Y, Armstrong TJ. A design checklist for control and prediction of cumulative trauma disorders in hand intensive manual jobs. In: Proceedings of the 30th Annual Meeting of Human Factors Society, September 29 October 3, Dayton, OH: Human Factors and Ergonomics Society. 1986, pp. 837–841.
39. Kemmert K. A method assigned for the identification of ergonomic hazards: PLIBEL. *Appl Ergon* 1995; 26:199–201.
40. International Labour Office. *Ergonomic checkpoints: practical and easy-to-implement solutions for improving safety, heath and working conditions*. Geneva: International Labour Office, 1996.
41. Washington Labor and Industries. 2014. Ergonomics Checklist. Available at: http://www.lni.wa.gov/Safety/SprainsStrains/AwkwardPostures/ErgonomicsChecklist.pdf. Accessed on June 15, 2016.
42. Latko WA, Armstrong TJ, Foulke JA, et al. Development and evaluation of an observational method for assessing repetition in hand tasks. *Am Ind Hyg Assoc J* 1997; 58:278–85.
43. Rodgers S. Job evaluation in worker fitness determination. In: Himmelstein JS, Pransky GS, eds. *Occupational medicine, state of the art reviews: worker fitness and risk evaluations*. Philadelphia, PA: Handley & Belfus, Inc., 1988:219–40.

44. Bao S, Howard N, Spielholz P, et al. Two posture analysis approaches and their application in a modified rapid upper limb assessment evaluation. *Ergonomics* 2007; 50(12):2118–36.
45. Moore JS, Garg A. The strain index: a proposed method to analyze jobs for risk of distal upper extremity disorders. *Am Ind Hyg Assoc J* 1995; 56:443–58.
46. McAtamney L, Corlett EN. RULA: a survey method for the investigation of work-related upper limb disorders. *Appl Ergon* 1993; 24:91–9.
47. Louhevaara V, Suurnakki T. *OWAS: a method for the evaluation of postural load during work.* Training Publication No. 11. Helsinki: Institute of Occupational Health, 1992.
48. American Conference of Governmental Industrial Hygienists (ACGIH). *2015 TLVs® and BEIs®: threshold limit values for chemical substances and physical agents and biological exposure indices.* Cincinnati, OH: ACGIH Signature Publications, 2015: 179–81.
49. Kapellusch JM, Garg A, Hegmann KT, et al. The Strain Index and ACGIH TLV for HAL: risk of trigger digit in the WISTAH prospective cohort. *Hum Factors* 2014; 56(1):98–111.
50. American Academy of Orthopedic Surgeons. *Joint motion: method of measuring and recording.* Edinburgh: Churchill Livingstone, 1965.
51. Eastman Kodak Company. *Ergonomic design for people at work,* Vol. 2. New York: Van Nostrand Reinhold Company, 1983: 255.
52. Occupational Safety and Health Administration. Ergonomics. Available at: https://www.osha.gov/SLTC/ergonomics/. Accessed on June 15, 2016.
53. California Department of Industrial Relations. General Industry Safety Orders: Ergonomics. Available at: http://www.dir.ca.gov/title8/5110.html. Accessed on June 15, 2016.
54. Armstrong RB. Mechanisms of exercise-induced delayed onset muscular soreness: a brief review. *Med Sci Sports Exerc* 1984; 16:529–36.
55. Armstrong RB. Initial events in exercise-induced muscular injury. *Med Sci Sports Exerc* 1990; 22:429–35.
56. Armstrong RB, Warren GL, Warren JA. Mechanisms of exercise-induced muscle fiber injury. *J Sports Med* 1991; 12:184–207.
57. Ashton-Miller JA. Soft tissue responses to physical stressors: muscles, tendons, and ligaments. In: National Resource Council, eds. *Work-related musculoskeletal disorders.* Washington, DC: National Academy Press, 1999:39–41.
58. Faulkner JA, Brooks SV. Muscle fatigue in old animals: unique aspects of fatigue in elderly humans. In: Gandevai S, Enoka R, McComas A, et al., eds. *Fatigue, neural and muscular mechanisms.* New York: Plenum Press, 1995:471–80.
59. Thelen DG, Ashton-Miller JA, Schultz AB, et al. Do neural factors underlie age differences in rapid ankle torque development? *J Am Geriatr Soc* 1996; 44:804–8.
60. Brooks SV, Faulkner JA. The magnitude of the initial injury induced by stretches of maximally activated muscle fibers of mice and rats increases in old age. *J Physiol* 1996; 497:573–80.
61. Brooks SV, Faulkner JA. Contraction-induced injury: recovery of skeletal muscles in young and old mice. *Am J Physiol* 1990; 258:C436–42.
62. Ashton-Miller JA. Response of muscle and tendon to injury and overuse. In: National Resource Council, eds. *Work-related musculoskeletal disorders.* Washington, DC: National Academy Press, 1999:73–97.
63. Gelberman R, Goldberg V, An K-N, et al. Tendon. In: Woo SL-Y, Buckwater JA, eds. *Injury and repair of the musculoskeletal soft tissues.* Park Ridge, IL: American Academy of Orthopedic Surgeons, 1988:5–40.
64. Schumacher HR Jr. Morphology and physiology of normal synovium and the effects of mechanical stimulation. In: Gordon SL, Blair SJ, Fine LJ, eds. *Repetitive motion disorders of the upper extremity.* Rosemont, IL: American Academy of Orthopedic Surgeons, 1995:263–76.
65. Vogel KG. Fibrocartilage in tendon: a response to compressive load. In: Gordon SL, Blair SJ, Fine LJ, eds. *Repetitive motion disorders of the upper extremity.* Rosemont, IL: American Academy of Orthopedic Surgeons, 1995:205–15.
66. Terzis JK, Smith KL, eds. *The peripheral nerve: structure, function, and reconstruction.* New York: Raven Press, 1990.
67. Wolfe F, Smythe HA, Yunus MB et al. The American College of Rheumatology 1990: criteria for the classification of fibromyalgia. *Arthritis Rheum* 1990; 33:160–72.
68. Grosshandler S, Burney R. The myofascial syndrome. *N C Med J* 1979; 40:562–5.
69. Johansson H, Sojka P. Pathophysiological mechanisms involved in genesis and spread of muscle tension in occupational muscle pain and in chronic musculoskeletal pain syndromes: a hypothesis. *Med Hypothesis* 1991; 35:196–203.
70. Lindman R, Eriksson A, Thornell LE. Fiber type composition of the human female trapezius muscle. *Am J Anat* 1991; 190:385–92.
71. Lindman R, Hagberg M, Angquist KA, et al. Changes in the muscle morphology in chronic trapezius myalgia. *Scand J Work Environ Health* 1991; 17:347–55.
72. Larsson SE, Bengtsson A, Bodegard L, et al. Muscle changes in work related chronic myalgia. *Acta Orthop Scand* 1988; 59:552–6.
73. Hagberg M, Angquist KA, Eriksson HE, et al. EMG-relationship in patients with occupational shoulder–neck myofascial pain. In: deGroot G, Hollander AP, Huijing PA, et al., eds. *Biomechanics XI-A.* Amsterdam: Free University Press, 1988:450–4.
74. Murthy G, Kahan NH, Hargens AR, et al. Forearm muscle oxygenation decreases with low levels of voluntary contraction. *J Orthop Res* 1997; 15:507–11.
75. Gandevai S, Enoka R, McComas AJ, et al., eds. *Fatigue, neural and muscular mechanisms.* New York: Plenum Press, 1995.
76. Clancy WG Jr. Tendon trauma and overuse injuries. In: Leadbetter W, Buckwater JA, Gordon SL, eds. *Sports-induced inflammation: clinical and basic science concepts.* Park Ridge, IL: American Academy of Orthopedic Surgeons, 1990:609–18.
77. Hart DA, Frank CB, Bray RC. Inflammatory processes in repetitive motion and overuse syndromes: potential role of neurogenic mechanisms in tendon and ligaments. In: Gordon SL, Blair SJ, Fine LJ, eds. *Repetitive motion disorders of the upper extremity.* Rosemont, IL: American Academy of Orthopedic Surgeons, 1995:249.

78. Fu FH, Harner CD, Klein AH. Shoulder impingement syndrome: a critical review. *Clin Orthop* 1991; 269:162–73.
79. Jarvholm U, Palmerud G, Styf J, et al. Intramuscular pressure in the supraspinatus muscle. *J Orthop Res* 1988; 6:230–8.
80. Hagberg M, Silverstein B, Wells R, et al. Evidence of the association between work and selected tendon disorders: shoulder tendinitis, epicondylitis, de Quervain's tendinitis, Dupuytren's contracture, Achilles tendinitis. In: Kuorinka I, Forcier L, eds. *Work related musculoskeletal disorders (WMSDs): a reference book for prevention.* London: Taylor & Francis, 1995:55–6.
81. Nirschl RP, Ashman ES. Elbow tendinopathy: tennis elbow. *Clin Sports Med* 2003; 22:813–36.
82. Moore JS. De Quervain's tenosynovitis. Stenosing tenosynovitis of the first dorsal compartment. *J Occup Environ Med* 1997; 39:990–1002.
83. Moore JS. Flexor tendon entrapment of the digits (trigger finger and trigger thumb). *J Occup Environ Med* 2000; 42:526–45.
84. Leadbetter WB. Cell-matrix response in tendon injury. *Clin Sports Med* 1992; 11:533–78.
85. Rempel D. Musculoskeletal loading and carpal tunnel syndrome. In: Gordon SL, Blair SJ, Fine LJ, eds. *Repetitive motion disorders of the upper extremity.* Rosemont, IL: American Academy of Orthopedic Surgeons, 1995:123–32.
86. Rempel D, Dahlin L, Lundborg G. Biological response of peripheral nerves to loading: pathophysiology of nerve compression syndromes and vibration induced neuropathy. In: National Resource Council, eds. *Work-related musculoskeletal disorders.* Washington, DC: National Academy Press, 1999:98–115.
87. Gelbermann RH, Rydevik BL, Pess GM, et al. Carpal tunnel syndrome: a scientific basis for clinical care. *Orthop Clin North Am* 1988; 19:115–24.
88. Fuchs PC, Nathan PA, Meyers LD. Synovial histology in carpal tunnel syndrome. *J Hand Surg* 1991; 16A:753–8.
89. Gelbermann RH, Szabo RM, Williamson RV, et al. Tissue pressure threshold for peripheral nerve viability. *Clin Orthop* 1983; 178:285–91.
90. Weiss N, Gordon L, Bloom T, et al. Wrist position of lowest carpal tunnel pressure: implication for splint design. *J Bone Joint Surg* 1995; 77:1695–9.
91. Rempel D, Smutz WP, So Y, et al. Effect of fingertip loading on carpal tunnel pressure. *Trans Orthop Res Soc* 1994; 19:698.
92. Rempel D, Manojlovic R, Levinsohn D, et al. The effect of wearing a flexible wrist splint on carpal tunnel pressure during repetitive hand activity. *J Hand Surg* 1994; 19A:106–10.
93. Keyserling WM. Workplace risk factors and occupational musculoskeletal disorders. Part 2: A review of biomechanical and psychophysical research on risk factors associated with upper extremity disorders. *Am Ind Hyg Assoc J* 2000; 61:31–43.
94. Smith MJ, Carayon P. Work organization, stress, and cumulative trauma disorders. In: Moon SL, Sauter SD, eds. *Beyond biomechanics: psychosocial aspects of musculoskeletal disorders in office work.* New York: Taylor & Francis, 1996.
95. Melin B, Lundberg U. A biopsychosocial approach to work-stress and musculoskeletal disorders. *J Psychophysiol* 1997; 11:238–47.
96. Sauter SL, Swanson NG. An ecological model of musculoskeletal disorders in office work. In: Moon SL, Sauter SD, eds. *Beyond biomechanics: psychosocial aspects of musculoskeletal disorders in office work.* New York: Taylor & Francis, 1996.
97. Feuerstein M. Workstyle: definition, empirical support, and implications for prevention, evaluation and rehabilitation of occupational upper-extremity disorders. In: Moon SL, Sauter SD, eds. *Beyond biomechanics: psychosocial aspects of musculoskeletal disorders in office work.* New York: Taylor & Francis, 1996.
98. Hauke A, Flintrop J, Brun E, et al. The impact of work-related psychosocial stressors on the onset of musculoskeletal disorders in specific body regions: a review and meta-analysis of 54 longitudinal studies. *Work Stress* 2011; 25(3):243–56.
99. Bongers PM, Ijmker S, van den Heuvel S, et al. Epidemiology of work related neck and upper limb problems: psychosocial and personal risk factors (Part I) and effective interventions from a bio behavioural perspective (Part II). *J Occup Rehabil* 2006; 16:272–95.
100. Visser B, van Dieën JH. Pathophysiology of upper extremity muscle disorders. *J of Electromyogr Kinesiol,* 2006; 16(1):1–16.
101. Theorell T, Hasselhorn HM. Endocrinological and immunological variables sensitive to psychosocial factors of possible relevance to work-related musculoskeletal disorders. *Work Stress* 2002; 16:154–65.
102. Chrousos GP, Gold PW. The concepts of stress and stress system disorders: overview of physical and behavioral homeostasis. *J Am Med Assoc* 1992; 267(9):1244–52.
103. Bailey TS, Dollard MF, McLinton SS, et al. Psychosocial safety climate, psychosocial and physical factors in the aetiology of musculoskeletal disorder symptoms and workplace injury compensation claims. *Work Stress* 2015; 29(2):190–211.
104. Moshe S, Izhaki R, Chodick G, et al. Predictors of return to work with upper limb disorders. *Occup Med* 2015; 65:564–9.
105. Goldman RH, Peters JM. The occupational and environmental health history. *J Am Med Assoc* 1981; 246:2831–6.
106. Piligian G, Herbert R, Hearns M, et al. Evaluation and management of chronic work-related musculoskeletal disorders of the distal upper extremity. *Am J Ind Med* 2000; 37:75–93.
107. Moore JS. Clinical determination of work-relatedness in carpal tunnel syndrome. *J Occup Rehabil* 1991; 1:145–58.
108. Kusnetz S, Hutchison MK, eds. *A guide to the work-relatedness of disease.* DHHS (NIOSH) Publication No. 79-116. Cincinnati, OH: Department of Health and Human Services, National Institute for Occupational Safety and Health, 1979.
109. McKenzie F, Storment J, Van Hook P, et al. A program for control of repetitive trauma disorders associated with hand tool operations in a telecommunications manufacturing facility. *Am Ind Hyg Assoc J* 1985; 46(11):674–8.
110. Lederman RJ, Calabrese LH. Overuse syndromes in instrumentalists. *Med Probl Perform Art* 1986; 1:7–11.
111. Kuhn JE, Dunn WR, Sanders R, et al. Effectiveness of physical therapy in treating atraumatic full-thickness rotator cuff tears: a multicenter prospective cohort study. *J Shoulder Elbow Surg* 2013; 22(10):1371–9.
112. Kohia M, Brackle J, Byrd K, et al. Effectiveness of physical therapy treatments on lateral epicondylitis. *J Sport Rehabil* 2008; 17:119–36.
113. Piper S, Shearer HM, Côté P, et al. The effectiveness of soft-tissue therapy for the management of musculoskeletal disorders and injuries of the upper and lower extremities: a systematic

review by the Ontario Protocol for Traffic Injury management (OPTIMa) collaboration. *Man Ther* 2016; 21:18–34.
114. Raman J, MacDermid JC, Grewal R. Effectiveness of different methods of resistance exercises in lateral epicondylosis: a systematic review. *J Hand Ther* 2012; 25:5–25.
115. Crawford JO, Laiou E. Conservative treatment of work-related upper limb disorders: a review. *Occup Med* 2007; 57(1):4–17.
116. Verhagen AP, Bierma-Zeinstra SM, Burdorf A, et al. Conservative interventions for treating work-related complaints of the arm, neck or shoulder in adults. *Cochrane Database Syst Rev* 2013;12:CD008742.
117. Long L, Briscoe S, Cooper C, et al. What is the clinical effectiveness and cost-effectiveness of conservative interventions for tendinopathy? An overview of systematic reviews of clinical effectiveness and systematic review of economic evaluations. *Health Technol Assess* 2015; 19(8):1–134.
118. National Institute for Health and Care Excellence (NICE). *Conservative interventions for elbow tendinopathy (HTA No 12/73)*. London: NICE, 2012.
119. Page MJ, O'Connor D, Pitt V, et al. Exercise and mobilisation interventions for carpal tunnel syndrome. *Cochrane Database Syst Rev* 2012; 13(6):CD009899.
120. Page MJ, O'Connor D, Pitt V, et al. Therapeutic ultrasound for carpal tunnel syndrome. *Cochrane Database Syst Rev* 2013; 28(3):CD009601.
121. Bisset L, Coombes B, Vicenzino B. Tennis elbow. *BMJ Clin Evid* 2011; 2011:1117.
122. Hales TR, Bertsche PA. Management of upper extremity cumulative trauma disorders. *Am Assoc Occup Health Nurses J* 1992; 40:118–28.
123. Verdugo RJ, Salinas RA, Castillo JL, et al. Surgical versus non-surgical treatment for carpal tunnel syndrome. *Cochrane Database Syst Rev* 2008; 8(4):CD001552.
124. Klauke DN, Buehler JW, Thacker SB, et al. Guidelines for evaluation of surveillance systems. *MMWR* 1988; 17(Suppl 5):1–18.
125. Halperin WE, Ratcliffe J, Frazier TM, et al. Medical screening in the workplace: proposed principles. *J Occup Med* 1986; 28:547–52.
126. Baker EL, Melius JM, Millar JD. Surveillance of occupational illness and injury in the United States: current perspectives and future directions. *J Public Health Policy* 1988; 9:188–221.
127. Ehrenberg RL. Use of direct survey in the surveillance of occupational illness and injury. *Am J Public Health* 1989; 79(Suppl):11–4.
128. Kuorinka I, Jonsson B, Kilbom A, et al. Standardized Nordic questionnaires for the analysis of musculoskeletal symptoms. *Appl Ergon* 1987; 18:233–7.
129. National Institute for Occupational Safety and Health (NIOSH). *Elements of ergonomics programs: a primer based on workplace evaluations of musculoskeletal disorders.* DHHS (NIOSH) Publication No. 97-117. Cincinnati, OH: NIOSH, 1997.
130. Burton AK, Kendall NAS, Pearce BG, et al. Management of work-relevant upper limb disorders: a review. *Occup Med* 2009; 59(1):44–52.
131. Van Eerd D, Munhall C, Irvin E, et al. Effectiveness of workplace interventions in the prevention of upper extremity musculoskeletal disorders and symptoms: an update of the evidence. *Occup Environ Med* 2016; 73:62–70. doi:10.1136/oemed-2015-102992
132. Galinsky TL, Swanson NG, Sauter SL, et al. A field study of supplementary rest breaks for data-entry operators. *Ergonomics* 2000; 43:622–38.
133. Feuerstein M, Fitzgerald TE. Biomechanical factors affecting upper extremity cumulative trauma disorders in sign language interpreters. *J Occup Med* 1992; 34:257–64.
134. Kilbom A, Persson J. Work technique and its consequences for task. *Ergonomics* 1987; 30:273–9.
135. Cole DC, Theberge N, Dixon SM, et al. Reflecting on a program of participatory ergonomics interventions: a multiple case study. *Work* 2009; 34(2):161–78.
136. Hoffmeister K, Gibbons A, Schwatka N, et al. Ergonomics climate assessment: a measure of operational performance and employee well-being. *Appl Ergon* 2015; 50:160–9.
137. Mehorn MJ, Wilkinson L, Gardner P, et al. An outcomes study of an occupational medicine intervention program for the reduction of musculoskeletal disorders and cumulative trauma disorders in the workplace. *J Occup Environ Med* 1999; 41:833–46.
138. Government Accounting Office 1997. Report to Congressional Requestors: Worker Protection, Private Sector, Ergonomics Programs Yield Positive Results, Report No. GAO/HEHS-97-163. Washington, DC: United States Government Accounting Office/Health, Education, and Human Services Division.
139. Lutz G, Hansford T. Cumulative trauma disorder controls: the ergonomics program at Ethicon; Inc. Part 2. *J Hand Surg* 1987; 12A(5):863–6.

Chapter 3

MANUAL MATERIALS HANDLING

Robert B. Dick, Stephen D. Hudock, Ming-Lun Lu, Thomas R. Waters†, and Vern Putz-Anderson*

Manual handling is defined by the International Organization for Standardization as any activity requiring the use of human force to lift, lower, carry or otherwise move, or restrain an object, including humans and animals.[1] Manual materials handling (MMH) excludes animate items as objects in such activities. The main risk factors associated with the development of injuries with MMH tasks are forceful exertions, awkward postures, repetitive motions, pressure points, and static postures.[2] Many different types of jobs or occupations require MMH activities including moving household goods, warehouse handling, truck unloading, inventory restocking, production line loading and unloading, baggage handling, container loading and emptying, transferring material goods, and delivering packages.

Nongovernmental researchers annually examine US Bureau of Labor Statistics (BLS) injury data to ascertain which workplace events caused an employee to miss six or more days of work. Those workplace events are then ranked by total workers' compensation costs. For the year 2012, the most recent information available, the leading cause of disabling injury was overexertion involving outside sources. This event category includes injuries resulting from lifting, pushing, pulling, holding, carrying, or throwing, all primarily MMH activities. These events cost US businesses $15.1 billion in direct costs and accounted for over 25% of the overall national disabling injury burden.[3]

In 2011 approximately 3 million workers in private industry and 821 000 in state and local government experienced a nonfatal occupational injury or illness with an estimated cost to the US economy of $200 billion annually.[4] High-risk occupations, which are defined as one where the days-away-from-work (DAFW) rate is at least twice the DAFW rate of 113.3 cases of injury and illnesses per 10000 full-time employees (FTEs), included two occupations with more than a million workers where MMH activities are involved. These were drivers (sales and trucks) and hand laborers (freight, stock, and material movers).[4] The estimated number of drivers was 2 721 000 and the DAFW rate was 329.4/10 000 FTE, and for hand laborers the number employed was 1 616 000 and the DAFW rate was 440.3/10 000 FTE.[4]

Musculoskeletal disorders (MSDs), as defined by the BLS, include soft tissue injuries to the trunk and upper and lower extremities and occurred at a rate of 35.8 days-away-from-work cases per 10 000 FTE in 2013 (including private and state and local government).[5] The MSD rate for transportation and warehousing was 80.3 cases per 10 000 full-time workers which was more than twice the MSD rate for all private industry sectors. MSDs accounted for 33% of all reported illness and injury cases in 2013. Nursing assistants and laborers and freight, stock, and material movers showed the greatest number of MSD cases in 2013.

This chapter provides an overview of the hazards associated with MMH. It will focus on muscular strains and sprains, primarily in the torso or extremities. The prevention of these injuries requires sufficient knowledge to both identify workplace hazards and implement changes in the job or process that will reduce or eliminate these hazards.

OCCUPATIONAL SETTING

MMH poses a risk of injury to many workers; injury is more likely to occur when workers perform tasks that exceed their physical capacities. In addition, the physical capacities

*The findings and conclusions in this chapter are those of the authors and do not necessarily represent the views of the National Institute for Occupational Safety and Health.
†Deceased October 29, 2014

Physical and Biological Hazards of the Workplace, Third Edition. Edited by Gregg M. Stave and Peter H. Wald.

of individual workers vary substantially. Because MMH hazards are present in many industrial and service operations, workers in a wide variety of industries are potentially at risk.

The BLS reports that in 2013 there were 1.6 million DAFW cases in private industry and state and local government.[5] The overall incidence rate of nonfatal occupational injury and illness cases requiring DAFW to recuperate was 109.4 cases per 10 000 FTE. When the BLS data is broken down by body part, the incidence rate for trunk injuries is 26.4 per 10000 FTE (20.0 for back). The rate for lower extremity (e.g., knee, ankle, foot) is 24.8 and upper extremity (e.g., shoulder, arm, wrist, hand) is 32.5.[5]

A recent report by the National Safety Council indicated that the total number of nonfatal private industry occupational injuries and illnesses involving DAFW was 905690 cases in 2012.[6] Overexertion was reported for 3 31 130 cases, and lifting and lowering activities accounted for 106210 of these cases. On a percentage basis, the parts of the body affected that related to MMH activities were trunk (including the back) (20.4%), upper extremities (shoulder, arm, wrist, hand) (23.9%), lower extremities (knee, ankle, foot, or toe) (18.1%), and neck (1.3%). Transportation and warehousing, the industry sector where MMH activities prevail, had 89260 nonfatal DAFW cases in 2012. On a percentage basis, the parts of the body affected that related to MMH activities were trunk (including the back) (21.7%), upper extremities (shoulder, arm, wrist, hand) (21.8%), lower extremities (knee, ankle, foot, or toe) (19.5%), and neck (2.1%). The average total workers' compensation incurred costs per claim for parts of the body associated with MMH activities for 2011–2012 reported by the National Safety Council were as follows: (i) arm/shoulders, $42742; (ii) lower back, $38492; (iii) upper back, $34297; (iv) multiple trunk/abdomen, $22361; and (v) hand/fingers/wrist, $21726.[6] The actual percentage of workers employed in jobs that require MMH tasks is difficult to determine, but three recent surveys of the American workforce indicate that a large percentage of workers are in jobs that require heavy lifting and/or hand movements, which are two indicators of MMH. The three surveys were conducted in 2002, 2006, and 2010 and roughly indicate that 48% of respondents had jobs that required these activities.[7–9]

NORMAL ANATOMY AND PHYSIOLOGY OF THE SPINE

The spine is a complex structure made up of bony, muscular, and ligamentous components. The spine can be divided into two major subsystems—the anterior and posterior spine. The **anterior spine** is mainly composed of the large bony vertebral bodies. These vertebral bodies rest atop one another and are separated by the cartilaginous intervertebral disks, which act as "shock absorbers." The vertebral bodies and disks are held together by two sets of ligaments. The **posterior spine** is made up of the additional bony structures of the vertebral peduncles and laminae, which together form the spinal canal. The facet joints, which join two adjacent vertebrae, and the lateral and posterior spinous processes also form part of the posterior spine. The spinous processes are the attachment points for muscles that move and support the spine.

The spine is dependent on both bony and nonbony support for stability. Bony support is provided by the intervertebral disks and the facet joints. Nonbony support comes from the ligaments and the attached musculature. As the bony structures and ligaments do not have enough strength to resist the forces generated during movement and lifting, the spine is dependent on the muscles of the back, abdomen, hip, and pelvis for stability. This principle explains why muscular fatigue is so important in the pathophysiology of back injury. The parts of the spine with the greatest degrees of movement are at highest risk. Because the thoracic and sacral vertebrae are fixed in place by the ribs and the pelvis, the lumbar vertebrae are the most common sites of injury. An excellent compilation of the anatomy and clinical conditions involving the lumbar spine is available in another publication.[10]

PATHOPHYSIOLOGY OF INJURY AND RISK FACTORS

The interpretation of the research linking work-related MSDs and MMH is problematic because of the high prevalence of certain disorders in the general population, such as low back pain (LBP), which have a frequent association with nonoccupational factors. In addition, the relationship is further obscured by the wide range of disorders, the nonspecific nature of the condition, and the general lack of objective data relating different risk factors to overexertion injury. In 1997, the National Institute for Occupational Safety and Health (NIOSH) published an extensive review of the epidemiologic literature that assessed the strength of the association between specific work factors and certain upper extremity and lower back MSDs. The NIOSH researchers identified more than 2000 studies, examined more than 600 epidemiologic studies, and published a comprehensive review of the epidemiologic studies of back and upper extremity MSDs and occupational exposures.[11]

A 2001 report by the National Research Council's Institute of Medicine summarized a review of the literature on work-related risk factors for low back pain by stating "there is a clear relationship between low back disorders and physical load imposed by manual material handling, frequent bending and twisting, physically heavy work and whole body vibration."[12] A systematic review conducted by da Costa and Vieira[13] examined 1761 studies published since the NIOSH[11] review on work-related musculoskeletal disorders and concluded that there was reasonable evidence for causal relationship with heavy physical work and MSDs. The most frequently reported risk factors for MSDs were excessive repetition, awkward postures, and heavy lifting, which are activities involved in MMH.[13]

It is generally recognized that musculoskeletal injuries are a function of a complex set of variables, including aspects of job design, work environment, and personal factors. Moreover, work-related MSDs may result from direct trauma, a single exertion (overexertion), or multiple exertions (repetitive trauma); typically, it is difficult to determine the specific nature of the causal mechanism. A variety of manual handling activities increase a worker's risk of developing an MSD, including jobs that involve a significant amount of manual lifting, pushing, pulling, or carrying and jobs requiring awkward postures, prolonged standing or sitting, or exposure to cyclic loading (whole-body vibration). In addition to these frequent patterns of usage, a variety of personal and environmental factors may affect the risk of developing an MSD. Risk factors increase the probability of occurrence of a disease or disorder, though they are not necessarily causal factors. For the problem of low back injuries involving MMH, four categories of risk factors have been identified by epidemiologic studies—personal, environmental, job related, and psychosocial/organizational:

- **Personal risk factors** are conditions or characteristics of the worker that affect the probability that an overexertion injury may occur. Personal risk factors include attributes such as age, gender, level of physical conditioning, strength, and medical history.
- **Environmental risk factors** are conditions or characteristics of the external surroundings that affect the probability that an overexertion injury may occur. Environmental risk factors include attributes such as temperature, lighting, noise, vibration, and friction at the floor.
- **Job-related risk factors** are conditions or characteristics of the MMH job that affect the probability that an overexertion injury may occur. Job-related risk factors include attributes such as the weight of the load being moved, the location of the load relative to the worker when it is being moved, the size and shape of the object moved, frequency of handling, and the grip forces required to move an object. These factors should be identified because they may be very important for preventing the magnitude of physical hazards to the worker.
- **Psychosocial/organizational factors** are factors related to the social or organizational environment such as occupational stress, job satisfaction, monotonous work, social support at work, perceived work demands, etc. The mechanisms for how these factors might increase the risk of MSDs is not fully understood, but research shows strong associations with increased physiological and biomechanical responses with MMH activities.[7–9]

A tabular presentation of risk factors is provided in Table 3.1.

Work-related MSDs attributed to MMH can result from a direct trauma, a single overexertion, or repetitive loading. It may not always be possible to determine the specific cause of the injury, and the pathology of many types of MSDs is poorly understood. Personal risk factors such as age, fatigue, and concomitant diseases can modify the way the body responds to stressful exertions. Therefore, an injury can occur at different loading levels for different workers. Even for an individual worker, a load may be tolerable one day and excessive on another day due to fluctuations in muscular strength and aerobic fitness.[14]

Our knowledge of job-related, environmental, and personal risk factors is far from complete, but more recent research has included a broader range of relevant risk factors in the studies. An important review conducted by Ferguson and Marras[15] led to increased attention to multiple risk factors, highlighting the role that psychosocial factors play in affecting the risk of injury that may be as important as the

TABLE 3.1 Risk factors associated with manual material handling injuries.

Personal factors	Environmental factors	Job-related factors	Psychosocial factors
Gender	Humidity	Location of load relative to the worker	Occupational stress
Anthropometry (height and weight)	Light	Distance object is moved	Job satisfaction
Physical fitness and training	Noise	Frequency and duration of handling activity	Monotonous work
Lumbar mobility	Vibration	Bending and twisting	Social support at work
Strength	Foot traction	Weight of object or force required to move object	High perceived work demands
Medical history		Coupling/grip strength required	Job control
Years of employment		Stability of the load	Time to complete work tasks
Smoking			
Anatomical abnormality			

actual physical demands of the job. The data showed that an increase in perceived job demands coupled with a decrease in workers' job control and social support led to an increase in physical changes in the worker such as increased muscle tension, lower endurance, and modified body mechanics. Recent systematic reviews[13,16,17] have helped sharpen the focus on which risk factors are the most likely to result inMSDs.

Work-related MSDs are a function of a complex set of variables that include aspects of personal, environmental, job-related, and psychosocial factors. Considerable evidence exists that MMH activities can lead to MSDs, especially to the low back, and MSDs are attributable to numerous risk factors that can be considered in the design of a safe workplace. The interpretation of research findings linking MSDs and MMH is not without problems because of the high prevalence of LBP in the general population and its frequent association with nonoccupational factors. Additionally, significant risk associations can be obscured by a wide range of nonspecific disorders and the general lack of objective data relating different risk factors to overexertion injuries.

MEASUREMENT ISSUES

According to the Occupational Safety and Health Administration (OSHA) a "worksite analysis involves a variety of worksite examinations to identify not only existing hazards, but also conditions and operations in which changes might create hazards. Effective management actively analyzes the work and the worksite, to anticipate and prevent harmful occurrences."[18] A variety of ergonomic measurement tools are available for the evaluation of MMH tasks, especially manual lifting tasks. These tools range in complexity from simple checklists, which are designed to provide a general indication of the physical stress associated with a particular MMH job, to complicated computer models that provide detailed information about specific risk factors. Excellent reviews have been published summarizing the various methods and tools used in ergonomic assessments that can help in choosing an assessment method that will best fit a worker population.[19–21] Table 3.2 summarizes a variety of ergonomic assessment tools and offers a brief description of their advantages and disadvantages.

Ergonomic measurement tools provide objective information about the physical demands of manual handling tasks that will help the user develop an effective prevention strategy. These tools are generally based on scientific studies that relate physical stress to the risk of musculoskeletal injury, particularly when those stressors exceed the physical capacity of the workers.[22–25] Any assessment of physical stress or human capacity is complicated by the influence of a variety of psychosocial factors, including work performance, motivation, expectation, and fatigue tolerance.

TABLE 3.2 Ergonomic assessment tools.

Assessment tool	Advantages	Disadvantages
Checklists	• Simple to use • Best suited for use as a preliminary assessment tool • Applicable to a wide range of manual handling jobs	• Do not provide detailed information about the specific risk factors • Do not quantify the extent of exposure to the risk factors
Biomechanical models	• Provide detailed estimates of mechanical forces on musculoskeletal components • Can identify specific body structures exposed to high physical stress	• Not applicable for estimating effects of repetitive activities • Difficult to verify accuracy of estimates • Rely on a number of simplifying assumptions
Psychophysical tables	• Provide population estimates of worker capacities that integrate biomechanical and physiologic stressors • Applicable to a wide range of manual handling activities	• Reflect more about what a worker will accept than what is safe • May over- or underestimate demands for infrequent or highly repetitive activities
Physiologic models	• Provide detailed estimates of physiologic demands for repetitive work as a function of duration • Applicable to a wide range of manual handling activities	• Not applicable for estimating effects of infrequent activities • Lack of strong link between physiologic fatigue and risk of injury
Integrated assessment models	• Simple to use • Use the most appropriate criterion for the specified task	• Require a significant number of assumptions • Limited range of applications
Videotape assessment	• Economic method of measuring postural kinematics • Can be used to analyze a large number of samples	• Labor-intensive analysis required • Limited to two-dimensional analysis
Exposure monitors	• Provide direct measures of posture and kinematics during manual handling • Applicable to a wide range of MMH activities	• Require the worker to wear a device on the body • Lack of data linking monitor output and risk of injury

Checklists

A checklist is an observational technique that is often the first choice for a rapid ergonomic assessment of a particular workplace. Checklists are designed to provide a general evaluation of the extent of a specific hazard that may be associated with an MMH task or job. A checklist usually consists of a series of questions about physical stressors such as frequent bending, heavy lifting, awkward or constrained postures, poor couplings at the hands or feet, and hazardous environmental conditions. Some checklists use a yes/no format; others use a numerical rating format. Checklists are easy to use, but they lack specificity and they are imprecise. Although there many examples of checklists available, checklists that are more relevant to MMH appear in Appendix B of the NIOSH publication *Ergonomic Guidelines for Manual Material Handling*.[2] An example of an MMH checklist is presented in Figure 3.1. For more information on checklists, see David[19], Dempsey et al.[21], or Eastman Kodak.[26]

Biomechanical models

Various biomechanical models have been developed starting in the early 1970s to estimate the stresses on body joints from external forces (i.e., loads) for a specific MMH task.[27] The ultimate goal was to understand the causes of MSDs associated with MMH tasks in order to develop effective interventions. Models can be two-dimensional (2D) or (3D) three-dimensional and either static or dynamic. The two-dimensional (2D) models were first developed, particularly

NIOSH hazard evaluation checklist for lifting, carrying, pushing, or pulling

Risk factors	Yes	No
1. General		
1.1 Does the load handled exceed 50 lbs.?	[ ]	[ ]
1.2 Is the object difficult to bring close to the body because of its size, bulk, or shape?	[ ]	[ ]
1.3 Is the load hard to handle because it lacks handles or cutouts for handles, or does it have slippery surfaces or sharp edges?	[ ]	[ ]
1.4 Is the footing unsafe? For example, are the floors slippery, inclined, or uneven?	[ ]	[ ]
1.5 Does the task require fast movement, such as throwing, swinging, or rapid walking?	[ ]	[ ]
1.6 Does the task require stressful body postures, such as stooping to the floor, twisting, reaching overhead, or excessive lateral bending?	[ ]	[ ]
1.7 Is most of the load handled by only one hand, arm, or shoulder?	[ ]	[ ]
1.8 Does the task require working in environmental hazards, such as extreme temperatures, noise, vibration, lighting, or airborne contaminants?	[ ]	[ ]
1.9 Does the task require working in a confined area?	[ ]	[ ]
2. Specific		
2.1 Does the lifting frequency exceed 5 lifts per minute?	[ ]	[ ]
2.2 Does the vertical lifting distance exceed 3 ft?	[ ]	[ ]
2.3 Do carries last longer than 1 minute?	[ ]	[ ]
2.4 Do tasks which require large sustained pushing or pulling forces exceed 30 seconds duration?	[ ]	[ ]
2.5 Do extended reach static holding tasks exceed 1 minute?	[ ]	[ ]

Comment: "Yes" responses are indicative of conditions that pose a risk of developing low back pain. The larger the percentage of "yes" responses, the greater the possible risk.

FIGURE 3.1 Manual material handling checklist.

for the purpose of estimating joint loading due to manual lifting. These early static models were based on Newtonian mechanics and did not consider muscle cocontractions for stabilizing the trunk musculature during movements. The 2D models were mainly concerned about body movements on the sagittal plane without taking into account body movements outside the plane. During the late 1970s and the early 1980s, three-dimensional (3D) static models were developed which improved the accuracy of estimating body joint loading. While the early 3D models were an improvement, they did not account for the influence of body motion on joint loading during dynamic loading. Dynamic models, which typically use electromyography (EMG) data from multiple muscle groups involved in MMH tasks, were developed to address this problem. Similar to the 2D and 3D static models, dynamic models were developed in 2D or 3D depending on the complexity of the model. Each model has its strengths and limitations. Chaffin et al.[28] and Rodrick and Karwowski[29] have published detailed discussions of biomechanical modeling that describes various 2D and 3D models.

A brief discussion of some of the models is presented. The Chaffin and Baker model consists of several links representing the major joints of the body and can estimate the torques imposed on each joint as well as spinal compression.[30] A model by Schultz and Andersson estimates both the compression and shear forces for static lifting conditions based on modeling 10 trunk muscles and introducing intra-abdominal pressure.[31] Dynamic models that can estimate loads from various lifting tasks in the workplace have been developed by several investigators, including Freivalds et al.[32]

Important findings that have supported the development dynamic biomechanical models are briefly summarized. McGill and Norman reported that the load on the lumbar spine under dynamic lifting conditions was at least 25% greater than had been estimated from the static models.[33] Marras and Sommerich noted that there was a functional constraint in attempts to evaluate dynamic spine loading that was caused by the "coactivity of the trunk muscles during motion."[34] They found that the pattern of muscle coactivity changes with changes in the velocity and acceleration of the trunk motion during lifting, which in turn influenced the loading on the lumbar spine. The Marras biomechanical research model was based on understanding mechanical aspects of muscle contraction and its influence on trunk motion and loading.

To adequately model lifting tasks that are found in the workplace requires understanding and assessing the role of trunk velocity, trunk asymmetry, and load level. In recent years 3D models have been successfully developed to address these parameters by adding modeling components that consider the effects of dynamic activity, multiple muscles, intra-abdominal pressure (IAP), muscular co-contraction, and posterior ligamentous structures. Marras and Sommerich, in their validation study, reported that their 3D motion model of the lumbar spine was able to calculate estimates of compression, shear, and torsion loading on the lumbar spine, as well as the torque production of the trunk throughout the lift.[35] Their estimates from the model were compared with actual measures of trunk torque and found to be "robust" and "reflected the action of the trunk" under controlled laboratory conditions. The 3D dynamic biomechanical model developed by Marras and colleagues is an example of a dynamic model that has been used in MMH research.[36]

In summary, static models have been used in both field studies and laboratory studies, whereas the dynamic models have been used more in laboratory studies. Researchers, however, have been able to develop methodology that has allowed for dynamic models to be used in field studies or in simulated workplace tasks.[37,38] Dynamic models are also better for assessing upper extremity risks in MMH activities because of the additional body components measured.

Although biomechanical models are typically used to help in the design of stressful activities with various levels of exertion, there is also the question of fatigue failure in the lumbar spine and sacrum from repetitive compression. Damage can occur at loads less than the ultimate compression strength of these structures and within the range of forces and repetitions in activities of daily living, work, and sporting activities.[10] Probability of failure is a function of the load applied and the number of repetitions. Increasing the load increases the probability of damage with fewer repetitions. Loads below 30% of maximal stress are less likely to result in damage even with multiple repetitions, but at loads of 50–60% maximum stress damage after 100 repetitions is 39% and at loads of 60–70% the probability rises to 63%.[39] Repetitions of 100 and up to 1000 are in the range for many occupational activities.[39]

Psychophysical tables

Psychophysics has been defined as the exact science of the functional relations between "body and mind."[40] Whenever the stimulus increases, the intensity of the sensation grows in accordance with a common basic principle which states that equal stimulus ratios produce equal subjective ratios. In every sense modality, sensation is a power function of stimulus. This is the basic principle that underlies the psychophysical law.[41] A worker's subjective determination is used to assess the synergistic effects of combined physiologic and biomechanical stress created by various MMH factors.

Although the vast majority of psychophysical research involving MMH activities has emphasized lifting tasks, the use of psychophysical techniques is not restricted to lifting. Psychophysics is also applicable to lowering, pushing, pulling, holding, and carrying activities. The use of psychophysical data to assess the physical demands of MMH is most appropriate for repetitive activities that are performed more often than once a minute. Databases are available that provide acceptable levels of MMH for various segments of the population. These levels of work are typically presented as

TABLE 3.3 Portion of a Psychophysical Table for maximum acceptable Push Force.

Maximum Acceptable Forces of Sustained Push (kg) for 15.2 m Push								
Height from floor to hands (cm)	Percentage of industrial Females	One push every:						
		6 s	12 s	1 min	2 min	5 min	30 min	8 h
89	90	5	6	6	7	7	8	10
	75	7	8	9	10	11	11	14
	50	9	11	13	13	14	15	19
	25	12	14	16	16	18	19	24
	10	14	17	19	19	21	23	28

Adapted from Snook SH, Ciriello VM. The design of manual-handling tasks: revised tables of maximum acceptable weights and forces. *Ergonomics* 1991; 34:1197–1213. Reprinted with permission.

acceptable weights for a lift or carry or acceptable forces of pushing and pulling.[42] A psychophysical table for acceptable forces for sustained push is shown in Table 3.3. For more information on psychophysical databases, see Snook,[24] Snook and Ciriello,[42] or Ayoub and Mital.[43]

Other psychophysical assessment methods have been developed to assess various MMH activities. For example, self-report measures such as rating of perceived exertion (RPE)[44] and body part discomfort (BPD)[45] have been used to assess a variety of lifting jobs. These assessment measures provide useful information about the worker's perception of the physical demands of the job. Moreover, RPE and BPD compare favorably with measures of physical demand.

Databases containing whole-body and segmental strength measures have also been developed for the design of MMH tasks. These include isometric, isokinetic, and isoinertial strength databases for whole-body activities, such as lifting, and various databases for the arms, legs, and back. For more information on strength measurement, see Ayoub and Mital[43] and Chaffin et al.[28]

It should be noted that psychophysics relies on self-reports from subjects; consequently, the perceived "acceptable" limit may differ from the actual "safe" limit. In addition, the psychophysical approach may not be valid for all tasks, such as high-frequency lifting.[25]

Physiological models

One of the goals in designing an MMH task is to avoid the accumulation of physical fatigue, which may contribute to an overexertion injury. This fatigue can affect specific muscles or groups of muscles, or it can affect the whole body by reducing the aerobic capacities available to sustain work. Two physiological factors that affect the suitability of an MMH task at the local muscle effort level include the **duration of force exertion** and the **frequency of exertions**. Local muscular fatigue will develop if a heavy effort is sustained for a long period. With heavy loads, the muscles need a substantially longer recovery period to return to their previous state. Small changes in workplace layout or handling heights, however, can often solve a local muscle fatigue problem through a reduction in holding duration. In addition, local muscle fatigue associated with maintaining awkward postures or constant bending can reduce the capacity of the muscles needed for lifting and therefore increase the potential for an overexertion injury to occur.

Local muscle fatigue may limit the acceptable workloads for MMH tasks that are performed for short but intensive periods during a work shift. However, it is the energy expenditure demands of repetitive tasks that have the most profound effect on what a worker is able to do over a longer period of time. Energy expenditure demands are dependent upon the extent of muscular exertion, frequency of activity, and duration of continuous work. A worker's limit for physiologic fatigue is often affected by a combination of discomfort in local muscle groups and more centralized (systemic) fatigue associated with oxygen demand and cardiovascular strain.[26,46]

To assess the cardiovascular demands of MMH tasks, physiologic parameters such as heart rate (HR), oxygen consumption, and ventilation rate may be used. In addition, electromyographic (EMG) assessments and blood lactate provide a relative measure of the instantaneous level of physiologic status and muscular fatigue. Assuming that fatigued workers are at a higher risk of overexertion injury, these measures can be used to help prevent overexertion injuries by predicting the limits of fatigue for repetitive handling tasks.

Physiological models provide a method for estimating the cardiovascular demands associated with a specific MMH activity. One such model, developed by Garg, allows the analyst to estimate the energy expenditure demands associated with a complex MMH job.[47] The first step is to separate the job into distinct elements or subtasks for which individual energy expenditure values can be predicted, such as standing and bending, walking, carrying, vertical lifting or lowering, and horizontal arm movement. The total energy expenditure requirements for the job are then estimated by summing the incremental expenditures of all of the subtasks.

Direct measure of oxygen consumption may provide the best estimate of physiological demand. For this type of measurement, it is assumed that the consumption of 1 L of oxygen per minute is approximately equal to an energy expenditure of 5 kcal/min.

HR is also useful in predicting physiologic demand, but it is less reliable than direct oxygen consumption measures, due to individual differences in the relationship between HR and energy expenditure. Portable monitors can be used to measure HR and oxygen consumption during MMH tasks. For information on assessing physiologic demands, see Astrand and Rodahl[48] or Eastman Kodak.[26]

Integrated assessment models

An integrated assessment model involves a unique approach that considers all three of the primary lifting criteria—biomechanics, physiology, and psychophysics. Such models can merge qualitative and quantitative methodologies and produce a single indicator of risk potential for an MSD. The integrated approach provides a measure of the relative magnitude of physical demand for a specific MMH task that relies on the most appropriate stress measure for that task. The result of the assessment is typically represented as a weight or force limit or as an index of relative severity. An integrated model considers the synergistic effects of the various task factors and uses the most appropriate stress measure to estimate the magnitude of hazard associated with each task factor.

One model, the NIOSH revised NIOSH lifting equation is described in more detail later in this chapter. Another method, which has been used in MMH assessments is Ayoub's job severity index (JSI).[29,49] Examples of integrated assessment models for MMH are briefly summarized and critiqued in a recent review article.[50]

Videotape assessment

Most ergonomic assessments include the use of videotape analysis, where a video camera is used to record the work activity for later analysis. Videotape recordings make it easy to stop or freeze the action so that body posture or workplace layout can be evaluated. Videotape analysis often consists of general observation by the analyst that results in subjective estimates of physical hazards, such as posture angle and task repetition. Ergonomic checklists can also be used to review physical hazards recorded in the video. Complex computerized video analysis programs are available that can capture and analyze individual frames from videotape recordings of workers performing MMH activities. These video frames can then be used to make more detailed assessments of spatial or dynamic biomechanical hazards that may not be apparent from the observational approach.

To use a computer program to assist in estimating body posture angles, videotape recordings need to be digitized in certain file formats that can be read by the computer program. With the advent of digital camcorder technology, digital camcorders are now available without taking the above analog-to-digital signal conversion step. Video recorded on digital media can be downloaded to the computer for the videotape assessment using a stand-alone media player or a third-party computer program with ergonomic assessment features. The Multimedia Video Task Analysis (MVTA™) program.[51] and the 3D Static Strength Prediction Program[52] (3DSSPP™) are two such third-party programs that are commercially available and commonly used by researchers and practitioners.

The MVTA video analysis software program was developed by the University of Wisconsin–Madison to help automate time and posture analyses of visually discerned activities through an interactive graphical user interface. User-defined arbitrary events are discerned by interactively identifying terminal breakpoints in the timed activity. Breakpoints are characteristic occurrences that define the start and end of an MMH activity. To determine the breakpoints, the video recording may be reviewed at any speed and in any sequence (real time, slow motion, fast motion, or frame by frame in either forward or reverse direction). The analyst can replay any event as much as desired in a continuous loop or direct the video to display an arbitrary event or point in time in stop action. The MVTA program computes the frequency and time duration of occurrence of each MMH activity and produces a conventional time study report.

The 3DSSPP was developed by the University of Michigan. A 3D biomechanical model is built in the program for estimating spinal loading variables according to input posture data, hand loads, and the worker's anthropometric information. To specify a body posture, 15 different body angles are required to be entered into the program. The 3DSSPP displays a humanlike mannequin to help identify the posture input by the user.

These computer programs are generally easy to use, and the output data are presented in a form that is easy to understand and apply. They may be limited, however, in their capability to analyze activities that occur outside of the camera focal plane (i.e., the plane parallel to the face of the camera lens). For example, when a body segment or group of segments moves outside the camera focal plane, the joint angles and positions measured from the digitized frame are distorted. The amount of distortion depends on the degree of displacement of the joint or segment from the focal plane. To improve the accuracy of specifying body posture angles, more than one camera may be used for recording the same work activity at the same time. Several studies have shown that more than one camera view can improve the accuracy of estimating body posture angles.[53–55] Compared to front- and back-viewing angles, side-viewing angle is associated with the least error for posture specification.[53,54] Estimation of larger body parts generally results in better reliability.[56]

Guidelines for videotape job analysis, which have been developed by NIOSH researchers, are provided in Table 3.4. For more information on videotape analysis and motion analysis, see Chaffin et al.[28] or Eastman Kodak.[26]

TABLE 3.4 Guidelines for recording work activities on videotape.

- If the video camera has the ability to record the time and date on the videotape, use these features to document when each job was observed and filmed. Recording time on videotape can be especially helpful if a detailed motion study will be performed at a later date (time should be recorded in seconds). Make sure that the time and date are set properly before videotaping begins
- If the video camera cannot record time directly on the film, it may be useful to position a clock or a stopwatch in the field of view
- At the beginning of each recording session, announce the name and location of the job being filmed so that it is recorded on the film's audio track. Restrict subsequent commentary to facts about the job or workstation
- For best accuracy, try to remain unobtrusive, that is, disturb the work process as little as possible while filming. Workers should not alter their work methods because of the videotaping process
- If the job is repetitive or cyclic in nature, film at least 10–15 cycles of the primary job task. If several workers perform the same job, film at least two or three different workers performing the job to capture differences in work method
- If necessary, film the worker from several angles or positions to capture all relevant postures and the activity of both hands. Initially, the worker's whole-body posture should be recorded (as well as the work surface or chair on which the worker is standing or sitting). Later, closeup shots of the hands should also be recorded if the work is manually intensive or extremely repetitive
- If possible, film jobs in the order in which they appear in the process. For example, if several jobs on an assembly line are being evaluated, begin by recording the first job on the line, followed by the second, third, etc.
- Avoid making jerky or fast movements with the camera while recording. Mounting the camera on a tripod may be useful for filming work activities at a fixed workstation where the worker does not move around much

Exposure monitors

Monitoring devices have been developed to measure various aspects of physical activity, such as position, velocity, and acceleration of movement. Some monitors can even measure three-dimensional joint angles in real time. These systems consist of mechanical sensors that are attached to various parts of the worker's body, such as the wrist, back, or knees. The mechanical sensors convert angular displacement (rotation) into voltage changes that can be displayed in real time or saved to a computer for later analysis. The position measures acquired from the sensors can then be differentiated to obtain rotational velocities and acceleration components. These movement characteristics may be used to estimate the extent of risk associated with a particular task and help to identify potential ergonomic solutions. Examples of positional monitoring equipment include potentiometer-based lumbar and wrist motion monitors, such as those developed at the Ohio State University[57,58] as well as strain gauge-based strip goniometers, such as those available from NexGen Ergonomics, Inc. (Pointe-Claire, Quebec, Canada, www.nexgenergo.com).

Another type of device that has been used to assess the extent of exposure to repetitive movement is an accelerometer-based motion recording system. In the early 1990s, an activity monitor was developed and used to track worker's activities throughout the work shift. The activity monitor consists of one or more accelerometers mounted within a small aluminum case that is connected with a Velcro strap to a worker's wrist, leg, or trunk. The accelerometers are sensitive to the movements of the body. They are capable of counting and recording rapid movements inherent in a specific task or activity. The data acquired from the activity monitor are typically plotted as a series of temporal histogram plots showing the extent of dynamic movement as a function of time. The greater the total dynamic activity, the greater the height of the sequential histogram bars and the greater the potential for injury.[59]

In recent years, a new type of inertial motion capture system has been developed for field applications. These motion capture systems determine a worker's movements by estimating body segment orientation and position changes using integration of gyroscope and accelerometer signals that are collected by sensors worn on multiple body segments. Based on some biomechanical models and computer algorithms, these systems can provide dynamic tracking of whole-body movements in real time.[60] With only one sensor worn on the lower back, the system can monitor trunk inclination (i.e., a physical hazard for MSDs) with a reasonable accuracy.[61]

It is important to note that the output from exposure monitors alone cannot provide all the information needed to assess the extent of physical demand required by an MMH task. It is also important to know the weight of the load and its position, velocity, and acceleration relative to the body during the task. This approach is best suited for repetitive or high-speed MMH tasks where the internal forces on the body may be affected more by extreme postures or rapid movements than by the weight or position of the external load.

GUIDELINES AND STANDARDS

Early attempts to prevent overexertion injuries associated with MMH focused on adopting arbitrary weight limits for lifting loads, hiring strong workers, or using training procedures that emphasized correct (but not necessarily safe) lifting techniques. None of these approaches, however, have proven to be effective in significantly reducing overexertion injuries.[62] Recently, industry leaders have started to recognize the risks associated with MMH. To reduce costs and increase productivity, these companies have implemented ergonomic programs or practices aimed at preventing these injuries. In many cases, these ergonomic programs rely on exposure guidelines or standards recommended by the federal government.

The National Institute for Occupational Safety and Health (NIOSH)

NIOSH, the leading federal research agency for ergonomic hazards, does not have a recommended exposure limit (REL) for general MMH activities, but has issued guidelines that can be used for prevention of MMH-related injuries. In 1981, NIOSH published its *Work Practices Guide for Manual Lifting* (WPG).[63] The 1981 WPG contained an equation for assessing certain manual lifting tasks. The 1981 NIOSH lifting equation provided a unique method for determining weight limits for selected two-handed manual lifts, but it was limited in its scope of application. It only applied to lifting tasks that occurred directly in front of the body (sagittal plane lifts) and had optimal hand-to-object couplings (i.e., handles).

Responding to the need for a guideline with a broader application and more flexibility, NIOSH revised the lifting equation. In addition to the four risk factors addressed by the 1981 equation (i.e., horizontal location, vertical height, vertical distance traveled, and frequency), the revised NIOSH lifting equation includes weight reduction factors for the assessment of asymmetric or nonsagittal plane lifts that begin or end to the side of the body and for lifting objects with less than optimal hand-to-object couplings (i.e., no handles).[25] The original 1981 NIOSH lifting equation was based on the concept that the overall physical stress for a specific lifting task is a function of the various task-related factors that define the lift, such as task geometry, load weight, and lifting frequency. This concept also forms the basis for the revised NIOSH lifting equation, which provides a practical method for determining the overall physical stress attributable to the various task-related factors.

The revised NIOSH lifting equation yields a unique set of evaluation parameters that include (i) intermediate task-related multipliers that define the extent of physical stress associated with individual task factors; (ii) the NIOSH recommended weight limit (RWL), a task-specific value that defines the load weight that is considered safe for nearly all healthy workers; and (iii) the NIOSH lifting index (LI), which provides a relative estimate of the overall physical stress associated with a specific manual lifting task.

The RWL is defined by the following equation:

$$\text{RWL} = \text{LC} \times \text{HM} \times \text{VM} \times \text{DM} \times \text{AM} \times \text{CM} \times \text{FM}$$

where the load constant (LC) is equal to 51 lb and the terms HM, VM, DM, AM, CM, and FM are task-specific multipliers within the equation that serve to reduce the recommended weight limit according to the specific task factor to which each multiplier applies. The magnitude of each multiplier will range in value between zero and one, depending on the value of the task factor to which the multiplier applies. The multipliers are defined as in Table 3.5.

TABLE 3.5 Formulas for individual multipliers.

Component	Metric	US customary
HM = horizontal multiplier	$(25/H)$	$(10/H)$
VM = vertical multiplier	$[1-(0.003\|V-75\|)]$	$[1-(0.0075\|V-30\|)]$
DM = distance multiplier	$[0.82+(4.5/D)]$	$[0.82+(1.8/D)]$
AM = asymmetric multiplier	$[1-(0.0032A)]$	$[1-(0.0032A)]$
FM = frequency multiplier	(Determined from Table 3.6)	
CM = coupling multiplier	(Determined from Table 3.7)	

In order to use the revised NIOSH lifting equation, you need to make the following measurements:

- L = weight of the load being lifted (lb or kg).
- H = horizontal location of the hands from midpoint between the ankles. Measure at the origin and the destination of the lift (cm or in).
- V = vertical location of the hands above the floor. Measure at the origin and destination of the lift (cm or in).
- D = vertical travel distance between the origin and the destination of the lift (cm or in).
- A = angle of asymmetry, defined as the angular displacement of the load from the sagittal plane when lifts are made to the side of the body. Measure at the origin and destination of the lift (°).
- F = average frequency rate of lifting measured in lifts/minute. Duration is defined as <1, 1–2, or 2–8 hours. Specific recovery allowances are required for each duration category (see Tables 3.6 and 3.7).

The LI provides a relative estimate of the level of physical stress associated with a particular manual lifting task. The estimate of the level of physical stress is defined by the relationship of the weight of the load lifted to the recommended weight limit. The LI is defined by the following equation:

$$\text{LI} = \frac{\text{load weight}}{\text{recommended weight limit}} = \frac{L}{\text{RWL}}$$

A detailed explanation of the use of the revised NIOSH lifting equation, including definitions of terms and procedures, is available in an applications manual for the revised NIOSH lifting equation.[64] The document can be

TABLE 3.6 Frequency multiplier (FM).

Frequency lifts/minute	Work duration					
	≤1 hour		≤2 hours		≤8 hours	
	$V<75$	$V\geq75$	$V<75$	$V\geq75$	$V<75$	$V\geq75$
0.2	1.00	1.00	0.95	0.95	0.85	0.85
0.5	0.97	0.97	0.92	0.92	0.81	0.81
1	0.94	0.94	0.88	0.88	0.75	0.75
2	0.91	0.91	0.84	0.84	0.65	0.65
3	0.88	0.88	0.79	0.79	0.55	0.55
4	0.84	0.84	0.72	0.72	0.45	0.45
5	0.80	0.80	0.60	0.60	0.35	0.35
6	0.75	0.75	0.50	0.50	0.27	0.27
7	0.70	0.70	0.42	0.42	0.22	0.22
8	0.60	0.60	0.35	0.35	0.18	0.18
9	0.52	0.52	0.30	0.30	0.00	0.15
10	0.45	0.45	0.26	0.26	0.00	0.13
11	0.41	0.41	0.00	0.23	0.00	0.00
12	0.37	0.37	0.00	0.21	0.00	0.00
13	0.00	0.34	0.00	0.00	0.00	0.00
14	0.00	0.31	0.00	0.00	0.00	0.00
15	0.00	0.28	0.00	0.00	0.00	0.00
>15	0.00	0.00	0.00	0.00	0.00	0.00

Values of V are in cm; 75 cm = 30 in.

TABLE 3.7 Coupling multiplier (CM).

Couplings	Coupling multipliers	
	$V<75$ cm (30 in.)	$V\geq75$ cm (30 in.)
Good	1.00	1.00
Fair	0.95	1.00
Poor	0.90	0.90

downloaded from the NIOSH website (https://www.cdc.gov/niosh/docs/94-110/).

The developers of the revised NIOSH lifting equation indicated that studies were needed to determine the effectiveness of the NIOSH lifting equation in identifying jobs with increased risk of lifting-related LBP for workers. Researchers at NIOSH published the results of a cross-sectional epidemiologic study designed to evaluate the effectiveness of the equation in identifying jobs with elevated risk of causing LBP.[65] In the NIOSH study, 50 jobs were evaluated using the revised NIOSH lifting equation. The LI values of the jobs were compared to information obtained about the LBP symptoms in people who worked in those jobs. Using logistic regression modeling, the odds ratio (OR) for LBP was determined for various categories of LI values compared to the unexposed control group. LBP was assessed with a symptom and occupational history questionnaire that was administered to 204 workers employed in lifting jobs and 80 workers employed in nonlifting jobs. Participation was 89–95% among the exposed workers and 82–100% among unexposed workers at four facilities. The authors found that as the LI increased from 1.0 to 3.0, the odds of LBP increased, with a peak and statistically significant OR occurring in the two to three LI category (unadjusted OR = 2.45; CI 1.29–4.85). For jobs with a LI higher than 3.0, however, the OR was lower (1.63; CI 0.66–3.95). The decrease in the OR for these highly exposed jobs is likely to result from a combination of worker selection and a survivor effect. This study also examined several confounding variables such as age, gender, body mass index, and psychosocial factors that were included in the multiple logistic models. The highest OR was in the two to three LI category (2.2; CI 1.01–4.96).

Additional studies conducted in the last decade have broadened the research base that has established the revised NIOSH lifting equation and confirmed its versatility as a valid methodology for estimating LBP risk. For jobs that require multiple MMH activities, a new index methodology has been developed. This method is based on the concept that a composite lifting index (CLI), which represents the collective demands of the job, is equal to the sum of the largest single task lifting index (STLI) and the incremental increases in the CLI as each subsequent task is added. The incremental increase in the CLI for a specific task is defined as the difference between the lifting index for that task at the cumulative frequency and the lifting index for that task at its actual frequency. Other studies have incorporated individual, psychosocial, and work organizational risk factors and their influence on self-reported back pain and the calculation of the lifting index. An excellent article which summarizes this recent research involving the revised NIOSH lifting equation is available from Lu et al.[66]

DIAGNOSIS AND TREATMENT

The primary objectives of the diagnosis and treatment of occupationally related musculoskeletal injuries are to (i) assist the recovery of workers and allow for a rapid return to work (ii) ensure that proper diagnostic tools are used so that an accurate assessment is made of the magnitude of the injury, and (iii) provide appropriate cost-effective treatments that avoid unnecessary surgery. The system should be based on an organized approach to evaluation, diagnosis, and treatment, which is essential for an early return to work and reduction in costs and lost work time. For example, a standardized diagnostic and treatment protocol has been shown to be effective in significantly and continuously reducing the number of incidents, days lost from work, low back surgery cases, and financial costs of LBP.[67]

Although a variety of MSDs result from MMH, the single most costly medical condition is work-related LBP, which affects millions of Americans. Experts have indicated that there is significant variation in assessment and treatment of LBP that results in inappropriate or at least less than optimal care for many patients with low back disorders. This issue was addressed in 1994 with the publication of the Clinical Practice Guideline for Acute Low Back Pain in Adults by the Agency for Health Care Policy and Research (AHCPR).[68] Copies of the guidelines are available on the agency's (now renamed the Agency for Healthcare Research and Quality) website at www.ahrq.gov. These guidelines, although somewhat outdated, focused on returning the patient to normal activity and were in widespread use. A more recent review, which is more inclusive and includes additional return to work (RTW) models, is provided by Schultz et al.[69]

Complicating the diagnosis and treatment of occupationally related LBP are the legal issues of disability and compensation and how they relate to pain and impairment. Pain and impairment, which are direct measures of the extent of injury, primarily depend on the severity of the injury. Disability and compensation, however, may depend more on the nature of the compensation system and laws than on the severity of the injury.[70]

Diagnosis

Acute low back pain is often nonspecific and it is often difficult to identify a specific cause. There are some estimates that more than 90% of all episodes of back pain are probably attributable to mechanical causes,[71] which could certainly be the case with MMH activities. Thus, the early diagnostic evaluation of back pain is designed to rule out systemic disease; grossly identify neurologic or anatomic abnormalities that may eventually require surgery such as fracture, tumor, infection, or cauda equina syndrome; and identify characteristic indicators of injury that may influence the selection of treatment. Clinical guidelines for diagnosis and treatment of low back pain are available from Chou et al.[72] and Casazza.[73]

The first step in diagnosing LBP, and perhaps the most important, is the medical history. The medical history should include an assessment of current symptoms, individual history of injury, functional status, and injury documentation and psychosocial risk factors which may be helpful in predicting risk for chronic back pain.[72,74–77] The second step in the diagnostic process is a complete physical examination. At minimum, this examination would include a check of the deep tendon reflexes at the Achilles and quadriceps tendons, bilateral straight-leg raising, palpation of the paraspinal muscles for spasm, and a screening motor and sensory neurologic examination. The neurologic examination should give special attention to the L5 and S1 nerve roots, since the vast majority of disk herniations occur at either the L4–L5 or L5–S1 interspaces. It has been suggested that a system should be developed to ensure that all parts of the examination are completed.[73,76]

Chou et al. proposes that after the medical history and physical examination patients be categorized into three broad categories: (i) nonspecific low back pain, (ii) back pain potentially associated with radiculopathy or spinal stenosis, or (iii) back pain potentially associated with another specific spinal cause.[72] Most back injuries from MMH activities would be categorized as nonspecific.

Imaging is not recommended for most patients without signs and symptoms of a serious condition.[73] Magnetic resonance imaging (MRI) may be most appropriate if a serious condition is suspected. Computed tomography can be used if an MRI is unavailable. However, these sophisticated imaging tests should not be ordered too early in the course of back pain, especially when there is an absence of clinical findings that suggest a need for surgical intervention. In general, these tests should be limited to workers who have continued neurologic findings after 4–6 weeks of conservative therapy. It is also important to remember that an abnormal imaging study and back pain are not necessarily related. The clinical course of each worker must be considered before this causal link is made and surgery is considered.

Treatment

Epidemiologic data suggest that the vast majority of low back injuries are not serious and that most workers return to work after a short time with only conservative therapy. In general, only a minority of all affected persons have back pain >2 weeks in duration. By contrast, 90% of patients return to work within 6 weeks of onset of the back pain.[76,78]

Recent recommendations for treatment on nonspecific back pain have changed from previous recommendations as evaluations of effective treatments improved. Practitioners treating LBP should review the ACOEM Practice Guidelines for low back disorders.[79]

Casazza[73] categorizes treatments into (i) recommended, (ii) acceptable, (iii) unsupported, and (iv) inadvisable, using an evidence-based approach. Recommended treatments are medications and patient education. Nonsteroidal anti-inflammatory drugs (NSAIDS) are effective for short-term relief as are benzodiazepine muscle relaxants. Opioids, while frequently prescribed, showed little evidence of benefit over NSAIDS in relief or return to work. Epidural steroid injections are frequently offered after 2–6 weeks of no relief from radicular pain with noninvasive treatment. However, the FDA has not approved corticosteroids for such use and has stated that the effectiveness and safety of the drugs for this use have not been established.[80] Patient education involves advising patients to stay active as much as the pain allows, avoid twisting and bending, and return to normal activities as soon as possible. Assurance that acute back pain is mostly benign in nature is important. Acceptable treatments are physical therapy, especially the McKenzie method,[81] and physical therapist-directed home exercise and application of ice or heat treatments in the first 5 days of pain occurrence. Unsupported treatments include oral steroids, acupuncture, exercise, lumbar support, massage, spinal manipulation, and traction. Many of the treatments in the unsupported list lack high-quality studies to evaluate intervention effectiveness but may be beneficial for some patients. An inadvisable treatment is bed rest, especially prolonged bed rest.

If conservative therapy fails and 4–6 weeks have passed, further diagnostic tests should be scheduled. These cases generally fall into one of the following four categories, by location of the patient's predominant complaints: (i) LBP, which is typically diagnosed as back sprain; (ii) leg pain radiating below the knee, which is commonly referred to as sciatica; (iii) posterior thigh pain; and (iv) anterior thigh pain. Each of these conditions warrants a unique treatment regimen.[67] It may also be appropriate to thoroughly investigate and address psychological and psychosocial issues that were not identified at the initial assessment. Failure to deal with a significant psychological or psychosocial issue will delay the patient's return to work and may ultimately result in treatment failure and continuing disability.

A few back pain patients have symptoms suggesting sciatica, which is usually the first clue to a herniated disk. It is estimated that only 5–10% of patients with persistent sciatica require surgery. In general, only patients diagnosed with cauda equina compression (CEC) or a similar mechanoanatomic problem require immediate surgical intervention. Other indications for referral include severe or progressive neurologic deficits, or persistent neurologic finding after 4–6 weeks of conservative therapy. In the overwhelming number of cases of LBP, the appropriate treatment includes a course of conservative therapy.[67,82]

There is general agreement among clinicians that treatment for LBP must be given according to a strict timetable so that patients do not develop a dependence that would prolong symptoms and functional limitations. Also, patient reassurance and education is an important aspect of therapy. The clinician must avoid labeling the patient. The patient should be reassured that the natural history of their condition is favorable and that he or she should be able to return to work in a short time.[82]

MEDICAL SURVEILLANCE

Active and passive surveillance of the workplace are necessary to establish a database of potential exposures in each workplace. Medical management of injuries and early reporting should be incorporated within the general medical treatment program at each work site. Since there is usually a short latency period between exposure and the onset of symptoms, medical surveillance per se is not helpful in eliminating MMH injuries. It may be useful to screen employees in order to match their physical abilities to the job. However, this is more of an exercise in prevention of injuries and is thus covered in the next section on "Prevention."

PREVENTION

Several strategies have been developed to try and prevent MSD injuries from MMH activities. These strategies can be categorized into three general types: (i) workplace directed, (ii) worker directed, and, (iii) employee screening.

Workplace-directed approaches

Automation

Workplace automation should be a top priority when the job has high physical demands, is highly repetitive, or is performed in a hazardous environment. Automation may consist of one or more machines or machine systems such as conveyors, automated handling lines, automated storage and retrieval systems, or robots. This approach is best suited for the design of new work processes or activities or for the redesign of highly stressful tasks. Because automation often requires large capital expenditures, this approach may be prohibitive for small companies with only a few workers.

Mechanical aids

In cases where the physical demands are high and automation is not practical, mechanical aids can be used to ameliorate the extent of those demands. Mechanical handling aids include machines or simple devices that provide a mechanical advantage during the MMH task, such as hand trucks, cranes, hoists, lift tables, powered mobile equipment and lift trucks, overhead handling and lifting equipment, and vacuum lift devices. As mentioned earlier, the publication *Ergonomic Guidelines for Manual Material Handling* illustrates a number of mechanical aids that would be useful in reducing exposure to MMH.[2] Additionally, a recent NIOSH publication *Ergonomic Solutions for Retailers* is specifically oriented for prevention of MMH injuries in the grocery sector and illustrates mechanical aids that are useful in the grocery sector.[83]

Ergonomic design changes

For MMH jobs, ergonomic design or redesign may be accomplished by modifying the job layout, engineering redesign, or incorporating procedures to reduce bending, twisting, horizontal extensions, heavy lifting, forceful exertions, and repetitive motions. The ergonomic approach is largely based on the assumption that work activities involving less weight, repetition, awkward postures, and applied force are less likely to cause injuries and disorders. The ergonomic approach is desirable because it seeks to eliminate potential sources of problems. Ergonomics also seeks to make safe work practices a natural result of the tool and work site design.

There are at least four advantages to adopting an ergonomic design/redesign strategy. First, an ergonomic approach does not depend on specific worker capabilities or learned behaviors, such as training. Second, human biological factors and their variations are accounted for in ergonomic approaches using design data that accommodate large segments of the populations. Third, ergonomic intervention is relatively permanent, since the workplace hazard is eliminated. Fourth, to the extent that sources of biomechanical stress at the work site are eliminated or significantly reduced, the difficult issues involving potential worker discrimination, lifestyle modifications, or attempts at changing behavioral patterns of workers will be of lesser practical significance. Many MMH jobs entail a variety of specialized tasks with overlapping sources of physical stress. Each MMH task may contribute in some unknown manner to the onset of an overexertion injury. Often, no single job modification is feasible or sufficient, and numerous adjustments may be required for each activity to minimize the hazard of overexertion injury. It is always important to develop an evaluation process to determine if the job modification is effective.

Worker-directed approaches

Training and education

According to OSHA,[84] the purpose of training and education is to "ensure that employees are sufficiently informed about the ergonomic hazards to which they may be exposed and thus are able to participate actively in their own protection." Training may include general instruction on the types of injuries that may occur, what risk factors may contribute to these injuries, how to recognize and report symptoms, and how to prevent these injuries. Training should also include job-specific instruction on the proper use of essential MMH equipment and techniques for the specific task or job. The training program should include the following individuals: (i) all exposed workers, (ii) supervisors, (iii) managers, (iv) engineers and maintenance personnel, and (v) health-care providers.

The term training has been used to describe two distinctly different approaches to injury prevention and control—instructional training in safe materials handling and fitness training (e.g., conditioning, strengthening, or work hardening). The basic premise of instructional training in MMH is that people can more safely handle greater loads when they perform the task correctly than if they perform the task incorrectly. Fitness training is based on the premise that people can safely handle greater loads when their strength or aerobic capacity is increased.

Instructional training

Instructional training is widely used and fundamentally sound, but there some potential problems when not appropriately used. First, care must be taken to ensure that appropriate, scientifically sound work practices are being taught. Today, with the advent of the Internet and the mass media exposure, work practices that have not been properly evaluated can be promoted that may be entirely unsound for MMH activities. For example, back belts were introduced as a device for preventing back injuries and were implemented in many workplaces with instructional techniques before scientific evaluation showed limited usefulness. NIOSH concluded, in 1994, that there was insufficient evidence to recommend the use of back belts as a back injury prevention measure.[85,86] Since then, NIOSH conducted a large epidemiologic study and two laboratory evaluations to determine more conclusively the effects of back belt use. The further research did not provide evidence to change NIOSH's earlier conclusions.[87–89]

Additionally, in some training courses, workers are taught to lift with a straight back and bent legs (i.e., squat posture) rather than bending over to pick up the load (i.e., stooped posture). This either/or approach does not always provide the "safest" lifting style.[62] Recent research has shown that the horizontal distance of the load from the worker may be more important in determining the spinal loading forces that the amount of forward bending. For this reason, NIOSH does not recommend a single, correct lifting style for all manual lifting tasks, but suggests that a free style lift, in which workers choose whatever style they prefer may be the most appropriate keeping in mind the avoidance of excessive bending and reach.

Secondly, the instructional approach relies on the worker's ability to comply with a set of recommended practices that may be forgotten or changed from time to time. To some extent, worker's ability to comply with safe practices can be improved with training materials that are more specifically directed at the MMH activities they are involved with. Two recent NIOSH documents such as *Simple Solutions for Home Building Workers*[90] and the aforementioned *Ergonomic Solutions for Retailers* are examples.[83]

Regardless of the potential problems associated with MMH training, all workers who perform MMH activities should receive basic instructional training in the recognition of hazardous tasks and should have a thorough knowledge of what to do when a hazardous task is identified. Furthermore, the instructional training should provide information to workers on how they can become involved in the process of preventing and controlling injuries on the job. An excellent review of the effectiveness of training and education programs for prevention of workplace injuries has recently been published.[91]

Fitness training

The basic premise of fitness training is that a worker's risk of injury would decrease if his or her strength or fitness increased. Although this seems intuitive, it is not clear whether an individual's strength and risk of injury performing MMH activities are related. Certainly, a worker's capacity to perform heavy work might be increased, but there is some controversy about the relationship between worker strength and risk of injury.[92] Moreover, it is not known how the soft tissues of the body respond to increased loads associated with stronger muscles, especially if the worker performs tasks requiring greater strength demands. Regardless of strength, an improper lifting technique could lead to an injury. Strength and fitness exercises are available in several documents and in preparation for MMH activities.[90]

Although both types of training programs have been used to prevent MMH injuries, the effectiveness of training in preventing or controlling injuries is unclear at the present time. Therefore, training programs should be used as a supplement to workplace-directed approaches.

Employee screening

Screening methodologies, which rely on the assessment of one or more physical characteristics of the worker to select specific workers for certain MMH jobs, have been advocated by some ergonomic experts. These methodologies are designed to identify workers (i) with a high risk of overexertion injury, such a history of previous physical injury that may limit MMH activities, and/or (ii) screen workers according to some preselected set of strength or endurance criteria in an attempt to match the capacity of the worker to the demands of the job.

Risk assessment screening

Attempts to identify workers with a high risk of overexertion injury have included such activities as spinal radiographs, psychological testing, and medical examinations, which are designed to provide an objective basis for excluding certain individuals from stressful MMH jobs. Few of these methodologies, however, have been shown to be reliable in predicting an individual's risk of overexertion injury. Although no psychological tests have been found to quantify a worker's risk of overexertion injury, psychological testing can provide an indication of how a worker might respond to a severe injury. Shaw et al. evaluated patients with recent onset of occupational low back pain and using a survey of potential disability risks and a questionnaire reporting symptoms of pain.[93] Return to work was more strongly related to employer factors (e.g., job tenure, physical work demands, and modified duty availability) and self-ratings of pain and mood than by health history or physical examination.

Physical capacity screening

Another type of screening approach that has been used to select workers for MMH tasks includes individual testing of physical characteristics, such as strength, aerobic capacity, or functional capability. The underlying basis for using tests such as these to screen workers for MMH jobs is the belief that the risk of injury is dependent on the relationship between the capacity of the worker and the demands of the job. When the physical demands of the job exceed the capacity of the worker, the worker is at risk for developing an MSD. Thus, the idea of this approach is that workers should be matched to jobs according to the demands of the work.

A number of studies have been conducted to develop databases of maximum strength capacities (i.e., population averages) that could be used to design MMH tasks and workstations. These studies, however, disagree as to which of the three principal testing methods—isometric, isokinetic, or isoinertial—is most useful for determining strength capacity guidelines. Some researchers argue that traditional isometric lifting strength measurements, by which thousands of workers have been tested, are limited in assessing what workers can do under dynamic task conditions. These researchers suggest that dynamic strength testing is more appropriate than static strength testing for determining strength capacity.[94] This assertion is based on how well the test replicates the job requirements. Other researchers, however, claim that isokinetic lifting strength measurements probably have no greater inferential power to predict risk of injury or job performance than any other form of testing.[95] Kroemer claims that isoinertial strength testing is the most appropriate lifting strength testing method because it matches actual lifting conditions.[96] Isoinertial methods, however, have not been generally validated in terms of their ability to predict risk of injury.[97]

Maximum isometric lifting strength (MILS) has been studied and reported extensively. It has a well-established testing procedure[98] and has been reported in field tests to predict risk of injury.[22,99] Extensive measures of MILS have been made for various work postures and activities. One study, for example, measured the isometric strength of 1239 workers in rubber, aluminum, and electronic component industries.[100] In another study that measured the standardized isometric strengths (i.e., arm lift, torso lift, and leg lift) of 2178 aircraft manufacturing workers,[90] the employees were followed for >4 years to document back pain complaints. The investigators found that worker height, weight, age, and gender are poor predictors of standardized isometric strength, a finding that agrees with other studies.[100] The investigators also found, however, that standardized isometric strength is a poor predictor of reported back pain, a finding that conflicts with the results of some other studies.[101]

Marras et al. published an ergonomic guide for assessing dynamic measures of low back performance.[102] This guide provides information on elements of dynamic performance, techniques to assess dynamic performance, and relationships between testing techniques and internal forces. Bos et al. reported the results of a systematic literature search to identify specific lifting, pushing, and pulling demands and evaluate reliable and valid tests for the assessment of acceptable load on an individual level.[103] Results showed that there was no universal strategy for the definition of occupational demand and none of the tests met the criteria of reliability and prognostic value for musculoskeletal complaints completely on an individual level. There was some epidemiological evidence that for the development of musculoskeletal complaints, there was prognostic value with strength capacity tests, but not for a maximum acceptable load at an individual level.

SUMMARY

In summary, MMH activities, such as excessive lifting, pushing, pulling, and carrying, represent a serious hazard for MSD for many workers. There are analytic tools to identify ergonomic hazards that may result in overexertion injury and prevention tools that are effective in reducing the potential for risk of work-related overexertion injury. Successful ergonomic programs require the full cooperation of management, labor organizations, government, and the workers themselves. The solution requires a team effort with a commitment to identify and eliminate hazardous material handling tasks from the workplace.

Current status and future concerns

There is clear evidence that many MMH activities that involve lifting, pushing, pulling, or carrying present risks to workers for developing MSDs. In the past MSDs were frequently just considered as a cost of doing business, but as health problems mounted coupled with the increased costs to employers and taxpayers to deal with these increases, attitudes have changed. Prevention of MSDs from MMH activities has come to the forefront.

In this chapter MMH activities and the potential MSDs that can result have been described, and strategies for preventing and identifying MSDs have been presented. Analytical tools have been presented to identify the physical hazards that lead to injuries and are effective in reducing the risks for developing work-related MSDs.

Evident in this chapter is that much knowledge has been gained in the assessment and treatment of MSDs related to MMH activities; there are also significant research gaps that also exist. Several of these gaps have been identified by national expert reviews.[11,12,104] A list of some identified research gaps is presented below.

1. How accurate are methods for measuring a worker's maximum safe capacity? Worker selection approaches assume that it is possible to accurately determine a worker's safe capacity. In addition, does the safe capacity change over time (e.g., hourly, daily, monthly, yearly)?
2. Ergonomic approaches, which are considered to be more effective in reducing MSDs, can also be expensive to implement. Do such approaches have reasonable cost/benefit advantages?
3. How accurate are the existing assessment tools (e.g., biomechanical, psychophysical, physiological, integrated) that are used to determine risks of injury from MMH tasks. Do these tools have limitations in predicting long-term risks?
4. Do performance standards or incentive programs increase a workers' risk of injury from MMH activities? If performance standards are used, are there criteria available to ensure safety for the worker?
5. What are the effects psychosocial risk factors (job stress, organizational, job safety, work time, work freedom, work schedule) on MMH activities that can influence the risk of MSDs?
6. The workforce is changing, with older workers and more women involved in MMH activities that require frequent lifting, pushing, and pulling. Are these changes being incorporated into job designs that can reduce the physical demands of many MMH activities?

These questions, although not exhaustive, identify the many issues that safety and health experts and industry leaders must address to reduce work-related injuries. Efforts to limit workplace hazards will only be successful with the full cooperation of management, government, and labor and, of course, the workers themselves. Solutions will require collaborative efforts of all involved.

References

1. International Organization for Standardization. ISO 11228-1:2003. Ergonomics: Manual Handling. Part 1: Lifting and Carrying. Geneva: ISO, 2003.
2. Cal/OSHA Consultation Service and the National Institute for Occupational Safety and Health. Ergonomic Guidelines for Manual Material Handling. DHHS (NIOSH) Publication No. 2007-131. Sacramento/Cincinnati, OH: Department of Industrial Relations/Department of Health and Human Services, Centers for Disease Control and Prevention, National Institute for Occupational Safety and Health, 2007.
3. Liberty Mutual Research Institute for Safety. 2014 Liberty Mutual Workplace Safety Index. Hopkinton, MA: Liberty Mutual Research Institute for Safety, 2014.
4. Centers for Disease Control and Prevention. Nonfatal Work-Related Injuries and Illnesses – United States, 2010. CDC Health Disparities and Inequalities Report—United States, 2013. *MMWR* 2013; 62 (Suppl 3):35–40.
5. Bureau of Labor Statistics. Nonfatal Occupational Injuries and Illnesses Requiring Days Away From Work, 2013. U.S. Department of Labor, USDL-14-2246, December 16, 2014.
6. National Safety Council. Injury Facts®. Itasca, IL: National Safety Council, 2015.
7. Waters TR, Dick RB, Davis-Barkley J, et al. A cross-sectional study of risk factors for musculoskeletal symptoms in the workplace using data from the general social survey (GSS). *J Occup Environ Med* 2007; 49(2):172–84.
8. Waters TR, Dick RB, Krieg EF. Trends in work-related musculoskeletal disorders: a comparison of risk factors for symptoms using quality of work life data from the 2002 and 2006 general social survey. *J Occup Environ Med* 2011; 53(9):1013–24.
9. Dick RD, Lowe BD, Lu M-L, et al. Further trends on work-related musculoskeletal disorders: a comparison of risk factors for symptoms using quality of work life data from the 2002, 2006 and 2010 general social survey. *J Occup Environ Med* 2015; 57(8):910–28.
10. Bogduk N. Clinical Anatomy of the Lumbar Spine and Sacrum, 5th ed. New York: Elsevier/Churchill Livingstone, 2012.
11. National Institute for Occupational Safety and Health. Musculoskeletal Disorders and Workplace Factors: A Critical Review of Epidemiological Evidence for Work-Related Musculoskeletal Disorders of the Neck, Upper Extremity, and Low Back. DHHS (NIOSH) Publication No. 97-141. Washington, DC: US Department of Health and Human Services, 1997.
12. National Research Council, Institute of Medicine—Panel on Musculoskeletal Disorders and the Workplace, Commission on Behavioral and Social Sciences and Education. Musculoskeletal Disorders and the Workplace: Low Back and Upper Extremities, Washington, DC: National Academy Press, 2001.
13. da Costa BR, Vieira RE. Risk factors for work-related musculoskeletal disorders: a systematic review of recent longitudinal studies. *Am J Ind Med* 2010; 53:285–323.
14. Pope MH, Frymoyer JW, Andersson G. Occupational Low Back Pain. New York: Praeger, 1984.
15. Ferguson SA, Marras WS. A literature review of low back disorder surveillance measures and risk factors. *Clin Biomech* 1997; 12(4):211–26.
16. Mayer J, Kraus T, Ochsmann E. Longitudinal evidence for the association between work-related physical exposures and neck and/or shoulder complaints: a systematic review. *Int Arch Occup Environ Health* 2012; 85:587–603.
17. Gallagher S, Heberger JR. Examining the interaction of force and repetition on musculoskeletal disorder risk: a systematic literature review. *Hum Factors* 2013; 55(1):108–24.
18. U.S. Department of Labor. Safety and Health Management Systems eTool website. Occupational Safety and Health Administration, http://www.osha.gov/SLTC/etools/safetyhealth/comp2.html accessed August 7, 2015.
19. David GC. Ergonomic methods for assessing exposure to risk factors for work-related musculoskeletal disorders. *Occup Med* 2005; 55:190–9. doi:1093/occmed/kqi082
20. van der Beek A, Frings-Dresen MHW. Assessment of mechanical exposure in ergonomic epidemiology. *Occup Environ Med* 1998; 55:190–5.
21. Dempsey PG, McGorry RW, Maynard WS. A survey of tools and methods used by certified professional ergonomists. *Appl Ergon*; 2005, 36:489–503.
22. Chaffin DB, Park KS. A longitudinal study of low-back pain as associated with occupational weight lifting factors. *Am Ind Hyg Assoc J* 1973; 34:513–25.
23. Frymoyer JW, Pope MH, Costanza MC, et al. Epidemiologic studies of low back pain. *Spine* 1980; 5:419–23.
24. Snook SH. The design of manual-handling tasks. *Ergonomics* 1978; 21:963–85.
25. Waters TR, Putz-Anderson V, Garg A, et al. Revised NIOSH equation for the design and evaluation of manual lifting tasks. *Ergonomics* 1993; 36(7):749–76.
26. The Eastman Kodak Company. Kodak's Ergonomic Design for People at Work, 2nd ed. Hoboken, NJ: John Wiley & Sons, Inc., 2004.
27. Ayoub MM, Mital A, Asfour SS, et al. Review, evaluation, and comparison of models for predicting lifting capacity. *Hum Factors* 1980; 22(3):257–69.
28. Chaffin DB, Andersson GBJ, Martin BJ. Occupational Biomechanics, 4th ed. New York: Wiley-Interscience, 2006.
29. Roderick D, Karwowski W. Manual materials handling. In: Salvendy G, ed. Handbook of Human Factors and Ergonomics, 3rd ed. Hoboken, NJ: John Wiley & Sons, Inc., 2006, pp. 818–54.
30. Chaffin DB, Baker WH. Biomechanical model for analysis of symmetric sagittal plane lifting. *AIIE Trans* 1970; 2(1): 16–27.
31. Schultz AB, Andersson GBJ. Analysis of loads on the lumbar spine. *Spine* 1981; 6(1):76–82.
32. Freivalds A, Chaffin DB, Garg A, et al. A dynamic biomechanical evaluation of lifting maximum acceptable loads. *J Biomech*, 1984; 17(4):251–62.

33. McGill SM, Norman RW. Dynamically and statically determined low back moments during lifting, *J Biomech* 1985; 18(12):877–85.
34. Marras WS Sommerich CM. A three-dimensional motion model of loads on the lumbar spine: I. Model structure. *Hum Factors* 1991; 33(2):123–37.
35. Marras WS, Sommerich CM. A three-dimensional motion model of Loads on the lumbar spine: II. Model validation. *Hum Factors* 1991; 33(2):139–49.
36. Marras WS, Allread WG, Burr DL, et al. Prospective validation of a low-back disorder risk model and assessment of ergonomic interventions associated with manual materials handling tasks. *Ergonomics* 2000; 43(11):1866–86.
37. Kerr MS, Frank JW, Shannon HS, et al. Biomechanical and psychosocial risk factors for low back pain at work. *Am J Pub Health* 2001; 91(7):1069–75.
38. Skotte JH, Essendrop M, Hansen AF, et al. A dynamic 3D biomechnical evaluation on the load on the low back during different patient-handling tasks. *J Biomech* 2002; 35:1357–66.
39. Brinckmann P, Johannleweling N, Hilweg D, et al. Fatigue fracture of human lumbar vertebrae: brief report. *Clin Biomech* 1987; 2(2):94–6.
40. Stevens SS. Mathematics, measurement, and psychophysics. In: Stevens SS, ed. Handbook of Experimental Psychology. New York: John Wiley & Sons, Inc., 1951.
41. Ayoub MM. Psychophysical Basis for Manual Lifting Guidelines. Purchase order number 88-79313, 1988, Preliminary report received 12/16/88. Final report received 2/1/89. Cincinnati, OH: National Institute for Occupational Safety and Health, 1988.
42. Snook SH, Ciriello VM. The design of manual-handling tasks: revised tables of maximum acceptable weights and forces. *Ergonomics* 1991; 34:1197–213.
43. Ayoub MM, Mital A. Manual Materials Handling. London: Taylor & Francis, 1989.
44. Borg G. Psychophysical scaling with applications in physical work and the perception of exertion. *Scand J Work Environ Health* 1990; 16(Suppl 1):55–8.
45. Corlett EN, Bishop RP. A technique for assessing postural discomfort. *Ergonomics*, 1976; 19:175–82.
46. Rodgers SH, Yates JW, Garg A. The Physiological Basis for Manual Lifting Guidelines. Report No. 91-227-330. Springfield, VA: National Technical Information Service, 1991.
47. Garg A. A Metabolic Rate Prediction Model for Manual Materials Handling Jobs. Ann Arbor, MI: University of Michigan, 1976.
48. Astrand PO, Rodahl K. Textbook of Work Physiology, 3rd ed., New York: McGraw-Hill, 1986.
49. Ayoub MM. Determination and Modeling of Lifting Capacity. Final report. Grant no. 5-ROI-OH-00545002. Cincinnati, OH: Department of Health and Human Services, National Institute for Occupational Safety and Health, 1978.
50. Takala EP, Pehkonen I, Forsman M, et al. Systematic evaluation of observational methods assessing biomechanical exposures at work. *Scand J Work Environ Health* 2010; 36(1):3–24.
51. University of Wisconsin-Madison, Multimedia Video Task Analysis™ (MVTA™). Madison, WI: Ergonomics Analysis and Design Consortium. http://mvta.engr.wisc.edu accessed April 27, 2016.
52. University of Michigan, 3D Static Strength Prediction Program™ (3DSSPP™). Ann Arbor, MI: Center for Ergonomics. http://c4e.engin.umich.edu/tools-services/3dsspp-software/ accessed April 27, 2016.
53. Liu Y, Zhang Z, Chaffin D. Perception and visualization of human information for computer-aided ergonomic analysis. *Ergonomics* 1997; 40:818–33.
54. Lu M, Waters T, Werren D. Development of human posture simulation method for assessing posture angles and spinal loads. *Hum Factors Ergon Manuf Serv Ind* 2015; 25(1): 123–36.
55. Lowe B. Accuracy of validity of observational estimated of wrist and forearm posture. *Ergonomics* 2004; 47(5):527–54.
56. Bao S, Howard N, Spielholz P, et al. Interrater reliability of posture observations. *Hum Factors* 2009; 51(3):292–309.
57. Marras WS, Fattalah F. Accuracy of three dimensional lumbar motion monitor for recording dynamic trunk motion characteristics. *Int J Ind Ergon* 1992; 9:75–87.
58. Marras WS, Schoenmarklin RW. Wrist motions in industry. *Ergonomics* 1993; 36:341–51.
59. Grant KA, Galinsky, TL, Johnson PW. Use of the actigraph for objective quantification of hand/wrist activity in repetitive work. In: Proceedings of the Human Factors and Ergonomics Society 37th Annual Meeting, Sheraton Seattle Hotel & Towers/Washington State Convention Center, October 11–15, 1993, Seattle WA. Santa Monica, CA: Human Factors and Ergonomics Society, pp. 720–4.
60. Roetenberg D, Slycke P. *Xsens MVN*: Full 6 DOF Human Motion Tracking using Miniature Inertial Sensors. Enschede: Xsens Technologies, 2013.
61. Fabers G, Kingma I, Bruijin S, et al. Optimal inertial sensor location for ambulatory measurement of trunk inclination. *J Biomech* 2009; 42:2406–9.
62. Garg A. What basis exists for training workers in correct lifting technique? In: Marras WS, Karwowski W, Smith JC, et al., eds. The Ergonomics of Manual Work. London: Taylor & Francis, 1993.
63. National Institute for Occupational Safety and Health. Work Practices Guide for Manual Lifting. NIOSH Technical Report No. 81-122. Cincinnati, OH: Department of Health and Human Services, National Institute for Occupational Safety and Health, 1981.
64. Waters TR, Putz-Anderson V, Garg A. Applications Manual for the Revised NIOSH Lifting Equation. DHHS (NIOSH) Publication No. 94-110. Cincinnati, OH: National Institute for Occupational Safety and Health, 1994.
65. Waters TR, Baron SL, Piacitelli LA, et al. Evaluation of the revised NIOSH lifting equation: a cross-sectional epidemiological study. *Spine* 1999; 24(4):386–95.
66. Lu M-L, Putz-Anderson V, Garg A, et al. Evaluation of the impact of the revised National Institute for Occupational Safety and Health Lifting Equation. Hum Factors 2016; 58(5):667–82.
67. Boden SD, Wiesel SW. Standardized approaches to the diagnosis and treatment of low back pain and multiply operated low back patients. In: Wiesel SW, Wiesel SW, eds. Industrial Low Back Pain: A Comprehensive Approach, 2nd ed. Charlottesville, VA: The Michie Company, 1989.

68. Bigos SJ, Bowyer OR, Braen RG, et al.. Acute Low Back Problems in Adults. Clinical Practice Guideline No. 14. AHCPR Publication No. 95-0642. Rockville, MD: Agency for Health Care Policy and Research, Public Health Service, U.S. Department of Health and Human Services, 1994.
69. Schultz IZ, Stowell AW, Feuerstein M, et al. Models of return to work for musculoskeletal disorders. *J Occup Rehabil* 2007; 17:327–52.
70. Andersson GBJ, Pope MH, Frymoyer JW, et al.. Epidemiology and cost. In: Pope MH, Andersson GBJ, Frymoyer JW, et al., eds. Occupational Low Back Pain: Assessment, Treatment, and Prevention. St. Louis, MO: Mosby-Year Book, 1991, pp. 95–113.
71. White AA, Gordon SL. Synopsis: workshop on idiopathic low back pain. *Spine* 1982; 7:141–9.
72. Chou R, Qaseem A, Snow V, et al. Diagnosis and treatment of low back pain: a joint clinical practice guideline from the American College of Physicians and the American Pain Society. *Ann Intern Med* 2007; 147(7):478–91.
73. Casazza BA. Diagnosis and treatment of acute low back pain. *Am Fam Physician* 2012; 85(4):343–50.
74. Silverstein BA, Fine U. Evaluation of Upper Extremity and Low Back Cumulative Trauma Disorders: A Screening Manual. Ann Arbor, MI: University of Michigan, School of Public Health, Occupational Health Program, 1984.
75. Frymoyer JW, Haldermans S. Evaluation of the worker with low back pain. In: Pope MH, Andersson GBJ, Frymoyer JW, et al., eds. Occupational Low Back Pain: Assessment, Treatment, and Prevention. St. Louis, MO: Mosby-Year Book, 1991, pp. 151–82.
76. Andersson GBJ, Frymoyer JW. Treatment of the acutely injured worker. In: Pope MH, Andersson GBJ, Frymoyer JW, et al., eds. Occupational Low Back Pain: Assessment, Treatment, and Prevention. St Louis, MO: Mosby-Year Book, 1991, pp. 183–94.
77. Putz-Anderson V. Cumulative Trauma Disorders: A Manual for Musculoskeletal Disorders of the Upper Limbs. London: Taylor & Francis, 1988.
78. Deyo RA. Non-operative treatment of low back disorders. In: Frymoyer JW, ed. The Adult Spine: Principles and Practice. New York: Raven, 1991.
79. Low back disorders. In: Hegmann KT, eds. Occupational Medicine Practice Guidelines. Evaluation and Management of Common Health Problems and Functional Recovery in Workers, 3rd ed. Elk Grove Village, IL: American College of Occupational and Environmental Medicine (ACOEM), 2011, pp. 333–796. http://www.guideline.gov/content.aspx?id=38438 accessed April 27, 2016.
80. Food and Drug Administration. FDA Drug Safety Communication: FDA Requires Label Changes to Warn of Rare but Serious Neurologic Problems after Epidural Corticosteroid Injections for Pain April 23, 2013. http://www.fda.gov/Drugs/DrugSafety/ucm394280.htm accessed April 27, 2016.
81. Machado L, de Souza M, Ferreira P, et al. The McKenzie method for low back pain. *Spine* 2006, 31(9):254–62.
82. Deyo RA, Loeser JD, Bigos SJ. Herniated lumbar intervertebral disk. *Ann Intern Med* 1990; 112:598–603.
83. Putz-Anderson V, NIOSH. Ergonomic Solutions for Retailers: Prevention of Material Handling Injuries in the Grocery Sector. DHHS (NIOSH) Publication No. 2015-100. Cincinnati, OH: U.S. Department of Health and Human Services, Centers for Disease Control and Prevention, National Institute for Occupational Safety and Health, 2014.
84. Occupational Safety and Health Administration. Ergonomics Program Management Guidelines for Meatpacking Plants. OSHA Document No. 3123. Washington, DC: Department of Labor, Occupational Safety and Health Administration, 1990.
85. NIOSH. Workplace Use of Back Belts. DHHS (NIOSH) Publication No. 94-122. Cincinnati, OH: U.S. Department of Health and Human Services, Centers for Disease Control and Prevention, National Institute for Occupational Safety and Health, 1994.
86. NIOSH. Back Belts: Do They Prevent Injury? DHHS (NIOSH) Publication No. 94-127, Cincinnati, OH: U.S. Department of Health and Human Services, Centers for Disease Control and Prevention, National Institute for Occupational Safety and Health, 1994.
87. Wassell JT, Gardner LI, Landsittel DP, et al. A prospective study of back belts for prevention of back pain and injury. *JAMA* 2000; 284(21):2727–32.
88. Bobick TG, Belard J-L, Hsaio H, et al. Physiological effects of back belt wearing during asymmetric lifting. *Appl Ergon* 2001; 32:541–7.
89. Giocelli RJ, Hughes RE, Wassell JT, et al. The effect of wearing a back belt on spine kinematics during asymmetric lifting of large and small boxes. *Spine* 2001; 26(16): 1794–8.
90. NIOSH. Simple Solutions for Home Building Workers: A Basic Guide for Preventing Manual Material Handling Injuries. DHHS (NIOSH) Publication No. 2013-111. Cincinnati, OH: U.S. Department of Health and Human Services, Centers for Disease Control and Prevention, National Institute for Occupational Safety and Health, 2013.
91. Robson L, Stephenson C, Schulte P, et al. A Systematic Review of the Effectiveness of Training and Education for the Protection of Workers. DHHS (NIOSH) Publication No. 2010-127. Toronto/Cincinnati, OH: Institute for Work & Health/National Institute for Occupational Safety and Health, 2010.
92. Battie MC, Bigos SJ, Fisher LD, et al. Isometric lifting strength as a predictor of industrial back pain reports. *Spine* 1989; 14(8):851–6.
93. Shaw WS, Pransky G, Patterson W, et al. Early disability factors for low back pain assessed at outpatient health clinics. *Spine* 2005; 30(5):572–80.
94. Kroemer KHE. Testing individual capability to lift material: repeatability of a dynamic test compared with static testing. *J Safety Res* 1985; 3:4–7.
95. Rothstein JM, Lamb RL, Mayhew TP. Clinical uses of isokinetic measurements. *Phys Ther* 1987; 67:1840–4.
96. Kroemer KHE. An isoinertial technique to assess individual lifting capacity. *Hum Factors* 1983; 25:493–506.
97. Kroemer KHE. Matching individuals to the job can reduce manual labor injuries. *Occup Safety Health News Dig* 1987; 3:4–7.
98. Chaffin DB. Ergonomics guide for the assessment of human static strength. *Am Ind Hyg Assoc J* 1975; 36:505–11.

99. Herrin GD, Jaraiedi M, Anderson CK. Prediction of overexertion injuries using biomechanical and psychophysical modes. *AIHA J* 1986; 47:322–30.
100. Herrin GD, Keyserling WM, Chaffin DB, et al. A Comparison of selected work muscle strength. In: Proceeding of the 22nd Annual Meeting of the Human Factors Society, October 1978, Detroit, MI. Santa Monica, CA: Human Factors and Ergonomics Society.
101. Keyserling WM, Herrin GD, Chaffin DB. Isometric strength testing as a means of controlling medical incidents on strenuous jobs. *J Occup Med* 1980; 22:332–6.
102. Marras WS, McGlothlin JD, McIntyre DR, et al. Dynamic Measures of Low Back Performance: An Ergonomics Guide. Fairfax, VA: American Industrial Hygiene Association, 1993.
103. Bos J, Kuijer PPFM, Frings-Dresen MHW. Definition and assessment of specific occupational demands concerning lifting, pushing, pulling based on systematic literature search. *Occup Environ Med* 2002; 59:800–6.
104. Marras WS, Cutlip RG, Burt SE, et al. National occupational research agenda (NORA) future direction in occupational musculoskeletal disorder health research, *Appl Ergon* 2009; 40:15–22.

Chapter 4

OCCUPATIONAL VIBRATION EXPOSURE

David G. Wilder and Donald E. Wasserman*

Human beings have been vibration exposed for thousands of years to nonhuman-generated cyclic or repetitive loading since the beginning of the use of tools, boats, airplanes, railway travel, or other transport methods using animals or platforms that could be dragged or placed on runners/skids, rollers, or wheels. Impact and vibration (acceleration) have been of increasing interest since the early 1900s. As early as 1918, there was intense interest in the medical effects of power tool vibration exposure on workers by the famous occupational medicine pioneer Dr. Alice Hamilton.[1]

OCCUPATIONAL SETTING

Vibration is the periodic motion of a body in alternately opposite directions from a position of rest. Vibration is present in most work settings where mechanical equipment is used. When vibration interacts with the human body, the coupling pathway in which it enters and moves through the body defines its path of travel. There are two major types of vibration that have human health concerns. Whole-body vibration (WBV) affects the entire body and is usually transmitted in a sitting or standing position from a vibrating seat or platform. Segmental or hand–arm vibration (HAV) affects one or both upper extremities and is usually transmitted to the hand and arm only from a motorized hand tool. WBV is generated by motor vehicle operation, including cars, trucks, buses, trains, marine craft, construction and agricultural (tractors, threshers, or combined) equipment, and heavy manufacturing equipment such as looms, large machine tools, and presses. HAV is generated by any powered hand tool, including chippers, jackhammers, chainsaws, trimmers and blowers, nut-tightening guns, polishers, grinders, and rivet guns. These tools have widespread use in industry and may be electric, pneumatic, hydraulic, or combustion engine powered. Workers using any powered hand tools have potential exposure.

In most cases, WBV and HAV exposures are distinctly separable. In a few cases there are crossover situations where simultaneously both intense WBV and HAV exposures occur simultaneously. For example,

- When a "pavement breaker tool" is operated at arm's length away from the operator's body, only the fingers, hands, and arms are involved, and this is clearly HAV exposure; but when this same tool is operated differently and the operator now also leans into this tool while working and places it against the stomach, in an attempt to damp the vibration, it results in HAV and WBV exposure because the vibration has a second pathway coupling it to the body (hands and stomach) with the distinct possibility of injury or disease to the operator's fingers and hands (HAV) and/or the omentum (WBV).
- The off-road operation of a motorcycle, resulting in intense HAV exposure to the hands via vibration coupled from the handlebars and simultaneously intense WBV exposure entering the spine via the operator's seat from the engine, ground conditions, and vehicle speed.

*Malcolm Pope and the late physicians Peter Pelmear, and William Taylor contributed to previous editions of this chapter.

Physical and Biological Hazards of the Workplace, Third Edition. Edited by Gregg M. Stave and Peter H. Wald.

The health consequences and effects of WBV and HAV exposures are mostly distinctly different as are the WBV and HAV protective standards, so they will be discussed separately in this chapter. The basic physics and vibration engineering terminology are the same for both WBV and HAV.

OCCUPATIONAL VIBRATION MEASUREMENTS

Vibration is a physical agent, expressed in terms of motion (acceleration), time, and frequency. It is not practical or acceptable to determine a dose of vibration that an individual receives inside the body as this would require an invasive measurement method such as fixing a motion sensor directly to a point of interest in or on the skeleton. Instead, surrogate motion information is used, measured at an interface between the individual and a vibration source.

Vibratory motion is by definition a mathematical "vector quantity" which simply means that it is described by both a direction and a magnitude. At each measurement point, the total motion is described using six possible vector directions; three so-called linear directions (up–down, side to side, front to back) with their magnitudes and three rotational directions (pitch, yaw, and roll) with their magnitudes. In most human vibration work, only the linear directions are measured, reported, and compared to health/safety standards.

Figure 4.1 shows the internationally accepted "biodynamic coordinate system" used for head to toe, or whole-body vibration (WBV), and similarly for segmental or hand–arm vibration (HAV). For WBV the sternum is the reference point of the measurements. The vibration intensity or magnitude quantity of choice for WBV (and HAV) vibration is "frequency-weighted acceleration (ms^{-2}), root mean squared (rms)." For WBV, by definition, motion in the "*Z*" direction is head to toe; the "*Y*" direction is side to side (shoulder to shoulder); and the "*X*" direction is front to rear. Each of these three linear directions and corresponding acceleration magnitudes is separately measured, digitally stored, and reported, and these data are used for exposure calculations.

For HAV measurements, the third metacarpal is the reference measurement point (Figure 4.2). The defined motion in the "*Z*" direction (axis) is along the long bones of the forearm; the "*Y*" direction is motion across the knuckles; and the "*X*" direction is motion through the palm. Many times it is not practical to place measurement accelerometers on the third metacarpal; thus measurements are obtained directly from the vibrating tool handle as the operator works and grips the handle. This tool handle frame of reference is called the "basicentric coordinate system." The reason for using these types of WBV and HAV coordinate systems is to establish uniform measurement methods worldwide and a means of easily comparing measurements to health and safety standards.

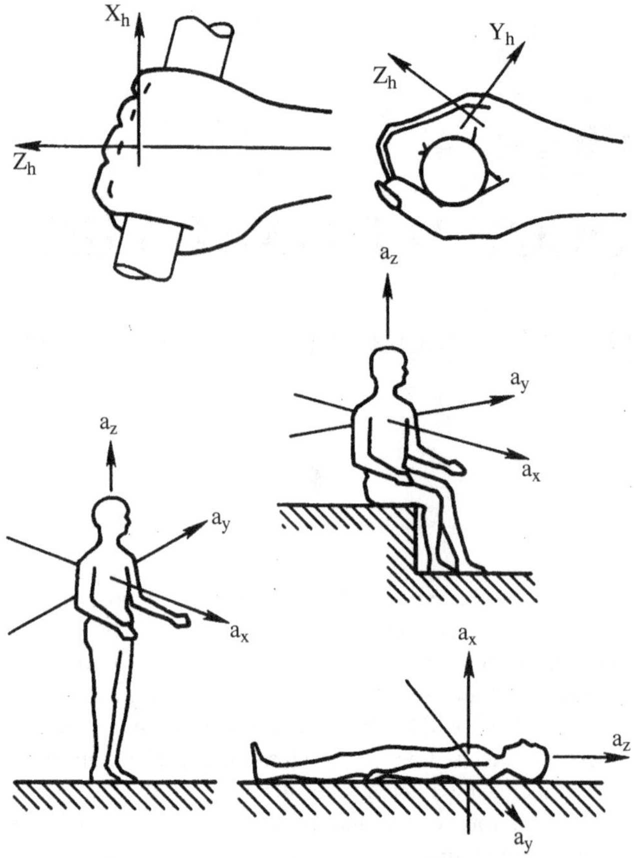

FIGURE 4.1 Hand-arm and whole-body biodynamic coordinate systems for human measurements. *Source*: Adapted from Figure 1 (page 87) from the *1991–1992 Threshold Limit Values for Chemical Substances and Physical Agents and Biological Exposure Indices* book; Figure 3 (page 130) from the *1996 TLVs® and BEIs®* book. From ACGIH®, *1991–1992 and 1996 TLVs® and BEIs®* Books. Copyrights 1991 and 1996 respectively. Reprinted with permission.

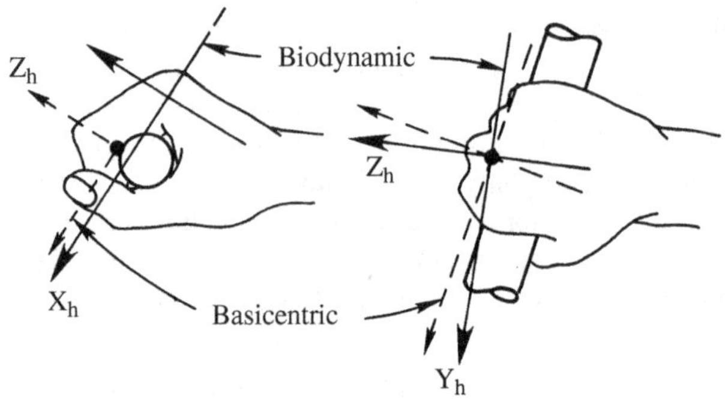

FIGURE 4.2 Biodynamic and basicentric coordinate systems for hand-arm vibration and power tool measurements. *Source*: From Figure 1 (page 87) from the ACGIH®, *1991-1992 TLVs® and BEIs®* Book. Copyright 1991. Reprinted with permission.

In practice, three-axis (triaxial) vibration measurements are simultaneously and individually obtained and recorded on separate channels (on a computer-based data acquisition system), and all triaxial data are saved. In both WBV and HAV measurements, what is usually sought is the total impinging vibration or the so-called weighted vector sum of all three measurement axes. Once this quantity has been determined, it is next compared to the appropriate health and safety standard(s) to determine if and by how much said standards have been exceeded. If these results indicate that the appropriate standard has been exceeded, then vibration controls are most likely needed.

The expression of vibration magnitude or intensity is "acceleration" expressed in units of meters/second/second root mean squared or ms^{-2}rms as, for example, 10 ms^{-2}rms, 20 ms^{-2}rms, etc. worldwide; in the United States we also use the familiar "gravitational *g* term for acceleration," where 1 g = 9.81 meters/second/second or ms^{-2}. Acceleration is defined as the time rate of change of velocity. Vibration frequency is expressed as hertz (Hz), defined as a complete vibration cycle occurring in a 1 second time span.

A unique concern is a vibration characteristic called "resonant or natural frequency vibration." Resonance or vibration at a resonant frequency is an undesirable condition when input vibration impinges on a structure or mechanism that in turn triggers an uncontrolled, undesirable, internally amplified, and exacerbated response from the structure or mechanism. This is of concern in occupational vibration because humans can respond at their human resonant frequencies when impinging vibration is uncontrollably amplified and exacerbated by the human body, depending on the impinging vibrations direction, magnitude, frequency, and exposure duration time. Human response to vibration at these resonant frequencies represents the "Achilles heel" of human vulnerability, impairment, and disease.

Because of resonance and other less well-understood factors, human response to vibration is not linear, meaning that our bodies do not respond to each vibration frequency the same way. At resonance, these special frequencies of vibration can unleash a large involuntary response from the body, sometimes with small triggering inputs, as compared to the results of nonresonant frequencies of vibration, being triggered with the same vibration intensities. Vibration axis/direction, exposure time, frequency, and intensity all matter. To take these factors into account when determining a dose–response relationship, we use a mathematical calculation called a "frequency weighting function," whose purpose is to mimic and attempt to characterize human WBV response across a wide frequency range or bandwidth and similarly use another weighting function for HAV response.

Another measure used to analyze vibration is Crest Factor, which is defined as the ratio of peak acceleration value divided by its corresponding root-mean-squared acceleration value. Since acceleration units are both in the numerator and denominator, the result is expressed as a pure number. This is of particular concern in WBV when vehicles are driven on bumpy roads or when trains experience "slack action" due to changes in their car couplers.

Most vibration measurements of WBV and HAV workplace environments contain complex vibration consisting of multiple vibration frequencies occurring simultaneously. Currently, a very few vibration standards data analysis require that a Fourier analysis be performed for each of the three axes of data. This analysis mathematically dissects and identifies each of the vibration frequencies from its complex overall vibration revealing its total elemental contents and its respective total amplitude contribution to the original overall complex vibration. Fourier analysis is a mathematical decomposition technique that is analogous to the well-known electro-optical "spectroscopy" technique used in chemical analysis when examining an unknown complex compound by determining and revealing the compound's elemental composition and the respective concentration for each element contained in the complex compound.

In this chapter the terms jolt, impact, and mechanical shock are used synonymously. "*Shock* is a somewhat loosely defined aspect of vibration wherein the excitation is nonperiodic, e.g., in the form of a pulse, a step, or transient vibration" rather than repeating itself over and over again. "The word 'shock' implies a degree of suddenness and severity."[2]

OCCUPATIONAL VIBRATION GUIDELINES USED IN THE UNITED STATES

Standards relating to human exposure to vibration have been evolving since the early 1970s. With the exception of the European Union's Directive 2002 where WBV and HAV standards both are binding law in member states, WBV and HAV standards used in the United States are voluntary nonbinding consensus standards/guidelines which are periodically reviewed and updated as new information becomes available.[3] Currently there are no official Occupational Safety and Health Administration (OSHA) vibration standards. However, OSHA can and does cite organizations for excessive WBV and/or HAV exposures under the general duty clause of the Occupational Safety and Health Act.

Currently four whole-body vibration standards/guidelines are used in the United States:

ACGIH standard for WBV,[4] ISO 2631 WBV standard,[5] ANSI S3.18-2002/ISO 2631-1:1997, Nationally Adopted International Standard,[6] and the EU Physical Agents Directive 2002.[3]

Three hand–arm vibration standards/guidelines are used in the United States:

ACGIH standard for HAV,[7] ANSI S2.70-2006,[8] and the National Institute for Occupational Safety and Health (NIOSH) Publication #89-106 criteria document for a standard for HAV.[9] There is also a 2014 antivibration/reduced HAV glove testing certification standard ISO 10819[10] used worldwide including the United States. The following discussion contains the basic elements of the occupational vibration standards.

Whole-body vibration standards

All WBV standards are generally used as follows:

1. Triaxial WBV acceleration measurements are made, digitized, and stored as separate channels in memory.
2. Signals from each of the three axes are weighted.
3. A weighted vector sum is calculated resulting in a single weighted total value. This is what is sought.
4. With knowledge of the worker's daily WBV exposure time, this resultant value is compared to the exposure range limits: from 0.5 to 1.15 ms^{-2} rms (for an 8-hour daily exposure).
5. If the calculated value for an 8-hour exposure is:
   ≤0.5 ms^{-2} rms, then the conditions require no treatment,
   >0.5 ms^{-2} rms but ≤1.15 ms^{-2} rms, conditions must be treated, and
   >1.15 ms^{-2} rms, exposure must be reduced.

ACGIH WBV STANDARD

The early 1970s saw the introduction of the world's first occupational WBV standard: the international standard ISO 2631.[4,11] The limits used were similar in the 1978 and 1985 versions and were the "gold standard" and were used extensively worldwide until 1997 when the ISO chose to significantly change and modify its standard to its current version.[5,12,13] Some strongly believed this early version of ISO 2631 should continue to be used because it was developed from some of the best and most pristine research conducted in this field to date. Eventually ACGIH agreed and adopted the best parts as their own.[14] Graphs of the values from the ACGIH WBV standard are given in Figures 4.3 and 4.4. In order to use this standard, all measured WBV acceleration per axis must first be Fourier analyzed and converted into spectra where Figure 4.3 is used to evaluate WBV measured spectra in the vertical "*Z*" direction. Figure 4.4 is used separately, twice, first to evaluate spectra in the side-to-side "*Y*" direction and separately again in the front-to-rear "*X*" direction.

Referring to Figure 4.3, the horizontal axis shows vibration frequency in so-called 1/3 octave bands, extending from 1 to 80 Hz. The vertical axis of Figure 4.3 gives a measure of the vibration "intensity" magnitude in root-mean-squared (rms) acceleration in either "*g*'s" or meters per second per second, where 1 *g* = 9.81 ms^{-2}. A "family" or set of parallel "U"-shaped daily exposure time curves is also shown in the graph. These U-shaped exposure curves are the frequency weighting function for *Z* axis WBV, where the trough of the curves is 4–8 Hz in which are found the resonant frequencies, and greatest sensitivities, mentioned above in the *Z* direction. What Figure 4.3 really shows are

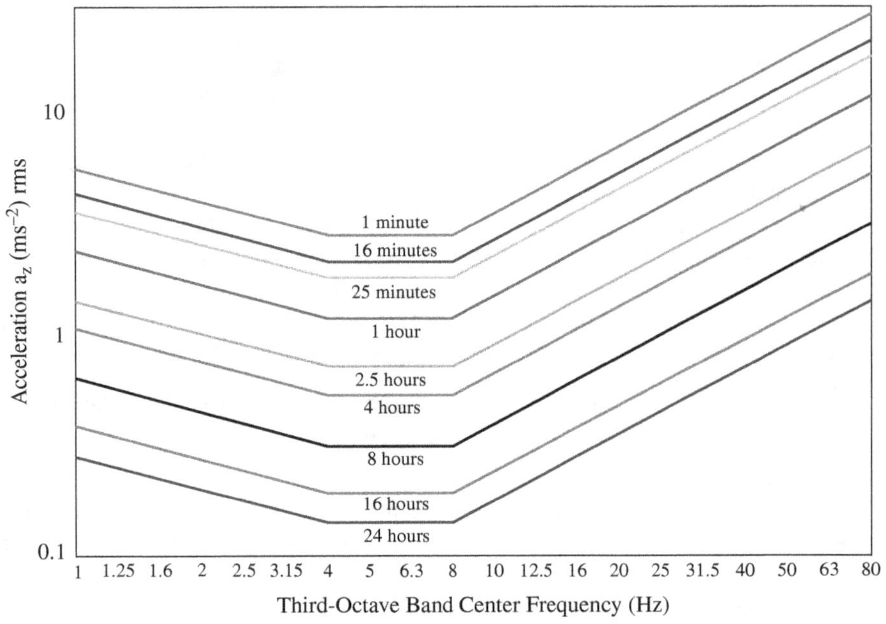

FIGURE 4.3 a_z acceleration limits as a function of vibration frequency and exposure time (compiled from data from ACGIH-WBV Standard, 1995–2015; ISO 2631 WBV Standard, 1979–1996).

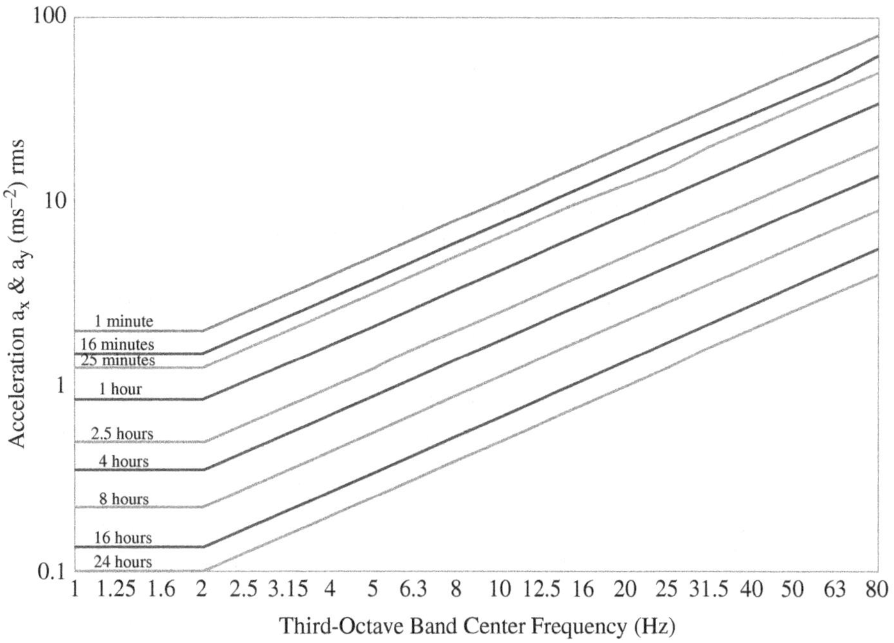

FIGURE 4.4 a_x and a_y acceleration limits as a function of vibration frequency and exposure time (compiled from data from ACGIH-WBV Standard, 1995–2015; ISO 2631 WBV Standard, 1979–1996).

vibration intensity (acceleration) limits as function of vibration frequency and daily worker exposure time for the "*Z*" axis. Figure 4.4 shows the same type of information for the "*Y*" and "*X*" axes where 1–2 Hz are the resonant frequencies. If one or more axis acceleration exceeds the standard, then the entire standard has been exceeded, and appropriate vibration control action is necessary.

ISO 2631 WBV STANDARD (ISO 2631-1:1997; 2010), (ISO 2631-2:2003; 2003), (ISO 2631-4:2001; 2001), (ISO 2631-5:2004; 2004)

ISO 2631 is a multipart standard related to human exposure to whole-body vibration that is promulgated and periodically updated by the International Standards Organization of Geneva, Switzerland. The latest version of ISO 2631 used in the United States includes several parts that address measurement and evaluation of human exposure to whole-body vibration. Current versions address possible health, comfort, perception, and motion sickness effects of vibration in general (2631-1)[5]; occupant comfort and annoyance in buildings (2631-2)[15]; passenger and crew comfort in fixed-guideway transport systems, such as trains (2631-4)[16]; and adverse health effects on the lumbar spine in conditions containing multiple shocks (2631-5).[17]

European Union (EU) Physical Agents (Vibration) Directive 2002

The EU Physical Agents Directive is legally binding in all EU member states.[3] The EU Physical Agents Directive established two levels of WBV exposure evaluation criteria:

- The daily exposure action value (DEAV) = 0.5 ms^{-2} rms for an 8 hours/day WBV exposure
- The daily exposure limit value (DELV) = 1.15 ms^{-2} rms for an 8 hours/day WBV exposure

When the DEAV 8 hours/day WBV exposure total weighted rms acceleration value is less than 0.5 ms^{-2} rms, it is considered a safe level. If the DEAV = 0.5 ms^{-2} rms or greater, then WBV corrective action is required to start reducing exposure levels.

When the DELV 8 hours/day WBV exposure total weighted rms acceleration limit value of 1.15 ms^{-2} rms or greater is reached, then all efforts must be made to reduce exposure levels even if the WBV source must be used for less than 8 hours/day, or added control measures must be used, all in an effort to reduce human WBV daily exposure to acceptable levels.

In practice, triaxial WBV acceleration measurements are made, usually at or near the point where vibration enters the body, for example, for seated postures, using a special rubber "pie plate" circular disk, containing the triaxial accelerometer buried at its center. This instrumented disk is placed between the seat top cushion and the operator's buttocks. Measurements are conducted under actual work conditions, digitized, converted to rms values, weighted, and stored in memory. The resultant total weighted rms acceleration value is calculated and compared to the DEAV and DELV EU values just described, and a determination is made as to the need for any corrective vibration control measures.

Fourier spectrum analysis is not needed to use this EU standard. If such analysis is later needed in order to correct a problem and reduce the WBV, then the so-called stored raw acceleration data for each vibration axis can be revisited with more analysis.

WBV STANDARD LIMITATIONS Jolt/impact or repetitive mechanical shock has emerged as an important problem. In the past, the standards/guidelines have tried to cope with jolt/impact by comparing its frequency-weighted peak to the frequency-weighted root mean square of the signal of which it is a part over a period of time. This ratio is known as the crest factor.[18] The current standards/guidelines emphasize that if the crest factor exceeds 9, then the exposed individual's risk is underestimated.[5] Jolt/impact can result from a combination of sinusoidal signals and are not fully addressed by the current standards. When Cohen, Wasserman, and Hornung studied human response to single versus combined sinusoidal whole-body vibration signals, it was clear that, for the same root-mean-squared acceleration levels, people (healthy firefighters) were more sensitive to the combined vibration.[19] The above standards are based on the assumption that workers at many jobs are exposed to sinusoidal vibration, but workers are also exposed to nonsinusoidal vibration. For these jobs, the above standards do not fully address and likely underestimate the risk. To address these issues, research by Fethke et al.[20], Morrison et al.[21], Robinson[22], Sandover[23], Gant et al.[24], and others has led to ISO and ANSI standard development efforts for exposure to nonsinusoidal vibration and repetitive mechanical jolt/impact.[17] Exposure to seated, whole-body, or nonsinusoidal vibration and jolt/impact or repetitive mechanical shock conditions can be found in, but are not limited to, the marine, agricultural, rail-guided, rail-constrained, forklift, mining/quarrying, over-the-highway trucking, and heavy equipment vehicle environments. In a recent "prospective study of professional drivers, measures of internal spinal load" via ISO/WD 2631-5[25] "were better predictors of the occurrence of sciatic pain than the measures of daily vibration exposure established by the EU Directive (2002).[3] Herniated lumbar disc, lumbar trauma and physical work load were also associated with sciatic pain."[26]

Hand–arm vibration standards

Most hand–arm vibration standards are focused at minimizing the probability that vibration white fingers (i.e., Raynaud's phenomenon) would occur.[27]

ACGIH HAND–ARM VIBRATION STANDARD

The first HAV standard to be introduced in the United States was the ACGIH standard for hand–arm vibration in 1984 and used thru 2013.[28] Weighted triaxial acceleration measurements are obtained over a 1/3 octave band vibration frequency range of 5.6–1250 Hz. A radically modified version of their old HAV standard was introduced in 2014, which emphasizes the need to determine overall total HAV-weighted rms acceleration from separately weighted totals for each of the three linear axes. A total weighted vector sum is next calculated with a numerical result, which should not exceed 5 ms^{-2} for a worker using the HAV tested tool for 8 hours/day.[7] If the total weighted overall rms acceleration has exceeded this limit, then protective measures must be established to reduce HAV exposure; as appropriate, this can also include operating said tool for less than 8 hours/day as calculated.

ANSI HAND-ARM VIBRATION STANDARD

A second HAV standard was promulgated in the United States in 1986 (ANSI HAV document S3.34, 1986) which was extensively used from 1986 to 2006.[8] This document is no longer used and has been replaced with a simpler-to-use and better standard in current use (ANSI S2.70, 2006). Figure 4.5 graphically identifies the HAV weighting function which is retained and used in virtually all HAV standards in use worldwide today and used for all three measurement axes. The vibration frequency range extends from 5.6 to 1250 Hz. Vibration gain is given in the vertical axis. The elbow shape of this weighting curve reveals that the emphasis from potential HAV-damaging vibration appears to be at the lower vibration frequencies, 5.6–16 Hz, and as frequencies increase, from 16 to 1250 Hz, the vibration damage potential is lessened. This assumed that frequency dependency is under investigation.

In 2002 the EU issued a directive/law to all its member countries and included WBV and HAV limits.[3] The HAV limits were established for "total weighted rms acceleration vector sum" at two levels for power tool operators who are exposed to 8 hours/day HAV. The Daily Exposure Action Value (DEAV) = 2.5 ms^{-2}. If the total weighted vector sum is <2.5 ms^{-2}, then the tool is safe and allowed to be operated for the full 8 hours/day; if the total weighted sum greater than 2.5 ms^{-2}, but less than 5.0 ms^{-2}, this falls in a cautionary range where HAV disease can occur and the employer needs to begin measures to reduce HAV exposures; the daily exposure limit value (DELV) = 5.0 ms^{-2} for an 8 hours/day; if the total weighted sum equals or is greater than 5 ms^{-2} where disease is highly likely to occur, then the employer must significantly reduce these high HAV exposures, not use the tested tool, or use it in a very restricted time limited way. The EU holds the employer responsible, not the employee. As a result of this EU directive, power tool manufacturers worldwide are developing new and better low vibration power tools.

In 2006, ANSI decided to help create a "level playing field" by using the EU HAV criteria worldwide and replaced its old S3.34 HAV standard with S2.70, adopting the EU HAV 8 hours/day exposure criteria for both DEAV = 2.5 ms^{-2} rms and DELV = 5 ms^{-2} rms. In 2014 ACGIH decided to adopt only the EU's DELV maximum exposure criteria = 5 ms^{-2} rms for 8 hours/day for HAV exposures.[7,8,29]

Finally, because of the high prevalence of HAV-related disease worldwide, special reduced vibration called "Anti-Vibration" (A/V) gloves have been designed and marketed

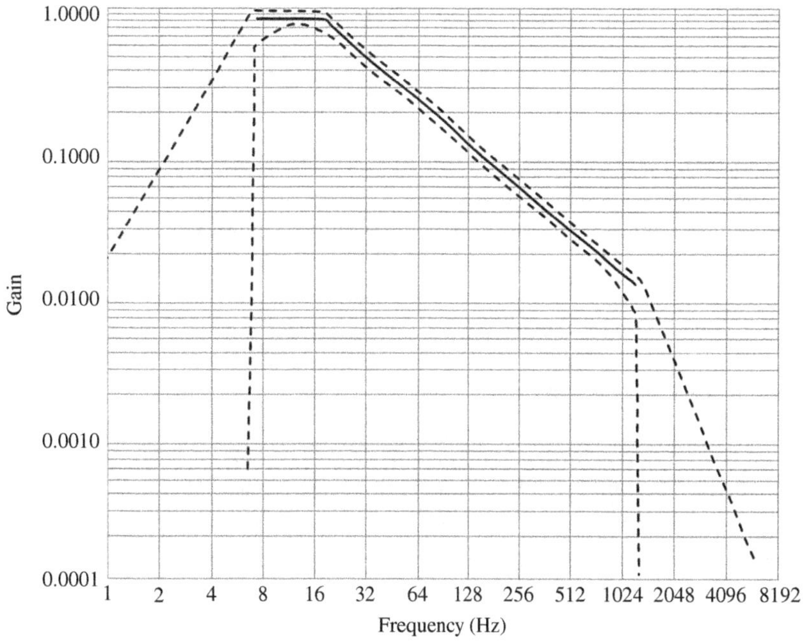

FIGURE 4.5 Nominal gain characteristics of the filter network used to frequency-weight acceleration components (continuous line). The dashed lines suggest filter tolerances. This is the basis of the ACGIH HAV TLV® (1984-present), ISO 5349-1 (2001-present), ANSI S3.34 (1986–2006), and ANSI S2.70 (2006-present). *Source*: From Figure 2 (page 189) from the ACGIH®, *2015 TLVs® and BEIs®* Book. Copyright 2015. Reprinted with permission.

worldwide. There was no uniform standard prior to 1996 certifying the efficacy of these products. The international A/V glove standard ISO 10819 was developed in 1996 and updated in 2013.[30,31] This standard has been used in the United States and elsewhere for many years; it is a difficult standard to meet but does serve the purpose of determining which products are most effective at reducing overall HAV exposure. The hallmark of this standard is to use, test, and certify only full-finger protected gloves which provide the best overall protection of fingers, palms, and hands and not test gloves where the fingers are exposed, because irreversible hand–arm vibration syndrome (HAVS) virtually always begins at the fingers working its way toward the palm.[27]

Limitations The NIOSH HAV criteria for an HAV standard #89-106[9] is old and different from ACGIH or ANSI documents. This was an interim, and rather more limited standard, since NIOSH has not chosen a maximum permissible acceleration value for HAV. It relies instead on medical monitoring and engineering controls and an extended high frequency range cutoff of 5000 Hz instead of 1200 Hz.[32–34]

The NIOSH document was developed in 1989 and is not often used. When it is cited, it is usually when NIOSH research discusses and compares HAV-unweighted acceleration data to identical weighted acceleration data, because a major current effort of NIOSH HAV research is to verify or challenge the current aforementioned HAV weighting functions adequacy and correctness, so that NIOSH can propose modifications if necessary. The heart of WBV and HAV measurements is the accuracy of the respective weighting functions of HAV and WBV, which form a critical link between vibration exposure dose and human response to vibration impinging daily. In the United States only NIOSH has the necessary legal mandate and resources to perform this very difficult and expensive task to protect workers.[32,35]

WHOLE-BODY VIBRATION AND LOW BACK PROBLEMS

Low back problems are a significant disabling health issue. Numerous epidemiologic studies have shown an association between back trouble and whole-body vibration.[36–41] Hoogendoorn et al.[42] and Bovenzi and Hulshof[43] reviewed the literature associating mechanical factors with back trouble and reported that there were sufficient properly executed studies to conclude that low back trouble is associated with WBV exposure. Demonstrating an association of WBV exposure with low back pain (LBP) is more challenging. A comprehensive review assessing the quality of the evidence conducted by the Evidence-Based Practice Workers' Compensation Board of British Columbia in 2002 and updated in 2008 identified methodological limitations, inconsistencies, and bias in the literature regarding a causal relationship between WBV and LBP. The reviews concluded that the available evidence did not satisfy the Bradford Hill criteria for causation of LBP.[44,45] This indicates that studies to date have not demonstrated that WBV causes LBP, although it does not exclude the possibility of causation. Low back pain is a common nonoccupational health problem, and workers who are exposed to WBV are also exposed to many other workplace hazards, making it difficult to sort out the cause of symptoms.

Possible etiologic factors for WBV and low back problems

Posture

The seated posture can be a mechanically extreme orientation for the lumbar intervertebral disk (i) increasing its internal pressure, (ii) increasing its anteroposterior shear flexibility, while (iii) decreasing its resistance to buckling instability and (iv) stressing the posterior region of the disk. Vibration and impact are additional mechanical stressors that can lead to large stresses and strains in the disk and subsequent mechanical fatigue of the disk material.[46,47]

Mechanical studies have been performed to evaluate the effect of WBV in seated, standing, and supine postures, in both single and multiple directions. The dynamic behavior of the human subject can be determined by two means: acceleration transmissibility and impedance. Using the former method, one compares the output acceleration resulting from the input or driving acceleration. At resonance, the ratio of output to input exceeds unity. For the impedance method, one computes the ratio of the force to move the body to its resulting velocity. This ratio, as a function of frequency, defines the mechanical response.

The degree to which an operator moves in a vibration environment is demonstrated by the magnitude of the acceleration transmissibility at the frequency of interest. Acceleration transmissibility is greatest at the resonant frequency, and many studies have shown transmissibilities greater than 1.0 for the first resonant frequency of the seated subject. Resonant frequencies were reported to occur between 4 and 6 Hz, which was usually attributed to the upper torso vibrating vertically with respect to the pelvis, and between 10 and 14 Hz, representing a bending vibration of the upper torso with respect to the lumbar spine.

It was also found that vibration response could be altered by posture.[48] Postures which are quite common in the occupational workplace (lateral bend and axial rotation) lead to greater transmission of vibrations. Typical examples would include the twisted posture of a tractor or forklift driver. We can gain considerable insight into the biomechanics and biodynamics of the spine if we look at the way it is assembled. The spine consists of a stack of boney elements that are each able to resist compression and bending. Between each of these relatively rigid elements is a softer intervertebral disk, an element that allows relative motion between adjacent vertebrae. A network of ligaments and muscles, tensile elements, stabilizes the stack overall and regionally. According to Levin,[49] the similarity of the spine to a structure that consists of discontinuous compressive elements and continuous tensile elements is striking and reminds us of the original insight and sculptures of Kenneth Snelson[50] and the subsequent structures of Buckminster Fuller. In those sculptures and structures, a locally applied load affected the entire structure. Hence, disturbance or change in any of the structure's individual tensile or compressive elements would have an effect on the rest of the structure. The additional complexity in the spine relates to the need for the coordinated action of agonist and antagonist muscles. Not only do the muscles and ligaments play a stabilizing role in static and quasistatic conditions, but they must also provide stability in dynamic environments. When muscles are needed as active stabilizers, we must also be aware that muscles may become less effective when the stimulus rate exceeds the muscles' ability to respond. Therefore, the posture held in the workplace can influence the spine, its network of support, the load path in the trunk, and the possibility of injury.

Using accelerometers and pins implanted in the lumbar region, Dupuis[51], Panjabi et al.[52], and Pope et al.[53] showed that the resonant frequency of the seated operator was 4.5 Hz. However, Panjabi found little or no relative motion between L1 and L3. In contrast though, by vibrating primates in seated postures, Quandieu and Pellieux[54] and Slonim[55,56] showed relative motion between lumbar levels and between the upper and lower spine via accelerometry. Coermann[57] also found relative motion between the pelvis and neck in humans by using mercury strain gauges between aluminum shells surrounding and tightly fitting the pelvic and chest regions. Zagorski et al.[58], using accelerometers taped to the back of human subjects, found greater acceleration at L3 than at the sacrum, in the 2–5 Hz frequency range. Wilder et al.[59,60] found relative motion on the surface of the lumbar region by means of filming seated subjects vibrating at their natural frequency as a moiré fringe pattern was projected on their backs. Pope et al.[53] found relative motion at the first natural frequency using transducers rigidly fixed to the lumbar spinous processes and fixed to the skin. Under local anesthesia, a threaded K-wire was threaded transcutaneously into the spinous process at L3. The greatest transmissibility was reported at 4 Hz, and substantial differences were noted between the vertical displacement of the pelvis and the adjacent LED marker and L3 and its adjacent LED marker. In a later study, using an intervertebral relative orientation sensing device, Pope et al.[61] found that greater rotations and translations occurred at 5 Hz compared to 8 Hz, again confirming the effect of the natural frequency.

Muscle response

Wilder et al. measured the electromyographic signals of the erector spinae and external obliques at each of three vibrational frequencies, first in the neutral position and then in varying body postures, as well as during the Valsalva maneuver.[59] Wide variations were observed in the EMG activity with respect to body posture. Increased activity of external obliques was found in rotation and lateral bend and during the Valsalva maneuver. A significantly increased myoelectric activity of the erector spinae was observed in women at the third resonant frequency. At this frequency, marked acceleration of the female breast mass could explain the increase in myoelectric activity.

Seroussi et al. measured the phasic activity of the erector spinae muscles in male subjects free of low back pain.[62] The ensemble-averaged EMG signals were converted to torque using an *in vivo* EMG–torque calibration technique. From these data, the phase relationship between the input signal to the platform and the resulting torque was established. Output data were the average, maximum, and minimum torque as a function of frequency. Higher average EMG levels, or muscle torques, were found for the vibration condition. The time lag between the input displacement and the peak torque varied from 30–100 ms at 3 Hz to 70–100 ms at 10 Hz. At 10 Hz, the muscle contraction tended to coincide with the input signal, or to be 360° out of phase. At all other frequencies, it was out of phase. Seidel and Heide have also monitored the timing of the back muscle response to a vibration stimulus and found that the muscles are not able to protect the spine from adverse loads.[63]

At the vertical natural frequency, Pope et al. found significantly greater erector spinae muscle activity without any foot support than with foot support. Pelvic rocking reduced with the aid of a foot support was shown to be an important factor in the reduction of the first natural frequency response of the seated individual.[64]

Magnusson et al. investigated the fatigue of the dorsal muscles under 5 Hz sinusoidal vibration.[65] To increase the response, the subjects wore pouches placed anteriorly over the ribs with 10 kg weights placed inside. The median frequency of the myoelectric power spectrum was used to establish the fatiguing effect of vibration as measured over the 30-minute time interval. Among seated, nonvibrated subjects, no change was observed over a 30-minute observation period. In contrast, among subjects vibrated over the 30-minute interval, a shift in the median frequency of erector spinae muscles was recorded in response to the vibratory input, suggesting muscle fatigue. In the industrial environment, especially those with awkward postures, vibration will lead to muscular fatigue.

Solomonow et al.[66–68] and Gedalia et al.[69] found that cyclic loading compromised the neuromuscular control system's reflexive ability to stabilize the lumbar spine. Wilder et al. also found that when subjects deliberately adopted an awkward posture in a well-configured, vertically vibrating seat, there was a significant increase in back muscle electrical activity.[70]

WBV: mechanical fatigue due to vibration loading

Although this chapter discusses soft tissue responses to impact/vibration, the review by Brinckmann et al. includes bony responses in the spine.[71]

Just as static postures produce intradiscal pressures unique to them, vibration also has an effect on the intradiscal pressure. Hansson et al. vibrated pigs longitudinally while they simultaneously obtained measurements of intradiscal pressure.[72] The vibration frequency used varied from 1 to 12 Hz. Intradiscal pressure was sensitive to frequency. Disk pressure peaked at 5 Hz and was 2.5 times than at 3 Hz, thus indicating a natural frequency similar to that of the seated human. If disk pressure in the human were similarly sensitive to vibration frequency, then vibrating at natural frequency would introduce time-varying disk pressure as a fatigue factor.

Adams and Hutton simulated a day of heavy flexion and torsion labor at a rate of 40 times/minute with loads based on the person's body weight.[73] Forty-one cadaveric lumbar motion segments in 12–57-year-olds showed plainly visible distortion as a precursor to a disk herniation tracking tear. In further work, Adams and Hutton produced disk prolapse in 6 of 29 specimens subjected to flexion and cyclic compression, which was increased at regular intervals.[74] Five of the prolapses occurred at the posterolateral "corner" of the disk, while the other occurred centrally.

Brown, Hansen, and Yorra produced a tear throughout the annulus (parallel to the endplate and to within 3.2 mm of the disk periphery) of a nondegenerated segment, with ligaments and posterior elements removed, as a result of 1000 cycles of 63.6 N compression load and 5° of forward flexion at a frequency of 1100 cycles/minute.[75]

Ten lumbar segments tested by Liu et al. experienced bony facet or vertebral body failures or disk annular or facet capsular ligament tears as a result of 0.5 Hz cyclic torque with a 445 N axial preload and testing until failure or elapsing of 10 000 cycles.[76] Torques applied were ±11.3, 22.6, or 33.9 N m. Generally failures occurred in segments subjected to more than 1.5° axial rotation. Other responses to this testing included discharge of synovial fluid from the articular facet joint capsule and joint "looseness" at the end of the test.

Wilder et al. found energy absorbed in the process of cyclically loading the spine.[77] Wilder, Pope, and Frymoyer also produced disk herniations as a result of combined vibration loading.[78] The herniations occurred in young calf disks subjected to a 9.5 Hz combined flexion and lateral bend cyclic loading with a constant superimposed axial rotation torque. Specimens were loaded from 6 000 to 37 740 times at a frequency of 9.5 Hz. A motion segment from a 68-year-old human male was also tested in a similar combined loading mode. After 2778 cycles, the specimen failed suddenly through the disk as a result of a tear beginning in the posterolateral portion of the disk, a point through which clinically observed herniations occur.

Forty lumbar motion segments tested either in simulated vibrating or nonvibrating, sitting environments showed mechanical sensitivity to load exposure history.[79] When subjected to 5 Hz vibration loads corresponding to accelerations occurring at physiologic levels (the 8-hour Fatigue, Decreased Proficiency Limit of ISO 2631[13]), significant mechanical changes were produced in the motion segments. In addition, the vibrated segments exhibited rapid, short-column buckling.

WBV: vertebral buckling instability

A long, slender, flexible column has the potential to buckle or give way suddenly. Buckling can lead to a rapidly occurring mechanical failure. When a buckling event occurs, the column's mode of resisting a vertical load applied coaxially changes so that it must resist that same load with the column in a bent shape. When the transition from straight to bent occurs, the bent column is less stiff than the straight column and the point at which the load is applied accelerates rapidly. The reader can observe this using a screen door spring or a bamboo shish-kabob skewer. Buckling can also occur in short columns. It is most easily understood in the case the load vector becomes directed outside the base of support. This is why a catamaran-type sailboat will continue to capsize or tip over once its center of gravity is outside either of its hulls or its base of support. The intervertebral motion segment can be considered a flexible short column, susceptible to buckling, especially if its disk has been compromised by injury, fatigue, disease, or degenerative processes, thereby decreasing the size of its effective base of support.

Wilder[60] and Wilder et al.[79] described experimental observations of short-column buckling in individual lumbar motion segments *in vitro*, in response to simulated exposure to a seated vertical vibration environment. Often, those segments buckled in a combination of flexion and lateral bend, placing the posterolateral aspect of the intervertebral disk at risk of experiencing a tensile impact load. Since then, 82 segments have been tested with the additional condition of maintaining a simulated laterally bent posture.[46] Of the normal segments, 79% buckled due to vibration exposure in a simulated awkward posture, while only 10% buckled due to the same awkward posture maintained in a static environment. Typically, the buckling of the motion segments occurred in less than 0.1 seconds. This raises important challenges for the neuromuscular control system in terms of its ability to sense and prevent or control a buckling event in the lumbar spine.

Many researchers have pointed out the importance of synchronized lumbar and trunk muscle activity in the active stabilization of the lumbar spine.[80–87] Quint et al. concluded that the lumbar spine is susceptible to injury during a buckling event occurring during movements associated with action as apparently simple as picking up a pencil[88]. McGill described an apparent buckling event and associated it with muscle activation behavior.[87]

WBV: impact as a sudden and unexpected load

If we consider that an impact event can be considered a suddenly, and in many cases an unexpectedly applied load, then we can apply another area of the literature to the understanding of the body's response to impact. Impact loading can come from many sources such as load shifting, slips, trips, stepping off a unexpected curb, mechanical slop in seats, sloshing of liquid in a tank trailer, and slack action in train or truck couplings (Figures 4.6, 4.7, and 4.8). The trunk musculature around the lumbar region responds differently to sudden loads depending on whether or not the load is expected. Whether standing or sitting, the pelvis acts as a foundation for the spine. The orientation of the pelvis is affected by leg muscles located below the pelvis along with trunk muscles located above the pelvis. Sudden events such as slips, trips, and falls affect actions of the leg and trunk muscles. In 1981, Manning and Shannon[89] and, in 1984, Manning et al.[90] showed that slip events are considered first events that lead to back injuries and expressed concern that this was a neglected research area in back injury etiology. Marras et al. showed that sudden, unexpected loads applied at the hands lead to large overcompensations in the back muscles, suggesting that the hazard is the body's excessive reaction or overcompensation to the applied load.[91] Mannion et al. predicted that disk overloading occurs as a result of sudden, unexpected loading.[92] These papers show that sudden

FIGURE 4.6 Typical trailer hook in a coupler ring on a dolly to allow a tractor-trailer set to pull a second trailer. The constraining latch is raised and the compression pad is retracted for visualization purposes. Without the compression pad in place, the space between the pin and tongue ring would be a potential source of slack-action impact. (Photograph by David Wilder).

FIGURE 4.7 Typical train car couplers. The space provides a source for potential slack-action impact. Couplers attached to their cars using a spring tend to reduce the impact from slack action at the coupler. (Insight based on personal communication with Robert Hitson) (Photograph by David Wilder).

FIGURE 4.8 Side view of a style of train-car coupler that has a design that tends to reduce slack-action impact at the coupler. (Insight based on personal communication with Robert Hitson) (Photograph by David Wilder).

loads and sudden movements of the body can place the back at risk of injury, as they each describe conditions that require the back muscles to respond rapidly to an imposed load or movement.

Responses to sudden loads applied at the hands are not necessarily symmetrical either. Pelvic orientation,[93] load application location,[94] and fatigue and hand dominance[95] can all affect the symmetry of the trunk muscle response.

WBV: triggering buckling of an unstable system

The work of Fethke et al.[20], Morrison et al.[21], and Robinson[22] show that the response of the trunk musculature of a seated subject to lateral impacts is asymmetrical in response time, duration, and amplitude. This condition raises the serious potential that asymmetrical muscle response could trigger a buckling event in the spine. This also corroborates the impression of a long time operator of rail-constrained vehicles who felt that the toughest part of the job was maintaining awkward postures while also trying to remain prepared for an unexpected horizontal impact.

PREVENTION

Low back pain can have a variety of occupational and non-occupational causes. Many industrial risk factors can be modified to reduce the rate of back disorders. Workers in occupations where vibration is present are also frequently exposed to lifting, pulling, or pushing, for example, truck drivers who also load and unload trucks. Regardless of the role of WBV in precipitating injury, many of the preventive measures that may mitigate vibration may also reduce other hazards.

Engineering redesign may include using vibration damping techniques (i.e., converting vibration into a small amount of heat due to the deformation of a viscoelastic material) and/or vibration isolation (i.e., intentionally mismatching the vibration pathway between the vibrating source and the worker receiving this vibration).

Epidemiologic data are supported by laboratory studies of spine changes that might produce back problems—for example, lumbar disk flattening, disk fiber strain and height increase, and intradiscal pressure. It is apparent from WBV data that the human spinal system has a characteristic response to vibration in a seated posture. Resonances occur at uniform frequencies for all of the subjects. The first vertical resonance occurs within a band of 4.5–5.5 Hz.

These studies indicate that maximum strain or stretching occurs in the seated operator's lumbar region at the first natural frequency. In addition, back muscles are not able to protect the spine from adverse loads. At many frequencies, the muscles' responses are so far out of phase that their forces are added to those of the stimulus. The fatigue that was found in muscles after vehicular vibration is indicative of the loads in the muscles. Thus, it would is advisable to walk around for a few minutes before bending over or lifting after vibration exposure (i.e., unloading a truck). One fuel company had their tank truck drivers take care of paperwork before handling hoses at a fuel delivery stop (P Wald, Personal communication). It would also be advisable for those exposed to prolonged vibration (i.e., long-distance driving) to take frequent breaks including walking around for a few minutes.

The field of mechanics provides an encouraging note for attempts to improve the whole-body vibration environment. If one considers that the damage to the spine due to vibration occurs from the work performed on the body by the kinetic energy from the vibration, two things become very clear. In the simplest form, the work performed on the body is equal to the kinetic energy applied to the body. The equation for that kinetic energy is $1/2\,mv^2$ (work = kinetic energy = $1/2\,mv^2$), where "m" is mass and "v" is velocity. Because velocity = acceleration × time ($v = at$), solving the energy equation in terms of acceleration and time yields an equation where kinetic energy equals $1/2m(a^2t^2)$. Using that formulation, it is then apparent that the work performed on the body from the kinetic energy of vibration depends on the square of the vibration acceleration and the square of the vibration exposure duration (time). This is important because it means a small reduction in acceleration and/or exposure time can lead to relatively large reductions in energy absorbed by the worker. For example, a 10% reduction in either vibration acceleration or time of exposure to the vibration can result in a 19% reduction in the energy applied to the body. Reducing by 10% both the vibration acceleration and time of exposure to the vibration can result in a 34% reduction in the energy or work applied to the body. Inexpensive accommodations therefore can have a big effect.

Jolt/impact is emerging as an important factor and is a subset of sudden/unexpected load conditions. More epidemiological studies in which relevant occupational exposures are quantified are needed. The relationship between intrinsically and extrinsically applied mechanical stresses, and the accompanying hard and soft tissue deformations, both acute and chronic, still needs greater definition. It is particularly important that seated vibration exposure not be used as a "work-hardening" treatment modality in trying to return someone with low back trouble to work.[96] It is also very important to realize that because the vibration standards/guidelines have limitations, it is critical to monitor the health and proficiency of people in the whole-body impact/vibration environment.

In a setting where vibration measurements have been made and the data have been evaluated with regard to the appropriate standard and it is determined that vibration control is necessary, usually a series of multiple control steps are taken depending on the problem as follows[96]:

1. Reduce vibration exposure by placing the worker away from vibrating surfaces by using remote controls, closed-circuit TV monitors, etc.
2. Reduce exposure time by modifying work organization, job sharing, etc.
3. If possible, mechanically isolate the vibrating surface, machine, etc.
4. Maintain mechanisms and replace worn-out mechanisms that contribute to production of jolt/impact and vibration.
5. In vehicles, use vibration isolating "suspended or air-ride" seats and cabs and replace vehicle suspension systems as necessary.
6. Once changes have been made, repeat vibration measurements and compare data to WBV standards as well as previous data.
7. Ask equipment users to comment on the effectiveness of the solutions.
8. Institute a surveillance program.

HAND–ARM VIBRATION MEDICAL EFFECTS

Adverse health effects from exposure to hand–arm vibration (HAV) have been recognized since 1911 when Loriga reported "dead fingers" among the Italian miners who used pneumatic tools.[97] Such tools had been introduced into the French mines in 1839 and were being extensively used by 1890. In the United States pneumatic tools were first introduced into the limestone quarries of Bedford, Indiana, about 1886. In 1918, Dr. Alice Hamilton and her colleagues subsequently investigated the health hazard from their use.[1] Since then there have been many reports of health hazards arising from the use of handheld vibratory tools in the literature from all over the world.[27]

It is now evident that adverse health effects can result from almost any vibrating source if the vibration is sufficiently intense over a wide frequency range for a significantly long period of time. The most important sources of HAV are pneumatically driven tools (air compressed and electrical), for example, grinders, drills, fettling tools, jackhammers, riveting guns, and chainsaws. Users of brush saws, hedge cutters, and speedway (dirt-track) motorbike riders are also at risk.

The predominant health effect of regular HAV exposure is now known as hand–arm vibration syndrome (HAVS), a generally irreversible disease entity with the following separate peripheral components[98,99]:

- Circulatory disturbances: cold-induced vasospasm with local finger blanching "white finger"
- Sensory and motor disturbances: numbness, loss of finger coordination and dexterity, clumsiness, and inability to perform intricate tasks
- Musculoskeletal disturbances: muscle, bone, and joint disorders

The vasospasm, also known as Raynaud's phenomenon, is precipitated by exposure to cold and/or damp conditions and

TABLE 4.1 The Stockholm Workshop Scale for the classification of cold-induced Raynaud's phenomenon in the Hand-arm Vibration Syndrome [vascular].[a]

Stage	Grade	Description
0		No attacks
1	Mild	Occasional attacks affecting only the tips of one or more fingers.
2	Moderate	Occasional attacks affecting the distal and middle fingers (rarely also proximal) phalanges of one or more fingers.
3	Severe	Frequent attacks affecting all phalanges of most fingers.
4	Very severe	As in stage 3, with trophic skin changes in the fingertips.

[a]The staging is made separately for each hand. In the evaluation of the subject, the grade of the disorder is indicated by the stages of both hands and the number of affected fingers on each hand; example: "2*L*(2)/1*R*(1)", "—/3*R*(4)", etc. From Gemme G, Pyykkö I, Taylor W, Pelmear PL. The Stockholm Workshop scale for the classification of cold-induced Raynaud's phenomenon in the hand-arm vibration syndrome (revision of the Taylor-Pelmear scale). *Scand J Work Environ Health* 1987; 13(4):277. Reprinted with permission.

TABLE 4.2 The Stockholm Workshop Scale for the classification of sensorineural affects of the Hand-arm Vibration Syndrome.[a]

Stages	Symptoms
OSN	Exposed to vibration but no symptoms
1SN	Intermittent numbness, with or without tingling.
2SN	Intermittent or persistent numbness, reduced sensory perception.
3SN	Intermittent or persistent numbness, reduced tactile discrimination and/or manipulative dexterity.

[a]The sensorineural stage is to be established for each hand. From Brammer AJ, Taylor W, Lundborg G. Sensorineural stage of hand-arm vibration syndrome. *Scand J Work Environ Health* 1987; 13(4):281. Reprinted with permission.

sometimes vibration exposure itself. The time period between first exposure to HAV and the onset of fingertip blanching is termed the "latent interval." It may range from 1 month to several years, depending on the intensity of vibration entering the hand and the susceptibility of the worker. The blanching is restricted initially to the tips of one or more fingers but progresses to the base of the fingers as the vibration exposure time increases. The thumbs are usually the last to be affected.

The blanching is accompanied by numbness, and as the circulation to the digits returns, there is usually tingling and pain. Tingling and paresthesia may precede the onset of blanching in many subjects. These sensory symptoms and signs may be the predominant complaint in some patients, and their recognition as a distinct entity led to the revision of the Taylor–Pelmear classification for assessment of HAVS devised in 1968.[100] It was replaced in 1985 by the Stockholm classification based on the subjective history supported by the extensive results of a battery of clinical tests to stage the severity of disease (Tables 4.1 and 4.2).[101,102] The vascular and sensorineural symptoms and signs are evaluated separately and for both hands individually.

In advanced cases the peripheral circulation becomes very sluggish, giving a cyanotic tinge to the skin of the digits, while in the very severe cases trophic skin changes (gangrene) will appear at the fingertips. The toes may be affected if directly subjected to vibration from a local source, that is, vibrating platforms, or they may be affected by reflex spasm in subjects with severe hand symptoms. Reflex sympathetic vasoconstriction may also account for the increased severity of noise-induced hearing loss in HAVS subjects.[103,104]

In addition to tactile, vibrotactile, and thermal threshold impairment, which may vary from subject to subject, impairment of grip strength is a common symptom in longer exposed workers.[105,106] Discomfort and pain in the upper limbs is also a common complaint. Bone cysts and vacuoles, although often reported, are more likely to be caused by biodynamic and ergonomic factors.[107,108]

Carpal tunnel syndrome, an entrapment neuropathy affecting the median nerve at the wrist, is often associated with HAVS.[109–111] Usually it is due to ergonomic stress factors including the constant repetitive nature of the work, grip force, and mechanical stresses, for example, torque and posture. When vibration is the primary cause of the median nerve neuropathy, the edematous reaction in the adjacent tissues and the nerve sheath compresses the central axon.[112] The median nerve is affected together with the ulnar in two thirds of the cases. Rarely the ulnar nerve may be affected alone.[113]

Whether smoking accelerates the onset of HAVS has not yet been proved conclusively, but this aggravating factor has been shown to increase the risk in several studies.[114,115]

DIAGNOSIS

The diagnosis of HAVS is based on a history of HAV exposure and the exclusion of other causes of Raynaud's phenomenon, that is, primary Raynaud's phenomenon

(Raynaud's disease or constitutional white finger), local trauma to the digital vessels, thoracic outlet syndrome, drugs, and peripheral vascular and collagen diseases, including scleroderma. The diagnosis of HAVS is confirmed and the severity assessed by stage from the results of a battery of laboratory tests.[113,116–118]

Vascular tests should include some or all of the following, that is, Doppler studies, plethysmography, finger systolic pressure measurement, and cold-water provocation tests to verify that vasospasm occurs on cold exposure. Subjective sensorineural tests should include depth sense and two-point discrimination, fingertip vibration threshold measurement, thermal hot/cold perception, and current perception threshold tests. Objective nerve conduction tests should be undertaken to confirm the presence and severity of the neuropathy which in HAVS normally affects both median and ulnar nerves. When the median nerve myelinated fibers are involved at the wrist level, there can be confusion with CTS nerve entrapment because the symptoms and signs are similar.

PATHOPHYSIOLOGY

The pathophysiology of HAVS has been well reviewed by Gemne.[119] The basic mechanism is not yet fully understood. Due to the mechanical stimulus, specific anatomical changes occur in the digital vessels, that is, vessel wall hypertrophy and endothelial cell damage. In the initial stages there is extrusion of fluid into the tissues. This edema, together with the subsequent spasmodic ischemia from the cold-included vasospasm, damages the mechanoreceptor nerve endings and nonmedullated fibers. Subsequently, a demyelinating neuropathy of the peripheral nerve trunks develops.

The vascular response to cold is complex because in addition to the diversity of receptor systems (adrenergic, cholinergic, purinergic, and serotonergic), there are several subtypes of specific receptors. The differential distribution and functional significance of the various receptor types is largely unknown. It is probable that the cold-induced pathological closure of the digital arteries and end vessels is mainly mediated by alpha-2 adrenoreceptors in the wall of the arterioles and veins. It has been demonstrated that the alpha-2 receptors are more receptive to the cold stimulus. In HAVS it is postulated that there is selective damage of alpha-1 receptors; hence the cold stimulus is more effective. While arterial spasm is necessary to stop the blood flow, vasospasm in the skin arterioles is essential to produce the blanching.

Cold, as well as vessel wall injury, causes platelet aggregation. The subsequent release of serotonin (5-hydroxytryptamine (5-HT)) promotes further release of 5-HT from the platelets, and the increased concentration stimulates smooth muscle to contract. Besides promoting contraction, serotonin may also contribute to vasodilation by inducing the release of endothelium-derived relaxing factor (ERDF) and prostacyclin from the endothelial cells. Acetylcholine and its agonist methacholine acting through the muscarinic receptors also release ERDF, while nitric oxide and its agonists, nitroprusside and nitroglycerine, release prostacyclin. The prostacyclin and ERDF so released, besides inhibiting platelet aggregation, stimulate the production of cyclic adenosine monophosphate (cAMP) and cyclic guanosine monophosphate (cGMP) in the smooth muscle cell. The latter substances inhibit calcium utilization by the smooth muscle cells so they do not contract. A delicate balance between smooth muscle contraction and relaxation is produced by these mechanisms interacting simultaneously in the normal person.

TREATMENT AND MANAGEMENT

To reduce the frequency of blanching attacks, the central body temperature must be maintained, and cold exposure must be avoided. When possible, mitts rather than gloves should be worn. Discontinuation of smoking and all nicotine is an essential requirement because of the adverse effect of nicotine and carbon monoxide on the digital arterial system.

To attempt to reverse the pathology and seek recovery, further vibration exposure must be avoided. If avoidance of HAV is not possible, a modified work routine should be practiced to reduce vibration exposure and slow the progression of this condition. This should also include using both antivibration power tools and antivibration full-finger protective gloves.

Recent advances in drug therapy have focused on three areas: (i) use of calcium channel antagonists to produce peripheral vasodilation, (ii) use of drugs to reduce platelet aggregation in combination with the above, and (iii) drugs to reduce blood viscosity and emboli formation. The preferred calcium channel antagonist is a slow-release product. In the more severe cases, additional medications are often prescribed for platelet deaggregation.

The necessity for drug therapy increases with the severity of the symptoms and the age of the subject. The results are encouraging for resolution of the vascular symptoms, particularly if vibration exposure is avoided. Unfortunately drug intolerance to the earlier calcium channel antagonists caused some young patients to abandon their use prematurely. Where this has happened, the newer products should be tried. Recovery from the sensorineural effects has not been reported. It remains to be seen whether an improvement in the peripheral circulation will result in reversal of the sensorineural symptomatology. For the most part, HAVS is irreversible; therefore prevention measures are the watchword in order to minimize the effects of HAV exposure.

When patients are thought to be suffering from HAVS, their employers should be advised to assess the work situation and introduce preventive procedures. All workers should be advised of the potential vibration hazard and receive training on the need to service their tools regularly; to grip the tools as lightly as possible within the bounds of safety; to use the protective clothing equipment provided; to attend for periodic medical surveillance; and to report all signs and symptoms of HAVS as soon as they develop.

HAND–ARM VIBRATION CONTROL

As with whole-body vibration control, once vibration measurements have been made and the data have been evaluated with regard to the appropriate standard and it is determined that vibration control is necessary, usually a series of multiple control steps are taken depending on the problem[9,27,33–35,120,121]:

1. Use only ergonomically correct antivibration (A/V) power tools wherever possible.
2. If possible do *not* use materials that wrap around tool handles and claim to significantly reduce HAV exposure. Usually they increase the tool handle diameter causing the worker to grip the tool handle more forcefully, thereby compressing the material and reducing their limited usefulness. It is far better to use a well-designed A/V tool.
3. Use only good fitting "full-finger protected" ISO 10819 certified A/V gloves[10]; do not use exposed finger gloves that only protect the palm. Hand–arm vibration syndrome (HAVS) nearly always begins at the fingertips advancing toward the root; only full-finger protected gloves offer the desired protection.
4. In factory situations, use suspended "tool balancers" to remove the weight of the tool from the operator.
5. Workers are advised to do the following as good work practices:
   a. Let the tool do the work, gripping it as lightly as possible consistent with safe tool handling.
   b. Operate the tool only when necessary and at reduced speeds if possible.
   c. Properly maintain hand tools and replace as necessary.
   d. Do not smoke, since nicotine, vibration, and cold all constrict the blood vessels.
   e. Keep your hands and body warm and dry.
   f. Consult a physician if signs of digit tingling, numbness, or blanching occur.
   g. Medical prescreening of workers is advised to minimize the risk to idiopathic primary Raynaud's disease sufferers and others from operating hand tools that could exacerbate such a preexisting condition.
   h. Workers and others need to be made aware of this problem and its signs and symptoms.
6. Recently some vibrating tool manufacturers have begun placing warning labels on their tools and in their instruction books.
7. Use HAV standards as critical adjuncts to control measures.
8. Institute a surveillance program.

Finally, for HAV exposures, these controls may need additional engineering redesign methods using vibration damping techniques (i.e., converting vibration into a small amount of heat due to the deformation of a viscoelastic material) and/or vibration isolation (i.e., intentionally mismatching the vibration pathway between the vibrating source and the worker receiving this vibration) to be effective.

In the most significant HAVS prevention projects to date, Geiger et al.[122] and Torelli et al.[123] have shown that it is possible to change the vibrating hand tool acquisition process and use the culture of a large organization, the US Department of Defense, in order to improve the well-being of tool users and reduce their risk of long-term, irreversible injury. Prior to this, protective antivibration hand tools were not available to the US federal workforce. The project provided a vital template for approaching such a large challenge. It made convincing arguments based on issues related to the EU Physical Agents Directive[3] (binding law), avoiding injury, life-cycle costs of tools, and making the tools available to the US federal workforce.

References

1. Hamilton A. A study of spastic anaemia in the hands of stonecutters. *Ind Accident Hyg Serv Bull* 1918; 236(19):53–66.
2. Harris CM, Crede CE. Introduction to the handbook. In: Harris CM, Crede CE, eds. *Shock and Vibration Handbook*, 2nd edition. McGraw-Hill Book Company, New York, 1976, pp. 1-1–2.
3. European Parliament, Council of the European Union. Directive 2002/44/EC of the European Parliament and of the Council of 25 June 2002 on the minimum health and safety requirements regarding the exposure of workers to the risks arising from physical agents (vibration) (sixteenth individual Directive within the meaning of Article 16(1) of Directive 89/391/EEC). *Off J Eur Communities L* 2002; 177:13–29.
4. American Conference of Governmental Industrial Hygienists (ACGIH). *Physical Agents: Threshold Limit Values: Ergonomic: Whole-Body Vibration, TLV®s and BEIs Threshold Limit Values for Chemical Substances and Physical Agents and Biological Exposure Indices*. ACGIH, Cincinnati, OH, 2015.
5. International Organization for Standardization, ISO 2631-1:1997. *Mechanical Vibration and Shock: Evaluation of Human Exposure to Whole Body Vibration. Part 1. General Requirements*, 2nd edition. International Organization for Standardization, Geneva, 1997; Corrected and reprinted, 1997-07-15; Amendment 1, 2010-07-01.

6. American National Standards Institute, Inc., ANSI S3.18-2002/ ISO 2631-1:1997. *Nationally Adopted International Standard (NAIS Standard): Mechanical Vibration and Shock—Evaluation of Human Exposure to Whole Body Vibration. Part 1. General Requirements.* American National Standards Institute, Inc., Standards Secretariat, Acoustical Society of America, Melville, NY, 2002.
7. American Conference of Governmental Industrial Hygienists (ACGIH). *Physical Agents: Threshold Limit Values: Ergonomic: Hand-Arm (Segmental) Vibration, TLVs and BEIs Threshold Limit Values for Chemical Substances and Physical Agents and Biological Exposure Indices.* ACGIH, Cincinnati, OH, 2015.
8. American National Standards Institute (ANSI). *ANSI: S2.70-2006: Guide for the Measurement and Evaluation of Human Exposure to Vibration Transmitted to the Hand.* ANSI, New York, 2006, (R2011).
9. National Institute for Occupational Safety and Health. *Criteria for a Recommended Standard: Occupational Exposure to Hand-Arm Vibration.* DHHS (NIOSH) Publication No. 89-106. U.S. Department of Health and Human Services, Public Health Service, Centers for Disease Control, National Institute for Occupational Safety and Health, Division of Standards Development and Technology Transfer, Cincinnati, OH, 1989.
10. American National Standards Institute (ANSI). *ANSI: S2.73/ ISO 10819: Mechanical Vibration and Shock: Measurement and Evaluation of the Vibration Transmissibility of Gloves at the Palm of the Hand.* ANSI, New York, 2014.
11. International Organization for Standardization (ISO), ISO 2631:1974. *Guide for the Evaluation of Human Exposure to Whole-Body Vibration.* ISO, Geneva, 1974.
12. International Organization for Standardization (ISO), ISO 2631:1978. *Guide for the Evaluation of Human Exposure to Whole-Body Vibration.* ISO, Geneva, 1978. Available at: http://www.iso.org/iso/rss.xml?csnumber=7608&rss=detail (accessed on April 30, 2016).
13. International Organization for Standardization (ISO), ISO 2631-1:1985. *Evaluation of Human Exposure to Whole-Body Vibration: Part 1: General Requirements.* ISO, Geneva, 1985.
14. American Conference of Governmental Industrial Hygienists ACGIH. *Notice of Intended Changes (for 1995–1996): Whole-Body Vibration, 1995–1996 Threshold Limit Values (TLVs®) for Chemical Substances and Physical Agents and Biological Exposure Indices (BEIs ®).* ACGIH, Cincinnati, OH, 1995, pp 123–31.
15. International Organization for Standardization (ISO), ISO 2631-2:2003. *Mechanical Vibration and Shock: Evaluation of Human Exposure to Whole-Body Vibration—Part 2: Vibration in Buildings (1 Hz to 80 Hz)* 2nd edition. ISO, Geneva, 2003.
16. International Organization for Standardization (ISO), ISO 2631-4:2001. *Mechanical Vibration and Shock: Evaluation of Human Exposure to Whole-Body Vibration—Part 4: Guidelines for the Evaluation of the Effects of Vibration and Rotational Motion on Passenger and Crew Comfort in Fixed-Guideway Transport Systems* 1st edition. ISO, Geneva, 2001.
17. International Organization for Standardization (ISO), ISO 2631-5:2004. *Mechanical Vibration and Shock: Evaluation of Human Exposure to Whole Body Vibration—Part 5: Method for Evaluation of Vibration Containing Multiple Shocks.* ISO, Geneva, 2004.
18. American National Standards Institute (ANSI). *ANSI:S3.18. Guide for the Evaluation of Human Exposure to Whole-Body Vibration.* ANSI, New York, 1979.
19. Cohen HH, Wasserman DE, Hornung RW. Human performance and transmissibility under sinusoidal and mixed vertical vibration. *Ergonomics* 1977; 20(3):207–16.
20. Fethke N, Wilder DG, Spratt K. Seated trunk-muscle response to impact. Presentation #51 at the International Society for the Study of the Lumbar Spine, Adelaide, Australia, April 9–13, 2000.
21. Morrison J, Robinson D, Roddan G, et al. Development of a standard for the health hazard assessment of mechanical shock and repeated impact in army vehicles: Phase 5. U.S. Army Aeromedical Research Laboratory, Fort Rucker, AL, Report CR-96-1, 1997.
22. Robinson DG. The dynamic response of the seated human to mechanical shock. PhD Dissertation, Simon Fraser University, Burnaby, BC, 1999.
23. Sandover J. High acceleration events: An introduction and review of expert opinion. *J Sound Vib* 1998; 215(4):927–45.
24. Gant LC, Wilder DG, Wasserman DE. Human response to single and combined sinusoidal vertical vibration: Revisited. *J Low Freq Noise Vib Active Control* 2012; 31(1)21–8.
25. International Organization for Standardization (ISO). *Mechanical Vibration and Shock: Evaluation of Human Exposure to Vibration: Part 5: Methods for Evaluation of Vibration Containing Multiple Shocks.* ISO/WD 2631-5. ISO/TC 108/SC 4/WG 15, N77. DIN, ISO, Geneva, 2013.
26. Bovenzi M, Schust M, Menzel G, et al. A cohort study of sciatic pain and measures of internal spinal load in professional drivers, *Ergonomics* 2015; 58(7):1088–102.
27. Pelmear P, Wasserman D (contributing editors). Hand-arm vibration. In: *A Comprehensive Guide for Occupational Health Professionals* 2nd edition, OEM Publishers, Beverly Farms, MA, 1998.
28. American Conference of Governmental Industrial Hygienists (ACGIH). *Ergonomics: Hand-Arm (Segmental) Vibration, 2013 TLVs® and BEIs ®, Based on the Documentation of the Threshold Limit Values for Chemical Substances and Physical Agents & Biological Exposure Indices.* ACGIH, Cincinnati, OH, 2013, pp. 185–8.
29. American National Standards Institute (ANSI). *ANSI: S3.34-1986: Guide for the Measurement and Evaluation of Human Exposure to Vibration Transmitted to the Hand*, ANSI, New York, 1986, (R1997).
30. International Organization for Standardization (ISO), ISO 10819:1996. *Mechanical Vibration and Shock: Hand-Arm Vibration—Method for the Measurement and Evaluation of the Vibration Transmissibility of Gloves at the Palm of the Hand.* ISO, Geneva, 1996.
31. International Organization for Standardization (ISO), ISO 10819:2013 *Mechanical Vibration and Shock: Hand-Arm Vibration—Method for the Measurement and Evaluation of the Vibration Transmissibility of Gloves at the Palm of the Hand.* ISO, Geneva, 2013.
32. Pelmear P, Leong D, Taylor W, et al. Measurement of vibration of hand-tools: Weighted or unweighted? *J Occup Med* 1989; 31:903.
33. Starck J, Pekkarinen J, Pyykko I. Physical characteristics of vibration in relations to vibration-induced white finger. *Am Ind Hyg Assoc J* 1990; 51:179.

34. Wasserman D. The control aspects of occupational hand-arm vibration. *Appl Ind Hyg* 1989; 4:22.
35. Wasserman D. To weight or not to weight...that is the question. *J Occup Med* 1989; 31:909.
36. Bovenzi M, Betta A. Low-back disorders in agricultural tractor drivers exposed to whole-body vibration and postural stress. *Appl Ergon* 1994; 25(4):231–41.
37. Bovenzi M, Zadini A. Self-reported low back symptoms in urban bus drivers exposed to whole-body vibration. *Spine* 1992; 17(9):1048–59.
38. Dupuis H, Zerlett G. *The Effects of Whole-Body Vibration*. Springer-Verlag, Berlin, 1986.
39. Magnusson M, Wilder DG, Pope MH, et al. Investigation of the long-term exposure to whole-body vibration: A 2-country study. Winner of the Vienna Award for Physical Medicine. *Eur J Phys Med Rehabil* 1993; 3(1):28–34.
40. Sandover J. Dynamic loading as a possible source of low-back disorders. *Spine* 1983; 8:652–8.
41. Bovenzi M. A longitudinal study of low back pain and daily vibration exposure in professional drivers. *Ind Health* 2010; 48(5):584–95.
42. Hoogendoorn WE, van Poppel MNM, Bongers PM, et al. Physical load during work and leisure time as risk factors for back pain. *Scand J Work Environ Health* 1999; 25(5):387–403.
43. Bovenzi M, Hulshof CTJ. An updated review of epidemiologic studies on the relationship between exposure to whole-body vibration and low back pain. *J Sound Vib* 1998; 215(4):595–611.
44. WorkSafeBC Evidence-Based Practice Group. Whole body vibration and low back pain, 2002. Available at: http://www.worksafebc.com/health_care_providers/Assets/PDF/whole_body_vibration_low_back_pain.pdf (accessed on April 30, 2016).
45. WorkSafeBC Evidence-Based Practice Group. Whole body vibration and low back pain: First update 2008. Available at: http://www.worksafebc.com/health_care_providers/Assets/PDF/whole_body_vibration_low_back_pain_first_update.pdf (accessed on April 30, 2016).
46. Wilder DG. The biomechanics of vibration and low back pain. *Am J Ind Med* 1993; 23(4):577–88.
47. Wilder DG, Pope MH. Epidemiological and etiological aspects of low back pain in vibration environments: An update. *Clin Biomech* 1996; 11(2):61–73.
48. Wilder DG, Woodworth BB, Frymoyer JW, et al. Vibration and the human spine. *Spine* 7(3):243–54, 1982.
49. Levin S. The icosahedron as a biological support system. International Society for the Study of the Lumbar Spine, Toronto, June 6–10, 1982. Society Administration Office, Toronto.
50. Snelson K. (1965) Continuous Tension, Discontinuous Compression Structures, US patent #3,169,611.
51. Dupuis H. Belastung durch mechanische Schwingungen und moegliche Gesundheitsschäedigungen im Bereich der Wirbelsaule. *Fortschr Med* 1974; 92(14):618–20.
52. Panjabi MM, Andersson GBJ, Jorneus L, et al. *In vivo* measurement of spinal column vibrations. *J Bone Joint Surg* 1986; 68A(5):695–703.
53. Pope MH, Svensson M, Broman H, et al. Mounting of the transducer in measurements of segmental motion of the spine. *J Biomech* 1986; 19(8):675–7.
54. Quandieu P, Pellieux L. Study *in situ et in vivo* of the acceleration of lumbar vertebrae of a primate exposed to vibration in the Z-axis. *J Biomech* 1982; 15:985–1006.
55. Slonim AR. Some vibration data on primates implanted with accelerometers on the upper and lumbar spine: Methodology and results in rhesus monkeys. Air Force Aerospace Medical Research Laboratory, Wright-Patterson Air Force Base, OH. Technical Report TR-81-153, 1983.
56. Slonim AR. Some vibration data on primates implanted with accelerometers on the upper thoracic and lower lumbar spine: Results in baboons. Air Force Aerospace Medical Research Laboratory, XX. Technical Report TR-83-091, 1984.
57. Coermann RR. Mechanical vibrations. Proceedings of a Symposium of Ergonomics and Physical Environmental Factors, Geneva, Switzerland, September 16–21, 1968. The International Labour Office, Geneva, pp. 17–41.
58. Zagorski J, Jakubowski R, Solecki L, et al. Studies on the transmissions of vibrations in human organism exposed to low-frequency whole-body vibration. *Acta Physiol Pol* 1976; 27:347–54.
59. Wilder DG, Frymoyer JW, Pope MH. The effect of vibration on the spine of the seated individual. *Automedica* 1985; 6:5–35.
60. Wilder DG. On loading of the human lumbar intervertebral motion segment. PhD Dissertation, University of Vermont, October 1985. Abstract: *Dissertation Abstracts International* 1986; 46(12): 4328-B, June 1986; Manuscript #DA8529728: University Microfilms International, Ann Arbor, MI, 1986.
61. Pope MH, Kaigle AM, Magnusson M, et al. Intervertebral motion during vibration. *Proc Instn Mech Eng J Eng in Med* 1991; (205):39–44.
62. Seroussi RE, Wilder DG, Pope MH. Trunk muscle electromyography and whole body vibration. *J Biomech* 1989; 22(3):219–29.
63. Seidel H, Heide R. Long-term effects of whole-body vibration: A critical survey of the literature. *Int Arch Occup Environ Health* 1986; 58:1–26.
64. Pope M, Wilder D, Seroussi R. Trunk muscle response to foot support and corset wearing during seated, whole-body vibration. *Trans Ortho Res Soc* 1988; 13:374.
65. Magnusson ML, Aleksiev A, Wilder DG, et al. European Spine Society: The AcroMed Prize for Spinal Research 1995. Unexpected load and asymmetric posture as etiologic factors in low back pain. *Eur Spine J* 1996; 5(1):23–35.
66. Solomonow M, Zhou B-H, Harris M, et al. The ligamento-muscular stabilizing system of the spine. *Spine* 1998; 23(23):2552–62.
67. Solomonow M, Zhou B-H, Baratta RV, et al. Biomechanics of increased exposure to lumbar injury caused by cyclic loading. Part 1. Loss of reflexive muscular stabilization. 1999 Volvo Award Winner in Biomechanical Studies. *Spine* 1999; 24(23):2426–34.
68. Solomonow M, Zhou B-H, Baratta RV, et al. Biexponential recovery model of lumbar viscoelastic laxity and reflexive muscular activity after prolonged cyclic loading. *Clin Biomech* 2000; 15:167–75.
69. Gedalia U, Solomonow M, Zhou B-H, et al. Biomechanics of increased exposure to lumbar injury caused by cyclic loading. Part 2. Recovery of reflexive muscular stability with rest. *Spine* 1999; 24(23):2461–7.

70. Wilder DG, Tranowski JP, Novotny JE, et al. Vehicle seat optimization for the lower back. International Society for the Study of the Lumbar Spine, Marseilles, June 15–19, 1993.
71. Brinckmann P, Wilder DG, Pope MH. Effects of repeated loads and vibrations. In: Weisel SW, Weinstein JN, Herkowitz H, Dvorak J, Bell G, eds. *The Lumbar Spine* 2nd edition. International Society for the Study of the Lumbar Spine, W.B. Saunders Co., Philadelphia, PA, 1996, pp. 181–202.
72. Hansson TH, Keller TS, Holm S. The load on the porcine lumbar spine during seated whole body vibrations. *Ortho Trans* 1988; 12(1):85.
73. Adams MA, Hutton WC. The effect of fatigue on the lumbar intervertebral disc. *Orthop Trans* 1983; 7(3):461.
74. Adams MA, Hutton WC. Gradual disc prolapse. *Spine* 1985; 10(6):524–31.
75. Brown T, Hansen RJ, Yorra AJ. Some mechanical tests on the lumbosacral spine with particular reference to the intervertebral discs: A preliminary report. *J Bone Joint Surg* 1957; 39A:1135–65.
76. Liu YK, Goel VK, DeJong A, et al. Torsional fatigue of the lumbar intervertebral joints. *Orthop Trans* 1983; 7(3):461.
77. Wilder DG, Woodworth BB, Frymoyer JW, et al. Energy absorption in the human spine. In: Paul I, ed. Proceedings of Eighth Northeast (New England), 1980 Bioengineering Conference, March 27–28, 1980, MIT, Cambridge, MA; 1980, pp. 443–5.
78. Wilder DG, Pope MH, Frymoyer JW. Cyclic loading of the intervertebral motion segment. In: Hansen Ew, ed. Proceedings of the Tenth Northeast Bioengineering Conference, March 15–16, 1982, Dartmouth College, Hanover. New York: IEEE; 1982, pp. 9–11.
79. Wilder DG, Pope MH, Frymoyer JW. The biomechanics of lumbar disc herniation and the effect of overload and instability. American Back Society Research Award. *J Spinal Disord* 1988; 1(1):16–32.
80. Cholewicki J, McGill SM. Mechanical stability of the *in vivo* lumbar spine: Implications for injury and chronic low back pain. New concepts and hypotheses. *Clin Biomech* 1996; 11(1):1–15.
81. Cholewicki J, Panjabi MM, Khachatryan A. Stabilizing function of trunk flexor-extensor muscles around a neutral spine posture. *Spine* 1997; 22(19):2207–12.
82. Cholewicki J, Juluru K, McGill SM. Intra-abdominal pressure mechanism for stabilizing the lumbar spine. *J Biomech* 1999; 32:13–17.
83. Crisco JJ, Panjabi MM. The intersegmental and multisegmental muscles of the lumbar spine. A biomechanical model comparing lateral stabilizing potential. *Spine* 1991; 16(7):793–9.
84. Crisco JJ, Panjabi MM, Yamamoto I, et al. Euler stability of the human ligamentous lumbar spine. Part II: Experiment. *Clin Biomech* 1992; 7:27–32.
85. Gardner-Morse MG, Stokes IAF, Laible JP. Role of muscles in lumbar spine stability in maximum extension efforts. *J Orthop Res* 1995; 13:802–8.
86. Gardner-Morse MG, Stokes IAF. The effects of abdominal muscle coactivation on lumbar spine stability. *Spine* 1998; 23(1):86–92.
87. McGill SM. The biomechanics of low back injury: Implications on current practice in industry and the clinic. ISB Keynote Lecture. *J Biomech* 1997; 30(5):465–75.
88. Quint U, Wilke H-J, Shirazi-Adl A, et al. Importance of the intersegmental trunk muscles for the stability of the lumbar spine. A biomechanical study *in vitro*. *Spine* 1998; 23(18):1937–45.
89. Manning DP, Shannon HS. Slipping accidents causing low-back pain in a gearbox factory. *Spine* 1981; 6(1):70–2.
90. Manning DP, Mitchell RG, Blanchfield LP. Body movements and events contributing to accidental and non accidental back injuries. *Spine* 1984; 9(7):734–9.
91. Marras WS, Rangarajulu SL, Lavender SA. Trunk loading and expectation. *Ergonomics* 1987; 30:551–62.
92. Mannion AF, Adams MA, Dolan P. Sudden and unexpected loading generates high forces on the lumbar spine. *Spine* 2000; 25(7):842–52.
93. Aleksiev A, Pope MH, Hooper D, et al. Pelvic unlevelness in chronic low back pain patients: Biomechanics and EMG time-frequency analyses. Recipient of the 1995 Vienna Award in Physical Medicine and Rehabilitation. *Eur J Phys Med Rehabil* 1996; 6(1):3–16.
94. Schumacher C, Wilder DG, Goel VK, et al. Back muscle response to sudden load with a modified lumbar support. 27th Annual Meeting International Society for the Study of the Lumbar Spine, Adelaide, Australia, April 9–13, 2000; Presentation #277. Society Administration Office, Toronto.
95. Wilder DG, Aleksiev A, Magnusson M, et al. Muscular response to sudden load: A tool to evaluate fatigue and rehabilitation. *Spine* 1996; 21(22):2628–39.
96. Wasserman D, Wilder D, Pope M, et al. Whole-body vibration exposure and occupational work hardening. *J Occup Environ Med* 1997; 39(5):403–7.
97. Loriga G. Il Lavoro Con i Martellie Pneumatici. *Boll Inspett Lovoro* 1911; 2:35–60.
98. Gemne G, Taylor W. eds. Foreword: Hand-arm vibration and the central nervous system. *J Low Freq Noise Vib* 1983; XI.
99. Brammer AJ, Taylor W, eds. *Vibration Effects on the Hand and Arm in Industry*. John Wiley & Sons, Inc., New York, 1982.
100. Taylor W, Pelmear PL, eds. *Vibration White Finger in Industry*. Academic Press, London, 1975, pp. XVII–XXII.
101. Brammer AJ, Taylor W, Lundborg G. Sensorineural stages of the hand-arm vibration syndrome. *Scand J Work Environ Health* 1987; 13:279–83.
102. Gemne G, Pyykkö I, Taylor W, et al. The Stockholm Workshop scale for the classification of cold-induced Raynaud's phenomenon in the hand-arm vibration syndrome (revision of the Taylor–Pelmear scale). *Scand J Work Environ Health* 1987; 13:279–83.
103. Iki M, Kurumantani N, Satoh M, et al. Hearing of forest workers with vibration induced white finger: A five year follow-up. *Int Arch Occup Environ Health* 1989; 61:437–42.
104. Pyykkö I, Starck J, Färkkilä M, et al. Hand-arm vibration in the etiology of hearing loss in lumberjacks. *Br J Ind Med* 1981; 38:281–9.
105. Färkkilä M. Grip force in vibration disease. *Scand J Work Environ Health* 1978; 4:159–66.
106. Färkkilä M, Aatola S, Stark J, et al. Hand-grip force in lumberjacks. Two year follow-up. *Int Arch Occup Environ Health* 1986; 58:203–8.

107. Gemne G, Saraste H. Bone and joint pathology in workers using hand-held vibratory tools: An overview. *Scand J Work Environ Health* 1987; 13:290–300.
108. James PB, Yates JR, Pearson JCG. An investigation of the prevalence of bone cysts in hands exposed to vibration. In: Taylor W, Pelmear PL, eds. *Vibration White Finger in Industry*. Academic Press, New York, 1975, pp. 43–51.
109. Färkkilä M, Koskimies K, Pyykkö I, et al. Carpal tunnel syndrome among forest workers. In: Okada A, Taylor W, Dupuis H, eds. *Hand-Arm Vibration*. Kyoei Press, Kanazawa, 1990:263–5.
110. Koskimies K, Färkkilä M, Pyykkö I et al. Carpal tunnel syndrome in vibration disease. *Br J Ind Med* 1990; 47:411–16.
111. Wieslander G, Norback D, Gothe CJ, et al. Carpal tunnel syndrome (CTS) and exposure to vibration, repetitive wrist movements, and heavy manual work: A case-referent study. *Br J Ind Med* 1989; 46:43–7.
112. Lundborg G, Dahlin LB, Danielsen N, et al. Intraneural edema following exposure to vibration. *Scand J Work Environ Health* 1987; 13(4 Special Issue):326–9.
113. Pelmear PL, Taylor W. Clinical evaluation. In: Pelmear PL, Taylor W, Wasserman DE, eds. *Hand-Arm Vibration: A Comprehensive Guide*. Van Nostrand Reinhold, New York, 1992:77–91.
114. Ekenvall L, Lindblad LE. Effect of tobacco use on vibration white finger disease. *J Occup Med* 1989; 31(1):13–16.
115. Virokannas H, Anttonen H, Pramila S. Combined effect of hand-arm vibration and smoking on white finger in different age groups. *Arch Complex Environ Stud* 1991; 3(1–2):7–12.
116. McGeoch KL, Taylor W, Gilmour WH. The use of objective tests as an aid to the assessment of hand-arm vibration syndrome by the Stockholm classification. In: Dupuis H, Christ E, Sandover J, et al., eds. Proceedings of the Sixth International Conference on Hand-Arm Vibration, Bonn, Germany. Essen: Druckzentrum Sutter & Partner GmbH; 1992, pp. 783–92.
117. Pelmear PL, Wong L, Dembek B. Laboratory tests for the evaluation of hand-arm vibration syndrome. In: Dupuis H, Christ E, Sandover J, et al., eds. Proceedings of the Sixth International Conference on Hand-Arm Vibration, Bonn, Germany. Essen: Druckzentrum Sutter & Partner GmbH; 1992.
118. Pelmear PL, Taylor W. Hand-arm vibration syndrome: Clinical evaluation. *J Occup Med* 1991; 33(11):1144–9.
119. Gemne G. Pathophysiology and pathogenesis of disorders in workers using hand-held vibratory tools. In: Pelmear PL, Taylor W, Wasserman DE, eds. *Hand-Arm Vibration: A Comprehensive Guide*. Van Nostrand Reinhold, New York, 1992, pp. 41–76.
120. Wasserman D. *Human Aspects of Occupational Vibration*. Elsevier, Amsterdam, 1987.
121. National Institute for Occupational Safety and Health. *Vibration Syndrome*, Current Intelligence Bulletin #38, DHHS/NIOSH Pub. #83-110. U.S. Department of Health and Human Services, Public Health Service, Centers for Disease Control, National Institute for Occupational Safety and Health, Cincinnati, OH 1983.
122. Geiger MB, Wasserman DE, Chervak SG, et al. Hand-arm vibration syndrome. Protecting powered hand tool operators. *Prof Saf* 2014; 59(11)34–42.
123. Torelli N, Rodriquez-Johnson E, Geiger M, et al. Hand-Arm Vibration Syndrome. 30 September 2011, Pentagon Brief, Defense Safety Oversight Committee, Acquisition and Technology Programs Task Force, Department of Defense, USA.

Chapter 5

MECHANICAL ENERGY

JAMES KUBALIK*

OCCUPATIONAL SETTING

Mechanical energy impacting on the human body is the most frequent cause of direct physical injuries (Table 5.1). Our workplaces and individual workspaces are dynamic and have both direct and indirect exposure potential. The management of these interactions will determine whether the outcome is efficient production and productivity or a failure resulting in an employee injury and costs to a company.

Our daily work activities are a series of interactions and physical contacts with equipment, work surfaces, chemicals, and materials. Viewing the workplace and an individual's workspace as a 360° environment that moves, bends, twists, and interacts with sources of mechanical energy provides a "graphic display" of the potential and magnitude of the exposure. With each direct physical contact and the subsequent transfer of sufficient mechanical energy, this potential can be manifested as a cut, bruise, strain, fracture, amputation, or other physical injury or illness.

The perception that eliminating manual tasks will also eliminate all exposures to mechanical energy is often erroneous. In many cases we are substituting one source for another. Replacing tasks with tools (i.e., pallet jacks, forklifts, punch presses, computers) has also exposed workers to equipment- or mechanical energy-related hazards. Often, the impacts of these changes are not proactively recognized, and the trailing consequences of these changes are injuries and illnesses.

An aging workforce further compounds the results of uncontrolled exposures. Workers are less able to endure the consequences of physical and repetitive tasks (i.e., lifting, twisting, repetitive motions) and contact with equipment.

There are many sources of occupational exposure to mechanical energy. Uncontrolled and unmanaged interactions with mechanical energy cause many serious injuries and are the most common reason for a worker to seek medical attention. These injuries and their causes are summarized in Table 5.1.

The magnitude and cost of these exposures are substantial. According to the National Safety Council, in 2012 there were 4 900 000 medically consulted work injuries with a cost of $198.2 billion.[1] Approximately one quarter of the total injuries were the result of worker contact with objects and equipment. Mechanical energy was most likely a factor in these incidents.

MEASUREMENT ISSUES

Mechanical hazards are measured in terms of the forces of kinematics and mechanics, which were reviewed in Chapter 1. In general, measurements are made during an accident investigation or while investigating a cluster of similar accidents. Examples of measurements include velocity, distance, acceleration, force, weight, temperature, pressure, and friction. These are all relatively simple measurements that do not require complex instruments.

There are no established "action levels" for workplace injuries. There are Occupational Safety and Health Administration (OSHA) regulatory standards that require control of hazards, many of which include exposure to mechanical energy. There are also general industry standards (i.e., ANSI) that address specific conditions and procedures

* Peter H. Wald and A.B. Barnes contributed to the previous edition of this chapter.

Physical and Biological Hazards of the Workplace, Third Edition. Edited by Gregg M. Stave and Peter H. Wald.

TABLE 5.1 Types of direct injuries and their causes.

Injury type	Causes/locations	Safeguard
Traffic accidents	Roadways Rail spurs	Bridges over crossings Signals Seat belts
Falls from heights	Platforms Walkways Aerial baskets Open structures Ladders	Railings Enclosures Fall protection systems
Slips, trips, and falls	Slippery surfaces	Abrasive surfaces
	Cluttered work site	Good housekeeping
Major crush injuries	Forklifts Cranes	Restraints Rollover protection Overhead guards Audible travel alarms Inspection Crane director Audible travel alarms
Explosions	Flammables/combustibles High-pressure steam/air/product systems	Fire protection program and systems Grounding No smoking Hydrotesting
Burns	Steam Hot surfaces Cryogens Open flame	Insulate process Insulating clothing Energy isolation (during maintenance)
	Electrical	Electrical safety program (see also explosions above)
Electrocutions	Any electrical processes	Electrical safety program (see Chapter 17)
	Power tools Electric moving equipment Extension cords	
Abrasions, lacerations, and contusions	Moving equipment—routine operation	Guard openings Two-handed "trip" operation
	Moving equipment—maintenance operations Power and hand tools	Enclosures Interlocks Automatic feeds "Presence" sensors Lockout/tagout
Musculoskeletal strain	Manual material handling	Maintain and repair Appropriate tool choice Hand and eye protection Solid work surface (see Chapter 3)

and a number of trailing indicators based on past occupational injury and illness experience such as the OSHA log or workers' compensation injury and illness/claim experience.

Workplace safety observation techniques can be useful in identifying and proactively correcting unsafe conditions and acts. This is a growing field, and proactive safety observations are based on observing employee work habits, specifically safe and unsafe work behavior. The ratio of unsafe acts and conditions observed to the total number of "safety-related" observations can be used as a predictive safety indicator and to develop an "action level." Once a baseline is established, any increase in the ratio is usually associated with an increase in workplace accidents. Likewise, a decreasing ratio often predicts a decrease in accidents. These observational techniques are a tool that can monitor real-time safety conditions and proactively target intervention.

EXPOSURE GUIDELINES

There are regulatory standards that indirectly cover injuries from mechanical energy. However, as stated above, there is no "action level" for workplace injuries. Injury statistics are used to calculate workers compensation insurance rates, and companies with higher injury rates will pay higher workers compensation premiums, so there is a direct financial incentive to keep these rates as low as possible. Increasingly, regulatory agencies will also target companies with high injury and illness rates.

Regulatory agency standards and recommended industry practices

There are a number of regulations that pertain to controlling worker access to moving parts of production equipment and the management of exposure to mechanical energy. The Code of Federal Regulations (CFR) contains a number of minimum safety and health standards and best practices that pertain to mechanical energy and overall employee safety and prevention of injuries. Many are contained in 29 CFR 1900–1910.999 (OSH Act), which are summarized in Table 5.2. Some states operate state OSHA programs, and their regulations may be more stringent than the federal OSHA regulations. Readers should refer to the pertinent state OSHA websites for further information.

Designing process and production methods to eliminate hazardous conditions is one of the best ways to prevent accidents. When these conditions cannot be eliminated by design or engineering or substitution, then machine guarding and other controls should be considered. Regulations are intended to address proper controls should the design, engineering, and substitution options not be feasible or possible.

Two standards of particular importance are the OSHA machine guarding standard (29 CFR 1910.211—Subpart O) and the OSHA lockout/tagout standard (29 CFR 1910.147—The Control of Hazardous Energy Lockout/Tagout).

These two standards outline the requirements for (i) guarding or controlling access to moving machine parts and (ii) the elimination or management of energy that can power machinery or exist within the equipment when primary sources of power are disconnected or interrupted. These regulations are excellent tools for managing mechanical energy and will be reviewed in some detail in the section on Prevention later in this chapter.

PATHOPHYSIOLOGY OF INJURY

Machines are designed to apply large amounts of force to wood, metal, or other materials. When applied to bone and tissue, that same force can produce disastrous results. Direct injuries can result from dramatically different types of accidents, which were reviewed in Table 5.1.

TABLE 5.2 Table of contents for 29 CFR Part 1910 occupational safety and health standards (https://www.osha.gov/pls/oshaweb/owadisp.show_document?p_table=STANDARDS&p_id=9696).

1910—Table of Contents
1910 Subpart A—General (1910.1 to 1910.8)
1910 Subpart B—Adoption and Extension of Established Federal Standards (1910.11 to 1910.19)
1910 Subpart C—Adoption and Extension of Established Federal Standards (1910 Subpart C)
1910 Subpart D—Walking–Working Surfaces (1910.21 to 1910.30)
1910 Subpart E—Means of Egress (1910.35 to 1910.38)
1910 Subpart F—Powered Platforms, Manlifts, and Vehicle-Mounted Work Platforms (1910.66 to 1910.68)
1910 Subpart G—Occupational Health and Environmental Control (1910.94 to 1910.98)
1910 Subpart H—Hazardous Materials (1910.101 to 1910.126)
1910 Subpart I—Personal Protective Equipment (1910.132 to 1910.139)
1910 Subpart J—General Environmental Controls (1910.141 to 1910.147 App A)
1910 Subpart K—Medical and First Aid (1910.151 to 1910.152)
1910 Subpart L—Fire Protection (1910.155 to 1910.165)
1910 Subpart M—Compressed Gas and Compressed Air Equipment (1910.166 to 1910.169)
1910 Subpart N—Materials Handling and Storage (1910.176 to 1910.184)
1910 Subpart O—Machinery and Machine Guarding (1910.211 to 1910.219)
1910 Subpart P—Hand and Portable Powered Tools and Other Hand-Held Equipment (1910.241 to 1910.244)
1910 Subpart Q—Welding, Cutting, and Brazing (1910.251 to 1910.255)
1910 Subpart R—Special Industries (1910.261 to 1910.272 App C)
1910 Subpart S—Electrical (1910.301 to 1910.399)
1910 Subpart T—Commercial Diving Operations (1910.401 to 1910.441)
1910 Subpart Z—Toxic and Hazardous Substances (1910.1000 to 1910.1450 App B)

Slips, trips, and falls are common in commerce and industry and result from the same primary and secondary causes as those in and around the home. In industry, however, there is always the potential for further trauma as a result of additional hazards in the environment, such as unguarded mechanisms, hot surfaces, or unprotected chemical processes. Electrical injuries and muscular strains due to material handling are covered in their own chapters in the physical hazards section.

TREATMENT

Medical personnel should be prepared to treat traumatic injury. The reader is advised to consult evidence-based guidelines and one of the many available texts on emergency treatment.

SURVEILLANCE PROGRAMS

Injury surveillance programs

Recognition and assessment of mechanical energy hazards are the first steps in successful management. Unfortunately, the first recognition of hazards is often realized after an injury, interruption of production, or an alarming "near miss." Reactive assessments and incident investigations are an important part of identifying, measuring, and improving. However, using these assessments or investigations as the primary methods to identify mechanical energy hazards is an ineffective and costly practice.

A basic review of injury and illness experience is the first step. In the United States four readily available sources of information are (i) the OSHA 300 Log, (ii) the Summary of Work-Related Injuries and Illnesses (OSHA Form 300A), (iii) the OSHA 301 or equivalent accident/injury investigation reports, and (iv) workers' compensation claim data.

These data will highlight your past experience, some of the causes associated with your experience and costs and your actions to prevent injuries and illnesses. After a thorough review of your loss experience, a series of inspections of production areas is highly recommended.

OSHA regulations require an employer with more than 10 employees (at any time during the calendar year immediately preceding the current calendar year) to keep and actively manage an OSHA 300 Log (Figure 5.1). Some industries are partially exempt from the recordkeeping requirement based on the industry classification (see https://www.osha.gov/recordkeeping/ppt1/RK1exempttable.html). All employers, including those partially exempted by reason of company size or industry classification, must report any workplace incident that results in a fatality within 8 hours to OSHA. In-patient hospitalization, amputation, or loss of an eye within 24 hours of the event. Inpatient hospitalizations of three or more employees must be reported to OSHA within 8 hours.

Each OSHA recordable injury and illness must be recorded on the log within six working days of notification of the incident. A good-faith effort is required to maintain and update pertinent OSHA 300 Log information. Injury and illness recording requirements are contained in OSHA 29 CFR Part 1904—Recording and Reporting Occupational Injuries, and additional guidance is available online.[2] You may also contact your local OSHA consultation office for additional information and can contact Federal OSHA to request copies of any letters of interpretation on questions submitted by industry or formally request them under the Freedom of Information Act. States with state OSHA programs may also have additional recording and reporting requirements.

Five years of updated OSHA logs must be readily available, so these documents are a good historical record of your health and safety performance. Keeping the OSHA log information up to date, especially after the change of a calendar year, is often a challenge but is required to meet regulatory requirements. A listing of commonly used safety statistics based on the OSHA 300 Log is included in Table 5.3.

Another source of information are the OSHA 301s and accident and injury investigation reports. These reports will give you valuable information, including data on the number and types of injuries resulting in employees being unable to work (lost time), restricted in their regular work activity (unable to perform all their normal duties), and injuries requiring medical treatment by a qualified provider. Severity information on injuries and illness based on the days lost/restricted activity will also be included.

Your accident investigation reports of employee injuries and noninjury accidents including property damage, equipment damage, and, if available, near-miss incidents will help identify high-hazard work areas, tasks, equipment, and overall accident/injury trends. Once trends have been identified, plans can be developed to address improvements.

An OSHA 301 document (Figure 5.2) or its equivalent (your accident/injury investigation report) is also required for each entry on the log. You can design your company incident investigation form to meet the OSHA 301 requirements.

Your insurance carrier/broker can provide you with the third recommended source of information, your workers' compensation (WC) loss/claim data. These data will provide general information on types of incidents, causes, body part injured, and costs incurred due to each reported WC claim. (Note: A number of state regulations may have a significant impact on the recording and reporting of WC claims. In addition, the recording and reporting requirements for OSHA log maintenance and WC claims are different. Some injuries may be recorded and reported in one system but not the other (i.e., OSHA recordkeeping requirements will capture restricted data activity cases, while workers' compensation will not).)

WC costs can have a "long tail," meaning that it often takes 5–8 years before they are fully realized. Thus costs reported at the close of a fiscal year will significantly increase

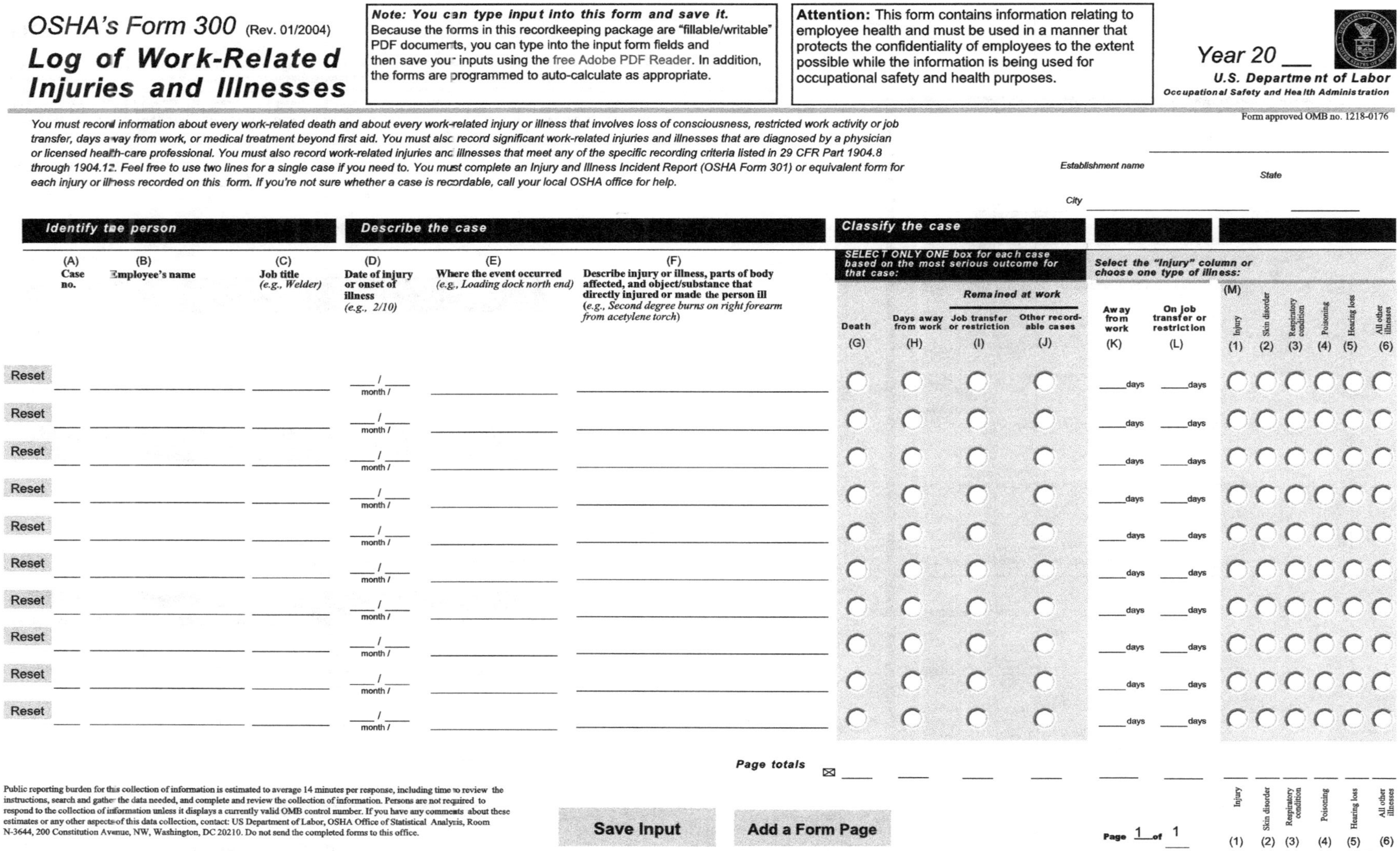

OSHA's Form 300 (Rev. 01/2004)

Log of Work-Related Injuries and Illnesses

Note: You can type input into this form and save it. Because the forms in this recordkeeping package are "fillable/writable" PDF documents, you can type into the input form fields and then save your inputs using the free Adobe PDF Reader. In addition, the forms are programmed to auto-calculate as appropriate.

Attention: This form contains information relating to employee health and must be used in a manner that protects the confidentiality of employees to the extent possible while the information is being used for occupational safety and health purposes.

Year 20 __

U.S. Department of Labor

Occupational Safety and Health Administration

Form approved OMB no. 1218-0176

You must record information about every work-related death and about every work-related injury or illness that involves loss of consciousness, restricted work activity or job transfer, days away from work, or medical treatment beyond first aid. You must also record significant work-related injuries and illnesses that are diagnosed by a physician or licensed health-care professional. You must also record work-related injuries and illnesses that meet any of the specific recording criteria listed in 29 CFR Part 1904.8 through 1904.12. Feel free to use two lines for a single case if you need to. You must complete an Injury and Illness Incident Report (OSHA Form 301) or equivalent form for each injury or illness recorded on this form. If you're not sure whether a case is recordable, call your local OSHA office for help.

Establishment name ____________

City ____________ State ______

Identify the person			Describe the case			Classify the case: SELECT ONLY ONE box for each case based on the most serious outcome for that case:		Remained at work				Select the "Injury" column or choose one type of illness: (M)					
(A) Case no.	(B) Employee's name	(C) Job title (*e.g., Welder*)	(D) Date of injury or onset of illness (*e.g., 2/10*)	(E) Where the event occurred (*e.g., Loading dock north end*)	(F) Describe injury or illness, parts of body affected, and object/substance that directly injured or made the person ill (*e.g., Second degree burns on right forearm from acetylene torch*)	Death (G)	Days away from work (H)	Job transfer or restriction (I)	Other record-able cases (J)	Away from work (K)	On job transfer or restriction (L)	Injury (1)	Skin disorder (2)	Respiratory condition (3)	Poisoning (4)	Hearing loss (5)	All other illnesses (6)
Reset			__/__ month /							___days	___days						
Reset			__/__ month /							___days	___days						
Reset			__/__ month /							___days	___days						
Reset			__/__ month /							___days	___days						
Reset			__/__ month /							___days	___days						
Reset			__/__ month /							___days	___days						
Reset			__/__ month /							___days	___days						
Reset			__/__ month /							___days	___days						
Reset			__/__ month /							___days	___days						
Reset			__/__ month /							___days	___days						
					Page totals	___	___	___	___	___	___	Injury (1)	Skin disorder (2)	Respiratory condition (3)	Poisoning (4)	Hearing loss (5)	All other illnesses (6)

Public reporting burden for this collection of information is estimated to average 14 minutes per response, including time to review the instructions, search and gather the data needed, and complete and review the collection of information. Persons are not required to respond to the collection of information unless it displays a currently valid OMB control number. If you have any comments about these estimates or any other aspects of this data collection, contact: US Department of Labor, OSHA Office of Statistical Analysis, Room N-3644, 200 Constitution Avenue, NW, Washington, DC 20210. Do not send the completed forms to this office.

Save Input | Add a Form Page

Page 1 of 1

FIGURE 5.1 OSHA 300 and 301 forms (US Department of Labor—Occupational Safety and Health Administration). Fillable PDF files are available at https://www.osha.gov/recordkeeping/ppt1/RK1exempttable.html

OSHA's Form 300A (Rev. 01/2004)

Summary of Work-Related Injuries and Illnesses

Note: You can type input into this form and save it. Because the forms in this recordkeeping package are "fillable/writable" PDF documents, you can type into the input form fields and then save your inputs using the free Adobe PDF Reader.

Year 20 ___

U.S. Department of Labor
Occupational Safety and Health Administration

Form approved OMB no. 1218-0176

All establishments covered by Part 1904 must complete this Summary page, even if no work-related injuries or illnesses occurred during the year. Remember to review the Log to verify that the entries are complete and accurate before completing this summary.

Using the Log, count the individual entries you made for each category. Then write the totals below, making sure you've added the entries from every page of the Log. If you had no cases, write "0."

Employees, former employees, and their representatives have the right to review the OSHA Form 300 in its entirety. They also have limited access to the OSHA Form 301 or its equivalent. See 29 CFR Part 1904.35, in OSHA's recordkeeping rule, for further details on the access provisions for these forms.

Number of Cases

Total number of deaths	Total number of cases with days away from work	Total number of cases with job transfer or restriction	Total number of other recordable cases
______	______	______	______
(G)	(H)	(I)	(J)

Number of Days

Total number of days away from work	Total number of days of job transfer or restriction
______	______
(K)	(L)

Injury and Illness Types

Total number of . . .
(M)

(1) Injuries	______	(4) Poisonings	______
(2) Skin disorders	______	(5) Hearing loss	______
(3) Respiratory conditions	______	(6) All other illnesses	______

Post this Summary page from February 1 to April 30 of the year following the year covered by the form.

Public reporting burden for this collection of information is estimated to average 50 minutes per response, including time to review the instructions, search and gather the data needed, and complete and review the collection of information. Persons are not required to respond to the collection of information unless it displays a currently valid OMB control number. If you have any comments about these estimates or any other aspects of this data collection, contact: US Department of Labor, OSHA Office of Statistical Analysis, Room N-3644, 200 Constitution Avenue, NW, Washington, DC 20210. Do not send the completed forms to this office.

Establishment information

Your establishment name ______________________

Street ______________________

City ______________ State ______ Zip ______

Industry description (*e.g., Manufacture of motor truck trailers*)

____

Standard Industrial Classification (SIC), if known (*e.g., 3715*)

______

OR

North American Industrial Classification (NAICS), if known (e.g., 336212)

______

Employment information (*If you don't have these figures, see the Worksheet on the next page to estimate.*)

Annual average number of employees ______

Total hours worked by all employees last year ______

Sign here

Knowingly falsifying this document may result in a fine.

I certify that I have examined this document and that to the best of my knowledge the entries are true, accurate, and complete.

______________________ Company executive ______________ Title

Phone ____-_____-______ Date ____/____/____

Save Input

FIGURE 5.1 (*Continued*)

TABLE 5.3 OSHA log-based safety performance metrics.

Lost workday case
An employee work-related injury or illness that is so severe that a worker is unable to come to work
Restricted activity case
An employee injury or illness that is severe; a worker, although able to come to work, is unable to perform all his/her regular job duties
Medical treatment cases
Employee injuries and illnesses that result in a worker being able to work; however, due to his/her work-related injury, treatment by a medical professional is required (i.e., stitches, prescriptions). The worker is able to return to full duties after treatment
Total employee hours worked
Represents the total hours worked (exposure to workplace hazards) by an employee population represented in the calculation.
Note: 200 000 hours represents 100 employees working 2000 hours over a 1-year period. The rates can be reported as lost workday = restricted activity days per 100 employees or per hours worked
OSHA lost workday case rate

$$\frac{\text{No. of lost work days and restricted activity cases} \times 200\,000}{\text{total employee hours worked}}$$

OSHA recordable rate

$$\frac{\text{No. of lost work days} + \text{restricted activity} + \text{medical treatment cases} \times 200\,000}{\text{total employee hours worked}}$$

over a 5-year period. As a general rule, your compensation costs at the close of a year will easily double within this 5-year period. Your insurance carrier may have loss control services, and their professionals can assist you in the prevention, management, and reporting of injuries and illnesses.

You may also use your WC experience modification rate (EX Mod) as an overall and general performance indicator. The EX Mod is based on a standard formula, and, with industry data, it considers but is not limited to your payroll, industry performance, and company WC performance for three previous and complete years (prior to your last year's experience). A modifier can be compared between companies. An Ex Mod of 1 means that your costs and experience are average for your industry/competitors. A rate below 1 means that your performance is better than your competitors'/industry average and your costs are lower. This would be a competitive advantage. If your rate is greater than 1, your costs are higher and you are paying more than your competitors. With 3 years of experience considered, 1 poor year will impact your costs for up to 3 years.

After a thorough review of your claim data, a walk/inspection of the active production areas, offices, and especially the areas with serious or frequent incidents is recommended. These real-time observations of employee work practices, production processes, and equipment operation will complement your experience review. Conducting this walk with a knowledgeable and experienced safety professional and production manager or foreman is highly recommended. Their observations and answers to your questions on production processes and safe work methods will be invaluable.

Observations should be focused on the immediate work of employees and include the workers' 360° environment. For instance, observe the activity and interaction of the employees in their workspace, with their equipment, materials, and equipment around them, to each side, overhead, underfoot, and behind.

Count the work actions involved in lifting, using tools, moving materials, and operating equipment. This will help you identify some sources of mechanical energy and the exposure potential. Your partner on the walk can assist you in identifying the work practices that are not consistent with safety requirements. Also count the pieces of powered equipment used in the areas you review. The more equipment and materials, the higher the potential for injuries.

These simple methods will help highlight the dynamics of your workplace, identify some of the sources of mechanical energy, and assist you in identifying a need for further assessment. Positive results from any of these basic injury surveillance program components suggest the need for further investigation and are summarized in Table 5.4.

Based on this initial assessment, a determination can be made of the need for further evaluation. Poor accident experience combined with an active production area, frequent employee equipment, and material interactions and high numbers of items of powered equipment indicate a greater exposure to mechanical energy and the potential for injury.

Safety surveillance

An effort to identify the need for an energy management system entails a comprehensive review, which is best achieved with a team approach. Experienced and knowledgeable personnel are required to identify equipment and sources of mechanical energy. The team should consist of members from production management: foreman, experienced workers, safety and industrial hygiene professionals, maintenance and plant engineers, and, where appropriate, equipment manufacturers.

OSHA's Form 301

Injury and Illness Incident Report

Note: *You can type input into this form and save it.* Because the forms in this recordkeeping package are "fillable/writable" PDF documents, you can type into the input form fields and then save your inputs using the free Adobe PDF Reader. In addition, the forms are programmed to auto-calculate as appropriate.

Attention: This form contains information relating to employee health and must be used in a manner that protects the confidentiality of employees to the extent possible while the information is being used for occupational safety and health purposes.

U.S. Department of Labor
Occupational Safety and Health Administration

Form approved OMB no. 1218-0176

This *Injury and Illness Incident Report* is one of the first forms you must fill out when a recordable work-related injury or illness has occurred. Together with the *Log of Work-Related Injuries and Illnesses* and the accompanying *Summary,* these forms help the employer and OSHA develop a picture of the extent and severity of work-related incidents.

Within 7 calendar days after you receive information that a recordable work-related injury or illness has occurred, you must fill out this form or an equivalent. Some state workers' compensation, insurance, or other reports may be acceptable substitutes. To be considered an equivalent form, any substitute must contain all the information asked for on this form.

According to Public Law 91-596 and 29 CFR 1904, OSHA's recordkeeping rule, you must keep this form on file for 5 years following the year to which it pertains.

If you need additional copies of this form, you may photocopy the printout or insert additional form pages in the PDF and then use as many as you need.

Completed by ____

Title ____

Phone ___ - ___ - ___ **Date** ___ ___ ___ (Month Day Year)

Information about the employee

1) Full name ____

2) Street ____

3) City ____ **State** ____ **ZIP** ____

4) Date of birth ___ ___ ___ (Month Day Year)

5) Date hired ___ ___ ___ (Month Day Year)

○ **Male**
○ **Female**

Information about the physician or other health-care professional

6) Name of physician or other health-care professional ____

7) If treatment was given away from the worksite, where was it given?

Facility ____

Street ____

City ____ **State** ____ **ZIP** ____

8) Was the employee treated in an emergency room?

○ **Yes**
○

9) Was the employee hospitalized overnight as an in-patient?

○ **Yes**
○ **No**

Information about the case

10) Case number from the *Log* ____ *(Transfer the case number from the Log after you record the case.)*

11) Date of injury or illness ___ ___ ___ (Month Day Year)

12) Time employee began work ____ ○ **AM** ○ **PM**

13) Time of event ____ ○ **AM** ○ **PM** ○ **Check if time cannot be determined**

14) *What was the employee doing just before the incident occurred?* Describe the activity, as well as the tools, equipment, or material the employee was using. Be specific. *Examples:* "climbing a ladder while carrying roofing materials," "spraying chlorine from hand sprayer," "daily computer key entry."

15) *What happened?* Tell us how the injury occurred. *Examples:* "When ladder slipped on wet floor, worker fell 20 feet"; "Worker was sprayed with chlorine when gasket broke during replacement"; "Worker developed soreness in wrist over time."

16) *What was the injury or illness?* Tell us the part of the body that was affected and how it was affected; be more specific than "hurt," "pain," or "sore." *Examples:* "strained back"; "chemical burn, hand"; "carpal tunnel syndrome."

17) *What object or substance directly harmed the employee? Examples:* "concrete floor"; "chlorine"; "radial arm saw." *If this question does not apply to the incident, leave it blank.*

18) *If the employee died, when did death occur?* Date of death ___ ___ ___ (Month Day Year)

Page 1 **of** 1

Save Input **Add a Form Page** **Reset**

Public reporting burden for this collection of information is estimated to average 22 minutes per response, including time for reviewing instructions, searching existing data sources, gathering and maintaining the data needed, and completing and reviewing the collection of information. Persons are not required to respond to the collection of information unless it displays a current valid OMB control number. If you have any comments about this estimate or any other aspects of this data collection, including suggestions for reducing this burden, contact: US Department of Labor, OSHA Office of Statistical Analysis, Room N-3644, 200 Constitution Avenue, NW, Washington, DC 20210. Do not send the completed forms to this office.

FIGURE 5.2 The OSHA 301 form is used to complete the OSHA 300 Log. It must be kept on file for 5 years with the OSHA 300 record (US Department of Labor—Occupational Safety and Health Administration). Fillable PDF files are available at https://www.osha.gov/recordkeeping/RKforms.html

TABLE 5.4 Indicators of a need for further assessment.

A high number of injuries and illnesses recorded on your OSHA 300 Logs (especially if the incidents are in the areas where you have identified powered equipment and work habits that are not consistent with company work practices)
A high percentage of workers using powered equipment and working near such equipment
A high number of workers' compensation claims and costs. (An experience modification rate can be used as an indicator, i.e., an experience modification rate of 1.25 or higher)
A frequent number of employee actions requiring interaction with or touching equipment or materials (i.e., lifting, cutting, activating powered equipment, feeding raw materials into powered equipment) per unit of time (half-hour increments recommended)
Difficult environmental conditions, such as slippery or steep surfaces

Select a team leader (preferably an operations manager/director), keep the number of members to a minimum, and assign individuals specific project tasks. Using project management techniques, milestones, and timelines will keep the process on schedule. If there are other plants with the same equipment, divide the tasks among plants.

Table 5.5 contains the key elements for conducting a more detailed assessment. This list should be used as a reference and sections used only as appropriate. It is not all encompassing or complete; however, by researching your workplace needs and using parts of this list and adding others, you will customize the criteria for assessment. The assessments will identify the needs. Choosing corrective action and a long-term plan to implement and maintain a system will require the same team effort, commitment, and participation needed for the assessment.

PREVENTION

The key to reducing injuries is preventing accidents. An effective prevention program presumes a surveillance component to identify high-risk potential hazard exposures and processes and then targets interventions at these identified items. The surveillance program is a "leading indicator," because it predicts likely accidents before they happen and allows us to prevent them.[3,4]

Johnson has defined an accident as an unwanted transfer of energy because of lack of barriers or controls, which produces injury to persons, property, or processes and which is preceded by sequences of planning and operation errors which (i) failed to adjust to changes in physical or human factors and (ii) produced unsafe conditions and/or unsafe acts.[5]

If there is a hazard present, the results of this sequence may be an injury. In the work environment, and to a certain extent any environment with which people interact, there are basically three categories of hazard:

1. **Inherent properties** of the mechanism, process, or other environmental variables, such as walking or working surface, electrical hazards, and the hazards of mechanical motion, weight, or radiation which are most likely to affect the end result of the sequence
2. **Failures** of materials, equipment, safeguards, or the human element that may be the proximate cause of an accident
3. **Environmental stresses**, which are contributory to the sequence of events such as natural, ergonomic, thermal, physical energy or electromagnetic stresses

Accident prevention depends on correcting both **unsafe conditions and unsafe acts**. Unsafe conditions can be corrected with process substitution, engineering design changes, and "guarding." Guarding in the safety profession refers to (i) machine guarding (which is really an engineering control) and (ii) personnel guarding, or the use of personal protective equipment. Good guard design shields the hazard and anticipates how workers might disregard or inactivate the guarding.

Unsafe acts can be divided into (i) lack of knowledge, which can be prevented by training; (ii) inattention; and (iii) deliberate acts. Some workers do not recognize hazards, some will defeat controls and guards to save time, and most people are not, and cannot be, 100% alert at all times. Training workers in the consequences of unsafe acts is the main method to reduce accidents from these acts. Deliberate acts such as bypassing guards and using "shortcuts" are common because of the human tendency to find the easiest or quickest (but not necessarily the safest) way to do a job.

The systems approach

Prevention of injuries and accidents utilizes various levels of safeguards to shield workers from hazards. The systems approach is a modern safety technique. In brief terms, it looks at the total system—people, materials, facilities, equipment, procedures, and process—to determine what could go wrong and in what way, what could cause it to go wrong, and what to do to keep it from going wrong. The approach should be and has been applied proactively from the design phase through the life of the system.

Systems are comprehensive and interrelated activities which, when acted upon in a consistent sequence, serve to complement each other to achieve continuous improvement. There are a number of approaches to developing a safety system, and the hallmarks include, but are not limited to:

1. Leadership—Management plays an active and visible role in establishing a team approach to developing, implementing, and maintaining continuous measurable improvement.

TABLE 5.5 Identifying pertinent equipment, operations, and procedures.

1. Conduct a physical plant equipment and energy survey/inventory
    i. Identify all powered equipment
    ii. Identify all sources of equipment and building power supplies. Identify the common sources of equipment energy, including but not limited to:
        (a) Electrical
        (b) Hydraulic
        (c) Mechanical
        (d) Pneumatic
        (e) Others
    iii. Evaluate individual equipment to identify all power sources. Potential sources of power, both external and within the equipment (including the above sources):
        (a) Compressed (i.e., spring)
        (b) Gravity actuated
        (c) Partially cycled equipment
        (d) Energized capacitors
        (e) Any other source that might cause unexpected movement
    iv. Identify type and points of potential contact between the worker and equipment, materials, power sources, and moving parts (refer to Table 5.6 and Figure 5.1)
    v. Evaluate equipment operating software and procedures. Many equipment operations are directed by software. The potential for unplanned equipment activation/action during maintenance, power interruptions and surges, and changes due to software upgrades is increasing the possibility of injuries)
2. Inventory work areas to identify high to low employee work activity and describe the types of work conducted:
    i. Materials and movement due to production processes (i.e., lifting, raw materials)
    ii. Interaction with equipment
    iii. Climbing and walking
    iv. Repetitive actions
    v. Manual processes
    vi. Automated processes
    vii. Work surfaces and walkways
3. Work procedures and processes
    i. Manuals, manufacturers, and company-recommended procedures
    ii. Modifications of equipment and procedures
    iii. Foreign-manufactured equipment and operating/maintenance procedures
    iv. Equipment deenergizing and reenergizing procedures
4. Observe and evaluate employees' work activities
    i. Work areas to identify sources of exposure to mechanical energy
    ii. Equipment and work procedures
    iii. Maintenance of equipment
    iv. Pertinent employee training
    v. Job safety/hazard analysis (JSA/JHA) by trained professionals
    vi. Controls
5. Detailed assessment of injury, illness, and incident experience
    i. OSHA 300 Logs and Metrics (see Tables 5.1 and 5.2)
    ii. Workers' compensation experience and specific claims
    iii. Incident reports (i.e., injuries, illnesses, near misses, serious potential incidents, production interruptions)
    iv. Trend analysis of incidents (refer to #3 above)
    v. Regulatory citations and notices
6. Worker evaluations of work activity and work place hazards
7. Identification and "evaluation" of work place hazards
    i. Control of hazards
    ii. Evaluation of work procedures
    iii. Worker training
    iv. Changes in process and equipment
    v. Unsafe acts and conditions
    vi. Regulations and industry best practices
    vii. OSHA regulations on managing workplace safety and mechanical energy (i.e., 29 CFR 1910 and 29 CFR 1910.147—Control of Hazardous Energy Lockout/Tagout)
    viii. Industry and best practices (i.e., American National Standards Institute (ANSI), Chemical Manufacturers of America (CMA), American Petroleum Institute (API), National Fire Protection Association (NFPA), International Standards Organization (ISO))

2. Self-assessment and performance management—Comprehensive self-assessments identify risks and liabilities and opportunities to manage these risks and limit potential losses. Metrics are established to accurately measure and report performance with the goal of continuous measurable improvement.
3. Personnel development—Organizational needs are identified and personnel selected to match both immediate and future needs. Personnel are developed to grow and change with the needs of the organization and business climate and as opportunities develop.
4. Design and operational integrity—Production and operational standards are identified, and immediate and long-range design, implementation, and operational standards are implemented.
5. Planning and change management—Strategic plans are developed to meet organizational and operational needs and ensure high-caliber performance in a competitive business environment. Change is an anticipated and integral part of the planning process. Contingencies are designed and readily implemented to maximize opportunities and ensure best practices.
6. Incident investigation—System feedback mechanisms are established to promptly assess success and failures. Lessons learned are promptly communicated and leveraged throughout the organization.
7. Audits—Self-evaluations are periodically conducted to compare performance to standards and implementation of system best practices. Results are reported and improvements monitored until fully implemented.

Injury and illness prevention programs have evolved beyond "initial behavioral" safety programs (e.g., OSHA Voluntary Protection Program (VPP), DuPont STOP Program) and include the next generation of worker behavior modification systems (e.g., Behavioral Safety Technology and Predictive Solutions), which has a predictive element and supports "unsupervised real-time behavioral change."

Two specific examples of systems of prevention applied to mechanical energy are machine guarding and energy control.

Machine guarding

Mechanical energy and machine guarding are part of the greater system of engineering controls. The system consists of a single unit of equipment and the associated production process, including raw materials, manufacturing process to modify materials, employee interaction in the process, and the general work environment. All these factors are interdependent. The work area must be assessed as a 360° environment, with special emphasis on exposure to mechanical energy to help identify mechanical exposure and guarding/control. Failures in the system can result in contact with mechanical energy and ultimately injuries. A list of commonly encountered machines is included in Table 5.6, and definitions used in machine guarding are listed in Table 5.7.

TABLE 5.6 Examples of machines that usually require point of operation guarding.

Guillotine cutters
Shears
Alligator shears
Power presses
Milling machines
Power saws
Jointers
Portable power tools
Forming rolls and calendars

Source: Obtained from Federal OSHA 29 CFR 1910.212.

One or more methods of machine guarding are required to protect employees.[6] Machine guarding regulatory requirements have been published and applied to broad classes of equipment. Common hazards have been recognized through industrial experience and are roughly classified as point of operation, ingoing nip points, rotating parts, flying chips, and sparks.

Point of operation presents the most common and generally obvious hazard potential. These hazards occur where materials are modified, and the employee is usually in close proximity to moving equipment and raw materials.[7] Table 5.8 lists regulatory requirements for specific controls designed to eliminate access to the points of operation, specifically the power press design criteria for the distance of guards from a "danger zone" or point regulatory design criteria. The guarding device shall be in conformity with any appropriate standards therefore, or, in the absence of applicable specific standards, shall be so designed and constructed as to prevent the operator from having any part of his body in the "danger zone" during the operating cycle.

One or more types of controls, machine guarding, and protective methods must be employed to properly protect employees and meet regulatory requirements. The most common mechanical hazards are present at the point of operation (see definitions), during power transmission, and at a number of locations as materials are processed with equipment. Aside from traditional physical machine guards, there are a number of presence-sensing devices, hand removal or restraint devices, and two-hand trip devices that help prevent contact with moving machine parts and tools. The most common devices are:

1. Presence-sensing devices—Prevent normal press operation if the operator's hands are inadvertently within the point of operation, prevent the initiation of a stroke, or stop a stroke in progress, if a body part, material, or

TABLE 5.7 Definitions for machine guarding.

Point of operation—the point at which cutting, shaping, boring, or forming is accomplished upon production materials. This is where exposure to mechanical energy is greatest and production material (or a body part) is actually positioned and work is performed.

Pinch point—any point other than the point of operation in which it is possible for a part of the body to be caught between the moving parts of a press or auxiliary equipment, or between moving and stationary parts of a press or auxiliary equipment or between the material and moving part or parts of the press or auxiliary equipment.

Safety system—the integrated system, including the pertinent elements of equipment, the controls, the safeguarding and any required supplemental safeguarding, and their interfaces with the operator, and the environment, designed, constructed and arranged to operate together as a unit, in such a way that one error will not cause injury to personnel due to point of operation hazards.

Nip-point belt and pulley guard—devices that enclose pulleys and are provided with rounded or rolled edge slots through which the belt passes.

Authorized person—an individual who has the authority and responsibility to perform a specific assignment and has been given authority by the employer.

Fixed barrier guard—a barrier guard fixed to a press frame, which is "unmovable" and restricts the operator's access to moving machine parts such as the point of operation.

Interlocked press barrier guards—barrier guards attached to the press frame and interlocked so that the press stroke cannot be started under most conditions unless the guard itself, or its hinged or movable sections, are in place and enclose the point of operation or other hazardous machine parts.

Adjustable barrier guard—a barrier requiring adjustment for each job or die setup.

Source: Adapted from Federal OSHA 29 CFR 1910.212.

TABLE 5.8 OSHA required openings in inches to guard a power punch press (CFR 1910.217(g)).

Distance of opening from point of operation hazard	Maximum width of opening
½ to 1½	1/4
1½ to 2½	3/8
2½ to 3½	1/2
3½ to 5½	5/8
5½ to 6½	3/4
6½ to 7½	7/8
7½ to 12½	1¼
12½ to 15½	1½
15½ to 17½	1⅞
17½ to 31½	2⅛

device passes into an electronic sensing field. Presence-sensing devices like machine guards have specific regulatory requirements (see 29 CFR—1910 Appendix A to 1910.217). The employer is ultimately responsible for ensuring (in this case) that the regulatory certification/validation requirements are met.

2. Hand removal devices—Automatically withdraw the operator's hands if the operator's hands are inadvertently within the point of operation.
3. Two-hand trip/control devices—Require two hands to trip/operate the equipment. For instance, the operator must simultaneously depress two buttons located on each side of the operator. Thus the operator must have his hands out of the danger zone to activate the equipment. Design criteria for the use of two-hand control devices and calculation of the distance from the sensing field to the point of operation are essential elements in a safety design.

Other common devices for preventing operator contact with mechanical energy include the following:

1. Special hand tools for placing and removing material into equipment points of operation. These tools are designed to permit easy handling of material without the operator placing a hand in the danger zone. Such tools shall not be in lieu of other guarding required by this section, but can only be used to supplement protection provided.
2. Gates or movable barrier devices are barriers arranged to enclose the point of operation. The gates must be in place before the press/equipment can start a part of or a new production/machine cycle.
3. Holdout or restraint devices are mechanisms that prevent an operator's hands from entering the point of operation. The operator's hands are secured with cables (i.e., cord, metal cable), and the cables are adjusted and secured to ensure that the operator cannot reach into a point of operation.
4. Pullout devices are mechanisms attached to the operator's hands and to a movable equipment part. These must be adjusted to each worker's unique physical characteristics; they pull the operator's hands away from the point of operation or danger zone. They operate inversely to the movement of the equipment. As a point of operation is reaching its most hazardous point (i.e., mechanical energy is transferred as work is performed), this device pulls the worker's hands back to a point beyond the danger zone.
5. Sweep devices are arms or bars that move the operator's hands to a safe position as the equipment cycles. They will literally sweep the operator's hands away from the danger zone as the machine cycles.

Managing hazardous energy

Lockout/tagout programs are one of the most common methods to control mechanical energy during equipment maintenance, adjustment, or repair. The standard 29 CFR 1910.147—The Control of Hazardous Energy Lockout/Tagout—applies during maintenance, repair, or equipment adjustment and does not apply to normal production operations. Important definitions of the standard are listed in Table 5.9. The primary objective of the standard is to eliminate or manage the unexpected "energization" or start-up of the machines or equipment or release of stored energy that could cause injury to employees.

Managing hazardous energy requires employers to establish programs and utilize procedures to control or eliminate uncontrolled hazardous energy. This program must include procedures, practices, and training to ensure that contact with energized equipment or equipment components is managed and includes appropriate lockout and tagout devices that isolate and/or disable machines or equipment to prevent unexpected energization, start-up, or release of stored energy in order to prevent injury to employees.[8]

TABLE 5.9 Definitions for lockout/tagout.[8]

Affected employee—an employee whose job requires him/her to operate or use a machine or equipment on which servicing or maintenance is being performed under lockout or tagout or whose job requires him/her to work in an area in which such servicing or maintenance is being performed

Authorized employee—a person who locks out or tags out machines or equipment in order to perform servicing or maintenance on that machine or equipment. An affected employee becomes an authorized employee when that employee's duties include performing servicing or maintenance covered under this section

Energized—connected to an energy source or containing residual or stored energy. The source of energy can be electrical, mechanical, hydraulic, pneumatic, chemical, thermal, or other energy

Energy-isolating devices—are devices that physically prevent the transmission or release of energy, including but not limited to the following: a manually operated electrical circuit breaker; a disconnect switch; a manually operated switch by which the conductors of a circuit can be disconnected from all ungrounded supply conductors and, in addition, no pole can be operated independently; a line valve; a block; and any similar device used to block or isolate energy. Push buttons, selector switches, and other control circuit-type devices are not energy-isolating devices

Lockout/tagout devices—are devices placed in accordance with an established procedure, ensuring that the energy-isolating devices and tags and the equipment being controlled cannot be operated until the lockout device is removed

Energy control programs—are employer-implemented programs consisting of energy control procedures, training, periodic inspections, and lockout/tagout devices to ensure that before any employee performs any servicing or maintenance on a machine or equipment where the unexpected energizing, start-up, or release of stored energy could occur and cause injury, the machine or equipment shall be isolated from the energy source and rendered inoperative

Employee protection and energy elimination and isolation are the goals of lockout/tagout. This requires that the employer demonstrate that the tagout program will provide a level of safety necessary to ensure no contact occurs with energized equipment or machinery. The employer shall demonstrate full compliance with all tagout-related provisions of the standard. Additional means of protection must be considered including periodic inspection of the lockout tagout program.

Procedures need to be developed, documented, and utilized for the control of potentially hazardous energy when employees are engaged in lockout/tagout activities.

The procedures need to clearly and specifically outline the scope, purpose, authorization, rules, and techniques to be utilized for the control of hazardous energy, and the means to enforce compliance include, but not limited to, the following:

1. A specific statement of the intended use of the procedure
2. Shutting down, isolating, blocking, and securing machines or equipment
3. Placement, removal, and transfer of lockout/tagout devices:
    a. Lockout/tagout devices are locks, tags, chains, wedges, key blocks, adapter pins, self-locking fasteners, or other hardware for isolating, securing, or blocking of machines or equipment from energy sources and must be provided. These tags must be standardized, easily identifiable, and only used for controlling energy. Lockout devices shall be substantial enough to prevent removal without the use of excessive force or unusual techniques, such as with the use of bolt cutters or other metal-cutting tools.
    b. Tagout devices, including their means of attachment, shall be substantial enough to prevent inadvertent or accidental removal.
4. Responsibility for ensuring that lockout/tagout devices are utilized and who will ensure that this is done
5. Testing a machine or equipment to determine and verify the effectiveness of lockout devices, tagout devices, blocks, and other energy control equipment
6. Procedures for reenergizing equipment and removing lockout/target devices

Training and communication

Employers should provide training to ensure that employees understand the purpose and function of all energy control programs and that the knowledge and skills required for the safe application, usage, and removal of the energy controls are acquired by employees. Examples of the training

TABLE 5.10 Training and communication for lockout/tagout as a paradigm for hazards control training.

Each authorized employee shall receive training in the recognition of applicable hazardous energy sources, the type and magnitude of the energy available in the workplace, and the methods and means necessary for energy isolation and control
Each affected employee shall be instructed in the purpose and use of the energy control procedure
All other employees whose work operations are or may be in an area where energy control procedures may be utilized shall be instructed about the procedure and about the prohibition relating to attempts to restart or reenergize machines or equipment which are locked out or tagged out
Retraining shall be provided for all authorized and affected employees whenever there is a change in their job assignments, machines, equipment, or processes that present a new hazard or a change in the energy control procedures
Affected employees shall be notified by the employer or authorized employee of the application and removal of lockout devices or tagout devices. Notification shall be given before the controls are applied and after they are removed from the machine or equipment

recommendations for the lockout/tagout regulations are included in Table 5.10; they are generalizable to training for all control technologies.

Periodic inspection

The final component of the systems approach is to inspect existing controls to ensure continued optimum operation. The employer shall conduct a periodic inspection of the energy control procedure at least annually to ensure that all equipment lockout/tagout procedures and the requirements of this standard are fully implemented.

The periodic inspection shall be performed by an authorized employee other than the ones(s) utilizing the energy control procedure being inspected and shall be conducted to correct any deviations and or areas requiring improvement.

Where tagout is used for energy control, the periodic inspection shall include a review, between the inspector and each authorized and affected employee, of that employee's responsibilities under the energy control procedure being inspected.

The employer shall certify that the periodic inspections have been performed. The certification shall identify the machine or equipment on which the energy control procedure was being utilized, the date of the inspection, the employees included in the inspection, and the person performing the inspection.

Utilization of as many of these principles of prevention as possible will maximize production and safety and minimize injuries from mechanical hazards in the workplace.

SUMMARY

Mechanical energy is an ever-present exposure in our industrial environment. Our changing workplaces and work activities present numerous challenges for effectively identifying and managing exposure to this energy. OSHA regulations provide guidance; however, this is not a substitute for an effective systems approach to identifying and actively managing exposures. This chapter provides general guidance and basic tools needed to begin the process of systematically managing mechanical energy, starting with recognition and quantification and ultimately leading to elimination and control.

References

1. National Safety Council. Injury Facts. Chicago: National Safety Council, 2015.
2. US Department of Labor. Detailed Guidance for OSHA's Injury and Illness Recordkeeping Rule. https://www.osha.gov/recordkeeping/entryfaq.html (accessed April 28, 2016).
3. Hagan PE, Montgomery JF, O'Reilly JT. Accident prevention manual for business and industry: engineering and technology, 13th edn. Chicago: National Safety Council Press, 2009.
4. CoVan J. Safety engineering. New York: John Wiley & Sons, Inc., 1995.
5. Johnson WG. MORT safety assurance systems. New York: Marcel Dekker, 1980.
6. Occupational Safety and Health Administration. The principles and techniques of mechanical safeguarding, Bulletin No. 197. Washington, DC: US Department of Labor, 1973.
7. Wadden RA, Scheff PA. Engineering design for the control of workplace hazards. New York: McGraw-Hill, 1987.
8. American National Standards Institute/American Society of Safety Engineers. Control of hazardous energy lockout/tagout and alternative methods Z 244.1–2003 (R2014). New York: ANSI, 2014.

II The Physical Work Environment

Chapter 6

HOT ENVIRONMENTS*

David W. DeGroot and Laura A. Pacha**

Exposure to high ambient temperatures while working in hot indoor environments or while working outdoors in hot weather is a common and potentially fatal occupational hazard. Unlike cold exposure (Chapter 7), which can be almost completely mitigated with proper clothing and equipment, it is impossible to fully mitigate the effects of a hot environment. Normally, body core temperature is maintained within a very narrow range. In the occupational setting, heat stress from the combined effects of environmental heat, metabolic heat, and often the use of impervious clothing can strain the ability to maintain heat balance, and core temperature may begin to rise, potentially leading to an exertional heat illness. Workplace heat exposure, in addition to causing heat-related illness, has been found to decrease productivity and to increase job-related accidents.[1]

OCCUPATIONAL SETTING

No recent estimate of the number of workers exposed to hot environments has been published. In 1986, the National Institute for Occupational Safety and Health (NIOSH) estimated that 5–10 million Americans worked in jobs where heat stress was an occupational health hazard, and this range is probably still a reasonable estimate.[2] Table 6.1 lists some common work sites where workers may be expected to experience heat stress.[3] Heat-related deaths are common in the United States. From 1999 to 2009, there was an average of 658 heat-related deaths annually.[4] Most of the heat-related deaths reported by the Centers for Disease Control and Prevention (CDC) were in nonworking populations. Occupational- or exertional-related heat deaths, while known to occur, are infrequently reported in the medical literature.[5] Thirteen workplace heat-related deaths were cited for federal enforcement by the Occupational Safety and Health Administration (OSHA) in 2012–2013.[5] On the other hand, heat-related hospitalizations are relatively common, with over 28 000 cases in a 10-year period.[6] How many of these cases were associated with occupational exposure to heat stress is not known. A case series of heat-related casualties in the mining industry reported the incidence of heat exhaustion in the summer months as 43 cases/million man-hours worked.[7] Military personnel, who are often required to achieve very high levels of work output in hot environments, have a long history of high risk for heat stress.[8] Reports from the military continue to provide information on the occurrence of heat injury in healthy, young individuals.[9,10]

In any hot environment, whether indoors or outdoors, high humidity or heavy manual labor increases the workers' risk for heat strain. Specific occupational groups, such as firefighters, are at exceptionally high risk for heat stress. Not only are they exposed to extremely high temperatures while fighting fires but they also must perform demanding physical tasks while wearing fire-resistant protective clothing that attenuates heat dissipation to the environment. The hazardous waste cleanup industry is another field where heat strain is a potentially significant occupational hazard, due to required respiratory protection and full protective clothing.

* **DISCLAIMER:** The views expressed in this chapter are those of the authors and do not necessarily reflect the official policy of the Department of Defense, Department of the Army, or the US Army Medical Department.

** The authors gratefully acknowledge Gail M. Gullickson, the author of the previous edition of this chapter, for her contributions.

Physical and Biological Hazards of the Workplace, Third Edition. Edited by Gregg M. Stave and Peter H. Wald.

TABLE 6.1 Work sites with heat exposure.

Iron and steel foundries
Non-ferrous foundries
Brick-firing and ceramics plants
Glass products facilities
Rubber products factories
Electrical utilities
Bakeries
Confectioneries
Commercial kitchens
Laundries
Food canneries
Chemical plants
Mining sites
Smelters
Steam tunnels
Fires (firefighting)
Outdoor operations
Surface mines
Agriculture sites
Construction sites
Merchant marine ships
Hazardous waste sites
Military training sites
Athletic competitions

Source: Department of Labor. OSHA Technical Manual In: OSHA, ed. Washington, DC 1999.

These requirements limit the use of conventional heat stress monitoring guidelines and place an additional stress on the hazardous waste cleanup worker.

MEASUREMENT ISSUES

Environmental heat

Ambient or environmental heat affecting the worker and the worker's ability to transfer body heat to the environment is determined by four environmental factors: (i) air or dry-bulb temperature (T_{db}), (ii) air moisture content or wet-bulb temperature (T_{wb}), (iii) air velocity, and (iv) radiant heat (solar and infrared). Microwave radiation may also be a source of environmental heat in some work situations (see Chapter 15). Various measures of ambient heat load, reflecting the factors influencing heat transfer, are available for use in the industrial setting.[11] The most commonly used instruments for quantifying the thermal environment are described in Table 6.2.

Metabolic heat

Heat production in the body is the by-product of normal basal metabolism and occurs mostly in the liver, brain, heart, and skeletal muscles. During exercise or work, skeletal muscle activity greatly increases metabolic heat production. A measurement of metabolic heat production or, at least, an estimate of metabolic heat is essential in determining a worker's total heat stress and in calculating workplace heat exposure limits. Direct or indirect measurement of each worker's level of work and quantity of heat produced in the occupational setting is not practical. Therefore, tables of the workload or energy cost of various tasks have been developed and are used to estimate metabolic heat (Table 6.3).

EXPOSURE GUIDELINES

Heat stress indexes

The guidelines currently used for worker exposure to heat stress are based on indexes developed through subjective and objective testing of workers or from combinations of external heat measurements. Earlier examples include the effective temperature index, resultant temperature, equivalent temperature, the Oxford index, the Botsball, the heat stress index, and the wet globe temperature index. It is important to note that some of these are indicators of subjective thermal comfort and not of heat stress.[12] Today, the most commonly used index is the Wet-Bulb Globe Temperature (WBGT) index, which was originally developed in an effort to reduce the incidence of heat illness during Marine Corps basic training at Parris Island.[13]

The WBGT index is recommended by OSHA, NIOSH, and American Conference of Governmental Industrial Hygienists (ACGIH) and required by the US Armed Forces.[14,15]

The WBGT index is calculated from measurements of the natural wet-bulb (T_{wb}), the black globe (T_g), and the dry-bulb (T_{db}) temperatures.

For outdoor environments with a solar heat source, the WBGT formula is

$$\mathrm{WBGT} = 0.7T_{wb}\mathrm{NWB} - 0.2T_g + 0.1T_{db}$$

For indoor use or for outdoor settings without a solar load, the formula is

$$\mathrm{WBGT} = 0.7T_{wb} + 0.3T_g$$

The necessary measurements require relatively simple instrumentation and can be easily obtained in an industrial environment. Automated heat stress monitors that measure all three temperatures and calculate the WBGT index temperature are also available. An example of a manual WBGT monitor is shown in Figure 6.1. The WBGT index was developed for men exercising outdoors in military fatigues. Therefore, if different types of clothing are worn, correction factors are required. This index is not applicable in settings where sweat-impermeable clothing is required and may not be as effective as other indices in preventing heat casualties in extreme heat stress conditions.

TABLE 6.2 Measures of external heat.[2,11]

Measure	Device	Comments
Dry bulb	Liquid-in-glass thermometer; thermocouple; resistance thermometer	Measures ambient air temperature and is useful in determining comfort zone for lightly clothed sedentary workers. Does not measure effect of humidity, radiant heat, or air movement on temperature
Wet bulb	Thermometer bulb or sensor covered by a wet cotton wick that is exposed to air movement	Measures effect of humidity on evaporation and effect of air movement on ambient temperature. Natural wet-bulb temperature is the term used if the wet bulb is exposed to prevailing natural air movement; may be a useful guide in preventing heat stress in hot, humid, and still environments where radiant heat does not contribute to heat load, such as underground mines
Globe temperature	Black globe—temperature sensor in the center of a 15 cm hollow copper sphere painted flat black Wet globe—temperature sensor in the center of a 3 in. copper sphere covered by a wet black cloth	Measures effect of radiant heat. Wet globe thermometer (Botsball) also reflects the effect of humidity and supposedly exchanges heat with the environment similarly to a nude man with totally wet skin
Air velocity	Vane and cup, hot-wire, pulsed-wire, ultrasonic, and Laser doppler anemometers	Vane and cup and hot-wire anemometers are directional devices, remaining anemometers are insensitive to flow direction and may be more suitable for use in locations with varying air flow direction

TABLE 6.3 Metabolic rate categories and example activities.

Category	Metabolic rate (W)*	Examples
Resting	115	Sitting
Light	180	Sitting with light manual work with hands or hands and arms, and driving. Standing with some light arm work and occasional walking.
Moderate	300	Sustained moderate hand and arm work, moderate arm and leg work, moderate arm and trunk work, or light pushing and pulling. Normal Walking.
Heavy	415	Intense arm and trunk work, carrying, shoveling, manual sawing; pushing and pulling heavy loads; and walking at a fast pace.
Very Heavy	520	Very intense activity at fast to maximum pace.

*The effect of body weight on the estimated metabolic rate can be accounted for by multiplying the estimated rate by the ratio of actual body weight divided by 70 kg (154 lb). From ACGIH®, *2015 TLVs® and BEIs®* Book. Copyright 2015. Reprinted with permission.

Heat strain indicators

Under identical heat stress conditions, the heat strain experienced by different individuals may vary widely, which may require estimation of the physiologic responses to environmental heat stress. Historically, the cumbersome nature of monitoring equipment and its lack of durability have limited our ability to make real-time measurements of a workers' core temperature or heart rate. As technology improves and durable and convenient monitors and sensors become available, the physiologic measures of heat strain may take on increasing importance, especially for workers in heavy protective clothing.

Heart rate is easily measured and may be a reliable indicator of overall heat strain, rising with both increasing workload and increasing core temperature. Utilizing this physiologic response to heat, a method of measuring oral temperature, postwork heart rate, and recovery heart rate to monitor for heat strain has been developed. Heat stress exposure should be discontinued if (i) sustained heart rate is in excess of 180 bpm minus the individual's age in years or (ii) recovery heart rate at 1 minute after peak work effort is greater than 120 bpm.[14] The availability of inexpensive electronic pulse monitors and timers has made this method easier, and it can be a valuable indicator of heat strain in certain settings.

Body core temperature appears to be a more reliable indicator of heat stress than heart rate. The World Health Organization recommends that body core temperature should not, under circumstances of prolonged daily work and heat, be permitted to exceed 38°C (100.4°F) rectally or 37.5°C (99.5°F) orally.[2] Though this heat strain index may seem ideal, monitoring internal or core body temperature with rectal or esophageal probes is not acceptable to many workers. Oral temperatures, while easy to obtain, may be inaccurate measures of core temperatures because of mouth breathing or drinking hot or cold

FIGURE 6.1 Manual, analog WBGT monitor; automated WBGT monitors are available from a variety of manufacturers. Photo courtesy of the authors.

liquids immediately before using a thermometer. An ingestible capsule containing a temperature sensor and a device producing a telemetry signal is available to monitor real-time internal temperatures as the capsule passes through the gastrointestinal tract. Validation studies suggest that given appropriate time to exit the stomach, this method is an acceptable surrogate for rectal or esophageal temperature.[16,17] Tympanic membrane temperature monitors may also provide an acceptable measurement of core temperature, but use of these monitors is uncommon because of ear discomfort and the need for a good seal in the ear canal. Mathematical models to predict core temperature and/or heat strain risk from noninvasive inputs have been developed; further validation and packaging in an end-user-friendly interface are necessary before they can be utilized in the occupational setting.[18]

Estimation of fluid loss or hydration status by regular weight measurements prior to work and throughout the day has also been proposed as an indicator of heat strain. Based upon weight measurements (assuming that the worker was fully hydrated before beginning work), the heat-exposed worker can be encouraged to drink liquids to maintain hydration and constant body weight throughout the day. Urine specific gravity measurements before, during, and after work have also been used to assess hydration status in workers. Weight loss of greater than 1.5% of body weight over the course of a normal shift may indicate that an individual is at greater risk of heat illness.[14]

ACGIH GUIDELINES

In the United States, the ACGIH guidelines are frequently used by industry to determine acceptable heat exposure for employees. These guidelines or threshold limit values (TLVs®) permit working conditions "that nearly all adequately hydrated, unmedicated, healthy workers may be repeatedly exposed without adverse health effects".[14] The International Organization for Standardization also provides guidance for use of the WBGT index and for estimating thermal strain using physiological measurements.[19,20]

The 2006 revision of the ACGIH heat stress and strain guidelines includes a decision tree for the assessment of a worker's risk of heat stress (Figure 6.2). The previous ACGIH guidelines were similar to the NIOSH guidelines and recommended that workers in hot environments rest for a portion of each hour, with the amount of rest based upon WBGT index, level of activity, and acclimation status of the worker. The current ACGIH recommendations still use a "work–rest regimen" but expand the guidelines for work situations where heat stress conditions (e.g., use of encapsulating clothing) exceed the typical work–rest cycle recommendations.

In general, the ACGIH decision tree incorporates an assessment of the type of clothing worn, the level of environmental heat or WBGT index, an estimate of the worker's physical activity, and an analysis of the work and work site. Table 6.4 provides work–rest cycles based upon the worksite WBGT index, work intensity, and worker acclimatization status. If the temperature index or work exceeds the limits outlined in Table 6.4, a detailed analysis of the heat stress potential of the work site, using a rationale model of core temperature, is required.[20] If the detailed analysis reveals excessive heat stress, physiologic monitoring of the workers is needed. The reader should consult an experienced health professional and review the specific guideline documentation before attempting to implement these guidelines in the workplace.[14]

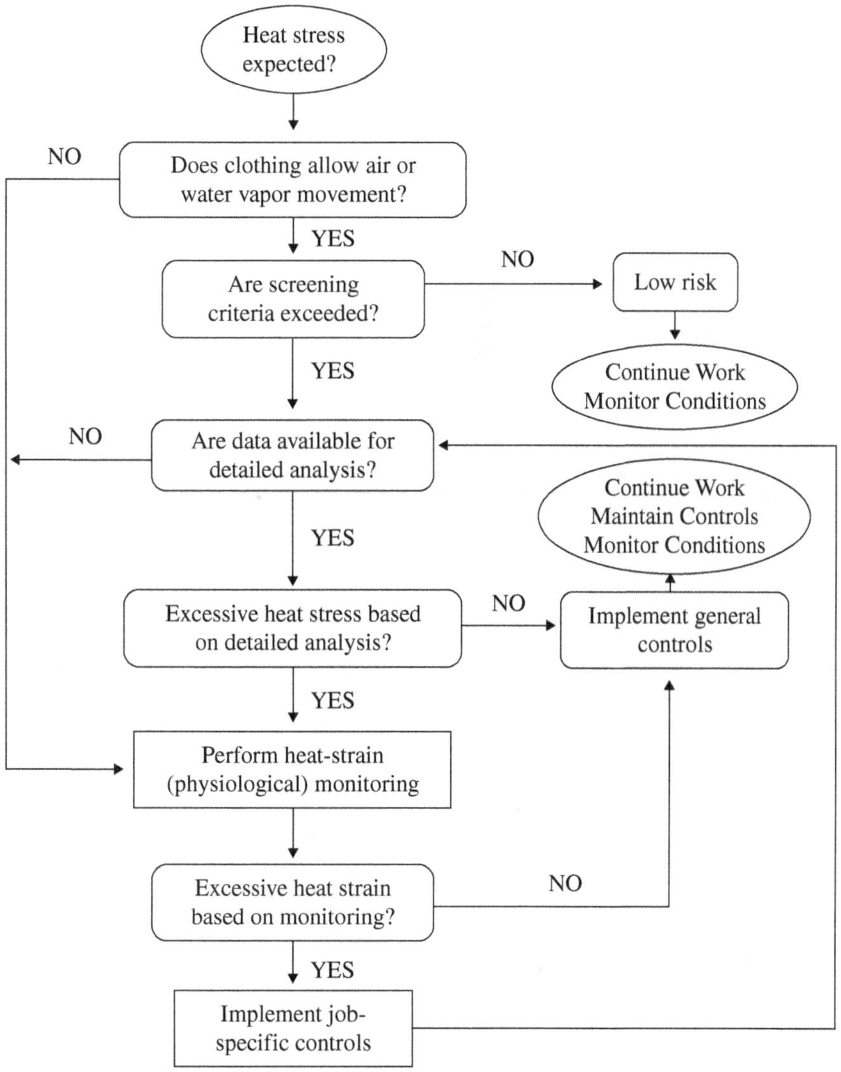

FIGURE 6.2 Heat stress decision tree. From American Conference of Governmental Industrial Hygienists, (ACGIH®), 2015 Threshold Limit Values (TLVs®) for Chemical Substances and Physical Agents and Biological Exposure Indices (BEIs®). Reprinted with permission.

TABLE 6.4 Examples of permissible heat exposure threshold limit values: screening criteria for TLV and action limit for heat stress exposure.

Allocation of work in a cycle or work and recovery	TLV (WBGT values in °C)				Action limit (WBGT values in °C)			
	Light	Moderate	Heavy	Very heavy	Light	Moderate	Heavy	Very heavy
75–100%	31.0	28.0	–	–	28.0	25.0	–	–
50–75%	31.0	29.0	27.5	–	28.5	26.0	24.0	–
25–50%	32.0	30.0	29.0	28.0	29.5	27.0	25.5	24.5
0–25%	32.5	31.5	30.5	30.0	30.0	29.0	28.0	27.0

See Table 6.3 for work demand categories.

WBGT values are expressed in °C and represent thresholds near the upper limit of the metabolic rate category.

If work and rest environments are different, hourly time-weighted averages (TWA) should be calculated and used. TWAs for work rates should also be used when the work demands vary within the hour.

Because of the physiologic strain associated with heavy and heavy work among less fit workers regardless of WBGT, criteria values are not provided for continuous work and for up to 25% rest in an hour. The screening criteria are not recommended, and a detailed analysis and/or physiologic monitoring should be used. From ACGIH®, *2015 TLVs® and BEIs®* Book. Copyright 2015. Reprinted with permission.

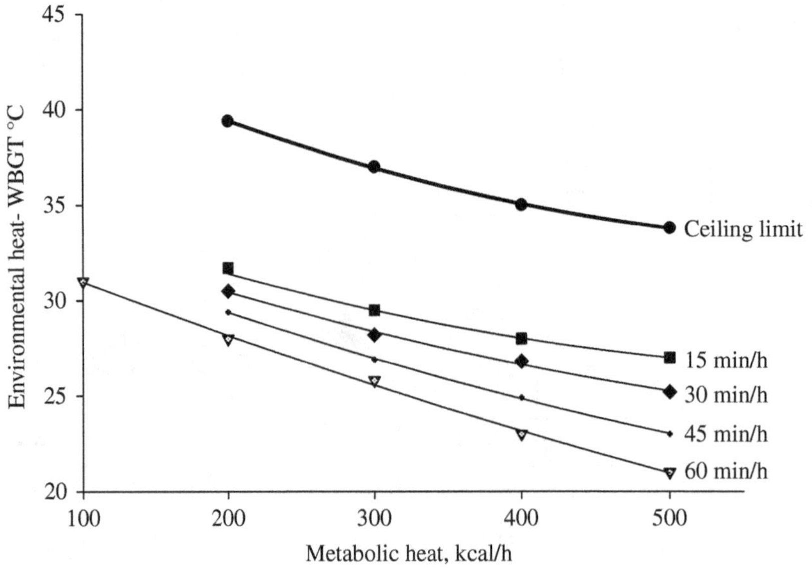

FIGURE 6.3 Recommended heat stress alert limits for heat-unacclimatized workers. C = ceiling limit; RAL = Recommended Alert Limit. Calculations are for a standard worker of 70 kg (154 lb) body weight and 1.8 m² (19.4 ft²) body surface. Reproduced from National Institute for Occupational Safety and Health. Criteria for a recommended standard. Occupational exposure to hot environments. Revised criteria. DHHS (NIOSH) publication no. 86-113. Washington, DC: US Government Printing Office, 1986.

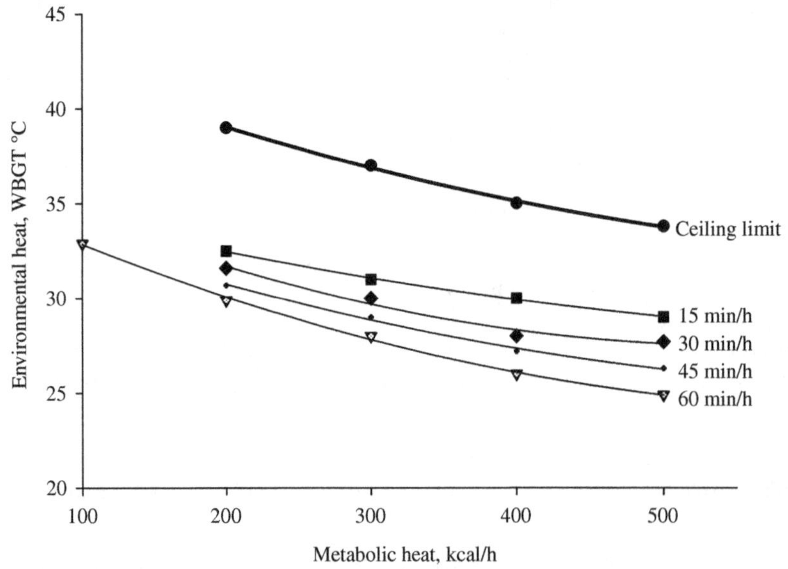

FIGURE 6.4 Recommended heat stress exposure limit for heat-acclimatized workers. C = ceiling limit; REL = Recommended Exposure Limit. Calculations are for a standard worker of 70 kg (154 lb) body weight and 1.8 m² (19.4 ft²) body surface. Reproduced from National Institute for Occupational Safety and Health. Criteria for a recommended standard. Occupational exposure to hot environments. Revised criteria. DHHS (NIOSH) publication no. 86–113. Washington, DC: US Government Printing Office, 1986.

NIOSH GUIDELINES

NIOSH has published a draft update to the 1986 "Criteria for a Recommended Standard Occupational Exposure to Hot Environments".[2] A system of work–rest cycles, similar to the ACGIH guidelines in Table 6.4, to prevent heat-related illnesses are presented. Work–rest time curves based upon WBGT index measurements and metabolic heat estimates were developed for unacclimatized and acclimatized workers; they are called recommended alert limits (RALs) and recommended exposure limits (RELs), respectively (Figures 6.3 and 6.4). The NIOSH RALs and RELs were developed for healthy workers who are physically and medically fit for their level of activity and who are wearing customary work clothes (i.e., a long-sleeved shirt and trousers).

NORMAL PHYSIOLOGY

Body core temperature is normally maintained in the range of 36.7–37°C (98–98.6°F) by oral measurement and 37.3–37.6°C (99–99.6°F) by rectal measurement. This narrow range of temperature control is maintained through the balance of heat production and conservation of metabolic heat in cold ambient conditions or the dissipation of metabolic heat to the environment in hot ambient conditions. When core temperature is elevated, active vasodilation of the cutaneous blood vessels and thermoregulatory sweating are initiated in order to facilitate heat dissipation. In humans, neurons in the preoptic area of the anterior hypothalamus, along with deep body temperature sensors, detect small changes in blood temperature. As the blood temperature rises, neurons in the hypothalamus activate the autonomic nervous system, which initiates circulatory, endocrine, and eccrine or sweat gland, responses to rid the body of excess heat.

Heat stress places demands on many body systems, primarily the cardiovascular and thermoregulatory systems, though the pulmonary, renal, and endocrine systems are also involved in the heat stress response. The cardiovascular system plays a prominent role in heat dissipation. As core temperature rises, the cutaneous active vasodilator mechanism initiates an increase in skin blood flow, which necessitates an increase in cardiac output. Cardiac output is redistributed so that less blood flow goes to internal organs and reducing the overall cardiovascular strain.[21] To maintain blood pressure and blood flow to exercising muscles and vital organs while blood is being shunted to the skin, adequate hydration is essential. Sweat evaporation is the body's primary method of heat loss. For each gram of sweat that evaporates from the skin, 0.58 kcal of heat is lost.[22] Under heat strain conditions, the sweat glands can be stimulated to produce up to 2 liters per hour (L/h) in an acclimatized individual, and sweat rates of 1 L/h would not be unusual in industrial workers.[2] Sweat rates this high cause significant loss of body water. Sweat rate is reduced in proportion to the magnitude of hypohydration, illustrating the importance of adequate fluid replacement during heat stress.[23] Sweat also contains sodium, chloride, and potassium and can account for measurable losses of these electrolytes from the body.[24]

After repeated exposures to heat, physiological adaptations occur that improve heat dissipation and tolerance. This adaptation to heat stress, or heat acclimatization, occurs over 5–14 days with daily exercise in hot ambient conditions and of sufficient duration and level of exertion to raise body temperature and initiate vigorous sweating.[22] Once acclimatized, a worker exposed to the same levels of heat will have a lower core temperature, a lower heart rate, and an increased volume of sweat that contains a lower concentration of sodium. The reduced sweat sodium concentration is the result of aldosterone-mediated reabsorption of sodium and chloride ions in the sweat glands. With acclimatization, sweat rates can increase from 0.6 to 2 L/h, and sodium chloride loss can decrease from 15–30 grams per day (g/day) to 3–5 g/day.[22] Overall, the sodium loss is reduced to <5 g/day after acclimatization. The average American diet provides 8–14 g of salt each day, which is adequate for the acclimatized worker. Before acclimatization, while salt deficits may occur in heavily sweating workers, increasing dietary salt is rarely required.

Generally, physically fit workers are better able to tolerate heat, and they achieve acclimatization more rapidly than nonphysically fit workers. Physical fitness, independent of heat acclimatization status, enhances sweat secretion. The adaptations of heat acclimatization are rapidly lost when the worker leaves the heat stress environment. Workers who return to hot environments after more than a few days away from the job—especially those who return after an illness—should be allowed to reacclimatize to the hot environment before resuming full-time work, though evidence suggests that reacclimatization occurs more rapidly than the initial acclimatization.

Heat exchange

Heat exchange between the body and the environment is influenced by air temperature and humidity, skin temperature, air velocity, evaporation of sweat, radiant temperature, and the clothing worn.[25] The heat balance equation describes the major modes of heat exchange or loss by the body (Table 6.5).

Calculated heat loss by radiation and convection is typically estimated from the skin-air thermal gradient. Consequently as T_{db} approaches skin temperature, heat loss via $R+C$ is diminished, and there is an increased reliance on evaporative heat loss. If T_{db} exceeds skin temperature, heat will be gained rather than lost. Conductive heat exchange, where there is direct transfer of heat to air or objects in contact with the body, is rarely an important source of heat gain or loss in clothed workers in most occupational settings. However, conductive heat exchange may be an important mode of heat transfer for some workers, such as divers working in hot or cold water.

In addition to elevated ambient temperature, an increase in the water vapor content of the air and clothing will limit heat dissipation to the environment. Relative humidity is the

TABLE 6.5 Heat balance equation.

$S=(M\pm W)\pm(R+C)\pm K-E$

where:

S = storage, the amount of heat gained or lost by the body

$(M\pm W)$ − metabolism ± external mechanical work performed

$(R+C)$ = combined radiative and convective heat gain or loss between the skin and the ambient air

K = heat gain or loss from conductive heat transfer

E = evaporative heat loss as sweat evaporates from the skin surface most common way of expressing water vapor content of the air, but comparisons between different levels of humidity are appropriate only when the air temperature is the same. The evaporation of sweat is dependent on the vapor pressure gradient between the skin and the air, which is temperature dependent. As an example, the vapor pressure of air at 30°C, 40%RH, is approximately equal to that at 40°C, 20%RH. Clothing insulates the body from heat loss via radiation, and convection will also serve as a barrier to the evaporation of sweat. Protective clothing, especially sweat-impermeable clothing, which effectively eliminates any body cooling from sweat evaporation, places workers at significantly greater risk for heat strain and heat-related illnesses.

Numerous acute and chronic medical conditions, medications, and individual characteristics may diminish the body's ability to tolerate heat stress and dissipate internal heat (Table 6.6).[26] Dehydration from any cause, whether due to increased sweating or associated with underlying illness, fever, vomiting, or diarrhea, increases the risk of hyperthermia. Many medications, especially antihypertensive or cardiac medications that affect the cardiovascular or renal systems or medications with anticholinergic effects on sweat glands, may alter a worker's physiologic responses to heat and therefore increase the risk of heat illness. Previous heat-related illness, especially heatstroke, has historically been considered an indicator of both heat intolerance and a possible underlying defect in the individual's thermoregulatory system. One investigation of 10 individuals with previous exertional heatstroke found that nine of them readily acclimatized to heat while one displayed evidence of persistent heat intolerance for almost 1 year following heatstroke. The investigators concluded that a small percentage of individuals with previous heatstroke may be heat intolerant.[27]

TABLE 6.6 Risk factors predisposing to heat disorders.

Individual factors
Increased age
Obesity
Lack of acclimatization
Use of water-impermeable/heavy clothing
Fatigue/sleep deprivation
Underlying medical conditions/states
Dehydration
Infection/fever/recent immunization
Overuse of ethanol
Diseases
Cardiovascular disease
Renal disease
Hyperthyroidism
Diabetes mellitus
Parkinson's disease
Skin conditions limiting sweating, including sunburn
Previous heat disorder
Pesticide poisoning
Drugs
Medications with anticholinergic effects
Antispasmodics
Tricyclic antidepressants
Psychotropics
Antihistamines
Antihypertensive medications
Diuretics
Stimulants (decongestants, amphetamine, cocaine)

PATHOPHYSIOLOGY OF ILLNESS AND TREATMENT

The spectrum of heat-related disorders ranges from relatively harmless illnesses such as various skin conditions and heat cramps to serious heat illnesses, which represent a continuum of severity from heat exhaustion to potentially fatal heatstroke.[26]

Heat-related skin conditions

Heat rash, or miliaria, is caused by sweat duct obstruction and resultant sweat retention within the sweat gland. Obstruction of the sweat duct leads to duct rupture and an inflammatory reaction surrounding the duct. Because the rupture of the sweat ducts may occur within different layers of the skin, three forms of miliaria—crystalline, rubra, and profunda—are described.[28]

Miliaria crystallina is a mild, asymptomatic skin condition consisting of small, clear vesicles resulting from sweat duct rupture within the surface layers of skin. Small erythematous macules resulting from sweat duct rupture within the middle layers of the skin and associated with burning and itching is known as miliaria rubra or "prickly heat." Miliaria rubra commonly affects the skin of the trunk and intertriginous areas of the body. Extensive cases of miliaria rubra involving large numbers of sweat glands can impede body heat dissipation and contribute to more severe heat-related illness.[29] Miliaria profunda results from sweat duct rupture deep within the skin. The lesions, which usually appear only after prolonged periods of miliaria rubra, are small, white to flesh-colored papules and occur most commonly on the trunk. Sunburned skin and occlusive clothing that precludes the free evaporation of sweat increase the risk of all forms of miliaria. Treatment involves reducing sweating in the affected individual and keeping the skin cool and dry. In workers whose jobs require sweat-impermeable protective clothing, the treatment of miliaria may include a temporary transfer to a job not requiring protective clothing or into an air-conditioned workspace. Miliaria rubra usually resolves within a week. Complete resolution of miliaria profunda may take several weeks in a cool environment.

Heat edema

Heat edema is often not considered a true heat-related disorder. It is a rather a common condition in which the extremities swell during the first 7–10 days of exposure to higher temperatures. Typically it is found in unacclimatized individuals, often women or military trainees, who stand or sit for long periods during hot weather and is not associated with cardiac or renal impairment. The etiology of heat edema is uncertain, but it may involve local vasomotor changes or be associated with changes in aldosterone activity.[24] It is sometimes prominent in pregnancy, when heat can aggravate the underlying condition of pregnancy-associated edema. Heat edema usually resolves spontaneously within a few days as the individual acclimatizes. Diuretic therapy does not provide significant relief and is not indicated. Symptomatic treatment—elevation of the legs, compression stockings, and gradual exposure to heat—is generally all that is required.

Heat cramps

Heat cramps are painful muscle spasms that occur during or following intense physical exercise in hot environments. The affected individual is usually acclimatized to heat and gives a history of heavy exertion in the heat, profuse sweating, drinking large quantities of water, and minimal salt or electrolyte replacement. Inadequate electrolyte replacement is probably associated with the underlying mechanism responsible for the muscle spasm.[24,30]

The muscles involved in the spasm are usually the same muscles used during the preceding exercise, such as the abdominal muscles or the large muscles of the thigh. Heat cramps may be heralded by fasciculations, and while multiple muscle spasms may occur simultaneously, usually only a small section of the muscle is involved. Individual heat-induced muscle spasms last less than a minute; but if untreated, attacks of intermittent heat cramps may last for 4–8 hours.

Heat cramps respond to rest in a cool place and ingestion of 0.1% saline solution (one teaspoon of salt in a quart of water) or fluids containing electrolytes. Salt tablets should not be given. If nausea and vomiting preclude administration of oral solutions, intravenous electrolyte solutions may be necessary. Heat cramps can be prevented by ensuring that workers, especially acclimatized workers, maintain adequate dietary salt intake in addition to adequate fluid replacement.

Heat syncope

Heat syncope occurs in individuals who stand for prolonged periods, who make sudden postural changes, or who exercise strenuously in the heat. The underlying mechanism of heat syncope is similar to that of orthostatic syncope. Venous return to the heart is reduced by pooling of blood in dependent extremities or in dilated peripheral vessels, and cardiac output is inadequate to maintain cerebral circulation and consciousness. Heat syncope is not associated with elevated body temperature, and the syncope victim may remember a typical prodrome of nausea, sweating, and dimming of vision before loss of consciousness. Following syncope and falling to a recumbent position, consciousness returns rapidly. Heat syncope is generally a benign "faint." The primary health concern is the potential for falling and injury, especially for workers on roofs and scaffolding.

Following an episode of heat syncope, the worker should be allowed to recover in a cool area. To ensure that he or she has not been injured by the fall, a medical examination should be performed. The employer should make sure that the worker is properly hydrated and acclimatized before returning to a job that requires heavy exertion or standing in a hot environment.

Heat exhaustion

Heat exhaustion is a mild to moderate illness characterized by an inability to sustain cardiac output during strenuous work or exercise in the heat that is necessary to meet the combined demands of blood flow for metabolic and thermoregulatory requirements.[15] The signs and symptoms include weakness, ataxia, fatigue, headache, nausea, hypotension, tachycardia, and transient alterations in mental status. Since the symptoms of heat exhaustion are similar to those of early heatstroke, all heat exhaustion victims must be evaluated to eliminate the diagnosis of heatstroke (see Heat Stroke below). Heat exhaustion occurs more commonly in workers who are unacclimatized to heat and who are without adequate water and salt replacement. Treatment of heat exhaustion must be individualized based upon the severity of the presenting symptoms and the underlying cause. Mild heat exhaustion may be treated by having the worker rest in a cool area and providing oral fluid and salt replacement. Severe cases of heat exhaustion—especially severe water-depletion type requiring intravenous fluids—need to be referred to an emergency room for careful replacement of body water. Timing of return to work after heat exhaustion has not been fully studied, but it seems prudent to allow at least 24–72 hours for full rehydration and correction of electrolyte abnormalities before the individual returns to work.

Hyponatremia occurs when electrolyte replacement is insufficient and/or fluid replacement is overly aggressive and can lead to potentially serious and sometimes fatal outcomes. Exercise- or work-related hyponatremia is due to overdrinking, both in absolute (related to total volume consumed) and relative (related to sodium loss via sweating) terms.[31] In the occupational setting, as hyponatremia rarely occurs, specific prevention measures beyond encouraging appropriate fluid replacement and consumption of salt-containing foods are not necessary.

Heat injury and heatstroke

Heat injury is a moderate to severe illness characterized by tissue (muscle and gut) and organ (liver, renal) injury resulting from work or exercise during heat exposure.[26] Individuals with exertional heat injury do not demonstrate sufficient neurological abnormalities to meet the usual definition of heatstroke. Organ and tissue damage may not manifest early in the presentation of heat injury, making it difficult to distinguish it from heat exhaustion, and a thorough clinical evaluation, including assessment of possible organ damage, is necessary.[15]

Heatstroke is a life-threatening medical condition. Symptoms include altered mental status and rectal temperatures are often but not always >40°C (104°F).[26] However, core temperature >40°C should not be used as a universal diagnostic criteria, as well-conditioned individuals can have core temperatures above this threshold yet remain asymptomatic of heat illness.[32,33] During physical work in the heat, skin blood flow is high, reducing perfusion of the intestines and other viscera, which can result in ischemia, endotoxemia, systemic inflammatory response, and multiorgan damage.

Heatstroke requires immediate, aggressive cooling of the affected individual. Classic heatstroke occurs during summer heat waves predominantly in infants or elderly individuals. The condition is most common in poor and/or elderly individuals who take medications for underlying medical conditions and live in poorly ventilated housing without air conditioning. After days of hot, humid environmental conditions, their ability to maintain body heat balance fails, and heatstroke occurs. A second form of heatstroke, exertional heatstroke, occurs in workers, athletes, or military recruits who perform vigorous exercise in hot, humid conditions and is the focus of this section. Mental status changes are the predominant initial presenting symptom. Heatstroke victims may present with any form of mental status change ranging from irrational behavior, poor judgment, and confusion to delirium, seizures, and coma. Sweating, characteristically absent in classic heatstroke, may be present in exertional heatstroke. The rectal temperature is often >40°C (104°F) and may be much higher; the pulse is elevated; the blood pressure is normal or low; and hyperventilation is common. Nausea, vomiting, and diarrhea may be present.

Initial laboratory findings often include proteinuria, with red blood cells and granular casts also present in the urine, an elevated white blood cell count, and a decreased platelet count. Serum electrolyte levels will vary with level of hydration, acid–base status, and underlying tissue damage. Serum enzymes—lactic dehydrogenase, creatinine phosphokinase, aspartate aminotransferase, and alanine aminotransferase are characteristically elevated. With hyperventilation, respiratory alkalosis may be present; but in exertional heatstroke, lactic acidosis may be the presenting acid–base abnormality. Abnormalities of blood clotting consistent with disseminated intravascular coagulation—including decreased fibrinogen, prolonged prothrombin time and partial thromboplastin time, and elevated levels of fibrin split products—may be present on initial evaluation.

The treatment for all heatstroke victims is immediate initiation of cooling and appropriate resuscitation. The method of cooling used in military settings is immersion of the victim in ice water or wrapping the victim in wet sheets. Even though this type of cooling will cause cutaneous vasoconstriction and shivering, thereby decreasing conductive heat loss and increasing metabolic heat formation, it is still probably the most effective method of cooling outside of a healthcare facility. Rapid cooling to a core temperature <38.3°C (101°F) has been associated with no mortality and an attenuation of organ damage.[34,35] Other rapid cooling methods used for heatstroke include packing the individual in ice or applying ice packs and/or spraying with cold water. Once cooling has been initiated, the victim should be transferred immediately to a hospital for continuous monitoring of core temperature and definitive care. Endotracheal intubation and invasive cardiovascular monitoring are often needed during resuscitation and follow-up care.

The severity of sequelae and mortality in heatstroke is determined by the degree and duration of the elevated temperature. After initial resuscitation of an individual with heatstroke, failure or disruption of the function of multiple organ systems is common and should be expected. Frequent sequelae to heatstroke include liver function abnormalities, disseminated intravascular coagulation, rhabdomyolysis, and acute renal failure.[24] Other reported complications of heatstroke include pancreatitis, pulmonary edema, myocardial infarction, and central and peripheral nervous system damage.

MEDICAL SURVEILLANCE

Medical surveillance of heat-exposed workers is one aspect of the overall prevention of heat-related illness. Preplacement and periodic medical examinations of heat-exposed workers should ensure that they can meet the total demands and stresses of the hot job environment without putting their safety and health or that of fellow workers in jeopardy.[2] The components of the preplacement and periodic medical examinations recommended by NIOSH are provided in Table 6.7. There are no strict guidelines to identify which heat-exposed workers require medical surveillance. An initial medical history may be taken on all heat-exposed workers to identify those at risk for heat illness. The initial history should elicit information about underlying medical conditions, use of prescription and over-the-counter medications, and previous episodes of heat-related illness. Workers may not require regular medical surveillance if they (i) do not have medical conditions that increase their risk of heat injury, (ii) are not exposed to heat above recommended guidelines, and (iii) are properly trained in the prevention of

TABLE 6.7 NIOSH recommended medical surveillance for heat-exposed workers.

Pre-placement medical evaluation		Periodic medical evaluation	
Component	**Special emphasis on**	**Component**	**Special emphasis on**
History		History	
Occupational	Previous heat exposure jobs Use of personal protective equipment	Occupational	Changes in job or personal protective equipment
Medical	Diseases of the following systems: cardiovascular, respiratory, endocrine, gastrointestinal, dermatologic, renal, neurologic, hematologic, reproductive	Medical	Change in health status Symptoms of heat strain
Personal habits	Alcohol and drug use	Personal habits	Update
Medications	Prescription and over the counter	Medications	Update
Characteristics	Height, weight, gender, age	Characteristics	Update
Direct evaluation		Direct evaluation	
Physical examination	Cardiovascular, respiratory, nervous and musculoskeletal systems Skin	Physical examination	Systems emphasized in pre-placement examination
Blood pressure		Blood pressure	
Clinical chemistry tests	Fasting blood glucose Blood urea Serum creatinine Serum electrolytes Hemoglobin Urinary sugar and protein	Clinical chemistry tests	Fasting blood glucose Blood urea Serum creatinine Serum electrolytes Hemoglobin Urinary sugar and protein
Mental status	Assessment of worker's ability to understand heat, Communicate and respond to emergencies	Mental status	Reassessment of ability to understand heat
Detailed medical evaluation	Cardiovascular disease Pulmonary disease Medication use which might interfere with heat tolerance or acclimatization Hypertension Need to use respiratory protection History of skin disease that may impair sweating Obesity Women with childbearing potential	Detailed medical evaluation	Based upon changes in health

Source: National Institute for Occupational Safety and Health. Criteria for a Recommended Standard: Occupational Exposure to Heat and Hot Environments: Revised Criteria 2013. Centers for Disease Control and Prevention; 2013:1–184.

heat-related illness. Special categories of workers likely to be exposed to extreme heat stress conditions, such as hazardous waste site workers who wear heavy protective clothing and respirators, should be screened carefully before placement and should receive periodic medical surveillance.

Return to Work Considerations

Return to work/play guidelines are often "common sense" recommendations based on a return to normal laboratory parameters and an asymptomatic state.[36] Recent data indicates that these criteria may not be appropriate, as tissue damage and/or impaired organ function may persist after the individual appears to have recovered. Biological markers of exertion heat illness (EHI) severity are lacking; however, recent evidence from a rat model of heatstroke suggests that there are novel patterns of core temperature responses to mild, moderate, and severe pathophysiology that may assist in the differential diagnosis. Additionally, associations between hemodynamic changes during heat exposure, high circulating cardiac troponin I levels, and histopathological changes in cardiac muscle during recovery were identified.[37] Whether similar or other biomarkers of EHI severity and/or recovery exist in humans remains to be determined.

The return-to-play recommendations from the American College of Sports Medicine are to refrain from exercise for 1 week after release from medical care, with follow-up examination, testing, and imaging as appropriate prior to resuming exercise. Work intensity, duration, and heat stress should be gradually increased until tolerance to vigorous exercise in the heat has been demonstrated. Development of clinical aids and/or biomarkers of EHI injury severity and recovery are necessary, as well as determination of the long-term health consequences, if any, of exertional heatstroke.

HEAT EXPOSURE AND REPRODUCTION

Physiologic and hormonal changes during early pregnancy are associated with a slight increase in maternal resting core temperature; however, there is no evidence that a pregnant woman's ability to eliminate excess heat is diminished. In fact, physiologic adaptations during pregnancy—increase in blood volume, increase in cardiac output and resting pulse rate, and increase in cutaneous blood flow—appear to offset the increased metabolic heat load associated with pregnancy.[38,39] In one study, physically fit pregnant women appeared to maintain thermoregulation during exercise throughout their pregnancy at least as well as nonpregnant women.[38] During the late third trimester of pregnancy, decreased venous return, due to the size of the uterus, may compromise cardiac output in the pregnant worker and impair heat tolerance.

In experimental animal studies on a variety of species, hyperthermia has been found to be teratogenic.[2] Early in the gestation, excessive heat exposure in animals is associated with structural defects, predominantly of the central nervous system and skeleton, and embryo death. Heat stress in animals later in gestation has been associated with retarded fetal growth and postnatal neurobehavioral defects. Studies of the effect of hyperthermia on the developing human fetus have been primarily concerned with the effect of illness-related fever during pregnancy. The results of these studies have been inconsistent. Some investigators have reported an association between maternal fevers of 38.9°C (102°F) and abnormal fetal development, whereas others found no association between maternal hyperthermia and adverse pregnancy outcomes. Heat exposure, in addition to its possible effects on female reproduction, is associated with decreases in sperm count and motility in male workers.[40] Considering the potential for reproductive effects due to heat in men and women, preventive measures to limit excessive heat exposures are essential for all workers.

PREVENTION

The prevention of heat-related illnesses and conditions of unacceptable heat stress can be categorized into four basic areas of control: (i) engineering controls, (ii) administrative controls, (iii) work practices controls, and (iv) protective clothing and devices.[2] Table 6.8 summarizes these prevention measures. Engineering controls, such as shielding and increased air movement, are the most desirable preventive measures. However, they are not effective in many outdoor work sites or work sites requiring full protective clothing. In these cases, administrative and work practices controls—such as reducing heat exposure, reducing work rates, enhancing fitness and heat tolerance, and offering special heat safety training—should be utilized to the fullest extent

TABLE 6.8 Heat strain/heat-related illness prevention.

Engineering controls
Decrease convection heat gain by worker
Cool air temperature to below mean skin temperature
Increase air movement (if ambient temperature ≤ skin temperature)
Decrease radiant heat gain by worker
Insulate hot surfaces
Use shielding between worker and heat source
Increase evaporative heat loss by worker
Eliminate humidity sources (steam leaks, standing water)
Decrease air humidity (ambient water vapor pressure)
Increase air movement
Administrative and work practices controls
Limit workers' exposure to hot working environment
Use appropriate environmental monitoring
Work during cool parts of day or in the shade
Schedule hot work for cool seasons
Provide cool rest areas
Increase the number of workers for a given job
Use recommended work/rest regimens
Decrease the metabolic heat load
Mechanize heavy work when possible
Rotate heavy work over entire workforce or increase workforce
Decrease shift time; allow liberal work breaks; restrict overtime
Enhance tolerance to heat
Encourage physical fitness in workers
Require minimum level of fitness in certain jobs
Use heat acclimatization program for new workers or workers returning from vacations, layoffs, or illness
Encourage regular fluid and salt replacement
Avoid conditions increasing risk to heat strain
Health and safety training for supervisors and workers
Recognize signs and symptoms of heat intolerance
Emphasize acclimatization, fluid and salt replacement
Avoid conditions increasing risk of heat strain
Use control methods to prevent heat strain
Use protective clothing
Use buddy system, if applicable
Medical screening of workers with heat intolerance
Establish heat alert program
Establish heat alert committee
Reverse plant winterization measures
Ensure water sources, fans and air conditioners are working
Ensure that medical department is prepared to treat heat casualties
Establish criteria for heat alerts
Take all appropriate preventive measures during heat alerts
Post signs identifying heat hazard areas
Protective clothing and auxiliary body cooling
Water-cooled garments
Air-cooled garments
Ice-packet vests
Wetted overgarments
Aluminized overgarments

Source: National Institute for Occupational Safety and Health. Criteria for a Recommended Standard: Occupational Exposure to Heat and Hot Environments: Revised Criteria 2013. Centers for Disease Control and Prevention; 2013:1-184.

possible. Adequate fluid replacement in the heat-exposed worker is critical. The sensation of thirst has been proven to be inadequate to prevent hypohydration in heavily sweating individuals. Therefore, workers need to be encouraged to drink adequate amounts of liquid to replace sweat losses. Any water or beverage provided for workers should be cool (10–15°C or 50–59°F) and should be consumed in small volumes. In the acclimatized worker, sweat rates of 1 L/h are possible, and fluid replacement with approximately 5–7 oz every 15–20 minutes approximates fluid losses. Maximum gastric emptying in exercising individuals is 1–1.5 L/h; therefore, larger volumes of fluid are not effective.[41] Overhydration and resultant hyponatremia, while rare, have been reported in marathon runners.[42] Numerous carbohydrate and electrolyte solutions are marketed for fluid replacement in athletes and workers. Studies have shown that the addition of carbohydrates and electrolytes to fluids may be beneficial for athletes who exercise strenuously for long periods.[41] For acclimatized workers whose diet includes sufficient calories and salt, water alone should provide adequate fluid replacement. Because of their diuretic effect, caffeinated beverages should be discouraged as a primary source of fluid replacement.

Personal protective equipment can be very effective in prolonging intervals of heat exposure. Ice-packet vests are the least cumbersome items to wear because they require no connection to an external device to provide cooling air or water. At work sites where radiant heat sources are the primary exposure problem, aluminized suits are a good choice because they reflect heat.

References

1. Dukes-Dobos FN. Hazards of heat exposure: a review. *Scand J Work Environ Health* 1981;7(2):73–83.
2. Jacklitsch B, Williams WJ, Musolin K, et al. 2016. NIOSH criteria for a recommended standard: occupational exposure to heat and hot environments. Cincinnati, OH: U.S. Department of Health and Human Services, Centers for Disease Control and Prevention, National Institute for Occupational Safety and Health, DHHS (NIOSH) Publication 2016-106. Available at: https://www.cdc.gov/niosh/docs/2016-106/pdfs/2016-106.pdf (accessed on June 28, 2016).
3. Department of Labor. *OSHA Technical Manual* OSHA, ed. Washington, DC: Occupational Safety & Health Administration; 1999.
4. Centers for Disease Control and Prevention. Heat-related deaths after an extreme heat event—four states, 2012, and United States, 1999–2009. *Morb Mortal Wkly Rep* 2013;62(22):433–6.
5. Arbury S, Jacklitsch B, Farquah O, et al. Heat illness and death among workers—United States, 2012–2013. *Morb Mortal Wkly Rep* 2014;63(31):661–5.
6. Choudhary E, Vaidyanathan A. Heat stress illness hospitalizations: environmental public health tracking program, 20 states, 2001–2010. *MMWR Surveill Summ* 2014;63(13):1–10.
7. Donoghue AM, Sinclair MJ, Bates GP. Heat exhaustion in a deep underground metalliferous mine. *Occup Environ Med* 2000;57(3):165–74.
8. Minard D, Belding HS, Kingston JR. Prevention of heat casualties. *J Am Med Assoc* 1957;165(14):1813–8.
9. Carter R, III, Cheuvront SN, Williams JO, et al. Epidemiology of hospitalizations and deaths from heat illness in soldiers. *Med Sci Sports Exerc* 2005;37(8):1338–44.
10. Wallace RF, Kriebel D, Punnett L, et al. Risk factors for recruit exertional heat illness by gender and training period. *Aviat Space Environ Med* 2006;77(4):415–21.
11. International Organization for Standardization (ISO). *ISO 7726, 1998 Ergonomics of the Thermal Environment: Instruments for Measuring Physical Quantities*. Geneva: ISO; 1998.
12. Parsons K. *Human Thermal Environments*. London and New York: Taylor & Francis; 2003.
13. Yaglou CP, Minard D. Control of heat casualties at military training centers. *AMA Arch Ind Health* 1957;16(4):302–16.
14. American Conference of Governmental Industrial Hygienists. *2014 Threshold Limit Values for Chemical Substances and Physical Agents and Biological Exposure Indices*. Cincinnati, OH: American Conference of Governmental Industrial Hygienists; 2014:206–24.
15. Department of the Army. TB MED 507 Heat stress control and heat casualty management. Washington, DC: Department of the Army and Air Force; 2003. TB MED 507/AFPAM 48–152.
16. Byrne C, Lim CL. The ingestible telemetric body core temperature sensor: a review of validity and exercise applications. *Br J Sports Med* 2007;41:126–33.
17. Goodman DA, Kenefick RW, Cadarette BS, et al. Influence of sensor ingestion timing on consistency of temperature measures. *Med Sci Sports Exerc* 2009;41(3):597–602.
18. Buller MJ, Latzka WA, Yokota M, et al. A real-time heat strain risk classifier using heart rate and skin temperature. *Physiol Meas* 2008;29(12):N79–85.
19. International Organization for Standardization (ISO). *ISO 7243, 2003, Hot Environments: Estimation of the Heat Stress on Working Man, Based on the WBGT-Index (Wet Bulb Globe Temperature)*. Geneva: ISO; 2003.
20. International Organization for Standardization (ISO). *ISO 9886, 2004, Ergonomics: Evaluation of Thermal Strain by Physiological Measurements*. Geneva: ISO; 2004.
21. Rowell LB. *Human Circulation: Regulation During Physical Stress*. New York: Oxford University Press; 1986.
22. Sawka MN, Young AJ. Physiological systems and their responses to conditions of heat and cold. In: Tipton CM, ed. *ACSM's Advanced Exercise Physiology*. Baltimore, MD: Lippincott Williams and Wilkins; 2006:535–63.
23. Sawka MN, Young AJ, Francesconi RP, et al. Thermoregulatory and blood responses during exercise at graded hypohydration levels. *J Appl Physiol* 1985;59(5):1394–401.
24. Knochel JP. Heat stroke and related heat stress disorders. *Dis Mon* 1989;35(5):301–77.
25. Gagge AP, Gonzalez RR. Mechanisms of heat exchange: biophysics and physiology. In: *Handbook of Physiology: Environmental Physiology*. Vol 1. New York: Oxford University Press; 1996:45–84.
26. Winkenwerder W, Sawka MN, Goldman L, et al. Disorders due to heat and cold. In: *Cecil Medicine*. Vol 2, 3rd ed. Philadelphia, PA: Saunders Elsevier; 2007:763–7.

27. Armstrong LE, De Luca JP, Hubbard RW. Time course of recovery and heat acclimation ability of prior exertional heat-stroke patients. *Med Sci Sports Exerc* 1990;22(1):36–48.
28. Kanerva L. Physical causes and radiation effects. In: Adams R, ed. *Occupational Skin Diseases*. 3rd ed. Philadelphia, PA: WB Saunders; 1999:47–8.
29. Pandolf KB, Griffin TB, Munro EH, et al. Heat intolerance as a function of percent of body surface involved in miliaria rubra. *Am J Physiol Regul Integr Comp Physiol* 1980;239:R233–40.
30. Eichner ER. Heat cramps in sports. *Curr Sports Med Rep* 2008;7(4):178–9.
31. Sawka MN, Leon LR, Montain SJ, et al. Integrated physiological mechanisms of exercise performance, adaptation, and maladaptation to heat stress. *Compr Physiol* 2011;1:1883–928.
32. Ely BR, Ely MR, Cheuvront SN, et al. Evidence against a 40°C core temperature threshold for fatigue in humans. *J Appl Physiol* 2009;107(5):1519–25.
33. Byrne C, Lee JK, Chew SA, et al. Continuous thermoregulatory responses to mass-participation distance running in heat. *Med Sci Sports Exerc* 2006;38(5):803–10.
34. O'Brien KK. Case studies of exertional heat injury/stroke in military populations. *Med Sci Sports Exerc* 2003;35:S3.
35. Gaffin SL, Gardner JW, Flinn SD. Cooling methods for heat-stroke victims. *Ann Intern Med* 2000;132(8):678.
36. O'Connor FG, Casa DJ, Bergeron MF, et al. American College of Sports Medicine Roundtable on exertional heat stroke–return to duty/return to play: conference proceedings. *Curr Sports Med Rep* 2010;9(5):314–21.
37. Quinn CM, Duran RM, Audet GN, et al. Cardiovascular and thermoregulatory biomarkers of heat stroke severity in a conscious rat model. *J Appl Physiol (1985)* 2014;117(9):971–8.
38. Jones RL, Botti JJ, Anderson WM, et al. Thermoregulation during aerobic exercise in pregnancy. *Obstet Gynecol* 1985;65(3):340–5.
39. Vaha-Eskeli K, Erkkola R, Seppanen A. Is the heat dissipating ability enhanced during pregnancy? *Eur J Obstet Gynecol Reprod Biol* 1991;39(3):169–74.
40. Thonneau P, Bujan L, Multigner L, et al. Occupational heat exposure and male fertility: a review. *Hum Reprod* 1998;13(8):2122–5.
41. Gisolfi CV, Duchman SM. Guidelines for optimal replacement beverages for different athletic events. *Med Sci Sports Exerc* 1992;24(6):679–87.
42. Frizzell RT, Lang GH, Lowance DC, et al. Hyponatremia and ultramarathon running. *JAMA* 1986;255(6):772–4.

Chapter 7

COLD ENVIRONMENTS*

David W. DeGroot and Laura A. Pacha**

OCCUPATIONAL SETTING

Cold is a physical hazard that may affect workers, both indoors and outdoors, virtually anywhere in the world. Workers at greater risk include construction workers, farmers, fishermen, utility workers, lumberjacks, soldiers, petroleum workers, police, firefighters, postal workers, butchers, and cold storage workers. For the purpose of this chapter, an extreme cold environment exists when ambient temperature is <0°C/32°F.[1]

Cold injuries may be either freezing (frostbite) or nonfreezing (trench/immersion foot and hypothermia) and localized or systemic. While they occur sporadically in civilian populations, in both occupational and recreational settings, cold injuries have been a significant problem in military campaigns throughout history. The armies of Xenophon (400 BC), Hannibal (218 BC), and Napoleon (1812–1813) experienced significant numbers of cold injuries.[2] Despite numerous advances in our understanding of thermoregulation during cold stress, and in the development of protective clothing, cold injuries continued to affect military operations in the twentieth century, notably in the trench warfare of World War I (hence the immersion foot synonym trench foot), the German experience in Russia during World War II, and the British experience in the Falkland Islands in the 1980s.[3] The incidence or frequency of cold injury in the civilian sector is much more difficult to determine; unlike the military, hypothermia, frostbite, and other cold injuries are not reported to health authorities for surveillance purposes. Cold weather is not a barrier to safe and successful work performance, and it is important to remember that "a man in the cold is not necessarily a cold man."[4]

*DISCLAIMER:** The views expressed in this chapter are those of the authors and do not necessarily reflect the official policy of the Department of Defense, Department of the Army, or the US Army Medical Department.

** The authors gratefully acknowledge Erik T. Evenson, the author of the previous edition of this chapter, for his contributions.

MEASUREMENT ISSUES

Two important concepts for evaluating cold exposures are the body core temperature and the wind chill index. Body core temperature is measured by a low-reading rectal thermometer. In the hospital, an esophageal temperature probe is the preferred instrument to monitor core temperature. Measurement at other anatomical sites, including sublingual, temporal, and axillary, will result in an inaccurate reading of body core temperature and is not recommended.[5] Ambient temperature should be measured by thermometers capable of measuring temperatures down to at least −40°C (−40°F). The wind chill index is used to determine the risk of cold injury by estimating the relative cooling ability of a combination of air temperature and wind velocity. The wind chill index is measured as the equivalent chill temperature (Table 7.1). In outdoor work situations, wind speed should be measured and recorded, together with air temperature, whenever the air temperature is below −1°C (30.2°F). The equivalent chill temperature should be

Physical and Biological Hazards of the Workplace, Third Edition. Edited by Gregg M. Stave and Peter H. Wald.

TABLE 7.1 Cooling power of the wind on exposed flesh expressed as equivalent temperature (under calm conditions).

Estimated wind speed (in miles/hour)	Actual temperature reading (°F)											
	50	40	30	20	10	0	−10	−20	−30	−40	−50	−60
	Equivalent chill temperature (°F)											
Calm	50	40	30	20	10	0	−10	−20	−30	−40	−50	−60
5	48	37	27	16	6	−5	−15	−26	−36	−47	−57	−68
10	40	28	16	4	−9	−24	−33	−46	−58	−70	−83	−95
15	36	22	9	−5	−18	−32	−45	−58	−72	−85	−99	−112
20	32	18	4	−10	−25	−39	−53	−67	−82	−96	−110	−121
25	30	16	0	−15	−29	−44	−59	−74	−88	−104	−118	−133
30	28	13	−2	−18	−33	−48	−63	−79	−94	−109	−125	−140
35	27	11	−4	−20	−35	−51	−67	−82	−98	−113	−129	−145
40	26	10	−6	−21	−37	−55	−69	−83	−100	−116	−132	−148
(Wind speeds > 40 mph have little additional effect)	**Little danger** In <1 hour with dry skin Maximum danger of false sense of security				**Increasing danger** Danger from freezing of exposed flesh within 1 minute			**Great danger** Flesh may freeze within 30 seconds				
	Non-Freezing Cold Injury may occur at any time.											

recorded with these data whenever the equivalent chill temperature is below −7°C (19.4°F).[6]

Threshold limit values (TLVs®) for cold stress are based on the wind chill index, and they require workplace temperature monitoring.[6] Suitable thermometry should be available at any workplace where the environmental temperature is below 16°C (60.8°F), and whenever the air temperature at a workplace falls below −1°C (30.2°F), the dry bulb temperature should be measured and recorded at least every 4 hours.

EXPOSURE GUIDELINES

Cold stress TLVs® are intended to protect workers from the severest effects of cold stress (hypothermia) and cold injury (frostbite) and to define exposures to cold working conditions under which it is believed that nearly all workers can be repeatedly exposed without adverse health effects.[6] The objectives of TLVs® are to prevent the body core temperature from falling below 36°C (96.8°F) and to prevent cold injury of body extremities. For a single, occasional exposure to a cold environment, a drop in core temperature to no lower than 35°C (95°F) should be permissible. However, some clinical signs and symptoms of cold injury would be expected at that temperature. Continuous, vigorous shivering should be taken as a danger sign and cold exposure should be terminated immediately. Practically speaking, useful physical or mental work is limited once severe shivering occurs.

Whole-body protection, in the form of adequate, insulated, and dry clothing, should be provided if work is performed in air temperatures below 4°C (39.2°F). Nomograms for the determination of required clothing insulation are available.[7] For unprotected skin, continuous exposure should not be permitted when the equivalent chill temperature is below −32°C (−26.5°F).

Provisions for additional total body protection, such as shielding the work area and wearing wind-resistant and water-repellent outer clothing, are required if work is performed at or below 4°C (39.2°F). If work is performed continuously in the cold at an equivalent chill temperature below −7°C (19.4°F), heated warming shelters (e.g., tents, cabins) should be available nearby and used at regular intervals (Table 7.2). Heavy shivering, frostnip, a distant gaze, a feeling of excessive fatigue, drowsiness, irritability, or euphoria are indications for immediate return to the shelter.

NORMAL PHYSIOLOGY

Body temperature is the sum of heat produced internally, plus heat gain and loss from the environment, as described by the heat balance equation presented in Chapter 6. Briefly, body core temperature is maintained when the sum of the terms of the equation equals zero. Heat is produced via metabolism and shivering, while heat is lost through radiation, conduction, convection, and evaporation. While heat can be gained via radiation, conduction, or convection, this is unlikely in a cold environment. The most significant heat loss in the cold occurs with cold water immersion or with exposure to low air temperature and strong winds while in wet clothing.

TABLE 7.2 Threshold limit values® for work/warm-up schedule for 4-h shift.

Air temperature-sunny skies		No wind		5 mile/h wind		10 mile/h wind		15 mile/h wind		20 mile/h wind	
°F	°C	Max work period	No. of breaks	Max work period	No. of breaks	Max work period	No. of breaks	Max work period	No. of breaks	Max work period	No. of breaks
−15 to −19	−26 to −28	Normal shifts	1	Normal shifts	1	75 min	2	55 min	3	40 min	4
−20 to −24	−29 to −31	Normal shifts	1	75 min	2	55 min	3	40 min	4	30 min	5
−25 to −29	−32 to −34	75 min	2	55 min	3	40 min	4	30 min	5		
−30 to −34	−35 to −37	55 min	3	40 min	4	30 min	5				
−35 to −39	−38 to −39	40 min	4	30 min	5						
−40 to −44	40 to −42	30 min	5			**Emergency work only**					
≤−45	≤−43										

Source: From ACGIH®, *2015 TLVs® and BEIs®* Book. Copyright 2015. Reprinted with permission.

The preoptic area of the anterior hypothalamus controls body core temperature in response to both heat and cold. The hypothalamus is responsible for initiating the two main defenses against cold: **peripheral vasoconstriction** and **shivering**. Body core temperature is maintained by decreasing heat loss (peripheral vasoconstriction) and increasing heat production (shivering). Increasing physical activity also increases heat production. Peripheral vasoconstriction is the initial response to reduced skin temperature. Vasoconstriction directs blood away from the surface of the body to the core, thereby increasing tissue insulation and conserving heat. Shivering is produced by involuntary muscle contraction and results in increased metabolic heat production, which replaces heat being lost. There is an associated increase in the respiratory rate and heart rate. Shivering may increase the metabolic rate two- to fivefold.[7] However, if body core temperature is reduced with continued cooling, the metabolic, respiratory, and heart rates decrease.

With continued cold exposure, cold-induced vasodilation (CIVD) alternates with peripheral vasoconstriction to conserve body heat and, at the same time, intermittently conserve dexterity and function of the extremities via periodic rewarming.[8] Blood shunting from the skin to the body core due to vasoconstriction results in cold diuresis and decreased fluid volume. Specialized thermoreceptors in the skin detect the sensation of cold and provide the individual with the sensation of thermal comfort. However, when the skin freezes as during the development of frostbite, the sensation of cold is lost and the individual may be unaware, which underscores the importance of buddy checks.

There is some evidence that humans are capable of minor physiologic acclimatization to the cold, which manifest in three patterns: habituation, metabolic acclimatization, and insulative acclimatization, depending on the nature of the exposure.[9] The physiological adjustments to chronic cold exposure are slower to develop, less pronounced, and less practical in relieving thermal strain and preventing cold injury, when compared to the effects of heat acclimatization. Workers in a cold environment require 10–15% more calories, as additional heat is expended working in heavy, protective clothing, and caloric expenditure increases in order to maintain normal body core temperature.[10]

The importance of behavioral thermoregulation, or how an individual reacts in response to cold stress, on the prevention of cold weather injuries cannot be overstated. A cursory glance at the widely divergent environments in which humans can live and work illustrates the importance of behavioral thermoregulation. When an individual is from a climate where cold stress is minimal or absent, or is enduring his or her first season in extreme cold, the risk of suffering a cold injury is increased.[11,12] Such individuals lack the experience of prior cold exposure to guide their decision-making regarding the proper selection, wear, and use of protective clothing and equipment. Most cold weather injuries are preventable, provided that the appropriate clothing and equipment is available and utilized.

PATHOPHYSIOLOGY OF INJURY

The three major cold weather injuries are freezing injury (frostbite), hypothermia, and nonfreezing injury (formerly trench/immersion foot or pernio). Additionally, there are cold-associated problems, which include accidental injury, sunburn, snow blindness, carbon monoxide poisoning, dehydration, and cold urticaria. In the following sections, we will address the pathophysiology, diagnosis, and treatment of the major cold weather injuries.

Frostbite

Mild frostbite, sometimes called frostnip, is a reversible, superficial freezing cold injury that leads to no loss of tissue. Frostbite is the localized freezing of tissue, with formation of ice crystals and disruption of the cells. The capillary walls, particularly the endothelial cells, of the frostbitten area are damaged, increasing cell wall permeability. Fluid is released into the tissues and is accompanied by local inflammation. The most peripheral parts of the body, such as the toes, fingers, nose, ears, and cheeks, are the most common sites of freezing cold injury. The incidence of freezing cold injuries in the United States is unknown.

Hypothermia

Hypothermia is the lowering of core body temperature below 35°C (95°F). Hypothermia results when sustained heat loss to the environment exceeds heat production. Below-freezing temperatures are not necessary for the development of hypothermia, as the effects of wind chill, water immersion, rain, and/or sweating dampening the innermost layer of clothing can contribute. Air temperature as "warm" as 18.3°C (65°F) or at water temperatures up to 22.2°C (72°F) may be sufficient to cause hypothermia, especially if the duration of exposure is prolonged.

While the incidence of hypothermia is unknown, approximately 700 deaths in the United States each year are attributed to hypothermia, with most occurring in persons aged 60 years or older.[13] Death due to cold exposure occurs more frequently in men than in women. Hypothermia is associated with prolonged environmental exposure to the cold, physical exertion, wind, and skin wettedness. The insulating capability of wet clothing is reduced because the layers of trapped, dead-space air are lost. Since the thermal conductivity of water is 25 times that of air, hypothermia occurs rapidly in cold water (Table 7.3).[14]

TABLE 7.3 Survival times in cold water.

Water Temperature, °C (°F)	Exhaustion or Unconsciousness	Survival Time
>21–27 (70–80)	3–12 hrs	3 hrs – indefinite
16–21 (60–70)	2–7 hrs	2 hrs – indefinite
10–16 (50–60)	1–2 hrs	1–6 hrs
4–10 (40–50)	30–60 min	1–3 hrs
0–4.0 (32–40)	15–30 min	30–90 min
<0 (32)	<15 min	15–45 min

Source: Adapted from United States Search and Rescue Task Force, Cold Water Survival Guide (http://www.ussartf.org/cold_water_survival.htm). The time estimate variability is due to difficulty in predicting survival time, which is influenced by numerous variables.

Nonfreezing cold injury

Previously referred to as trench or immersion foot, nonfreezing cold injury (NFCI) is due to prolonged exposure to cold water. It usually occurs in dependent parts of the lower extremities in relatively immobile workers who are partially immersed in cold water. Minor trench/immersion foot injuries occur after 3–12 hours of exposure; significant tissue damage occurs after 12 hours to 3 days of exposure; and severe amputation-type injuries occur beyond 3 days of exposure. Prevention of NFCI is most important, because treatment of this injury is relatively ineffective. As NFCI is relatively infrequent in non-military scenarios, the interested reader is directed elsewhere for more in-depth discussion.[15]

Intense and prolonged cold-induced peripheral vasoconstriction is understood to initiate NFCI, in which the neuro-endothelio-muscular components of the walls of blood vessels are affected. Depending on the severity of the injury, clinically, there are four stages to the injury and recovery process. The first stage occurs during cold exposure; the key diagnostic criterion is complete local anesthesia and loss of proprioception. The intense ischemic vasoconstriction is responsible for the visual appearance of affected tissue, which initially appears bright red but often changes to a paler color before becoming almost completely white. The second stage follows cold exposure and typically only lasts for several hours. During this stage tissue is reperfused, and skin color changes to a mottled pale blue, while anesthesia to pain, touch, and/or temperature remain. The third stage is characterized by hyperemia, persistence of pain in the affected area, and anhydrosis. Pain relief using conventional analgesics and anti-inflammatory agents is usually completely unsuccessful. This stage may last for a few days or several months in more severe cases. The fourth and final stage, once hyperemia has resolved, may last for weeks, months, or even for the remainder of their lifetime in some individuals. This stage is often characterized by long-lasting cold sensitivity, hyperhydrosis, and related increased risk of recurrent fungal infection.

Cold-associated problems

In addition to injuries directly attributable to cold exposure, there are several problems associated with cold, such as cold urticaria, carbon monoxide poisoning, sunburn and snow blindness, accidental injury, and dehydration. Cold urticaria, in susceptible individuals, is characterized by local or systemic formation of wheals, with redness, swelling, and edema of the skin associated with cold exposure. The severity of cold urticaria is proportional to the rate of skin cooling and not to the absolute temperature,[16] and treatment is limited to antihistamines for relief of swelling.

Carbon monoxide (CO) poisoning is another common event in cold environments because of the use of unvented combustion heaters. CO poisoning should be considered in all cold, unresponsive patients. Symptoms include an initial headache, followed by confusion, dizziness, and somnolence.

Snow, ice, and reduced visibility increase the incidence of accidents in the cold, especially slips, trips, and falls. Snow blindness and sunburn may occur when the skin and eyes are unprotected from the ultraviolet rays of the sun and their reflection off the snow.

Body fluid requirements are no different in cold than in temperate environments. Because individual perception of thirst and the need to drink is suppressed in the cold, dehydration occurs when fluid intake is reduced. Individuals may also voluntarily dehydrate in an effort to avoid urination, which may present a nuisance when warm restroom facilities are not available. Dehydration results in decreased mental alertness, impaired cognitive ability, and reduced work capacity, and, therefore, the importance of fluid replacement should not be overlooked during cold exposure.

DIAGNOSIS

Hypothermia is insidious in onset and may be difficult to identify. In the setting of exposure to cold, hypothermia should be suspected. An appropriate clinical history, physical examination, and rectal temperature must be performed to make the diagnosis. The signs and symptoms of hypothermia may begin to appear at a body temperature of 36.1°C (97°F). Maximum shivering occurs at 35°C (95°F). Progressive decrease in core temperature results in confusion, unusual behavior, impaired coordination, slurred speech, drowsiness, weakness, lethargy, disorientation, and unconsciousness. There is slowing of the heart rate and respiratory rate. The pulse is weak, and blood pressure is decreased. Movements are slow, and deep tendon reflexes are reduced.

With body core temperatures from 32.2°C (90°F) to 35°C (95°F), peripheral vasoconstriction and shivering occur. Between 25°C (77°F) and 32.2°C (90°F), shivering is diminished, and peripheral vasoconstriction is lost. Below 25°C (77°F), there is a failure of all heat regulatory and heat conservation mechanisms.[16] Loss of consciousness occurs between 30°C (86°F) and 32°C (89.6°F). Loss of central nervous system function and coma occur below 28°C (82.4°F). Apnea occurs below 27°C (80.6°F) and asystole below 22°C (71.6°F).

TREATMENT

The degree of the cold and the duration of exposure are the two most important factors in determining the extent of a frostbite injury.[17] The keys to frostbite treatment are to protect the tissue from further injury and to increase blood flow to the interface between injured and uninjured tissue.

Frostbite may be classified as superficial or deep. Skin freezes at about −2°C (28°F). Superficial frostbite involves the skin and subcutaneous tissue, and there is no loss of tissue. The skin is gray–white, dry, and hard, with loss of sensation. With rewarming, there is pain, redness, and swelling of the skin, and blisters containing clear fluid may form. Deep frostbite involves the skin and subcutaneous tissue, as well as deeper tissues, including muscle, and bone. With deep frostbite, the affected area is pale, cold, and solid, and there is eventual loss of tissue. Formation of deep hemorrhagic blisters, death of tissue (necrosis), and ulceration occurs. Dry gangrene may develop, with autoamputation of the dead tissue. Superficial frostbite corresponds to first and second degree frostbite, under older classification systems, while third and fourth degree frostbite together are classified as deep frostbite. Frostnip and mild frostbite may be rewarmed in the workplace by placing the injured part in the armpits or the groin. In more severe injuries, if there is absolutely no possibility of the tissue refreezing, the frozen tissue may be rewarmed outside the hospital. The rewarmed part should be insulated and the patient transported to the hospital. Care must be taken not to apply excessive heat to rewarm frozen tissue, since this may produce a devastating secondary burn injury. Ideally, thawing of frozen tissue should occur in the hospital. Tissue should be rapidly rewarmed in a controlled-temperature water bath (40–42.2°C (104–108°F)). Attention should also be given to warming the whole body.

Hypothermia is a medical emergency. Only conscious patients with mild hypothermia (above 32.2°C (90°F)) should be rewarmed in the workplace.[18] For conscious patients their core temperature should be determined, and the patients should be prevented from losing additional body heat by insulating them and then rewarming them with passive external rewarming. They may be placed in a sleeping bag, wrapped in blankets, or exposed to a radiant heat source. Shivering and voluntary physical activity, such as walking, should be encouraged to generate body heat. Warm, decaffeinated, and nonalcoholic drinks should be provided to rewarm the body and replace lost fluids.

Severely hypothermic (below 28°C (82.4°F)) patients and unconscious victims of hypothermia are in a life-threatening situation. They should be handled gently, insulated, provided with intravenous fluids (5% dextrose), and transported to definitive medical care for physiologic monitoring, controlled rewarming, and management of sequelae. Peripheral pulses should be checked before initiating CPR, in order to avoid disturbing a faint but otherwise normal cardiac rhythm.[19] Attempts to rewarm these hypothermia victims in the workplace should be avoided.[19]

In severely hypothermic individuals, a body core temperature of less than 25–26.1°C (77–79°F) is a poor prognostic sign. At these temperatures, the myocardium is easily irritated and ventricular fibrillation or asystole are significant risks. Resuscitative measures such as active external rewarming, active core rewarming, cardiopulmonary resuscitation, or defibrillation should not be performed unless cardiac monitoring capability is available.[19] Aggressive rewarming with extracorporeal blood warming has been associated with good long-term prognosis in a group of patients suffering severe hypothermia with cardiac arrest.[20] Patients should be transported to a definitive care facility as quickly as possible. If cardiopulmonary resuscitation is started, it must be continued until the patient has been warmed to 36°C (96.8°F). Since metabolic processes are slowed with hypothermia, asystolic survival times are prolonged. Drowning may occur with sudden immersion in cold water, and resuscitation efforts should continue for the same reasons. Expected survival times in cold water, depending on the water temperature, are presented in Table 7.3.[21]

For nonfreezing cold injury (NFCI), slow, gradual rewarming of affected tissues is required, though treatment of coincident hypothermia may take precedence. Pain treatment is very challenging in NFCI, and while surgical or pharmacological sympathectomy may provide short-term relief in severe cases, pain often returns after several months, contraindicating this approach. Unfortunately there are no known medications that are appropriate for treatment of NFCI, beyond those used for symptomatic relief.

MEDICAL SURVEILLANCE

Identification of workers who will work in a cold environment requires: (i) determination of the physical and mental qualifications appropriate to the specific job, (ii) medical evaluation of the individual's physical and psychological ability to work in the cold, and (iii) identification of specific medical conditions which may be contraindications to working in the cold.[1] Conditions that may preclude work in the cold include exertional angina, previous cold injury, asthma, peripheral vascular disease, coronary artery disease, alcohol abuse, use of tranquilizers, and thermoregulatory disorders.

Workers should be excluded from the workplace at −1°C (30.2°F) or below if they are suffering from any of these medical conditions or if they are taking medication which either interferes with normal body temperature regulation or reduces tolerance to work in cold environments.[2] Workers who are routinely exposed to temperatures below −24°C (−11.2°F) with wind speeds less than 5 miles/hour, or air temperatures below −18°C (0°F) with wind speeds above 5 miles/hour, should be certified as medically cleared for such exposures.

Risk factors for cold injury are associated with the agent (cold), the host (the individual), or the environment (wind chill, humidity, duration of exposure, amount of activity, and protective clothing). A decrease in the equivalent chill temperature and working with cold metal objects or super-cooled volatile liquid fuels, which conduct heat away from the skin very rapidly, increase the risk of cold injury to unprotected skin. There is significant variation in individual susceptibility to cold injury. Several individual risk factors for cold injury have been identified.[13,21] While there is a lack of data in civilian populations, data from the US military indicates that women and African Americans may be at increased risk.[22] Poor physical condition, fatigue, age (the very young and the very old), inadequate caloric intake, acute or chronic illness (e.g., angina or cardiovascular disease), and a previous cold injury are additional risk factors associated with an increased risk of cold injury. Frostbite is the most frequently occurring cold injury, occurring in 40–50% of all cases.[23]

Alcohol, stimulants, and prescription drugs also affect the body's cold adaption mechanisms. Alcohol impairs judgment and reduces awareness of the signs and symptoms of cold injury. It produces peripheral vasodilation, which interferes with peripheral vasoconstriction, increasing body heat loss. Alcohol also increases urine output, exacerbating dehydration. Caffeine may have similar effects on blood vessels and urine production. Nicotine increases the risk of a peripheral cold injury by increasing the degree of peripheral vasoconstriction, which increases the rate of skin cooling and heat loss to the environment. The use of major tranquilizers (e.g., phenothiazines) also increases the risk of cold injury. Chlorpromazine suppresses peripheral shivering and produces vasodilation.[19]

PREVENTION

Cold injuries may be prevented by properly protecting workers from a cold environment through the use of appropriate protective clothing and shelter. All workers should be trained in the proper use of protective clothing. The selection of the proper clothing system for a cold environment is based on the principles of insulation, layering, and ventilation.[10] Insulation depends on clothing thickness, the properties of the material, and the amount of dead-space air trapped within the garment.[22] The inner layer of clothing, such as polypropylene, should wick moisture to the outer layers. The intermediate layers, such as wool or Thinsulate, provide insulation and may be increased or decreased for appropriate warmth. The outer layer, such as Gor-Tex, should be wind resistant and water repellent and allow water vapor and moisture, generated as perspiration, to pass through the layer. The outer layer should also be easily vented to release body heat and to prevent sweating and to allow the evaporation of moisture if sweating does occur.

A similar layering system should be used to protect the head, hands, and feet. Gloves should be worn. If fine manual dexterity is required, thin inner gloves may be worn under heavier outer gloves or mittens. The outer gloves may be temporarily removed as needed. The head should be protected, since 30% of body heat is lost through the head. Workers at increased risk of cold injury may require additional clothing.

When wearing cold weather clothing, the mnemonic "COLD" should be used to guide appropriate clothing maintenance:

- Keep clothing **C**lean to ensure maximum insulation.
- Avoid **O**verheating by adding or removing insulating layers, as appropriate.
- Wear **L**oose clothing and in **L**ayers to allow free blood circulation, trap dead-space air, and adapt to changes in the workload or environment.
- Keep clothing **D**ry to ensure maximum insulation.

Workers should wear anticontact gloves to prevent contact frostbite. For work performed at temperatures below −7°C (19.4°F), a warning or safety briefing should be given at least daily to each worker by the supervisor to prevent inadvertent contact of bare skin with cold surfaces. If the air temperature is −17.5°C (0°F) or less, the hands should be protected by mittens. Workers handling evaporative liquids (gasoline, alcohol, or cleaning fluids) at air temperatures below 4°C (39.2°F) should take special precautions. Workers handling liquefied gases (liquid natural gas, liquid oxygen, and liquid nitrogen) must also take special precautions, particularly in the event of a spill.

Special protection of the hands is required to maintain manual dexterity. If fine work is to be performed with bare hands for more than 10–20 minutes in an environment below 16°C (60.8°F), special provisions should be established for keeping the workers' hands warm. Metal handles should be covered by thermal insulating material at temperatures below −1°C (30.2°F). In the absence of extenuating circumstances, cold injury to body parts other than the hands, feet, and/or head is not likely to occur without showing the initial signs and symptoms of hypothermia.

Prevention of non-freezing cold injury centers on proper foot care, including keeping the feet dry and allowing proper circulation. Vapor barrier boots, which are very effective in keeping moisture out, present a double-edged sword, as they also retain fluid from sweat or that accumulated inside the boot due to leaks or dripping. Socks should be changed whenever they become wet, and feet should not remain motionless.

If available clothing does not provide adequate protection against the cold, work should be modified or suspended until adequate clothing is available or until weather conditions improve. If clothing is wet, workers should immediately change into dry clothing. If work is done at normal temperatures prior to entering a cold area, it may be necessary for the worker to change clothing that is damp due to sweating.

Shelter is used to reduce exposure to the cold, when periodic rewarming breaks must be taken. Workers must be encouraged to drink fluids at regularly scheduled times to avoid dehydration. Fluid intake should increase with decreasing temperature and increasing levels of exertion. Warm, sweet drinks, and soups provide calories and fluid volume. Caffeine has a minor diuretic effect,[23] so coffee intake should be moderated, though the warming and stimulant effects may be desirable, and alcohol use is contraindicated.

Protective eyewear should be used to protect against blowing snow and ice crystals, airborne particulates, ultraviolet radiation, and glare. Sunscreen will prevent sunburn, and moisturizers will reduce the effects of dry cold on the skin, lips, and nose. Self-aid and "buddy" aid, with frequent checks, should be used to identify early signs and symptoms of cold injury.

Additional workplace requirements exist for refrigerator rooms, working with toxic substances, and exposure to vibration.

Health and safety education on the recognition and treatment of cold injuries should be provided to workers. Training should be performed on how to wear and work in cold weather clothing. Workers should also be instructed in safety and health procedures such as proper rewarming methods, appropriate first aid treatment, good clothing practices, eating and drinking requirements, recognition of impending frostbite and hypothermia, and safe work practices. New employees should be allowed to become accustomed to working conditions in the cold and use of the required protective clothing. The weight and bulkiness of clothing should be included in estimating the required work performance. For work at or below −12°C (10.4°F) equivalent chill temperature, there should be constant protective observation ("buddy" system or direct supervision). The work rate should not be so high as to cause heavy sweating that will result in wet clothing, and there should be frequent rest periods. Most important of all, the worker must respect and use common sense in dealing with the cold.

References

1. International Organization for Standardization. *Ergonomics of the Thermal Environment: Medical Supervision of Individuals Exposed to Extreme Hot or Cold Environments.* Geneva: ISO; 2001.
2. Whayne TF, DeBakey ME. Cold injury, ground type. Washington, DC: Office of the Surgeon General, Department of the Army; 1958.
3. Paton BC, Pandolf KB, Burr RE. Cold, casualties, and conquests: the effects of cold on warfare. In: Pandolf KB, Burr RE, editors. *Medical Aspects of Harsh Environments* Volume 1. Falls Church, VA: Office of the Surgeon General; 2001. p. 313–49. Available at: http://www.shtfinfo.com/shtffiles/medical_remedies/Harsh_Environments_Vol_1.pdf (accessed on June 28, 2016).
4. Bass DE. Metabolic and energy balances of man in a cold environment. In: Horvath SM, editor. *Cold Injury.* Montpelier, VT: Capital City Press; 1958. p. 317–38.
5. Bagley JR, Judelson DA, Spiering BA, et al. Validity of field expedient devices to assess core temperature during exercise in the cold. *Aviat Space Environ Med* 2011;82:1098–103.
6. American Conference of Governmental Industrial Hygienists. *2014 Threshold Limit Values for Chemical Substances and Physical Agents and Biological Exposure Indices.* Cincinnati, OH: American Conference of Governmental Industrial Hygienists; 2014.
7. Eyolfson DA, Tikuisis P, Xu X, et al. Measurement and prediction of peak shivering intensity in humans. *Eur J Appl Physiol* 2001;84:100–6.
8. Hamlet M. Human cold injuries. In: Pandolf K, Swaka M, Gonzales R, editors. *Human Performance Physiology and Environmental Medicine at Terrestrial Extremes.* Indianapolis, IN: Benchmark Press, Inc.; 1988. p. 435–66.
9. Sawka MN, Young AJ. Physiological systems and their responses to conditions of heat and cold. In: Tipton CM, editor. *ACSM's Advanced Exercise Physiology.* Baltimore, MD: Lippincott Williams and Wilkins; 2006. p. 535–63.
10. Department of the Army. *Prevention and Management of Cold-Weather Injuries.* Washington, DC: Headquarters, DA; 2005.
11. Miller D, Bjornson DR. An investigation of cold injured soldiers in Alaska. *Mil Med* 1962;127:247–52.
12. Candler WH, Ivey H. Cold weather injuries among US soldiers in Alaska: a five-year review. *Mil Med* 1997;162:788–91.
13. Fallico F, Nolte K, Siciliano L, et al. Hypothermia-related deaths—United States, 2003–2004. *Morb Mortal Wkly Rep.* 2005;54:173–5 (in English).
14. United States Coast Guard (USCG). *National Search and Rescue Manual.* Washington, DC: USCG; 1970.
15. Thomas JR, Oakley HN. Nonfreezing cold injury. In: Pandolf KB, Burr RE, editors. *Textbooks of Military Medicine: Medical Aspects of Harsh Environments.* Falls Church, VA: Office of the Surgeon General, US Army; 2001. p. 467–90. Available at: http://www.shtfinfo.com/shtffiles/medical_remedies/Harsh_Environments_Vol_1.pdf (accessed on June 28, 2016).
16. Smith D, Robson M, Heggers J. Frostbite and other cold induced injuries. In: Auerbach P, Geehr E, Lewis E, editors. *Wilderness Medicine: Management of Wilderness and Environmental Emergencies.* New York: Macmillan; 1995. p. 101–18.
17. Mills WJ. Clinical aspects of freezing cold injury. In: Pandolf KB, Burr RE, editors. *Textbooks of Military Medicine: Medical Aspects of Harsh Environments.* Falls Church, VA: Office of the Surgeon General, Department of the Army; 2001. p. 429–66. Available at: http://www.shtfinfo.com/shtffiles/medical_remedies/Harsh_Environments_Vol_1.pdf (accessed on June 28, 2016).

18. Lazar HL. The treatment of hypothermia. *N Engl J Med* 1997;337:1545–7. (in eng).
19. Reed G, Anderson R. Accidental hypothermia. In: Wolcott B, Rund D, editors. *Emergency Medicine Annual*. Norwalk, CT: Appleton, Crofts; 1984. p. 93–124.
20. Walpoth BH, Walpoth-Aslan BN, Mattle HP, et al. Outcome of survivors of accidental deep hypothermia and circulatory arrest treated with extracorporeal blood warming. *N Engl J Med* 1997;337:1500–5.
21. DeGroot DW, Castellani JW, Williams JO, et al. Epidemiology of US Army cold weather injuries, 1980–1999. *Aviat Space Environ Med* 2003;74:564–70.
22. Young A, Roberts D, Scott D, et al. Sustaining health and performance in tho cold: A pocket guide to environmental medicine aspects of cold-weather operations. DTIC Document; 1992.
23. Zhang Y, Coca A, Casa DJ, et al. Caffeine and diuresis during rest and exercise: A meta-analysis. *J Sci Med Sport* 2015; 18(5):569–74.

Chapter 8

HIGH-PRESSURE ENVIRONMENTS

TONY L. ALLEMAN AND JOSEPH R. SERIO*

Humans function well only within a narrow range of barometric pressures. Outside this range, they are subject to major physiologic stresses that occasionally result in disease. On land, workers are exposed to hyperbaric environments (i.e., increased barometric pressure) during tunneling projects that require the use of compressed air or when caissons are used to work in ground saturated with water. In addition, hyperbaric chamber support staff members are routinely exposed when they treat patients in hyperbaric medical treatment facilities. In the water, occupational exposures are diverse. Examples of exposures include breath-hold divers, such as the ama pearl divers of Japan, and compressed gas divers, ranging from instructors of recreational SCUBA students, who breathe compressed air, to saturation divers supporting offshore oil exploration, who dive in excess of up to 1000 ft of seawater (fsw), depending on the scope of the project, while breathing artificial gas mixtures. Divers can also be found inland. These divers are involved in such jobs as inspecting dams and reservoirs, cleaning filters, maintaining fish farms, placing underwater demolitions, and conducting police searches.

The first practical method developed for conducting useful work underwater was the diving bell, which is essentially an upside-down cone. Invented by Smeaton in 1778, the diving bell was the forerunner of the modern caisson (caisse, in French, means "box").[1] This method is still used today in the construction of tunnels and bridge footings. In 1819 in England, Augustus Siebe invented the first practical diving dress; it consisted of a copper helmet bolted to a leather coverall. This apparatus gave humans the ability to walk and function relatively unencumbered underwater without holding their breath. In 1837, Siebe introduced an improved version of this gear, which served as the basic diving dress for deep-sea diving until the early 1980s and is still used by some divers today.[2] Modern surface-supplied equipment has incorporated various advances in technology, such as helmets made of lightweight composites with increased fields of vision and improved gas delivery systems that reduce breathing resistance.

Along with these advances, which have enabled divers and caisson workers to work at greater pressures or depths for longer periods of time, have come associated medical problems. The first descriptions of decompression-related disorders were made by Triger in 1841 among pressurized tunnel workers[3] and by Alfonse Galin in 1872 with Greek sponge divers.[4] By the early 1900s, it was recognized that decompression-related symptoms were due to inert gas and could be relieved by returning the individual to pressure. However, there was still no method for controlling the exposure to prevent the disease. The Royal Navy enlisted the help of the eminent physiologist J. S. Haldane to develop such a method. Haldane published his first set of decompression tables in 1908, based on experiments using sheep.[3] The tables incorporated delays on ascent (called stops) to allow time for excess inert gas dissolved in body tissues to be eliminated (off-gassing). Haldanian principles form the basis for most decompression tables used today.

OCCUPATIONAL SETTING

Diving

Commercial divers are often involved in construction, inspection, and repair of pipelines, vessels, aquariums, tunnels, and marine platforms and typically spend a significant part of

* The authors wish to acknowledge the contribution of David J. Smith who authored this chapter in the prior edition of this book.

Physical and Biological Hazards of the Workplace, Third Edition. Edited by Gregg M. Stave and Peter H. Wald.

their time in diving operations. Scientific divers are devoted to the study of the marine environment and may dive only a few days a month as part of their job. Public safety divers become involved primarily with search and rescue operations and may only dive a few days a year. Engineers that design and oversee construction and maintenance of bridges often dive to inspect the underwater structures.

Diving operations can be classified into three basic categories: air, mixed gas, and saturation. Air diving is restricted to relatively shallow depths, due to the increasing narcotic effect of the nitrogen component of air as depth increases (the Occupational Safety and Health Administration (OSHA) restricts exposures on air to <190 fsw but will permit surface-supplied air for dives with bottom times of 30 minutes or less for depths of 220 ft). Mixed-gas diving uses a breathing mixture other than air; it is employed principally when deeper working depths are required. It generally specifies the use of helium as the inert gas constituent; however, nitrogen–oxygen mixtures (nitrox) are becoming more common for relatively shallow applications, and hydrogen has been tested for deep commercial applications. Although greater depths are permitted in standard mixed-gas diving, they are also limited by the amount of decompression required. For a constant bottom time, the required decompression time increases significantly with depth, prolonging exposure to the environment and limiting the useful work period. To overcome this problem, saturation diving methods have been developed.

Saturation diving, a specialized extension of standard decompression diving, is based on the principle that at a constant depth tissues will on-gas until the tissue partial pressure reaches equilibrium with the ambient pressure (see Henry's law, p. 116). Once tissue equilibrium is reached, no net uptake will occur (unless the pressure is increased), resulting in no further increase in decompression time. At this point, the bottom time may be increased without additional decompression penalty. The divers are housed at "depth" for up to a month in a dry, pressurized chamber on the surface and transported under pressure to and from the work site via a diver transport capsule (Figure 8.1). Saturation diving techniques represent a cost-effective alternative to standard diving when deep or prolonged bottom times are required, since the need for repetitive, long in-water decompressions is avoided. The other advantage of saturation diving is that only one decompression is required, thereby reducing the risk of decompression-related incidents.

FIGURE 8.1 A saturation system diving bell (Courtesy of D. R. Chandler).

TABLE 8.1 High-pressure and diving environments.

Class	Equipment type	Type of air supplied
SCUBA	Open circuit	Air
	Closed circuit	
	Semiclosed circuit	
Surface supplied	Lightweight	Air
	Deep sea	Mixed gas
	Saturation	Mixed gas
Caisson	Shirt sleeve	Pressurized air

Equipment

There are two basic classes of equipment employed by divers today: self-contained underwater breathing apparatus (SCUBA) and surface-supplied equipment. Choice of equipment depends primarily on the job requirements and the employer's capabilities. Equipment generally used in high-pressure and diving environments is summarized in Table 8.1.

Modern SCUBA has evolved from equipment originally developed during World War II by Cousteau and Gagnon.[2] The advent of SCUBA ushered in a new era of underwater mobility. SCUBA equipment is designed either to be open circuit, closed circuit, or semiclosed (a hybrid between open circuit and closed circuit). Modern open-circuit SCUBA (Figure 8.2) is the principal equipment used in

FIGURE 8.2 SCUBA equipment (Courtesy of US Navy).

FIGURE 8.3 Diver using Draeger LAR V closed-circuit SCUBA equipment (Courtesy of T.J. Doubt).

recreational diving; it is also used for many commercial applications. Closed-circuit SCUBA (rebreathers) (Figure 8.3) removes the exhaled carbon dioxide from the breathing gas prior to returning the "scrubbed" gas to the diver. Oxygen is added as needed. Some closed-circuit rigs maintain a constant partial pressure of oxygen in the breathing gas, increasing depth capabilities. Closed-circuit equipment is generally more complex and expensive, but it has the advantage of less gas consumption without the generation of bubbles. The absence of bubbles is useful in marine research, marine photography, and various military applications.[5]

Many advances have been made in surface-supplied diving since the 1960s, including specialized materials for helmets and suits, hot water heating for suits, and advanced communication systems. Surface-supplied diving equipment can generally be divided into two types: lightweight and deep sea. Deep-sea equipment can be further subdivided into air, mixed gas, and saturation capable. Lightweight equipment, such as a "band mask," is customarily used for work where surface communications are required, but significant diver protection is not necessary. Deep-sea diving equipment (Figure 8.4) affords increased protection, generally consisting of a "hard hat" offering maximum head/neck protection along with surface-to-diver communication video capability, protective gloves and boots, weights, an emergency gas supply, and varying levels of thermal protection, as complex as suits that circulate hot water supplied from the surface.

As in most occupational settings, the choice of equipment used in diving generally depends on the characteristics of the environment and the work to be completed, along with the personal preferences and the capabilities of the individual

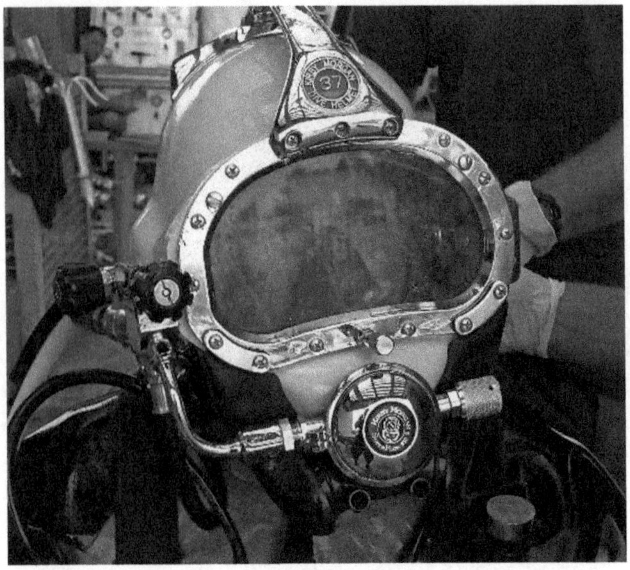

FIGURE 8.4 Deep sea diving equipment (*Source*: https://commons.wikimedia.org/wiki/File:US_Navy_051026-N-0000X-001_Electronics_Technician_1st_Class_Matthew_Ammons,_a_diver_assigned_to_Mobile_Diving_and_Salvage_Unit_Two_(MDSU-2),_is_fitted_with_a_Kirby_Morgan_37_Dive_Helmet.jpg).

diver. The principal advantages of SCUBA diving are its ease of use, transportability, and freedom of movement. Its disadvantages include limited gas supply, depth restrictions, minimal head protection, lack of communication, and inability to function safely in strong currents. Surface-supplied diving overcomes many of the disadvantages of SCUBA by providing increased head protection, thus reducing the chance of drowning if the diver becomes unconscious; "unlimited" gas supply; better buoyancy control, thus enabling the use of heavy construction techniques; hardwired communications; and greater depth capabilities (Figures 8.5, 8.6, 8.7, and 8.8). However, surface-supplied diving is more complex and costly, because it requires significantly more surface support in both equipment and personnel. In addition, it decreases the diver's comfort on the surface due to weight, decreases mobility in the water, carries a risk of entanglement, and can present a significant noise hazard. Essentially all of the commercial diving in the offshore environment is done with surface-supplied air or gas or in a saturation environment.

Caissons

A caisson is a watertight retaining structure that is used for engineering projects where water or waterlogged soil precludes standard construction techniques. It can be envisioned as an inverted cone or diving bell that rests on the bottom. It is pressurized to exclude water and to allow a relatively dry working environment. Similar principles are

FIGURE 8.5 Surface-supplied diving: Diver using a grinding tool. Copyright 2016 Oceaneering International, Inc. Used with permission of Oceaneering International, Inc.

also applied in the construction of tunnels through waterlogged or unstable "mucky" earth. Workers enter and exit the workplace through a pressurized lock system where decompression may be carried out on exit. At depth, the workers may labor for up to 8 hours at pressure. Decompression is dependent on the exposure time and the pressure. The length of the decompression is determined from tables developed specifically for this purpose. Fire safety is of concern in pressurized environments, due to the higher partial pressures of oxygen; however, modern systems are constructed from steel, minimizing flammable materials.[6] OSHA provides guidelines for safety in caisson work (29 CFR 1926.801).

The need for caisson techniques has been gradually declining as other engineering methods, such as pressure-balanced shields and unmanned excavating systems, have been developed to avoid the cost and complexity of caisson work. Currently, there is limited caisson work being done in the United States. Noncaisson construction is less costly since it eliminates both high-pressure equipment and the extra labor costs related to high-pressure work. However, with automated equipment, men occasionally need to enter a

FIGURE 8.6 Surface-supplied diving: Diver water gouging. Copyright 2016 Oceaneering International, Inc. Used with permission of Oceaneering International, Inc.

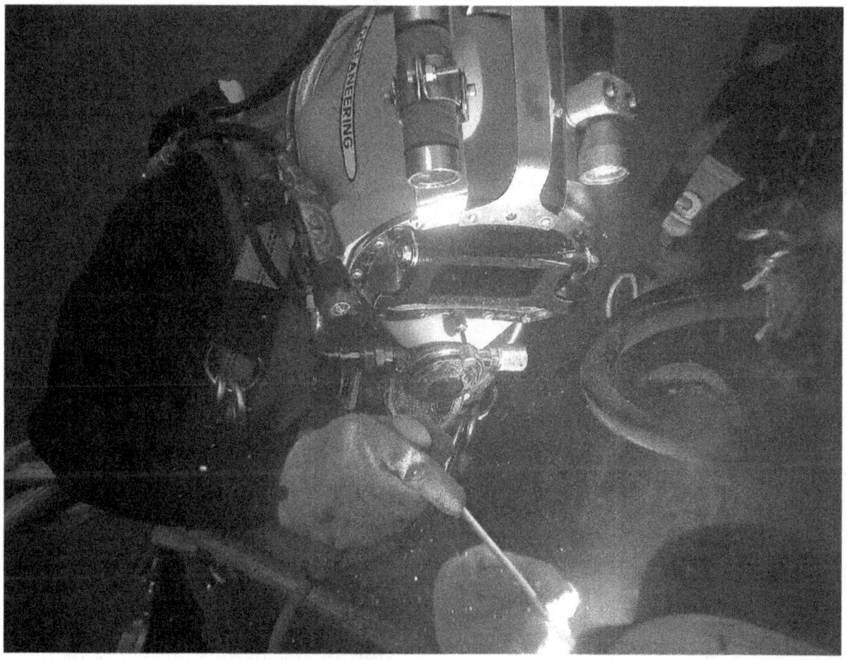

FIGURE 8.7 Surface-supplied diving: Diver welding. Copyright 2016 Oceaneering International, Inc. Used with permission of Oceaneering International, Inc.

high-pressure compressed air environment to repair or maintain the equipment.[7]

The majority of hyperbaric exposures today occur in the undersea environment. Since caisson work is very limited, the remainder of this chapter will focus on undersea hyperbaric exposures. The pathophysiology, diagnosis, and treatment of hyperbaric exposures and their sequelae are similar in caisson and undersea workers.

FIGURE 8.8 Surface-supplied diving: Diver working on Habitat with Bell. Copyright 2016 Oceaneering International, Inc. Used with permission of Oceaneering International, Inc.

MEASUREMENT ISSUES AND PHYSICS OF PRESSURE

Pressure is a measurement of force per unit area. Air pressure is the force exerted by the column of atmosphere above a particular point. The height of the column determines the pressure. Water pressure, measured by a gauge, is defined as the force exerted by the column of water above the submerged object. Pressure increases linearly with depth. Absolute pressure measures the force per unit area of the combined air and water column. The absolute pressure at 33 fsw is 2 ata (atmospheres absolute) since 1 atm of pressure is contributed by both the air column (weight of atmosphere) and the water column (weight of water). Table 8.2 defines some useful measurements of pressure in diving.

An understanding of the physiologic effects of pressure requires knowledge of basic physics. The relationship between pressure, volume, and temperature is defined by the ideal gas law, which states

$$PV = nRT$$

where P is absolute pressure, V is volume, n is the number of moles of the gas, R is the universal gas constant, and T is the absolute temperature.

Boyle's law states that at constant temperature, the volume is inversely proportional to the absolute pressure—that is, no gas enters or exits the system ($P_1V_1 = P_2V_2$).[8] For example, if a balloon is filled with 2.0L of compressed gas at a depth of 99 ft (4.0 ata) and brought to 33 ft (2.0 ata), the gas volume will have expanded to 4.0L ($V_2 = P_1V_1/P_2 = 4\times2/2 = 4$). If taken to the surface (1.0 atm), the volume will have expanded to 8.0L ($V_S = P_1V_1/P_S = 4\times2/1 = 8$). It is important to note that the proportional change in volume per depth change increases toward the surface.

TABLE 8.2 Units of pressure used in hyperbaric environments.

10.00 m of seawater (msw)[a] = 32.646 ft of seawater (fsw)[b]
10.00 msw = 1 bar = 100 kilopascals (kPa)[c]
1 atmosphere (atm) = 1.013 bar
1 atm = 760 Torr (mm of mercury)
= 1033 cm of water
= 14.69 lb per in.^{2}
= 33.08 fsw

[a] The definition of fsw (feet seawater) assumes a density (weight per unit volume) for seawater of 1.025 at 40°C.
[b] The definition of msw (meters seawater) assumes a density for seawater of 1.020 at 40°C.
[c] The unit for the pascal is defined as a Newton per meter squared.

The law of partial pressures (**Dalton's law**) states that, in a mixture of gases, the total gas pressure is the same as the sum of the partial pressures of the individual gases in the mixture ($P = p_1 + p_2 \cdots p_n$). As a result, the partial pressure of the gas can be calculated by knowing the total pressure of the mixture and the percentage makeup of a particular component. As the total pressure increases, the volume of the individual gases becomes more significant, and eventually Dalton's law no longer applies.[8]

Henry's law states that the amount of gas that will dissolve into a solution is directly proportional to the partial pressure of that gas and inversely proportional to the absolute temperature.[8] Therefore, as the pressure increases (depth increases), the amount of gas dissolved in tissues will increase. Once at a constant depth, gas will continue to increase in the tissues until the equilibrium or saturation is attained (i.e., when the partial pressure of the gas in tissue equals the ambient pressure). Gases such as oxygen may be metabolized, whereas inert gases will not be. Subsequently, when the ambient pressure is decreased, inert gases will come out of solution. This law explains why nitrogen bubbles may form when you surface from an air dive and why a can of carbonated beverage fizzes when it is opened.

Archimedes' principle, which describes buoyancy, states that an object immersed in a fluid is buoyed up by a force equal to the weight of the volume of fluid that the object displaces. The effect of this principle is seen throughout diving. For example, inhaling a large breath will increase the diver's buoyancy (due to greater displacement of fluid).

EXPOSURE GUIDELINES

Commercial diving in the United States is covered by various regulations, including OSHA Standard Title 29 Code of Federal Regulations 1910 Subpart T, Commercial Diving Operations, and Title 46 Code of Federal Regulations, Subchapter V, Marine Occupational Safety and Health Standards, Subpart B, Commercial Diving Operations. Diving contractors and operators generally have developed local standard operating procedures that define procedures and safety guidelines in more detail.

Caisson work, and other high-pressure construction work, is covered by OSHA Standard Title 29 CFR 1926 Subpart S—Underground Construction, Caisson, Cofferdams, and Compressed Air.

SPECIAL UNDERWATER STRESSORS

The worker in the undersea environment is exposed to a number of unique environmental stressors that may be enhanced by pressure. These environmental factors include sensory input modifications (sound, visual, proprioceptive), thermal challenges, and gas effects. Professionals generally do not have the luxury of waiting for ideal environmental conditions. As a result, most operations are conducted when at least one element is not ideal. Medical planning for diving operations should take this into consideration.

Sensory changes

Divers must tolerate seriously reduced sensory input while working, forcing increased alertness and vigilance. As water temperature decreases, there can be an associated decrease in sensory perception. If the temperature drops significantly enough, there will be difficulty in motor control as well as sensory perception.[9]

Sound

Sound travels approximately four times faster in water than in air; as a result, it travels further. Because of this increased speed, the normal delay cues required to place sound in three dimensions are in practice lost, making the localization of sound extremely difficult. Moreover, the increased density of air and increased middle-ear impedance inhibit air conduction, causing a ≤40 dB loss.[10] Sound does not transmit across an air–water interface efficiently. As a result, vocal communications through the water are effectively ruled out, thus adding to the sensory deprivation already present secondary to other effects.[10] In addition, many surface-supplied helmets are noisy secondary to their design, which compounds hearing difficulties. Divers may sustain aural trauma from underwater blasts or activation of sonar devices while submerged.

Vision

The underwater environment seriously affects vision in a number of ways. First, water transmits light poorly. This is caused principally by turbidity, that is, suspended particles that obstruct light. At 10 m in clear water, ~60% of the light is filtered out, leaving only 40% visible light.[11] However, in highly turbid waters, like those found in many ports and inland waters, light may be reduced substantially more than this.

In addition, color vision is lost. As depth increases, long wavelengths (red end of the spectrum) are filtered initially, followed by the blues. In the absence of artificial light, perception of red is generally eliminated by 10 m, followed by yellow by ~20 m.

As the diver descends, the eyes must adapt from day to night light levels. However, the working diver generally descends faster than the retina can compensate, further compounding the visual adjustment.[8] Objects appear 25% larger and closer underwater because the refractive index of light is increased ~1.3 times over that in air. This causes problems with determination of distances and eye–hand coordination that improve somewhat, although not completely, with experience. Finally, if the cornea comes into direct contact with water—for example, in the absence of the air interface provided by a mask or helmet—a substantial hyperoptic refractive error of 40–50 diopters (D) results.[12]

Proprioception

Proprioceptive inputs for orientation to the diver are reduced by buoyancy and the diver's protective clothing. Additionally, divers receive fewer visual cues. On land, workers rely on visual, proprioceptive, and vestibular inputs to maintain balance and orientation in space. This capacity is significantly impaired by the underwater environment.[11]

Thermal transfer

Maintaining a constant body temperature while submerged is a challenge, even in relatively warm water. Heat is lost during water immersion, primarily through convection and conductance from the skin and through the respiratory tract. Water has a high coefficient of thermal conductivity (~25 times greater than air), and submersion effectively eliminates the normal protective air insulation layer. Water cools four to five times faster than an unprotected individual at the same temperature in air.[13] As a result, an unprotected diver in 27°C (80°F) water loses heat at the same rate as an unprotected subject in 6°C (42°F) air. A water temperature of ~35°C (95°F) is required to keep a resting, unprotected immersed diver thermoneutral.[14] To protect from this loss, passive insulation, such as neoprene, is used for mild to moderate cold shallow applications. Wet suits are generally made of neoprene, a material that is compressible and as a result loses insulation capacity with depth. For deeper, colder dives, dry suits (passive) and hot water suits (active) are used. Dry suits consist of an impermeable outer barrier and an insulating inner garment made of a variety of different materials. Dry suits can be inflated with pressurized gas as depth increases to maintain the insulation barrier. Active heating garments, such as hot water suits, which bathe the diver in warm water, are employed in very cold and deep diving, but they require more technical support.

At greater depths (>100 m), heat loss from the respiratory tract becomes significant secondary to the high heat capacity of the dense gas. This respiratory heat loss may not be sensed and therefore can cause asymptomatic hypothermia. Breathing gas is heated to prevent this phenomenon. Occasionally, in unique circumstances, the diver may be exposed to increased water temperatures. Such dives require special planning and measures to protect against hyperthermia, since the diver will be unable to self-regulate body heat by the evaporation of sweat. Hypothermia and cold exposure are discussed in detail in Chapter 7.

Gas effects

Table 8.3 summarizes the toxic effects of gases on divers.

Inert gas narcosis

Inert gas narcosis is defined as the progressive development of symptoms of intoxication/anesthesia with increasing partial pressures of the gas. Inert gases vary substantially in their ability to induce narcosis at a given partial pressure of exposure. Xenon is anesthetic at 1 atm, whereas helium has little narcotic effect even at very great depths.[15] The narcotic effect of a specific inert gas is related to its lipid solubility; however, the precise pathophysiologic mechanism of narcosis is not well understood.[15]

As a practical point, the depth or pressure to which a diver may descend while breathing air is restricted principally by nitrogen narcosis. Significant individual variability in response to the effect exists; but on average, at 100–200 ft (30–60 m), an individual feels lightheaded and euphoric (similar to alcohol intoxication) and experiences decreased reasoning capability, reaction time, and manual dexterity.[16] At 200–300 ft, reflexes slow, paresthesias may develop, and decrements in judgment and reasoning may produce dangerous overconfidence.[11] At 300–400 ft, marked impairment of judgment,[15] anesthesia, progressive depression of the sensorium with auditory and visual hallucinations, and amnesia precede syncope. Finally, at depths >400 ft on air, the diver becomes unconscious.[11]

Primary prevention involves limiting the depth of exposure and substituting a less narcotic gas, such as helium. Individuals do not acclimatize to narcosis; however, repeated exposures help a diver adjust to this effect.

Oxygen toxicity

Because oxygen is a mandatory constituent of breathing gas, it would be logical to postulate that diving on 100% oxygen (O_2) would solve the problems associated with inert gas during hyperbaric exposures. Applications of 100% O_2 breathing are restricted, however, because oxygen in the hyperbaric environment becomes increasingly toxic to the central nervous system (CNS) as the partial pressure of oxygen increases. Breathing oxygen with a partial pressure as low as ~1.3 atm abs (equivalent on the surface to 130%) may cause acute CNS manifestations. The risk of CNS oxygen toxicity increases as the oxygen partial pressure increases. The most serious CNS manifestation secondary to oxygen is a tonic–clonic seizure, which may occur without prodromal symptoms. If a seizure occurs underwater, it may cause drowning. Prodromal manifestations include[11]:

- *V* = Visual disturbances, such as tunnel vision
- *E* = Ear problems, including tinnitus or decreased acuity
- *N* = Nausea and vomiting
- *T* = Twitching
- *I* = Irritability and restlessness
- *D* = Dizziness and vertigo

Manifestations of CNS O_2 toxicity are treated by decreasing the partial pressure of oxygen. Divers try to prevent toxicity by limiting their exposure time and the partial pressure of oxygen breathed. Factors that are thought to predispose to oxygen convulsions include exercise, carbon dioxide reten-

TABLE 8.3 Toxic effects of gases on divers.

Gas	Causes/sources	Symptoms
Inert gas narcosis	Increased partial pressure of inert gas	100–200 fsw Lightheaded Decreased complex reasoning Loss of fine discrimination Euphoria 200–300 fsw Poor judgment/reasoning Slowed reflexes Paresthesia Dangerous marine life 300–400 fsw Progressive depression Auditory/visual hallucinations Loss of memory >400 fsw Unconsciousness
Oxygen toxicity Central nervous system (acute)	>1.3 atm abs partial pressure of oxygen Risks increase with increase in partial pressure	Visual disturbance Tinnitus/decreased visual acuity Nausea/vomiting Twitching Irritability/restlessness Vertigo Convulsions (may occur with no prodromal symptoms from above; no long-term consequences)
Pulmonary (chronic)	Prolonged exposure to >0.5 atm abs oxygen	Chest pain with inspiration Inspiratory irritation Coughing Progressive shortness of breath
Hypoxia	Hyperventilation prior to breath-hold diving Improper gas mixture	Unconsciousness (without symptoms)
Carbon dioxide toxicity	Equipment design Increased gas density Increased partial pressure of oxygen blunts response to CO_2 CO_2 contamination	Increased breathing rate Shortness of breath Headache Unconsciousness
Carbon monoxide toxicity	Use of improper compressor lubricants (flashable) Compressor failure Improper intake placement (e.g., next to combustion engine)	Headache Nausea/vomiting Unconsciousness

tion, some medications, and water immersion itself. Long-term health consequences from CNS oxygen toxicity have been demonstrated. Oxygen seizures per se are of less clinical importance than other seizures because the recipient is well oxygenated prior to their occurrence.

Prolonged exposures to partial pressures of oxygen >0.5 atm abs may cause clinically apparent pulmonary toxicity. Usually, only saturation diving scenarios and hyperbaric oxygen therapy, for example, a prolonged therapy associated with decompression illness (DCI), are sufficiently prolonged to cause clinically apparent pulmonary toxicity. However, in rare cases, pulmonary oxygen toxicity may become a limiting factor in repetitive deep bounce dives, which require prolonged decompressions. Manifestations of pulmonary oxygen toxicity are the same as those of patients on ventilators exposed for prolonged periods to increased partial pressures of oxygen. These include (i) substernal chest pain, which begins as irritation on inspiration and progresses with continued exposure to severe burning chest pain during both inspiration and expiration; (ii) coughing, which gradually increases in frequency and duration with exposure; and (iii) progressive shortness of breath.[11]

Physical examination is generally unremarkable and chest radiographs are clear except in severe cases. In practice, within a few hours after removal from exposure, symptoms begin to resolve, and they generally resolve completely within 24–48 hours (although if severe, the forced vital capacity may take substantially longer to recover). Limiting exposures is the only preventive measure available.

Inspiration of high partial pressures of oxygen can also cause ocular toxicity. Conditions associated with ocular oxygen toxicity include myopia, cataracts, keratoconus, and retinopathy.[17] Myopia caused by hyperbaric oxygen requires multiple exposures and is generally reversible. Cataract formation may be dependent upon the partial pressure of oxygen exposure. The retina is susceptible to oxidative stress, the cause for development of retinopathy in hyperbaric conditions.

Hypoxia

Hypoxia is particularly hazardous, because it can cause unconsciousness without warning. Although hypoxia rarely occurs in air diving, it may occur in mixed-gas diving secondary to procedural errors, such as improper gas mix or mechanical failures. Individuals are also prone to hypoxia during breath-hold diving, which is preceded by hyperventilation. This is a result of decreased predive carbon dioxide (CO_2) levels secondary to hyperventilation. Since elevated CO_2 levels drive the need to breathe, low initial CO_2 levels permit divers to comfortably overstay their permissible dive time at depth (secondary to elevated partial pressure of oxygen at depth). On ascent, the individual becomes hypoxic as the partial pressure of oxygen decreases below safe levels.

Carbon dioxide toxicity

Carbon dioxide buildup or retention can occur commonly in the hyperbaric environment. Causes include:

- Increased work of breathing caused by increased gas density (Boyle's law), increased breathing resistance, and dead space as a result of the use of breathing equipment. Added dead space and breathing resistance are dependent on the design of the breathing equipment.
- Increased partial pressure of oxygen, which blunts the body's response to elevated partial pressures of CO_2.
- CO_2 in the breathing gas secondary to failure of CO_2 absorbent, poor gas analysis, or contaminated breathing gas.
- Deliberate reduction of ventilation. With some breathing equipment designs, ventilation to the helmet causes a great deal of noise, which interferes with communications. Divers sometimes reduce helmet ventilation to reduce noise levels and facilitate communications, thus predisposing to CO_2 buildup.

The early manifestations of CO_2 toxicity are shortness of breath, anxiety, and increased heart rate. These symptoms may go unnoticed when a diver is performing hard work. As the exposure increases, the worker may experience a headache (which presents as a mild to moderate throbbing during exposure but may increase in severity post–post exposure). With higher levels, the diver becomes progressively confused, eventually losing consciousness. Eliminating the CO_2 buildup through ventilation or surfacing from the dive (i.e., removal from the work site) is the treatment of choice. Symptoms resolve rapidly after removal of exposure, with the possible exception of the headache.

CO_2 retention reduces the divers' exercise tolerance, increases the manifestations of N_2 narcosis,[18] and may increase the risk of oxygen toxicity and decompression illness (since cerebral vasodilation facilitates on-gassing).

Carbon monoxide toxicity

The principle source of carbon monoxide (CO) in diving is contaminated breathing gas. For example, CO can be drawn into the compressor intake (e.g., from an idling truck in the vicinity of an intake) or via improper compressor maintenance (e.g., use of flashable lubricants or a failure of compressor rings). OSHA requires that breathing gas be tested every 6 months and as needed to prevent unrecognized elevations of CO, oil mist, and CO_2.[19] The effects of CO may be masked in the diver while breathing contaminated gas because of the high partial pressure of oxygen in the breathing media. Unless the CO concentration is high, the symptoms of CO exposure will be delayed until the partial pressure of oxygen decreases—that is, during ascent/decompression or after surfacing. However, if there is not a high degree of suspicion, manifestations may be confused with other diving-related disorders. In addition, the symptoms associated with carbon monoxide poisoning such as headaches, dizziness, confusion, disorientation, nausea, vomiting, weakness, visual disturbances, loss of consciousness, cardiac arrhythmia, and infarction may be attributed to other diseases rather than effects of carbon monoxide.[13]

OTHER HAZARDS IN THE DIVING ENVIRONMENT

The elements of the professional diving environment are varied and diverse, but its hazards are similar to the occupational hazards commonly encountered on land. These hazards are summarized in Table 8.4, and most of them are covered elsewhere in the book.

In addition, it should be remembered that many divers do not dive full time. They often have primary jobs that expose

TABLE 8.4 General industrial hazards that may be encountered in the underwater work environment.

- **Manual materials handling**. While underwater, the diver is frequently involved in moving heavy loads. The normal complexity of safely handling materials is compounded, even when buoyancy aids and hoists are used. Lifting characteristics of materials are affected by their buoyancy. In addition, the diver's protective equipment is designed to be negatively buoyant in the water and is extremely heavy when carried out of the water by the diver. As a result, undue stress is often placed on the diver's shoulders and back by poor ergonomic design
- **Welding, cutting, and brazing** are frequently performed underwater, requiring both standard and environment-specific precautions. Underwater explosions secondary to buildup of explosive gases during these operations may occur
- **Electrical hazards** may be compounded by the aquatic environment
- **Ionizing radiation**. Divers should be issued personal dosimetric devices when working around ionizing radiation sources
- **Underwater blasts**. Blast injuries may be compounded by the effects of pressure wave propagation
- **Vibration**. Dives are frequently conducted in cold water, potentially compounding the effects of vibration
- **Marine animal hazards** are frequently encountered in the work area
- **Chemical exposures**, for example, to hydrogen sulfides in sunken ships, may occur in enclosed spaces underwater. Many foams used in salvage contain isocyanates
- **Pollution**. Many divers routinely work in contaminated waters. Therefore, meticulous ear and skin care is important. Preventive measures, such as vaccinations, may also be recommended in some circumstances

them to more traditional hazards, such as welding, degreasing, paint removal, or other chemical, physical, or biological hazards.

PATHOPHYSIOLOGY OF DIRECT PRESSURE INJURY

The health effects of direct pressure injury can be divided into two major groups: barotrauma and decompression illness. Barotrauma results from the expansion or contraction of gases in anatomic spaces, causing trauma. Decompression illness results when bubbles of inert gas form in body tissues. Body fluids become supersaturated with inert gases at high pressures, and this gas comes out of solution when the ambient pressure is decreased. A summary of direct pressure effects can be found in Table 8.5.

Barotrauma

Barotrauma may occur in any gas-filled space in the body. To sustain barotrauma, commonly referred to as a "squeeze," a pressure change must occur in an enclosed gas-filled space in that consequences of this pressure change will depend on the location of the space, the magnitude of the pressure change, and the physiologic response mounted. Common locations for barotrauma include the ear (middle most commonly, outer and inner depending on circumstances), sinuses, teeth, gastrointestinal tract, and the lung. In addition, the skin may be traumatized by pressure changes in gas pockets found in a suit or under a face mask.

TABLE 8.5 Human health effects of environmental pressure change.

Barotrauma site
Ear
Sinus
Lung
Skin
Teeth
Gastrointestinal tract
Manifestations of decompression illness
Pain
Neurologic effects
Pulmonary effects
Cutaneous effects
Lymphatic effects
Constitutional effects

Middle-ear squeeze is the most common form of barotrauma seen in diving. It occurs when the middle-ear space is not vented properly by the Eustachian tube. When an individual descends in a pressure column, whether to a dive site or in an airplane, the gas within the middle ear compresses in accordance with Boyle's law. To counterbalance this effect, the individual must introduce air into this space via the Eustachian tube, which joins the middle ear to the pharynx. This process, known as clearing the ears, may be accomplished by various methods such as yawning, swallowing, moving the jaw around, or a Valsalva maneuver. The ears will not clear if the Eustachian tube is blocked.

If descent continues without clearing, the diver will initially experience a sense of fullness and pressure, followed by sharp pain.[20] If descent is not stopped, the eardrum may rupture, allowing immediate equalization of pressure difference, or the middle-ear space may fill with blood, a noncompressible fluid to equalize the pressure. Middle-ear barotrauma occurs most frequently within the first 10–20 ft of descent, which is the period of greatest proportional pressure change. Following the dive, residual symptoms may include pain, a sensation of fullness in the ear, or a temporary, mild conductive hearing loss across all frequencies, generally <20 dB (or greater if an ossicular rupture has occurred).[11] In a small proportion of cases, blood may be visible in the mouth or nose. Treatment depends on the amount of damage sustained. It may range from a mild squeeze, requiring no diving for 48–72 hours, to severe barotrauma, requiring diving restrictions up to 6 weeks or longer.

To prevent this form of barotrauma, an individual should not dive when the ears do not clear properly, such as during periods of significant upper respiratory congestion. The absence of predive symptoms does not guarantee adequate Eustachian tube function. In addition, descent should be stopped at the first sign of difficulty in equalizing (clearing ears).

Sinus barotrauma may occur during descent when the openings that vent the sinuses into the nasal cavity are obstructed. It presents as increasing pain over the effected sinus(es). On descent, if the sinus does not properly equalize, pressure in the sinuses decreases relative to ambient pressure. As a result, edema and hemorrhage of the mucosal lining of the sinus may occur. If a sinus opening subsequently becomes blocked (secondary to edema/hemorrhage) during a dive, sinus barotrauma may present during the ascent phase of a dive. In such cases, the sinus pain is caused by a relative increase in pressure within the sinus. This pain may continue for a number of hours after the dive. Relief may be accompanied by a discharge and often a high-pitched sound as gas leaves the sinus.

Pulmonary barotrauma is a very serious form of this disorder. Gas present in the lung expands during ascent. If the lung is allowed to overpressurize by as little as 90–110 cm H_2O (1 m of seawater (msw), or 3 ft), the lung may rupture.[21,22] This displaced gas causes a variety of sequelae that present individually or in combination. These include mediastinal/subcutaneous emphysema, pneumothorax, and arterial gas embolism.

After rupture, gas may migrate along the bronchial tree to the mediastinum. The result is mediastinal emphysema, which may remain asymptomatic or present as substernal burning chest pain. Mediastinal emphysema is thought to be the most common manifestation of pulmonary barotrauma. From the mediastinum, extrapulmonary gas may track into the neck, presenting as subcutaneous emphysema or occasionally as a voice change secondary to pressure directly on the larynx or the recurrent laryngeal nerve. Alternatively, the gas may be driven into the retroperitoneal region. The rupture may expel gas into the intrapleural space, causing a pneumothorax.[23] If this occurs at depth, the damage will be further exacerbated by ascent. Finally, the overpressurization may force gas into the pulmonary veins, causing an arterial gas embolism that presents as neurologic sequelae. Any type of cerebral neurologic manifestation is possible secondary to arterial embolization; symptoms range from subtle neurologic findings to hemiplegia, convulsions, coma, and death.

Any or all of these sequelae may occur simultaneously. Therefore, it is critical to perform a complete neurologic examination to rule out arterial embolization whenever pulmonary overinflation is suspected or detected. Individuals are at an increased risk of lung rupture during (i) diver training courses, particularly during underwater removal and donning of gear (ditch and don); (ii) buoyant ascent training (diver may not exhale completely due to loss of buoyancy); (iii) diving with predisposing lung pathology that impedes gas flow; and (iv) panic/emergency/uncontrolled ascents. It is important to remember that lung rupture can occur after surfacing from a compressed gas dive from as little as 3–4 ft of water (1 msw), even when normal procedures are followed. This should not happen in breath-hold diving, since the volume of gas present on ascent will not exceed the original breath taken on the surface.

Barotrauma may also occur in a tooth, causing implosion or expulsion of amalgam/dental material. Gas can also expand in the gastrointestinal tract; however, this rarely causes more than mild discomfort, unless there is a hernia present. Trapped gas within an intestinal loop may cause incarceration.

Decompression illness

In accordance with Henry's law, hyperbaric workers will take up or onload gas while breathing compressed gas at elevated pressures. Subsequently, when the diver ascends in the water column or the caisson worker leaves the work site, inert gas already present in the body expands and must be off-gassed. Decompression tables have been developed that provide rules for both ascent rates and stops in the pressure water column, thus allowing time to asymptomatically off-gas the inert components on ascent. Rigorous adherence to these tables is a critical component to prevent decompression illness. The tables and diving computers used today are based principally on perfusion-limited theories (Haldanian principles), which have been modified as needed to reflect human experience. However, even when the tables are rigorously adhered to, decompression illness (DCI) can still occur. Decompression illness is obviously more common if the tables are disregarded. Examples of commonly used tables in the United States today include the US Navy standard air tables[24] and the DCIEM air tables.[25] These tables are widely available in the sports and commercial diving community. There is an average predicted range of incidence of DCI on commonly used tables of <1 to 5%, but specific dive profiles within each table vary greatly.[26,27]

The overall incidence of DCI is virtually impossible to measure,[28] since records of the total numbers of divers (the denominator) are not routinely maintained. It has been estimated that the operational incidence in US Navy divers is <5 cases per 10 000 dives (0.05%) and the incidence within the US sports diving community has been estimated at <1 per 10 000.[29–31] US Navy analysis of dives not requiring decompression completed between 21 and 55 fsw from 1991 to 1994 showed an incidence of 2.9 cases/10 000 dives (0.029%), with the incidence increasing with depth.[32] Shields et al. have documented an overall incidence of decompression sickness in the commercial UK sector of the North Sea over the period 1982–1986 of 0.31% and an incidence of 0.10% after the implementation of depth–time restrictions.[33]

Luby reported an incidence of approximately 4.0 cases/10 000 dives (0.04%) shallower than 100 ft and 10 cases/10 000 dives (0.1%) deeper than 100 ft in commercial diving in the Middle East.[34]

During and after ascent, bubbles may form in tissues. If bubbles form, they may either remain asymptomatic or cause clinically apparent damage. It is generally believed that bubbles form in the tissues, causing damage locally (autochthonous bubbles),[35] or that they are distributed widely by the venous system, principally filtered at the lungs. However, this filter may become overloaded with large amounts of venous gas, causing the filter to leak, particularly with an increased load. Alternatively, venous bubbles may traverse a patent foramen ovale, atrial septal defect, or even an extracardiac shunt. Once in the arterial system, the bubbles probably distribute commensurate with the organ's blood flow, as with the bubbles associated with pulmonary barotrauma.[36]

Generally, any new pain or neurologic manifestation presenting shortly after a hyperbaric exposure must be considered DCI until ruled out. The range of manifestations of bubble disease is seemingly infinite, but the manifestations may be grouped into the following categories: pain, neurologic, pulmonary, cutaneous, lymphatic, and constitutional.[37] Any combination of manifestations from these categories may be present. Limb pain is believed to be the most common manifestation.[38–40]

DCI pain generally presents as a deep, toothache-like periarticular pain that is not affected by movement. Girdle pain, a distinct DCI pain syndrome, is characterized by a poorly localized, constricting sensation radiating from the back and often heralds the onset of severe neurologic manifestations. A wide variety of neurologic manifestations may be seen, including alterations in consciousness, higher-function abnormalities, derangements in sensory modalities, strength deficits, problems with special senses (audiovestibular particularly, including vertigo, tinnitus, and hearing loss), and loss of sphincter control (particularly bladder function).[37] Cutaneous presentations (cutis marmorata) generally start with intense itching, most commonly on the torso, which progresses to an erythematous rash and may continue on to cyanotic marbling (mottling). Lymphatic disease presents as a painful swelling of an individual lymph node or group of lymph nodes, which on rare occasions may be accompanied by swelling/edema, presumably due to obstruction. Pulmonary manifestations (chokes) include shortness of breath, cough, chest pain, or cyanosis. Most of these pulmonary manifestations are quite rare unless substantial decompression has been omitted. Constitutional symptoms, including fatigue, nausea, and anorexia, may accompany any of these manifestations.

A descriptive system of nomenclature for DCI is presented in Table 8.6. The evolutionary and clinical manifestation terms are used to form the label—for example, acute progressive neurologic DCI or acute static neurologic and limb pain DCI. "Acute" is used to discriminate the case from potential long-term health effects related to decompression.[37]

Traditionally, manifestations of arterial gas embolism have been distinguished from those of DCI, despite probable overlapping pathophysiology. As a result of experience in caring for the caisson workers digging the Dartford tunnel below the Thames River in London, Golding et al. further divided DCI into two categories based on presumed severity and disease location.[41] Pain, cutaneous effects, and lymphatic manifestations were designated as Type I decompression sickness, whereas neurologic and pulmonary manifestations alone or with any other combination of manifestations were called type Type II decompression sickness. Type II was believed to represent serious disease. This classification quickly became the standard, based on 35 cases of Type II disease. It is still in frequent use today. However, this system has been demonstrated to give inconsistent diagnoses.[42–44] Furthermore, as can be seen from their clinical description and our present concepts of the pathophysiology, cerebral arterial gas emboli (CAGE) and DCI probably cannot be distinguished except in a few isolated circumstances.[37]

DCI can begin on ascent or after surfacing. When Francis et al. reviewed 1070 well-documented cases of neurologic DCI, excluding all cases with histories thought to predispose to arterial gas embolism, they found that 50% presented within 10 minutes of surfacing, >85% of cases presented within 1 hour of surfacing, and >95% presented within 6 hours of surfacing.[45]

However, symptoms attributable to DCI may present at 24–48 hours or more after a dive.[43] The classical arterial gas embolic mechanism secondary to barotrauma may occur as stated above when compressed gas is inhaled deeper than 3 ft and generally presents on surfacing or quickly thereafter. The inert gas mechanism, on the other hand, requires the diver to stay for a minimum time at depth to acquire an adequate gas burden. With normal diving, depths >33 fsw are generally required.[46]

A number of proposed predisposing factors for decompression illness have been identified in addition to the dive profile. They include individual susceptibility[26] (which is most likely multifactorial), patent foramen ovale,[47–49] atrial septal defect, obesity,[50–52] exercise at depth and during decompression,[26,53] dehydration,[54] and low ambient air temperature/wind chill,[55] hot water suits, hot showers in cold divers, oral contraceptives, rapid ascent, heavy work and exertion, residual deficit from previous DCI, and obstructive lung disease.[56] Data are lacking to translate these associations, such as obesity and patent foramen ovale, into specific individual recommendations. Patent foramen ovale, for example, has a high prevalence in normal populations, including divers, despite a low incidence of resulting DCI. Therefore, most experts do not recommend screening divers for a patent foramen ovale unless they have experienced

TABLE 8.6 A matrix for describing decompression illness.

Acute decompression illness
The five following terms are used to describe a case of DCI adequately: 1. Evolution. Used to summarize the development of a case from onset to the present moment, prior to recompression. Terms used to describe the evolution include: Progressive Static Spontaneously improving Relapsing 2. Manifestation. Used to describe the organ systems or parts of the body that are affected. Unlike "presenting" symptoms, which describe the initial manifestation of a medical condition, these terms are used to describe the disease complex at the time of the report. Manifestation terms include: • Pain Limb pain Girdle pain • Cutaneous • Neurologic Audiovestibular • Pulmonary • Lymphatic • Constitutional 3. Time of onset. For each manifestation, record the time that has elapsed between surfacing from the dive and the onset of each principal manifestation. 4. Gas burden. An estimate of residual inert gas load present after a pressure exposure; gives some indication of the "sensitivity" of the exposure until a standard is established. An accurate record of the dive profile is probably the most useful measure at present. 5. Evidence of barotrauma. Clinical or radiographic evidence of barotrauma should be documented. When barotrauma cannot be diagnosed definitively, DCI terminology should be used; otherwise, the barotrau-mata are diagnosed as before.
Format for decompression illness "label" Since lengthy descriptions are unwieldy for communication purposes, a specific abbreviated "label" is needed for each case.
The general form of the proposed label is as follows: Acute [Evolution term], [Manifestation term(s)], decompression illness (see text).

Source: Adapted from Francis TJR, Smith DJ, Sykes JJW. The prevention and management of diving accidents. INM Technical Report R92004.[20] Reprinted with permission.

recurrent episodes of "unexplained" neurologic DCI.[57] In addition, age, poor physical fitness, recent tissue injury, previous DCI, and dehydration have been suggested as predisposing factors; however, little epidemiologic evidence exists to support these suggestions, and the studies are conflicting.

Diving while pregnant is a controversial issue, due to the paucity of available data, which are also conflicting. However, most hyperbaric authorities agree that a woman should not dive while pregnant.[58–61] Safe depth–time profiles have not been established.[62] The fetus may be susceptible to intravascular bubble formation, and these bubbles may have a deleterious effect on the nervous system of the fetus with a patent foramen ovale and ductus arteriosus. Moreover, there is concern about the potential effects of hypoxia during an unanticipated emergency while diving. The military and most commercial diving operations prohibit women from diving while pregnant.

Unique problems of saturation diving

Saturation diving techniques represent a cost-effective alternative to standard diving when deep or prolonged bottom times are required, since the need for frequent and long in-water decompressions is avoided. The worker essentially completes one decompression after a prolonged exposure. However, at greater depths, unique medical problems arise.

Rapid compression at deep depths (>150m) is associated with high-pressure nervous syndrome (HPNS) and compression arthralgia. The symptoms of HPNS frequently include a 5–8-Hz tremor, dizziness, nausea and associated vomiting, decreased mental alertness, and microsleep.[63] The manifestations are related to depth and rate of compression. Compression arthralgia comprises pains or ill-defined discomforts in joints on moving during and immediately after compression. The knees, hips, and wrists are most commonly involved.[11] To avoid both HPNS and compression arthralgia, the compression rates are slowed in comparison to conventional diving, and nitrogen is added to the helium and oxygen mixture (heliox) to produce trimix (helium, oxygen, and nitrogen).[63]

Maintenance of thermal balance is a significant problem at depth. At 300 m, the comfort range varies by less than 2°C. Additionally, while the diver is lying down, the exposed surfaces may become cold secondary to thermal conduction, and the surfaces in contact with the mattress may become too warm. While the diver is actually working in the water, thermal balance is maintained with hot water suits, and the breathing media are heated.

Because the saturation environment is humid and warm, pathogens grow well. Meticulous housekeeping is required to keep all divers within this closed community healthy. Otitis externa has been a particular problem, forcing the early completion of some dives. A preventive regimen using 2% acetic acid in aluminum acetate eardrops has been developed; this regimen has effectively controlled the problem in most situations.[64]

Communication is extremely difficult in a helium environment. Electronic "unscramblers" are mandatory for adequate communications between the diver and support personnel. All materials are transferred to depth and back via "medical locks," which are small pressure-transfer chambers with doors on the outside and inside of the chamber. These locks can be pressurized to allow transfer of food, medical supplies, mail, and any important personal objects. Greater depths degrade the taste of food, requiring increased flavoring and spices. The heat of compression must be anticipated, because the additional heat generated will further cook food on descent.

Once they enter the chamber, the physician no longer has ready access to the divers; therefore, medical screening for deep saturation diving must be rigorous. The lack of access is due to the slow compression times required and subsequent decompression obligations incurred by the medical attendant, making access unfeasible. Depending on the depth of storage of the saturation system, decompression obligations for saturation divers may take up to a week or more. Predive physicals should pay particular attention to the ears, skin, and respiratory tract. To be better prepared for emergencies, divers should be trained in various first aid and life-support techniques. It is standard practice in commercial diving for one or more of the saturation dive team members to be trained as a diving medical technician (DMT).[65]

LONG-TERM HEALTH EFFECTS

Dysbaric osteonecrosis, or aseptic bone necrosis, is a well-recognized, relatively uncommon long-term occupational hazard associated with compressed air work. The bone necrosis lesions occur principally in the femur, tibia, and humerus.[66] As a rule, shaft lesions and lesions that occur away from articular joints do not produce clinical symptoms. Juxta-articular lesions, on the other hand, can progress and produce debilitating disease. Juxta-articular lesions are more commonly found in compressed air workers than in divers. Necrosis is associated with increasing depth and duration of exposure, increasing age (although age may only be a surrogate measure of exposure), and a history of DCI, though not related to the site of DCI.[66] The lesions of dysbaric osteonecrosis are indistinguishable from other causes of aseptic necrosis. However, as McCallum and Harrison note, "in men with a history of work in compressed air or diving, the probability of bone necrosis being due to compressed air is very high".[66] The pathogenesis of the dysbaric form is not well understood. An excellent review of dysbaric osteonecrosis was completed by Jones and Neuman.[66]

Divers are at risk for sensorineural hearing loss secondary due to barotrauma, DCI, and potentially long-term noise exposure. Conductive hearing loss is also a possibility with middle-ear barotrauma and exposure to underwater explosions. Although the hearing threshold is increased underwater, a number of studies have shown that measured sound intensity levels of some diving helmets when pressurized are as high as 90–120 dB(A), depending on the application.[67,68] In addition, hyperbaric chambers during compression and some chambers during ventilation have high noise levels. Some investigators have documented standard threshold shifts with "normal" dive profiles.[68–71] The principal cause of the high noise level is the volume of gas flow required to sufficiently ventilate helmets to prevent CO_2 buildup. The frequency range of noise is 800–3000 Hz, which is within normal communication ranges. The redesign has significantly improved the noise levels of modern helmets.[69] Some epidemiologic studies have shown little difference in hearing acuity between professional divers and matched controls when corrected for age,[70] whereas others[71–74] have noted significant differences in divers. One recent age-adjusted prospective study in SCUBA divers demonstrated increased hearing loss in low frequencies only, suggesting a compression/decompression effect.[74] Although some of the high-frequency hearing loss seen in divers is probably due to other exposures, the studies imply that some of the losses may be due to workplace noise exposure.[75] Therefore, engineering controls and hearing protection should be instituted wherever possible.

Sequelae from episodes of acute DCI, particularly neurologic ones, are well recognized. However, there may be less obvious but potentially serious long-term health effects from diving, based primarily on descriptive and anecdotal evidence. This possibility has generated a number of hypotheses, which currently lack good epidemiologic support. Subsequent studies have frequently not supported the hypothesized findings.

FIGURE 8.9 Hyperbaric recompression chamber (Courtesy of U.S. Navy).

These effects include neuropsychiatric deficits, such as short-term memory deficits and emotional lability,[76–79] electroencephalographic abnormalities[80,81] (principally slow waves and spikes), and retinoangiography aberrations, including pigment and capillary changes in the retina.[82] Pulmonary function changes have been documented in groups of divers; these include increased vital capacities, which may be adaptive.[83] The clinical relevance and validity of these findings are still being investigated. An excellent review of the relevant literature can be found in Elliott and Moon.[84] This review concludes that "in the absence of a history of acute decompression illness, the possibility of a clinical syndrome among divers or ex-divers remains unproven. If it exists, the prevalence is unknown, and probably low."

TREATMENT OF BUBBLE-RELATED DISEASE

Because symptoms and signs can progress, a diagnosis of DCI should be treated urgently. The patient should be placed on oxygen at as high a partial pressure as reasonably feasible. Additionally, fluids (generally oral, but intravenous or intraosseous fluids may be administered as appropriate for the condition of the patient) are pushed, and the individual is placed in a recumbent position. A hyperbaric chamber should be found and the patient transferred for recompression therapy (Figure 8.9). An excellent source of emergency information and referral is the Divers Alert Network.[#]

[#]The Peter B. Bennett Center, 6 West Colony Place, Durham, NC 27705; Telephone: (919) 684-8111; http://www.diversalertnetwork.org

Many recompression tables exist; however, the US Navy Treatment Table 6 (Figure 8.10 and Table 8.7) is the principal therapeutic table used for the treatment of decompression illness. The patient is given oxygen at depth (60 fsw) initially, with intermittent air breaks to reduce the incidence of oxygen toxicity. Response to therapy is generally excellent if recompression therapy is instituted promptly. Depending on the response to therapy, table modifications and follow-on hyperbaric treatments may be required if manifestations do not resolve or recur. Deeper tables include the Comex 30 Treatment Table using 50% oxygen and nitrogen mixture (100 fsw), US Navy Treatment Table 6A using 50% oxygen and nitrogen mixture (165 fsw), and the Lamberston/SOSI Treatment Table 7A to depths greater than 165 fsw. Rarely saturation tables such as US Navy Treatment Table 7 will be used if the diver does not respond to treatment based on the other treatment tables.

MEDICAL SURVEILLANCE

Healthcare personnel must understand that few other workers experience the magnitude of physiologic stresses imposed routinely on divers and caisson workers. Medical surveillance requires a thorough understanding of the physiologic aspects of diving, along with the specific workplace hazards that may be encountered by the worker. In general, different classes of workers are exposed to varying levels or types of hazards, requiring a tailored approach. Most professional divers, including military, commercial, and scientific divers, as well as caisson workers and hyperbaric chamber attendants are provided with specific guidance and standards by their employer. For example, the Association of Diving Contractors International periodically publishes

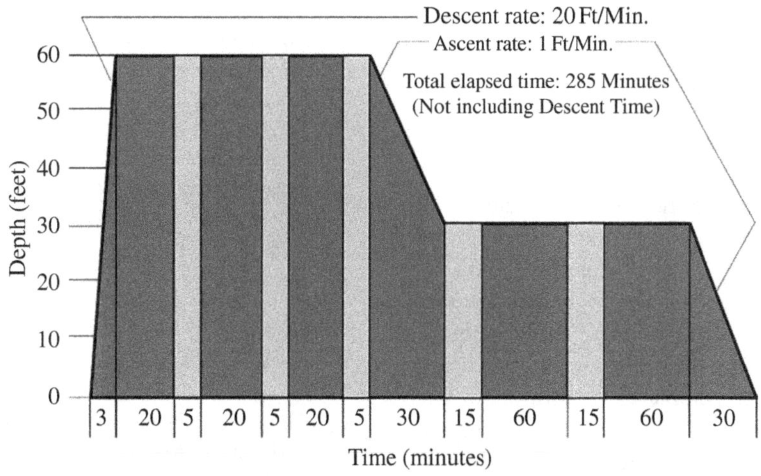

FIGURE 8.10 U.S. Navy Treatment Table 6. *Source*: U.S. Navy Diving Manual, Chapter 21. NAVSEA 0910-LP-0708-8000, Washington, DC: U.S. Government Printing Office, 1999.

TABLE 8.7 US Navy Treatment Table 6: Oxygen Treatment of Type II Decompression Sickness[a,b].

Depth (ft)[c]	Time (min)	Breathing media[d]	Total elapsed time (h:min)
60	20	O_2[e]	0:20[f]
60	5	Air	0:25
60	20	O_2	0:45
60	5	Air	0:50
60	20	O_2	1:10
60	5	Air	1:15
60 to 30	30	O_2	1:45
0	15	Air	2:00
30	60	O_2	3:00
30	15	Air	3:15
30	60	O_2	4:15
30 to 0	30	O_2	4:45

[a]Treatment of Type II or Type I DCI when symptoms are not relieved within 10 minutes at 60 ft.

[b]Extensions to Table 6: Table 6 can be lengthened up to two additional 25-minute oxygen-breathing periods at 60 ft (20 minutes on oxygen and 5 minutes on air) or up to two additional 75 minutes oxygen-breathing periods at 30 ft (15 minutes on air and 60 minutes on oxygen) or both. If Table 6 is extended only once at either 60 or 30 ft, the tender breathes oxygen during the ascent from 30 ft to the surface. If more than one extension is done, the care-giver begins oxygen breathing for the last hour at 30 ft during ascent to the surface.

[c]Descent rate—25 ft/min. Ascent rate—1 ft/min. Do not compensate for slower ascent rates. Compensate for faster rates by halting the ascent.

[d]Caregiver breathes air throughout unless he has had a hyperbaric exposure within the past 12 hours, in which case he breathes oxygen at 30 ft.

[e]If oxygen must be interrupted because of adverse reaction, allow 15 minutes after the reaction has entirely subsided and resume schedule at point of interruption.

[f]Time at 60 ft begins on arrival at 60 ft.

Source: Adapted from the US Navy Diving Manual, Revision 4, chapter 21.

medical requirements for their members. Title 29 CFR 1910 Subpart T Appendix A provides examples of conditions that restrict or limit exposure to hyperbaric conditions.[19] Other sources provide more detailed guidance.[57,85–89] Examination frequency varies depending on type of exposure and the regulations being followed. Reexamination must be completed after any significant illness or injury, particularly exposure-related injuries, diving with any neurological involvement, gas embolus, or pulmonary or ear barotrauma.

Healthcare providers who conduct medical surveillance on divers should have formal instruction in diving/hyperbaric medicine. This training enables them to correlate various medical conditions to the unique hyperbaric environment. One source for information on diving medicine courses, scientific meetings, general information on diving or hyperbaric medicine, or addresses of practitioners with an interest in diving or hyperbaric medicine is the Undersea and Hyperbaric Medical Society.[##]

Medical history

The medical history is of primary importance in hyperbaric medical surveillance. It must be remembered that in the underwater environment, any condition that incapacitates an individual, even temporarily (e.g., seizure, fainting), may cause drowning in addition to the standard sequelae. Moreover, divers for the most part rely on the buddy system, which means that the divers must be able to help their buddies when they are in distress and must not endanger their buddies secondary to their own medical condition.

[##]631 US Highway 1, Suite 307, North Palm Beach FL 33408, United States; telephone: 919-490-5149 or 877-533-8467; http://www.uhms.org

In addition, a careful occupational history—including type of dives, number of dives, maximum depth obtained, and any untoward events—is uniquely important.

Physical examination

A complete physical examination should be completed, with an emphasis on ear, nose, and throat, pulmonary, cardiovascular, skeletal, and neurologic systems. Caisson and diving workers require a significant amount of cardiovascular reserve and aerobic work capacity, along with adequate dexterity and strength. During each examination, a neurologic examination should be completed to, at a minimum, document preexisting deficits. A well-documented neurologic examination may prevent confusion during evaluation of symptoms and signs postdiving.

PREVENTION

The only certain method to prevent hyperbaric injuries is to simply avoid all high-pressure and diving work. Thanks to modern engineering techniques, the need for caisson work has been significantly decreased. However, preventing diving injuries by avoiding diving altogether is generally regarded as unfeasible. Therefore, the principal methods used to prevent diving injuries are (i) extensive training, including both an academic understanding of principles and job-specific practical training; (ii) dive planning, to include emergency procedures and appropriate use of tables; (iii) maintenance of a high level of fitness; (iv) meticulous care of equipment, along with ongoing improvement in equipment design; and (v) a healthy respect for the indigenous hazards of the profession. These same principles also apply to caisson work. Current research is directed toward the refinement of decompression models and tables, improving equipment design, enhancing treatment methods, and understanding the pathophysiology and associated risk factors of diving disorders.

References

1. Elliott DH. Raised barometric pressure. In: Baxter PJ, Adams PH, Aw T-C, Cockcroft A, Harrington JM, eds. Hunter's diseases of the occupations 9th ed. London: Arnold, 2000:343–60.
2. Bachrach AJ. A short history of man in the sea. In: Bennett PB, Elliott DH, eds. The physiology and medicine of diving, 3rd edn. San Pedro, CA: Best Publishing, 1982:1–14.
3. Kindwall EP. A short history of diving and diving medicine. In: Bove AA ed. Bove and Davis' diving medicine, 4th edn. Philadelphia, PA: Elsevier, 2004:1–9.
4. Gal A. Des dangers du travail dans l'air comprimé et des moyens de les prévenir. In: Bert P, ed. Barometric pressure—researches in experimental physiology. Columbus, OH: College Book Company, 1943:398.
5. Butler FK Jr., Smith DJ. U.S. Navy diving equipment and techniques. In: Bove AA ed. Bove and Davis' diving medicine, 4th edn. Philadelphia, PA: Elsevier, 2004:547–71.
6. Kindwall EP. Compressed air work. In: Brubakk AO, Neuman TS, eds. Bennett and Elliott's physiology and medicine of diving, 5th edn. Philadelphia, PA: WB Saunders, 2003:17–28.
7. Kindwall EP. Compressed air tunneling and caisson work decompression procedures: development, problems, and solutions. *Undersea Hyperb Med* 1997; 24(4):337–45.
8. Taylor LH. Diving physics. In: Bove AA ed. Bove and Davis' diving medicine, 4th edn. Philadelphia, PA: Elsevier, 2004:11–35.
9. Bookspan J. Diving in cold and heat. In: Bookspan J ed. Diving physiology in plain English. Dunkirk, MD: Underwater and Hyperbaric Medical Society, 2006.
10. Farmer JC. Vestibular and auditory function. In: Shilling CXV, Carlston CB, Mathias RA, eds. The physician's guide to diving medicine. New York: Plenum, 1984:192–8.
11. Flynn ET, Catron PW, Bayne CG. Diving medical officer's student guide. Memphis, TN: Naval Technical Training Command, 1981.
12. Kinney JAS. Physical factors in underwater seeing. In: Shilling CXV, Carlston CB, Mathias RA, eds. The physician's guide to diving medicine. New York: Plenum, 1984:199–205.
13. Tipton MJ, Mekjavic IB, Golden FSC. Hypothermia. In: Bove AA ed. Bove and Davis' diving medicine, 4th edn. Philadelphia, PA: Elsevier, 2004:261–73.
14. Craig AB, Dvorak M. Thermal regulation during water immersion. *J Appl Physiol* 1966; 21:1577–85.
15. Bennett PB, Rostain JC. Inert gas narcosis. In: Brubakk AO, Neuman TS, eds. Bennett and Elliott's physiology and medicine of diving, 5th edn. Philadelphia, PA: WB Saunders, 2003:300–22.
16. Kiessling RJ, Maag CH. Performance impairment as a function of nitrogen narcosis. *J Appl Psychol* 1962; 46:91–5.
17. McMonnies CW. Hyperbaric oxygen therapy and the possibility of ocular complications or contraindications. *Clin Exp Optom* 2015; 98:122–5.
18. Hesser CM, Fagraeus L, Adolfson J. Roles of nitrogen, oxygen and carbon dioxide in compressed air narcosis. *Undersea Biomed Res* 1978; 5:391–400.
19. Code of Federal Regulations 29, part 1910, subpart T—Commercial diving operations. 1, Washington: US Government Printing Offices, 1992 July.
20. Francis TJR, Smith DJ, Sykes JJW. The prevention and management of diving accidents. INM Technical Report R92004. Alverstoke: Institute of Naval Medicine, 1992.
21. Malhotra MC, Wright HC. The effects of a raised intrapulmonary pressure on the lungs of fresh unchilled cadavers. *J Pathol Bacteriol* 1961; 82:198–202.
22. Vann RD, Butler FK, Mitchell SJ, et al. Decompression illness. *Lancet* 2011; 377:153–64.
23. Broome JR, Smith DJ. Pneumothorax as a complication of recompression therapy for cerebral arterial gas embolism. *Undersea Biomed Res* 1992; 19:447–55.
24. US Navy Diving Manual; rev. 6. U. S. Department of the Navy, Naval Sea Systems Command 2008. Available at: http://www.navsea.navy.mil/Portals/103/Documents/SUPSALV/Diving/Dive%20Manual%20Rev%206%20with%20Chg%20A.pdf?ver=2016-02-26-123349-523 (accessed on September 3, 2016).

25. Lauckner GR, Nishi RY. Decompression tables and procedures for compressed air diving based on the DCIEM 1983 decompression model. No. 84-R-74. Toronto: DCIEM, 1984.
26. Vann RD, Thalmann ED. Decompression physiology and practice. In: Bennett PB, Elliott DH, eds. The physiology and medicine of diving, 4th edn. Philadelphia, PA: WB Saunders, 1993:376–432.
27. Weathersby PK, Survanshi SS, Homer LD, et al. Statistically based decompression tables. I. Analysis of standard air dives: 1950–1970. NMRI Report 85-16. Bethesda, MD: Naval Medical Research Institute, 1985.
28. Sykes JJW. Is the pattern of acute decompression sickness changing? *J R Nav Med Serv* 1989; 75:69–73.
29. Dembert ML. Individual factors affecting decompression sickness. In: Vann RD, ed. The physiological basis of decompression. Proceedings of the 38th Undersea and Hyperbaric Medical Society Workshop, Duke University Medical Center, Durham, NC, June. Bethesda, MD: Undersea and Hyperbaric Medical Society, 1989:355–67.
30. Wilmshurst P, Allen C, Parish T. Incidence of decompression illness in amateur SCUBA divers. *Health Trends* 1994–1995; 26(4):116–8.
31. Arness MK. Scuba decompression illness and diving fatalities in an overseas military community. *Aviat Space Environ Med* 1997; 68:325–33.
32. Flynn ET, Parker EC, Ball R. Risk of decompression sickness in shallow no-stop air diving: an analysis of US Navy experience 1990–94. In: Proceedings of the 14th meeting of United States–Japan Cooperative Program in Natural Resources (UJNR), Panel on Diving Physiology, Panama City, FL, USA, September 16–17, 1997. Spring, MD: U.S. Department of Commerce, National Oceanic and Atmospheric Administration, National Undersea Research Program, 1998:23–38. Available at: https://searchworks.stanford.ed (accessed on June 28, 2016).
33. Shields TG, Duff PM, Wilcock SE, et al. Decompression sickness from commercial offshore air-diving operations on the UK continental shelf during 1982 to 1988. In: Subtech'89. Fitness for Purpose, Vol. 23. Amsterdam: Society for Underwater Technology, 1990:259–77. Available at: https://www.onepetro.org/conference-paper/SUT-AUTOE-v23-259?sort=&start=0&q=isbn%3A%28%220-7923-0742-9%22%29&fromSearchResults=true&rows=50# (accessed on June 28, 2016).
34. Luby J. A study of decompression sickness after commercial air diving in the northern Arabian gulf: 1993–95. *Occup Med* 1999; 49(5):279–83.
35. Francis TJR, Dutka AJ, Flynn ET. Experimental determination of latency, severity, and outcome in CNS decompression sickness. *Undersea Biomed Res* 1988; 15:419–27.
36. Francis TJR. A current view of the pathogenesis of spinal cord decompression sickness in a historical perspective. In: Vann RD, ed. The physiological basis of decompression. Proceedings of the 38th Undersea and Hyperbaric Medical Society Workshop, Duke University Medical Center, Durham, NC, June. Bethesda, MD: Undersea and Hyperbaric Medical Society, 1989:241–79.
37. Francis TJR, Smith DJ, eds. Describing decompression illness. Proceedings of the 42nd Undersea and Hyperbaric Medical Society Workshop, Institute of Naval Medicine, Alverstoke, Gosport, Hampshire, UK, October 9–10, 1990. Bethesda, MD: Undersea and Hyperbaric Medical Society, 1991.
38. Rivera JC. Decompression sickness among divers: an analysis of 935 cases. *Mil Med* 1964; 129:314–34.
39. Kelleher PC, Francis TJR, Smith DJ, et al. INM diving accident database: analysis of cases reported in 1991 and 1992. *Undersea Biomed Res* 1993; 20(suppl):13 (abstract).
40. Denoble P, Vann RD, de L Dear G. Describing decompression illness in recreational diving. *Undersea Biomed Res* 1993; 20(suppl):14 (abstract).
41. Golding FC, Griffiths P, Hemplemen HV, et al. Decompression sickness during the construction of the Dartford tunnel. *Br J Ind Med* 1960; 17:167–80.
42. Kemper GB, Stegmann BJ, Pilmanis AA. Inconsistent classification and treatment of type I/type II decompression sickness. *Aviat Space Environ Med* 1992; 63:153 (abstract).
43. Smith DJ, Francis TJR, Pethybridge RJ, et al. Concordance: a problem with the current classification of diving disorders. *Undersea Biomed Res* 1992; 19(suppl):47 (abstract).
44. Smith DJ, Francis TJR, Pethybridge RJ, et al. An evaluation of the classification of decompression disorders. *Undersea Hyperbaric Med* 1993; 20(suppl):11 (abstract).
45. Francis TJR, Pearson RR, Robertson AG, et al. Central nervous system decompression sickness: latency of 1070 human cases. *Undersea Biomed Res* 1988; 15:403–17.
46. Elliott DH, Moon RE. Manifestations of the decompression disorders. In: Bennett PB, Elliott DH, eds. The physiology and medicine of diving, 4th edn. Philadelphia, PA: WB Saunders, 1993:492.
47. Moon RE, Camporesi EM, Kisslo JA. Patent foramen ovale and decompression sickness in divers. *Lancet* 1989; I:513–4.
48. Wilmshurst P, Byrne JC, Webb-Peploe MM. Relation between interatrial shunts and decompression sickness in divers. *Lancet* 1989; II:1302–6.
49. Gernompré P, Dendale P, Unger P, et al. Patent foramen ovale and decompression sickness in sports divers. *J Appl Physiol* 1998; 84(5):1622–6.
50. Medical Research Council Decompression Central Registry, University of Newcastle-upon-Tyne. Decompression sickness and aseptic necrosis of bone. Investigations carried out during and after the construction of the Tyne Road Tunnel (1962–66). *Br J Ind Med* 1971; 28:1–21.
51. Lam TH, Yau KP. Analysis of some individual risk factors for decompression sickness in Hong Kong. *Undersea Biomed Res* 1989; 16:283–92.
52. Dembert ML, Jekel JF, Mooney LW. Health risk factors for DCS. *Undersea Biomed Res* 1984; 11:395–406.
53. Van der Aue OE, Kellar RJ, Brinton ES. The effect of exercise during decompression from increased barometric pressures on the incidence of decompression sickness on man. Report no. 8–49. Panama City, FL: United States Navy Experimental Diving Unit, 1949.
54. Suzuki N, Yagishita K, Togawa S, et al. Risk factors for decompression sickness. *Undersea Hyperb Med* 2014; 41(6):521–30.
55. Broome JR. Climatic and environmental factors in the aetiology of DCI in divers. *Undersea Biomed Res* 1992; 19(suppl):17 (abstract).
56. Moon RE. Treatment of decompression illness. In: Bove AA ed. Bove and Davis' diving medicine, 4th edn. Philadelphia, PA: Elsevier, 2004:195–223.

57. Elliott DH. Medical evaluation of working divers. In: Bove AA ed. Bove and Davis' diving medicine, 4th edn. Philadelphia, PA: Elsevier, 2004:533–45.
58. Fife WP, ed. Effects of diving on pregnancy. Proceedings of the 19th Undersea Medical Society Workshop. Bethesda, MD: Undersea Medical Society, 1978:15–9.
59. Fife WP, ed. Women in diving. Proceedings of the 35th Undersea and Hyperbaric Medical Society Workshop, Bethesda, MD, May 21–22. Bethesda, MD: Undersea and Hyperbaric Medical Society, 1986:3–10.
60. Vorosmarti J, ed. Fitness to dive. Proceedings of the 34th Undersea and Hyperbaric Medical Society Workshop, Bethesda, MD, May 15–16. Bethesda, MD: Undersea and Hyperbaric Medical Society, 1987:101–2. Available at: http://rubicon-foundation.org/uhms-workshops/ (accessed on June 28, 2016).
61. Hill RK. Pregnancy and travel. *JAMA* 1989; 262:498.
62. Cresswell JE, St Leger-Dowse M. Women and scuba diving. *Br Med J* 1991; 302:1590–1.
63. Bennett PB, Rostain JC. The high pressure nervous syndrome. In: Brubakk AO, Neuman TS, eds. Bennett and Elliott's physiology and medicine of diving, 5th edn. Philadelphia, PA: WB Saunders, 2003:323–57.
64. Thalmann ED. A prophylactic program for the prevention of otitis externa in saturation divers. Report no. 10-74. Washington, DC: Navy Experimental Diving Unit, 1974.
65. International Marine Contractors Association, International Code of Practice for Offshore Diving, 2007. International Consensus Standards for Commercial Diving and Underwater Operations, 6.1 edn. London: Association of Diving Contractors International, 2014.
66. Jones JP Jr., Neuman TS. Dysbaric osteonecrosis: aseptic necrosis of the bone. In: Brubakk AO, Neuman TS, eds. Bennett and Elliott's physiology and medicine of diving, 5th edn. Philadelphia, PA: WB Saunders, 2003:659–99.
67. Summitt JK, Reimers SD. Noise: a hazard to divers and hyperbaric chamber personnel. *Aerosp. Med* 1971; 42:1173–7.
68. Curley MD, Knafelc ME. Evaluation of noise within the MKl2 SSDS helmet and its effect on divers' hearing. *Undersea Biomed Res* 1987; 14:187–204.
69. Molvaer OI, Gjestland T. Hearing damage to divers operating noisy tools under water. *Scand J Work Environ Health* 1981; 7:263–70.
70. Brady JI, Summitt JK, Berghage TE. An audiometric survey of navy divers. *Undersea Biomed Res* 1976; 3:41–7.
71. Edmonds C. Hearing loss with frequent diving (deaf divers). *Undersea Biomed Res* 1985; 12:315–9.
72. Molvær OI, Lehmann EH. Hearing acuity in professional divers. *Undersea Biomed Res* 1985; 12:333–49.
73. Molvaer OI, Albrektsen G. Hearing deterioration in professional divers: an epidemiologic study. *Undersea Biomed Res* 1990; 17(23):1–46.
74. Haraguchi H, Ohgaki T, Okubo J, et al. Progressive sensorineural hearing impairment in professional fishery divers. *Ann Otol Rhinol Laryngol* 1999; 108:1165–9.
75. Molvær O. Otorhinolaryngological aspects of diving. In: Brubakk AO, Neuman TS, eds. Bennett and Elliott's physiology and medicine of diving, 5th edn. Philadelphia, PA: WB Saunders, 2003:227–64.
76. Rózsahegyi I. Late consequences of the neurological forms of decompression sickness. *Br J Ind Med* 1959; 16:311–7.
77. Edmonds C, Boughton J. Intellectual deterioration with excessive diving. *Undersea Biomed Res* 1985; 12:321–6.
78. Edmonds C, Hayward C. Intellectual impairment in diving: a review. In: Bove AA, Bachrach AJ, Greenbaum U, eds. Proceedings of the 9th International Symposium on Underwater and Hyperbaric Physiology, September 16–20, 1986. Kobe, Japan. Bethesda, MD: Undersea and Hyperbaric Medical Society, 1987:877–86.
79. Curley MD. US Navy saturation diving and diver neuropsychologic status. *Undersea Biomed Res* 1988; 15:39–50.
80. Todnem K, Nyland H, Skiedsvoll H, et al. Neurological long term consequences of deep diving. *Br J Ind Med* 1991; 48: 258–66.
81. Murrison AW. The contribution of neurophysiologic techniques to the investigation of diving-related illness. *Undersea Biomed Res* 1993; 20(4):347–73.
82. Polkinghome PJ, Sebmi K, Cross MR, et al. Ocular fundus lesions in divers. *Lancet* 1988; 2:1381–3.
83. Crosbie WA, Reed JW, Clarke MC. Function characteristics of the large lungs found in commercial divers. *J Appl Physiol* 1979; 46:639–45.
84. Elliott DH, Moon RE. Long-term health effects of diving. In: Bennett PB, Elliott DH, eds. The physiology and medicine of diving, 4th edn. Philadelphia, PA: WB Saunders, 1993: 585–604.
85. Davis JC, ed. Medical examination of sports scuba divers, 2nd edn. San Antonio, TX: Medical Seminars, 1986.
86. Bove AA. Fitness to dive. In: Brubakk AO, Neuman TS, eds. Bennett and Elliott's physiology and medicine of diving, 5th edn. Philadelphia, PA: WB Saunders, 2003:700–17.
87. Health and Safety Executive. The medical examination of divers (MA 1), revision 4. London: Health and Safety Executive, 2015:1–27. Available at: http://www.hse.gov.uk/pubns/ma1.pdf (accessed on May 2, 2016).
88. Elliott DH, ed. Medical assessment of fitness to dive. Proceedings of an International Conference, Edinburgh Conference Center, Edinburgh, UK, March 8–11, 1994. Ewell: Biomedical Seminars, 1995. Available at: https://catalog.hathitrust.org/Record/008331599 (accessed on June 28, 2016).
89. Wendling J, Elliott DH, Nome T, eds. Medical assessment of working divers. Hyperbaric Editions, CH 2501. Biel-Bienne, Switzerland, 2004.

Chapter 9

LOW-PRESSURE and HIGH-ALTITUDE ENVIRONMENTS

WORTHE S. HOLT*

The discussion of high-pressure environments in Chapter 8 noted that humans function well only within a narrow range of barometric pressures. Ascent to altitude places workers in an adverse environment and exposes them to multiple stressors—decreased barometric pressure, reduced oxygen levels (hypoxia), ionizing and nonionizing radiation, and low temperatures. Most workers are acclimated to sea-level or near-sea-level pressures; reduced barometric pressures and oxygen levels will produce a range of symptoms ranging from mild discomfort to severe disease or even death.

OCCUPATIONAL SETTING

It was estimated that more than 140 million people worldwide permanently resided above 2440 m (8000 ft). Tourism to mountainous regions of the Western United States exposed an estimated 35 000 000 people to the hypobaric environment.[1] Occupationally, pilots and flight attendants have the greatest potential exposure, although actual incidents in commercial aviation are infrequent due to cabin pressurization. Inside observers in hypobaric pressure chambers and scientists at research laboratories located at high altitudes routinely perform duties at decreased barometric pressures. Individuals who travel to mountainous regions as employees of the construction or travel industries are also at risk, depending on the altitude reached and the time taken to reach that altitude. Rescue workers and the military may also be at risk.

LOW-PRESSURE ENVIRONMENTS

Measurement issues

Barometric pressure is a measurement of the weight of the atmosphere at any given point. To ensure uniformity in the calibration of altimeters used in the aviation industry, the concept of standard atmosphere was universally accepted in the 1920s.[2] Standard atmosphere is defined as a sea-level pressure of 760 mmHg at a temperature of +15°C and a linear decrease in temperature as one ascends of 6.5°C/km (Table 9.1). Equivalent units of standard atmosphere include 1 atmosphere (atm), 29.92 inHg, 14.7 lb/in.^{2}, 760 Torr, and 1013.2 mbar. Atmospheric pressure is denser at the lower altitudes, as the weight of the atmosphere above compresses the air below. On a standard day, 5486 m (18 000 ft) marks the midpoint of atmospheric pressure.

Exposure guidelines

There are no formal guidelines for hypobaric exposures. As the agency responsible for aviation safety in the United States, the Federal Aviation Administration (FAA) establishes requirements for supplemental oxygen use in both pressurized and unpressurized aircraft as detailed in US Code of Federal Regulations (CFR), Title 14, Part 91.211.

* Glenn Merchant and Roy DeHart contributed to previous editions of this chapter.

Physical and Biological Hazards of the Workplace, Third Edition. Edited by Gregg M. Stave and Peter H. Wald.

TABLE 9.1 Altitude–pessure–temperature relationships.

Altitude (m)	Pressure		Temperature (°C)
	(mbar)	(Torr)	
Sea level	1013	760	15.00
100	1001	751	14.35
200	989	742	13.70
300	977	733	13.05
400	966	724	12.40
500	954	716	11.75
1000	898	674	8.50
2000	795	596	2.00
3000	701	525	−4.49
4000	616	462	−10.98
5000	540	405	−17.47
10000	264	198	−49.90
15000	121	90	−56.50
20000	55	41	−56.50
25000	25	19	−51.60
30000	11	8	−46.64
40000	2	2	−22.80
50000	0.8	0.6	−2.5

Physiology and the physics of gases

The behavior and impact of gases in the body are largely the result of the three well-described gas laws previously discussed (Chapter 8): Boyle's law, Dalton's law, and Henry's law. Boyle's law established the inverse relationship of the pressure and volume of a gas in a closed system at a constant temperature. Pressure reduction results in directly proportional expansion of a fixed mass of a gas. Dalton's law, or the law of partial pressures, states that each gas in a mixture exerts pressure independent of the other gases present. Approximately 80% of the Earth's atmosphere is composed of nitrogen. Oxygen fills the remaining 20%. Finally, Henry's law describes the behavior of gases dissolved in a liquid under pressure. If the pressure is increased, the amount of gas dissolved in a liquid will increase; conversely, if the pressure acting on the liquid's surface is reduced, gases will exit the solution.

Pathophysiology, diagnosis, and treatment

Physiologically, hypoxia is the greatest threat to workers' survival in low-pressure environments. Hypoxia is covered on its own in the next section of this chapter. This section will focus on the deleterious behavior of gases contained in the body during exposure to reduced pressures.

Barotrauma (trapped gases)

As an individual ascends to altitude, gases present in the body respond according to Boyle's law, that is, gases expand inversely to the pressure acting on them. The middle ear, lungs, gastrointestinal tract, and paranasal sinuses are gas-containing cavities and normally vent expanding volumes through physiologic openings such as the Eustachian tubes, mouth or rectum, or paranasal ostia. Preexisting disease may interfere with the passage of expanding gases, creating a trapped gas syndrome. Trapped gases produce a range of symptoms, from mild discomfort to pain of such intensity as to interfere with job performance. Workers usually experience trapped gas symptoms on ascent. Exceptions include ear or sinus blocks, which occur on descent. Gas in the middle ear normally vents through the Eustachian tube on ascent. The same is true for air found in the paranasal sinuses, even in the presence of preexisting disease, for example, an upper respiratory infection. On descent, the volume in the middle-ear space or sinus is recompressed, and unless one is able to equalize the space to ambient conditions, barotrauma may result from the negative pressure in the middle ear/sinus, with actual tearing of the mucosal lining.

Diagnosis Most trapped gas symptoms appearing during routine flight resolve on return to the surface, with the exception of ear or sinus blocks associated with descent. Patients presenting with flight-related symptoms should be carefully evaluated for preexisting disease, especially those of the upper respiratory tract. In the case of descent-related ear block, direct visualization of the ear will reveal a retracted tympanic membrane with increased vascular marking in mild cases and middle-ear effusion, hemotympanum, or perforation in more severe cases. Patients experiencing sinus barotrauma will typically complain of sharp localized pain over the affected region, 80% of the time involving the frontal sinuses. Additional findings include epistaxis in 15% of patients.[3] A sinus series may reveal clouding or fluid levels in the affected sinus.

Treatment Symptomatic treatment of persistent symptoms includes temporary removal from flying duties, oral or nasal decongestants, and analgesics. Otitic barotrauma symptoms usually resolve in a matter of days. Barosinusitis sequelae may persist for weeks. As it is impossible to directly measure the ability to equilibrate sinuses, an acceptable indirect measure of normal function is when the patient can comfortably equalize the ears through the Valsalva maneuver (forced expiration with the lips closed and the nostrils compressed). Until such time, workers should be removed from flying duties.

Aviation decompression illness

The mechanisms of altitude- and diving-related decompression illness are identical—nitrogen present in the tissues leaves solution as the ambient pressure acting on the body is reduced on ascent. Decompression illness is more common in aviators operating unpressurized aircraft above 5500 m, although Voge reported a case occurring at 4268 m (14 000 ft).[4] Commercial aircraft protect occupants from decompression illness by maintaining cabin pressures at or below 2400 m (8000 ft). Risk factors for aviation-related decompression illness include altitude, duration and rate of exposure, physical exertion, low temperatures, age greater than 40 years, female gender, and recent exposure to increased pressure, for example, SCUBA diving.[5,6] Neurological decompression sickness has been documented in the high-altitude aviation environment and effectively addressed by reducing the frequency and duration of exposure as well as increasing the pressure differential in the aircraft. A safe interval between diving and flying may be as short as 2 hours or as long as 48 hours, depending on the number of dives and whether decompression was required. Divers should consult the Navy Dive Tables (discussed in Chapter 8).

RAPID DECOMPRESSION Sudden decompression occurs when a pressurized aircraft suffers a structural or mechanical failure that results in loss of internal pressurization. Rapid changes in ambient pressure greatly increase the risk of decompression illnesses. The effect on passengers and crew is dependent on multiple factors, including the length of time of equilibration from the aircraft's internal pressure to the ambient external pressure and the differential in pressures. Extremely rapid decompression is termed an "explosive" decompression and can produce an instantaneous overexpansion of the lung with resultant pulmonary trauma, leading to a pneumothorax, subcutaneous emphysema, or air emboli entering the vascular system. Fortunately, such decompressions are limited to aircraft with small cockpit spaces flying at higher altitudes and are relatively rare.

DIAGNOSIS AND TREATMENT Decompression illness symptoms might develop in flight, on descent, shortly after landing or be delayed for several hours.[5] All workers experiencing a rapid decompression should be thoroughly evaluated for signs and symptoms of decompression illness. In a study of 447 cases of altitude-related decompression illness, Ryles observed that 83.2% had musculoskeletal involvement, 70% of the time involving the knees. Approximately 3% experienced pulmonary symptoms, 10.8% developed paresthesias, and 0.5% had frank neurologic findings.[7] Any person with signs or symptoms suggesting postflight decompression illness should be referred immediately to a hyperbaric chamber, as immediate recompression is the only appropriate therapy. Maintaining the greatest possible ambient pressure during transport is critical to avoid further sequelae. If aeromedical evacuation to a distant recompression chamber is necessary, the flight should be at the lowest possible altitude, preferably in an aircraft pressurized to sea level.

Prevention

Clearly, primary prevention is the method of choice for avoiding decompression sickness. Those exposed to a changing pressure environment should receive education and training. Workers must be aware of the risks associated with exposure to differing atmospheric pressures, as well as the signs and symptoms that are the first manifestations of decompression illness. All workers exposed to changing atmospheric pressure must follow the guidelines established for safe entry, work, egress, and emergencies.

HYPOXIA

Most of the life-threatening effects of altitude are due to hypoxia. The response of humans to hypoxia is complex and is heavily dependent on the severity and rate of exposure. The aviator who experiences sudden loss of cabin pressure at altitude has a different physiologic response from a traveler who has traveled for weeks on the ground to reach the same altitude. Broadly speaking, acute hypoxia occurs over seconds to an hour or two; chronic hypoxia occurs from many hours to days.

Pathophysiology of acute hypoxia

Ascent to altitude reduces both the ambient pressure and the oxygen content available for gas exchange (Table 9.2). At 2438 m, there is a 25% reduction in the partial pressure of oxygen entering the lungs; by 5500 m it is reduced by half. Commercial and military aviators are at increased risk of hypoxia, as jet aircraft routinely operate at altitudes exceeding 7315 m (24 000 ft). Loss of pressurization or

TABLE 9.2 Atmospheric pressure and oxygen levels at altitude.

Altitude		Pressure		Ambient	
(m)	(ft)	(PSIA)	(mmHg)	PO_2 (mmHg)	PaO_2 (mmHg)
Sea level		14.69	759	159	103
610	2 000	13.66	706	148	93.8
1219	4 000	12.69	656	137	85.1
1829	6 000	11.77	609	127	76.8
2438	8 000	10.91	564	118	68.9
3048	10 000	10.10	522	109	61.2
3658	12 000	9.34	483	101	54.3
4267	14 000	8.63	446	93	47.9
4877	16 000	7.96	411	86	42.0
5486	18 000	7.34	379	79	37.8
6096	20 000	6.76	349	73	34.3
6706	22 000	6.21	321	67	32.8
7315	24 000	5.70	294	61	31.2

PSIA, pounds per square inch atmospheric.

failure of personal breathing equipment can result in loss of consciousness in a matter of minutes. In some environments requiring an increase in the work of breathing, reduced tidal volume and resultant hyperventilation may mimic hypoxia.

As the partial pressure of oxygen decreases, the body's ability to maintain adequate oxyhemoglobin saturation is impaired. Table 9.3 outlines the impact of hypoxia on oxygen availability to the arterial circulation on ascent to 6706 m (22 000 ft).

TABLE 9.3 Oxyhemoglobin saturation at selected altitudes.

Tissue level	Altitude			
	Sea Level	3048 m (10 000 ft)	5486 m (18 000 ft)	6706 m (22 000 ft)
Alveolus PO_2 (mmHg)	100	60	38	30
Arterial PO_2 (mmHg)	100	60	38	30
Venous PO_2 (mmHg)	40	31	26	22
A–a gradient (mmHg)	60	29	12	8
Oxyhemoglobin saturation (%)	98	87	72	60

Physiological response to acute hypoxia

Acute hypoxia occurs in a series, of stages, progressively affecting those tissues with the greatest requirement for oxygen, particularly the nervous system.

Indifferent stage (surface to 3000 m)

Early symptoms begin to appear but are frequently unnoticed by the individual. Vision will be mildly impaired, especially night vision and color vision. Cognitive functioning is normal, with slight decrements in novel task performance. Respiratory rate and depth of inspiration and cardiac output begin to increase. Oxygen saturation is maintained at 90–98%.

Compensatory stage (3000–4500 m)

Oxygen saturation falls below 90%. Errors in skilled task performance appear along with euphoria and impaired judgment, although workers are frequently unaware of any deficiencies. Prolonged exposure produces a generalized headache.

Disturbance stage (4500–6100 m)

Oxygen saturation falls below 80%. Cerebral functions are severely impaired. Mental calculations become unreliable. Headaches increase in severity, and neuromuscular control is greatly diminished. Tunnel vision frequently occurs. Personality and emotional changes appear and range from elation or euphoria to belligerence. Increased respiratory drive leads to hyperventilation and hypocapnia. Paresthesias of the extremities and lips are followed in severe cases by tetany and carpopedal or facial spasms. The chances of recovery are poor, due to serious deficiencies in judgment and loss of muscular coordination.

Critical stage (above 6100 m)

Oxygen saturation is below 70%. Comprehension and mental performance decline rapidly, and unconsciousness occurs within minutes, often without warning (Table 9.4).

TABLE 9.4 Duration of useful consciousness.

Altitude		Duration of useful consciousness
(m)	(ft)	
5 486	18 000	20–30 minutes
6 706	22 000	10 minutes
7 620	25 000	3–5 minutes
9 144	30 000	1–2 minutes
10 668	35 000	30–60 seconds
12 192	40 000	14–20 seconds
13 106	43 000	9–12 seconds

Treatment and prevention of acute hypoxia

Initial management of hypoxia is with the immediate use of 100% oxygen. Aircraft lacking an oxygen system or experiencing a depressurization should begin an immediate emergency descent. Recovery usually occurs within seconds of supplemental oxygen use, although a transient worsening of symptoms may occur for 15–60 seconds.

Hypoxia prevention requires adequate oxygen during flight, through either individual oxygen systems or aircraft pressurization. In commercial aircraft, the high-flying passenger jet provides a pressurized cabin that rarely exceeds an altitude of 2400 m (8000 ft). In unpressurized aircraft, supplemental oxygen is required for the pilot at a cabin altitude of 14 000 ft or higher. If flight is maintained for >30 minutes at altitudes between 12 500 and 14 000 ft, the pilot must use oxygen. When flying above 15 000 ft, all occupants must have supplemental oxygen.[8] At an altitude of 10 363 m (34 000 ft), 100% oxygen is required in order to maintain adequate oxygenation, equivalent to sea level. In some aircraft, particularly in the military high-altitude environment, regulators to increase the partial pressure of oxygen and/or provide positive pressure ventilation are necessary. Sustained flight at ambient altitudes of 13 700 m (45 000 ft) or higher requires use of pressure suits.

Prolonged elevated partial pressures of oxygen can introduce a new set of challenges, including alveolar atelectasis and chronic cough.

Medical surveillance and education

Commercial airline transport pilots are required to undergo FAA-approved physical examinations every 6 months to maintain their medical certification.[9] Annual flight physicals are mandatory for military aviators, who receive excellent medical surveillance, given the high ratio of flight surgeons to aircrew and robust prevention programs in place throughout the armed services. Preemployment and periodic examinations pay great attention to otorhinolaryngeal conditions that might predispose aircrew to otitic or sinus barotrauma.

Primary prevention is the goal. Aircrews receive extensive training in the physiology of operating in hypobaric environments. The US Code of Federal Regulations (CFR), Title 14, Part 61.31 (g)(2)(i) indicates that "no person may act as pilot in command of a pressurized airplane that has a service ceiling or maximum operating altitude, whichever is lower, above 25 000 ft MSL unless that person has completed ground training that includes instruction on respiration; effects, symptoms, and causes of hypoxia and any other high altitude sicknesses; duration of consciousness without supplemental oxygen; effects of prolonged usage of supplemental oxygen; causes and effects of gas expansion and gas bubble formations; preventive measures for eliminating gas expansion, gas bubble formations, and high altitude sicknesses; physical phenomena and incidents of decompression; and any other physiological aspects of high altitude flight."

The FAA coordinates low-pressure chamber "flights" at US Air Force bases to allow civilian aviators to experience hypoxia in a controlled environment—information for such courses is available on the World Wide Web at hhttps://www.faa.gov/pilots/training/airman_education/aerospace_physiology/cami_enrollment/. The US armed services maintain a large network of hypobaric chambers, as military aviators are required to complete low-pressure training every 3–4 years. Aviators learn early in their career to refrain from flying when congested. Foods that cause excessive intestinal gas—beans, cabbage, cauliflower, carbonated beverages, or peas—are also soon avoided.

The risk of decompression illness in flight can be minimized through several strategies. Use of 100% oxygen for 30 minutes before flight will reduce the body's nitrogen load, as will use of oxygen throughout a flight. Limiting the altitude and duration of exposure will significantly reduce the incidence. SCUBA divers should allow a sufficient interval between diving and flight. A minimum of 12 hours should elapse following a no-compression dive with less than 2 hours total bottom time; 24–48 hours is recommended after more complex dive profiles.[5]

HIGH-ALTITUDE ACCLIMATIZATION AND ILLNESS

Acute exposure to high altitude is fatal to unprotected workers in a matter of minutes (Table 9.4), yet men have climbed the tallest peaks in the world with nothing more than thermal protection. The difference lies in a man's ability to adapt to severely hypoxic conditions through progressive acclimatization over extended time periods (days to months).

Acute mountain sickness (AMS)

Acute mountain sickness is the most common altitude-related illness affecting travelers to high altitude. Chinese authors first described AMS in 32 BC. Jose de Acosta, a Jesuit priest living in Peru in the sixteenth century, provided a more complete description based on his experiences in the Andes.[10] AMS consists of a group of symptoms occurring 6–48 hours after rapid ascent. At elevations over 3000 m, 25% of travelers will have mild symptoms, including headache, fatigue, nausea, malaise, loss of appetite, and disturbed sleep.[11] Symptoms are so nonspecific that patients often fail to recognize their condition. The most important reason to diagnosis AMS is that it is often unrecognized; it may progress to high-altitude cerebral edema (HACE) or high-altitude pulmonary edema (HAPE) (see below) if there is further increase in altitude without adequate time for acclimatization. A greater incidence has been reported in individuals in their early 20s.[12]

Pathophysiology of AMS

No etiology has been firmly established for AMS. There is evidence of cerebral vasodilation with increased blood flow and leakage of proteins and fluid across the blood–brain barrier, but the exact mechanism remains elusive.[13]

Diagnosis and therapy

Complete history and physical examination are usually adequate to rule out conditions with similar symptoms such as viral illnesses, exhaustion, dehydration, or hangover. Mild AMS is limited to the symptoms described above and is best treated with arrest of ascent or a slight descent to allow time for acclimatization—usually 1–3 days is sufficient. Acetazolamide (Diamox) 125–250 mg twice a day has been effective in reducing symptoms[14] but may result in diffuse paresthesias. Acetaminophen or nonsteroidal anti-inflammatory drugs are indicated to manage headaches. Theophylline reduced AMS symptoms in 14 subjects given 375 mg oral slow-release theophylline twice a day at simulated altitudes of 3454 m.[15]

Symptoms of moderate AMS include severe headache not relieved by medication, nausea and vomiting, progressive weakness and fatigue, shortness of breath, and loss of coordination. Moderate AMS should be treated with descent. When descent is delayed, oxygen and acetazolamide should be considered.

High-altitude cerebral edema (HACE)

Pathophysiology

HACE is a potentially fatal metabolic encephalopathy believed to be of vasogenic etiology, with leaking of protein and water across the blood–brain barrier.[16]

Diagnosis

HACE occurs in 2–3% of trekkers at altitudes of 5500 m, although HACE can appear in individuals above 2500 m.[14] Symptoms include severe headache, nausea, vomiting, ataxia, disorientation, hallucinations, seizures, stupor, and coma. Mild AMS can progress to HACE in 12–72 hours.

Treatment

HACE is life-threatening—definitive therapy is immediate descent with close supervision as symptoms may worsen while descending. When descent is delayed due to the situation or patient's condition, oxygen and dexamethasone, 10 mg intravenously and then 4 mg intramuscularly every 6 hours, are indicated.[11] Prognosis is poor once the patient becomes comatose.

High-altitude pulmonary edema (HAPE)

Pathophysiology

HAPE is the most frequent cause of death of the altitude illnesses. It frequently develops on the second night of exposure at altitudes above 2500 m. In Colorado, 1 in 10 000 skiers will develop HAPE, with a higher incidence in younger men. HAPE victims develop substantial increases in pulmonary artery pressures, with increased vascular permeability. Fluid increases in the lungs, reducing oxygen exchange. Respiratory alkalosis and severe hypoxemia follow, with mean oxygen saturations of 56%.[14]

Diagnosis

Early symptoms include dyspnea on exertion, fatigue, weakness, and dry cough. Signs typically are tachycardia, tachypnea, rales, pink-tinged frothy sputum, and cyanosis. Radiographs show patchy peripheral infiltrates, which may be unilateral or bilateral.[14]

Treatment

Immediate descent to lower altitudes with close monitoring is required. Patients should be kept warm and minimize exertions. If descent is impossible due to conditions, nifedipine (10 mg every 4 hours)[17] and oxygen (4–6 L/min) have been shown to improve patients. Descent remains the key to therapy.

Prevention of altitude illnesses

The rate of ascent and individual susceptibility are the primary determinants of altitude illnesses. Physical fitness is not protective. The key to prevention is a gradual ascent, when feasible. Workers traveling from altitudes below 1200 m to altitudes above 2500 m should spend at least one night at 1500–2200 m or alternatively 2 or 3 nights at 2800–3000 m before proceeding higher.[18]

There is a diversity of recommendations regarding further ascents and only two prospective studies to support the guidance.[19,20] The Himalayan Rescue Association recommends ascending no more than 300 m/day with a rest day (no ascent in sleeping altitude) for every additional 600–900 m and no single day gain greater than 800 m.[21] The Wilderness Medical Society recommends limiting ascent to 500 m/day with a rest day every 3–4 days.[22]

Returning to lower altitudes at night to sleep enhances acclimatization, possibly because of the relative hypoxemia during sleep.[18] Sleep hypnotics and alcohol should be avoided, as they suppress breathing during sleep, worsening cerebral oxygenation. High-carbohydrate diets and avoidance of dehydration or overexertion have also been widely reported as helping to prevent high-altitude illnesses, although precise mechanisms are unknown.[23]

Acetazolamide is the drug of choice to prevent or limit the severity of AMS and HACE and accelerates the rate of acclimatization. Acetazolamide 125 mg should be given twice a day beginning the day before ascent and continued until 2 days at the highest sleeping altitude or until descent begins.[18] Acetazolamide is a nonantibiotic sulfonamide with low cross-reactivity with sulfa antibiotics, but patients with a history of drug allergy should be assessed before receiving acetazolamide. While the most common allergic reaction is a rash, anaphylaxis has been reported.[18]

Dexamethasone can mask symptoms and can be started on the day of ascent. The recommended dose is 4 mg every 12 hours for passive (sedentary) ascent and 4 mg every 6 hours for active ascent.[18] It can be stopped after 2–3 days at the highest sleeping altitude or when starting descent, but should not be taken for more than 10 days to prevent adrenal suppression.[18]

In military or rescue operations requiring rapid ascent above 3500 m without time for acclimatization, acetazolamide and dexamethasone can be used together.[18]

For workers with a history of HAPE, nifedipine is recommended to prevent recurrence.[24] Salmeterol may be a helpful addition for those at the highest risk.[22]

Workers with underlying medical conditions that can be worsened by hypoxia (including lung disease, heart disease, sickle cell disease, and others) should be evaluated to determine whether work restrictions, additional medications, or additional oxygen is needed.

OTHER ALTITUDE-RELATED CONDITIONS

Vision

Beck Weathers' experience on an ill-fated Mt Everest expedition in 1996 heightened the public's awareness of high-altitude effects on patients who have undergone eye surgery. Weathers, a Texas pathologist, suffered severe hyperopia at altitude following his radial keratotomy (RK), effectively disabling him.[24] Ng et al. experimentally demonstrated reversible hyperopic changes in RK subjects exposed to altitudes of 4300 m.[25] Studies of postphotorefractive keractomy (PRK) and laser in situ keratomileusis (LASIK) patients revealed no change in PRK patients and a small but statistically significant myopic change in LASIK patients.[26] The mechanism in all cases appears to be hypoxia-induced corneal hydration.[27]

Extreme altitudes have also been implicated in high-altitude retinopathy (HAR). Weidman and Tabor examined 40 climbers climbing Mt Everest. Fourteen of 19 climbers who ascended to altitudes between 4880 and 7620 m developed HAR, and 19 of the 21 who exceeded 7620 m (25 000 ft) developed HAR.[28] Most patients were asymptomatic; descent was not required.

Pregnancy

Altitude has been implicated in a number of complications of pregnancy. Ali et al. and Niermeyer have suggested in independent studies that there is an increased incidence in preterm labor among pregnant high-altitude travelers.[29,30] Palmer et al. reported a 16% incidence of preeclampsia at 3100 m compared to a 3% rate at 1260 m at high and low altitudes in Colorado. Birth weight averaged 285 g less in those deliveries at 3100 m.[31]

Radiation exposure

It has long been known that high altitude exposes workers to elevated levels of cosmic radiation, especially in higher latitudes.[32] Aircrews on polar routes have significantly higher exposures than those flying equatorial routes, but the long-term impact on health is currently unknown. Gundestrup and Storm reported increased acute myeloid leukemia, malignant melanoma, and skin cancer rates in Danish male jet cockpit crew members. The melanomas and skin cancers were attributed to sun exposure during vacations rather than occupational exposure at altitude.[33] Other studies have shown individual exposures to be well within current international recommended exposures,[34–38] although a pregnant flight attendant would have to change routes to remain under the exposure limits.[38]

References

1. Hultgren HN. High altitude medicine. Stanford: Hultgren Publications, 1997:10–1.
2. Ward MP, Milledge JS, West JB. High altitude medicine and physiology. London: Chapman & Hall Medical, 1995:32–7.
3. O'Reilly BJ. Otorhinolaryngology. In: Ernsting J, Nicholson AN, Rainford DJ, eds. Aviation medicine. 3rd edn. Oxford: Butterworth-Heinemann, 1999:319–36.
4. Voge VM. Probable bends at 14,000 feet: a case report. *Aviat Space Environ Med* 1989; 60(11):1102–3.
5. Heimbach RD, Sheffield PJ. Decompression sickness and pulmonary overpressure accidents. In: DeHart RL, ed. Fundamentals of aerospace medicine. 2nd edn. Baltimore: Williams & Wilkins, 1996:131–61.
6. Weien RW, Baumgartner N. Altitude decompression sickness: hyperbaric results in 528 cases. *Aviat Space Environ Med* 1990; 61(9):833–6.
7. Ryles MT, Pilmanis AA. The initial signs and symptoms of altitude decompression sickness. *Aviat Space Environ Med* 1996; 67(10):983–9.
8. Code of Federal Regulations 14, part 91, subpart C, section 91.211 Supplemental oxygen. April 25, 2000. Government Printing Office, Washington, DC.
9. Code of Federal Regulations 14, part 61, subpart A, section 61.23 Medical certification and duration. October 10, 2000. Government Printing Office, Washington, DC.
10. Hultgren HN. High altitude medicine. Stanford: Hultgren Publications, 1997:213–4.
11. Hultgren HN. High altitude medicine. Stanford: Hultgren Publications, 1997:212–48.
12. Hackett PH, Rennle D. The incidence, importance, and prophylaxis of acute mountain sickness. *Lancet* 1976; 2:1149–55.
13. Hackett PH. The cerebral etiology of high-altitude cerebral edema and acute mountain sickness. *Wilderness Environ Med* 1999; 10(2):97–109.
14. Kloche DL, Decker WW, Stepanek J. Altitude-related illnesses. *Mayo Clin Proc* 1998; 73(10):988–93.
15. Fischer R, Lang SM, Steiner U, et al. Theophylline improves acute mountain sickness. *Eur Respir J* 2000; 15(1):123–7.
16. Hackett PH. High altitude cerebral edema and acute mountain sickness. A pathophysiology update. *Adv Exp Med Biol* 1999; 474:23–45.
17. Oelz O, Maggiorini M, Ritter M, et al. Nifedipine for high altitude pulmonary oedema. *Lancet* 1989; 2(8674):1241–4.
18. Zafren K. Prevention of high altitude illness. *Travel Med Infect Dis* 2014; 12(1):29–39.

19. Bloch KE, Turk AJ, Maggiorini M, et al. Effect of ascent protocol on acute mountain sickness and success at Muztagh Ata, 7546 m. *High Alt Med Biol* 2009; 10(1):25e32.
20. Beidleman BA, Fulco CS, Muza SR, et al. Effect of six days of staging on physiologic adjustments and acute mountain sickness during ascent to 4300 meters. *High Alt Med Biol* 2009; 10(3):253e60.
21. Zafren K, Honigman B. High-altitude medicine. *Emerg Med Clin North Am* 1997; 15(1):191e222.
22. Luks AM, McIntosh SE, Grissom CK, et al. Wilderness Medical Society consensus guidelines for the prevention and treatment of acute altitude illness. *Wilderness Environ Med* 2010; 21(2):146e55.
23. Bartsch P, Maggiorini M, Ritter M, et al. Prevention of high-altitude pulmonary edema by nifedipine. *N Engl J Med* 1991; 325(18):1284e9.
24. Krakauer J. Into thin air. New York: Anchor Books, 1997:246–9.
25. Ng JD, White LJ, Parmley VC, et al. Effects of simulated high altitude on patients who have had radical keratotomy. *Ophthalmology* 1996; 103(3):452–7.
26. White LJ, Mader TH. Refractive changes at high altitude after LASIK. *Ophthalmology* 2000; 107(12):2118.
27. Mader TH, Blanton CL, Gilbert BN, et al. Refractive changes during 72-hour exposure to high altitude after refractive surgery. *Ophthalmology* 1996; 103(8):1188–95.
28. Wiedman M, Tabin GC. High-altitude retinopathy and altitude illness. *Ophthalmology* 1999; 106(10):1924–6; discussion 1927.
29. Ali KZ, Ali ME, Khalid ME. High altitude and spontaneous preterm birth. *Int J Gynaecol Obstet* 1996; 54(1):11–5.
30. Niermeyer S. The pregnant altitude visitor. *Adv Exp Med Biol* 1999; 474:65–77.
31. Palmer SK, Moore LG, Young D, et al. Altered blood pressure during normal pregnancy and increased preeclampsia at high altitude (3100 meters) in Colorado. *Am J Obstet Gynecol* 1999; 180(5):1161–8.
32. Mohr G. The future perspective. In: DeHart RL, ed. Fundamentals of aerospace medicine. 2nd edn. Baltimore: Williams & Wilkins, 1996:37–55.
33. Gundestrup M, Storm HH. Radiation-induced acute myeloid leukaemia and other cancers in commercial jet cockpit crew: a population-based cohort study. *Lancet* 1999; 354(9195): 2029–31.
34. Bagshaw M, Irvine D, Davies DM. Exposure to cosmic radiation of British Airways flying crew on ultra-longhaul routes. *Occup Environ Med* 1996; 53(7):495–8.
35. Oksanen PJ. Estimated individual annual cosmic radiation doses for flight crews. *Aviat Space Environ Med* 1998; 69(7):621–5.
36. Tume P, Lewis BJ, Bennett LG, et al. Assessment of cosmic radiation exposure on Canadian-based routes. *Health Phys* 2000; 79(5):568–75.
37. Bagshaw M. Cosmic radiation in commercial aviation. *Travel Med Infect Dis* 2008; 6(3):125–7.
38. Waters M, Bloom TF, Grajewski B. The NIOSH/FAA Working Women's Health Study: evaluation of the cosmic-radiation exposures of flight attendants. *Health Phys* 2000; 79(5):553–9.

Chapter 10

SHIFT WORK

Allene J. Scott

Shift work refers to hours of work occurring outside the regular daytime schedule, that is, work schedules not falling between 6a.m. and 6p.m. According to the May 2004 *Work Schedules and Work at Home* supplement to the *Current Population Survey by the US Bureau of Labor Statistics*, considering only the primary job of full-time workers, 19% of male and 16% of female workers are shift workers including 3–4% each reporting working "irregular schedules" (varying with the needs of the business), rotating shifts, or night shifts. Over one half of protective service workers, for example, police and firefighters, are shift workers. Similarly the majority of employees in eating and drinking establishments work nondaytime schedules. One third of transportation workers are shift workers. Industries with 5% or less of employees working shift work schedules include construction, education, finance, and insurance.

Alertness and performance of shift workers, particularly when working schedules involving the night (graveyard) shift, may be compromised. Night workers and rotating shift workers are at increased risk of falling asleep on the job and of falling asleep when driving home after work.[1,2] Public safety has been compromised by catastrophes involving chemical plant, nuclear power plant, and transportation accidents attributed to work schedule-related drowsiness.[3,4]

Shift work also takes a toll on individuals and their families, due to conflicts between the work schedule and domestic responsibilities. Shift work has been identified as an emerging risk factor for several medical and mental health conditions, which receive extra attention in this chapter.

OCCUPATIONAL SETTING

There are many different types of shift work schedules. The type of schedule is often determined by traditional practices of the industry. For example, the Southern-Swing schedule, a weekly, backward rotation, has been commonly used for generations in the steel industry. Worker preferences, labor union interests, and job process needs are common factors influencing schedule design.

Common shift work schedule designs include the following:

1. Fixed ("permanent")—Each individual employee works one shift. The number of days worked in a row and the number of days off vary between industries. "Permanent" night work is, in one sense, a misnomer, because almost all workers return to daytime activity on their days off and thus still switch back and forth between night work and daytime activity.
2. Rotating—An individual employee regularly changes the shift worked. Rotating schedules vary in the direction and speed of rotation:
   - Direction of rotation: This may be clockwise (forward rotation, phase delay) or counterclockwise (backward rotation, phase advance).
   - Speed of rotation: This may be slow (greater than weekly), weekly, or rapid. Examples of rapidly rotating systems include the European metropolitan (two identical consecutive shifts are worked consecutively, followed by two of the next shift until

Physical and Biological Hazards of the Workplace, Third Edition. Edited by Gregg M. Stave and Peter H. Wald.

all three shifts have been worked, 2-2-2) and the continental (same as the metropolitan, except three of one of the shift times are worked, 2-2-3) rotation schedules.

3. Split shifts—A hiatus of a few hours separates the work hours performed on the same day. For example, this may be done in the restaurant or transportation industries to cover peak business hours.
4. Alternative rotations—An increasingly popular 4-day "compressed workweek" with 10–12 hours shifts used in a single-, two-, or three-shift operation. The 8-day week with 4 10-hours work days followed by 4 days off is used primarily in firms operating for 10 hours/day, 7 days/week or operating 20 hours/day in two shifts. Twelve-hour shift schedules are common in hospital nursing schedules. Schedules with "flexi-time" give employees some choice in designing their own personal daily work hours to meet weekly requirements.

MEASUREMENT GUIDELINES

Several factors affect tolerance to night work and rotating shift work, including (i) individual differences in susceptibility to performance and alertness deficits related to altered sleep–wake cycles and to symptomatology from disruption of biological circadian rhythms, (ii) differences in social/family responsibilities and supportiveness, (iii) effectiveness of off-the-job and on-the-job coping strategies for maximizing sleep and minimizing performance deficits, (iv) the schedule design, and the predictability and flexibility of time off.

In order to evaluate the appropriateness of a particular work schedule for a specific industrial site, information should be gathered from human resources, from safety officers, and directly from the workers. The general categories of necessary information include (i) demographics of the workforce, (ii) frequency of cases of shift work intolerance from medical surveillance programs and/or medical claims, (iii) rates of on-the-job fatigue-related accidents or performance deficits, and (iv) information from confidential worker surveys concerning sleepiness on the job, motor vehicle accidents or near misses driving home from each shift, general well-being, and worker satisfaction with the schedule.

Tepas et al. have described a survey method for use in designing and evaluating shift work systems tailored to specific workers and plants.[5] Their work–sleep survey has demonstrated that there are significant differences between industrial plants with respect to demographics, worker habits, and worker preferences and that "Before recommending a new work-system shift scheme to a plant, the complexity of shift-work issues must be recognized by assessing a wide range of personal, social, and health issues." Education of workers with respect to the reasons for the selection and expectations of how the scheduling system will work is also important. Follow-up surveys, conducted after the system has been in operation for some time, are recommended to evaluate the effects on the workers' well-being and performance. Kogi made similar recommendations based on an extensive survey of male and female shift workers in Japan working various rotating schedules.[6] These recommendations included considering other factors, in addition to the shift rotation, when determining ways to minimize the detrimental effects of shift work on the worker and his/her family, such as social and family life responsibilities, worker commuting time, and possibilities for anchor sleep (regularly scheduled sleep periods) during night shifts.

Direct comparisons of studies looking at performance or health outcomes of shift workers are often difficult, due to differences in schedule design. Pertinent shift work exposure parameters include, in addition to the number of years employed as a shift worker, schedule design variables such the direction and frequency of shift rotation, the number of nights worked in a row, the number of days off after night shift assignment, the amount of overtime, and the length of shifts. National data collection is limited in the United States. The United States Bureau of Labor Statistics does annually collect and tabulate occupational injury data available from employers regarding the time of day the injury event occurred and hours on the job worked before the event.[7] The Federal Occupational Safety and Health Administration's (OSHA) Form 301 Incident Report (Rev 01/2004) which employers are required to complete and maintain in their files for 5 years also request this information, if available. The BLS includes this data in some of its published estimates.[8] However, OSHA only requires employers to record work site-wide annual average employment and total hours worked by all payroll employees, and BLS injury statistics do not include data regarding the composition of work force by shift, so shift work hour-based rates are not determinable from the national labor statistics (Bureau of Labor Statistics, personal E-mail communication, December 31, 2014).[8] Limited opportunities therefore exist to look at US national data concerning on-the-job accidents and scheduling factors.

Despite the inherent design problems in shift work research and the limited national data, field and laboratory studies of 24-hour sleep and performance rhythms, and of health outcomes in shift workers, have provided information that has been used to make reasonable recommendations concerning shift work scheduling. These are presented in the following section.

EXPOSURE GUIDELINES (SCHEDULE DESIGN)

The various combinations of individual susceptibility factors and scheduling designs make it difficult to assess the effective "exposure" to shift work and thus the "risk" that different shift systems may have on health and safety outcomes. Comparisons of groups must account for the type of shift schedule as well as years spent in shift work. Wedderburn has proposed criteria for assessing shift work schedules for potential risk for health and well-being including the perturbation of the circadian rhythm, performance at work, health, and social life.[9] The "Rota Risk Profile Analysis," developed by Jansen and Kroon utilizes several physiological and psychosocial risk factors associated with the schedule design including regularity of shift timetable, periodicity (the degree to which the "biological clock" is disturbed), shift load (the average length of shifts) and week load (the average length of the work week), opportunities for nighttime sleep (between 11 p.m. and 7 a.m.) and constancy in night rest (variation in the week), predictability of the shift cycles, and opportunities and constancy for household and family tasks, for evening recreation, and for weekend recreation.[10]

As is true of threshold limit values® (TLVs®) for chemical exposures, which are developed to protect nearly all healthy workers, with the recognition that a small percentage of workers are susceptible to lower levels of exposure,[11] recommendations for shift work scheduling design, which control exposure to night work, are applicable to most workers. Shift work scheduling recommendations may not be adequate for workers with greater individual susceptibility to shift work-related biological rhythm disruption or with family/social situations limiting daytime sleep opportunities. In addition, with respect to performance on the job, the appropriateness of recommendations for the length of shift varies with the type of industry involved and related productivity versus safety and alertness demands.

Chronotoxicologic considerations

OSHA Permissible Exposure Limits (PELs) for occupational exposures to chemicals are determined for an 8-hour time-weighted average (TWA). TLVs® calculated by the American Conference of Governmental Industrial Hygienists (ACGIH) for airborne concentrations of chemical exposures are also based on an 8-hour workday.[11] Neither exposure limit adjusts for nocturnal differences in metabolism when working at night. With the exception of the extended shift reduction factor requirement included in the OSHA lead standard, OSHA PELs do not require measurement adjustment for shifts over 8 hours.[12] The ACGIH[11] suggests that industrial hygienists follow the Brief and Scala model[13] for adjusting TLVs® for work shifts over 8 hours. The Brief and Scala model considers the increased uptake due to longer work exposure and adds a second component to account for the decreased metabolic clearance during the shorter time away from the workplace exposure.

In addition, the susceptibility to the adverse affects of the toxic materials may be greater when exposure occurs at night, just as response to medications varies with the time of administration. Smolensky and Reinberg[14] have prepared a detailed discussion of chronotoxicology as it relates to biological monitoring of workplace exposures.

Musculoskeletal considerations

Work involving heavy physical labor and/or repetitive motion may need to be adjusted for production speed and number of breaks. Job assignments may need to be rotated to prevent repetitive musculoskeletal stress/injury. The number of hours between shifts worked and the number of sequential days off may be significant factors in the management of acute and chronic strain/sprain injuries. A model for special provisions for night workers who are also exposed to physical and/or toxic stress developed by the Austrian "Night Shift/Heavy Work Law" has been summarized by Koller.[15]

General scheduling considerations

Ideally, shift work schedules should be designed to minimize the potential negative effects of night work on worker sleep, health, and performance. However, medical and performance/safety considerations are not the only driving force in scheduling design. Economic ramifications including staffing levels and production needs, labor union issues, and worker preferences are important factors in influencing the final outcome. In order to help balance all of these considerations, shift work scheduling consultants are increasingly being utilized to assist companies in evaluating and redesigning schedules.

Advantages and disadvantages of different schedules

Rotating schedules

Entrainment considerations Differences in recommendations concerning whether permanent night shift schedules are preferable to rotating schedules, and if a rotating schedule is used, the appropriate speed of schedule rotation, reflect differing opinions about the effect of the schedule design on nocturnal adjustment by the night worker and the desirability of circadian re-entrainment to nocturnal work.[16] In order to maximize night work performance, rapid re-entrainment is desirable. However, minimizing circadian system disruption in order to avoid related deleterious effects on health and well-being conflicts with this goal and the goal of entraining the night worker to the inverted day–night cycle is difficult to reach. It has been demonstrated that more than a week of consecutively worked nights is needed before complete adjustment of the circadian system begins to occur.[17,18] Most workers revert to day activity on nonwork days, and because considerable re-entrainment to diurnal activity occurs over only a couple of days off, biological rhythm desynchronization can be expected to occur each time the worker starts on the night shift period.[19] Weekly and rapidly rotating schedules do not allow enough time for full adjustment to night orientation. When no more than two or three nights in a row are worked, little diurnal shifting of the circadian system will have occurred.[20] If the degree of adjustment to night work has been overestimated as has been suggested,[21] the rationale is weakened for using "permanent" or slowly rotating shift systems in order to maximize adjustment of the circadian system.

Studies of long-term tolerance to shift work have suggested that shift workers with large amplitudes in circadian temperature rhythms (indicating resistance to adjustment) have better long-term tolerance to shift work.[22–24] Reinberg et al. have therefore suggested that if tolerance to shift work over the long term is associated with large circadian amplitude of the temperature rhythm and therefore slow adjustment, rapid rotation is preferable to weekly rotation.[22] Mills et al. suggested that if re-entrainment were accomplished, the worker would have trouble functioning in the diurnal world on days off.[25]

Circadian and sleep debt considerations The forward (clockwise, phase-delay) direction for schedule changes is usually considered preferable, since the biological clock is easier to set back than ahead. Jet-lag and shift work laboratory studies have demonstrated that it is easier to adjust to a phase-delay than to a phase-advance time shift.[25–28] Shift work field studies also have supported a clockwise rotation over a counterclockwise direction.[25,29–31] Folkard has pointed out that when there is not a day off between shift changes, systems that rotate forward allow a break of 24 hours, while many backward rotations have a "quick return" after only an 8-hour break.[32]

With respect to the speed of rotation, Vidacek et al., in their study of weekly rotating shift workers, found, for the night shift only, a day-of-the-week effect, with productivity increasing through the third day and then falling over the last two days of the 5-day shift period, but not to the first-day low.[33] They suggested that a weekly rotation might capitalize on circadian adjustment while minimizing effects of sleep deprivation in comparison to rapidly rotating or more slowly rotating schedules. Wilkinson et al. found, for a slowly rotating system, that compared to the first night worked, performance on tests sensitive to the effects of sleep deprivation was significantly poorer on the seventh work night, but not on the fourth night.[34] Other studies have shown evidence of sleep deprivation after only two consecutive nights on the job[35,36] and that, for rotating workers, sleep is most disturbed at the beginning of the night shift.[37] Czeisler et al. reported that most weekly rotators on a phase-advance schedule need 2–4 days for their sleep schedule to adjust to shift changes.[38]

Chronic sleep deprivation has been demonstrated after several nights worked in succession in rotating systems.[39–41] Williamson and Sanderson evaluated the effects of switching from a clockwise, slowly rotating (seven straight shifts) to a clockwise, rapidly rotating system with no more than three nights worked in a row.[42] After the switch, workers reported improved sleep, and no longer complained of feeling tired and irritable at work. Smith et al. compared a slowly rotating continuous 8-hour shift system involving seven shifts worked in succession to rapidly rotating continuous 8- and 12-hour shifts.[43] After switching to the rapid rotations, day sleep was reported to be improved, fatigue decreased, home and social life improved, and symptoms of circadian disruptions decreased.

Eastman has described a protocol involving worker's controlling bright light exposure before bed to achieve minimal circadian misalignment when working a three-shift system with 2-week slowly rotating schedule in the delaying direction.[44] Smith and Eastman have concluded that rapidly rotating shift schedules that include day and night work should be abolished, because the circadian biological clock cannot phase-shift fast enough to reduce circadian rhythm misalignment (however an effective sleep and light schedule can be applied if the rotation is between evening and night shifts), and that rotation between day and evening shifts also produces less circadian disruption than switching between day and night work.[45]

Maasen et al. found that good sleep was obtained by workers on a weekly rotating system; however, the schedule provided 6 days off following the week of night work, and the morning shift did not start until 0800.[46] Based on health, safety, and productivity evaluations before and after the introduction of a new shift schedule, Moore-Ede's group has been successful with a 21-day slowly rotating schedule which requires workers to maintain nocturnal orientation on their days off during the 3 weeks of night work.[47]

There are other variables to be considered when contrasting rotating schedules with different speeds of rotation. Providing adequate time off after night shifts is necessary to prevent chronic sleep deprivation. Dahlgren has concluded that "the speed of rotation in itself, without consideration of how the free days are organized is an insufficient criteria for judging the relative merits of different shift systems."[48] Others have recommended that there should not be many night shifts in succession and that workers should have at least 24 hours off after each night shift.[35,49] With respect to worker preference, several studies have found rapid rotations to be preferred by workers who have experienced them over weekly and permanent night shifts.[39,41,50–53]

Permanent shifts

For the permanent night worker, sleep deprivation appears to be persistent. Some research indicates that the sleep reduction associated with night work is greater for rotating workers; however, sleep deprivation performance deficits can be reversed during nonnight shift rotation time.[54] Tepas et al. found measures of vigilance to be poorer for permanent night workers than for slowly rotating workers when on the night shift; other evidence of chronic sleep deprivation was found for permanent night workers but not for the rotators.[55] The total average duration of sleep appears to be shorter for permanent night workers than for rotating ones, although sleep when on the night shift may be shorter for rotators.[56]

Length of shifts: 8- versus 12-hour shifts

In addition to the type of rotation, the length of the shift may affect fatigue-related performance parameters. Several studies have found a negative impact of 12-hour shifts on performance and alertness parameters. Rosa et al., in a simulated work study, measured several performance parameters and collected worker reports of drowsiness and fatigue; the results were interpreted to support that 12-hour/4-day "compressed" weeks were more fatiguing than 8-hour/6-day weeks.[57] Using the same fatigue test battery in a work site assessment, an overall decrease in performance and alertness as well in total sleep time was observed after a switch from an 8-hour three-rotating work shift schedule to a 12-hour shift schedule.[58]

There have been reports of fatigue effects on performance toward the end of 10- and 12-hour shifts.[59–62] Of special concern is that an increase in fatigue and sleepiness may occur in the last hours of a long shift coinciding with time of the circadian nadir of alertness. In a case–control study of large truck crashes on interstate highways, driving for over 8 hours was associated with nearly a doubling of the risk of crash involvement compared to drivers who had driven fewer hours.[63] Lisper et al. found, in a 12-hour observed driving test, that most episodes of falling asleep occurred after 8 hours of driving.[64] Several studies, however, comparing 8- and 12-hour systems have reported improvement or minimal deleterious effects on health, performance, and alertness or sleep parameters with the introduction of a 12-hour shift.[65–70] Mitchell and Williamson[71] reported for a sample of electrical power station workers, changing from an 8-hour slowly backward rotating system to a forward rapidly rotating 12-hour system resulted in improvements in perceived sleep quality, with less broken sleep patterns being reported, and less use of alcohol as a sleep aid; these workers also reported feeling fresher at the beginning and end of the 12-hour shifts. In this study, while there were also no ill effects of 12-hour shifts on four out of five performance measures, performance on a vigilance task declined over 12-hour shifts, but not over 8-hour shifts; injury data was similar for the 8- and 12-hour systems (two injuries being reported during the 8-hour schedule compared to the one during the 12-hour schedule). In his extensive review of the available research, Tucker concluded that there was little direct evidence of an increase in risk to safety from switching to a compressed workweek; however he also pointed out evidence from field studies suggesting that fatigue increases toward the end of extended shifts in some work settings.[72] This observation is consistent with increasing evidence that accident risk increases with the duration of the work shift, and the increase in risk may be dramatic toward the end of shifts longer than 8 hours.

Using compressed workweeks with 12-hour shifts to cover 24-hour operations has been recommended by shift work scheduling consultants.[73] Although compressed workweeks require working long shifts, they have become popular because they provide several days off in a row allowing workers to have blocks of time off for family/social activity. In addition, some switches to 12-hour schedules come with less potential for required overtime and working back-to-back shifts.

Compared to the estimated 5–7% of US workers in general who work more than one job, about 25% of 12-hour shift workers report moonlighting. Concern has been raised that these workers may be returning to work, possibly at night, already tired from other work activities.[62,74,75]

Multifactorial issues contribute to the advantageous and disadvantageous outcomes reported when 12-hour shifts have been introduced into various industries and worker populations[76]; and limitations and inconsistencies in the research data published on this topic have been noted.[62,76–78] Reasons for conflicting findings and recommendations include differences in the demographic characteristics of workforces, differences in priorities of employees versus employers, differences in production and safety issues for various types of industries, and the inconsistencies in the types of work schedules examined across studies, for example, fixed versus rotating schedules, speed and direction of rotation, number of hours worked per week, number of consecutive days worked, and number of rest days and weekends off. Sirois and Moore-Ede have pointed out that workplace staffing levels rather than the shift schedule determines the amount of overtime worked and may be the primary determinant of the actual length of shifts, the time employees have off between shifts, and the number of consecutive days worked.[79]

When considering changing to a 12-hour shift schedule, in order to avoid fatigue-related problems, particularly when public safety is a factor, Knauth has suggested that shifts over 8 hours be used only if (i) the nature of work and the workload are suitable, (ii) sufficient time for breaks are provided, (iii) the design of the shift system minimizes the accumulation of fatigue, (iv) coverage for absent workers is provided and overtime is not involved, (v) toxic stressors are limited, and (vi) a complete recovery is possible after the shift.[80]

Recommendations by Rosa and Colligan for shift work scheduling design involving 12-hour shifts include the following: only two to three shifts be worked in a row, for night work two is probably best, and a day or two off should follow night shifts.[81]

Tucker recommended the following fatigue countermeasures for compressed workweek schedules based on several previously published guidelines and additional studies reviewed[72]:

- Overtime, moonlighting, and long commutes should be avoided where it impacts on recovery.
- Utilize fatigue countermeasures (e.g., liberal rest breaks, job rotation) to minimize the impact of extended shifts to avoid boredom.
- Allow shorter, more frequent breaks within shifts, rather than fewer long ones.
- Ensure adequate recovery between shifts, for example, three successive rest days for long sequences of long shifts and extended recovery periods (3–4 days) after periods of night work due to the likelihood of circadian disruption.
- Take account of changes in risk outside the workplace, such as the risk of fatigue while commuting after a long shift and domestic care responsibilities.
- Readjust margins of error inherent in the work design (e.g., overtime policies) to account for the effects of extended shift length and/or quick returns.
- Ensure availability of sufficient personnel to cover all shifts to avoid increased risk associated with overtime on longer shifts.
- Redistribute workloads to be low at times of high fatigue/low alertness, for example, midafternoon and last few hours of the shift, especially at night.

Other factors related to scheduling decisions

Early morning starting times Sleep deprivation and related fatigue have been associated with early starting times (before 7–8 a.m.) for the morning shift. In addition to being sleep deprived when working night shifts, rotating shift workers may have their sleep cut short when working first shifts requiring very early rising times; and permanent day workers on early starting 12-hour shifts have been reported to be more sleepy than their night-working counterparts.[82,83] Day workers, who must begin work early in the morning, may have job/social-bound sleep restrictions with significant consequences.[84,85] Knauth et al. have advised that morning shifts should not begin too early to avoid an accumulation of sleep deficits.[35]

Schedule predictability and flexibility One reason for worker preference for a 12-hour shift system may be that there is less variation in work times compared to 8-hour three-shift rotations. In order to maximize participation in family and social activities, work schedule predictability is necessary for making plans and keeping commitments. The Centers for Disease Control and Prevention (CDC) has recommended that rotating shift schedules be stable and predictable.[86] Reorganization of the on-call shifts may minimize the negative effects on social and family well-being.[87] A recent systematic review of 10 controlled before-and-after studies of the effects of different types of flexible working arrangement found evidence that shift scheduling interventions, which allow for flexibility in working patterns and give workers more control over working time, improve health and/or well-being.[88] In their recent analysis of sickness absence days in Finnish workers, Nätti et al. found a higher incidence of sickness absence in shift work compared with day work primarily due to less working time control, supporting the conclusion that increased worker control of their work scheduling may counteract the negative health effects of shift work.[89]

Approaches to making schedule changes

The actual shift system design chosen is not the only factor which will determine the success of a schedule change. The manner in which the decision was made may be equally important. A schedule system is more apt to be well accepted by the workers if their desires with respect to free time and family/social responsibilities are considered in its design.[90,91] The acceptance of an ergonomically "good" schedule may reflect whether workers or management initiated the introduction of the new schedule.[92] According to Kogi, the trend in scheduling design is to use a participatory process.[93] Recommended steps to follow when making shift schedule changes are as follows: (i) carry out a group study of operational needs, worker preferences, health and tolerance issues of the workforce, and potential options; (ii) utilize joint planning to make plans for feasible options and specific measures; (iii) provide for feedback and dialogue to build a consensus and allow for adjustment and training; (iv) implement jointly the new work organization, in a progressive fashion if appropriate; and (v) jointly evaluate the change, taking further action as needed.

The following general guidelines for shift system design, not specific to 12-hour/compressed workweek schedules, were published by Costa et al.[94] and included in Tucker's 2006 recommendations[72]:

- Minimize night work.
- Favor quick shift rotation and clockwise rotation.
- Avoid permanent night work except for safety critical conditions, where complete adjustment is necessary.
- Avoid early morning changeovers.

- Maximize regularity, free weekends, minimum of two consecutive days off, and promote flexible work practices.
- Assess working environment for hazards and stressors.
- Assess individual workers for health status and individual differences impacting shift work tolerance (e.g., morningness, rigid sleeping habits) and social situation (responsibility for young children, long-distance commutes)
- Include provision for medical surveillance.

NORMAL PHYSIOLOGY

Numerous psychological and physiologic variables have been documented to have a demonstrable 24-hour, circadian (Latin: *circa* = about, and dies = a day)[95] rhythm, for example, body temperature, the sleep–wake cycle, cardiovascular parameters, cognitive performance, hormonal and immunologic factors, metabolic responses including to medications and toxins, and psychological variables.[96,97]

Circadian rhythms do not merely reflected responses to external time cues but also have an endogenous component, creating significant consequences for shift workers. The existence of a biological clock in humans was initially demonstrated over 50 years ago in temporal isolation studies, in which subjects were separated from all environmental and social time cues.[98] Under normal nychthermal conditions (daytime activity and nighttime sleep), the circadian system is synchronized with the 24-hour solar day by external triggers to which the biological clock is responsive. The normal phase relationships of the multitude of biological rhythms are achieved by an orchestrated response to the internal pacemaker.

The time cues, which are capable of entraining the biological clock to an external periodicity, have been termed "zeitgebers" (German: time giver). Zeitgebers allow the biological clock, which typically runs slightly slower than the 24-hour day, to be reset and entrained to the 24-hour day.[99,100] Various agents have been shown to act as zeitgebers, including light, social factors, and behavioral patterns such as eating schedules and sleep–wake schedules.[101,102] Social cues, the sleep–wake schedule and the rest–activity cycle, are relatively weak zeitgebers in comparison to sunlight (or electrical lighting of at least 7000–13000 lx).[103]

The phase-shifting effect of light on the circadian timing system is secondary to its suppressing action on melatonin secretion by the pineal gland; melatonin induces sleep and depresses the core body temperature.[104] Melatonin release is controlled by the suprachiasmatic nuclei (SCN) in the hypothalamus; the SCN is the primary endogenous circadian pacemaker/master biological clock.[105,106]

The scientific research regarding the molecular level of circadian rhythms has advanced rapidly over the last decade. Genetic researchers have identified clock genes and proteins in mammals that control the intracellular circadian clock via transcriptional modulators that allow the cellular recognition of time of day and preparation for expected stimuli. Circadian rhythm of electrical activity in SCN neurons reflects the expression of clock genes that are ultimately responsible for circadian biological rhythms. The circadian system is complex involving intercellular events and transcriptional–translational feedback loops. The expression of clock genes is not limited to the SCN, having been identified in several organs, for example, vasoactive intestinal peptide-expressing cells have been shown to play a role in entrainment by light; circadian clocks identified in cardiomyocytes and vascular smooth muscle cells allow the cardiovascular system to anticipate diurnal variations in stimuli.[107–114]

PATHOBIOLOGY (CIRCADIAN RHYTHMS AND SHIFT WORK)

During night shift work, activity is out of phase with the circadian body temperature and other coupled rhythms. In addition, because individual biological rhythms re-entrain to a time shift at different rates, each time the work schedule rotates after the time shift, the circadian system will be in a desynchronized state for a period of time. For example, sodium and potassium excretions are closely linked in a stable rhythmic environment but have significant differences in their rate of re-entrainment to phase shifts.[115] The sleep–wake rhythm adjusts faster than the body temperature rhythm, and activity re-entrains faster than many physiologic functions.[116] This desynchronization of re-entrainment of individual biological rhythms is a reflection of differences in response times of the multiple clock genes involved in maintaining the circadian system; for example, circadian oscillators in the anterior section of the SCN adapt faster than those in the posterior region[117]; oscillators in the hypothalamus adjust faster than those in peripheral tissues—resulting in temporary loss, after imposed time shifts, of the normal central control of the circadian system.[109]

Circadian rhythms are more easily re-entrained after a time shift if all the important zeitgebers, including the light–dark cycle and activity, are synchronously shifted, such as occurs with transmeridian flights. For shift workers, zeitgebers are shifted in a nonsynchronized manner. Knauth and Rutenfranz failed to find complete inversion of the body temperature rhythm in shift workers even after 21 consecutively worked night shifts and concluded that the circadian system never fully adapts to night work.[17] Other field studies of shift workers have also found adaptation to night work to be incomplete.[118–120]

The circadian system is responsible for maintaining the internal sequencing and normal relationship of physiologic events and metabolism. Biological processes are thus coordinated for optional functioning of the organism. In animal studies, circadian system disruption has been shown to result in metabolic dysregulation and misalignment of physiologic parameters and biomarkers and increase progression or susceptibility to disease.[121–128] The critical restorative functions of sleep are maximized during the nighttime hours by the normal phase relationship of biological rhythms. Overnight activity results in desynchrony between the circadian biological clock and the sleep–wake cycle. Studies in humans have provided increasing evidence of manifestations of circadian misalignment of sleep and wake activities during night work.[129–131]

Workers on night shifts experience circadian misalignment of metabolic rhythms and chronic sleep loss; both contribute to the increased risk of symptoms and diseases reported. The International Agency for Research on Cancer (IARC) has recently classified shift work as a Group 2A carcinogen—"probably carcinogenic to humans."[132,133] Other long-term health risks have been associated with the desynchronization of circadian oscillators (biological clocks) in both central nervous and peripheral tissues, which are experienced by night and rotating shift workers. Increased risk for various medical syndromes and diseases and possible pathophysiologic mechanisms are discussed in later sections in this chapter.

Night work and sleep deprivation

Regular night work is associated with chronic sleep deprivation with sleep after the night shift typically being shortened by 2–4 hours.[134,135] The sleep length of night workers is 15–20% that of day and afternoon workers, averaging 4–6 hours compared to 7–9 hours, respectively.[85,136] In addition to being shorter than nighttime sleep, day sleep is poor in quality due to frequent awakenings and disruptions of the normal REM/non-REM sleep stage pattern.[137–139]

The etiology of the sleep problem of night workers is multifactorial. A major determinant of sleep duration and quality is the endogenous circadian system.[139–142] Job schedule requirements, domestic responsibilities, and environmental conditions may also significantly contribute to the sleep problems of night workers.[143–145]

Sleep deprivation is associated with increased irritability and generalized fatigue that can compromise social and domestic interactions.[146–148] In addition to decreasing quality of life and general well-being, inadequate sleep has been shown to have deleterious effects on metabolism and hormonal functions. Short sleep duration has also been associated with increased risk of obesity, diabetes, cardiovascular disease, and overall mortality.[129,149–153]

Sleep deprivation associated with night work negatively impacts alertness and job performance. Sleepiness and falling asleep on the job are reported by workers and have been documented by objective measures. In safety-sensitive industries such as transportation, sleepiness related to work scheduling has had catastrophic consequences.[4,154–156]

DIAGNOSIS

Jet lag versus shift lag

The signs and symptoms of jet lag are an example of desynchronosis due to desynchronization of the normal phase relationships between biological rhythms within the circadian system and to the external desynchronization between the circadian system and the 24-hour solar day–night cycle to which the biological clock is normally synchronized (entrained).

Symptoms of jet lag include (in order of frequency as typically reported by frequent jet travelers) daytime sleepiness and fatigue, difficulty sleeping at night, poor concentration, slow physical reflexes, irritability, digestive system complaints, and feelings of depression. Not surprisingly, studies of shift workers have demonstrated that shift workers experience very similar symptoms,[36,157] so symptoms of "shift lag"[158] are essentially the same as those of jet lag. However, symptoms of shift work-related desynchronosis, which often go unrecognized, have more significance for long-term health. Unlike jet-lag symptoms, which are limited to a few days following travel, shift work-related desynchronosis is chronic; and the ongoing malalignment between the night work and the predominantly day-oriented social/business schedule opposes achieving circadian re-entrainment.

Shift work intolerance

For most shift workers, shift-lag symptoms are not debilitating, but for a significant minority, the symptoms of desynchronosis are significant. Surveys of former shift workers indicate that, for some, health complaints increase with continued shift work and become severe enough to cause the worker to give up a job, often following medical advice.[159,160]

Up to 20% of night or rotating shift workers have a disproportionate amount of symptoms of illness when assigned to chronobiologically poorly designed schedules.[96,161] Clinical intolerance to night work has been defined by the presence and intensity of the following set of medical complaints: (i) sleep alterations, (ii) persistent fatigue (not disappearing after time off to rest), (iii) changes in behavior, (iv) digestive system problems, and (v) the regular use of sleeping pills (near pathognomonic of shift work intolerance).[162,163] Askenazi et al. consider the presence of the symptoms in categories (i), (ii), and (v) to be essential to classify a worker as shift work intolerant.[163]

The term shift work maladaptation syndrome (SMS) has been used to refer to the typical constellation of signs and symptoms seen in shift work-intolerant workers. Symptoms are pronounced and worsen with continued exposure to shift work. The longer the worker stays on shift work, the worse the symptoms become, and eventually the worker may be fired, quit his/her job, or be involved in an accident. Inability to adjust family/social life to the work schedule and poor schedule design may significantly contribute to the degree of intolerance.[164]

Circadian rhythm sleep disorders recognized in the International Classification of Sleep Disorders (ICSD-2) include shift work disorder and jet-lag disorder.[165,166] Essential features of these two disorders include a misalignment between endogenous circadian rhythm and exogenous factors affecting the sleep period timing or duration. Criteria for the diagnosis of shift work sleep disorder (SWSD), also referred to as shift work disorder (SWD), are:

1. Complaints of insomnia or excessive sleepiness temporally associated with a recurring work schedule that overlaps the usual sleep period
2. Symptoms associated with the shift work schedule for at least 1 month
3. Circadian and sleep-time misalignment as demonstrated by sleep log or actigraphy for at least 7 days
4. Sleep disturbance that is not explained by another sleep disorder, medical or mental disorder, or medication or substance abuse disorder

Estimates of the prevalence of shift workers with SWD are similar to that reported for shift work intolerance and SMS. Using study instruments to diagnose SWD in accordance with the ICSD-2 criteria, Waage et al. found that the prevalence of SWD in offshore oil riggers was 23%.[167] Significant differences found in subjective health complaints between workers with and without SWD included, but were not limited to, gastrointestinal (GI) complaints and depression.

Some individuals are apparently relatively asymptomatic during circadian misalignment.[166] Individual factors that predispose to shift work intolerance are not fully understood. In general, age over 40–50 years, extreme morningness, and rigid sleep requirements are characteristics that have been associated with decreased tolerance for night work.[22,168–171] A review of research assessing circadian factors and shift work tolerance[16] suggests that (i) individuals with small amplitudes of certain circadian rhythms, for example, body temperature, may be more prone to desynchronization of rhythms when subjected to time shifts; (ii) some individuals are more likely to experience desynchronization of biological circadian rhythms unrelated to zeitgeber manipulation; and (iii) certain individuals are particularly sensitive to rhythm desynchronization manifesting clinically significant symptomatology. Roden et al.[172] found that night workers with high work satisfaction tended not to lose diurnal orientation of melatonin rhythms and suggested factors other than resynchronization of the circadian systems may be important for shift work tolerance.

In addition to individual biological susceptibility, factors affecting shift work tolerance include social and family situations, working conditions, and shift work schedule arrangements. Support at home and from coworkers and supervisors at work facilitate adjustment and tolerance to shift work.[173,174] Shift work disorders must be assessed in the complex framework of interrelationships of these factors.

Shift work and specific medical disorders

Gastrointestinal (GI) disorders

GI dysfunction is common in shift workers.[175–177] Gastritis or other digestive disorders have been an explanation frequently given by shift workers for absenteeism and for switching to day work for health reasons.[178] While some studies have not found an increased incidence of peptic ulcer disease (PUD) in shift workers, the majority of studies investigating PUD have found shift workers to at greater risk for the disease than day workers.[179,180]

The etiology of GI symptoms and disorders in shift workers is probably multifactorial, involving dietary factors, psychosocial stress, sleep loss, as well as circadian disruption. Night workers' mealtimes are in conflict with the circadian rhythms of gastric acidity and gastric emptying.[181,182] Shift workers may alter their diet due to lack of eating facilities available during the night shift.[183–185] In their recent review of GI conditions in shift workers, Knuttson and Bogglid discuss studies suggesting that disturbance of gastrin/pepsin secretion and decreased resistance to *Helicobacter pylori* infection are contributing factors to PUD in shift workers.[180]

Metabolic syndrome, cardiovascular disease, and diabetes mellitus

Shift work involving night work has been found to be a risk factor for development of metabolic syndrome (obesity together with dyslipidemia, hypertension, and often impaired glucose tolerance due to insulin resistance). There is also evidence that abnormalities seen in metabolic syndrome are risk factors for development of both cardiovascular disease (CVD) and Type 2 diabetes mellitus (DM).[186–189]

Results from a cross-sectional study of 226 female hospital nurses and 134 male workers at a manufacturing firm by Ha and Park[190] were consistent with there being an association between metabolic risk factors for CVD or metabolic syndrome and shift work. Regression analyses revealed an association between shift work duration and the metabolic risk factors for cardiovascular disease. Duration of shift work was significantly associated with systolic blood pressure or cholesterol level among male workers aged 30 or more. Body mass index (BMI) was nonsignificantly associated with the duration of shift work in both male workers and female nurses ≥30 years of age. Recently Scheer et al. conducted a laboratory protocol study of metabolic measurements during inverted sleep–wake cycles.[191] This allowed control of mealtimes. Circadian misalignment, when subjects ate and slept 12 hours out of phase from their habitual times, systematically decreased leptin, increased glucose (despite increased insulin), reversed the daily cortisol rhythm, and increased mean arterial blood pressure. In three of eight subjects, during circadian misalignment, postprandial glucose responses were in the prediabetic range. Several studies have found increased risk of hypertension, undesirable lipoprotein profiles, elevated triglyceride levels, and obesity in shift workers compared to day workers.[192–200]

CARDIOVASCULAR DISEASE Well-designed epidemiological studies have found an increased risk of CVD associated with shift work of, on average, around 40%.[201] Increased risk of stroke has also been reported in shift workers. Analysis of cohort study data from the Nurses' Health Study looked at the risk of ischemic stroke related to the total number of years nurses had worked rotating night shifts.[202] After adjusting for multiple vascular risk factors, of the 80 108 study participants available for analysis, 60% reported working at least 1 year of rotating night work. Rotating night shift work was associated with a 4% increase in risk of ischemic stroke for every 5 years of shift work exposure (hazard ratio, 1.04; 95% CI: 1.01–1.07; $p_{trend} = 0.01$). The increased risk may be limited to women with a history of shift work for ≥ 15 years.

Prospective and historical prospective studies have found a dose–response relationship between shift work exposure and cardiovascular disease. In a small historical prospective study of rotating shift workers, Knutsson et al. found an increased risk of CVD in shift workers.[203] The relative risk of ischemic heart disease (IHD) increased with increasing years of exposure to shift work (6–10 years, relative risk (RR) = 2.0; 11–15 years, RR = 2.2; 16–20 years, RR = 2.8; combined RR = 1.4). This dose–response relationship continued for up to 20 years of exposure ($p < 0.05$). (Twenty-one or more years of shift work had a RR = 0.4, attributed to the healthy worker effect). Based on multiple logistic regression analysis, the association between shift work and an increased risk of IHD was independent of age and smoking habits. Subsequently, analysis of data from the ongoing prospective Nurses' Health Study during 4 years of follow-up of 79 109 nurses between 1988 and 1992 found evidence that ≥ 6 years of night shift work exposure may increase the RR of CHD, consistent with the previous findings in the paper mill workers.[204] This large study allowed for control of more confounders, in addition to age and smoking, including body mass index (BMI), hypertension, diabetes, hypercholesterolemia, physical activity, and alcohol use. The multivariate adjusted RR was 1.21 for women reporting < 6 years and 1.5 for women reporting ≥ 6 years of rotating night work. In 2005, Karlsson et al. reported findings for a total and cause-specific mortality in pulp and paper workers in Sweden in a historical cohort between 1952 and 2001.[205] Plant records provided accurate shift work exposure information to rotating night work shifts; mortality data were obtained from the national cause of death register. A longer duration of shift work was associated with increased risk of CHD. Shift workers with > 30 years of shift work exposure had the highest risk (SRR, 1.24, 95% CI: 1.04–1.49).

Additional analysis of data from the Helsinki Heart Study by Tenkanen et al. demonstrated an interaction of shift work exposure with lifestyle factors known to increase the risk of coronary heart disease (CHD).[206] For shift workers, the relative risk of CHD rose gradually with increasing numbers of adverse lifestyle factors, but for day workers, no clear dose–response pattern was found. Overall the results of this study were consistent with the notion that shift work may have a triggering effect on other lifestyle factors that can increase the risk of CHD, and that active preventive medicine intervention is particularly important for shift workers.

There is evidence that chronobiological factors independently contribute to the increased risk of CVD reported in shift workers. Young has provided a discussion of evidence supporting that notion that disruption of the normal diurnal sleep–wake activity which occurs during night work results in desynchrony of circadian clocks within the cardiovascular system interfering with its ability to anticipate the variations in neurohormonal stimuli, which may contribute to the development of cardiovascular disease in shift workers.[207] Shift work has been shown in studies controlled for differences in work demands and work stress to be an independent risk factor for CVD.[208]

Peter et al. studied the association between shift work and cardiovascular risk factors of hypertension and elevated blood lipids, in the context of the psychosocial work environment.[209] In addition to finding direct effects of shift work on cardiovascular risk, mediating effects of psychosocial work factors (effort–reward imbalances) were found, supportive of the hypothesis that a stressful work environment can act as a mediator of adverse effects of shift work on hypertension and partly on atherogenic lipid levels.

Boggild and Knutsson[201] have previously presented evidence for a complex mechanistic model involving interdependent pathways involving circadian system disruption, behavioral lifestyle changes, disruption of social interactions, as well as changes in biomarkers of CVD, for example, development of dyslipidemia and hypertension. Most recently, Puttonen et al.[210] reviewed current research information regarding mechanisms between shift work and CVD finding sufficient epidemiological evidence for several possible disease mechanisms. They described an updated model of interrelated pathways stemming from "circadian" stress brought on by shift work. The model describes mechanisms whereby the interaction of psychosocial, behavioral, and physiologic stress factors may contribute to cardiovascular disease outcomes (see Figure 10.1).

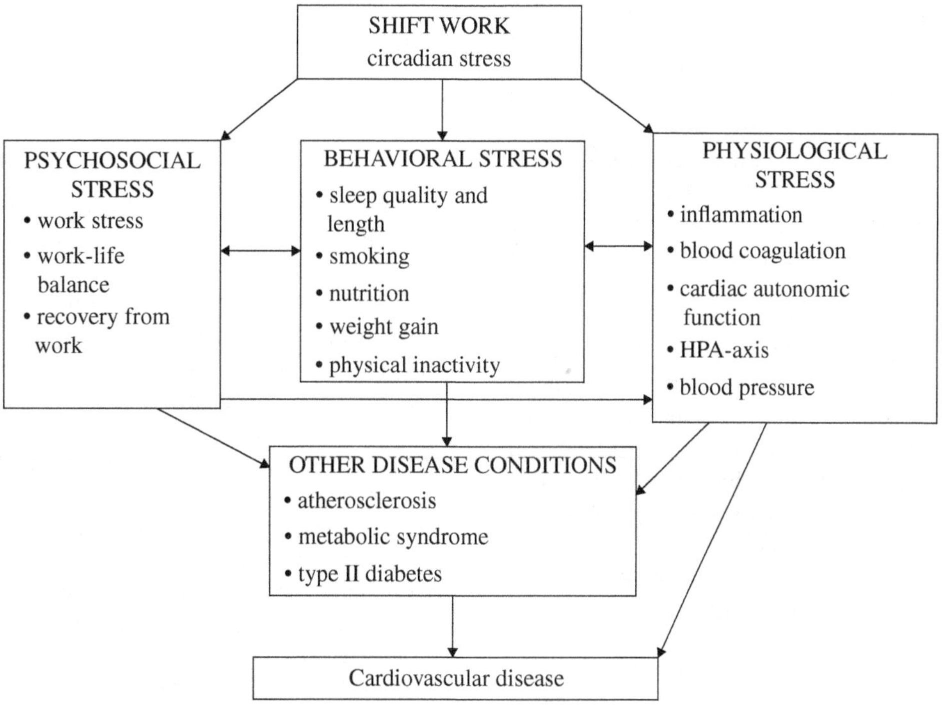

FIGURE 10.1 Model of interrelated pathways from shift work "circadian stress" to cardiovascular disease outcomes. Used with permission from Puttonen S, Hãrmã M, Hublin C. Shift work and cardiovascular disease – pathways from circadian stress to morbidity. *Scan J Work Environ Health* 2010; 36(2):96–108.[210]

Diabetes mellitus (DM) Type 2 As noted above, changes in glucose tolerance with postprandial glucose levels in prediabetic ranges have been demonstrated in laboratory studies of circadian misalignment of sleep–wake cycle. Several studies have supported an association of night work and an increased risk of obesity and metabolic syndrome, conditions known to be related to Type 2 DM.[187,188,211–214]

In addition to the increased risk of CHD reported in 2005 by Karlsson et al., the risk of diabetes was also increased as the number of years of shift work exposure increased.[205] Kroenke et al. analyzed data over 6 years regarding work characteristics and incidence of Type 2 diabetes mellitus (DM), from the prospective Nurses' Health Study (NHS) II.[215] They found a positive association ($p_{trend}<0.001$) between years in rotating night shift work and diabetes, mediated by BMI. Subsequently, Pan et al. performed an updated analysis of cohort data from the Nurses' Health Study I (1988–2008, 69 269 women) and Nurses' Health Study II (1989–2007, 107 915) studies; statistical analysis adjusted for diabetes risk factors revealed a monotonic association with duration of rotating night shift work and increased risk of Type 2 DM in both cohorts.[216] Compared to women with no shift work exposure, the hazards ratio (95% CI) for women with 1–2 years of shift work was 1.05 (1.00–1.11), for 3–9 years 1.20 (1.14–1.26), for 10–19 years 1.40 (1.30–1.51), and for ≥20 years 1.58 (1.43–1.74), $p_{trend}<0.001$. After adjustment for updated BMI, there was attenuation in the association. These findings were consistent with an increased risk of Type 2 DM in women, in part related to body weight; women with >20 years of shift work exposure still had a 44% increased risk of developing diabetes after the adjustment for BMI. In 2005, Japanese researchers reported the results from a longitudinal study of male blue-collar and white-collar workers over 8 years using data from annual health examinations and glycosylated hemoglobin levels.[217] Relative risk of DM for two-shift workers and three-shift workers, compared with fixed daytime workers, was 1.73 and 1.33, respectively, after adjustment for confounding factors, but these increases were not statistically significant. Using white-collar workers as the reference group, there was a significantly increased risk of DM for the two-shift workers (RR = 2.01), but not for three-shift rotators, or daytime blue-collar workers. Another Japanese longitudinal study, a year later, of 5629 male steelworkers compared onset of DM in day workers with alternating shift workers over a 10-year period using data from annual health examinations.[218] Analysis of this data revealed an increased odds ratio (OR) for alternating shift workers compared to day workers of 1.35 (95% CI: 1.05–1.75); OR for BMI was also increased 1.28 (1.23–1.33).

Reproductive health

Preterm delivery (PTD), small for gestational age (SGA), and low birth weight (LBW) McDonald and colleagues conducted a large cross-sectional study in Montreal, in which 22761 live births were assessed for maternal employment risk factors for LBW and preterm delivery (PTD).[219] Observed-to-expected (*O/E*) ratios for LBW and PTD were increased significantly for long hours of work (46 hours/week or more), with ratios of 1.34 ($p<0.01$) and 1.24 ($p=0.03$), respectively. Rotating shift work was also associated with LBW, $O/E=1.38$ ($p<0.01$). A significant association was found in the services sector between rotating shift work and PTD, $O/E=1.88$ ($p<0.01$). Analysis of prematurity factors from the Montreal study, in which gestational age was allowed for in the analysis of birth weight data, done by Armstrong et al. suggested that shift work might slow fetal growth and increase the risk of PTD.[220]

In a retrospective cohort study, Axelsson et al. looked at the effect of night work, evening work, working irregular hours outside the period of 0645–1745, and rotating shift work compared to permanent day work on pregnancy outcome.[221] They found an association between irregular working hours and an increased risk of LBW. For nonsmoking mothers with infants of birth order 2+, an increased risk of LBW was associated with irregular hours ($p<0.01$) or rotating shift work ($p<0.05$).

Xu and colleagues found higher proportions of preterm birth and LBW for rotating shift workers compared to regular day workers, with adjusted OR of 2.0 for PTD and 2.1 for LBW.[222] Subsequently Fortier et al. reported the results of a Canadian study of 4390 women who had recently delivered live-born singleton infants.[223] No deleterious effect was found for those doing shift work (defined as occasional evening or night work), OR = 1.03. No increase in the risk of delivering a small-for-gestational-age (SGA) baby was found related to working shift work or regular evenings or nights. Interestingly, long work hours (40 or more) were also not related to PTD or SGA outcomes.

Nurimen found a small association between shift work (varying types of nonregular day work including two- and three-shift rotations) and SGA infants in a study comparing 1475 mothers of infants with certain structural malformations to mothers of babies without malformations.[224] Mothers who had worked shift work schedules during most of their pregnancy had a slightly increased risk of having SGA infants compared to day-working mothers (adjusted rate ratio = 1.4).

Research data is limited for permanent night work. Saurel-Cubizolles and Kaminski did not find an association between night work and preterm delivery or low birth weight.[225] However, only 4% of the female hospital employees interviewed were night workers, and some of them worked nights only occasionally. In another study, no increased risks were found for pregnancy outcome of permanent night workers; however, 91% of the night workers questioned were part-time employees.[221] Fortier et al. reported an increased OR of 1.45 for delivering preterm for women working regular evening or night shifts who continued to do so after 23 weeks of pregnancy.[223]

Croteau et al. conducted a case–control study by evaluating the relationship of occupational conditions during pregnancy and the increased risk of delivering a small-for-gestational-age (SGA) infant and whether taking measures to eliminate these conditions decreases that risk.[226] 1536 cases and 4441 controls were selected from 43898 women who had single live births between January 1997 and March 1999 in Québec. The women were interviewed by telephone after delivery. The risk of having an SGA infant increased with an irregular or shift work schedule alone; and there was a cumulative index of at least two of the following occupational conditions: night work, irregular or shift work schedule, standing, lifting loads, noise, and high job strain with low social support. As the number of job conditions that were not eliminated during pregnancy increased, the risk increased ($p_{trend}=0.004$; ORs = 1.00, 1.08, 1.28, 1.43, and 2.29 for 0, 1, 2, 3, and 4–6 conditions, respectively). Elimination of the conditions before 24 weeks of pregnancy brought the risks close to those of unexposed women.

Bonzini et al. reported the results of a systematic review of the epidemiological research between 1966 and February 2010 relating shift work and pregnancy outcomes including PTD, LBW, small for gestational age (SGA), and preeclampsia.[227] Twenty-three relevant studies were retrieved. Preterm delivery was consistently defined as birth at <37 weeks of gestation. LBW was defined as a birth weight <2500 g, except for one study that reported risk estimates for birth weight <3000 g. SGA was defined as birth weight <10th percentile for gender and gestational age, with one study including infants <5th percentile. Pooled estimates of relative risk (RR) were calculated in random-effects meta-analyses. Their search identified 17 original studies that investigated the association between shift work and PTD: 10 studies concerning SGA and 6 studies concerning LBW. The pooled estimate of RR (16 studies) for PTD was 1.16 (95% CI: 1.00–1.33); when five reports of poorer methodological quality were excluded, there was no longer statistical significance. Increased RRs for LBW were also observed (RR 1.27, 95% CI: 0.93–1.74) and for SGA (RR 1.12, 95% CI: 1.03–1.22). Estimates of risk for PTD tended to reduce with inclusion of more recent studies in meta-analyses. The authors pointed out that one explanation of this time trend could be that in most countries, precautionary legislation had been introduced, allowing pregnant workers to be assigned to nonshift work schedules, or to take earlier prenatal leave, so that fewer women continued shift work during the later stages of pregnancy. Night shifts may have been voluntarily suspended if women suspected that it could be detrimental to their pregnancy, and thus risks in women who continue to work shifts late in pregnancy might be underestimated due to a healthy pregnant worker effect (healthier women with uncomplicated pregnancies being less likely to change work schedules). Two studies were noted to support the notion that the risk of PTD associated with shift work was higher in women whose work conditions did not change during the pregnancy[226,228]; and a study comparing European countries found that significant associations of PTD with shift work were mainly observed in countries where long prenatal leaves were infrequent and legislative support for preventive measures was weaker.[229] Ten studies were analyzed looking at the association between shift work and the risk of delivering an SGA baby, including five prospective cohort investigations. With the exception of one, the studies tended to rule out a more than moderate effect (RRs ranging from 0.8 to 1.5). The pooled risk estimate was 1.12 (95% CI: 1.03–1.22, test for heterogeneity $p=0.39$). When a poor quality study was removed from the analysis, the RR was 1.10 (95% CI: 1.00–1.20). Six studies were analyzed looking at the association between shift work and risk of delivering an LBW infant. One of the three cohort studies showed a significantly elevated risk. In the pooled meta-analysis, the combined risk estimate was 1.27 (95% CI: 0.93–1.74, test for heterogeneity $p=0.39$). Overall, the authors concluded their findings suggest that the risk of preterm delivery, low birth weight, or small for gestational age from working shift work in pregnancy is small; and the available data does not make a compelling case for mandatory restrictions on shift working in pregnancy. Pooled estimates of risk for specific patterns of work schedules were not calculated because of the small number of studies with sufficient information. The authors point out that further studies are needed to address the question of whether adverse birth outcomes are related to different types of rotating work schedules or to fixed night work and suggest, as circumstances permit, pregnant women who wish to reduce their exposure to shift and night work be allowed to do so.

Fetal loss and spontaneous abortion In a 1988 cross-sectional analysis of 22613 previous pregnancies, in 56067 Montreal women, McDonald et al. found a small but significant increase in spontaneous abortion associated with working 46 hours or more per week (*O/E* 1.19, $p<0.01$) and for rotating shift work (*O/E* 1.25, $p<0.01$).[230] Axelsson et al. studied the relation between shift work and spontaneous abortion in a cross-sectional study of 3358 Swedish midwives.[231] Hours of work information was divided into day workers, permanent night work, and two- or three-shift rotators. The OR was increased for night work and three-shift work (OR = 1.63 and 1.49, respectively) for women who worked during the first trimester. After restricting analysis to first pregnancies, a significant increased risk was also found for night work (OR = 6.89). The OR was significantly elevated for two-shift workers (OR = 2.70), but not significantly increased for three-shift schedules. When analysis was done separately for early and late spontaneous abortions, night work, but not rotating shift work, was associated with an elevated risk of late (beyond the 12th week of pregnancy) abortions (OR = 3.33).

In a case–control study, Infante-Rivard et al. compared the work schedules of 331 women who had experienced pregnancy loss to 993 pregnant women matched for gestational age. For fixed evening schedules, the adjusted OR was substantially elevated at 4.17, and for fixed nights the adjusted OR was also elevated, but to a lesser degree (OR = 2.68).[232] Axelsson et al. reported the results from their retrospective cohort study included a slight, but not statistically significant, increase in the risk for miscarriage (RR = 1.44; 95% CI: 0.83–2.51) associated with irregular working hours including rotating shift work.[221] Axelsson and Molin found an increased miscarriage rate of borderline significance in shift workers (OR = 2.07; 95% CI: 0.98–4.34).[233] In a retrospective cohort study of laboratory workers, a significantly increased risk of miscarriage was reported for shift workers (RR = 3.2; 95% CI: 1.36–7.47).[234] Two other studies, as reviewed by Nurimen, also found an increased risk of miscarriage in shift workers.[235–237]

Zhu et al. reported the results of a prospective cohort study using the Danish National Birth Cohort. Fixed night work was associated with fetal loss.[238] Over 33000 pregnancies were identified in day workers and over 8000 pregnancies in shift workers between 1998 and 2001. Fetal loss included spontaneous abortion and stillbirth. The hazard ratios for fetal loss compared to day workers, adjusted for potential confounders, was increased for fixed night shift work: for all fetal loss, HR = 1.85 (CI: 1.00–3.42); for spontaneous abortion, HR = 1.81 (0.88–3.72); and for stillbirth, HR = 1.92 (0.59–6.24). No high risk was found for rotating shifts.

Two case–control studies of spontaneous abortion and shift work did not find an increased risk.[239,240] As pointed out by Nurimen, in these two negative studies, rotating shifts were part of broad-based exposure categories and were not analyzed explicitly.[237]

Bonde et al. recently conducted a systematic review and meta-analysis of published studies between 1996 and 2012, which looked at the risk of miscarriage related to shift work, working hours, and physical stressors.[241] Working fixed nights was associated with a moderately increased risk of miscarriage (pooled RR 1.51; 95% CI: 1.27–1.78), as determined by the fixed model OR in five better quality studies reporting RRs for fixed night work compared with day work. Pooled fixed meta-OR for studies ($n=7$) reporting risk of miscarriage for three-shift schedules or evening/night shifts compared to day or two-shift workers was associated with a small increased risk (OR 1.12; 95% CI: 0.96–1.42).

SHIFT WORK AND SUBFECUNDITY Ahlborg et al. studied subfertility in Swedish midwives in a survey sent to 3985 midwives (84% response rate).[242] Those who worked rotating shifts or permanent nights had decreased fertility compared to day workers. Bisanti et al. surveyed women during prenatal clinic visits or after giving birth to determine the time of unprotected intercourse prior to conception.[243] Rotating shift work was associated with an increased risk of subfecundity (OR=2.0). Spinelli et al. interviewed new mothers who had delivered within the preceding week. After adjustment for confounders, fertility was not significantly reduced in shift-working mothers [fecundability (conception rate) ratio=0.9; $p>0.10$].[244] A Japanese questionnaire study on working conditions found pregnancy rates lower for women doing shift work (10.0%) compared to day workers (18.1%) ($p<0.01$).[236]

Tuntiseranee et al. recently reported no association between shift work and subfecundity, but did find long hours of work to increase time to pregnancy.[245] This effect was greatest when both partners were working over 70 hours/week (OR=4.1 in primigravid women and 2.0 for all pregnant subjects). There was no increased OR when only the male partner worked long hours.

Zhu et al. analyzed data from the Danish National Birth Cohort (DNBC), including 39 913 pregnant women who were enrolled from March 1, 1998, to May 1, 2000, to examine whether shift work was associated with reduced fecundity as estimated by time to pregnancy (TTP).[246] Data on job characteristics and TTP (0–2, 3–5, 6–12, and >12 months) were used for 17 531 daytime workers and 3 907 shift workers who had planned pregnancies. Fecundity odds ratios were calculated (OR>1 indicating a shorter TTP). Fixed evening workers and fixed night workers had a longer TTP. Compared with daytime workers, the adjusted ORs were 0.80 (95% CI: 0.70–0.92) for fixed evening workers, 0.80 (95% CI: 0.63–1.00) for fixed night workers, 0.99 (95% CI: 0.91–1.07) for rotating shift (without night) workers, and 1.05 (95% CI: 0.97–1.14) for rotating shift (with night) workers. The proportions of unplanned pregnancies and contraceptive failures were higher among fixed evening and fixed night workers. The researchers concluded that the slightly reduced fecundity among fixed evening workers and fixed night workers may be mediated by pregnancy planning bias or differential options for sexual contacts; and there was no unequivocal evidence of a causal association between shift work and subfecundity.

CANCER

Taylor and Pokock were the first to report an increased risk of cancer in shift workers in a 1972 mortality study.[247] A 1990 mortality study by Raffnsson and Gunnarsdottir also reported an increased risk of cancer associated with night work.[248]

A 1996 Norwegian nested case–control study of breast cancer incidence in radio and telegraph operators, working at sea, followed 2619 mostly postmenopausal women for about 30 years.[249] Cancer cases were identified from National Norwegian Cancer Registry. Job histories of work on ships were collected for shift work, as well as travel through time zones classified for each ship mentioned in the job histories to define shift work. After controlling for duration of employment and after adjustment for age and year of birth of first child, there was an excess risk of breast cancer associated with exposure to night shift work (SIR, 1.5; 95% CI: 1.1–2.0). Breast cancer risk was highest in women with the highest cumulative shift work history compared to no shift work, RR=6.1(95% CI: 1.5–24.2). There appeared to be an increased risk of breast cancer in women ≥50 years of age with increasing cumulative exposure to shift work compared to no shift work (low exposure 0–3.1 years, adjusted for duration of employment, RR=3.2, 95% CI: 0.6–17.3; high exposure 3.1–20.7 years, adjusted for duration of employment, RR=4.3, 95% CI: 0.7–26.0; $p_{trend}=0.13$).

Subsequent studies looking at the risk of breast cancer in shift workers followed between 2001 and 2007 includes the following:

- A 2001 case–control study by Davis et al. looked at sleep patterns and habits, lighting in bedrooms, and shift work history in 813 cancer cases (identified by the Cancer Surveillance System of Seattle) and 793 age-matched controls.[250] Information about sleeping habits, light-at-night exposure, and lifetime occupational history was obtained from in-person interviews. Night work was defined as at least one "graveyard" (beginning after 19:00 and ending before 09:00) shift/week in the 10 year prior to diagnosis. After adjusting for parity, family history of breast cancer, contraceptive use, and history of hormone replacement therapy, there were an increased risk for breast cancer associated with graveyard shift work (OR 1.6, 95% CI: 1.0–2.5), a 13% increase in risk for additional year of having worked at least one such shift per week (OR 1.13; 95% CI: 1.01–1.27), and a 14% increase risk in women reporting, for any reason, frequently not sleeping during the middle of the night (*when melatonin levels are typically highest*) (OR, 1.14; 95% CI: 1.01–1.28, per night/week; OR 1.7 for the group with at least 2.6 nights/week of interrupted sleep).
- A population-based case–control study nested in a cohort of all female employees in Denmark based on national pension fund data found that women who worked at night for at least 6 months had a 50% greater risk for breast cancer compared to age-matched controls (OR, 1.5; 95% CI: 1.2–1.7).[251] A positive trend was noted with increasing duration of work at night. 7035 women with incident breast cancer were identified in

the Danish Cancer Registry. Individual employment histories were reconstructed using information from the pension fund. Night work definition was based on information from the nationwide interview survey of living and working environment conditions; trades with at least 60% of female responders working at night were considered to have a predominant nighttime schedule; trades with <40% reporting nighttime schedules were classified as day workers. The RR of breast cancer was 1.5 (95% CI: 1.3–1.7; 434 cases) among women who worked in night work trades at least a half year, at least 5 years before diagnosis. Analysis was done after controlling for age, social class, parity, and age at birth of first child and last child. For the subgroup of women with more than 6 years predominantly working at night, the RR was 1.7 (95% CI: 1.3–1.7; 117 cases). In a subanalysis of nurses (41% having predominantly night work), the RR of breast cancer was also significantly evaluated (RR, 1.3; 95% CI: 1.1–1.4).[252]

- Two large prospective cohort studies of night shift work and breast cancer risk used data from the Nurses' Health Study cohorts (Nurses' Health Study I and Nurses' Health Study II).[253,254] The Nurses' Health Study began in 1976, when 121 701 registered nurses in the United States, 30–55 years of age, were enrolled and completed a questionnaires about their health status, medical history, and known or suspected risk factors for cancer. Since baseline, questionnaires have been mailed biannually with the exception of lifetime history of night work (total number years working rotating night shifts at least three nights per month), which was assessed in 1988. In 2001 Schernhammer et al. reported the results of follow-up of 78 562 women in the Nurses' Health Study who answered the 1988 questionnaire and were cancer-free at baseline over 10 years (1988–1998).[253] 2441 incident breast cancer cases were documented during that time. After adjustment for known breast cancer risk factors, their analysis revealed a significantly increased risk of breast cancer in women who had worked ≥30 years of rotating night work compared to women who never worked rotating night shifts (RR 1.36; 95% CI: 1.04–1.78). The risk increased with numbers of years of shift work (test for trend, $p=0.02$). The similarly designed Nurses' Health Study II study was established in 1989, when the prospective cohort of 116 671 registered female nurses (no overlap with Nurses' Health Study) were enrolled and completed the baseline questionnaire, including questions on total months worked on rotating night shifts at least three nights per month. Information was updated in 1991, 1993, 1997, and 2001. In 2006, Schernhammer et al. reported, in 115 022 mostly premenopausal women, an elevated breast cancer risk of 1.79 (95% CI: 1.06–3.01; $p=0.65$) among those who worked ≥20 years of rotating night shift work compared with women with no work history of rotating night shifts.[254]
- In 2006, Lie et al.[255] reported the results of a nested case–control study within a cohort of 44 835 Norwegian nurses. Using the nationwide cancer registry, 537 breast cancer diagnoses during 1960–1982 were identified. Work history was reconstructed from the nurses' registry (self-report of work history periodically updated) and census information. It was assumed that nurses doing clinical work at infirmaries worked at nights (with the exception of physiotherapy and outpatient departments); it was assumed that work sites other than infirmaries involved day work only. An association was found between duration of night work and breast cancer risk ($p_{trend}=0.01$). The RR associated with >30 years of night work was 2.21.
- In 2006, O'Leary et al. published results of their case–control study in New York—the Electromagnetic Fields and Breast Cancer on Long Island Study (EBCLIS).[256] They did not observe an association between night work and breast cancer risk. To evaluate the effects of electromagnetic frequency, women from within this case–control study were selected according to their degree of residential stability (EBCLIS component). EBCLIS comprised 576 breast cancer cases and 585 matched population-based controls. Occupational history was obtained in personal interviews as well as history of residential light-at-night exposures. Shift work was defined as "ever" working in at least one job during the past 15 years that included evening shifts (could start in the afternoon and end as late as 02:00) and overnight shifts (starting as early as 19:00 and continuing until the next morning). Results were adjusted for age, parity, education, family history of breast cancer, and history of benign breast disease.
- Results of a retrospective registry-based cohort study were reported in 2007 comprising all 1 148 661 Swedish women active in the workforce per 1960 and 1970 census reports.[257] Workers were followed up for breast cancer morbidity using the Swedish Cancer Registry. Annual surveys of living conditions among 46 438 randomly selected subjects who participated in a personal interview were used for assessing night work. Shift workers were defined as those who reported their workplace had a rotating schedule with at least three possible shifts per day or had any night work hours (between 01:00 and 04:00) at least 1 day during the week preceding the interview. Participants working in job titles and industry combinations (from census data regarding workers' industry and socioeconomic status) with at least 40% shift work were classified as shift workers.

> The reference group comprised subjects in occupation–industry combinations in which <30% reported being shift workers. In analyses using the census information for the definition of exposure, no increase in risk was found for women classified as shift workers.

In October 2007, a panel of 24 scientists who reviewed the epidemiological, experimental, and mechanistic research for International Agency for research on Cancer (IARC) concluded that there is "limited evidence in humans for the carcinogenicity of shift work that involves night work"; "sufficient evidence in experimental animals for the carcinogenicity of light during the daily dark period (biological night)"; and "shiftwork that involves circadian disruption is probably carcinogenic to humans (*Group 2A*)."[132] The research included in the IARC panel analysis is reviewed in the subsequent IARC Monograph published in 2010.[258] The epidemiological data was primarily based on the seven breast cancer studies reviewed above.[249–257] The IARC panel discussion noted that each of the studies used different definitions of shift work; that six of the eight breast cancer studies, including the two prospective cohort studies from the Nurses' Health Study, "consistently pointed towards a modestly increased risk of breast cancer among long-term employees who performed night shiftwork, defined in different ways"; and that one of the two negative studies had important limitations in design. Other points included the following: there were a relatively limited number of studies, most focusing on the single profession of nursing; there was some potential for confounding by unknown risk factors, and inconsistent and inaccurate exposure assessments of shift work, which may have biased the results toward the null; the evidence for an association with breast cancer and night work was consistent in the studies specifically designed to address this question; and studies of the incidence of breast cancer in female flight cabin crews provided additional support.

The conclusions of the IARC working group drew attention to the evidence of an association of night work with risk for developing cancer, prompting additional reviews, meta-analysis of the existing data of the breast cancer studies, and new research.

Costa et al.[259] reviewed the epidemiologic studies included in the IARC report and a recent case–control study, published in 2010, using data from the German Gene–Environment Interaction and Breast Cancer (GENICA) study[260] and concluded that the loose definitions of exposure to night work in the published studies did not allow for proper assessment of the risk of cancer associated with circadian disruption. In the GENICA study, shift work exposure (defined as working the full period between 24:00 and 05:00) was based on personal interview information; cases experience about 800 night shifts compared to 300 among controls. Long-term night work, ≥20 years, was associated with a statistically nonsignificant increase in breast cancer risk, OR 2.48 (95% CI: 0.62–9.99). Limitations of the GENICA study include low prevalence of night work, especially long-term exposure, and the retrospective assessment of work history.

In 2011 Hansen and Stevens published results from a Danish case–control study of breast cancer in nurses (identified from the national cancer registry) and collected detailed information on lifetime shift work history.[261] Overall, nurses working rotating shifts after midnight had significantly increased risk of developing breast cancer compared to permanent day-working nurses (OR 1.8, 95% CI: 1.2–2.8). The greatest increased risk was seen associated with long-term day–night rotating shift schedules (OR 2.6, 95% CI: 1.8–3.8).

Three meta-analysis of breast cancer studies in shift workers were published in 2013. Kamdar et al. identified 15 studies published by 2012 that met their inclusion criteria.[262] Using a random-effects model, they found the pooled RR for women with ever night shift exposure to be 1.21 (95% CI: 1.00–1.27, $p=0.056$). For long-term night work, defined as ≥8 years, RR was 1.04 (95% CI: 0.92–1.18). Subgroup analysis for nurses with long-term exposure was reported to suggest an increased risk of breast cancer. Jia et al. reported, in their analysis of 13 studies through 2012, a pooled adjusted RR of 1.20 (95% CI: 1.08–1.33) for risk of breast cancer associated with ever working versus never working night shift work.[263] Later, in August 2013, Wang et al. published results of their meta-analysis of cohort and case–control studies published in May 2013.[264] They summarized evidence by frequency and duration of cumulative exposure to night shift work and used a dose–response regression model to evaluate the relationship between exposure to night work and risk of breast cancer. Ten studies were included. The pooled adjusted RR for association between "ever exposed to night shift work" and breast cancer was 1.19 (95% CI: 1.05–1.35). Analysis of the dose–response relationship revealed a 3% increase in risk of breast cancer for every 5 years of exposure to night work (pooled RR 1.03 (95% CI: 1.01–1.05) $p_{heterogeneity} < 0.001$). This meta-analysis also suggested that an increase of 500-night shifts increases the RR 13%.

Also, in August 2013, results of an Australian case–control study were published by Fritschi et al. Women diagnosed with breast cancer between 2009 and 2011 were identified from the Western Australian Cancer Registry (mandatory reporting of invasive cancer); randomly selected controls were from the Western Australian electoral roll (compulsory adult enrollment). 1202 cases and 1785 controls completed the study.[265] Questionnaires followed by telephone interviews were used to obtain work history. A statistically significant association was found between breast cancer and phase shift caused by night work (OR 1.22, 95% CI: 1.01–1.47). There was evidence of a dose–response relationship, but no duration–response relationship.

In 2014, Hansen looked at the influence of night shift work on the survival of breast cancer.[266] Women participating in two separate Danish case–control studies of breast cancer were interviewed by phone to obtaining information including night shift work history and known risk factors for breast cancer. The national cause of death register was used to follow up for death. Time-to-event analyses were done using Cox proportional hazards models and Kaplan–Meier survival plots. There was a significant tendency of decreasing survival in fixed and rotating night shift workers compared to day workers and by increasing years of prior nondaytime work ($p = 0.04$).

The current research provides support for biological plausibility and potential mechanisms for an association of long-term exposure to night shift work with increased risk of cancer. For example, animal studies have demonstrated that the development of neoplasias has been associated with disruption of the rhythmicity of the circadian period 2 gene.[267] Additional findings have been reviewed in detail by Costa et al.[259] and in the IARC monograph[258] including alteration of endocrine rhythms and immune system function due to sleep deprivation, nocturnal suppression of melatonin by light at night (LAN), and circadian disruption at the cellular and molecular level. The phase shifts experienced by night and rotating shift workers result in molecular level desynchronization in circadian oscillators in peripheral tissues, as well as the CNS, which provides a pathophysiological basis for the reported increased risk for shift workers of developing cancers. Immune responses may be suppressed via circadian disruption of the central oscillator in the SCN. Immune suppression reflects multiple events including decreased number of natural killer (NK) cells, cytotoxic lymphocytes, and proinflammatory cytokines, interferon, and tumor necrosis factor (TNF). Suppression of melatonin may lead to estrogen elevation with clinical significance in endocrine-dependent tumors; melatonin counteracts the enhancing effects of estradiol on breast cancer cell activity; melatonin can act as a free radical scavenger by activation of antioxidative pathways and has antiproliferative effects on human cancer cells *in vitro*. Melatonin has been shown to be oncostatic in certain tumor cells; exogenous melatonin has been shown to have anti-initiating and oncostatic activity on chemically induced cancers.[268–272] Clinical trials in humans have also demonstrated favorable responses to melatonin in cancer patients, alone or in combination with standard treatment regimens.[273] Chronobiologists have also noted the effects of loss of normal diurnal rhythm in cortisol levels associated with sleep disruption which support a mechanistic shift work role in the observed association between shift work and increased risk of breast cancer.[274] The circadian clock may act as a tumor suppressor at systemic and molecular levels via clock-controlled genes involved in controlling cell cycles and tumor suppressor genes. Processes relevant to carcinogenesis, for example, DNA repair and cellular proliferation, have circadian variation.

In the very recent study of molecular mechanisms, Liu et al. followed up on previous findings indicating that long-term exposure to LAN in night workers may result in dysregulated patterns of methylation, inducing alteration of microRNAs relevant to cancer.[275] They found a 49% increase in miR-34b promotor methylation in shift workers—with results suggesting that long-term shift work may increase the risk of breast cancer via methylation-based suppression of miR-34b, with a related reduction in immune-mediated antitumor capacity.

In another recent study, Monsees et al. hypothesized that circadian genes influence breast cancer risk in women and investigated gene–environment interactions in nurses working rotating shift work.[276] Using blood samples collected from >29 000 cancer-free participants in the Nurses' Health Study II study between 1996 and 1999, the researchers looked at variants of genes relevant to the circadian system in women in the Nurses' Health Study II cohort. They tested for associations between certain genotypes and breast cancer risk and potential interactions between genotype and rotating shift work in a subset. Data was available for cumulative exposure to rotating shift work prior to the last blood sample collection on the subset (438 cases; 880 controls). Analysis of interactions between clock genes and shift work was limited to incident cases as diagnosis of cancer may influence work schedule. The researchers tagged genes with known key roles in circadian regulation. The results provided evidence that a coding single-nucleotide polymorphism (SNP) in NPAS2 (Ala394Thr;rs2305160) may modify the influence of shift work on risk of breast cancer. None of the selected genetic variants were significantly associated with breast cancer risk; however rs23051560 (Ala394Thr), in the largest circadian gene, was most strongly associated with breast cancer risk ($p = 0.0005$). Among women with minimal (<2 years) exposure to rotating night work, NPAS2Ala394Thr variant Thr genotypes (Ala/Thr and Thr/Thr) were associated with significantly less risk of breast cancer compared to the Ala/Ala genotype. Women homozygous for the minor allele, Thr/Thr genotype, who had ≥2 years of rotating shift work exposure had 2.83 times higher risk for breast cancer compared to women of the same genotype with <2 years exposure (95% CI: 1.47–5.56). In their discussion of their findings, the authors pointed out that Asian populations have lower frequencies of Thr compared to Europeans, and if the interaction observed is causal, a weaker effect of shift work would be expected in populations with a lower prevalence of Thr genotype at Ala394Thr. They also noted that the prospective Shanghai Women's Health Study observed no association between lifetime history of shift work and risk of breast cancer and that genetic interactions modifying the effects of night work may explain the lack of association of shift work with risk of breast cancer observed in the Asian population.[277]

Overall there are increasing epidemiologic support and molecular research evidence that a history of working several years of shift work involving nights increases a woman's risk of developing breast cancer and the magnitude of the risk is similar to other known risk factors for breast cancer. There is also some epidemiological evidence of increased risk of developing other cancers in women. In the Nurses' Health Study, the RR for colorectal cancer in women working rotating nights 15 years or more compared to women who had never worked rotating night shifts was 1.35 (95% CI: 1.03–1.77; $p_{trend}=0.04$).[278] 602 incident cases of colorectal cancer were documented among 78586 women followed from 1988 to 1998. Study participants were asked how many years in total they had worked in rotating night shifts, at least three nights per month in addition to working days or evenings. Working rotating night shifts for 20 or more years has also been found to be associated with a significant increase in risk for endometrial cancer, especially in obese women (multivariate relative risk =1.47 (95% CI: 1.03–1.14)).[279] Stratified analysis of obese rotating night workers found a doubled risk of endometrial cancer (MVRR=2.09, 95% CI: 1.24–3.52). A nonsignificant increase was found in obese women who did not do rotating night work.

There are also epidemiological studies implicating that shift work is a risk factor for cancer in men. Limitations of the epidemiological studies available in 2010, looking at shift work-related circadian disruption and the risk of cancer in men and women, were critically reviewed by Costa et al.[259] The issues highlighted were mostly related to the diversity of shift work ascertainment methods, varying exposure durations, different definitions of exposure windows, assessment of changes in night shift schedules during the working lifetime, and adjustment for confounders. Parent et al. specifically addressed some of these issues in their recent report of a population-based case–control study of night work and cancer in men conducted in Canada, in the 1980s, by (i) including the period between 1:00 and 2:00 a.m. in the definition of night work precluding evening shift work being considered night work exposure; (ii) calculating a cumulative index of night work exposure; and (iii) incorporating adequate analytic control for several potential confounders.[280] Cases were male patients, 35–70 years old, diagnosed at any of the 18 major Montreal hospitals with incident, pathologically confirmed cancer. Participation of all large hospitals in this area ensured a nearly complete (97%) population-based identification of cases. Between 1979 and 1985, 4576 eligible cancer patients were accrued; 3730 patients (82%) were successfully interviewed. Controls were recruited from the general population using electoral lists (a nearly complete list of voting age citizens in Quebec). Interviews were conducted from 1979 to 1986. For each job held, history of shift work was obtained including the start and finish times of shifts. The results suggest that night work may increase cancer risk at several anatomic sites in men. Compared with men who never worked at night, the adjusted ORs among men who ever worked at night were 1.76 (95% confidence interval (CI): 1.25, 2.47) for lung cancer, 2.03 (95% CI: 1.43, 2.89) for colon cancer, 1.74 (95% CI: 1.22, 2.49) for bladder cancer, 2.77 (95% CI: 1.96, 3.92) for prostate cancer, 2.09 (95% CI: 1.40, 3.14) for rectal cancer, 2.27 (95% CI: 1.24, 4.15) for pancreatic cancer, and 2.31 (95% CI: 1.48, 3.61) for non-Hodgkin's lymphoma. The significant increased risks of cancer in the prostate and colon were irrespective of the timing of night work. There was an effect of long-term night work (>10 years) for cancers of the prostate, colon, and bladder and for non-Hodgkin's lymphoma; the evidence was weaker for other cancer types. There were too few men in the study who had worked in jobs involving night work for over 20 years to conduct analyses for longer exposure. As the authors point out, the findings of increased risks across a wide array of cancer types suggest a common underlying mechanism, with the anticancer effects of melatonin being the most often evoked theory.

Lahti et al. reported the results of a retrospective cohort study in 2008 showing a modest increased risk of non-Hodgkin's lymphoma in men with high shift work exposure based on a job-exposure matrix.[281] Cancer cases were identified from the Finnish Cancer Registry. 6307 cases were identified diagnosed between 1971 and 1995. Exposure to night work for 10 years was associated with an RR of 1.10 (95% CI: 1.03–1.19).

Three of the four studies looking at the risk of prostate cancer in shift workers have found an elevated risk for prostate cancer associated with shift work exposure. In 2006, Kubo et al. reported the results of a prospective cohort study of Japanese shift workers.[282] Individual questionnaires were used to collect shift work history. The RR for fixed night work was 2.3 (95% CI: 0.6–9.2) and 3.0 for rotating shifts (95% CI: 1.2–7.7). The following year, Conlon et al. reported the results of a Canadian case–control study.[283] Study participants completed a questionnaire including what was their "usual work time (daytime, evening/nightshift, rotating shift, other)." OR for rotating shift workers with ≤7 years' shift work exposure was 1.44 (95% CI: 1.10–1.87); for 7.1–22 years, 1.14 (95% CI: 0.86–1.52); for 22.1–34 years, 0.93 (95% CI: 0.70–1.23); and for >34 years, 1.30 (95% CI: 0.97–1.74). The Parent et al. study, discussed above, also found a significantly increased risk of prostate cancer.[281] A retrospective cohort study by Schwartzbaum et al., which included job sectors with 40% rotating shift workers, did not identify a statistically significant elevated risk, reporting an SIR of 1.04 for shift work in 1970 (CI: 0.99–1.10) and 1.02 for shift work in 1960 and 1970 (CI: 0.95–1.10).[257]

Flynn-Evans et al. reported a strong positive association with shift work and elevated prostate-specific antigen (PSA) level, supporting the hypotheses that sleep or circadian disruption is associated with elevated PSA and that shift-working men are at some increased risk of developing prostate cancer.[284] Shift work and PSA test data was obtained as part of the National Health and Nutrition Examination Survey (NHANES) study. Data was combined from three *NHANES surveys* (2005–2010) to obtain current work schedule among employed men aged 40–65 years with no prior history of cancer (except nonmelanoma skin cancer). Men who reported working regular night shifts or rotating shifts were considered shift workers. Using multivariable logistic regression models, PSA levels were compared among current shift workers and nonshift workers. Statistical analysis revealed a statistically significant age-adjusted association between current shift work and elevated PSA $\geq$4.0 ng/mL (odds ratio = 2.48, 95% CI: 1.08–5.70; $p = 0.03$). The confounder-adjusted odds ratio was 2.62 (95% CI: 1.16–5.95; $p = 0.02$). The confounder-adjusted odds ratio for those with total PSA $\geq$4.0 ng/mL and free PSA $\leq$25% was 3.13 (95% CI: 1.38–7.09; $p = 0.01$).

Aggravation or exacerbation of medical disorders

Circadian variation in individual physiological parameters is discussed in the preceding section "Normal Physiology." Symptom manifestation of numerous medical conditions and response to prescribed medications also follow circadian rhythms and may be altered by the reversal of the sleep-activity period experienced when doing night work and the associated phase shifts in various metabolic parameters. Differences in the rate of absorption, distribution, and elimination of a drug can be seen depending on the timing of the drug administration. There are a physiological coupling between the circadian activity–rest cycle and the chronopharmacological mechanism and a related predictable time-of-day variation with respect to best time for administration of some medications. Chronotherapeutic schedules used in the treatment of human cancer are based on predictable circadian rhythms in tolerance to chemotherapy and tumor responsiveness; and changing the time of day of ingestion of certain blood pressure-lowering medications alters the effectiveness on blood pressure control.[285–287] Thorough reviews of these medical chronobiology and chronopharmacology issues have been provided in reviews by Smolensky and D'ALonzo[288] and Smolensky et al.[289]

Diabetes mellitus

Regularity of timing of meals and administration of antiglycemic medication are important in the management of diabetes mellitus. The type and amount of food eaten are also essential elements in the control of glucose levels. Plasma glucose levels in diabetics are higher in the morning than at night, and studies have shown that insulin response to a glycemic stimulus follows a circadian rhythm and that blood glucose control may vary with the time of day. Diurnal variation has been demonstrated in the effect of the type of meal (simple vs. carbohydrates) on blood glucose control in insulin-dependent diabetics. Gastric emptying response limits the rate of digestion and absorption of nutrients, and there is a circadian rhythm to the gastric emptying response; diets that slow the gastrointestinal digestion of carbohydrates have been used successfully in controlling glucose levels in diabetic patients. Studies in rats suggest that the timing of meals, relative to the gastric emptying response, may induce changes in the number and affinity of insulin receptors and thus affect responsiveness to insulin.[182,290–294]

Shift work may interfere with the timing and type of meals eaten by workers, which make compliance with dietary recommendations more difficult. Circadian rhythm disruption inherent in night work may alter the response to pharmacological control of blood glucose levels. Provision for counseling and medical surveillance in cooperation with recommendations from treating providers is essential for night workers with diabetes. However an a priori restriction for diabetics from working shift work is not supported by the available medical evidence. A study by Poole et al. of automobile workers found that insulin-dependent DM is not necessarily a contraindication for night work, with the availability of newer insulin regimens and glucose monitoring meters.[295] However the type of rotation did adversely impact control; very rapid rotations were not used. Compared to day workers, glucose control for insulin-dependent diabetic workers was not significantly better for day workers than for shift workers. More slowly rotating shifts (2-week rotations) were associated with better control than weekly rotating shifts.

Epilepsy

There are circadian time-of-day patterns in the frequency of epileptic seizure occurrences. Partial seizures have been found to demonstrate different peaks in frequency depending on what lobe the seizure arises from. Temporal lobe seizures are more likely to occur in the late afternoon and early evening.[296,297]

The onset of seizures has also been related to changes in corticosteroid levels, which become desynchronized with phase inversions of the sleep–wake cycle.[298] Recently van Campen et al. studied the relationship between circadian rhythm of cortisol and time of epileptic seizure occurrence.[299] They used a systematic literature search to identify relevant reports with 24-hour data of seizure occurrence and then combined and related the data to a standard circadian cortisol rhythm. The occurrence of generalized seizures and focal seizures originating from the parietal lobe in particular followed the circadian rhythm of cortisol with a sharp rise in the early morning, followed by a gradual decline. The results support the hypothesis that changes in cortisol (stress hormone) level influence the occurrence of epileptic seizures.

Sleep deprivation techniques are used to provoke epileptic electroencephalographic discharges in patients under diagnostic evaluation for epilepsy. Sleep deprivation of 24–26 hours can cause electroencephalogram (EEG) activation in epileptics. In patients in whom this has been observed, seizure episodes have been related to loss of sleep.[300,301] Bergonzi et al. found that REM sleep stage deprivation activates EEG epileptic activity and sometimes clinical seizures in persons with generalized or focal epilepsy.[302]

Tarp's review concluded that a lack of sleep leads to an increased frequency of seizures in some epileptics.[303] Kendris et al. recently found that primary generalized epilepsy patients were five times more likely to have a late (evening-type) chronotype compared to healthy controls and generalized epilepsy patients were more likely to be evening types compared to study participants with focal epilepsy or those without epilepsy.[304] These researches point out that late chronotype is a risk factor for circadian misalignment which can impact seizure control in patients with epilepsy and that preventing sleep deprivation and integrating chronotype evaluations and chronotherapy are important in comprehensive care of patients with epilepsy.

Asthma

Circadian rhythms of airway resistance have been demonstrated by pulmonary function testing in both normal and asthmatic subjects. Dyspnea in symptomatic asthmatics is greatest, and peak expiratory flow rate (PEFR) lowest, in the early morning, with the reverse occurring in the early afternoon. A threefold difference in amplitude has been found for dyspnea, and the amplitude for PEFR was found to be 20% of the mesor.[305–307]

Nonnocturnal asthmatic subjects showed an increase frequency and severity of response to environmental exposures when challenged in the evening in comparison to a morning exposure. A relationship between the nocturnal propensity for respiratory symptoms in asthmatics and circadian rhythms is demonstrated by (i) coincidentally decreased levels of circulating cortisol or epinephrine, or urinary levels of adrenocorticosteroid hormones and catecholamines; (ii) decreased dynamic lung compliance; (iii) decreased airway patency; and (iv) increased bronchial reactivity to allergen triggers such as house dust and to histamine and acetylcholine.[308–310]

The pharmacokinetics and effectiveness of bronchodilators vary with the time of administration in a circadian fashion. Chronobiological considerations have been used to optimize the effectiveness of bronchodilators and steroids in controlling asthma.[311–313] The time of administration of steroidal anti-inflammatory and beta-agonist (bronchodilator) medications can be critical in achieving control of respiratory symptoms and minimizing side effects. Treatment schedules are however based on regular diurnal activity by the patient. For regular night workers, it may be possible to adjust the recommended dosing schedule to fit the reversal in the activity–sleep cycle, if the worker is able to maintain a consistent activity–rest pattern. Chronotherapeutic applications may be limited and consistent control of symptoms difficult in an individual engaged in rotating shift work with irregular sleep–wake schedules. On the other hand, it has been suggested that for unmanageable asthmatics, irregular schedules might actually reduce bronchospastic episodes by decreasing the amplitude of the PEFR rhythm.[314] It remains to be determined to what extent the severity of asthma is affected by being awake and active through the night and sleeping during the day. Research is needed for studying the effects of shift work on the control of asthmatic symptoms and on effectiveness of chronotherapeutic interventions for asthmatic shift workers.

Psychosocial disruption and depression

Shift work increases social and family stress. Night work, evening work, and irregular schedules often conflict with family life and social events. Social and family life factors may interfere with good sleep hygiene and other chronobiological coping strategies. This disruption may contribute significantly to the shift work intolerance seen in individuals with desynchronosis which often includes depressive symptomatology.[96,161,315,316]

Shift systems that usually keep the worker away from home during the afternoon and evening are most disruptive for family interaction.[317] The significance of the stress that shift work may impose on marriages is reflected in the results of studies reporting higher divorce rates for shift workers and for shift work dropouts than for day workers.[318–320]

Smith and Folkard surveyed the wives of nuclear power plant operators concerning the impact of their husbands' shift work on themselves and their family.[321] Over 70% believed there were occasional or frequent marital conflicts related to shift work. Approximately one third of the spouses had tried to persuade their husbands to quit shift work. Wyatt and Marriott reported that night workers blamed their shift schedule for broken or strained marriages.[322] These research findings are consistent with the observation of Mott and colleagues who pointed out in their extensive studies of the effects of shift work on families that "… the schedules of husbands and wives are so closely interwoven that it would be a serious mistake to consider the effects of shift work upon only one of the marriage partners."[323]

Although it is recognized that shift workers report decreased well-being related to shift work schedules, the role shift work may have in causing *specific* psychiatric illnesses is unclear.[324] However, chronobiological observations raise concern about the mental health of shift workers—particularly about their risk of developing depression. As discussed in the preceding section "Jet Lag versus Shift Lag," time shifts imposed by transmeridian jet travel or by shift work produce internal rhythm desynchronization. Related symptoms of desynchronosis can include psychophysiological disturbances of well-being, resembling symptoms of affective disorders.[315,324–326] In addition, changes in mood have been shown to be associated with irregularity of sleep patterns and changes in attitude with sleep deprivation.[327,328]

Studies of patients diagnosed with depressive disorder have demonstrated that these patients typically have a dysfunction of the circadian system. This may reflect an abnormal functioning of the endogenous biological clock and/or an abnormal response to external zeitgebers.[329–331]

Hypotheses proposed to explain the abnormalities in circadian rhythms seen in depression have been reviewed by Monk.[332] Despite the uncertainties, it is clear that clinical depression is often associated with circadian dysfunction. This dysfunction has most consistently been evidenced by reduced amplitudes of circadian rhythms in depressed patients and phase changes, such as the early morning awakening characteristic of this disorder. The typical early morning awakening seen in endogenous depression suggests phase advancement of the circadian rhythm, and it has been suggested that depressed patients suffer from a phase disruption of the sleep–wake and "awakening–readiness" rhythms. Certain antidepressant medications, such as tricyclic antidepressants, monoamine oxidase inhibitors, and lithium, cause a phase delay of several circadian rhythms.[333,334] In temporal isolation studies, shifts in the light–dark cycle were associated with an increase in depressive symptoms.[335] Not surprisingly appropriately timed bright light therapy can be an effective antidepressant.[336]

Depressive patients have also been found to have a shorter REM sleep latency and to experience more REM sleep in the first third of the night and less in the last third than controls.[335–337] Phase advancing the time of sleep with respect to the REM–temperature–cortisol circadian rhythm leads to a remission of symptoms in some depressed patients.[338]

The converse, that is, delaying the time of sleep would precipitate depression in susceptible persons, may have been observed. Two weeks following a study where four subjects underwent a 12-hour sleep delay, one of the subjects committed suicide. Retrospectively, it was observed that the subject's circadian temperature rhythm was phase-advanced with respect to the other three subjects' and did not re-entrain to the schedule shift, but remained advanced relative to the shift.[337] There are also studies which support the notion that imposed time shifts may lead to the development of clinical depression in vulnerable individuals. Healthy subjects were observed to have an increase in depressive symptoms after experiencing a phase-delay time shift.[339] In bipolar patients, episodes of mania have been triggered by time zone changes and by a night of sleep deprivation and have then been successfully managed by regularization the manic patient's schedule.[340] A review of psychiatric incidents at London's Heathrow Airport revealed the direction of air travel to predict the type of affective disorder experienced, with mania primarily occurring after eastbound flights (phase advance) and depression mainly after westbound travel (phase delay).[341]

A model proposed by Ehlers and colleagues to explain the relationship between imposed circadian rhythm disruption and the occurrence of depression involves a cascade of effects linking adverse life events with the onset or recurrence of depression in vulnerable individuals.[342,343] The life event is proposed to result in a change in the individual's daily social routines (analogous to the effect of shift work schedules), leading to circadian rhythm maladaptation. The resulting desynchronosis includes depressive symptomatology, and in vulnerable individuals, a major depressive episode may result. The term *zeitstörers* (German: time disrupter) was coined to describe these agents/events, including shift work, that disrupt the circadian system.[343]

As a group, shift workers report excessive symptoms of depressive illness, suggesting that shift work may predispose vulnerable individuals to affective disorders. Increased psychological symptoms and increased scores for depression on mood profiles have been reported in nurses during their first months of working starting shift work.[344,345] Costa et al. found shift workers to have a 5–15% increase in a "neurotic disorders" category, which included depression.[346] In a comparative study of retired day workers and retired shift workers, cases of depression as identified by a neuropsychiatrist were more frequent in the retired shift workers.[347] Results from a pilot study of the prevalence of major depressive disorder (MDD), defined by the Structured Clinical Interview for DSM-III-R criteria, in 100 current and former shift workers found a monotonic trend of increasing prevalence of MDD as years of exposure to shift work increased up to 20 years of exposure.[348] The rate decreased after 20 years, likely reflecting a healthy worker effect. Overall lifetime prevalence of MDD was 15%, compared with the prevalence in the general population, which is estimated to be around 10%.[349]

TREATMENT (COUNTERMEASURES)

In order for individual coping strategies to be effective, families must be involved. The shift worker needs to be aware of the toll that the shift work schedules may take on the family, and the family to be aware of the effect of the shift work schedule on the worker. The provision of educational programs for both the worker and family is essential for employees to successfully cope with shift work schedules in terms of performance at work, responsibilities at home, and health considerations. Educational materials addressing shift work issues, including countermeasures published in laymen's terms, are available which will assist employers and employees in this endeavor.[81,350,351]

Diet and exercise

Good dietary habits and regular exercise are recommended in general for preventive health reasons. As noted in the above section "Shift Work and Specific Medical Disorders," shift working is associated with metabolic disorders, particularly related to hyperlipidemia, glucose intolerance, and metabolic syndrome. Dietary habits may be even more important for shift workers' preventive health to reduce the risk of risk of coronary artery disease, diabetes, and obesity than for workers in general.[206] Shift work and the time of the shift worked appear to affect the amount and quality of food eaten and the energy distribution over the day. Lowden et al., in their comprehensive review of the available studies of shift work-related dietary issues, observed that factors such as time availability and social context are important in determining food intake at work, particularly at night, and that a case can be made that shift workers need to be provided with both the opportunity and the appropriate facilities to maintain healthy eating habits in the workplace.[352]

Sleep deprivation and circadian disruption can affect the endogenous signals and disrupt the homeostatic control of food intake; and moderate sleep deprivation has been shown to be associated with an increase in consumption of energy from snacks with a higher carbohydrate content.[353,354] It has been proposed that disruption of these peripheral circadian oscillators may be involved in the development of obesity, Type 2 diabetes, and metabolic syndrome[355]; if this is so, it supports the argument that circadian disruption should be minimized when working at nights by keeping the same mealtimes across the shift cycle to maintain a relatively diurnal dietary rhythm[356] and avoiding eating, or restricting energy intake, between midnight and 0600 hours.

Carbohydrate-rich meals produce greater decrements in mental performance (in contrast to physical performance) and increase sleepiness as compared to fat-rich meals; although compared to circadian effects on sleep and performance, the effects of meal differences and the postprandial response are relatively small. Regarding whether night workers should fast or feed during the night shift, laboratory studies have shown subjective ratings of sleepiness and energy levels to be lower at night in the fasting condition.[357–360]

In their review, Lowden et al. discuss the limitations of study designs and contradictory findings and the complexity of eating habits of shift workers that limit conclusions regarding dietary recommendations for shift workers.[352] They note in particular that additional research is needed to identify when shift workers eat, with respect to their work hours and circadian rhythms, in order to answer the question of when and what night workers should eat to avoid inducing metabolic disturbances and optimize wakefulness and performance. Acknowledging the gaps in the current research, they have identified some broad guidelines that may be included in nutrition management strategies—noting that these guidelines are to be considered in parallel with appropriate fatigue management strategies, are targeted directly at the individual or the employer, and are appropriate not only for shift workers but also other populations:

General guidelines

- Avoid eating, or at least restrict energy intake, between midnight and 0600 hours.
- Try to eat at the beginning and end of the shift.
- Avoid "large meals" (>20% of daily energy intake) 1–2 hours before the main daily sleep episode.
- Provide a variety of food choices: complete or vegetarian meals and high-quality snacks are recommended.
- Avoid foods and beverages classified as low-quality snacks.
- Provide appropriate dining facilities in as pleasant a surrounding as possible for works to eat with coworkers, or allow meals to be eaten away from the workplace.
- Maintain a healthy lifestyle with exercise, regular mealtimes, and good sleep hygiene when not working.

Specific guidelines for shift work

- Eat breakfast before day sleep to avoid wakening due to hunger.
- Stick as closely as possible to a normal day and night pattern of food intake. See, for example, the Nordic Nutrition Recommendations (*Nordic Council of Ministers. Nordic Nutrition recommendations. Copenhagen: Nordic Council of Ministers; 2004*).
- Divide the 24-hour intake into eating events with three satiating meals each contributing 20–35% of 24-hour intakes. Higher energy needs require more frequent eating.
- Avoid overreliance on (high-energy content) convenience foods and high-carbohydrate foods during the shift. Select vegetable soups, salads, fruit, yogurt, whole grain sandwiches, cheese or cottage cheese (topped with fruit), boiled egg, nuts, and green tea (for antioxidant activity)—noting that this may not be palatable for some workers.
- Design shift schedules which allow adequate time between shifts for sleep and meal preparation, avoid quick returns.
- Avoid sugar-rich products such as soft drinks, bakery items, sweets, and nonfiber carbohydrate foods (high glycemic load), for example, white bread.

Physical exercise has been demonstrated to reduce general fatigue in shift workers and sleepiness at work, increase sleep duration and the quantity of slow-wave sleep which is vital to the restorative functions of sleep, and decrease musculoskeletal symptoms. Physical exercise can cause circadian rhythm phase advances and delays. Overall the research supports that there is benefit for shift workers from appropriately timed, regular physical exercise on sleep and performance outcomes. Recommendations for exercise for shift workers include the following: (i) moderate physical exercise is preferred over intensive training, (ii) exercise should be done a few hours before the main sleep period, and (iii) for morning or day shifts, the best exercise time is after the shift. After night shifts, the exercise should be done before an evening nap.[169,361–363]

Maximizing sleep

The most significant factor interfering with sleep for night workers is daytime noise. Even if the worker is not aware of actually being awakened by noise, sleep quality may be compromised.[364] Actions should be taken to soundproof the bedroom as much as possible. In addition to utilizing sound damping materials, for example, ceiling tiles and carpeting, white noise from a fan or air conditioner may be helpful. Family and neighbor cooperation may be needed to control noisy activities near the night worker's sleeping quarters. The phone and doorbell should not be audible in the bedroom. Comfortable earplugs can also be used to attenuate noise.

Light exposure should be limited to as close to nighttime conditions as possible. Lined drapery and window blinds or dark room shades are suggested. Eyeshades are another option for decreasing light exposure.

Applying "sleep hygiene," a technique initially developed to help patients with insomnia, is also a recommended coping strategy for shift workers. Sleep hygiene is a program applying regular procedures and following behavioral rules that enhance the ability to fall asleep and stay asleep.[365]

Although the regular use of sleeping pills is contraindicated, short-acting hypnotics such as triazolam have been shown to improve quality and duration of daytime sleep. Intermittent use for a day or two, under a physician's care, may be useful when beginning a run of night shifts or following a transmeridian flight. However, caution must be exercised regarding the timing of administration of even short-acting hypnotics, as impaired cognition may linger 8 hours after administration.[366–368]

Caffeine and other alertness-enhancing drugs

Caffeine belongs to the xanthine class of drugs, which have been shown to cause phase shifts of the temperature rhythm in animals. Caffeine is an effective countermeasure for night workers due to its stimulant effect in counteracting sleepiness and to its ability to delay sleep onset at night. Caffeine has been shown to have beneficial effects on alertness and performance and to decrease sleep tendency as measured by multiple sleep latency tests.[369–372]

The dose of caffeine should be limited to avoid undesirable side effects, such as heart palpitations. Caffeine's effect in increasing alertness is most apparent after a time of abstinence, and with repeated doses, and the effect may diminish with repeated doses. Caffeine disrupts daytime sleep more than nocturnal sleep, and consumption should be avoided closer than around 5 hours before bedtime.[373–375] Shift workers should limit use to the first half of night or evening shifts. It is important to avoid caffeine during the last half of the evening shift or night shift, since the worker's bedtime will come soon after getting home. Fruit juice is good alternative drink for the second half of the shift.

Amphetamines and stimulant diet pills should never be used to treat shift work-related sleepiness due to adverse side effects and potential for abuse. Newer alertness-enhancing drugs may have some usefulness for occasional alertness promotion.[376] Research on the wake-promoting medication, modafinil, has demonstrated its ability to improve performance and decrease extreme sleepiness in night workers.[377] Modafinil has been approved by the US Food and Drug Administration for use in increasing alertness in night workers; however it may also increase insomnia and is not a substitute for adequate sleep. As pointed out by the American Academy of Sleep Medicine (AASM), caffeine is a readily available, inexpensive alternative.[378] The AASM guidelines do include modafinil as an indicated medication to enhance alertness during the night shift for patients diagnosed with SWD (shift work disorder, i.e., shift work-related sleep disturbances and impairment of waking alertness and performance). However, ethical concerns are raised with the use of medication rather than changing the work schedules for workers who have been determined to be shift work intolerant.[376,379]

Bright light and melatonin

In addition to its sleep-inducing property, animal and human laboratory studies have shown that melatonin effects phase shifts of circadian rhythms when administered with appropriate timing.[380,381] Several field studies have demonstrated melatonin to be useful for ameliorating jet-lag symptoms.[104,382] Five milligrams daily, taken orally, is the typical dose used in research protocols, although lower doses may also be effective.[383] Specific instructions (and side effect warnings) for taking the hormone for eastbound and westbound flights, as given to subjects participating in jet-lag studies, are included in the review by Arendt and Deacon.[104] The American Academy of Sleep Medicine (AASM) has recommended the use of melatonin at appropriate times to reduce the symptoms of jet lag and improve sleep following transmeridian flights and suggests that immediate-release formulations in doses from 0.5 to 5 mg may be effective.[378]

Shift workers have anecdotally reported benefits from using over-the-counter preparations of melatonin for shift-lag symptoms.[384] There is limited research available on the use of melatonin in real shift work situations. Beneficial effects on sleep and alertness have been reported associated with bedtime administration, but some performance measures may be adversely affected.[385] Sharkey et al. found that in laboratory-simulated night shifts, melatonin was effective in preventing decreased sleep time during daytime sleep only on the first day of administration and had no effect on alertness (assessed with the multiple sleep latency test) or on performance measures or mood during the night shift.[384] Inappropriately timed administration may be dangerous due to its sedative effect. Effects of long-term, regular usage of melatonin are not known. In addition to its sleep-inducing role, melatonin may influence blood pressure regulation, immune modulation, control of tumor growth, and antioxidant action on free radicals.[105]

The AASM has recommended the administration of melatonin as a guideline, prior to day sleep, for shift workers diagnosed with SWD. The report also points out, however, that there is mixed evidence supporting the use of melatonin, that it is difficult to draw firm conclusions from the current research due to variability in shift schedules and dosage and timing, and that subjects have seldom been diagnosed according to SWD criteria.[378]

Exposure to bright light, 7000–12000lx (comparable to sunlight), has been demonstrated to result in phase shifts of the circadian timing system. The timing of the exposure determines the direction of the shifts, that is, either a phase-advance or phase-delay response.[44,386,387] Although appropriately timed bright light exposure can enhance adjustment to night shifts, practical application for shift workers is a different matter. Not only is the timing of the light exposure critical, but also prevention of outdoor sunlight exposure at times is necessary (e.g., on the commute home after dawn). In addition, there is considerable individual variation in the degree of phase-shift response. The American Academy of Sleep Medicine report points out that although circadian realignment has been achieved with light exposure in simulated shift work situations, larger studies are needed to determine the clinical utility of timed light therapy for the treatment of SWD.[378]

The use of both bright light and melatonin together for readaptation from night work to a daytime schedule has been studied. Specific protocols have been described in detail for the timing of light and medication following long-term and short-term night work by Pallesen et al.[388] Assistance from a chronobiologist is probably needed to make practical, understandable schedules for a worker. The goal is to provide predictability of the shift-work schedule without unexpected overtime, and commitment from the worker and employer for successful application. A final consideration is the concern that has been raised regarding use of bright light at night related to the possible oncogenic risk for estrogen-sensitive breast cancers associated with long-term bright light exposure during the night shift; as discussed in the preceding "Cancer," section, melatonin suppresses estrogens, and light exposure decreases melatonin secretion.

Naps

About one-third of night workers take a nap for about an hour in the late afternoon before night shifts.[389,390] Although not usually allowed in the United States, provision for on-the-job naps during night shifts is not uncommon in Japan. Scheduled nap times during the first night shift are effective in counteracting the extreme decrease during the early morning circadian trough in alertness.[391] Field study evidence that scheduled napping at work improves performance has been reported in aviation studies of 30minutes cockpit naps.[392]. Other studies have shown that naps taken during the night shift can increase alertness and performance.[393,394] Napping in on call rooms is a standard practice for medical interns. Napping during the night shift has been shown to be beneficial for counteracting some effects of sleep deprivation in nurses.[395] Overall, it appears that naps can be an effective countermeasure against on-the-job sleepiness. Allowance should be planned for the initial 5–15minutes period of sleep inertia after awakening. Short naps of less than 15minutes do not appear to have significant risk of sleep inertia.

MEDICAL SURVEILLANCE

Although there are no US federally mandated requirements for medical evaluations for night work exposure, the International Labor Organization (ILO) 1990 Convention (No. 171) includes provisions for a health assessment for workers before beginning their night work and health assessments of night workers at regular intervals as well as for work-related problems that may be secondary to the work schedule.[396] In addition, the European Directive No. 93/104/EC, "Concerning certain aspects of the organization of working time," also considers it a right of workers to have a free health assessment before beginning their first assignment to the night shift.[397]

Recommendations for medical evaluations of workers before they begin night work assignments have been made by occupational medicine practitioners and by chronobiologists who have studied health effects of shift work. Identification of individual characteristics that are associated with poor tolerance of night work is recommended, not with the goal of disqualifying workers for night work but with the recognition that, for some, night work may medically not be advisable. In most situations, the preplacement examination will provide an opportunity to make susceptible workers aware of their individual risks and plan appropriate medical supervision and develop coping strategies.[15,96,97,173,398,399]

The frequency of medical surveillance examinations is somewhat arbitrary. However, recommendations are consistent in advising evaluation during the first few months after beginning shift work and at regular but less frequent intervals, depending on the work schedule and the age of the worker. A reasonable schedule has been outlined by Harma which includes the following: the first follow-up health check scheduled no later than 2 months after night/shift work has begun; subsequently, for workers between 25 and 45 years of age, 3–5 year intervals; for those under 25 or over 45 years of age, 2-year intervals; and for those over 60, 1-year intervals are advised.[169] More frequent evaluations may be needed for individuals with underlying conditions that may be aggravated by shift work. Costa has made more general recommendations for the first medical surveillance health check to be during the first year of shift or night work, for those under 45 years successive evaluations to be at least every 3 years, and for those over 45 successive checkups to be every 2 years.[400] Ongoing medical surveillance programs, including periodic medical screening examinations and appropriate laboratory testing, have been recommended for rotating and permanent night workers. In addition, follow-up evaluations of day workers who have left shift work for medical reasons have also been advised.[15,173]

Common conditions that may be exacerbated by shift work have already been reviewed in this chapter. Smolensky et al. have recently reviewed numerous other medical and psychiatric conditions exhibiting circadian fluctuation in symptoms and response to external temporal triggers and pharmacological treatments.[289]

Potential contraindications to working shift work involving night shifts are summarized in Table 10.1. For example, due to the recognized increase in likelihood of seizure events in epileptic individuals associated with circadian rhythm disruption and sleep deprivation, clearance from the neurologist managing a potential shift worker with epilepsy before initial assignment to a night work schedule is a reasonable requirement. For asthmatics, medical surveillance with involvement of the treating physician is essential for monitoring any changes in frequency of bronchoconstriction and response to prescribed medications. Both rotating and permanent night workers with diabetes should be monitored for changes in response to dietary and pharmacological management of glucose control. Circadian disruption in glucose tolerance has been noted. Studies of permanent night workers have shown only partial adjustment of glucose and insulin rhythms after 2 years of regular night work.[401] Sleep restriction can also lower glucose tolerance.[402,403] Changes in the timing and quality of meals and metabolism related to working nights also necessitate monitoring for increased levels of undesirable triglycerides and lipoproteins.

Based on the available epidemiological evidence, medical conditions exacerbated by shift work may be absolute or relative contraindications to shift work.[96,173,402] Identification of individual characteristics that are associated with poor tolerance of night work is recommended not with the goal of disqualifying workers for night shifts but providing appropriate medical counseling and medical surveillance for those for whom night work may not be medically advisable. Depending on the severity of the condition and stability of treatment needed, the individual's overall tolerance to shift work, and the particular shift work schedule involved, temporary or permanent restriction from night work may be in the best medical interest of the worker. Occupational health physicians should remember that shift work intolerance is a manifestation of complex medical and psychosocial interactions and individual worker responses will vary in terms of severity and timing of onset of clinical signs and symptoms.[403]

While medical surveillance programs for shift workers are important for early detection of shift work-related health problems, Kogi has pointed out that medical surveillance examinations alone cannot adequately meet the health needs of shift workers.[93] The medical surveillance program for shift workers should include educational/counseling opportunities related to the assessment and optimization of shift work coping strategies.[169,173] Joint efforts by occupational health and safety teams, in conjunction with working with supervisors and workers, are necessary to provide necessary preventive medicine programs and address scheduling considerations. Adjustments may be needed in medical surveillance schedules for chemical exposures to account for quick turnover times or extended hours of work.

PREVENTION AND ADMINISTRATIVE CONTROLS

Scheduling decisions

Scheduling decisions should be made with the goal of minimizing the potential negative impact of shift work on worker sleep, health, and performance. Although scheduling designs understandably reflect business needs and employee preferences, sleep and health considerations should not be secondary concerns. Before making schedule changes, the demographics of the workforce, including lifestyles, sleep habits, common medical problems, and shift scheduling preferences, as well as the type of work and the environment in which it is performed, should be assessed.[86,404]

Recognizing that "tailor-made" shift systems need to involve compromises between conflicting interests of employees and employers and ergonomic considerations, Knauth has recently provided detailed practical recommendations for achieving ergonomically sound shift schedule systems.[405] Four general categories of factors important in the evaluation of the degree of shift system compliance with ergonomic recommendations are reviewed: (i) the sequence

TABLE 10.1 Potential contraindications for working night or rotating shifts.

Condition	Examples	Comments
Asthma	Poor control or increasing use of rescue inhalers after starting shift work. The time of administration of steroidal anti-inflammatory and beta-agonist (bronchodilator) medications can be critical in achieving control of respiratory symptoms and minimizing side effects	Medical surveillance indicated to monitor for changes in frequency of bronchoconstriction and response to prescribed medications
Cancer	Endocrine-sensitive cancers Other cancers in treatment involving chronotherapeutic schedules	Avoid work schedules interfering with chronotherapy considerations of timing of chemotherapy or radiation treatments or associated with sleep deprivation
Cardiovascular disease	Ischemic heart disease; poorly controlled hypertension; high-risk/multiple risk factors for acute myocardial infarction	Shift work has been shown to be an independent risk factor for CVD. Shift work may have a triggering effect on lifestyle factors that can increase the risk of CHD; active preventive medicine intervention is important for shift workers. Medical surveillance lab work should include lab work for monitoring for increased levels of undesirable triglycerides and lipoproteins
Diabetes mellitus	Poorly controlled diabetes; changes in control on rotating shifts. Regularity of timing of meals and administration of insulin and antiglycemic medication are important in the management of diabetes mellitus	Monitor for changes in response to dietary and pharmacological management of glucose control. Medical surveillance programs should include lab work for monitoring for prediabetes
Epilepsy	Generalized or partial seizures. Clearance should be obtained from the neurologist managing a potential shift worker with epilepsy before initial assignment to a night work schedule	Sleep deprivation associated with night work and circadian rhythm misalignment may increase frequency of seizures in some epileptics
Gastrointestinal disorders	Chronic peptic ulcer disease; symptomatic inflammatory bowel diseases	Uncontrolled with standard treatments; history of exacerbations related to work schedule changes; work schedules not allowing for regular timing of meal break
Psychiatric disorders	Bipolar disorder—irregular schedule may trigger manic behavior Other diagnoses as determined by the treating psychiatrist	Treatment involving timed light therapy may not be compatible with the shift work schedule. Shift work intolerance seen in individuals with desynchronosis often includes depressive symptomatology
Pregnancy	Increased risk or history of preterm delivery, miscarriage, or low birth weight while working a shift work schedule	Pregnant shift workers should advise their obstetrician regarding their hours of work. Do not increase shift length or initiate shift work during a pregnancy
Prescription medications	If time of dosing affects drug effectiveness. For example, changing the time of day of ingestion of certain blood pressure-lowering medications reduces effectiveness	Differences in the rate of absorption, distribution, and elimination of a drug can be seen depending on the timing of the drug administration
Sleep disorders	Narcolepsy, shift work disorder, and other circadian rhythm sleep disorders, long or rigid sleep requirements, uncontrolled sleep apnea	Treatment involving timed light, melatonin, and/or planned sleep schedule exposure may not be compatible with the shift work schedule

of shifts, including the speed and direction of rotation and special cases; (ii) the duration and distribution of working time, including the number of consecutive working days, shift duration, and time off; (iii) the position of the working time, including the start of the morning shift and the end of the evening and night shift and number of free weekends; and (iv) short-term deviations from the established shift schedule resulting from wishes of the employees or from requirements of the employer.

It is clear from the above discussions that preparing schedule designs for shift systems is a complex matter. In addition, there may be individual workers with medical restrictions that need to be worked into particular rosters. Fortunately, computer software programs have been developed to assist in the process.[406,407]

Occupational health programs

Workplace facilities and environmental conditions can impact on tolerance to shift work and shift worker performance. In order to assist employees in dietary countermeasures, equivalent canteen/eating facilities should be provided for night workers as for day workers. At a minimum, a microwave, refrigerator, and vending machines with low-fat nutritious foods should be available, including dairy products and fruit juices.

Other environmental factors should be assessed which can increase alertness on the job and help prevent episodes of falling asleep. For instance, bright, uniform lighting will enhance alertness. Nonvariable background noise which promotes boredom may be replaced with judiciously selected music and, if appropriate, social interactions between workers. Keeping room temperatures below 70°F and providing opportunities for physical activity have been recommended to maximize alertness on the night shift.[350,408]

Educational programs should be provided for workers and their families that provide information for shift work coping strategies. In addition, workers should be advised of the increased risk of motor vehicle accidents on the drive home when working night shifts. Provision of sleeping facilities for workers who need to sleep before driving home should be considered. The degree of driving risk for the individual workforce should be assessed, and the aggressiveness of preventive measures taken based on the findings. Some proactive companies, having recognized the difficulty that night workers with families face in obtaining childcare, have established 24-hour childcare facilities for their workers.[409] Monk and Folkard have recommended that employers develop a "Shift Work Awareness Program" for coordinating educational and social support programs.[351]

References

1. Coleman RM, Demant WC. Falling asleep at work: a problem for continuous operations. *Sleep Res* 1986; 15:265.
2. Novak RD, Novak SE. Focus group evaluation of night nurse shiftwork difficulties and coping strategies. *Chronobiol Int* 1996;13(6):457–63.
3. Scott AJ. Chronobiological considerations in shiftworker sleep and performance and shiftwork scheduling. *Hum Perform* 1994; 7(3):207–33.
4. Scott AJ. Sleepiness and fatigue: risks for the transportation industry. *Clin Occup Environ Med* 2003; 3:81–108.
5. Tepas DI, Armstrong DR, Carlson ML, et al. Changing industry to continuous operations: different strokes for different plants. *Behav Res Methods Instrum Comput* 1985; 17(6):670–6.
6. Kogi K. Comparison of resting conditions between various shift rotation systems for industrial workers. In: Reinberg A, Vieux N, Andlauer P, eds. *Night and shift work: biological and social aspects*. Oxford: Pergamon Press, 1981:417–24.
7. U.S. Department of Labor, Bureau of Labor Statistics. Survey of Occupational Injuries and Illnesses, 2014 OMB No. 1220-0045. Page 6 of the survey instrument. Available at http://www.bls.gov/respondents/iif/forms/soii2014.pdf (accessed May 3, 2016).
8. Bureau of Labor Statistics. Case and Demographic Characteristics for Work-related Injuries and Illnesses Involving Days Away From Work Resource Table Categories: Calendar Year 2013 Survey Results Tables R76 and R83. Available at http://www.bls.gov/iif/oshcdnew2013.htm#13 first link (accessed May 3, 2016).
9. Wedderburn A. *Instruments for designing, implementing and assessing working time arrangements*. Bulletin of European Studies on Time, 7. Luxembourg: Office for Official Publications of the European Communities, 1994
10. Jansen B, Kroon H. Rota-risk-profile-analysis. *Work Stress* 1995; 9:245–55.
11. American Conference of Governmental Industrial Hygienists. *2014 TLVs and BEIs, threshold limit values for chemical and physical agents biological exposure indices*. Cincinnati: ACGIH, 2014.
12. Occupational Safety and Health Administration (OSHA). Standard Interpretation 11/10/1999 [corrected 6/21/2007]: OSHA policy regarding PEL adjustments for extended work shifts. Available at https://www.osha.gov/pls/oshaweb/owadisp.show_document?p_table=INTERPRETATIONS&p_id=22818 (accessed May 3, 2016).
13. Brief RS, Scala RA. Occupational exposure limits for novel work schedules. *Am Ind Hyg Assoc J* 1975 36:467–9.
14. Smolensky M, Reinberg A. Clinical chronobiology: relevance and applications to the practice of occupational medicine. *Shiftwork Occup Med State Art Rev* 1990; 5(2):239–72.
15. Koller M. Occupational health services for shift and night workers. *Appl Ergon*, 1996; 27(1):31–7.
16. Monk TH, Folkard S. Circadian rhythms and shift work. In: Hockey GRS, ed. *Stress and fatigue in human performance*. Chichester: John Wiley & Sons, Ltd, 1983:97–121.
17. Knauth P, Rutenfranz J. Experimental shift work studies of permanent night and rapidly rotating shift systems: 1. Circadian rhythm of body temperature and re-entrainment at shift change. *Int Arch Occup Environ Health* 1976; 37:125–37.
18. Knauth P, Emde E, Rutenfranz J, et al. Re-entrainment of body temperature in field studies of shiftwork. *Int Arch Occup Environ Health* 1981; 49:137–49.
19. Van Loon JH. Diurnal body temperature curves in shiftworkers. *Ergonomics* 1963; 6:267–73.
20. Costa G, Ghirlanda G, Tarondi G, et al. Evaluation of a rapidly rotating shift system for tolerance of nurses to nightwork. *Int Arch Occup Environ Health* 1994; 65:305–11.
21. Folkard S. The pragmatic approach to masking. *Chronobiol Int* 1989; 6(1):55–64.
22. Reinberg A, Vieux N, Andlauer P, et al. Tolerance of shift work, amplitude of circadian rhythms, and aging. In: Reinberg A, Vieux N Andlauer P, eds. *Night and shift work: biological and social aspects*. Oxford: Pergamon Press, 1981:341–54.
23. Leonard R. Night- and shift-work. In: Reinberg A, Vieux N, Andlauer P, eds. *Night and shift work: biological and social aspects*. Oxford: Pergamon Press, 1981:323–9.
24. Reinberg A, Motohashi Y, Bourdeleau P, et al. Alteration of period and amplitude of circadian rhythms in shift workers. *Eur J Appl Physiol* 1988; 57:5–25.
25. Mills JN, Minors DS, Waterhouse JM. Exogenous and endogenous influences on rhythms after sudden time shift. *Ergonomics* 1978; 21:755–61.

26. Klein KE, Wegman HM, Hunt BI. Desynchronization as a function of body temperature and performance circadian rhythm as a result of outgoing and homecoming transmeridian flights. *Aerosp Med* 1972; 43:119–32.
27. Klein KE, Herrmann R, Kuklinski P, et al. Circadian performance rhythms: experimental studies in air operations. In: Mackie R, ed. *Vigilance: theory, operational performance, and physiological correlates*. New York: Plenum Press, 1977: 111–32.
28. Bodanowitz M. *The change of circadian rhythms of psychomotor performance after transmeridian flights*. Translation of DLR-FB 73-52, *Die veranderung tagesperiodischer schwankungen der psychomotorischen leistung nach transmeridian flugen*. Bonn: Deutsche Forschungs- und Versuchsanstalt fur Luft- und Raumfahrt, Institut fur Flugmedizin, 1973.
29. Orth-Gomer K. Intervention on coronary risk factors by changing working conditions of Swedish policemen. In: Harvath M, Frankth E, eds. *Psychophysiologic risk factors of cardiovascular diseases*. International Symposium, Suppl. 3. Prague: Avicenum-Czechoslovak Medical Press, 1982.
30. Orth-Gomer K. Intervention on coronary risk factors by adapting a shiftwork schedule to biologic rhythmicity. *Psychosom Med* 1983; 45:407–15.
31. Lavie P, Tzischinsky O, Epstein R, et al. Sleep–wake cycle in shift workers on a "clockwise" and "counter-clockwise" rotation system. *Isr J Med Sci* 1992; 28(8–9):636–44.
32. Folkard S. Shift work: a growing occupational hazard. *Occup Health* 1989; 41:182–6.
33. Vidacek S, Kaliterna L, Radosevic-Vidack B, et al. Productivity on a weekly rotating shift system: circadian adjustment and sleep deprivation effects? *Ergonomics* 1986; 29(12):1583–90.
34. Wilkinson R, Allison S, Feeney M, et al. Alertness of night nurses: two shift systems compared. *Ergonomics* 1989; 32(3):281–92.
35. Knauth P, Landau K, Droge C, et al. Duration of sleep depending on the type of shift work. *Int Arch Occup Environ Health* 1980; 46:167–77.
36. Wojtczak-Jaroszowa J. Circadian rhythm of biological functions and night work. In: Wojtczak-Jaroszowa J, ed. *Physiological and psychological aspects of night and shift work*. Cincinnati: National Institute for Occupational Safety and Health, 1977:3–12.
37. Dahlgren K. Adjustment of circadian rhythms and EEG sleep functions to day and night sleep among permanent nightworkers and rotating shiftworkers. *Psychophysiology* 1981; 18(4):381–91.
38. Czeisler CA, Weitzman ED, Moore-Ede MC, et al. Human sleep: its duration and organization depend on its circadian phase. *Science* 1980; 210:1254–67.
39. Walker J. Frequent alternation of shifts on continuous work. *Occup Psychol* 1966; 40:215–25.
40. Kiesswetter E, Knauth P, Schwarzenau P. Daytime sleep adjustment of shiftworkers. In: Koella WP, Ruther E, Schulz H, eds. *Sleep '84*. New York: Gustav Fischer Verlag, 1985:273–5.
41. Kogi K. Estimation of sleep deficit during a period of shift rotation as a basis for evaluating various shift systems. *Ergonomics* 1978; 21(10):861–74.
42. Williamson AM, Sanderson JW. Changing the speed of shift rotation: a field study. *Ergonomics* 1986; 29(9):1085–96.
43. Smith PA, Wright BM, Mackey RW, et al. Change from slowly rotating 8-hour shifts to rapidly rotating 8-hour and 12-hour shifts using participative shift roster design. *Scand J Work Enviorn Health* 1998; 24(S3):55–6.
44. Eastman CI. Circadian rhythms and bright light: recommendations for shift work. *Work Stress* 1990; 4:245–60.
45. Smith MR, Eastman CI. Shift work: health, performance and safety problems, traditional countermeasures, and innovative management strategies to reduce circadian misalignment. *Nat Sci Sleep* 2012; 4:111–32.
46. Maasen A, Meers A, Verhagen P. Quantitative and qualitative aspects of sleep in four shift workers. *Ergonomics* 1978; 21(10): 861–74 (abstract).
47. Circadian Technologies, Inc. Improving human performance and health in round-the-clock operations. Seminar, Pittsburgh, PA, November 1987.
48. Dahlgren K. Adjustment of circadian rhythms to rapidly rotating shift work: a field study of two shift systems. In: Reinberg A, Vieux N, Andlauer P, eds. *Night and shift work: biological and social aspects*. Oxford: Pergamon Press, 1981:357–65.
49. Rutenfranz J, Knauth P. Hours of work and shiftwork. *Ergonomics* 1976; 19(3):331–40.
50. Ghata J, Reinberg A, Vieux N, et al. Adjustment of the circadian rhythm of urinary 17-OHCS, 5-HIAA, catecholamines, and electrolytes in oil refinery operators to a rapidly rotating shift system. *Ergonomics* 1978; 21(10):61–874 (abstract).
51. Wedderbrun AAI. How important are the social effects of shiftwork? In: Johnson LC, Tepas DI, Colquhoun WP, Colligan MJ, eds. *Biological rhythms, sleep, and shift work*. Advances in Sleep Research, Vol 7. New York: Spectrum Publications, 1981:257–69.
52. Conroy RT, Mills JN. *Human circadian rhythms*. London: J & A Churchill, 1970.
53. Reinberg A. Clinical chronopharmacology. In: Reinberg A, Smolensky M, eds. *Biological rhythms and medicine: cellular, metabolic, physiopathologic, and pharmacologic aspects*. New York: Springer-Verlag, 1983:211–57.
54. Tepas DI, Monk TH. Work schedules. In: Salvendy G, ed. *Handbook of human factors*. New York: John Wiley & Sons, Inc., 1987:819–43.
55. Tepas DI, Walsh JK, Moss PD, et al. Polysomnographic correlates of shiftworker performance in the laboratory. In: Reinberg A, Vieux N, Andlauer P, eds. *Night and shift work: biological and social aspects*. Oxford: Pergamon Press, 1981:179–86.
56. Tepas DI, Walsh JK, Armstrong DR. Comprehensive study of the sleep of shift workers. In: Johnson LC, Tepas DI, Colquhoun WP, Colligan MJ, eds. *Biological rhythms, sleep and shift work*. Advances in Sleep Research, Vol 7. New York: SP Medical & Scientific Books, 1981:347–55.
57. Rosa RR, Wheeler DD, Warm JS, et al. Extended workdays: effects on performance and ratings of fatigue and alertness. *Behav Res Methods Instrum Comput* 1985; 17(1):6–15.
58. Rosa RR, Colligan MJ, Lewis P. Extended workdays: effects of 8-hour and 12-hour rotating shift schedules on performance, subjective alertness, sleep patterns, and psychosocial variables. *Work Stress* 1989; 3(1):21–32.
59. Mills DQ. Does organized labor want the 4-Day week? In: Poor R, ed. *4 days, 40 hours*. Cambridge, MA: Bursk and Poor Publishing, Inc., 1970:61–9.

60. Steele JL, Poor R. Work and leisure: the reactions of people at 4-day firms. In: Poor R, ed. *4 days, 40 hours*. Cambridge, MA: Bursk and Poor Publishing, Inc., 1970:105–22.
61. Brief RS, Scala RA. Occupational health aspects of unusual work schedules: a review of Exxon's experiences. *Am Ind Hyg Assoc J* 1986; 47(4):199–202.
62. Colligan MS, Tepas DI. The stress of hours of work. *Am Ind Hyg Assoc J* 1986; 47:686–95.
63. Jones IS, Stein HS. *Effect of driver hours of service on tractor-trailer crash involvement*. Arlington: Insurance Institute for Highway Safety, 1987.
64. Lisper HO, Laurell H, van Loon J. Relation between time to falling asleep behind the wheel on a closed track and changes in subsidiary reaction time during prolonged driving on a motorway. *Ergonomics* 1986; 29(3):445–53.
65. Laundry BR, Lees RE. Industrial accident experience of one company on 8-and 12-hour shift systems. *J Occup Med* 1991; 33(8):903–6.
66. Tucker P, Smith L, MacDonald I, et al. Shift length as a determinant of retrospective on-shift alertness. *Scand J Work Environ Health* 1998; 24(S3):49–54.
67. Smith PA, Wright BM, Mackey RW, et al. Change from slowly rotating 8-hour shifts to rapidly rotating 8-hour and 12-hour shifts using participative shift roster design. *Scand J Work Environ Health* 1998; 24(S3):55–61.
68. Lowden A, Kecklund G, Aselsson J, et al. Change from an 8-hour shift to a 12-hour shift, attitudes, sleep, sleepiness, and performance. *Scand J Work Environ Health* 1998; 24(S3):69–75.
69. Williamson AM, Gower CG, Clarke BC. Changing the hours of shiftwork: a comparison of 8- and 12-hour shift rosters in a group of computer operators. *Ergonomics* 1994; 37(2):287–98.
70. Johnson MD, Sharit J. Impact of a change from an 8-h to a 12-h shift schedule on workers and occupational injury rates. *Int J Ind Ergon*, 2001; 27(5):303–19.
71. Mitchell RJ, Williamson AM. Evaluation of an 8-hour versus a 12-hour shift roster on employees at a power station. *Appl Ergon* 2000; 31(1):83–93.
72. Tucker, P. *Compressed working weeks*. Conditions of Work and Employment Series, No. 12. Geneva: International Labour Organization, 2006. Available at http://www.ilo.org/wcmsp5/groups/public/—ed_protect/—protrav/—travail/documents/publication/wcms_travail_pub_12.pdf (accessed May 4, 2016).
73. Mardon S, ed. *Shiftwork alert, 2(7) and (8) circadian information*. Cambridge, MA: Shiftwork Newsletter, 1997.
74. Finn P. The effects of shift work on the lives of employees. *Monthly Labor Rev* 1981; 104(10):31–5.
75. Krell, E. Second jobs: blessing or curse? *HR Mag* 2010; 55(3):57.
76. Smith L, Folkard S, Tucker P, et al. Work shift duration: a review comparing eight hour and 12 hour shift systems. *Occup Environ Med* 1998; 55:217–29
77. Smith L, Hammond T, Macdonald I, et al. 12-hr shifts are popular but are they a solution? *Int J Ind Ergon* 1998; 21:323–31.
78. Caruso CC, Hitchcock EM, Dick RB, et al. *Overtime and extended work shifts: Recent findings on illnesses, injuries, and health behaviors*. DHHS (NIOSH) Publication, No. 2004-143. Cincinnati: Department of Health and Human Services, Centers for Disease Control and Prevention, 2004.
79. Sirois WG, Moore-Ede M. Staffing Levels: Key to managing risk in 24/7 operations. Circadian White Paper. LP 2013. Circadian Information, Stoneham.
80. Knauth P. Extended work periods. *Ind Health* 2007; 45:125–36.
81. Rosa R, Colligan M. *Plain language about shiftwork*. NIOSH Publication, No. 97–145. Cincinnati: U.S. Department of Health and Human Services, National Institute for Occupational Safety and Health, 1997.
82. Folkard S, Arendt J, Clark M. Sleep and mood on a "weekly" rotating (7-7-7) shift system: some preliminary results. In Costa G, Cesana G, Kogi K, Wedderburn A, eds. *Shiftwork: health, sleep, performance*. Frankfurt: Peter Lang, 1989: 484–9.
83. Gillberg M. Subjective alertness and sleep quality in connection with permanent 12-hour day and night shifts. *Scand J Work Environ Health* 1998; 24(S3):76–81.
84. Folkard S, Barton J. Does the 'forbidden zone' for sleep onset influence morning shift sleep duration? *Ergonomics* 1993; 36:85–9.
85. Akerstedt T. Work schedules and sleep. *Experientia* 1984 40:417–22.
86. Center for Disease Control. Leading work-related diseases and injuries. *MMWR* 1986; 35(39):613–4; 619–21.
87. Imbernon E, Warret G, Roitg C, et al. Effects on health and social well-being of on-call shifts: an epidemiologic study in the french national electricity and gas supply company. *J Occup Med* 1993; 35(11):1131–7.
88. Joyce K, Pabayo R, Critchley JA, et al. Flexible working conditions and their effects on employee health and wellbeing. *Cochrane Database Syst Rev* 2010; (2):CD008009. Available at http://summaries.cochrane.org/CD008009/PUBHLTH_flexible-working-conditions-and-their-effects-on-employee-health-and-wellbeing#sthash.LiB1suFd.dpuf (accessed May 4, 2016). 10.1002/14651858.CD008009.pub2.
89. Nätti J, Oinas T, Härmä M, et al. Combined effects of shiftwork and individual working time control on long-term sickness absence: a prospective study of Finnish employees. *J Occup Environ Med* 2014; 56(7):732–8.
90. Ernst G, Rutenfranz J. Flexibility in shiftwork—some suggestions. *Ergonomics* 1978; 21(10):861–74 (abstracts).
91. Smith P. A study of weekly and rapidly rotating shift workers. *Ergonomics* 1978; 21(10):861–74 (abstracts).
92. Northrup HR, Wilson JT, Rose KM. The twelve-hour shift in the petroleum and chemical industries. *Ind Labor Relat Rev* 1979; 32(3):312–26.
93. Kogi K. Improving shift workers' health and tolerance to shiftwork: recent advances. *Appl Ergon* 1996; 27(1):5–8.
94. Costa G, Folkard S, Harrington JM. Shift work and extended hours of work. In Baxter PJ, Adams PH, Caw T-C, et al., eds. *Hunter's diseases of occupation*, 9th edn. London: Arnold, 2000:581–9.
95. Halberg F, Halberg E, Barnum CP, et al. Circadian rhythm: coined. In: Withrow RB, ed. *Photoperiodism and related phenomenon in plants and animals*. Washington, DC: AAAS, 1959:803–78.
96. Scott AJ, Ladou J. Shiftwork: effects on sleep and health. In: Scott AJ, ed. *Shiftwork: State of the art reviews*. Occupational Medicine, State of the Art Reviews, 5(2). Philadelphia: Hanley & Belfus, 1990:273–99.

97. Scott AJ. Shift work and health: primary care, clinics in office practice. *Occup Environ Med* 2000; 27(4):1057–78.
98. Wever R. Man in temporal isolation: basic principles of the circadian system. In: Folkard S, Monk TH, eds. *Hours of work: temporal factors in work scheduling*. Chichester: John Wiley & Sons, Ltd, 1985:15–28.
99. Aschoff J. Circadian rhythms in man. *Science* 1965; 148:1427–32.
100. Czeisler CA, Duffy JF, Shanahan TL, et al. Stability, precision, and near 24-hour period of the human circadian pacemaker. *Science* 1999; 284(5423):2177–81.
101. Vernibos-Danelles J, Winget CN. The importance of light, postural, and social cues in the regulation of the plasma cortisol rhythm in man. In: Reinberg A, Halberg F, eds. *Chronopharmacology: Proceedings of the 7th International Congress of Pharmacology, Paris 1978*. New York: Pergamon Press, 1979:101–6.
102. Webb WB, Agnew HW Jr. The effects of a chronic limitation of sleep length. *Psychophysiology* 1974; 11(5):265–74.
103. Duffy JF, Kronauer RE, Czeisler CA. Phase-shifting human circadian rhythms: influence of sleep timing, social contact and light exposure. *J Physiol* 1996; 95(pt 1):289–97.
104. Arendt J, Deacon S. Treatment of circadian rhythm disorders: melatonin. *Chronobiol Int* 1997; 14(2):185–204.
105. Benarroch, EE. Suprachiasmatic nucleus and melatonin: reciprocal interactions and clinical correlations. *Neurology* 2008; 71:594–8.
106. Vela-Bueno A, Olavarrieta-Bernardino S, Fernández-Mendoza J, et al. Melatonin, sleep, and sleep disorders. *Sleep Med Clin* 2007; 2:303–8.
107. Gekakis N, Staknis D, Nguyen HB, et al. Role of the CLOCK protein in the mammalian circadian mechanism. *Science* 1998; 280(5369):1564–8.
108. Darlington TK, Wager-Smith K, Ceriani MF, et al. Closing the circadian loop: CLOCK-induced transcription of its own inhibitors per and time. *Science* 1998; 280(5369): 1599–603.
109. Reppert SM, Weaver DR. Coordination of circadian timing in mammals. *Nature* 2002; 418:935–41.
110. Young ME. The circadian clock within the heart: potential influence on myocardial gene expression, metabolism, and function. Invited review. *Am J Physiol Heart Circ Physiol* 2006; 290:H1–16.
111. Davidson AJ, London B, Block GD, et al. Cardiovascular tissues contain independent circadian clocks. *Clin Exp Hypertens* 2005; 27:307–11.
112. Durgan DJ, Hotze MA, Tomlin TM, et al. The intrinsic circadian clock within the cardiomyocyte. *Am J Physiol Heart Circ Physiol* 2005; 289:H1530–41.
113. McNarmara P, Seo SP, Rudic RD, et al. Regulation of CLOCK and MOP4 by nuclear hormone receptors in the vasculature: a hormonal mechanism to reset a peripheral clock. *Cell* 2001; 105:877–89.
114. Nonaka Y, Emoto N, Ikeda K, et al. Angiotensin II induces circadian gene expression of clock genes in cultured vascular smooth muscle cells. *Circulation* 2001; 104:1746–8.
115. Webb WB. Sleep and biological rhythms. In: Webb WB, ed. *Biological rhythms, sleep, and performance*. Chichester: John Wiley & Sons, Ltd, 1982:87–141.
116. Wever R. Phase shifts of human circadian rhythms due to shifts of artificial zeitgebers. *Chronobiologia* 1980; 7:303–27.
117. Nagano M, Adachi A, Nakaham K, et al. An abrupt shift in the day/night cycle causes desynchrony in the mammalian circadian center. *J Neurosci* 2003; 23:6141–51.
118. Folkard S, Monk TH, Lobban MC. Short and long-term adjustment of circadian rhythms in permanent night nurses. *Ergonomics* 1978; 21:785–99.
119. Akerstedt T, Patkai P, Dahlgren K. Field studies of shift work: II. Patterns in psychophysiological activation in workers alternating between night and day work. *Ergonomics* 1977; 20:849–56.
120. Reinberg A, Andlauer P, DePrins J, et al. Desynchronization of the oral temperature circadian rhythm and intolerance to shiftwork. *Nature* 1984; 308:272–4.
121. Martino TA, Tata N, Belsham DD, et al. Disturbed diurnal rhythm alters gene expression and exacerbates cardiovascular disease with rescue by resynchronization. *Hypertension* 2007; 49:1104–13.
122. Martino TA, Oudit GY, Herzenberg AM, et al. Circadian rhythm disorganization produces profound cardiovascular and renal disease in hamsters. *Am J Physiol Regul Integr Comp Physiol* 2008; 294:R1675–83.
123. Penev PD, Kolker DE, Zee PC, et al. Chronic circadian desynchronization decreases the survival of animals with cardiomyopathic heart disease. *Am J Physiol* 1998; 275:H2334–7.
124. Karatsoreos IN, Bhagat S, Bloss EB, et al. Disruption of circadian clocks has ramifications for metabolism, brain, and behavior. *Proc Natl Acad Sci U S A* 2011; 108:1657–62.
125. Varcoe TJ, Wight N, Voultsios A, et al. Chronic phase shifts of the photoperiod throughout pregnancy programs glucose intolerance and insulin resistance in the rat. *PLoS One* 2011; 6:e18504.
126. Fonken LK, Workman JL, Walton JC, et al. Light at night increases body mass by shifting the time of food intake. *Proc Natl Acad Sci U S A* 2010; 107:18664–9.
127. Lee S, Donehower LA, Herron AJ, et al. Disrupting circadian homeostasis of sympathetic signaling promotes tumor development in mice. *PLoS One* 2010; 5:e10995.
128. Logan RW, Zhang C, Murugan S, et al. Chronic shift-lag alters the circadian clock of NK cells and promotes lung cancer growth in rats. *J Immunol* 2012; 188:2583–91.
129. Buxton OM, Cain SW, O'Connor SP, et al. Adverse metabolic consequences in humans of prolonged sleep restriction combined with circadian disruption. *Sci Transl Med* 2012; 4:129ra43. 10.1126/scitranslmed.3003200.
130. Bass J, Takahashi JS. Circadian integration of metabolism and energetics. *Science* 2010; 330(6009):1349–54.
131. McHill AW, Melanson EL, Higgins J, et al. Impact of circadian misalignment on energy metabolism during simulated night-shift work. *Proc Natl Acad Sci U S A* 2014; 111(48):17302–7.
132. Straif K, Baan R, Grosse Y, et al. Carcinogenicity of shift-work, painting, and fire-fighting. *Lancet Oncol* 2007; 8:1065–6.
133. International Agency for Research on Cancer. *Painting, fire-fighting, and shiftwork*. IARC Monographs on the Evaluation of Carcinogenic Risks to Humans, Vol 98. Lyon Cedex: WHO, Published by the International Agency for Research on Cancer, 2010.

134. Akerstedt T. Shift work and disturbed sleep/wakefulness. *Occup Med* 2003; 53:89–94.
135. Tepas DI, Maham RP. The many meanings of sleep. *Work Stress* 1989; 3:93–102.
136. Tepas DI, Stock CG, Maltese JW, et al. Reported sleep of shift workers: a preliminary report. In: Chase M, Mitler, M, Walter P, eds. *Sleep research*, Vol. 7. Los Angeles: Brain Information Research Institute, University of California, 1978.
137. Smith MJ, Colligan MJ, Tasto DL. Health and safety consequences of shift work in the food processing industry. *Ergonomics* 1982; 25(2):133–44.
138. Tilley AJ, Wilkinson RT, Drud M. Night and day shifts compared in terms of the quality and quantity of sleep recorded in the home and performance measured at work: a pilot study. In: Reinberg A, Vieux N, Andlauer P, eds. *Night and shift work: biological and social aspects*. Oxford: Pergamon Press, 1981:187–96.
139. Weitzman ED, Godmacher D, Kripke D, et al. Reversal of sleep–waking cycle: effect on sleep stage pattern and certain neuroendocrine rhythms. *Trans Am Neurol Assoc* 1968; 93:153–7.
140. Akerstedt T, Gillberg M. Sleep disturbances and shiftwork. In: Reinberg A, Vieux N, Andlauer P, eds. *Night and shift work: biological and social aspects*. Oxford: Pergamon Press, 1981:127–37.
141. Gillberg M, Akerstedt T. Body temperature and sleep at different times of day. *Sleep* 1982; 5(4):378–88.
142. Walsh JK, Tepas DI, Moss PD. The EEG sleep of night and rotating shiftworkers. In: Johnson LC, Tepas DI, Colquhoun WP, Colligan MJ, eds. *The twenty-four hour workday: Proceedings of a Symposium on Variations in Work–Sleep Schedules*. Washington, DC: US Government Print Office, 1981:81–127.
143. Tepas DI, Sullivan PJ. Does body temperature predict sleep length, sleepiness, and mood in a lab-bound population? *Sleep Res* 1982; II:42.
144. Knauth P, Rutenfranz J, Schulz H. Experimental shift work studies of permanent night and rapidly rotating shift systems, II. Behavior of various characteristics of sleep. *Int Arch Occup Environ Health* 1980; 46:111–25.
145. Gadbois C. Women on night shift: interdependence of sleep and off-the-job activities. In: Reinberg A, Vieux N, Andlauer P, eds. *Night and shift work: biological and social aspects*. Oxford: Pergamon Press, 1981:223–7.
146. Johnson LC, MacLeod WL. Sleep and awake behavior during gradual sleep reduction. *Percept Mot Skills* 1973; 36:87–97.
147. Grandjean E. *Fitting the task to the man, an ergonomic approach*. London: Taylor & Francis, 1982.
148. Cameron C. A theory of fatigue. *Ergonomics* 1973; 16(5):633–48.
149. Speigel K, Leproult R, Van Cauter E. Impact of sleep debt on metabolic and endocrine function. *Lancet* 1999; 354:1435–9.
150. Grandner MA, Hale L, Moore M, et al. Mortality associated with short sleep duration: the evidence, the possible mechanisms, and the future. *Sleep Med Rev* 2010; 14(3):191–203.
151. Ayas NT, Whit DP, Mason JE, et al. A prospective study of sleep duration and coronary heart disease in women. *Arch Intern Med* 2003; 27(3):205–9.
152. Lyytikainen P, Rakonen O, Lahelma E, et al. Association if sleep duration with weight and weight gain: a prospective follow-up study. *J Sleep Res* 2011; 20:298–302.
153. Morselli, LL. Sleep and metabolic function. *Pflugers Arch* 2012; 463(1):139–60.
154. Torsvall L, Akerstedt T, Gillander K, et al. Sleep on the night shift: 24-hour EEG monitoring of spontaneous sleep/wake behavior. *Psychophysiology* 1989; 26(3):352–8.
155. Kecklund G, Akerstedt T. Sleepiness in long distance truck driving, an ambulatory EEG study of night driving. *Ergonomics* 1993; 36:1007–17.
156. Winget CM, DeRoshia CW, Markley CL, et al. A review of human physiological and performance changes associated with desynchronosis of biological rhythms. *Aviat Space Environ Med* 1984; 55(12):1085–93.
157. Kogi K. Introduction to the problems of shiftwork. In: Folkard S, Monk TH, eds. *Hours of work*. New York: John Wiley & Sons, Inc., 1985:165–84.
158. Smith MJ, Colligan MJ, Hurrell JJ Jr. A review of the psychological stress research carried out by NIOSH: 1971–1976. In: *New developments in occupational stress*. Cincinnati: US. Department of Health and Human Services, 1980:1–9.
159. Frese M, Okonek K. Reasons to leave shiftwork and psychological and psychosomatic complaints of former shiftworkers. *J Appl Psychol* 1984; 69(3):509–14.
160. Verhaegan P, Maasen A, Meers A. Health problems in shiftworkers. In: Johnson L, Colquhoun WP, Tepas D, eds. *Biological rhythms, sleep, and shift work*. Advances in Sleep Research, Vol 7. New York: Spectrum Publications, 1981:271–87.
161. Moore-Ede MC, Richardson GS. Medical implications for shift work. *Annu Rev Med* 1985; 36:607–17.
162. Reinberg A, Motohashi Y, Bourdeleau P, et al. Internal desynchronization of circadian rhythms and tolerance of shiftwork. *Chronobiologia* 1989; 16:21–34.
163. Askenazi IE, Reinberg AE, Motohashi Y. Interindividual differences in the flexibility of human temporal organization: pertinence to jet lag and shiftwork. *Chronobiol Int* 1997; 14(2):99–113.
164. Coleman RM. Shiftwork scheduling for the 1990s. *Personnel* 1989; 66(1):10–5.
165. American Academy of Sleep Medicine. *The international classification of sleep disorders: diagnostic and coding manual*, 2nd ed. Westchester: American Academy of Sleep Medicine, 2005.
166. Sack RL, Auckley D, Auger R, et al. Circadian rhythm sleep disorders: Part I. basic principles, shift work, and jet lag disorders: an American Academy of Sleep Medicine review. *Sleep* 2007; 30(11):1460–83.
167. Waage JS, Moen BE, Pallesen S, et al. Shift work disorder among oil rig workers in the North Sea. *Sleep* 2009; 32(4):558–65.
168. Graeber RC, Lauber JK, Connel LJ, et al. International aircrew sleep and wakefulness after multiple time-zone flights: a cooperative study. *Aviat Space Environ Med* 1986; 57(12):B3–9.
169. Härmä M. Aging, physical fitness and shiftwork tolerance. *Appl Ergon* 1996; 27(1):25–9.
170. Akerstedt T, Froberg J. Interindividual differences in circadian patterns of catecholamine excretion, body temperature, performance and subjective arousal. *Biol Psychol* 1976; 4:277–92.

171. Costa G, Lievore F, Casaletti G, et al. Circadian characteristics influencing interindividual differences in tolerance and adjustment to shiftwork. *Ergonomics* 1989; 32:373–85.
172. Roden KM, Koller M, Pirich K, et al. The circadian melatonin and cortisol secretion pattern in permanent night shift workers. *Am J Physiol* 1993; 265:R261–7.
173. Costa G. Guidelines for the medical surveillance of shiftworkers. *Scand J Work Environ Health* 1998; 24(S3):151–5.
174. Loudon R, Bohle P. Work/non-work conflict and health in shiftwork: relationships with family status and social support. *Int J Occup Environ Health* 1997; 3:S71–7.
175. Angersbach D, Knauth P, Loskant H, et al. A retrospective cohort study comparing complaints and diseases in day and shift workers. *Int Arch Occup Environ Health* 1980; 45:127–40.
176. Minors DS, Scott AR, Waterhouse JM. Circadian arrhythmia: shiftwork, travel, and health. *J Soc Occup Med* 1986; 36(2):39–44.
177. Colligan MJ, Frock IJ, Tasto D. Shift work: the incidence of medication use and physical complaints as a function of shift. In: *Occupational and health symposia, 1978*. NIOSH Publication, No. 80-105. Cincinnati: US Department of Health, Education, and Welfare, 1980:47–57.
178. Walker J, De la Mare G. Absence from work in relation to length and distribution of shift hours. *Br J Ind Med* 1971; 28:36.
179. Costa G. The impact of shift and night work on health. *Appl Ergon* 1994; 27(1):9–16.
180. Knutsson A, Bogglid H. Gastrointestinal disorders among shift workers. *Scand J Work Environ Health* 2010; 36(2):85–95.
181. Moore JC, Englert E. Circadian rhythms of gastric acid secretion in man. *Nature* 1970; 226:1261–2.
182. Goo RH, Moore JG, Greenberg E, et al. Circadian variation in gastric emptying of meals in humans. *Gastroenterology* 1987; 93:515–8.
183. Reinberg A, Migraine A, Apfelbaum C. Circadian and ultradian rhythms in the eating behavior and nutrient intake of oil refinery operators (Study 2). *Chronobiologia* 1979; 6(suppl 1): 89–102.
184. Stewart AJ, Wahlquist ML. Effect of shiftwork on canteen food purchase. *J Occup Med* 1985, 27(8):552-4.
185. Tepas DI. Do eating and drinking habits interact with work schedule variables? *Work Stress* 1990; 4(3):203–11.
186. Knutson A, Boggild H. Shiftwork, risk factors and cardiovascular disease: review of disease mechanisms. *Rev Environ Health* 2000; 15:359–72.
187. De Bacquer D, Van Risseghem M, Clays E, et al. Rotating shift work and the metabolic syndrome: a prospective study. *Int J Epidemiol* 2009; 38:848–54.
188. Lin YC, Hsiao TJ, Chen PC Persistent rotating shift-work exposure accelerates development of metabolic syndrome among middle-aged female employees: a five-year follow-up. *Chronobiol Int* 2009; 26:740–55.
189. Haffner SM. Epidemiology of insulin resistance and its relation to coronary artery disease. *Am J Cardiol* 1999; 84:11J–4.
190. Ha M, Park J. Shiftwork and metabolic risk factors of cardiovascular disease. *J Occup Health* 2005; 47(2):89–95.
191. Sheer FA, Hilton MF, Mantzoros CS, et al. Adverse metabolic and cardiovascular consequences of circadian misalignment. *Proc Natl Acad Sci U S A* 2009; 106 (11):4453–8.
192. Orth-Gomér K. Intervention on coronary risk factors by changing working conditions of Swedish policemen. In: Harvath M., Frankth E., eds. *Psychophysiologic risk factors of cardiovascular diseases*, International Symposium, Suppl. 3, Actio. Nerv. Sup. (Praha) Avicenum-Czechoslovak Med. Prague: Avicenum Czechoslovak Medical Press, 1982.
193. DeBacker G, Kornitzer M, Peters H, et al. Relation between work rhythm and coronary risk factors. *Eur Heart J* 1984; 5(suppl 1):307 (abstract).
194. Knutsson A. Relationships between serum triglycerides and gamma-glutamyltransferase among shift and day workers. *J Intern Med* 1989; 226:337–9.
195. Lavie P, Chillag N, Epstein R, et al. Sleep disturbances in shift-workers: marker for maladaptation syndrome. *Work Stress* 1989; 3(1):33–40.
196. Harenstam A, Theorell T, Orth-Gomer K, et al. Shiftwork, decision latitude and ventricular ectopic activity: a study of 24-hour electrocardiograms in Swedish prison personnel. *Work Stress* 1987; 1(4):341–50.
197. Knutsson A, Anderson H, Berglund U, Serum lipoproteins in day and shift workers: a prospective study. *Br J Ind Med* 1990; 47:132–4.
198. Yamasaki F, Schwartz JE, Gerber LM, et al. Impact of shift work and race/ethnicity on the diurnal rhythm of blood pressure and catecholamines. *Hypertension* 1998; 32(3):417–23.
199. Morikawa Y, Nakagawa H, Miura K. Relationship between shiftwork and onset of hypertension in a cohort of manual workers. *Scand J Work Environ Health* 1999; 25(2):100–4.
200. Karlsson BH, Knutsson AK, Lindahl BO, et al. Metabolic disturbances in male workers with rotating 3 shiftwork. Results of the WOLF study. *Int Arch Occup Environ Health* 2003; 76:424–30.
201. Boggild H, Knutsson A. Shiftwork, risk factors and cardiovascular disease. *Scand J Work Environ Health* 1999; 25(2):85–99.
202. Brown DL, Feskanich D, Sánchez BN, et al. Rotating night shift work and risk of ischemic stroke. *Am J Epidemiol* 2009; 169(11):1370–7.
203. Knutsson A, Akerstedt T, Jonsson BG, et al. Increased risk of ischemic heart disease in shift workers. *Lancet* 1986; 12(2):89–92.
204. Kawachi I, Colditz GA, Stampfer MJ, et al. Prospective study of shiftworkers and risk of coronary heart disease in women. *Circulation* 1995; 92 (11):3178–82.
205. Karlsson B, Alfredsson L, Knutsson A, et al. Total mortality and cause-specific mortality of Swedish shift- and day workers in the pulp and paper industry in 1952–2001. *Scand J Work Environ Health* 2005; 31(1):30–5.
206. Tenkanen L, Sjoblom T, Hãrmã M. Joint effect of shiftwork and adverse life-style factors on the risk of coronary heart disease. *Scand J Work Environ Health* 1998; 24(5):351–7.
207. Young ME. The circadian clock within the heart: potential influence on myocardial gene expression, metabolism, and function. *Am J Phys Heart Circ Phys* 2006; 290:H1–16. 10.1152/ajpheart.00582.2205.

208. Knutsson A, Hallquist J, Reuterwall C, et al. Shiftwork and myocardial infarction: a case–control study. *Occup Environ Med* 1999; 56:46–50.
209. Peter R, Alfredsson L, Knuttsson A, et al. Does a stressful psychosocial work environment mediate the effects of shift work on cardiovascular risk factors? *Scand J Work Environ Health* 1999; 25(4):376–81.
210. Puttonen S, Hãrmã M, Hublin C. Shift work and cardiovascular disease: pathways from circadian stress to morbidity. *Scan J Work Environ Health* 2010; 36(2):96–108.
211. Barbadoro P, Santarelli L, Croce N, et al. Rotating shiftwork as an independent risk factor for overweight Italian workers: a cross-sectional study. *PLoS One* 2013; 8(5):e63289. 10.1371/journal.pone.0063289.
212. Antunes IC, Levandovksi R, Dantas G, et al. Obesity and shift work: chronobiological aspects. *Nutr Res Rev* 2010; 23:155–68.
213. Suwazono Y, Dochi M, Sakata K, et al. Longitudinal study on the effect of shift work on weight gain in male Japanese workers. *Obesity* 2008; 16:1887–93.
214. Zhao I, Bogossian F, Turner C. Does maintaining of changing shift types affect BMI? *J Occup Environ Med* 2012; 54(5):525–31.
215. Kroenke CH, Spiegelman, D, Manson J, et al. Work characteristics and incidence of type 2 diabetes in women. *Am J Epidemiol* 2007; 165(2):175–83.
216. Pan A, Schernhammer ES, Sun Q, et al. Rotating night shift work and risk of type 2 diabetes: two prospective cohort studies in women. *PLoS Med* 2011; 8(12):e1001141. 10.1371/jornal.pmed.1001141.
217. Morikawa Y, Nakagawa H, Miura K, et al. Shift work and the risk of diabetes mellitus among Japanese male factory workers. *Scand J Work Environ Health* 2005; 31(3):179–83.
218. Suwazono Y, Sakata K, Okubo Y, et al. Long term longitudinal study on the relationship between alternating shiftwork and the onset of diabetes mellitus in male Japanese workers. *J Occup Envion Med* 2006; 48:455–61.
219. McDonald A, McDonald J, Armstrong B, et al. Prematurity and work in pregnancy. *Br J Ind Med* 1988; 45:56–62.
220. Armstrong G, Nolin A, McDonald A. Work in pregnancy and birth weight for gestational age. *Br J Ind Med* 1989; 46:196–9.
221. Axelsson G, Rylander R, Molin I. Outcome of pregnancy in relation to irregular and inconvenient work schedules. *Br J Ind Med* 1989; 46:393–8.
222. Xu X, Ding M, Li B, et al. Association of rotating shiftwork with preterm births and low birth weight among never smoking women textile workers in China. *Occup Environ Med* 1994; 51(7):470–4.
223. Fortier I, Marcoux S, Brisson J. Maternal work during pregnancy and the risks of delivering a small-for-gestational-age or preterm infant. *Scand J Work Environ Health* 1995; 21(6):412–8.
224. Nurimen T. Shift work, fetal development and course of pregnancy. *Scand J Work Environ Health* 1989; 15: 395–403.
225. Saurel-Cubizolles M, Kaminski M. Pregnant women's working conditions and their changes during pregnancy: a national study in France. *Br J Ind Med* 1987; 44:236–43.
226. Croteau A, Marcoux S, Brisson C. Work activity in pregnancy, preventive measures, and the risk of delivering a small-for-gestational-age infant. *Am J Public Health* 2006; 96(5):846–55.
227. Bonzini M, Palmer KT, Coggon D, et al. Shift work and pregnancy outcomes: a systematic review with meta-analysis of currently available epidemiological studies. *Br J Obstet Gynaecol* 2011; 118(12):1429–37.
228. Bonzini M, Coggon D, Godfrey K, et al. Occupational physical activities, working hours and outcome of pregnancy: findings from the Southampton Women's Survey. *Occup Environ Med* 2009; 66(10):685–90.
229. Saurel-Cubizolles MJ, Zeitlin J, Lelong N, et al. Employment working conditions, and preterm birth: results from the Europop case–control survey. *J Epidemiol Community Health* 2004; 58(5):395–401.
230. McDonald A, McDonald J, Armstrong B, et al. Fetal death and work in pregnancy. *Br J Ind Med* 1988; 45:148–57.
231. Axelsson G, Ahlborg G Jr., Bodin L. Shift work, nitrous oxide exposure, and spontaneous abortion among Swedish midwives. *Occup Environ Med* 1996; 53:374–8.
232. Infante-Rivard C, David M, Gauthier R, et al. Pregnancy loss and work schedule during pregnancy. *Epidemiology* 1993; 4(1):73–5.
233. Axelson G, Molin I. Outcome of pregnancy among women living near petrochemical industries in Sweden. *Int J Epidemiol* 1988; 17(2):363–9.
234. Axelsson G, Lutz C, Rylander R. Exposure to solvents and outcome of pregnancy in university laboratory employees. *Br J Ind Med* 1994; 41:305–12.
235. Uehata T, Sasakawa N. The fatigue and maternity disturbances of night workwomen. *J Hum Ergol (Tokyo)* 1982; 11:465–74.
236. Hemminki K, Kyyronen P, Lindbohm ML. Spontaneous abortions and malformations in the offspring of nurses exposed to anesthetic gases, cytostatic drugs, and other potential hazards in hospitals, based on registered information of outcome. *J Epidemiol Community Health* 1985; 39:141–7.
237. Nurminen T. Shift work and reproductive health. *Scand J Work Environ Health* 1998; 24(suppl 3):28–34.
238. Zhu JL, Hjollund NH, Andersen AM, et al. Shift work, job stress and late fetal loss: the national birth cohort in Denmark. *J Occup Environ Med* 2004; 24(11):1144–9.
239. Eskenazi B, Fenster L, Wright S, et al. Physical exertion as a risk factor for spontaneous abortion. *Epidemiology* 1994; 5(1):6–13.
240. Bryant E, Love EJ. Effect of employment and its correlates on spontaneous abortion risk. *Soc Sci Med* 1991; 33:795–800.
241. Bonde JR, Jorgensine KT, Bonzinie M, et al. Risk of miscarriage and occupational activity: a systematic review and meta-analysis regarding shift work, working hours, lifting, standing, and physical work load. *Scan J Work Environ Health* 2013; 39(4):325–34.
242. Ahlborg G Jr., Axelsson G, Bodin L. Shift work, nitrous oxide exposure and subfertility among Swedish midwives. *Int J Epidemiol* 1996; 25(4):783–90.
243. Bisanti L, Olsen J, Basso O, et al. Shift work and subfecundity: a European multicenter study. *J Occup Environ Med* 1996; 38(4):352–8.

244. Spinelli A., Figà-Talamanca I, Osborn J. Time to pregnancy and occupation in a group of Italian women. *Int J Epidemiol* 1997; 26(3):601–9.
245. Tuntiseranee P, Olsen J, Geater A, et al. Are long working hours and shiftwork risk factors for subfecundity? A study among couples from southern Thailand. *Occup Environ Med* 1998; 55(2):99–105.
246. Zhu JL, Hjollund NH, Boggild H, et al. Shift work and subfecundity: a causal link or an artefact? *Occup Environ Med* 2003; 60(9):E12.
247. Taylor P, Pocock S. Mortality of shift and day workers 1956–68. *Br J Ind Med* 1972; 29:201–7.
248. Rafnsson V, Gunnarsdottir H. Mortality study of fertilizer manufacturers in Iceland. *Br J Ind Med* 1990; 47:721–5.
249. Tynes T, Hannevik M, Andersen A, et al. Incidence of breast cancer in Norwegian female radio and telegraph operators. *Cancer Causes Control* 1996; 7(2):197–204.
250. Davis S, Mirick DK, Stevens RG. Night shift work, light at night, and risk of breast cancer. *J Natl Cancer Inst* 2001; 93(20):1557–62.
251. Hansen J. Increased breast cancer risk among women who work predominantly at night. *Epidemiology* 2001; 12:74–7.
252. Hansen J. Breast cancer among women who work at night. *Epidemiology* 2001; 12:588–99.
253. Schernhammer ES, Laden F, Speizer FE, et al. Rotating night shifts and risk of breast cancer in women participating in the nurses' health study. *J Natl Cancer Inst* 2001; 93:1563–8.
254. Schernhammer ES, Kroenke CH, Laden F, et al. Night work and risk of breast cancer. *Epidemiology* 2006; 17:108–11.
255. Lie JA, Roessink J, Kjaerheim K. Breast cancer and night work among Norwegian nurses. *Cancer Causes Control* 2006; 17:39–44.
256. O'Leary ES, Schoenfeld ER, Stevens RG, et al. Electromagnetic fields and breast cancer on long Island Study Group. Shift work, light at night, and breast cancer on Long Island, New York. *Am J Epidemiol* 2006; 164:358–66.
257. Schwartzbaum J, Ahlbom A, Feychting M. Cohort study of cancer risk among male and female shift workers. *Scand J Work Environ Health* 2007; 33:336–43.
258. International Agency for Research on Cancer (IARC). *Painting, firefighting and shiftwork*, IARC Monographs on the Evaluation of Carcinogenic Risk to Humans, Vol 98. Lyon: IARC, 2010.
259. Costa G, Hause E, Stevens R. Shift work and cancer-considerations on rationale, mechanisms, and epidemiology. *Scand J Work Environ Health* 2010; 36(2):163–79.
260. Pesch B, Harth V, Rabstein S, et al. Night work and breast cancer-results from the German GENICA study. *Scand J Work Environ Health* 2010; 36(2):134–41.
261. Hansen J, Stevens RG. Case-control study of shift-work and breast cancer risk in Danish nurses: impact of shift systems. *Eur J Cancer* 2011; 48:1722–9.
262. Kamdar BB, Tergas AI, Mateen FJ, et al. Night-shift work and risk of breast cancer: a systematic review and meta-analysis. *Breast Cancer Res Treat* 2013; 138(1):291–301. 10.1007/s10549-013-2433-1.
263. Jia Y, Lu Y, Wu K, et al. Does night work increase the risk of breast cancer? A systematic review and meta-analysis of epidemiological studies. *Cancer Epidemiol* 2013; 37:197–203.
264. Wang F, Yeung KL, Chan WC, et al. A meta-analysis on dose–response relationship between night shift work and the risk of breast cancer. *Ann Oncol* 2013; 24:2724–32.
265. Fritschi L, Erren TC, Glass DC, et al. The association between different night shiftwork factors and breast cancer: a case–control study. *Br J Cancer* 2013; 109:2472–80.
266. Hansen J. Night shiftwork and breast cancer survival in Danish women. *Occup Environ Med* 2014; 71(21):A26.
267. Fu L, Pelicano H, Liu J, et al. The circadian gene Period 2 plays an important role in the tumor suppression and DNA damage response in vivo. *Cell* 2002; 111:41–50.
268. Cos S, Fernandez R, Guezmes A, et al. Influence of melatonin on invasive and metastatic properties of MCF-7 human breast cancer cells. *Cancer Res* 1998; 58:4383–90.
269. Hill SM, Blask DE. Fracture of the pineal hormone melatonin on the proliferation and morphological characteristics of human breast cancer cells (MCF-seven) and culture. *Cancer Res* 1998; 48:6121–6.
270. Petranka J, Baldwin W, Biermann J, et al. The non-prostatic action of melatonin in an ovarian carcinoma cell line. *J Pineal Res* 1999; 26:129–36.
271. Anisimov VN, Popovich IG, Zabezhinski MA. Melatonin and colon carcinogenesis: I. Inhibitory effect of melatonin on development of intestinal tumors induced by 1,2-dimethylhydrazine in rats. *Carcinogenesis* 1997; 18:1549–53.
272. Cini G, Coronnello M, Mini E, et al. Melatonin has growth-inhibitory effect on hepatoma AHI30 in the rat. *Cancer Lett* 1998; 125:51–9.
273. Vijayalaxmi T, Thomas CR Jr., Reiter RJ, et al. Melatonin: from basic research to cancer treatment clinics. *J Clin Oncol* 2002; 20:2575–601.
274. Haus E, Smolensky M. Biological clocks and shiftwork: circadian dysregulation and the potential long term effects. *Cancer Causes Control* 2006; 17:489–500.
275. Liu R, Jacob DI, Hansen J, et al. Aberrant methylation of miR-34b is associated with long-term shiftwork: a potential mechanism for increased breast cancer susceptibility. *Cancer Causes Control* 2015; 26(2):171–8.
276. Monsees GM, Kraft P, Hankinson S, et al. Circadian genes and breast cancer susceptibility in rotating shift workers. *Int J Cancer* 2012; 131(11):2547–52.
277. Pronk A, Ji BT, Shu XO, et al. Night-shift work and breast cancer risk in a cohort of Chinese women. *Am J Epidemiol* 2010; 171:953–9.
278. Schernhammer ES, Laden F, Speizer FE, et al. Night shift work and risk of colorectal cancer in nurses' health study. *J Natl Cancer Inst* 2003; 95:825–8.
279. Viswanathan AN, Hankinson SE, Schernhammer ES. Night shift work and the risk of endometrial cancer. *Cancer Res* 2007; 67:10618–22.
280. Parent M-E, El-Zein M, Rousseau M-C, et al. Night work and the risk of cancer among men. *Am J Epidemiol* 2012; 176(9):751–9.
281. Lahti TA, Partonen T, Kyyronen P, et al. Night-time work predisposes to non-Hodgkin lymphoma. *Int J Cancer* 2008; 123:2148–51.
282. Kubo T, Ozasa K, Mikami K, et al. Prospective cohort study of the risk of prostate cancer among rotating-shift workers: findings from the Japan Collaborative Cohort Study. *Am J Epidemiol* 2006; 164(6):549–55.

283. Conlon M, Lightfoot N, Kreiger N. Rotating shift work and risk of prostate cancer. *Epidemiology* 2007; 18(1):182–3.
284. Flynn-Evans EE, Mucci L, Stevens RG, et al. Shiftwork and prostate-specific antigen in the National Health and Nutrition Examination Survey. *J Natl Cancer Inst* 2013; 105(17):1292–7.
285. Hrushesky WJ, Bjarnason GA. Circadian cancer therapy. *J Clin Oncol* 1993; 11(7):1403–17.
286. Smolensky MH, Hermida RC, Ayala DE, et al. Administration-time-dependent effect of blood pressure-lowering medications: basis for chronotherapy of hypertension. *Blood Press Monit* 2010; 15:173–80.
287. Hermida RC, Ayala DE, Fernandez JR, et al. Circadian rhythms in blood pressure regulation and optimization of hypertension treatment with ACE inhibitor and ARB medications. *Am J Hypertens* 2011; 24:383–91.
288. Smolensky MH, D'Alonzo GE. Medical chronobiology: concepts and applications. *Am Rev Respir Dis* 1993; 147:S2–19.
289. Smolensky MH, Portaluppi F, Manfredini R, et al. Diurnal and twenty-four hour patterning of human diseases: acute and chronic common and uncommon medical conditions. *Sleep Med Rev* 2014; 21:12–22. 10.1016/j.smrv.2014.06.005.
290. Faiman C, Moorehouse JA. Diurnal variation in the levels of glucose and related substances in health and diabetic subjects during starvation. *Science* 1967; 32:111.
291. Mejean L, Bicakova-Rocher A, Kolopp M. Circadian and ultradian rhythms in blood glucose and plasma insulin of healthy adults. *Chronobiol Int* 1988; 5:227–36.
292. Sensi S, Capani F, Bertini M, et al. Circadian variation in insulin response to tolbutamide and related hormonal levels in fasting healthy subjects. *Int J Chronobiol* 1976; 3:141–53.
293. Jenkins DJA. Lente carbohydrate: a newer approach to the dietary management of diabetes. *Diabetes Care* 1982; 5:634–41.
294. Trout DL, Bhathena SJ, Mobarak M, et al. Evidence that therapeutic alterations of a circadian rhythm for gastric emptying response may be possible. *Prog Clin Biol Res* 1990; 341B:155–65.
295. Poole CJ, Wright AD, Nattrass M. Control of diabetes mellitus in shift workers. *Br J Ind Med* 1992; 49:513–5.
296. Durazzo TS, Spencer SS, Duckrow RB, et al. Temporal distributions of seizure occurrence from various epileptogenic regions. *Neurology* 2008; 70(15):1265–71.
297. Pavolva MK, Shea SA, Bromfield EB. Daylight patterns of focal seizures. *Epilepsy Behav* 2004; 5(1):44–9.
298. Engle R, Halberg F, Gully RJ. The diurnal rhythm in EEG discharge and in circulating eosinophils in certain types of epilepsy. *Electroencephalogr Clin Neurophysiol* 1952; 4:115–6.
299. van Campen JS, Valentijn FA, Jansen FE, et al. Seizure occurrence and the circadian rhythm of cortisol: a systematic review. *Epilepsy Behav* 2015; 47:132–7. 10.1016/j.yebeh.2015.04.071.
300. Ellingson RJ, Wilken K, Bennett DR. Efficacy of sleep deprivation as an activation procedure in epilepsy patients. *J Clin Neurophysiol* 1984; 1:83–101.
301. Pratt KL, Mattson RH, Wubers NJ, et al. EEG activation of epileptics following sleep deprivation: a prospective study of 114 cases. *Electroencephalogr Clin Neurophysiol* 1968; 24:11–5.
302. Bergonzi P, Chuunvilla C, Tempesta E. Selective deprivation of sleep stages in epileptics. In: Koella WP, Levi P, eds. *Sleep: physiology, biochemistry, psychology, pharmacology, clinical implications, 1st European Congress on Sleep Research, Basel, 1972.* Basil: Kayer, 1973.
303. Tarp B. Epilepsy and sleep. In: Guilleminault C, ed. *Sleeping and waking disorders: indications and techniques.* Menlo Park: Addison-Wesley Publishing Company, 1982:373–81.
304. Kendis H, Baron K, Schuele SU, et al. Chronotypes in patients with epilepsy: does the type of epilepsy make a difference? *Behav Neurol* 2015; 2105:941354. 10.1155/2015/941354.
305. Guberan E, Williams MK, Walford J, et al. Circadian variation of FEV in shift workers. *Br J Ind Med* 1969; 26:121–5.
306. Reinberg A, Guillet P, Gervais P, et al. One-month chronocorticotherapy (Dutimelan 8–15 mite). Control of the asthmatic condition without adrenal suppression and circadian rhythm alteration. *Chronobiologia* 1977; 4:295–312.
307. Smolensky MH. Aspects of human chronopathology. In: Reinberg A, Smolensky MH, eds. *Biological rhythms and medicine: cellular, metabolic, physiopathologic, and pharmacologic aspects.* New York: Springer-Verlag, 1983:131–209.
308. De Vries, Goei JT, Booy-Noord H, et al. Changes during 24 hours in lung function and histamine hyperreactivity of the bronchial tree in asthmatic and bronchitic patients. *Int Arch Allergy Appl Immunol* 1962; 20:93–101.
309. Barnes P, FitzGerald G, Brown M, et al. Nocturnal asthma and changes in circulating epinephrine, histamine, and cortisol. *N Engl J Med* 1980; 303:263–7.
310. Mohiuddin A. Martin RJ. Circadian basis of late asthmatic response. *Am Rev Respir Dis* 1990; 142:1153–7.
311. Smolensky M. The chronobiology and chronotherapy of asthma. Paper presented at Clinical Applied Chronobiology Conference, NIH, Bethesda, MD, June 20, 1989.
312. Pincus DJ, Szefler SJ, Ackerson LM, et al. Chronotherapy of asthma with inhaled steroids: the effect of dosage timing on drug efficacy. *J Allergy Clin Immunol* 1995; 95:1172–8.
313. Beam WR, Weiner DE, Martin RJ. Timing of prednisone and alterations of airways inflammation in nocturnal asthma. *Am Rev Respir Dis* 1992; 146(6):1524–30.
314. Hetzel MR, Clark TJ. The clinical importance of circadian factors in severe asthma. In: Reinberg A, Halberg F. eds. *Chronopharmacology: Proceedings of the Seventh International Congress of Pharmacology*, New York: Pergamon Press, 1982:213–21.
315. Reinberg A., Vieux N, Andlauer P, et al. Tolerance to shift work: a chronobiological approach. *Adv Biol Psychiatry* 1983; 11:35–47.
316. Colligan MJ, Rosa RR. Shiftwork: effects of social and family life. In: Scott AJ, ed. *Shiftwork: occupational medicine.* State of the Art Reviews, Vol 5(2). Philadelphia: Hanley & Belfus, 1990:315–22.
317. Nilsson C. Social consequences of the scheduling of working hours. In: Reinberg A, Vieux N, Andlauer P, eds. *Night and shiftwork: Biological and social aspects.* Oxford: Pergamon Press, 1981:187–96.
318. Carpentier J, Cazamian P. *Night work: Its effects on the health and welfare of the worker.* Geneva: ILO, 1977.
319. Koller M, Kundi M, Cervinka R. Field studies of shift work at an Austrian oil refinery: I. Health and psychosocial

well-being of workers who drop out of shiftwork. *Ergonomics* 1978; 21(10):835–47.
320. Simon BL. Impact of shift work on individual and families. *Fam Soc* 1990; 71:342–8.
321. Smith L, Folkard S. The impact of shiftwork upon shiftworkers' partners. *Proceedings of the 10th International Symposium on Night and Shiftwork.* Sheffield, September 18–22, 1991, International Commission on Occupational Health.
322. Wyatt S, Marriott R. Night work and shift changes. *Br J Ind Med* 1953; 10:164–72.
323. Mott PE. *Shiftwork: the social, psychological, and physical consequences.* Ann Arbor: University of Michigan Press, 1965.
324. Åkerstedt T, Torsvall L. Experimental changes in shift schedules: their effects on well-being. *Ergonomics* 1978; 21(10):849–56.
325. Meer A, Maasen A, Verhaegaen P. Subjective health after six months and after four years of shift work. *Ergonomics* 1978; 21(10):857–9.
326. Holley DC, Winget CM, DeRoshia CM. *Effects of circadian rhythm phase alteration on physiological and psychological variables: Implications to pilot performance.* NASA Technical Memorandum, 81277. Washington, DC: National Aeronautics and Space Administration, 1981.
327. Friedman RC, Bigger JT, Kornfeld DS. The intern and sleep loss. *N Engl J Med* 1971; 285(4):201–3.
328. Boivin D, Czeisler CA, Dijk DJ. Complex interaction of the sleep–wake cycle and circadian phase modulates mood in healthy subjects. *Arch Gen Psychiatry* 1997; 54:145–52.
329. Healy D, Waterhouse JM. The circadian system and affective disorders: clocks or rhythms? *Chronobiol Int* 1990; 7(11):5–10.
330. Healy D, Waterhouse JM. Reactive rhythms and endogenous clocks (Editorial). *Psychol Med* 1991; 21:557–64.
331. Wehr TA. Reply to D. Healy and J. M. Waterhouse, The circadian system and affective disorders: clocks or rhythms? *Chronobiol Int* 1990; 7(11):11–4.
332. Monk TH. Biological rhythms and depressive disorders. In: Mann JJ, Kupfer DJ, eds. *The biology of depressive disorders.* New York: Plenum Press, 1993:19–122.
333. Wehr TA, Goodwin FK, Biological rhythms and psychiatry. In: Arieti S, Brodie K, eds. *American handbook of psychiatry,* Vol 7, 2nd ed. New York: Basic Books, 1981:46–107.
334. Wirz-Justice A. Antidepressant drugs: effects on the circadian system. In: Wehr TA, Goodwin FK, eds. *Circadian rhythms in psychiatry.* Los Angeles: Boxwood Press, 1983:235–43.
335. Rockwell DA, Hodgson MG, Beljan JR. Psychologic and psychophysiologic responses to 105 days of social isolation. *Aviat Space Environ Med* 1976; 10:1087–93.
336. Lewy AJ, Sack RL, Miller LS, et al. Antidepressant and circadian phase-shifting effects of light. *Science* 1987; 135:352–4.
337. Rockwell DA, Winget CM, Rosenblatt LS, et al. Biological aspects of suicide: circadian disorganization. *J Nerv Ment Dis* 1978; 166(12):851–9.
338. Sack DA, Nurnberger J, Rosenthal NE, et al. Potentiation of antidepressant medications by phase advance of the sleep–wake cycle. *Am J Psychiatry* 1985; 142(5):606–8.
339. David MM, MacLean AW, Knowles JB, et al. Rapid eye movement latency and mood following a delay of bedtime in healthy subjects: do the effects mimic changes in depressive illness? *Acta Psychiatr Scand* 1991; 84(1):33–9.
340. Frank E, Kupfer DJ, Ehlers CL, et al. Interpersonal and social rhythm therapy for bipolar disorder: integrating interpersonal and behavioral approaches. *Behav Ther* 1995; 17:144–9.
341. Jauhar P, Weller MPI. Psychiatric morbidity and time zone changes: a study of patients from Heathrow Airport. *Br J Psychiatry* 1982; 140:231–5.
342. Ehlers CL, Frank E, Kupfer DJ. Social zeitgebers and biological rhythms: a unified approach to understanding. *Arch Gen Psychiatry* 1988; 45:948–52.
343. Ehlers CL, Kupfer DJ, Frank E, et al. Biological rhythms and depression: the role of zeitgebers and zeitstorers. *Depression* 1993; 1:285–93.
344. Tasto DL, Colligan MJ, Polly SJ. *Health consequences of shiftwork.* NIOSH. Cincinnati: U.S. Department of Health Education and Welfare, 1978.
345. Bohle P, Tilley AJ. The impact of night work on psychological well-being. *Ergonomics* 1989; 3:1089–99.
346. Costa G., Apostali P, d'Andrea F, et al. Gastrointestinal and neurotic disorders in textile shift workers. In: Reinberg A, Vieux N, Andlauer P, eds. *Night and shift work: Biological and social aspects.* Oxford: Pergamon Press, 1981:187–96.
347. Michael-Briand C., Chopard JL, Guiot A, et al. The pathological consequences of shiftwork in retired workers. In: Reinberg A, Vieux N, Andlauer P, eds. *Night and shiftwork: Biological and social aspects.* Oxford: Pergamon Press, 1981:399–406.
348. Scott AJ, Monk TH, Brink L. Shiftwork as a risk factor for depression. *Int J Occup Environ Health* 1997; 3(3):S3–9.
349. Spitzer RL, Endicott J, Robins E. Research diagnostic criteria: rational and reliability. *Arch Gen Psychiatry* 1978; 35:733–82.
350. Monk TH. *How to make shift work safe and productive.* Des Plains: American Society of Safety Engineers, 1988.
351. Monk TH, Folkard S. Strategies for the employer. In: *Making shiftwork tolerable.* London: Taylor & Francis, 1992:69–75.
352. Lowden A, Moreno C, Holmbäck U, et al. Eating and shift work: effects on habits, metabolism, and performance. *Scand J Work Environ Health* 2010; 36(2):150–62.
353. Rohleder N, Kirschbaum C. Effects of nutrition on neuroendocrine stress responses. *Curr Opin Clin Nutr Metab Care* 2007; 10(4):504–10.
354. Nedeltcheva AV, Kilkus JM, Imperial J, et al. Sleep curtailment is accompanied by increased intake of calories from snacks. *Am J Clin Nutr* 2009; 89(1):126–33.
355. Zvonic S, Floyd ZE, Mynatt RL, et al. Circadian rhythms and the regulation of metabolic tissue function and energy homeostasis. *Obesity (Silver Spring)* 2007; 15(3):539–43.
356. Duchon JC, Keran CM. Relationships among shiftworker eating habits, eating satisfaction, and self-reported health in a population of US miners. *Work Stress* 1990; 4(2):111–20.
357. Holmbäck U, Lowden A, Åkerfeldt T, et al. The human body may buffer small differences in meal size and timing during a 24-h wake period provided energy balance is maintained. *J Nutr* 2003; 133(9):2748–55.
358. Dye L, Lluch A, Blundell JE. Macronutrients and mental performance. *Nutrition* 2000; 16(10):1021–34.

359. Lowden A, Holmbäck U, Åkerstedt T, et al. Performance and sleepiness during a 24 h wake in constant conditions are affected by diet. *Biol Psychol* 2004; 65:251–63.
360. Landström U, Knutsson A, Lennernäs M, et al. Onset of drowsiness and satiation after meals with different energy contents. *Nutr Health* 2001; 15(2):87–95.
361. Hãrmã MI, Ilmarinen J. Physical training intervention in female shift workers: I. The effects of intervention on fitness, fatigue, sleep, and psychosomatic symptoms. *Ergonomics* 1988; 31(1):39–50.
362. Mistlberger RE, Skene DJ. Nonphotic entrainment in humans? *J Biol Rhythms* 2005; 20:339–52.
363. Uroponen H, Vuori I, Partine M. Self-evaluations of factors promoting and disturbing sleep: an epidemiological survey in Finland. *Soc Sci Med* 1988; 26:443–50.
364. Vallet M, Mouret J. Sleep disturbance due to transportation noise: ear plugs vs. oral drugs. *Experientia* 1984; 40:429–36.
365. Doghramji K, Fredman S. Clinical frontiers in the sleep/psychiatry interface. Satellite Symposium of the 1999 American Psychiatric Association Annual Meeting, Golden, CO. Golden: Medical Education Collaborative, 1999.
366. Walsh JK, Muehlbach MJ, Walker PK. Acute administration of triazolam for the daytime sleep of rotating shiftworkers. *Sleep* 1984; 7:223–9.
367. Penetar DM, Belenky G, Garrigan JJ, et al. Triazolam impairs learning and fails to improve sleep in a long-range aerial deployment. *Aviat Space Environ Med* 1989; 60:594–8.
368. Walsh JK, Muehlbach MJ, Schweitzer PK. Hypnotics and caffeine as countermeasures for shiftwork-related sleepiness and sleep disturbance. *J Sleep Res* 1995; 4:80–3.
369. Ehret CF, Potter VR, Dobia KW. Chronobiological action of theophylline and of pentobarbital as circadian zeitgebers in the rat. *Science* 1975; 188:1212–4.
370. Dews PB, ed. Behavioral effects of caffeine. In: *Caffeine: perspectives from recent research*. New York: Springer-Verlag, 1984:86–103.
371. Bonnet MH, Arand DL. The use of prophylactic naps and caffeine to maintain performance during a continuous operation. *Ergonomics* 1994; 37:1009–20.
372. Muehlbach MH, Walsh JK. The effects of caffeine on simulated night-shift work and subsequent daytime sleep. *Sleep* 1995; 18(1):22–9.
373. Smith AP, Brockman R, Flynn A, et al. Investigation of the effects of coffee on alertness and performance during the day and night. *Neuropsychology* 1993; 27:217–23.
374. Kelly T, Gomez S, Engelland S, et al. Repeated administration of caffeine during sleep deprivation does not affect cognitive performance. *Sleep Res* 1993; 22:336.
375. Carrier J, Fenanndez-Bolanos M, Robillard R, et al. Effects of caffeine are more marked on daytime recovery sleep than on nocturnal sleep. *Neuropsychopharmacology* 2007; 32:964–72.
376. Akerstedt T, Ficca G. Alertness-enhancing drugs as a countermeasure to fatigue in irregular work hours. *Chronobiol Int* 1997; 14(2):145–58.
377. Cziesler CA, Walsh JK, Roth T, et al. Modafinil for excessive sleepiness associated with shift-work sleep disorder. *N Engl J Med* 2005; 353:476–86.
378. Morgenthaler T, Lee-Chiong T, Alessi C, et al. Practice parameters for the clinical evaluation and treatment of circadian rhythm sleep disorders. An American Academy of Sleep Medicine Report. *Sleep* 2007; 30(11):1445–59.
379. Basner R. Shift-work sleep disorder: the glass is more than half empty. *N Engl J Med* 2005; 353(5):520–1.
380. Redman J, Armstrong S, Ng KT. Free-running activity rhythms in the rat: entrainment by melatonin. *Science* 1983; 219:1081–9.
381. Arendt J, Aldhous M, English J, et al. Some effects of jet lag and their alleviation by melatonin. *Ergonomics* 1987; 30: 1379–93.
382. Comperatore CA, Krueger GP. Circadian rhythm desynchronosis, jet lag, shift lag, and coping strategies. *Occup Med* 1990; 5(2):323–42.
383. Suhner A, Schlagenhauf P, Johnson R, et al. Comparative study to determine the optimal melatonin dosage form for the alleviation of jet lag. *Chronobiol Int* 1998; 15(6): 655–66.
384. Sharkey KM, Foff LF, Eastman CI. Effects of administration on daytime sleep after simulated night shift work. *J Sleep Res* 2001; 10(3):181–9.
385. Folkard S, Arendt J, Clark M. Can melatonin improve shift workers' tolerance of the night shift? Some preliminary findings. *Chronobiol Int* 1993; 10(5):315–20.
386. Czeisler CA, Kronauer RE, Allan JS, et al. Bright light induction of strong (type 0) resetting of the human circadian pacemaker. *Science* 1989; 244:1328–33.
387. Czeisler CA, Johnson MP, Duffy JF, et al. Exposure to bright light and darkness to treat physiologic maladaptation to night work. *N Engl J Med* 1990; 322(18):1253–9.
388. Pallesen S, Bjorvatn B, Magerøy N, et al. Measures to counteract the negative effects of night work. *Scand J Work Environ Health* 2010; 36(2):109–20.
389. Åkerstedt T, Kecklun G, Knuttson A. Spectral analysis of sleep electroencephalography in rotating three-shift worker. *Scand J Work Environ Health* 1991; 17:330–6.
390. Tepas DI. Shiftworker sleep strategies. *J Hum Ergol* 1982; 11:325–36.
391. Dinges DF, Orne MT, Whithouse WG, et al. Temporal Placement of a nap for alertness: contribution of circadian phase and prior wakefulness. *Sleep* 1987; 10:313–29.
392. Rosekind MR, Graber RC, Dinges DF, et al. *Crew factors in flight operations. IX. Effects of planned cockpit rests on crew performance and alertness in long haul operations*. Technical Memorandum, A-94134, Moffet Field: NASA, 1995.
393. Kogi K. Should shift workers nap? Spread roles and effects of on-duty napping. In: Hornberger S, Knauth P, Costa G, Folkard S., eds. *Shift work in the 21st century*. Frankfurt: Peter Lang, 2000:31–6.
394. Tepas D. Should a general recommendation to nap be made to workers? In: Hornberger S, Knauth P, Costa G, Folkard S., eds. *Shift work in the 21st century*. Frankfurt: Peter Lang, 2000:25–30.
395. Ribeiro-Silva F, Rotenberg L, Soares RES, et al. Sleep on the job partially compensates for sleep loss in night-shift nurses. *Chronobiol Int* 2006; 26(3):1389 99.
396. International Labour Office (ILO). *Night work convention No. 171*. Geneva: ILO, 1990. Available at http://www.ilo.org/dyn/normlex/en/f?p=NORMLEXPUB:12100:0::NO::P12100_INSTRUMENT_ID:312316 (accessed on June 28, 2016).

397. European Council. Concerning certain aspects of working time. *Off J Eur Community* 1993; L307:18–24 (Council Directive 93/104/EC).
398. Rutenfranz J, Knauth P, Angerback D. Shiftwork research issues. In: Johnson LC, ed. *Advances in sleep research.* New York: Medical and Scientific Books, 1985:165–96.
399. Rutnfranz J, Haider M, Koller M. Occupational health measures for night workers and shift workers. In: Folkard S, Monk TH, eds. *Hours of work: temporal factors in work scheduling*. Chichester: John Wiley & Sons, Ltd, 1985:199–210.
400. Costa G. Shift work and occupational medicine. *Occup Med* 2003; 53:83–8.
401. Simon C, Weibel L, Brandenberger G. Twenty-four-hour rhythms of plasma glucose and insulin secretion rate in regular night workers. *Am J Physiol Endocrinol Metab* 2000; 278(3):E413–20.
402. Knuttsson A. Health disorders of shiftworkers. *Occup Med* 2003; 53:103–8.
403. Haider M, Cervinka R, Koller M. A destabilization theory of shiftworkers effects. In: Hekkens JJM, Kerkhof GA, Rietveld WJ, eds. *Trends in chronobiology*. Oxford: Pergamon Press, 1988:209–17.
404. Siwolop S, Therrien L, Oneal M, et al. Helping workers stay awake at the switch. *Business Week*; December 8, 1986, p. 108.
405. Knauth P. Changing scheduling: shiftwork. *Chronobiol Int* 1997; 14(2):159–71.
406. Gartner J, Wahl S. Design tools for shift schedules: empowering assistance for skilled designers & groups. *Int J Ind Ergon* 1998; 21:221–32.
407. Schwarenau P, Knauth P, Keisswetter E, et al. Algorithms for the computerized construction of shift systems which meet ergonomic criteria. *Appl Ergon* 1986; 17:169–76.
408. Circadian Technologies, Inc. Control room operator alertness and health in nuclear power plants. Prepared for Electric Power Research Institute, EPRI Rep. No. NP-6748—project 2184-7. Palo Alto CA, 1990.
409. Mardon S, ed. Some companies meet shiftworkers' family needs with 24-hour child care. Should yours? *Shift Work Alert* 1996; 1(2):9–11.

III Energy and Electromagnetic Radiation

Chapter 11

IONIZING RADIATION

James P. Seward*

Although most physicians and other healthcare professionals do not often encounter individuals injured by ionizing radiation in their practices, they can anticipate questions about radiation exposure and its potential health effects. In the rare event that a patient does present following radiation exposure or contamination with radioactive materials, the following information will be of assistance in the individual's case management. Important information on resources for emergency information and expert advice are included at the end of this chapter.

BACKGROUND RADIATION

Although this chapter is generally concerned with unusual high exposures to ionizing radiation, routine low-level exposure to radiation is unavoidable. Background radiation is the primary source of most individuals' exposure during their lifetimes. Background radiation has both a natural and an artificial ("human-made") component. In the United States the population average for total natural and human-made background is approximately 6.2 mSv/year (620 mrem). Worldwide the range of natural background radiation varies considerably from 1 to 10 mSv with an average of about 2.4 mSv. Radon accounts for about one half of the dose globally. In the United States natural background is typically somewhat higher at about 3.1 mSv due to higher radon levels. Cosmic radiation, terrestrial sources (e.g., thorium and uranium), and internally deposited radionuclides (e.g., potassium-40 and carbon-14) make up the other principal sources of natural background dose. These background exposures vary with geography and altitude[1,2] (http://www.nrc.gov/about-nrc/radiation/around-us/doses-daily-lives.html).

Human-made radiation usually comes principally from medical sources. This component has been growing in many parts of the world, particularly in the United States, where human-made background is about 3.1 mSv (equivalent to natural background). Other artificial sources include consumer products (0.1 mSv/0.01 rem) and a very small component (<0.01 mSv/0.001 rem) due to occupational exposure, nuclear power, or radioactive fallout. Clearly, personal medical factors, choice of residence, and occupation can be significant factors in determining this component of background exposure (see Figure 11.1).

OCCUPATIONAL EXPOSURES

Exposure to ionizing radiation can occur in a number of different industries and industrial settings. Subsequent sections of this chapter will discuss the potential acute and chronic health effects. For many occupational groups there is a substantial body of epidemiologic literature that is definitive in a few situations where doses are high but is more often inconclusive. Accurate estimation of lifetime occupational dose is often challenging. It is beyond the scope of this chapter to review industry-specific findings in detail. Examples of occupational groups in which radiation-induced carcinogenesis has been demonstrated epidemiologically would include uranium miners (in the United States), nuclear remediation workers (at Chernobyl), and some other nuclear industry workers (e.g., at Mayak in Russia). Reviews of occupational exposures in relation to cancer causation may

* Bryce D. Breitenstein MD and John H. Spickard MD contributed to previous editions of this chapter.

Physical and Biological Hazards of the Workplace, Third Edition. Edited by Gregg M. Stave and Peter H. Wald.

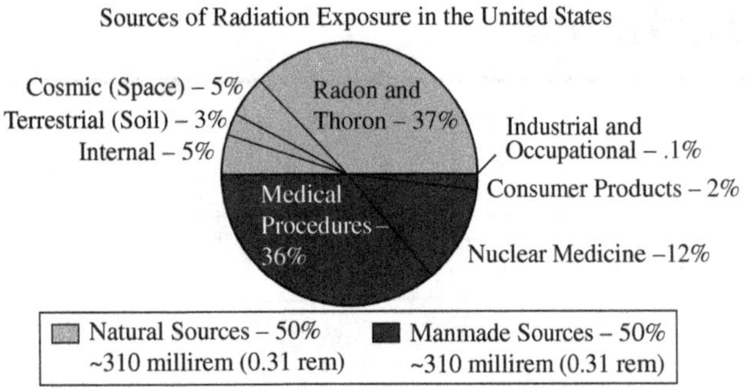

FIGURE 11.1 Sources of Natural and Manmade Background Radiation. *Source*: NCRP Report No. 160(2009). Full report is available on the NCRP web site at www.NCRPpublications.org.

TABLE 11.1 Examples of potential occupational exposure to ionizing radiation beyond US EPA 1 mSv limit set for general public[a].

Airline pilots and crew
Food irradiation facilities
Hazardous waste management
Manufacture of consumer products (e.g., luminous dials, as mantles)
Nuclear power/fuel cycle operators
Nuclear medicine
Research involving ionizing radiation sources/radionuclides
Uranium mining and milling
Use of X-ray equipment
Dentists and dental technicians
Industrial radiographers
Radiologists and X-ray technicians

[a]Listing does not necessarily imply documented health effects for these groups (see text).

TABLE 11.2 Estimates of Effective Dose from Common Single X-rays.

Estimates of the dose an individual might receive from one x ray.	
Single Radiograph	**Effective Dose, mrem (mSv)**
Skull (PA or AP)[1]	3 (0.03)
Skull (lateral)[1]	1 (0.01)
Chest (PA)[1]	2 (0.02)
Chest (lateral)[1]	4 (0.04)
Chest (PA and lateral)[2]	6 (0.06)
Thoracic spine (AP)[1]	40 (0.4)
Thoracic spine (lateral)[1]	30 (0.3)
Lumbar spine (AP)[1]	70 (0.7)
Lumbar spine (lateral)[1]	30 (0.3)
Abdomen (AP)[1]	70 (0.7)
Abdomen[3]	53 (0.53)
Pelvis (AP)[1]	70 (0.7)
Pelvis or hips[3]	83 (0.83)
Bitewing dental film[3]	0.4 (0.004)
Limbs and joints[3]	6 (0.06)

Source: Reproduced with permission of Health Physics Society Outreach, http://hps.org/physicians/documents/Doses_from_Medical_X-Ray_Procedures.pdf

[1]Wall BF, Hart D. Revised radiation doses for typical x-ray examinations. The British Journal of Radiology 70:437–439; 1997 (5,000 patient dose measurements from 375 hospitals).

[2]National Council on Radiation Protection and Measurements. Sources and magnitude of occupational and public exposures from nuclear medicine procedures. Bethesda, MD: National Council on Radiation Protection and Measurements; NCRP Report 124; 1996.

[3]United Nations Scientific Committee on the Effects of Atomic Radiation. Sources and effects of ionizing radiation, Vol. 1: Sources. New York, NY: United Nations Publishing; 2000.

be found in References 1 and 3. Table 11.1 provides a partial listing of occupations where ionizing radiation exposure may be present.

DIAGNOSTIC MEDICAL EXPOSURES

Patient exposure has been growing with increased reliance medical procedures that utilize ionizing radiation, including both diagnostic and therapeutic uses of radiation sources. Between 1982 and 2006 the average per capita dose from medical exposure in the United States increased by 600% from 0.54 to about 3.0 mSv. The increase was largely due to CT scanning.[4] Tables 11.2 and 11.3 illustrate examples of selected diagnostic radiology procedures and provide an estimate of the typical associated radiation dose. Epidemiologic studies of groups that have received high doses in the past (e.g., radiotherapy, ankylosing spondylitis cohorts) have shown the potential for medical therapy using ionizing radiation to induce cancers. There is concern that patients receiving multiple diagnostic procedures with X-ray may be at increased risk for cancer. In the absence of definitive studies, this concern is based primarily on the assumption of linear no-threshold dose–response for cancer (see Pathophysiology and Health Effects, p. 181). As a protective response, many professional medical organizations are placing increased emphasis on "Choosing Wisely" to weigh the benefit and potential harm of diagnostic procedures before ordering them. Helpful information for patients and providers related to many imaging studies can be found at http://www.choosingwisely.org. Improvements in technology and procedural technique have been lowering the doses received in many specific diagnostic procedures in recent years.

MEASUREMENT ISSUES AND THE PHYSICS OF IONIZING RADIATION

Exposure to ionizing radiation can be delivered either by electromagnetic radiation (X-rays and gamma rays) or by moving particles (alpha and beta particles, neutrons, and

TABLE 11.3 Estimates of Effective Dose from Complete X-ray Procedures.

Estimates of the dose an individual might receive if undergoing an entire procedure (e.g., a lumbar spine series typically consists of five films).	
Complete Exams	**Effective Dose, mrem (mSv)**
Intravenous pyelogram (kidneys, 6 films)[1]	250 (2.5)
Barium swallow (24 images, 106 sec fluoroscopy)[1]	150 (1.5)
Barium meal (11 images, 121 sec fluoroscopy)[1]	300 (3.0)
Barium follow-up (4 images, 78 sec fluoroscopy)[1]	300 (3.0)
Barium enema (10 images, 137 sec fluoroscopy)[1]	700 (7.0)
CT head[1]	200 (2.0)
CT chest[1]	800 (8.0)
CT abdomen[1]	1,000 (10)
CT pelvis[1]	1,000 (10)
CT (head or chest)[2]	1,110 (11.1)
PTCA (heart study)[3]	750–5,700 (7.5–57)
Coronary angiogram[3]	460–1,580 (4.6–15.8)
Mammogram[3]	13 (0.13)
Lumbar spine series[3]	180 (1.8)
Thoracic spine series[3]	140 (1.4)
Cervical spine series[3]	27 (0.27)

Source: Provided Courtesy of the Health Physics Society, http://hps.org/physicians/documents/Doses_from_Medical_X-Ray_Procedures.pdf

[1]Wall BF, Hart D. Revised radiation doses for typical x-ray examinations. The British Journal of Radiology 70:437–439; 1997 (5,000 patient dose measurements from 375 hospitals).

[2]National Council on Radiation Protection and Measurements. Sources and magnitude of occupational and public exposures from nuclear medicine procedures. Bethesda, MD: National Council on Radiation Protection and Measurements; NCRP Report 124; 1996.

[3]United Nations Scientific Committee on the Effects of Atomic Radiation. Sources and effects of ionizing radiation, Vol. 1: Sources. New York, NY: United Nations Publishing; 2000.

TABLE 11.4 Types of ionizing radiation important to radiologic health.

Source	Symbol	Character	Mass[a]	Charge	Example
X-ray	X	Electromagnetic energy	0	0	X-ray tube
Gamma	γ	Electromagnetic energy	0	0	^{60}Co ^{192}Ir
Alpha	α	Particulate (helium nucleus)	4	++	^{239}Pu ^{212}Po
Beta	β	Particulate (electron)	1/2000	–	^{90}Sr ^{3}H
Neutron	n	Nucleus particle	1	0	^{235}U fission
Proton	p	Nucleus particle	1	+	Proton beam

[a]The atomic mass unit (AMU) is chosen so that a neutral carbon-12 atom has a relative mass exactly equal to 12. This is equal to ~1 AMU for both a proton and a neutron.

protons). The physical characteristics of these types of radiation are reviewed in Table 11.4.

The basic unit of radiation exposure is the gray (Gy) given in international system (SI) units or the rad (given in traditional units). Since some types of ionizing radiation are more effective at causing ionization than others, a second measure called the dose equivalent is used to adjust for the tissue damage caused by the exposure. The unit is the sievert (Sv) in SI units or rem in traditional units. The dose equivalent is obtained by multiplying the radiation dose by a quality factor (also called weighting factor). For X-rays, gamma rays, and electrons, the quality factor equals 1, and 1 rad equals 1 rem. For alpha particles, the quality factor is 20; so, 1 Gy of alpha dose equals 20 Sv. Table 11.5 shows the units of radiation and radioactivity. There is an ongoing effort on the part of the scientific community to regularly use SI units of becquerel, gray, and sievert.[5] Scientific publications commonly require that units be given in SI units. As with conversion of measurements to metric terms, use of the older terminology persists. Both units will be used here to facilitate familiarity with them.

TABLE 11.5 Units of radiation and radioactivity.

Unit description	Unit name	Symbol	Definition
Activity	Curie	Ci	3.7×10^{10} disintegrations/s
	Becquerel	Bq	1 disintegration/s
Exposure	Roentgen	R	2.58×10^{-4} C/kg
Absorbed dose	Rad[a]	rad	100 ergs/g of absorbing material
	Gray	Gy	100 rad
Dose equivalent	Rem[b]	rem	rad × Q (quality factor)
	Sievert	Sv	100 rem

Measurement terminology: 1000 millicuries in 1 curie; 1 000 000 microcuries in 1 curie
1 000 000 000 nanocuries in 1 curie; 1 000 000 000 000 picocuries in 1 curie
1 000 000 000 000 000 femtocuries in 1 curie
1 curie = approximate disintegration rate of 1 g of ^{226}Ra or 3.7×10^{10} disintegrations per second
[a]Radiation absorbed dose.
[b]Roentgen equivalent man.

The penetration of radioactive dose in the body depends on the energy and, in the case of particles, the mass; the more massive the particle, the less distance it travels in the body. Since the relatively massive, positively charged alpha particles do not penetrate the skin or clothing, they are primarily a concern if deposited internally.

Beta particles (electrons) can penetrate the skin to the germinal layer of the epidermis, and beta-emitting contaminants that are allowed to reside on the skin for significant periods can cause dermal injury. Beta radiation can travel several feet in air. Gamma radiation and X radiation are highly penetrating to human tissue and require dense shielding materials. Uncharged neutron particles can be relatively penetrating and damaging and, therefore, have a quality factor up to 20 depending on their energy. It is important to recognize that radioactive sources often emit more than one type of ionizing radiation.

Ionizing radiation can be detected by a number of devices, but personal dosimeters are the instrument of choice to record individual exposures. Film badges have been almost completely replaced by thermoluminescent dosimeters (TLDs) (Figure 11.2). TLDs use materials (most commonly lithium fluoride) that glow on heating after exposure to ionizing radiation. TLDs are useful to detect beta and gamma radiation. Polycarbonate "foils" are often incorporated into TLDs to detect neutrons. Neutrons disrupt the structure of the polycarbonate, and these trails can be visually counted under a microscope after "etch" development.

Alpha particles cannot be counted directly with a dosimeter, but they can be identified by their associated low-energy gamma emissions. Special thin-window Geiger–Müller counters can directly measure surface alpha contamination. Whole-body counters are very sensitive gamma cameras that identify the low-energy gamma coemissions from internally deposited alpha emitters.

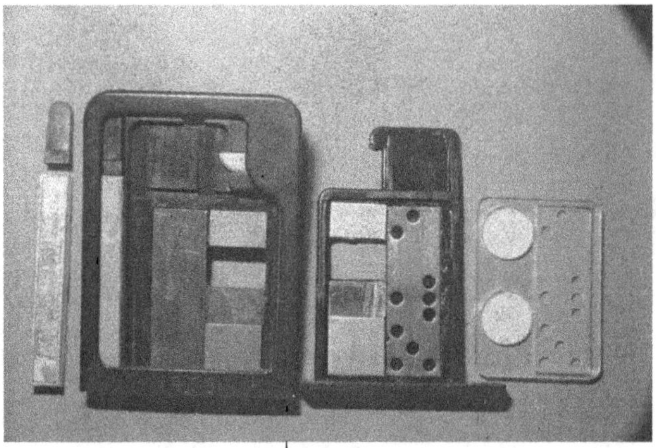

FIGURE 11.2 Interior of a thermoluminescent dosimeter (TLD), showing areas for beta, gamma, and neutron detection.

EXPOSURE GUIDELINES

Dose limits for ionizing radiation have been established for occupationally exposed individuals and for the general public. These limits are based on recommendations of the National Council on Radiation Protection and Measurements (NCRP).[6] The dose limits that have been established by the Nuclear Regulatory Commission (NRC) and the US Department of Energy (DOE) reflect the findings and recommendations of this advisory group. The resultant dose limit guidelines are outlined in Table 11.6. Whole-body exposure limits have been established to reduce the risk of cancer. The annual risk of a fatal radiation-induced cancer in a population from exposure at the annual allowable limit of 50 mSv (5 rem) is approximately 1 in 10 000.

The NCRP also recommends a cumulative dose limit of 10 mSv (1 rem) × age. Workers exposed at this average limit would have a cumulative lifetime risk of a fatal

TABLE 11.6 Ionizing radiation exposure guidelines for the United States.

Category	Dose limit guidance (NRC, DOE, NCRP)	
	mSv	rem
Occupational exposure (annual)	50	5
Lens of the eye	150	15
Skin, other organs/tissues	500	50
Unborn child of worker	5	0.5
Members of the public (annual) from a licensed nuclear operation	1	0.1

DOE, Department of Energy (see 10 CFR 835); NCRP, National Council on Radiation Protection (see Report 116); NRC, Nuclear Regulatory Commission (see 10 CFR 20).

TABLE 11.7 US federal OSHA radiation exposure limits under 29 CFR 1910.1096.[a]

	Rem per calendar quarter
Whole body: head and trunk; active blood-forming organs; lens of eyes; or gonads	1 1/4[b]
Hands and forearms; feet and ankles	18 3/4
Skin of whole body	7½

[a]Activities carried out by Department of Energy facilities or under license by the Nuclear Regulatory Commission have separate regulations.
[b]An employer may permit an individual in a radiation restricted area to receive doses to the whole body greater than those above so long as:

- During any calendar quarter the dose to the whole body shall not exceed 3 rem.
- The dose to the whole body, when added to the accumulated occupational dose to the whole body, shall not exceed $5(N-18)$ rem, where "N" equals the individual's age in years at his last birthday.

radiation-induced cancer between 1 per 1000 and 1 per 10 000. The International Commission on Radiological Protection (ICRP) has recommended lowering the current recommended standards on the basis of the reevaluation of the atomic bomb survivor cohorts from Japan.[7] Their recommendation calls for a maximum of 100 mSv (10 rem) exposure over 5 years, with a yearly average exposure of 20 mSv (2 rem). These recommendations have not yet been incorporated into US NRC or DOE exposure standards.

To reduce the potential for *in utero* developmental effects on the fetus, both NRC and DOE standards require that occupational exposure of a pregnant worker must not exceed 5 mSv (0.5 rem) during the entire gestation period. There are also separate dose limits for exposure to the skin, extremities, the lens of the eye, and any specific body organ in order to prevent dose-related deterministic effects.

In the United States an Occupational Safety and Health Administration (OSHA) standard for ionizing radiation use in general industry is found in 29 CFR 1910.1096, and the exposure limits are summarized in Table 11.7. OSHA incorporates applicable NRC standards for protection against radiation exposure in its Construction (29 CFR 1926.53) and Shipyard Employment (29 CFR 1915.57) Standards.

Exposure standards for underground miners have been developed to limit exposure to alpha-emitting radon progeny. Exposure is measured in working level months (WLM). A *working level* (WL) is any combination of short-lived radon progeny (for radon-222: polonium-218, lead-214, bismuth-214, and polonium-214; and for radon-220: polonium-216, lead-212, bismuth-212, and polonium-212) in 1 L of air that will result in the ultimate emission of 1.3×10^5 MeV (megaelectron volts) of potential alpha particle energy. A WLM is 170 hours of exposure at one WL. Under US federal standards enforced by the Mine Safety and Health Administration, the annual exposure limit is 4 WLM for an individual worker.

A general principle for all potential exposure situations is that every effort must be made to limit exposure to ionizing radiation to a level that is as low as reasonably achievable (ALARA). This means that radiation exposures must be kept as low as possible while still allowing workers to get their jobs done.

PATHOPHYSIOLOGY AND HEALTH EFFECTS

Ionizing radiation consists of electromagnetic waves or moving particles that carry sufficient energy to produce ions in matter. Ionization occurs when enough radiation energy transfers to atoms in the material through which it is passing to displace an orbital electron, thus leaving these atoms as electrically charged ions. In tissue, ionization can cause cellular injury, leading to apoptosis, mutagenesis, carcinogenesis, and cytogenetic changes. DNA damage may include single- and double-strand breaks. Mechanisms of injury may also include the generation of reactive oxygen species and the bystander effect in which agents or signals from irradiated cells reduce survival in surrounding cells.

The spectrum of health effects from ionizing radiation can be divided into the stochastic and nonstochastic (also called deterministic) effects. While most deterministic effects occur within days or weeks of exposure, a few, such as increases in cardiovascular disease, are delayed. The nonstochastic effects are those that appear predictably as a function of the dose received and worsen with increasing dose. The stochastic effects, including cancer and the theoretical risk of birth defects, are "probabilistic." They may or may not occur in an individual as a result of an exposure, but the risk in a population of similarly exposed individuals increases with the dose.

Nonstochastic (deterministic) effects

All organs can undergo nonstochastic, dose-dependent effects from radiation exposure. In addition to acute effects, many organs may be susceptible to delayed effects that develop over months or years. Fibrosis is the most common cause of delayed organ failure. The reader is referred to additional references for detailed discussions of specific organ effects and thresholds as well as acute versus fractionated dose thresholds for organ damage.[8,9]

The acute radiation syndrome (ARS)

Large whole-body doses of ionizing radiation above 1 Gy (100 rad) can cause the acute radiation syndrome that is characterized by a sequence of dose-dependent organ effects. ARS may progress through four phases: prodrome, latent phase, manifest illness, and recovery or death. The time course of these phases and the need for hospitalization vary with the dose (Table 11.8).

During the prodromal period the exposed individual may experience dose-dependent nausea, vomiting, diarrhea, headache, loss of consciousness, and fever. Individuals exposed at or below 1 Gy (100 rad) may not manifest any of these symptoms, while individuals with whole-body exposures above 8 Gy (800 rad) are likely to have them all. In the absence of good information on the amount of exposure, the time to onset of vomiting can be used to assess the approximate radiation dose received. Other early clinical indicators that can be used to estimate dose are the drop in the lymphocyte count and the dicentric chromosome assay. Most clinical laboratories can quickly perform the total lymphocyte count, whereas the dicentric chromosome assay is a specialized test that is only available in a few reference laboratories and requires multiple days' turnaround time (see section "Assessment of External Ionizing Radiation Whole-Body Exposure" and also http://www.remm.nlm.gov/ars_timephases1.htm).

Prodromal symptoms usually resolve within approximately 2 days. The latent phase begins as prodromal symptoms subside and ends when (if) manifest symptoms of ARS develop. The latency period is dose dependent and may last 30 days or more with lesser levels of exposure. Patients with high levels of exposure (> 15 Gy, 1500 rad) may move quickly from prodrome to manifest illness with no latent period. During the latent phase patients may experience dose-dependent skin erythema, epilation, and a gradual reduction in hematopoietic indices.

The illness phase of ARS is often characterized by four subsyndromes affecting the hematopoietic, gastrointestinal, dermatologic, and central nervous systems (CNS) that manifest some of the most predictable health effects. However, many body organs can be affected by radiation exposure, and clinical findings are not limited to these four systems.

Radiation causes death of hematopoietic cells in the bone marrow with doses as low as 1 Gy (100 rad), resulting in a decline in white blood cell and platelet counts and delayed tissue healing. Peripheral lymphocyte counts may drop progressively in the first 24 hours, well in advance of the decline in granulocytes and thrombocytes that typically reach their nadir in 30 days. Red blood cell counts decline more slowly. Consequences of the hematopoietic system effects include increased risk of bleeding, immune dysfunction, infection, and delayed wound healing.

The gastrointestinal epithelium, with its high rate of turnover, is often affected by whole-body radiation beginning at about 6 Gy (600 rad) of exposure. A syndrome of vomiting, diarrhea, hematochezia, and malabsorption may result within hours to several days on a dose-dependent basis. Patients may have an ileus as well as fluid and electrolyte imbalances.

The skin reacts to local and whole-body exposures with a dose-dependent progression of signs including erythema, epilation, edema, dry and moist desquamation, blistering, ulceration, and necrosis; epilation may occur at doses above 3 Gy/300 rad manifesting after about 2 weeks. There may be both an early-phase and a late-phase skin erythema if the dose has been sufficient. The hours to onset of skin erythema in the first 24 hours correlates with the dose received. Dose-dependent secondary skin erythema may occur days later. Local skin exposure is one of the most frequent types of radiation injury and often results from inadvertent exposure to ionizing radiation beams.

Nonspecific central nervous system effects such as nausea and vomiting, fatigue, anorexia, and mild headache can manifest at the lower range of exposure in the ARS. More devastating CNS findings occur at high exposure levels (> 20 Gy/2000 rad) and may include early onset of severe headache and hypotension, cognitive impairment, cerebral edema, ataxia, convulsions, coma, and death within several days.

Gonadal effects

The testes (spermatogonia) are particularly sensitive to radiation; decreases in sperm counts can occur at low doses around 150 mGy (15 rad). Exposures in the range of 3–4 Gy (300–400 rad) can result in permanent sterility in men. The organ dose to create sterility in women is upward of 2–3 Gy (200–300 rad) with the follicles nearest to ovulation being the most sensitive. Lower doses can temporarily impair fertility.[8]

In utero *developmental effects*

Radiation exposure of > 0.5 Gy (50 rad) to the fetus can cause growth retardation and congenital malformations. A dose-dependent relationship has been found between radiation exposure and reduced IQ in children of Japanese atomic bomb survivors who were irradiated between 8 and 15 weeks of gestation and to a lesser extent between 15 and 25 weeks. Children who were irradiated while *in utero* are also at higher risk from the stochastic effect of leukemia and solid cancers.

TABLE 11.8 Signs and Symptoms of Acute Radiation Syndrome in the Three Phases after Exposure. Adapted from Diagnosis and treatment of radiation injuries. Safety series no. 2. Vienna/International Atomic Energy Agency 1998. Reprinted with Permission.

PRODROMAL PHASE OF ACUTE RADIATION SYNDROME (ARS) (WBE = WHOLE BODY EXPOSURE)					
	Degree of ARS and approximate dose of acute WBE (Gy)				
Symptoms and medical response	**Mild (1–2 Gy)**	**Moderate (2–4 Gy)**	**Severe (4–6 Gy)**	**Very severe (6–8 Gy)**	**Lethal[a] (>8 Gy)**
Vomiting					
Onset	2 hrs after exposure or later	1–2 hrs after exposure	Earlier than 1 hr after exposure	Earlier than 30 min after exposure	Earlier than 10 min after exposure
% of incidence	10–50	70–90	100	100	100
Diarrhea	None	None	Mild	Heavy	Heavy
Onset	—	—	3–8 hrs	1–3 hrs	Within minutes or 1 hr
% of incidence	—	—	<10	>10	Almost 100
Headache	Slight	Mild	Moderate	Severe	Severe
Onset	—	—	4–24 hrs	3–4 hrs	1–2 hrs
% of incidence	—	—	50	80	80–90
Consciousness	Unaffected	Unaffected	Unaffected	May be altered	Unconsciousness (may last seconds/minutes)
Onset	—	—	—	—	Seconds/minutes
% of incidence	—	—	—	—	100 (at >50 Gy)
Body temperature	Normal	Increased	Fever	High fever	High fever
Onset	—	1–3 hrs	1–2 hrs	<1 hr	<1 hr
% of incidence	—	10–80	80–100	100	100
Medical response	Outpatient observation	Observation in general hospital, treatment in specialized hospital if needed	Treatment in specialized hospital	Treatment in specialized hospital	Palliative treatment (symptomatic only)

LATENT PHASE OF ACUTE RADIATION SYNDROME					
	Degree of ARS and approximate dose of acute WBE (Gy)				
	Mild (1–2 Gy)	**Moderate (2–4 Gy)**	**Severe (4–6 Gy)**	**Very severe (6–8 Gy)**	**Lethal (>8 Gy)**
Lymphocytes (G/L) (days 3–6)	0.8–1.5	0.5–0.8	0.3–0.5	0.1–0.3	0.0–0.1
Granulocytes (G/L)	>2.0	1.5–2.0	1.0–1.5	≤0.5	≤0.1
Diarrhea	None	None	Rare	Appears on days 6–9	Appears on days 4–5
Epilation	None	Moderate, beginning on day 15 or later	Moderate or complete on days 11–21	Complete earlier than day 11	Complete earlier than day 10
Latency period (days)	21–35	18–28	8–18	7 or less	None
Medical response	Hospitalization not necessary	Hospitalization recommended	Hospitalization necessary	Hospitalization urgently necessary	Symptomatic treatment only

(*Continued*)

TABLE 11.8 (Continued)

	CRITICAL PHASE OF ACUTE RADIATION SYNDROME				
	Degree of ARS and approximate dose of acute WBE (Gy)				
	Mild (1–2 Gy)	**Moderate (2–4 Gy)**	**Severe (4–6 Gy)**	**Very severe (6–8 Gy)**	**Lethal (>8 Gy)**
Onset of symptoms	>30 days	18–28 days	8–18 days	<7 days	<3 days
Lymphocytes (G/L)	0.8–1.5	0.5–0.8	0.3–0.5	0.1–0.3	0–0.1
Platelets (G/L)	60–100	30–60	25–35	15–25	<20
	10–25%	25–40%	40–80%	60–80%	80–100%[b]
Clinical manifestations	Fatigue, weakness	Fever, infections, bleeding, weakness, epilation	High fever, infections bleeding, epilation	High fever, diarrhoea, vomiting, dizziness and disorientation, hypotension	High fever, diarrhoea, unconsciousness
Lethality (%)	0	0–50	20–70	50–100	100
		Onset 6–8 weeks	Onset 4–8 weeks	Onset 1–2 weeks	1–2 weeks
Medical response	Prophylactic	Special prophylactic treatment from days 14–20; isolation from days 10–20	Special prophylactic treatment from days 7–10; isolation from the beginning	Special treatment from the first day; isolation from the beginning	Symptomatic only

[a]With appropriate supportive therapy, individuals may survive whole body doses as high as 12 Gy.
[b]In very severe cases, with a dose >50 Gy, death precedes cytopenia.

Cataracts

Ionizing radiation can cause cataracts of the lens of the eye from acute exposures or significant smaller exposures over time. The type of lens opacity induced is usually a posterior subcapsular cataract. There is increasing evidence of cataract formation among those who conduct interventional medical radiography studies with insufficient eye protection with a threshold of approximately 0.5 Gy (50 rad) that varies with the type of radiation (neutrons more potent) and dose rate (fractionated slower).

Other organ effects

Ionizing radiation can cause also pneumonitis that may progress into fibrosis and can be a limiting factor in long-term survival. Long-term follow-up of Japanese atomic bomb survivors and patients who received chest or head radiotherapy has demonstrated an association between cardiovascular disease and radiation exposure at doses even below 1 Sv (100 rem). Deterministic radiation effects on the vascular endothelium are a potential explanation for this increased risk. Ionizing radiation may cause acute renal failure and nephrosclerosis as well as liver injury. Both thyroid nodules and hyperparathyroidism are more common in individuals who have received radiation to the neck.

Stochastic effects

The stochastic health effects of ionizing radiation are related to mutagenic and carcinogenic events in the cell. The key demonstrated stochastic effect in humans is malignancy. Radiation-induced cancers have no special features differentiating them from other cancers. The expression of hereditary genetic effects in humans has not been found in human observational studies. However, since they have been seen in mice and other organisms, there is a possibility that human hereditary genetic effects could occur.

Knowledge about the dose–response curve for cancer comes from the study of exposed cohorts. The most intensive ongoing study is the Long-Term Survivor Study (LSS) of the Hiroshima and Nagasaki bombing victims. Examples of additional studies include research on fluoroscoped tuberculosis patients as well as patients receiving radiation therapy for ankylosing spondylitis, cervical cancer, and *tinea capitis*. There has been much scientific discussion on the shape of the dose–response curve, particularly at low doses (Figure 11.3). There is good evidence that the curve is linear for solid tumors at higher doses (above 100 mSv/10 rem).

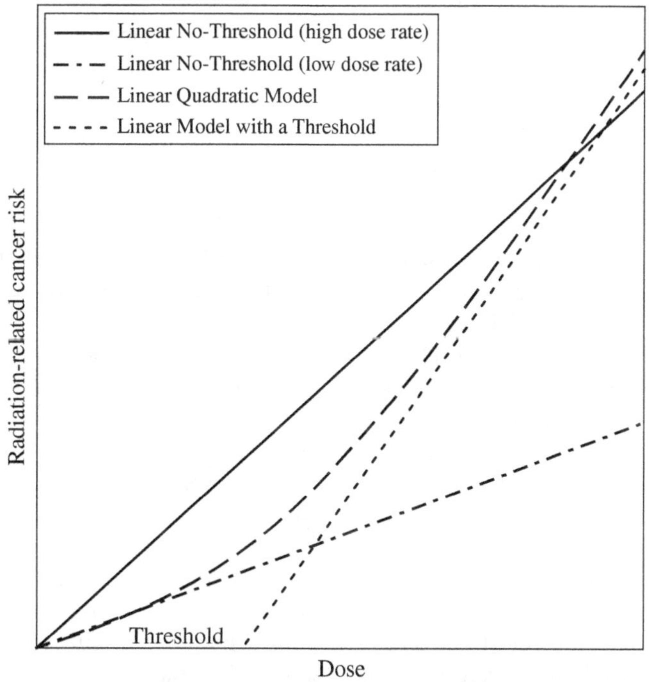

FIGURE 11.3 Different conceptual models for cancer risk from ionizing radiation dose. The National Research Council Biological Effects of Ionizing Radiation VII Phase 2 report adopts the Linear No-Threshold model (noted above for high and low dose rates) as more scientifically plausible than the threshold model. The linear quadratic model was adopted for leukemias. *Source*: Brenner, DJ, R. R, Goodhead DT, Hall EJ, Land CE, Little JB, Lubin JH, Preston DL, Preston RJ, Puskin JS, Ron E, Sachs RK, Samet JM, Setlow RB, Zaider M. 2003. Cancer risks attributable to low doses of ionizing radiation: Assessing what we really know. P Natl Acad Sci USA 100: 13761–13766. Copyright (2003) National Academy of Sciences, U.S.A. Reprinted with permission.

The Committee on the Biological Effects of Ionizing Radiation VII (BEIR VII) of the US National Research Council has concluded that there is a linear dose–response relationship between exposure to ionizing radiation and the development of radiation-induced solid tumors. BEIR VII assessed the health risks from exposure to *low doses* below 100 mSv (10 rem) and concluded that it is unlikely that a threshold exists for the induction of cancers, although the occurrence of cancers at low doses will be small.[1]

Recent epidemiologic evidence from the LSS of cancer incidence in those atomic bomb survivors who received low doses offers additional support for the linear no-threshold model.[10,11]

With respect to leukemias, the evidence supports a linear quadratic dose–response model (Figure 11.3). Leukemia was the earliest cancer attributed to radiation exposure in the LSS with a similar dose–response curve for various leukemia subtypes that were seen.[2] Most sources continue to classify chronic lymphocytic leukemia as nonradiogenic based on the paucity of epidemiologic support. Leukemia can have a relatively short latency period with indications that excess cases may have occurred in atomic bomb survivors in less than 5 years.

The BEIR VII committee discounted the arguments for hormesis, the theory that low doses of radiation are beneficial in humans.

Radiation exposure has been demonstrated to increase the risk of most types of solid tumors, although there is considerable uncertainty in quantifying risk and statistical significance has not been reached for some tumors in the Japanese LSS. Thyroid, breast, bladder, colon, lung, ovary, and skin cancers are examples of solid tumors with significant excess relative risk. Excellent discussions of specific tumor sites can be found in the existing literature.[1,3]

Carcinogenic effects may occur from internally deposited radionuclides. Examples include lung cancer in uranium miners and osteogenic sarcomas in radium dial painters. Underground miners of uranium and other minerals have an increased incidence of lung cancer as a result of chronic exposure to radiation from inhaled radon progeny. Radon is a decay product of naturally occurring radium and uranium. Its decay in the air results in "radon progeny" or radionuclide products that emit alpha particles, which are inhaled. Because they are solid, they attach themselves to dust particles in the air that can be inhaled. The primary hazardous radiation dose from the radon progeny is due to the alpha particles that are deposited in the lungs. These particles can induce metaplasia and atypical cell growth in the tracheobronchial epithelium that may subsequently develop into bronchial carcinoma. In recognition of the demonstrated lung cancer risk among uranium miners, the Congress passed the Radiation Exposure Compensation Act in 1990 to compensate exposed uranium miners with lung cancer.

Because cigarette smoking has a synergistic effect with this radiation exposure, smokers have an increased risk of cancer, along with a decreased latency period from time of original exposure to the expression of disease.

Monitoring of ionizing radiation exposure from radon progeny in miners is based on the concept of a Working Level Month (as described in "Exposure Guidelines").

Radon and radon progeny also pose an internal radiation hazard to the general public. Radon exposure is estimated to be the second leading cause of lung cancer in the United States after tobacco smoking. The risk of cancer in the general population is dependent on the amount of cumulative radon progeny exposure, the age distribution of the population, the time since the start of exposure, and the extent of cigarette smoking.

Radon is found in soils and trapped in basements and the lower floors of housing where significant human exposure can occur. While there is substantial geographic variation in the presence of radon, geography alone is not a sufficient predictor (Figure 11.4). State public health agencies have developed programs to encourage the public to test their homes for the presence of radon. A variety of short- and longer-term detection devices are commercially available; longer-term monitoring is preferable due to greater accuracy and seasonal variation. These screening devices are usually placed in the basement or the area of greatest risk. The EPA has developed an exposure guideline of 4 picocuries/liter of air as a threshold for remediation measures. This level is based on technical feasibility for remediation, not "zero risk" to inhabitants.[12] Smokers living in homes with elevated levels of radon should be strongly supported in efforts to quit.

DIAGNOSIS AND TREATMENT

External ionizing radiation exposure, radionuclide contamination, and internal deposition

In the initial patient evaluation, it is important to decide whether the individual has been exposed to an energy source (irradiated) or contaminated with radioactive materials that emit ionizing radiation.

Exposure to X-ray and gamma (photon) radiation is called external exposure, because this form of radiation lacks mass and is entirely composed of electromagnetic waves. Although it may produce ionization, it does not make matter or tissue radioactive. Therefore, rescue workers dealing with these victims do not need to be concerned about receiving radiation exposure once the victim has been removed from the field of ionizing radiation exposure. However, appropriate measures must be taken to ensure emergency responder safety if the victim is in an active radiation field or contaminated with radionuclides.

Radionuclides, such as radioactive iodine (^{131}I) or cesium (^{137}Cs), have a mass that is radioactive and can become

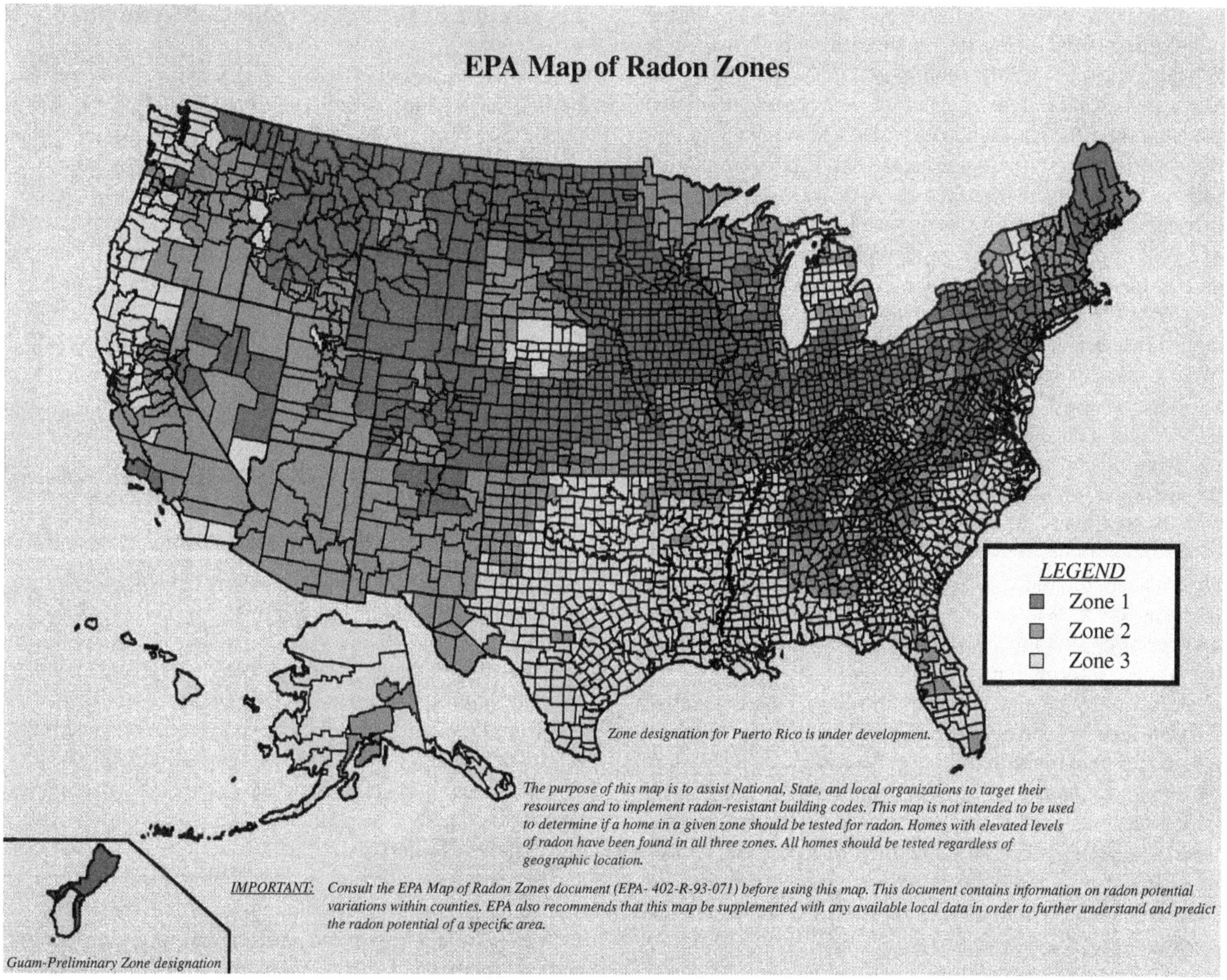

FIGURE 11.4 EPA Map of Radon Zones. *Source*: https://www.epa.gov/sites/production/files/2015-07/documents/zonemapcolor.pdf.

distributed on the body surface (contamination) or internally deposited in the body by way of inhalation or ingestion and through the skin by puncture, laceration, abrasion, or burn. The contaminating or internally deposited material continues to emit radiation types (e.g., alpha, beta) and amounts that are characteristic of the specific radionuclide. The radionuclide may also contaminate the surrounding area and other individuals in that area.

Assessment of external ionizing radiation whole-body exposure

Rescue workers responding to a radiation incident should remove the victim quickly from the radiation field and limit medical care for injuries to lifesaving procedures until both the rescuers and the patient are removed from the radiation exposure.

Extremely high whole-body exposures can result in an acute radiation syndrome, as described in Section The Acute Radiation Syndrome, p. 182, including the development of signs and symptoms that indicate the severity of the exposure. Prodromal acute radiation syndrome findings are summarized in Table 11.8.

It is important to gather information about the amount of exposure, either from a health physicist or from any other person who is knowledgeable about the exposure circumstances. Radiation dosimetry, if available, will also help in the management of the case. If the patient has been irradiated, the clinician should attempt to establish whether the exposure was limited to an extremity or a specific body area or whether it could be described as whole-body exposure. In addition to information gathered from experts on the scene, there are clinical indicators that can provide information about the likely dose. The timing of onset of nausea and vomiting after exposure corresponds with the dose. The amount and timing of a decline in the peripheral lymphocyte count (lymphocyte depletion kinetics) and the neutrophil/lymphocyte (N/L) ratio are also good indicators of exposure. A baseline differential white blood cell count should be performed immediately and then approximately every 4–6 hours. The USDHS Radiation Emergency Medical Management website has calculators that can convert vomiting latency and lymphocyte kinetics into dose estimates (http://www.remm.nlm.gov/ars_wbd.htm), and additional guidance on use of the N/L ratio in early assessment can be found at http://orise.orau.gov/files/reacts/medical-aspects-of-radiation-incidents.pdf.

Bioassay cytogenetic studies are another way to assess the dose received, particularly when the level of exposure is uncertain. In acute exposure situations the preferred test is the dicentric chromosome assay (Figure 11.5). This test can be performed on blood drawn as soon as 1 hour after exposure, but optimally after 24 hours if there is a sufficient pool of surviving lymphocytes. The sample should be sent to an experienced reference laboratory; it may detect exposure levels as low as 0.1–0.2 Gy (10–20 rad) and is both sensitive and specific. This procedure takes 4–5 days, but it provides useful dose information.

A second technique of chromosome painting called fluorescence *in situ* hybridization (FISH) is the preferred method for retrospective (>6 months) assessment of exposure. FISH uses chromosome-specific probes in peripheral blood lymphocytes to evaluate translocations. Since chromosome translocations are relatively stable, their frequency can be used to estimate past radiation exposures. Another technique called premature chromosome condensation (PCC) may be useful for exposures over 3.5 Gy (350 rad).[13]

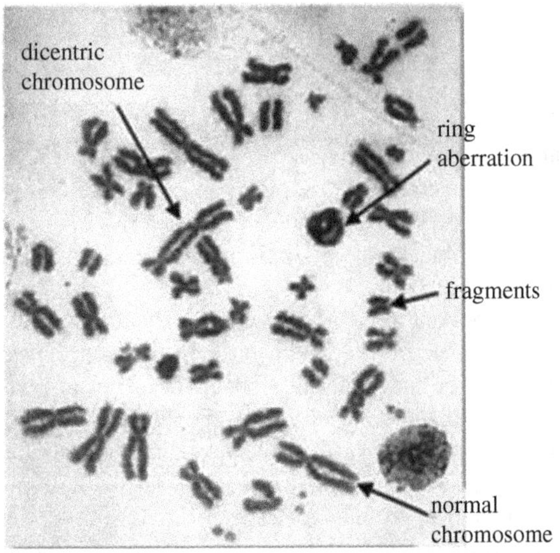

FIGURE 11.5 Cytogenetic Assay for Dicentric Chromosomes. *Source*: https://orise.orau.gov/files/reacts/medical-aspects-of-radiation-incidents.pdf, page 42.

Treatment of the acute radiation syndrome

Appropriate treatment of ARS depends on assessing the dose and the stage of the patient's condition based on the typical sequence of the syndrome. Support of the hematopoietic system and prevention of infection are key goals to promote recovery.

Useful laboratory studies include[14]:

- Complete blood count with differential (every 4–6 hours) (see above)
- Serum amylase (baseline and serially every 24 hours) (amylase increases with dose assuming salivary glands irradiated)
- Plasma FLT-3 ligands (correlates with radiation bone marrow injury)
- Blood citrulline (decreases with radiation-induced bowel mucosal atrophy)
- Interleukin-6 (indicator of high radiation dose)
- C-reactive protein (increases with radiation dose)
- Cytogenetic studies (see above)

Clinical interventions are tailored to the indicators of radiation injury. They may include management of nausea and vomiting with antiemetics, hospitalization, and administration of hematopoietic cell line colony-stimulating factors. The effectiveness of granulocyte and granulocyte-macrophage stimulating factors is greatest when given early—within 24–72 hours of injury.

Prevention of infection is a key objective. In addition to maintaining adequate circulating granulocytes, measures such as skin and bowel decontamination, reverse isolation, and environmental controls should be considered according to standard protocols for immune-impaired patients. Support with transfusions and antibiotics may be necessary.

Donor bone marrow transplantation (BMT) was performed on a number of the Chernobyl incident victims; however, most of these individuals died of other radiation complications or of graft-versus-host disease. Two BMT survivors were later determined to have regenerated blood cells from their own marrow, although the donor cells may have helped them to survive for this to occur.

Lung effects may be seen with doses above 8 Gy (800 rad) and include acute effects on the alveoli with cytokine-mediated inflammation. Over time pulmonary fibrosis can result.

In ARS situations with trauma or when surgery or wound closure is required for a patient, consideration should be given to performing the operation early in the course. Over time the patient's wound healing abilities are likely to decline with the fall in hematologic cell lines.

The survival of ARS patients depends significantly on medical support. The LD50 for patients with medical support is approximately 5–6 Gy (500–600 rad) as opposed to 3.5–4 Gy (350–400 rad) without support.

Treatment of cutaneous irradiation injury

Irradiation to specific body regions often manifests as skin injury, and the effects are typically delayed in nature. Such injuries frequently result from exposure to contained sources or to X-ray or gamma ray or beta-generating devices. The higher the dose, the more quickly the physical changes are seen. For example, a dose of 20 Gy (2000 rad) may produce reddening of the skin in 2–3 hours, whereas a dose of 6–8 Gy (600–800 rad) might result in the development of erythema after 2–3 days. The extent of injury is dependent upon the energy of the radiation type that affects penetration, the dose, and the dose rate. After an accident, it may not be possible to immediately reconstruct the exposure circumstances. Local radiation injury may be observed concurrent with significant whole-body exposure, and the patient should be evaluated for ARS as discussed above. Careful clinical evaluation and observation for progressing signs and symptoms are helpful in managing cutaneous injury. Consultation with a specialist in plastic or reconstructive surgery is often indicated.

Usually, local radiation exposure is not an acute medical emergency. Symptoms and effects are frequently delayed, so there is adequate time for evaluation, supportive treatment, and consultation. Careful observation and documentation, particularly with the use of serial color photographs, also help in evaluating the progress of local injuries. Therapeutic goals include limiting the extent of tissue damage, preventing/treating infection, and managing pain.

As seen in Table 11.9, local injury may proceed through a sequence of stages, depending on the significance of the initial exposure. Stages may include erythema, epilation, moist desquamation, dry desquamation, necrosis, and healing with fibrosis/sclerosis. Skin grafting or amputation may be necessary, and there will be an increased risk of skin cancer in irradiated areas.

Therapeutic modalities that may be employed to manage the extent of the tissue damage and recovery include[14]:

- Topical corticosteroids
- Hyperbaric oxygen
- Pentoxifylline with vitamin E
- State-of-the-art wound care, including microvasculature imaging techniques

External and internal contamination with radionuclides

There are two ways in which a person may be exposed to ionizing radiation through direct contact with radionuclides—by external contamination (i.e., contamination of the skin or exposed body parts only) or internal deposition (i.e., by inhalation, ingestion, wounds, or burns). Both types of exposure can occur together.

External contamination

Contamination incidents require prompt removal and containment of the radionuclides and measures to limit the dose received. However, life- and limb-threatening conditions take precedence over contamination evaluation and control. Early decontamination facilitates the reduction of exposure from external irradiation and reduces the risk of cross-contaminating other individuals and areas. It may also help to reduce the risk of internal deposition of radionuclides.

Most healthcare facilities have established emergency plans addressing chemical and radiologic hazards as required by the Joint Commission on Accreditation of Healthcare Organizations. Preplanning for management of patients contaminated with radioactive materials addresses the need for prompt decontamination and care of the patient as well as protection of medical personnel and facilities. Detailed guidance for this preplanning effort is available from a number of sources.[15–17] None of the required supplies are highly specialized, except for the instruments used for radiation detection. These devices are operated by a radiation safety officer (RSO), who should be formally designated at all healthcare facilities where radiotherapy devices are used or nuclear medicine is practiced. If no RSO has been named based on these activities, one should be designated in the facility's emergency plan.

The RSO should survey the patient as soon as the patient is medically stabilized to assess the location, type, and intensity of radiation contamination. The RSO or other trained individuals use specific radioactivity-detecting devices for this purpose. In most instances, radiation contamination does not pose an external radiation threat to the healthcare providers. However, they must be protected both from becoming contaminated and from internal deposition; this can usually be achieved with surgical gowns, gloves, caps, and masks. Decontamination of patients can usually be performed readily using soap or detergents and water; in the case of wounds, copious irrigation and debridement are necessary. Collection of irrigation fluids and tissues and control of instruments, drapes, and dressings also help to constrain the spread of contamination.

External radionuclide contamination alone (i.e., with no injury) requires identification of the involved skin surfaces to control and prevent spread of the contamination and to determine which areas to decontaminate. If the skin contamination is identified and removed promptly, the radionuclide is unlikely to cause damage to the skin or deeper structures.

If the skin is broken or burned, there is potential for absorption of the radionuclide through the wound. In some cases particulate matter may be found in an open lesion or be injected subcutaneously. In these cases care should be taken to remove as much of the material as is reasonably possible. On occasion the excision of wound margins, the use of a

TABLE 11.9 Stages of Cutaneous Radiation Injury.

	Grades of cutaneous radiation injury						
Grade	Skin dose*	Prodromal stage	Latent stage	Manifest illness stage	Third wave of erythema†	Recovery	Late effects
I	>2 Gy (200 rads)‡	1–2 days postexposure or not seen	no injury evident for 2–5 weeks postexposure§	• 2–5 weeks postexposure, lasting 20–30 days: redness of skin, slight edema, possible increased pigmentation • 6–7 weeks postexposure, dry desquamation	not seen	complete healing expected 28–40 days after dry desquamation (3–6 months postexposure)	• possible slight skin astrophy • possible skin cancer decades after exposure
II	>15 Gy (1500 rads)	6–24 hours postexposure with immediate sensation of heat lasting 1–2 days	no injury evident for 1–3 weeks postexposure	• 1–3 weeks postexposure; redness of skin, sense of heat, edema, skin may turn brown • 5–6 weeks postexposure, edema of subcutaneous tissues and blisters with moist desquamation • possible epithelialization later	• 10–16 weeks postexposure, injury of blood vessels, edema, and increasing pain • epilation may subside, but new ulcers and necrotic changes are possible	healing depends on size of injury and the possibility of more cycles of erythema	• possible skin astrophy or ulcer recurrence • possible telangiectasia (up to 10 years postexposure) • possible skin cancer decades after exposure
III	>40 Gy (4000 rads)	4–24 hours postexposure, with immediate pain or tingling lasting 1–2 days	none or less than 2 weeks	• 1–2 weeks postexposure: redness of skin, blisters, sense of heat, slight edema, possible increased pigmentation • followed by erosions and ulceration as well as severe pain	• 10–16 weeks postexposure: injury of blood vessels, edema, new ulcers, and increasing pain • possible necrosis	can involve ulcers that are extremely difficult to treat and that can require months to years to heat fully	• possible skin atrophy, depigmentation, constant ulcer recurrence, or deformity • possible occlusion of small vessels with subsequent disturbances in the blood supply, destruction of the lymphatic network, regional lymphostasis, and increasing fibrosis and sclerosis of the connective tissue • possible telangiectasia • possible skin cancer decades after exposure
IV	>550 Gy (55,000 rads)	occurs minutes to hours postexposure, with immediate pain or tingling, accompanied by swelling	none	• 1–4 days postexposure accompanied by blisters • early ischemia (tissue turns white, then dark blue or black with substantial pain) in most severe cases • tissue becomes necrotic within 2 weeks following exposure, accompanied by substantial pain	does not occur due to necrosis of skin in the affected area	recovery possible following amputation of severely affected areas and possible skin grafts	• continued plastic surgery may be required over several years • possible skin cancer decades after exposure

*Absorbed dose to at least 10 cm^2 of the basal cell layer of the skin.
†Especially with beta exposure.
‡The Gray (Gy) is a unit of absorbed dose and reflects an amount of energy deposited in a mass of tissue (1 Gy = 100 rads).
§Skin of the face, chest, and neck will a shorter latent phase than the skin of the palms of the hands and the skin of the feet.
Source: Centers for Disease Control and Prevention http://emergency.cdc.gov/radiation/pdf/criphysicianfactsheet.pdf pp 3-5.

dermatologic punch, or other surgical intervention may be necessary. In such situations the benefits of removal must be weighed against the risks of the procedure.

Internal contamination

Internally deposited radionuclides require treatment based on the chemical properties of the specific radionuclide and its deposition kinetics in the body. The advice of a physician trained in treating this type of medical situation should be sought (see resources at end of chapter). Treatment of the patient with internal radionuclide contamination begins with any needed decontamination. Assessment of the route and estimates of amount of radioactive material involved and its chemical form are necessary to evaluate absorption and distribution in the body. Specialized assessment procedures such as use of nasal swabs, use of a Geiger–Müller counter on skin surfaces, and thyroid and lung and whole-body counting may be required as well as capabilities for measuring radioactive material in excreta. Thresholds for intervention with medication and other therapies depend on the potential long-term dose to the patient from decay of internally deposited nuclides.

A variety of decorporation treatments can be used for internally deposited radionuclides. These include chelation, competitive binding, enhanced urinary excretion, and enhanced intestinal excretion (Table 11.10). For example, diethylenetriaminepentaacetate (DTPA) is used to chelate internally deposited plutonium and enhance its excretion through the kidneys. This is similar to treating lead poisoning with dimercaptosuccinic acid (DMSA, also referred to as succimer). Chelation or other treatments for internal contamination must be administered as soon as possible to reduce internal deposition in the target organs. Because this is a highly specialized activity, prompt consultation with experts in the management of internal contamination cases is critical.

Administration of iodine tablets or supersaturated potassium iodide (SSKI) solution is an important prophylactic measure that can reduce the absorption of the radioactive iodine into the thyroid gland. This measure, used after fission accidents such as Chernobyl and Fukushima, reduces radioiodine deposition and subsequent risk of thyroid cancer. Because of the high affinity of the thyroid for iodine, it is important to provide the iodine doses as soon as possible after exposure (or ideally before) to saturate the iodine receptors in the thyroid gland.

The approach for radiocesium or thallium is to administer Prussian blue (ferric hexacyanoferrate/Radiogardase) orally so that ion exchange takes place in the gut during enterohepatic circulation of the radionuclide.

For tritium (heavy water) exposure the guidance is to provide fluids and potentially diuretics to hasten the excretion through the kidneys.

A comprehensive discussion of specific isotopes and interventions to minimize internal deposition can be found in NCRP Report No. 161.[18] Examples of decorporation interventions may be found in Table 11.10. Doses of potassium iodide for thyroid prophylaxis are given in Table 11.11.[19]

Healthcare providers who manage radionuclide-contaminated victims should be properly trained in radiation protection procedures in order to avoid becoming contaminated or incurring an internal radionuclide deposition.

Psychological aspects of radiation exposure incidents

Psychological considerations are important in all accidents, but they are even more critical for victims of radiation incidents. Public fears of radiation exposure and the difficulty

TABLE 11.10 Examples of decorporation therapies for internal contamination.

Radionuclides	Therapy	Comment
Americium, curium, plutonium, and some other transuranic radionuclides and rare earths	CaDTPA or ZnDTPA	May be obtained from REAC/TS (see end of chapter)
Cobalt	DTPA, DMSA, EDTA, *N*-acetyl cysteine (all off label)	DTPA preferred
Cesium, thallium	Prussian blue	Prussian blue is given orally and binds radionuclide in the intestine, reducing enterohepatic recirculation
Iodine	KI or SSKI	Success depends on early administration after exposure. See doses in Table 11.11
Strontium	Aluminum hydroxide	Barium sulfate and alginates are alternatives to block gastrointestinal uptake
Tritium	Forced water diuresis	To promote excretion
Uranium	Sodium bicarbonate	Alkalinize urine to promote excretion and prevent acute tubular necrosis

TABLE 11.11 Thyroid prophylaxis doses in radiation incident involving radioiodine.

Threshold Thyroid Radioactive Exposures and Recommended Doses of KI for Different Risk Groups					
	Predicted Thyroid gland exposure (cGy)	**KI dose (mg)**	**Number or fraction of 130 mg tablets**	**Number or fraction of 65 mg tablets**	**Milliliters (mL) of oral solution, 65 mg/mL*****
Adults over 40 years	≥500	130	1	2	2 mL
Adults over 18 through 40 years	≥10	130	1	2	2 mL
Pregnant or Lactating Women	≥5	130	1	2	2 mL
Adolescents, 12 through 18 years*	≥5	65	½	1	1 mL
Children over 3 years through 12 years	≥5	65	½	1	1 mL
Children 1 month through 3 years	≥5	32	Use KI oral solution**	½	0.5 mL
Infants birth through 1 month	≥5	16	Use KI oral solution**	Use KI oral solution**	0.25 mL

*Adolescents approaching adult size (≥150 lbs) should receive the full adult dose (130 mg).
** Potassium iodide oral solution is supplied in 1 oz (30 mL) bottles with a dropper marked for 1, 0.5, and 0.25 mL dosing, each mL contains 65 mg potassium iodide.
*** See the home preparation procedure for emergency administration of potassium iodide tablets to infants and small children.
Source: US Food and Drug Administration, http://www.fda.gov/Drugs/EmergencyPreparedness/BioterrorismandDrugPreparedness/ucm072248.htm

individuals have in assessing their level of exposure may generate anxiety or panic. Individuals at increased risk of psychological issues in a radiation incident include individuals with children, children, first responders, evacuees, and individuals with limited social support, prior trauma, or mental illness.

Healthcare personnel should actively engage patients using attentive listening skills and promoting a sense of calmness, safety, and willingness to help without false reassurances. Information about exposure and potential health consequences needs to be communicated in clear terms that convey the level of risk appropriately. Clinicians should assess the patient's understanding of the information provided. If the victim is not adequately informed, the adverse psychological stress can be the most serious consequence of the radiation exposure or contamination.

Social stigma is also an issue that some radiation-exposed individuals have encountered. Communication with family members is important so that they do not unnecessarily perceive risk to themselves from contact with the patient. Support from mental health professionals can often be valuable for exposure victims, especially given the long-term uncertainties that may result from exposure. Radiation incidents often generate substantial interest from the media, and it may be worthwhile preparing patients for this possibility.

Nuclear power plant incidents, the public health response, and ongoing environmental exposures

Since the beginning of the nuclear age, there have been numerous situations involving unplanned human exposure to ionizing radiation, but two events that have particularly drawn attention to radiation's public health impacts include the nuclear power plant meltdowns at Chernobyl, Ukraine, in 1986 and Fukushima, Japan, in 2011. Both events involved the release of fission products containing iodine-131 and other radionuclides. Substantial preventable exposures to I-131 occurred around Chernobyl because of delays in notification, failure to distribute iodine tablets in a timely way, and continued consumption of contaminated foods, such as dairy products. As a result there have been many cases of papillary cancer of the thyroid in the surrounding region in individuals whose exposure to the gland exceeded 0.05 Gy (5 rem). Thyroid cancer predominantly affected children and was essentially limited to people who were under 40 years at the time of exposure. Although leukemias have been seen in remediation workers at the plant, thyroid cancer is the only malignancy that has been definitively attributed to the general public to date as a result of the nuclear meltdown. The World Health Organization estimates that some children in the areas around Fukushima have also received elevated radioiodine doses and may have an increased risk of thyroid cancer. Table 11.11 shows recommended iodine dose by age, usually administered one time as soon as possible after exposure. Iodine is most likely to be effective if given as soon as possible, but there is substantial protection even 3–4 hours after exposure.[19] Due to its short half-life of 7 days, I-131 is essentially gone from the environment after 90 days.

Environmental contamination from other radionuclides during a nuclear power plant meltdown can cause additional serious public health issues. In the Chernobyl and Fukushima disasters, the principal contaminants of concern, apart from radio-iodines, are cesium-137 and -134 with respective half-lives of 30.2 and 2.06 years. Strontium-90, with a half-life of 28.8 years is also a concern. Animals, seafood, and plant products take up cesium and strontium systemically. Projected exposures from external and internal radionuclides can result in unacceptable levels of risk to residents and render some districts unsafe for habitation for many years until exposure levels drop. This has occurred around both Chernobyl and Fukushima. In some areas near the Fukushima Nuclear Power Plant, washing down of surfaces, removal or remediation of soil, and other decontamination measures have reduced risk to a level deemed acceptable for rehabilitation by public health authorities. Other areas with higher levels of contaminants will remain uninhabitable for many years. Another aspect of the Fukushima disaster is the need to monitor potentially contaminated food products to keep exposure levels within regulatory limits. Some agricultural lands and fisheries have been taken out of production until safe radiation levels are reached.

PREVENTION

The principles for prevention of exposure to radiation are similar in many ways to the hierarchy of controls in other areas of occupational health including engineering measures, administrative controls, and personal protective equipment. Key protective concepts in radiation protection include use of shielding, reduction of exposure time, and distance from the source. For point sources, intensity of the exposure is inversely related to the square of the distance. Health professionals must understand the nature of the radiation source, its strength, and its capabilities to cause exposure and penetrate human tissue as well as other matter.

Regulatory exposure standards (discussed above) provide guidance to health professionals in establishing safety programs. In the United States commercial and research nuclear reactors are under the jurisdiction of the Nuclear Regulatory Commission. Other facilities using radiation or radioactive materials may be covered by regulations of the Department of Energy, OSHA, or state agencies. It is a good practice—and required under some agencies—for an institution to establish a radiation safety committee and to appoint a radiation safety officer (RSO) who is ideally a health physics professional. The RSO is responsible for the overall radiation protection program including (i) educating users about safety procedures, (ii) monitoring environmental and personal exposure, and (iii) ensuring that all recommended radiation safety policies, procedures, and controls are followed.

The use of engineering methods to protect workers against exposure is the most reliable approach. Engineering controls such as lead shielding, high-density concrete walls, and safety interlocks are used to control external radiation exposure. Glove boxes and ventilated hoods are used to prevent the exposure of individuals who handle radionuclides. Special shields might be used to protect the eyes, arms, or other body parts from exposure.

Administrative controls must be built into standard safe radiation practices. Work practices should be used that effectively manage the unseen hazards of radiation and radioactive materials. Administrative controls may reduce time of exposure in radioactive environments and limit the proximity to sources. Other practices may include the following injunctions: (i) work only in designated areas, (ii) use spill paper under radionuclide operations, (iii) never mouth-pipette; and (iv) handle all materials to prevent secondary contamination of work areas, equipment, and personnel. The institutional RSO and the radiation safety committee enforce operational policies and procedures required for licensing.

Personal protective equipment designed to avoid radionuclide contamination or internal deposition includes special clothing and respirators with appropriate filters for radionuclides that are worn only in the work area and then discarded or collected for cleaning when the worker leaves the area. The clothing to prevent contamination may include a hood, coveralls, gloves, and booties. Plastic tape is usually used to seal the areas between the gloves and the sleeves or other gaps in protective clothing. In heavily contaminated areas, a supplied air respirator or Scott air pack may be used.

Personal protective equipment for external exposures is widely used in industrial radiography and medical radiology. This equipment consists primarily of lead aprons, gloves, thyroid shields, and leaded glasses. Personal protective equipment for medical and research use of radionuclides typically includes gloves, laboratory coats, and eye/face protection.

Employees who work regularly around radiation or have the possibility of significant exposure should be monitored for exposure levels. A standard practice is for dosimeter (e.g., TLD) to be worn in the chest area of each employee who may be at risk. The dosimeters are collected and read at set intervals, that is, weekly for higher-risk exposures and monthly or quarterly for lower-risk exposures. Employees are informed of the results, and an RSO or radiation protection specialist investigates exposures greater than expected. Any worker who has reached a threshold exposure (action level) can be removed from potential exposure.

Environmental monitoring for external hazards is usually performed with fixed equipment. TLDs, continuous air monitors, or other monitoring instruments are placed outside the shielded area to ensure integrity of engineering controls. Equipment that generates X-rays must be kept in calibration. For radionuclide use, swipe samples can be taken at work areas to check for contamination.

For individuals who work with radionuclides, periodic thyroid, chest, or whole-body scanning, as appropriate, for detection of internal deposition of the radioactive material is common practice. For those who work with radionuclides that can be excreted in the urine or feces, collection and radionuclide analysis of these excretions are done on a scheduled basis. The frequency of these examinations is dependent on the type and activity of the materials used; therefore, an expert in surveillance of these materials must be consulted to set up an appropriate schedule.

Ionizing radiation should also be covered under an institutional reproductive health policy and education provided to all workers. All pregnant employees, or employees considering a pregnancy, should be voluntarily evaluated. A reproductive health assessment should be obtained as soon as possible to determine the types and levels of potential exposure. This assessment will direct any job accommodations that might be necessary. In addition, some employers increase exposure monitoring during gestation to assure that exposures are ALARA and within gestational exposure standards. In the absence of anticipated exposures above the NCRP recommended limit of 5 mSv (500 mrem) during pregnancy, it is generally not necessary to restrict workers from their usual job duties.

There is no standard practice regarding medical surveillance of workers who have radiation exposure. The decision regarding preplacement examinations, periodic evaluations, and evaluations at employment termination or transfer depends not only on the anticipated levels of exposure but also on the other tasks and risks inherent in the work. When performed, medical surveillance should be tailored to the exposures as much as possible. Risks of both deterministic and stochastic effects are very low as long as exposures are within permissible exposure limits and ALARA principles are followed. However, in some operations with significant exposure potential, medical surveillance may provide insights as to whether proper protections are being followed and detect unanticipated consequences such as skin or lens changes. Routine screening for cancer in workers with low levels of ionizing radiation exposure is not warranted other than the age- and gender-specific screenings offered to the general public. However, the NRC has adopted guidelines developed by the American National Standards Institute and the American Nuclear Society for the medical evaluation of nuclear power reactor operators to help assure their physical abilities to perform the work.[20]

EMERGENCY INFORMATION AND EXPERT ADVICE

The Oak Ridge Institute for Science and Education (ORISE) operates the Radiation Emergency Assistance Center/Training Site (REAC/TS). The US Department of Energy funds this facility. Expert advice and information on managing radiation accidents are available on a 24 hours/7-day-per-week basis by calling 865-576-1005. The REAC/TS also has a website at www.orau.gov/reacts/. REAC/TS provides training courses in radiation protection, as well as courses that prepare hospital and emergency personnel to manage medical response in radiation accidents. Information about these courses can be obtained on the website.

State public health departments usually have a radiation protection division that can provide guidance in setting up radiation protection programs. These professionals also help in dealing with ionizing radiation problems in industry, healthcare settings, or research and development projects. In most states, the health department radiation division is involved with licensing, auditing, inspecting, or accrediting hospitals and other facilities that use ionizing radiation equipment and materials and are not covered by the NRC. Some hospitals have a radiation safety expert on staff that may be available to answer questions or provide assistance.

References

1. National Research Council. *Health risks from exposure to low levels of ionizing radiation: BEIR VII Phase 2*. Washington, DC: National Academy of Sciences, 2006:189–206.
2. US Nuclear Regulatory Commission. Sources of Radiation. http://www.nrc.gov/about-nrc/radiation/around-us/doses-daily-lives.html (accessed June 20, 2016).
3. International Agency for Research on Cancer. *Radiation* Vol 100D. Geneva: World Health Organization Press, 2012:132–5. http://monographs.iarc.fr/ENG/Monographs/vol100D/mono100D.pdf (accessed June 20, 2016).
4. Mettler FA. Medical radiation exposure in the US in 2006: preliminary results. *Health Phys* 2008; 95(5):502–7.
5. National Council on Radiation Protection and Measurements. *SI units in radiation protection and measurements*. NCRP Report no. 82. Washington, DC: National Council on Radiation Protection and Measurements, 1985.
6. National Council on Radiation Protection and Measurements. *Recommendations on limits for exposure to ionizing radiation*. NCRP Report no. 116. Washington, DC: National Council on Radiation Protection and Measurements, 1993.
7. International Commission on Radiologic Protection. Recommendations of the International Commission on Radiologic Protection. ICRP Publication no. 103. *Ann ICRP* 2007; 37(2–4):1–332.
8. International Commission on Radiation Protection. Non-stochastic effects of ionizing radiation. ICRP Report no. 41. *Ann ICRP* 1984; 14(3).

9. Anno GH, Baum SJ, Withers HR, et al. Symptomatology of acute radiation effects in humans after exposure to doses of 0.5–30 Gy. *Health Phys* 1989; 56(6):821–38.
10. Preston DL, Ron E, Tokuoka S, et al. Solid cancer incidence in atomic bomb survivors: 1958–1998. *Radiat Res* 2007; 168:1–64.
11. Shore RE. Low-dose radiation epidemiology studies: Status and issues. *Health Phys* 2009; 97(5):481–6.
12. US Environmental Protection Agency. Radon: A Physician's Guide. 1993. http://www.epa.gov/radon/pubs/physic.html#SOL6 (accessed May 6, 2016).
13. International Atomic Energy Agency. *Cytogenetic analysis for radiation dose assessment*. Technical Report Series no. 405. Vienna: International Atomic Energy Agency, 2001.
14. Sugarman SL, Goans RE Garrett AS, et al. The Medical Aspects of Radiation Incidents. Radiation Emergency Assistance Center/Training Site ORISE Revised September 25, 2013. http://orise.orau.gov/files/reacts/medical-aspects-of-radiation-incidents.pdf (accessed May 6, 2016).
15. Ricks RC. *Hospital emergency department management of radiation accidents*. ORAU no. 224. Washington, DC: Federal Emergency Management Agency, 1984
16. Radiation Emergency Assistance Center/Training Site (REAC/TS). Guidance for Radiation Accident Management. http://orise.orau.gov/reacts/guide/procedures.htm (accessed May 6, 2016).
17. Centers for Disease Control and Prevention. Radiation Emergency Training and Education. http://emergency.cdc.gov/radiation/training.asp (accessed May 6, 2016).
18. National Council on Radiation Protection and Measurements. *Management of persons accidentally contaminated with radionuclides*. NCRP Report no. 161. Washington, DC: National Council on Radiation Protection and Measurements, 2008.
19. US Food and Drug Administration, Center for Drug Evaluation and Research (CDER). Guidance: Potassium Iodide as a Thyroid Blocking Agent in Radiation Emergencies. Rockville, MD: US Department of Health and Human Services, 2001. http://www.psr.org/assets/pdfs/fda-iodine-guidelines.pdf (accessed September 4, 2016).
20. American National Standards Institute/American Nuclear Society. Medical Certification and Monitoring of Personnel Requiring Operator Licenses for Nuclear Power Plants. ANSI 3.4.-2013 (Adopted by the US Nuclear Regulatory Commission).

Chapter 12

ULTRAVIOLET RADIATION

JAMES A. HATHAWAY AND DAVID H. SLINEY

Ultraviolet radiation is that portion of the electromagnetic spectrum between visible light (about 400 nm) and the lower limit of ionizing radiation (about 100 nm). The energy of ultraviolet radiation photons increases as the wavelength decreases. The ultraviolet spectrum is divided into the following three bands: from 315 to 400 nm, it is called UV-A; from 280 to 315 nm, it is designated UV-B; and from 100 to 280 nm, it is referred to as UV-C.[1]

OCCUPATIONAL SETTING

Natural sunlight includes biologically significant amounts of energy in the UV-A and UV-B bands. The upper atmosphere filters out the UV-C radiation, although there is concern about UV-C exposure in areas where the ozone layer is absent. Employees who work in the natural environment incur the greatest occupational exposure to ultraviolet radiation. Examples of such occupations include farmers and other agricultural and forestry workers, fishermen, outdoor construction workers, landscapers, and lifeguards.

Altitude, latitude, time of year, time of day, and the thickness of the ozone layer affect UV strength. Levels are higher at higher altitudes, closer to the equator, and during summer. For latitudes between the Tropic of Cancer and the Tropic of Capricorn, levels are moderate to extreme year-round. UV strength is also increased between 10 a.m. and 4 p.m. Levels are also higher where the ozone layer is thinner. While exposure is highest under clear skies, cloud cover provides only a limited reduction in exposure. Exposure is also increased when ultraviolet radiation reflects off water, snow, or sand.

The most common exposure to significant levels of nonnatural ultraviolet radiation occurs among welders. Other workers may receive exposure from sources such as gas discharge lamps and carbon arcs.[2] Low-pressure mercury vapor lamps are used to control microorganisms in operating rooms, to control bacterial growth in meat, to prevent contamination in biological laboratories, to reduce airborne bacterial levels in air ducts, and to eliminate coliform bacteria in drinking water. High-pressure mercury vapor lamps are used for photochemical reactions and to identify minerals. High-pressure xenon arcs and carbon arcs have a broad spectrum of radiation, including visible and ultraviolet radiant energy. They are used as high-intensity light sources, such as searchlights, as well as in the printing industry.[3]

Workers with occupational exposure also have nonoccupational exposure, including exposure during outdoor activities and possibly through artificial (indoor) tanning.

MEASUREMENT ISSUES

Measurement is not a significant issue when using many of these exposure sources since it is already known that they emit harmful levels of ultraviolet radiation. In these cases, skin and eye protection is required. However, if direct measurement of ultraviolet radiation is needed, several UV meters are commercially available. Care should be taken to ensure that the UV meter is effective in the wavelength range of the source. UV detection devices include photodiodes, certain photomultipliers, and vacuum photodiodes.

Physical and Biological Hazards of the Workplace, Third Edition. Edited by Gregg M. Stave and Peter H. Wald.

Selective filters are frequently used to isolate the part of the UV spectrum under study. Frequent calibration of meters is often necessary with heavy use. No single instrument perfectly matches the biological hazard action spectrum of ultraviolet radiation; therefore, it is necessary to calibrate meters to several wavelengths to evaluate broad-spectrum sources of UV.[4] Some recently developed detectors are remarkably well matched to the action spectrum.

EXPOSURE GUIDELINES

The American Conference of Governmental Industrial Hygienists (ACGIH) has published exposure limits called threshold limit values (TLVs®) for ultraviolet radiation.[5] To protect against photokeratitis effects on unprotected eyes from UV radiation in the 320–400 nm range, total irradiance should not exceed 1.0 mW/cm^2 for periods >10 seconds (~16 mm) or >1.0 J/cm^2 for exposures less than ~10 seconds. Exposure limits for other wavelengths vary significantly by wavelength and duration of exposure. The tables in the ACGIH TLV® guide should be used to determine exposure limits for a particular set of conditions. The International Commission on Non-Ionizing Radiation Protection (ICNIRP) also adopted these limits with minor modifications in the UV-A region.[6]

NORMAL PHYSIOLOGY

Ultraviolet irradiation of 7-dehydrocholesterol in the skin produces previtamin D3, which is converted to vitamin D3 (cholecalciferol). This vitamin and some vitamin D3 from the diet are converted to circulating vitamin D.[7] Vitamin D is essential for the regulation of metabolism of bone minerals. Some level of ambient ultraviolet radiation below the TLV® appears to be necessary to maintain good health. Low levels of vitamin D result in increased risks of numerous health problems, including rickets and osteomalacia.[7] Several studies also suggest that ultraviolet radiation can reduce the risks of some types of cancer, although this has not been shown in all studies.[8-14]

Another normal physiologic reaction to ultraviolet radiation is the tanning of the skin. Tanning provides some protection against UV exposure, with a sun protection factor (SPF) of 1.5–4.[15] (SPF is a measure of UV-B protection, the factor by which the time required for unprotected skin to become sunburned is increased.) Individuals with darker skin may also have greater protection than those with fair skin.[15] Exposure limits can be increased for tanned individuals.[5] However, tanning will not result from exposures below the TLV®, and growing evidence suggests that a risk of skin cancer still exists even with careful tanning.[6]

PATHOPHYSIOLOGY OF INJURY

The penetration of ultraviolet radiation into human tissue is very limited. As a consequence, adverse health effects have historically been thought to be limited to acute and chronic skin and eye damage. The response of biologic tissue to UV radiation is highly dependent on the depth of absorption, which in turn is wavelength specific. The wavelengths of concern will be discussed along with each type of tissue damage.[16] Some studies have demonstrated that UV radiation can suppress the immune response in the skin and may cause some systemic immunosuppression.[17,18] These effects may play a role in the development of skin cancer, and it has been speculated that they could alter host response to infectious diseases.

Acute effects on the eye

The typical acute condition of the eye caused by UV exposure is photokeratitis of the cornea, commonly called welder's flash, arc eye, or flash burn. It is caused by UV radiation below 315 nm, usually from unprotected exposure to a welding arc. Between 2 and 24 hours after exposure, the worker experiences severe pain, redness, photophobia, and spasm of the eyelids if the exposure was severe. The condition usually clears up in 1–5 days, depending on the severity of the exposure. Healing is usually complete, and there is no residual injury.[19]

Acute effects on the skin

Ultraviolet radiation, especially in the UV-C and UV-B bands, produces erythema of the skin (sunburn). If the exposure is more severe, edema and blistering will result. Although UV radiation above 315 nm is less efficient at producing erythema, there is sufficient energy in the UV-A band from tropical sunlight to produce sunburn even if the UV-B is filtered out. The maximum effective wavelength for producing sunburn is 300–307 nm in sunlight and at shorter wavelengths in artificial light. The National Institute for the Occupational Safety and Health (NIOSH) criteria document on ultraviolet radiation includes an extensive discussion of the histological and cytological changes in the skin induced by acute UV exposure.[20] These effects include an inflammatory response with edema, lymphocytic infiltrates, capillary leakage, and evidence of local dermal cell damage. The response to UV radiation is photochemical. A short-duration phase occurs in 1–2 hours; a later phase appears after 2–10 hours and may last for several days. The intensity and duration of the burn are proportional to the dose and wavelength of UV radiation; it is also related to the individual's skin pigmentation. The latent period to erythema becomes shorter with more intense exposures.[21]

An additional concern with acute exposure is the potential for photosensitivity reactions. Photosensitizing agents may have biological action spectra in the UV-A range, and they can act either systemically or locally. A number of medications (systemic photosensitizers) can predispose an individual to photosensitivity. Drugs such as sulfonamides, sulfonylureas, chlorothiazides, phenothiazines, and tetracyclines are also well-known sensitizers. Occupational exposure to certain chemicals that may remain on the skin (local, or contact, sensitizers) can also act synergistically with UV radiation to produce erythema at much lower dose of UV radiation than would ordinarily be required. Coal tar products are well known for their photosensitizing properties. When combined with UV exposure, severe irritation and blistering can result.[22] Individuals with certain underlying diseases or a genetic predisposition may be unusually sensitive to the acute effects of UV radiation. Examples of conditions caused or aggravated by acute exposure to UV radiation include solar urticaria, polymorphous light eruption, the porphyrias, and systemic lupus erythematosus. Numerous other skin conditions can also be aggravated by the UV radiation from sunlight or artificial sources.

Chronic effects on the eye

Ultraviolet radiation in the 295–400 nm range can cause photochemically induced opacities of the lens of the eye. The most effective wavelengths are 295–325 nm.[23] Radiation above 315 nm also causes cataracts in experimental animals. Some authorities have long theorized that ambient exposure to ultraviolet radiation is the primary cause of cataract development in older persons, and this has been demonstrated in some epidemiology studies.[24-26] Potential effects on other ocular structures are not as well understood, but UV exposure may contribute to macular and retinal degeneration.[27,28] UV-B radiation may also be a risk factor for the development of pterygium.[29]

Chronic effects on the skin

Chronic exposure to ultraviolet radiation results in accelerated aging of the skin and an increased risk of skin cancer.[2,30] Prolonged exposure causes loss of elasticity, resulting in wrinkles. Actinic keratoses may form in the epidermis and are precursors to squamous cell carcinoma.

Three types of skin cancer have been associated with UV radiation—basal cell carcinoma, squamous cell carcinoma, and melanoma. Squamous cell carcinomas are related to the cumulative dose of ultraviolet radiation to the skin. They occur primarily on sun-exposed areas of the body, and it has been known for many years that persons in outdoor occupations are at a higher risk for these cancers.[31,32] Basal cell carcinoma risk is more dependent on exposure patterns, and melanoma risk is even more so with intermittent exposure being the most dangerous.[33] Although UV radiation is important in the pathogenesis of melanoma, cumulative exposure to UV does not appear to be the primary factor. Nor has occupational exposure to UV radiation been correlated with the incidence of melanoma. Several studies have found that childhood exposures to sunlight that resulted in severe blistering sunburns were predictive of malignant melanoma risk.[34,35] The fact that melanoma is more common on the trunk than on commonly sun-exposed areas such as the face, back of the neck, and hands supports this theory. Melanoma is also rare on the buttocks and on women's breasts, which are typically covered during sunbathing.

Fair-skinned people have a higher risk of developing skin cancers than darker ones. Individuals of Celtic origin are also at greater risk. Genetic factors that affect DNA repair, such as xeroderma pigmentosum, can greatly increase the risk of skin cancer, and immunosuppression may also be an important factor. Kidney transplant patients have been identified as being at greater risk of developing skin cancer.[36]

Immunosuppression

Research on the effects of UV radiation on the immune system has led to a new field of research called photoimmunology. In animal studies it was found that many UV-B-induced cancers were highly antigenic and were rejected when transplanted into normal syngenic animals. The UV radiation not only induced the tumors but also lead to systemic T-lymphocyte-mediated immunosuppression, which reduced the host animal's ability to reject the tumor cells. This has raised the possibility that UV radiation-induced immunosuppression might also lead to a reduced response to infectious diseases.[24,25]

DIAGNOSIS AND TREATMENT

Detailed discussion of the treatment of chronic effects of UV radiation is beyond the scope of this book. The most significant acute effects are UV burns of the eye and skin (i.e., sunburn).

Ultraviolet burns of the eye (photokeratitis) should be examined with a slit lamp using fluorescein stain. Diffuse punctate staining of the corneas will be seen within the palpebral fissure. Both eyes should be patched, and cycloplegic agents should be used. Local anesthetics should not be prescribed. Recovery is usually complete in 24 hours.[37]

Sunburn can range from mild to severe. Aspirin or other nonsteroidal anti-inflammatory agents may be useful for fever and pain. Corticosteroids may be required for severe reactions.

MEDICAL SURVEILLANCE

Medical surveillance has generally not been recommended for workers who are exposed to ultraviolet radiation. It would not be useful for acute effects on the skin or eye and has not been demonstrated to be of particular value for chronic effects. Nonmelanotic skin cancer and ocular cataracts both have relatively long latencies (time from exposure to the earliest manifestations of disease) in most people. Most of the diseases attributable to occupational exposure would probably not be seen until after retirement; even if detected sooner, it would still be many years after the initial exposure. At best, medical surveillance might be expected to help detect some basal and squamous cell carcinomas when they are small and therefore more easily treated. It would also be possible to treat precursor lesions, such as actinic keratoses.

Periodic examinations of the skin have been suggested for high-risk groups such as coal tar workers with concurrent outdoor exposure. Skin cancers that are hard to see, such as those on the back of the neck or ears, could be detected earlier through this type of examination.

Early detection of cataracts for workers with ultraviolet exposure has not been specifically investigated. Studies on workers exposed to lasers or microwave radiation have attempted to assess early changes in the lens of the eye. These studies did not identify early changes that would be useful from a medical surveillance perspective.[38,39]

PREVENTION

Prevention of exposure to man-made sources of ultraviolet radiation is accomplished through a combination of engineering controls and personal protective equipment. Typical controls include the use of opaque shields or curtains when welding to eliminate exposure to coworkers. UV radiation used for germicidal purposes can usually be installed in ducts or recessed areas so that exposure to individuals in the same room is eliminated. Individuals performing operations such as welding use welding helmets and protective clothing.[40]

For outdoor work, simple measures such as long-sleeved shirts, broad-brimmed hats, canopies, and awnings provide significant protection. Sunscreens with an adequate SPF and UV-A protection should be used on exposed parts of the body. The wearing of tinted glasses is also recommended for protection from UV exposure, and is especially important for occupations near water, sand, or snow, where ambient exposures are amplified.[27,41] Eyeglasses with lenses made of glass normally provide substantial UV-B protection without special tinting, but tinted plastic lenses do not necessarily stop UV exposure. Plastic lenses usually state what degree of UV protection they provide, and this should be specifically checked. Sunglasses that block wavelengths below 400 nm are widely available. Lenses should meet ANSI Z80.3 blocking requirements. Tinted lenses should also be of a wraparound design or have side shields or large temples to block peripheral exposure of the eye.[27,41] Contact lenses that absorb significant amounts of UV radiation may provide additional protection.[27] Polarized lenses may be helpful to reduce reflected glare, but do not enhance UV protection.

Outside of work, individuals should also follow safe sun practices and avoid indoor tanning. Behavioral counseling is recommended by the US Preventive Services Task Force (USPSTF).[42]

References

1. Commission Internationale de l'Eclairage (International Commission on Illumination). *International lighting vocabulary*. Publication, no. 17. 3rd ed. Paris: Commission Internationale de líEclairage, 1970.
2. Yost MG. Occupational health effects of nonionizing radiation. In: Shusterman DJ, Blanc PD, eds. *Occupational medicine: state of the art reviews*. Philadelphia: Hanley & Belfus, 1992:543–66.
3. Zenz C. Ultraviolet exposures. In: Zenz C, ed. *Occupational medicine*. 2nd ed. Chicago: Yearbook Medical Publishers, 1988:463–7.
4. Wilkening GM. Nonionizing radiation. In: Clayton GD, Clayton FE, eds. *Patty's industrial hygiene and toxicology; vol 1, Pt B*. 4th ed. New York: John Wiley & Sons, Inc., 1991.
5. American Conference of Governmental Industrial Hygienists. *Threshold limit values*. Cincinnati: American Conference of Governmental Industrial Hygienists, 2015.
6. International Commission on Non-Ionizing Radiation Protection (ICNIRP). Guidelines on UV radiation exposure limits. *Health Phys* 1996; 71:978–82.
7. Federman DD. Parathyroid. In: Rubenstein E, Federman DD, eds. *Scientific American medicine; section 3, VI*. New York: Scientific American, 1992:1.
8. Grant WB, Strange RC, Garland CF. Sunshine is good medicine. The health benefits of ultraviolet-B induced vitamin D production. *J Cosmet Dermatol* 2003; 2:86–98.
9. Lin SW, Wheeler DC, Park Y, et al. Prospective study of ultraviolet radiation exposure and risk of cancer in the United States. *Int J Cancer* 2012; 131:1015–23.
10. Holick MF. Sunlight, ultraviolet radiation, vitamin D and skin cancer: how much sunlight do we need? *Adv Exp Med Biol* 2014, 810:1–16.
11. Monnereau A, Glaser SL, Schupp CW, et al. Exposure to UV radiation and risk of Hodgkin lymphoma: a pooled analysis. *Blood* 2013; 122:3492–9.
12. Tran B, Lucas R, Whiteman D, et al. Association between ambient ultraviolet radiation and risk of esophageal cancer. *Am J Gastroenterol* 2012; 107:1803–13.
13. Berwick M. Can UV exposure reduce mortality? *Cancer Epidemiol Biomarkers Prev* 2011; 20:582–4.
14. Lin S, Wheeler DC, Park Y, et al. Prospective study of ultraviolet radiation exposure and mortality risk in the United States. *Am J Epidemiol* 2013; 178:521–33.

15. Brenner M, Hearing VJ. The protective role of melanin against UV damage in human skin. *Photochem Photobiol* 2008; 84(3):539–49.
16. World Health Organization (WHO). *Ultraviolet radiation*, Environmental health Criteria, 160. Joint Publication of the United Nations Environmental Program, The International Radiation Protection Association and the World Health Organization. Geneva: WHO, 1994.
17. Kripke ML. Ultraviolet radiation and immunology: something new under the sun—presidential address. *Cancer Res* 1994; 54:6102–5.
18. Beissert S, Scharz T. Mechanisms involved in ultraviolet light-induced immunosuppression. *J Investig Dermatol Symp Proc* 1999; 4:61–4.
19. Pitts DG, Tredici TJ. The effects of ultraviolet on the eye. *Am Ind Hyg Assoc J* 1971; 32:235–46.
20. U.S. Department of Health, Education and Welfare. *A recommended standard for occupational exposure to ultraviolet radiation*, HSM publication, no. 73-11009. Rockville: National Institute of Occupational Safety and Health, 1977.
21. CIE. *Erythema reference action spectrum and standard erythemal dose*, CIE standard, S007-1998. Vienna: CIE, 1998; also available as ISO 17166:1999.
22. Harber LC, Bickers DR. Drug induced photosensitivity. In: *Photosensitivity diseases: principles of diagnosis and treatment*. Philadelphia: WB Saunders, 1981:121–53.
23. Pitts DG, Cullen AP. *Ocular effects from 295 nm to 335 nm in the rabbit eye*, DHEW (NIOSH) publication, no. 177-30. Washington, DC: National Institute of Occupational Safety and Health, 1976.
24. Duke-Elder S. The pathological action of light upon the eye. Part II (continued)-action upon the lens: theory of the genesis of cataract. *Lancet* 1926; 1:1250–4.
25. Hanna C. Cataract of toxic etiology. In: Bellows JG, ed. *Cataract and abnormalities of the lens*. New York: Grune & Stratton, 1975:217–24.
26. Wang Y, Yu J, Gao G, et al. The relationship between the disability prevalence of cataracts and ambient erythemal ultraviolet radiation in China. *PLoS One* 2012; 7:e51137 (published online).
27. Roberts JE. Ultraviolet radiation as a risk factor for cataract and macular degeneration. *Eye Contact Lens* 2011; 37(4):246–9.
28. Sliney DH. Photoprotection of the eye: UV radiation and sunglasses. *J Photochem Photobiol B* 2001; 64:166–75.
29. Saw SM, Tan D. Pterygium: prevalence, demography and risk factors. *Ophthalmic Epidemiol* 1999; 6:219–28.
30. International Agency for Research on Cancer (IARC). *Solar and ultraviolet radiation*, Monograph on the evaluation of carcinogenic risk to humans. Vol. 55. Lyon: IARC, 1992.
31. Belisario JC. Effects of sunlight on the incidence of carcinomas and malignant melanoblastomas in the tropical and subtropical areas of Australia. *Dermatol Trop* 1962; 1:127–36.
32. Nicolan SG, Balus S. Chronic actinic cheilitis and cancer of the lower lip. *Br J Dermatol* 1964; 76:278–84.
33. Moan J, Grigalaviccius M, Baturaite Z, et al. The relationship between UV exposure and incidence of skin cancer. *Photodermatol Photoimmunol Photomed* 2014; 31(1):26–35.
34. Shore RE. Nonionizing radiation. In: Rom WN, ed. *Environmental and occupational medicine*. Boston: Little, Brown and Company, 1992:1093–108.
35. Sober AJ, Lew RA, Kob HK, et al. Epidemiology of cutaneous melanoma. *Dermatol Clin* 1991; 9:617–29.
36. Walder BK, Robertson MR, Jeremy D. Skin cancer and immunosuppression. *Lancet* 1971; 2:1282–90.
37. Riordan-Eva P, Vaughan DG. Ultraviolet keratitis. In: Schroeder SA, ed. *Current medical diagnosis and treatment*. Norwalk: Lange Medical Books, 1990:120.
38. Friedman Al. The ophthalmic screening of laser workers. *Ann Occup Hyg* 1978; 21:277–9.
39. Hathaway JA, Stern N, Soles EM, et al. Ocular medical surveillance on microwave and laser workers. *J Occup Med* 1977; 19:683–8.
40. Tenkate TD. Optical radiation hazards of welding arcs. *Rev Environ Health* 1998; 13:131–46.
41. Sliney DH. Eye protective techniques for bright light. *Ophthalmology* 1983; 90:937–44.
42. Lin JS, Eder M, Weinmann S. Behavioral counseling to prevent skin cancer: A systematic review for the U.S. Preventive Services Task Force. *Ann Intern Med* 2011; 154:190–201.

Chapter 13

VISIBLE LIGHT and INFRARED RADIATION

JAMES A. HATHAWAY AND DAVID H. SLINEY

Visible light is generally defined as that portion of the electromagnetic spectrum between approximately 380–400 nm and approximately 760 nm.[1] Some reference sources list the upper limit of the visible light band as 780 or 800 nm.[2,3] Within the visible light spectrum, blue light (400–500 nm) is of particular importance. Infrared radiation is divided into the following three bands: IR-A is between 760 and 1400 nm, IR-B is between 1.4 μm (1400 nm) and 3 μm, and IR-C is between 3 and 1000 μm (1 mm). This ABC notation is sometimes referred to as near, middle, and far IR.

OCCUPATIONAL SETTING

Visible light, along with the adjacent portions of the ultraviolet and infrared bands of radiation, makes up much of the solar radiation reaching the surface of the Earth. Outdoor occupations naturally have greater exposure to visible light and IR radiation. Visible light reflecting off sand and snow can create hazardous conditions that require eye protection. Ambient IR radiation can contribute to heat load, particularly in persons who work outdoors while wearing impervious clothing. Issues related to heat stress are covered in Chapter 6. Man-made sources of broad-spectrum intense visible light include arc welding or cutting, arc lamps, spotlights, gas and vapor discharge tubes, flash lamps, open flames, and explosions.[4] Even though ultraviolet radiation is the main concern with many of these exposures, the potential for visible light-induced damage cannot be ignored. More recent concerns have been raised regarding potential blue light hazards from crystal glass-blowing, use of LED dental illumination applications, and newer types of theater projectors.[5–7]

Infrared radiation is emitted by many sources besides the sun. Man-made sources include heated metals, molten glass, home electrical appliances, incandescent bulbs, radiant heaters, furnaces, welding arcs, and plasma torches. Glassblowing and working in glass and steel plants are considered potentially hazardous due to excessive IR radiation.[8,9]

MEASUREMENT ISSUES

Among the adaptive responses to intense visible light are constriction of the pupil, light adaptation of the retina, squinting, and blinking. Intense light causes a natural aversion response, including shutting the eyes and turning away from the source of exposure. Measurement of continuous visible light emissions is usually not necessary to determine if the level of exposure is excessive or not, because the human eye itself provides adequate warning. Pulsed sources of visible light and sources that are turned on suddenly may present problems if the intensity of light is high enough to cause damage before an aversion response can take place. Usually such sources will be labeled with appropriate warnings; specific measurement of output levels will not be necessary.

Unfortunately, there are virtually no instruments designed as optical safety meters; so when measurements are necessary, a scientist experienced in radiometry may have to be consulted. A variety of instruments that use photodiodes or thermal detectors may be required for the measurement of visible light levels. These devices detect optical energy and convert the optical radiation to a measurable electrical signal. Similar instruments are also available to measure infrared radiation. IR radiation is most frequently measured using

Physical and Biological Hazards of the Workplace, Third Edition. Edited by Gregg M. Stave and Peter H. Wald.

thermal detectors such as thermopiles or disk calorimeters.[10,11] These detectors measure heat from absorbed energy; they are suitable for the entire range of IR radiation, although the response time is slow. Lamp safety standards require the lamp manufacturer to perform detailed radiometric measurements of the optical radiation hazards and to group the lamp in one of four risk groups.

EXPOSURE GUIDELINES

Visible light and the near portion of the IR spectrum have threshold limit values (TLVs®) developed by the American Conference of Governmental Industrial Hygienists (ACGIH).[12] These TLVs® are for visible and near-infrared radiation between 400 and 3000 nm. The TLVs® apply to 8 hours exposures and require knowledge of the spectral radiance and total irradiance of the source as measured at the eyes of the worker. Moderately complex formulas and reference tables are required to calculate the TLV® for each exposure situation; these calculations are beyond the scope of this book. TLVs® can be calculated for three types of injury—retinal thermal injury from exposure to 400–3000-nm radiation, retinal photochemical injury from chronic blue light (400–500-nm) exposure, and possible delayed effects leading to cataract formation from exposure to 770–1400-nm radiation. There are additional calculations for persons who have had a lens removed (cataract surgery) and not had a UV-absorbing intraocular lens surgically inserted. Such persons (although now rare) are at increased risk for photochemical retinal injury. The International Commission on Non-Ionizing Radiation Protection (ICNIRP) has published guidelines for human exposure and these are available at no cost from the ICNIRP web site (http://www.icnirp.org).[13,14]

NORMAL PHYSIOLOGY

Life on Earth would not be possible without visible light and infrared radiation. Infrared radiation provides warmth, allowing a climate where life is possible; visible light provides the energy upon which life is based. Plant life uses chlorophyll to acquire energy from visible light. Using this energy, it converts carbon dioxide and water to carbohydrates in a process called photosynthesis. Plant-eating animals ingest stored carbohydrates in plants, and meat-eating animals acquire photosynthetically produced energy directly by feeding on plant-eating animals. The energy in all food consumed by humans is ultimately derived from visible light reaching the Earth's surface. Most animal species have a sense of vision that is responsive in the near-UV, visible, or near-IR portion of the electromagnetic spectrum. For humans, visual response defines the relatively narrow band of radiation called visible light (400–760 nm). Photons of light enter the eye and are focused by the cornea and lens onto the retina. In the retina, there are two types of photoreceptors: the cones, which are responsible for color vision and detailed visual acuity, and the rods, which allow peripheral vision and are responsive to lower light levels. The macula is an area on the retina that is densely packed with cones; it is the site of maximum visual acuity.

Photochemical reactions take place in both the cones and the rods. This stimulus results in a neurosensory transmission to the brain where visual images are perceived. The retina and its photoreceptors are able to adapt to a wide range of light intensities. Adaptation to brighter levels of light typically occurs rapidly in a period of a few seconds. Adaptation to darkness requires many minutes, and in some cases more than an hour, to achieve maximum adaptation. In most circumstances, visible light is not hazardous. In addition to light adaptation, other normal protective mechanisms such as pupillary constriction, squinting, and blinking occur rapidly when bright light is encountered. When exposed suddenly to a highly intense visible light source, most people exhibit an aversion response that includes blinking and turning the head. This response typically occurs within 0.25 second; this time period is used to calculate exposure limits for radiation in the visible spectrum. The eyes are also naturally shaded from ambient sunlight by the eyebrows and the periorbital socket ridge.

Under some circumstances, visible light can be harmful—for example, when it is presented suddenly, as in a flash or explosion, or when equipment is first turned on. If the intensity is high enough to cause damage in <0.25 second, the natural protective mechanisms will be insufficient. It is also possible to create a hazardous situation by suppressing the aversion response and staring directly at a high-intensity light source such as the sun (solar maculopathy or eclipse photoretinitis) or a welding arc (welding-arc photoretinitis).

PATHOPHYSIOLOGY OF INJURY

Potential adverse health effects from overexposure to visible light or infrared radiation occur primarily in either the eye or the skin. Systemic effects of infrared radiation from general body heating are considered in Chapter 6. Adverse effects can result from acute and chronic exposure. In the case of the eye, injury can result in different structures depending on the wavelength of radiation.

Acute chorioretinal injury

Visible light and near-infrared radiation from 400 to 1400 nm can be focused on the retina.

Sudden exposures to high-intensity sources of such radiation can cause adverse effects ranging from temporary flash blindness and afterimages to chorioretinal burns that produce scotomas (i.e., blind spots) in the field of vision. Retinal burns from gazing at the sun or observing a solar eclipse have been described throughout history. Man-made sources of luminance comparable to the sun have been developed in more recent decades. Even so, there have been fewer incidents of chorioretinal burns from man-made sources such as electric arcs, explosions, and nuclear fireballs than from directly viewing the sun.[4] Measurement of the spectral radiance of the sun has shown that the exposure limits for blue light can easily be exceeded when viewing the sun. The exposures increase with solar elevation. Viewing the sun directly can be very hazardous and should be avoided.[15]

Several factors are important in determining the exposure to the retina. These include (i) pupil size, (ii) spectral transmission through the ocular media, (iii) spectral absorption by the retina and choroid, and (iv) the size and quality of the image. A dark-adapted pupil may be as large as 7 mm, as compared to a normal pupil size of 2–3 mm in outdoor sunlight. The area of a 7-mm pupil is about 12 times greater than a 2-mm pupil; thus, it allows that much more radiation to enter the eye. Although some radiation from 400 to 1400 nm can reach the retina, absorption in the ocular media (cornea, aqueous humor, lens, and vitreous humor) varies by wavelength. Optical transmission is greater from 500 to 900 nm, dropping about 50% to 1000 nm, rising again to 1100 nm, and dropping to low levels by 1200–1400 nm. Absorption of energy by the choroid and retina peaks around 500–700 nm, dropping gradually as the wavelength increases to 1000 nm, with a small rise peaking at about 1100 nm and with very little absorption past 1200 nm. The more energy that reaches and is absorbed by the retina and choroid, the greater the potential damage. For large, uniform images, the total absorbed dose per area on the retina is a good predictor of damage. Small images or images with "hot spots" blur as they are focused on the retina due to diffraction and therefore produce reduced peak retinal irradiance. Involuntary eye movements also spread the radiant energy over larger retinal areas.[16]

The mechanism of injury from accidental exposure to arc lamps or the sun was once thought to be primarily thermal, resulting in protein denaturation and enzyme inactivation. Today we know that most retinal injuries from staring at the sun or at a welding arc actually result from photochemical reactions that dominate particularly with exposure to wavelengths of visible light between 400 and 500 nm.[17] However, thermal effects are still important. The threshold for thermal injury is dependent on light absorption, heat flow, and duration of exposure. Thermal injury is a rate-dependent process, so there is no single critical temperature that results in damage. In general, shorter exposures require higher temperatures to produce the same degree of damage.[4]

The degree of impairment caused by an acute chorioretinal injury depends on the size of the lesion in the retina and its location. If the source of exposure was directly viewed, as in gazing at a solar eclipse or looking at an explosion, the macula of the eye will be involved. Since fine visual acuity is dependent on intact macular function, damage in this area typically causes significant impairment of visual acuity. Injury to peripheral regions of the retina produces scotomas in the visual fields, but in many cases, peripheral lesions have minimal effect on overall visual function. Obviously, larger lesions cause more impairment than small lesions in the same location.

Chronic blue light-induced retinal injury

Whereas thermal effects of visible and near-infrared radiation on the retina are acute phenomena, the photochemical effects of blue light photoretinitis are additive over time periods of seconds to hours and are probably partially additive even over many years. Exposure to light capable of causing thermal injury that does not cause actual injury is virtually not additive with subsequent exposures. In contrast, blue light, especially 400–500 nm radiation, can cause subclinical changes, which with repeated exposures can result in observable retinal damage. Subacute retinal injury due to photochemical mechanisms can occur at thresholds well below those of thermal injury. This threshold is only slightly higher than normal exposures to sunlight in outdoor work environments.[18] A number of mechanisms have been proposed to explain these effects, including photooxidative membrane damage, toxic chemical production in the outer retina, and metabolic disruption from extended overbleaching of retinal pigments. Ophthalmic examinations of experimental animals show both edema and pigmentary changes.

Some researchers believe that even typical or "normal" outdoor exposure to sunlight can result in damage to the retina over a period of many years. They believe that macular degeneration, which is an important cause of blindness in older persons, is the result of lifelong exposure to the blue light portion or possibly the entire visible spectrum of ambient sunlight. Many of these researchers regularly wear amber or red-tinted glasses to reduce blue light exposure. Even though the link between macular degeneration and chronic blue light or visible exposure must still be considered hypothetical, the results of subacute experiments provide support for the theory.[19]

Near-infrared exposure and cataracts

Near-infrared radiation is capable of producing cataracts; such damage has been noted historically in glassblowers and furnace men. Radiation between 800 and 1200 nm is most likely responsible for temperature increases in the lens itself because of its spectral-absorption characteristics. Visible wavelengths may also contribute to the problem, since the heat absorbed by the iris could result in heat transfer to the lens.[20] Other structures of the eye, such as the cornea, absorb at longer wavelengths beyond 1200 nm and may also conduct thermal energy to the lens. Both mechanisms probably play a role in the relative importance of each being dependent on the wavelength characteristics of the exposure.[21]

Clinically, glassblowers' cataract has been described as a well-defined opacity in the outer layers of the axial posterior cortex of the lens, appearing as an irregular latticework with a cobweb appearance.[22] If exposure to IR between 700 and 800 nm or between 1200 and 1400 nm is a more significant factor, the cataracts are more likely to occur in the periphery of the lens.[23]

Acute skin, cornea, and iris injury

Both the skin and the cornea of the eye are opaque to wavelengths >1400 nm. IR radiation in this region produces injury through thermal mechanisms, with absorbed radiation being converted to heat. Injury to the cornea is described as a gray appearance detectable by slit lamp that is caused by energy just above the threshold for injury.[24] Larger amounts of energy can produce extensive opacification of the cornea or even more severe injury. Focused sources of energy can create localized burns to the skin that resemble those caused by other sources of heat. There is some transmission of energy into the skin for radiation between 750 and 1300 nm, with maximum transmission at 1100 nm. At this wavelength, 20% of the energy will reach a depth of 5 mm. The nature of the injury will still be thermal. IR radiation below 3000 nm will penetrate into different depths of the cornea to varying degrees, depending on the specific wavelength. The iris of the eye can absorb energy and play a role only at wavelengths below approximately 1300 nm.

Solar urticaria and drug-induced photosensitivity

Although photosensitivity *per se* is primarily due to ultraviolet radiation, solar urticaria is often the result of visible light radiation, while drug-induced photosensitivity may be caused by visible light in the blue region, depending on the action spectrum of the specific drug. Solar urticaria is manifested by urticaria lesions on sun- or light-exposed areas of the body. Typically, the reaction begins as reddened skin; mild to moderate itching develops rapidly into urticaria lesions with edema.[25] The lesions resolve over several hours. Different parts of the body have variable degrees of susceptibility. Typically, chronically sun-exposed areas such as the face and arms are more tolerant to light exposure.

Some patients react only to ultraviolet radiation in the 320–400 nm region, whereas others have an action spectrum of 400–500 nm. Still other patients have a broad-action spectrum of 280–600 nm. The mechanism of action is believed to be immunologic; in some cases, sensitivity can be transferred by a patient's serum. An antigen may be formed in the skin of susceptible individuals following exposure to light, leading to an antigen–antibody reaction that produces the urticaria.[26] In some individuals, a nonimmunologic mechanism may be present where light causes the production of a substance that causes the urticaria directly.

Drug-induced photosensitivity may be caused by exposure to ultraviolet radiation or visible light in the blue region, depending on the action spectrum of the particular substance. For example, the action spectrum for coal tar pitch is 340–430 nm and for dimethylchlorotetracycline, it is 350–450 nm. Both of these are examples where visible blue light as well as near UV can cause reactions. The clinical presentation may vary greatly. Lesions are usually on the light-exposed areas of the body, such as the face, the "V" of the neck, the back of the hands, and the extensor surfaces of the arms. Various degrees of redness, edema, and vesicle formation may occur. In chronic cases, scaling and lichenification may occur.

Either phototoxicity (most common) or photoallergy may cause drug-induced photosensitivity. The former can be photodynamic, requiring oxygen, or it can be oxygen independent. Phototoxicity is usually targeted at nuclear DNA or cell membranes. Photodynamic sensitizers interact with oxygen to form phototoxic compounds in the presence of UV or blue light. Oxygen-independent photosensitizers form toxic photoproducts even in the absence of oxygen. Photoallergy is the result of an immunologic response. The drug or chemical absorbs a photon of UV or blue light and is converted to a photoproduct that binds to a soluble or membrane protein to form an antigen.[27]

Porphyrias

While a number of conditions—such as systemic lupus erythematosus, atopic dermatitis, acne vulgaris, and herpes simplex—can be aggravated by exposure to UV radiation, the porphyrias are the result of blue light interaction with porphyrins produced by aberrations in the enzymatic control of heme synthesis. The action spectrum is most predominant between 400 and 410 nm. Photons in this narrow wavelength band cause porphyrins to go to an "excited" state. Reactions with oxygen lead to peroxide formation, which in turn damages vital components of cell membranes, leading to cell death. The porphyrias may be due to either hereditary or acquired abnormalities in heme synthesis. They are classified into hepatic or erythropoietic categories, depending on the site of excess porphyrin. Most of the porphyrias are due to autosomal dominant defects in the enzymes responsible for heme synthesis. Porphyria cutanea tarda is the most common form and photosensitivity is the major finding. The disease, which usually manifests itself in middle age, may be triggered by exposure to certain medications such as barbiturates, phenytoin, and tolbutamide.

TREATMENT

Thermal burns to the skin from visible or infrared radiation are treated like any thermally caused burn. If minor in nature, burns to the cornea are evaluated using fluorescein stain and slit lamp. Treatment focuses on the prevention of infection during healing. Typically it includes the use of cycloplegic agents and antibiotics in addition to patching of the eyes. Injuries to the deeper structures of the eye do not lend themselves well to specific treatment; they often result in permanent damage. Visual impairment depends on the extent and location of the injury.

Photosensitivity reactions can be treated with nonsteroidal anti-inflammatory agents to control fever and pain. Corticosteroids may be needed for severe reactions and can be used both topically and systemically.

MEDICAL SURVEILLANCE

Medical surveillance has not been generally recommended for individuals exposed to intense levels of visible or infrared radiation. Surveillance would not be appropriate for the acute effects of visible or IR radiation. Nor have specific subclinical effects been identified that would be useful for the surveillance of individuals with chronic exposures. Also, the magnitude of most occupational exposure is dwarfed by the contribution from ambient sunlight. Examination of the lens of the eye by slit lamp has shown a far greater prevalence of opacities of the lens in IR-exposed individuals than in controls.[26] However, it was not possible to demonstrate a dose–response in these studies, and the changes noted were indistinguishable from naturally occurring cataracts. Although this type of examination has been worthwhile in epidemiological study, it is doubtful that it would be useful for individual medical surveillance.

PREVENTION

Exposure to man-made sources of visible and infrared radiation can be prevented through engineering controls and protective equipment. Typical controls include barriers and reflectors or opaque shields to eliminate exposure to individuals. Viewing windows or ports can be equipped with glass or plastic with appropriate tinting materials to block the radiation. For visible light, neutral density filters are commonly used. When eye exposure is the major concern, tinted glasses, goggles, or face shields can be used.[28] Reflective suits can help reduce thermal loading from exposures to the entire body and prevent burns.

To prevent photosensitivity reactions, exposure to sources of bright light including sunlight should be minimized. Simple measures include wearing long-sleeved shirts and broad-brimmed hats and using canopies or awnings. Sunscreen agents also offer some protection from blue light photosensitivity reactions. Beta-carotene in doses of 60–80 mg/day can help to prevent photosensitivity reactions in persons with porphyria.

References

1. Sliney DH, Moss E, Miller CG, et al. Semitransparent curtains for control of optical radiation hazards. *Appl Optics* 1981; 20:2352–66.
2. Yost MG. Occupational health effects of nonionizing radiation. In: Shusterman DJ, Blanc PD, eds. *Occupational medicine: state of the art reviews*. Philadelphia: Hanley & Belfus, 1992:543–66.
3. Harber LC, Bickers DR, Kocherer I. Introduction to ultraviolet and visible radiation. In: *Photosensitivity diseases*. Philadelphia: WB Saunders, 1981:13–23.
4. Sliney DH, Freasier BC. Evaluation of optical radiation hazards. *Appl Optics* 1973; 12:1–23.
5. Okuno T, Ueno S, Kobayashi Y, et al. Blue light hazards associated with crystal glassware production. *Sangyo Eiseigaku Zasshi* 2013; 55:85–9.
6. Stamatacos C, Harison JL. The possible ocular hazards of LED dental illumination applications. *J Tenn Dent Assoc* 2013; 93:25–9.
7. Sliney DH, Stack C, Schnuelle D, et al. Optical safety of comparative theater projectors. *Health Phys* 2014; 106:353–64.
8. Goldman H. The genesis of the cataract of the glass blower. *Am J Ophthalmol* 1935; 18:590–1.
9. Wallace J, Sweetnam PM, Warner CG, et al. An epidemiologic study of lens opacities among steel workers. *Br J Ind Med* 1971; 28:265–71.
10. Wilkening GM. Nonionizing radiation. In: Clayton GD, Clayton FE, eds. *Patty's industrial hygiene and toxicology: vol 1, Pt B*. 4th ed. New York: John Wiley & Sons, Inc., 1991:657–742.
11. Sliney DH, Wolbarsht ML. *Safety with lasers and other optical sources*. New York: Plenum Press, 1980.
12. American Conference of Governmental Industrial Hygienists. *Threshold limit values*. Cincinnati: American Conference of Governmental Industrial Hygienists, 2015.
13. Sliney DH. Risks of occupational exposure to optical radiation. *Med Lav* 2006; 97:215–20.
14. International Commission on Non-Ionizing Radiation Protection. ICNIRP Guidelines on limits of exposure to incoherent visible and infrared radiation. *Health Phys* 2013; 105:74–96.
15. Okuno T. Hazards of solar blue light. *Appl Optics* 2008; 47:2988–92.
16. Ness, JW, Zwick, H, Stuck BE, et al. Retinal image motion during deliberate fixation: implications to laser safety for long duration viewing. *Health Phys* 2000, 78(2):131–42.
17. Ham WT, Mueller HA, Sliney DH. Retinal sensitivity to damage from short wavelength light. *Nature* 1976; 260:155–7.
18. Ham AT, Mueller HA, Williams RC, et al. Ocular hazards from viewing the sun unprotected through various windows and filters. *Appl Optics* 1973; 12:2122–9.
19. Mainster MA. Light and macular aging [Editorial]. *Lasers Light Ophthalmol* 1993; 5:117–9.

20. Goldman H. Genesis of the heat cataract. *Arch Ophthalmol* 1933; 9:314.
21. Lydahl E. Infrared radiation and cataract. *Acta Ophthalmol* 1984; 166(suppl):1–63.
22. Dunn KL. Cataracts from infrared rays (glass worker's cataracts). *Arch Ind Hyg Occup Med* 1950; 1:166–80.
23. Langley RK, Martimer CB, McCulloch C. The experimental production of cataracts by exposure to heat and light. *Arch Ophthalmol* 1960; 63:473–88.
24. Leibowitz HM, Peacock GR. Corneal injury: produced by carbon dioxide laser radiation. *Arch Ophthalmol* 1969; 81:713–21.
25. Botcherby PK, Gianelli F, Magnus I, et al. UV-A induced damage in skin cells from actinic reticuloid and normal individuals. In: Cronly-Dillon J, Rosen ES, Marshall J, eds. *Hazards of light.* Oxford: Pergamon Press, 1973:95–9.
26. Horio T, Minami K. Solar urticaria: photoallergen in a patient's serum. *Arch Dermatol* 1977; 113:157–60.
27. Harber LC, Bickers DR. Drug-induced photosensitivity. In: *Photosensitivity diseases.* Philadelphia: XVB Saunders, 1981:120–53.
28. American National Standards Institute. *Standard for occupational and educational eye and face protection.* ANSI Z 87.1. Washington, DC: American National Standards Institute, 1991.

Chapter 14

LASER RADIATION

David H. Sliney and James A. Hathaway

Lasers are devices that produce an intense, coherent, directional beam of light by stimulating electronic or molecular transitions to lower energy levels.[1] The beam of radiation emitted by lasers in common use may have a wavelength anywhere from the ultraviolet (UV) region of the electromagnetic (EM) spectrum to the far-infrared (FIR) region. This includes numerous lasers operating in the visible light portion of the EM spectrum. Lasers vary widely in the intensity of their outputs; they may generate brief bursts or pulses of energy or operate continuously. The potential hazard of laser radiation depends on all of these factors.

OCCUPATIONAL SETTING

The use of lasers in industry, construction, research, medicine, and the military is widespread and increasing. Lasers are used in alignment, welding, trimming, spectrophotometry, range-finding, interferometry, flash photolysis, fiber-optic communication systems, and surgical removal or repair procedures.[2,3] Low-power lasers are also widely used in commercial activities and consumer applications, including supermarket checkout counters, detection of motor vehicle speed, as pointers for presentations, in CD-ROM drives for computers, and in CD, DVD, and laser disk players for home entertainment. Specific occupational titles may not be particularly helpful in identifying where lasers may be used. In industries using high-technology processes, various craftsmen, operators, and service workers may be expected to use lasers.[4] Lasers are used for range-finding in advanced weapon systems by military personnel. Maintenance personnel may actually be at higher risk of accidental exposure than the operators because they may need to remove protective shielding and interlocks to repair the equipment. Similarly, the nature of laboratory research often precludes the use of engineering safeguards and may increase the risk of accidental exposure.[2] Medical uses usually require lasers with sufficient power to damage tissue. Accidental exposures have the potential to injure operating room personnel as well as patients.

MEASUREMENT ISSUES AND CLASSIFICATION OF LASER POWER

In general, measurements of laser radiation are not necessary. The laser classification scheme described in the following paragraph was designed to minimize the need for measurements. It is the responsibility of laser manufacturers to perform measurements and classify their products. The classification system allows the user to determine potential risks and provide for the necessary safeguards, procedures, and personal protective equipment. Measurements are required only when information from a manufacturer is not available or when a laser system has been modified. Detailed information on measurement can be found in Section 9 of the ANSI Z136.1 standard[1] on the safe use of lasers. Appendix H4 of the same document provides a listing of catalogs on commercially available laser-measuring instruments.

The primary hazard from laser radiation is from exposure to the eye and, to a lesser extent, the skin. Therefore, the classification is based on the laser's capability of injuring the eye or the skin. Lasers manufactured in the United States are classified in accordance with the Federal Laser Product Performance Standard.[5,6] The actual process is somewhat complex, because numerous types of lasers have

Physical and Biological Hazards of the Workplace, Third Edition. Edited by Gregg M. Stave and Peter H. Wald.

been developed that operate at different wavelengths. The threshold for biological injury varies with the wavelength of radiation. It is also dependent on the operating conditions of the laser—that is, on whether the radiation is continuous or pulsed. If it is pulsed, the duration and repetition rate of the pulse must also be considered. Details of the classification scheme are described in ANSI Z136.1. The following outline provides a somewhat simplified view of the classification scheme:

- Class 1 laser—Will not produce injury even if the direct beam is looked at for the maximum possible duration inherent in the design of the laser. For many lasers, this essentially amounts to an unlimited viewing time.
- Class 1M laser—Will not produce injury except potentially when viewed by binoculars or telescopes from within the beam.
- Class 2 laser—Will not produce injury if the direct beam is viewed for 0.25 second, the time period necessary for a protective aversion response. Class 2 lasers are limited to lasers emitting visible light on a continuous basis.
- Class 2M laser—Is equivalent to class 2 unless viewed from within the beam with a telescope.
- Class 3 laser—Can produce eye damage if the direct beam is viewed, but would almost never pose a risk to the skin. This classification is subdivided into classes 3R and 3B. Class 3R, which is limited to the lower accessible outputs of this class, is believed to present less risk of actual injury from a practical standpoint. Class 3B represents those class 3 lasers with higher outputs where the risk of real ocular injury from even momentary viewing of the direct beam is high.
- Class 4 laser—Even the diffuse reflection of some pulsed lasers with this level of power output can produce biological damage to the eye. The direct laser beam can injure the skin or pose a fire hazard.

Control measures apply primarily to lasers in class 3B or 4. Limited precautions such as product labeling apply to classes 2 and 3R lasers.

EXPOSURE GUIDELINES

Exposure guidelines have been developed by the American Conference of Governmental Industrial Hygienists (ACGIH) and by the American National Standards Institute (ANSI) Z136 Committee on the Safe Use of Lasers.[1,7] Both of these organizations have issued guidelines for safe laser use. The Occupational Safety and Health Administration (OSHA) does not specifically regulate laser radiation, although the ANSI standard would be consulted in cases where the OSHA "general duty" clause is applied.

The ACGIH standards are called threshold limit values (TLVs®). They vary, depending on the wavelength of the laser radiation and depending on whether the radiation is pulsed or continuous. Certain assumptions are also made regarding aversion time (0.25 second) and the size of the pupil under various exposure conditions. ACGIH Tables 2 and 3 list the TLVs® for either eye or skin exposure by wavelength and exposure time. The output of pulsed lasers is described in terms of energy (joules), while the output from continuous wave lasers is described in terms of power (watts). The TLVs® are expressed as radiant exposure in joules per square centimeter (J/cm^2) or as irradiance in watts per square centimeter (W/cm^2). The ANSI Z136 committee has labeled their exposure limits as maximum permissible exposures (MPEs). The MPEs for various conditions and types of lasers are listed according to wavelength in Table 5-7 of the ANSI Z136.1 standard.[1] The complexity of these tables and those of the ACGIH preclude them from being summarized here.

PATHOPHYSIOLOGY OF INJURY

Research on the biological effects of laser radiation has been directed toward determination of the thresholds for tissue damage. The threshold for identifying damage has typically been grossly apparent findings or findings observable using instruments such as microscopes, slit lamps, and ophthalmoscopes. The ANSI Z136 committee used these data to determine exposure levels that produce damage 50% of the time. This would be analogous to an ED_{50}, or a dose that produces an adverse effect 50% of the time in experimental animals exposed to a chemical. A factor of 10 below the 50% damage level was then typically used to arrive at the MPE level, where the probability of damage was negligible. Actual regression lines were used to determine the slope of the dose–response curve where possible; when this slope was very steep, a factor <10 was used.[8] The principal biological hazards associated with laser radiation occur with acute short-term or intermittent exposures. Chronic effects are theoretically possible based on results of exposures to experimental animals or based on analogy to the chronic effects produced by ambient or artificial sources of UV, visible, or IR radiation. Chronic exposure to laser radiation of sufficient power to be of concern is rare in occupational settings, because laser beams have very limited spatial extent. Therefore, we will focus on the acute biological effects of laser radiation.

Corneal damage from the infrared region (1400 nm to 1.0 mm)

Depending on the power level of the laser, tissue damage to the cornea from acute IR laser radiation can range from a minimal lesion involving only the epithelium, which appears as a small white area, to massive destruction of the cornea with severe burns to adjacent structures of the eye such as the conjunctiva and lids. Damage results from absorption of energy by tears and tissue water in the cornea. The heat is diffusely absorbed; a simple heat flow model is believed to explain the observed effects adequately.[1] Minor damage may heal completely within 48 hours; more severe damage will have permanent sequelae.

Corneal damage from the ultraviolet region (100–400 nm)

Biological damage from UV laser radiation is similar to that caused by other artificial or ambient sources of UV. Corneal effects following acute or subacute exposures include epithelial stippling, granules, haze, debris, exfoliation, and stromal haze and opacities.[8] Clinical symptoms and findings may include photophobia, tearing, conjunctival discharge, and redness. The damage caused by UV radiation is not due to heating effects but rather to photochemical denaturation of proteins and other macromolecules, such as DNA and RNA. Thermal damage can occur from pulsed lasers at some wavelengths necessitating dual exposure limits.

Retinal damage from the visible and near-infrared region (400–1400 nm)

The cornea, lens, and ocular media are mostly transparent to visible light in the 400–700-nm wavelength range. Nearly all of the visible energy reaches the retina. Near-IR radiation in the 700–1400-nm range also reaches the retina in significant amounts and produces damage similar to that caused by visible radiation. Clinically, the minimal lesion is a small white patch apparently caused by the coagulation of protein. It may be asymptomatic. It is visible within 24 hours of exposure. More significant exposures may produce immediate symptoms, such as loss of vision in the visual fields, spots (scotomata) in the field of vision, or persistent afterimages.[2] More severe exposure can cause substantial damage, including significant hemorrhage from the retina into the vitreous humor.

Laser radiation in the visible and near-IR can cause damage by a variety of mechanisms, depending on the type of laser. Damage has been attributed to thermal, thermoacoustic, and photochemical phenomena.[3] Lasers with short-pulse durations of $<10^{-9}$ seconds may cause "blast" damage through nonlinear mechanisms such as ultrasonic resonance and acoustic shock waves.[9] Most of the radiation is absorbed in the melanin granules of the retina in the retinal pigment epithelium and choroid. This structure underlies the cones and rods. When damage is caused by a thermal mechanism, it is due to protein denaturation. Damage caused by heat or photochemical mechanisms is similar to what would be expected from equally intense doses of noncoherent light.

Examination of an injured individual typically reveals a blind spot (scotoma) or spots in one or both eyes. Visual acuity may or may not be decreased, depending on the proximity of the injury to the macula. Obviously, injuries to or near the macula produce greater functional loss than injuries in the periphery of the retina. Fundoscopic examination may show retinal or subretinal hemorrhages and hemorrhage into the vitreous. More minor injuries may not be immediately obvious on fundoscopy, or they may present as minor retinal burns with edema. Healing takes place over a course of weeks. Some improvement in visual acuity may occur as the edema subsides. Generally, a blind spot remains in the visual field. The extent of functional loss depends on the size and location of the injury.

There are two reports on series of patients injured from exposure primarily to Nd:YAG lasers operating at 1064 nm.[10,11] In one report, 8 of 12 patients had macular lesions. Visual loss ranged from minimal to severe. There was no improvement in vision over time, in spite of vasoprotective and corticosteroid treatments. In two cases, the extent of the injury was progressive. In one case, there was hemorrhage in the vitreous humor. In the other report, 25 of 31 eye injuries resulted in macular damage. Macular damage was progressive over a 1-week time period in seven cases. Ten eye injuries were followed for 4–10 years. The extent of injury remained stable over this time for nine of the eye injuries. Another report indicates that the presence of a hemorrhage, which may initially severely impair vision, does not preclude the possibility of a return to normal vision. When a short-pulse laser produces a microscopic retinal hole, recovery can be remarkable.[12]

Other ocular damage

Radiation in the near-UV zone and radiation in the zone between near-IR and IR have absorption characteristics such that significant levels of energy may be absorbed in structures of the eye between the cornea and the retina, including the lens and the iris. Acute damage to these structures would be expected from very high-energy lasers (e.g., at 1315 nm, iodine laser wavelength). Concurrent damage to the lens or the retina (depending on the wavelength) would also be expected. Chronic exposure to noncoherent sources of UV or IR radiation in these wavelength regions causes lenticular damage leading to cataracts. Theoretically, the same damage could be incurred from coherent laser radiation at similar wavelengths. In actual occupational settings, chronic exposure of unprotected workers is unlikely. Chronic effects on the lens have not been studied.

Skin damage

Laser radiation can cause injury to the skin. Higher levels of energy are required to produce skin damage than for eye injury. The focusing power of the cornea and lens of the eye increases the energy density reaching the retina, thus allowing lower levels of total energy to produce localized injuries. However, UV lasers can cause photochemical damage to the skin similar to acute sunburn. Visible and IR lasers can produce thermal burns from acute exposures. The power output of the laser determines whether accidental exposure to the skin produces a minor injury or a more severe one. Theoretically, chronic exposure to UV lasers would have the same risk of causing premature aging of the skin and increase the risk of skin cancer; however, exposure conditions that would result in chronic exposures are unlikely, given the current uses of lasers.

TREATMENT

Individuals with suspected injuries to the retina should be referred to an ophthalmologist. In many cases, no treatment is required, but continued follow-up is important to evaluate functional loss (both visual acuity and blind spots in visual fields). Complications such as growth of new blood vessels in the vicinity of the injury may require treatment. More severe retinal injuries from very high-power lasers require immediate evaluation by an ophthalmologist.

Minor UV injuries to the cornea can be treated in the same way as photokeratitis from other sources of UV radiation. Patching the eye and using cycloplegics are recommended. Anesthetic drops should not be used. Complete recovery takes about 48 hours. More severe corneal injuries from either UV or IR radiation require specialized treatment from an ophthalmologist.

The following precautions should be taken when dealing with eye injuries that require referral to an ophthalmologist:[13]

- Eye ointments should never be used because they make clear visualization of the retina very difficult.
- Topical anesthetics should not be used to relieve pain from a UV injury.
- Prolonged use of these anesthetics can cause corneal breakdown and lead to blindness.
- Topical steroids should never be used unless prescribed by an ophthalmologist.
- If in doubt about the seriousness of an injury, err on the side of caution and refer the patient to an ophthalmologist.
- Keep in mind that some suspected laser-induced ocular injuries may not actually originate from laser exposure.[14]

Megadose intravenous methylprednisolone has been used in studies with cynomolgus monkeys to determine if it might improve healing of retinal laser burns caused by visible or near-IR laser radiation. An overall beneficial effect was noted. The authors indicated that the effect might be ascribed to the anti-inflammatory action, protection of microcirculation, and antilipid peroxidation effects.[15]

Skin injuries from visible or IR radiation can be treated in the same way as localized thermal burns. Intramuscular vitamin E and/or use of vitamin E and an occlusive dressing have been reported to improve wound healing in miniature swine following exposure to IR lasers.[16] UV radiation can produce a localized injury equivalent to sunburn; it should be treated accordingly.

MEDICAL SURVEILLANCE

At one time, medical surveillance requirements were included in the ANSI Z136.1 standard.[1] They were required only for individuals working with class 3B or 4 lasers. Medical surveillance was never required for use of class 1, 2, or 3R lasers. Examinations were required before work with lasers and after suspected injuries. No periodic examinations were required. At this time the ANSI Z136.1 standard requires no medical surveillance.

Although some wavelengths of laser radiation have the theoretical potential to produce chronic effects, the nature of their use makes this unlikely. A few studies have looked for possible chronic effects in laser workers, but no evidence of adverse chronic effects has been found.[17–20]

At this time, the preplacement evaluation of laser workers may be used to establish a baseline against which the effects of accidental injury could be compared. Some employers or institutions may require this for potential legal protection against claims for preexisting damage to parts of the eye.

Medical examinations should be required following accidental exposures to the eye or skin. It is recommended that an ophthalmologist conduct the exam if there are injuries to the retina.

PREVENTION

Protecting the skin, and particularly the eye, from high-power laser radiation is critical because permanent damage—including blindness—can result. There are a number of excellent references on laser safety that can be consulted. Other references provide considerably more details on preventive measures than are appropriate here.[1,21–26]

In their review of reported accidental exposures to laser radiation, the ANSI Z136 committee specifically noted the following important causes of the incidents:[1]

- Unanticipated eye exposure during alignment
- Available eye protection not used
- Equipment malfunction
- Intentional exposure of unprotected persons
- Operators unfamiliar with laser equipment
- Improper restoration of equipment following service

The ANSI committee also noted that several serious accidents were traceable to ancillary hazards such as electric shock, toxic gas exposure, and vaporized tissue exposure from medical procedures. These topics are covered in several reports.[1,22,25,27]

Obviously, the preferred method of prevention is to incorporate engineering control measures that limit access to laser radiation. Indeed, many laser systems are designed to embed more powerful class 3B and 4 lasers within shields or enclosures. This safeguard eliminates the risk of accidental operator exposure, but it does not eliminate the risk to persons servicing the equipment. For many applications of lasers, however, it is not feasible to rely on enclosure; other methods of engineering control must be used, along with training, administrative procedures, personal protective equipment, and warning systems.

The ANSI Z136.1 standard requires the appointment of a laser safety officer (LSO) to monitor and enforce the control of laser hazards. This task may involve training requirements, administrative procedures, standard operating procedures, and selection of engineering control measures.

Depending on how the laser is used, numerous engineering control measures may be necessary. Examples include protective housing, interlocks on protective housings, interlocked service access panels, master switches that are disabled when the laser is out of use, interlocks, filters or attenuators for viewing portals and display screens and collecting optics, enclosed beam paths, remote interlock connectors, beam stops or attenuators, emission delay systems, and remote firing and monitoring.

Under certain circumstances, some of these engineering controls may not be feasible and alternative methods will be necessary. One important control measure is the establishment of what is called a laser controlled area. Access to the area is limited to personnel who have been specially trained. These workers must have appropriate protective equipment, and they must follow all applicable administrative and procedural controls. Controlled areas need to be posted with warning signs. They must have limited access, be operated by qualified and authorized personnel, and be under the supervision of someone specially trained in laser safety. They should use beam stops of appropriate material, diffuse reflecting materials where feasible, and appropriate eye or skin protection. Furthermore, these areas limit the beam path to above or below eye level except as required for medical use; they eliminate the possibility of transmission of laser radiation through doors, windows, and they include a system that can disable the laser to prevent unauthorized use. Class 4 laser controlled areas require safety controls to allow rapid egress, emergency alarms, nondefeatable area/entry controls where feasible, and other controls for particular operations. Inherent in the controlled area concept is the need for rigorous compliance with training, administrative, and procedural requirements. Protective equipment is mandatory whenever it is needed.

Industrial employers have generally complied well with the ANSI Z136.1 consensus standard. This has not always been the case with research laboratories, particularly those in university settings. One article described a number of injuries in university research laboratories where persons using lasers had not received proper training and appropriate protective eyewear was not used.[2] The authors proposed a registration system for research lasers that would ensure that laser personnel receive proper training and that appropriate protective equipment is available. In recent years, a number of other application-specific safety standards were developed in the ANSI Z136.1 series. These include ANSI Z136.5 (educational institutions), ANSI Z136.6 (outdoor use), ANSI Z136.8 (research, development, and testing), and ANSI Z136.9 (manufacturing environments).

References

1. Laser Institute of America. *American national standard for the safe use of lasers*. ANSI Z 136.1. Orlando, FL: Laser Institute of America, 2014.
2. Barbanel CS, Ducatman AM, Garston MJ, et al. Laser hazards in research laboratories. *J Occup Med* 1993; 35:369–74.
3. Wilkening GM. Nonionizing radiation. In: Clayton GD, Clayton FE, eds. *Patty's industrial hygiene and toxicology, Vol. 1, Part B*, 4th edn. New York: John Wiley & Sons, Inc., 1991:657–742.
4. US Department of Labor. *Guidelines for laser safety and hazard assessment*. OSHA instruction PUB 8-1.7. Washington, DC: US Government Printing Office, 1991.
5. Code of Federal Regulations, Title 21, Subchapter J, Part 1040. Laser product performance standard. Washington, DC: US Government Printing Office, 1999.
6. Code of Federal Regulations, Title 21, Parts 1000 and 1040. Laser products: amendments to performance standard. Washington, DC: US Government Printing Office, 2015.
7. American Conference of Governmental Industrial Hygienists. *Threshold limit values, 2015*. Cincinnati, OH: American Conference of Governmental Industrial Hygienists, 2015.
8. Sliney DH, Mellerio J, Gabel VP, et al. What is the meaning of threshold in laser injury experiments? Implications for human exposure limits. *Health Phys* 2002; 82(3):335–47.
9. Ham WT, Williams RC, Mueller HA, et al. Effects of laser radiation on the mammalian eye. *Trans N Y Acad Sci* 1966; 28:517–26.
10. Pariselle J, Sastourne JC, Bidaux F, et al. Eye injuries caused by lasers in military and industrial environment. *J Fr Ophtalmol* 1998; 21:661–9.
11. Lui HF, Gao GH, Wu DC, et al. Ocular injuries from accidental laser injuries. *Health Phys* 1989; 56:711–6.
12. Hirsch DR, Booth DG, Schockett S, et al. Recovery from pulsed dye laser retinal injury. *Arch Ophthalmol* 1992; 110:6188.
13. Vinger PF, Sliney DH. Eye disorders. In: Levy BS, Wegman DH, eds. *Occupational health*, 2nd edn. Boston, MA: Little, Brown, 1988:387–97.

14. Mainster MA, Sliney DH, Marshall J, et al. But is it really light damage? *Ophthalmology* 1997; 104:179–80.
15. Takahashi K, Lam TT, Tso MO. The effect of high dose methylprednisolone on laser-induced retinal injury in primates: an electron microscopic study. *Graefes Arch Clin Exp Ophthalmol* 1997; 253:723–32.
16. Simon GA, Scmid P, Reifenrath WG, et al. Wound healing after laser injury to skin—the effect of occlusion and vitamin E. *J Pharm Sci* 1994; 83:1101–6.
17. Wolbarsht WL, Sliney DH. Historical development of the ANSI laser safety standard. *J Laser Appl* 1991; 3:5–11.
18. Hathaway JA, Stein N, Soles EM, et al. Ocular medical surveillance on microwave and laser workers. *J Occup Med* 1977; 19:683–8.
19. Friedman Al. The ophthalmic screening of laser workers. *Ann Occup Hyg* 1978; 21:277–9.
20. Hathaway JA. The needs for medical surveillance of laser and microwave workers. In: Tengroth B, ed. *Current concepts in ergophthalmology*. Stockholm: Societas Ergophthalmologica Internationalis, 1978:139–60.
21. Laser Institute of America. *Laser safety guide*. Cincinnati, OH: Laser Institute of America, 2015.
22. Sliney DH, Wolbarsht ML. *Safety with lasers and other optical sources: a comprehensive handbook*. New York: Plenum Press, 1980.
23. Sliney DH, LeBodo H. Laser eye protectors. *J Laser Appl* 1990; 2:9–13.
24. Laser Institute of America. *American national standard for the safe use of optical fiber communication systems utilizing laser diode and LEP sources*. ANSI Z 136.2. Orlando, FL: Laser Institute of America, 2014.
25. Laser Institute of America. *American national standard for the safe use of lasers in health care facilities*. ANSI Z 136.3. Cincinnati, OH: Laser Institute of America, 2005.
26. Thach AB. Laser injuries of the eye. *Int Ophthalmol Clin* 1999; 39:13–27.
27. Sliney DH, Clapham T. Safety of medical excimer laser with an emphasis on compressed gases. *Ophthalmol Technol* 1991; 1423:157–62.

Chapter 15

MICROWAVE, RADIOFREQUENCY, and EXTREMELY LOW-FREQUENCY ENERGY

RICHARD COHEN AND PETER H. WALD

Microwaves (MW) include that portion of the electromagnetic spectrum between 300 megahertz (MHz) and 300 gigahertz (GHz). Radiofrequency (RF) radiation comprises that portion of the electromagnetic energy spectrum in which wave frequency varies from 3 kilohertz (kHz) to 300 MHz. Extremely low-frequency (ELF) radiation includes frequencies <3 kHz; it commonly refers to radiation associated with electric power generation and transmission. This chapter will address MW, RF, and ELF radiation.

MICROWAVE AND RADIOFREQUENCY RADIATION

Occupational setting

The four types of devices that generate RF and MW energy are power grid tubes, linear beam tubes (klystrons), crossed-field devices, and solid-state devices. Sources of RF and MW energy can operate in three modes: continuous, intermittent, and pulsed. The continuous mode is used in some communication devices, the intermittent mode is used in heating devices, while the pulsed mode is used in radar and digital communication.

MW energy can be transmitted from the generating device through a wave guide or through a transmission line to an applicator or antenna. Microwaves are used to transmit signals in telecommunications, navigation, radar, and broadcasting (i.e., radio and television); they can also be used to produce heat in industrial and home microwave ovens and dielectric heaters (i.e., heaters used to heat electrically nonconductive materials by means of a rapidly alternating electromagnetic field).

Cellular telephones operate at frequencies between 800 and 2200 MHz. Home microwave ovens use a microwave frequency of 2.45 GHz. Dielectric heaters are used in the manufacture of automobiles, furniture, glass fiber, paper products, rubber products, and textiles. RF dielectric heater applications include sealing and molding plastics; drying glues after manufacturing; drying textiles, paper, plastic, and leather; and curing materials such as epoxy resins, polymers, and rubber. Video display terminals (VDT) can generate RF radiation (at ~10–30 kHz) because they have a cathode ray tube, which is a source of electrons (measured levels have been extremely low). Industrial welding also generates RF radiation, typically ~400 kHz. RF radiation is also used for diathermy (deep-tissue heating) applications in medical treatment.

Measurement issues

An electromagnetic wave results from the combination of electric and magnetic field vectors, each perpendicular to the other, that produce a wave that is propagated perpendicular to the first two vectors (see Chapter 1, Figure 1.1). The power density or energy of the wave is derived from the measured intensity of the electric and magnetic field vectors. The total energy of the wave is expressed in milliwatts per square centimeter (mW/cm^2). The individual field strengths can be measured; the electric field strength is measured in volts per meter (V/m), and the magnetic field strength is measured in amperes per meter (A/m). Whole body absorption is also measured/estimated as specific absorption rate (SAR) and expressed as watts per kilogram (W/kg).

Physical and Biological Hazards of the Workplace, Third Edition. Edited by Gregg M. Stave and Peter H. Wald.

Measurements are most often made using meters with frequency ranges from 2 kHz to 40 GHz. These meters yield point/spot measurements of the strength of either the electric or the magnetic fields. From there, the power density in mW/cm^2 is calculated. The instruments do not directly measure power density but have sensing probes that measure voltages or currents. These are usually displayed in V/m or A/m, which are then converted to power density as W/m^2 or mW/cm^2.

At frequencies above 300 MHz, the primary measurement is the electric field. Below 300 MHz, separate electric and magnetic field measurements must be made and combined. The reasoning behind this methodology is complicated, but it is related to whether the measurement is taken in the "near" field or the "far" field. In the far field (greater than one wavelength from the emitter), the ratio between the magnetic and electric fields is constant. When measured in the near field (less than one wavelength from the emitter), the ratio between the electric and magnetic field varies and both need to be measured. At 300 MHz, the corresponding wavelength is 1 m. As the frequency decreases, the wavelength (and the length of the near field) increases. This increases the chance that measurements are made in the near field. Personal dosimeters are also available to measure exposures.

Exposure guidelines

The exposure standards for RF and MW radiation are based on the assumption that the primary way that energy is absorbed for these frequencies is by heat deposition. Power absorption and heat deposition are affected by the following factors:

- **Frequency** of the radiation
- Body **position** relative to wave direction
- **Distance** between body (target) and source (MW energy generally decreases with the inverse of the square of the distance from the energy source)
- Exposure **environment** (surrounding objects may reflect, resonate, or modify incident waves)
- **Electrical properties** of the tissue (conductivity and dielectric constant).

The dielectric constant is a measure of the "permittivity" of the tissue; it measures the ratio of the amount of electric current that will flow in a specific medium versus the amount that will flow in a vacuum. Tissue electrical properties are constant and depend on water content. Higher energy absorption occurs in tissues with higher water (higher dielectric constant) contents, such as brain, muscle, and skin; lower energy absorption occurs in tissues with lower water (lower dielectric constants) contents, such as bone and fat.

The boundaries between tissues can reflect the energy waves differently, resulting in "hot spots" that can cause localized injury. Changes in position can change the amount of energy absorbed, because the body acts as an antenna to receive the MW or RF energy.

The Occupational Safety and Health Administration (OSHA)'s Nonionizing radiation regulation (29 CFR 1910.97)[1] specifies an exposure limit of 10 mW/cm^2 over a 6-min period (or longer) for frequencies between 10 MHz and 100 GHz.[2] There is also a consensus ACGIH standard (Table 15.1) that applies to a much broader frequency range. The ACGIH standards are derived on the basis of an SAR equivalent to 4 W/kg to which a safety factor of 10 is applied. This standard is based on data that show no effects from exposures at 4 W/kg over the 6-min period. Accordingly, the resulting standard is based on an SAR of 0.4 W/kg. This relates to comparison energy values of 1 W/kg for an individual at rest and 5 W/kg when the person is exercising.

Because it is difficult to measure SAR directly, the standard is expressed in measurable quantities such as power density in mW/cm^2 or field strengths in V/m (or V^2/m^2) or A/m (or A^2/m^2). These power densities and field strengths represent allowable exposures that do not exceed the SAR. The standard, in effect, allows an equivalent power density of 1 mW/cm^2 for frequencies

TABLE 15.1 Radiofrequency and microwave radiation standards.

Agency	Frequency	Power density, S (mW/cm^2)	Electric field strength, E (V/m)	Magnetic field strength, H (A/m)	Averaging time E^2, H^2, or S (minutes)
ACGIH	30–100 kHz		1842	163	6
	100 kHz–1 MHz		1842	$16.3/f$	6
	1–30 MHz		$1842/f$	$16.3/f$	6
	30–100 MHz		61.4	$16.3/f$	6
	100–300 MHz	10	61.4	0.163	6
	300 MHz–3 GHz	$f/30$			6
	3–30 GHz	10			$33\,878.2/f^{1.079}$
	30–300 GHz	10			$67.62/f^{0.476}$
OSHA	10 MHz–100 GHz	10			

f = frequency in MHz, V/m = volt/meter, A/m = ampere/meter, mW/cm^2 = milliwatts/square centimeter, ACGIH = American Conference of Governmental Industrial Hygienists, OSHA = Occupational Safety and Health Administration.

between 100 and 300 MHz. The exposures are averaged over 6 min; and the electric and magnetic fields must be measured separately below 300 MHz. Exposures at frequencies from 300 MHz to 3 GHz are calculated according to the formula $P=f/300$ (allowable power density in mW/cm^2 equals the frequency in MHz divided by 300).

The penetration of energy is a function of frequency with penetration depth of ~1.7 cm at 2.45 GHz in comparison with 2–4 cm of penetration at 0.915 GHz. Although there is greater penetration depth at lower frequencies, the resulting heating decreases with frequency. Below 10 MHz, the body is essentially transparent to RF radiation, and little heating takes place. The resonance frequency, or frequency that generates the greatest energy deposition, occurs at ~70 MHz.

Pathophysiology of injury

Biological tissues respond to MW and RF radiation exposure with the induction of their own electric and magnetic fields. Depending on the polarity of the biological molecules, rotation and agitation of molecules can occur, resulting in heat generation. Thermal injury can occur as a result of RF/MW exposure when exposures are in excess of 10 mW/cm^2. Because internal hot spots may result from internal resonance due to differences in dielectric properties or radiation reflection, there may be localized increases in energy absorption and heating. The net heating of the body is related to the amount of energy absorbed minus the amount lost through the usual heat-dissipating mechanisms (blood flow, evaporation, radiation, convection, and conduction). The phenomenon of MW clicking ("hearing" MW radiation), originally thought to be a nonthermal response, appears to be caused by thermal elastic expansion and contraction of the cochlea.

The least likely tissues and organs to be affected are those with greater thermal regulatory ability, usually due to increased blood flow and greater heat-dissipation potential. RF burns, which are the most frequently encountered industrial effects, usually involve the skin. Subcutaneous tissue heating usually occurs simultaneously with skin exposure. Where full-thickness skin burns occur and subcutaneous tissues are involved, healing may be prolonged due to the lack of base granulation tissue. Skin burns appear similar to sunburn. The patient may initially present with a feeling of warmth, as if the skin or exposed portion were being heated. Within hours, redness and slight induration can occur. The course is usually characteristic of any thermal burn, but it may include vesiculation and ulceration. Similarly, thermogenic exposures have been associated with cataract formation in exposed workers, but cataracts have not occurred following exposures below recommended limits. There are case reports of massive exposures to MW and other RF sources that resulted not only in eye or skin burns but also in symptoms of neurasthenia or posttraumatic stress-like disorders (recurrent headache, malaise, fatigue, depression). Hypertension and/or peripheral neuropathy has also been reported.[2,3]

The proliferation of cellular telephones and their potential MW emission when held within a few centimeters from the brain have resulted in concern and investigation of possible health effects. Cell phones are now used by over 90% of American adults. The conclusion of Corle's (2012) review and analysis of studies of cell phone use and glioma risk summarizes the current knowledge relating to cell phones and brain cancer risk:

> Despite the results pointing to an association in one direction or another, it is clear that there is no definite answer to the question of whether cell phone use is associated with increased brain cancer risk. Notwithstanding the inconsistencies in the epidemiological studies, a few of the human studies do suggest an association between cell phone use and brain tumors for a 10 year or greater induction period and/or a high number of cumulative call hours.[4]

There have been anecdotal reports of nonthermal effects from MW and RF exposures. These include carcinogenic, reproductive, hematopoietic, immunologic, neurologic, neuroendocrine, and psychological effects. These bioeffects have been reviewed extensively, but a specific nonthermal mechanism has not been identified. Because of inconsistent and conflicting animal and human data, genotoxicity, carcinogenicity, reproductive toxicity, and other systemic/organ effects have not been clearly linked to nonthermal exposures.[5,6] Similarly, extensive research of reproductive effects in relation to VDT use has not found an association.[7] However, continued positive findings in some cancer studies have led IARC to designate radio frequency radiation as a category "2-B" carcinogen. Specifically, IARC concluded: "Radiofrequency electromagnetic fields are *possibly carcinogenic to humans* (Group 2B)."[8]

Treatment

The medical response to MW/RF radiation exposure should involve (i) removal from exposure, (ii) determination of radiation frequency and exposure intensity, and (iii) medical treatment for thermal injury to skin or subcutaneous tissues. High-intensity exposures can lead to deep-tissue injury. Localized subcutaneous hot spots and deeper-penetration heating may make such thermal injury more difficult to evaluate. There have been reports of burning of the skin with undamaged subcutaneous fat and burned muscle tissue below the fat layer. If a high-intensity exposure is suspected, tests for deeper-tissue injury, such as creatine phosphokinase (CPK) to evaluate muscle injury, can be performed. Tests of specific organ function can also be ordered if injury is suspected. Routine burn management can be followed for superficial burns. Deeper or more serious burns should be referred to a burn specialist for specific medical treatment and follow-up.

Medical surveillance

Because no effects have been consistently demonstrated following long-term low-intensity exposures, periodic examination would not yield findings that would indicate a need for preventive actions. Given the inconsistencies and lack of scientific consensus regarding nonthermal effects, there is no basis for any periodic monitoring. Following an acute high-intensity exposure that results in thermal injury, appropriate follow-up should be instituted. Other than that predicted based on the thermal tissue effects, sequelae to that injury would not be expected.[9]

Prevention

Identification of RF/MW exposure in excess of recommended levels should be accomplished using available instrumentation that creates a plot of the potential fields and intensities. These measurements should occur at the time of initial equipment use and following any equipment changes thereafter.

Engineering controls include partial enclosure and elimination of leakage. Enclosing an area with wire mesh and sealing the seams with copper tape is a common engineering measure. Care must be taken to ensure that enclosures do not allow leakage. Where enclosures are not sufficient to reduce potential exposures, identification of the distance necessary for adequate energy dissipation can be effective. For example, hazard zones can be clearly marked surrounding the MW source. Although some personal protective equipment, such as eyewear and clothing, has been developed, its effectiveness is controversial, and it is not usually recommended.

There are no specific pregnancy-related recommendations for RF and MW exposures. The current recommendation of a SAR of 0.4 V/kg limits exposures to levels below those that cause significant thermal effects. Studies of the reproductive effects of MW and RF radiation have shown fetal loss and teratogenesis postexposure. However, most positive studies used exposures in the >100-V/kg range and were associated with internal temperature increases of 2–10°C.[10]

Heat has adverse effects on the testis and can also cause decreased spermatogenesis.[11] Exposures incapable of thermal effects are not considered reproductive hazards.

EXTREMELY LOW-FREQUENCY ELECTROMAGNETIC RADIATION: MAGNETIC FIELDS

Occupational setting

ELF energy is generated from electric power transmission and most household appliances. Workers with potential ELF exposure include electrical and electronic engineers and technicians; electric power line, telephone, and cable workers; electric arc welders; electricians; television and radio repair workers; power station operators; and motion picture projectionists. Magnetic fields and ELF electromagnetic radiation form the lowest end of the electromagnetic spectrum and include radiation at frequencies from 0 to 300 Hz. Concern regarding health effects in this portion of the electromagnetic spectrum centers around the power frequencies of 50 or 60 Hz (wavelength of 5000 km). Although it is composed of an electric and a magnetic field, the electric field does not have significant human tissue penetration.

Measurement issues

A magnetic field is formed whenever there is a flow of electric current. A static magnetic field occurs as a result of direct voltage or direct current (DC), while a time-varying magnetic field results from alternating current (AC). Magnetic field intensities are measured with a unit of magnetic flux called tesla or gauss (10 000 gauss = 1 T, 10 gauss = 1 mT). Magnetic fields freely penetrate many materials, including biological systems, and are very difficult to shield. Time-varying magnetic fields are of greatest interest because of their suggested association with molecular biological perturbation and health effects. Sources of time-varying magnetic fields include any device that utilizes an AC energy source, such as appliances, electrical equipment, and high-power transmission lines.

Two types of instruments are available for the measurement of magnetic fields: one is a multiturn loop used with a portable voltmeter and the other is a gauss meter. Otherwise, the measurement issues are similar to those discussed in the MW/RF section. Because cell membranes are relatively poor electrical conductors, the internal electric field is reduced between 10^6- and 10^8-fold compared to the external field. External fields of 1 million V/m would be needed to produce an internal field on the order of the existing transmembrane potential. Because of the tremendous attenuation of the electric field, it is not measured when evaluating ELF exposures. Personal dosimeters (EMDEX models A, B, and C; Electric Field Measurements, West Stockton, MA, United States) are also available to measure exposures in the 35–300-Hz band of frequencies.

Exposure guidelines

The ACGIH has recommended threshold limit values for electric and magnetic fields in the 1–30-kHz range. At exposures between 1 and 300 Hz, whole body exposure limits are determined by the formula $B = 60/f$ where B is the limit in millitesla (mT) and f is frequency. At 60 Hz, this would result in a 1-mT exposure (10 gauss). For frequencies between 300 Hz and 30 kHz, the whole body exposure limit is 0.2 mT (2 gauss) (1 gauss = 0.1 mT).

TABLE 15.2 Electrical and magnetic field strengths at ground level near high-tension transmission lines.[12]

	Distance from line (ft)				
	0	50	65	100	200
500-kV line					
Electric field (kV = m)	7	3		1	0.3
Magnetic field (mG)	70	25		12	3
230-kV line					
Electric field (kV = m)	2	1.5		0.3	0.05
Magnetic field (mG)	35	15	5	1	
115-kV line					
Electric field (kV = m)	1	0.5		0.07	0.01
Magnetic field (mG)	20	5		1	0.3

TABLE 15.3 Typical electric and magnetic fields.

Field source	Electric field (V/m)	Magnetic field (mG)
Home wiring	1–10	1–5
Electrical appliances	30–300	5–3000
Neighborhood distribution lines	10–60	1–10
Electrified railroad cars	–	10–200
High voltage transmission lines	1000–7000	25–100

Source: Adapted from Oak Ridge Associated Universities. Health effects of low frequency electric and magnetic fields. NTIS publication ORAU 92/F8, 1992. The highest appliance fields are recorded at the center of a spiral hot plate.

The recommended standard for the electric field varies with frequency and is designed to both prevent induced internal currents and eliminate spark discharges and other safety hazards that take place at field strengths >5–7 kV/m.

Natural background exposure from the static Earth's field is ~450 mG, and it may change by as much as 0.5 mG per day due to changes in solar activity. The Earth's electric field is ~120 V/m, which is comparable to the field found under a typical 12-kV urban power distribution line. Disturbances in the Earth's local electric field are commonly found in the form of lightning, which needs at least 3 million V/m to ionize the air.

Electric and magnetic fields at the edges of a restricted right-of-way have been characterized by many public utilities. Typical electric and magnetic fields from the Bonneville Power Administration are reviewed in Table 15.2.[12] Table 15.3 summarizes some of the other typical electric and magnetic fields that might be encountered.

Pathophysiology and health effects

Many biological effects have been proposed to be associated with ELF radiation. Savitz and Calle first aroused interest in this area with their 1979 review of the incidence of leukemia in workers exposed to high electromagnetic fields.[13] Thereafter, this interest was heightened when a two to fourfold increase in childhood leukemia in the Denver area was attributed to ELF exposures in a report, also published in 1979, by Wertheimer and Leeper.[14] Their study linked high childhood cancer rates to ELF exposures by wiring code configurations (WCC), which were used as a surrogate for ELF exposure.

Good examples of childhood residential studies have been published by Savitz et al. (Denver) in 1988 and London et al. (Los Angeles) in 1991.[15,16] In the Denver study, spot magnetic field measurements were used in addition to WCC. In the Los Angeles study, spot measurements and 24-hour magnetic field measurements were recorded. Based on WCC data, the relative risk for leukemia in the high- versus low-current classification was 1.54 (95% confidence interval 0.9–2.63) in the Denver study and 1.73 (95% confidence interval 0.82–3.66) in the Los Angeles study. Both these studies show a rise in the relative risk with increasing WCC. However, a significantly increased risk was not demonstrated in either study when the risk was assessed in relation to measured magnetic fields.

Since 1979, over 1000 articles have been published in the area of ELF effects. Research papers can generally be divided into the following areas: (i) human epidemiologic studies of cancer, focusing on childhood and adult residential exposures and on adult occupational exposures; (ii) effects on growth control; (iii) neurobehavioral effects; and (iv) other physiologic effects.

Overall, research has yielded conflicting results about the presence (or absence) of an association between ELF and health effects. For example, residential/nonoccupational studies of adults have a number of epidemiologic shortcomings centered on the confounding factors of occupational versus home/other exposures, and they have not consistently shown an increased risk with measured ELF exposures. Nonoccupational exposures include exposures from home wiring, electrical distribution lines and substations, transportation equipment (i.e., electric trains and buses), and home appliances (i.e., hot plates, refrigerators, hair dryers, electric blankets, and any other electrically powered device). Many of the studies negative for cancer were performed with field strengths well above what should have been considered typical for an occupational or residential exposure. The two effects that have been reported at typical residential exposure levels (Table 15.4) are changes in calcium ion flux and inhibition of melatonin secretion.[17,18]

Epidemiologic studies suggest a weak association between ELF fields and childhood leukemia. Kheifets et al. reviewed childhood leukemia ELF research and found small overall elevations in relative risks (<1.5).[19] Studies of ELF association with other cancers (e.g., breast) have been negative or inconclusive.[20,21] The effect of ELF on reproduction has also been studied. Huuskonen et al.'s review

TABLE 15.4 Biological effects reported with ELF EMF fields.

Enhanced RNA synthesis in insect salivary gland culture
Decreased cell growth in slime molds
Enhanced DNA synthesis in mammalian cell culture in certain frequency windows
Lack of evidence for altering DNA structure
Promoting repair of non-union fractures
Epidemiologic evidence of cancer in humans
Changes in calcium ion flux from chick brain and embryo culture
Behavioral and EEG changes in mammals
Inhibition of melatonin secretion from the pineal gland

Source: Adapted from Oak Ridge Associated Universities. Health effects of low frequency electric and magnetic fields. NTIS publication ORAU 92/F8, 1992.

concluded that "the epidemiologic evidence does not, taken as a whole, suggest strong associations between exposure to ELF magnetic fields and adverse reproductive outcome. An effect at high levels of exposure cannot be excluded, however."[22] Other conditions, including multiple sclerosis and Alzheimer's disease, have not been shown to be associated with ELF magnetic field exposure.[23] The biological plausibility of ELF effects rests on the ability of the magnetic field to interact with the body at the cellular level. Although the induced ELF fields are weak, the evidences of changes in calcium ion fluxes, increased rates of bone healing, and changes in melatonin secretion suggest that magnetic fields are biologically active.

IARC's most recent (2002) review of the available ELF and cancer data concluded as follows:

> There is *limited evidence* in humans for the carcinogenicity of extremely lowfrequency magnetic fields in relation to childhood leukaemia. There is *inadequate evidence* in humans for the carcinogenicity of extremely lowfrequency magnetic fields in relation to all other cancers. There is *inadequate evidence* in humans for the carcinogenicity of static electric or magnetic fields and extremely low-frequency electric fields. There is *inadequate evidence* in experimental animals for the carcinogenicity of extremely low-frequency magnetic fields. No data relevant to the carcinogenicity of static electric or magnetic fields and extremely low-frequency electric fields in experimental animals were available.
>
> **Overall evaluation**
>
> Extremely low-frequency magnetic fields are *possibly carcinogenic to humans (Group 2B).*
>
> Static electric and magnetic fields and extremely low-frequency electric fields are *not classifiable as to their carcinogenicity to humans (Group 3).*[24]

The 2007 report published by the World Health Organization reflects the current knowledge regarding the biological effects of ELF magnetic fields.[25] The summary of the WHO report concludes:

> Although a causal relationship between magnetic field exposure and childhood leukaemia has not been established, the possible public health impact has been calculated assuming causality in order to provide a potentially useful input into policy. However, these calculations are highly dependent on the exposure distributions and other assumptions, and are therefore very imprecise. Assuming that the association is causal, the number of cases of childhood leukaemia worldwide that might be attributable to exposure can be estimated to range from 100 to 2400 cases per year. However, this represents 0.2 to 4.9% of the total annual incidence of leukaemia cases, estimated to be 49 000 worldwide in 2000. Thus, in a global context, the impact on public health, if any, would be limited and uncertain.
>
> A number of other diseases have been investigated for possible association with ELF magnetic field exposure. These include cancers in both children and adults, depression, suicide, reproductive dysfunction, developmental disorders, immunological modifications and neurological disease. The scientific evidence supporting a linkage between ELF magnetic fields and any of these diseases is much weaker than for childhood leukaemia and in some cases (for example, for cardiovascular disease or breast cancer) the evidence is sufficient to give confidence that magnetic fields do not cause the disease.

In their review published in 2015, the European Commission's Scientific Committee on Emerging and Newly Identified Health Risks (SCENIHR) incorporated newer research in their analysis and reached similar conclusions.[26] The main findings of their report are[27]:

> Overall, the epidemiological studies on radiofrequency EMF exposure do not show an increased risk of brain tumours. Furthermore, they do not indicate an increased risk for other cancers of the head and neck region.
>
> Previous studies also suggested an association of EMF with an increased risk of Alzheimer's disease. New studies on that subject did not confirm this link.
>
> Epidemiological studies associate exposure to Extremely Low Frequency (ELF) fields, from long-term living in close proximity to power lines to a higher rate of childhood leukaemia. No mechanisms have been identified and no support from experimental studies could explain these findings, which, together with shortcomings of the epidemiological studies prevent a causal interpretation.
>
> Concerning EMF hypersensitivity (idiopathic environmental intolerance attributed to EMF), research consistently shows that there is no causal link between self-reported symptoms and EMF exposure.

Prevention

Even though there are no proven adverse health effects related to EMF and ELF, a number of simple steps may be taken to reduce exposure without significant expense. This is known as the strategy of "prudent avoidance." Identifying high-voltage transmission equipment at the worksite will help to focus on the kinds of monitoring that may be necessary. It would be prudent to reduce exposures to below the recommended levels. Distance is the best control since the field strength falls inversely with the square of the distance. Shielding is more difficult because magnetic fields react differently to different metals and different metal configurations (e.g., screen, mesh, sheet metal). In some cases, rewiring to oppose adjoining field polarities may diminish exposure, but this procedure can often be prohibitively expensive.

References

1. Occupational Safety and Health Administration. 29 CFR1910.97. Available at https://www.osha.gov/pls/oshaweb/owadisp.show_document?p_table=STANDARDS&p_id=9745 (accessed June 22, 2016).
2. Forman SA, Holmes CK, McManamon TV, et al. Psychological symptoms and intermittent hypertension following acute microwave exposure. *J Occup Med* 1982; 24:932–4.
3. Schilling CJ. Effects of exposure to very high frequency radiofrequency radiation on six antenna engineers in two separate incidents. *Occup Med* 2000; 50:49–56.
4. Corle C, Makale M, Kesari S. Cell phones and glioma risk: a review of the evidence. *J Neurooncol* 2012; 106:1–13.
5. Jauchem JR. Effects of low-level radio-frequency (3 kHz to 300 GHz) energy on human cardiovascular, reproductive, immune, and other systems: a review of the recent literature. *Int J Hyg Environ Health* 2008; 211:1–29.
6. Morgan RW, Kelsh MA, Zhao K, et al. Radiofrequency exposure and mortality from cancer of the brain and lymphatic/hematopoietic systems. *Epidemiology* 2000; 11(2):118–27.
7. COMAR Reports. Biological and health effects of electric and magnetic fields from video display terminals. A technical information statement. *IEEE Eng Med Biol Mag* 1997; 16(3): 87–92.
8. IARC Working Group on the Evaluation of Carcinogenic Risks to Humans, World Health Organization, International Agency for Research on Cancer. *Non-ionizing radiation, Part 2: Radiofrequency electromagnetic fields*. IARC monographs on the evaluation of carcinogenic risks to humans, volume 102. Lyon: IARC/WHO, 2013. Available at http://monographs.iarc.fr/ENG/Monographs/vol102/mono102.pdf (accessed June 22, 2016).
9. Reeves GI. Review of extensive workups of 34 patients overexposed to radiofrequency radiation. *Aviat Space Environ Med* 2000; 71(3):206–15.
10. Polk C, Postow E. *Handbook of biological effects of electromagnetic fields*, 2nd edn. Boca Raton, FL: CRC Press, 1996.
11. van Demark VVR, Free JR. Temperature effects. In: Johnson AD, Gomes WR, van Derman ML, eds. *The testis*, 3rd edn. New York: Academic Press, 1973.
12. Bonneville Power Administration. *Electric power lines*, 6th edn. DOE/BP-961. Portland, OR: Bonneville Power Administration, 1990:4.
13. Savitz DA, Calle EE. Leukemia and occupational exposure to electromagnetic fields: a review of epidemiology studies. *J Occup Med* 1979; 29:47–51.
14. Wertheimer N, Leeper E. Electrical wiring configuration and childhood cancer. *Am J Epidemiol* 1979; 109:273–84.
15. Savitz DA, Wachtel H, Barnes FA, et al. Case-control study of childhood cancer and exposure to 60-Hz magnetic fields. *Am J Epidemiol* 1988; 128:21–38.
16. London SJ, Thomas DC, Bowman JD, et al. Exposure to residential electric and magnetic fields and risk of childhood leukemias. *Am J Epidemiol* 1991; 134:923–37.
17. Blackman CF, Benane SG, Elliott DJ, et al. Influence of electromagnetic fields on the efflux of calcium ions from brain tissue in vitro: a three-model analysis consistent with the frequency response up to 510 Hz. *Bioelectromagnetics* 1988; 9:215–27.
18. Burch JB, Reif JS, Noonan CW, et al. Melatonin metabolite levels in workers exposed to 60-Hz magnetic fields: work in substations and with 3-phase conductors. *J Occup Environ Med* 2000; 42(2):136–42.
19. Kheifets LI, Ahlbom A, Crespi CM, et al. Pooled analysis of recent studies on magnetic fields and childhood leukaemia. *Br J Cancer* (2010) 103, 1128–35.
20. Sorahan T, Hamilton L, Gardiner K, et al. Maternal occupational exposure to electromagnetic fields before, during, and after pregnancy in relation to risks of childhood cancers: findings from the Oxford Survey of Childhood Cancers, 1953–1981 deaths. *Am J Ind Med* 1999; 35(4):348–57.
21. Caplan LS, Schoenfeld ER, O'Leary ES, et al. Breast cancer and electromagnetic fields: a review. *Ann Epidemiol* 2000; 10(1):31–44.
22. Huuskonen H, Linbohm ML, Juutilainen J. Teratogenic and reproductive effects of low-frequency magnetic fields. *Mutat Res* 1998; 410:167–83.
23. Mattsson MO, Simkó M. Is there a relation between extremely low frequency magnetic field exposure, inflammation and neurodegenerative diseases? A review of in vivo and in vitro experimental evidence. *Toxicology* 2012; 301(1–3):1–12.
24. International Agency for Research on Cancer, World Health Organization. *Non-ionizing radiation, Part 1: Static and extremely low-frequency (ELF) electric and magnetic fields*. IARC monographs on the evaluation of carcinogenic risks to humans, volume 80. Geneva: IARC/WHO, 2002.
25. World Health Organization, International Labour Organization, International Commission on Non-Ionizing Radiation Protection. *Extremely low frequency fields*. Environmental health criteria monograph, volume 238. Geneva: WHO, 2007. Available at http://www.who.int/peh-emf/publications/elf_ehc/en/ (accessed June 11, 2016).
26. Scientific Committee on Emerging and Newly Identified Health Risks. Opinion on potential health effects of exposure to electromagnetic fields (EMF), European Commission, January 27, 2015. Available at http://ec.europa.eu/health/scientific_committees/emerging/docs/scenihr_o_041.pdf (accessed June 11, 2016).
27. European Commission. Final opinion on potential health effects of exposure to electromagnetic fields (EMF). Available at http://ec.europa.eu/health/scientific_committees/consultations/public_consultations/scenihr_consultation_19_en.htm (accessed June 11, 2016).

Chapter 16

NOISE

Robert A. Dobie

About five to nine million Americans had potentially hazardous noise exposures in the workplace in the 1980s;[1,2] massive job reductions in manufacturing in the United States since then have surely reduced the number at risk. Noise can be annoying and distracting. It interferes with spoken communication and masks the warning signals necessary for safety and productivity. As one of several generalized stressors, noise may contribute to cardiovascular disorders. However, the most important and best-characterized effect of excessive noise exposure is hearing loss. It is widely believed that reducing noise to a level low enough to prevent hearing loss will also prevent its other harmful effects (some important exceptions to this rule will be noted later). Therefore, this chapter will primarily deal with noise-induced hearing loss (NIHL), with special emphasis on risk assessment and prevention.

OCCUPATIONAL SETTING

Excessive noise is produced by an almost infinite variety of processes—anything that cuts, grinds, collides, explodes, or just moves (itself, another object, or a gas or liquid) will make noise. The industries listed in Table 16.1 are responsible for much of the hazardous occupational noise exposures in the United States. In some industry groups such as fishing, forestry, construction, transportation, trade, and services, fewer than half of workers receive hazardous exposures. According to surveys, >50% of workers in such industries as textiles, lumber and wood, and mining receive hazardous exposures.[1]

Although these survey results are interesting, they are of limited use to the occupational physician. No industrial sector is completely free of hazardous noise exposures. Only assessment of individual workplaces—or, even better, individual workers' exposures—can accurately identify persons at risk for NIHL.

MEASUREMENT ISSUES

Fortunately, risk to hearing (unlike annoyance, sleep deprivation, and some other effects of noise) is relatively well predicted by three measurable physical properties of sound: **frequency**, **intensity**, and **time**. In general, more hazardous sounds are louder, longer, and more concentrated in the frequency range where we hear best.

Frequency

A vibrating object moves air molecules back and forth, creating a sound wave that propagates outward. The number of complete cycles or oscillations per second is the frequency of the sound, measured in hertz (Hz) or kilohertz (1 kHz = 1000 Hz). Normal young people can hear sounds ranging in frequency from about 20 Hz to 20 kHz, but our best hearing is in the 1–5-kHz region.

Only artificial objects like tuning forks and electronic oscillators put out pure tones, that is, sounds having energy at only one frequency. Natural sounds like speech contain many frequencies simultaneously; indeed, it is the relative intensities of these different frequencies, or harmonics, that permit us to recognize speech sounds or to distinguish one musical instrument from another.

Physical and Biological Hazards of the Workplace, Third Edition. Edited by Gregg M. Stave and Peter H. Wald.

TABLE 16.1 Some industries with risks of hazardous noise exposure.[3]

Construction
Manufacturing
Ship building
Mining
Agriculture
Textiles
Utilities
Paper
Machining
Forestry
Landscaping
Paper mills
Transportation
Stone cutting
Furniture
Oil and gas
Airports

Intensity

Sound intensity (energy flow per unit area per unit time) is difficult to measure directly but is directly proportional to the square of sound pressure, which is easily measured and thus much more commonly reported, in units of pascals (Pa) or micropascals (1 Pa = 10^6 μPa). The ratio between the softest audible sound pressure (20 μPa) and the loudest tolerable sound pressure (20 Pa) is a million to one. To avoid having to either switch units or use too many zeroes, this pressure range is compressed by using logarithms, just as when converting hydrogen ion concentration to pH.

Specifically, the decibel (dB) is defined as

$$\text{dB} = 20\log_{10}\left[\frac{P}{P_0}\right]$$

For general-purpose sound level measurements, P_0 is set at 20 μPa, a barely audible level for the best-heard frequencies (inaudible for higher and lower frequencies). Decibels measured with $P_0 = 20$ μPa are identified as "sound pressure level" (SPL). Thus, a sound pressure (P) of 20 μPa would be 0-dB SPL.

$$\text{dB} = 20\log_{10}\left(\frac{20}{20}\right) = 20\log_{10}(1) = 0\text{-dB SPL}$$

Sounds less intense than 20 μPa would represent negative values in dB SPL. In other words, 0 dB does not represent the absence of sound but a sound whose pressure equals the reference pressure level. A very intense sound of 20 Pa (barely tolerable) would have a sound pressure level of 120-dB SPL:

$$\text{dB} = 20\log_{10}\left(\frac{20\ 000\ 000}{20}\right) = 20(6) = 120\text{-dB SPL}$$

Since decibels are logarithmic, they combine in ways that may seem surprising. For example, if two sound sources which individually produce 80-dB SPL are turned on simultaneously, the result is only a 3-dB increase, to 83-dB SPL, rather than a doubling, as might be expected. Since doubling the distance from a sound source reduces the acoustic energy flow per unit area by a factor of 4 (area is proportional to distance^2), sound pressure level decreases by 6 dB under ideal circumstances. However, this only holds true outdoors; because of reverberation, sound levels change much less in most indoor workplaces as one moves away from a sound source.

Sound pressure level is not enough to specify hazard to hearing: 120-dB SPL would be extremely loud (and hazardous) at a well-heard frequency such as 2 kHz, but it would be inaudible (and harmless) at 50 kHz. One could measure the sound pressure in each of several audible octave bands (an octave comprises a 2 : 1 frequency ratio, e.g., 500–1000 Hz), but this would be too cumbersome. Instead, hearing conservation professionals universally use decibels on the A-scale to combine frequency and intensity for a single-number measurement of potential hazard. A sound level meter operating in the A-scale mode uses electronic filters to cut out inaudible frequencies altogether, to partially remove poorly heard frequencies, and to give full weight to the best-heard frequencies (1–5 kHz). Thus, two sounds with identical sound pressure levels could have different A-scale readings; the one with most energy in the 1–5-kHz range would be higher, more accurately reflecting the risk of NIHL (Table 16.2).

Sound level meters can be deceptively easy to use. However, microphone care, selection, calibration, and placement, timing of measurements, and proper interpretation of data are just a few of the variables that can affect the reliability and validity of sound level measurements. Use of these instruments is normally best delegated to acoustic engineers, industrial hygienists, audiologists, or others who have been trained in their use.

Time

Even an A-scale reading is not enough; 5 minutes/day at 100 dBA is much less hazardous than 8 hours/day at 90 dBA. Using the logarithmic decibel scale, a 3-dB change is equivalent to a doubling (or halving) of sound energy per unit time. If the hazard were proportional to the total sound energy received by the ear, factoring in time would be simple;

TABLE 16.2 Sound pressure level versus A-scale.

Source	dB SPL	dBA
Jet	90	92
Diesel	90	85

TABLE 16.3 Time–intensity trading.

Level (dBA)	Duration (hours)
90	8
95	4
100	2
102	1.5
105	1
110	0.5
115	0.25 or less

Source: Adapted from OSHA General Industry Standard 29 CFR 1910.95, Table G-16.

a twofold increase in exposure time would be equivalent to a 3-dB increase in sound pressure level (dBA). Indeed, many experts support the use of a 3-dB exchange rate to relate time and level to overall hazard. However, there is considerable evidence that intermittent exposures are less hazardous than continuous exposures at the same level and total duration.[4] Since high-level exposures are often interrupted, the Occupational Safety and Health Administration[3] (OSHA) has adopted a 5-dB exchange rate in an attempt to incorporate the protective effects of intermittency. Under OSHA regulations, for example, a 90-dBA exposure for 8 hours is considered as hazardous as a 95-dBA exposure for 4 hours (Table 16.3).

Time-weighted average (TWA) is a useful single number characterizing a day's exposure in terms of frequency, intensity, and time. This is the level (in dBA) that, if present for 8 hours, would present a hazard equal to that of the exposure in question. Both of the previously mentioned exposures (90 dBA/8 hours and 95 dBA/4 hours) would be described as 90-dBA TWA.

Calculating TWA for varying exposure times (e.g., 6 hours, 37 minutes) and combinations of exposures at different levels and durations is mathematically straightforward but cumbersome. Fortunately, most acoustic hazard assessment now uses noise dosimeters, wearable devices that use built-in microprocessors to automatically calculate TWA for a day's exposure.

EXPOSURE GUIDELINES

The Occupational Safety and Health Act,[5] modified by the OSHA Hearing Conservation Amendment,[3] established extensive regulations for industries in the United States. The mining and railroad industries are subject to separate federal regulation by the Mine Safety and Health Administration and the Federal Railroad Administration, respectively.[6]

OSHA defines the permissible exposure limit (PEL) as 90-dBA TWA. Higher exposures must be reduced by engineering or administrative controls or by the use of hearing protection devices (HPDs). OSHA also recognizes a borderline or low-risk range of exposures at 85–90-dBA TWA. Workers with daily exposures >85-dBA TWA must be covered by hearing conservation programs (HCPs), which will be discussed later.

Exposures are sometimes described in terms of noise dose, where the PEL (90-dBA TWA) is equivalent to a noise dose of 1.0 (or 100%). Exposures of 85-dBA TWA and 95-dBA TWA would be described as noise doses of 0.5 (50%) and 2.0 (200%), respectively. TWA and noise dose are interchangeable descriptions of noise exposure hazard, and most dosimeters will read out whichever the user prefers. OSHA requires that dosimeters incorporate all continuous, intermittent, and impulsive sounds between 80 and 130 dBA into their calculations.

These guidelines are intended to protect almost all covered workers from substantial occupationally related NIHL. However, a few highly susceptible workers may incur mild NIHL at exposure levels below the PEL. Some workers are not covered by OSHA regulations; others may increase their risk by moonlighting at a second noisy job. Nonoccupational exposures, especially hunting and target shooting, are often more hazardous to hearing than workplace exposures.

Observance of OSHA noise exposure guidelines will not always protect against safety hazards other than NIHL. For example, a brief period of intense noise (1 hour per day at 95 dBA) would neither exceed the PEL nor require an HCP. However, during that period, speech communication would be severely disrupted, and warning signals could become inaudible.

Speech interference can be a useful clue to a potentially hazardous noise environment. People raise their voices to be heard over loud background noises. Most will need to shout to converse at arm's length at sound levels >85 dBA; this is also the level at which OSHA requires hearing conservation programs for workers with 8-hour exposures.

NORMAL PHYSIOLOGY

The **outer ear** includes the pinna or auricle (the visible part) and the ear canal, a skin-lined tube leading to the eardrum; together, these structures provide resonances that enhance transmission of certain frequencies (around 3 kHz) and impair others. These effects are rather small; traumatic or surgical alterations of outer-ear structures produce only minimal hearing changes. Complete blockage of the ear canal with earwax or a foreign body will cause mild to moderate hearing loss.

The **middle ear** is separated from the ear canal by the eardrum, a thin membrane connected to a chain of three tiny bones, or ossicles: the malleus, incus, and stapes. The footplate of the stapes, which transmits the vibrations of the eardrum and ossicular chain into the inner ear, is much smaller in area than the eardrum. Thus, the pressure exerted on the inner-ear fluids is increased or amplified, much as the difference in cross sections amplifies the force in a hydraulic system. Without this ingenious mechanical arrangement, most of the sound energy reaching the inner ear would be reflected back into the air. Two small muscles attach to the malleus and stapes and contract in response to loud sounds. This *acoustic reflex* stiffens the ossicular chain, impairing transmission of low-frequency sounds and slightly enhancing transmission of sounds above 2 kHz; the reflex offers some protection against NIHL, at least for low frequencies (1 kHz and lower). The middle ear is an air-containing space that receives a regular air supply via the Eustachian tube. Disruptions of the middle-ear mechanism (perforated eardrum, fixed or disconnected ossicles, and fluid-filled middle-ear space) can cause mild to moderately severe hearing loss.

Hearing loss caused by outer-ear or middle-ear disorders is called conductive, because it interferes with normal air conduction of sound to the inner ear. Sounds presented by an oscillator held directly against the skull reach the inner ear by bone conduction and are heard normally by people with conductive hearing loss.

The **inner ear** contains organs of both hearing and balance; the latter will not be discussed here. The hearing organ, or cochlea, includes a spiral basilar membrane encased in a snail-shaped cavity in the temporal bone; on that membrane are spirally arranged hair cells, which change mechanical fluid vibrations into nerve impulses traveling to the brain. The hair cells of the base of the cochlear spiral respond best to high frequencies, whereas those at the apex respond only to low frequencies. At any given point along the spiral, there are three rows of outer hair cells, essential for hearing soft sounds, and one row of inner hair cells, which connect to almost all the nerves carrying sound to the brain.

Inner-ear hearing losses are usually lumped together as *sensorineural*, although the vast majority affect the hair cells (sensory) and very few directly affect the nerve cells (neural). Outer hair cells are more vulnerable than inner hair cells to most diseases and injuries causing sensorineural hearing loss, including age-related hearing loss, NIHL, and ototoxic drugs.

PATHOPHYSIOLOGY OF NOISE-INDUCED HEARING LOSS

Most hazardous noise exposures cause reversible inner-ear injury at first. Hair cells may lose their normal ability to respond to sound, accompanied by a temporary elevation of threshold (the softest sound that can be heard), or temporary threshold shift (TTS), lasting hours to days. Individuals often experience this as a muffling of sound, together with tinnitus (i.e., ringing in the ears) and fullness. After repeated TTS-inducing exposures, permanent hair cell loss and noise-induced permanent threshold shift (NIPTS) occur. At least in some animals, exposures that cause very large TTSs can cause immediate and permanent loss of auditory nerve cell function without measurable threshold shifts.[7] Some very brief but intense exposures, especially involving impulsive sounds such as gunfire, can cause immediate permanent threshold shift (PTS) without intervening TTS; this is called *acoustic trauma*, as distinct from ordinary NIHL.

We might expect NIHL to affect hearing for the same frequencies contained in the offending sound, and this is true, up to a point. However, most occupational and recreational noises contain a broad spectrum of frequencies; thus, the frequency pattern of NIHL is determined more by the sensitivity of the ear than by the frequency content of the noise. Other factors, such as the acoustic reflex, probably also play a role; the result is that for almost all cases of NIHL and acoustic trauma, the first and most severe effects are seen in the 3–6-kHz region.

Hearing loss is usually represented using an audiogram, a graph of hearing sensitivity (the softest sounds a patient can hear) as a function of frequency. The vertical axis plots thresholds in dB hearing level (HL). Recall that dB SPL implies a reference level of 20 μPa and that dBA represents a weighted sum across the audible range of frequencies, suitable for noise hazard assessment. In contrast, dB HL implies a reference level of normal human (young adult) hearing; for each audiometric frequency, 0-dB HL is average normal hearing, whereas thresholds above 15-dB HL are abnormal for young adults (although not necessarily handicapping or even noticeable).

Figure 16.1 shows median audiograms for a group of retired jute mill workers with noise exposure above 100 dB and an age- and gender-matched control group.[8] The well-known 4-kHz dip is evident in the noise-exposed audiogram. NIPTS is simply the decibel difference between audiometric thresholds for noise-exposed and non-noise-exposed populations.

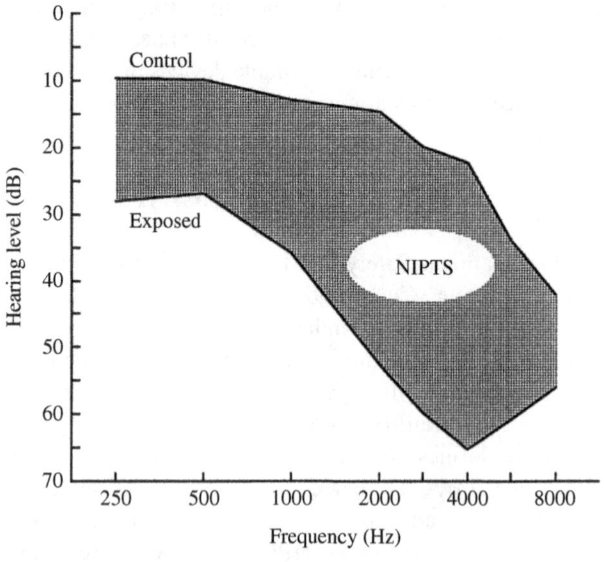

FIGURE 16.1 Median audiograms for noise-exposed and control subjects.

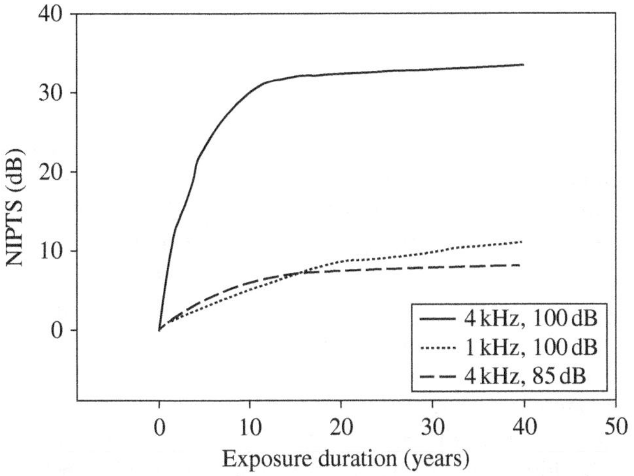

FIGURE 16.2 Noise-induced permanent threshold shift as a function of time for different frequencies and exposure levels.

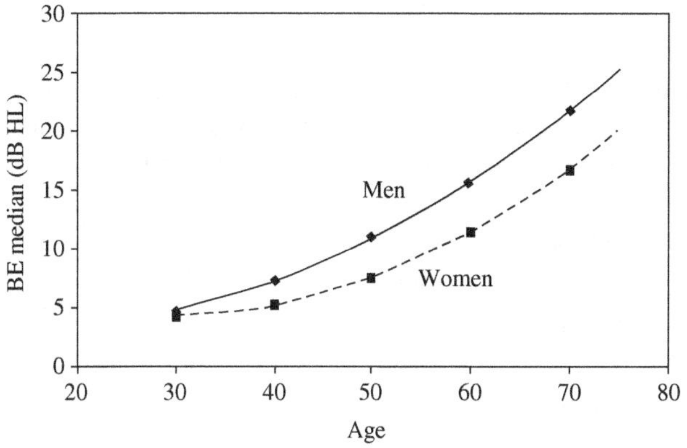

FIGURE 16.3 Speech frequency age-related permanent threshold shift as a function of age and gender.

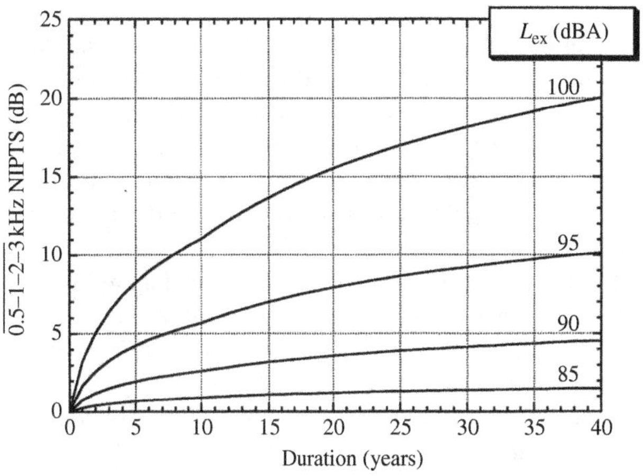

FIGURE 16.4 Speech frequency noise-induced permanent threshold shift as a function of time for different exposure levels.

For repeated exposures to occupational noise, NIPTS grows gradually. After about 10 years, the growth decelerates markedly, and NIPTS approaches a plateau. The International Organization for Standardization has published tables and formulas (ISO 1999[9]) that describe the growth of NIPTS over time for different frequencies and exposure levels. Some selected curves from ISO 1999 are reproduced in Figure 16.2. Note that all three curves show (i) the plateau effect, (ii) that the predicted NIPTS is much greater for 4 kHz than for 1 kHz, and (iii) that 40 years of exposure at 85 dBA produces only slight changes at 4 kHz (and none at 1 kHz). These curves represent median NIPTS; that is, half of exposed persons would be expected to show greater losses, and half would show lesser losses.

With prolonged exposure (or immediately in cases of acoustic trauma), frequencies below 3 kHz may demonstrate substantial losses. The AMA and most workers' compensation programs consider the speech frequencies to be 0.5, 1, 2, and 3 kHz, because losses for these frequencies interfere with everyday speech communication, in quiet or noisy backgrounds.[10] Most of the acoustic energy in normal speech is concentrated below 1 kHz, especially in vowel sounds. However, most of the information content is above 1 kHz, where the consonants have their peak energies. Thus, people with NIHL (or any other high-frequency sensorineural hearing loss) will often complain that they can hear speech but cannot understand it. This can lead to social isolation and depression.

NIPTS, as plotted in Figures 16.1 and 16.2, has not been directly measured in human epidemiologic studies. Rather, the hearing levels of a group of noise-exposed workers (known intensity and duration) are compared to hearing levels for non-noise-exposed workers with similar age, gender, and other characteristics. The differences between these two groups are reported as NIPTS. Implicit in this approach is the fact that people lose hearing as they age (high frequency > low frequency, men > women) and the assumption that age-related permanent threshold shift (ARPTS) and NIPTS are additive in decibels.

Figure 16.3 shows median ARPTS curves for men and women, from the US database included in ISO 1999, averaged across the speech frequencies. Note that ARPTS is an accelerating process. Contrast this with Figure 16.4, which shows median speech frequency NIPTS for exposures ranging from 85 to 100 dBA. NIPTS decelerates; approximately 60% of the 40-year total is present after 10 years.

Susceptibility to both age-related hearing loss and noise-induced hearing loss varies widely. It seems likely that genetic differences underlie much of this variation,[11] but when those with worse hearing are found to have particular genetic markers, it is difficult to know whether these are markers for age effects or noise effects.[12] To date, genetic tests have not been shown to assist in the diagnosis or prevention of NIHL.

DIAGNOSIS

The American College of Occupational and Environmental Medicine[13] has enumerated a series of criteria for the diagnosis of occupational NIHL, which they have described as:

1. Sensorineural
2. Bilateral and symmetrical
3. Not profound
4. Not progressive after cessation of noise exposure
5. Decelerating
6. Greatest in 3–6-kHz range
7. Stable after 10–15 years in high frequencies
8. Less severe after interrupted exposures than after continuous daily exposures of the same level and duration

Occupational NIHL is too frequently diagnosed casually, without adequate attention to the patient's noise exposure history, both occupational and nonoccupational. Employers and HCP managers can improve the quality of medical reports by providing examining physicians with noise exposure data, especially dosimetry, relevant to the individuals being evaluated.

Except in very young workers, noise and aging must be considered together. At retirement age, most workers exposed at levels below 100 dBA (the vast majority of noise-exposed workers) will have more age-related loss than noise-induced loss. Statistical methods for estimating the relative contributions of aging, various periods of occupational exposure, nonoccupational exposure, and other otologic disorders have been developed.[14]

TREATMENT

No medical or surgical treatment has been shown to be helpful for NIHL in humans, and none is usually offered. Some authorities have recommended treatments aimed at improving cochlear blood supply in cases of acoustic trauma, and multiple antioxidants, vitamins, and minerals have been shown to partially prevent hearing loss in animals when given prior to noise exposure,[15] but NIHL remains a disorder without medical treatment of proven efficacy.

NIHL can be palliated with hearing aids and assistive listening devices, such as "TV ears." However, hearing aids rarely if ever compensate completely for sensorineural hearing loss; they can amplify inaudible sounds into the patient's audible range, but distortions usually persist, especially with loud sounds.

The lack of available medical treatment and the inadequacy of hearing aids combine to emphasize the role of prevention. A well-organized HCP can dramatically reduce the risk of NIHL and acoustic trauma for motivated workers.

MEDICAL SURVEILLANCE

The Hearing Conservation Amendment[3] requires HCPs for all workers with exposures exceeding 85-dBA TWA. An HCP must include baseline and annual pure-tone air-conduction audiometry (0.5, 1, 2, 3, 4, and 6 kHz), with monitoring for standard threshold shifts (STS), defined as a 10-dB or greater change for the worse in either ear for the average of 2, 3, and 4 kHz. In determining whether an STS has occurred, age correction is optional. Audiometry within the HCP (usually on-site or in a mobile audiometric van) does not require bone conduction, speech, or middle-ear tympanometric tests.

Otoacoustic emissions (OAEs) are soft sounds emitted by the healthy cochlea in response to acoustic stimulation (they were initially called "cochlear echoes"). People with weak OAEs appear to be more susceptible to NIHL,[16] perhaps because they have already suffered subclinical noise-induced cochlear damage, but OAEs have not been shown to be useful in HCPs, either to replace or to supplement conventional pure-tone audiometry.

OSHA requires every HCP to be supervised by an otolaryngologist or other physician, or by an audiologist. If an STS occurs, the professional supervisor may elect either to retest within 30 days or to accept the STS without confirmation. An accepted or confirmed STS requires the initiation of earplug or earmuff use for all workers in the 85–90-dBA TWA zone who were not previously required to use HPDs, or the refitting of HPDs for those previously using them. Employee notification and counseling are required for all STSs. When the 2-, 3-, and 4-kHz average is 25 dB or greater after an STS, OSHA requires that the STS be recorded on Form 300, unless it is determined not to be work related.[17]

The National Institute for Occupational Safety and Health (NIOSH) published the revised *Criteria for a Recommended Standard: Occupational Noise Exposure* in 1998,[18] reaffirming their support for an 85-dBA recommended exposure limit. With a 40-year lifetime exposure at 85 dBA, NIOSH estimates an excess risk of developing "material hearing impairment" of 8%—considerably lower than the 25% excess risk they calculate at the 90-dBA PEL currently enforced by OSHA. However, NIOSH changed their definition of material hearing impairment, to give more weight to high frequencies and less weight to low frequencies, prior to calculating excess risk. When excess risk is calculated using the definition most widely used in the United States (that of the AMA[10]), the corresponding figures are 3% for 85 dBA and 8% for 90 dBA.[6]

NIOSH now recommends an exchange rate (ER) of 3 dB for the calculation of TWA exposures to noise. The 5-dB exchange rate remains in the OSHA regulations and appears to be more accurate than the 3-dB exchange rate for intermittent and fluctuating exposures,[4] but the most appropriate exchange rate remains controversial.[4] NIOSH also recommends a new criterion for significant threshold shift—an increase of 15 dB in the hearing threshold level (HTL) at 500, 1000, 2000, 3000, 4000, or 6000 Hz in either ear, as determined by two consecutive audiometric tests. OSHA has not adopted this recommendation. NIOSH no longer recommends age correction on individual audiograms. This practice is felt by some not to be scientifically valid; on the good side, it can decrease false positives (STSs not due to noise), but on the bad side it can increase false negatives, delaying intervention to prevent further hearing losses in workers whose HTLs have increased because of occupational noise exposure. OSHA still allows age correction as an option.

Referral is not required for STSs, but it is required when the program supervisor suspects a medical problem affecting the ears or believes that the audiometric data are questionable. The American Academy of Otolaryngology—Head and Neck Surgery[19] has published criteria for otologic referral from HCPs that have been adopted by many companies. These criteria recommend referral based on otologic symptoms, baseline audiometry (with special emphasis on substantial asymmetries, which often indicate serious otologic disease), and changes seen on periodic audiometry (including the lower frequencies, where communicatively significant or medically serious hearing losses are more likely to be seen).

PREVENTION

The OSHA Hearing Conservation Amendment requires more than periodic audiometry and medical or audiologic surveillance. After all, audiograms do not prevent hearing loss; they only measure it. Health professionals responsible for an HCP should carefully review the entire standard. The additional important elements of an HCP include:

1. Risk assessment using sound level meters or dosimetry
2. Education and motivation of workers
3. Exposure reduction through engineering control, administrative controls, and use of HPDs
4. Recordkeeping

Education and motivation are important for two reasons. First, if the HCP includes the use of HPDs, there may be significant worker resistance and poor compliance unless they believe that hearing is really at risk, is worth saving, and can effectively be saved. Second, many—perhaps most—industrial workers have potentially hazardous nonoccupational noise exposures; if they fail to reduce their recreational exposures, NIHL will continue to accrue despite reduction of on-the-job exposures to safe levels.

Engineering controls are obviously the most desirable way to reduce occupational noise exposures. If all workplaces can be brought below 85 dBA and no workers spend >8 hours per day on the job, the employer no longer needs an HCP at all. Lesser reductions are also valuable if they reduce the number of workers who must be enrolled in the HCP, the number who must use HPDs, or the HPD performance requirements. Cost is sometimes an issue, but newer equipment is often designed to reduce noise output.

Administrative controls involve reducing the individual's exposure to hazardous noise. Most experts in HCP design and administration have found that these changes are rarely feasible without unacceptable disruption of work routines.

In practice, then, most HCPs rely heavily on the use of HPDs to reduce exposures. Earplugs and earmuffs can both be effective if properly fitted and used. OSHA requires that employers offer their workers a variety of HPDs, recognizing that individuals and their jobs vary too much to permit specification of a single HPD for all. HPDs are rated by a *noise reduction rating* (NRR), which estimates the number of dB of attenuation obtained by proper use. However, most authorities believe that NRR numbers are usually too generous to apply to real-world situations. For conservative application, many recommend using half the NRR as a guide. For example, workers in a 100-dBA environment who use earplugs with a 30-dB NRR should be considered to be exposed to ~85 dBA (100 – 15 dB). Simple observation by supervisors is not enough to ensure proper use of HPDs. Earplugs are sometimes trimmed or perforated by uncooperative workers, and earmuffs can be stretched to reduce the spring force holding the muffs to the head, making them more comfortable but less protective. In either case, the worker would pass a cursory inspection.

It is now possible to measure HPD attenuation, as well as daily effective ("behind-the-HPD") exposures, for individual workers in field situations; this technology could help supervisors to detect and remediate cases where HPDs are inadequately fit. However, in a recent nonrandomized trial, workers using a device that gave them and their management daily feedback on their effective daily exposures did not have significantly less future threshold shift than control group workers, after appropriate statistical controls.[20]

Workers should be urged to use HPDs on and off the job whenever they are exposed to loud noises (generally, when it is necessary to raise the voice to converse at arm's length). For many workers, nonoccupational exposures—especially shooting—are more hazardous than occupational exposures.

The speech interference and safety risks (difficulty in hearing warning signals) of loud noise have already been mentioned. Simply making the desired signal louder than the background noise, for example, by shouting, can alleviate this problem. Once this has been done, the use of HPDs will not further degrade listening performance in normally hearing people. Both the signal and the noise will be reduced in intensity by the HPD, but as long as both are audible, the signal-to-noise ratio will be the same as without the HPD.

Persons with preexisting high-frequency hearing loss may have poorer detection and discrimination of speech and warning signals when using HPDs. This occurs because the HPD may actually make high-frequency signals (including the high-frequency portions of speech) inaudible.

Intermittent noise poses special safety hazards because what is protective in noisy environments may disrupt communication in quiet ones. Workers who wear HPDs during quiet periods may have difficulty in hearing and understanding normal speech. HPDs can make the worker's own voice seem louder (try plugging your ears with your fingers while speaking at a constant level); thus, workers may speak less loudly than normal when wearing HPDs in quiet environments.

These risks may be managed by implementing special communication strategies, such as those often used in radio communication: using a restricted message set to increase redundancy, spelling out words, and having important messages repeated back by the listener, to mention a few. Communication headsets are available that combine an earplug or earmuff, for attenuation of ambient noise, with a built-in radio receiver. In addition, active HPDs are available as earmuffs with built-in microphone–amplifier circuits. In quiet environments, these units amplify sound to overcome the attenuation of the earmuff, while in high levels of noise, the amplifier is automatically disabled. These units are particularly helpful for workers with hearing loss, who may need amplification in quiet places, but still need protective attenuation in noisy ones. Provision of this type of equipment may constitute the sort of accommodation necessary to comply with the Americans with Disabilities Act (ADA). Earmuffs with active noise reduction circuits are also available. These headsets use noise cancellation technology, actively decrease low-frequency exposures, and are particularly helpful for pilots in general aviation (small planes) where cabin noise is dominated by low frequencies.

References

1. Franks R. Number of workers exposed to occupational noise. *Semin Hear* 1988; 9:287–97.
2. Dobie RA. The burdens of age-related and occupational noise-induced hearing loss in the United States. *Ear Hear* 2008; 29:565–77.
3. US Department of Labor, Occupational Safety and Health Administration. Occupational noise exposure: hearing conservation amendment final rule. *Fed Regist* 1983; 48:9738–84.
4. Dobie RA, Clark WW. Exchange rates for intermittent and fluctuating noise exposures: a systematic review of studies of human permanent threshold shift. *Ear Hear* 2014; 35:86–96.
5. US Department of Labor, Occupational Safety and Health Administration. Occupational safety and health standards, national consensus standards and established federal standards. *Fed Regist* 1971; 36:10518.
6. Jayne TR. Legal remedies for hearing loss. In: Dobie RA, ed. *Medical-legal evaluation of hearing loss*. 3rd edn. San Diego, CA: Plural Publishing, 2015. 243–64.
7. Kujawa SG, Liberman MC. Adding insult to injury: cochlear nerve degeneration after "temporary" noise-induced hearing loss. *J Neurosci* 2009; 29:14077–85.
8. Taylor W, Pearson J, Mair A, et al. Study of noise and hearing in jute weaving. *J Acoust Soc Am* 1965; 38:113–20.
9. International Organization for Standardization. *Acoustics: determination of occupational noise exposure and estimation of noise-induced hearing impairment*. ISO-1999. Geneva: International Organization for Standardization, 2013.
10. Dobie RA. The AMA method of estimation of hearing disability: a validation study. *Ear Hear* 2011; 32:732–40.
11. Konings A, Van Laer L, Van Camp G. Genetic studies on noise-induced hearing loss: a review. *Ear Hear* 2009; 30:151–9.
12. Kowalski TJ, Pawelczyk M, Rajkowska E, et al. Genetic variants of CDH23 associated with noise-induced hearing loss. *Otol Neurotol* 2014; 35:358–65.
13. Kirchner DB, Evenson E, Dobie RA, et al. ACOEM guidance statement: occupational noise-induced hearing loss. *J Occup Environ Med* 2012; 54:106–8.
14. Dobie RA. Diagnosis and allocation. In: Dobie RA, ed. *Medical-legal evaluation of hearing loss*. 3rd edn. San Diego: Singular Thomson Learning, 2015.
15. Oishi N, Schacht J. Emerging treatments for noise-induced hearing loss. *Expert Opin Emerg Drugs* 2011; 16:235–45.
16. Marshall L, Lapsley-Miller JA, Heller LM. Detecting incipient inner ear damage from impulse noise with otoacoustic emissions. *J Acoust Soc Am* 2009; 125:995–1013.
17. Dobie RA. Is this STS work-related? ISO 1999 predictions as an adjunct to clinical judgment. *Am J Ind Med* 2015; 58:1311–18.
18. National Institute for Occupational Safety and Health. *Criteria for a recommended standard: occupational noise exposure. Revised criteria*. DHHS (NIOSH) publication no. 98–126. Washington, DC: US Government Printing Office, 1998.
19. American Academy of Otolaryngology, Head and Neck Surgery, Medical Aspects of Noise Subcommittee. *Otologic referral criteria for occupational hearing conservation programs*. Alexandria, VA: American Academy of Otolaryngology, Head and Neck Surgery Foundation, 1997.
20. Rabinowitz PM, Galusha D, Kirsche SR, et al. Effect of daily noise exposure monitoring in annual rates of hearing loss in industrial workers. *Occup Environ Med* 2011; 68:414–18.

Chapter 17

ELECTRICAL POWER and ELECTRICAL INJURIES

Jeffrey R. Jones*

Electrical hazards constitute a narrow but ubiquitous class of occupational physical hazards. Since the first report of death by electrocution in 1879, when a stage carpenter was killed after exposure to a 250 V AC generator, electric current has been responsible for a significant number of accidents that result in severe injury or death. Approximately 1 500 cases of electrocution occur annually in the United States, including about 100 lightning-caused deaths.[1–3] Approximately 300–400 of these occur at work, and the vast majority could be prevented if currently mandated safeguards were used and proper procedures followed. Although we now know a great deal about how to prevent electrical injuries, electrical fatalities still account for a significant number of fatalities in the workplace. The majority of injuries and deaths occur while workers are performing duties they normally undertake in the course of their job, suggesting inadequate training or an underestimation of the risks inherent to working around electricity.[4] Because lightning injuries differ in significant ways from other electrical injuries, they will be discussed in a separate section in the second part of this chapter. While the focus of this chapter is electrical injuries, it should be recognized that the injured workers might also have exposure to chemical hazards from fires and explosions, including ozone and PCBs.

OCCUPATIONAL SETTING

Electrocution and electric shock

Electrical Injury Epidemiology

The National Institute for Occupational Safety and Health (NIOSH) examined death certificates over a 13-year period to determine deaths at work from electrical causes (5 348 total deaths, average of 411 deaths per year).[1] A subset of those deaths (224 cases) was examined in greater detail to determine actions and risk factors that underlay the electrocutions. Electrocutions accounted for 7% of traumatic fatal injuries each year. Forty percent of the deaths were in construction and 60% occurred in workers less than 35 years old. Ninety-nine percent of the electrocutions were among men; 86% were Caucasian, probably reflecting the demographics of the workforce. Other industries with high rates included transportation/communication/public utilities (16%), manufacturing (12%), and agriculture/forestry/fishing (11%).

When NIOSH examined some of the electrocutions in detail, several trends were noted. Of the total, 33% involved low voltages (less than 600 V) and 66% involved high voltage. Most of the high-voltage deaths involved voltages of 7 200–13 800 (distribution voltages). Most of the low-voltage deaths involved 110–120 V. Utility linemen, who would be expected to have the most safety training, had the highest number of fatalities. In 55% of the deaths evaluated, there had been a failure to use required personal protective equipment (gloves, sleeves, mats, blankets, etc.). The risk greatly increases when repairs are conducted under conditions of widespread damage to electrical transmission and distribution systems, as in the aftermath of a natural disaster such as a hurricane.[3] Laborers, who would be expected to have substantially less electrical safety training, had slightly fewer fatalities. In 35% of the incidents, there was no safety program or written safety procedures. When safety procedures did exist, there was a lack of enforcement or supervisory intervention. Supervisors were present at

* The author wishes to thank Gary Pasternak, the author of this chapter in the first edition, for providing a strong foundation on which to build.

Physical and Biological Hazards of the Workplace, Third Edition. Edited by Gregg M. Stave and Peter H. Wald.

53% of the incidents, and 17% of the victims were supervisors. Forty-one percent of the victims had been on the job less than a year.

In the construction industry, electrocution is the second leading cause of death. Painters seem to be at particular risk, most likely from working near energized lines using ladders and other potentially conductive equipment.[5] The constantly changing nature of the construction work site exposes workers to temporary wiring, which may be substandard, and to harsh or unanticipated conditions that damage tool power cords and temporary power supplies. Portable arc-welding equipment is also responsible for numerous accidents, often as a result of improper grounding.

The greatest number of electrical injuries occurs in white men in the 25–34-year age group. Most injuries occur during the summer months, probably related to increased outdoor activity and the use of electrical equipment and machinery during this time of year. Increased sweating during the warmer months may also increase the severity of the electrical shock.[6] Alcohol and drugs have not been consistently found to be significant contributing factors in most work-related cases; however, widespread postaccident testing has not been routinely performed.[7,8]

Risk factors for electrocution

NIOSH described five scenarios that accounted for the deaths they investigated:[1]

1. Direct worker contact with an energized line (28%)
2. Direct worker contact with energized equipment (21%)
3. Boomed vehicle contact with an energized power line (18%)
4. Improperly installed or damaged equipment (17%), typically involving improper grounding
5. Conductive equipment, such as an aluminum ladder, contacting energized power lines (16%)

Risk factors for injury also include contact with moisture. When electric tools are handled with wet or perspiration-covered hands, resistance to electricity passing through the body is lowered. Risk is also heightened when workers wear clothing inadequate to protect them from electric arc and flash. Working on damp ground or where water is an integral part of a process, such as in irrigation, concrete work, and slaughtering, predisposes the victim to electric shock. Aerial power lines are a well-described hazard in the vicinity of irrigation pipes. In some agricultural regions, irrigation pipes have been the most common source of fatal human contact with electric lines.[9]

Exposure guidelines

The Occupational Safety and Health Administration (OSHA) has comprehensive standards covering electrical safety in both high and low voltages intended to prevent exposure. The primary required electric safety standards can be found in the following OSHA regulations:

- Subpart S, 29 CFR 1910.302 through 1910.399, which applies to all electrical installations and related equipment regardless of when they were installed
- Subpart K, 29 CFR 1926.402 through 1926.408 of the Construction Safety and Health Standards apply to installation of and work on electrical equipment, including equipment to supply power and light to job sites

Other consensus standards, such as National Fire Protection Association (NFPA) Standard 70E, may be incorporated by reference into some state standards.[10]

Specific sections address the need for lockout/tagout procedures (also covered in a specific lockout/tagout standard, 29 CFR 1910.147—see Chapter 5), proper or assured grounding of equipment, use of protective equipment, work on overhead power lines, and use of portable extension cords and cables. Improper lockout/tagout is one of the most common citations issued by OSHA. Protective equipment, such as rubber gloves and nonconductive sticks, must be adequate for the voltage levels encountered and must meet specific standards established by the American National Standards Institute (ANSI). Insulating gloves and other nonconductive personal protective equipment must be regularly inspected and periodically tested to ensure that they afford adequate protection. OSHA requires the use of nonconductive ladders where the employee or the ladder could contact exposed electrical conductors; it further requires that all metal ladders be prominently marked with a warning label. Clothing must be fire resistant when workers might be exposed to arc or flash, since injuries typically include burns that may be life threatening. NFPA Standard 70E addresses prevention of these injuries.

Pathophysiology of injury

Basic concepts of electrical energy

In order to grasp the consequences of human exposure to electricity, it is important to understand certain basic concepts of electrical energy. The tissue damage produced by electricity is proportional to the intensity of the current that passes through the body, as stated in Ohm's law:

$$\text{Resistance}\ (\Omega) = \frac{\text{voltage}}{\text{amperage}}$$

Voltage (tension or potential) is the electromotive force of the system, amperage (intensity) is a measure of current flow per unit time, and an ohm is the unit used to measure the resistance to conduction of electricity.

Two types of current are encountered in the workplace. Voltage is supplied either as a continuous source (direct current or DC), with electron flow moving in one direction, or as a cyclic source (alternating current or AC), with a cyclic reversal of electron flow. Nearly all injuries are the result of contact with the much more common AC. The exception is lightning, which can be thought of as a high-voltage DC shock.

AC is generated by large electromagnetic devices and generating stations, and it is transmitted many miles at very high voltages, typically greater than 100 000 V. Transformers reduce this voltage to 7 620 V, the usual voltage feeding residential and industrial distribution lines. At the local level and at residences, the voltage is again decreased to 110–220 V for domestic use.[11] The frequency of most commercial AC current in the United States is 60 cycles per second or hertz (Hz); 50 Hz in much of the rest of the world. Sixty hertz means that the flow of electrons changes direction 60 times per second. The use of a 50–60-Hz frequency has evolved because it is optimal for the transmission and utilization of electricity and because it has advantages in terms of generation.

Pathophysiology of electrical injury

Whether or not death or injury occurs is directly related to the following related factors:[12,13]

- Type of current (AC vs. DC)
- Frequency of current
- Voltage
- Amperage
- Duration of exposure
- Exposure pathway through the body
- Area of contact
- Resistance at points of electrical contact and in tissues

AC is significantly more dangerous than DC at similar voltages. AC can produce tetanic contractions that freeze the victim to the source of the current (an inability to "let go" of energized elements), and it can interfere with the respiratory and cardiovascular centers. Victims do not freeze to DC as with AC, but they can be injured when hurled from the point of electrical contact by the shock or when burned.

The frequency of AC is of importance physiologically. The most critical frequencies are in the range 20–150 Hz. Human muscular tissue responds well to frequencies between 40 and 150 Hz. Sixty-cycle household current lies directly within this range. If the shock spans the vulnerable period of the cardiac cycle, ventricular fibrillation may result. As the frequency increases beyond 150 Hz, human tissue response is decreased, and the current is generally less dangerous.

Although the divisions are somewhat arbitrary, harmful effects from electrical current can be subdivided into the effects of high-voltage current, greater than 600 V, and the effects of low-voltage current, less than 600 V. While most people recognize high voltage as dangerous, such as with overhead power lines, low-voltage current can prove equally deadly. Injury or death can occur with currents of very low voltage; instances of death have been reported with AC of 46 and 60 V. It has been claimed that any current greater than 25 V should be considered potentially lethal, although reports of fatalities from common electrical equipment operating below 50 V are rare.[14] Ventricular fibrillation is the most common cause of immediate death in high- and low-voltage injuries. Severe burns are more commonly associated with high-voltage currents.

The amount of current flow, expressed in amperes (A), is the single most important factor in injuries from electricity. Low-power alternating currents may be characterized by the thresholds of perception, "let-go", and ventricular fibrillation. The **threshold of perception** is the minimum current that causes any sensation in the human body. The **let-go threshold** is the maximum level of current flowing through an energized pathway at which a person taking hold of an element of the circuit is still able to release it. Currents greater than the let-go threshold are especially dangerous, because the body receiving the shock is unable to respond to it and break away. The International Electrotechnical Commission (IEC) has derived general values for the thresholds of perception and let-go. The threshold of perception is 0.5 milliamperes (mA), and the let-go threshold is 10 mA for signal frequencies in the range 15–100 Hz.

The threshold for an adverse effect is clearly related to the duration of exposure to the electrical energy. The IEC has derived values for the **threshold of ventricular fibrillation** by adapting results from animal experiments. For a 50/60-Hz signal, a current of 500 mA may cause fibrillation for a shock duration of 10 ms: likewise, 400 mA for 100 ms, 50 mA for 1 second, and 40 mA for greater than 3 seconds. The likelihood of ventricular fibrillation increases if the shock continues for a complete cardiac cycle.

The pathway of the current is critical to injury. Current passing through the head or chest is more likely to produce immediate death by affecting the central respiratory center, the respiratory muscles, or the heart. Arm-to-arm conduction can pass through the heart, and head-to-foot conduction can affect the central nervous system and the heart. Whereas a small current through the chest can be fatal, a large current passing through a single extremity may have little effect. The above values for the threshold of ventricular fibrillation are for current paths through the whole body, but fibrillation is only triggered by that level of current directly affecting the heart.[15] Nonfatal tissue injury is also determined by the current path. However, prediction of injuries from observation of the tissue path can be unreliable. Burn injuries have been shown to be more severe when the electrical energy traverses the long axis of the body.

The effect of electric current on the body depends on the resistance of the tissues involved. Intact dry skin has a relatively high resistance, but resistance falls dramatically where there is moisture, cuts, or abrasions. A callused palm has a resistance of 1 000 000 Ω, dry skin may have 5 000 Ω, and wet skin may have less than 1 000 Ω. The resistance of individual tissues is thought to differ considerably and should play a role in differential injury to tissues. Resistance to current flow occurs in decreasing order through the bone, fat, tendon, skin, muscle, blood vessels, and nerves. However, the theory that electric current always passes through the body along lines of least resistance may not be true; the path may largely depend on the voltage.

Tissue damage in electrical injury is caused by at least two mechanisms. Electrical energy may be converted to thermal energy in tissues and cause damage by heating, including protein denaturation and coagulation. The heat produced is a function of the current strength, duration of contact, and tissue resistance.[11,16] The other mechanism is the creation of pores in cell membranes by means of electrical current, known as electroporation.[16,17] Electroporation disrupts cell membranes and leads to cell death without significant heating. This occurs when high electrical field strengths are applied.[18] It is possible that tissue damage may result from other undescribed electrical effects.[11]

Health effects

Electric current can cause a spectrum of acute and chronic health effects and often results in multiorgan system manifestations.[11,19–21] Injury is most often the result of a direct effect on the heart, nervous system, skin, and deep tissues. Ventricular fibrillation, respiratory arrest, and burns are the most common causes of immediate death. Burns may be minor, requiring minimal debridement, or may cause devastating destruction, requiring aggressive resuscitation and major limb amputation. Other health effects include vascular damage (such as thrombosis, rhabdomyolysis, and subsequent renal failure), fractures due to high temperature or tetanic contractions, cataracts, neuropsychological changes, degenerative neurologic syndromes, and associated trauma (such as from falling off a ladder).

Electrical injury, with the exception of arc flash, resembles a crush injury more than a thermal burn. The extent of internal damage is often more severe than the cutaneous wound makes it appear. Small entry and exit wounds give little useful indication of the extent of underlying tissue damage and may not even be present in brief low-voltage injuries. In addition, there may be a direct electrical effect on the heart, nervous system, and skeletal muscles. Immediate death results from ventricular fibrillation, with or without asphyxia, from paralysis of the respiratory centers, or from prolonged tetany of respiratory muscles. Low-voltage injuries more commonly produce ventricular fibrillation rather than severe burns. Current passing through lower-resistance tissues causes necrosis of muscles, vessels, nerves, and subcutaneous tissues; it can also cause thrombosis and vascular insufficiency.

Cutaneous and deep-tissue effects Cutaneous injuries result from induced thermal burns, flame, and arc flash burns. An arc flash occurs when powerful, high-amperage currents travel, or arc, through the air between ungrounded conductors or between ungrounded conductors and grounded conductors.[10,22] This results in an instant release of tremendous amounts of energy. Temperatures as high as 36 000 F have been recorded.[22] Arc injuries result from current coursing external to the body, jumping from its source to the victim, or arcing between different sites on an extremity. The flexor surface of the forearm is a frequent site of arcing injury. Arcing may cause significant flame burns from ignition of clothing or other materials or from radiant heat. Frequently, an entrance and exit wound can be identified, but there may be severe deep-tissue damage with minimal cutaneous involvement. In severe burns, there may be extensive limb damage, requiring extensive debridement, fasciotomy, and limb amputation.

Arc blast results from the explosive expansion of air and metal in the arc path. When copper turns from a solid to a vapor, it expands 67 000 times.[10] This produces high pressures (hundreds to thousands of pounds per square foot), intense sound (exceeding 160 dB), and shrapnel (traveling at speeds >700 mph).[10] The pressure can knock workers off ladders, rupture eardrums, and collapse lungs.[10]

Cardiac effects The heart is particularly susceptible to electrical injury. A wide range of abnormalities may be seen, from arrhythmias to structural damage.[19] Ventricular fibrillation is a common cause of death in electric shocks. In nonfatal cases, arrhythmias are an important complication. The most common EKG abnormalities noted are sinus tachycardia and nonspecific ST-T wave changes. Ventricular and atrial ectopy, atrial fibrillation, bundle branch blocks, and ventricular tachycardia have all been reported. Generally, these abnormalities do not persist. Patients exposed to low voltages that are asymptomatic and initially have a normal EKG do not typically develop arrhythmias.[20]

High-voltage injury can cause myocardial necrosis, although the diagnosis can be difficult to make because of the absence of typical chest pain and EKG changes.[23] Evaluation of myocardial injury is based on measuring the cardiac fraction of the creatine kinase (CK) enzyme. Recent evidence suggests that a raised creatine kinase MB isoenzyme (CK-MB) level, which is elevated in acute myocardial infarction, is not necessarily indicative of myocardial damage.[24] Myocardial infarction has been reported as a rare complication of electrical injury; it may be the result of electrically induced coronary vasospasm.[25] Extensive burns and entrance and exit wounds in the upper and lower parts of the body may predict patients at risk of myocardial involvement and thus warrant intensive monitoring.[23]

NEUROLOGIC EFFECTS Acute central nervous system complications include respiratory center arrest or depression, seizures, mental status changes, coma, localized paresis, and amnesia. In mild cases, the patient may experience headaches, irritability, dizziness, and trouble in concentrating. These symptoms usually resolve in a few days. Peripheral nerve injuries are seen most often with extensive limb burns. Peripheral neuropathy may be seen following exposure to high current loads, even in the absence of extensive burns. Spinal cord damage is the most common permanent neurologic problem. It may cause progressive muscular atrophy or illness simulating amyotrophic lateral sclerosis or transverse myelitis. Symptoms characteristically appear after a latency period, with no (or minimal) neurologic symptoms in the acute stage. Some investigators have described a stereotypical generalized cerebral dysfunction leading to depression, divorce, and unemployment, as well as a high incidence of atypical seizures.[26] Complex regional pain syndrome Types I and II (formerly RSD and causalgia, respectively) have been reported following electrical injury.[19]

RENAL EFFECTS Electrical injuries produce a higher incidence of renal damage than other burns. Factors include shock, direct damage to the kidneys by high-voltage current, and the release of toxic products from the breakdown of damaged muscles. Myoglobinuria is commonly present and is proportional to the amount of muscle injury. In a retrospective case series, myoglobinuria was associated with high-voltage exposure, prehospital cardiac arrest, full-thickness burns, and compartment syndrome.[27] Timely administration of fluids and diuretics is important in the prevention of pigment-induced acute renal failure.[28] The development of acute renal failure in electrical injury correlates poorly with the extent of surface burns, given that the volume of tissue destroyed is often much greater than observed in a surface burn.[29] The therapeutic implication is that the formula for estimating fluid replacement in surface burns may seriously underestimate the fluid required in patients with electrical burns.

VASCULAR EFFECTS Vascular complications have included immediate and delayed major vessel hemorrhage, arterial thrombosis, abdominal aortic aneurysms, and deep-vein thrombosis. This large- and small-tissue damage, typically leading to vascular insufficiency, may be responsible for the tissue damage in electrical injury that is not immediately apparent.[11]

OTHER ORGAN SYSTEMS Musculoskeletal injuries are often initially overlooked. They may include multiple fractures and dislocations. Shoulder and scapular fractures are frequently reported. Delayed diagnosis of femoral neck fractures has been reported. Injury may result from falls, being hurled from the electrical source, or simply abrupt muscle contraction. Cataracts, conjunctival burns, and corneal burns can occur. Cataracts usually form 2–6 months after the shock, but they can appear immediately or many years later. Intra-abdominal injury should be expected in patients having burns of the abdominal wall. Stress ulcers of the duodenum (Curling's ulcer) can occur following severe burns.[19] Where arcing or burning has occurred, especially in vaults and other enclosed spaces, inhalation injury should be considered, since metals, dielectric fluids, and a variety of other materials may have been vaporized.

PREGNANCY

Electric shock during pregnancy may have serious health consequences for the mother and fetus, which may be even more vulnerable. The uterus and amniotic fluid are excellent conductors and fetal skin has a lower resistance to current.[30] Reported adverse fetal outcomes include spontaneous abortion, placental abruption, cardiac arrhythmias, fetal burn, and intrauterine fetal death.[30,31] When the pregnancy has continued, decreased fetal movements, asphyxia, pathological fetal heart patterns, intrauterine growth retardation, damage to the fetal central nervous system, and oligohydramnios have occurred.[30,31]

Diagnosis

The diagnosis of electrical injury can usually be made based on history from the patient or witnesses. Even with the available history, an assessment of the extent of trauma, burns, and related complications may be challenging due to related injuries that appear to explain signs and symptoms and other factors.[11] Assessment of the patient requires a thorough physical examination, laboratory testing, electrocardiogram (EKG) and cardiac monitoring, and possibly radiology studies. In general, the greater the amount of energy that was absorbed, the greater the underlying tissue damage. Appropriate testing to assess end-organ damage and rhabdomyolysis may include complete blood count (CBC), electrolytes, creatinine, urinalysis, myoglobin, creatine kinase (CK) component levels, and troponin.

Treatment

Treatment in the field involves rapidly and safely removing the victim from the source of the current, immediate and prolonged cardiopulmonary resuscitation (CPR), attention to other life-threatening injuries such as cervical spine injury, and initiation of advanced cardiac life support, if indicated. A rescuer can inadvertently become an electrocution victim. Do not touch victims who could still be in contact with the power source. Before touching the victim, the rescuer should attempt to turn off the electrical source. The victim can be separated from the power by using nonconductive materials such as rubber, a wooden tool handle, a mat, or heavy blankets. If it is not known whether the victim was thrown or fell, a cervical collar should be applied. Efforts should be made to revive victims who appear dead, since there is a good chance that some of them will respond to prolonged resuscitation attempts.[11,32]

Ideally, fluid resuscitation should begin in the field, especially in high-voltage injuries, which can result in significant volume depletion secondary to exudation and sequestration of fluids in burned and damaged areas. Fluid requirements frequently are much greater than those recommended by formulas that predict fluid needs from the area of cutaneous burns (note: this is not true for lightning injuries). Because deep-tissue damage can occur with limited surface burns, fluid should be administered in sufficient volume to maintain a urine output of 50–100 mL/hour to prevent renal insufficiency from myoglobin deposition. Mannitol may help to ensure an adequate urine flow. Tetanus immunization should be administered if indicated.

There are no clear-cut criteria for the hospital admission of less severely injured patients. Inpatient observation is advisable for patients who show evidence of cardiac dysfunction, symptoms of neurologic impairment, presence of significant surface burns, suspicion of deep-tissue damage, or laboratory evidence of acidosis or myoglobinuria. The evolution of tissue injury and vascular necrosis is usually complete within 8–10 days postexposure. Spies[16] has published a comprehensive summary of electrical injuries and proposed treatment approaches. Criteria for cardiac monitoring after electric injury were suggested by Fish[20] following a review of the available literature:

1. Loss of consciousness
2. Cardiac dysrhythmias
3. Abnormal 12-lead EKG
4. Abnormality on mental status or physical examination
5. Burns or tissue damage that would be expected to cause hemodynamic instability or electrolyte imbalance

Fish[20] further summarized factors to be considered in a patient that may have been exposed to significant electricity (Table 17.1)

TABLE 17.1 Factors in the Physical Examination and Work-Up Suggesting a High-Risk Electric Injury.

1. Factors suggesting significant effects on the patient
    A. Evidence of inhalation injury
    B. Dysrhythmia
    C. Confusion (do a non-contrast CT scan of the head)
    D. Abnormal physical examination (neurologic, orthopedic, vascular)
    E. Abnormal laboratory examination (Urinalysis, EKG, CK, CK-MB, or troponins)*
    F. Significant burn or other condition requiring treatment
2. Signs of deep tissue injury, especially in extremities
    A. Edema
    B. Ischemic changes
    C. Sensory or motor loss
    D. Full-thickness skin injury along the path of current without flame burns in the same area
    E. Persistent flexion deformity

*EKG = Electrocardiogram; CK = Creatine kinase; CK-MB = creatine kinase MB isoenzyme.

Source: Fish R. Electric injury Part III: Cardiac monitoring indications, the pregnant patient, and lightning. *J Emerg Med* 2000; 18(2):186. Reprinted with permission.

Arc flash injuries should be treated as trauma following established emergency burn protocols.[10,22,33,34]

Medical surveillance

No specific medical surveillance is required for workers engaged in electrical work, although they may require surveillance for exposure to other workplace hazards. Injured workers should be evaluated for delayed complications, including cataracts. Injured pregnant workers require ongoing obstetrical monitoring.

Prevention

Occupational electrical injuries are generally preventable. Each electrical injury or fatality should be viewed as a sentinel health event. The incident should prompt an analysis of the work site with the intent of preventing any further injuries. Any injury suggests the need for a job-specific electrical safety analysis and provides an opportunity for preventive intervention in the workplace. The majority of these injuries occur either from lack of education concerning the specific hazards of the work or from failure to follow safe work practices.

Primary prevention of electrical injuries is the ultimate goal. Prevention can be accomplished through the use of engineering and administrative controls, personal protective equipment, and training. Lockout/tagout (disabling machinery or equipment to prevent the release of hazardous energy), the use of appropriate personal protective equipment, and an understanding of the hazards are key to injury prevention. Lockout/tagout and other administrative controls require commitment from management and workers and should be part of a written safety program. Their effectiveness depends on rigorous implementation and enforcement in the workplace. An example of an administrative control would be a scheduled inspection and preventive maintenance program for all power tools and electrical cords.

Training

Persons exposed to electrical risk fall into two general categories: (i) those with training and experience in electrical work as a craft, such as electricians and linemen, and (ii) those not engaged in "electrical work" who nonetheless use or work near equipment with potential electrical hazards. Injury prevention requires both education regarding hazards and attention to safe work practices. Many electrical injuries occur among nonutility/electrical workers, and these employees should be specifically targeted for primary prevention activities. Work site electrical safety education should focus on the recognition of potential electrical hazards and how to avoid exposure to live electrical circuits.[28]

Proactive safety programs are needed in all occupations that have a high risk of electrical injury. NIOSH and OSHA have produced numerous documents that contain elements of effective safety training, including electrical safety. Some state OSHA programs have stricter and/or more detailed requirements than the federal program. In general, comprehensive safety programs should include written rules and safe work procedures for dealing with electrical hazards.[1,35]

Workers must be educated about the potential dangers of low voltage, for example, 120 V. They should be instructed to always ground hand tools properly, especially portable powered hand tools. Extension cords used to supply power to portable tools need particular attention, especially at temporary work sites, because grounding can fail. Using battery-powered or double-insulated power tools plugged into ground fault circuit interrupters (GFCIs) can help prevent electrical shock, even under adverse conditions.

Engineering controls

Many options are available to reduce electrical hazards. The single most effective is to de-energize and lock out any active circuit or equipment that could be contacted during the work activity. Failing that, risk can be reduced through measures such as enhancement with visual markers, sleeving to insulate lines in high-risk areas, using ladders made of nonconducting materials, and using procedures to stabilize and prevent equipment from moving into and contacting power lines. "Lookout" workers can help guide aerial equipment. A minimum of 10 ft of clearance is required between operating aerial equipment and a power line.

Many accidents occur during the use of electrically powered machinery or portable powered hand tools. Repairing damaged power cords, maintaining proper grounding, and using GFCIs can prevent the majority of these injuries. GFCI-protected circuits are required at construction sites, but can have much broader application, even at fixed locations. GFCIs represent a simple engineering control that could save dozens of lives each year.

LIGHTNING INJURIES

According to the National Weather Service, 100 000 thunderstorms annually produce approximately 30 million lightning strikes in the United States. These kill, on average, 82 people and injure 1 000–2 000 people per year. It is estimated that 25% of these cases are occupationally related.[36,37]

Occupational setting

The Centers for Disease Control and Prevention (CDC) summarized lightning-caused deaths from 1980 to 1995.[21] During this period, there were 1 318 deaths (82 deaths per year, range 53–100 deaths). Occupationally, many of these accidents occurred in the agricultural and construction industries, but they also occurred in other jobs where people work outdoors, such as wild land firefighting, sailing, and golfing; other activities involving open, exposed situations; and with less frequency, indoors. CDC reported 350 military personnel injured by lightning, with one death, from 1998 to 2001.[38] Most victims are injured during the summer months, when thunderstorms are most frequent. The states with the greatest number of deaths were Florida and Texas, but the highest death rates were in New Mexico, Arizona, Arkansas, and Mississippi.

Working (and thus getting paid) may be an incentive to risk exposure. Workers may feel compelled or be forced to continue working when thunderstorms are near for fear of losing their jobs, thus prolonging their potential exposure to lightning. The hazard of lightning is often not realized by those exposed: lightning has struck 10 miles away from the rain of a thunderstorm.

Pathophysiology of injury

Lightning kills 30% of its victims, and 74% of survivors experience a permanent disability.[36] Sixty-three percent of deaths occur within an hour of injury. The most common cause of death is immediate cardiopulmonary arrest. Lightning is dangerous due to high voltage, heat generation, and explosive force. Lightning may also injure indirectly by starting forest and house fires, causing explosions, or by felling objects such as trees.

TABLE 17.2 Lightning versus high-voltage electrical injury.[39]

Factor	Lightning injury	High-voltage injury
Energy level	Very high voltage and amperage	Lower
Duration of exposure	Brief instantaneous	Prolonged
Pathway	Flash over	Deep, internal
Burns	Superficial	Deep, major injury
Cardiac	Asystole more common	Ventricular fibrillation
Renal	Rare myoglobinuria	Myoglobinuric renal failure common
Fasciotomy and amputation	Very rare	Common, may be extensive
Blunt trauma	Explosive thunder effect	Falls, being thrown

Source: Adapted from Cooper M. Lightning injuries. This article was published in: Auerbach PS, Geehr EC, eds. *Management of wilderness and environmental emergencies*. 2nd ed. St. Louis: Mosby-Year Book, 1989: 171–93. Copyright Elsevier 1989. Printed with permission.

There are significant differences between injuries from electric current and injuries from lightning (Table 17.2).[39] The factor that seems to be most important in distinguishing lightning from electric current injuries is the duration of exposure to the current. Exposure to lightning current is nearly instantaneous, so that prolonged contact does not occur. The energy generally travels superficially over the surface of the body. Distinct entry and exit wounds are rare, and deep burns are infrequent. The explosive force of the lightning strike may cause significant blunt trauma if the victim is thrown by a direct strike or by the shock wave created by the flash. Clothing and shoes may be literally blown from the body. Injuries are classified from minor to severe. As with electrical injuries, several organ systems may be affected, and both acute and chronic effects are seen.[21]

With minor injury, patients experience confusion and temporary amnesia. They rarely have significant burns but may complain of paresthesias and muscular pain. These patients usually recover completely. A ruptured tympanic membrane is a frequent finding, due to the explosive force of the lightning shock wave. With moderate injury, patients may be disoriented, combative, or unconscious. Motor paralysis, more often of the lower extremities, may be seen, along with mottling of the skin and diminished or absent pulses due to arterial spasm and sympathetic instability. Hypotension should be ruled out and, if found, should prompt a search for inapparent blunt trauma and fractures. Victims may have experienced temporary cardiopulmonary arrest at the time of the strike. Respiratory arrest may be prolonged and lead to cardiac arrest from hypoxia. Seizures may also occur. Ruptured tympanic membranes are commonly found, and minor burns may become apparent after a few hours. Patients generally improve, although they may experience chronic symptoms such as sleep disturbances, weakness, paresthesias, and psychomotor abnormalities. As with electrical shock injuries, rare cases of spinal paralysis have been reported.

Severely injured patients may present in cardiac arrest, with either asystole or ventricular fibrillation. In fact, victims are unlikely to die unless cardiopulmonary arrest occurred at the time of the lightning strike.[40] Because persons struck by lightning have a better chance of survival than persons suffering cardiopulmonary arrest from other causes, resuscitation should be started immediately.[20,36] Resuscitation may not be successful if there was a significant delay in initiating CPR. Direct brain damage may have occurred. Findings of blunt trauma suggest a direct strike. Long-term sequelae in survivors also include visual and hearing deficits, due most often to damaged tympanic membranes and cataracts. Neurological sequelae are described in Table 17.3.

TABLE 17.3 Neurological Sequelae of Lightning Strike.

I. Early (and sometimes spontaneously reversible) conditions
   A. Loss of consciousness
   B. Confusion
   C. Abnormalities of motor and sensory function of one or more limbs
   D. Paraplegia, quadriplegia, and focal paralysis, often resolves within hours to days
II. Conditions that can be persistent
   A. Retrograde amnesia
   B. Late-developing hemiplegia and aphasia
   C. Later-occurring neuritis and painful neuralgia in extremities
   D. Peripheral (e.g., median) neuropathy
   E. Neuropsychiatric disorders with normal brain CT scan and EEG
   F. Coma
   G. Cerebral edema
   H. Inappropriate secretion of antidiuretic hormone (SIADH)
   I. Seizures
   J. Cerebellar ataxia
   K. Painful sensory disturbances

Source: Fish R. Electric injury Part III: Cardiac monitoring indications, the pregnant patient, and lightning. *J Emerg Med* 2000; 18(2):185. Reprinted with permission.

Lightning injury during pregnancy has resulted in fetal death *in utero*, abortion, stillbirth, and neonatal death.[30]

Diagnosis

The diagnosis of injury from lightning can usually be made based on history from the patient or witnesses. Since injury can affect many organ systems, an assessment of the extent of trauma, burns, and related complications requires a thorough physical examination and testing. Initial assessment and follow-up should include an evaluation of neurological and neuropsychological symptoms.

Treatment

Initial treatment includes rapid attention to CPR and support, if necessary. In general, fluids should be restricted unless there is evidence of hypotension. Although minor burns may be present, vigorous fluid therapy and mannitol diuresis are not indicated unless myoglobinuria is found. Overhydration with resultant cerebral edema has probably killed more lightning victims than pigment-induced renal failure.[39] Fasciotomy is rarely needed in lightning injuries, since most burns are superficial.

Medical Surveillance

No specific medical surveillance is indicated. Follow-up assessment for injured workers should include evaluation of neurological and neuropsychological sequelae.

Prevention

Workers involved in activities in areas where lightning strikes are possible should be familiar with the preventive measures recommended by CDC (Table 17.4).[37]

TABLE 17.4 Recommendations for preventing injuries from lightning strike.

1. During a storm, take shelter inside a home, large building, or vehicle
2. If outside and unable to reach shelter, do not stand under or near a tall tree or other structure in an open area. Go into a ravine or gully. Assume a squatting posture on the balls of your feet with your head down and your hands over your ears (the lightning crouch) to minimize the chance of exposure to lightning. Do not lie flat
3. Get out of and away from open water
4. Get away from metal equipment or objects such as tractors, antennas, drainpipes, and metal stairs
5. Put down any objects that might conduct electricity (shovels, rakes, ladders, etc.)
6. In forested areas, seek shelter in a low-lying area under a thick growth of small trees
7. When indoors, avoid using electrical appliances or telephones

Source: Adapted from Center for Disease Control. Lightning-caused deaths 1980–1995. *MMWR* 1998; 47:391.[36]

References

1. National Institute for Occupational Safety and Health (NIOSH). *Worker Deaths by Electrocution: A Summary of NIOSH Surveillance and Investigative Findings*. DHHS (NIOSH) Pub. No. 98–118. NIOSH, Cincinnati, OH, 1998.
2. Mellen PE, Weed VW, Kao G. Electrocution: a review of 155 cases with emphasis on human factors. *J Forens Sci* 1992; 37:1016–22.
3. Cawley, JC, Homce, GT. Occupational electrical injuries in the United States, 1992–1998, and recommendations for safety research, *J Safety Res* 2003; 34:241–8.
4. Harvey-Sutton PL, Driscoll TR, Frommer MS, et al. Work-related electrical fatalities in Australia, 1982–1984. *Scand J Work Environ Health* 1992; 18:293–7.
5. Centers for Disease Control. Up-date: work-related electrocutions associated with Hurricane Hugo—Puerto Rico. *MMWR* 1989; 38:718–20.
6. Suruda AJ. Work-related deaths in construction painting. *Scand J Work Environ Health* 1992; 18:30–3.
7. Baker SP, Samkoff JS, Fisher RS, et al. Fatal occupational injuries. *JAMA* 1982; 248:692–7.
8. Berkelman RL, Herndon JL, Callaway JL, et al. Fatal injuries and alcohol. *Am J Prev Med* 1985;1(6):21–8.
9. Helgerson SD. Farm workers electrocuted when irrigation pipes contact power lines. *Public Health Rep* 1985; 100:325–8.
10. National Fire Protection Association. NFPA 70E Standard for Electrical Safety in the Workplace (2015). Available at: http://www.nfpa.org/codes-and-standards/all-codes-and-standards/list-of-codes-and-standards?mode=code&code=70e (accessed on June 14, 2016).
11. Fish R. Electric injury, Part I: Treatment priorities, subtle diagnostic factors, and burns. *J Emerg Med* 1999; 17(6):977–83.
12. Koumbourlis AC. Electrical injuries. *Crit Care Med* 2002; 30(11 Suppl):S424–30.
13. Cooper A., Price TG. Electrical and Lightning Injuries. Available at: http://lightninginjury.lab.uic.edu/Electr&Ltn.pdf (accessed on June 23, 2016). University of Illinois at Chicago, Chicago, IL.
14. Bikson, M. A Review of Hazards Associated with Exposure to Low Voltages, Department of Biomedical Engineering, City College of New York, New York. Available at: http://bme.ccny.cuny.edu/faculty/mbikson/BiksonMSafeVoltageReview.pdf (accessed on June 14, 2016).
15. Robinson MN, Brooks CG, Renshaw GD. Electric shock devices and their effects on the human body. *Med Sci Law* 1990; 30:285–300.
16. Spies C, Trohman RG. Narrative review: electrocution and life-threatening electrical injuries. *Ann Intern Med* 2006; 145:531–7.
17. Block TA, Aarsvold JN, Matthews KL 2nd, et al. The 1995 Lindberg Award. Nonthermally mediated muscle injury and necrosis in electrical trauma. *J Burn Care Rehabil* 1995; 16:581–8.
18. Lee RC, Gaylor DC, Bhatt D, et al. Role of cell membrane rupture in the pathogenesis of electrical trauma. *J Surg Res* 1988; 44:709–19.
19. Fish R. Electric injury: Part II: Specific injuries. *J Emerg Med* 2000; 18(1):27–34.

20. Fish R. Electric injury: Part III: Cardiac monitoring indications, the pregnant patient, and lightning. *J Emerg Med* 2000; 18(2):181–7.
21. Sanford A, Gamelli RL. Lightning and thermal injuries. *Handb Clin Neuro* 2014, 120:981–6.
22. Occupational Health and Safety Administration (OSHA). Understanding "Arc Flash." Available at: https://www.osha.gov/dte/grant_materials/fy07/sh-16615-07/arc_flash_handout.pdf (accessed on June 14, 2016), 2007.
23. Chandra NC, Siu CC, Munster AM. Clinical predictors of myocardial damage after high voltage electrical injury. *Crit Care Med* 1990; 18:293–7.
24. McBride JW, Labrosse KR, McCoy HG, et al. Is serum creatine kinase-MB in electrically injured patients predictive of myocardial injury? *JAMA* 1986;255(6):764–8.
25. Xenopoulos N, Movahed A, Hudson P, et al. Myocardial injury in electrocution. *Am Heart J* 1991; 122:1481–4.
26. Hooshmand H, Radfar F, Beckner E. The neurophysiological aspects of electrical injuries. *Clin Electroencephalogr* 1989; 20:111–20.
27. Rosen CL, Adler JN, Rabban JT, et al. Early predictors of myoglobinuria and acute renal failure following electrical injury. *J Emerg Med* 1999; 17(5):783–9.
28. Gupta KL, Kumar R, Sekhar S, et al. Myoglobinuric acute renal failure following electrical injury. *Renal Failure* 1991; 13: 23–5.
29. DiVincenti FC, Moncrief JA, Pruitt BA. Electrical injuries: a review of 65 cases. *J Trauma* 1969; 9:497–507.
30. Fatovich DM. Electric shock in pregnancy. *J Emerg Med* 1993; 11:175–7.
31. Sparić R, Malvasi A, Nejković L, et al. Electric shock in pregnancy: a review. *J Matern Fetal Neonatal Med* 2015 25:1–7.
32. Kearns, RD, Rich PB, Cairns CB, et al. Electrical Injury and Burn Care: A Review of Best Practices, September 1, 2014. Available at: http://www.emsworld.com/article/11621404/electrical-injury-electrocution-and-burn-care (accessed on June 14, 2016).
33. Washington State Department of Labor and Industries. Burn Injury Facts: Arc Flash/Blast, Report # 86-1-2006, April 2006.
34. Curwick, CC. *Hospitalized Work-Related Burns in Washington State*, Technical Report Number 86-2-2006, Washington State Department of Labor and Industries, 2006.
35. National Institute of Occupational Safety and Health (NIOSH). *Preventing Falls and Electrocutions during Tree Trimming*. DHHS (NIOSH) publication No. 92-106. NIOSH, Cincinnati, OH, 1992.
36. Centers for Disease Control. Lightning-caused deaths 1980–1995. *MMWR* 1998; 47:391.
37. Duclos PJ, Sanderson LM, Klontz KC. Lightning-related mortality and morbidity in Florida. *Public Health Rep* 1990; 105:276.
38. Centers of Disease Control. Lightning-associated injuries and deaths among military personnel – United States, 1998–2001, *MMWR* 2002 51(38); 859–62.
39. Cooper MA. Lightning injuries. In: Auerbach PS, Geehr EC, eds. *Management of Wilderness and Environmental Emergencies*, 2nd edn. St. Louis, IL: Mosby-Year Book, 1989:173–93.
40. Cooper MA. Lightning injuries: prognostic signs for death. *Ann Emerg Med* 1980; 9:134–8.

Part II

BIOLOGICAL HAZARDS

Chapter 18

GENERAL PRINCIPLES of MICROBIOLOGY and INFECTIOUS DISEASE

WOODHALL STOPFORD*

Common occupational and environmental biological hazards include microorganisms (viruses, rickettsia, chlamydiae, bacteria, fungi, and parasites), allergens of biological origin (e.g., the aeroallergenic fungi and animal dander), and the by-products of microbial growth (e.g., the endotoxins and mycotoxins). Because of their invisible and frequently undetectable nature, biohazards are considered "silent hazards." Among the occupations associated with biohazards are the healthcare industry, agriculture, science and technology, livestock management, fish and shellfish processing, forestry, waste management, and recreation management.

Microorganisms are found everywhere in nature. They inhabit all environmental niches from the polar icecap to the tropics and deserts. Microorganisms are intimately associated with all living species. Many forms are present as the normal flora of the skin and body orifices, whereas others may cause disease. Most of the microorganisms found on earth, including most of the human and animal pathogens, belong to the mesophilic species, which survive best at ambient temperatures of 20–400°C. Microorganisms that require elevated temperatures for growth belong to the thermophilic species and those that thrive at lower temperatures belong to the psychrophilic species.

The human host is constantly exposed by a variety of routes to biological materials, including living microbes and their products. Humans are also the source of many microorganisms through the shedding process, whereby thousands of organisms are released continuously from the skin, mucous membranes, and body orifices. The great majority of these microorganisms are harmless to us and to our ecosystem. Most are saprophytic organisms that live on inanimate substrates and represent normal human and environmental flora. Nevertheless, pathogenic microorganisms have had a tremendous impact on humankind throughout history, with devastating pandemics of smallpox, yellow fever, influenza, AIDS, plague, tuberculosis, and malaria.

There is also a potential for new infectious diseases of occupational significance to emerge, such as hantaviruses and coronoaviruses. Such outbreaks could result from environmental changes, microbial adaptation, or population movements. Awareness of workplace biohazards is critical to the prevention of occupational disease; complacency can result in serious and life-threatening illness.

Microorganisms found in the workplace range from extremely small viruses and single-celled bacteria to multicellular fungi and parasites. To understand how these organisms produce disease, it is necessary to know how they function, how disease is transmitted, and what factors influence whether infection will develop.

ETIOLOGY OF DISEASE

When Robert Koch discovered the causative microbe of anthrax in 1876, he formulated criteria to establish that a specific microorganism was the cause of a clinically discernible disease. These criteria, known as Koch's postulates, stipulate that:

1. The specific organism must be found in diseased animals and not in healthy animals.
2. The specific organisms must be isolated from the diseased animal and grown in pure culture.

* The author wishes to thank Jerry J. Tulis for his contribution to an earlier version of this chapter.

Physical and Biological Hazards of the Workplace, Third Edition. Edited by Gregg M. Stave and Peter H. Wald.

3. The identical disease must be produced upon inoculation of healthy susceptible animals with a pure culture of the originally isolated organism.
4. The identical organism must be isolated from the experimentally infected animal.

Microorganisms that can produce disease are classified into categories including viruses, rickettsia, chlamydia, bacteria, fungi, and parasites.

Viruses

Viruses represent the smallest etiologic agent of human diseases (measuring 20–300 nm). They are responsible for the great majority of human infections, especially through inhalation. The classification of viruses depends on the following criteria: (i) morphology, (ii) the presence of envelopes surrounding the viral capsid, (iii) the type of genetic material (RNA or DNA), (iv) organs and tissues preferentially infected, and (v) the nature of disease caused. All viruses are obligate intracellular parasites. Viruses cannot multiply outside the host cell. Their survival as naked particles in the environment is limited, ranging from several hours to a few weeks. Viruses can only infect cells where appropriate receptors are present. When a virus infects a host cell, it utilizes the metabolic machinery of that cell to replicate itself.

Viruses may contain either deoxyribonucleic acid (DNA) or ribonucleic acid (RNA) as their genetic material. If the virus contains DNA, this nucleic acid is a recognizable substrate for the host cell's DNA polymerase enzymes. These enzymes translate the DNA into messenger RNA (mRNA). The mRNA is then transcribed by the host cell's ribosomes into viral proteins. By contrast, the RNA in RNA viruses is not recognized by the host cell's ribosomes. These viruses, classified as retroviruses, contain a unique enzyme known as reverse transcriptase. This enzyme allows for transcription of infecting viral RNA to host cell DNA within infected cells. Once this DNA is created, the genetically programmed viral replication proceeds in a manner similar to that for DNA viruses. The replication process results in the production of many additional copies of the infecting virus. These are subsequently liberated into the surrounding milieu to infect other target cells. Some viruses, termed cytopathogenic viruses, cause the destruction of the host cell during the replication process.

Rickettsia, Coxiella, Ehrlichia, and Anaplasma

The *Rickettsia, Coxiella, Ehrlichia*, and *Anaplasma* are primarily intracellular parasites, although they are considerably more complex than the viruses. They are coccobacillary in morphology, contain both RNA and DNA, and resemble the Gram-negative bacteria. As intracellular parasites, the rickettsiae multiply through the process of binary fission. They are completely independent of host cell metabolic activity. Most rickettsial agents are transmitted to humans through arthropod vectors, for example, the transmission of Rocky Mountain spotted fever by the Dermacentor tick. However, Q fever, caused by *Coxiella burnetii*, is readily transmitted through contaminated aerosols and has been responsible for numerous laboratory-acquired infections. Outdoor sites represent the greatest risk of rickettsial infection for workers, including those employed in agriculture, forestry, and construction.

Chlamydiae

Chlamydiae are usually classified as belonging to the domain of the bacteria, although many authors separate them for purposes of discussion. Like viruses, chlamydiae are obligate intracellular parasites; however, they differ from viruses by being susceptible to antibiotics. Among the vertebrate host range of the chlamydiae are birds, mammals, and humans. The leading sexually transmitted disease in the United States is trachoma, caused by *Chlamydia trachomatis*; worldwide, it is the primary cause of human blindness. The most important occupational disease caused by the chlamydiae is psittacosis, a zoonotic disease caused by *Chlamydophila psittaci*. The primary reservoirs of *C. psittaci* are the psittacine birds (e.g., parrots) and domestic chickens and turkeys. Psittacosis is readily transmitted by aerosol, and the organisms remain stable in dried form for extended periods.

Bacteria

Bacteria are single-celled organisms. They have semirigid cell walls and a cell nucleus containing DNA that is not membrane bound. They reproduce through the process of binary fission. Bacteria include thousands of species, encompassing numerous genera. In addition to genus and species classification, bacteria can be categorized in several ways.

Morphologically, bacteria are categorized as cocci, bacilli, and spirilla. The cocci, which are round, spheroidal or ovoid in shape, are found as single cells, doublets, tetrads, clusters, and chains. The bacilli, or rod-shaped bacteria, occur as coccobacilli, square-ended bacilli, round-ended bacilli, club-shaped bacilli, and fusiform bacilli. The spirilla include the corkscrew and comma-shaped organisms (e.g., the vibrios and spirochetes).

Bacteria may also be differentiated based on the results of a commonly used differential specimen stain, the Gram stain. Organisms detected microscopically with this technique are described as Gram positive or Gram negative. Approximately 67% of the cocci and 50% of the bacilli are Gram positive. All spirilla are Gram negative. The mycobacteria (organisms that cause tuberculosis and atypical tuberculosis) have a waxy envelope and do not stain readily with the Gram stain. To detect mycobacteria, a special acid-fast staining procedure is required.

Bacteria may also be differentiated based on biochemical characteristics. Organisms may exhibit characteristic patterns of sugar fermentation and metabolic product formation. They may also require specific substrates or nutrients for growth.

Other techniques used to classify bacteria include bacteriophage typing (classification of bacteria based on susceptibility to different strains or types of species-specific viruses) and bacterial chromosome analysis using restriction endonucleases.

All Gram-negative bacteria possess a lipopolysaccharide component of the cell wall that displays toxic properties. This is referred to as endotoxin. Although the potencies of endotoxins produced by different bacterial species vary, all endotoxins possess pyrogenic (fever-inducing) properties. In the occupational setting, exposure to aerosols contaminated with polluted water, animal feces, or soil can result in human exposure to endotoxins.

Fungi

Fungi are composed of molds and yeasts; some species exhibit dimorphic properties, growing as either molds or yeasts depending on the substrate and temperature. Although thousands of fungal species are found in nature, <100 species are responsible for all human and animal diseases, and less than a dozen species are responsible for the majority of human mycotic infections. Fungal diseases are classified as mycoses, mycotoxicoses, and allergies.

The mycoses can be localized or systemic. The occupational mycoses transmitted by the respiratory route include blastomycosis, cryptococcosis, histoplasmosis, and coccidioidomycosis, which have been implicated as a etiologic agents in laboratory-acquired infections. All of these fungi are natural inhabitants of the soil and become aerosolized when the soil is disturbed, as occurs during construction, demolition, and other earth-moving activities. *Histoplasma capsulatum* and *Cryptococcus neoformans*, the causative agents of histoplasmosis and cryptococcosis, have a predilection for growth in soils contaminated with bird droppings; they are often found in the vicinity of poultry houses and bird-roosting areas.

Mycotoxicoses are intoxications resulting from exposure to fungal toxins (mycotoxins). Although numerous mycotoxins have been identified, the best studied are the aflatoxins, elaborated by species of *Aspergillus*. Besides possessing mutagenic, carcinogenic, and teratogenic properties, the aflatoxins are acute toxins affecting various body organs. The substrates for these molds are extensive and include most agricultural products, for example, corn and peanuts in the United States. Rigid standards have been imposed by the US Department of Agriculture to control aflatoxin levels of these commodities.

Fungal allergies are represented by clinical cases of allergic rhinitis, hypersensitivity pneumonitis, and asthma. The allergic manifestation is not an infection *per se*, and fungal viability is not required to induce allergic disease, since mycelial fragments, dead spores, and other fungal debris can elicit a host response. Common aeroallergenic fungal genera include *Cladosporium*, *Penicillium*, and *Aspergillus*, among others. Clinical and laboratory findings indicate that building-related illness is sometimes due to the inhalation of mold-contaminated air, but so-called sick-building syndrome remains a poorly defined illness of unknown etiology.

Parasites

Parasites may be involved as etiologic agents of occupational infections as a result of travel to, or work in, endemic areas. Illnesses such as giardiasis and amebic infections can result from workplace exposure to contaminated water. Human parasites can be classified as the protozoa and the helminths. The protozoa are single-celled organisms composed of the amoebae, the ciliated protozoa, the flagellated protozoa, the malarial parasites, *Toxoplasma gondii*, and *Pneumocystis carinii*. The helminths are multicellular parasitic worms.

Parasites may have complex life cycles that involve sexual and asexual reproductive states. Parasites may also infect both intermediate and definitive hosts. For different organisms, humans may serve the role as intermediate host, definitive host, or both.

TRANSMISSIBILITY OF DISEASE

The recognized routes of human exposure to etiologic agents of disease include (i) the respiratory route, (ii) the oral route, (iii) the contact route, (iv) the parenteral route, and (v) transmission through arthropod vectors.

Etiologic agents of occupational disease are most frequently transmitted through the respiratory route. Aerosolized particles (bioaerosols) composed of infectious or airborne allergenic (aeroallergenic) agents are difficult to detect or control. Exposure may result in sporadic and multiperson exposure in indoor and outdoor environments. The great majority of documented laboratory-acquired infections have resulted from apparent respiratory exposure, including disease agents not normally transmitted through aerosols (e.g., Rocky Mountain spotted fever and rabies). Bioaerosol particles, which measure 0.5–5.0 µm in aerodynamic size, can readily penetrate deep into the respiratory tract, reaching the alveolar spaces.

Occupational exposure to infectious agents through the oral route occurs by the following mechanisms: sprays and splatters, ingestion while mouth pipetting, consumption of contaminated foods, or touching the nose or mouth with contaminated hands. Since mouth pipetting has been prohibited in most laboratories, infection by accidental ingestion has been reduced significantly. Occupational infection by enteric pathogens, hepatitis A virus (HAV), listeria, and other agents continues to occur.

Contact exposure in the workplace has resulted in a variety of occupational infections, including tularemia, Newcastle disease, hepatitis B virus (HBV), human immunodeficiency virus (HIV), brucellosis, anthrax, glanders, erysipeloid, herpes, and leptospirosis. Transmission of disease organisms has occurred by contact with contaminated surfaces or fomites, exposure of mucous membranes and skin surfaces (including nonintact skin), and exposure to spatters and sprays of infectious agents. Routine handwashing practices and the use of gloves and other protective apparel can significantly reduce the spread of infectious organisms by contact.

Parenteral exposure to infectious organisms in the workplace results primarily from accidental needlestick or other penetrating trauma, such as skin puncture with sharp instruments or animal bites and scratches. Most workplace infections with HBV or HIV are the result of accidental needlestick or sharps injury.

Arthropods may serve as vectors in transmitting occupational infections. Examples of vector-borne diseases include the mosquito in malaria and the encephalitides, the flea in plague and tularemia, and the tick in Rocky Mountain spotted fever and Lyme disease. This route of infection is primarily associated with outdoor work, including forestry management and lumbering, agriculture, construction, and recreation management. Employees involved in outdoor activities and fieldwork need to be cognizant of vector-borne diseases, especially in endemic areas.

INFECTIVITY OF DISEASE

Following exposure to etiologic agents of disease, the infective process depends on a number of factors—namely, the resistance or susceptibility of the host, the exposure route and dose, and the virulence of the specific pathogen. Although host susceptibility is difficult to document, certain factors are recognized as being contributory, including age, race, gender, health status, underlying disease, pregnancy, vaccination status, and immunosuppression. The infectious dose varies significantly for different diseases, ranging from a single cell to millions of organisms. Moreover, the infectious dose differs by many orders of magnitude when exposure to the identical disease agent occurs through different routes of exposure. Exposure of nonhuman primates to *Francisella tularensis*, the causative agent of tularemia, results in disease with respiratory exposure to ~10 organisms. In excess of 100 000 are required to initiate disease when exposure occurs by ingestion. Additionally, although the infective dose of the blood-borne pathogens HBV and HIV are unknown, the significantly higher concentration of HBV in body fluids may be associated with the documented higher workplace infection rate of HBV. From the viewpoint of risk management, those pathogens that possess low infectious doses (e.g., tuberculosis, which has an infective dose of one tubercle bacillus) require a considerably higher level of infection control practices in the workplace.

Following exposure to disease-causing organisms, the infective process may lead to clinical, subclinical, or asymptomatic disease, which occurs after an incubation period of several days to several months. Clinical disease, associated with hallmark signs and symptoms, often begins abruptly with elevated temperature and general malaise, whereas subclinical infections are generally milder, of shorter duration, and associated with fleeting symptoms. However, with most diseases, the majority of those infected experience asymptomatic disease. These persons are completely devoid of clinical symptoms and any outward appearance of illness. The diagnosis of clinical disease is aided by the presence of clinical findings, whereas asymptomatic disease is usually only recognized through specific serologic tests. The process of seroconversion and elevation in specific antibody titer represents important criteria in the screening of employees for occupational exposures to infectious organisms. Workplace monitoring of employees for asymptomatic disease has provided invaluable information on infections such as tuberculosis and the hemorrhagic fevers.

Although many infectious diseases are transmitted to humans by a primary route, some are transmitted by several routes. Thus, it is generally recognized that tuberculosis is transmitted via aerosol, HAV by ingestion, erysipeloid by contact, rabies by penetration, and Lyme disease by a tick. However, in some occupational settings, especially in diagnostic, research, and production facilities, employees may be exposed to pathogens by abnormal routes, thereby leading to infectious diseases with puzzling clinical symptoms. Moreover, because large concentrations of etiologic agents are grown and manipulated in these workplaces, the opportunity exists for doses far exceeding community exposures. Thus, in workplaces where large quantities of infectious agents are being used, vigilance must be exercised to prevent human exposures via abnormal routes or with an overwhelming exposure dose.

Opportunistic infections occur in individuals, whose normal resistance to infection has been compromised, thereby making them susceptible to microorganisms that would not ordinarily cause disease. Those at higher risk include employees undergoing drug or steroid therapy resulting in transient immunosuppression and those with underlying disease associated with a permanent state of immunosuppression. *Pneumocystis carinii* infection in HIV-infected individuals and *Aspergillus fumigatus* infection in bone marrow transplant recipients are examples of opportunistic infections.

Zoonotic infections result from human exposure to animal diseases. There are more than 200 recognized zoonoses. Data on laboratory-acquired infections have demonstrated that many were zoonotic in nature and represented all classes of infectious agents. The transmission of zoonotic agents can occur in numerous occupations, including veterinary practice, agriculture, animal husbandry, and forest management. It also occurs in such workplaces as animal-holding areas, abattoirs, research laboratories, field operations, commercial fishing, and pet operations. It is imperative that specific infection control practices be instituted to protect workers from zoonotic infections, including the use of prophylactic vaccination, quarantine of feral animals, containment procedures, serologic screening, animal husbandry practices, vector management, and the use of personal protection equipment.

CLASSIFICATION OF MICROORGANISMS FOR LABORATORY WORK

The Centers for Disease Control and Prevention and National Institutes of Health (CDC/NIH) classification of biosafety levels for infectious agents is based on a combination of pathogenicity and transmissibility. Combinations of engineering controls, work practices, and personal protective equipment are recommended for each of the four biosafety levels (i.e., BSL1, BSL2, BSL3, and BSL4). The hierarchy of levels is based on the transmissibility of infectious agents by the aerosol route. For example, commonly used laboratory strains of *E. coli*, which are not virulent and not easily transmitted, are classified in BSL1. The blood-borne pathogens (e.g., HBV and HIV) are classified as BSL2 agents because they are not easily transmitted as aerosols but cause more serious and life-threatening disease. (For work with large quantities of these agents, such as viral cultures, they are classified as BSL3.) The aerosol-spread Venezuelan equine encephalomyelitis virus and the yellow fever virus, for example, are classified as BSL3 agents.

A related system of animal biosafety levels (ABSL-1 to ABSL-4) exists for work with research animals.

Further Reading

Bennett JE, Dolin R, Blaser M, eds. Mandell, Douglas, and Bennett's Principles and Practice of Infectious Disease, 8th edn. Philadelphia: Elsevier Saunders, 2015.

Chosewood LC, Wilson DE, eds. Biosafety in Microbiological and Biomedical Laboratories, 5th edn. HHS publication no. (CDC) 21-1112. Washington, DC: US Government Printing Office, 2009. Available at: http://www.cdc.gov/biosafety/publications/bmbl5/bmbl.pdf (accessed June 24, 2016).

Willey JM, Sherwood LM, Woolverton CJ. Prescott's Microbiology, 9th edn. New York: McGraw-Hill, 2014.

Chapter 19

CLINICAL RECOGNITION of OCCUPATIONAL EXPOSURE and HEALTH CONSEQUENCES

GARY N. GREENBERG AND GREGG M. STAVE

Occupational and environmental health effects associated with exposure to biological hazards are mediated primarily by two distinct mechanisms. These mechanisms are infection by intact organisms and immunologic reaction to materials from biological sources. This chapter will describe these basic disease processes and their associated patterns of illness. Practical guidance will be offered as to when particular illnesses should be suspected and how they can be confirmed.

INFECTION

Infection versus colonization

Infection results when living microorganisms (viruses, bacteria, fungi, or parasites) establish an active and growing presence within the human host. This situation creates characteristic pictures of illness. Some disease elements are created by the damage caused directly by the invading pathogens. Others result from the host's response to the organisms. The detection of disease requires knowledge of the microbiology of the attacking microorganism and an understanding of the human body's reactions.

Although bacteria are the most commonly isolated source of infectious illness, only some interactions between mammalian organisms and bacteria produce disease. *Infection*, an event with important medical consequences, must be distinguished from *colonization*, a term used for the harmless or adventitious presence of the microorganism in contact with human tissue.

The human host provides many microenvironments that act as sites for a complex microbiological ecology inhabited by diverse strains of bacteria. For example, the colon augments the breakdown of food wastes with innumerable bacterial species, termed coliforms. Coliforms represent harmless symbiotes as long as they remain in their usual habitat. However, these same bacteria can be the source of critical illness when opportunities place them at other biological locations, including the spaces outside the intestinal wall, within the urinary system, or within the bloodstream.

Bacteriologic evaluation of patient specimens can be extremely useful in distinguishing between infection and colonization. Awareness of the source of the specimen is important for the selection of the appropriate test and interpretation of the results. Samples from many sites are commonly contaminated, including saliva, stool, vaginal secretions, and skin swabs. In these sites, the presence of bacteria (and often fungi) need not be interpreted as evidence of disease unless the organism is not part of the usual biology at that location. Clinicians must recognize the species of normal flora specific to each body site and to distinguish them from the harmful organisms recognized as pathogens.

Systemic infection versus localized infection

When illness results from the presence of microorganisms within the host's tissues, infection is most easily diagnosed by the evaluation of the affected area. Some infections are localized and superficial, such as those involving the skin (e.g., cellulitis) or a mucous surface (e.g., streptococcal pharyngitis or strep throat). For other infections, illness is diffuse, resulting either from a total body invasion by the organism (e.g., the spread of rickettsial organisms in Rocky Mountain spotted fever) or from the body's global reaction to infection.

Physical and Biological Hazards of the Workplace, Third Edition. Edited by Gregg M. Stave and Peter H. Wald.

Clinical manifestations of infections are based on both local and systemic mechanisms, mediated by the immune system and result from activated defensive cells. As the body recognizes the assault of foreign organisms, immediate reactions at the site are involved in a complex process known as inflammation. Redness and warmth arise from the stimulated local circulation, causing increased blood arrival through dilated capillaries. Swelling results from increased blood vessel permeability, permitting the escape of antibacterial proteins and plasma into the surrounding area. Tenderness and limited local function are due to the presence of the offending pathogen and to the effect of local mediators released from the host's activated protector cells.

In addition to the local effects of low molecular weight chemical signals, mediators circulate throughout the body and produce systemic signs of illness. Fever is mediated by the brain's hypothalamus and shows that infection and inflammation have triggered a systemic response. Although it is unclear what advantage is gained by raising the body temperature, fever is one of the earliest and most common responses of infected mammals. When stimulated by the arrival of infectious debris or cellular activating proteins, regional lymph nodes (collection and production sites for immune cells) activate and enlarge. Another manifestation of infection is a more dynamic circulation, resulting in increased heart rate, reduced vascular resistance to blood flow, and lowered blood pressure. Generalized muscular aches and stiffness occur in many areas not directly involved by the infection.

The next elements in the local infection process are much more common for bacterial organisms than for viruses or fungi. When the body's immune response fails to eradicate a local invasion, the process can result in a closed-space infection. Within this abscess are active and dead defensive cells and countless foreign organisms that are "walled off" from nearby tissue. An abscess may either open spontaneously or require surgical drainage. When infections are especially severe, widespread bodily invasion via the circulatory system, or sepsis, can occur. This process, which allows seeding of distant tissues, is one of the most dangerous late stages of bacterial illness.

Sometimes, the cumulative volume of circulating infectious material and the massive release of immune mediators combine to produce the syndrome of septic shock. This dangerous situation of thready, weak pulse, and poor circulatory perfusion results in deteriorating vital organs' function, including the heart, brain, and kidneys. Septic shock requires prompt diagnosis followed by aggressive and intensive treatment and may often be fatal.

SPECIFIC CLINICAL DISEASES

Different bacteria, fungi, and viruses may invade each component of the human host. For many body elements, the disease syndromes that result are clinically similar regardless of the attacking infections. The characteristic symptoms for these diagnoses are described here to provide a basis for the ensuing chapters, which will describe the consequences of infection due to specific organisms with such terms as meningitis, hepatitis, or pneumonia.

Upper respiratory infections

Although the most common of infections is often called a "cold" and is usually viral in origin, respiratory infections are not much different when caused by other invading organisms. Patients develop irritation of all respiratory surfaces, including the nose, throat, middle ear, and facial sinuses. These lining tissues become swollen and moist and may obstruct the passage of air. When lymphoid structures (tonsils, adenoids, and lymph nodes) are stimulated to respond protectively to the infection, they enlarge and cause additional symptoms of obstruction and pain. When a corridor for mucus drainage becomes persistently obstructed, immune mechanisms are rendered less effective, and bacteria that are not ordinarily pathogenic can become successful sources of infection and can cause complications. This results in secondary infections in areas where drainage is blocked, including facial sinuses and the middle ear.

Many of the symptoms of the respiratory infection are nonspecific and result from circulating immune activators and foreign proteins. Fever, muscular aches, stiffness, chills, and headaches are common symptoms for any infection. They are especially linked with respiratory infection in most patients' minds only because colds and flu are such common forms of infection.

Bronchitis and pneumonia

Bronchitis and pneumonia are diagnoses that represent infection of lower respiratory structures. The clinical picture characteristically involves a productive cough, chest pain, shortness of breath, and fever. Bronchitis is associated with excess mucus (usually containing infected material) arising from within the chest. Wheezing (or asthma) represents an inappropriate muscular reflex of the contractile elements surrounding the air passage, which narrows these passages and obstructs air exchange. Pneumonia represents infection of the surface of the lung where oxygen transfers to the passing circulation. Many pneumonias involve "consolidation" or filling of the usually empty spongy lung tissue with the combined debris of the attacking organisms, reactive immune cells, and fluids. Pneumonia patients, with illness in one or more of the lungs' lobes, suffer obliteration of the respiratory surface in addition to the consequences of reduced airflow into the lungs, causing respiratory insufficiency. This causes poor oxygenation (manifested by the desaturated blue color of circulating blood), weakness, shortness of breath, and risk of death.

Hepatitis

Direct infection of the liver is often subtle, only occasionally manifested by local tenderness over the liver itself (in the upper right corner of the abdomen). More universal among patients with liver disease is prominent and disabling fatigue, with loss of appetite, and the onset of jaundice. This last sign represents the escape of bilirubin (a breakdown product of normal red blood cell turnover) from the liver's usual metabolic machinery. This displaced pigment accounts for the yellow skin and dark urine of hepatitis patients, and its absence from feces may be reported as "clay-colored" stools.

Fibrous scarring from repeated and widespread liver infection leads to cirrhosis. This situation can occur even when cellular damage has quieted but causes dangerous disruption in blood reaching the liver and pressure alterations in the abdominal cavity.

Liver failure is a rare but potentially fatal consequence of infection in this organ. Reduction in the metabolic and synthetic functions of the liver can lead to the development of clotting disorders, abdominal distension with extracellular fluid, deteriorating mental function, and unregulated blood sugar.

Dermatitis

Since skin is the outermost layer of the host, infectious rashes are the most easily noted of the body's reactions to invading organisms. Unfortunately, for many common patterns of illness, the skin's pattern of reaction is nonspecific, and noninfectious immune responses can confuse the patient and clinician regarding their origin. Furthermore, some rashes may be associated with infectious disease affecting body elements elsewhere, including systemic illnesses like measles, Lyme disease, and meningococcal meningitis.

Cellulitis is a primary infection of the skin appearing as a spreading area of redness. It usually originates with local mechanical injury. Other obvious and direct skin infections may involve specific skin structures, such as hair follicles, sweat glands, and nail beds. Again, primary injury often triggers local infection at these sites. The organisms responsible are usually those already present at the skin's surface. Infection may also be caused by organisms introduced by the agent of mechanical injury (e.g., an animal tooth in a bite wound). Remote and systemic disease is also possible from skin infection. Rheumatic fever and toxic shock syndrome are both consequences of local infection at the skin. Tetanus organisms cause systemic illness unrelated to their direct skin effects. Illness results from the remote effects of tetanus toxin, released by the organism after successful infection of anaerobic spaces beneath the skin.

Central nervous system infections

The brain and spinal cord can be attacked by microorganisms in three distinct patterns of illness—meningitis, encephalitis, and abscess. The most common of these, meningitis, is an inflammation of the brain's lining and suspending fluids. Patients have severe headaches and are unwilling to stretch or fold these covering membranes by such activities as bending their necks or even turning their eyes. The brain itself shows normal function until late in the course of infection, when there is possible damage to the nervous tissue, affecting thought, movement, and behavior. Making the diagnosis requires finding evidence of infection in samples of spinal fluid.

Encephalitis involves a direct attack of a microorganism into the nervous tissue of the brain. It usually occurs in a diffuse pattern throughout the brain's substance. Some patients suffer damage in only one cerebral area, with symptoms relating to the specific brain structures involved. Since the brain is a fragile and shielded structure, diagnosis usually requires indirect testing, such as evaluation of cerebrospinal fluid or computerized brain imaging.

An abscess in the brain represents a circumscribed collection of infected material that not only causes local damage but also, as it expands, causes compression of the brain as a whole, trapped in the skull's rigid compartment. This is a rare illness that requires specific treatment, possibly including drainage to empty the area where the infection is localized.

Gastroenteritis and dysentery

The presence of disruptive organisms in the upper gastrointestinal tract can result in painful abdominal distension and discomfort, caused by irritation of the gastric and esophageal lining. It can also cause reflex reactions from the brain, manifested as nausea, vomiting, and loss of appetite. In the lower intestinal tract, the presence of infection by pathogenic microorganisms can lead to different forms of diarrhea. If the organisms—or, more commonly, their secreted toxins—merely interfere with the intestine's ability to resorb liquid, then the patient suffers a watery diarrhea. Dehydration and electrolyte imbalance may result. This type of infection is often difficult to confirm, because the organism is not easily available for laboratory identification. If the infection actually attacks the intestine's wall, the patient loses more than the usual colonic contents. Dysentery is recognized by the presence of blood, inflammatory cells, and mucus in the stool. In this form of illness, invasive microorganisms are more likely to be successfully identified in laboratory evaluation.

"Flu-like illness"

Many illnesses with important consequences initially manifest themselves with widespread symptoms that are nonspecific and not easily diagnosed. Even though the clinical picture may be different from the specific illness of influenza, whenever patients have prominent respiratory symptoms, the phrase flu-like illness will be applied to almost any fever-associated syndrome. The symptoms most often recognized to be "flu like" are body aches, chills, stiffness, mild to moderate fever, headache, and fatigue. The term is rarely applied to illness with only respiratory symptoms, such as cough, nasal congestion, or sore throat.

IMMUNE MECHANISMS AND HYPERSENSITIVITY DISORDERS

The host immune system triggers a cascade of cellular, antibody, and chemical activity in response to the presence of foreign materials. Though this system performs an essential protective function, its inappropriate activation by otherwise benign materials can sometimes lead to deleterious consequences. These responses may manifest themselves as allergy, asthma, arthritis, or other hypersensitivity disorders.

Classical allergy, such as hay fever and laboratory animal allergy, results from antibodies of the immunoglobulin E (IgE) class. These antibodies are adherent to mast cells, an inflammatory cell present in many tissues. When specific target proteins arrive at the cellular surface, IgE antibodies bind to their molecular targets and activate the mast cell's response. Mast cell products released include histamine, a small circulating compound that is responsible for many of the clinical manifestations of allergy. The propensity to develop an allergic response as a result of a specific exposure is not distributed uniformly in the population. A personal or family history of allergy, asthma, eczema, or sinusitis (known collectively as atopy) is associated with an increased risk of developing allergy. However, history of atopy alone cannot predict whether a worker will develop symptoms from work exposures. Previously nonatopic individuals can also develop allergic illness.

Not all immune reactions are mediated through IgE. In hypersensitivity pneumonitis, specific antibodies of the immunoglobulin G (IgG) class recognize foreign airborne antigens. The resultant antibody-triggered cascade of immune events can produce a devastating reaction in surrounding lung tissue.

Cellular-mediated immune mechanisms are slower and less well-characterized than those initiated by circulating antibodies. The application of specific proteins to sensitized tissues causes activation of local immune cells, resulting in the activation of monocytes and the migration of macrophages to the area. These arriving cells cause changes in skin thickness and firmness.

SPECIFIC CLINICAL SYNDROMES

Upper respiratory allergy

Allergy to ragweed pollen ("hay fever") is a common environmental respiratory allergy. Symptoms of this disorder are identical to those of allergy caused by other airborne proteins and result from direct contact of inhaled particles with the respiratory mucosa. Manifestations of exposure include increased production of tears and a continuous clear nasal discharge. The lining of the nasal passages may swell, resulting in obstruction. Patients commonly experience sneezing, along with itching of the eyes and throat.

Lower respiratory allergy

Asthma is an episodic illness. Symptoms result from constriction of muscle-lined air passages in the lungs. The ability to exhale air is limited. Patients experience spells of shortness of breath, chest tightness, and possibly cough. Wheezing may be audible or noted only with a stethoscope or if measured with a flow meter. The inability to achieve adequate air exchange results in a diminished blood level of oxygen.

Patients with acute hypersensitivity pneumonitis also commonly experience shortness of breath and cough. Unlike asthma, symptoms include chills, high fever, and muscle aches, which are signs of widespread reaction to active inflammation. Symptoms generally resolve after a brief illness lasting several hours, but some patients experience milder symptoms for several days. This illness may be confused with an infectious pneumonia caused by organisms actually present in the lung. Careful evaluation and a high degree of clinical suspicion are required to arrive at the correct diagnosis.

Skin reactions

Immunologic skin reactions are extremely varied and can often be confusing. However, the classic lesion associated with allergic contact dermatitis is the rash associated with poison ivy and poison oak. This red, itchy rash contains small, clear, fluid-filled raised blisters. For other allergic skin eruptions, the clinical findings are frequently much less specific, with evidence of only redness and possibly scaling.

The skin may also be active as part of a systemic response to ingestion or injection of antigens. One such manifestation may be widespread edema (swelling of soft tissues as a result of excess fluid accumulation) and diffuse hives accompanied by pale swelling and itching. In severe cases, this reaction may be associated with leakage of circulating blood plasma into peripheral tissues and consequent shock. Emergency medical treatment is required in these cases.

Another skin response is similar to that seen with the intradermal injection of purified protein derivative (PPD) used diagnostically for tuberculosis skin testing. The skin becomes firm and raised as a result of local cellular infiltration responding to a secondary exposure to this foreign protein. The specific reaction does not appear for 2–3 days following re-exposure.

Irritations

Nonimmunologic individual variation in the response to other biological stimuli is also widely recognized. Wood and tobacco smoke and other irritants usually do not act as specific allergens and do not provoke the illness mechanisms described above. Nonetheless, variation among the doses tolerated by the human population is considerable. Even though irritation occurs without specific antibodies or immunologic mechanisms, there are sensitive individuals who react to many stimuli (including odors) with more severe symptoms than the general population, sometimes with easily recognized objective clinical findings.

LABORATORY CONFIRMATION OF INFECTIOUS AND HYPERSENSITIVITY DISEASES

Several technologies are available for the direct identification of microorganisms in human tissue. Each has specific advantages and they are commonly used in combination. Although microscopic evaluation of stained smears of collected liquids can be performed rapidly, it lacks sensitivity to small numbers of organisms, and there is poor precision regarding the microbe's identity. Culturing of the organism with specific nutrient media permits recognition of even small numbers of pathogenic germs. It may be a slow process, dependent on the growth of the microbes sought. A variety of tests are available to detect infecting viruses, usually based on measurement of DNA, RNA, or the patient's antibody response.

Microscopic visualization of the organism

The most rapid means for disease recognition is direct microbial identification in stained biological specimens. In pneumonia, for example, sputum samples are smeared onto a glass slide and allowed to dry. Specific stains and selective rinses are then applied before careful microscopic examination. Expert evaluation can reveal the nature and number of the organisms present and even indicate whether the microorganisms are pathogens or colonizers, based on whether they are seen engulfed by the host's defensive cells. This technique not only provides a glimpse into the nature of the disease but also discloses the possibility of mixed infection by several organisms and indicates the intensity of the battle between the microbes and the host cells.

For general bacterial evaluation, the Gram stain is used. This technique provides useful information regarding the organism's shape and the type of cell membrane of the microbes present. The Gram stain procedure is designed for a primary pigment to be selectively removed from the interior of certain Gram-negative bacteria, permitting the loss of the dark color and permitting staining only by the last stage, a light pink universal stain. The darkly stained bacteria are recognized as Gram positive, whereas the paler organisms are Gram negative. The organism's shape offers additional diagnostic clues. Rectangular and elongated bacteria are characterized as rods; circles, coffee bean shapes, and clusters are cocci.

Special stains are required for certain organisms and situations. The acid-fast reaction is a special stain that provides the classic means to recognize mycobacteria (e.g., tuberculosis). Silver-based staining is used to identify microscopic protozoa, fungi, and spirochetal bacteria (e.g., syphilis and pneumocystis). Viruses, which are much smaller and necessarily intracellular, are more difficult to visualize by direct microscopic evaluation. Usually, the identity of the exact virus must be inferred from the clinical circumstances and the source tissue being evaluated. Viral organisms can sometimes be confirmed with special staining techniques or electron microscopy.

The most targeted stains use specifically created antibodies against individual organisms. Once these antibodies attach to their targets, they are linked to fluorescent compounds that allow microscopic detection. This technique, called immunofluorescence, allows precise confirmation of actual microbial identity without the delay required for cultures to grow or the host's own antibodies to emerge.

Growth and identification of microbial colonies

The organism's ability to multiply in the host tissue is the most common disease mechanism. Culture techniques use this same capability to identify organisms that would otherwise be missed by amplifying their numbers in specific artificial environments called media. Once isolated and flourishing in the microbiological laboratory, pathogenic colonies can be tested for additional information regarding their precise speciation, antibiotic sensitivity, and biochemical activity. Molecular markers may be useful for epidemiologic evaluation.

The greatest weakness of culture as investigative tool is the delay before the organisms are sufficiently numerous to be detected. For bacteria, the lag is usually 24 hours. Most viruses fail to grow in laboratory settings, but even where it is possible, there is a similar delay. For fungi and mycobacteria, although a positive sample might be reported earlier, a sample cannot be considered "no growth" until it has been incubated for a full 6 weeks.

In addition, there are many occasions when a sample will fail to yield any growth, even in the presence of infection confirmed by other means. To protect the microorganisms they contain, samples must be spared any risk of heat, cold, or drying. If the patient has taken antibiotics, if the sample is mishandled, or if nonpathogenic organisms are present that inhibit the pathogen's *in vitro* behavior, then the culture will not only be slow to diagnose the illness but will also yield false-negative results.

After collecting a specimen with viable microbes, the next step in microbiological culture preparation is selection of the appropriate growth medium. This requires specific broths or agar gels with nutrients and cofactors designed to encourage the growth of even particularly fastidious organisms. In addition, when specimens are collected from a source known to be contaminated with selective nonpathogenic organisms (e.g., from the throat), antibacterial chemicals must be included to suppress their competitive growth.

Finally, the environment for culture growth must be selected. When the organism sought is a tissue-invading bacterial pathogen, 37°C is used to simulate the human host. However, when an allergen is cultivated, its disease-producing biology is different, since the mechanism of disease involves the organism's growth at ambient temperature and the subsequent release into the environment. Environmental samples should be cultured at a temperature that accurately reflects the biology of the area where they were collected.

The results from cultures, especially environmental, do not necessarily constitute a clinical diagnosis. For each biological sample, there are guidelines for culture isolates. For sputum, culture interpretation requires knowledge of whether the specimen included host cells that prove its origin from lung tissue (as opposed to mere oral saliva). For urinary cultures, microscopic identification of cells from the vagina makes diagnosis from culture results suspect. Because most infections are caused by only a single species, growth of even small numbers of organisms from just one species constitutes stronger support for the presence of an important microbiological presence.

Even when a pure sample is achieved, the concentration of viable colonies can be an important interpretative fact. Rare stray organisms found in urine specimens do not constitute proof of disease. Useful culture reports must indicate, for example, that the culture showed 100 000 colonies of *Escherichia coli* per milliliter of sampled fluid. Clearly, there are many times when the mere identification of an organism in a sample does constitute a conclusive diagnosis of disease. Where a material is ordinarily sterile, the presence of a single colony is persuasive. Such samples include spinal fluid, liver biopsy material, and urine obtained via sterile catheter. In those cases, every microorganism must be considered a pathogen.

In other circumstances, the presence of any microbial colonies is sufficient proof for a firm diagnosis where the infecting agent is never part of the benign normal flora. The mere isolation of any of these organisms constitutes a firm diagnosis, regardless of concentration, coexisting organisms, and whether the specimen was otherwise contaminated by other body materials. These agents include *Neisseria gonorrhoeae* (the gonorrhea organism), *Mycobacterium tuberculosis*, and the herpes simplex virus.

Diagnostic evaluation by immunity testing

For many infections, the responsible organisms cannot be identified by direct visualization or by culture because they are too fastidious or too few, or because they are located in unreachable anatomic sites. In these cases, including those caused by many viral organisms and several atypical bacteria, the best way to identify an illness is to monitor the patient's immune response to the microbe as indirect evidence of its presence.

A basic concept in biology is that each individual's immunity is developed as a consequence of its own exposure and infection experience. The immune system serves as a data bank for the aggregate history of the host's exposures. Each foreign protein and macromolecule serves as a unique immune stimulus (called an antigen). Each exposure to a new antigen constitutes a new immunization and results in a uniquely responsive set of precise cellular and antibody reactions. These reactions create permanent changes in the way that the host reacts to the antigen on any future exposure. The sum of these learned reactions is retained in the organism's immunologic "memory." This represents a catalog of identifiable exposures, each of which resulted from a prior exposure and each of which can be tested and identified for proof of prior contact. As an example, when a patient is effectively exposed to the mumps virus, a characteristic set of targeted proteins called antibodies are produced that react specifically to the viral proteins. Whether the exposure arrives as a vaccine or is the consequence of an actual infection, the result is the lifelong presence of identifiable circulating antibodies specifically reactive to this microbe.

Most immune testing evaluates the existence, quantity, and type of circulating antibodies. The presence of each specific antibody proves prior exposure and infection. Because the production of immunoglobulins of differing types occurs in predictable sequential fashion, this pattern of response can reveal the timing of an infection. The classes of antibody are named with single-letter suffixes, for example, immunoglobulin M (IgM). IgM is an antibody subtype with a transitory role, lasting only a few weeks until immunoglobulin G

(IgG) is synthesized. Because of its transient presence, the identification of IgM against a microbe is itself proof of recent disease. Evidence of recent immune activation may also be obtained by demonstrating a recent rise in antibody concentration. Antibody levels are usually expressed as titers, referring to serial dilutions of serum that still demonstrate a positive reaction. For example, if four twofold dilutions of a sample produced a positive test result but the fifth did not, the result would be reported as positive at 1 : 16. Two samples collected 6 weeks apart (labeled acute and convalescent) constitute a matched set for analytic purposes. The customary threshold for a significant antibody elevation is a fourfold titer increase. Since the clinical consequences are likely to have run their course, the intervening delay is usually too long for clinical decisions concerning an affected patient. However, it may provide important documentation for workplace or public health use.

Immunoglobulins are also utilized in other diagnostic tests. Fluorescent compounds chemically linked to antibodies can be used to stain specific proteins for immunohistologic microscopic examination. This technique of linking a detectable moiety to an antibody is also used in enzyme-linked immunosorbent assay (ELISA). When an antigen from the biological substrate of interest binds to the ELISA antibody, the linked enzyme generates a measurable product. This product may be a fluorescent compound or another easily measured chemical.

Although ELISA provides a rapid and relatively simple technique for antigen detection, results are not highly specific. A positive ELISA may need to be confirmed by antibody electrophoresis testing (e.g., the Western blot assay). This more expensive and time-consuming assay increases specificity by recognition of the size and the electric charge characteristic of the detected antibody.

Polymerase chain reaction testing

Biotechnology techniques have provided new opportunities to specifically identify microorganisms. Based on the same means used for gene cloning and synthesis, methods now exist to identify segments of either DNA or RNA that represent specific identifiable microbial genes. Identification is accomplished by a process that amplifies any genetic material present. This technique can prove the presence of infecting organisms without relying on the immune system's response and without requiring successful growth of the organism outside the infected host. It replaces whole-pathogen cultures with the detection of recognizable species-specific genes and gene segments. There are several advantages to this elegant laboratory tool. Confirming the presence of microbial DNA or RNA, even when the organism itself is too weak or too slow growing to be cultured, enables the identification of organisms whose culture has never been possible. It can also provide results more rapidly than other diagnostic methods. However, there is concern about the possibility of genetic contamination, where random bits of genetic material erroneously suggest that a particular microbe is present. Clinical experience is rapidly accumulating to where this technique might achieve the reliability and standardization now available with traditional culture or antibody testing.

CLINICAL TESTING FOR HYPERSENSITIVITY

There are several mechanisms by which patients may become ill from biological sources without direct microbial infection. Each has a specific evaluative technology for recognition and diagnosis. None of these techniques to evaluate hypersensitivity is as well validated or standardized as the diagnostic tools for evaluation of infection. Testing techniques used for allergy and other forms of hypersensitivity require considerable judgment in their application. Consultation with an appropriate clinical specialist is usually required.

Allergy testing with skin prick tests

The mechanism by which traditional allergy occurs involves mast cell activation by the binding of a high-molecular-weight compound to the cell-attached IgE antibodies. The classical allergy test involves placing a drop of a dilute antigenic solution onto the skin, pricking the skin shallowly with a clean pin, and awaiting an immediate reaction. The response, when one occurs, shows the effects of the activated mast cell's release of histamine and other mediators, causing a localized reaction of swelling, reddening, and notable itching.

When properly administered, this testing procedure is useful and reliable. Results are skewed by varying circumstances, including nonspecific skin reactions, antihistamine use, and many complex issues involving the applied solution. An experienced clinician should perform this test.

IgE evaluations with RAST testing

Because the actual mechanism of classical allergic symptoms involves IgE, diagnostic tools have focused on measuring circulating levels of this molecular class. Measurement of total IgE may be useful in some cases (e.g., allergic bronchopulmonary aspergillosis). The concentration of IgE antibody directed against specific allergens can also be quantitated.

The technology most commonly used is called radioallergosorbent test (RAST) (because the test involves radioactive methods in the laboratory). These tests are capable of accurately measuring very small concentrations of IgE antibodies in the serum of allergic patients. The resulting information has proven highly comparable with the results from skin testing. While both tests correlate with clinical diagnoses of allergy, evaluations can be positive without clinical meaning.

Patch testing and intradermal skin testing

Cellular-mediated immunity is also called delayed hypersensitivity, because it represents a slower mechanism of response. Although the recognition of the foreign agent still depends on the lymphocytes' molecular memory, the response utilizes an entirely different class of activated cells, and the disease is manifested by different mechanisms.

Because delayed hypersensitivity involves several cellular classes working together, testing currently requires measurement of the response by the intact host rather than any cellular extract or circulating antibody. Thus, the diagnostic tests require applying the potential offending allergen directly onto the patient's skin and waiting 48 hours for cellular infiltration. The response considered to be diagnostic results from the arrival of enough immune-activated cells to produce a circular area of irritation and stiffening (induration).

Patch testing is performed to evaluate suspected cases of contact dermatitis. An extremely dilute antigen solution is applied to a gauze pad and held in place with a shallow aluminum protector during the test's incubation. To ascertain the reactions' specificity, skin tests are always done simultaneously with control solutions. These include antigens known to produce positive results in the population at large as well as the saline preservative solution used for the allergen's dilution to reveal nonspecific responses. These skin test batteries thus often require an entire grid of applied patches, sometimes covering the patient's entire back for the 2-day waiting period.

When the tissue response in question involves organs other than the skin (including possible inapparent tuberculosis), more invasive dosing is performed, depositing 0.1 mL of the solution directly into the skin with a tiny needle (intradermal testing). In this case, the number of applied solutions is limited by the patient's discomfort, but control solutions may still be used. Many patients with suppressed immune response (including corticosteroid treatment, HIV infection, or even overwhelming systemic infection) will be unable to mount any cellular response, thus producing a false-negative result termed anergy. By including simultaneous doses of antigens with universal response (mumps, *Trichophyton*, *Candida*), the skin test battery can be self-validating.

For intradermal skin tests, the puncture site becomes the center of a spreading firm area. Reading the test simply involves measurement of the firm region's greatest diameter after a delay of 48–72 hours. The medical interpretation of skin testing requires additional consideration of the setting and the patient. Even the most common of intradermal skin tests, the purified protein derivative (PPD) for tuberculosis, can be called positive at 5, 10, or 15 mm of induration, depending upon the clinical setting (see Chapter 23).

Exposure challenge testing

In cases where the clinician tries to evaluate environmental disease potentially explained by mechanisms of hypersensitivity, objective measures of dose may be totally misleading. Responses by allergic individuals are frequently many orders of magnitude more sensitive than the best industrial hygiene techniques, especially when others who share the exposure show no symptoms at all. Investigators of potentially allergenic environments must therefore cope with an obvious temptation—direct patient challenge.

There is a significant danger in sending potentially affected individuals into situations where they are suspected to be allergic to an airborne agent. Depending on the patient's prior reactions, such experiments must be done with ample opportunity for rescue and medical attention. They should only be done when the prior illness has been mild (e.g., skin rash) and when the symptoms are easily reversible. The exposure testing must progress in a stepwise fashion, where earliest exposures should be chosen to produce no response at all.

Exposure testing should be used only when no other means of evaluation is available and only with both medical guidance and the fully informed permission of the patient. It should be considered the choice of last resort in diagnostic techniques.

WHEN TO SUSPECT OCCUPATIONAL ILLNESS OF BIOLOGICAL ORIGIN

Most illness due to infection or allergy results from nonoccupational exposures. However, certain settings or specific illnesses increase the likelihood that a medical problem has an occupational origin.

Unusual job activities

The health consequences of many workplaces can be predicted and prevented with planning and conscientious concern, but despite such measures, some employee populations remain at increased risk. Healthcare workers have direct contact with patients with transmissible illnesses. For organisms spread by airborne contagion (e.g., influenza or tuberculosis), the risk of contracting disease is greater for these workers than for the population as a whole, because the disease is more concentrated among their clients. For other illnesses, healthcare workers are uniquely susceptible where exposure requires deposit or liberation of an infective organism. Infection usually occurs as an untoward consequence of an invasive procedure (e.g., exposure to bloodborne pathogens).

In these settings, clinicians and safety professionals must remain aware of the opportunity for illness to transform care providers into patients. Routine preventive measures, including vaccination against expected exposures (e.g., influenza and hepatitis), universal precautions with blood and body fluids, and routine hand washing, are essential.

Animal workers

Unfortunately, workers with direct contact with other species are at risk for both allergy and infection. The proteins released from animal urine, skin, and other tissues can easily become airborne, resulting in rashes, hives, allergic nasal and ocular symptoms, and even asthma. This form of hypersensitivity usually occurs immediately after exposure, facilitating the diagnosis. However, conditions such as asthma may be delayed. The lack of an immediate response does not automatically exclude an occupational association.

For infections, the parameters of risk attribution are reversed. Infection resulting from other species is rare and poorly identified. Clinicians often fail to recognize the nature of illness and may not even know what microbial agents to suspect or what treatment is needed. When an agent is identified by culture or serologic means, it is not hard to determine that this unusual pathogen must have arisen as a consequence of work exposure. Populations at risk include workers at abattoirs, zoological parks, and veterinary clinics and those involved in biology research. Pet owners also have large exposures to the possibility for allergy and for infection. Since many animal workers are also pet owners, both occupational and home environmental exposures need to be investigated when evaluating suspicious illnesses.

Workers handling waste and sewage

Before proper sanitation and sterilized water supplies, numerous infections associated with human waste posed community-wide dangers. The risk for diseases prevented by these techniques are now concentrated among the workers with potential exposure, usually in municipal water treatment centers. These agents include both viral organisms and bacteria. Recognition of disease in these workers is important to provide proper treatment and to minimize identified exposures for their coworkers.

Travelers

Geographic dislocation may result in environmental illnesses through a variety of mechanisms. Many areas of the world, including regions in the United States, contain unique pathogens in such high environmental quantities that the rate of pediatric infection is universal while the danger to adults exists only among new arrivals to the community. Usually fungal in nature, examples include histoplasmosis and coccidioidomycosis, endemic in the Ohio River valley and in the American southwest, respectively. Even "traveler's diarrhea" can sometimes be explained by organisms that produce no symptoms among local inhabitants because they were naturally immunized years earlier.

Many illnesses are climate specific and thus are seen in industrialized societies only among returning travelers, new immigrants, and visitors. Malaria, yellow fever, and Zika infection represent important concerns for travelers to tropical areas, where insects act as potent vectors for disease. For these diagnoses, the link to foreign travel is essential. In addition to the area of the world, some consideration is required for the traveler's choice of accommodation. Urban life, with its air-conditioned hotels and restaurant meals, represents a drastically reduced risk for tropical disease compared to traveling to remote villages and spending long, unprotected hours in the wilderness.

Sanitation and public health measures in other cultures are often less thorough than the norms in European and American societies. Water and food supplies are often the source of infection for adventurous travelers who sample local edibles contaminated with viable organisms that would not be present in their home food markets. Vibrio organisms and *Shigella* are much more common in settings where food and water regulations are lax for reasons of poverty, societal disruption, or crowding.

Travelers may also contract contagious illness from their new human contacts. The geographic migration of many illnesses (including measles, Ebola, HIV, and resistant gonorrhea) requires migrating human hosts. Thus, an infection may result just as easily from contact with a newly arrived immigrant to the domestic environment as from visits to foreign regions.

Hypersensitivity is not usually related to travel. A period of several weeks to years is required for exposure to produce the necessary antibodies that create allergy, and travelers have by then become residents. Additionally, when travelers develop hypersensitivity-related illness, the best therapy is removal from exposure, which is easily accomplished when a visit is short term by its nature.

Unusual clusters of disease events

Even among workers with unremarkable job activities, some evaluation is required in response to what appears to be an outbreak or disease cluster. Even for illnesses that are common in our society, there is a poorly defined threshold when an investigation is required to explain the simultaneous development of numerous similar medical problems within a worker community. The rarity and nature of the illness, its prevalence within the at-risk workforce, and the pattern of its occurrence and spread are important criteria in evaluating these situations.

In the office setting, occupational health professionals may be asked to evaluate health complaints that the occupants have ascribed to "sick building syndrome." Symptoms reported commonly include headache; irritation of the eyes, nose, and throat; fatigue; and sensitivity to odors. Employees may report that symptoms occur only while they are in the building. Although controversy persists as to the most common etiology of this syndrome, it appears unlikely that a specific biological organism causes such nonspecific symptoms. Extensive searches for biological sources of illness are usually not warranted. A systematic review of employee complaints is an important first step in understanding the problem. If indicated, an evaluation of the ventilation system should be undertaken. Efforts should be directed toward providing adequate airflow and air exchanges, as well as appropriate regulation of temperature and humidity. Altering these physical aspects of the ambient environment may reduce occupant complaints. Psychosocial factors, including a variety of workplace stressors, may also contribute to health complaints. An evaluation of the role of psychosocial issues should be conducted contemporaneously with the rest of the evaluation.

By contrast, "building-related illness" describes a situation where building occupants suffer specific clinically diagnosable illnesses, such as hypersensitivity pneumonitis. Biological organisms may either cause or contribute to these illnesses. A thorough evaluation for potential sources of contamination should be pursued. Molds (usually fungal colonies and spores) are commonplace. Although they are found even in well-maintained office settings, their environmentally released dose is greatly magnified wherever imperfect ventilation and filtration are permitted. Clusters of symptomatic workers with "hay fever" symptoms (nasal congestion, tearing, sneezing, and coughing) should prompt an assessment of the air purity for potential contamination with invisible microbes and proteins. Evidence of either water condensation or prior flooding makes it even more likely that the symptoms can be explained by occupational exposures. These factors suggest a need for special testing for environmental flora that may be present either on surfaces or in the air.

Situations involving hypersensitivity pneumonitis (such as "farmer's lung") present a rarer but more critical problem. Because of the delayed onset of their illness, the sensitive workers do not develop immediate symptoms; therefore, they may not notice an association with any particular activity. Several episodes of illness, either in just a single worker or among a work team, may occur before a connection can be made to the work environment. In these cases, the evaluation of the environment and the patient should be coordinated, with open communication between the clinician and the environmental health professional.

There are occasions when work-related controls must be considered for an epidemic of infection as a result of a common source illness. Food-related illnesses can be introduced into the workplace by any common eating opportunity, including vending machines, in-house cafeterias, or popular neighborhood restaurants. Direct contagion must be considered for outbreaks of conjunctivitis ("pinkeye") among workers using shared optical devices, such as microscopes.

Rare or severe diseases

In some cases, the patient's diagnosis is sufficiently unusual on its own that the occupational environment must be considered to explain the source of disease. Just as with chemical exposures, where the development of peripheral neuropathy or bladder cancer requires a thoughtful assessment of the potential for environmentally triggered illness, there are certain diagnoses where occupational and environmental causes must be conscientiously sought, even without explicit hints or leads.

Recurrent asthma and respiratory compromise, even in just one worker, represent such a commonly environmentally mediated danger that an investigation of workplace environment is a reasonable supplement to medical management. An inspection for potential organic contaminants or dusts is a prudent adjunct to the treatment of this hazardous and progressive condition. A diagnosis of Legionnaires' disease should prompt a consideration of where the pathogen was acquired.

Tuberculosis in a worker should prompt an assessment of coworkers who might have provided or received the organism in the work setting. This public health response is similar for those with a shared home environment and is usually performed by the same governmental prevention specialists. The need to identify those with recent exposure and early infection is very important, both to the individual and to the rest of the work community, since curative treatment abolishes the risk of further exposure in only a few days.

EVALUATION OF SUSPECTED OCCUPATIONAL ILLNESS

Clinical suspicion of a possible occupational illness should be heightened whenever a patient is a member of a group at increased risk for exposure to biological hazards or belongs in one of the other categories described above. Specific evaluation will vary greatly, depending upon the clinical presentation and differential diagnosis. When the illness is suspected to be occupational in origin, a detailed history should be taken to establish how the exposure occurred. In addition, a walk-through evaluation of the work site should be considered. Worksite visits should allow for a thorough understanding of job functions and work practices. Depending on circumstances and available resources, the walk-through team may include occupational physicians, occupational health nurses, industrial hygienists, or biohazard scientists.

The visit should result in a determination of the need for further action, possibly including additional diagnostic testing and a trial of early work site remediation. Environmental sampling, commonly used for the evaluation of chemical exposure, should only be used with caution in the evaluation of biological hazards. Although it may be useful in some situations, the ubiquity of microbes and the lack of "normal" values renders interpretation difficult.

Certain occupational diseases must be reported to the state health department as a matter of law. Many health departments are staffed by experts who can assist with the evaluation of occupational and environmental illness. Additionally, they may coordinate relevant public health measures to protect other workers and the community. Other government resources include the Centers for Disease Control and Prevention (CDC) and the National Institute for Occupational Safety and Health (NIOSH). CDC and the state public health department can be particularly helpful where there is an opportunity for prevention (or research) or when specialty evaluations would contribute to the resolution of the situation or crisis. The role of NIOSH includes the evaluation of how the job contributed to the illness. This agency is also interested in studying potential new disease mechanisms in order to develop health and safety standards that will prevent future occurrences.

Further Reading

Bennett JE, Dolin R, Blaser MJ, eds. Mandell, Douglas, and Bennett's Principles and Practice of Infectious Diseases, 8th edn. Philadelphia, PA: Saunders, 2015.

Kasper DL, Fauci AS, Longo DL, et al., eds. Harrison's Principles of Internal Medicine, 19th edn. New York: McGraw-Hill, 2015.

Chapter 20

PREVENTION of ILLNESS from BIOLOGICAL HAZARDS

Gregg M. Stave*

Once we become aware of the potential biological hazards in a particular work setting, we can develop an effective plan to prevent occupational illness. Prevention of illness from biological hazards is accomplished by a combination of the three classic prevention strategies—primary, secondary, and tertiary prevention.

Primary prevention aims to prevent illness before the disease process begins. Strategies include vaccination and measures to limit potentially hazardous exposure to biological agents and organisms. Exposures can be limited by using engineering controls (including ventilation and containment systems), proper work practices, and personal protective equipment (such as gloves, uniforms, laboratory coats, safety glasses, and respirators). Environmental monitoring may be useful to determine if controls are effective in reducing potential exposures.

Secondary prevention entails intervention when the physiologic changes that precede illness are recognized or when subclinical illness develops. Secondary prevention is most effective when a surveillance system detects these events systematically. Medical screening must therefore focus on both the results for individuals and those for the group (epidemiologic evaluation).

Tertiary prevention is directed at limiting the consequences of clinical illness once it has occurred. It may involve medical treatment, work restrictions, and/or removal of the worker from further potential exposure. Specific preventive practices vary, depending on the work setting and the level of hazard.

* The author wishes to thank Linda M. Frazier and Jerry J. Tulis for their contributions to an earlier version of this chapter.

OCCUPATIONS WITH POTENTIAL BIOLOGICAL HAZARDS

Occupational biological hazards are those encountered when the workplace has greater risk of exposure than the surrounding community. Thus, the common cold is not usually considered to be an occupational biological hazard even though one employee can contract a cold from another, because cold viruses are ubiquitous in the community at large.

Biological hazards may be found in diverse work settings. In some settings, such as research laboratories conducting studies on specific biological agents or organisms, the hazards are clearly identified. On farms and at zoos, specific zoonoses (animal infections that may be transmitted to humans) may be a risk. However, in most settings, the hazard is an indirect consequence of the work or a risk that arises in the work environment. Before we develop a prevention program, we need to evaluate the setting for reasonably anticipated hazards.

THE OSHA BLOODBORNE PATHOGENS STANDARD

The Occupational Safety and Health Administration (OSHA) has issued only one standard to date that addresses biological hazards. The Bloodborne Pathogens Standard (29 CFR 1910.1030) was issued in December 1991 and became effective in March 1992.[1] The standard applies to all employers

Physical and Biological Hazards of the Workplace, Third Edition. Edited by Gregg M. Stave and Peter H. Wald.

with one or more employees where employees may have exposure to blood-borne pathogens. Blood-borne pathogens are defined as pathogenic microorganisms that are present in human blood and can cause disease. These pathogens include, but are not limited to, hepatitis B virus (HBV) and human immunodeficiency virus (HIV). The standard applies not only to hospitals and doctors' offices but also to many other work settings, such as clinical and research laboratories, mortuaries, emergency response teams, lifeguarding, and medical equipment maintenance.

The first requirement of the Bloodborne Pathogens Standard is the performance of an Exposure Determination. Employers must evaluate the potential for employees to be exposed to blood-borne pathogens. Occupational exposure means reasonably anticipated skin, eye, mucous membrane, or parenteral contact with potentially infectious materials on the job. Even employees who use personal protective equipment, such as gloves, are considered to be potentially exposed. If employees have the potential for occupational exposure to blood-borne pathogens, then the employer must develop a written Exposure Control Plan. Blood-borne pathogen exposure can occur from handling substances other than blood. Also, body fluids from deceased individuals can be infectious.

An employee is at risk of exposure if he or she handles these substances. Exposure risk is negligible for personnel who work in healthcare settings but who do not handle body substances—for example, telephone repair personnel. In contrast, many employers recruit work site first aid teams that include employees whose usual work does not involve contact with human body fluids. If the emergency response duties of these volunteers may lead to contact with blood or body fluids of injured coworkers, the employer should provide appropriate training and personal protective equipment and offer the hepatitis B vaccine. (OSHA issued a ruling in June 1993 that permits employers to delay the vaccination of first aid providers in specific situations. Employers considering this option should carefully review the practical implications of this policy.)

Although not specifically addressed by the OSHA standard, tissues or cell lines derived from human or primate sources are also potentially infectious. HIV does not replicate in cells outside the host, so cell lines of relatively recent origin are potentially infectious, whereas later generations will not be infectious because of a dilution effect. HBV can persist in cell lines.

When workers can become exposed to blood-borne pathogens occupationally, the employer's written Exposure Control Plan must include certain critical elements that are described in detail in the standard. Issues to be addressed include engineering controls, personal protective equipment, and work practice controls that focus on universal precautions. Universal precautions means that all blood and body fluids should be regarded as potentially infectious; it is not sufficient to use precautions with some samples and not with others. The employer must provide training (and annual retraining) in the use of personal protective equipment, safe storage and transport of body fluids, safe disposal of potentially infectious wastes, effective decontamination of contaminated work surfaces, and prohibition of storage or consumption of food and drink in areas where there is a reasonable likelihood of exposure. Employers must provide hepatitis B vaccine promptly to employees at reasonable risk of exposure at no cost. Employees who refuse the vaccine should sign an OSHA-specified declination form. A procedure to evaluate employees who have had an exposure to determine the potential infectivity of the source and to provide appropriate medical care for the exposed worker is also required. The employer must keep records documenting training, vaccination (or declination), and postexposure evaluation.

In 2000, several states and the US Congress considered laws and regulations to encourage or require the use of newer needlestick and sharps injury prevention technology. The federal Needlestick Safety and Prevention Act was passed in November 2000 and went into effect on April 18, 2001.[2] The law required OSHA to modify the Bloodborne Pathogens Standard. Legal requirements include the use of needleless systems and other engineering approaches that effectively reduce the risk of an exposure incident. The standard requires that frontline employees who are using the equipment have the opportunity for input into purchasing decisions. The needlestick log helps both employees and employers track all needlesticks to help identify problem areas or operations. When OSHA transitioned to the OSHA 300 Log for reporting occupational injuries and illnesses in 2002, it incorporated the requirement to record needlesticks and cuts from sharp objects that are contaminated with another person's blood or other potentially infectious material as an injury. To protect the employee's privacy, the employee's name is not entered on the OSHA 300 Log.[3] If an employee develops a blood-borne illness, such as HIV, hepatitis B, or hepatitis C, as a result of exposure, that needs to be recorded as an illness.

THE HISTORY OF OSHA GUIDELINES FOR TUBERCULOSIS

In late 1993, OSHA issued mandatory guidelines (revised February 1996) for an enforcement policy intended to protect workers from tuberculosis (TB). These guidelines were based primarily on the Centers for Disease Control and Prevention (CDC) 1990 *Guidelines for Preventing the Transmission of Tuberculosis in Health-Care Settings, with Special Focus on HIV-Related Issues*.[4] The guidelines pertain to employers in settings where workers are at increased risk of exposure, such as healthcare facilities, correctional institutions, homeless shelters, drug treatment centers, and long-term care facilities.

OSHA guidelines required employee training and information on the signs and symptoms of tuberculosis, hazards of transmission, medical surveillance, and site-specific controls. Employers must institute a program of early identification of suspected cases. Medical surveillance should include preplacement evaluation, periodic Mantoux testing, and management of persons with positive test results. Persons who are infectious must be treated in respiratory isolation rooms under negative pressure. A formal OSHA compliance directive was planned for after CDC completed the second edition of its guidelines. A draft of the guidelines included a requirement to create a TB infection control plan. Elements of the plan include a risk assessment, administrative controls, engineering controls, use of respiratory protection, employee education and training, and medical surveillance.

In 1997, OSHA proposed a comprehensive standard for preventing TB transmission among healthcare workers.[5] It differed from the CDC guidance in the areas of risk assessment, medical surveillance, and Respiratory Protection. It also contained medical removal protection for employees. However, controversy over the standard and a precipitous drop in TB in acute care hospitals resulted in OSHA withdrawing the proposed rule at the end of 2003.[6] This also resulted in OSHA withdrawing the TB-specific respirator protection standard and instead relying on the general Respiratory Protection standard. The US Congress next stepped in and restricted OSHA from enforcing provisions that require annual fit testing of respirators for occupational exposure to tuberculosis. OSHA was to take no further action until the CDC issued their revised guidelines. In December of 2005, the CDC issued *Guidelines for Preventing the Transmission of Mycobacterium tuberculosis in Health-Care Settings*, 2005, which recommended periodic fit testing.[7] When Congress passed the 2008 appropriations bill, it did not contain the restriction on OSHA enforcement. On January 2, 2008, OSHA resumed full enforcement of the Respiratory Protection Standard, including the requirements for fit testing of respirators when they are used for protection against TB.[8]

PROPOSED OSHA INFECTIOUS DISEASE STANDARD

In 2010, OSHA published a request for information (RFI) on occupational exposure to infectious agents in settings where healthcare is provided and healthcare-related settings.[9] The standard could require the development of a comprehensive infection control plan. The focus would be on protecting workers from infections transmitted by contact, droplet, and aerosol routes—areas not addressed by the Bloodborne Pathogens Standard. This could include TB, influenza, MRSA, Ebola, and emerging infectious diseases.

The RFI is the first step in a process that may lead to a new OSHA standard. Since releasing the RFI, OSHA held stakeholder meetings in 2011 and initiated the Small Business Advocacy Review Panel process in 2014 as required by the Small Business Regulatory Enforcement Fairness Act of 1996. As of mid 2016, OSHA has not published a notice of proposed rulemaking for this standard.

PREVENTION OF EXPOSURE TO BIOLOGICAL AGENTS

Hazardous exposures to biological agents occur mainly through inhalation and ingestion, although skin contact (or penetration) can cause illness with some agents. These routes of exposure can be eliminated through engineering, administrative or work practice controls, and the use of personal protective equipment.

Engineering controls

The preferred preventive measure for prolonged or highly hazardous potential exposures is the use of engineering controls. Workplace controls are intended to contain biohazards at their source, reduce their airborne concentration, and limit their movement through the work site. Heating, ventilation, and air conditioning (HVAC) systems must also be appropriately designed and maintained to prevent contamination by fungi and bacteria (including *Legionella pneumophila*). For indoor settings, such as medical or research facilities, room ventilation can be engineered to provide directional and single-pass airflow. In hospitals, air exhausted from high-risk infectious disease isolation rooms can be further decontaminated by filtration. Use of ultraviolet light to treat exhausted air is under study. In research and clinical laboratories, handling infectious agents in a biological safety cabinet (BSC) can prevent inhalation exposures. For bioaerosol control, the correct type of unit must be used.

Class I cabinets provide personnel protection but little or no product protection. Room air flows into this open cabinet and is ducted through a high-efficiency particulate air (HEPA) filter. HEPA filters clean air supplied to the work zone, providing product protection; and the HEPA filtration of exhaust air provides environmental protection. This filtration system traps all microorganisms, including viruses, with 99.97% efficiency at the 0.3 μm particle size and essentially 100% capture of particles larger than 0.3 μm. The class I cabinet is designed for work with low- to moderate-risk biological agents. It can be used to house various aerosol-generating equipment, including blenders, centrifuges, and mixers. Since the cabinet work zone is not protected from external contamination by the inward flow of unfiltered laboratory air, the cabinet should not be used for work that requires aseptic conditions.

Class II laminar flow cabinets are the most commonly used laboratory containment devices. An air barrier at the front opening of the cabinet provides personnel protection.

The air circulating in the workspace is HEPA filtered, providing protection from contamination for the biological material inside the cabinet. The exhaust is also passed through a HEPA filter and either returned to the room or ducted outside. Class II cabinets are classified as A or B, based on design, airflow, and exhaust. The class II type A cabinet is used for work with biological agents in the absence of volatile or toxic chemicals and radioisotopes, since cabinet air is recirculated within the work zone. These cabinets may be exhausted to the room or externally via ductwork. Class II type B cabinets are ducted directly to the exhaust system; the plena remain under negative pressure.

Class III cabinets are totally enclosed gastight ventilated chambers. They are used in laboratories for work with organisms that are highly infectious through the airborne route.

Clean benches are not considered BSCs. They are designed only to protect the product from contamination by providing positive pressure airflow. Using a clean bench to handle an infectious organism would cause the organism to be exhausted onto the user.

Other engineering controls include special containers for waste and sharps disposal, needleless systems, and devices such as self-resheathing needles.

Administrative controls

Administrative control focuses on maintaining good work habits to minimize exposures due to spills, accidental releases, or other causes. Hands should be washed frequently, work surfaces should be decontaminated properly, and under no circumstances should food, beverages, or tobacco products be consumed in the same work area as biohazardous agents. Access to biohazard work areas should be restricted to employees who have had appropriate safety training and who have the necessary personal protective equipment. In laboratories, mouth pipetting should be prohibited.

Personal protective equipment

The use of personal protective equipment (PPE) is indicated whenever the hazards cannot be eliminated through the use of facility design and other engineering controls. Gloves should always be worn when handling infectious agents or secretions from potentially infectious patients or animals. Protective clothing is desirable in many instances, including use of reinforced hand and arm wear (using leather or steel mesh) for certain animal handling tasks where there is a risk of bite or laceration. Eye protection is important when working with certain airborne biological hazards. Instead of ordinary safety glasses, goggles or face shields should be employed when potentially infectious particulates may arise, such as when performing dental or surgical procedures on potentially infectious patients or animals.

Protection from inhalation exposure of biologicals can be accomplished by wearing a respirator. Because even the most lightweight respirators can be somewhat uncomfortable after prolonged periods of use, engineering controls are preferred except for short-term control. Surgical masks only protect the patient, animal, or product from exposure to the worker's exhaled organisms. The worker breathes unfiltered air that enters the airway from around the sides of the mask. To protect the worker from biological hazards in the environment, one of many varieties of certified respirators must be used, such as a HEPA filter mask or dust/mist respirator. Respirator selection should be specific for the hazard and work situation. Employees using respirators are required by OSHA regulation to obtain medical clearance and to attend a training program. Training must include a fit test for the specific type of respirator being worn. These requirements apply to all respirators, including the simple dust mask.

Waste handling

The proper handling, decontamination or containment, and disposal of biological waste are important infection control measures in all work settings. In medical facilities and laboratories, wastes that are potentially infectious must be initially segregated from other wastes and placed in identifiable biohazard storage bags, affixed with the international biohazard symbol. All sharps must be placed in hard-walled, leakproof, and secure containers. Contaminated needles should not be cut or recapped prior to disposal.

Decontamination can be accomplished by means of sterilization, disinfection, sanitization, or antisepsis. Sterilization means the eradication of all living microorganisms and spores. Disinfection means elimination of most biological organisms, although hardier organisms and spores may survive. Sanitization is the lowest level of disinfection and removes most pathogenic organisms. Antisepsis means reducing bacterial counts by applying compounds to skin or other body tissues.

Disinfectants have been classified according to chemical composition and level of activity. Commercially available disinfectants often combine one or more agents. The high-level disinfectants possess a broad spectrum of antimicrobial properties and are recommended for use in the destruction of mycobacteria and the blood-borne pathogens; they are often referred to as mycobactericidal or tuberculocidal. The low-level disinfectants are recommended for use in sanitation and other public health applications. The resistance of microorganisms to chemical disinfection, from most resistant to most sensitive, is as follows: bacterial spores, tubercle bacilli, fungal spores, hydrophilic viruses, mycelial fungi, lipophilic viruses, Gram-negative vegetative bacteria, and Gram-positive vegetative bacteria. Common disinfectants and their uses are listed in Table 20.1.

Sterilization can be accomplished by several techniques. Steam autoclaving is a commonly used method to sterilize

TABLE 20.1 Disinfectants and their uses.

Disinfectant	Antimicrobial activity	Use
Chlorine-liberating halogens	Hypochlorites at 1–5% aqueous concentration possess wide spectrum of activity against microbials, including HIV and HBV	A 1 : 1000 dilution of household bleach recommended for use against blood-borne pathogens. Chlorination of potable water conducted at 0.2 ppm available chlorine
Formaldehyde	Bactericidal, tuberculocidal, and virucidal; hours of exposure required for destruction of bacterial spores; aqueous formaldehyde (formalin) is 37% formaldehyde with 10–15% methanol in water	Usefulness limited by toxicity and odor. Formalin is used as a spray and surface disinfectant. Vapor-phase formaldehyde is used to routinely disinfect biological safety cabinets and other enclosures
Alcohols (ethanol, propanol, and isopropanol)	Bactericidal, tuberculocidal, and virucidal; devoid of sporicidal activity; most effective concentration is 70% in water	General disinfection of surfaces and equipment. Rapid destruction (in seconds) of vegetative bacteria, fungi, and certain viruses. Leaves little to no residue
Glutaraldehyde	Broad spectrum of antimicrobial activity; hours of exposure required for destruction of bacterial spores; used as 2% alkaline glutaraldehyde	Excellent high-level disinfectant for inhalation therapy equipment and other devices. Residues need to be removed with sterile water wash
Iodophors	Not effective as a disinfectant	Primarily used as an antiseptic
Phenolics	Antimicrobial properties of phenolics vary considerably	Used primarily for housekeeping and sanitizing applications (e.g., Lysol)
Quaternary ammonium compounds (e.g., benzalkonium chloride (monoalkyldimethyl benzyl ammonium salt))	Possess detergent and surfactant properties; antimicrobial properties are questionable	Popular for sanitizing

cultures and stocks of microorganisms, laboratory ware, and contaminated devices and instruments. Dry heat sterilization (i.e., 160°C for 1–2 hours) can be used to sterilize glassware and metallic instruments when corrosive effects of steam on sharps and cutting edges are undesirable. However, the penetrability and killing effects of dry heat are poorer than those of steam autoclaving or gaseous sterilization. Ethylene oxide, a commonly used gaseous sterilant with high penetrability, is found in most commercial and hospital sterile processing units. Excellent containment is required for these sterilization machines to avoid worker exposures. Ionizing radiation is gaining worldwide acceptance as a commercially feasible sterilization procedure. Attributes of radiation sterilization include penetrability, final package processing, and lack of toxic residues.

SURVEILLANCE

Surveillance for infectious organisms has become a common practice in hospital settings since Semmelweis and others began promoting hand washing and aseptic surgical technique in the nineteenth century. Hospital infection control programs were initiated in the 1950s in response to the first epidemics of antibiotic-resistant staphylococcal infection among hospitalized patients. An initial enthusiasm for routine environmental culturing has been replaced by monitoring programs that tabulate rates of nosocomial infection among patients.

Target infections are usually surgical wound infections, urinary tract infections, bacteremias, and pneumonias. Now that tuberculosis rates are rising again in the United States, some hospitals are also attempting to determine if nosocomial tuberculosis infection is occurring, especially among patients with the acquired immunodeficiency syndrome (AIDS). When rates are found to be elevated, patient care practices are reviewed to correct deficiencies in urinary catheter care, intravenous equipment care, respiratory therapy, surgical care, patient isolation, or hand washing. In agriculture, animal breeding, veterinary practices, and related settings, infected animals should be segregated and promptly diagnosed. To prevent zoonoses among workers, animals should be treated or killed and disposed of properly.

In industrial settings, occupational disease surveillance generally has a different target group. Workers themselves are monitored to detect disease caused by a work exposure, such as development of elevated blood lead levels among battery-manufacturing workers. Surveillance of only a few infectious diseases (e.g., tuberculosis) is conducted in this manner.

Tuberculosis monitoring through surveillance of workers by tuberculin skin testing has three goals. First, certain workers who convert from skin test negative to positive can be treated with antimicrobial therapy to prevent development of active TB in the worker. For workers whose previous skin test reactivity is unknown and who experience an acute exposure to tuberculosis occupationally, a skin test should be done immediately and then again in 6–12 weeks to check for tuberculin test conversion. Second, early identification of potentially infectious healthcare, food service, or other workers with extensive contact with the public can prevent the infection of patients or others. Third, skin test conversion rates provide a measure of the quality of infection control procedures in healthcare workplaces.

HBV surveillance is generally limited to workers who have been exposed to a patient's blood or body fluids, such as from a needlestick. Surveillance of healthcare workers for antibody conversion could be used to assess the quality of infection control procedures. However, this is usually not done because of its expense, because of the sometimes difficult task of determining if a conversion is work related, and because surveillance for occupational exposures such as needlesticks is more efficient.

When an exposure occurs, a targeted postexposure follow-up and treatment protocol should be initiated for the employee. The protocol includes a baseline visit in which acute treatment is based on the source's HBV status and the employee's vaccination status. The employee and source should also be evaluated for HIV at baseline. During follow-up over 3–6 months, the employee is then assessed for seroconversion from either virus.

Mandatory screening of physicians, dentists, and other healthcare workers for HIV has been hotly debated. Surveillance has not been required or recommended to date, because the risk for transmitting the virus to patients is very low. However, professionals who are infected should not perform exposure-prone invasive procedures and should refrain from patient contact when they have open skin lesions. There is no justification for HIV screening of workers who do not have patient contact.

In addition to healthcare workers, laboratory workers and agricultural workers can also be at risk of contracting infectious diseases occupationally. Infections that could occur from processing human tissue specimens include HBV, HIV/tuberculosis, and the following bacterial and fungal pathogens: *Brucella* species, *Francisella tularensis*, *Shigella*, *Salmonella*, *Coccidioides immitis*, *Blastomyces dermatitidis*, and *Histoplasma capsulatum*.

Laboratory animals can potentially transmit infections to humans, including HBV, simian immunodeficiency virus, rabies, plague, and tuberculosis. Wild animals and farm animals or their products can potentially transmit anthrax, brucellosis, erysipeloid, leptospirosis, plague tularemia, candidiasis, coccidioidomycosis, dermatophytoses, histoplasmosis, hookworm, toxoplasmosis, ornithosis, Q fever, Rocky Mountain spotted fever, viral encephalitis, hantavirus, paramyxovirus, rabies, or parasitic infections. Safe handling procedures and vaccination are recommended for preventing these occupational infections rather than surveillance among workers.

Universal precautions when handling macaque monkeys are essential, including physically and chemically restraining the monkeys and use of goggles and arm-length-reinforced leather gloves. Relying on periodic serologic testing in the colony to determine which monkeys need to be handled cautiously is hazardous, because monkeys may seroconvert between testing but may appear clinically free from infection.

VACCINATION

Vaccinations are given in occupational settings for four common indications. First, employees may be vaccinated to protect them from an infectious organism such as tetanus or hepatitis B when they are at increased risk of being exposed in the workplace. Some vaccines are given as part of a post-exposure protocol. Second, healthcare workers may be vaccinated against agents such as rubella or influenza to prevent them from inadvertently passing the infection to patients. Third, some company health units, as part of a corporate wellness program, may provide vaccinations against community-acquired infections such as tetanus to employees who are not at increased occupational risk of the infection. Fourth, vaccinations against agents such as yellow fever may be provided to prepare employees for international travel.

Vaccines commonly administered in occupational settings are listed in Table 20.2. For laboratory workers who handle unusual organisms, consult the chapters on specific organisms later in this book. The Infectious Diseases sections of CDC can provide helpful information. For information on dosage, administration, and contraindications, consult the package insert for each vaccine. For further information about commonly used vaccines, including those that generally do not need to be administered for occupational indications, consult the most recent Adult Immunization Schedule-United States[10] and the American College of Physicians *Guide for Adult Immunization*.[11] Vaccines used before international travel are listed separately in Table 20.3.

SPECIAL SITUATIONS

Immunocompromised workers

Many individuals remain in the workforce even after developing health problems that may lead to immune system compromise. The most highly publicized group comprises those with HIV infections, but other conditions can also confer some degree of immune dysfunction. Individuals with diabetes mellitus, splenic disorders, renal dysfunction, alcoholism, or cirrhosis do not generally require special work restrictions or special work-related immunizations, but they should be followed closely by a personal physician, who may administer vaccines against pneumococcus, influenza, or other agents. Conditions in which special occupational considerations may be warranted are listed in Table 20.4.

TABLE 20.2 Vaccines and immunobiologicals commonly administered in occupational settings.

Vaccine or immunobiological	Type	Worker groups	Immunocompromised employees	Comments
Hepatitis A vaccine	Inactivated whole virus	Institutional workers (caring for developmentally challenged), child care workers, laboratory workers handling hepatitis A virus, primate handlers working with animals that may harbor HAV*	OK to give	
Hepatitis B vaccine	Recombinant DNA vaccine	Healthcare workers, other workers handling human blood or body fluids	OK to give	No risk of acquiring HIV from vaccine
Hepatitis B immune globulin	Pooled human antiserum	Postexposure, hepatitis B	OK to give	Also known as HBIG
Immune serum globulin	Pooled human antiserum	Postexposure, hepatitis A	OK to give	
Measles, mumps, rubella	Attenuated live viruses	Healthcare workers, day care workers	Do not give	Can test employee for rubella immunity in lieu of vaccination
Polio vaccines	Oral = attenuated live virus Parenteral = killed virus	Laboratory workers handling polio cultures	Do not give the oral or live vaccine	
Rabies vaccine	Inactivated virus	Workers handling animals which may have contracted rabies in the wild	Do not give	
Rabies immune serum globulin	Human antiserum	Postexposure	OK to give	
Tetanus–diphtheria vaccine	Killed bacteria and toxoid	Animal handlers, postexposure, corporate wellness program	OK to give	Should have booster every 10 years
Tetanus toxoid	Human antiserum	Postexposure	OK to give	
Varicella vaccine	Attenuated live virus	Healthcare workers	Do not give	
Vaccinia vaccine	Attenuated live virus	Workers handling vaccinia cultures	Do not give	

*HAV, hepatitis A virus.

Individuals with potentially immunosuppressive conditions should receive training in techniques to prevent exposure to infectious agents in the workplace, including the proper use of personal protective equipment. The need for special vaccines or surveillance should be reviewed, bearing in mind that certain vaccines—especially live virus vaccines—may be contraindicated. Consideration should be given to restricting employees from high-risk work exposures for which protective vaccinations are contraindicated. Although some employees may request to avoid certain other infectious agents and some employers may be able to accommodate these requests, standards generally do not exist for when immunocompromised employees absolutely must be restricted from working with specific infectious agents.

Pregnant workers

Some infectious diseases acquired during pregnancy can cause direct harm to the fetus or substantial maternal morbidity with indirect consequences for the developing fetus. Common agents of concern from the occupational standpoint are rubella, human parvovirus B19, cytomegalovirus, varicella zoster, hepatitis B, coccidioidomycosis, and toxoplasmosis. Prenatal screening can determine if a pregnant woman is susceptible to contracting any of these agents during pregnancy. Prenatal infection with Lyme disease, malaria, Zika, or viral encephalitis has also been associated with adverse fetal outcomes, whereas perinatal transmission has not been demonstrated for polio, rabies, or influenza.

Standards are evolving for restricting occupational exposures among pregnant workers. Some authorities focus on educating workers about optimal work practices, because it has not been demonstrated that pregnant workers are any more likely than nonpregnant workers to contract the infections of concern, even though the health consequences of becoming infected may be serious. Other authorities recommend restricting susceptible pregnant employees from working with patients in acute aplastic crisis (human parvovirus B19), adult patients or children shedding cytomegalovirus, or persons with chickenpox or herpes zoster (varicella zoster virus).

TABLE 20.3 Vaccines and immunobiologicals commonly administered for international travel.

Vaccine or immunobiological	Type	Required versus recommended	Immunocompromised employees	Comments
Yellow fever	Live attenuated virus	May be required	Do not give	
Cholera	Live attenuated virus	May be required; physician statement contraindicating use for specific patient may be accepted	Do not give	Approved in 2016 for adults age 18–64, targets *Vibrio cholerae* serogroup O1
Typhoid				
Killed bacteria	Not required but highly recommended for certain high-risk locales	OK to give killed vaccine only	Live attenuated vaccine under development	
Polio	Live attenuated virus (oral) or killed (parenteral)	Recommended for certain high-risk locales	Do not give live attenuated	If not fully immunized previously, complete primary series
Tetanus and diphtheria	Killed bacteria and toxoid	Recommended	OK to give	
Immune serum globulin	Pooled human antiserum	Recommended for certain high-risk locales	OK to give	Do not administer at same time as some live virus vaccines; may suppress immune response to them
Hepatitis A vaccine	Inactivated whole virus	Recommended for certain high-risk locales	OK to give	May be administered concomitantly with immune globulin if needed
Hepatitis B vaccine	Recombinant DNA vaccine	Recommended for certain high-risk locales	OK to give	6 months required for full series
Rabies vaccine	Inactivated virus	Recommended if high-risk animal contact is likely	Do not give	
Measles, mumps, rubella	Attenuated live viruses	Recommended for certain high-risk locales	Do not give	
Meningococcal vaccine	Mixed polysaccharides	Recommended for certain high-risk locales	OK to give	
Malaria	Chemoprophylaxis, not a vaccine	Recommended for certain high-risk locales	OK to give	

TABLE 20.4 Immunocompromising conditions with potential occupational significance.

Condition	Occupational significance
HIV 1 infection	Review work practices Do not administer live vaccines If employee is a healthcare provider, restrict from performing exposure-prone invasive procedures
Organ transplantation, receiving immunosuppressive drugs	Review work practices Consult transplant physician before administering vaccines to avoid nonspecific immunologic response that may trigger allograft rejection
High-dose chronic corticosteroid therapy	Review work practices Consult oncologist to determine if employee is immunosuppressed
Malignant disease, receiving immunosuppressive chemotherapy	Review work practices Consult oncologist before administering live vaccines
Congenital immunodeficiency diseases	Review work practices Consult treating physician before giving vaccines Do not give immune globulin to persons with selective IgA deficiency

The potential reproductive risks of uncommon infectious agents that could be encountered in occupational settings should be evaluated individually. The American College of Obstetricians and Gynecologists maintains a resource center in Washington, DC, that can provide technical bulletins on perinatal care. Although female employees are often the focus of concern about occupational reproductive issues, men are also susceptible to reproductive hazards. The best-known infectious reproductive hazard for men is mumps; mumps can cause orchitis, which may lead to sterility.

Workers concerned about contracting disease from coworkers

Serious illness in an employee can generate substantial concern among coworkers. If an employee looks sick but the diagnosis is not known, coworkers may contact company health or safety personnel. Ethical issues about the confidentiality of the employee suspected of illness must then be addressed. HIV infection is an apt example. Even if coworkers suspect that an individual has the infection, there is no reason to invade his or her privacy to confirm or allay fears of infection through casual contact. The virus is not transmitted through handshakes, work surfaces, or telephones. While the virus has been isolated in very small quantities from saliva, there has never been a documented case of salivary transmission through food or eating utensils, or even through kissing on the lips. There is no evidence that the virus will replicate in insects, let alone be transmitted to humans from insects. There is no reason to restrict HIV-infected individuals from engaging in their normal work in order to allay unrealistic fears of contagion among coworkers. Coworker concerns should instead be addressed rapidly and decisively through education.

In general, employees with common viral respiratory infections are not restricted from work, because such viruses are endemic in the community. Employees with suggestive symptoms should be evaluated to rule out tuberculosis. If tuberculosis is confirmed, the individual should be restricted from work until his or her sputum becomes free of acid-fast bacilli. The local health department should be notified, coworkers should be skin tested, and follow-up should be provided as appropriate.

Occasionally, employees may become concerned that a coworker with a rash could have an infectious condition, such as scabies. Coworker anxiety may be accompanied by itching. Although scabies transmission is unlikely in the absence of personal contact, it may be helpful for medical personnel to determine the specific diagnosis in the "index" employee, to inform the work unit if treatment of coworkers is indicated, and to provide education on the myriad causes of rash that are not contagious.

International travel

International travel has become commonplace for business purposes. To prevent unnecessary illness or injury, a preventive health review before travel is strongly advised. Information on health and safety conditions for each country to be visited can be obtained from CDC. CDC maintains an international travel hotline and travel website. It issues a publication entitled *Health Information for International Travel* and has a database, accessible by computer, which is updated monthly. A printout can be obtained for each country that includes advice on infectious disease hazards, required and recommended immunizations, malaria prophylaxis, and food and water safety; it also details any recent incidents of civil unrest. This list underscores the importance of taking health precautions beyond vaccinations required for visa purposes. Vaccination requirements can also be obtained from embassies, the World Health Organization (Albany, NY), and sometimes from local health departments.

Travel can be categorized as high or low risk based on the country to be visited and on the person's itinerary during his or her stay. Travel to Westernized countries or to first-class hotels in some developing countries carries a lower risk than travel to rural areas of developing countries. Immunizations can be separated into those that may be required for visa purposes (e.g., yellow fever and cholera vaccines) and those that are recommended for personal protection (e.g., diphtheria–tetanus, hepatitis A vaccine). The more commonly used vaccines are listed in Table 20.3. For information on contraindications, dosage, and timing of vaccine administration, see the product information in the *Physicians' Desk Reference (PDR)*. The American College of Physicians *Guide for Adult Immunization* provides a good overview of major aspects of preventive health during foreign travel.

Along with vaccinations and chemoprophylactic drugs, travelers should take additional precautions to reduce the risk of vector-borne diseases. No vaccines or medications are available to prevent many mosquito-borne diseases (including dengue, chikungunya, Zika, and West Nile) and tick-borne diseases (including Lyme disease). Travelers to endemic areas should take precautions including avoiding being outdoors at times of peak exposure, wearing appropriate clothing to cover exposed skin, and using repellents.

For all international business travelers, especially persons with chronic health problems, it is also advisable to investigate health insurance coverage for foreign travel, to ensure that medications are carried in their original prescription containers, and to obtain the location and telephone numbers of the US embassies in each country to be visited.

Required reporting

Certain issues related to infectious diseases require reporting by health professionals, including infections that are diagnosed in occupational health units. Each state can provide lists of infections that must be reported to local health departments. Physicians and other healthcare providers must also maintain permanent records of immunization and report certain adverse effects of vaccination to the US Department of Health and Human Services.

Business continuity and pandemic planning

Business continuity plans are developed to ensure that business interruptions have the smallest possible impact on the ability of the business to function and can recover as quickly as possible. Business continuity planning is a formal exercise that is usually connected to emergency and disaster planning. Along with the development of roles and responsibilities, and a determination of essential personnel, organizations often conduct tabletop exercises and/or drills to practice the response to varied scenarios. These exercises also provide opportunities to identify the need to modify the plan.

Potential causes of business interruption include natural disasters, fires, and accidents. Community-based infectious diseases, such as pandemic influenza, may also cause business interruption. Business continuity plans should include a response to pandemics.

Pandemics are characterized by rapid spread that can overload the healthcare system. Approaches that may mitigate impact on workplaces include social distancing (including allowing nonessential personnel to work from home when feasible), vaccination programs, and facilitating access to medications that can reduce the duration or severity of illness. Medical facilities need additional procedures that reduce the likelihood that staff, visitors, or other patients will be exposed.

OSHA developed *Guidance on Preparing Workplaces for an Influenza Pandemic* in 2007.[12] The guidance is advisory and does not create new obligations. However, businesses are still obligated under the General Duty Clause to provide workplaces that "are free from recognized hazards that are causing or are likely to cause death or serious physical harm."[13]

The guidance recommends a risk stratification approach for employers:

Very high exposure risk occupations are those with high potential exposure to high concentrations of known or suspected sources of pandemic influenza during specific medical or laboratory procedures.

High exposure risk occupations are those with high potential for exposure to known or suspected sources of pandemic influenza virus.

Medium exposure risk occupations include jobs that require frequent close contact (within 6 ft) exposures to known or suspected sources of pandemic influenza virus such as coworkers, the general public, outpatients, schoolchildren, or other such individuals or groups.

Lower exposure risk (caution) occupations are those that do not require contact with people known to be infected with the pandemic virus, nor frequent close contact (within 6 ft) with the public. Even at lower risk levels, however, employers should be cautious and develop preparedness plans to minimize employee infections.

Based upon risk level, the guidance includes recommendations for planning, work practice controls, engineering controls, administrative controls, and personal protective equipment.

Concern about bioterrorism

Concerns about biological warfare and bioterrorism have existed for several decades and have been heightened by the horrific events of September 11, 2001, and their aftermath.

Several countries are known to have had biological warfare programs and stocks of agents are known to exist. At least 35 agents and organisms have been classified as possible

TABLE 20.5 Potential Biological Warfare Agents

Disease	Incubation	Symptoms	Signs	Diagnostic tests	Transmission and Precautions	Treatment (Adult dosage)	Prophylaxis
Inhaled Anthrax	2–6 days Range: 2 days to 8 weeks	Flu-like symptoms Respiratory distress	Widened mediastinum on chest X-ray (from adenopathy) Atypical pneumonia Flu-like illness followed by abrupt onset of respiratory failure	Gram stain ("boxcar" shape) Gram positive bacilli in blood culture ELISA for toxin antibodies to help confirm	Aerosol inhalation No person-to-person transmission Standard precautions	Mechanical ventilation Antibiotic therapy Ciprofloxacin 400 mg iv q 8–12 hours Doxycycline 200 mg iv initial, then 100 mg iv q 8–12 hours Penicillin 2 mil units iv q 2 hours – possibly add gentamicin	Ciprofloxacin 500 mg po bid or doxycycline 100 mg po bid for ~8 weeks (shorter with anthrax vaccine) FDA-approved vaccine: administer after exposure if available
Botulism	12–72 hours Range: 2 hours – 8 days	Difficulty swallowing or speaking (symmetrical cranial neuropathies) Symmetric descending weakness Respiratory dysfunction No sensory dysfunction No fever	Dilated or un-reactive pupils Drooping eyelids (ptosis) Double vision (diplopia) Slurred speech (dysarthria) Descending flaccid paralysis Intact mental state	Mouse bioassay in public health laboratories (5–7 days to conduct) ELISA for toxin	Aerosol inhalation Food ingestion No person-to-person transmission Standard precautions	Mechanical ventilation Parenteral nutrition Trivalent botulinum antitoxin available from State Health Departments and CDC	Experimental vaccine has been used in laboratory workers
Plague	1–3 days by inhalation	Sudden onset of fever, chills, headache, myalgia Pneumonic: cough, chest pain, hemoptysis Bubonic: painful lymph nodes	Pneumonic: Hemoptysis; radiographic pneumonia – patchy, cavities, confluent consolidation Bubonic: typically painful, enlarged lymph nodes in groin, axilla, and neck	Gram negative coccobacilli and bacilli in sputum, blood, CSF, or bubo aspirates (bipolar, closed "safety pin" shape on Wright, Wayson's stains) ELISA, DFA, PCR	Person-to-person transmission in pneumonic forms Droplet precautions until patient treated for at least three days	Streptomycin 30 mg=kg=day in two divided doses × 10 days Gentamicin 1–1.75 mg=kg iv/im q 8 hours Tetracycline 2–4 g per day	Asymptomatic contacts; or potentially exposed Doxycycline 100 mg po q 12 hours × 7 days Ciprofloxacin 500 mg po Tetracycline 250 mg po q 6 hours × 7 days Vaccine production discontinued

(*Continued*)

TABLE 20.5 (Continued)

Disease	Incubation	Symptoms	Signs	Diagnostic tests	Transmission and Precautions	Treatment (Adult dosage)	Prophylaxis
Tularemia "pneumonic'	2–5 days Range: 1–21 days	Fever, cough, chest tightness, pleuritic pain Hemoptysis rare	Community-acquired, atypical pneumonia Radiographic: bilateral patchy pneumonia with hilar adenopathy (pleural effusions like TB) Diffuse, varied skin rash May be rapidly fatal	Gram negative bacilli in blood culture on BYCE (Legionella) cysteine- or S-H-enhanced media Serologic testing to confirm: ELISA, microhemagglutination DFA for sputum or local discharge	Inhalation of agents No person-to-person transmission but laboratory personnel at risk Standard precautions	Streptomycin 30 mg = kg = day IM divided bid for 10–14 days Gentamicin 3–5 mg = kg = day iv in equal divided shoulders × 10–14 days Ciprofloxacin possibly effective 400 mg iv q 12 hours (change to po after clinical improvement) × 10–14 days	Ciprofloxacin 500 mg po q 12 hours × 2 weeks Doxycycline 100 mg po q 12 hours × 2 weeks Tetracycline 250 mg po q 6 hours Experimental live vaccine
Smallpox	12–14 days Range: 7–17 days	High fever and myalgia; itching; abdominal pain; delirium Rash on face, extremities, hands, feet; confused with chickenpox which has less uniform rash	Maculopapular then vesicular rash – first on extremities (face, arms, palms, soles, oral mucosa) Rash is synchronous on various segments of the body	Electron microscopy of pustule content PCR Public health lab for confirmation	Person-to-person transmission Airborne precautions Negative pressure Clothing and surface decontamination	Supportive care Vaccinate care givers	Vaccination (vaccine available from CDC)

Courtesy of Michael Hodgson, M.D., Office of Public Health and Environmental Hazards, Veterans Health Administration, Washington, D.C.

bioterrorism concerns. The most widely discussed are anthrax, botulism, plague, tularemia, and smallpox (Table 20.5). An epidemic of anthrax occurred in the former Soviet Union following an accidental release from a military facility in Sverdlovsk in 1979. In 2001, there were exposures to anthrax sent through the mail in the United States, resulting in several cases of cutaneous anthrax and three fatalities due to inhalation anthrax. In 2008, the Department of Justice and FBI officials released documents and information showing that charges were about to be brought against an Army scientist at Ft. Detrick, Maryland, who committed suicide before those charges could be filed. These episodes have led to heightened fears about the possibility of further small- and also large-scale bioterrorism activities.

While publicized incidents of bioterrorism led to significant fears, it should be kept in mind that the actual risk of being involved in an event is extremely small. The dissemination of large quantities of bioterrorism agents and organisms fortunately has many significant technical challenges.

A prudent response to concerns involves different actions for individuals, professionals, and organizations. In general, people should be reassured that their personal risk is very low and that they should take reasonable precautions in everyday life. This includes not handling suspicious looking mail and packages and accessing their local emergency response system as needed. The practices of hoarding antibiotics or using antibiotics without a medical diagnosis should be discouraged.

For the medical and emergency response community, there is a need to learn to recognize the signs and symptoms of bioterrorism agents and organisms (Table 20.5). This is of special concern because many of these diseases are otherwise uncommon or, as in the case of smallpox, have not been seen for decades. Significant improvements in the public health infrastructure are needed to continue to enhance readiness for bioterrorism.

The animal care community can provide an early warning, since some of the organisms involved are animal pathogens. Veterinarians, farmers, and others who work with and care for animals need to recognize potential public health implications of certain problems seen in animals.

Organizations and companies that handle mail and packages need to develop prudent handling procedures to recognize and isolate suspicious items. Medical, maintenance, safety, and security staffs and emergency response personnel should receive appropriate training.

Finally, concerns about bioterrorism should be kept in perspective in the context of the everyday risks. Most preventable morbidity and mortality are due to addictions (especially tobacco), modifiable lifestyle factors, and treatable diseases and risk factors. Individuals, healthcare personnel, and health systems should maintain and increase the focus on these more mundane issues, as they will ultimately have the greatest impact on life and health.

References

1. Occupational Safety and Health Administration. Occupational exposure to bloodborne pathogens: final rule. 29 CFR Part 1910.1030. Washington, DC: Department of Labor, 1991. https://www.osha.gov/pls/oshaweb/owadisp.show_document?p_id=10051&p_table=STANDARDS (accessed June 14, 2016).
2. Needlestick Safety and Prevention Act. An act to require changes in the bloodborne pathogens standard in effect under the Occupational Safety and Health Act of 1970. Pub. L. 106-430, 114 Stat. 1901 (2000). https://history.nih.gov/research/downloads/PL106-430.pdf (accessed June 23, 2016).
3. Occupational Safety and Health Administration. Recording criteria for needlestick and sharps injuries. 29 CFR 1904.8. Washington, DC: Department of Labor, 2001. https://www.osha.gov/pls/oshaweb/owadisp.show_document?p_table=STANDARDS&p_id=9639 (accessed June 14, 2016).
4. Dooley SW Jr, Castro KG, Hutton MD, et al. Guidelines for preventing the transmission of tuberculosis in health-care settings, with special focus on HIV-related issues. *MMWR* 1990; 39(RR-17):1–29
5. Occupational Safety and Health Administration. Occupational exposure to tuberculosis; proposed rule. *Fed Regist* 1997; 62:54159–308.
6. Occupational Safety and Health Administration. Occupational exposure to tuberculosis; proposed rule; termination of rulemaking respiratory protection for M. tuberculosis; final rule; revocation. 29 CFR 1910, Docket No. H-371. Washington, DC: Department of Labor, 2003. https://www.osha.gov/pls/oshaweb/owadisp.show_document?p_table=FEDERAL_REGISTER&p_id=18050 (accessed June 14, 2016).
7. Jensen PA, Lambert LA, Iademarco MF, et al. Guidelines for preventing the transmission of *Mycobacterium tuberculosis* in health-care settings, 2005, *MMWR* 2005; 54(RR-17): 1–141.
8. Occupational Safety and Health Administration. Tuberculosis and respiratory protection enforcement. 29 CFR 1910.134; 1910.134(f)(2). Washington, DC: Department of Labor, 2008. https://www.osha.gov/pls/oshaweb/owadisp.show_document?p_table=Interpretations&p_id=26013 (accessed June 14, 2016).
9. Occupational Safety and Health Administration. Infectious diseases. Request for information. *Fed Regist* 2010 75(87):24835–44. https://www.osha.gov/pls/oshaweb/owadisp.show_document?p_table=FEDERAL_REGISTER&p_id=21497 (accessed June 23, 2016).
10. Centers for Disease Control and Prevention. Adult Immunization Schedule: United States, 2016. http://www.cdc.gov/vaccines/schedules/hcp/adult.html (accessed June 23, 2016).
11. American College of Physicians. Guide to Adult Immunization. http://immunization.acponline.org (accessed June 14, 2016).
12. OSHA. Guidance on Preparing Workplaces for an Influenza Pandemic. https://www.osha.gov/Publications/influenza_pandemic.html (accessed June 14, 2016).
13. Occupational Safety and Health Act of 1970. 29 USC 654 Sec. 5(a)1 https://www.osha.gov/pls/oshaweb/owadisp.show_document?p_id=3359&p_table=oshact (accessed June 14, 2016).

Further Reading

American College of Obstetricians and Gynecologists. Perinatal viral and parasitic infections. ACOG practice bulletin 151. Cytomegalovirus, Parvovirus B19, Varicella Zoster, and Toxoplasmosis in Pregnancy (June 2015). *Obstet Gynecol* 2015; 125(6):1510–25.

Babcock HM, Woeltje KF. The development of infection surveillance and control programs. In: Bennett JV, Brachman PS, eds. Hospital infections, 6th edn. Philadelphia, PA: Lippincott, Williams & Wilkins, 2013:57–62.

Burge HA, Feeley JC. Indoor air pollution and infectious diseases. In: Samet JM, Spengler JD, eds. Indoor air pollution: a health perspective. Baltimore, MD: The Johns Hopkins University Press, 1991:273–84.

Centers for Disease Control and Prevention, National Institutes of Health. Biosafety in microbiological and biomedical laboratories, 5th edn., HHS publication no. (CDC) 21-1112. Washington, DC: U.S. Department of Health and Human Services, Public Health Service, Centers for Disease Control and Prevention, National Institutes of Health, 2009. http://www.cdc.gov/biosafety/publications/bmbl5/bmbl.pdf (accessed June 23, 2016).

Committee on Hazardous Biologic Substances in the Laboratory, National Research Council. Biosafety in the laboratory: prudent practices for the handling and disposal of infectious materials. Washington, DC: National Academy Press, 1989.

Fleming DO, Hunt DL, eds. Biological safety: principles and practices, 3rd edn. Washington, DC: American Society of Microbiology Press, 2000.

Henderson DA, O'Toole T, Inglesby TV. Bioterrorism: guidelines for medical and public health management. Chicago, IL: American Medical Association, 2002.

Livingston EG. Infectious agents and non-infectious biologic products. In: Frazier LM, Hage ML, eds. Reproductive hazards of the workplace. New York: John Wiley & Sons, Inc., 1998:463–505.

North Carolina Department of Labor, Division of Occupational Safety and Health. Farm safety, NC-OSHA industry guide no.10. Raleigh, NC: North Carolina Department of Labor, 2008. http://www.unctv.org/content/sites/default/files/0000011508-NCDOL%20farm%20work.pdf (accessed June 23, 2016).

Sears SD, James NW. International medicine: care of travelers and foreign-born patients. In: Fiebach NH, Kern DE, Barker LR, et al., eds. Principles of ambulatory medicine, 7th edn. Philadelphia, PA: Lippincott, Williams & Wilkins, 2006:609–44.

Welch LS, Blodgett DW. Occupational and environmental disease and bioterrorism. In: Fiebach NH, Kern DE, Barker LR, Ziegelstein RC, Zieve PD, Thomas PA, eds. Principles of ambulatory medicine, 7th edn. Philadelphia, PA: Lippincott, Williams & Wilkins, 2006:118–36.

Wilson ML, Reller LB. Clinical laboratory-acquired infections. In: Bennett JV, Brachman PS, eds. Hospital infections, 6th edn. Philadelphia, PA: Lippincott, Williams & Wilkins, 2013:320–8.

Suggested websites

Centers for Disease Control and Prevention, http://www.cdc.gov (accessed June 14, 2016).

National Institutes of Health, http://www.nih.gov (accessed June 14, 2016).

National Institutes of Health. Biodefense and Bioterrorism, http://www.nlm.nih.gov/medlineplus/biodefenseandbioterrorism.html (accessed June 14, 2016).

Occupational Safety and Health Administration, http://www.osha.gov (accessed June 14, 2016).

Bioterrorism websites

CDC. Bioterrorism, http://emergency.cdc.gov/bioterrorism/ (accessed June 14, 2016).

Dembek ZF, Alves DA, U.S. Army Medical Research Institute of Infectious Diseases. Medical management of biological casualties handbook, 7th edn. Fort Detrick, MD: U.S. Army Medical Research Institute of Infectious Diseases, 2011. http://www.usamriid.army.mil/education/bluebookpdf/USAMRIID%20BlueBook%207th%20Edition%20-%20Sep%202011.pdf (accessed June 23, 2016).

Pandemic planning websites

OSHA. Guidance on Preparing Workplaces for an Influenza Pandemic, https://www.osha.gov/Publications/influenza_pandemic.html (accessed June 14, 2016).

U.S. Department of Health & Human Services Influenza Pandemic, http://www.flu.gov/pandemic/ (accessed June 14, 2016).

Chapter 21

VIRUSES

Manijeh Berenji*

ARBOVIRUSES

Common names for disease: Yellow fever, Dengue fever, Chikungunya, Zika fever, Japanese encephalitis, St Louis encephalitis, West Nile encephalitis, Eastern equine encephalitis, Western equine encephalitis, La Crosse encephalitis, Colorado tick fever

Classification: Family—Flaviviridae, Togaviridae, Bunyaviridae, Reoviridae

Occupational setting

Travelers to and workers in Africa, tropical North and South America, and Asia are most at risk for contracting arboviral diseases. Workers exposed to frequent mosquito bites are at risk for the encephalitides in various parts of the United States and around the world. Locations of arboviral outbreaks continue to change through time. For example, West Nile virus (WNV), previously localized to the Caribbean, was first identified in the United States in 1999. WNV is now the number one cause of domestically acquired arboviral disease in the United States.[1] In 2011, there were 712 cases of WNV reported through the national surveillance system ArboNET, of which 68% were identified as being neuroinvasive (Figure 21.1).[2] Workers in and travelers to the western United States and western Canada in locations between 4000 and 10000 ft above sea level are at risk for Colorado tick fever, also referred to as mountain tick fever and mountain fever (Figure 21.2).[3] Workers and travelers to the Caribbean and Central and South America are at risk for Zika fever, after an epidemic began in Brazil in 2015.[4] Although it has not occurred at the time of this writing, it is anticipated that infection may occur in the United States as soon as the summer of 2016 due to the range of the primary host vector, the *Aedes aegypti* mosquito, which may be present in up to 30 states.[5]

* The author wishes to thank George W. Jackson for his contribution to previous versions of this chapter.

Exposure (route)

The arboviruses are mainly transmitted to vertebrate hosts by arthropod bites (mainly via mosquitoes and ticks). Yellow fever, dengue fever, chikungunya, and Zika are primarily associated with domestic mosquitoes (including *A. aegypti* and *Aedes albopictus*).[6,7] For Zika virus, *A. aegypti* is the major vector, but transmission has also been reported through blood transfusion, perinatally, and through sexual contact.[8]

Sandflies have been identified in the transmission of arboviruses in US military personnel during recent operations in the Middle East.[9,10] There has been a case report of WNV infection in laboratory workers without other known risk factors who acquired infection through percutaneous inoculation.[11] Colorado tick fever (which is distinct from Rocky Mountain spotted fever) is usually spread by the bite of the infected Rocky Mountain wood tick (*Dermacentor andersoni*).[12] Cases have also been reported in California following exposure to a different tick vector, *Dermacentor variabilis*.[13]

Pathobiology

The flaviviruses, togaviruses, and bunyaviruses are enveloped, single-stranded RNA viruses that are sensitive to heat and detergents.[6] Reoviruses (which cause Colorado tick fever and other diseases) are double-stranded RNA viruses.

Physical and Biological Hazards of the Workplace, Third Edition. Edited by Gregg M. Stave and Peter H. Wald.

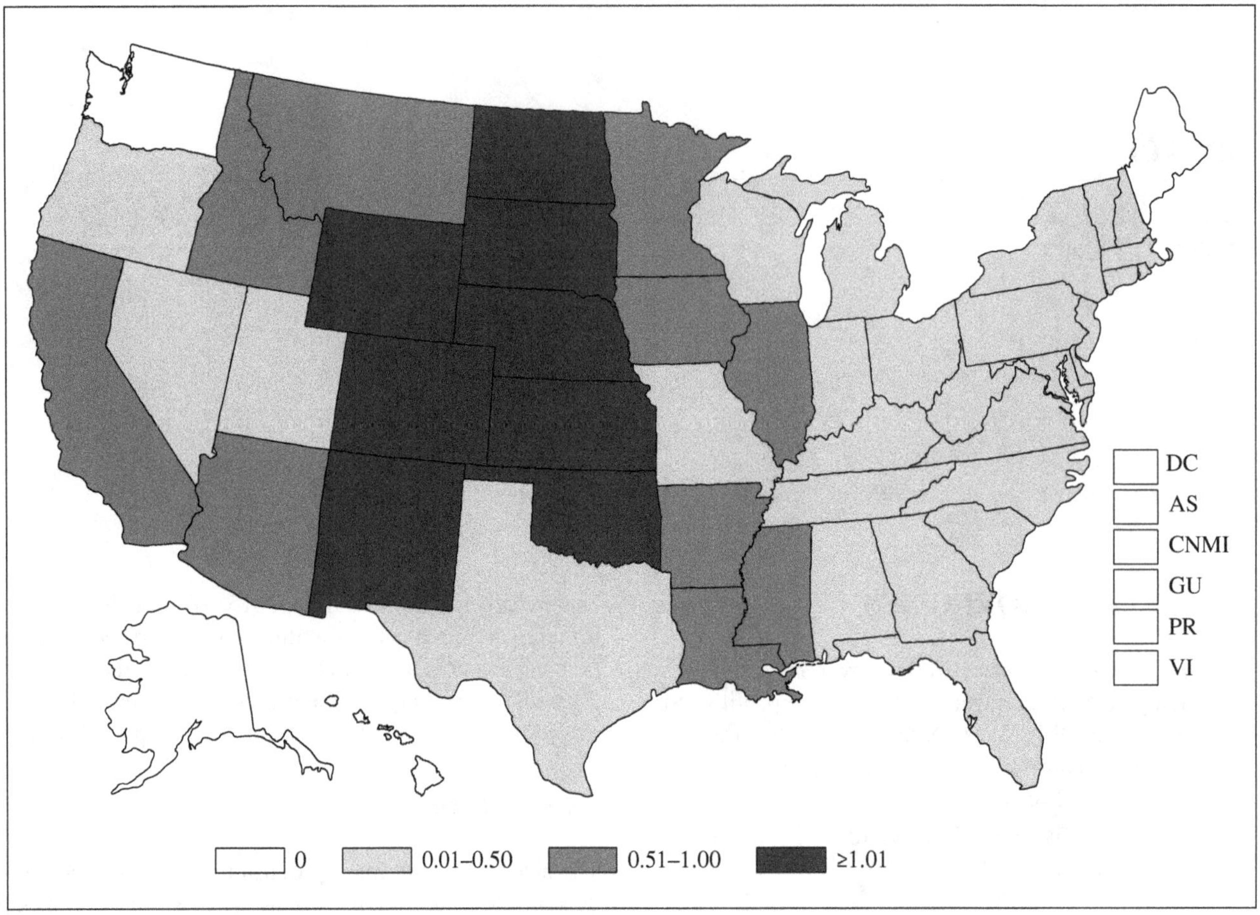

FIGURE 21.1 West Nile Virus Incidence per 100 000 of reported cases of neuroinvasive disease — United States and U.S. territories, 2013. Data from the Division of Vector-Borne Diseases (DVBD), National Center for Emerging and Zoonotic Infectious Diseases (ArboNET Surveillance). In 2013, eight states reported an incidence of West Nile virus (WNV) neuroinvasive disease >1 case per 100,000; the four states with the highest reported incidence were North Dakota (8.9), South Dakota (6.8), Nebraska (2.9), and Wyoming (2.8). Six states reported approximately half of the WNV neuroinvasive disease cases: California (237 cases), Texas (113), Colorado (90), Illinois (86), North Dakota (64), and Oklahoma (60). *Source*: MMWR 2015; 62(53):56.

Yellow fever is a flavivirus infection. The incubation period for yellow fever is 3–6 days, followed by the abrupt onset of fever, chills, headache, malaise, nausea, and vomiting. Yellow fever virus causes damage to the liver, kidneys, heart, and gastrointestinal tract. Degeneration of these organs results in the jaundice, hematemesis, and cardiovascular changes characteristic of yellow fever. The case fatality rate is approximately 5% overall and 20% in patients who develop jaundice. Yellow fever is endemic only in tropical South America and Africa.[14] In 2013, there were an estimated 130 000 cases of yellow fever and 78 000 deaths in the African continent, taking into account yellow fever occurrence data, vaccination rates, and new statistical analyses.[15]

Dengue fever is a flavivirus with an incubation period of 5–8 days. The virus is endemic and epidemic in the tropical Americas, Africa, and Asia. There are four serotypes (DENV 1–4). According to the World Health Organization, there were approximately 2.35 million cases of dengue reported in the Americas in 2013, including 773 travel-related and 49 locally transmitted cases identified in the continental United States (Figure 21.3).[7,16] Dengue is characterized by fever, chills, headache, prostration, nausea, vomiting, cutaneous hyperesthesia, and altered taste. The fever is commonly bimodal; a maculopapular rash accompanies the second episode. Hemorrhagic manifestations occur in some cases. A variant of classic dengue is dengue hemorrhagic fever, which is similar to the viral hemorrhagic fevers (e.g., Lassa, Ebola). Dengue is most commonly a self-limited illness requiring only limited supportive measures, but in rare cases, it can result in severe hemorrhage and fatal shock.[17]

Chikungunya virus, an arbovirus in the Togavirus family, was initially identified in what is now known as Tanzania in the early 1950s. The word chikungunya means "to walk bent over" in Kimakonde, a language spoken by the Makonde ethnic group in Tanzania. The incubation period for chikungunya ranges from 3 to 7 days, with the most common symptoms

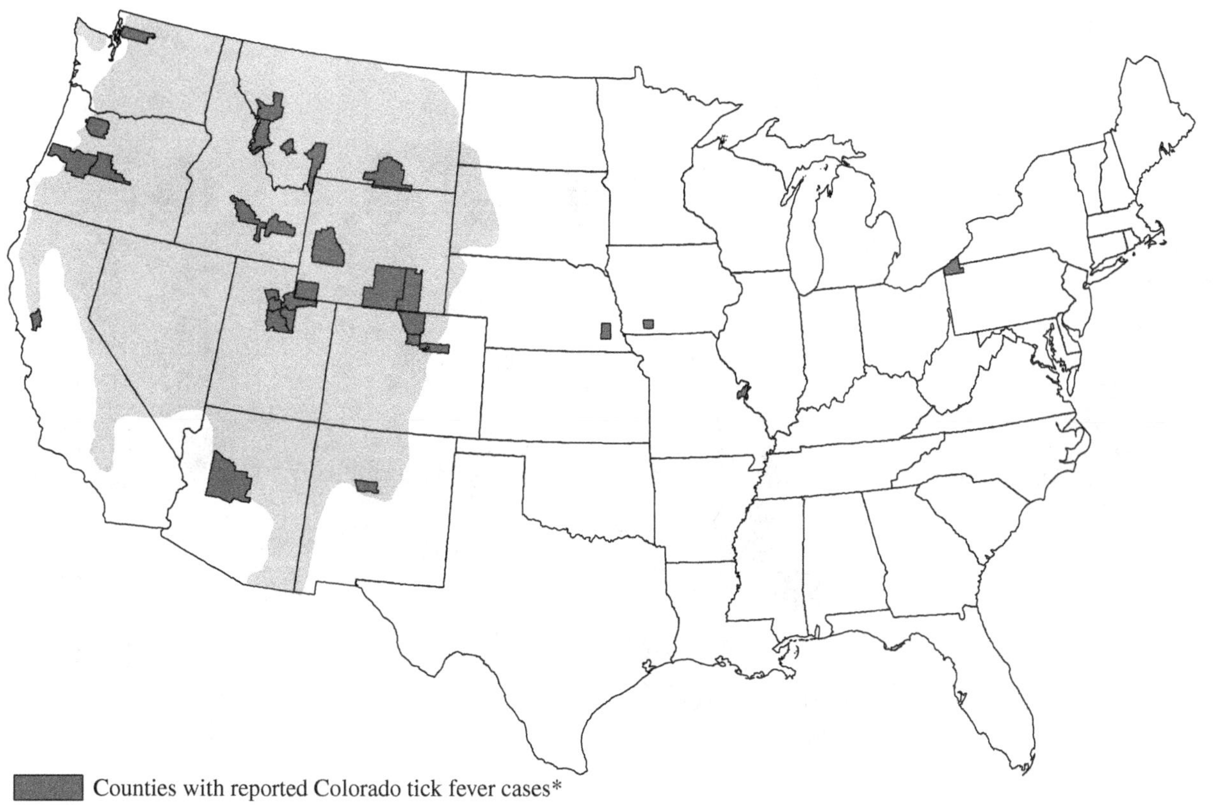

FIGURE 21.2 Approximate geographic distribution of *Dermacentor andersoni* ticks and counties of residence for confirmed and probable Colorado tick fever (CTF) virus disease cases, United States, 2002–2012. *All cases were acquired in states where local transmission of CTF virus has been reported previously. Two additional cases were reported from Colorado with unknown county of residence. **Derived from James AM, Freier JE, Keirans JE, Durden LA, et al. Distribution, seasonality, and hosts of the Rocky Mountain wood tick in the United States. *J Med Entomol* 2006; 43:17–24. *Source*: http://www.cdc.gov/coloradotickfever/pdfs/ctfmappdf.pdf.

reported including fever, muscle and joint pains, and a nonspecific maculopapular rash.[16,17] While most of these infections have occurred in continental Africa and Asia, there have been recent outbreaks in Comoros Islands and adjacent islands in the Indian Ocean (mid 2000s), as well as in the Americas (the Caribbean and Central America; beginning in late 2013).[7,18–20] As of early August 2014, there were over 575000 suspected and laboratory-confirmed chikungunya cases in the Americas and over 700 cases of imported chikungunya into the United States.[7,18,19] While most cases were in travelers, a few cases were acquired locally in the continental United States.[18] It should be noted that *Aedes* species of mosquitoes can harbor and transmit both dengue and chikungunya and there has been a reported case of simultaneous coinfection.[21]

Zika virus is a flavivirus that was first identified in a caged febrile rhesus monkey in Uganda in 1947.[6,22] The incubation period is 2–7 days and the course is similar to a dengue fever-like illness without hemorrhagic features.[6,23] Symptoms may include rash, fever, arthralgia, conjunctivitis, and headache.[6] The illness lasts about 2–7 days.[23] However, it is estimated that 80% of infections are asymptomatic.[24] The largest outbreak prior to 2013 occurred in Micronesia in 2007 included 59 probable cases.[6] A much larger outbreak occurred in French Polynesia in 2013–2014. Beginning in 2015, Zika virus spread through tropical regions of the Americas, possibly beginning in Brazil. By early 2016, infections were reported in more than 30 countries.[23] In 2015, it was estimated that in Brazil there may have been between 440000 and 1300000 cases.[24] A substantial increase in cases of microcephaly in newborns has been associated with this epidemic.[25] Infants born with microcephaly may also have vision-threatening findings including macular and perimacular lesions, optic nerve abnormalities, and other ocular abnormalities.[26] Zika virus infections have also been associated with an increased incidence of Guillain–Barré syndrome.[25] Zika has also been associated with acute disseminated encephalomyelitis (ADEM), which can mimic multiple sclerosis.[27] In the United States, by early 2016, hundreds of cases of Zika virus

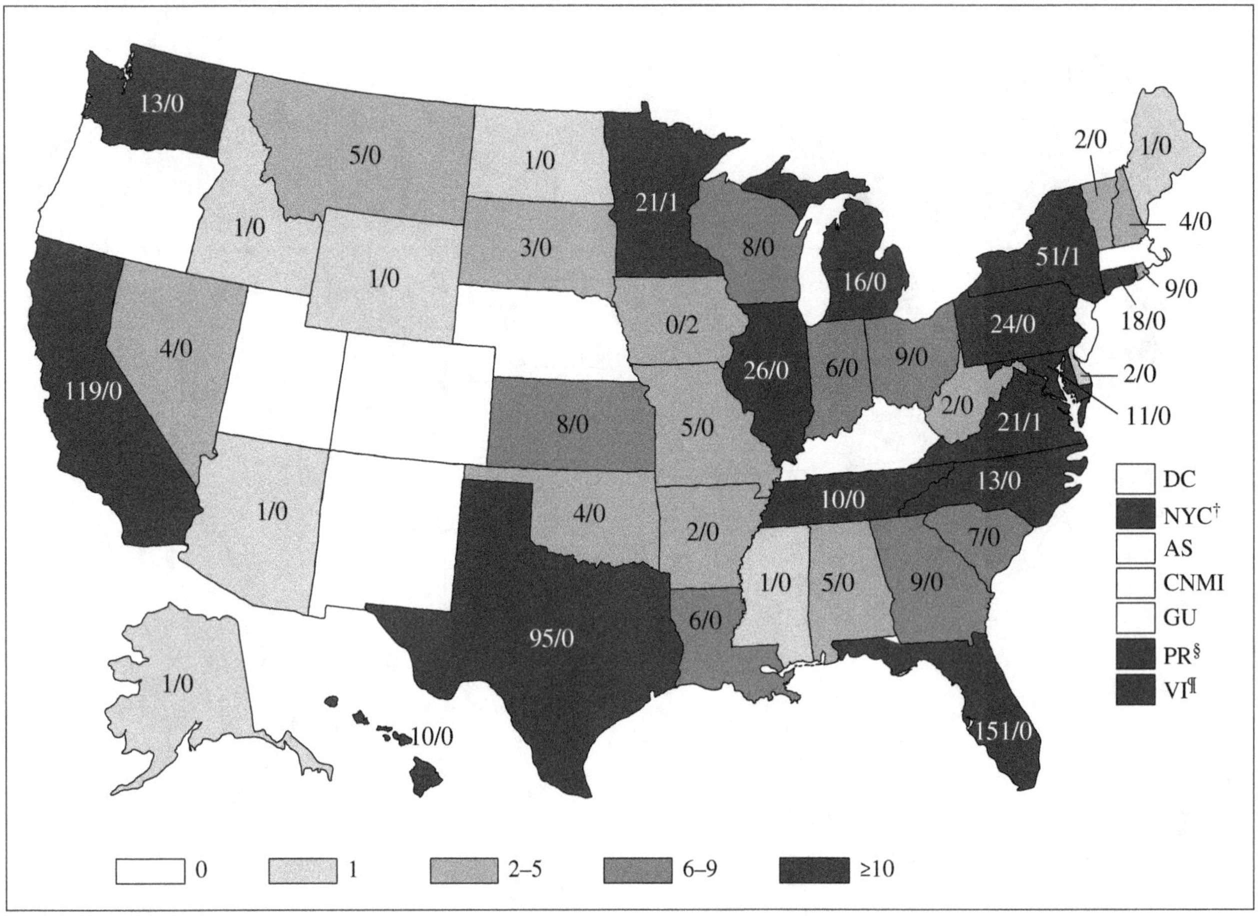

FIGURE 21.3 Dengue fever and dengue hemorrhagic fever. Number of reported cases, by location of residence — United States and U.S. territories, 2013. Number of Dengue fever cases/number of Dengue Hemmorrhagic fever. Data from the Division of Vector-Borne Diseases (DVBD), National Center for Emerging and Zoonotic Infectious Diseases (ArboNET Surveillance). †New York City reported cases 131/1. §Puerto Rico locally acquired cases 9 557/153. ¶Virgin Islands reported cases 169/5. *Source*: MMWR 2015; 62(53):65.

infection were identified that were most likely acquired during travel.[28] The first domestically acquired mosquito borne cases of Zika infection occurred in Florida in the summer of 2016. The range of the mosquito vector *A. aegypti* includes many southern states and stretches up the eastern seaboard.

The encephalitides may present primarily as meningitis or encephalitis or both. They are characterized by fever and headache and then signs of meningeal irritation and/or altered levels of consciousness. Lethargy and drowsiness is followed by evidence of abnormal mental status and, in severe cases, focal and generalized seizures. Cerebrospinal fluid usually reveals a lymphocytic pleocytosis with a slight increase in protein and a normal glucose level. The initial step in evaluation is to exclude nonviral causes of meningitis/encephalitis. Herpes virus infection must also be excluded.

Coltiviruses, members of the reovirus family, cause Colorado tick fever (CTF). CTF is not common, but it is likely to be underreported. The illness occurs primarily in spring and summer, with most cases occurring in April and May at lower altitudes and in June and July at higher elevations. Symptoms usually begin 3–6 days following a tick bite and the illness commonly presents with fever, chills, body aches, and fatigue. Patients may also have sore throat, vomiting, abdominal pain, or a maculopapular or petechial rash. About half of patients have a biphasic fever with several days of fever followed by a period of improvement and then a second shorter period of fever and illness.[12] Life-threatening illness and death are rare, but patients may develop more severe illness that includes meningitis, encephalitis, hepatitis, epididymo-orchitis, pericarditis, myocarditis, pneumonia, hemorrhagic fever, and coma.[3,29] Patients who are immunocompromised or who have undergone a splenectomy are at increased risk for severe complications.[30]

Diagnosis

Yellow fever must be differentiated from malaria, hepatitis, typhoid, and the viral hemorrhagic fevers. Specific diagnosis is based on viral isolation or detection of the viral antigen by a specific ELISA test.

Dengue must be differentiated from malaria, typhus, influenza, and a variety of other viral syndromes (including chikungunya). The hemorrhagic variant is similar to the other hemorrhagic fevers. The specific diagnosis can be obtained by viral isolation or a serologic test.

Chikungunya must be differentiated from leptospirosis, malaria, rickettsia, group A streptococcus, rubella, measles, parvovirus, enteroviruses, adenovirus, other alphavirus infections, postinfection arthritis, and rheumatologic conditions.[7] The specific diagnosis can be made based on PCR testing for chikungunya virus and identification of IgM antibodies.

Zika virus disease can be diagnosed during the early phase of illness using reverse transcription polymerase chain reaction (RT-PCR).[31] Virus-specific IgM and neutralizing antibodies typically develop toward the end of the first week of illness, but cross-reaction with dengue and yellow fever viruses is common.[31] Plaque-reduction neutralization testing can measure virus-specific neutralizing antibodies and discriminate between cross-reacting antibodies in primary flavivirus infections.[31] Zika virus testing is performed at the Centers for Disease Control and Prevention (CDC) Arbovirus Diagnostic Laboratory and by a few state health departments.[31] A commercial test became available in May 2016. CDC recommends serum testing for Zika virus using RT-PCR if specimens are collected within 7 days of the onset of symptoms.[32] In addition, CDC recommends that urine be tested for Zika virus by RT-PCR if the urine is collected within 2 weeks of symptom onset, since viral RNA can be detected in the urine for a longer period.[32] A positive result of either tests suggests a diagnosis of Zika virus infection.[32]

Colorado tick fever (CTF) needs to be differentiated from Q fever, Lyme disease, ehrlichiosis, tularemia, Rocky Mountain spotted fever, and relapsing fever. Laboratory diagnosis of CTF is generally accomplished by testing of serum to detect viral RNA or virus-specific IgM and neutralizing antibodies. Antibody production can be delayed with CTF, so tests that measure antibodies may not be positive for 14–21 days after the onset of symptoms.[3] RT-PCR (reverse-transcriptase polymerase chain reaction) is a more sensitive test early in the course of disease.[33] CTF testing is available at some commercial and state health department laboratories and at CDC. IgG antibodies can be found in campers who regularly visit endemic areas, so an elevated IgG does not necessarily indicate acute infection. An ELISA for antibodies is available.[34]

Treatment

Supportive care should be provided, including antipyretics, intravenous fluids, and analgesics. No specific treatment is available. For Colorado tick fever, aspirin and nonsteroidal anti-inflammatory drugs should be avoided due to their exacerbation of platelet dysfunction and theoretical increased risk of coagulopathy.[27]

Medical surveillance

No specific medical surveillance is recommended. Cases should be reported to state health departments or CDC as required.

Prevention

Protection from arthropod bites in endemic areas is currently the only method to prevent transmission and is a requisite for travelers and workers in those environments. This includes the use of long sleeve shirts and pants, and using insect repellants, typically a solution containing at least 20% DEET (*N*,*N*-diethyl-meta-toluamide). A thorough check for ticks should be performed after work outdoors. Ticks should be removed safely.

As a result of concerns about microcephaly following Zika virus infection, in 2016 CDC issued *Interim Guidelines for Pregnant Women During a Zika Virus Outbreak*.[24] The guidelines recommend that women who are pregnant consider postponing travel to any area where Zika transmission is ongoing and women who are trying to become pregnant discuss their travel plans with their doctors.[24] Travel exclusively to areas above 2000 m may carry only a minimal risk of transmission.[35] Pregnant women who must travel are encouraged to follow strict precautions to prevent mosquito bites.[24] An update to the CDC interim guidelines recommends offering serologic testing to asymptomatic pregnant women who have traveled to areas with ongoing Zika virus transmission 2–12 weeks after returning from travel.[36] For pregnant women with symptoms consistent with Zika virus disease, testing is recommended during the first week of illness.[36] For asymptomatic pregnant women residing in areas with ongoing Zika virus transmission, testing is recommended at the initiation of prenatal care and mid-second trimester.[36] Given the risk of sexual transmission, men who reside in or travel to areas of Zika virus transmission should be counseled, especially if their partner is pregnant and as part of family planning.[37]

In April 2016, CDC and the Occupational Safety and Health Administration (OSHA) issued new guidance and information for protecting workers from occupational exposure to Zika virus in anticipation that Zika virus may spread in the United States. The recommendations are comprised of:

1. Interim guidance for outdoor workers, healthcare and laboratory workers, mosquito control workers, and business travelers to protect against occupational exposure to Zika virus
2. Interim guidance and recommendations for employers to use to protect their workers
3. Interim guidance and recommendations for workers to consider to protect themselves from mosquito bites and exposure to an infected person's blood or other body fluids[38]

Specific recommendations for employers include worker training on risks and protection, providing insect repellent, providing protective clothing including hats with mosquito netting, removing sources of standing water, and to consider reassigning workers who indicate they are or may become pregnant, or male workers who have a sexual partner who is or may become pregnant, to indoor tasks to reduce their risk of mosquito bites.[39] Outdoor workers are encouraged to use insect repellant, wear clothing that covers exposed skin, remove sources of standing water, and talk with supervisors about reducing exposure if you are or may become pregnant or, for male workers, if your sexual partner is or may become pregnant.[39] For healthcare workers, bloodborne pathogen prevention practices should also reduce the risk of Zika virus transmission (at the time of this writing, there have not been any reported cases of Zika transmission in the healthcare setting).[39] Mosquito control workers should follow the same precautions as other outdoor workers and may also need additional protective clothing if handling insecticides.[39]

Vaccines

A vaccine is available for yellow fever that provides long-term (10 years) immunity. The CDC's Advisory Committee on Immunization Practices (ACIP) states that a single dose of yellow fever vaccine can provide long-lasting immunity for most travelers but a 10-year booster is advised for those individuals planning to work in a high risk setting (such as planning to spend a prolonged period in endemic areas or those traveling to highly endemic areas such as rural West Africa during peak transmission season).[40] In addition to travelers and workers in endemic areas, researchers and other laboratory personnel working with yellow fever virus should also be vaccinated. Workers who routinely handle wild-type yellow fever virus should have antibody titers measured at least every 10 years and be given booster yellow fever vaccine if titers are too low for sufficient immunity.[39] Adverse reactions to the vaccine are infrequent, but the most serious ones include a severe nervous system reaction and organ failure. Immediate hypersensitivity has been seen primarily in persons with egg allergy. Individuals with attenuated immune states should not be vaccinated.[14,41]

Research continues on developing a dengue vaccine that can protect against all four serotypes of dengue (DENV 1–4). A trial of a tetravalent live attenuated vaccine against dengue virus resulted in a vaccine efficacy of 30.2%, with almost no efficacy against serotype DENV 2.[42] Recent analysis of antibodies from dengue-infected patients has revealed a new dengue virus envelope epitope that has cross-reactivity across all four dengue subtypes.[43]

No vaccine is available for Zika virus although a vaccine is being developed.

No vaccine is available for Colorado tick fever. As of January 2015, CTF was specifically reportable to local public health authorities in six states: Arizona, Colorado, Montana, Oregon, Utah, and Wyoming.[3]

Laboratory work

In the laboratory, Zika and dengue viruses are classified as biological safety level 2 (BSL-2). Chikungunya virus is classified as a BSL-3 agent. For work with Zika virus, laboratory personnel should also follow Biosafety Guidance for Transportation of Specimens and for Work with Zika Virus in the Laboratory, developed by CDC.[44]

References

1. Reimann CA, Hayes EB, DiGuiseppi C, et al. Epidemiology of neuroinvasive arboviral disease in the United States, 1999–2007. *Am J Trop Med Hyg* 2008;79:974–9.
2. Lindsey NP, Lehman JA, Campbell GL, et al. West nile virus disease and other arboviral disease – United States, 2011. *MMWR Morb Mortal Wkly Rep* 2012;61(27):510–4.
3. Centers for Disease Control and Prevention. Colorado Tick Fever. Available at: http://www.cdc.gov/coloradotickfever/index.html (accessed on June 17, 2016).
4. Campos GS, Bandeira AC, Sardi SI. Zika Virus Outbreak, Bahia, Brazil. *Emerg Infect Dis* 2015;21(10):1885–6.
5. Centers for Disease Control and Prevention. Surveillance and Control of *Aedes aegypti* and *Aedes albopictus* in the United States. Available at: http://www.cdc.gov/chikungunya/resources/vector-control.html (accessed on June 17, 2016).
6. Thomas SJ, Endy TP, Rothman AL, et al. Flaviviruses (Dengue, Yellow Fever, Japanese Encephalitis, West Nile Encephalitis, St Louis Encephalitis, Tick-Borne Encephalitis, Kyasanur Forest Disease, Alkhurma Hemmorrhagic Fever, Zika). In: Bennett JE, Dolin R, Blaser M. eds. *Mandell, Douglas, and Bennett's Principles and Practice of Infectious Disease*, 8th edn. Philadelphia, PA: Elsevier Saunders, 2015:1881–903.
7. Staples JE, Fischer M. Chikungunya virus in the Americas—What a vector-borne pathogen can do. *N Engl J Med* 2014;10:887–9.
8. Musso D, Roche C, Robin E, et al. Potential sexual transmission of zika virus. *Emerg Infect Dis* 2015;21(2):359–61.
9. Richards AL, Hyams KC, Merrell MS. Medical aspects of Operation Desert Storm [Letter]. *N Engl J Med* 1991;325:970.
10. Riddle MS, Althoff JM, Earhart K, et al. Serological evidence of arboviral infection and self-reported febrile illness among U.S. troops deployed to Al Asad, Iraq. *Epidemiol Infect* 2008;136(5):665–9.
11. Campbell G, Lanciotti R. Laboratory-acquired West Nile Virus infections – United States, 2002. *MMWR Morb Mortal Wkly Rep* 2002;51(50):1133–5.
12. Goodpasture HC, Poland JD, Francy DB, et al. Colorado tick fever: clinical, epidemiologic, and laboratory aspects of 228 cases in Colorado in 1973–1974. *Ann Intern Med* 1978;88:303–10.
13. Attoui H, Mohd JF, Biagini P, et al. Genus Coltivirus (family Reoviridae):genomic and morphologic characterization of Old World and New World viruses. *Arch Virol* 2002;147:533–61.

14. Centers for Disease Control and Prevention. Yellow fever vaccine. *MMWR Morb Mortal Wkly Rep* 1990;39(RR-6):1–6.
15. Garske T, Van Kerkhove MD, Yactayo S, et al. Yellow fever in Africa: estimating the burden of disease and impact of mass vaccination from outbreak and serological data. *PLoS Med* 2014;11(5):e1001638.
16. World Health Organization. Dengue and Severe Dengue. Available at: http://www.who.int/mediacentre/factsheets/fs117/en/ (accessed on June 17, 2016).
17. Hayes EB, Bugler DJ. Dengue and dengue hemorrhage. *Pediatr Infect Dis* 1992;11:311–7.
18. Centers for Disease Control and Prevention. Infectious Disease Related to Travel – "Chikungunya." In *Traveler's Health.* Available at: http://wwwnc.cdc.gov/travel/yellowbook/2014/chapter-3-infectious-diseases-related-to-travel/chikungunya (accessed on June 17, 2016).
19. Morens DM, Fauci AS. Chikungunya at the door – deja vu all over again. *N Engl J Med* 2014;10:885–7.
20. Hochendz P, Jaureguiberry S, Debruyne M, et al. Chikungunya infection in travelers. *Emerg Infect Dis* 2006;12(10):1565–7.
21. Myers RM, Carey DE. Concurrent isolation from patient of two arboviruses, chikungunya and dengue type 2. *Science* 1967;157:1307–8.
22. Dick GW. Zika virus, II. Pathogenicity and physical properties. *Trans R Soc Trop Med Hyg* 1952;46:521–34.
23. Pan American Health Organization. Zika Virus Infection. Available at: http://www.paho.org/hq/index.php?option=com_topics&view=article&id=427&Itemid=41484 (accessed on June 17, 2016).
24. Petersen EE, Staples JE, Meaney-Delman D, et al. Interim Guidelines for Pregnant Women during a Zika Virus Outbreak—United States, 2016. *MMWR* 2016;65:2:30–3. Available at: http://www.cdc.gov/mmwr/volumes/65/wr/mm6502e1.htm?s_cid=mm6502e1_w (accessed on June 17, 2016).
25. European Centre for Disease Prevention and Control (ECDC). Rapid Risk Assessment: Zika Virus Epidemic in the Americas: Potential Association with Microcephaly and Guillain-Barré syndrome, December 10, 2015. Stockholm: ECDC; 2015. Available at: http://ecdc.europa.eu/en/publications/Publications/zika-virus-americas-association-with-microcephaly-rapid-risk-assessment.pdf (accessed on June 17, 2016).
26. de Paula Freitas B, de Oliveira Dias J, Prazeres J, et al. Ocular findings in infants with microcephaly associated with presumed zika virus congenital infection in Salvador, Brazil. *JAMA Ophthalmol* 2016. 10.1001/jamaophthalmol.2016.0267 [Epub ahead of print].
27. Steenhuysen J. Brazilian scientists find new Zika-linked brain disorder in adults. *Reuters*. April 10, 2016. Available at: http://www.reuters.com/article/us-health-zika-brain-idUSKCN0X70VP (accessed on June 17, 2016).
28. Armstrong P, Hennessey M, Adams M, et al. Travel-associated zika virus disease cases among U.S. residents—United States, January 2015–February 2016. *MMWR Morb Mortal Wkly Rep* 2016;65:286–9. http://dx.doi.org/10.15585/mmwr.mm6511e1.
29. Davis LE, Beckham JD, Tyler KL. North American encephalitic arboviruses. *Neurol Clin* 2008;26(3):727, ix. 10.1016/j.ncl.2008.03.012.
30. Bratton RL, Corey R. Tick-borne disease. *Am Fam Physician*. 2005;71(12):2323–30.
31. Centers for Disease Control and Prevention, Division of Vector-Borne Diseases, Arboviral Diseases and Dengue Branches. Memorandum: Updated Diagnostic Testing for Zika, Chikungunya, and Dengue Viruses in US Public Health Laboratories, January 13, 2016. Available at: http://www.cdc.gov/zika/pdfs/denvchikvzikv-testing-algorithm.pdf (accessed on June 17, 2016).
32. Centers for Disease Control and Prevention. Interim guidance for Zika virus testing of urine—United States, 2016. *MMWR Morb Mortal Wkly Rep* 2016; 65(18):474. Available at: http://www.cdc.gov/mmwr/volumes/65/wr/mm6518e1.htm (accessed on June 27, 1016). http://dx.doi.org/10.15585/mmwr.mm6518e1.
33. Lambert AJ, Kosoy O, Velez JO, et al. Detection of Colorado tick fever viral RNA in acute human serum samples by a quantitative real-time RT-PCR assay. *J Virol Methods* 2007;140 (1–2):43–8.
34. Mohd Jaafar F, Attoui H, Gallian P, et al. Recombinant VP7-based enzyme-linked immunosorbent assay for detection of immunoglobulin G antibodies to Colorado tick fever virus. *J Clin Microbiol* 2003;41(5):2102–5.
35. Cetron M. Revision to CDC's zika travel notices: minimal likelihood for mosquito-borne zika virus transmission at elevations above 2,000 meters. *MMWR Morb Mortal Wkly Rep* 2016;65:267–8. http://dx.doi.org/10.15585/mmwr.mm6510e1.
36. Oduyebo T, Petersen EE, Rasmussen SA, et al. Update: interim guidelines for health care providers caring for pregnant women and women of reproductive age with possible zika virus exposure—United States, 2016. *MMWR Morb Mortal Wkly Rep* 2016;65:1–6. http://dx.doi.org/10.15585/mmwr.mm6505e2er.
37. Oster AM, Brooks JT, Stryker JE, et al. Interim guidelines for prevention of sexual transmission of zika virus—United States, 2016. *MMWR Morb Mortal Wkly Rep* 2016;65:120–1. http://dx.doi.org/10.15585/mmwr.mm6505e1.
38. Centers for Disease Control and Prevention. CDC and OSHA Issue Interim Guidance for Protecting Workers from Occupational Exposure to Zika Virus. Media Release April 22, 2016. Available at: http://www.cdc.gov/media/releases/2016/s0422-interim-guidance-zika.html (accessed on June 17, 2016).
39. Occupational Safety and Health Administration. OSHA Fact Sheet: Interim Guidance for Protecting Workers from Occupational Exposure to Zika Virus. Available at: http://www.cdc.gov/niosh/topics/outdoor/mosquito-borne/pdfs/osha-niosh_fs-3855_zika_virus_04-2016.pdf (accessed on June 17, 2016).
40. Staples JE, Bocchini JA Jr, Rubin L, et al. Yellow fever vaccine booster doses: recommendations of the Advisory Committee on Immunization Practices, 2015. *MMWR* 2015;64(23):647–50.
41. Lange WR, Beall B, Deny SC. Dengue fever: a resurgent risk for the international traveler. *Am Fam Phys* 1992;45:1161–8.
42. Sabchareon A, Wallace D, Sirivichayakul C, et al. Protective efficacy of the recombinant, live-attenuated, CYD tetravalent dengue vaccine in Thai schoolchildren: a randomised, controlled phase 2b trial. *Lancet* 2012;380:1559–67.
43. Dejnirattisai W, Wongwiwat W, Supasa S, et al. A new class of highly potent, broadly neutralizing antibodies isolated from viremic patients infected with dengue virus. *Nat Immunol* 2015;16(2):170–7. 10.1038/ni.3058.

44. Centers for Disease Control and Prevention. Biosafety Guidance for Transportation of Specimens and for Work with Zika Virus in the Laboratory. Available at: http://www.cdc.gov/zika/state-labs/biosafety-guidance.html (accessed on June 17, 2016).

ARENAVIRUSES

Common names for disease: LCMV (lymphocytic choriomeningitis virus), Dandenong virus, Lassa virus, Junin virus (Argentine hemorrhagic fever), Machupo virus (Bolivian hemorrhagic fever), Guanarito virus (Venezuelan hemorrhagic fever), Sabia-associated virus (Brazilian hemorrhagic fever), Chapare virus (Chapare hemorrhagic fever), Lujo virus (Lujo hemorrhagic fever)

Classification: Family—Arenaviridae

Occupational setting

These illnesses can occur in travelers going through areas with endemic infection, including Europe, Africa, and the Americas. Researchers and other laboratory personnel working with arenaviruses have become infected.[1] More recently, healthcare workers attending to an infected patient in southern Africa contracted Lujo virus, a newly identified arenavirus.[2]

Exposure (route)

Direct contact with rodents in the form of rodent bites is the mode of transmission in endemic areas. Aerosolized virus from rodent excreta (including urine and feces) can result in the transmission from the rodent reservoir to humans.[3] The virus can be transmitted from person to person via direct contact with an infected person's blood or bodily fluids, through mucous membrane, or through sexual contact, and patients are not believed to be infectious before the onset of symptoms.[4] Person-to-person contact via aerosols is less than likely to occur. Transmission of LCMV and an LCMV-like arenavirus through solid organ transplantation has been reported in the literature.[5–7] Lassa, Machupo, and Lujo viruses have been associated with nosocomial transmission via blood and bodily fluids.[8] Contact with objects contaminated with arenaviruses (i.e., medical equipment) has also been identified as a mode of arenavirus transmission.[8]

Pathobiology

The arenaviruses are enveloped, single-stranded RNA structures with a natural reservoir in rodents. The various agents in this group exhibit specificity for rodent species: LCMV infects *Mus musculus*, Lassa infects *Mastomys natalensis*, Junin virus infects *Calomys musculinus*, Machupo virus infects *Calomys callosus*, and Guanarito virus infects *Zygodontomys brevicauda*.[8,9] These viruses are inactivated at 56°C, pH < 5.5 or >8.5, or by exposure to UV and/or gamma radiation.[10]

Lymphocytic choriomeningitis virus (LCMV) infection is primarily seen in Europe and the Americas. A spotty distribution of infected mice has been seen when studies have been conducted in urban settings. Human infection occurs through aerosol spread, direct rodent contact, and rodent bites. The incubation period is highly variable but commonly ranges from 5 to 10 days.

LCMV begins as a low-grade fever with headache and myalgias. Lymphadenopathy and a maculopapular rash may develop during the incubatory period. A subsequent decrease in fever is followed by several days of recurrent high fever and worsening headache. A minority of patients develop clinical meningitis during this second phase. Rare complications include encephalitis, orchitis, pericarditis, and arthritis.[11] Congenital LCMV infection is proposed as a cause of central nervous system (CNS) disease in infants.[12]

The Lassa virus causes an illness characterized by an insidious onset with fever, sore throat, and malaise. Anorexia, vomiting, severe abdominal pain, and chest pains frequently follow. The incubation period is approximately 5–21 days.[10] Hemorrhage occurs in up to 20% of cases and can significantly increase mortality. Neurologic complications include deafness, meningitis, encephalitis, and global encephalopathy.

The South American arenaviruses have an average incubation period from 5 to 19 days.[9,10] The onset of disease is gradual, but these viruses have a tendency to progress rapidly. The hemorrhagic and neurologic symptoms associated with the South American arenaviruses can be severe and can lead to significant morbidity and mortality.

Diagnosis

Diagnosis of these illnesses is generally made by clinical and epidemiologic parameters.

Specific diagnosis is by the detection of viral RNA via RT-PCR from patient's blood, urine, throat washings, and other tissues. Immunofluorescent antibody assay and ELISA are techniques used to detect viral antibodies.[10]

Treatment

Most infections with arenaviruses result in recovery without treatment. In severe cases, intensive supportive therapy reduces mortality. Ribavirin has been found to be most effective in treating Lassa fever if given within the first 6 days of illness. Ribavirin has also been utilized for off-label use in patients infected with the South American arenaviruses but patient responses have been mixed and severe toxicity has been reported.[13]

There is evidence that demonstrates a significant reduction in case fatality from Junin virus with administration of immune serum therapy (from 30% down to 1%).[14]

Experimental studies suggest that the immune plasma neutralizes the Junin virus.[9]

Medical surveillance

Surveillance is indicated in hospital or research workers with possible exposure. Employees with a febrile illness should be fully evaluated. Workers exposed to these viruses where they are endemic should be evaluated thoroughly if a febrile illness occurs within the incubation period.

Prevention

Careful hand washing and barrier personal protective equipment are indicated for healthcare employees. Initial reports of high levels of person-to-person transmission of Lassa virus in hospitals have not recurred in more sophisticated healthcare settings. Patient isolation for blood and body fluids and respiratory transmission is indicated. All secretions and contaminated materials should be treated as biological hazards. High-level biosafety containment (BSL-3 or BSL-4) is indicated for laboratory research and viral isolation with these viruses, because aerosol infectivity is high. Candid #1 vaccine, a live attenuated vaccine, is currently used to prevent Argentinian hemorrhagic fever caused by the Junin virus.[15]

References

1. Barry M, Russi M, Armstrong L, et al. Treatment of a laboratory acquired Sabia virus infection. *N Engl J Med* 1995;333:294–296.
2. Paweska JT, Sewlall NH, Ksiazek TG, et al. Nosocomial outbreak of novel arenavirus infection, *Southern Africa. Emerg Infect Dis* 2009;15(10):1598–1602.
3. Howard CR, Simpson DIH. The biology of arenaviruses. *Gen Virol* 1980;51:1.
4. Centers for Disease Control and Prevention. Lassa Fever Confirmed in Death of U.S. Traveler Returning from Liberia, 2015. Available at: http://www.cdc.gov/media/releases/2015/p0525-lassa.html (accessed on May 3, 2016).
5. Fischer SA, Graham MB, Kuehnert MJ, et al. Transmission of lymphocytic choriomeningitis virus by organ transplantation. *N Engl J Med* 2006;354:2235–2249.
6. Barry A, Gunn J, Tormey P, et al. Brief report: lymphocytic choriomeningitis virus transmitted through solid organ transplantation—Massachusetts, 2008. *MMWR Morb Mortal Wkly Rep* 2008;57(29):799–801.
7. Palacios G, Druce J, Du L, et al. A new arenavirus in a cluster of fatal transplant-associated diseases. *N Engl J Med* 2008; 358:991–998.
8. Centers for Disease Control and Prevention. Arenaviridae. 2013. Available at: http://www.cdc.gov/vhf/virus-families/arenaviridae.html (accessed on May 4, 2016).
9. Seregin A, Yun N, Paessler S. Lymphocytic choriomeningitis, Lassa virus and the South American hemorrhagic fevers (arenaviruses). In: Bennett JE, Dolin R, Blaser M. eds. *Mandell, Douglas, and Bennett's Principles and Practice of Infectious Disease*, 8th edn. Philadelphia, PA: Elsevier Saunders, 2015:2031–2038.
10. St. Georgiev, V. *Viral Hemorrhagic Fever – Arenaviruses. National Institute of Allergy and Infectious Diseases, NIH – Volume 2: Impact on Global Health.* Bethesda, MD: Humana Press–Springer Science & Business Media, 2009, pp. 260–264.
11. Lehmann-Grube E. *Lymphocytic choriomeningitis virus.* New York: Springer, 1971.
12. Barton LL, Peters CJ, Ksiazek TG. Lymphocytic choriomeningitis virus: an unrecognized teratogenic pathogen. *Emerg Infect Dis* 1995;1(4):152–153.
13. Enria DA, Briggiler AM, Sánchez Z. Treatment of Argentine hemorrhagic fever. *Antiviral Res* 2008;78(1):132–139.
14. Enria DA, Briggiler AM, Fernandez NJ, et al. Importance of dose of neutralising antibodies in treatment of Argentine haemorrhagic fever with immune plasma. *Lancet* 1984;2(8397):255–256.
15. Ambrosio A, Saavedra M, Mariani M, et al. Argentine hemorrhagic fever vaccines. *Hum Vaccin* 2011;7(6):694–700.

CORONAVIRUS

Common names for disease: Severe acute respiratory syndrome coronavirus (SARS-CoV), Middle East respiratory syndrome coronavirus (MERS-CoV)

Classification: Alpha coronaviruses, beta coronaviruses, gamma coronaviruses, delta coronaviruses

Occupational setting

Severe acute respiratory syndrome coronavirus (SARS-CoV)

Those individuals caring for SARS-infected individuals in a healthcare facility (including inpatient wards and outpatient clinics) are most susceptible for contracting the virus. When SARS presented in Toronto Canada in 2003, up to 50% of those workers infected were healthcare workers.[1] Laboratory personnel are also vulnerable to acquiring SARS if the sample the worker is handling contains live SARS-CoV.[2]

Middle East respiratory syndrome coronavirus (MERS-CoV)

Preliminary investigation in Qatar has demonstrated that people working closely with camels (such as farm workers, slaughterhouse workers, and veterinarians) may be at higher risk of MERS-CoV infection than people who do not have regular close contacts with camels.[3] A camel herder working with ill camels came down with MERS-CoV and extensive genomic sequencing identified the camels as the source of the herder's disease.[4] Nosocomial infections have also occurred primarily in the Arabian Peninsula where the majority of these patients have been treated, namely, because of poor infection control practices. There is currently evidence demonstrating sustained human-to-human transmission of MERS-CoV in South Korea, where an outbreak occurred in 2015 in a healthcare setting (including outpatient clinics and hospitals) in Seoul.[5]

Exposure (route)

SARS

In addition to humans, SARS-CoV has been found to infect animals, including monkeys, Himalayan palm civets, raccoon dogs, cats, dogs, and rodents.[6] It is primarily spread via airborne droplets shed from the respiratory secretions of those infected.[6,7] SARS-CoV has been found in the stool samples of infected patients, suggesting that oral–fecal spread may occur.[7]

MERS

MERS-CoV has been detected in camels and bats. Mechanisms behind exposure (both from animal to human and human to human) continue to be elucidated. It is postulated that MERS-CoV is transmitted by one of two ways: (i) via airborne droplets spread by coughing and sneezing or (ii) via close personal contact (such as touching or shaking hands).[6]

Pathobiology

Coronaviruses are enveloped, positive single-stranded RNA viruses. Human coronaviruses cause mainly mild respiratory illness although in some instances symptoms can progress rapidly, leading to severe respiratory infection.[7] These coronaviruses can survive on hard surfaces for up to 3 hours.[8]

SARS

The SARS coronavirus was first identified in Asia in late 2002. The index case was a healthcare worker from Guangdong province, China, who visited Hong Kong and became ill, and succumbed to the infection soon after being hospitalized.[9,10] By means of international travel, SARS started to disseminate on a global scale, resulting in more than 8000 infections and almost 800 deaths between the winter of 2002 and spring of 2003.[11,12] Adjoining Asian countries soon were affected. SARS spread to Toronto, Canada, in early 2003, where a subsequent epidemic resulted in almost 400 probable and suspect cases and 44 deaths between March and July 2003, making Toronto the most affected center outside of Asia.[11,12]

SARS is characterized largely by constitutional symptoms, including high fever (>104°F), headache, and muscle aches.[13] Diarrhea can also be present in 10–20% of cases. With respect to lower respiratory tract impact, affected individuals can develop nonproductive cough and dyspnea 2–7 days after initial exposure, which can progress to pneumonia and ultimately respiratory failure requiring intubation and mechanical ventilation.[14]

MERS

MERS-CoV disease was first described in a patient living in Saudi Arabia in mid-2012.[15] Retrospective investigations revealed that the first cases of the disease had occurred previously in a cluster of hospital-associated cases in Jordan earlier that year.[16] A marked increase in the number of cases of MERS-CoV infection occurred in Jeddah, Saudi Arabia, in early 2014. An epidemiologic investigation has identified 255 patients in Jeddah in 2014 who had laboratory-confirmed MERS-CoV infection, with 78 of these infections (including asymptomatic and symptomatic cases) occurring in healthcare personnel.[17] Additional analyses have determined that among the 191 symptomatic patients who were not healthcare personnel, 112 (74.2%) had data that could be assessed, and 109 of these 112 patients (97.3%) had been found to have had contact with a healthcare facility, a person with a confirmed case of MERS-CoV infection, or someone with severe respiratory illness before the onset of illness.[17] There have also been travel-associated cases identified in both Europe and the United States, but in all of these cases, infection had occurred prior to the travelers' departure from the Arabic Peninsula.

An outbreak of MERS occurred in South Korea in 2015. The index case, a 68-year-old national of the Republic of Korea with a recent history of travel to four countries in the Middle East, had been asymptomatic during his return flight to the Republic of Korea on May 4, 2015, but developed symptoms a week later. He sought care at two outpatient clinics and two hospitals, where multiple healthcare workers, patients, family members, and visitors were infected. The Ministry of Health has reported that two additional confirmed cases represent a third generation of transmission—from the index case, to someone exposed to that case, to a third person with no direct exposure to the index case.[5] The epidemic was declared over at the end of July 2015.[18] A total of 186 MERS-CoV cases, including 36 deaths, have been reported in the South Korea outbreak including the index case, healthcare workers caring for the index case, patients who were being cared for at the same clinics or hospitals, and family members and visitors to the hospital.[18,19] Globally, as of late June 2015, there have been 1357 laboratory-confirmed cases of MERS-CoV infection, including 351 deaths (Figure 21.4).[20]

The incubation period for MERS-CoV ranges from 2 to 14 days. Those presenting with MERS-CoV infection most commonly have fever, cough, and shortness of breath. Gastrointestinal symptoms (nausea, vomiting, diarrhea) have also been reported. All confirmed cases have

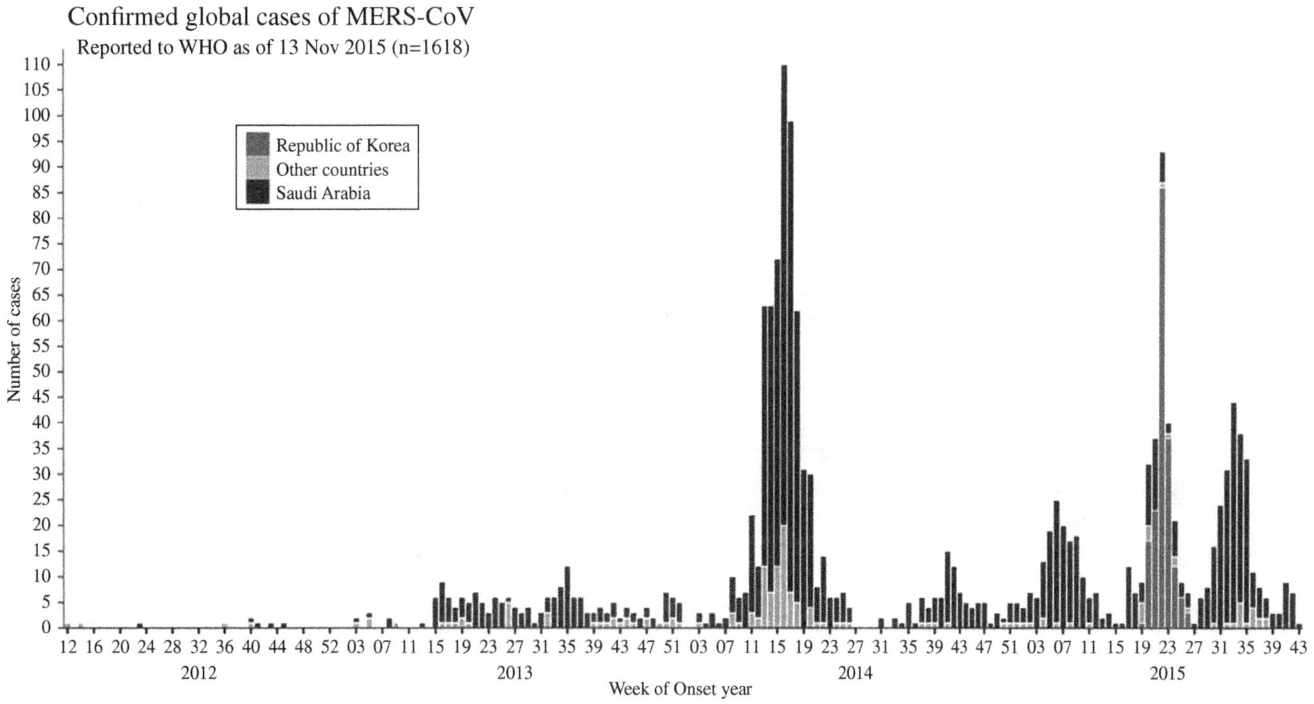

Other countries: Algeria, Austria, China, Egypt, France, Germany, Greece, Iran, Italy, Jordan, Kuwait, Lebanon, Malaysia, Netherlands, Omen, Philippines, Qatar, Thailand, Tunisia, Turkey, United Arab Emirates, United Kingdom, United States of America, Yeman

Please note that the underlying data is subject to change as the investigations around cases are ongoing. Onset date estimated if not available.

FIGURE 21.4 Confirmed global cases of MERS-CoV. *Source*: World Health Organization http://www.who.int/csr/disease/coronavirus_infections/maps-epicurves/en/ Reprinted with permission

experienced significant respiratory disease. Complications include pneumonia with respiratory failure requiring mechanical ventilation, acute respiratory distress syndrome (ARDS) with multi-organ failure, renal failure requiring dialysis, consumptive coagulopathy, and pericarditis.[16]

Diagnosis

SARS

There have been no reported SARS cases since 2004. Based on the most recent SARS experience, guidelines have been developed on how to diagnose SARS. In individuals admitted to the hospital for radiologically confirmed pneumonia, SARS must be ruled out in the following scenarios: (i) travel history is consistent with recent SARS cases (previously travel from China, Hong Kong, or Taiwan) within 10 days of symptom onset, (ii) occupation with high exposure risk for contracting SARS (i.e., a healthcare worker), and/or (iii) the patient's clinical presentation is similar to other recently admitted patients with similar constellation of symptoms.[2] Laboratory confirmation of SARS-CoV involves detection of serum antibodies to SARS-CoV in a single serum specimen. Detection of SARS-CoV RNA by RT-PCR validated by CDC is also acceptable, with subsequent confirmation in a reference laboratory.[21]

MERS

Performing RT-PCR on respiratory swab samples can identify the presence of MERS-CoV RNA.[15]

Treatment

Currently there are no therapeutic agents developed to combat SARS or MERS. Supportive treatment (including ventilatory support) is the mainstay of care.

Surveillance

SARS

During the peak of the SARS epidemic, the Frontlines of Medicine (an organization consisting of emergency medicine, public health, and informatics professionals) created the SARS Surveillance Project (SARS-SP). The primary goal of this workgroup was to develop, disseminate, and update a practical screening (case-finding) form for patients presenting to the emergency room via a robust syndromic surveillance electronic system in the United States.[22] The initial test site was in Milwaukee, WI, and three other metropolitan sites also were involved (Denver, CO; Akron, OH; and Fort Worth, TX). The surveillance system only detected one person meeting the CDC definition of SARS and no confirmed cases were reported; thus sensitivity and specificity could not be calculated.

MERS

As the pathobiology behind MERS continues to evolve, understanding the current epidemiology of the disease (including analysis of specimens collected from all patients tested, even from those that test negative) is critical in order to develop robust surveillance systems.[22] In 2012, the United Kingdom developed a surveillance system to identify cases of MERS-CoV infection among travelers to England from the Middle East. After 12 months of surveillance, 77 individuals had met the case definition for MERS (including having symptoms of acute respiratory illness) and were tested for virus. Infection was confirmed in two of these travelers and two of their contacts.[23]

Prevention

SARS

Three activities—case detection, patient isolation, and contact tracing—can reduce the number of people exposed to each infectious case and eventually break the chain of transmission.[24] In a case control study in five Hong Kong hospitals involving 241 noninfected and 13 infected staff with documented exposures to 11 index patients, no infection was observed among 69 healthcare workers who reported the use of mask, gloves, gowns, and hand washing.[25] N95 masks provided the best protection for exposed healthcare workers, whereas paper masks did not significantly reduce the risk of infection. For laboratory personnel working with whole blood, serum, plasma, and urine samples from infected patients, standard precautions should be employed.[26] This includes the use of gloves, gown, mask, and eye protection.[26] Any procedure capable of generating aerosols (i.e., vortexing or sonication of specimens in an open tube) should be performed in a biosafety cabinet.[26]

MERS

Currently there are no vaccines available against MERS-CoV. Healthcare workers are advised to adhere to recommended infection control measures (including standard, contact, and airborne precautions).[19] Camel farm and slaughterhouse workers should practice hand hygiene (soap and water), use face shields, and wear protective clothing (which should be removed after work and washed daily).[27] Workers should also avoid exposing family members to soiled work clothing, shoes, or other items that may have come into contact with camels' secretions or excretions. In cases where a particular animal has been tested and confirmed positive for MERS-CoV, people should avoid direct contact with these animals.

References

1. Grace SL, Hershenfield S, Robertson E, et al. Factors affecting perceived risk of contracting severe acute respiratory syndrome among academic physician. *Infect Control Hosp Epidemiol* 2004;25:1111–1113.
2. Centers for Disease Control and Prevention. Clinical Guidance on the Identification and Evaluation of Possible SARS-CoV Disease among Persons Presenting with Community-Acquired Illness. 2012. Available at: http://www.cdc.gov/sars/clinical/guidance.html (accessed on May 4, 2016).
3. World Health Organization. Update on MERS-CoV Transmission from Animals to Humans, and Interim Recommendations for At-Risk Groups. 2014. Available at: http://www.who.int/csr/disease/coronavirus_infections/MERS_CoV_RA_20140613.pdf?ua=1 (accessed on May 4, 2016).
4. Memish ZA, Cotten M, Meyer B, et al. Human infection with MERS coronavirus after exposure to infected camels, Saudi Arabia, 2013. *Emerg Infect Dis* 2014;20(6):1012–1015.
5. World Health Organization. Middle East Respiratory Syndrome Coronavirus (MERS-CoV) in the Republic of Korea – Situation Assessment. 2015. Available at: http://www.who.int/mediacentre/news/situation-assessments/2-june-2015-south-korea/en/ (accessed on May 3, 2016).
6. Centers for Disease Control and Prevention. About Coronaviruses. 2014. Available at: http://www.cdc.gov/coronavirus/about/index.html (accessed on May 3, 2016).
7. Kamps BS, Hoffmann C. eds. Transmission. SARS Reference – 10/2003. Available at: http://www.sarsreference.com/sarsreference.pdf (accessed on May 3, 2016).
8. Sizun J, Yu MW, Talbot PJ. Survival of human coronaviruses 229E and OC43 in suspension after drying on surfaces: a possible source of hospital-acquired infections. *J Hosp Infect* 2000;46:55–60.
9. Ksiazek TG, Erdman D, Goldsmith CS, et al. A novel coronavirus associated with severe acute respiratory syndrome. *N Engl J Med* 2003;348:1953–1966.
10. Drosten C, Gunther S, Preiser W, et al. Identification of a novel coronavirus in patients with severe acute respiratory syndrome. *N Engl J Med* 2003;348:1967–1976.
11. Loutfy MR, Blatt LM, Siminovitch KA, et al. Interferon alfacon-1 plus corticosteroids in severe acute respiratory syndrome: a preliminary study. *JAMA* 2003;290:3222–3228.
12. Farcas GA, Poutanen SM, Mazzulli T, et al. Fatal severe acute respiratory syndrome is associated with multiorgan involvement by coronavirus. *J Infect Dis* 2005;191(2):193–197.
13. Centers for Disease Control and Prevention. Frequently Asked Questions about SARS. 2012. Available at: http://www.cdc.gov/sars/about/faq.html (accessed on May 4, 2016).
14. Hwang DM, Chamberlain DW, Poutanen SM, et al. Pulmonary pathology of severe acute respiratory syndrome in Toronto. *Mod Pathol* 2005;18:1–10.
15. Zaki AM, van Boheemen S, Bestebroer TM, et al. Isolation of a novel coronavirus from a man with pneumonia in Saudi Arabia. *N Engl J Med* 2012;367:1814–1820.
16. World Health Organization. Interim Surveillance Recommendations for Human Infection with Middle East Respiratory Syndrome Coronavirus. 2013. Available at: http://www.who.

int/csr/disease/coronavirus_infections/InterimRevised SurveillanceRecommendations_nCoVinfection_27Jun13.pdf (accessed on May 4, 2016).
17. Oboho IK, Tomczyk SM, Al-Asmari AM, et al. 2014 MERS-CoV outbreak in Jeddah—a link to health care facilities. *N Engl J Med* 2015;372:846–854.
18. San-Hun C. South Korea: Government Declares End to MERS Outbreak. New York Times, July 27, 2015. http://www.nytimes.com/2015/07/28/world/asia/south-korea-government-declares-end-to-mers-outbreak.html?_r=5 (accessed on May 3, 2016).
19. World Health Organization. Middle East respiratory syndrome coronavirus – MERS-CoV Republic of Korea. Disease Outbreak News, June 26, 2015. Available at: http://www.who.int/csr/don/26-june-2015-mers-korea/en/ (accessed on May 3, 2016).
20. World Health Organization. Middle East respiratory syndrome coronavirus – MERS-CoV Saudi Arabia. Disease Outbreak News, June 23, 2015. Available at: http://www.who.int/csr/don/23-june-2015-mers-saudi-arabia/en/ (accessed on May 3, 2016).
21. Centers for Disease Control and Prevention. Appendix F8 – Guidelines for Laboratory Diagnosis of SARS-CoV Infection, 2005. Available at: http://www.cdc.gov/sars/guidance/F-lab/app8.html (accessed on May 3, 2016).
22. Foldy SL, Barthell E, Silva J, et al. SARS surveillance project – internet-enabled multiregion surveillance for rapidly emerging disease. *MMWR Morb Mortal Wkly Rep* 2004;53S:215–220. Available at: http://www.cdc.gov/mmwr/preview/mmwrhtml/su5301a39.htm (accessed on May 3, 2016).
23. Thomas HL, Zhao H, Green HK, et al. Enhanced MERS coronavirus surveillance of travelers from the Middle East to England. *Emerg Infect Dis* 2014;20(9):1562–1564.
24. World Health Organization. SARS. Weekly Epidemiological Record, May 16, 2003. Available at: http://www.who.int/docstore/wer/pdf/2003/wer7820.pdf (accessed on May 3, 2016).
25. Seto WH, Tsang D, Yung R, et al. Effectiveness of precautions against droplets and contact in prevention of nosocomial transmission of severe acute respiratory syndrome (SARS). *Lancet* 2003;361(9368):1519–1520.
26. Centers for Disease Control and Prevention. Section VIII-E: Viral agents. *Biosafety in Microbiological and Biomedical Laboratories*, 5th edn. In Wilson DE, Chosewood LC. eds. Available at: http://www.cdc.gov/biosafety/publications/bmbl5/BMBL5_sect_VIII_e.pdf (accessed on May 4, 2016).
27. World Health Organization. Middle East respiratory syndrome coronavirus Joint Kingdom of Saudi Arabia/WHO mission, Riyadh, June 4–9, 2013. Available at: http://www.who.int/csr/disease/coronavirus_infections/MERSCov_WHO_KSA_Mission_Jun13_.pdf (accessed on May 3, 2016).

CYTOMEGALOVIRUS (CMV)

Common names for disease: CMV mononucleosis, heterophile-negative mononucleosis
Classification: Family—Herpesviridae

Occupational setting

Child day-care workers are at greater risk of exposure to CMV compared to other occupational groups because of the high prevalence of CMV infection in children (specifically in those under the age of 24 months) as well as nature of the work itself (i.e., diaper changing, cleaning of contaminated surfaces).[1–3] Studies examining the occupational risk of CMV acquisition in day-care educators in industrialized countries have estimated seroprevalence rates between 38 and 67% and seroconversion rates between 12 and 23.8%.[3] Although the potential for occupational exposure to CMV among healthcare workers has been a concern in the past, several studies have found that the risk of CMV acquisition among healthcare workers is not higher than that of the general population.[4–6] Moreover, pregnant healthcare workers are not at increased risk of acquiring CMV compared to nonpregnant healthcare workers.[7]

Exposure (route)

Transmission requires intimate contact with secretions or body fluids of an infected individual.

The virus has been isolated from blood, saliva, urine, tears, breast milk, cervical secretions, and semen, suggesting multiple possible modes of transmission. CMV has not been shown to be transmitted via respiratory secretions or aerosolized virus.[8] For the pregnant woman, contact with urine or saliva of young children (their children in particular) and sexual contact are the most common exposures to CMV.[9] Less than one in five parents of children who are shedding CMV become infected over the course of a year.[9] Congenital infection is well documented.[6]

Pathobiology

CMV is a member of the herpes virus family. It is a double-stranded DNA virus. Infection is usually asymptomatic; only 10% of adults develop a mononucleosis syndrome. This may include fever, malaise, hepatitis, lymphadenopathy, or splenomegaly. Infrequent manifestations of CMV infection include Guillain–Barré syndrome, meningoencephalitis, myocarditis, thrombocytopenia, hemolytic anemia, and granulomatous hepatitis. After primary infection, the virus becomes latent, but it can reactivate with production of infectious virions. In the United States, between 1 and 4% of women who have never been infected with CMV have a primary CMV infection during pregnancy, with a third of these women passing the infection to their infant.[9] Out of 1000 live births, less than 1% will have congenital CMV infection and only 0.1% of these infants will have permanent CMV-related morbidity.[10]

Diagnosis

The diagnosis of CMV infection requires laboratory confirmation, but it should be suspected in cases of mononucleosis where tests for Epstein–Barr virus (EBV) are negative (e.g., monospot, heterophile). A rise in CMV-specific IgG titer over time or the presence of IgM antibody usually reflects active CMV infection. The enzyme-linked immunosorbent assay (ELISA) is the most commonly available serologic test for measuring antibody to CMV. Recently IgG avidity assays (which measure antibody maturity) have been shown to reliably detect recent primary CMV infection, with low CMV IgG avidity suggesting a primary CMV infection has occurred within the past 2–4 months.[11] Active infection with CMV can also be diagnosed by PCR from urine, saliva, and throat swab specimens.

Treatment

CMV infection is usually mild in the immunocompetent host and is treated with rest and supportive care. Specific treatment is usually not indicated. Antivirals are recommended in those patients with depressed immunity. Limited data suggests that ganciclovir can be used off-label to treat infants with CMV infection with concurrent central nervous system involvement to reduce adverse developmental outcomes.[12]

Medical surveillance

Screening programs to identify susceptible workers are not recommended by CDC.[9,13] For pregnant healthcare workers whose antibody status is negative or unknown, there are no data to indicate that job reassignment or modification alters the risk of infection.[6]

Prevention

The only known effective method of reducing the risk of CMV infection is to stress careful hand washing and strict adherence to universal infection precautions.[8,13,14]

References

1. Stagno S, Cloud GA. Working parents: the impact of day care and breast-feeding on cytomegalovirus infections in offspring. *Proc Natl Acad Sci USA* 1994;91:2384–2389.
2. Murph JR, Bale JF Jr, Murray JC, et al. Cytomegalovirus transmission in a Midwest day care center: possible relationship to child care practices. *J Pediatr* 1986;109:35–39.
3. Joseph SA, Beliveau C, Muecke CJ, et al. Cytomegalovirus as an occupational risk in daycare educators. *Paediatr Child Health* 2006;11(7):401–407.
4. Balcarek K, Bagley R, Cloud GA, et al. CMV infection among employees of a children's hospital. *JAMA* 1990;263:840–844.
5. Sepkowitz K. Occupationally acquired infections in health care workers: Part II. *Ann Intern Med* 1996;125(11):917–928.
6. Pomeroy C, Englund J. CMV: epidemiology and infection control. *Am J Infect Control* 1987;15:107–118.
7. Wicker S. Viral infections – occupational risk for pregnant health-care personnel? *Proc Vaccinol* 2012;6:156–158.
8. Cannon MJ, Davis KF. Washing our hands of the congenital cytomegalovirus disease epidemic. *BMC Pub Health* 2005;5:70.
9. Centers for Disease Control and Prevention. CMV transmission. 2010. Available at: http://www.cdc.gov/cmv/transmission.html (accessed on May 4, 2016).
10. Cannon MJ. Congenital cytomegalovirus (CMV) epidemiology and awareness. *J Clin Virol* 2009;46(Suppl 4):S6–S10.
11. Centers for Disease Control and Prevention. CMV – Interpretation of Laboratory Results. 2010. Available at: http://www.cdc.gov/cmv/clinical/lab-tests.html (accessed on May 3, 2016).
12. Centers for Disease Control and Prevention. CMV – Clinical Diagnosis and Treatment. 2010. Available at: http://www.cdc.gov/cmv/clinical/diagnosis-treatment.html (accessed on May 3, 2016).
13. Williams W. CDC guidelines for the prevention and control of nosocomial infections. *Am Infect Control* 1984;12:34–63.
14. Centers for Disease Control and Prevention. CMV – Prevention. 2010. Available at: http://www.cdc.gov/cmv/prevention.html (accessed on May 3, 2016).

FILOVIRUSES (EBOLA AND MARBURG VIRUSES)

Common name for disease: Viral hemorrhagic fever
Classification: Family—Filoviridae

Occupational setting

Ebola

With the Ebola epidemic that began in 2014 (Figure 21.5), many workers are susceptible to contracting the virus. Humanitarian aid, healthcare, medical air, laboratory, mortuary, airline, and waste management workers are particularly at risk. Healthcare providers are the most susceptible for acquiring the virus. In the United States, two healthcare workers contracted the virus as they were treating an imported case of Ebola in Dallas, Texas. CDC has issued guidelines for exposure mitigation and infection control practices.[1–7] Researchers studying the viruses and primate handlers are also at risk from exposure.

Marburg

The first reported outbreaks of Marburg virus occurred in the occupational setting, specifically among workers in Germany (Marburg and Frankfurt) and the former Yugoslavia (Belgrade) in 1967. These simultaneous outbreaks led to 25

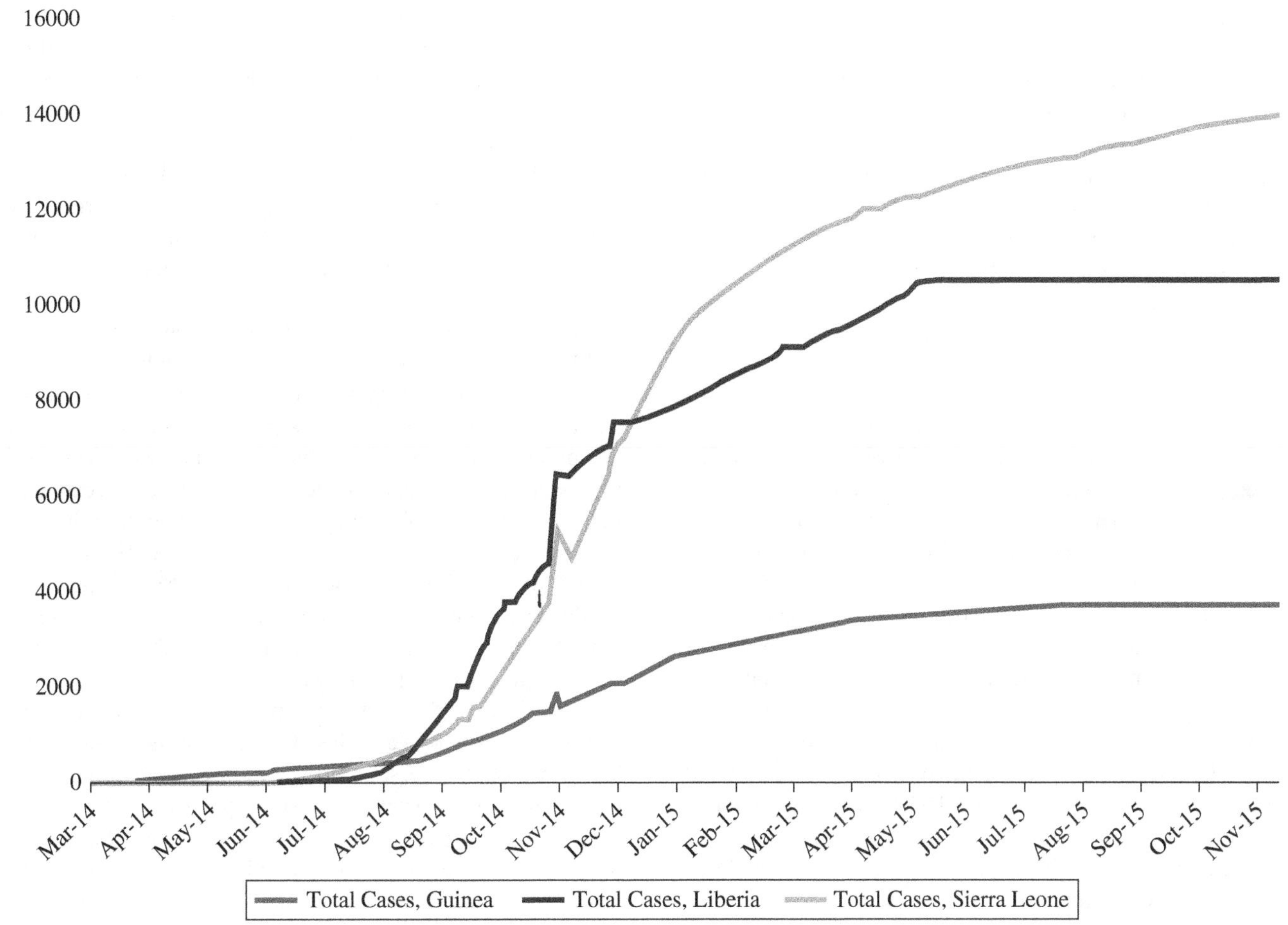

FIGURE 21.5 Total reported suspected, probable, and confirmed cases in Guinea, Liberia, and Sierra Leone beginning March 25, 2014 through November 8, 2015. *Source*: http://www.cdc.gov/vhf/ebola/outbreaks/2014-west-africa/cumulative-cases-graphs.html

primary infections (in laboratory staff handling African green monkeys (*Cercopithecus aethiops*) or their tissues), 6 secondary infections (in healthcare personnel via direct contact with primary cases), and 7 deaths.[8]

Exposure (route)

Ebola

The exact mechanism by which Ebola virus is transmitted from the natural host to humans is unknown. The hammer-headed fruit bat (*Hypsignathus monstrosus*), Franquet's epauletted fruit bat (*Epomops franqueti*), and the little collared bat (*Myonycteris torquata*) have been implicated in the Ebola outbreak of 2014.[9]

Transmission of Ebola virus from person to person requires direct contact with an infected individual's blood or bodily fluids (i.e., sputum, urine, feces), via abrasions in the skin, splashes to unprotected mucosal tissues, or parenteral exposure. Accidental injuries by sharps (such as scalpel blades or needles that are contaminated with infectious material) must also be considered. Other transmission routes (via direct contact with medical equipment contaminated with infected fluids from Ebola patients) have also been implicated in the spread of the disease, including unintentional transfer of contaminated materials to mucous membranes.[10]

Exposure may also occur through ingestion, via the consumption of infected primates, bats, and bushmeat. Although filoviruses such as Ebola display some capability of infection through small-particle aerosols, airborne spread among humans has not been clearly demonstrated.[10]

Marburg

The natural reservoirs for Marburg virus have been found in the African fruit bat (*Rousettus aegyptiacus*).[10,11] The route of acquisition of natural infection is unclear, but as with the Ebola virus, person-to-person transmission occurs by direct contact with blood and other body fluids. In the laboratory, these viruses have been shown to be highly infectious via the respiratory route.

Pathobiology

Each filovirus virion (or complete virus particle) is made up of one molecule of single-stranded, nonsegmented, negative-sense RNA. The virions are arranged in filaments (either branched or U-shaped) and enveloped in a lipid bilayer coat that protects the virus genome and allows easy entry into host cells.[10] Filoviruses are classified as Category A potential bioterrorism agents by CDC. Because of their highly hazardous nature, these viruses are investigated in high-level (BSL-4) microbiological containment facilities.[12,13]

Ebola

Ebola virus was first identified in 1976 when two outbreaks occurred: northern Zaire (now the Democratic Republic of Congo) and southern Sudan. The case fatality rate was 90% for the Zairian outbreak and 50% for the Sudanese outbreak. Five species of Ebola virus have been identified to cause disease: Taï Forest (formerly Ivory Coast), Sudan, Zaire, Reston, and Bundibugyo. The Ebola Reston strain affects nonhuman primates and more recently swine in Southeast Asia, but does not appear to cause clinical infection in humans. Between 1977 and 2012, there have been four large outbreaks and a number of smaller outbreaks in Central and East Africa, with approximately 1600 deaths attributed to the virus.[14]

Beginning in late 2013, the Zairian strain of the Ebola virus emerged in the West African country of Guinea, expanding into adjacent Liberia and Sierra Leone in early 2014. The World Health Organization declared an Ebola outbreak in March 2014. These three countries have been the main epicenter of the outbreak, with vigilant containment efforts helping to keep the Ebola virus from spreading in Nigeria and Senegal. Imported cases were reported in Spain and the United States. From March 25, 2014, through January 17, 2016, there have been approximately 28 638 infections (including probable, suspected, and laboratory-confirmed cases) and 11 316 deaths attributed to the Ebola outbreak in Sierra Leone, Liberia, and Guinea.[15] The cumulative case fatality rate is 70% among the three affected countries.[16] The outbreak in Liberia was declared over on May 9, 2015, and WHO determined that Ebola virus transmission ended in Sierra Leone on November 7, 2015.[15] No new cases were reported in Guinea in the week prior to November 7, 2015.[15] However, subsequent sporadic cases suggest the ongoing possibility of flare-ups and small outbreaks in the affected countries since the virus may persist in survivors.

The Ebola virus implicated in the 2014–2015 outbreak shares 97% homology with the Zairian strain. Additional sequencing analyses have demonstrated that there has been a rapid accumulation of mutations (~400) since transmission from the natural reservoir to humans.[13]

Ebola patients transferred from West Africa to the United States were given primarily supportive care, including aggressive fluid resuscitation and electrolyte replacement.[17] They were also administered experimental drug therapies (including ZMapp—see "Treatment" section below).

There has been one documented case of imported Ebola disease in the United States. A 42-year-old man, originally from Liberia, had been in contact with an Ebola-infected individual in Monrovia, Liberia, on September 15, 2014.[18] He did not exhibit fever as he made his passage from Monrovia, Liberia, to Brussels, Belgium, on September 19, 2014. He entered the United States through Washington Dulles International Airport before finally arriving in Dallas, Texas, on September 20, 2014. He first began to exhibit symptoms (including fever) on September 24, 2014, and thus went to a hospital on September 25, 2014, for treatment, but was sent home. Given his worsening condition, he was transported back to the hospital via ambulance. He was confirmed to have Ebola on September 30, 2014. The Texas Health Department began a contact investigation, identifying 10 people at highest risk of contracting Ebola (including the four persons the patient was living with, as well as the three medical workers who transported him to the hospital) and 40 others who were classified as low risk. During his hospital stay, two nurses who were taking care of him contracted the virus. They were subsequently transferred to Emory University Medical Center and the National Institutes of Health for treatment where they both made a full recovery. The Liberian patient succumbed to the disease on October 8, 2014.

The incubation period for Ebola virus ranges from 2 to 21 days, with an average incubation of 5–7 days. The symptoms associated with initial infection include fever, chills, myalgia, and malaise. These symptoms overlap with malaria or dengue in tropical climates, thus making early detection difficult. Ebola virus RNA is detectable in blood once fever starts to manifest itself but may not be reliably detectable in some patients during the first 3 days of illness.[19,20] Ebola virus RNA levels in the blood have been shown to increase logarithmically during the acute phase of illness.[19] The initial period is followed by flu-like and gastrointestinal symptoms, with more severe cases demonstrating maculopapular rash, petechiae, conjunctival hemorrhage, epistaxis, melena, hematemesis, shock, and encephalopathy.[13] Based on studies conducted in nonhuman primates, the virus replicates in many types of cells (including macrophages, dendritic cells, and monocytes). The virus disables the innate, humoral, and cell-mediated immune response, permitting unchecked replication of virus. As viral loads increase in the infected patient and the patient gets more ill, the risk of transmitting the Ebola virus grows. Based on prior epidemiologic analyses of risk factors and exposure histories, the risk of a noninfected person acquiring Ebola is highest when the patient is severely ill, since bodily fluids are copiously emitted in the later phases of disease.[21] A higher viral load in the infected patient is also associated with higher fatality rates.[22] Thus the bodies of deceased Ebola-infected persons are highly infectious. Among patients who survive, the Ebola virus RNA levels in the blood decrease during clinical recovery.[19] A clinical report suggests that viable Ebola virus was detected in the aqueous humor of the inflamed eye of an Ebola patient 14 weeks after the onset of disease and 9 weeks after the clearance of the viremia.[23] Ebola virus RNA has been detected in semen up to 9 months after recovery in some patients.[24] Ebola virus RNA has also been detected in vaginal fluid from one woman 33 days after symptom onset, although live virus has not been isolated from vaginal fluids. It is not known for how long virus typically persists in vaginal fluids or whether it can be sexually transmitted from females to males.[25]

Marburg

Since the initial presentation in the late 1960s, there have been five outbreaks of Marburg virus, all occurring in Africa.[11,26] As with Ebola virus, contracting Marburg virus requires close contact with the source patient. Infection results from contact with blood or other body fluids (feces, vomitus, urine, saliva, and respiratory secretions) with high virus concentration, especially when these fluids contain blood.[11] The incubation period for Marburg virus is 3–9 days. Initial symptoms include headache and malaise. High fever and gastrointestinal symptoms (including nausea, vomiting, diarrhea) then ensue. Hemorrhagic (bleeding from the gums, nose, reproductive tract) and neurological sequelae (confusion, irritability, aggression) indicate serious disease. Mortality is associated with severe hemodynamic instability.

Marburg virus may also be transmitted via semen, as suggested by a single instance of heterosexual transmission from a male survivor to a female partner reported during an outbreak in 1967.[25]

Diagnosis

Ebola

Assays used to test for Ebola virus in the acute period include real-time quantitative PCR, IgM ELISA, and Vero E6 cell lines.[13] Later during the course of disease IgG ELISA can be used. The ELISA has been standardized by CDC for the detection of Ebola-specific antibodies and is capable of detecting antibodies in the sera of humans exposed 10 years previously to Ebola.[13] Additional laboratory tests demonstrate elevated aminotransferase levels, hematologic effects (including lymphocytopenia and thrombocytopenia), and coagulopathic complications.[14] New diagnostic tests are currently in development (spearheaded by WHO) for expedient and accurate diagnosis of Ebola in the field.[27] Healthcare facilities working up a patient with suspected Ebola are advised to contact the state and local public health departments for testing logistics. The respective health departments can contact the CDC Emergency Operations Center at 770-488-7100 for additional consultation and testing.[28]

Marburg

Because of the rarity of these illnesses, the clinical diagnosis is difficult and generally missed except in those cases where there is a high index of suspicion due to known exposure. A rapidly progressing, severe viral illness in a primate handler or exposed healthcare worker should be evaluated for possible hemorrhagic fever.

Medical surveillance

Surveillance protocols for healthcare workers returning from the three affected West African countries have been devised, in accordance with CDC recommendations. It must be stressed to healthcare workers prior to their departure to the epidemic zone the potential for risk that may arise from unprotected contact with asymptomatic, or individuals assumed ill with another disease, who develop disease a few days later, even with appropriate use of PPE. These workers, even those with no known Ebola exposure, will be under medical surveillance in conjunction with local public health authorities upon their return to the United States. The paradigm is as follows: (i) daily symptom review and twice-daily temperature check for 21 days after return and (ii) daily contact with the worker's respective occupational health department and local public health department on the presentation of any symptoms.

Healthcare workers caring for patients with Ebola should self-monitor for fever and symptoms and regularly report the results.

Treatment

Ebola

Treatment should include aggressive fluid resuscitation (3–5 L or more of intravenous fluids per day along with oral rehydration as tolerated) and electrolyte replacement as early as possible in the disease course.[17] However, excessive fluid administration can lead to substantial pooling of fluid in the third space, causing pleural effusions and significant lower extremity swelling among other effects, which will need to be managed.[17] The volume of intravenous fluids needed should be determined based on the degree of dehydration and evidence of hypovolemic shock, as well as fluid losses (volume of diarrhea or vomitus or both). Large volumes of fluid replacement (>10 L/day) may be needed in febrile patients with active diarrhea.[29] Hemodynamic monitoring is key in assessing these patients. The use of a new therapeutic agent ZMapp (an experimental cocktail of three Ebola virus glycoprotein-specific monoclonal antibodies derived from tobacco plants) in treating two healthcare workers transported from West Africa to the United States has been documented, although there is no data on the safety or efficacy of ZMapp.[17] Moreover, current stockpiles of ZMapp have run out. Current clinical trials of convalescent plasma obtained from Ebola survivors are underway in West Africa in an attempt to combat the epidemic. Results using an in vivo murine Ebola virus infection model have shown two drugs, bepridil and sertraline, to be protective against the Ebola virus, blocking a late stage of viral entry.[30]

Marburg

There is no known effective treatment beyond supportive care. The same paradigm for treatment of Ebola virus applies for the treatment of Marburg virus.

Prevention

Infection control is the main strategy by which to halt the spread of the Ebola and Marburg viruses in the occupational setting. For the Ebola epidemic of 2014, CDC and OSHA developed extensive protocols, algorithms, and guidelines for specific worker populations who may have had contact with an Ebola patient. Vigilant airport screening procedures are of paramount importance in halting the spread of disease across international borders. Screening of travelers as they exit the respective West African countries included (i) temperature reading to check for fever and (ii) questions pertaining to Ebola exposure.[31] For those travelers coming to the United States, five airports had been designated for entry screening (New York John F. Kennedy (JFK) International Airport, Washington Dulles International Airport, Newark Liberty Airport (NJ), Chicago O'Hare International Airport, and Atlanta Hartsfield-Jackson International Airport).[32] The entry screening process included review by Department of Homeland Security (including temperature reading), provision of Check and Report Ebola (CARE) kit (which comes with thermometer, health log, and cell phone with 21 days of service to ensure communication between the individual and the local public health department).[30] However, screening will only detect those who are acutely ill at the time of screening and will not necessarily identify those with subclinical infection. Healthcare facilities need to ensure that procedures have been thoroughly reviewed and systems are in place in advance of providing patient care for suspected Ebola-infected patient. All portals of entry in the healthcare system where potential patients harboring Ebola virus seek medical attention must be identified and respective protocols developed to safely quarantine and transport the patient under investigation. Isolation of patients, use of protective clothing, and disinfection procedures (together called viral hemorrhagic fever isolation precautions or barrier nursing) are critical to prevent further transmission.[10] Medical first responders, including paramedics and other emergency medical service (EMS) personnel, are at risk for exposure, given the precarious nature of working in the field. The 911 call centers have a key role in identifying potentially Ebola-infected patients, activating the protocol, communicating this information to the paramedics, and mediating the transport process to the appropriate healthcare facility in a safe and expedient manner. CDC has developed a guidance document for medical first responders, with the primary goal of minimizing exposure among this worker cohort.[33] It is recommended that initial screening be performed at a minimum of 3 ft distance from the patient and only by one EMS provider. If this EMS provider finds probable cause that the patient may have Ebola, PPE must be readily available for the EMS provider to put on so he/she can approach the patient. Other EMS personnel should be advised to keep away and to assist the primary EMS provider as indicated in the transport process.

Since ill patients may first seek care in the Emergency Department, staff (including healthcare, clerical, and other ancillary workers) should be properly educated on how to screen recent travelers from countries with active cases. CDC has devised an algorithm to systematically evaluate and manage the patient with suspected Ebola virus, with the underlying goal that all staff members be ready to (i) identify, (ii) isolate, and (iii) inform.[34] Taking a detailed travel history is imperative, including where a patient has been in the prior 4 weeks. Any indication that a patient may have been in one of the countries with active cases and exhibits a fever should be immediately quarantined and an Ebola activation response team should be triggered. Detailed protocols focusing on the proper utilization of PPE have been developed. Every organization should conduct a risk assessment to determine who is at greatest risk and the level of protection they need to prevent infection, taking into account factors that influence both the magnitude and probability of exposure.[21] CDC has placed great emphasis that PPE must cover all exposed skin surfaces and provide maximum Respiratory Protection. An impermeable gown and two pairs of gloves are recommended. Since a half-mask negative pressure respirator may not offer adequate protection from aerosols, a powered air-purifying respirator (PAPR) is the preferred respirator for emergency responders, ED triage personnel, and inpatient providers directly working with Ebola patients. CDC has promoted rigorous training, practice, competence, and observation of healthcare workers in correct donning and doffing of PPE to minimize occupational Ebola exposure.[35,36] Those healthcare personnel must have received repeated training and demonstrated proficiency in performing all Ebola-related infection control practices and procedures and specifically in donning/doffing proper PPE.[37] OSHA and the National Institute of Occupational Safety and Health (NIOSH) have also developed guidelines for workers handling the waste material of infected patients (including sharps, dressings used in patient care, laboratory supplies used for diagnostics, disinfection of hospital rooms, transport vehicles, and other locations housing Ebola patients, and PPE used by personnel) to minimize exposures to this cohort.[38]

The American College of Occupational and Environmental Medicine (ACOEM) developed guidelines for the medical clearance of designated Ebola caregivers.[39] Facilities that may care for patients with Ebola should identify potential members of the treatment team and conduct a proactive clearance process. The process should address the physical, medical, and psychological ability to utilize the required PPE for 2 hours at a time and fully participate in care without placing themselves or others at risk. If there are changes in health status following initial clearance, these should be assessed to determine whether continued participation is appropriate.

At least 15 vaccines are under development vaccines with 4 in advance stages of clinical trials. The lead candidates are (i) Chimpanzee adenovirus 3-based candidate vaccine expressing the glycoprotein of a *Zaire ebolavirus* (ChAd3-ZEBOV), developed by GlaxoSmithKline in collaboration with the National Institute of Allergy and Infectious Diseases, and (ii) recombinant, replication-competent vesicular stomatitis virus-based candidate vaccine expressing the glycoprotein of a *Zaire ebolavirus* (rVSV-ZEBOV), developed by NewLink Genetics and Merck Vaccines, in conjunction with the Public Health Agency of Canada.[40] Interim results of the phase 3 trials through July 2015 suggested that the rVSV-ZEBOV vaccine may be highly efficacious and safe.[41]

References

1. Centers for Disease Control and Prevention. Advice for Humanitarian Aid Workers Traveling to Guinea, Liberia, or Sierra Leone during the Ebola Outbreak. 2014. Available at: http://wwwnc.cdc.gov/travel/page/humanitarian-workers-ebola (accessed on May 3, 2016).
2. Centers for Disease Control and Prevention. Information for Healthcare Workers and Settings. 2014. Available at: http://www.cdc.gov/vhf/ebola/hcp/ (accessed on May 3, 2016).
3. Centers for Disease Control and Prevention. Guidance on Air Medical Transport for Patients with Ebola Virus Disease. 2014. Available at: http://www.cdc.gov/vhf/ebola/hcp/guidance-air-medical-transport-patients.html (accessed on May 3, 2016).
4. Centers for Disease Control and Prevention. Interim Guidance for Specimen Collection, Transport, Testing, and Submission for Persons Under Investigation for Ebola Virus Disease in the United States. 2014. Available at: http://www.cdc.gov/vhf/ebola/hcp/interim-guidance-specimen-collection-submission-patients-suspected-infection-ebola.html (accessed on May 3, 2016).
5. Centers for Disease Control and Prevention. Guidance for Safe Handling of Human Remains of Ebola Patients in U. S. Hospitals and Mortuaries. 2014. Available at: http://www.cdc.gov/vhf/ebola/hcp/guidance-safe-handling-human-remains-ebola-patients-us-hospitals-mortuaries.html (accessed on May 3, 2016).
6. Centers for Disease Control and Prevention. Ebola Guidance for Airlines – Interim Guidance about Ebola Infection for Airline Crews, Cleaning Personnel, and Cargo Personnel. 2014. Available at: http://www.cdc.gov/quarantine/air/managing-sick-travelers/ebola-guidance-airlines.html (accessed on May 3, 2016).
7. Centers for Disease Control and Prevention. Interim Guidance for Managers and Workers Handling Untreated Sewage from Individuals with Ebola in the United States. 2014. Available at: http://www.cdc.gov/vhf/ebola/prevention/handling-sewage.html (accessed on May 3, 2016).
8. World Health Organization. Marburg Hemorrhagic Fever – Fact Sheet. 2005. Available at: http://www.who.int/csr/disease/marburg/factsheet/en// (accessed on May 3, 2016).
9. Gatherer D. The 2014 Ebola virus disease outbreak in West Africa. *J Gen Virol* 2014;95:1619–1624.
10. Centers for Disease Control and Prevention. 2014. Filoviridae. Available at: http://www.cdc.gov/vhf/virus-families/filoviridae.html (accessed on May 3, 2016).
11. Centers for Disease Control and Prevention. Marburg Hemorrhagic Fever (Marburg HF). 2014. Available at: http://www.cdc.gov/vhf/marburg/ (accessed on May 4, 2016).
12. Sanchez A, Kiley MP. Identification and analysis of Ebola virus proteins. *Virology* 1987;157:414.
13. Ansari AA. Clinical features and pathobiology of Ebolavirus infection. *J Autoimmun* 2014;55C:1–9.
14. Fauci AS. Ebola – underscoring the global disparities in health care resources. *N Engl J Med* 2014;371:1084e6.
15. World Health Organization. Ebola Virus Disease Outbreak. Available at: http://www.who.int/csr/disease/ebola/en/ (accessed on May 4, 2016).
16. World Health Organization. Ebola Response Roadmap – Situation Report. 2014. Available at: http://www.who.int/csr/disease/ebola/situation-reports/en/ (accessed on May 3, 2016).
17. Lyon GM, Mehta AK, Varkey JB, et al. Clinical care of two patients with Ebola virus disease in the United States. *N Engl J Med* 2014;371(25):2402–2409.
18. Buchanan L, Copeland B, Yourish K, et al. Retracing the Steps of the Dallas Ebola Patient. New York Times, October. 8, 2014. Available at: http://www.nytimes.com/interactive/2014/10/01/us/retracing-the-steps-of-the-dallas-ebola-patient.html?_r=0 (accessed on June 27, 2016).
19. Towner JS, Rollin PE, Bausch DG, et al. Rapid diagnosis of Ebola hemorrhagic fever by reverse transcription-PCR in an outbreak setting and assessment of patient viral load as a predictor of outcome. *J Virol* 2004;78(8):4330–4341.
20. Ksiazek TG, Rollin PE, Williams AJ, et al. Clinical virology of Ebola hemorrhagic fever (EHF): virus, virus antigen, and IgG and IgM antibody findings among EHF patients in Kikwit, Democratic Republic of the Congo, 1995. *J Infect Dis* 1999;179(Suppl 1):S177–S187.
21. Jones RM, Brosseau LM. Commentary: Ebola virus transmission via contact and aerosol – a new paradigm. 2014. Available at: http://www.cidrap.umn.edu/news-perspective/

2014/11/commentary-ebola-virus-transmission-contact-and-aerosol-new-paradigm (accessed on May 3, 2016).
22. Schieffelin JS, Shaffer JG, Goba A, et al. Clinical illness and outcomes in patients with Ebola in Sierra Leone. *N Engl J Med* 2014;371(22):2092–2100.
23. Varkey JB, Shantha JG, Crozier I, et al. Persistence of Ebola Virus in ocular fluid during convalescence. *N Engl J Med* 2015; 372(25):2423–7. Available at: http://www.nejm.org/doi/full/10.1056/NEJMoa1500306 (accessed on June 27, 2016).
24. Deen GF, Knust B, Broutet N, et al. Ebola RNA persistence in semen of Ebola virus disease survivors – preliminary report. *N Eng J Med* 2015. Available at: http://www.nejm.org/doi/full/10.1056/NEJMoa1511410 (accessed on June 27, 2016). 10.1056/NEJMoa1511410.
25. World Health Organization. Interim Advice on the Sexual Transmission of the Ebola Virus Disease. Available at: http://www.who.int/reproductivehealth/topics/rtis/ebola-virus-semen/en/ (accessed on May 3, 2016).
26. World Health Organization. Marburg Virus Disease. Available at: http://www.who.int/csr/disease/marburg/en/ (accessed on May 3, 2016).
27. World Health Organization. Joint WHO/FIND Meeting on Diagnostics and Ebola Control. Available at: http://www.who.int/medicines/ebola-treatment/meetings/2015-0123_EbolaDxMtg_reportDec2014_Final.pdf (accessed on June 27, 2016).
28. Centers for Disease Control and Prevention. Ebola Virus Disease Information for Clinicians in U.S. Healthcare Settings. 2015. Available at: http://www.cdc.gov/vhf/ebola/healthcare-us/preparing/clinicians.html (accessed on May 3, 2016).
29. Kreuels B, Wichmann D, Emmerich P, et al. A case of severe Ebola virus infection complicated by gram-negative septicemia. *N Engl J Med* 2014;371(25):2394–2401.
30. Johansen LM, DeWald LE, Shoemaker CJ, et al. A screen of approved drugs and molecular probes identifies therapeutics with anti-Ebola virus activity. *Sci Transl Med* 2015;7(290):290ra89.
31. Centers for Disease Control. Fact Sheet: Screening and Monitoring Travelers to Prevent Spread of Ebola. 2015. Available at: http://www.cdc.gov/vhf/ebola/travelers/ebola-screening-factsheet.html (accessed on May 3, 2016).
32. Centers for Disease Control. Fact Sheet: Screening and Monitoring Travelers to Prevent Spread of Ebola. 2015. Available at: http://www.cdc.gov/vhf/ebola/travelers/ebola-screening-factsheet.html (accessed on May 3, 2016).
33. Centers for Disease Control and Prevention. Interim Guidance for Emergency Medical Services (EMS) Systems and 9-1-1 Public Safety Answering Points (PSAPs) for Management of Patients Under Investigation (PUIs) for Ebola Virus Disease (EVD) in the United States. 2015. Available at: http://www.cdc.gov/vhf/ebola/healthcare-us/emergency-services/ems-systems.html (accessed on May 3, 2016).
34. Centers for Disease Control and Prevention. Identify, Isolate, Inform: Emergency Department Evaluation and Management of Patients with Possible Ebola Virus Disease. Available at: http://www.cdc.gov/vhf/ebola/pdf/ed-algorithm-management-patients-possible-ebola.pdf (accessed on May 3, 2016).
35. Occupational Safety and Health Administration. Ebola – Control and Prevention. Available at: https://www.osha.gov/SLTC/ebola/control_prevention.html (accessed on May 3, 2016).
36. Centers for Disease Control and Prevention. Guidance on Personal Protective Equipment To Be Used by Healthcare Workers During Management of Patients with Ebola Virus Disease in U.S. Hospitals, Including Procedures for Putting On (Donning) and Removing (Doffing). 2014. Available at: http://www.cdc.gov/vhf/ebola/hcp/procedures-for-ppe.html (accessed on May 3, 2016).
37. Centers for Disease Control and Prevention. Guidance on Personal Protective Equipment To Be Used by Healthcare Workers During Management of Patients with Ebola Virus Disease in U.S. Hospitals, Including Procedures for Putting On (Donning) and Removing (Doffing). 2014. Available at: http://www.cdc.gov/vhf/ebola/healthcare-us/ppe/guidance.html (accessed on May 3, 2016).
38. Occupational Safety and Health Administration. OSHA Fact Sheet – Cleaning and Decontamination of Ebola on Surfaces – Guidance for Workers and Employers in Non-Healthcare/Non-Laboratory Settings. Available at: https://www.osha.gov/Publications/OSHA_FS-3756.pdf (accessed on May 4, 2016).
39. Swift M, Hudson TW, Behrman A, et al. American College of Occupational and Environmental Medicine Guidelines for the Medical Clearance of Designated Ebola Caregivers in US Hospitals. Available at: http://www.acoem.org/uploadedFiles/Public_Affairs/Policies_And_Position_Statements/Guidelines/Position_Statements/Medical%20Clearance%20Guidelines%20for%20Ebola%20Care%20Teams%20in%20US%20Hospitals.pdf (accessed on June 27, 2016).
40. World Health Organization. Essential Medicines and Health Products: Vaccines. Available at: http://www.who.int/medicines/ebola-treatment/emp_ebola_vaccines/en/ (accessed on May 3, 2016).
41. Henao-Restrepo AM, Longini IM, Egger M, et al. Efficacy and effectiveness of an rVSV-vectored vaccine expressing Ebola surface glycoprotein: interim results from the Guinea ring vaccination cluster-randomised trial. *Lancet* 2015. Available at: http://www.thelancet.com/pdfs/journals/lancet/PIIS0140-6736(15)61117-5.pdf (accessed on June 27, 2016).

HANTAVIRUSES

Common names for disease: Hemorrhagic fever with renal syndrome, nephropathia epidemica, hantavirus pulmonary syndrome (HPS)

Classification: Family—Bunyaviridae; genera—*Bunyavirus, Phlebovirus, Nairovirus, Tospovirus, Hantavirus*

Occupational setting

Outdoor workers in endemic areas may be at risk from exposure. The risk may be greater for pest/rodent control workers as well as construction and utility workers. Laboratory workers have also become infected while working with infected wild and laboratory rodents.[1–3] In 2012, there was an outbreak of hantavirus in Yosemite National Park where 10 visitors contracted the virus while staying in the tent cabins. The staff maintaining the tent cabins (including laborers, electricians, hospitality, and housekeeping workers) was evaluated for exposures but there were no confirmed occupational cases.[4]

Exposure (route)

Rodents identified as harboring hantavirus include deer mouse, white-footed mouse, striped field mouse, yellow-necked field mouse, cotton rat, rice rat, brown or Norway rat, bank vole, and meadow vole.[5–7] Most of the rodent reservoirs are prevalent in rural settings.[1] Infection occurs when the rodent urine, saliva, vomit, or feces are stirred up, creating tiny droplets that become airborne. As a result, there is either direct aerosolization of the infected rodent droppings or secondary aerosolization of dried rodent excreta. Rodent bites have resulted in disease transmission. Person-to-person transmission is rare.[1,8]

Pathobiology

Hantaviruses are negative-sense, single-stranded, enveloped RNA viruses whose genome is made up of three RNA segments: S (small), M (medium), and L (large).

In general, the hantaviruses fall into two groups: old world and new world. There are numerous subgroups. The old-world group is primarily associated with hemorrhagic fever with renal syndrome (HFRS), while the new-world group express as the hantavirus pulmonary syndrome (HPS).

Hemorrhagic fever with renal syndrome (HFRS)

Hemorrhagic fever with renal syndrome (HFRS) is a group of clinically similar illnesses including diseases such as Korean hemorrhagic fever, epidemic hemorrhagic fever, and nephropathia epidemica.[7] The viruses that cause HFRS include Hantaan, Dobrava, Saaremaa, Seoul, and Puumala. Hantaan virus is widely distributed in eastern Asia, particularly in China, Russia, and Korea.[7] Puumala virus is found in Scandinavia, western Europe, and western Russia. Dobrava virus is found primarily in the Balkans, and Seoul virus is found worldwide. Saaremaa is found in central Europe and Scandinavia.[7] The striped field mouse (*Apodemus agrarius*) is the reservoir for both the Saaremaa and Hantaan virus, the brown or Norway rat (*Rattus norvegicus*) is the reservoir for Seoul virus, the bank vole (*Clethrionomys glareolus*) is the reservoir for Puumala virus, and the yellow-necked field mouse (*Apodemus flavicollis*) carries Dobrava virus.[7]

The incubation period for HFRS is 12–21 days, with a mild, nonspecific prodrome followed by the febrile, hypotensive, oliguric, diuretic, and convalescent phases. The febrile phase may include chills, high fever, lethargy, dizziness, headache, photophobia, myalgia, abdominal and back pain, anorexia, nausea, and vomiting. Most patients experience a slow, uneventful recovery. Others go on to develop the classical symptoms of hypotension, hemorrhage, and renal failure. Estimates of the case fatality rate range from 1 to 15%.[7] Vascular dysfunction with impaired vascular tone and increased vascular permeability produce hypotension and shock. Hemorrhage may result from disseminated intravascular coagulopathy. Renal dysfunction results from antibody complex deposition, as opposed to a direct effect of the virus.[9] In parts of Europe (including Sweden), the Puumala hantavirus causes nephropathia epidemica (NE), a milder form of HRFS. The clinical course of NE begins with sudden onset with fever, headache, back pain, and gastrointestinal symptoms, with severe complications and even death being reported.[10] The case fatality rate is 0.1–1%.[11]

Hantavirus pulmonary syndrome (HPS)

Hantavirus disease has been identified in various sites in the Americas. Before 1993, no cases of human disease due to hantaviruses had been reported in the United States. In 1993, an outbreak of an acute illness occurred initially in the Four Corners region of the southwestern United States. It was characterized by the abrupt onset of fever, myalgias, headache, and cough, followed by the rapid development of respiratory failure. This syndrome, now known as hantavirus pulmonary syndrome (HPS), was caused by the Sin Nombre virus (SNV) and is now known to be the predominant cause of HPS in the United States.[12] SNV is found in the urine, feces, and saliva of the deer mouse (*Peromyscus maniculatus*), predominant in the central and western part of the United States. HPS has a 1–5-week incubation. The onset is nonspecific with fever and myalgias. Patients may also experience headaches, nausea, vomiting, and diarrhea. Cough, tachypnea, and tachycardia do not usually begin until day 7.[13] Rapid progression to dyspnea, pulmonary edema, shock, and death may occur. Hemorrhage and renal failure are not usual components of the syndrome.[4,13,14]

The New York hantavirus is harbored in the white-footed mouse, which is found in the northeastern United States. The natural host for the Black Creek hantavirus is the cotton rat, prevalent in the southeastern United States.

HPS is more common in South America than in North America with cases identified in Argentina, Chile, Uruguay, Paraguay, Brazil, and Bolivia. Andes virus causes HPS in Argentina and Chile and is the only hantavirus known to have been transmitted from person to person.[15] Andes, Bermejo, Hu39694, Lechiguanas, Maciel, Oran, and Pergamino viruses have been linked to HPS cases in Argentina. Bermejo and Laguna Negra viruses cause HPS in Bolivia, and Laguna Negra virus is also linked to HPS in Paraguay. Araraquara, Castelo dos Sonhos, and Juquitiba viruses have been associated with HPS in Brazil.[15]

A 1999 outbreak in Panama marked the first cases of HPS identified in Central America. This outbreak led to the identification of another hantavirus, Choclo virus, which is associated with the rodent host *Oligoryzomys fulvescens*. The broad geographic distribution of sigmodontine rodents suggests that human cases of hantavirus infections will eventually be identified from all countries in the Americas.[15]

Diagnosis

During the prodrome, determination of hantavirus infection is difficult, but soon after presentation, shortness of breath and a cough may be present. The index of suspicion is raised by myalgias of large muscle groups plus gastrointestinal symptoms such as nausea, vomiting, and abdominal pain. Interstitial edema is frequently present on chest X-ray. Bilateral alveolar edema develops later, and pleural effusions may be seen. Specific diagnosis is made with IgM testing of acute-phase serum. Sin Nombre virus antigen, a common subgroup, will cross-react with other hantaviruses that cause HPS in the Americas.[16]

Treatment

Treatment consists primarily of supportive care. This may include mechanical ventilation, renal dialysis, and transfusions, as appropriate. Ribavirin has been used to treat HRFS, although initial studies have not shown a dramatic response.[4,17] Extracorporeal mechanical ventilation has been implemented to treat HPS patients.[18]

Prevention

Control of rodent populations in endemic areas is helpful but may not be practicable. Rodent nests and dead rodents can be wetted with disinfectant prior to removal. Activities that may result in contact with rodents or aerosolization of rodent excreta should be avoided. For workers involved in animal control activities in endemic areas, work practices and personal protective equipment should protect from exposure by direct skin contact and inhalation. In the healthcare setting, universal precautions are considered adequate to prevent exposure. In the research setting, utilizing BSL-2 and BSL-3 facilities is critical. There is no vaccine available at this time.

References

1. Hart CA, Bennett M. Hantavirus infections: epidemiology and pathogenesis. *Microbes Infect* 1999;1:1229–1237.
2. Desmyter J, LeDuc JW, Johnson KM, et al. Laboratory rat associated outbreak of haemorrhagic fever with renal syndrome due to Hantaan-like virus in Belgium. *Lancet* 1983;2:1445–1448.
3. Lloyd G, Jones M. Infection of laboratory workers with hantavirus acquired from immunocytomas propagated in laboratory rats. *J Infect* 1986;12:117–125.
4. California Department of Public Health, Occupational Health Branch. Work-Related Hantavirus Exposures at Yosemite National Park: Key Findings and Recommendations. 2013. Available at: http://www.cdph.ca.gov/programs/ohb/Documents/HantaRept.pdf (accessed on May 3, 2016).
5. Centers for Disease Control and Prevention. Facts about Hantavirus – What You Need to Know to Prevent the Disease Hantavirus Pulmonary Syndrome. Available at: http://www.cdc.gov/hantavirus/pdf/hps_brochure.pdf (accessed on May 3, 2016).
6. Centers for Disease Control and Prevention. Virology – Hantaviruses. 2012. Available at: http://www.cdc.gov/hantavirus/technical/hanta/virology.html (accessed on May 3, 2016).
7. Centers for Disease Control and Prevention. Hemorrhagic Fever with Renal Syndrome. 2011. Available at: http://www.cdc.gov/hantavirus/hfrs/index.html (accessed on May 4, 2016).
8. Centers for Disease Control. Outbreak of acute illness—Southwestern United States, 1993. *MMWR* 1993;42:421–424.
9. Cosgriff TM. Mechanism of disease in hantavirus infection: pathophysiology of hemorrhagic fever with renal syndrome. *Rev Infect Dis* 1991;13:97–107.
10. Hjertqvist M, Klein SL, Ahlm C, et al. Mortality rate patterns for hemorrhagic fever with renal syndrome caused by Puumala virus. *Emerg Infect Dis* 2010;16(10):1584–1586.
11. Vapalahti O, Mustonen J, Lundkvist A, et al. Hantavirus infections in Europe. *Lancet Infect Dis* 2003;3(10):653–661.
12. Centers for Disease Control and Prevention. How People Get Hantavirus Pulmonary Syndrome. 2012. Available at: http://www.cdc.gov/hantavirus/hps/transmission.html (accessed on May 3, 2016).
13. Centers for Disease Control and Prevention. HPS Technical/Clinical Information: Clinical Manifestation. Available at: http://www.cdc.gov/hantavirus/technical/hps/clinical-manifestation.html (accessed on May 3, 2016).
14. Graziano KL, Tempest B. Hantavirus pulmonary syndrome: a zebra worth knowing. *Am Fam Phys* 2002;66(6):1015–1021.
15. Centers for Disease Control and Prevention. HPS Technical/Clinical Information: Ecology. Available at: http://www.cdc.gov/hantavirus/technical/hps/ecology.html (accessed on May 4, 2016).
16. Peters CJ, Simpson GL, Levy H. Spectrum of hantavirus infection: hemorrhagic fever with renal syndrome and hantaviruses pulmonary syndrome. *Ann Rev Med* 1999;50:531–545.
17. Huggins JW, Hsiang CM, Cosgriff TM, et al. Prospective, double-blind, concurrent, placebo controlled clinical trial of intravenous ribavirin therapy for hemorrhagic fever with renal syndrome. *J Infect Dis* 1991;164:119–127.
18. Crowley MR, Katz RW, Kessler R, et al. Successful treatment of adults with severe hantavirus pulmonary syndrome with extracorporeal membrane oxygenation. *Crit Care Med* 1998;26(2):409–414.

HEPATITIS A VIRUS (HAV)

Common name for disease: Infectious hepatitis
Classification: Family—Picornaviridae; genus—*Enterovirus*

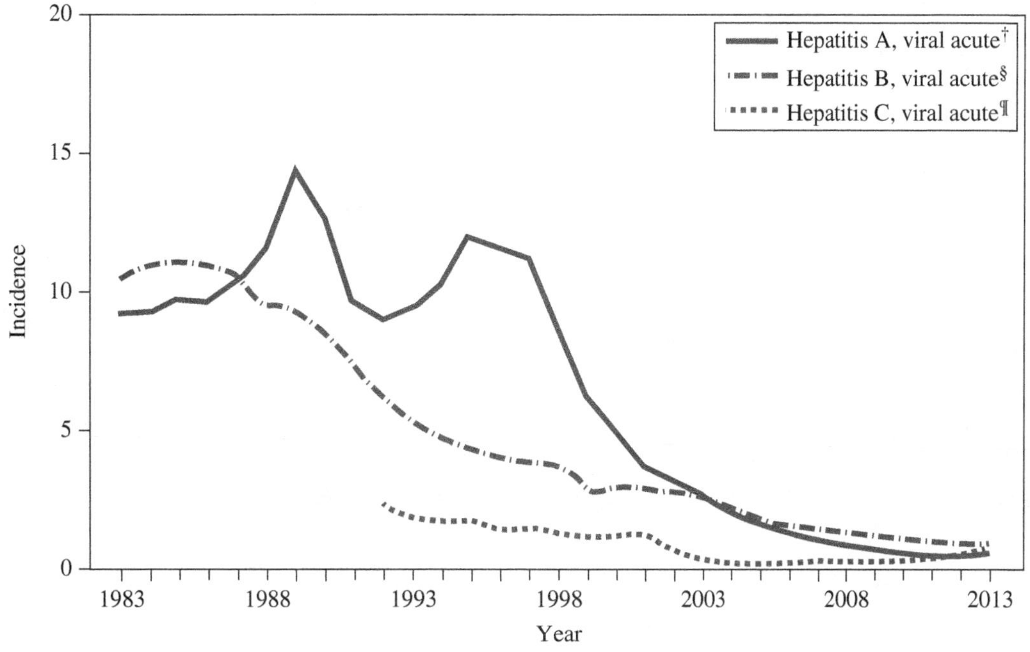

FIGURE 21.6 Viral Hepatitis Incidence per 100 000 population by year — United States, 1983–2013. †Hepatitis A vaccine was first licensed in 1995. §Hepatitis B vaccine was first licensed in June 1982. ¶An anti-hepatitis C virus (HCV) antibody test first became available in May 1990. Hepatitis A incidence declined during 1998–2011 and increased in 2012 and 2013. The hepatitis A vaccine became available in 1995, the last year a peak in incidence of acute, symptomatic hepatitis A was observed. Coinciding with the implementation of the national vaccination strategy to eliminate hepatitis B infections, the incidence of acute hepatitis B has declined since 1987. Acute hepatitis B incidence has remained stable since 2008. The incidence of acute hepatitis C remained fairly stable during 1992–2000, declined in 2001 and 2002, remained stable during 2003–2005, and increased in 2011, 2012, and 2013. Recent investigations suggest this increase is largely driven by acute infections in nonurban young persons who start injecting drugs after habituation to oral prescription opioid drugs such as "OxyContin" and oxycodone. *Source*: MMWR 2015; 62(53):77.

Occupational setting

Workers in food handling, day-care centers, and healthcare institutions, while they may be implicated in localized outbreaks given the nature of their jobs, do not display increased prevalence rates of HAV infections.[1,2] Healthcare workers do not have an increased prevalence of HAV infections, but inpatient outbreaks have been observed in neonatal intensive care units and in association with adult fecal incontinence.[3] Epidemiologic studies in healthcare workers suggest that younger workers are susceptible to HAV infection.[2] An increased risk for workers exposed to sewage has been suggested by some earlier studies, but not confirmed.[4] However, given the low overall anti-HAV antibody seroprevalence among sewage and wastewater workers (including younger workers) and the inherent fecal exposure that occurs during the course of work suggest that sewage workers are at risk for occupational exposure to HAV.[2]

Work with nonhuman primates carries an increased risk. Travelers to areas of high endemicity may also have an increased risk of infection, primarily through ingestion of contaminated food or water.

Exposure (route)

Infection is transmitted almost exclusively by the fecal–oral route. Transmission by saliva or blood transfusion is possible but uncommon.

Pathobiology

HAV is a nonenveloped RNA picornavirus that is stable in most environments but can be inactivated by high temperature (185°F (85°C) or higher), formalin, and chlorine.[3] Pathogenesis begins with entry into the mouth, where the virus goes to the liver to begin replication. The incubation period of hepatitis A is approximately 28 days (range 15–50 days). The symptoms of illness include fever,

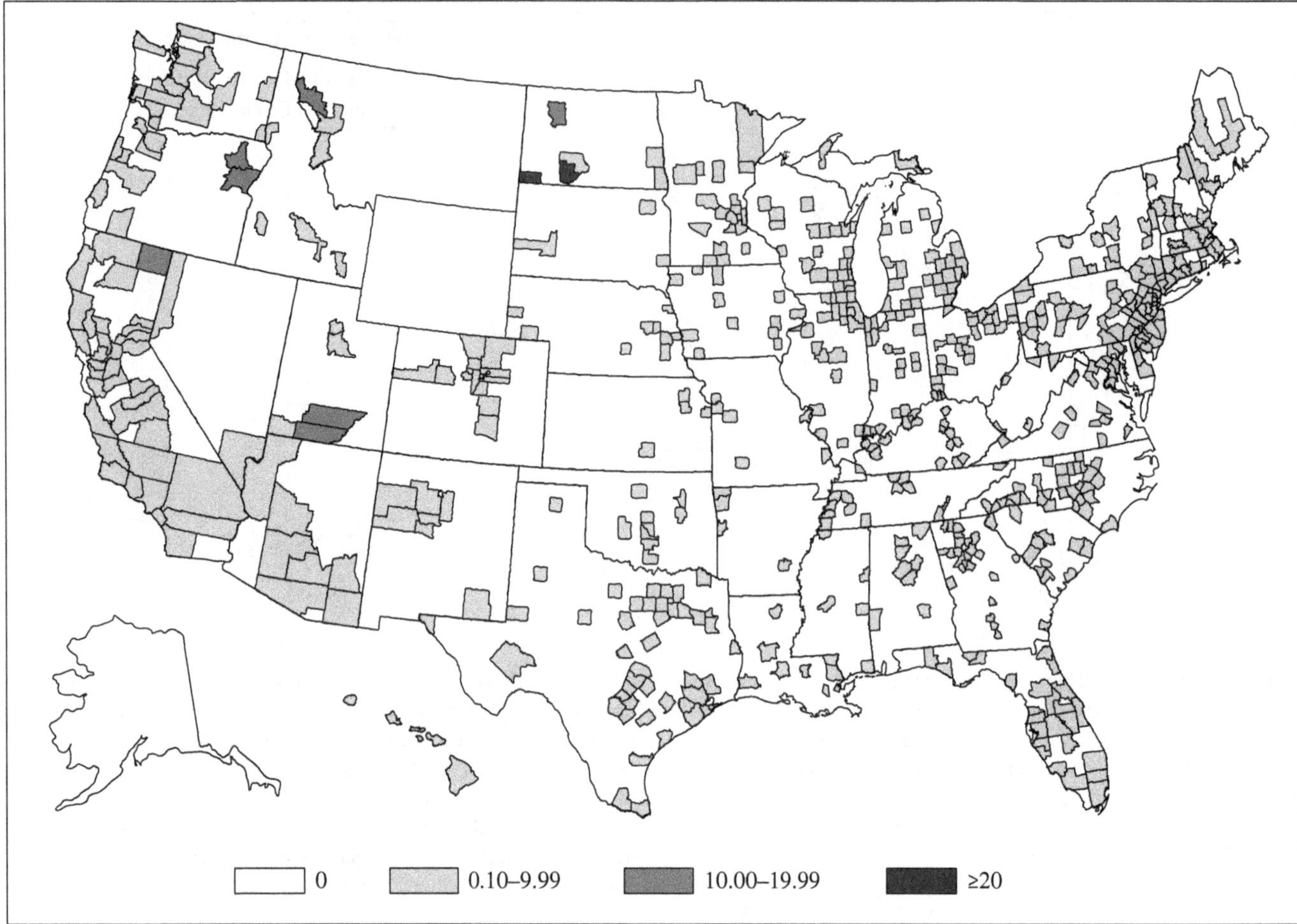

FIGURE 21.7 Hepatitis A Incidence by county — United States, 2013. Although effective vaccines to prevent Hepatitis A virus infections have been available in the United States since 1995, cases still occur in almost every state. In 2013, a total of 1,781 cases were reported and 17 counties in 13 states reported incidence rates of >10 cases per 100,000 population. Two of these counties in one state reported incidence rates of >20 cases. *Source*: MMWR 2015; 62(53):784.

malaise, anorexia, nausea, abdominal discomfort, dark urine, and jaundice. After 10–12 days, HAV is present in the bloodstream, with peak titers occurring during the 2 weeks before illness onset.[3] HAV is then excreted via the biliary system into the stool. Transmissibility is greatest 1–2 weeks before the onset of illness, when HAV concentration in stool is highest. Viral shedding in feces continues for up to 3 weeks after symptoms resolve. Fulminant disease is rare, with an estimated case fatality rate of 0.3% (1.8% for adults over 50).[4] There is occasional recurrence, but no recognized chronic or carrier state.

The incidence of hepatitis A has declined by 95% with the advent of the hepatitis A vaccine in the mid-1990s (Figure 21.6) (see "Prevention" section below for more information on hepatitis A vaccine).[5] In 2013, there were 1781 symptomatic cases of hepatitis A reported and 3473 total estimated cases (including adjustment for asymptomatic infection and underreporting).[6] Even with this decrease, outbreaks continue to occur in almost all states across the United States (Figure 21.7).

Diagnosis

Clinical features plus elevated alkaline aminotransferase (ALT), aspartate aminotransferase (AST), and bilirubin support the diagnosis. Elevated anti-HAV IgM antibody confirms acute infection, is reliably present at the onset of symptoms, and can persist for up to 6 months.

Treatment

Supportive care includes fluid replacement (oral and in severe dehydration cases intravenous) and electrolyte replacement if indicated.

Medical surveillance

National surveillance for acute hepatitis A is through the CDC's National Notifiable Diseases Surveillance System. At the local/institutional level, surveillance of exposed susceptible workers may be indicated during the incubation period.

Prevention

There are two single-antigen hepatitis A vaccines (Havrix and Vaqta) and one combined hepatitis A–B vaccine (Twinrix) currently licensed in the United States. All three of these vaccines are inactivated. Havrix and Vaqta are both a two-dose vaccine, with the second vaccination recommended at 6–12 months for added immunity. Twinrix is offered on either a three-dose (at 1 month, 6 months, and 12 months) or expedited four-dose (0, 7 days, 21–30 days, and 12 months) schedule. Based on a recent panel evaluation, the protective levels of antibody to HAV could last for at least 25 years in adults and at least 14–20 years in children; thus checking postvaccination titers for hepatitis A is not indicated.[5] CDC recommends hepatitis A vaccine for persons at increased risk of exposure (i.e., during a community-wide hepatitis A outbreak). CDC recommends vaccination for those who work with HAV-infected primates or in a laboratory conducting HAV research.

Business travelers or other individuals planning to work for extended periods of time in developing countries where sanitation status may be of concern should avoid local water (for drinking or food preparation), milk, shellfish, and uncooked fruits and vegetables. In all settings, thorough hand washing with running water and single-use towels are essential after potentially infectious contact and before handling food. Susceptible travelers or expatriates to locations of intermediate or high endemicity should receive a first dose of hepatitis A vaccine at least 2–4 weeks prior to departure. The Advisory Committee on Immunization Practices (ACIP) has recently amended its guidelines for hepatitis A vaccination for travelers, emphasizing that one dose of single-antigen hepatitis A vaccine administered at any time before departure may provide adequate protection for most healthy persons under the age of 40.[5] However, those individuals with preexisting chronic medical conditions above the age of 40 who are planning to depart in less than 2 weeks to a developing country should receive the initial dose of vaccine and immune globulin (IG) (0.02 mL/kg) at a separate anatomic injection site.

Comprehensive sanitation and hygiene measures remain of great importance. Enteric infection control procedures are crucial for occupations at risk, since transmission occurs primarily during the presymptomatic incubation phase. HAV retains its infectivity on surfaces from days to months depending on the environmental conditions. Moreover, HAV is effectively transmitted on human hands and resists many common disinfectants.[7] Effective decontaminating agents for environmental surfaces include quaternary ammonium compounds with 23% HCl, 2% glutaraldehyde, or hypochlorite with >5000 ppm free chlorine.[8]

For susceptible persons, postexposure prophylaxis may be up to 85% effective in preventing clinical hepatitis A if administered up to 2 weeks after exposure. There is no apparent harm in administering hepatitis A vaccine or IG to persons with existing immunity (e.g., from prior undiagnosed infection). Therefore, any decisions about serologic testing before immunization should be based on projected cost reduction. The vaccine should not be used without IG for postexposure prophylaxis. Previously unvaccinated households and sexual contacts of persons with serologically confirmed hepatitis A are candidates for postexposure prophylaxis.

References

1. Forbes A, Williams R. Changing epidemiology and clinical aspects of hepatitis A. *Br Med Bull* 1990;46:303–318.
2. Keeffe EB. Occupational risk for hepatitis A – a literature-based analysis. *J Clin Gastroenterol* 2004;38(5):440–448.
3. Centers for Disease Control and Prevention. *Hepatitis A: The Pink Book*, 12th edn, 2012. Available at: http://www.cdc.gov/vaccines/pubs/pinkbook/downloads/hepa.pdf (accessed on May 3, 2016).
4. Bell BP, Wasley A, Shapiro CN, et al. Prevention of hepatitis A through active or passive immunization: recommendations of the Advisory Committee on Immunization Practices. *MMWR* 1999;48:1–37.
5. Centers for Disease Control and Prevention. Hepatitis A FAQs for health professionals. 2015. Available at: http://www.cdc.gov/hepatitis/HAV/HAVfaq.htm (accessed on May 4, 2016).
6. Centers for Disease Control and Prevention. Viral Hepatitis: Statistics and Surveillance. 2015. Available at: http://www.cdc.gov/hepatitis/statistics/2013surveillance/index.htm (accessed on May 4, 2016).
7. Mbithi JN, Springthorpe VS, Boulet JR, et al. Survival of hepatitis A on human hands and its transfer on contact with animate and inanimate surfaces. *J Clin Microbiol* 1992;30:757–763.
8. Mbithi JN, Springthorpe VS, Satter SA. Chemical disinfection of hepatitis A virus on environmental surfaces. *Appl Environ Microbiol* 1990;56:3601–3604.

HEPATITIS B VIRUS (HBV)

Common name for disease: Serum hepatitis
Classification: Family—Hepadnaviridae; genus—Orthohepadnavirus Type 1

Occupational setting

Healthcare workers are at increased risk of infection if they are exposed to blood products, especially surgeons, pathologists, dentists and dental hygienists, personnel in emergency departments, personnel in labor and delivery, laboratory technicians and researchers, and hemodialysis staff. Other occupations with increased risk include hospital custodial staff, morticians, and employees of institutions that care for the mentally disabled. From 1983 to 2010, the number of HBV infections among healthcare workers has been estimated to have declined from approximately 17 000 infections to 263 acute HBV infections, a result of routine pre-exposure hepatitis B vaccination required by the OSHA Bloodborne Pathogen standard and improvements in infection control practices.[1,2] From 2008 to 2013, there have been 20 identified healthcare-associated hepatitis B outbreaks, with 15 of the 20 occurring in long-term care facilities and 13 of these 15 associated with infection control protocol breaches.[3] The prevalence of HBV surface antigen (HBsAg; an antigenic determinant found on the surface of the virus) in the serum of healthcare workers with no or infrequent blood contact is 0.3% (the same prevalence rate as in general population) and in those healthcare workers with frequent blood contact it is 1–2%.[4] In the staff of mental health institutions, the prevalence of HBsAg is 1%.[4] In the general US population, reported clinical cases of acute HBV infection have decreased from a peak of 26 611 cases in 1985 to 3050 cases in 2013 (Figure 21.6).[5] However, the prevalence of chronic HBV in the United States has remained stable over the last decade at 3 per 100 000 cases, with an estimated 700 000–1.4 million persons having chronic HBV.[6]

Worldwide, more than one third of the population has been infected with the hepatitis B virus. About 5% are chronic carriers, and nearly 25% of all carriers develop serious liver diseases including chronic hepatitis, cirrhosis, and primary hepatocellular carcinoma.[7]

Exposure (route)

Transmission occurs by parenteral, sexual, and perinatal routes. Blood, serum, saliva, semen, and vaginal fluid are known to carry HBV. The risk of HBV infection following a needlestick to both HBsAg-positive and hepatitis B e antigen (HbeAg; a marker of HBV infectivity and viral load)-positive blood was estimated to be 22–31%, while the risk for HBV infection following a needlestick to HBsAg-positive, HBeAg-negative blood was 1–6%.[8] Contact of infectious material with skin breaks or mucous membranes, including the eye, can result in infection. There is no evidence for fecal–oral transmission. Transmission via oral secretions is rare (in one recorded case, disease was transmitted via a bite from an infected patient).[9] Indirect transmission of HBV infection can occur from blood-contaminated instruments and other objects.[10] Transmission of HBV from healthcare workers (usually chronic carriers) to patients is rare, with suspected or known cases having been associated with invasive procedures, either those with high risk of sharps injury or when failure to wear gloves on hands led to some compromise of skin barrier integrity (cuts, dermatoses, bleeding warts). Although transmission risk is low with proper infection control technique, it has been documented in two cases where transmission from infected surgeons to patients occurred despite double gloving and elimination of certain high-risk procedures.[11,12] Chronic hepatitis B infection should not prevent the practice or study of medicine, surgery, dentistry, or any other health profession.[13] Healthcare workers who are HBsAg positive must be carefully evaluated prior to starting employment, including clear identification of expected job tasks. These workers should be counseled on how to prevent HBV transmission to patients. Those who perform exposure-prone procedures should be advised regarding the procedures they can perform safely, based on the latest CDC recommendations for the management of HBsAg-positive healthcare providers and students.[13]

Pathobiology

HBV is a double-shelled DNA virus. HBV contains numerous antigenic components, including HBsAg, HBeAg, and hepatitis B core antigen (HBcAg; the nucleocapsid protein core of HBV, is undetectable in serum but present in liver tissue in those with acute or chronic HBV infection). All three antigens are immunologically distinct.[4,14]

Following infection, the incubation period ranges from 45 to 160 days with an average incubation of 120 days.[4] There are three phases in the acute stage: (i) preicteric, (ii) icteric, and (iii) recovery. The preicteric phase is characterized by malaise, anorexia, nausea, vomiting, right upper quadrant abdominal pain, fever, headache, myalgia, skin rashes, arthralgia and arthritis, and dark urine. It can last from 3 to 10 days. The serum transaminases AST (SGOT) and ALT (SGPT) start to rise in this phase.

The icteric phase is variable but usually lasts from 1 to 3 weeks and is characterized by jaundice, gray stools, hepatic tenderness, and hepatomegaly. Only 20–50% of cases of acute HBV infections manifest clinical signs of jaundice. The characteristic hepatic pathology of acute infection includes panlobular, lymphocytic, hepatocellular necrosis, and cholestasis. During recovery, HBsAg is eliminated from the blood and the production of surface antibody (anti-HBs) ensues, generating lasting immunity against HBV. Malaise and fatigue may persist for weeks or months, while jaundice, anorexia, and other symptoms disappear. Fulminant hepatitis (defined as the rapid development of impaired liver function and progression to liver failure) occurs in about 1–2% of acutely infected HBV patients. In the United States, 200–300 Americans die of fulminant HBV disease each year (case fatality rate 6–93%).[4] Acute HBV infection progresses to chronic disease in 1–12% of adults.[15] The carrier state increases the risk of hepatocellular carcinoma.

Hepatitis delta virus (HDV) may be seen as a coinfection or superinfection with HBV infection and generally increases the severity of disease. HDV is found only in the presence of HBsAg, which it requires for its replication. Prevalence of HDV is low in the US general population but higher in groups with multiple parenteral exposures, such as intravenous drug users and people with multiple blood transfusions.

Diagnosis

Diagnosis of HBV infection is based on serologic testing. While epidemiology, clinical presentation, and laboratory features direct the differential diagnosis to include HBV infection, many cases are subclinical and in some instances occur with no elevation of serum transaminase activity.[14] HBsAg is the first serologic marker in acute infection, with HBeAg appearing shortly after. Anti-HBc (core antibody) develops in all HBV infections after HBsAg appears, suggesting HBV infection at some undefined time in the past. IgM anti-HBc antibody is detectable 4–6 months after onset of illness and is the best serologic marker of acute HBV infection.[4] The presence of anti-HBs following acute HBV infection means that the respective individual has recovered and has developed immunity against reinfection. Such an individual is classified as no longer being infectious only after anti-HBs appears and HBsAg is not detectable. HBsAg should be checked every 1–2 months following an acute infection until it disappears. HBsAg persisting 6 months after the acute infection indicates that a carrier state has developed. Five to ten percent of patients with acute infections become carriers; HbsAg and HBeAg become negative in carriers at rates of 2 and 10% per year, respectively.[14] IgG anti-HBc in the presence of HBsAg and HBeAg is diagnostic of chronic infection or the carrier state. A small but significant percentage of chronic carriers are HBsAg negative and can transmit HBV in donated blood screened for HBsAg.[14] Table 21.1 shows patterns that may be seen with a standard HBV serology screen.

Treatment

There is currently no specific treatment for acute HBV infection. General supportive care is indicated. For chronic HBV infection, guidelines developed by the American Association for the Study of Liver Diseases (AASLD) provide recommendations for evaluation, treatment, and follow-up, as well as which patients are candidates for drug treatment since therapy is suppressive, but does not eradicate HBV.[16] Drugs include antivirals (lamivudine, adefovir, tenofovir, entecavir, telbivudine, Truvada) and interferon (IFN-α or pegylated IFN-α) alone or in combination.[16] Special consideration is needed for patients coinfected with HIV. Patients with chronic HBV who are not already immune should receive hepatitis A vaccine.[16]

Medical surveillance

National surveillance for acute and chronic hepatitis B is done through the CDC's National Notifiable Diseases Surveillance System. At the local/institutional level, individual outbreak investigations are conducted by the respective health agency. Healthcare, emergency medical, and public safety workers who sustain a percutaneous, mucosal, or nonintact skin exposure to HBV-positive blood should be assessed for HBV immunity. Those who are not immune to HBV should be identified quickly and given hepatitis B immune globulin (HBIG) immediately. The hepatitis B vaccine series should be offered if not done previously. The workers should be followed closely per the respective institutional protocol and monitored by occupational health personnel for seroconversion. Systems and databases could ideally be electronically linked to employee health records and should include results that identify hepatitis B vaccine responders and nonresponders and the nature and HBV status of the exposure source and postexposure management.[1]

Prevention

HBV vaccination must, per 29 CFR 1910.1030 of the OSHA statute, be made available to at-risk employees within 10 days of employment unless they are already immune or the vaccine is medically contraindicated.[17] All healthcare workers whose activities involve considerable risk for exposure to blood or body fluids should be vaccinated with the complete hepatitis B three-dose vaccine series.[1] Adults 20 years of age and older should receive either Recombivax HB (Merck) or Engerix-B (GlaxoSmithKline). The usual schedule for adults is two doses separated by no less than 4 weeks and a third dose 4–6 months after the second dose. If an accelerated schedule is needed, the minimum interval between the first two doses is 4 weeks, and the minimum interval between the second and third doses is 8 weeks.[4]

CDC guidelines recommend that employees be tested for adequate titer response after vaccination. Employees who are nonresponders should be revaccinated and retested. Prophylaxis must be considered when a significant exposure has occurred. This may be defined as blood or other body fluid contamination via the percutaneous route (e.g., needlestick or incised, bitten, broken, or otherwise compromised skin) or the permucosal route (splash into eye or mouth). The decision on whether to proceed with prophylaxis will depend on the availability of the source of the blood or body fluid, the HbsAg status of the source person, and the HBV vaccination and vaccine-response status of the exposed person.[1] Table 21.2 summarizes recommendations for specific prophylaxis based on these considerations.

A hepatitis B exposure prevention plan is required to address the potential risk of HBV transmission from patients

TABLE 21.1 Hepatitis B Serologic Patterns.

HBsAg	Anti-HBs	HBeAg	Anti-HBe	Anti-HBc	Interpretation
+	–	–	–	–	Late incubation or early acute HBV, infective
+	–	+	–	–	Early acute HBV, infective
+	–	+	–	IgM	Acute HBV, infective
–	–	–	–	IgM	Serologic window(late acute HBV), infective
–	–	–	+	+	Late acute HBV, low infectivity
–	+	–	+	IgG	Recovery from HBV, not infective, immune
–	+	–	–	IgG	Recovery from HBV not infective, immune
+	–	+	–	IgG	Chronic HBV (carrier), infective
+	–	–	+	IgG	Late acute HBV or chronic HBV, low infectivity
–	–	–	–	IgG	Low-level HBsAg carrier (possibly infective)
–	+	–	–	–	Active immunization (vaccination) or passive immunization (post-HB immunoglobulin infusion) or remote past infection

to workers. The OSHA Bloodborne Pathogen standard (29 CFR 1910.1030) mandates that universal blood and body fluid precautions be implemented as part of a written Exposure Control Plan by any employer having employees with "any reasonably anticipated" occupational exposure.[17] Occupational exposure is defined as "skin, eye, mucous membrane, or parenteral contact with blood or other potentially infectious materials that may result from the performance of an employee's duties." Other potentially infectious material includes semen, vaginal secretions, saliva where dental procedures are involved, unfixed tissues or organs, and cerebrospinal, synovial, pleural, pericardial, and amniotic fluids. When a fluid is contaminated with blood, or where it is difficult or impossible to differentiate between body fluids, the material should be considered potentially infectious. The current available data do not justify automatic exclusion of HBV carriers from healthcare duties, because transmission requires direct parenteral or mucosal membrane exposure to HBV. Although HBV carriers who are HBeAg negative are considered less infectious than those who are HBeAg positive, there have been well-documented cases of transmission during surgery.[18] Thus all healthcare workers who are HbsAg positive and potentially perform exposure-prone procedures must be evaluated regarding risk. Testing for viral load of those who are HbeAg negative can provide useful information.

The efficacy of universal precautions, where all human blood and certain body fluids are treated as if known to be infectious for HIV, HBV, and other bloodborne pathogens, is limited by compliance. Workers should receive training in the implementation of work practice controls, such as puncture precautions, hand washing and skin washing, environmental controls, and standard infection control practices, and in the proper use of barrier precautions (personal protective equipment, or PPE). Gloves, masks, protective eyewear or face shields, barrier gowns or aprons, surgical caps or hoods, and shoe covers should be used appropriately.

All healthcare workers should take precautions to prevent injuries: caused by needles, scalpels, and other sharp instruments or devices used during procedures; when cleaning used instruments; during disposal of used needles; and when handling sharp instruments after procedures. To prevent needlestick injuries, needles should not be recapped, purposely bent or broken by hand, removed from disposable syringes, or otherwise manipulated by hand. After they are used, disposable syringes and needles, scalpel blades, and other sharp items should be placed in puncture-resistant containers for disposal; the puncture-resistant containers should be located as close as practical to the use area. Large-bore reusable needles should be placed in a puncture-resistant container for transport to the reprocessing area.

Instruments and devices used in patient care that enter the patient's vascular system or other normally sterile areas of the body, or touch nonintact skin or mucous membranes, should be sterilized in accordance with normal sterilization procedures before being used for each patient. Those that at most touch intact mucous membranes should be sterilized when possible. They should undergo high-level disinfection if they cannot be sterilized before being used for each patient. Those that do not touch the patient or that only touch intact skin of the patient need only be cleaned with a detergent or as indicated by the manufacturer.[19] Environmental surfaces contacted by blood or other potentially infectious materials should be cleaned of visible material and disinfected immediately or after the procedures. All surfaces that may have been so contaminated should be disinfected at the end of each shift. Medical and other equipment surfaces should undergo intermediate-level disinfection, as should any spill of blood or other patient material. A low-level disinfectant may be used for routine housekeeping surfaces (countertops, floors, sinks, etc.). Intermediate-level disinfection should be used for routine decontamination in serology and microbiology sections of clinical and research laboratories.[20]

TABLE 21.2 Post-exposure management of healthcare personnel after occupational percutaneous and mucosal exposure to blood and body fluids, by healthcare personnel Hepatitis B vaccination and response status.

Health-care personnel (HCP) status	Post-exposure testing		Post-exposure prophylaxis		Post-vaccination serologic testing[†]
	Source patient (HBsAg)	HCP testing (anti-HBs)	HBIG*	Vaccination	
Documented responder[§] after complete series (≥3 doses)	No action needed				
Documented nonresponder[¶] after 6 doses	Positive/unknown	—**	HBIG x2 separated by 1 month	—	No
	Negative	No action needed			
Response unknown after 3 doses	Positive/unknown	<10 mIU/mL**	HBIG x1	Initiate revaccination	Yes
	Negative	<10 mIU/mL	None		
	Any result	≥10 mIU/mL	No action needed		
Unvaccinated/incompletely vaccinated or vaccine refusers	Positive/unknown	—**	HBIG x 1	Complete vaccination	Yes
	Negative	—	None	Complete vaccination	Yes

Abbreviations: HCP = health-care personnel; HBsAg = hepatitis B surface antigen; anti-HBs = antibody to hepatitis B surface antigen; HBIG = hepatitis B immune globulin.

*HBIG should be administered intramuscularly as soon as possible after exposure when indicated. The effectiveness of HBIG when administered >7 days after percutaneous, mucosal, or nonintact skin exposures is unknown. HBIG dosage is 0.06 mL/kg.

[†]Should be performed 1–2 months after the last dose of the HepB vaccine series (and 4–6 months after administration of HBIG to avoid detection of passively administered anti-HBs) using a quantitative method that allows detection of the protective concentration of anti-HBs (≥10 mIU/mL).

[§]A responder is defined as a person with anti-HBs ≥10 mIU/mL after ≥3 doses of HepB vaccine.

[¶]A nonresponder is defined as a person with anti-HBs <10 mIU/mL after ≥6 doses of HepB vaccine.

**HCP who have anti-HBs <10 mIU/mL, or who are unvaccinated or incompletely vaccinated, and sustain an exposure to a source patient who is HBsAg-positive or has unknown HBsAg status, should undergo baseline testing for HBV infection as soon as possible after exposure, and follow-up testing approximately 6 months later. Initial baseline tests consist of total anti-HBc; testing at approximately 6 months consists of HBsAg and total anti-HBc.

Source: MMWR 2013; 62(RR10);14 http://www.cdc.gov/mmwr/preview/mmwrhtml/rr6210a1.htm

References

1. Schille S, Murphy TV, Sawyer M, et al. CDC guidance for evaluating health-care personnel administering postexposure management for hepatitis B virus protection and for administering postexposure management. *MMWR* 2013;62(10):1–24.
2. Centers for Disease Control and Prevention. Viral Hepatitis: Statistics and Surveillance. Available at: http://www.cdc.gov/hepatitis/statistics/index.htm (accessed on May 3, 2016).
3. Centers for Disease Control and Prevention. Healthcare-Associated Hepatitis B and C Outbreaks Reported to the Centers for Disease Control and Prevention (CDC) in 2008–2013. 2014. Available at: http://www.cdc.gov/hepatitis/Statistics/HealthcareOutbreakTable.htm (accessed on May 3, 2016).
4. Centers for Disease Control and Prevention. *Hepatitis B: The Pink Book*, 12th edn, 2012. Available at: http://www.cdc.gov/vaccines/pubs/pinkbook/downloads/hepb.pdf (accessed on May 3, 2016).
5. Centers for Disease Control and Prevention. Viral Hepatitis: Statistics and Surveillance. 2015. Available at: http://www.cdc.gov/hepatitis/statistics/2013surveillance/index.htm (accessed on May 3, 2016).
6. Wasley A, Kruszon-Moran D, Kuhnert W, et al. The prevalence of hepatitis B virus infection in the United States in the era of vaccination. *J Infect Dis* 2010;202:192–201.
7. World Health Organization. Hepatitis B. Available at: http://www.who.int/csr/disease/hepatitis/whocdscsrlyo20022/en/index4.html (accessed on May 3, 2016).
8. Werner BG, Grady GF. Accidental hepatitis-B surface-antigen-positive inoculations: use of e antigen to estimate infectivity. *Ann Intern Med* 1982;97:367–369.
9. Centers for Disease Control and Prevention. Hepatitis-B transmitted by human bite. *MMWR* 1974;23:45.
10. Weber DJ, Hoffmann KK, Rutala WA. Management of health care worker infected with human immunodeficiency virus: lessons from nosocomial transmission of hepatitis B virus. *Infect Control Hosp Epidemiol* 1991;12:625–630.
11. LaBrecque DR, Muhs JM, Lutwick LI. The risk of hepatitis B transmission from health care workers to patients in a hospital setting—a prospective study. *Hepatology* 1986;12:205–208.
12. Centers for Disease Control and Prevention. Recommendations for preventing transmission of human immunodeficiency virus and hepatitis B virus to patients during exposure-prone invasive procedures. *MMWR* 1991;40(RR-8):1–8.
13. Centers for Disease Control and Prevention. Updated CDC recommendations for the management of hepatitis B virus-infected health-care providers and students. *MMWR* 2012;61(RR-3):1–12.
14. Dienstag J, Delemos, AS. Viral hepatitis. In: Bennett JE, Dolin R, Blaser M. eds. *Mandell, Douglas, and Bennett's Principles and Practice of Infectious Disease*, 8th edn. Philadelphia, PA: Elsevier Saunders, 2015:1439–1468.
15. Hyams KC. Risks of chronicity following acute hepatitis B virus infection: a review. *Clin Infect Dis* 1995;20:992–1000.
16. Lok ASF, McMahon BJ. AASLD Practice Guidelines: Chronic Hepatitis B: Update 2009. *Hepatology* 2009;50(3):1–36. Available at: http://www.aasld.org/sites/default/files/guideline_documents/ChronicHepatitisB2009.pdf (accessed on May 3, 2016).
17. US Department of Labor. Bloodborne Pathogens, 29 CFR1910.1030. Available at: https://www.osha.gov/pls/oshaweb/owadisp.show_document?p_table=standards&p_id=10051 (accessed on May 3, 2016).
18. The Incident Investigation Teams. Transmission of hepatitis B from four infected surgeons without hepatitis B e antigen. *N Engl J Med* 1997;336:178–183.
19. Centers for Disease Control and Prevention. *Guidelines for infection control in hospital personnel*, GPO no. 544-436, 24441. Atlanta, GA: Public Health Service, 1985, pp. 1–20.
20. Favero MS, Bond WW. Sterilization, disinfection and antisepsis in the hospital. In: Balows A, Hausler WJ (eds). *Manual of clinical microbiology*. Washington, DC: American Society for Microbiology, 1991, p. 192.

HEPATITIS C VIRUS (HCV)

Common name for disease: Hepatitis C
Classification: Family—Flaviviridae

Occupational setting

Hepatitis C virus (HCV) infection is the most common bloodborne chronic infection in the United States, with an estimated 3.2 million people living with chronic HCV disease.[1] While the number of reported cases of acute hepatitis C has declined rapidly from the early 1990s through the 2000s (Figure 21.6), there has been a recent uptick in the number of acute hepatitis C cases. In 2013, there were 2138 reported cases of acute HCV from 41 states.[2] After adjusting under-ascertainment and underreporting, an estimated 29718 acute hepatitis C cases occurred in 2013. Twelve of the 41 reporting states—California, Florida, Indiana, Kentucky, Massachusetts, Michigan, New Jersey, New York, North Carolina, Ohio, Pennsylvania, and Tennessee—accounted for 68.6% of acute cases reported in 2013.[2] Healthcare workers with an exposure to blood or blood-tinged bodily fluids are considered at risk for contracting HCV.

Exposure (route)

Transmission of HCV occurs primarily by the parenteral route, although sexual and perinatal exposures do occur.

Pathobiology

HCV is an enveloped RNA virus with a diameter of about 50 nm. Eleven genotypes plus numerous subtypes and strains have been identified. Research has demonstrated that HCV can undergo frequent mutation (especially in the segment of the genome coding for the HCV envelope glycoproteins), allowing for continuous evasion of the host's immune system.[3] As a consequence, most HCV-infected people develop chronic infection. After an incubation period of 6–10 weeks, HCV causes an illness that is not clinically different from other forms of viral hepatitis. Symptoms are nonspecific and may include fatigue, nausea, anorexia, and low-grade fever. More specific findings, such as hepatomegaly and jaundice, are infrequent. There is also elevation in liver enzymes, specifically ALT. It has been estimated that up to 90% of infected HCV patients fail to clear the virus during the acute phase of the disease and become chronic carriers, with progression to chronic liver disease occurring in over 60% of cases.[4,5]

Diagnosis

The diagnosis of hepatitis C is made by detection of HCV antibody (anti-HCV) in serum or plasma via enzyme immunoassays (EIAs). However, a single positive anti-HCV result cannot distinguish between acute and chronic HCV infections. After an acute exposure, HCV RNA is identified as early as 2 weeks following exposure, whereas anti-HCV is generally not detectable before 8–12 weeks.[6] Thus testing for both anti-HCV and HCV RNA is critical in differentiating between acute and chronic diseases. Genotyping of the HCV RNA helps the clinician in optimizing the medical treatment for the patient. Genotype 1 (subtypes 1a and 1b) is the most common in the United States, followed by genotypes 2 and 3.[6]

Medical surveillance

National surveillance for hepatitis C is done through the CDC's National Notifiable Diseases Surveillance System. In 2012, CDC recommended one-time screening for HCV infection among all those born during 1945–1965, based on estimates that persons born during these years had a greater prevalence of HCV antibodies compared to other age cohorts.[7]

Healthcare, emergency medical, and public safety workers who sustain a percutaneous, mucosal, or nonintact skin exposure to HCV-positive blood should be identified by the respective occupational health personnel and monitored for seroconversion over the course of 6 months, with periodic testing for HCV RNA, anti-HCV, and liver function tests (depending on the institution's hepatitis C exposure protocol). If infection occurs, seroconversion (based on positive anti-HCV test) occurs in 50% of patients within 9 weeks of exposure, in 80% of patients within 15 weeks of exposure, and in at least 97% of patients within 6 months of exposure.[8] There is currently no postexposure prophylaxis for hepatitis C. Neither immunoglobulin nor antiviral agents are recommended for HCV postexposure prophylaxis.[8]

Healthcare workers with HCV infection should be evaluated by an occupational medicine physician, in conjunction with infectious disease specialist, to consider risk of HCV transmission from worker to patient. Although the likelihood of such an instance is very low, one report has described the transmission during an exposure-prone surgical procedure to five patients.[9]

Treatment

In the case that a healthcare worker seroconverts and develops acute hepatitis C, referral to a hepatologist or other specialist prepared to offer comprehensive management and treatment is indicated. Treatment decisions should be individualized based on viral genotype, as well as severity of liver disease, the potential for serious side effects, the likelihood of treatment response, the presence of comorbid conditions, and the patient's readiness for treatment.[6] Earlier poorly tolerated pegylated interferon alfa and ribavirin hepatitis C viral (HCV) treatments have now been largely replaced by safer, highly efficacious oral direct acting antivirals which treat genotype-specific HCV including sofosbuvir (an HCV RNA polymerase nucleotide analog inhibitor), ledipasvir (viral phosphoprotein inhibitor that blocks HCV viral replication, assembly, and secretion), simeprevir (NS3/4A protease inhibitor), ribavirin (nucleoside analogue indicated as an HCV combination treatment), and ombitasvir/paritaprevir/ritonavir + dasabuvir (otherwise known as the Viekira Pak). The Viekira Pak consists of (i) NS5A inhibitor ombitasvir, (ii) NS3/4A protease inhibitor paritaprevir, and (iii) HIV protease inhibitor ritonavir in combination with the nonnucleoside NS5B palm polymerase inhibitor dasabuvir.[10] As new antiviral treatments continue to be approved, current guidelines should be reviewed prior to initiating treatment.

Prevention

There is currently no hepatitis C vaccine commercially available. Vaccine development is challenging because HCV rapidly mutates into quasispecies. However, preliminary results from a recent clinical trial using a two-tier vaccine approach (first vaccine "primes" an initial immune response to HCV, second vaccine "boosts" this immune response) have demonstrated a robust T-cell response in healthy volunteer study participants which continued over the 6-month study period.[11,12]

The OSHA Bloodborne Pathogen standard (29 CFR 1910.1030) mandates that universal blood and body fluid precautions be implemented as part of a written Exposure Control Plan by an employer having employees with "any reasonably anticipated" occupational exposure. Occupational exposure is defined as "skin, eye, mucous membrane, or parenteral contact with blood or other potentially infectious materials that may result from the performance of an employee's duties." Other potentially infectious materials include semen, vaginal secretions, saliva where dental procedures are involved, unfixed tissues or organs, and cerebrospinal, synovial, pleural, pericardial, and amniotic fluids. When a fluid is contaminated with blood, or where it is difficult or impossible to differentiate between body fluids, the material should be considered potentially infectious. The efficacy of universal precautions, where all human blood and certain body fluids are treated as if known to be infectious for HIV, HBV, HCV, and other bloodborne pathogens, is limited by compliance. Workers should understand the importance of, and be well trained in, the implementation of work practice controls, such as puncture precautions, hand washing and skin washing, environmental controls, and standard infection control practices, and in the proper use of barrier precautions and personal protective equipment (PPE). Gloves, masks, protective eyewear or face shields, barrier gowns or aprons, surgical caps or hoods, and shoe covers should be used appropriately. All healthcare workers should take precautions to prevent injuries caused by needles, scalpels, and other sharp instruments or devices during procedures; when cleaning used instruments; during disposal of used needles; and when handling sharp instruments after procedures. To prevent needlestick injuries, needles should not be recapped, purposely bent or broken by hand, removed from disposable syringes, or otherwise manipulated by hand. After they are used, disposable syringes and needles, scalpel blades, and other sharp items should be placed in puncture-resistant containers for disposal; the puncture-resistant containers should be located as close as practical to the use area. Large-bore reusable needles should be placed in a puncture-resistant container for transport to the reprocessing area. Instruments and devices used in patient care that enter the patient's vascular system or other normally sterile areas of the body, or touch nonintact skin or mucous membranes, should be sterilized in accordance with normal sterilization procedures before being used for each patient. Those that at most touch intact mucous membranes should be sterilized when possible. They should undergo high-level disinfection if they cannot be sterilized before being used for each patient. Those that do not touch the patient or that only touch intact skin of the patient need only be cleaned with a detergent or as indicated by the manufacturer. Environmental surfaces contacted by blood or other potentially infectious materials should be cleaned of visible material and disinfected immediately or after the procedures. All surfaces that may have been so contaminated should be disinfected at the end of each shift. Medical and other equipment surfaces should undergo intermediate-level disinfection, as should any spill of blood or other patient material. A low-level disinfectant may be used for routine housekeeping surfaces (countertops, floors, sinks, etc.). Intermediate-level disinfection should be used for routine decontamination in serology and microbiology sections of clinical and research laboratories.

References

1. Denniston MM, Jiles RB, Drobeniuc J, et al. Chronic hepatitis C virus infection in the United States, National Health and Nutrition Examination Survey 2003 to 2010. *Ann Intern Med* 2014;160:293–300.
2. Centers for Disease Control and Prevention. Viral Hepatitis: Statistics and Surveillance. 2015. Available at: http://www.cdc.gov/hepatitis/statistics/2013surveillance/index.htm (accessed on May 3, 2016).
3. Burke KP, Cox AL. Hepatitis C Virus Evasion of Adaptive Immune Responses – A Model for Viral Persistence. *Immunol Res* 2010;47(1–3):216–227. Available at: http://www.ncbi.nlm.nih.gov/pmc/articles/PMC2910517/pdf/nihms215817.pdf (accessed on June 28, 2016).
4. Hsu HH, Greenberg HB. Hepatitis C. In: Hoeprich PD, Jordan MC, Ronald AR. eds. *Infectious Diseases: A Treatise of Infectious Processes*, 5th edn. Philadelphia, PA: JB Lippincott Co., 1994:820–825.
5. Ray SC, Thomas DL. Hepatitis C. In: Bennett JE, Dolin R, Blaser M. eds. *Mandell, Douglas, and Bennett's Principles and Practice of Infectious Disease*, 8th edn. Philadelphia, PA: Elsevier Saunders, 2015:1904–1927.
6. Ghany MG, Strader DB, Thomas DL, et al. Diagnosis, management, and treatment of hepatitis C: an update. *Hepatology* 2009;49(4):1335–1374.
7. Centers for Disease Control and Prevention. Recommendations for the identification of chronic hepatitis C virus infection among persons born during 1945–1965. *MMWR* 2012;61(RR-04):1–36.
8. Centers for Disease Control and Prevention. Updated US public health service guidelines for the management of occupation and exposure to HBV, HCV and HIV and recommendations for postexposure prophylaxis. *MMWR* 2001;50(RR-11):1–53.
9. Esteban JL, Gómez J, Martell M, et al. Transmission of hepatitis C virus by a cardiac surgeon. *N Engl J Med* 1996;334:555–560.
10. AASLD-IDSA. Recommendations for testing, managing, and treating hepatitis C. Available at: http://www.hcvguidelines.org (accessed on June 28, 2016). 10.1002/hep.27950.
11. Swadling L, Capone S, Antrobus RD, et al. A human vaccine strategy based on chimpanzee adenoviral and MVA vectors that primes, boosts, and sustains functional HCV-specific T cell memory. *Sci Transl Med* 2014;6(261):261ra153.
12. Hepatitis C Vaccine Shows Promise in Early Clinical Trial. Medical News Today. November 6, 2014. Available at: http://www.medicalnewstoday.com/articles/285027.php (accessed on June 28, 2016).

HERPES B VIRUS

Common names for disease: Monkey B virus, B virus
Classification: Family—Herpesviridae; genus/species—*Herpesvirus simiae*

Occupational setting

Persons at risk include zoo attendants, nonhuman primate handlers, veterinarians, and researchers working with macaque monkeys (a genus of Old-World monkeys that serve as the natural host) or with infected tissues or fluids originating from infected macaques. While herpes B virus infection has been reported most commonly in the rhesus and cynomolgus macaque (*Macaca fascicularis*), it has also been identified in the stump-tail (*Macaca arctoides*), pig-tailed (*Macaca nemestrina*), Japanese (*Macaca fuscata*), bonnet (*Macaca radiata*), and Taiwan (*Macaca cyclopis*) macaque.[1]

Exposure (route)

Route of transmission of herpes B virus include (i) bite or scratch from an infected animal, (ii) needlestick from a contaminated syringe, (iii) scratch or cut from a contaminated cage or other sharp-edged surface, and (iv) exposure to nervous tissue or skull of an infected animal.[2] B virus may survive for hours on the surface of objects, particularly on surfaces that are moist. The injury need not be severe for infection to occur, although nonpenetrating wounds appear to carry a lower risk of transmission.[2] A case of infection from mucocutaneous exposure has occurred[3] as well as a case of human-to-human transmission.[4]

Pathobiology

Herpes B virus is a large, double-stranded DNA virus which shares almost 80% amino acid sequence identity with herpes simplex virus 1 and 2.[5] Primary infection in monkeys frequently appears as buccal mucosal lesions. Subsequently, the virus remains latent in the host and may reactivate spontaneously or in times of stress. This may result in shedding of virus in saliva or genital secretions. The disease is characterized by a variety of symptoms, most of which develop within 1 month of exposure.[6] Among the symptoms are vesicular skin lesions at or near the site of inoculation. Neurologic symptoms, such as dysesthesias and numbness, often develop at the site of inoculation. Subsequent symptoms often include myalgias, headaches, and dizziness. Symptoms may progress rapidly, leading to permanent neurologic impairment or death.

There have only been 23 cases of symptomatic primate-to-human human infection described in the literature, with 18 of the 23 dying from encephalitis.[7] Hundreds of macaque bites and scratches are reported in animal handlers and other personnel working in primate facilities in the United States, but contracting herpes B virus is rare. In a study of more than 300 animal care workers, with 166 of them reporting possible transmission risk exposures to macaques, none of the workers were found to be B virus positive.[2]

In 1998, the first case of occupational herpes B virus infection from primate to human following a mucocutaneous exposure was reported. A 22-year-old primate worker sustained an ocular splash while moving an animal within a cage. She was first evaluated for eye symptoms 10 days after exposure and was hospitalized 5 days later. Despite institution of intravenous antiviral therapy, she developed acute demyelinating encephalomyelitis and died.[3]

There has only been one reported case of human-to-human transmission of herpes B virus that occurred when an animal worker contracted a B virus infection and his wife used an ointment to treat his skin lesions. She subsequently used ointment from the same container on herself to treat contact dermatitis. She seroconverted to B virus but never developed symptoms.[4]

Diagnosis

Wounds resulting from potential sources of B virus should be cultured. Lesions on mucous membranes of monkeys should also be cultured.[8] If vesicles are present, a Tzanck smear can be performed to look for multinucleated giant cells. A simple, rapid test based on restriction endonuclease analysis of labeled infected cell DNA is available. Serial serum samples should also be obtained from individuals inadvertently exposed to see if a rise in antibody titer occurs. Laboratory assistance may be obtained from the National B Virus Resource Center in Atlanta Georgia, which provides around-the-clock diagnostic testing, educational resources, and emergency information at 404-413-6550 (website: http://www2.gsu.edu/~wwwvir/).

Treatment

Wounds and exposed mucosa should be thoroughly cleaned immediately with soap and water for at least 15 minutes. After the 1998 incident, updated guidelines were issued, highlighting best practices and providing information on recommended postexposure prophylaxis.[9]. Postexposure prophylaxis with acyclovir should be strongly considered for a high-risk source and definitive route of exposure.[9]

Medical surveillance

Serum samples from all individuals working with macaques (including animal handlers, veterinarians, researchers) should be obtained and frozen for future analysis starting from the pre-hire phase and repeatedly annually.

Prevention

Primate handlers should wear long-sleeved garments and leather gloves.[4,7] In addition, a face mask and safety glasses should be worn. Individuals should be trained in the proper use of mechanical and chemical restraints and squeeze cages. Research laboratories should evaluate the feasibility of acquiring and maintaining a B virus-free colony of monkeys. Routine screening of macaques for B virus is not recommended. In situations where laboratory studies may lead to immunosuppression, the investigator may want to determine the infection status of the animal to be used, since viral shedding may be enhanced in such situations. Animal cages should be thoroughly cleaned and free of sharp edges at corners. Access to animal quarters should be restricted to individuals properly trained in procedures to avoid risk of infection. Workers must be educated to notify their supervisor immediately in the case of an animal bite, scratch, or mucous membrane exposure. There is no vaccine currently available to prevent herpes B virus infection.

References

1. Weigler BJ. Biology of B virus in macaque and human host: a review. *Clin Infect Dis* 1992;14:555–567.
2. Centers for Disease Control and Prevention. B Virus (Herpes B, Monkey B Virus, Herpesvirus Simiae, Herpesvirus B) – Transmission. 2014. Available at: http://www.cdc.gov/herpesbvirus/transmission.html (accessed on May 3, 2016).
3. Centers for Disease Control and Prevention. Fatal Cercopithecine Herpesvirus 1 (B Virus) Infection Following a Mucocutaneous Exposure and Interim Recommendations for Worker Protection. *MMWR* 1998;47:1073–1083.
4. Centers for Disease Control and Prevention. B Virus Infection in Humans—Pensacola, Florida. *MMWR* 1987;36:289–296.
5. Huff JL, Barry PA. B-virus (Cercopithecine herpesvirus 1) infection in humans and macaques: potential for zoonotic disease. *Emerg Infect Dis* 2003;9(2):246–250.
6. Centers for Disease Control and Prevention. B Virus Infection in Humans—Michigan. *MMWR* 1989;38:453–454.
7. Palmer AE. B-virus, herpesvirus simiae: historical perspective. *J Med Primatol* 1987;16:99–130.
8. Hilliard JK, Munoz RA, Lipper SL, et al. Rapid identification of herpesvirus simiae (B virus) DNA from clinical isolates in nonhuman primate colonies. *J Virol Methods* 1986;13:52–62.
9. Cohen JI, Davenport DS, Stewart JA, et al. Recommendations for prevention of and therapy for exposure to B virus (cercopithecine herpesvirus 1). *Clin Infect Dis.* 2002;35(10):1191–1203.

HERPES SIMPLEX VIRUS (HSV)

Common names for disease: Cold sores, fever blisters, genital herpes, herpetic whitlow

Classification: Family—Herpesviridae, subfamily—Alphaherpesvirinae, genus—*Simplexvirus*

Occupational setting

Healthcare workers, particularly dental, nursing, and respiratory care personnel, appear to beat greatest risk of occupational exposure to HSV. Transmission of HSV from patients to healthcare workers is well documented and appears to occur more frequently than transmission from workers to susceptible patients.[1–4]

Exposure (route)

The principal mode of spread is by direct contact with infected secretions. The virus enters the body via mucosal surfaces or abraded skin.

Pathobiology

HSV is a linear, double-stranded DNA virus enclosed within an icosahedral capsid, surrounded by a phospholipid-rich envelope. There are two subtypes, HSV-l and HSV-2. HSV-1 is most commonly associated with orofacial lesions and HSV-2 with genital lesions; however, this distinction is becoming increasingly blurred. Humans are the primary reservoir.

HSV-1 has a seroprevalence of up to 85% in adults in the United States and developed countries.[5] Primary infection with HSV-1 is frequently asymptomatic, but it may present as gingivostomatitis and pharyngitis in children.[6] After the primary HSV-1 lesion appears, it progresses to a papule, then to a vesicle, and finally to a full lesion. Typical herpetic lesions are small vesicles on an erythematous base that subsequently ulcerate and then crust over. The lesions heal in approximately 8–10 days. Recurrent lesions due to HSV-1 (referred to as cold sores or fever blisters) occur mainly over the vermillion border of the lip and occur in up to 40% of cases.[4]

Primary infection with HSV-2 is unusual before puberty, though perinatal transmission does occur. Most HSV-2 infections are transmitted from infected persons to others through direct contact with a lesion or infected body fluids (including vesicular exudates and genital fluids). Direct tactile contact with freshly contaminated articles and environmental surfaces can also occur but is rare.[7] HSV-2 is characterized by formation of multiple, bilateral, painful, and extensive genital ulcers, with fever, malaise, lymphadenopathy, and myalgia. Recurrent genital herpes disease is mild with no systemic symptoms and is shorter duration.

Herpetic whitlow, a painful lesion on a finger or thumb caused by HSV infection, is of particular occupational importance in healthcare personnel. The transmission of herpetic whitlow can occur via direct contact with an active herpetic lesion with the finger and subsequent inoculation of nonintact skin.[8] Herpetic whitlow has been found to occur more frequently in dentists than in the general population.[9] There are also case reports of herpetic whitlow in dental workers.[4] Whitlow presents with pain, erythema, swelling, and pustular lesions that are frequently difficult to distinguish from pyogenic bacterial infections. Diagnosis is usually made clinically based on the classic presentation of a vesicular lesion on an erythematous base.

Ophthalmic herpetic keratitis is the most frequent cause of corneal blindness in the United States and the most common source of infectious blindness in the Western world.[10] In the occupational setting, exposure can occur if a dental worker touches their eye after directly touching a patient's active HSV-1-infected lesion.[4] The infection primarily affects the epithelial layer of the cornea. Characteristics of infection include eye pain, swelling, redness, watery discharge, and photophobia. Clinical diagnosis based on characteristic features of the corneal lesion, with laboratory confirmation via viral culture, PCR, Papanicolaou stain (showing intranuclear eosinophilic inclusion bodies), and/or

Giemsa stain (scrapings of the cornea showing multinucleated giant cells).[10–13]

Diagnosis

Diagnosis is often made clinically on the basis of the typical vesicular lesions on an erythematous base. Tzanck smears can identify HSV-related changes (including multinucleated giant cells and epithelial cells containing eosinophilic intranuclear inclusion bodies) but cannot differentiate between HSV-1 and HSV-2. Because of sero-cross-reactivity, antibodies to HSV-1 and HSV-2 are not generally distinguishable unless a glycoprotein G antibody assay is available.[14] Moreover, antibody titer increases generally do not occur during HSV recurrences and thus antibody testing cannot be used for mucocutaneous HSV relapse.[14] The gold standard for diagnosis is via viral culture, with immunofluorescent staining of the tissue culture cells able to uniquely identify HSV-1 and HSV-2. However, the sensitivity of viral culture is low, especially for recurrent lesions, and declines rapidly as lesions begin to heal.[15] PCR assays for HSV DNA are more sensitive and are used to diagnose such conditions as HSV encephalitis and recurrent meningitis as well as identify asymptomatic viral shedding.[15]

Treatment

Antivirals including acyclovir, famciclovir, and valacyclovir are effective for episodic treatment of HSV-1 and HSV-2 disease.[15] However, the location of the lesions and the chronicity (primary or reactivation) of the infection should guide the clinician to dose the antiviral accordingly.[14]

Prevention

Since transmission occurs through direct contact, institution of universal precautions minimizes the risk of HSV transmission. Evidence-based occupational guidelines have been developed for dental workers for risk mitigation.[4,16] Some larger institutions recommend that routine treatment be delayed for patients with active lesions. However, for the majority of dental practices, the dentist proprietor will have the ultimate responsibility for setting policies that protect the safety of patients and the dental team. It is advisable to limit the treatment of patients with active lesions to urgent care only.[4] For elective cases, active lesions should be treated prior to dental work to reduce potential HSV exposure to dental workers.[4] As HSV can be potentially spread by aerosol, strict adherence to PPE may not completely eliminate the risk of aerosol transmission (since aerosols can be inhaled via leaks in the mask and permeate around safety glasses).[17] Reducing aerosol exposure can be achieved by the use of high-volume evacuators (which capture 95% of aerosols), safety glasses (with side shields and a face shield), and goggles.[17] There is currently a phase one/two clinical trial of an investigational therapeutic HSV-2 vaccine designed to prevent or reduce genital lesion recurrences and prevent transmission of this herpes virus to uninfected individuals.

References

1. Schiffer, JT, Corey L. Herpes simplex virus. In: Bennett JE, Dolin R, Blaser M. eds. *Mandell, Douglas, and Bennett's Principles and Practice of Infectious Disease*, 8th edn. Philadelphia, PA: Elsevier Saunders, 2015:1713–1731.
2. Perl TM, Haugen TH, Pfaller MA, et al. Transmission of herpes simplex virus type 1 infection in an intensive care unit. *Ann Intern Med* 1992;117:584–586.
3. Adams G, Stover BH, Keenlyside RA, et al. Nosocomial herpetic infections in a pediatric intensive care unit. *Am J Epidemiol* 1981;113:126–132.
4. Browning WD and McCarthy JP. A case series: herpes simplex virus as an occupational hazard. *J Esth Res Dent* 2012;24(1):61–66.
5. Seigel M. Diagnosis and management of recurrent herpes simplex infections. *J Am Dent Assoc* 2002;133:1245–1249.
6. Turner R, Shehab Z, Osborne K, et al. Shedding and survival of herpes simplex virus from 'fever blisters'. *Pediatrics* 1982;70:547–549.
7. Huber MA, Terezhalmy GT. HSV and VZV: Infection Control/Exposure Control Issues for Oral Healthcare Workers. 2013. Available at: http://www.dentalcare.com/media/en-US/education/ce323/ce323.pdf (accessed on May 3, 2016).
8. Leggat P, Kedjarune U, Smith D. Occupational health problems in modern dentistry: a review. *Ind Health* 2007;45:611–621.
9. Lewis MA. Herpes simplex virus: an occupational hazard in dentistry. *Int Dent* 2004;J54(2):103–111.
10. Liesegang TJ. Herpes simplex virus epidemiology and ocular importance. *Cornea* 2001;20(1):1–13.
11. Wilhelmus KR. Diagnosis and management of herpes simplex stromal keratitis. *Cornea* 1987;6(4):286–291.
12. Zaher SS, Sandinha T, Roberts F, et al. Herpes simplex keratitis misdiagnosed as rheumatoid arthritis-related peripheral ulcerative keratitis. *Cornea* 2005;24(8):1015–1017.
13. Seitzman GD, Cevallos V, Margolis TP. Rose bengal and lissamine green inhibit detection of herpes simplex virus by PCR. *Am J Ophthalmol* 2006;141(4):756–758.
14. Schiffer JT, Corey L. Herpes simplex virus. In: Bennett JE, Dolin R, Blaser M. eds. *Mandell, Douglas, and Bennett's Principles and Practice of Infectious Disease*, 8th edn. Philadelphia, PA: Elsevier Saunders, 2015:1713–1730.
15. Centers for Disease Control and Prevention. 2010 STD Treatment Guidelines – Diseases Characterized by Genital, Anal, or Perianal Ulcers. 2014. Available at: http://www.cdc.gov/std/treatment/2010/genital-ulcers.htm (accessed on May 3, 2016).
16. MacLean C. *Infection Control Manual. Appendix B: Management of Patients with Herpetic Lesions*. Halifax: Dalhousie University School of Dentistry. 2013. Available at: http://www.dal.ca/content/dam/dalhousie/pdf/dentistry/IC%20Manual%2713.pdf (accessed on May 3, 2016).

17. Harrel SK, Molinari J. Aerosols and splatter in dentistry: a brief review of the literature and infection control implications. *J Am Dent Assoc* 2004;135:429–437.

HUMAN IMMUNODEFICIENCY VIRUS (HIV-1)

Common names for disease: Acquired immunodeficiency syndrome (AIDS), AIDS-related complex
Classification: Family—Retroviridae; genus—*Lentivirus*

Occupational setting

Healthcare workers are at increased risk of infection if they are exposed to blood products, especially surgeons, pathologists, dentists and dental hygienists, personnel in emergency departments, personnel in labor and delivery, laboratory technicians and researchers, and hemodialysis staff. Other occupations with increased risk include hospital custodial staff, laboratory workers, morticians, and emergency response personnel. Research scientists conducting HIV research are also at risk.

Exposure (route)

The virus can be spread through blood, semen, or other infected body fluids via percutaneous exposure, mucous membrane exposure, exposure to abraded or otherwise nonintact skin, breastfeeding, perinatal routes from an infected mother, or sexual activity. Nonbloody saliva has not been identified as an effective mode of transmission. In the healthcare setting, the risk of infection following a percutaneous exposure has been estimated at 0.3%, while that for a mucous membrane exposure is approximately 0.09%.[1,2] The risk for HIV transmission via nonintact skin exposure, which has been reported, has not been precisely quantified but is estimated to be less than the risk for mucous membrane exposures.[3] Feces, nasal secretions, saliva, sputum, sweat, tears, urine, and vomitus are not considered potentially infectious unless they are visibly bloody.[1] The likelihood of acquiring an occupational HIV infection increases if the exposure occurs via:

- A medical instrument or device that is visibly contaminated with the patient's blood
- A procedure that involved a needle being placed directly in patient's vein or artery
- A deep injury to the healthcare worker[4]

Moreover, the larger the viral inoculum (i.e., from a source patient with acute HIV infection, untreated HIV, or terminal disease), the greater the transmission risk.[4]

From 1985 through 2013, there were 58 confirmed and 150 possible cases of occupationally acquired HIV infection among healthcare workers in the United States (Table 21.3), with only one confirmed case since 1999 (Figure 21.8).[3]

Pathobiology

HIV is a retrovirus whose inner core contains two copies of a single strand of RNA. This group of viruses uses reverse transcriptase to encode DNA, which is then incorporated in the host cell.

HIV infection may be associated with the sudden onset of an acute retroviral illness. The illness may last from 3 to 14 days

TABLE 21.3 Number of confirmed or possible cases of occupationally acquired HIV infection among health care workers reported to CDC — United States, 1985–2013.

Occupation	Confirmed (N=58)		Possible (N=150)	
	No.	(%)	No.	(%)
Nurse	24	(41.4)	37	(24.7)
Laboratory technician, clinical	16	(27.6)	21	(14.0)
Physician, nonsurgical	6	(10.3)	13	(8.7)
Laboratory technician, nonclinical	4	(6.9)	—	—
Housekeeper/maintenance	2	(3.4)	14	(9.3)
Technician, surgical	2	(3.4)	2	(1.3)
Embalmer/morgue technician	1	(1.7)	2	(1.3)
Hospice caregiver/attendant	1	(1.7)	16	(10.7)
Respiratory therapist	1	(1.7)	2	(1.3)
Technician, dialysis	1	(1.7)	3	(2.0)
Dental worker, including dentist	—	—	6	(4.0)
Emergency medical technician/paramedic	—	—	13	(8.7)
Physician, surgical	—	—	6	(4.0)
Technician/therapist, other	—	—	9	(6.0)
Other health care occupations	—	—	6	(4.0)

Source: MMWR 63(53);1245–1246

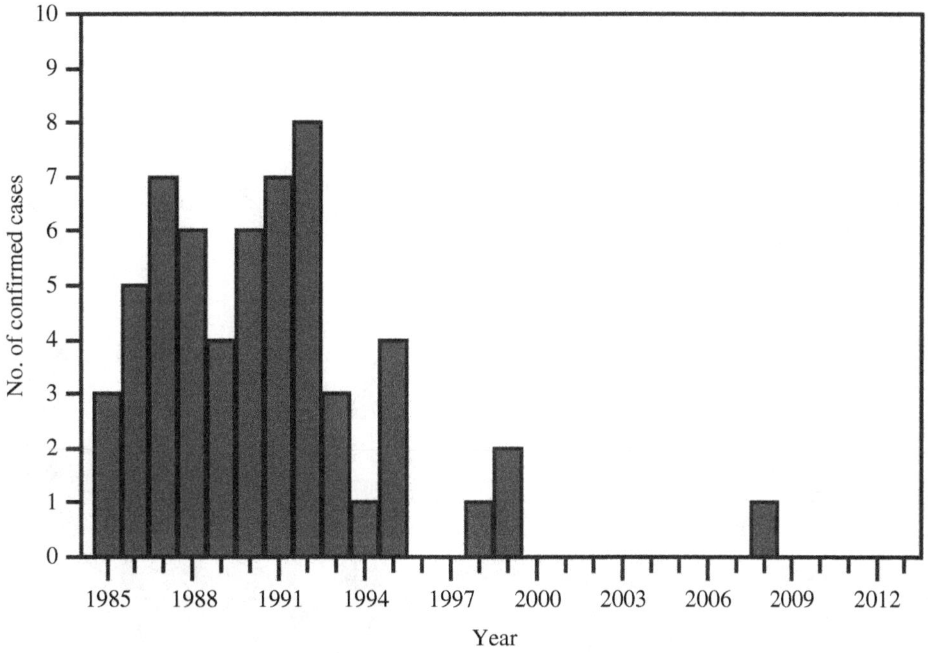

FIGURE 21.8 Number of confirmed cases (N = 58) of occupationally acquired HIV infection among health care workers reported to CDC — United States, 1985–2013. *Source*: MMWR 2015; 63(53):1245-1246.

and is associated with fevers, sweats, malaise, lethargy, anorexia, nausea, myalgia, arthralgia, headaches, sore throat, diarrhea, generalized lymphadenopathy, a macular erythematous truncal eruption, and thrombocytopenia.[5] Symptoms have been reported following an incubation period of 6–56 days.[5] Comparison of T-cell subsets before and after the acute illness may show inversion of the T4 : T8 ratio in some patients, due to increased numbers of circulating T8+ cells.[5]

HIV is initially widely disseminated in the lymphoid organs. HIV primarily infects CD4+ T cells (T-helper cells), interfering with normal function or resulting in cell destruction. HIV also disrupts the network of signaling molecules that normally regulates the immune response and damages lymph nodes and related immunologic organs. The resulting immunosuppression makes patients vulnerable to multiple opportunistic infections.[6]

HIV invariably escapes the host immune system due in large part to the high rate of mutations that occur during the process of HIV replication. In addition, the killer T cells that recognize HIV may be depleted or become dysfunctional. Early in the course of HIV infection, people may lose HIV-specific CD4+ T-cell responses that normally slow the replication of viruses including the secretion of interferons and other antiviral factors and the orchestration of CD8+ T cells. The virus may also hide within the chromosomes of an infected cell as a provirus and be shielded from surveillance by the immune system. The provirus is not targeted by current antiretroviral therapy (ART), which is active again replicating virions.[6]

The median time to develop AIDS following HIV infection in the absence of ART is 10–12 years, with a very wide range. Most infecting strains of HIV use the coreceptor CCR5, in addition to CD4, to enter certain of its target cells. HIV-infected people with a specific mutation in one of their two copies of the gene coding for CCR5 have a slower disease course than people with two normal copies of the gene. Rare individuals with two mutant copies of the CCR5 gene appear, in most cases, to be completely protected from HIV infection. Mutations in the gene for other HIV coreceptors may also influence the rate of disease progression.[6]

The level of HIV in an untreated person's plasma 6 months to a year after infection, termed viral "set point," is highly predictive of the rate of disease progression. Potent combinations of three or more anti-HIV drugs known as highly active antiretroviral therapy (HAART) can reduce the viral burden to very low levels and in many cases delay the progression of HIV disease for prolonged periods.[6]

Along with the risk of opportunistic infections and some cancers in patients with HIV/AIDS, neurotic manifestations are seen in half or patients. These include:

- Cognitive symptoms, including impaired short-term memory, reduced concentration, and mental slowing
- Motor symptoms such as fine motor clumsiness or slowness, tremor, and leg weakness
- Behavioral symptoms including apathy, social withdrawal, irritability, depression, and personality change

The neurologic symptoms appear to be the result of HIV that reaches the brain through circulating monocytes and macrophages that are not killed by the virus. While nerve cells do not become infected with HIV, astrocytes and

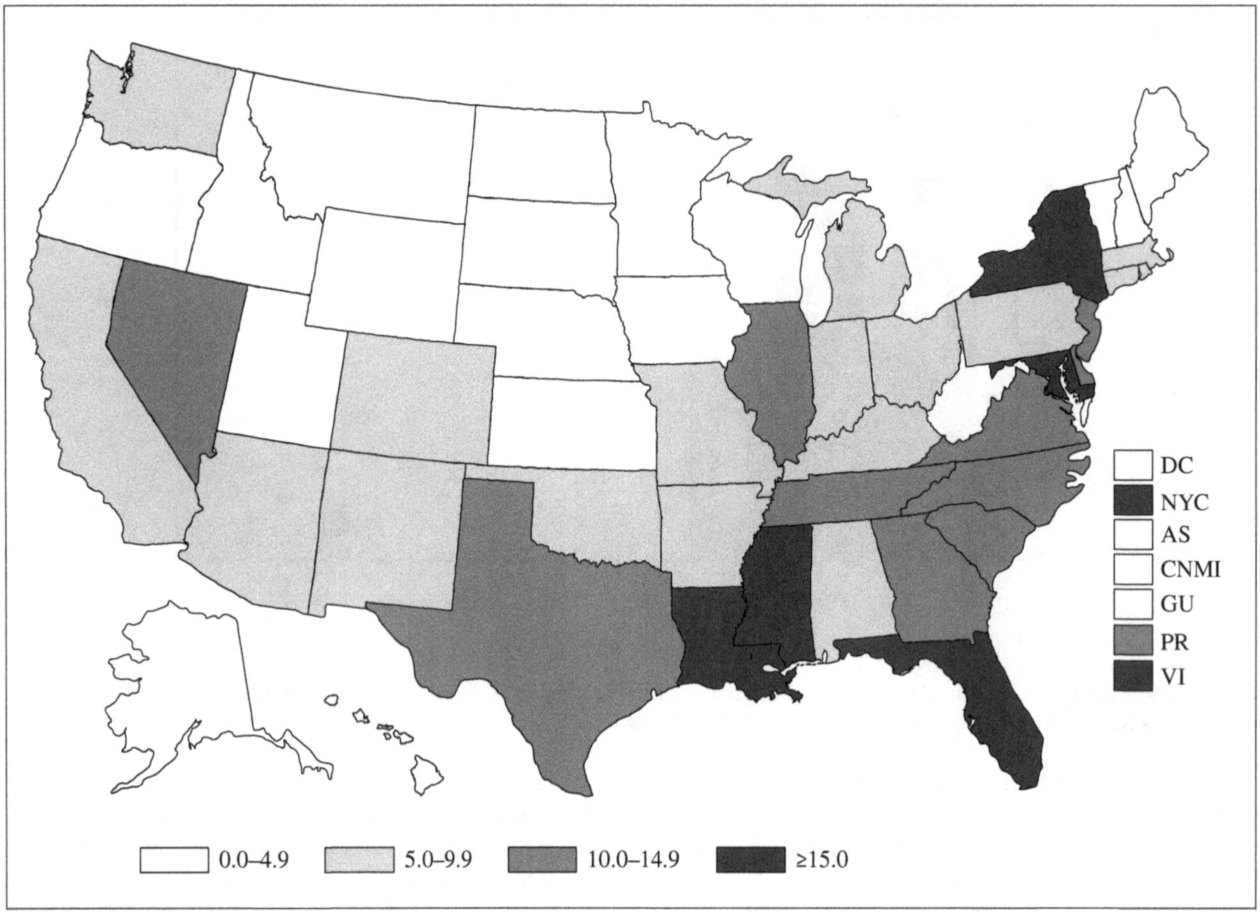

FIGURE 21.9 HIV Diagnoses Per 100 000 population – United States and U.S. territories, 2013. *Source*: MMWR 2015; 62(53):80.

microglia (as well as monocyte/macrophages that have migrated to the brain) can be infected with the virus. Infection may cause a disruption of normal neurologic functions by altering cytokine levels, delivering aberrant signals, and causing the release of toxic products in the brain.[6]

The global epidemic of HIV infection involves an estimated 35 million individuals including an estimated 1.3 million in the United States. The highest rates (≥15 diagnoses per 100 000 population) were observed in certain states in the Southeast, Northeast, and Washington, DC (Figure 21.9).

Diagnosis

HIV viremia does not develop until approximately 10–15 days after infection. Early HIV infection is characterized by markedly elevated HIV RNA levels. RT-PCR is a highly sensitive and specific modality in quantifying HIV RNA. Confirmation testing for HIV infection is via an FDA-approved antigen/antibody combination immunoassay that detects HIV-1 and HIV-2 antibodies and HIV-1 p24 antigen (a viral core protein that appears in the blood as the viral RNA level rises following HIV infection).[7] In 2015, FDA approved a test that can distinguish between HIV-1 and HIV-2 antibodies.[8]

Treatment

Since the mid-1990s, CDC has been recommended postexposure prophylaxis (PEP) with antiretrovirals for workers exposed to HIV-infected blood or bloody fluids. With the advent of newer antiretrovirals, the US Public Health Service issued its latest update to the HIV PEP guidelines for occupational exposures in 2013.[9] The basic principles guiding HIV PEP have remained the same:

- The HIV status of the source patient should be determined to help assess the need for PEP, but this inquiry should not delay treatment.
- PEP medication regimens should be started as soon as possible after occupational exposure to HIV (if the source is determined to be HIV negative after initiation of PEP, PEP can be discontinued without further follow-up).

- Medications should continue for a 4-week duration.
- Close follow-up should be provided that includes counseling, baseline and follow-up HIV testing, and monitoring for drug toxicity.

The main change from the previous 2005 guideline is that PEP medication regimens should now contain three or more antiretroviral drugs for all occupational exposures to HIV.[6] The impetus for this change comes from nonoccupational studies which demonstrate the treatment efficacy of three versus two antiretroviral drugs in reducing viral load in HIV-infected individuals, source patient drug resistance to commonly used HIV PEP medications, and reported better side effect profile with newer-generation antiretrovirals. The 2013 guidelines recommend emtricitabine (FTC) plus tenofovir (TDF) (these two agents may be dispensed as Truvada®, a fixed-dose combination tablet) plus raltegravir (RAL) as HIV PEP for occupational exposures (Table 21.4). Several drugs may be used as alternatives to FTC plus TDF plus RAL. TDF has been associated with renal toxicity, and an alternative should be sought in the presence of underlying renal disease. Zidovudine (ZDV) could be used as an alternative to TDF and could be conveniently prescribed in combination with lamivudine (3TC), to replace both TDF and FTC, as Combivir®. Alternatives to RAL include darunavir (DRV) plus ritonavir (RTV), etravirine (ETV), rilpivirine (RPV), atazanavir (ATV) plus RTV, and lopinavir (LPV) plus RTV. Preparation of this PEP regimen in single-dose "starter packets" may facilitate timely initiation of PEP. The preferred HIV PEP regimen recommended in the guideline should be reevaluated and modified whenever additional information is obtained concerning the source of the occupational exposure (e.g., possible treatment history or antiretroviral drug resistance) or if expert consultants recommend the modification. Whenever possible, consultation with an infectious disease specialist or another physician who is an expert in the administration of antiretroviral agents is recommended. Such consultation should not, however, delay timely initiation of PEP. Staff should be

TABLE 21.4 Preferred HIV PEP Regimen.

Raltegravir (Isentress; RAL) 400 mg PO twice daily Plus Truvada, 1 PO once daily (Tenofovir DF [Viread; TDF] 300 mg + emtricitabine [Emtriva; FTC] 200 mg)

Alternative Regimens

(May combine 1 drug or drug pair from the left column with 1 pair of nucleoside/nucleotide reverse-transcriptase inhibitors from the right column; prescribers unfamiliar with these agents/regimens should consult physicians familiar with the agents and their toxicities)[a]

Raltegravir (Isentress; RAL)	Tenofovir DF (Viread; TDF) + emtricitabine (Emtriva; FTC); available as Truvada
Darunavir (Prezista; DRV) + ritonavir (Norvir; RTV)	Tenofovir DF (Viread; TDF) + lamivudine (Epivir; 3TC)
Etravirine (Intelence; ETR)	Zidovudine (Retrovir; ZDV; AZT) + lamivudine (Epivir; 3TC); available as Combivir
Rilpivirine (Edurant; RPV)	Zidovudine (Retrovir; ZDV; AZT) + emtricitabine (Emtriva; FTC)
Atazanavir (Reyataz; ATV) + ritonavir (Norvir; RTV)	
Lopinavir/ritonavir (Kaletra; LPV/RTV)	

The following alternative is a complete fixed-dose combination regimen, and no additional antiretrovirals are needed: Stribild (elvitegravir, cobicistat, tenofovir DF, emtricitabine)

Alternative Antiretroviral Agents for Use as PEP Only with Expert Consultation[b]

Abacavir (Ziagen; ABC)
Efavirenz (Sustiva; EFV)
Enfuvirtide (Fuzeon; T20)
Fosamprenavir (Lexiva; FOSAPV)
Maraviroc (Selzentry; MVC)
Saquinavir (Invirase; SQV)
Stavudine (Zerit; d4T)

Antiretroviral Agents Generally Not Recommended for Use as PEP

Didanosine (Videx EC; ddI)
Nelfinavir (Viracept; NFV)
Tipranavir (Aptivus; TPV)

Antiretroviral Agents Contraindicated as PEP

Nevirapine (Viramune; NVP)

[a]Listed in order of preference, however, other alternatives may be reasonable based upon patient and clinician preference.

Source: Kuhar DT, Henderson DK, Struble KA, Heneine W, Thomas V, Cheever LW, Gomaa A, Panlilio AL; US Public Health Service Working Group. Updated US Public Health Service guidelines for the management of occupational exposures to human immunodeficiency virus and recommendations for postexposure prophylaxis. *Infect Control Hosp Epidemiol.* 2013;34(9):875-92, Appendix A.

trained and protocols should be in place to ensure prompt reporting and rapid initiation of treatment when indicated.[9]

Workers with possible occupational exposure to HIV should be seen in follow-up within 72 hours of the exposure regardless of whether they are taking PEP. This visit should address risk of infection, benefits of PEP and importance of adherence, side effects associated with PEP, any needed modification of PEP, prevention of secondary HIV transmission, and psychological counseling. Workers taking PEP should also be seen 2 weeks after initiating therapy to address any possible drug toxicity, including measuring blood counts, renal function, and hepatic testing.[9]

HIV testing should be used to monitor exposed workers for seroconversion after occupational exposure. Baseline testing should be performed at the time of exposure, with follow-up testing at 6 weeks, 12 weeks, and 6 months. If a fourth-generation combination HIV Ag/Ab test is used, HIV follow-up testing could be concluded as early as 4 months after exposure. Follow-up at 12 months is recommended for workers that develop HCV infection following exposure to patients coinfected with HIV and HCV and may be considered in other cases, including anxiety of the exposed worker. HIV tests should also be performed on any exposed person with an illness compatible with an acute retroviral syndrome, regardless of the interval since exposure. Cases of occupationally acquired HIV should be reported to CDC.[9]

Medical surveillance

CDC does not recommend routine screening for HIV infection at most worksites.

Screening of workers who handle concentrated sources of HIV, such as those in research laboratories, may be appropriate. This is only useful if done periodically as well as at the time of hire. Healthcare workers who are identified as HIV infected must be evaluated if they do exposure-prone invasive procedures. An expert panel should evaluate the risk to patients and considers any restrictions on the healthcare worker's activities. However, the risk of transmission appears to be extremely low with very few documented cases. A dentist in Florida may have transmitted HIV to six patients, although some researchers have questioned the conclusion.[10–12] An orthopedic surgeon in France, who had contracted HIV from a patient, infected a patient during hip replacement surgery.[13] A review of his work practices indicated several practices that could have transmitted the infection.[13] A nurse in France apparently transmitted HIV to a patient, despite lack of evidence of exposure to blood.[14] An obstetrician in Spain infected a patient during an emergency C-section.[15] By contrast, several large studies have looked back at hundreds or thousands of patients cared for by surgeons with HIV infection and did not identify any cases of transmission.[16–21]

HIV-infected workers who are immunocompromised need to be evaluated in the context of risk to their health when exposed to tuberculosis in patient care or animal care settings.

Prevention

The approach to prevention of the transmission of HIV and other bloodborne pathogens is specified in the OSHA Bloodborne Pathogen standard (29 CFR 1910.1030). In addition, CDC has released guidelines regarding precautions to reduce the spread of HIV infection. All workers should routinely use appropriate barrier precautions to prevent skin and mucous membrane exposure when contact is anticipated with blood or other body fluids of any patient (or other worker). Gloves should be worn when touching blood and body fluids, mucous membranes, or nonintact skin of all patients, when handling items or surfaces soiled with blood or body fluids, and when performing venipuncture and other vascular access procedures. Gloves should be changed after contact with each patient. Masks and protective eyewear or face shields should be worn during procedures that are likely to generate droplets of blood or other body fluids to prevent exposure of mucous membranes of the mouth, nose, and eyes. Gowns or aprons should be worn during procedures that are likely to generate splashes of blood or other body fluids. Hands and other skin surfaces should be washed immediately and thoroughly if contaminated with blood or other body fluids. Hands should be washed immediately after gloves are removed. All healthcare workers should take precautions to prevent injuries caused: by needles, scalpels, and other sharp instruments or devices during procedures; when cleaning used instruments; during disposal of used needles; and when handling sharp instruments after procedures. To prevent needlestick injuries, needles should not be recapped, purposely bent or broken by hand, removed from disposable syringes, or otherwise manipulated by hand. After they are used, disposable syringes and needles, scalpel blades, and other sharp items should be placed in puncture-resistant containers for disposal; the puncture-resistant containers should be located as close as practical to the use area. Large-bore reusable needles should be placed in a puncture-resistant container for transport to the reprocessing area. Although saliva has not been implicated in HIV transmission, to minimize the need for mouth-to-mouth resuscitation, mouthpieces, resuscitation bags, or other ventilation devices should be available for use in areas where the need for resuscitation is predictable. Healthcare workers who have exudative lesions on their hands should refrain from all direct patient care and from handling patient care equipment until the condition resolves. As with most safety issues, engineering controls are more effective than PPE. The introduction of various safety devices such as needleless intervention systems has reduced blood/body fluid exposures in many healthcare worksites.[22] Thorough training is necessary when introducing new devices to prevent a flare of injuries related to the new device. Medical/surgical/dental organizations and institutions should identify exposure-prone procedures where the procedures are performed. Healthcare workers who perform exposure-prone procedures should know their HIV antibody status. Healthcare workers who are infected with HIV should not perform exposure-prone procedures unless they have sought counsel from an expert review panel and been advised under what circumstances, if any, they may continue to perform these procedures. Such circumstances could include notifying prospective patients of the healthcare workers' seropositivity before they undergo exposure-prone invasive procedures.[23] There have been no documented cases of HIV transmission involving environmental surfaces. HIV is rapidly inactivated via contact with many common germicides at low concentrations. Household bleach (sodium hypochlorite) diluted with water at concentrations of one part bleach to 10 parts water can inactivate HIV. Spills or patient-derived infected material can be cleaned by removing visible material before chemical decontamination. Spills of concentrated research laboratory solutions (which may contain HIV in concentrations thousands of times higher than clinically encountered) should be chemically decontaminated, cleaned, and then redecontaminated. Cleaning spills of infective material requires the use of personal protective devices.

There is currently no HIV vaccine available. There have been clinical trials conducted in the past decade moderated through the US Military HIV Research Program (MHRP), an international collaborative based at the Walter Reed Army Institute of Research with affiliated sites in Kenya, Tanzania, Uganda, and Thailand. The Thai HIV vaccine efficacy trial, known as RV144, tested the "prime-boost" combination of two vaccines: (i) a canarypoxbased vaccine called ALVAC® HIV (which is the "priming" vaccine) and AIDSVAX® B/E vaccine (which is the "boosting" vaccine). Results of this trial showed that the prime-boost combination of ALVAC® HIV and AIDSVAX® B/E lowered the rate of HIV infection by 31.2% compared to placebo after 42 months of follow-up, with protection against HIV evident at 12 months postvaccination (60% vaccine efficacy at this time point).[24] There is a current clinical trial sponsored by the Pox-Protein Private-Public Partnership (including the National Institute of Allergy and Infectious Diseases) being conducted in South Africa to test HVTN 100. This is an HIV vaccine regimen consisting of two experimental vaccines: (i) ALVAC® HIV and a gp120 protein subunit vaccine with an adjuvant that enhances the body's immune response.[25] Both ALVAC® HIV (supplied by Sanofi Pasteur) and the protein vaccine (supplied by Novartis Vaccines) have been modified from RV144 to be specific to HIV subtype C, the predominant HIV subtype in southern Africa.

References

1. Bell DM. Occupational risk of human immunodeficiency virus infection in healthcare workers: an overview. *Am J Med* 1997;102(5B):9–15.
2. Ippolito G, Puro V, De Carli G, et al. The risk of occupational human immunodeficiency virus infection in health care workers: Italian multicenter study. *Arch Intern Med* 1993;153(12): 1451–1458.
3. Joyce MP, Kuhar D, Brooks JT. Notes from the Field: Occupationally Acquired HIV Infection among Health Care Workers—United States, 1985–2013. *MMWR* 2015;63(53): 1245–1246.
4. Cardo DM, Culver DH, Ciesielski CA, et al. A case–control study of HIV seroconversion in health care workers after percutaneous exposure. *N Engl J Med* 1997;337(21):1485–1490.
5. Cooper DA, Gold J, Maclean P, et al. Acute AIDS retrovirus infection. Definition of a clinical illness associated with seroconversion. *Lancet* 1985;1(8428):537–540.
6. National Institute of Allergy and Infectious Diseases. How HIV Causes AIDS. Available at: http://www.niaid.nih.gov/topics/HIVAIDS/Understanding/howHIVCausesAIDS/Pages/howhiv.aspx (accessed on May 3, 2016).
7. Centers for Disease Control and Prevention. Laboratory Testing for the Diagnosis of HIV Infection. 2014. Available at: http://www.cdc.gov/hiv/pdf/HIVtestingAlgorithmRecommendation-Final.pdf (accessed on May 3, 2016).
8. Food and Drug Administration. FDA News Release: FDA Approves Diagnostic Test to Differentiate Between Types of HIV Infection. 2015. Available at: http://www.fda.gov/NewsEvents/Newsroom/PressAnnouncements/ucm455813.htm (accessed on June 28, 2016).
9. Kuhar DT, Henderson DK, Struble KA, et al. Updated US public health service guidelines for the management of occupational exposures to human immunodeficiency virus and recommendations for postexposure prophylaxis. *Infect Control Hosp Epidemiol* 2013;34(9):875–892.
10. Weiss SH, Resnick L, DeBry R, et al. Analysis of reported HIV transmission in a dental practice. Ninth International Conference on AIDS, abstract PO-A11-0186, June 6–11, 1993, International Congress Center, Berlin, Germany. International AIDS Society. Available at: http://quod.lib.umich.edu/c/cohenaids/5571095.0067.025/167?view=image&size=200 (accessed June 28, 2016).
11. Barr S The 1990 Florida dental investigation: is the case really closed? *Ann Intern Med* 1996;124(2):250–254.
12. Brown D The 1990 Florida dental investigation: theory and fact. *Ann Intern Med* 1996;124(2):255–256.
13. Lot F, Séguier JC, Fégueux S, et al. Probable transmission of HIV from an orthopedic surgeon to a patient in France. *Ann Intern Med* 1999;130:1–6.
14. Goujon CP, Schneider VM, Grofti J, et al. Phylogenetic analyses indicate an atypical nurse-to-patient transmission of human immunodeficiency virus type 1. *J Virol* 2000;74(6): 2525–2532.
15. Mallolas J, Gatell JM, Bruguera M. Transmission of HIV-1 from an obstetrician to a patient during a caesarean section. *AIDS* 2006;20(2):285–287.

16. Crawshaw SC, West RJ. HIV transmission during surgery. *BMJ*, 1991;303:580;
17. Lowenfels AB, Wormser G. Risk of HIV transmission from surgeon to patient. *N Engl J Med*1991;325:888–889.
18. Porter JD, Cruickshank JG, Gentle PH, et al. Management of patients treated by a surgeon with HIV infection. *Lancet* 1990;335:113–114.
19. Von Reyn CF, Gilbert TT, Shaw FE, et al. Absence of HIV transmission from an infected orthopedic surgeon. A 13-year look-back study. *JAMA* 1993;269(14):1807–1811.
20. Rogers AS, Froggatt JW 3rd, Townsend T, et al. Investigation of potential HIV transmission to the patients of an HIV-infected surgeon. *JAMA* 1993;269(14):1795–1801.
21. Schwaber MJ, Sereti I. Investigation of patients treated by an HIV-infected cardiothoracic surgeon – Israel, 2007. *MMWR Morb Mortal Wkly Rep* 2009; 57(53):1413–1415.
22. Younger B, Hunt EH, Robinson C, et al. Impact of a shielded safety syringe on needlestick injuries. *Infect Control Epidemiol* 1992;13:349–353.
23. Centers for Disease Control and Prevention. Recommendations for preventing transmission of human immunodeficiency virus and hepatitis B virus to patients during exposure-prone invasive procedures. *MMWR* 1991;40:1–9.
24. Rerks-Ngarm S, Pitisuttithum P, Nitayaphan S et al. Vaccination with ALVAC and AIDSVAX to prevent HIV-1 infection in Thailand. *N Engl J Med* 2009;361:1–12.
25. National Institute of Allergy and Infectious Diseases. NIH-Sponsored HIV Vaccine Trial Launches in South Africa – Early Stage Trial Aims to Build on RV144 Results. 2015. Available at: http://www.niaid.nih.gov/news/newsreleases/2015/Pages/HVTN100.aspx (accessed on May 3, 2016).

HUMAN T-CELL LYMPHOTROPHIC VIRUS

Common names for disease: HTLV-1—adult T-cell leukemia, HTLV-1 myelopathy, tropical spastic paraparesis; HTLV-2—HTLV-2 myelopathy; HTLV-3 and HTLV-4—no disease association currently identified
Classification: Family—Retroviridae; genus—*Deltaretrovirus*

Occupational setting

Healthcare workers, laboratory workers, and researchers with blood and body fluid exposure potential are all at risk of exposure to HTLV. One healthcare worker who unintentionally inoculated himself with blood from an adult T-cell leukemia/lymphoma patient (with underlying HTLV-1 infection) has been reported to have seroconverted.[1] A case series identifying 31 laboratory and healthcare workers exposed to HTLV-1 via puncture wounds found no positive seroconversions.[2] A retrospective analysis of hospital occupational exposure events in an HTLV-1-endemic area of Central Australia found no seroconversions.[3] Percutaneous occupational transmission of HTLV-2 occurred as a result of a hospital laboratory worker recapping a syringe after collecting material for arterial blood gas analysis.[4]

Exposure (route)

The transmission of HTLV-1 and HTLV-2 occurs through mechanisms similar to HIV (formerly known as HTLV-3 or lymphadenopathy-associated virus, but was subsequently reclassified as a lentivirus). Infection may occur through transfusion of blood products (whole blood, red blood cells, and platelets), organ transplantation, sharing of contaminated needles (among intravenous drug users), sexual transmission, and mother-to-child transmission during breastfeeding. HTLV-3 and HTLV-4, discovered in 2005, are believed to be transmitted through direct human contact with primates, but data demonstrating this mechanism is currently lacking.[5]

Pathobiology

HTLV-1 and HTLV-2 are RNA retroviruses. They use reverse transcriptase to generate DNA that then becomes incorporated into the host's DNA. Among US blood donors, there has been a general decline in HTLV-1 and HTLV-2 infections since the 1990s.[6] In the United States, the overall prevalence of HTLV-1 and HTLV-2 is 22 per 100 000 population, with HTLV-2 more common than HTLV-1.[5]

Six HTLV-1 subclasses exist, and each subtype is endemic to a particular region: Subtype A (cosmopolitan subtype), Japan; Subtypes B, D, and F, Central Africa; Subtype C, Melanesia (a subregion of Oceania, including Papua New Guinea); and Subtype E, South and Central Africa.[7] HTLV-1 preferentially attacks helper (CD4) T-lymphocytes but is capable of infecting other types of cells. HTLV-1 can cause two types of disease: (i) adult T-cell leukemia/lymphoma and (ii) HTLV-1-associated myelopathy (an inflammation of the nerves in the spinal cord that can cause lower extremity weakness and bladder incontinence). HTLV-1 can also cause inflammation of the eye (uveitis), joints (arthritis), muscles (myositis), lung (alveolitis), and skin (dermatitis).[8]

HTLV-2 is endemic in native populations living in the central area of Africa and America and has recently spread among intravenous drug users in the United States and Europe.[9] HTLV-2 predominantly affects killer (CD8) T-lymphocytes. In the vast majority of cases (up to 99%), most individuals do not develop any physical manifestation of HTLV-2 disease but it has been associated with myelopathy.[10] HTLV-2 is also associated with milder neurologic disorders and chronic pulmonary infections.[5]

The novel HTLV-3 and HTLV-4 have been isolated only in a few cases with no disease yet been associated with these viruses.[5]

Diagnosis

Initial screening with an ELISA will not differentiate between HTLV-1 and HTLV-2.

Confirmation with Western blot, immunofluorescence assay (IFA), or PCR is required. PCR also quantifies the proviral load, which is used as a marker for progression in HTLV-1-associated myelopathy/tropical spastic paraparesis.[5]

Treatment

There is no treatment for these viral infections. All patients with HTLV-1 or HTLV-2 infection should be counseled extensively on the lifelong implications of their infection.[11,12] Treatment of HTLV-related disease (such as adult T-cell leukemia) is the same regardless of the presence or absence of HTLV infection.[5]

Medical surveillance

Surveillance for seroconversion is indicated for individuals with a known parenteral or mucous membrane exposure to infected blood or body fluids. There is no basis for routine testing of workers. There is no current standard regarding healthcare workers who are known to be infected with HTLV-1/HTLV-2.

Prevention

Control of exposure is the only effective approach. Because HTLV-1 and HTLV-2 are bloodborne pathogens, the requirements of OSHA regulation 29 CFR 1910.1030 are applicable. However, these agents are not specifically discussed in the regulations. Approaches that are protective of HIV transmission will provide protection from these viruses. Disinfection is accomplished with 10% sodium hypochlorite solution and other common germicides.

References

1. Kataoka R, Takehara N, Iwahara Y, et al. Transmission of HTLV-I by blood transfusion and its prevention by passive immunization in rabbits. *Blood* 1990;76:1657–1661.
2. Amin RM, Jones B, Rubert M, et al. Risk of retroviral infection among retrovirology and health care workers. American Society for Microbiology, 92nd General Meeting, New Orleans, LA, USA, May 26–30, 1992 (abstract T-20).
3. Hewagama S, Krishnaswamy S, King L, et al. Human T-cell lymphotropic virus type 1 exposures following blood-borne virus incidents in central Australia, 2002–2012. *Clin Infect Dis* 2014;59(1):85–87.
4. Menna-Barreto M.. HTLV-II transmission to a health care worker. *Am J Inf Control* 2006;34(3):158–160.
5. Murphy EL, Bruhn RL. Human T-lymphotropic virus (HTLV). In: Bennett JE, Dolin R, Blaser M. eds. *Mandell, Douglas, and Bennett's Principles and Practice of Infectious Disease*, 8th edn. Philadelphia, PA: Elsevier Saunders, 2015:2038–2053.
6. Cook LB, Taylor GP. HTLV-1 and HTLV-2 prevalence in the United States. *J Infect Dis* 2014;209(4):486–487.
7. Proietti FA, Carneiro-Proietti AB, Catalan-Soares BC, et al. Global epidemiology of HTLV-I infection and associated diseases. *Oncogene* 2005;24(39):6058–6068.
8. National Centre for Retrovirology. HTLV-1. 2008. Available at: http://www.htlv1.eu/htlv_one.html (accessed on May 3, 2016).
9. Vrielink H, Zaaijer HL, Reesink HW. The clinical relevance of HTLV type I and II in transfusion medicine. *Transfus Med Rev* 1997;11:173–179.
10. National Centre for Retrovirology. HTLV-2. 2006. Available at: http://www.htlv1.eu/htlv_two.html (accessed on May 3, 2016).
11. Centers for Disease Control and Prevention. Recommendations for counseling persons infected with human T-lymphotrophic virus, types I and II. *MMWR* 1993;42(RR-9):1–13.
12. Ramos JC, Lossos IS. Newly emerging therapies targeting viral-related lymphomas. *Curr Oncol Rep* 2011;13(5):416–426.

INFLUENZA VIRUS

Common names for disease: Influenza, flu, grippe
Classification: Family—Orthomyxoviridae; genera—Influenza virus A, Influenza virus B, Influenza virus C

Occupational setting

Healthcare, long-term care, and childcare workers are most susceptible to contracting the flu, given the fact that they may interact with individuals at higher risk of developing influenza and severe influenza-related complications (including adults greater than 65 years of age, pregnant women, immunocompromised adults, and children less than 2 years of age). Pig farmers, handlers, and pork producers can contract swine influenza as a result of having direct exposure to the animal or close contact with pork products.

Exposure (route)

Large-particle droplet transmission of influenza virus requires close contact between source and recipient persons (less than 1 m). Airborne transmission of evaporated droplets (less than 5 μm) suspended in air for extended periods of time is possible but limited data supports this mechanism. Contact with respiratory-droplet contaminated surfaces can also lead to infection.[1]

Pathobiology

There are three influenza types in humans: types A, B, and C. Influenza type A is the main source of flu in humans and causes seasonal disease every winter season in the United

States. Between 1976 and 2007, flu-associated deaths in the United States have ranged from a low of about 3 000 to a high of about 49 000 people.[2] Frequent changes in the virus's antigenicity are an important feature of influenza type A. Influenza A viruses are divided into subtypes based on two proteins on the surface of the virus: the hemagglutinin (H) and the neuraminidase (N). There are 18 different hemagglutinin subtypes and 11 different neuraminidase subtypes. H1N1 and H3N2 are the most common circulating strains, with H3N2 causing the greatest morbidity and mortality.[3] In any given flu season, there can be alterations to the H and N surface protein makeup. Such genetic variations are referred to as antigenic drift (small, minor nucleotide changes in the influenza viral genome that happen continually over time as the virus replicates) or antigenic shift (large, significant change in antigenic type resulting in new hemagglutinin and neuraminidase proteins).[4]

The most recent occurrence of antigenic shift happened in 2009 with the advent of a new H1N1 influenza strain. This virus had two genes from flu viruses that normally circulate in pigs in Europe and Asia, three genes that normally circulate in North American pigs, and genes from flu viruses from birds and humans.[5] The 2009 H1N1 was first detected in the United States in two isolated pediatric patients in California beginning in early April 2009. Laboratory testing found that the influenza viruses obtained from these two patients were similar to each other but were different from any previous strains on record. By late April 2009, the WHO had declared the 2009 H1N1 outbreak a "Public Health Emergency of International Concern" and recommended that countries establish rigorous influenza surveillance systems to identify positive cases. Over 200 countries reported laboratory-confirmed cases of pandemic influenza H1N1 2009 through early 2010. Globally, there were an estimated 200 000 respiratory-associated deaths with an additional 83 000 cardiovascular deaths associated with the 2009 H1N1 pandemic, with 80% of these deaths occurring in those under 65 years of age.[6] In the United States, there were approximately 60.8 million cases, 274 304 hospitalizations, and 12 469 deaths attributed to H1N1 between April 2009 and April 2010.[7] The most impacted populations that were affected by H1N1 include children, young adults, pregnant women, persons with underlying chronic medical conditions, and indigenous populations.[3] The H1N1 virus that caused the 2009 pandemic is now considered a prevalent human flu virus and continues to circulate seasonally worldwide.

In 2005 and 2006, three cases of infection with influenza A viruses that normally circulate in swine ("variant viruses") were reported in humans. Beginning in 2007, about three to four cases were reported per year in the United States, due to new regulations mandating reporting of human infection with a nonhuman influenza virus.[5]

Other influenza A viruses with pandemic potential include avian H5N1 and influenza A H7N9, which are two different "bird flu" viruses. Human infections with these viruses have occurred rarely but have been reported.[8,9] In 2014–2015, a large outbreak of avian influenza A (including H5N2, H5N8, and H5N1) occurred among birds in the United States with no identified disease in humans.[10]

Influenza B viruses can cause severe disease in humans. They are categorized by lineages and strains. Currently circulating influenza B viruses belong to one of two lineages: B/Yamagata and B/Victoria. Influenza type C virus causes a mild respiratory illness and is not thought to cause epidemics.

The typical incubation period for influenza is 1–4 days. Uncomplicated influenza illness is characterized by fever, myalgia, headache, malaise, nonproductive cough, sore throat, and rhinitis. Uncomplicated influenza illness typically resolves after 3–7 days, but it can cause primary influenza pneumonia as well as contribute to coinfections with other viral or bacterial pathogens.[11] Influenza viruses cause disease among persons in all age groups, but rates of infection are highest among children. The risks for complications (including hospitalizations and deaths) from influenza are usually higher among persons 65 and older, young children, and persons of any age who have medical conditions that place them at increased risk for complications from influenza.

Diagnosis

Tests do not need to be done on all patients, but can be helpful in contact investigations to identify cause of a respiratory illness outbreak. Rapid influenza diagnostic tests provide results within 15 minutes and are 50–70% sensitive and 90% specific for detecting influenza.[12] Confirmatory diagnostic tests available for influenza include viral culture, serology, rapid antigen testing, PCR, immunofluorescence assays, and rapid molecular assays.

Treatment

Neuraminidase inhibitors have activity against both influenza A and B viruses. There are currently 3 influenza antiviral medications recommended for use: oral oseltamivir (Tamiflu®), inhaled zanamivir (Relenza®), and intravenous peramivir (Rapivab®). For hospitalized patients and patients with severe or complicated illness, treatment with oral or enterically administered oseltamivir is recommended.[13] There has been reported antiviral resistance among the influenza strains in circulation. Antiviral resistance can also develop in those with weakened immune systems, either before or after treatment. Adamantanes are medications that are active against influenza A viruses, but not influenza B viruses. Amantadine and rimantadine are the two main antiviral drugs in this class. There continue to be high levels of resistance (>99%) to adamantanes among H3N2 and 2009 H1N1 viruses.[13]

Medical surveillance

The CDC's Advisory Committee on Immunization Practices (ACIP) has recommended that all healthcare workers (including health professional students and residents) be vaccinated annually against influenza.[14] To reduce transmission of influenza in the healthcare setting, workers with fever and respiratory symptoms should not report to work until 24 hours after their fever ends (100°F/37.8°C or lower) without the use of medication.[15] Additional surveillance measures that can be incorporated into the healthcare workplace include screening workers for flu symptoms before their shift starts and tracking absences of workers who care for flu patients, with contingency plans in place should a worker have to leave his/her shift early due to illness.[15]

Prevention

Because of the small size of respirable aerosols, surgical masks do not significantly reduce the transmission rate between individuals. Thus the administration of flu vaccine is the best option for reducing the chance of acquiring seasonal flu and spreading it to others. When more people get vaccinated against the flu, the chance for flu spreading in the workplace and the greater community is reduced.

To determine what flu strains will be in the following year's seasonal flu vaccine formulation, the WHO consults with experts to review surveillance data worldwide and make recommendations for the composition of the seasonal influenza vaccine for the Northern Hemisphere. In the United States, the Vaccines and Related Biological Products Advisory Committee considers the WHO recommendations and makes a final decision regarding composition of seasonal flu vaccine for the United States.

For the 2014–2015 flu year, there are two flu vaccine formulations: trivalent and quadrivalent. Trivalent flu vaccine protects against two influenza A viruses (primarily H1N1 and an H3N2) and an influenza B virus. The following trivalent flu vaccines are available: (i) standard-dose intramuscular shot developed in eggs for people 6 months of age and older, (ii) an intramuscular shot developed in cell culture for those 18 years and older, (iii) an intradermal shot approved for people 18 through 64 years of age, (iv) a high-dose shot for people 65 and older, and (v) a recombinant egg-free shot for people 18 years and older.[2] Quadrivalent flu vaccine protects against two influenza A viruses and two influenza B viruses. There are two quadrivalent vaccines available: (i) an intramuscular shot that can be given to people 6 months of age and older and (ii) a live attenuated nasal spray that is approved for people aged 2–49 who do not have any contraindication to receiving live virus (i.e., immunocompromised persons).[2]

While the risk of acquiring flu in the community is appreciable, worker populations are particularly susceptible to contracting and spreading influenza. Influenza affects 5–10% of the workforce annually, with up to 111 million workdays lost. These lost workdays cost an estimated $7 billion/year in sick days and lost productivity.[16] Influenza vaccination for healthy working adults has been found to be cost-saving for employers, associated with a marginal cost-effectiveness ratio of $113 per quality-adjusted day gained or $41 000 per quality-adjusted life-year saved compared with antiviral treatment.[17] Along with encouraging or providing vaccination, employers should develop business continuity plans that address the possibility of an influenza pandemic.

The premise behind vaccinating healthcare providers and other healthcare employees in close contact with high-risk individuals is to reduce transmission of the virus from employees to patients. Educating healthcare workers on the benefits and risks of influenza vaccination and providing vaccinations in the workplace at convenient locations and times at no cost are effective strategies to increase coverage among healthcare workers.[18] Universal precautions still apply for healthcare personnel, which include practicing effective hand hygiene (washing hands frequent with soap and running water before and after working with patients) and using appropriate personal protective equipment (including gloves, masks that cover your mouth and nose).

With respect to pig farmers and pork producers, they should get annual seasonal influenza vaccines. Although vaccination of people with seasonal influenza vaccine probably will not protect against infection from swine-specific influenza viruses (since they are substantially different from human influenza A viruses), flu vaccination of these workers is still imperative to reduce the risk of transmitting seasonal influenza A viruses from ill people to other people and to pigs. Universal precautions apply for these workers (including hand hygiene and proper PPE).

References

1. Centers for Disease Control and Prevention. Clinical Signs and Symptoms of Influenza. 2015. Available at: http://www.cdc.gov/flu/professionals/acip/clinical.htm (accessed on May 4, 2016).
2. Centers for Disease Control and Prevention. Key Facts about Seasonal Flu Vaccine. 2014. Available at: http://www.cdc.gov/flu/protect/keyfacts.htm (accessed on May 4, 2016).
3. Centers for Disease Control and Prevention. H1N1: Overview of a Pandemic. 2009. Available at: http://www.cdc.gov/h1n1flu/yearinreview/yir1.htm (accessed on May 4, 2016).
4. Centers for Disease Control and Prevention. How the Flu Virus Can Change: 'Drift' and 'Shift'. 2014. Available at: http://www.cdc.gov/flu/about/viruses/change.htm (accessed on May 4, 2016).
5. Centers for Disease Control and Prevention. What People Who Raise Pigs Need To Know About Influenza (Flu). Available at: http://www.cdc.gov/flu/swineflu/people-raise-pigs-flu.htm (accessed on May 4, 2016).

6. Dawood FS, Iuliano AD, Reed C, et al. Estimated global mortality associated with the first 12 months of 2009 pandemic influenza A H1N1 virus circulation: a modelling study. *Lancet Infect Dis* 2012;12(9):687–695.
7. Shrestha SS, Swerdlow DL, Borse RH, et al. Estimating the burden of 2009 pandemic influenza A (H1N1) in the United States (April 2009–April 2010). *Clin Infect Dis* 2011;52(Suppl 1):S75–S82.
8. Uyeki TM. Human infection with highly pathogenic avian influenza A (H5N1) virus: review of clinical issues. *Clin Infect Dis* 2009;49:279–290.
9. Gao HN, Lu HZ, Cao B et al. Clinical findings in 111 cases of influenza A (H7N9) virus infection. *N Engl J Med* 2013;368: 2277–2285.
10. Jhung MJ, Nelson DI. Outbreaks of avian influenza A (H5N2), (H5N8), and (H5N1) among birds—United States, December 2014–January 2015. *MMWR* 2015;64(4):111.
11. Centers for Disease Control and Prevention. Clinical Signs and Symptoms of Influenza. 2015. Available at: http://www.cdc.gov/flu/professionals/acip/clinical.htm (accessed on May 4, 2016).
12. Centers for Disease Control and Prevention. Influenza Symptoms and the Role of Laboratory Diagnostics. 2014. Available at: http://www.cdc.gov/flu/professionals/diagnosis/labrolesprocedures.htm (accessed on May 4, 2016).
13. Centers for Disease Control and Prevention. Influenza Antiviral Medications: Summary for Clinicians. 2015. Available at: http://www.cdc.gov/flu/professionals/antivirals/summary-clinicians.htm (accessed on May 4, 2016).
14. Centers for Disease Control and Prevention. Prevention and control of seasonal influenza with vaccines: recommendations of the Advisory Committee on Immunization Practices (ACIP). *MMWR* 2009;58:1–52.
15. Occupational Safety and Health Administration. Employer Guidance – Reducing Healthcare Workers' Exposures to Seasonal Flu Virus. Available at: https://www.osha.gov/dts/guidance/flu/healthcare.html (accessed on May 4, 2016).
16. Molinari NA, Ortega-Sanchez IR, Messonnier ML, et al. The annual impact of seasonal influenza in the US: measuring disease burden and costs. *Vaccine* 2007;25(27):5086–5096.
17. Rothberg MB, David DN. Vaccination versus treatment of influenza in working adults: a cost-effectiveness analysis. *Am J Med* 2005;118:68–77.
18. Centers for Disease Control and Prevention. Influenza Vaccination Information for Health Care Workers. 2014. Available at: http://www.cdc.gov/flu/healthcareworkers.htm (accessed on May 4, 2016).

MEASLES VIRUS

Common names for disease: Measles, red measles, rubeola
Classification: Family—Paramyxoviridae; genus—*Morbillivirus*

Occupational setting

Employees in the healthcare setting (particularly pediatrics personnel), childcare workers, and school employees are at increased risk of exposure. A study conducted in 1996 in medical facilities in a county in Washington State indicated that healthcare workers were 19 times more likely to develop measles than other adults.[1]

Exposure (route)

Measles is transmitted through the respiratory route via aerosol droplets. It can spread to others through coughing and sneezing. It is one of the most communicable of the infectious diseases: if one person has measles, 90% of nonmeasles immune people in close contact with that source person will contract the virus. Measles virus can live for up to 2 hours on a surface or in an airspace where the infected person coughed or sneezed.[2]

Pathobiology

The measles virus is 100–200 nm in diameter, with a core of single-stranded RNA. Two membrane envelope proteins are important in pathogenesis: (i) F (fusion) protein, which is responsible for fusion of virus and host cell membranes, viral penetration, and hemolysis, and (ii) H (hemagglutinin) protein, which is responsible for adsorption of virus to cells.[3] There is only one antigenic type of measles virus. Its natural hosts are humans.

The incubation period is 10–14 days, after which a prodromal phase of malaise, fever, anorexia, cough, and coryza begins. A few days of prodrome are followed by the development of Koplik's spots (small, red, irregular lesions with blue–white centers) in the mouth and an erythematous, maculopapular rash primarily on the face and trunk. The illness lasts 7–10 days in most cases. Although viral shedding is greatest during the late prodrome due to coughing, the contagious period extends from several days before the development of the rash to several days after.

Most common measles complications include diarrhea, otitis media, and pneumonia. Complications typically appear in children younger than 5 years of age and adults 20 years of age and older. Measles during pregnancy results in increased rates of premature labor, spontaneous abortion, and low birth weight, but it is not associated with congenital abnormalities. Acute encephalitis is seen in one of every 1000 cases. Subacute sclerosing panencephalitis (SSPE) is caused by persistent measles virus infection in the brain, with onset occurring an average of 7 years after measles illness. SSPE incidence is 5–10 cases per million reported measles cases. Death occurs in two of every 1000 reported cases with pneumonia accounting for about 60% of deaths. The most common causes of death are pneumonia in children and acute encephalitis in adults.[3]

While measles is still common in many parts of the world, it has recently had a resurgence in the United States (Figure 21.10). The disease was declared eliminated from the United States in 2000. Starting in 2008, there have been an increasing number of domestic measles cases which have largely been attributed

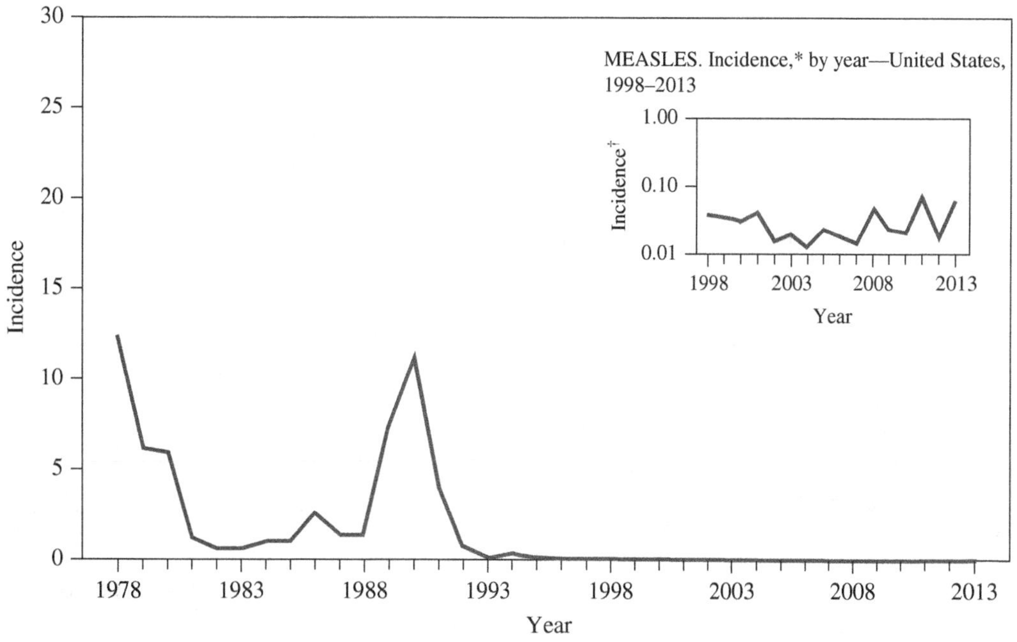

FIGURE 21.10 Measles Incidence per 100 000 by year — United States, 1978–2013. †In the inset figure, the *Y* axis is a log scale. Measles vaccine was licensed in 1963. Endemic measles was declared eliminated from the United States in 2000. *Source*: MMWR 2015; 62(53):85.

to parents choosing not to have their children vaccinated based on the paper from a British physician (which has since been negated by the publisher) that children who received MMR vaccine had increased risk of developing autism.[4] A series of measles outbreaks have ensued even though subsequent studies have demonstrated no connection.[4,5]

A measles epidemic beginning at Disneyland in Anaheim, California, resulted in occupational exposure to measles among theme park employees. The suspected source is believed to be a traveler who became infected overseas with measles, then visited the amusement park while infectious.[6] From December 28, 2014, through June 26, 2015, there were178 people from 24 states and the District of Columbia reported to have contracted measles with 117 linked to the California outbreak.[6] Analysis by CDC scientists showed that the measles virus in this outbreak is identical to the virus type that caused a measles outbreak in the Philippines in 2014.

Diagnosis

Measles is a clinical diagnosis based on the presence of the maculopapular rash and Koplik's spots, which are pathognomonic for disease. While isolation of measles virus is not recommended as a routine method to diagnose measles, it can be important especially in outbreaks. Measles virus can be isolated from urine, nasopharyngeal aspirates, heparinized blood, or throat swabs. If confirmatory testing is needed, specimens for virus culture should be obtained from every person with a clinically suspected case of measles and shipped to the state public health laboratory or CDC. Acute and convalescent serologic titer change can also be used to establish the diagnosis, via enzyme-linked immunoassay.

Treatment

There is no specific treatment other than supportive therapy for fever and dehydration.

Medical surveillance

All healthcare workers should observe airborne precautions in caring for patients with measles. If measles exposures occur in a healthcare facility, all contacts should be checked for measles immunity. Those healthcare workers without evidence of immunity should be offered the first dose of MMR vaccine, provided with postexposure immune globulin, and excluded from work from days 5 to 21 following exposure.[7] Those with documentation of one MMR vaccine dose may remain at work and should receive the second dose.[8] If a healthcare worker who has a history of measles immunity but contracts measles from an actively infected patient, that worker should be excluded from work for at least ≥4 days following rash onset.[8]

Prevention

The only measles virus vaccine available in the United States is formulated based on a live attenuated strain. This particular measles vaccine has been in circulation since 1968 because it causes less adverse side effects compared to its predecessors. Two doses of measles vaccine, as a combination measles–mumps–rubella (MMR) vaccine separated by a minimum of 4 weeks, are routinely recommended for all children. The CDC recommended schedule for MMR vaccination in children is to administer the first dose at age 12 through 15 months of age and the second dose at 4 through 6 years of age.

Documentation of vaccination for children entering day care and primary school provides protection for education personnel in these settings; however, they should be encouraged to ensure their own immunity for personal protection. Colleges and other post-high school educational institutions are potential high-risk areas for measles and thus prematriculation vaccination requirements for measles immunity have been shown to significantly decrease the risk of measles outbreaks in these settings. Students at post-high school educational institutions without evidence of measles immunity need two doses of MMR vaccine, with the second dose administered no earlier than 28 days after the first dose.

All persons who work in healthcare facilities should have evidence of immunity to measles. Such evidence includes any of the following: (i) written documentation of vaccination with 2 doses of live measles or MMR vaccine administered at least 28 days apart, (ii) laboratory evidence of immunity, (iii) laboratory confirmation of disease, or (iv) birth before 1957.[8] Serologic testing is appropriate if the immunity status cannot be verified. Those individuals who do not have evidence of immunity against measles should get at least one dose of MMR vaccine. The MMR vaccinations and serologic titers should be tracked to encourage compliance.

Contraindications to vaccination include altered immunocompetence, recent administration of immune globulin, and severe febrile illness. Individuals with a history of anaphylactic reaction to eggs can be vaccinated, since such reaction has been found not to be associated with hypersensitivity to egg antigens but to other components of the vaccines (such as gelatin).[3] Pregnancy is a theoretical contraindication with the monovalent measles vaccine but a necessary contraindication if administered together with mumps and rubella vaccines. Women should be advised not to become pregnant for at least 30 days after vaccination.[3]

References

1. Steingart KR, Thomas AR, Dykewicz CA, et al. Transmission of measles virus in healthcare settings during a communitywide outbreak. *Infect Control Hosp Epidemiol* 1999;20:115–119.
2. Centers for Disease Control and Prevention. Transmission of Measles. 2014. Available at: http://www.cdc.gov/measles/about/transmission.html (accessed on May 4, 2016).
3. Centers for Disease Control and Prevention. *Measles. The Pink Book: Course Textbook*, 12th edn, 2012. Available at: http://www.cdc.gov/vaccines/pubs/pinkbook/meas.html (accessed on May 4, 2016).
4. Taylor B, Miller E, Farrington CP, et al. Autism and measles, mumps and rubella vaccine: no epidemiological evidence for a causal association. *Lancet* 1999;353:2026–2029.
5. Jain A, Marshall J, Buikema A, et al.Autism occurrence by MMR vaccine status among US children with older siblings with and without autism. *JAMA* 2015;313(15):1534–1540. 10.1001/jama.2015.3077.
6. Centers for Disease Control and Prevention. Measles Cases and Outbreaks. Available at: http://www.cdc.gov/measles/cases-outbreaks.html (accessed on May 4, 2016).
7. Centers for Disease Control and Prevention. Measles, mumps, and rubella—vaccine use and strategies for elimination of measles, rubella, and congenital rubella syndrome and control of mumps: recommendations of the Advisory Committee on Immunization Practices (ACIP). *MMWR* 1998;47(RR-8):i–57.
8. Centers for Disease Control and Prevention. Immunization of Health-Care Personnel: Recommendations from the Advisory Committee of Immunization Practices. *MMWR* 2011;60(7):1–45.

MUMPS VIRUS

Common name for disease: Mumps
Classification: Family—Paramyxoviridae; genus—*Paramyxovirus*

Occupational setting

Persons working in healthcare facilities, day-care centers, schools, and clinical and research laboratories are at risk. Although healthcare-associated transmission of mumps is rare, it may be underreported because of the high percentage (up to 40%) of infected persons who might be asymptomatic.[1] There have been a number of healthcare facility-associated mumps outbreaks in the United States (including Tennessee 1986–1987; Utah 1994; Chicago, IL, 2006).[1] Exposures to mumps in healthcare settings can result in added economic costs (such as closure of inpatient units as well as on placement of work restrictions on essential personnel).[2] In the last 10 years, there have been several large mumps outbreaks in the United States (Figure 21.11) and around the world among fully vaccinated school-age children and young adults in high-contact settings.[3]

Exposure (route)

Mumps virus is primarily spread via airborne transmission of droplets of saliva or mucus from the mouth, nose, or throat of an infected person when he/she coughs, sneezes, or talks.[4] Direct contact with secretions or contact with fomites can also lead to transmission of virus. The greatest transmission of virus happens before the salivary glands begin to swell and up to 5 days after the swelling begins.[5] Mumps virus is rapidly inactivated by formalin, ether, chloroform, heat, and ultraviolet light.[6]

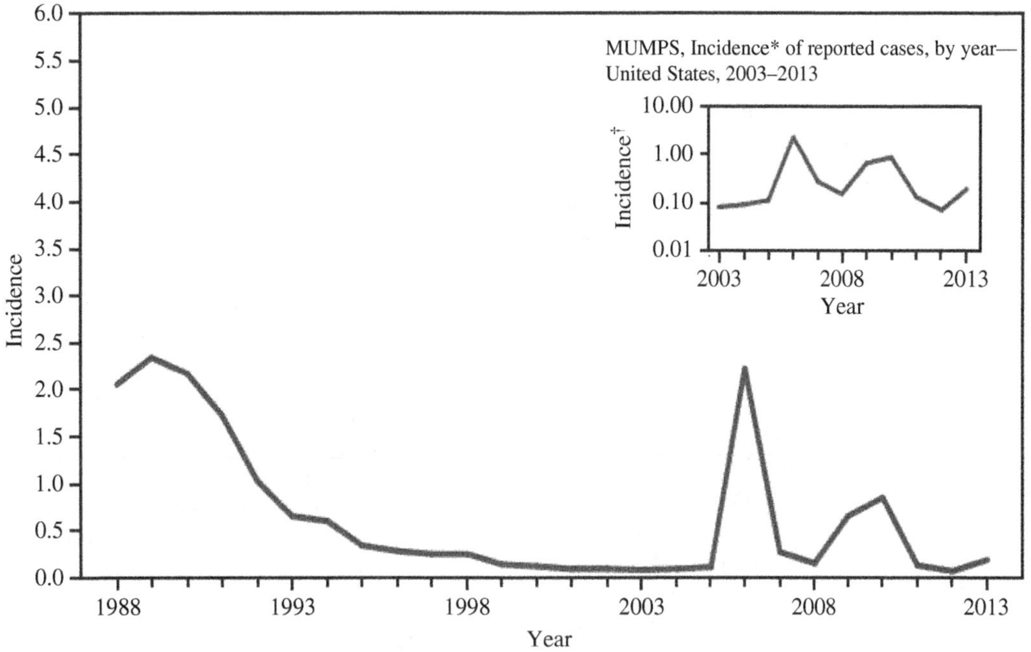

FIGURE 21.11 MUMPS Incidence per 100 000 by year — United States, 1988–2013. †In the inset figure, the *Y* axis is a log scale. The widespread use of a second dose of mumps vaccine beginning in 1989 was followed by historically low morbidity until 2006, when the United States experienced the largest mumps outbreak in two decades. The 2006 outbreak of approximately 6,000 cases primarily affected college students aged 18–24 years in the Midwest. A second large outbreak occurred during 2009–2010 and affected Orthodox Jewish communities in the Northeast. *Source*: MMWR 2015; 62(53):86.

Pathobiology

Mumps is caused by *Rubulavirus*, a single-stranded negative-sense RNA virus surrounded by a glycoprotein envelope.[7] Humans are the only natural host. A carrier state is not known. The incubation period is 14–18 days, with appearance of myalgia, anorexia, malaise, headache, and low-grade fever. Parotitis is the most common manifestation and occurs in 30–40% of infected persons, first presenting as an earache with accompanying tenderness on palpation of the angle of the jaw. The virus replicates in the nasopharynx and regional lymph nodes, with viremia occurring 12–25 days after initial infection. This viremia lasts from 3 to 5 days, during which time the virus spreads to multiple tissues (including the meninges, salivary glands, pancreas, testes, and ovaries). Central nervous system involvement is the most common extra-salivary complication of mumps, with aseptic meningitis being the prevalent manifestation.[7]

Symptomatic meningitis (including headache, stiff neck) appears in up to 15% of patients and resolves without sequelae in 3–10 days.[6] Adults are at higher risk for this complication than are children. Encephalitis is rare (less than 2 per 100 000 mumps cases).[6] Individuals who develop either meningitis or encephalitis usually recover fully. Orchitis (testicular inflammation) is the most common complication in postpubertal males, occurring in up to 50% of postpubertal males after parotitis. Oophoritis (ovarian inflammation) occurs in 5–7% of postpubertal females. Mumps can also lead to acute pancreatitis, which may develop without accompanying parotitis. Deafness attributed to the mumps virus happens in 0.5–5 per 100 000 cases, oftentimes with sudden onset.[7] In rare instances, this can lead to permanent unilateral hearing loss.[7]

Diagnosis

Mumps infection should be suspected when a patient presents with parotitis, acute salivary gland swelling, orchitis, or oophoritis that cannot be explained by other causes. Acute mumps infection can be detected by the presence of serum mumps IgM, a significant rise in IgG antibody titer in acute- and convalescent-phase serum specimens, IgG seroconversion, positive mumps virus culture, or detection of virus by RT-PCR.[8] Mumps virus can be detected from fluid collected from the parotid duct, other affected salivary gland ducts, the throat, from urine, and from cerebrospinal fluid (CSF). Parotid duct swabs yield the best viral sample.

Treatment

There is no therapy other than supportive measures, which are dependent on the type of glandular involvement.

Medical surveillance

Surveillance of workers exposed to mumps should be conducted during the incubation period to monitor for active signs of disease. Exposed workers in whom immunity cannot be established must be excluded during the lengthy incubation period.[9] Employers should report to public health authorities and follow up on cases of suspected mumps among workers.[10]

Prevention

The mumps vaccine is combined with measles and rubella, otherwise known as MMR vaccine. This vaccine is a live attenuated virus commercially produced since 1967. The vaccination schedule in children occurs at 12–15 months and 4–6 years of age. The vaccine produces an inapparent, mild, noncommunicable infection with antibody titer development in 95% of individuals.[11] Adverse reactions are rare but include parotitis and low-grade fever. Documentation of mumps disease, or laboratory evidence of mumps immunity, is an alternative to vaccination.

Workers who are potentially exposed to individuals with mumps should be immunized against mumps. In 2006, the Advisory Committee on Immunization Practices (ACIP) recommended an additional dose of MMR vaccine for adults at high risk of mumps exposure such as healthcare personnel.[6] A dose of mumps-containing vaccine is also administered in some non-outbreak settings. Healthcare personnel who may have been vaccinated as children but who lack documentation are routinely given an additional dose, which is often the third dose.[3] Healthcare personnel should be informed that mumps outbreaks have occurred in highly vaccinated populations in high-transmission settings and thus mumps should not be ruled out on the assumption they are immune because of vaccination. To protect day-care and school workers, children entering these facilities should show evidence of immunization. Day-care and school personnel may wish to ensure their immunity with revaccination.

References

1. Centers for Disease Control and Prevention. Immunization of Health-Care Personnel: Recommendations of the Advisory Committee on Immunization Practices (ACIP) Recommendations and Reports. *MMWR* 2011;60(RR07):1–45.
2. Centers for Disease Control and Prevention. Updated recommendations of the Advisory Committee on Immunization Practices (ACIP) for the control and elimination of mumps. *MMWR* 2006;55:629–630.
3. Fiebelkorn AP, Coleman LA, Belongia EA, et al. Mumps antibody response in young adults after a third dose of measles–mumps–rubella vaccine. *Open Forum Infect Dis* 2014;1(3):ofu094.
4. Centers for Disease Control and Prevention. Mumps Fast Facts. 2010. Available at: http://www.cdc.gov/mumps/about/mumps-facts.html (accessed on May 4, 2016).
5. Centers for Disease Control and Prevention. Transmission of Mumps. 2010. Available at: http://www.cdc.gov/mumps/about/transmission.html (accessed on May 4, 2016).
6. Epidemiology and Prevention of Vaccine-Preventable Diseases. The Pink Book: Course Textbook, 12th edn, 2012. Available at: http://www.cdc.gov/vaccines/pubs/pinkbook/mumps.html (accessed on May 4, 2016).
7. Defendi GL, Demirci CS, Abuhammour W, et al. Mumps. 2014. Available at: http://reference.medscape.com/article/966678-overview (accessed on May 4, 2016).
8. Fiebelkorn AP, Barskey A, Hickman C, et al. Mumps. In: *Manual for the Surveillance of Vaccine-Preventable Diseases*, 5th edn, 2012. Available at: http://www.cdc.gov/vaccines/pubs/surv-manual/chpt09-mumps.html (accessed on May 4, 2016).
9. Weber DJ, Rutala WA, Orenstein WA. Prevention of MMR among hospital personnel. *Pediatrics* 1991;119:322–325.
10. Kaplan KM, Marder DC, Cochi SL, et al. Mumps in the workplace. *JAMA* 1988;260:1434–1438.
11. Buynak EB, Hilleman MR. Live attenuated mumps virus vaccine. *Proc Soc Exp Biol Med* 1966;123:768.

NOROVIRUS (FORMERLY NORWALK VIRUS) AND OTHER ENTERIC VIRUSES

Common name for disease: Acute viral gastroenteritis
Classification: Family—Caliciviridae

Occupational setting

Workers employed in healthcare settings with limited sanitary facilities are at risk. Outbreaks of norovirus illness have afflicted personnel in nursing homes, hospitals, restaurants, cruise ships, schools, banquet halls, summer camps, and workplace cafeterias where food preparation occurs.[1]

Exposure (route)

Norovirus is found in the feces and vomit of infected people and is spread via the fecal–oral route. This virus is very contagious, with only a nominal number of viral particles (as few as 18) needed for transmission.[2] As a result, norovirus can disseminate rapidly. Personnel can become infected with the virus by consuming contaminated food or by (i) direct contact with another person who is infected (a healthcare worker, visitor, or another patient) or (ii) touching surfaces or objects contaminated with norovirus and then touching one's mouth or other food items.[3]

Pathobiology

Norovirus is a nonenveloped, single-stranded RNA virus. There are six norovirus genogroups, with three of the six (GI, GII, and GIV) impacting humans. More than 25 different genotypes have been identified within these three genogroups.

Since 2002, variants of the GII.4 genotype have been the most common cause of norovirus outbreaks.[2] The virus has been shown to cause structural changes in the lining cells of the stomach and small intestine. Following an incubation period of 12–48 hours, an illness characterized by nausea, vomiting, and watery nonbloody diarrhea develops, which can last for up to 72 hours. Abdominal pain is accompanied by fever and an elevated white count. Depending on the duration and severity of diarrhea and vomiting, dehydration may be of concern, especially in light of excessive fluid losses. In the healthy adult, this illness rarely results in mortality.

Diagnosis

Real-time reverse transcriptase-polymerase chain reaction (RT-PCR) is the most widely used diagnostic assay for detecting norovirus. The ideal method for testing for norovirus is to obtain stool specimens within 48–72 hours of a person developing acute symptoms. There are currently a few enzyme immunoassays (EIAs) available for detecting norovirus in stool samples, but they are not a sensitive test for diagnosing individual cases.[4]

Treatment

There is no treatment other than supportive care, which may include fluid and electrolyte replacement.

Medical surveillance

No routine surveillance is indicated. In outbreak situations, surveillance of staff for infection by symptom reporting may facilitate control of nosocomial spread. Suspected foodborne illness outbreaks in the workplace should be investigated.

Prevention

Although the possibility of vaccine development exists, it is currently in the preliminary stages of discussion. Excellent sanitation and personal hygiene is the principal means of control. Workers with symptoms of viral gastroenteritis should be evaluated by a healthcare provider and treated accordingly. If norovirus or other enteric virus is suspected, the worker should remain out of work for at least 48 hours after symptoms resolve to avoid spread of infection.[5] Precautions against secondary spread to family members should also be taken at home. In a healthcare facility, patients with suspected norovirus may be placed in private rooms or share rooms with other patients with the same infection. When caring for patients who are symptomatic with norovirus, strict hand hygiene and use of PPE (including gowns and gloves) are required.[3] Contaminated surfaces should be cleaned and disinfected with a chlorine bleach solution with a concentration of 1000–5000 ppm (5–25 tablespoons of household bleach (5.25%) per gallon of water) or other disinfectant registered as effective against norovirus by the Environmental Protection Agency (EPA).[6,7]

References

1. Centers for Disease Control and Prevention. Norovirus – For Food Workers. 2014. Available at: http://www.cdc.gov/norovirus/food-handlers/work-with-food.html (accessed on May 4, 2016).
2. Centers for Disease Control and Prevention. Norovirus – Clinical Overview. 2013. Available at: http://www.cdc.gov/norovirus/hcp/clinical-overview.html (accessed on May 4, 2016).
3. Centers for Disease Control and Prevention. Norovirus in Healthcare Settings. 2013. Available at: http://www.cdc.gov/HAI/organisms/norovirus.html (accessed on May 4, 2016).
4. Centers for Disease Control and Prevention. Norovirus – Laboratory Diagnosis and Treatment. 2012. Available at: http://www.cdc.gov/norovirus/hcp/diagnosis-treatment.html (accessed on May 4, 2016).
5. Centers for Disease Control and Prevention. Key Infection Control Recommendations for the Control of Norovirus Outbreaks in the Healthcare Setting. 2014. Available at: http://www.cdc.gov/hai/pdfs/norovirus/229110A-NorovirusControlRecomm508A.pdf (accessed on May 4, 2016).
6. Centers for Disease Control and Prevention. Preventing Norovirus Infection. Available at: http://www.cdc.gov/norovirus/preventing-infection.html (accessed on May 4, 2016).
7. US Environmental Protection Agency – Office of Pesticide Programs. List G: EPA Registered Hospital Disinfectants Effective Against Norovirus (Norwalk-like virus). 2014. Available at: https://www.epa.gov/sites/production/files/2016-06/documents/list_g_norovirus.pdf (accessed on June 28, 2016).

PARVOVIRUS B19

Common names for disease: Fifth disease, erythema infectiosum

Classification: Family—Parvoviridae; genus—*Erythrovirus*

Occupational setting

Employees in healthcare facilities, day-care centers, primary and secondary schools, and research laboratories may be at risk from exposure. There have been documented incidences of parvovirus transmission between infected source patient and healthcare workers, as well as between members of hospital staff.[1,2] Infection can be an occupational risk for school and childcare personnel, where up to 20% of susceptible workers can contract the virus from affected children.[3]

Exposure (route)

Transmission of parvovirus occurs most often by direct contact with large-droplet respiratory secretions (such as saliva, sputum, or nasal mucus) when an infected person coughs or sneezes.[4] Exposure can also occur via percutaneous exposure to blood, transfusion of blood products, and vertical transmission (from mother to fetus).[5]

Pathobiology

Parvovirus is a single-stranded, nonenveloped DNA virus. Three genotypes of the genus *Erythrovirus* are now recognized. Parvovirus B19, the first discovered human parvovirus, is the prototype of genotype 1 and is responsible for the majority of human infections worldwide.[6,7] Two additional human parvoviruses have been recently identified: (i) PARV4 (an orphan virus) and (ii) bocavirus (associated with respiratory infections).[8] The fact that this virus is not encapsulated within a lipid envelope makes it resistant to physical inactivation with heat or detergents.[9] The virus only replicates in human erythrocyte precursors.

The incubation period from acquisition of parvovirus B19 to onset of initial symptoms usually is between 4 and 14 days but can be as long as 21 days. The most frequent manifestation of parvovirus B19 infection is erythema infectiosum (EI). EI occurs worldwide, year-round, and most commonly in the school-age years. It typically starts as a mild, nonspecific illness consisting of fever, malaise, myalgia, and headache, and then after 7–10 days, the "slapped-cheek" facial rash appears. A symmetric, maculopapular, lacelike, and often pruritic rash then develops on the trunk, moving peripherally to involve the arms, buttocks, and thighs.[3] Parvovirus B19 also can cause other physiological manifestations including (i) asymptomatic infection, (ii) a mild respiratory tract illness with no rash, (iii) a rash atypical for EI that may be rubelliform or petechial, (iv) papulopurpuric gloves-and-socks syndrome (PPGSS; painful and pruritic papules, petechiae, and purpura of hands and feet, often with fever and enanthem), and polyarthropathy syndrome (arthralgia and arthritis in adults in the absence of other manifestations of EI), (v) chronic erythroid hypoplasia with severe anemia in immunodeficient patients, and (vi) transient aplastic crisis lasting 7–10 days in patients with hemolytic anemias.[3]

Parvovirus infection during pregnancy can potentially impact the fetus. The risk of fetal morbidity (including fetal hydrops—edema in at least two fetal compartments, intrauterine growth retardation, isolated pleural and pericardial effusions) and mortality is increased if maternal infection occurs during the first two trimesters of pregnancy.[10] The neonatal complications of maternal parvovirus B19 infection have been reported, including hepatic insufficiency, myocarditis, transfusion, dependent anemia, and central nervous system abnormalities.[10]

Diagnosis

In nonpregnant persons, detection of serum parvovirus B19-specific IgM antibody is the preferred diagnostic test for acute infection, as it can identify the presence of infection within the previous 4 months. Anti-parvovirus B19 IgG antibody appears by approximately day 7 of EI and persists for life, thus making it a less than reliable marker of acute infection. The optimal method for detecting chronic infection in the immunocompromised patient is demonstration of virus by PCR, as parvovirus B19 DNA can be detected for up to 9 months after the acute viremic phase.[3] For immunocompromised patients with severe anemia associated with chronic infection, dot blot hybridization of serum specimens may have adequate sensitivity.[3]

A pregnant woman who is exposed to or develops signs or symptoms of parvovirus B19 infection should have both parvovirus B19-specific IgG and IgM antibodies drawn. If a pregnant woman has presence of B19-specific IgG and absence of parvovirus B19-specific IgM, she can be considered immune and can thus be reassured because only the primary infection in pregnancy may cause fetal harm.[10,11] Measurement of parvovirus B19 IgM is a highly sensitive and specific marker for acute B19 infection.[12,13] The presence of parvovirus B19 IgM antibodies with the absence of parvovirus B19 IgG antibodies suggests either a very recent infection or a false-positive result and so the pregnant woman must repeat the parvovirus B19 IgG and IgM in 1–2 weeks. If the IgG is positive, it suggests a recent infection. If both parvovirus B19 IgG and IgM are negative, the woman is not immune so she is therefore susceptible to infection.[11] If she has had a recent exposure to the virus and may be incubating the infection, it is suggested that the IgG and IgM tests be repeated after 2–4 weeks. If repeat blood work reveals an increasing parvovirus B19 IgG titer, that means that recent infection has occurred. However, as maternal IgM antibodies may be undetectable at the time of fetal sampling for nonimmune fetal hydrops, recent studies have shown that PCR analysis of maternal blood samples appears to identify B19 infection with greater diagnostic sensitivity.[12]

Treatment

For most patients, supportive treatment is only indicated. Patients with aplastic crisis will need transfusions. Intravenous immune globulin therapy may speed recovery in some immunocompromised patients. If it is found that a pregnant woman has developed a recent infection, it is recommended that these women be referred to a maternal–fetal medicine specialist and that these women have serial ultrasounds to detect evidence of hydrops for 8–12 weeks after infection.[10] Some cases of parvovirus B19 infection concurrent with hydrops fetalis have been treated successfully with intrauterine blood transfusions.[3]

Medical surveillance

While women who are exposed to children at home or at work (e.g., teachers or childcare providers) are at increased risk of infection with parvovirus B19, given the fact that there may be concurrent spread of parvovirus B19 in the community, one cannot pinpoint the workplace as the primary source of infection. Thus routine exclusion of pregnant women from the workplace where EI is occurring is not recommended.[3] Women of childbearing age who are concerned can undergo serologic testing for IgG antibody to parvovirus B19 to determine their susceptibility to infection.[14] Determination of immune status has been recommended for parvovirus B19 research workers.[15]

Prevention

There is currently no vaccine or medicine that can prevent parvovirus B19 infection. During an outbreak, parents of preschool and school children as well as employees in the respective settings should be informed of the risk of infection and its management. In addition to standard precautions, droplet precautions are recommended for hospitalized children with aplastic crises, children with PPGSS, or immunosuppressed patients with chronic infection and anemia for the duration of hospitalization for up to 7 days.[3]

Pregnant healthcare professionals should be informed of the potential risks to the fetus from parvovirus B19 infections and about preventive measures that may decrease these risks (including adherence to infection control procedures such as hand hygiene and proper disposal of used facial tissues as well and not caring for immunocompromised patients with chronic parvovirus infection or patients with parvovirus B19-associated aplastic crises).[3] Each pregnant woman should be counseled about her individual risk, based on her risk of infection, gestational age, and other obstetrical considerations. As there is no evidence that susceptible women reduce their risk of infection by leaving work, routine exclusion from work is not recommended.[10,16,17]

References

1. Farr RW, Hutzel D, D'Aurora R, et al. Parvovirus B19 outbreak in a rehabilitation hospital. *Arch Phys Med Rehabil* 1996;77(2):208–210.
2. Miyamoto K, Ogami M, Takahashi Y, et al. Outbreak of human parvovirus B19 in hospital workers. *J Hosp Infect* 2000;45(3): 238–241.
3. American Academy of Pediatrics Committee on Infectious Diseases. Parvovirus B19. In: Pickering LK, Baker CJ, Kimberlin DW, et al. eds. 2009 *Red Book: Report of the Committee on Infectious Diseases*, 28th edn. Elk Grove Village, IL: American Academy of Pediatrics, 2009:491–493.
4. Centers for Disease Control and Prevention. Fifth Disease. 2012. Available at: http://www.cdc.gov/parvovirusB19/fifth-disease.html (accessed on May 4, 2016).
5. Broliden K, Tolfvenstam T, Norbeck O. Clinical aspects of parvovirus B19 infection. *J Intern Med* 2006;260:285–304.
6. Cossart YE, Field AM, Cant B, et al. Parvovirus-like particles in human sera. *Lancet* 1975;1(7898):72–73.
7. Servant A, Laperche S, Lallemand F, et al. Genetic diversity within human erythroviruses: identification of three genotypes. *J Virol* 2002;76(18):9124–9134.
8. Corcoran C, Hardie D, Yeats J, et al. Genetic variants of human parvovirus B19 in South Africa: cocirculation of three genotypes and identification of a novel subtype of genotype 1. *J Clin Microbiol* 2010;48(1):137–142.
9. Young NS, Brown KE. Parvovirus B19. *N Engl J Med* 2004;350:586–597.
10. Giorgio E, De Oronzo MA, Iozza I, et al. Parvovirus B19 during pregnancy: a review. *J Prenatal Med* 2010;4(4):63–66.
11. Rodis JF. Parvovirus infection. *Clin Obstet Gynecol* 1999;42:107–120.
12. Enders M, Weidner A, Rosenthal T, et al. Improved diagnosis of gestational parvovirus B19 infection at the time of nonimmune fetal hydrops. *J Infect Dis* 2008;197(1):58–62.
13. Beersma MF, Claas EC, Sopaheluakan T, et al. Parvovirus B19 viral loads in relation to VP1 and VP2 antibody responses in diagnostic blood samples. *J Clin Virol* 2005;34:71–75.
14. Levy R, Weissman A, Blomberg G, et al. Infection by parvovirus B19 during pregnancy: a review. *Obstet Gynecol Survey* 1997;52:254–259.
15. Cohen RJ, Brown KE. Laboratory infection with human parvovirus. *J Infect Dis* 1992;24:113–114.
16. Centers for Disease Control and Prevention. Risks associated with human parvovirus B19 infection. *MMWR* 1989;38:81–97.
17. Gillespie SM, Cartter ML, Asch S, et al. Occupational risk of human parvovirus B19 infection for school and day-care personnel during an outbreak of erythema infectiosum. *J Am Med Assoc* 1990;263:2061–2065.

RABIES VIRUS[1]

Common name for disease: Rabies
Classification: Family—Rhabdoviridae

Occupational setting

Persons at risk include veterinarians, animal handlers, laboratory workers, foreign travelers to endemic areas, recreational hunters, animal control officers, outdoor workers, and recreational enthusiasts who are potentially in contact with rabid animals (Figure 21.12). Physicians, nurses, therapists, and laboratory workers are also at potential risk if they provide care for patients with rabies.

Exposure (route)

Rabies is almost always transmitted to humans from an animal bite inoculation with virus-laden saliva. Nonbite exposures to saliva or other potentially infectious materials, such as brain tissue from a rabid animal, may occur through

[1]Rabies virus was written by Dennis J. Darcey.

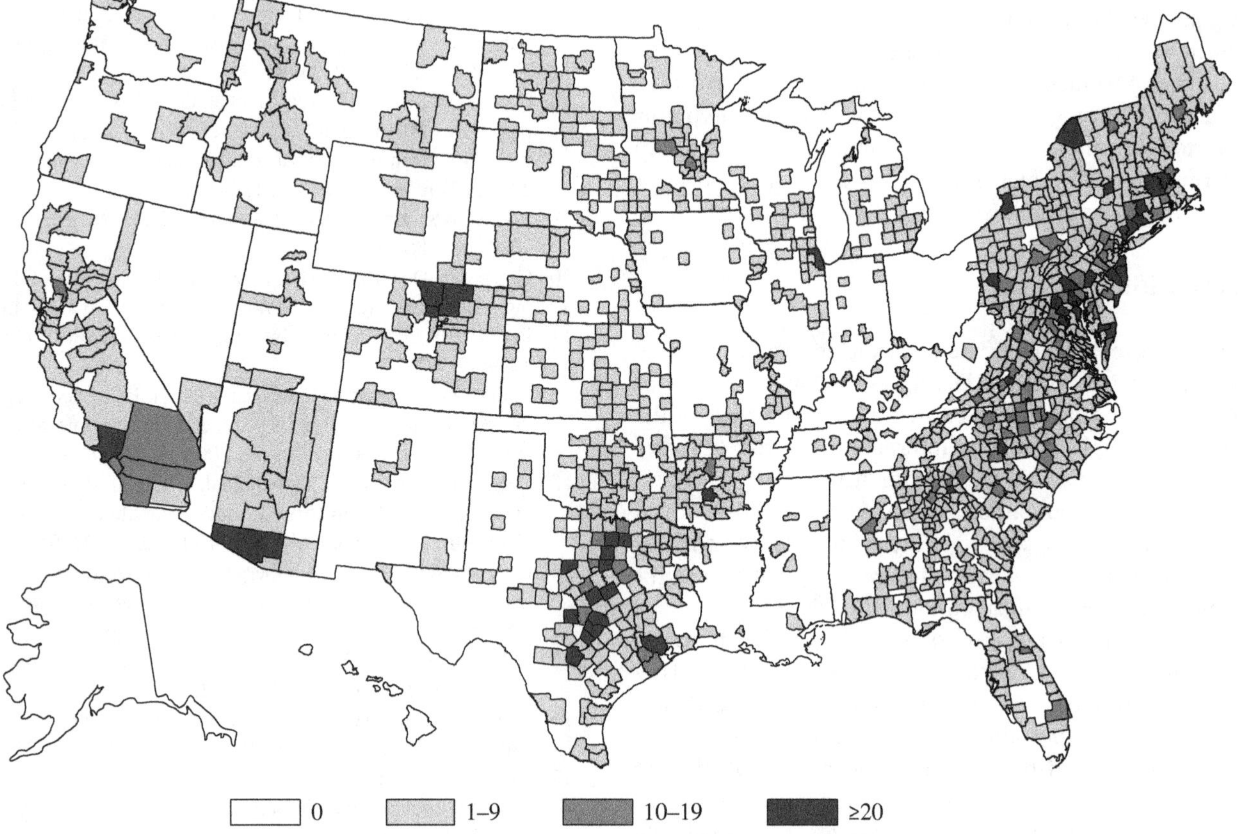

FIGURE 21.12 Animal Rabies reported cases, by county* — United States, 2013. *Data from the National Center for Emerging and Zoonotic Infectious Diseases, Division of High-consequence Pathogens and Pathology. Several rabies virus variants associated with distinct reservoir species have been identified in the United States. The circulation of rabies virus variants associated with raccoons (eastern United States), skunks (central United States and California), and foxes (Texas, Arizona, and Alaska) occur over defined geographic areas. Several distinct rabies virus variants associated with different bat species are broadly distributed across the contiguous United States. Hawaii is the only state considered free of rabies. *Source*: MMWR 2015; 62(53):89.

abraded skin, open wounds, or mucous membrane inoculation. Although nonbite exposures constitute a sufficient reason to initiate postexposure prophylaxis in some circumstances, these exposures rarely cause rabies.[1] Rare cases have been attributed to aerosol transmission in rabies research laboratories.[2] Two cases of rabies have also been attributed to probable airborne exposures in a bat-infested cave.[3] Rabies virus is present in a variety of human fluids and tissues during the first 5 weeks of illness, but there are only eight well-documented cases of human-to-human transmission, all in corneal transplant recipients.[4] Although it has never been documented, human-to-human transmission of rabies following saliva exposure remains a theoretical possibility. Thus, in these circumstances, adherence to contact isolation and use of personal protective equipment PPE standard precautions will minimize the risk of exposure.[5]

Pathobiology

Rabies is an RNA-containing virus. Virions measure 180×75 nanometers; they are cylindrical with one round or conical end and one flat end, giving each particle the shape of a bullet. Five proteins have been identified from purified rabies virus, including glycoprotein (G), nucleocapsid (N) protein, viral polymerase (L), and two smaller proteins (NS (P) and M).[6] Monoclonal antibodies directed against both nucleocapsid and glycoprotein antigens have been developed. The virus is rapidly inactivated by desiccation, ultraviolet and X-radiation, sunlight, trypsin, beta-propiolactone, ether, and detergents. Following inoculation, the virus replicates locally and then spreads via peripheral nerves to the CNS, where it causes encephalitis. The incubation period is usually 20–90 days; the incubation period is shorter when the bite is on the head or face rather than on an extremity. During the incubation period, the infected individual is usually well except for symptoms related to local wound healing.

After a latent period lasting from days to several weeks, months, or (in some cases) years, the virus spreads via peripheral nerves to the spinal cord and CNS, especially the limbic system. The virus is normally present in the CNS in high titer before the development of systemic symptoms. When symptoms occur, the virus has already traveled down efferent nerves to nearly every organ and tissue, including, most important for the life cycle of the virus, the salivary glands.

Clinical rabies usually begins with generalized, nonspecific constitutional symptoms, including malaise, fatigue, headache, anorexia, and fever. In 50% of patients, pain or paresthesia occurs at the site of the exposure and may be the first rabies-specific symptom.[6] Other nonspecific symptoms, including sore throat, cough, abdominal pain, nausea and vomiting, diarrhea, or chills, have also been associated with the prodrome, which normally lasts 2–10 days. This prodrome is followed by the neurologic phase, which may include intermittent hyperactivity, hallucinations, disorientation, bizarre behavior, seizures, nuchal rigidity, or paralysis. The hyperactive episodes may occur spontaneously, or they may be precipitated by a variety of tactile, auditory, visual, or other stimuli. Diaphoretic spasm often leads to the classical hydrophobia. Other symptoms seen during the acute neurologic phase include fever, muscle fasciculation, hyperventilation, hypersalivation, and convulsions.

The acute neurologic phase usually lasts 2–7 days, with longer duration in the paralytic forms. In patients in the United States who do not receive intensive supportive care, the average duration of illness until death is 7 days. For those who receive intensive support, the duration of illness averages 25 days. To date fewer than 10 documented cases of human survival from clinical rabies have been reported and only two have not had a history of pre- or postexposure prophylaxis.[7] Therefore, emphasis is placed on prevention rather than treatment. In the United States and other parts of the developed world, animal vaccination programs have greatly reduced the incidence of rabies among domestic dogs and cats. Wild animals now constitute the most important potential source of infection for both humans and domestic animals in the United States. Most of the animal rabies cases in the continental United States occur in skunks, foxes, raccoons, and bats. Only Hawaii remains consistently rabies-free. Bats are increasingly implicated in human rabies transmission. Transmission of rabies virus can occur from minor or unrecognized bites from bats. The dog remains the major source of human exposure to rabies outside the United States and is a concern for foreign travelers. Rodents, such as squirrels, hamsters, guinea pigs, gerbils, chipmunks, rats, mice, and rabbits, are rarely infected with the rabies virus and have not been known to transmit rabies to humans. The marked decrease of rabies cases among domestic animals in the United States has drastically decreased human infections. However, outside the United States and Europe, dogs remain the most common source of infection for humans.

Diagnosis

No tests are currently available to diagnose rabies in humans before the onset of clinical disease. The virus is immunologically protected and does not usually stimulate antibody production until after invasion of the CNS. Antemortem studies have shown that rabies virus can be isolated from human saliva, brain tissues, CSF, urine sediment, and tracheal secretions. Collection of samples for diagnosis of rabies in the United States should be performed only after consultation with state health department or with the rabies program at CDC (www.cdc.gov/rabies). All samples should be considered potentially infectious. The RT-PCR test is emerging as the diagnostic procedure of choice in suspected rabies cases and can be performed on saliva, CSF, or tissue.[7]

Treatment

There is no specific treatment once clinical rabies is established. Treatment consists of respiratory and cardiovascular support. Passive rabies immune globulin has been used in several cases with no clear benefit. To prevent secondary bacterial infection of the patient and to prevent exposure of hospital staff to rabies virus, the patient should be isolated. To avoid exposure, hospital personnel should wear face masks, gloves, and gowns. The administration of vaccine after the onset of clinical illness has not been successful.

Prevention

Human rabies prevention consists of pre-exposure vaccination and postexposure therapy and prophylaxis. There are currently two types of rabies-immunizing products available in the United States. Human rabies immune globulin (HRIG) provides rapid short-term passive immunity (half-life, 21 days). Rabies vaccines include the human diploid cell vaccine (HDCV), the purified chick embryo cell vaccine (PCEC), and the rabies vaccine adsorbed (RVA). In the United States, only the HDCV and PCEC are available (see Table 21.5). These vaccines induce an active immune response with the production of neutralizing antibodies, usually within 7–10 days, which persists for at least 2 years. All types of rabies vaccines are considered equally efficacious and safe when used as indicated. HDCV, PCEC, and RVA should be administered through the intramuscular route. Adverse reactions to HDCV, RVA, and PCEC are far less common and serious than reactions to previous vaccines. They include mild local reactions, such as pain, erythema, and swelling at the injection site. Systemic reactions such as

TABLE 21.5 Currently available rabies biologics.

Human rabies vaccine	Product name	Manufacturer	Dose	Route	Indications
Human diploid cell vaccine	Imovax® Rabies*	Sanofi Pasteur Phone: 800-822-2463 Website: http://www.vaccineplace.com/products/	1 mL	Intramuscular	Pre-exposure or post exposure[†]
Purified chick embryo cell vaccine	RabAvert®	Novartis Vaccines and Diagnostics Phone: 800-244-7668 Website: http://www.rabavert.com	1 mL	Intramuscular	Pre-exposure or post exposure[†]
Rabies immune globulin	Imogam® Rabies-HT	Sanofi Pasteur Phone: 800-822-2463 Website: http://www.vaccineplace.com/products/	20 IU/kg	Local[§]	Post exposure only
	HyperRab™ S/D	Talecris Biotherapeutics Bayer Biological Products Phone: 800-243-4153 Website: http://www.talecris-pi.info	20 IU/kg	Local[§]	Post exposure only

*Imovaxrabies I.D., administered intradermally, is no longer available in the United States.

[†]For postexposure prophylaxis, the vaccine is administered on days 0, 3, 7, 14 and 28 in patients who have not been previously vaccinated and on days 0 and 3 in patients who have been previously vaccinated. For pre-exposure prophylaxis, the vaccine is administered on days 0, 7 and 21 or 28.

[§]As much of the product as is anatomically feasible should be infiltrated in to and around the wound. Any remaining product should be administered intramuscularly in the deltoid or quadriceps (ata location other than that used for vaccine inoculation to minimize potential interference).

Source: Centers for Disease Control. Human Rabies Prevention—United States, 2008, Recommendations of the Advisory Committee on Immunization Practices. *MMWR Early Release* 2008;57(RR-3):4.

TABLE 21.6 Rabies post-exposure prophylaxis guide.

Animal type	Evaluation and disposition of animal	Post exposure prophylaxis recommendations
Dogs, cats, and ferrets	Healthy and available for 10 days observation Rabid or suspected rabid Unknown (e.g., escaped)	Persons should not begin prophylaxis unless animaldevelopsclinicalsignsofrabies* Immediately begin prophylaxis. Consult public health officials.
Skunks, raccoons, foxes, and most other carnivores; bats[†]	Regarded as rabid unless animal proven negative by laboratory tests[§]	Consider immediate prophylaxis.
Livestock, small rodents (rabbits and hares), large rodents (wood chucks and beavers), and other mammals	Consider individually	Consult public health officials. Bites from squirrels, hamsters, guinea pigs, gerbils, chipmunks, rats, mice, other small rodents, rabbits, and hares almost never require antirabies postexposure prophylaxis.

*During the 10-day observation period, begin postexposure prophylaxis at the first sign of rabies in a dog, cat, or ferret that has bitten someone. If the animal exhibits clinical signs of rabies, it should be euthanized immediately and tested.

[†]Postexposure prophylaxis should be initiated as soon as possible following exposure to such wild life unless the animal is available for testing and public health authorities are facilitating expeditious laboratory testing or it is already known that brain material from the animal hastested negative. Other factors that might influence the urgency of decision-making regarding initiation of postexposure prophylaxis before diagnostic results are known include the species of the animal, the general appearance and behavior of the animal, whether the encounter was provoked by the presence of a human, and the severity and location of bites. Discontinue vaccine if appropriate laboratory diagnostic test (i.e., the direct fluorescent antibody test) is negative.

[§]The animal should be euthanized and tested as soon as possible. Holding for observation is not recommended.

Source: Centers for Disease Control. Human Rabies Prevention—United States, 2008, Recommendations of the Advisory Committee on Immunization Practices, *MMWR* 2008; 57(No. RR-3):12.

headache, nausea, abdominal pain, muscle aches, and dizziness have been reported in 5–40% of recipients. Three cases of neurologic illness resembling Guillain–Barré syndrome that resolved without sequelae have been reported.[8] An immune complex-like reaction has been reported in 6% of persons receiving booster injections of HDCV. Symptoms include generalized urticaria, sometimes accompanied by arthralgia, myalgia, vomiting, fever, and malaise, but in no cases were these reactions life-threatening.[9] Corticosteroids, other immunosuppressive agents, antimalarial medications, and immunosuppressive illness can interfere with the antibody response, so it is especially important that these individuals be tested for effective antibody titers. Pregnancy is not considered a contraindication to postexposure or pre-exposure prophylaxis in high-risk individuals.

Pre-exposure vaccination should be offered to high-risk groups such as veterinarians, animal handlers, laboratory workers, and foreign travelers on extended visits to foreign countries where canine rabies is endemic. Pre-exposure prophylaxis (PEP) may provide protection to persons with inapparent exposures to rabies; it can also provide protection when postexposure therapy is delayed. However, pre-exposure vaccination does not eliminate the need for additional therapy after a known rabies exposure. PEP begins with effective local wound cleansing and, in most cases, administration of both HRIG and rabies vaccine. Table 21.6 shows the CDC recommendations on decisions to give PEP based on animal exposure risk.[10] Table 21.7 shows the updated CDC guidelines for rabies PEP in previously vaccinated and unvaccinated workers.[11] Table 21.8 contains CDC recommendations for PEP including primary and booster vaccinations.[10] Table 21.9 shows the CDC recommendations for determining which workers should receive PEP based on exposure risk.[10].

For laboratory personnel handling rabies virus, BSL-3 containment is recommended. Although rabies has the highest case fatality rate of any known human infection, it can almost always be prevented if exposures are recognized and PEP is initiated. Each year, 20000 people receive antirabies PEP in the United States. Even though most of these individuals have not had significant rabies exposure, there have been no postexposure vaccine failures in the United States since HDCV was licensed.[9]

References

1. Afshar A. A review of non-bite transmission of rabies virus infection. *Br Vet J* 1979;135:142–148.
2. Winkler XVG, Fashinell TR, Leffingwell L, et al. Airborne rabies transmission in a laboratory worker. *JAMA* 1973;226: 1219–1221.
3. Centers for Disease Control. Rabies prevention United States, 1991. *MMWR* 1991;40:1–14.
4. Centers for Disease Control. Human rabies prevention—United States, 1999. *MMWR* 1999;48(RR-1):1–17.
5. Garner JS, The Hospital Infection Control Practices Advisory Committee. Guideline for isolation precautions in hospitals. *Infect Control Hosp Epidemiol* 1996;17:54–80.
6. Singh K, Rupprecht CE, Bleck TP. Rabies (Rhabdoviruses). In: Bennett JE, Dolin R, Blaser M. eds. *Mandell, Douglas, and*

TABLE 21.7 Rabies post-exposure prophylaxis (PEP) schedule.

Vaccination status	Intervention	Regimen*
Not previously vaccinated	Wound cleansing	All PEP should begin with immediate thorough cleansing of all wounds with soap and water. If available, a virucidal agent (e.g., povidine-iodine solution) should be used to irrigate the wounds.
	Human rabies immune globulin (HRIG)	Administer 20 IU/kg body weight. If anatomically feasible, the full dose should be infiltrated around and into the wound(s), and any remaining volume should be administered at an anatomical site (intramuscular [IM]) distant from vaccine administration. Also, HRIG should not be administered in the same syringe as vaccine. Because RIG might partially suppress active production of rabies virus antibody, no more than the recommended dose should be administered.
	Vaccine	Human diploid cell vaccine (HDCV) or purified chick embryo cell vaccine (PCECV) 1.0 mL, IM (deltoid area†), 1 each on days 0,§ 3, 7 and 14.¶
Previously vaccinated**	Wound cleansing	All PEP should begin with immediate thorough cleansing of all wounds with soap and water. If available, a virucidal agent such as povidine-iodine solution should be used to irrigate the wounds.
	HRIG	HRIG should not be administered.
	Vaccine	HDCV or PCECV 1.0 mL, IM (deltoid area†), 1 each on days 0§ and 3.

*These regimens are applicable for persons in all age groups, including children

†The deltoid area is the only acceptable site of vaccination for adults and older children. For younger children, the outer aspect of the thigh may be used. Vaccine should never be administered in the gluteal area.

§Day 0 is the day dose 1 of vaccine is administered.

¶For persons with immunosuppression, rabies PEP should be administered using all 5 doses of vaccine on days 0, 3, 7, 14, and 28.

**Any person with a history of pre-exposure vaccination with HDCV, PCECV, or rabies vaccine adsorbed (RVA); prior PEP with HDCV, PCECV or RVA; or previous vaccination with any other type of rabies vaccine and a documented history of antibody response to the prior vaccination.

Source: Use of a Reduced (4-Dose) Vaccine Schedule for Postexposure Prophylaxis to Prevent Human Rabies: Recommendations of the Advisory Committee on Immunization Practices, *MMWR*, March 19, 2010; 59(RR02):6.

TABLE 21.8 Rabies pre-exposure prophylaxis schedule.

Type of vaccination	Route	Regimen
Primary	Intramuscular	Human diploid cell vaccine (HDCV) or purified chick embryo cell vaccine (PCECV); 1.0 mL (deltoid area), one each on days 0,* 7, and 21 or 28
Booster†	Intramuscular	

* Day 0 is the day the first dose of vaccine is administered.

† Persons in the continuous-risk category should have a serum sample tested for rabies virus neutralizing antibody every 6 months, and persons in the frequent-risk category should be tested every 2 years. An intramuscular booster dose of vaccine should be administered if the serum titer falls to maintain a value of atleast complete neutralization at a 1:5 serum dilution by rapid fluorescent focus inhibition test.

Source: Centers for Disease Control. Human Rabies Prevention —United States, 2008, Recommendations of the Advisory Committee on Immunization Practices, *MMWR* 2008; 57(RR-3):18.

Bennett's Principles and Practice of Infectious Disease, 8th edn. Philadelphia, PA: Elsevier Saunders, 2015:1984–1995.

7. Centers for Disease Control. Rabies. Available at: http://www.cdc.gov/rabies/symptoms/index.html (accessed on May 4, 2016).
8. Bernard KW, Smith PW, Kader FJ, et al. Neuroparalytic illness and human diploid cell rabies vaccine. *JAMA* 1982;248:3136–3138.
9. Centers for Disease Control. Systemic allergic reactions following immunization with human diploid cell rabies vaccine. *MMWR* 1984;33:185–187.
10. Centers for Disease Control. Human rabies prevention—United States, 2008, recommendations of the Advisory Committee on Immunization Practices. *MMWR* 2008;57(RR-3):1–28.
11. Rupprecht CE, Briggs D, Brown CM, et al. Use of a reduced (4-dose) vaccine schedule for postexposure prophylaxis to prevent human rabies: recommendations of the advisory committee on immunization practices. *MMWR Recomm Rep* 2010; 59(RR-02):1–9. Available at: http://www.cdc.gov/mmwr/preview/mmwrhtml/rr5902a1.htm (accessed on June 29, 2016)..

TABLE 21.9 Rabies pre-exposure prophylaxis guide — United States, 2008.

Risk category	Nature of risk	Typical populations	Pre-exposure recommendations
Continuous	Virus present continuously, often in high concentrations. Specific exposures likely to go unrecognized. Bite, nonbite, or aerosol exposure.	Rabies research laboratory workers; rabies biologics production workers.	Primary course. Serologic testing every 6 months; booster vaccination if antibody titer is below acceptable level.*
Frequent	Exposure usually episodic, with source recognized, but exposure also might be unrecognized. Bite, nonbite or aerosol exposure	Rabies diagnostic laboratory workers, cavers, veterinarians and staff, and animal-control and wildlife workers in areas where rabies is enzootic. All persons who frequently handle bats.	Primary course. Serologic testing every 2 years; booster vaccination if antibody titer is below acceptable level.*
Infrequent (greater than population at large)	Exposure nearly always episodic with source recognized. Primary course. Bite or nonbite exposure.	Veterinarians and animal-control staff working with terrestrial animals in areas where rabies is uncommon to rare. Veterinary students. Travelers visiting areas where rabies is enzootic and immediate access to appropriate medical care including biologics is limited.	Primary course. No serologic testing or booster vaccination.
Rare (population at large)	Exposure always episodic with source recognized. Bite or nonbite exposure.	U.S. population at large, including persons in areas where rabies is epizootic.	No vaccination necessary.

* Minimum acceptable antibody level is complete virus neutralization at a 1:5 serum dilution by the rapid fluorescent focus inhibition test. A booster dose should be administered if the titer falls below this level.
Source: MMWR 2008; 57(RR-3):19.

RESPIRATORY SYNCYTIAL VIRUS (RSV)

Common name for disease: Respiratory syncytial virus pneumonia
Classification: Family—Paramyxoviridae; subfamily—Pneumovirinae

Occupational setting

Healthcare and childcare workers as well as military personnel may be at risk of contracting respiratory syncytial virus (RSV). In one report, nearly half of the nurses, residents, and medical students on children's hospital wards developed RSV infections during an RSV outbreak.[1] An outbreak of acute febrile illness caused by RSV occurred in the Republic of Korea Air Force boot camp from May to July 2011, the first documented outbreak to have occurred in a healthy young adult group.[2]

Exposure (route)

Infection is transmitted by the respiratory route. Large-droplet inoculation is the primary mode. Small-particle aerosols may also transmit infection. Transmission can also occur through contact with contaminated fomites on skin and soft tissues (where the virus can survive up to an hour) or on environmental surfaces (where the virus can survive up to 30 hours on countertops and up to 1 hour on cloth or paper).[3–5]

Pathobiology

RSV is a single-stranded, enveloped RNA virus measuring 100–350 nm but may be as long as 10 μm. On electron microscopy, the virus is pleomorphic with spherical and filamentous forms. There are two strains, subgroup A and B. There are ten viral genes that are transcribed sequentially into separate mRNAs by the viral polymerase. The virus enters the body usually through the eye or nose, rarely through the mouth. The virus then spreads along the epithelium of the respiratory tract, mostly by cell-to-cell transfer.

Almost all children will have had an RSV infection by their second birthday. The incubation period is 3–6 days. Infants with a lower respiratory tract infection typically have a runny nose and a decrease in appetite before any other symptoms appear with cough developing 1–3 days later. Soon after the cough develops, sneezing, fever, and wheezing may occur. In very young infants, irritability, decreased activity, and apnea may be the only symptoms of infection. In children, the primary infection is manifested as pneumonia, bronchiolitis, tracheobronchitis, or upper respiratory tract illness with fever and otitis media.

Infection in otherwise healthy adults usually presents as upper respiratory tract illness with nasal congestion and cough. Incapacitation is seen in about half of adults with RSV. In some adults, pulmonary function change and airway hyperreactivity can be seen 8 weeks after the illness.[6] Residential crowding has been identified as a risk factor for RSV infections.[7]

Repeat infections with RSV are common, and they affect all age groups. Subsequent infection after the primary one generally results in less severe illness.

Diagnosis

Diagnosis of RSV is clinical and epidemiologic. Community evidence of RSV disease in the infant population associated with lower respiratory findings provides a good presumptive diagnosis.

Specific diagnostic tests for confirming RSV infection include culture, antigen-revealing techniques, PCR assay, and use of molecular probes.[8]

Treatment

Most healthy adults require no treatment for this self-limited illness. Infants, young children, or elderly individuals may require supportive therapy. Bronchodilators and the antiviral ribavirin are used in severe high-risk cases.[8] However, ribavirin treatment does not lead to better outcomes in some instances.[9]

Prevention

There is currently no RSV vaccine available, but active research continues.[10] Recent technological advances using a structure-based approach to vaccine development are showing promising RSV vaccine candidates.[11] Palivizumab is a monoclonal antibody recommended for the prevention of severe RSV infection in some infants and young children who are considered to be at high risk (although it does not treat infection).[5]

Effective control of transmission in the hospital or childcare settings is dependent on the interruption of hand carriage of virus from one individual to another and the interruption of self-inoculation of the eyes or nose. Frequent hand washing with soap and water (up to 20 seconds) before and after contact can help to reduce transmission. Using protective clothing and eye–nose goggles for close contact work may be helpful.[12–14] Standard laboratory practice should prevent transmission in the clinical laboratory. RSV is sensitive to temperature and pH changes and the virus is inactivated by a variety of detergents so vigilant decontamination of infected surfaces is imperative.

References

1. Hall GB, Geiman JM, Douglas RG. Control of respiratory syncytial viral infections. *Pediatrics* 1978;62:730.
2. Park W-J, Yoo S-J, Lee S-H, et al. Respiratory syncytial virus outbreak in the basic military training camp of the Republic of Korea Air Force. *J Prev Med Public Health* 2015;48(1):10–17.
3. Hall CB. The nosocomial spread of respiratory syncytial viral infections. *Annu Rev Med* 1983;34:311–319.
4. Hall GB, Geiman JM, Douglas RG. Possible transmission by fomites of respiratory syncytial virus. *Infect Dis* 1980;141: 98–102.
5. Centers for Disease Control and Prevention. RSV – Transmission and Prevention. 2014. Available at: http://www.cdc.gov/rsv/about/transmission.html (accessed on May 4, 2016).
6. Hall WJ, Hall CB, Speers DM. Respiratory syncytial virus infections in adults. *Ann Intern Med* 1978;88:203.
7. Colosia AD, Masaquel A, Hall CB, et al. Residential crowding and severe respiratory syncytial virus disease among infants and young children: a systematic literature review.*BMC Infect Dis.* 2012;12:95.
8. Walsh EW, Hall CB. Respiratory syncytial virus. In: Bennett JE, Dolin R, Blaser M. eds. *Mandell, Douglas, and Bennett's Principles and Practice of Infectious Disease*, 8th edn. Philadelphia, PA: Elsevier Saunders, 2015:1948–1966.
9. Ventre K, Randolph AG. Ribavirin for respiratory syncytial virus infection of the lower respiratory tract in infants and young children. *Cochrane Database Syst Rev* 2007;24(1):CD000181.
10. Hurwitz JL. Respiratory syncytial virus vaccine development. *Exp Rev Vac* 2011;10(10):1415–1433.
11. Graham BS, Modjarrad K, McLellan JS. Novel antigens for RSV vaccines. *Curr Opin Immunol* 2015;35:30–38.
12. Leclair JM, Freeman J, Sullivan BF, et al. Prevention of nosocomial respiratory syncytial virus infections through compliance with glove and gown isolation precautions. *N Engl J Med* 1987;317(6):329–334.
13. Graman PS, Hall CB. Epidemiology and control of nosocomial viral infections. *Infect Dis Clin North Am* 1989;3:815–823.
14. Simoes E. Respiratory syncytial virus infection. *Lancet* 1999;354:847–852.

RUBELLA VIRUS

Common names for disease: German measles, rubella, third disease

Classification: Family—Togaviridae; genus—*Rubivirus*

Occupational setting

Healthcare workers, childcare workers, and school teachers may have an increased risk of acquiring rubella if not fully immunized. Pregnant workers who become infected can transmit the virus *in utero*, which can result in miscarriages, stillbirths, therapeutic abortions, and congenital rubella syndrome (CRS), a constellation of birth defects that develop in the first trimester.[1] Rubella was declared eliminated from the United States in 2004.[2] To date there has been no documented transmission of rubella to healthcare workers in US healthcare facilities since 2004. On April 29, 2015, the Pan-American Regional Office of the World Health Organization (PAHO/WHO) announced that the Americas region had become the world's first region to eliminate rubella and CRS.[3]

Exposure (route)

Rubella is transmitted almost entirely by infected airborne droplets. The virus is spread by affected individuals via contact with nasopharyngeal secretions. Up to 50% of all rubella virus infections occur by individuals with subclinical or asymptomatic presentations.[4] There has been no documented transmission from animals or insects to humans.

Pathobiology

Rubella virus is a 60–70-nanometers spheroidal, enveloped, single-stranded RNA virus with a single antigenic type that does not cross-react with other members of its class. Rubella virus is inactivated by lipid solvents, trypsin, formalin, ultraviolet light, low pH, heat, and amantadine.[4]

The primary illness of rubella is preceded by an incubation period of 14–21 days and is commonly asymptomatic.[4,5] For those with recognized illness, early symptoms of fever and malaise often precede lymph node enlargement and eventual rash. The rash is very similar to rubeola (measles), commonly beginning on the face and spreading downward during the few days of its duration. The period of greatest risk for contagion is closest to the time that the rash is evident. Experimental studies have shown that the virus can be isolated from normal individuals several days before the occurrence of the rash and for up to 2 weeks after the onset of symptoms. Adults and children should be considered contagious until at least 4 days after the rash develops. Complications from rubella infection include arthralgias (with the fingers, wrists, and knees most often affected), hemorrhagic complications (including thrombocytopenic purpura being the most common manifestation), encephalitis (occurring in about 1 in 6000 persons), orchitis, neuritis, and panencephalitis.[4]

While the rubella virus causes a disease with usually trivial symptoms for the primary host, it produces catastrophic consequences *in utero*. When epidemics of the disease occurred in the past (prior to the licensing in 1969 of the MMR vaccine), many thousands of simultaneous birth defects were found among affected pregnancies. Early gestational exposure leads most commonly to spontaneous abortion. Somewhat later infections cause recognizable structural damage. Heart problems occur in up to 30–35% of fetuses infected during the third month of pregnancy. Late infection (after the 20th week) causes less structural damage but it carries a 10% risk of producing functional damage to the CNS, especially deafness. Fetal injury does represent a true infection with the virus, since the virus can be isolated from affected tissues. CRS produces fetal and neonatal damage, with recognizable cardiac, ocular, and CNS malformations. Newborns with congenital infection may remain infectious for months after birth. Damage may continue to be discovered through childhood with juvenile diabetes mellitus and progressive encephalopathy being observed in some older children with CRS.[4]

Diagnosis

Postnatal diagnosis is achieved by recognition of the rash and confirmed by testing. Confirmation of rubella disease is via positive viral culture for rubella or detection of rubella virus by polymerase chain reaction (PCR), the presence of rubella-specific IgM antibody, or demonstration of a significant rise in IgG antibody from paired acute- and convalescent-phase sera.[4]

Treatment

No antiviral therapy for rubella is available. Supportive therapy is recommended for adult patients with uncomplicated rubella symptoms. For arthralgias, nonsteroidal anti-inflammatory drugs (NSAIDs) may be helpful, but corticosteroids are not indicated. Patients with encephalitis should receive supportive care with adequate fluid and electrolyte maintenance.[6]

Medical surveillance

Documentation of adequate rubella vaccination is universally required by public health agencies and clinical credentialing bodies for all personnel in healthcare facilities.[1,7,8] Presumptive evidence of immunity to rubella for persons who work in healthcare facilities includes any of the following: written documentation of vaccination with 1 dose of live rubella or MMR vaccine, laboratory evidence of immunity, laboratory confirmation of rubella infection or disease, or birth before 1957.[1] Vaccinations should be routinely provided for workers with inadequate medical documentation and low serum antibody titers. Yet even in successfully vaccinated individuals, elevations in antibody titer can be seen following viral exposure, sometimes with isolated live virus. This implies that antibody-mediated immunity is incomplete and that those with normal immune function and successful vaccination can reacquire the virus. Five cases of post-rubella syndrome have been noted among "immune" patients whose antibody rise followed viral exposure during pregnancy. Thus obstetric monitoring of rubella antibody levels must be quantitative rather than qualitative. More broadly, the need to isolate infected individuals remains relevant even when the potentially exposed population has been shown to have antibody.

Prevention

The rubella vaccine is a live attenuated virus (part of the MMR vaccine) that confers excellent resistance against subsequent clinical illness and greatly reduces the risk of incurring and transmitting the illness. Even though the target population for protection is women of childbearing years, all children should be vaccinated and all healthcare, childcare, and school workers should be required to document immunity. Because the vaccine is a live virus, theoretical considerations require that it not be administered to individuals already or soon to be pregnant. Despite these protective concerns, inadvertent episodes of vaccination during pregnancy have occurred without subsequent CRS.[9]

References

1. Centers for Disease Control and Prevention. Immunization of health-care personnel: recommendations of the Advisory Committee on Immunization Practices (ACIP). *MMWR* 2011; 60(RR07):1–45.
2. Centers for Disease Control and Prevention. Achievements in public health: elimination of rubella and congenital rubella syndrome - United States, 1969–2004. *MMWR* 2005;54(11): 279–282.
3. Centers for Disease Control and Prevention. Rubella. 2015. Available at: http://www.cdc.gov/rubella/ (accessed on May 4, 2016).
4. Centers for Disease Control and Prevention. Rubella: Pink Book. Available at: http://www.cdc.gov/vaccines/pubs/pinkbook/downloads/rubella.pdf (accessed on May 4, 2016).
5. Gershon AA. Viral diseases: rubella virus (German measles). In: Bennett JE, Dolin R, Blaser M. eds. *Mandell, Douglas, and Bennett's Principles and Practice of Infectious Disease*, 8th edn. Philadelphia, PA: Elsevier Saunders, 2015:1875–1880.
6. Tunkel AR, Glaser CA, Bloch KC, et al. The management of encephalitis: clinical practice guidelines by the Infectious Diseases Society of America. *Clin Infect Dis* 2008;47(3): 303–327.
7. Centers for Disease Control and Prevention. *Immunization Recommendations for Health-Care Workers*. Atlanta, GA: Centers for Disease Control, Division of Immunization, Center for Prevention Services, 1989.
8. Centers for Disease Control and Prevention. Recommendations of the immunization practices advisory committee: update on adult immunization. *MMWR* 1991;40(RR-12):1.
9. Mann JM, Preblud SR, Hoffman RE, et al. Assessing risks of rubella infection during pregnancy: a standardized approach. *JAMA* 1981;245:1647.

SIMIAN RETROVIRUSES

Common names for disease: Simian lymphoproliferative disease, simian T-cell lymphoma, simian AIDS

Classification: Family—Retroviridae; genus—Lentivirus—Subfamily Orthoretrovirus—simian immunodeficiency virus (SIV)

Subfamily: Oncornavirus—simian T-lymphotropic virus (STLV), simian type D retrovirus (SRV); Subfamily Spumavirus—simian foamy virus (SFV)

Occupational setting

Veterinarians, nonhuman primate (NHP) handlers, animal caretakers, and virologists and associate researchers are at risk from exposure.

Exposure (route)

The simian retroviruses are transmitted via needlesticks, bites and scratches, and mucous membrane contact with NHP body fluids. Serologic surveys indicate that numerous species of wild and captive NHPs are infected with simian immunodeficiency virus (SIV), simian T-lymphotropic virus (STLV), simian type D retrovirus (SRV), and/or simian foamy virus (SFV). The viral culture material and any equipment that comes into contact with these materials must be considered infectious.

Pathobiology

SIV is closely related to HIV Types I (originating from chimpanzees, *Pan troglodytes*) and II (originating from sooty mangabeys, *Cercocebus atys*), but infection in these natural hosts is most always subclinical and rarely causes disease.[1,2] Once introduced into macaque populations (either experimentally as an animal model for AIDS-related research or acquired in captivity), SIV can cause severe immunodeficiency disease (including diarrhea, thymic atrophy, decrease in CD4 cell counts, lymphomas, encephalitis) and increase susceptibility to opportunistic infections.[2]

There has been evidence of seroconversion in a laboratory worker who was exposed by a needlestick to blood from an SIV-infected macaque.[3] Despite immediate scrubbing of the wound, inflammation and swelling developed at the puncture site and persisted for several weeks. Seroactivity to SIV developed within 3 months, peaked between the 3rd and 5th months, and declined afterward. Attempts to isolate the virus and to find SIV provirus by PCR were unsuccessful. Also, attempts to transmit SIV by inoculation of a macaque with the worker's blood were unsuccessful. It was concluded that the worker had not become permanently infected with SIV.

In 1994, there was a first report of an SIV laboratory worker who developed actual infection with SIV.[4] This researcher developed a severe dermatitis of the forearms and hands and continued to handle clinical specimens from infected monkeys without wearing gloves. SIV was successfully isolated from the worker's peripheral blood mononuclear cells. However, two monkeys inoculated with the researcher's blood remained seronegative. The researcher's infection had not resulted in clinical illness at the time of the report.

A survey of workers with reported occupational exposure to SIV (either via direct NHP contact or handling of NHP tissues) revealed that 2 of 550 workers (0.4%) had developed antibodies to SIV.[5]

STLV was identified in the early 1980s, with the corresponding human counterparts categorized as human T-lymphotropic virus (HTLV).[6] STLV-1 has been isolated from a wide variety of old-world monkeys in Asia and Africa (including macaques, baboons, African green monkeys, guenons, mangabeys, orangutans, and chimpanzees), STLV-2 has been identified in captive bonobos (*Pan paniscus*) from the Democratic Republic of Congo, and STLV-3 has been isolated from hamadryas baboons (*Papio hamadryas*) from East and West Africa and red-capped mangabeys (*Cercocebus torquatus*) and greater spot-nosed monkeys (*Cercopithecus nictitans*) from Cameroon.[6] Only a very small proportion of infected animals develop T-cell lymphoma or lymphoproliferative disease (LPD) and usually only after a prolonged period of infection.[2]

A study on HTLV-1 transmission through animal bites raised the question of whether STLV-1 could be similarly transferred.[7] In a cohort of Central African hunters bitten by gorillas, chimpanzees, or small monkeys during hunting activities, 8.6% had positive serology for HTLV-1 compared with 1.5% of controls.[7] Screenings of sera from 187 workers handling NHPs in zoos and research facilities were all found to be negative for antibodies to STLV-1, indicating that infection with STLV-1 in the workplace may be low because of lower prevalence of this virus in captive NHPs.[1]

SRVs are a group of closely related viruses that is present in up to 90% of wild and captive macaques. SRVs include five different serotypes (types 1–5).[8] SRV can elicit a broad spectrum of clinical and pathologic manifestations, ranging from subclinical state to rapidly fatal immunosuppressive disease.[2] As part of ongoing voluntary prospective surveillance for human infections with simian retroviruses among workers occupationally exposed to NHPs or their tissues, 2 of 231 (0.9%) of workers tested for presence of SRV antibodies were found to be strongly seropositive, showing reactivity against multiple SRV antigens.[9] Additional serological analyses of macaque workers found 2/481 (0.48%) of these individuals had developed antibody response to SRV, with one of the two workers having long-standing seropositivity.[8] The fact that neither of these individuals developed clinical disease and the inability to isolate virus from these individuals implies that there were lower levels of circulating virus to begin with.

Most New-World and Old-World monkeys as well as apes harbor SFV.[1] A study of workers employed at North American zoos and primate centers found that 14 of 418 workers (3.35%) were seroreactive to SFV; the infections originated from African green monkeys ($n=1$), baboons ($n=4$), and chimpanzees ($n=9$).[10] A subsequent surveillance study for simian retroviruses at research centers and zoos handling NHPs found that 10 of 187 persons (5.3%) tested positive for SFV antibodies.[1]

Diagnosis

Diagnosis of SIV infection in NHPs and exposed humans is typically made using a combination of serologic and molecular assays, including ELISA tests (using HIV-1 and HIV-2 as antigens) and assays using SIV-specific synthetic peptides.[8,11] Western blot (WB) testing utilizing SIV and HIV-1 or HIV-2 antigens, either alone or in combination, is used for confirmation. Specimens showing WB reactivity to both Env and Gag proteins are considered seropositive.[8]

Screening for STLV infection is performed by using ELISA or particle agglutination containing HTLV-1 and/or HTLV-2 viral lysates and/or with IFA (by using HTLV-infected cells).[8] Confirmation of infection is done using HTLV-1 WB assays spiked with recombinant Env proteins (GD21) common to both HTLV-1 and HTLV-2 and with peptides specific for HTLV-1 and HTLV-2.[8]

Serologic screening as well as virologic screening by culture or PCR testing of peripheral blood mononuclear cells is indicated to identify the presence of SRV. Criteria for WB positivity include reactivity to at least one Gag protein and one Env protein.[8]

Serologic WB testing for SFV antibodies requires the use of two tests: (i) one that contains antigen from a monkey and (ii) one that contains antigen from an ape to allow detection of respective antibodies.[8] There is an assay that combines both ape and monkey SFV antigens into a single WB assay, thus eliminating the need for two WB tests on each sample.[12] ELISA, IFA, and radioimmunoprecipitation assays (RIPA) have also been used for the detection of SFV antibody.[8]

Treatment

Procedures for management of NHP exposures should be developed institutionally in conjunction with specialists in infectious disease and occupational medicine, veterinarians, research personnel, and safety specialists. If there is a life-threatening injury, the worker should be transported immediately to the ED with open communication between ED and occupational medicine to facilitate care in accordance with respective NHP exposure policy. All NHP bite wounds or other skin exposures to NHP blood or body fluids should be cleansed with soap and water for a minimum of 15 minutes. If eyes or mucous membranes have been exposed, copious irrigation with flowing water should be initiated for a minimum of 10 minutes. There are currently no standard treatment guidelines for PEP in simian retrovirus exposures. The use of antiretrovirals for PEP as is used for HIV exposures should be initiated for SIV exposures.[13] The use of postexposure prophylaxis for STLV and SRV exposures has not been fully evaluated, but may be used. As SFV has not been documented to cause active disease in nonhuman primates nor in humans, postexposure prophylaxis is not recommended in these exposures.

Medical surveillance

Serum should be collected and stored at 6-month intervals for individuals performing research with SIV.[14] Serum banking should also be undertaken on individuals whose work entails exposure to nonhuman primates. If a worker is inadvertently exposed to SIV-contaminated material, he or she should undergo a medical evaluation and serum examination for antibody against SIV. Seronegative workers should be retested at 6 weeks, 12 weeks, and 6 months. They should seek medical attention for any acute illness that develops within 12 weeks of exposure.

Prevention

During work with clinical specimens, laboratory coats, gowns, or uniforms should be worn along with protective eyewear, face masks, and gloves.[15] If simian retroviruses are being propagated in research laboratories or procedures are performed that may generate aerosols, activity should be performed in BSL-2 facilities, with additional practices and containment equipment recommended for BSL-3. Activities involving large volume production should be conducted in BSL-3 facilities. BSL-2 standards are recommended when handling infected animals and when performing activities involving clinical specimens.[15] Work surfaces should be decontaminated and hands washed immediately after handling infectious material, even when gloves have been worn.

References

1. Switzer WM, Bhullar V, Shanmugam V, et al. Frequent simian foamy virus infection in persons occupationally exposed to nonhuman primates. *J Virol* 2004;78(6):2780–2789.
2. Lerche NW, Osborn KG. Simian retrovirus infections: potential confounding variables in primate toxicology studies. *Toxicol Pathol* 2003;31(Suppl):103–110.
3. Khabbaz RF, Rowe T, Murphey Corb M, et al. Simian immunodeficiency virus needlestick accident in a laboratory worker. *Lancet* 1992;340:271–273.
4. Khabbaz RF, Heneine XV, George JR, et al. Brief report: infection of a laboratory worker with simian immunodeficiency virus. *N Engl J Med* 1994;330:172–177.
5. Sotir M, Switzer W, Schable C, et al. Risk of occupational exposure to potentially infectious nonhuman primate materials and to simian immunodeficiency virus. *J Med Primatol* 1997;26(5):233–240.
6. Courgnaud V, Van Dooren S, Liegeois F, et al. Simian T-cell leukemia virus (STLV) infection in wild primate populations in Cameroon: evidence for dual STLV type 1 and type 3 infection in agile mangabeys (Cercocebus agilis). *J Virol* 2004; 78(9):4700–4709.
7. Filippone C, Betsem E, Tortevoye P, et al. A severe bite from a nonhuman primate is a major risk factor for HTLV-1 infection in hunters from Central Africa. *Clin Infect Dis* 2015;60(11):1667–1676.
8. Murphy HW, Miller M, Ramer J, et al. Implications of simian retroviruses for captive primate population management and the occupational safety of primate handlers. *J Zoo Wildl Med* 2006;37(3):219–233.
9. Lerche NW, Switzer WM, Yee JL, et al. Evidence of infection with simian type D retrovirus in persons occupationally exposed to nonhuman primates. *J Virol* 2001;75(4):1783–1789.
10. Heneine W, Switzer WM, Sandstrom P, et al. Identification of a human population infected with simian foamy viruses. *Nat Med* 1998;4:403–407.
11. National Research Council. *Occupational Health and Safety in the Care and Use of Nonhuman Primates*. Washington, DC: The National Academic Press, 2003 Available at: http://www.nap.edu/openbook.php?record_id=10713 (accessed on May 4, 2016).
12. Hussain AI, Shanmugam V, Bhullar VB, et al. Screening for simian foamy virus infection by using a combined antigen Western blot assay: evidence for a wide distribution among Old World primates and identification of four new divergent viruses. *Virology* 2003;309:248–257.
13. Murphy HW, Switzer WM. Occupational exposure to zoonotic simian retroviruses – health and safety implication for persons working with non human primates. In: Fowler ME, Eric Miller R (eds.) *Zoo and Wild Animal Medicine: Current Therapy*, 6th edn. London: Elsevier Health Sciences, 2007, pp. 251–264.
14. Centers for Disease Control. Guidelines to prevent simian immunodeficiency virus infection in laboratory workers and animal handlers. *MMWR* 1988;37:693–703.
15. Centers for Disease Control. Anonymous survey for Simian immunodeficiency virus (SIV) seropositivity in SIV laboratory researchers—United States. *MMWR* 1992;41:814–815.

VACCINIA

Common name for disease: None
Classification: Family—Poxviridae; genus—*Orthopoxvirus*

Occupational setting

Vaccinia is related to, but not the same as, cowpox, a rare zoonosis from domesticated animals. Vaccinia is also a highly effective immunizing agent against the closely related and far more dangerous smallpox virus (variola). Since global eradication of naturally acquired smallpox occurred in 1977, there is no longer a routine requirement for population immunization. Vaccinia immunization is now rare, except for selected researchers, first responders for bioterrorism preparedness, some healthcare personnel, and for military troops. Smallpox is considered to be a potential biological warfare threat; therefore, military forces may stockpile vaccinia immunizations.

Researchers and vaccine developers can potentially become infected with vaccinia virus in the course of their work. There have been a number of case reports in the literature of laboratory-acquired vaccinia virus infections in the United States over the past decade, with fewer than 5 per year reported to CDC.[1–5] Most of these exposures have been accidental inoculations. While most of the research using vaccinia virus is with highly attenuated virus (including recombinant forms), a number of research labs do utilize less-attenuated virus and thus estimating the incidence of vaccinia infection among at-risk laboratory workers is difficult.[3] Healthcare workers can also be potentially exposed in the course of treating a vaccinia-infected patient. Military personnel and other first responders are also at risk. From 2002 to 2009, the reported rate of vaccinia virus infection via contact transmission (via accidental spread of vaccinia from the vaccinated person to an unvaccinated person) in US military personnel was 5 cases per 100 000 persons, with intimate and sports-related contact being the most commonly cited exposures.[6,7]

Cattle and dairy workers have been found to have contracted vaccinia-like viruses (such as Cantagalo virus, Araçatuba virus) during the course of their employment.[8,9] Over the past 15 years, several outbreaks affecting dairy and cattle workers have been reported in Brazil as a result of exposures to these bovine-originating vaccinia-like viruses.[9]

Exposure (route)

Transmission of vaccinia virus in the occupational setting occurs via percutaneous or mucocutaneous exposure. A person can also become infected with direct contact to the smallpox vaccination site. As vaccinia virus is used in smallpox vaccination, inadvertent exposure to the injection site can result in contact transmission or inadvertent autoinoculation.[10]

Pathobiology

Vaccinia is a double-stranded DNA virus consisting of <200 kilobits per second encoding for some 260 proteins.[11] It replicates exclusively in the cytoplasm of infected cells.[12] Given its large genome and the numerous viral proteins it can produce, vaccinia has the ability to suppress the antiviral activity of its host at all stages of its replication cycle.[13] Vaccinia virus is able to stimulate actin polymerization at the plasma membrane, thus playing an important role in cell-to-cell dissemination of the virus.[11] Research is continuing to better elucidate the mechanism of pathogenesis of vaccinia in animal models.

Diagnosis

Diagnosis requires clinical suspicion and recognition of the typical lesion and its spread. Suspicion should be heightened for vaccinia research workers, recently immunized workers, and contacts of those recently immunized. History of exposure is the critical diagnostic clue. In addition, the virus can be isolated in scrapings and vesicle fluids of lesions, where samples can be obtained by unroofing vesicles, collecting the tissue, performing slide touch preps of the unearthed base of each vesicle, and obtaining viral swabs by using pox collection kits.[7] Recent infection can be inferred from sequential serologic testing.

Treatment

Treatment is supportive. In case of ocular exposure, timely irrigation of eyes is imperative. Vaccinia immune globulin, developed from pooled sera collected from vaccinated patients in the 1960s, can be used in patients with generalized vaccinia and eczema vaccinatum or those at high risk for developing complications following vaccination with vaccinia. However, vaccinia immune globulin is less successful in treating progressive vaccinia disease or vaccinia infection with CNS complications.[14] Cidofovir and adefovir are currently being investigated to evaluate the clinical effect and outcomes as a secondary treatment of vaccinia-related complications that do not respond to VIG treatment.[14]

Prevention

Routine smallpox vaccination for the general population stopped in 1972 after disease eradication in the United States. In 1976, smallpox vaccination of healthcare workers was discontinued. Since 1980, the CDC's Advisory Committee on Immunization Practices (ACIP) has recommended smallpox vaccine for laboratory workers who directly handle cultures or animals contaminated or infected with non-highly attenuated vaccinia virus.[15] In 1991, ACIP further expanded smallpox vaccination recommendations to include healthcare workers involved in clinical trials using recombinant vaccinia virus vaccines.[15] With heightened concerns about terrorism and bioterrorism after September 11, 2001, the ACIP recommended smallpox vaccination for healthcare workers providing direct medical care to smallpox patients as well as to those healthcare workers administering the vaccines in preparation for a possible bioterrorist attack involving smallpox.[16,17] Based on 2003 ACIP supplemental recommendations for using smallpox vaccine in a pre-event vaccination program, through December 2003, approximately 40 000 civilian personnel in the United States received licensed smallpox vaccine as part of state and local smallpox preparedness programs.[18] The majority were healthcare and public health response personnel, but the program also included law enforcement, firefighters, and EMTs.[18]

In 2007, the Dryvax vaccinia vaccine was discontinued and was replaced with the ACAM2000 smallpox vaccine.[19] ACAM2000 is a vaccinia virus vaccine derived from a plaque-purified clone of the same New York City Board of Health strain that was used to manufacture Dryvax vaccine.[19] CDC is the only vaccine source for civilians.[19] Routine vaccination with ACAM2000 is recommended for laboratory personnel who directly handle cultures or animals contaminated or infected with replication-competent vaccinia virus, recombinant vaccinia viruses derived from replication-competent vaccinia strains (i.e., those that are capable of causing clinical infection and producing infectious virus in humans), or other orthopoxviruses that infect humans (e.g., monkeypox, cowpox, and variola).[19] Healthcare personnel who currently treat or anticipate treating patients with vaccinia virus infections and whose contact with replication-competent vaccinia viruses is limited to contaminated materials (including dressings) and persons administering ACAM2000 vaccine who adhere to appropriate infection prevention measures can be offered vaccination with ACAM2000. The smallpox vaccine contains the live vaccinia virus. Smallpox vaccination provides high-level immunity for 3–5 years.[18] Persons should not receive the vaccine if they have a history or presence of atopic dermatitis or have other active exfoliative skin conditions (e.g., eczema, burns, impetigo, varicella zoster virus infection, herpes simplex virus infection, severe acne, severe diaper dermatitis with extensive areas of denuded skin, psoriasis, or keratosis follicularis (Darier disease).[19] It is also contraindicated for persons with immunosuppression (including human immunodeficiency virus (HIV) infection or acquired immune deficiency syndrome (AIDS), leukemia, lymphoma, generalized malignancy, solid organ transplantation, or therapy with alkylating agents, antimetabolites, radiation, tumor necrosis factor (TNF) inhibitors, or high-dose corticosteroids (≥2 mg/kg body weight or ≥20 mg/day of prednisone or its equivalent for ≥2 weeks), hematopoietic stem cell transplant recipients <24 months posttransplant or ≥24 months, but who have graft-versus-host disease or disease relapse, or other autoimmune disease, such as systemic lupus erythematosus, with immunodeficiency as a clinical component.[19] Other contraindications include women who are pregnant or breastfeeding; persons with a serious allergy to any component of ACAM2000; persons with known underlying heart disease with or without symptoms (e.g., coronary artery disease or cardiomyopathy); and primary vaccinees with three or more known major cardiac risk factors (i.e., hypertension, diabetes, hypercholesterolemia, heart disease at the age of 50 years in a first-degree relative, and smoking).[19]

The vaccine should be administered only by trained individuals and is administered percutaneously by scarification using 15 jabs of a bifurcated needle.[20] Healthy individuals may respond to primary vaccinia challenge with a mild fever and lymphadenopathy. A red papule forms at the site of inoculation 3–5 days after exposure. This papule vesiculates several days later and becomes pustular by the 9th or 11th day. The pustule dries after 2 weeks and drops off when encrusted, usually about 3 weeks after vaccination. The axillary lymphadenopathy often associated with the process may be persistent.[21,22] Serious complications resulting from immunization are rare. They include systemic vaccinia (also called progressive vaccinia or vaccinia gangrenosum). Systemic vaccinia leads to underlying destruction of skin, subcutaneous tissues, and even viscera of immunosuppressed individuals. Patients with agammaglobulinemia or T-cell deficiencies and those receiving immunosuppressive therapy are at greatest risk. The disease is usually fatal. Vaccinia infection of eczematous skin or other problem skin (eczema vaccinatum) is also a serious complication. Accidental inoculation of eyelids, vulva, or other normal mucous membrane is another rare skin complication. Unintentional exposures via transmission from infected individuals or from laboratory exposures are rare but may be serious if they involve eyes or if the exposed individual is immunosuppressed. Postvaccinal encephalomyelitis is the most dreaded complication of immunization of patients with normal immune systems. Its incidence is still debated; it is said to occur in 1 : 10 000–1 : 25 000 vaccinations or perhaps as infrequently as 1 : 300 000 vaccinations with at least 25% mortality and an equal risk of other severe neurological sequelae.

Personnel whose only occupational exposure to orthopoxviruses is through administering smallpox vaccine to others should be revaccinated every 10 years.[18] During a smallpox outbreak, individuals likely to be administering vaccine in larger vaccination efforts should receive revaccination, without regard to interval from last vaccination.[18] CDC recommends revaccination of volunteer responders from the pre-event smallpox program on an as-needed basis only after there is determination of a credible smallpox threat to public health and prior to engaging in activities involving a risk for exposure to smallpox virus. This applies to first responders who had been vaccinated as part of the US Civilian Smallpox Preparedness and Response and had a documented vaccine "take."[18] Specific guidance has not been provided for the research laboratory setting; however, CDC has indicated that the risk for those administering vaccine may be comparable to the risk for laboratory workers handling non-highly attenuated vaccinia strains.[18] This would suggest a revaccination interval of 10 years for this group. However, in 2003, ACIP recommended that laboratory workers handling orthopoxviruses at proposed variola testing sites would need to be revaccinated every 3 years to maintain optimal immunity or protection.[23]

A 2010 study assessing knowledge, attitudes, and beliefs of laboratory workers found a lack of adherence to current recommendations of the ACIP, a need for greater training of laboratory workers who handle vaccinia virus, and the importance of having robust institutional policies to ensure the safety of laboratory workers through smallpox vaccination.[24] In the vaccinia research laboratory, prevention consists of administrative controls, engineering controls, and safe work practices. Pregnant, immunosuppressed, or those with chronic dermatologic conditions with active lesions are advised not to work with vaccinia virus. Work practices are standard for the BSL-2 laboratory. Personal protective equipment must be used at all time when working with vaccinia virus. The primary engineering control of the recombinant laboratory is the biosafety cabinet. Class I and II cabinets provide partial containment and should prevent most inoculation. Work with recombinant vaccinia virus, which may confer increased risk of infection or else increased virulence with infection, should be performed in a Class III biosafety cabinet (also known as a glove box) to provide full containment. Class III biosafety cabinets are also indicated if manipulations of highly concentrated cultures are required in the laboratory.

References

1. Loeb M, Zando I, Orvidas MC, et al. Laboratory-acquired vaccinia infection. *Can Commun Dis Rep* 2003;29(15):134–136.
2. Lewis FM, Chernak E, Goldman E, et al. Ocular vaccinia infection in laboratory worker, Philadelphia, 2004. *Emerg Infect Dis* 2006;12(1):134–137.
3. Centers for Disease Control and Prevention. Laboratory-acquired vaccinia exposures and infections—United States, 2005–2007. *MMWR* 2008;57(15):401–404.
4. Centers for Disease Control and Prevention. Laboratory-acquired vaccinia virus infection—Virginia, 2008. *MMWR* 2009;58(29):797–800.
5. Hsu CH, Farland J, Winters T, et al. Laboratory-acquired vaccinia virus infection in a recently immunized person—Massachusetts, 2013. *MMWR* 2015;64(16):435–438.
6. Vaccine Healthcare Centers Network. Smallpox Vaccine Contact Transmission Fact Sheet. Updated June17, 2016. Available at: http://www.vaccines.mil/VHC/ContactTransmissionFactSheet_SM.aspx (accessed on June 29, 2016).
7. Young GE, Hidalgo CM, Sullivan-Frohm A, et al. Secondary and tertiary transmission of vaccinia virus from US military service member. *Emerg Infect Dis* 2011;17(4):718–721.
8. Damaso CR, Esposito JJ, Condit RC, et al. An emergent poxvirus from humans and cattle in Rio de Janeiro State: Cantagalo virus may derive from Brazilian smallpox vaccine. *Virology* 2000;277(2):439–449.
9. de Souza Trindade G, da Fonseca FG, Marques JT, et al. Araçatuba virus: a vaccinialike virus associated with infection in humans and cattle. *Emerg Infect Dis* 2003;9(2):155–160.
10. Casey C, Vellozzi C, Mootrey GT, et al. Surveillance guidelines for smallpox vaccine (vaccinia) adverse reactions. *MMWR Recomm Rep* 2006;55(RR-1):1–16.
11. Leite F, Way M. (2015) The role of signalling and the cytoskeleton during vaccinia virus egress. *Virus Res.* Available at: http://ac.els-cdn.com/S0168170215000490/1-s2.0-S0168170215000490-main.pdf?_tid=3f8375f2-1cea-11e5-ae3e-00000aab0f26&acdnat=1435422991_8adccdada655dcfbc2c4831105aa1905 (accessed on June 17, 2016).
12. Moss B. Poxviridae: the viruses and their replication. In: Knipe DM, Howley PM. eds. *Fields Virology*, 5th edn. New York: Lippincott Williams and Wilkins, 2007:2905–2945.
13. Haller SL, Peng C, McFadden G, Rothenburg S. Poxviruses and the evolution of host range and virulence. *Infect Genet Evol* 2014;21:15–40.
14. Peterson BW, Damon IK. Orthopoxviruses: vaccinia (smallpox vaccine), variola (smallpox), monkeypox, and cowpox. In: Bennett JE, Dolin R, Blaser M. eds. *Mandell, Douglas, and Bennett's Principles and Practice of Infectious Disease*, 8th edn. Philadelphia, PA: Elsevier Saunders, 2015:1694–1702.
15. Rotz LD, Dotson DA, Damon IK, et al. Vaccinia (smallpox) vaccine: recommendations of the Advisory Committee on Immunization Practices (ACIP), 2001. *MMWR Recomm Rep* 2001;50(RR-10):1–25.
16. Wharton M, Strikas RA, Harpaz R, et al. Recommendations for using smallpox vaccine in a pre-event vaccination program. Supplemental recommendations of the Advisory Committee on Immunization Practices (ACIP) and the Healthcare Infection Control Practices Advisory Committee (HICPAC). *MMWR Recomm Rep* 2003;52(RR-7):1–16.
17. Centers for Disease Control and Prevention. CDC interim guidance for revaccination of eligible persons who participated in the US civilian smallpox preparedness and response program—October 2008. Available at: http://emergency.cdc.gov/agent/smallpox/revaxmemo.asp?s_cid=EmergencyPreparedness1e (accessed on June 17, 2016).
18. Centers for Disease Control and Prevention. CDC Interim Guidance for Revaccination of Eligible Persons who Participated in the US Civilian Smallpox Preparedness and Response Program—October 2008. Available at: http://emergency.cdc.gov/agent/smallpox/revaxmemo.asp#downloadpage (accessed on June 17, 2016).
19. Petersen BW, Harms TJ, Reynolds MG, et al. Use of Vaccinia virus smallpox vaccine in laboratory and health care personnel at risk for occupational exposure to orthopoxviruses—recommendations of the Advisory Committee on Immunization Practices (ACIP), 2015. *MMWR Morb Mortal Wkly Rep* 2016; 65:257–262. http://dx.doi.org/10.15585/mmwr.mm6510a2
20. Highlights of Prescribing Information for ACAM2000. Available at: http://emergency.cdc.gov/agent/smallpox/vaccination/pdf/ACAM2000PackageInsert-Version8-2007.pdf (accessed on June 17, 2016).
21. Benenson AS, ed. *Control of Communicable Disease in Man*, 15th edn. Washington, DC: American Public Health Association, 1990.
22. Ray CG. Smallpox, vaccinia, and cowpox. In: Petersdorf R, Adams RD, Braunwald E, et al. eds. *Harrison's Principles of Internal Medicine*, 10th edn. New York: McGraw-Hill, 1983:1118–1121.
23. Medcalf S, Bilek L, Hartman T, et al. Smallpox vaccination of laboratory workers at US variola testing sites. *Emerg Infect Dis* 2015;21(8):1437–1439. 10.3201/eid2108.140956.
24. Benzekri N, Goldman E, Lewis F, et al. Laboratory worker knowledge, attitudes and practices towards smallpox vaccine. *Occup Med* 2010;60(1):75–77.

VARICELLA ZOSTER VIRUS (VZV)

Common names for disease: Chickenpox (varicella), shingles (herpes zoster)
Classification: Family—Herpesviridae

Occupational setting

Infection with varicella zoster virus (VZV) may be encountered in any occupational setting, usually from a community-acquired infection, although rates of infections have been falling since the introduction of the varicella vaccine (Figure 21.13). Infection of susceptible healthcare workers is a known cause of nosocomial spread. Other high-risk settings include teachers of young children, childcare employees, residents and staff members in institutional or correctional settings, and military personnel. Primary infection in pregnant women is of particular concern because of the risks of congenital and neonatal infection. If a pregnant woman gets varicella in her first or early second trimester, there is up to a 2% risk of the baby being born with congenital varicella syndrome.[1]

Exposure (route)

VZV is transmitted from person to person by direct contact, droplet or aerosol from vesicular fluid of skin lesions, or by secretions from the respiratory tract. The virus enters the host via the respiratory tract. The virus is labile, and transmission by inanimate objects is unlikely to occur.[2,3]

Pathobiology

VZV is a member of the Herpesviridae family. It is a double-stranded DNA virus. After primary infection (chickenpox) or vaccination, the virus establishes latency in the dorsal root ganglia. Primary infection confers lifelong immunity to varicella, but reactivation of the virus causes herpes zoster (shingles) in about one third of the population. Vaccination with the live attenuated VZV appears to result in a lower risk of herpes zoster compared with people who were infected with wild-type VZV.[4]

The incubation period for varicella is 14–16 days (range 10–21 days) after exposure to a varicella or a herpes zoster rash. A mild prodrome of fever and malaise may occur 1–2 days before rash onset, particularly in adults.[5] The rash is

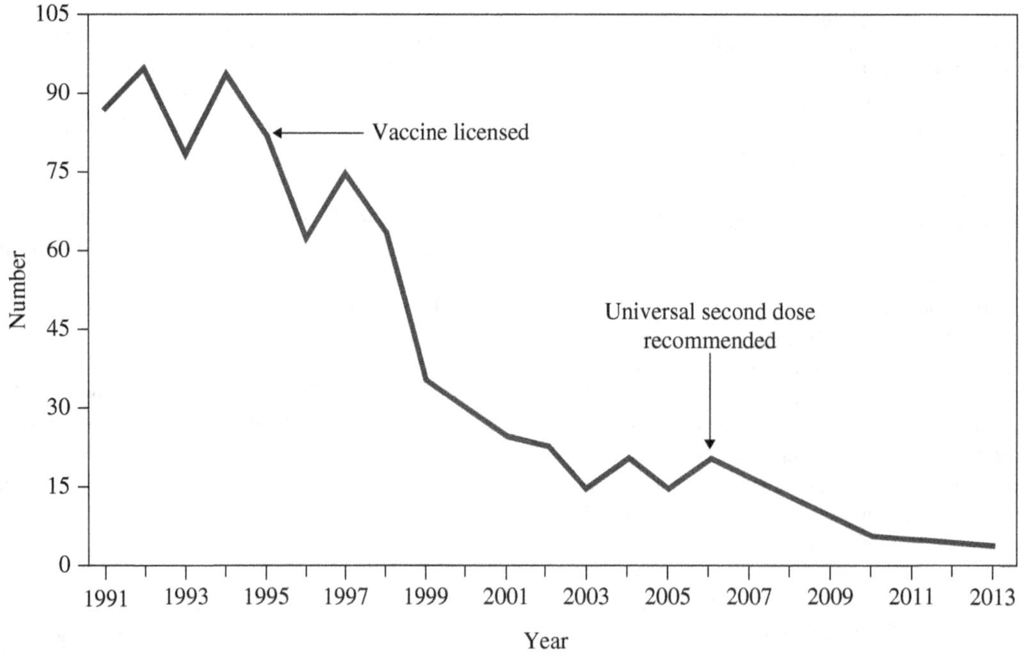

FIGURE 21.13 Varicella (Chickenpox). Number* of reported cases — Illinois, Michigan, Texas, and West Virginia, 1991–2013. *In thousands. Varicella was not nationally notifiable in 1996, when the first dose of the varicella vaccine was recommended in the United States. However, four states (Michigan, Illinois, Texas, and Virginia) were reporting varicella cases to CDC before the varicella vaccine was recommended and have continued reporting, providing consistent data to allow for monitoring of trends in varicella disease. In these four states, the number of cases reported in 2013 was 23% lower than 2012, 74% lower than the average annual number reported during the mature 1-dose varicella vaccination era of 2000–2006, and 96% lower than the average annual number reported during the prevaccine years of 1993–1995. *Source*: MMWR 2015; 62(53):104.

pruritic and progresses rapidly from macules to papules to vesicular lesions before crusting. The rash usually appears first on the head, chest, and back and then spreads to the rest of the body. The lesions are usually most concentrated on the chest and back.[5] The hallmark of the rash is the presence of lesions in all stages of development at the same time.

In healthy children, the illness is generally mild, with an itchy rash, malaise, and temperature up to 102 °F for 2–3 days. Adults may experience more severe disease and have a higher incidence of complications, including viral pneumonia. Less common complications include cerebellar ataxia, encephalitis, bacterial pneumonia, hemorrhagic conditions, septicemia, toxic shock syndrome, necrotizing fasciitis, osteomyelitis, and septic arthritis. More severe complications may occur in immunocompromised people, who are at risk of developing visceral dissemination leading to pneumonia, hepatitis, encephalitis, and disseminated intravascular coagulopathy.[1] They can have an atypical varicella rash with more lesions and can be sick longer than immunocompetent persons. The lesions may erupt for as long as 10 days, may appear on the palms and soles, and may be hemorrhagic.[1] Recurrence of varicella usually occurs only in people who are immunocompromised.[5]

Infection with wild-type VZV occurring in a vaccinated person more than 42 days after varicella vaccination is termed "breakthrough varicella".[5] The illness is usually mild. Patients are typically afebrile or have low fever and develop fewer than 50 skin lesions.[5] The illness is usually shorter and the rash is more likely to be predominantly maculopapular rather than vesicular. However, 25–30% of persons vaccinated with 1 dose who develop breakthrough varicella have clinical features typical of varicella in unvaccinated people.[5]

Varicella infection during pregnancy is of concern because of risks for the mother, fetus, and newborn. Pregnant women are at increased risk for developing pneumonia and, in some cases, may die as a result of varicella.[1] In less than 2% of cases, VZV is transmitted *in utero* during a primary maternal infection. If this occurs in the first 20 weeks of pregnancy, fetal demise, intrauterine growth restriction, hydrops, limb deformities, microcephaly, and other neurologic defects can result from the congenital infection (congenital varicella syndrome).[6] Infants with the highest risk of acquiring VZV are born to women with acute infection appearing between 5 days before and 2 days following delivery. This occurs because the infant is exposed to VZV at birth in the setting of limited or no placental VZV-specific IgG transfer.[6]

Varicella is extremely contagious; secondary attack rates among susceptible household contacts range as high as 90%. Patients are infectious for 48 hours prior to developing a rash and until the vesicles crust over.[5] It is typically a disease of childhood; 99.5% of adults over the age of 40 have immunity.[4] Although primary infection in adults is less common, the complication rate is higher.

Zoster can occur at any age, but the peak incidence is in the sixth decade. Zoster most commonly presents with a rash in one or two adjacent dermatomes (localized zoster), most commonly on the trunk along a thoracic dermatome. The rash does not usually cross the body's midline. Approximately 20% of people have a rash that overlaps adjacent dermatomes. A more widespread rash affecting three or more dermatomes (disseminated zoster) may occur in people with compromised or suppressed immune systems. Disseminated zoster can be difficult to distinguish from varicella.[4]

The rash is usually painful, itchy, or tingly. Symptoms may precede the onset of rash by days to weeks. Some people may also have headache, photophobia, and malaise in the prodromal phase. The rash develops into clusters of clear vesicles. New vesicles continue to form over 3–5 days and progressively dry and crust over. They usually heal in 2–4 weeks. There may be permanent pigmentation changes and scarring on the skin.[4]

Potentially serious complications can occur with involvement of the ophthalmic branch of the trigeminal nerve. Lesions on the tip of the nose often herald zoster ophthalmicus, which can be sight threatening. Other complications include bacterial superinfection of the lesions, usually due to *Staphylococcus aureus* and, less commonly, due to group A beta hemolytic streptococcus; cranial and peripheral nerve palsies; and visceral involvement, such as meningoencephalitis, pneumonitis, hepatitis, and acute retinal necrosis.[4] Postherpetic neuralgia is a frequent complication; 25% of patients over the age of 50 experience pain that persists for over a month. Herpes zoster is also associated with an increased risk of stroke.[7,8] Antiviral treatment may mitigate this risk.[8] Most people experience a single episode of zoster, but recurrences can occur.

Diagnosis

Diagnosis is generally made on the basis of history and physical examination. Specific laboratory testing is usually unnecessary. Laboratory testing is indicated to verify varicella infection in vaccinated people (which is often mild and atypical in presentation), as well as determining cause of outbreaks, establishing varicella infection as a cause of death, and determining a patient's susceptibility to contracting varicella.[9] PCR is the method of choice for rapid clinical diagnosis. Skin lesions and scabs from skin lesions are optimal specimen types for PCR detection. Whole infected cell ELISA is the most commonly used test to determine if a person has antibodies to VZV from past varicella disease, while the more sensitive purified glycoprotein ELISA or fluorescent antibody to membrane antigen (FAMA) tests have been used in research settings to detect seroconversion after vaccination.[9]

Treatment

Management of varicella in the immunocompetent patient involves supportive care and aggressive hygiene to avoid bacterial superinfection of skin lesions as well as to limit transmission. Pruritus can be treated with topical or systemic antipruritic drugs. Fever should be treated with acetaminophen; aspirin is contraindicated because of the risk of Reye's syndrome. Oral acyclovir, administered within 24 hours of the appearance of a rash, is recommended for patients with increased risk for moderate to severe disease, including:

- Healthy persons older than 12 years of age
- Persons with chronic cutaneous or pulmonary disorders
- Persons receiving long-term salicylate therapy
- Persons receiving short, intermittent, or aerosolized courses of corticosteroids[10]

Some experts recommend oral acyclovir or valacyclovir for pregnant women with varicella, especially during the second and third trimesters.[10] Intravenous acyclovir is recommended for the pregnant patient with serious complications of varicella.[10] Intravenous acyclovir therapy is also recommended for patients who are immunocompromised.[11]

Varicella zoster immune globulin (VariZIG) is indicated for postexposure prophylaxis of varicella for persons at high risk for severe disease who lack evidence of immunity to varicella and for whom varicella vaccine is contraindicated.[12] VariZIG should be administered within 10 days of exposure to:

- Immunocompromised patients without evidence of immunity
- Newborn infants whose mothers have signs and symptoms of varicella around the time of delivery (5 days before to 2 days after)
- Hospitalized premature infants born at ≥28 weeks of gestation whose mothers do not have evidence of immunity to varicella
- Hospitalized premature infants born at <28 weeks of gestation or who weigh ≤1000 g at birth, regardless of their mothers' evidence of immunity to varicella
- Pregnant women without evidence of immunity[12]

Herpes zoster is usually self-limiting, but antivirals have been shown to decrease the duration of the episode and reduce the likelihood of complications including postherpetic neuralgia. Antiviral agents, acyclovir, valacyclovir, or famciclovir, should be started within 72 hours of onset to reduce the severity of the infection, the duration of the eruptive phase, and the intensity of acute pain.[13] Treatment also includes analgesics for control of acute zoster pain and good skin care for healing and prevention of secondary bacterial infection (Bader). Patients who develop disseminated herpes zoster with multiple dermatomes affected and with CNS involvement should be treated immediately with antiretrovirals.[8] Postherpetic neuralgia is notoriously difficult to treat and may require judicious use of narcotic analgesics. Other treatment approaches include topical agents, such as lidocaine patches, and systemic agents, such as the anticonvulsants gabapentin and pregabalin.[13] Tricyclic antidepressants and phenothiazines may also be effective.

Medical surveillance

With introduction of varicella vaccine in the last 20 years, the number of varicella outbreaks has declined considerably. Since the varicella vaccines started to be administered, only eight vaccinated people have been documented as spreading vaccine virus to others with nine people (including household members and people in long-term care facilities, contracting the varicella virus.[14]

Outbreaks still do occur, with elementary schools now the most common sites for varicella outbreaks.[15] Among adult populations, outbreaks have been reported from correctional facilities, military training facilities, refugee centers, immigration detention facilities, homeless shelters, and cruise ships.[15] Nosocomial transmission of VZV and herpes zoster in hospitals and long-term healthcare facilities also do occur, with both immunocompetent and immunocompromised patients having been identified as sources.[16] Studies of VZV exposure in healthcare settings have documented that a single provider with unrecognized varicella can result in the exposure of >30 patients and >30 employees.[17] Airborne transmission of VZV from patients with either varicella or herpes zoster has resulted in varicella in healthcare workers and patients who had no direct contact with the index case-patient.[16,18] Immediate identification of susceptible patients and staff, medical management of susceptible exposed patients at risk for complications of varicella, and removal of susceptible exposed healthcare workers are indicated, with appropriate symptoms monitoring and follow-up of the healthcare worker to assess for presence of disease.[16] Unvaccinated VZV-susceptible healthcare workers are potentially contagious from days 8 to 21 after exposure and should be reassigned to nonpatient care activities. These individuals should receive postexposure vaccination within 3–5 days of exposure to reduce disease progression if infection did indeed occur but vaccination 6 or more days after exposure is still indicated because it induces protection against subsequent exposures.[19,20] For unvaccinated VZV-susceptible healthcare personnel at risk for severe disease and for whom varicella vaccination is contraindicated (such as pregnant healthcare personnel), treatment with VariZIG after the exposure is recommended.[19,20]

Prevention

The mainstay of prevention is surveillance and vaccination of susceptible workers in high-risk occupations who have no contraindications to vaccination. Varicella vaccine (Varivax) was licensed for use in the United States in 1995. It is an attenuated single-antigen varicella virus vaccine, which induces high antibody titers in 99% of adults after two doses. Vaccination provides 70–90% protection against infection and 95% protection against severe disease for 7–10 years after vaccination. In a randomized control study of health children who received Varivax over a 10-year observation period, the estimated vaccine efficacy of two doses was 98.3% compared with 94.4% for one dose.[19,20] There are currently no studies of Varivax efficacy or effectiveness in adults. In 2005, a combination varicella-containing vaccine, measles, mumps, rubella, varicella (MMRV), otherwise known as ProQuad, was licensed in 2005 for use in children 12 months through 12 years of age. This is also a live attenuated vaccine. At this time, it is not known how long a vaccinated person is protected against varicella when receiving either of these vaccinations.

Most healthcare institutions now implement assessment of varicella immune status as part of routine pre-employment assessment. Documentation includes two doses of varicella vaccine in adults or as school-age children/adolescents; laboratory evidence of immunity or laboratory confirmation of disease; or diagnosis or verification of a history of varicella or herpes zoster by a healthcare provider.[21]

With respect to the management of patients with varicella, CDC recommends following standard precautions plus airborne precautions (negative airflow rooms) and contact precautions until the VSV lesions are dry and crusted. If negative airflow rooms are not available, patients with varicella should be isolated in closed rooms and be cared for by staff with evidence of immunity.[19,20] CDC recommendations for prevention and managing clusters of VZV infection in hospitals include:

- Isolating patients who have varicella and other susceptible patients who are exposed to VZV
- Controlling airflow
- Using rapid serologic testing to determine susceptibility
- Furloughing exposed susceptible personnel or screening these persons daily for skin lesions, fever, and systemic symptoms
- Temporarily reassigning varicella-susceptible personnel to locations remote from patient areas

Effective prevention strategies must account for the fact that most varicella cases, even those among healthcare workers, result from community-acquired infection.[22]

People with active lesions from herpes zoster can spread VZV to susceptible people.[4] People with active lesions should avoid contact with susceptible people in their household and in occupational settings until their lesions dry and crusted.[4] Healthcare personnel with zoster may need to be restricted from work to avoid exposing patients.

Herpes zoster vaccine (Zostavax) is recommended for adults aged 60 or older who do not have contraindications, even if they have a history of zoster.[23] However, vaccine efficacy wanes quickly and cannot be demonstrated after 5 years.[23]

References

1. Centers for Disease Control and Prevention. Varicella – People at High Risk for Complications. 2014. Available at: http://www.cdc.gov/chickenpox/hcp/high-risk.html (accessed on May 5, 2016).
2. Centers for Disease Control and Prevention. Prevention of Varicella: Recommendations of the Advisory Committee on Immunization Practices. *MMWR* 1996;45(RR-11):1–36.
3. Centers for Disease Control and Prevention. Prevention of Varicella: Update Recommendations of the Advisory Committee on Immunization Practices. *MMWR* 1999; 48(RR-6):1–12.
4. Centers for Disease Control and Prevention. Shingles (Herpes Zoster): Clinical Overview. Available at: http://www.cdc.gov/shingles/hcp/clinical-overview.html (accessed on May 5, 2016).
5. Centers for Disease Control and Prevention. Chicken Pox (Varicella): Clinical Overview. Available at: http://www.cdc.gov/chickenpox/hcp/clinical-overview.html (accessed on May 5, 2016).
6. Pastuszak AL, Levy M, Schick B, et al. Outcome after maternal varicella infection in the first 20 weeks of pregnancy. *N Engl J Med* 1994;330:901–905.
7. Sreenivasan N, Basit S, Wohlfahrt J, et al. The short- and long-term risk of stroke after herpes zoster – a nationwide population-based cohort study. *PLoS One* 2013;8(7):e69156.
8. Langan SM, Minassian C, Smeeth L, Thomas SL. Risk of stroke following herpes zoster: a self-controlled case-series study. *Clin Infect Dis* 2014;58(11):1497–1503.
9. Centers for Disease Control and Prevention. Varicella – Interpreting Laboratory Tests. 2012. Available at: http://www.cdc.gov/chickenpox/hcp/lab-tests.html (accessed on May 5, 2016).
10. Centers for Disease Control and Prevention. Chicken Pox (Varicella): Managing People at Risk for Severe Varicella. Available at: http://www.cdc.gov/chickenpox/hcp/persons-risk.html (accessed on May 5, 2016).
11. Balfour HH Jr, McMonigal KA, Bean B. Acyclovir therapy of varicella-zoster virus infections in immunocompromised patients. *J Antimicrob Chemother* 1983;12(Suppl B):169–179.
12. Centers for Disease Control and Prevention. Updated Recommendations for Use of VariZIG—United States, 2013. *MMWR* 2013;62(28):574–576.
13. Bader MS. Herpes zoster: diagnostic, therapeutic, and preventive approaches. *Postgrad Med* 2013;125(5):78–91. 10.3810/pgm.2013.09.2703.

14. Centers for Disease Control and Prevention. Varicella Vaccination of Healthcare Personnel – Information for Healthcare Providers. 2012. Available at: http://www.cdc.gov/vaccines/vpd-vac/varicella/hcp-vacc.htm (accessed on May 5, 2016).
15. Centers for Disease Control and Prevention. Varicella. In *Manual for the Surveillance of Vaccine-Preventable Diseases*, 5th edn. 2001. Available at: http://www.cdc.gov/vaccines/pubs/surv-manual/chpt17-varicella.html (accessed on May 5, 2016).
16. Centers for Disease Control and Prevention. Immunization of Health-Care Personnel: Recommendations of the Advisory Committee on Immunization Practices (ACIP). *MMWR* 2011;60(RR07):1–45.
17. Haiduven-Griffiths D, Fecko H. Varicella in hospital personnel: a challenge for the infection control practitioner. *Am J Infect Control* 1987;15:207–211.
18. Lopez SA, Burnett-Hartman A, Nambiar R, et al. Transmission of a newly characterized strain of varicella-zoster virus from a patient with herpes zoster in a long-term-care facility, West Virginia, 2004. *J Infect Dis* 2008;197:646–653.
19. Centers for Disease Control and Prevention. Preventing Varicella-Zoster Virus (VZV) Transmission from Zoster in Healthcare Settings. 2014. Available at: (http://www.cdc.gov/shingles/hcp/HC-settings.html#healthcare-personnel (accessed on May 5, 2016).
20. Kuter B, Matthews H, Shinefield H, et al. Ten year follow-up of healthy children who received one or two injections of varicella vaccine. *Pediatr Infect Dis* 2004;23:132–137.
21. Centers for Disease Control and Prevention. Assessing Immunity to Varicella. 2013. Available at: http://www.cdc.gov/chickenpox/hcp/immunity.html (accessed on May 5, 2016).
22. Josephson A, Karanfil L, Gombert M. Strategies for the management of varicella-susceptible healthcare workers after a known exposure. *Infect Control Hosp Epidemiol* 1990;11(6):309–313.
23. Hales CM, Harpaz R, Ortega-Sanchez I, et al. Update on recommendations for use of herpes zoster vaccine. *MMWR Morb Mortal Wkly Rep* 2014;63(33):729–731.

Chapter 22

BACTERIA

Christopher J. Martin, Aletheia S. Donahue, and John D. Meyer

ACINETOBACTER SPECIES

Common name for disease: None

Occupational setting

Acinetobacter species are ubiquitous in nature. They are commonly isolated from work settings with moist environments and microenvironments (e.g., swine confinement buildings,[1] wastewater treatment plants,[2] composting plants,[3] poultry-processing plants,[4] cotton mills,[5] metal-working operations,[6] and bakeries.[7])

Exposure (route)

Inhalation is the main route of exposure in the occupational setting.

Pathobiology

Acinetobacter species, particularly *Acinetobacter baumannii* and the closely related species *A. pittii* and *A. nosocomialis*, are among the most common causes of healthcare-associated pneumonia and other infections, particularly in intensive care unit (ICU) settings and long-term care facilities.[8] Patient-to-patient transmission is frequent in such outbreaks, with isolated case reports of transmission to healthcare workers.[9] The number of isolates in healthcare settings increases during times of conflict and following natural disasters, with recent large outbreaks involving soft-tissue infections among previously healthy US soldiers wounded in Afghanistan and Iraq.[10]

These organisms rarely cause infection outside of the clinical setting. Community-acquired pneumonia has been described in persons with cancer and alcoholism, with a preponderance of case reports during warm, humid months and from Asia and Australia.[8]

An outbreak of pneumonia caused by *A. baumannii* has been reported among three individuals working in close proximity in a foundry.[11] Two of the cases were fatal, and an examination of the lung tissue identified evidence of a mixed dust pneumoconiosis with features compatible with siderosis in both. The concomitant presence of iron may increase the virulence of this microorganism.[12] *A. Iwoffii* has been implicated in an outbreak of hypersensitivity pneumonitis in workers in an automobile parts manufacturing plant using metalworking fluids.[6]

Exposure to water aerosols, such as metalworking fluids, from environments with contamination involving multiple microorganisms, including *Acinetobacter*, has been associated with a spectrum of respiratory diseases (asthma, hypersensitivity pneumonitis, bronchitis, humidifier fever) in several occupational settings, mostly involving automotive and aeronautical manufacturing.[13] Despite extensive investigation, the specific etiology of disease in these outbreaks remains elusive, and therefore the precise role of *Acinetobacter* is uncertain.

Diagnosis

Infections with this organism are diagnosed using standard isolation and culture methods of appropriately selected clinical specimens. Rapid molecular methods such as polymerase chain reaction-based assays are increasingly available both to detect *Acinetobacter* directly from patient specimens and to

Physical and Biological Hazards of the Workplace, Third Edition. Edited by Gregg M. Stave and Peter H. Wald.

identify the presence of specific antimicrobial resistance genes.[14] Genotyping can help to identify the source of the outbreak and guide infection control measures.

Treatment

Since many *Acinetobacter* strains have developed multidrug resistance, therapy depends on the clinical setting (healthcare-associated versus community) as well as results of susceptibility testing. There are a wide number of options including cefepime, imipenem, meropenem, ampicillin/sulbactam, tigecycline, colistin, and polymyxin B. *Acinetobacter* species with resistance to all routinely tested antibiotics have been described in healthcare-associated outbreaks.[15]

Medical surveillance

There are no recommended medical surveillance activities.

Prevention

Engineering controls and work practices should be aimed at reducing microbial contamination of water and other media. Aerosolized processes involving contaminated water are of particular concern. The use of air-purifying respirators may also be appropriate.

References

1. Cormier Y, Tremblay G, Meriaux A, et al. Airborne microbial contents in two types of swine confinement buildings in Quebec. *Am Ind Hyg Assoc J* 1990; 51:304–9.
2. Laitinen S, Kangas J, Kotimaa M, et al. Workers' exposure to airborne bacteria and endotoxins at industrial wastewater treatment plants. *Am Ind Hyg Assoc J* 1994; 55(11):1055–60.
3. Lundholm M and Rylander R. Occupational symptoms among compost workers. *J Occup Med* 1980; 22:256–7.
4. Fallschissel K, Klug K, Kämpfer P, et al. Detection of airborne bacteria in a German turkey house by cultivation-based and molecular methods. *Ann Occup Hyg* 2010; 54(8):934–43.
5. Delucca AJ and Shaffer GP. Factors influencing endotoxin concentrations on cotton grown in hot, humid environments: a two year study. *Br J Ind Med* 1989; 46:88–91.
6. Zacharisen MC, Kadambi AR, Schlueter DP, et al. The spectrum of respiratory disease associated with exposure to metal working fluids. *J Occup Environ Med* 1998; 40(7):640–7.
7. Domanska A and Stroszejn-Mrowca G. Endotoxin in the occupational environment of bakers: method of detection. *Int J Occup Med Environ Health* 1994; 7(2):125–34.
8. Munoz-Price LS and Weinstein RA. Acinetobacter infection. *N Engl J Med* 2008; 358(12):1271–81.
9. Whitman TJ, Qasba SS, Timpone JG, et al. Occupational transmission of *Acinetobacter baumannii* from a United States serviceman wounded in Iraq to a health care worker. *Clin Infect Dis* 2008; 47(4):439–43.
10. Centers for Disease Control and Prevention (CDC). *Acinetobacter baumannii* infections among patients at military medical facilities treating injured U.S. service members, 2002–2004. *MMWR* 2004; 53:1063–6.
11. Cordes LG, Brink EW, Checko PJ, et al. A cluster of Acinetobacter pneumonia in foundry workers. *Ann Intern Med* 1981; 95:688–93.
12. Phillips M. Acinetobacter species In: Bennett JE, Dolin R, and Blaser MJ (eds.), *Mandell, Douglas, and Bennett's Principles and Practice of Infectious Diseases*, 8th ed. Philadelphia: Elsevier Saunders, 2014.
13. Burton CM, Crook B, Scaife H, et al. Systematic review of respiratory outbreaks associated with exposure to water-based metalworking fluids. *Ann Occup Hyg* 2012; 56(4):374–88.
14. Denys GA and Relich RF. Antibiotic resistance in nosocomial respiratory infections. *Clin Lab Med* 2014; 34(2):257–70.
15. Pendleton, JN, Gorman, SP, and Gilmore, BF. Clinical relevance of the ESKAPE pathogens. *Exp Rev Anti Infect Ther* 2013; 11(3):297–308.

BACILLUS SPECIES

Common names for disease: Anthrax, woolsorter's disease, ragpicker's disease, splenic fever.

Occupational setting

Anthrax is an enzootic disease with a worldwide distribution transmitted to humans via contact with animals or animal products. In 2001, an outbreak of bioterrorism-related anthrax occurred from contaminated mail sent through the United States Postal Service.[1] Eventually, 23 confirmed or suspected cases were documented, including one in a laboratory worker who handled environmental samples from the outbreak.[2]

Anthrax is most commonly associated with herbivores, especially cattle, which acquire infection through ingestion of endospores on contaminated soil.[3] Potential sources of human exposure are raw wool or hair,[4] bone,[5] meat,[5] and hides or skins imported from areas where anthrax is enzootic, especially Africa and Asia.[6] Shepherds, farmers, craft workers, and workers in manufacturing plants using the above materials are at highest risk for occupational anthrax[7] and, in the past, textile mills that used these animal products presented a significant occupational hazard.[8] Rare cases continue to be reported in these settings.[9]

Exposure (route)

Naturally occurring anthrax is an extremely rare cause of human disease in the United States.[6] Transmission occurs via inhalation, cutaneous contact, or ingestion of endospores. Recent large outbreaks of disease among users of contaminated heroin in several European countries have led to the recognition of a distinct presentation from a fourth route of exposure termed "injectional" or "injection" anthrax.[10]

The portal of entry will also determine the clinical picture. Cutaneous anthrax cases are the most common and result from endospores being introduced through cuts or abrasions in the skin.[11] Endospores are within the respirable size range and are, therefore, deposited at the alveolar level following inhalation.[12] Ingestion of meat contaminated with endospores results in gastrointestinal anthrax. There are no known cases of person-to-person transmission of anthrax via the inhalation route. However, endospores are produced in cases of cutaneous anthrax, which have caused widespread contamination and secondary transmission of disease, including in a healthcare setting.[13]

Pathobiology

Anthrax *Bacilli* are Gram-positive, rod-shaped bacteria, which produce endospores that are not true spores since they are not the product of reproduction but a dormant form of the bacteria. Endospores are the infectious form of the disease, can persist for decades in the environment, and are resistant to environmental extremes of desiccation, heat, freezing, and ultraviolet light as well as many common disinfectants. The species of greatest concern is *B. anthracis*, although a recent outbreak of disease resembling cutaneous anthrax among shepherds has been attributed to *B. pumilus*.[14]

Upon penetration into the host, the endospores either germinate locally or are phagocytized and transported into the lymphatic system to regional lymph nodes with subsequent germination.[15] Within hours of germination, the bacilli produce potent exotoxins, which have multiple physiological effects, causing widespread inflammation, edema, necrosis, hypotension, hypoperfusion, congestion, and hemorrhage.[15]

Anthrax exists in three primary forms depending upon the route of entry: cutaneous, inhalational, and gastrointestinal. Cutaneous anthrax, which accounts for more than 95% of naturally occurring anthrax cases,[16] results from the introduction of endospores into the skin, most commonly on the head and neck or upper extremity, via a wound, penetrating animal fiber, or an insect bite. The incubation period is estimated to range from 1 to 12 days. The endospores germinate and multiply in the subcutaneous tissue, with production of exotoxin causing tissue necrosis. A slowly enlarging papule is first noticed, which then vesiculates, eventually bursting to form a black eschar around which smaller vesicles may appear. The lesion is generally painless and may be associated with impressive local edema, regional lymphadenopathy, and septicemia. The disease may be self-limited and mortality is less than 1% with appropriate therapy, although airway compromise from infections involving the face and neck can occur.[16] The diagnosis should be considered in any patient with a painless ulcer with vesicles and edema who has a history of exposure to animals or animal products.

Recently, a new type of anthrax infection, termed injection or injectional anthrax, has been described among European heroin users. No cases have been observed to date in North America. It is thought that contamination may occur because the heroin is produced in Afghanistan and transported through Iran and Turkey, all countries where anthrax is enzootic.[17] Unlike cutaneous anthrax, papules, vesicles, and eschars are generally not observed and there is an increased risk of shock and death, with mortality reported to be 37%.[10]

While inhalation anthrax is very rare in the natural setting, it accounted for 11 of the 22 cases in the bioterrorist attack of 2001.[17] Following inhalation and alveolar deposition, endospores are rapidly phagocytized in the terminal alveoli by macrophages and carried to mediastinal lymph nodes. There, the endospores germinate and multiply, producing large amounts of exotoxin. The result is a hemorrhagic, edematous mediastinitis, evidenced by the characteristic widening of the mediastinum on imaging studies. The initial prodromal phase of the illness follows an incubation period of 1–5 days and lasts 3–4 days with nonspecific, flu-like symptoms. Without prompt treatment, a second phase with septic shock, meningitis, and gastrointestinal involvement ensues. Death can occur within 24 hours of the onset of this phase. The mortality approaches 100% without treatment, but with early diagnosis, use of multiple antibiotics, and improvements in supportive care, mortality was 46% among the 11 inhalational cases in the outbreak in the United States.[16]

Gastrointestinal anthrax is extremely rare outside of enzootic regions and occurs after ingestion of contaminated meat. The incubation period has been estimated to be 42 hours.[3] The central feature of this form of disease is ulcers, which can occur anywhere along the gastrointestinal tract from the oral cavity to the cecum, depending upon where endospores are deposited. In the oropharyngeal form, symptoms and signs include fever, anorexia, cervical lymphadenopathy, and edema. In the intestinal form, there is mesenteric lymphadenitis with vomiting, anorexia, fever, abdominal pain, hematemesis, and bloody diarrhea. Ascites, septicemia, intestinal perforation, shock, and death may ensue. The case fatality ratio ranges from 25 to 60%.[16]

Although most commonly associated with inhalational anthrax, all forms of the disease can lead to shock, which is usually fatal despite aggressive supportive measures.[16]

Diagnosis

The Centers for Disease Control and Prevention (CDC) has provided definitions for suspected, probable, and confirmed cases of cutaneous, inhalational, gastrointestinal, oropharyngeal, and meningeal anthrax as well detailed guidance on diagnosis.[18] Bacilli can be cultured and identified from a variety of appropriately collected clinical specimens such as blood, drainage from cutaneous lesions, cerebrospinal fluid, sputum, pleural fluid, and feces. Several techniques are available to directly and rapidly identify the microorganism from samples, such as the polymerase chain reaction (PCR). A variety of immunological-based methods toward components of both the microorganism and the exotoxins have been developed. Isolates should be sent for confirmatory testing to the CDC's Laboratory Response Network.[19]

Treatment

Early antimicrobial therapy is essential. During the 2001 outbreak, all cases of inhalational anthrax treated during the prodromal phase survived, while all those treated later died.[17]

The approach to treating anthrax differs from that of other bacterial infections in several important ways.[16] Because of the potential for the persistence of endospores in the body, prolonged therapy (60 days) is often indicated. Since exotoxins rather than the bacilli mediate many of the effects of anthrax infection, antimicrobials that inhibit protein synthesis and exotoxin production (i.e., clindamycin or linezolid) may need to be added to bactericidal agents (fluoroquinolones).

The treatment of choice for cutaneous anthrax is doxycycline or an oral fluoroquinolone such as ciprofloxacin. Penicillin VK or amoxicillin may be used for penicillin-susceptible strains.[16] If systemic illness develops or there is a risk of airway compromise, patients should be treated like other forms of anthrax. Gastrointestinal, inhalational, or injectional anthrax should be treated with intravenous ciprofloxacin combined with either clindamycin or linezolid if meningitis has been excluded. If there is the possibility of meningeal involvement, three-drug therapy is recommended with ciprofloxacin, meropenem, and linezolid being the current antibiotics of choice.[16] Antitoxin therapy should be added if there is a high index of suspicion of systemic involvement.[16] In 2015, the FDA-approved Anthrasil—Anthrax Immune Globulin Intravenous, Human—to treat patients with inhalational anthrax in combination with appropriate antibacterial drugs.[20]

Medical surveillance

There are no validated measures to monitor individual's exposure to anthrax. While the results of serological studies and nasal swabs may be useful for epidemiologic purposes, they should not be used for medical surveillance purposes following suspected exposure of workers.[21] The Occupational Safety and Health Administration (OSHA) recommends baseline, periodic, and final evaluations of workers potentially exposed to anthrax, which includes an assessment of contraindications or adverse effects from vaccination or antibiotics.[22] Medical monitoring of workers with potential anthrax exposure should be performed within the context of a comprehensive occupational health and safety program which includes a risk assessment for various jobs, a health and safety plan, and on-site monitoring for heat, stress, fatigue, and adverse psychological effects associated with the response, including personal protective equipment, to this highly virulent agent.[23]

Anthrax is a nationally notifiable disease in the United States and a CDC Category A bioterrorism agent; therefore, suspected cases should be immediately reported to local public health authorities.

Prevention

The National Institute of Occupational Safety and Health (NIOSH) has provided detailed recommendations on personal protective equipment for biological agents, including anthrax, with an orientation toward terrorist-related events.[23] The specific level of respiratory protection varies depending upon the hazard and suspected level of exposure from a self-contained breathing apparatus to a full facepiece air-purifying respirator.

Historically in the industrial setting, measures such as the washing of potentially contaminated material, mechanization, improved ventilation, and hygiene of work areas have all been successful in greatly reducing occupational cases of anthrax infection.[24] OSHA provides detailed recommendations on the management of anthrax in the workplace including sampling, personal protective equipment, and decontamination measures.[22] Because of the well-known resistance of Bacillus endospores to many commonly used disinfection measures, careful attention must be paid to the correct choice of agent. Ethylene oxide, chlorine dioxide, paraformaldehyde, and irradiation have appropriate endosporicidal activity.[22] Alcohol, alcohol-based sanitizers,[25] and ultraviolet irradiation[26] are considered ineffective.

In the healthcare setting, standard precautions are recommended with contact precautions for cases of cutaneous anthrax with uncontained drainage.[27] In the agricultural setting, anthrax has been successfully mitigated through vaccination of livestock and epidemiologic measures to promptly identify, trace, and dispose of infected animals through incineration.[28] Since such measures are not uniformly applied throughout the world, added precautions should be taken when handling animal products imported from higher risk countries.

The Advisory Committee on Immunization Practices (ACIP)'s recommendations for prophylactic anthrax vaccination are summarized in Table 22.1.[29]

After exposure to anthrax, post-exposure prophylaxis consisting of ciprofloxacin or doxycycline as well as vaccination is recommended.[29] Anthrax immune globulin has also been approved for post-exposure prophylaxis.

TABLE 22.1 ACIP recommendations for anthrax vaccination.

Occupation/Group	Vaccine recommendation
General population	Not recommended prior to a bioterrorism event.
Medical personnel	Not recommended.
Persons who handle animals or animal products	Only recommended when other preventive measures deemed insufficient.
Persons who routinely have contact with animal hide drums or animal hides	Not recommended.
U.S. veterinarians and animal husbandry technicians	Not recommended, unless at higher risk due to potential exposures from work involving enzootic areas or research settings.
Laboratory workers	Only recommended for personnel with repeated exposure to endospores, especially with the potential for aerosolization.
Workers in postal processing facilities	Not recommended
Military personnel	Only when deemed to have a "calculable risk" of exposure to aerosolized endospores.
Environmental investigators and remediation workers	Recommended for those who repeatedly enter areas contaminated with endospores.
Emergency responders	Only for those whose response activities may involve exposure to endospores. Vaccination should be voluntary and incorporated within a broader occupational health and safety program.

Source: MMWR Recommendation Report 2010 Jul 23;59(RR-6):1–30.

References

1. Centers for Disease Control and Prevention. Investigation of bioterrorism-related anthrax and adverse events from antimicrobial prophylaxis. *JAMA* 2001; 286(20):2536–7.
2. Centers for Disease Control and Prevention. Public health dispatch: update: cutaneous anthrax in a laboratory worker-Texas, 2002. *JAMA* 2002; 288(4):444.
3. Sweeney DA, Hicks CW, Cui X, et al. Anthrax infection. *Am J Respir Crit Care Med* 2011; 184(12):1333–41.
4. Kissling E, Wattiau P, China B, et al. *B. anthracis* in a wool-processing factory: seroprevalence and occupational risk. *Epidemiol Infect* 2012; 140(5):879–86.
5. Brandes Ammann A and Brandl H. Anthrax in the canton of Zurich between 1878 and 2005. *Schweiz Arch Tierheilkd* 2007; 149(7):295–300.
6. Nguyen TQ, Clark N, and the 2006 NYC Anthrax Working Group. Public health and environmental response to the first case of naturally acquired inhalational anthrax in the United States in 30 years: infection of a New York City resident who worked with dried animal hides. *J Public Health Manag Pract* 2010; 16(3):189–200. doi:10.1097/PHH.0b013e3181ca64f2.
7. Shafazand S, Doyle R, Ruoss S, et al. Inhalational anthrax: epidemiology, diagnosis and management. *Chest* 1999; 116:1369–76.
8. Stone SE. Cases of malignant pustule. *Boston Med Surg J* 1868; I:19–21.
9. Winter H and Pfisterer RM. Inhalation anthrax in a textile worker: non-fatal course. *Schweiz Med Wochenschr* 1991; 121(22):832–5.
10. Berger T, Kassirer M, and Aran AA. Injectional anthrax – new presentation of an old disease. *Euro Surveill* 2014; 19(32):pii: 20877.
11. Goel AK. Anthrax: a disease of biowarfare and public health importance. *World J Clin Cases* 2015; 3(1):20–33.
12. Duncan EJ, Kournikakis B, Ho J, et al. Pulmonary deposition of aerosolized *Bacillus atrophaeus* in a Swine model due to exposure from a simulated anthrax letter incident. *Inhal Toxicol* 2009; 21(2):141–52.
13. Yakupogullari Y and Koroglu M. Nosocomial spread of *Bacillus anthracis*. *J Hosp Infect* 2007; 66(4):401–2.
14. Tena D, Martinez-Torres JA, Perez-Pomata MT, et al. Cutaneous infection due to *Bacillus pumilus*: report of 3 cases. *Clin Infect Dis* 2007; 44(4):e40–2.
15. Hendricks KA, Wright ME, Shadomy SV, et al. Centers for disease control and prevention expert panel meetings on prevention and treatment of anthrax in adults. *Emerg Infect Dis* 2014; 20(2):e130687. doi:10.3201/eid2002.130687.
16. Martin GJ and Friedlander AM. *Bacillus anthracis* (Anthrax). In Bennett JE, Dolin R, and Blaser MJ (eds.), *Mandell, Douglas, and Bennett's Principles and Practice of Infectious Diseases*, 8th ed. New York: Saunders, 2014.
17. Jernigan DB, Raghunathan PL, Bell BP, et al. Investigation of bioterrorism-related anthrax, United States, 2001: epidemiologic findings. *Emerg Infect Dis* 2002; 8(10):1019–28.
18. Centers for Disease Control and Prevention, National Notifiable Diseases Surveillance System, Anthrax (*Bacillus anthracis*): 2010 Case Definition. Available at: http://wwwn.cdc.gov/NNDSS/script/casedef.aspx?CondYrID=609&DatePub=1/1/2010%2012:00:00%20AM (accessed on June 1, 2016).
19. Centers for Disease Control and Prevention (CDC). The Laboratory Response Network Partners in Preparedness. Available at: http://www.bt.cdc.gov/lrn/ (accessed on June 1, 2016).
20. U.S. Food and Drug Administration. FDA Approves Treatment for Inhalation Anthrax. 2015. Available at: http://www.fda.gov/NewsEvents/Newsroom/PressAnnouncements/ucm439752.htm (accessed on June 1, 2016).
21. Centers for Disease Control and Prevention (CDC). Interim guidelines for investigation of and response to *Bacillus anthracis* exposures. *MMWR* 2001; 50(44):987–90.
22. Occupational Safety and Health Administration. eTools: Anthrax: How Should I Decontaminate During Response Actions? Available at: https://www.osha.gov/SLTC/etools/anthrax/decon.html (accessed on June 1, 2016).

23. National Institute for Occupational Safety and Health (NIOSH), Recommendations for the Selection and Use of Respirators and Protective Clothing for Protection against Biological Agents. 2009. DHHS (NIOSH) Publication Number 2009-132. Available at: http://www.cdc.gov/niosh/docs/2009-132/default.html (accessed on June 30, 2016).
24. Brachman PS. Inhalation anthrax. *Ann N Y Acad Sci* 1980; 353:83–93.
25. Weber DJ, Sickbert-Bennett E, Gergen MF, et al. Efficacy of selected hand hygiene agents used to remove *Bacillus atrophaeus* (a surrogate of *Bacillus anthracis*) from contaminated hands. *JAMA* 2003; 289(10):1274–7.
26. Spotts Whitney EA, Beatty ME, Taylor TH Jr, et al. Inactivation of *Bacillus anthracis* spores. *Emerg Infect Dis* 2003; 9(6): 623–7.
27. Siegel JD, Rhinehart E, Jackson M, et al. 2007 guideline for isolation precautions: preventing transmission of infectious agents in health care settings. *Am J Infect Control* 2007; 35(10 Suppl 2):S65–164.
28. Shadomy SV and Smith TL. Zoonosis update. Anthrax. *J Am Vet Med Assoc* 2008; 233(1):63–72.
29. Wright JG, Quinn CP, Shadomy S, et al. Use of anthrax vaccine in the United States: recommendations of the Advisory Committee on Immunization Practices (ACIP), 2009. *MMWR Recomm Rep* 2010; 59(RR-6):1–30.

BORRELIA SPECIES

Common names for disease: Lyme disease, Lyme borreliosis

Occupational setting

Lyme disease is a zoonosis and the most common vector-borne disease of humans in temperate regions of the Northern Hemisphere, including the United States. The number of confirmed cases in the United States peaked at 29,959 in 2009 and has been roughly stable over the past several years.[1] The disease is found worldwide, with important foci in forested areas of Europe, Asia, and North America. The distribution of the disease parallels the distribution of the vectors.[2] In the United States, the vast majority of cases occur within two main geographic regions: the northeastern coastal states from Virginia to Maine and the Upper Midwest, primarily Wisconsin and Minnesota (Figure 22.1).

All outdoor workers in endemic areas should be considered at an increased risk of Lyme disease. Specific occupational groups of concern include forestry and agricultural workers,[3] military recruits,[4] hunters,[5] and urban park workers.[6]

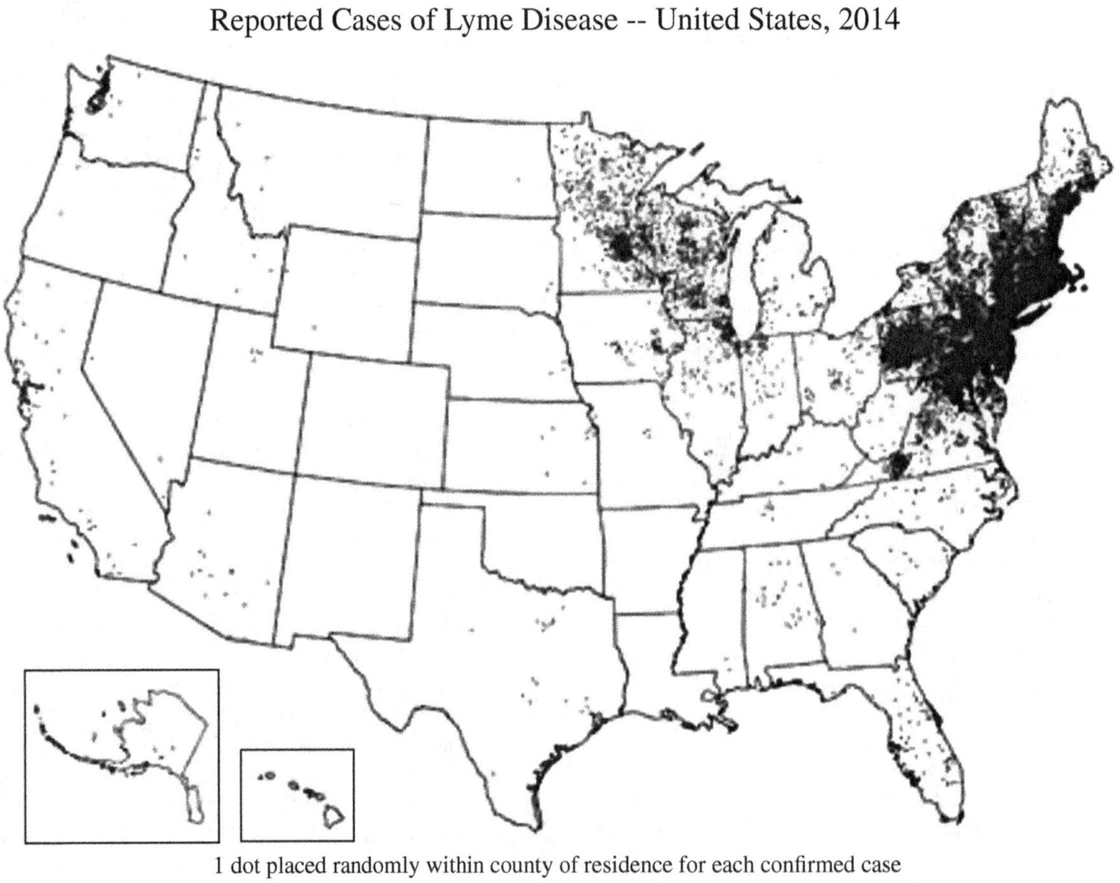

FIGURE 22.1 Reported cases of Lyme disease in USA, 2014. *Source*: http://www.cdc.gov/lyme/stats/maps.html.

Exposure (route)

Transmission to humans occurs through the bite of ticks of the *Ixodes* genus (hard-bodied ticks) infected by *Borrelia* bacteria. These include *I. scapularis* (deer tick or black-legged tick) in the northeastern and upper midwestern United States, *I. pacificus* (Western blacklegged tick) in the western United States, *I. ricinus* (castor bean tick) in Europe, and *I. persulcatus* (taiga tick) in Asia.

The tick vectors of Lyme disease have a 2-year life-span, which takes place in three stages: larva, nymph, and adult. Larval and nymphal ticks feed only once during the season before overwintering and emerging into the next stage the following year. Feeding activity for larvae peaks in the late summer/early fall, for nymphs during the summer, and for adults in the early spring and late fall. Although the subadult stages of *Ixodes* ticks are not species-specific in seeking hosts for their blood meals, the major reservoirs for *Borrelia* are small mammals such as chipmunks or mice as well as birds.[7] *Ixodes* ticks generally feed on their hosts for 3–5 days, increasing manyfold in size and weight during this time. Studies in rodents suggest that *B. burgdorferi* transmission does not occur when ticks are feeding on a host for less than 24 hours, with the highest risk being between 48 and 72 hours.[8]

Only nymphs and adult ticks are infected by *Borrelia*. However, as adults are much larger, they are usually discovered and removed before transmission can occur. As a result, the vast majority of Lyme disease in humans results from smaller nymph bites during summer months. Often, prolonged tick attachment occurs in less noticeable areas such as hair-covered parts of the body: the axilla and groin.

There is no evidence of person-to-person, airborne, waterborne, or foodborne transmission of disease. Although concern has been raised about the possibility of *in utero* transmission based on case reports, no convincing evidence of adverse effects in pregnancy among infected mothers has been demonstrated in large studies.[9]

Pathobiology

In the United States, Lyme disease is only caused by *Borrelia burgdorferi* sensu stricto, whereas in Europe and in Asia, *B. afzelii*, *B. garinii* as well as at least nine other species are additional causes of disease.[7] *Borrelia* are fastidious spirochetes that are difficult to culture from infected patients, and hence diagnosis is based most frequently on clinical and serologic criteria rather than isolation of the organism from cases.

After a tick bite that results in the transmission of spirochetes, the most common sign of Lyme disease is a distinctive rash at the site of attachment known as *erythema migrans*, which develops after 3–32 days.[7] Although this rash is classically described as having an erythematous border surrounding a clear center with the appearance of a "bull's-eye" in an archery target, most cases have either enhanced central or uniform erythema. The lesion is usually solitary and can expand to reach more than 60 cm in diameter. *Erythema migrans* may be asymptomatic, mildly pruritic, or painful.

This early, or localized, infection may be accompanied by more generalized symptoms of fever, headache, arthralgias, malaise, and fatigue. In untreated patients, *erythema migrans* can resolve within 3–4 weeks. Approximately 20% of patients will present with these nonspecific symptoms only without cutaneous findings.[10]

About 2–3% of patients will go on to develop disseminated disease within days or weeks.[10] The earliest sign of systemic involvement is multiple, smaller lesions of *erythema migrans*. Patients at this stage are generally ill with severe malaise, fatigue, headaches, and arthralgias. The manifestations of the disseminated stage can be protean, but several syndromes are of interest. Carditis usually presents as conduction abnormalities (the most characteristic being atrioventricular nodal block), which can be asymptomatic but can also result in sudden death.[11] The nervous system is also an important site of early disseminated infection, with meningitis, neuropathy of several different cranial nerves (most commonly a Bell's palsy), peripheral neuropathy, encephalitis, radiculoneuritis, and myelitis all being described. Arthritis, which most commonly affects the knee, is an additional complication of late-stage, disseminated disease. Without therapy, many of these manifestations can last weeks or months, recur, or become chronic.

In a small number of cases, subjective symptoms persist for months or years after completion of appropriate antimicrobial therapy and resolution of objective abnormalities. This condition has been labeled by some as "chronic" Lyme disease and has been an area of controversy. However, multiple, well-designed clinical trials have shown no benefit from prolonged antimicrobial treatment in ameliorating these subjective symptoms, which are highly nonspecific.[7]

Diagnosis

Lyme disease is a clinical diagnosis based on compatible symptoms and signs as well as a history of a possible bite by the appropriate type of tick.

As noted previously, *Borrelia* can be difficult to culture from patients, so serologic tests are the mainstay to confirm the diagnosis. CDC recommends a two-step approach with initial testing using enzyme immunoassay or immunofluorescence assay, followed by testing with the more specific Western blot test to confirm positive results or further test equivocal results as summarized in Figure 22.2.[12]

Diagnostic difficulties associated with Lyme disease have been well reported in the lay and scientific literatures. Although sensitivity increases over time following infection and with disseminated disease, it is poor throughout all phases, including convalescence.[7]

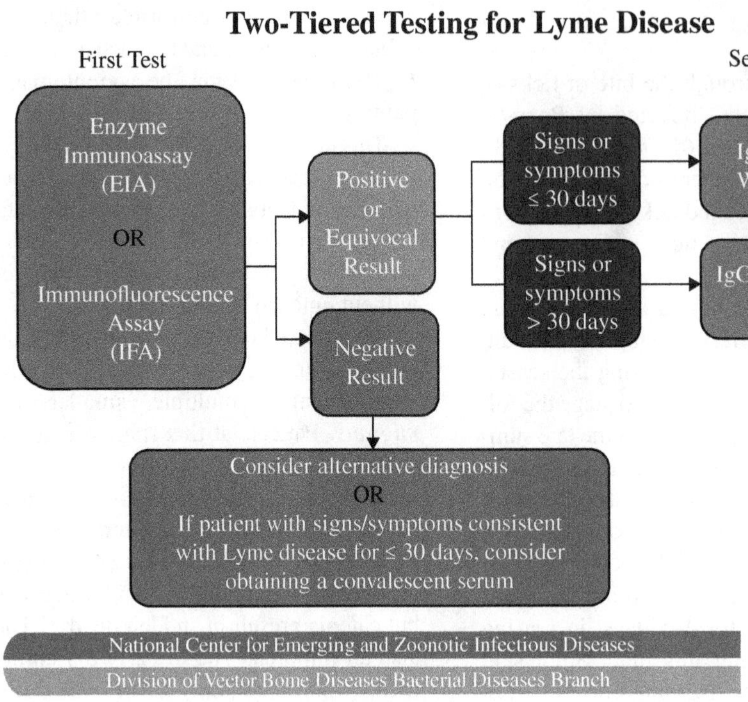

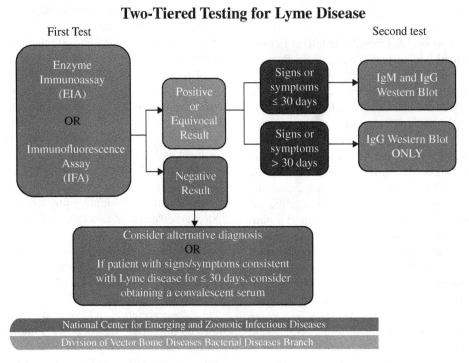

FIGURE 22.2 CDC diagnostic testing algorithm for Lyme disease. *Source*: MMWR 1995; 44:590–1.

Treatment

Early treatment is important to prevent the complications from disseminated disease. Doxycycline, amoxicillin, and cefuroxime are appropriate agents administered orally for 14 days for the treatment of *erythema migrans*, all with reported cure rates of approximately 90%.[13] This course should be extended to 28 days for cases with arthritis. Parental therapy with the same agents is indicated when carditis is present, while ceftriaxone or cefotaxime are recommended when there is meningeal involvement.

Approximately 15% of patients will develop worsening cutaneous findings and abrupt onset of arthralgias, myalgias, and increased fever usually within 24 hours of antimicrobial treatment as a result of endotoxins released from bacteriolysis (the Jarisch–Herxheimer reaction). This condition is usually self-limited within 48 hours and can be treated symptomatically with nonsteroidal anti-inflammatory drugs.[7] However, delayed onset and more severe reactions necessitating a discontinuation of antimicrobial therapy have also been reported.[14]

Medical surveillance

There are no currently recommended medical screening activities for Lyme disease. The use of anti-*Borrelia* antibody screening is not recommended both because of low sensitivity in detecting early infection and lack of specificity in distinguishing past exposure from current infection. Lyme disease is a nationally notifiable disease in the United States and cases must be reported to public health authorities.

Prevention

Tick populations can be controlled for a short time in the environment through the application of commercially available pesticides and removal of leaf litter.[15] Recent studies have shown reductions by about two-thirds in the prevalence of infected ticks following use of an oral vaccine targeting a mouse reservoir.[16] However, effectiveness in decreasing disease transmission to humans has not yet been demonstrated.

Personal preventive measures are the preferred means to reduce disease transmission. Workers should be advised to avoid contact with areas of heavy undergrowth and leaf litter wherever possible. Other preventive behaviors—wearing light-colored clothing for easier visualization of ticks, tucking trousers into socks, and wearing long-legged trousers and long-sleeved shirts—are recommended for their ease and low cost.

Insect repellents containing concentrations of at least 20% DEET (*N*,*N*-diethyl-*m*-toluamide) are effective after application to the skin. Clothing, tents, and footwear purchased pre-treated or treated with 0.5% permethrin provide protection from questing ticks even after several launderings because of binding of the insecticide to the fibers. Whereas DEET repels, permethrin rapidly kills ticks, and field studies have revealed decreases in the number of tick bites among outdoor workers who wear permethrin-treated clothing.[17] All gear, clothing, and any pets should be examined for the presence of ticks on a daily basis. Ticks on clothing can be killed in a laundry drier at high heat for 1 hour.

It is important to emphasize that studies in animals suggest that the spirochete is not transmitted from tick to host until after 24 hours of feeding.[8] Thus, workers should be advised to conduct a careful head-to-toe inspection of the body each day. Bathing or showering as soon as possible after work can both aid in identifying and washing away ticks.

Attached ticks should be removed with fine-tipped tweezers using slow, upward pressure, attempting to avoid breaking off the mouth parts.[18] After removal of the tick and any broken-off mouth parts, the area should be disinfected or washed with soap and water and the tick disposed by flushing down the toilet, submerging in alcohol, or sealing in plastic. The tick should not be crushed by hand.[18]

The use of prophylactic doxycycline for asymptomatic individuals with a history of tick bite in an endemic area remains controversial. Although effective in preventing Lyme disease, the low risk of acquiring infection after a single tick bite (from 1 to 3% in highly endemic regions) and high cure rate of symptomatic cases indicate to many that prophylactic therapy is not warranted.[7] The number needed to treat for people bitten by a deer tick has been calculated to be 50 to prevent one case of *erythema migrans*.[7] Vigilance for early symptoms and signs in persons who have recently sustained a tick bite or who have removed a tick is recommended in place of tick-bite prophylaxis in most instances. However, an exception may be made when there is evidence of prolonged tick attachment, such as a history of removal of an engorged nymphal tick.[7]

Manufacture of a Lyme disease vaccine for use in humans was discontinued in 2002 due to poor sales and perceptions of safety concerns. Since protection declines over time, workers reporting prior vaccination should not be regarded as immune. New vaccines are under development.[19]

References

1. Centers for Disease Control and Prevention. Lyme Disease: Lyme Disease Data. Available at: http://www.cdc.gov/lyme/stats/index.html (accessed on June 1, 2016).
2. Centers for Disease Control and Prevention. Appendix methods used for creating a national lyme disease risk map. *MMWR*, 1999; 48(RR07):21–24. Available at: http://www.cdc.gov/mmwr/preview/mmwrhtml/rr4807a2.htm (accessed on June 30, 2016).
3. Tokarska-Rodak M, Plewik D, Kozioł-Montewka M, et al. Risk of occupational infections caused by *Borrelia burgdorferi* among forestry workers and farmers. *Med Pr* 2014; 65(1):109–17.
4. Oksi J, Viljanen MK. Tick bites, clinical symptoms of Lyme borreliosis, and Borrelia antibody responses in Finnish army recruits training in an endemic region during summer. *Mil Med* 1995; 160(9):453–6.
5. Cetin E, Sotoudeh M, Auer H, et al. Paradigm Burgenland: risk of *Borrelia burgdorferi* sensu lato infection indicated by variable seroprevalence rates in hunters. *Wien Klin Wochenschr* 2006; 118(21–22):677–81.
6. Krstić M and Stajković N. Risk for infection by lyme disease cause in green surfaces maintenance workers in Belgrade. *Vojnosanit Pregl* 2007; 64(5):313–8.
7. Shapiro ED. Lyme disease. *N Engl J Med* 2014; 370:1724–31.
8. des Vignes F, Piesman J, Heffernan R, et al. Effect of tick removal on transmission of *Borrelia burgdorferi* and *Ehrlichia phagocytophila* by *Ixodes scapularis* nymphs. *J Infect Dis* 2001; 183(5):773–8.
9. Lakos A, Solymosi N. Maternal Lyme borreliosis and pregnancy outcome. *Int J Infect Dis* 2010; 14(6):e494–8.
10. Steere AC and Sikand VK. The presenting manifestations of Lyme disease and the outcomes of treatment. *N Engl J Med* 2003; 348:2472–4.
11. Centers for Disease Control and Prevention. Three sudden cardiac deaths associated with Lyme carditis – United States, November 2012–July 2013. *MMWR Morb Mortal Wkly Rep* 2013; 62(49):993–996. Available at: https://www.cdc.gov/mmwr/preview/mmwrhtml/mm6249a1.htm (accessed June 30, 2016).
12. Centers for Disease Control and Prevention. Recommendations for test performance and interpretation from the Second National Conference on Serologic Diagnosis of Lyme Disease. *MMWR* 1995; 44:590–1.
13. Wormser GP, Dattwyler RJ, Shapiro ED, et al. The clinical assessment, treatment, and prevention of lyme disease, human granulocytic anaplasmosis, and babesiosis: clinical practice guidelines by the Infectious Diseases Society of America. *Clin Infect Dis* 2006; 43(9):1089–134.
14. Kadam P, Gregory NA, Zelger B, et al. Delayed onset of the Jarisch-Herxheimer reaction in doxycycline-treated disease: a case report and review of its histopathology and implications for pathogenesis. *Am J Dermatopathol* 2015; 37(6):e68–74.
15. Hayes EB and Piesman J. How can we prevent Lyme disease? *N Engl J Med* 2003; 348:2424–30.
16. Richer LM, Brisson D, Melo R, Ostfeld RS, et al. Reservoir targeted vaccine against *Borrelia burgdorferi*: a new strategy to prevent Lyme disease transmission. *J Infect Dis* 2014; 209(12):1972–80.
17. Richards SL, Balanay JAG, and Harris JW. Effectiveness of permethrin-treated clothing to prevent tick exposure in foresters in the central Appalachian region of the USA. *Int J Environ Health Res* 2014; 7:1–10.
18. Centers for Disease Control and Prevention. Lyme Disease: Tick Removal. Available at: http://www.cdc.gov/lyme/removal/index.html (accessed on June 1, 2016).
19. Wressnigg N, Pöllabauer EM, Aichinger G, et al. Safety and immunogenicity of a novel multivalent OspA vaccine against Lyme borreliosis in healthy adults: a double-blind, randomised, dose-escalation phase 1/2 trial. *Lancet Infect Dis* 2013;13(8):680–9.

BRUCELLA SPECIES

Common names for the disease: Brucellosis, Bang's disease, Mediterranean fever, undulant fever, Neapolitan fever, Malta fever, Gibraltar fever, Cyprus fever.

Occupational setting

Brucellosis is the most common zoonotic infection globally and is caused by transmission of several species of the genus *Brucella* from different animals (Table 22.2).[1] As such, a high index of suspicion is warranted in travelers or military personnel returning from endemic areas, especially the Middle East.[2]

Species well recognized to cause disease in humans include *B. abortus*, *B. melitensis*, *B. canis*, and *B. suis*. Of these, *B. melitensis* is the most virulent species and causes the majority of cases diagnosed in humans globally. *B. abortus* is a common species implicated in occupational infections acquired in the agricultural setting, while *B. abortus* RB51 (RB51) and *B. abortus* S19 (S19) are attenuated strains used in vaccines for cattle, which can cause brucellosis in humans.[3] More recently identified species from marine mammals (*B. pinnipedialis* and *B. ceti*) have been identified very rarely as causes of disease in humans.[4,5]

Reported cases of brucellosis in humans in the United States have declined from a peak of 6341 in 1947 to the current plateau of about 100 per year,[6] primarily as a result of animal control methods, including vaccination, inspection, and prompt segregation of diseased animals (Figure 22.3). Brucellosis has been eradicated in all cattle herds in the United States apart from areas of Idaho, Wyoming, and Montana, adjacent to the Grand Teton National Park and Yellowstone National Park as a result of spillover from elk.[7]

TABLE 22.2 Host animals for Brucella species causing disease in humans (adapted from CDC: http://www.cdc.gov/brucellosis/veterinarians/host-animals.html)

Species	Main Host(s)	Less Common Host(s)
B. abortus	Cattle, water buffalo, bison	Pigs, elk, horses
B. melitensis	Goats, sheep, camel	
B. suis	Pigs, feral swine, boar	Cattle, horses, caribou, reindeer and hares
B. canis	Dogs	Foxes
B. pinnipedialis also known as *B. pinnipediae*	Pinnipeds (seals, sea lions, walruses)	
B. ceti also known as *B. cetaceae*	Cetaceans (dolphins, porpoises, whales)	

Brucellosis can be transmitted by consumption of unpasteurized milk or milk products such as cheese[8] as well as raw or undercooked meat.[9] This mode of transmission is still the most frequent source worldwide. Occupational activities with particularly high risk of exposure include animal slaughter, meat processing, meat-packing, hunting, and milking, or handling of semen, aborted animal fetal tissue, placentas, laboratory specimens, and *Brucella* vaccines. Occupations with the highest risk of exposure are livestock handling and slaughterhouse workers,[10] veterinarians,[11] meat-packers,[12] farmers, dairy workers,[13] and hunters.[14] Brucellosis is considered to be the most common laboratory-acquired infection[15] with outbreaks also described following exposure to attenuated vaccines.[16] Marine species of *Brucella* have been implicated in potential occupational exposures among university and laboratory employees performing a rescue and subsequent necropsy of an infected porpoise.[17]

Exposure (route)

Occupational infection usually occurs through direct contact or inhalation. Because the bacteria are easily aerosolized and have a low infectious dose, inhalational exposure is a significant concern, especially among slaughterhouse workers[18] and laboratory personnel as a result of mouth pipetting and sniffing of cultures.[19] The bacteria also enter the body by penetrating the mucosa of the mouth, or throat, or through the conjunctiva when infected material is splashed or sprayed. Transmission through the skin, particularly in slaughterhouse, abattoir workers, and veterinarians, may occur through cuts, abrasions, or percutaneously via injuries from sharps.

Ingestion is less likely in the occupational setting but remains an important route of exposure in cases transmitted by infected dairy products or meat products.[9] Transmission has rarely been reported from person to person, breastfeeding, sexual activity, blood transfusion, and tissue transplantation.

Pathobiology

Brucella organisms are small, nonmotile, aerobic, Gram-negative rods. They are facultative intracellular parasites, a property that allows these bacteria to evade the immune system. They can survive phagocytosis by neutrophils and macrophages and spread hematogenously throughout the host once within the bloodstream.

The clinical course is variable with an incubation period ranging from 1 week to several months. Although frequently described as having protean manifestations, fever, usually accompanied by chills, is always present.[20] Malodorous perspiration is regarded as being almost pathognomonic and additional

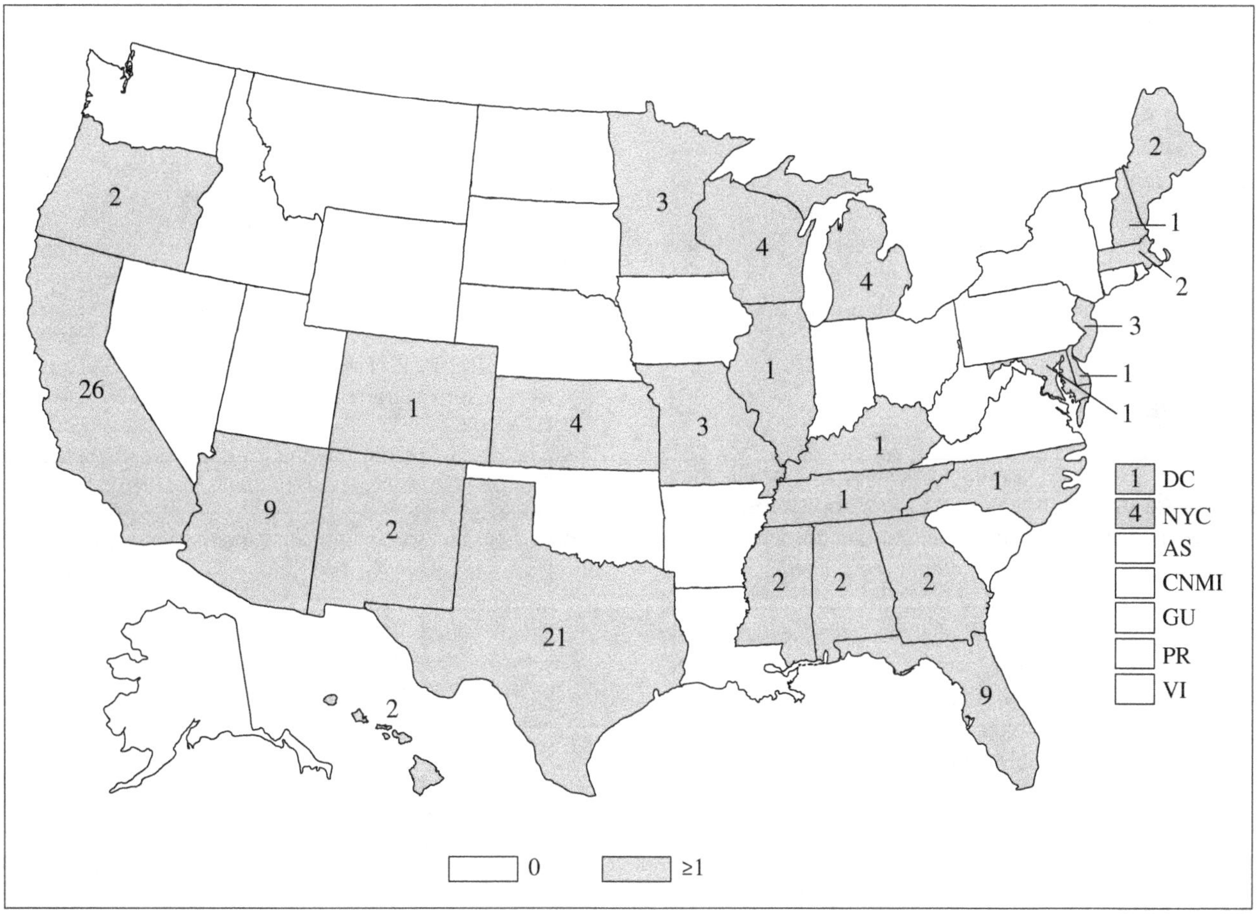

FIGURE 22.3 Number of reported cases of brucellosis – United States and U.S. territories, 2010. *Source*: http://www.cdc.gov/brucellosis/resources/surveillance.html.

constitutional symptoms are usually present. Any organ system can be involved, with the most common being the musculoskeletal (usually as arthritis of knees, hips, ankles, or wrists) reproductive system (causing loss of pregnancy in women or epididymo-orchitis in men), or liver (hepatitis with mild elevations in transaminases).[20] Marine species seem to have a predilection for central nervous system (CNS) involvement.[5] Although rare, endocarditis is the most common cause of death.

Approximately 10% of cases will relapse, often as a result of inadequate antimicrobial therapy. Such relapses are usually milder than the initial presentation and can be treated with appropriate antibiotics.[20] Overall, the prognosis with treatment is considered excellent, with mortality less than about 2–5%.

Diagnosis

Infection with *Brucella* species can be determined by standard laboratory methods. A definitive diagnosis is made by isolating *Brucella* organisms in cultures from blood or other specimens. Culturing is often difficult, given the slow growth and fastidious nature of these bacteria. As a result, the laboratory should be alerted when *Brucella* infection is suspected in order to prolong the length of culture time and ensure that laboratory staff takes appropriate protective measures.

Agglutination tests for *Brucella* antigen can detect infections arising from *B. abortus*, *B. melitensis*, and *B. suis*. Two serum samples are preferred for serological testing: the first drawn when the patient is acutely ill (within the first 7 days) and the second 2–4 weeks later. A fourfold or greater rise in antibodies is considered positive for brucellosis infection.

ELISA and polymerase chain reaction (PCR) tests are also available.

Treatment

Prolonged treatment is indicated, often to include an agent with intracellular activity (i.e., doxycycline). A variety of regimens have been recommended, which usually include a combination of doxycycline, rifampin, or trimethoprim-sulfamethoxazole, although it is not clear which regimen is most effective. A recent systematic review concluded that 6 weeks of doxycycline plus 2–3 weeks of streptomycin was more effective than a 6-week course of doxycycline plus rifampin.[21]

Medical surveillance

Brucellosis is a nationally notifiable disease in the United States, and *B. abortus*, *B. melitensis*, and *B. suis* are select agents requiring prompt reporting to the Federal Select Agent Program when isolation or release occurs.[22] Brucellosis toxin is a CDC Category B bioterrorism disease.

Laboratory workers exposed to *B. abortus* should have blood drawn for serological studies at 0, 6, 12, 18, and 24 weeks post exposure. Serological testing is not available for exposures to the RB51 vaccine.

Prevention

The most important preventive measures, which have resulted in a dramatic decline in human brucellosis in the United States, are the vaccination and careful inspection of animals at risk, along with immunologic testing of cows' milk and blood for evidence of *Brucella* infection. Diseased animals are segregated or slaughtered. Despite these eradication measures, work practice measures are essential for protection against remaining diseased animals. Kill floors should be isolated from other areas of the slaughterhouse and be under negative-pressure ventilation with entry restricted to essential personnel. All workers handling animal products, including milk, and especially placenta, uterine discharges, and blood, should wear heavy gloves, aprons, and goggles.[23] The use of high-top boots should be considered as well.

Areas where exposure is likely should be posted with information about brucellosis, including routes of exposure, disease symptoms, and preventive activities. Work sites should have accessible handwashing facilities, first aid kits for prompt treatment of wounds, and separate areas, isolated from animal work, for eating and drinking. Such activities should be prohibited in work areas.

CDC has provided detailed preventive recommendations for both laboratory personnel and those exposed to the RB51 vaccine.[24] Work with *Brucella* should be performed in a class II biosafety cabinet using biosafety level 3 precautions. Post-exposure prophylaxis of doxycycline and rifampin for at least 21 days is recommended following high-risk exposures.

Pasteurization of milk and thorough cooking of meat (especially from game) are important general preventive measures.

There is no vaccine available for use in humans.[25]

References

1. Centers for Disease Control and Prevention. Host Animals for *Brucella* Species. Available at: http://www.cdc.gov/brucellosis/veterinarians/host-animals.html (accessed on June 1, 2016).
2. Bechtol D, Carpenter LR, Mosites E, et al. *Brucella melitensis* infection following military duty in Iraq. *Zoonoses Public Health*. 2011; 58(7):489–92.
3. Ashford D, di Pietra J, Lingappa J, et al. Adverse events in humans associated with accidental exposure to the livestock brucellosis vaccine RB51. *Vaccine* 2004; 22:3435–9.
4. Sohn AH, Probert WS, Glaser CA, et al. Human neurobrucellosis with intracerebral granuloma caused by a marine mammal Brucella spp. *Emerg Infect Dis* 2003; 9:485–8.
5. Whatmore AM, Dawson CE, Groussaud P, et al. Marine mammal Brucella genotype associated with zoonotic infection. *Emerg Infect Dis* 2008; 14:517–8.
6. Centers for Disease Control and Prevention. Brucellosis Surveillance. Available at: http://www.cdc.gov/brucellosis/resources/surveillance.html (accessed on June 1, 2016).
7. Rhyan JC, Nol P, Quance C, et al. Transmission of brucellosis from elk to cattle and bison, Greater Yellowstone Area, USA, 2002–2012. *Emerg Infect Dis* 2013; 19(12):1992–5. Available at: http://www.ncbi.nlm.nih.gov/pmc/articles/PMC3840865/ (accessed on June 30, 2016). doi:10.3201/eid1912.130167.
8. Castell-Monsalve, J, Rullán JV, Peiró Callizo EF, et al. 1996. Epidemic outbreak of 81 cases of brucellosis following the consumption of fresh cheese without pasteurization. *Rev Esp Salud Publica*; 70(3):303–11.
9. Chen S, Zhang H, Liu X, et al. Increasing threat of brucellosis to low-risk persons in urban settings, China. *Emerg Infect Dis* 2014; 20(1):126–30.
10. Buchanan TM, Hendricks SL, Patton CM, et al. Brucellosis in the United States, 1960–1972: an abattoir-associated disease. *Medicine* 1974; 53:427–39.
11. Centers for Disease Control and Prevention (CDC). Human exposure to Brucella abortus strain RB51—Kansas, 1997. *MMWR Morb Mortal Wkly Rep* 1998; 47(9):172–5.
12. Landau Z, Green L. Chronic brucellosis in workers in a meat-packing plant. *Scand J Infect Dis* 1999; 31(5):511–2.
13. Trunnell TN, Waisman M, and Trunnell TL. Contact dermatitis caused by Brucella. *Cutis* 1985; 35(4):379–81.
14. Simoes EM and Justino JD. Brucellosis infection in a feral swine hunter. *Nurse Pract* 2013; 38(7):49–53.
15. Traxler RM, Guerra MA, Morrow MG, et al. Review of brucellosis cases from laboratory exposures in the United States in 2008 to 2011 and improved strategies for disease prevention. *J Clin Microbiol* 2013; 51(9):3132–6.
16. Wallach JC, Ferrero MC, Victoria Delpino M, et al. Occupational infection due to Brucella abortus S19 among workers involved in vaccine production in Argentina. *Clin Microbiol Infect* 2008; 14(8):805–7.
17. Centers for Disease Control and Prevention. Human exposures to marine Brucella isolated from a Harbor Porpoise – Maine, 2012. *MMWR* 2012; 61(25):461–3.
18. Trout D, Gomez TM, Bernard BP, et al. Outbreak of brucellosis at a United States pork packing plant. *J Occup Environ Med* 1995; 37(6):697–703.
19. Centers for Disease Control and Prevention. Overview of Laboratory Risks. Available at: http://www.cdc.gov/brucellosis/laboratories/risks.html (accessed on June 1, 2016).
20. Pappas G, Akritidis N, Bosilkovski M, et al. Brucellosis. *N Engl J Med* 2005; 352(22):2325–36.
21. Yousefi-Nooraie R, Mortaz-Hejri S, Mehrani M, et al. Antibiotics for treating human brucellosis. *Cochrane Database Syst Rev* 2012; 10:CD007179.

22. Centers for Disease Control and Prevention. Federal Select Agent Program. Available at: http://www.selectagents.gov (accessed on June 1, 2016).
23. Kligman EW, Peate WF, and Cordes DH. Occupational infections in farm workers. *Occup Med* 1991; 6(3):429–46.
24. Centers for Disease Control and Prevention. Brucellosis Homepage: Laboratory Personnel. Available at: http://www.cdc.gov/brucellosis/laboratories/index.html (accessed on June 1, 2016).
25. Oliveira SC, Giambartolomei GH, and Cassataro J. Confronting the barriers to develop novel vaccines against brucellosis. *Exp Rev Vac* 2011; 10(9):1291–305.

CAMPYLOBACTER SPECIES

Common name for disease: Campylobacteriosis

Occupational setting

Although *Campylobacter* is the most common bacterial cause of gastroenteritis, only a small proportion of cases are linked to outbreaks.[1] Point-source outbreaks in the workplace have been related to both food[2] and water[3] contamination. Direct, occupationally acquired infections have not only been described in poultry processing workers most frequently[4] but also farmworkers on dairy farms[5] and zoo workers.[6] Outbreaks have also been documented among occupants of child day care centers[7] and prisons,[8] raising the possibility of illness being acquired by workers in these sectors. As with other enteric pathogens, international travel is a well-known risk factor for illness from *Campylobacter*.[9]

Exposure (route)

Spread usually occurs through fecal–oral transmission. Since the infectious dose is very low, meat is easily contaminated during processing and cross-contamination creating other potential sources for infection is common. Poultry and, to a lesser extent, swine are the most common animal reservoirs, although an increasingly wide variety of animals are recognized as potential sources, including household pets such as dogs[10] as well as reptiles.[11] Animals infected with *Campylobacter* are not symptomatic.

Campylobacter is highly prevalent and widespread in poultry processing. The prevalence of colonization in flocks increases from 5–10% to 25–40% in the summer, with a corresponding seasonal increase in human infections.[12] Asymptomatic infections in humans are very common, among both experienced and new workers.[13] Several case reports have implicated airborne transmission of droplets as an additional route of infection, including an occupational setting in a poultry worker.[14]

In the general population, disease is most frequently caused when meat or dairy products are consumed following improper preparation and storage. Recent large outbreaks of disease have resulted from consumption of unpasteurized milk[15] and exposure to muddy surface water through participation in an obstacle course at a cattle ranch.[16] Person-to-person[17] transmission can occur but is considered uncommon and is only a risk from symptomatic cases, especially young children. Rare cases of sexual transmission have been reported.[18]

Pathobiology

Campylobacteria are motile, Gram-negative, curved rods. *C. jejenui* is the species responsible for most *Campylobacter* illness. The major reservoir for *C. jejuni* is poultry, particularly chickens. A closely related species more commonly found in swine, *C. coli*, produces an illness clinically indistinguishable from that produced by *C. jejuni*. *C. upsaliensis* is rarely implicated as a cause of gastroenteritis,[19] while *C. fetus* has additionally caused bacteremia and infections, such as endocarditis, abscesses, septic arthritis, and abortions, with pericarditis reported in a slaughterhouse worker.[20] Recently, several emerging *Campylobacter* species have been isolated from environmental, food, animal, or human clinical isolates: *C. hyointestinalis*, *C. lanienae*, *C. sputorum*, *C. concisus*, and *C. curvus*.[21]

Acute enterocolitis is caused by *C. jejuni*, in about 90–95% of cases, *C. coli* in about 5–10% of cases, and other species for less than 1% of cases. Following an incubation period of 1–10 days, the most common symptoms are diarrhea (with or without blood), abdominal pain, and fever. However, this represents the midpoint on a continuum of presentations, which may range from an asymptomatic carrier state to severe, prolonged diarrhea. Abdominal pain may be very prominent, mimicking a surgical abdomen or inflammatory bowel disease. Symptoms are typically self-limited and resolve within 2–5 days.

C. jejuni gastroenteritis may be followed by the development of Guillain–Barré syndrome, usually within 12 weeks of the infection, which has also been reported following large outbreaks.[22] This association may also exist for *C. coli*.[23] Reactive arthritis or Reiter syndrome, most frequently involving the knee, may also follow *Campylobacter* infection in less than 5% of those infected.[24]

Diagnosis

The presence of the bacteria with a characteristic "gull wing" appearance on microscopic examination of stool specimens supports the diagnosis, and isolation of the organism in culture confirms it. A variety of rapid testing methods based on PCR with improved sensitivity and specificity are under development that may allow more rapid and convenient testing.[25] Serologic testing is not recommended for routine clinical use.

Treatment

In the vast majority of cases, the disease is self-limited. In more severe cases of enteritis, supportive therapy consisting of fluid and electrolyte replacement is the primary consideration. Antibiotic therapy is generally not recommended, as it only shortens the duration of symptoms minimally and may result in further antibiotic resistance.[26] However, antibiotics may be indicated in select patients with high fever, bloody or profuse diarrhea, or protracted illness. Erythromycin and azithromycin are the antibiotics of choice. Ciprofloxacin and tetracycline resistance is now widespread as a result of use in the agriculture setting.[27] Antibiotic treatment eliminates excretion of the bacteria.

Medical surveillance

No specific surveillance measures are recommended. Campylobacteriosis is a nationally notifiable disease in the United States and cases must be reported to the local health authorities.

Prevention

Proper preparation and storage of food at the workplace will prevent this and other causes of infectious gastroenteritis. Work practices should emphasize good hygiene with strict handwashing after contact with potentially infected materials. Eating and smoking should not occur in work areas and gloves should be worn. In slaughterhouses, work practices and engineering controls should be directed toward minimizing fecal contamination.

While bacteria can be shed for days to weeks, the risk of person-to-person transmission is considered to be low. Employees in jobs with a high risk of transmission (e.g., food handlers, healthcare workers, and childcare workers) should be vigilant and promptly report any signs or symptoms compatible with gastroenteritis from any infectious cause to allow removal from high-risk activities while symptomatic. Travelers should take appropriate preventive precautions common to all enteric pathogens, with particular attention to avoiding undercooked poultry.

No vaccine is available.

References

1. Taylor EV, Herman KM, Ailes EC, et al. Common source outbreaks of Campylobacter infection in the USA, 1997–2008. *Epidemiol Infect* 2013; 141(5):987–96.
2. Murphy O, Gray J, Gordon S, and Bint AJ. An outbreak of campylobacter food poisoning in a health care setting. *J Hosp Infect* 1995; 30(3):225–8.
3. Rautelin H, Koota K, von Essen R, et al. Waterborne *Campylobacter jejuni* epidemic in a Finnish hospital for rheumatic diseases. *Scand J Infect Dis* 1990; 22(3):321–6.
4. de Perio MA, Niemeier RT, Levine SJ, et al. Campylobacter infection in poultry-processing workers, Virginia, USA, 2008–2011. *Emerg Infect Dis* 2013; 19(2):286–8.
5. Gilpin BJ, Scholes P, Robson B, et al. The transmission of thermotolerant Campylobacter spp. to people living or working on dairy farms in New Zealand. *Zoonoses Public Health* 2008; 55(7):352–60.
6. Forsyth MB, Morris AJ, Sinclair DA, et al. Investigation of zoonotic infections among Auckland Zoo staff: 1991–2010. *Zoonoses Public Health* 2012; 59(8):561–7.
7. Goosens H, Giesendorf BA, Vandamme P, et al. Investigation of an outbreak of *Campylobacter upsaliensis* in day care centers in Brussels: analysis of relationships among isolates by phenotypic and genotypic typing methods. *J Infect Dis* 1995; 172(5):1298–305.
8. Fernandez-Martin JI, Dronda F, Chaves F, et al. *Campylobacter jejuni* infections in a prison population coinfected with the human immunodeficiency virus. *Rev Clin Esp* 1996; 196(1):16–20.
9. Ricotta EE, Palmer A, Wymore K, et al. Epidemiology and antimicrobial resistance of international travel-associated Campylobacter infections in the United States, 2005–2011. *Am J Public Health* 2014; 104(7):e108–14.
10. Mughini Gras L, Smid JH, Wagenaar JA, et al. Increased risk for *Campylobacter jejuni* and *C. coli* infection of pet origin in dog owners and evidence for genetic association between strains causing infection in humans and their pets. *Epidemiol Infect* 2013; 141(12):2526–35.
11. Patrick ME, Gilbert MJ, Blaser MJ, et al. Human infections with new subspecies of *Campylobacter fetus*. *Emerg Infect Dis* 2013; 19(10):1678–80.
12. Jore S., Viljugrein H., Brun E., et al. Trends in Campylobacter incidence in broilers and humans in six European countries, 1997–2007. *Prev Vet Med* 2010; 93:33–41.
13. Ellström P, Hansson I, Söderström C, et al. A prospective follow-up study on transmission of campylobacter from poultry to abattoir workers. *Foodborne Pathog Dis* 2014; 11(9):684–8.
14. Wilson IG. Airborne Campylobacter infection in a poultry worker: case report and review of the literature. *Commun Dis Public Health* 2004; 7(4):349–53.
15. Centers for Disease Control and Prevention. Recurrent outbreak of *Campylobacter jejuni* infections associated with a raw milk dairy—Pennsylvania, April–May 2013. *MMWR Morb Mortal Wkly Rep* 2013; 62(34):702.
16. Zeigler M, Claar C, Rice D, et al. Outbreak of campylobacteriosis associated with a long-distance obstacle adventure race—Nevada, October 2012. *MMWR Morb Mortal Wkly Rep* 2014; 63(17):375–8.
17. Rotariu O, Smith-Palmer A, Cowden J, et al. Putative household outbreaks of campylobacteriosis typically comprise single MLST genotypes. *Epidemiol Infect* 2010; 138(12):1744–7.
18. Gaudreau C, Helferty M, Sylvestre JL, et al. *Campylobacter coli* outbreak in men who have sex with men, Quebec, Canada, 2010–2011. *Emerg Infect Dis*. 2013; 19(5):764–7.
19. Couturier BA, Hale DC, and Couturier MR. Association of *Campylobacter upsaliensis* with persistent bloody diarrhea. *J Clin Microbiol* 2012; 50(11):3792–4.

20. Ganeshram KN, Ross A, Cowell RP, et al. Recurring febrile illness in a slaughterhouse worker. *Postgrad Med J* 2000; 76(902):790–1.
21. Miller WG, Chapman MH, Yee E, et al. Multilocus sequence typing methods for the emerging Campylobacter species *C. hyointestinalis*, *C. lanienae*, *C. sputorum*, *C. concisus*, and *C. curvus*. *Front Cell Infect Microbiol* 2012; 2:45.
22. Jackson BR, Zegarra JA, López-Gatell H, et al. Binational outbreak of Guillain-Barré syndrome associated with *Campylobacter jejuni* infection, Mexico and USA, 2011. *Epidemiol Infect* 2014; 142(5):1089–99.
23. van Belkum A, Jacobs B, van Beek E, et al. Can *Campylobacter coli* induce Guillain-Barré syndrome? *Eur J Clin Microbiol Infect Dis* 2009; 28(5):557–60.
24. Porter CK, Choi D, Riddle MS. Pathogen-specific risk of reactive arthritis from bacterial causes of foodborne illness. *J Rheumatol* 2013; 40(5):712–4.
25. Van Lint P, De Witte E, De Henau H, et al. Evaluation of a real-time multiplex PCR for the simultaneous detection of *Campylobacter jejuni*, *Salmonella* spp., *Shigella* spp./EIEC, and *Yersinia enterocolitica* in fecal samples. *Eur J Clin Microbiol Infect Dis* 2015; 34(3):535–42.
26. Ternhag A, Asikainen T, Giesecke J, et al. A meta-analysis on the effects of antibiotic treatment on duration of symptoms caused by infection with Campylobacter species. *Clin Infect Dis* 2007; 44(5):696–700.
27. Melero B, Juntunen P, Hänninen ML, et al. Tracing *Campylobacter jejuni* strains along the poultry meat production chain from farm to retail by pulsed-field gel electrophoresis, and the antimicrobial resistance of isolates. *Food Microbiol* 2012; 32(1):124–8.

CLOSTRIDIUM BOTULINUM (INCLUDING *C. ARGENTINENSE, C. BARATII,* AND *C. BUTYRICUM*)

Common names for diseases: Botulism, infant botulism, wound botulism

Occupational setting

A toxin formed by the bacterium *Clostridium botulinum* (or, more rarely, *C. argentinense*, *C. butyricum*, and *C. baratii*) causes botulism. These organisms are ubiquitous in most soils, have also been found in agricultural products, and in a diverse array of animals, including marine animals.[1] Type A and B botulinum toxins are commercially available for cosmetic and therapeutic use, including the treatment of a variety of conditions involving involuntary muscle spasm, such as cervical dystonia, blepharospasm, and strabismus.

To date, only one report of nonfatal botulism acquired in the occupational setting has been documented among three veterinary lab workers.[7] They became ill 3 days after inhaling botulinum type A toxin while performing necropsies on guinea pigs and rabbits whose fur had been covered with aerosolized toxin.

Although botulism has been reported in patients receiving therapeutic injections[3] and intravenous drug users,[4] no cases have been reported among healthcare providers to date. In theory, laboratory workers in research or public health facilities or those involved in the manufacture of botulinum toxin are also at risk.

Exposure (route)

Several forms of botulism are recognized. Foodborne, wound, and intestinal botulism (which is further subdivided as infant or adult) are the natural forms of disease. Inhalational botulism requires aerosolization, while iatrogenic botulism occurs from overdosing of the injected toxin. In the United States in 2012, 160 laboratory-confirmed cases of botulism were reported to CDC. A total of 122 cases were of the infant form, 25 were foodborne, 8 were wounds, and 5 were cases of unknown or other etiology.[5]

Since the toxin is readily inactivated by heat, uncooked or improperly cooked foods are the source of disease. Although commonly associated with home-canned foods, almost any food can cause botulism and most cases in the United States involve vegetables. The bacteria cannot penetrate intact skin and person-to-person transmission has not been described.

Inhalational exposure is a significant concern in the context of bioterrorism,[6] although, as noted previously, only three human cases have been described from this route.[2]

Pathobiology

C. botulinum is a spore-forming, obligate anaerobic bacillus. Disease is caused by the toxin, which is regarded as the most toxic substance known. Foodborne, inhalational, and iatrogenic forms result from exposure to preformed, externally derived toxin, whereas the toxin in intestinal and wound botulism originates from *Clostridia* bacteria that have colonized these sites in the host.

In a conducive environment, such as the hypoxic atmosphere produced by canning, in a deep wound, or in the intestine, clostridial spores germinate. The growing bacterial colonies release a potent neurotoxin, which is taken up in the circulation and acts at peripheral cholinergic synapses to block the release of acetylcholine, causing multiple cranial nerve palsies and subsequent diffuse muscular weakness. In untreated cases, death is usually due to respiratory failure from paralysis of respiratory muscles. There are seven types of toxin, designated A–H, which may be elaborated by the bacillus, but most human cases are caused by types A, B, and E, with rare cases due to type F. In 2014, a novel-type H toxin was identified from a case of infant botulism.[7]

Diagnosis

A high level of clinical suspicion is needed to make this diagnosis, which is frequently missed.[8] The paralysis initially affects bulbar musculature and subsequently descends to a generalized weakness. Since disease results from intoxication rather than infection, constitutional signs and symptoms such as fever are not seen.

Often, extensive investigations are needed to rule out other neurological causes of paralysis. The diagnosis of botulism is supported by demonstrating the toxin in serum, stool, or in the suspected food source. *C. botulinum* can sometimes be cultured from the stool in cases among infants or those with intestinal anomalies. The presence of *C. botulinum* spores in the implicated food is less helpful than finding toxin, as the spores are ubiquitous and are not themselves harmful. In cases of suspected wound botulism, serum should be tested for toxin and the wound cultured for the organism. However, since the sensitivity of the toxin assay is only 33–44% and takes 4 days to yield results,[9] the diagnosis should be made on clinical grounds. The presence of bilateral cranial-nerve palsies with a subsequent descending paralysis should raise the suspicion of botulism, regardless of the exposure history.[10]

Treatment

After collection of serum for specific toxin identification, all suspected cases of botulism should be treated as soon as possible with antitoxin. The only available antitoxin in the United States is a heptavalent botulinum antitoxin, which covers toxin types A–G and can be obtained from CDC through referral from state health departments.[10] Since patient outcomes are much better with early antitoxin therapy (ideally within 24 hours), administration should not be withheld while awaiting laboratory confirmation of botulism.

Supportive treatment, which may include ventilation, is the second cornerstone of botulism management. Such care may be required for months, especially when antitoxin treatment has been delayed.

Cases of wound botulism should be treated with antitoxin as well as wound debridement or drainage, and antibiotics, with Penicillin G as the preferred agent.

Medical surveillance

There are no recommended medical screening activities for botulism. Botulism is a nationally notifiable disease in the United States and all cases, confirmed or suspected, must be reported immediately to the local public health authorities. Botulinum toxin is a CDC Category A bioterrorism agent.

Prevention

Appropriate food handling can prevent the majority of cases. Public notification and recall of tainted products are essential after identification of commercial food sources of poisoning. Tracing of others who may have consumed contaminated food is important when botulism has been identified in commercially prepared or distributed foods. The public should be educated about the risk of botulism being present in bulging containers, such as cans, but they should be aware that this sign of contamination is often absent. Those involved in home canning should be educated about the proper time, temperature, and pressure needed to destroy spores. Uneviscerated fish products should be avoided because of the risk of contamination. Prompt cleaning of wounds and careful attention (including irrigation or debridement) to wounds that are not healing may prevent wound botulism. Botulism associated with toxin manufacture and use is, to date, only a theoretical risk and should be preventable by maintaining strict containment procedures in handling or production of the toxin.

No vaccine is available. A toxoid vaccine was withdrawn in the United States in 2011 due to concerns about declining immunogenicity and adverse local reactions from boosters.[11]

References

1. From the Centers for Disease Control and Prevention. Outbreak of botulism type E associated with eating a beached whale—western Alaska, July 2002. *JAMA* 2003; 289(7):836–8.
2. Holzer VE. Botulism from inhalation [in German]. *Med Klin* 1962; 57:1735–8.
3. Coban A, Matur Z, Hanagasi HA, et al. Iatrogenic botulism after botulinum toxin type A injections. *Clin Neuropharmacol* 2010; 33(3):158–60.
4. Yuan J, Inami G, Mohle-Boetani J, et al. Recurrent wound botulism among injection drug users in California. *Clin Infect Dis* 2011; 52(7):862–6.
5. Centers for Disease Control and Prevention. Botulism Annual Summary, 2012. Atlanta, GA: US Department of Health and Human Services, CDC, 2014.
6. Arnon SS, Schechter R, Inglesby TV, et al. Botulinum toxin as a biological weapon: medical and public health management. *JAMA* 2001; 285(8):1059–70.
7. Barash JR and Arnon SS. A novel strain of *Clostridium botulinum* that produces type B and type H botulinum toxins. *J Infect Dis* 2014; 209:183–91.
8. St Louis ME, Peck SH, Bowering D, et al. Botulism from chopped garlic: delayed recognition of a major outbreak. *Ann Intern Med* 1988; 108(3):363–8.
9. Vasa M, Baudendistel TE, Ohikhuare CE, Clinical problem-solving. The eyes have it. *N Engl J Med* 2012; 367(10): 938–43.

10. Rao AK, Jackson KA, and Mahon BE. The eyes have it. *N Engl J Med* 2013; 368(4):392.
11. Centers for Disease Control and Prevention. Notice of CDC's Discontinuation of Investigational Pentavalent (ABCDE) Botulinum Toxoid Vaccine for Workers at Risk for Occupational Exposure to Botulinum Toxins. *MMWR* 2011; 60(42):1454–5.

CLOSTRIDIUM DIFFICILE

Common names for disease: C. *difficile* colitis, pseudomembranous colitis, antibiotic-associated colitis

Occupational setting

Although generally associated with individual antibiotic use, healthcare-associated infection from *Clostridium difficile* has been documented in nurses[1] and laboratory workers.[2] However, such reports are rare, and, in some cases, the workers had been prescribed antibiotics prior to acquiring the infection.[3]

Exposure (route)

C. difficile is a ubiquitous bacterium, widely found in soil and forming part of the normal colonic flora in many healthy adults.

Person-to-person transmission of *C. difficile* spores occurs through the fecal–oral route. Spores are resistant to gastric acid and subsequently germinate upon reaching the large intestine. Most infections are healthcare-associated, although there is increasing evidence that the sources of infection are more complex than previously recognized. One large study reported that only 35% of cases in a hospital could be traced to a symptomatic patient.[4] Asymptomatic carriers and spore-contaminated surfaces may represent additional sources of infection.[5]

In addition, community-acquired infections may now account for at least 20% of infections.[6] There is evidence that *C. difficile* is a zoonotic infection implicating new modes of transmission such as foodborne.[6]

Pathobiology

C. difficile is an anaerobic, spore-forming, Gram-positive bacillus, which produces a variety of toxins, the most important of which are denoted as toxins A and B.

Colitis due to *C. difficile* is more common in elderly and debilitated patients, the immunocompromised, and those taking antibiotics. Antibiotic-associated colitis occurs when alteration of the normal intestinal flora disrupts competitive inhibition and allows overgrowth of *C. difficile* with elaboration of toxins into the intestinal lumen. Clinical effects arise from the toxins, as the bacteria themselves are rarely invasive.

There is a wide range of symptoms arising from *C. difficile* infection, ranging from an asymptomatic carrier state to life-threatening colitis with the characteristic "pseudomembrane" of yellowish exudate. In antibiotic-associated cases, symptoms usually occur during treatment or within 1–2 weeks of completion but can begin as long as 12 weeks after therapy. A typical patient with *C. difficile* colitis presents with profuse, foul-smelling diarrhea, which may be watery or green and mucoid. There is usually crampy abdominal pain, with fever and abdominal tenderness on examination. Reactive arthritis may develop after *C. difficile* infection.

Diagnosis

The diagnosis should be suspected in anyone with three or more diarrheal stools within 1 day who received antibiotics within the previous 12 weeks.

Laboratory confirmation of *C. difficile* colitis can be challenging. The most widely available test is to detect toxins A and B in the stool using an enzyme immunoassay (EIA). However, not all labs test for both toxins and the sensitivity is limited. A variety of more rapid tests with higher sensitivity are increasingly available, some based on PCR and others on EIA.

Routine stool cultures have poor specificity because of the widespread presence of nonpathogenic strains of *C. difficile*. However, a culture followed by an assay for toxin performed on isolates, known as a toxigenic culture, solves this problem. Although currently regarded as the gold standard, this method is both labor intensive and requires 2–4 days to provide results. Therefore, authorities currently recommend either a PCR test for toxigenic strains or a two-step approach using a rapid EIA-based test followed by confirmatory testing.[7]

Empiric treatment should be given when clinically appropriate while awaiting test results. Testing should only be performed on diarrheal stool and repeat testing is discouraged, as it may yield falsely positive test results due to asymptomatic colonization, which can occur for a prolonged period following infection.

Treatment

Mild cases of antibiotic-associated colitis may be treated with discontinuation of antibiotics and supportive therapy only. Metronidazole is the first-line antibiotic for mild to moderate cases, and vancomycin for severe cases.[7] A recent meta-analysis has concluded that short-term use of probiotics in conjunction with antibiotics is effective and safe.[8] Antiperistaltic agents should not be used, since they may cause toxin retention and obscure symptoms. Recurrence occurs in 20–30% of patients.

Colectomy may be indicated for severe cases. Fecal microbiota transplantation shows considerable promise as a safe new therapy, although randomized controlled clinical trials are limited and this approach is not yet fully standardized.[9]

Medical surveillance

There are no recommended medical surveillance activities beyond those for healthcare-associated infections.

Prevention

Some preventive measures are common to many bacterial infections and include judicious use of antibiotics, prompt identification of cases, and contact precautions for cases of diarrhea.

A significant challenge in controlling *C. difficile* infection is that the spores are resistant to alcohol, handwashing, and usual cleaning measures. Therefore, universal glove use, enhanced cleaning measures using known sporicidal agents, and avoidance of shared equipment should be considered in high-risk healthcare areas.[10]

References

1. Strimling MO, Sacho H, and Berkowitz I. *Clostridium difficile* infection in health-care workers. *Lancet* 1989; 2(8667):866–7.
2. Bouza E, Martin A, Van den Berg RJ, et al. Laboratory-acquired *Clostridium difficile* polymerase chain reaction ribotype 027: a new risk for laboratory workers? *Clin Infect Dis* 2008; 47(11):1493–4.
3. Hell M, Indra A, Huhulescu S, et al. *Clostridium difficile* infection in a health care worker. *Clin Infect Dis* 2009; 48(9):1329.
4. Eyre DW, Wilcox MH, and Walker AS. Diverse sources of *C. difficile* infection. *N Engl J Med* 2014; 370(2):183–4.
5. Ali S, Manuel R, and Wilson P. Diverse sources of *C. difficile* infection. *N Engl J Med* 2014; 370(2):182.
6. Hoover DG and Rodriguez-Palacios A. Transmission of *Clostridium difficile* in foods. *Infect Dis Clin North Am* 2013; 27(3):675–85.
7. Cohen SH, Gerding DN, Johnson S, et al. Clinical practice guidelines for *Clostridium difficile* infection in adults: 2010 update by the Society for Healthcare Epidemiology of America (SHEA) and the Infectious Diseases Society of America (IDSA). *Infect Control Hosp Epidemiol* 2010; 31(5):431–55.
8. Goldenberg JZ, Ma SSY, Saxton JD, et al. Probiotics for the prevention of *Clostridium difficile*-associated diarrhea in adults and children. *Cochrane Datab Syst Rev* 2013, 5, art. no.: CD006095.
9. Kapel N, Thomas M, Corcos O, et al. Practical implementation of faecal transplantation. *Clin Microbiol Infect* 2014; 20(11):1098–105. Available at: http://www.clinicalmicrobiologyandinfection.com/article/S1198-743X(14)65300-3/fulltext (accessed on June 30, 2016).
10. Hsu J, Abad C, Dinh M, et al. Prevention of endemic healthcare-associated *Clostridium difficile* infection: reviewing the evidence. *Am J Gastroenterol* 2010; 105(11):2327–39.

CLOSTRIDIUM PERFRINGENS (ALSO *C. SEPTICUM, C. NOVI*)

Common names for diseases: Gas gangrene, myonecrosis, *enteritis necroticans*, *necrotic enteritis*, Darmbrand, pigbel.

Occupational setting

As is the case with other *Clostridium* species, *Clostridium perfringens* bacteria are ubiquitous. *C. perfringens* is considered to be one of the most widely distributed disease-causing bacteria, commonly found in soils and as an intestinal inhabitant in both animals and humans.

C. perfringens causes both a wide spectrum of gastrointestinal disease and a necrotic infection of tissue. Outbreaks of the common foodborne illnesses (food poisoning and diarrhea) occur most commonly in restaurants.[1] However, large outbreaks have also been reported in institutional settings, which included staff who ate at the cafeteria.[2] Some of these cases were fatal, due to a much rarer foodborne illness, variously known as necrotic enteritis, necrotizing colitis, *enteritis necroticans*, or "pigbel."[2]

C. perfringens can also cause myonecrosis or gas gangrene, which usually occurs after traumatic injury or surgery. Other clostridial species less commonly associated with gas gangrene include *C. septicum*[3] and *C. novyi*.[4]

While case reports are rare, occupations at risk are those with high risk of traumatic injury with the potential for contamination, such as agricultural workers[5] and military personnel.[6]

Exposure (route)

In cases of foodborne illness, the route of infection is ingestion. In gas gangrene caused by *C. perfringens*, bacteria are introduced into a wound from an external source such as soil, or the wound may be seeded from the patient's own colonic flora such as with penetrating abdominal injuries. *C. novyi* has only been implicated in cases of gas gangrene among injection drug users[4], while *C. septicum* can cause spontaneous gas gangrene without antecedent trauma, thought to arise from hematogenous seeding from the patient's own gastrointestinal flora.[3]

Person-to-person transmission has not been described.

Pathobiology

C. perfringens is an anaerobic, spore-forming, Gram-positive bacillus. It is also considered to be the most rapidly growing microorganism, with a generation time of less than 7 minutes under ideal conditions. Five different strains are recognized according to the toxin produced, denoted as types A–E. Only types A and C have been associated with disease in humans. While a variety of toxins are produced, the three clinically relevant forms are alpha, which causes gas gangrene; beta, which causes necrotic enteritis; and *C. perfringens* enterotoxin (CPE), which causes food poisoning.

Diseases caused by *C. perfringens* result from the effects of toxins produced after germination and replication of the spores in a hospitable anaerobic environment. The most common, but least serious, syndrome is food poisoning, caused by *C. perfringens* type A. When spore-contaminated food (usually meat) is inadequately heated, or inadequately reheated after slow cooling, spores germinate and produce CPE, which is then ingested with the contaminated food. The circumstances under which the toxin is produced are common in institutional kitchens preparing food in large batches. After an average incubation period of 8–12 hours, diarrhea and abdominal cramps develop. Fever and vomiting are uncommon. Symptoms are self-limited and usually resolve within 24 hours.

Enteritis necroticans is caused by the beta toxin produced by *C. perfringens* type C and is most frequently associated with the ingestion of undercooked pork.[7] The bacteria multiply in the small intestine and release beta toxin, which causes intestinal necrosis. Segmental intestinal gangrene and other severe complications may occur. This very serious disease is rare in industrialized countries. Children with protein malnutrition, diabetics, or those who have had pancreatic or gastric resection are at increased risk, possibly due to deficiencies in pancreatic proteases, which break down beta toxin.[7]

Gas gangrene is a rare but devastating syndrome caused by alpha toxin, which is produced by *C. perfringens* as well as *C. novyi* and *C. septicum*. It usually occurs after traumatic injury or surgery, but spontaneous cases have been described, especially as a result of *C. septicum*.[3] *C. perfringens* is a common contaminant of open wounds. Factors that are thought to promote clostridial replication in a wound are foreign bodies, vascular insufficiency, and concurrent infection with other bacteria. The incubation period is usually 2–3 days. The first symptom is usually sudden pain at the wound site, which may be pale, edematous, and tender. Crepitus may be palpated and gas from bacterial metabolism may be seen on radiographic studies. The skin color progresses from pale to magenta or bronze, and hemorrhagic bullae with a thin brown serosanguinous discharge, which has a characteristic offensive, sweet odor, may develop. Necrosis of muscle is an associated finding.

Other soft-tissue infections due to *C. perfringens* include uncomplicated polymicrobial abscesses, crepitant cellulitis, suppurative myositis, emphysematous cholecystitis, anaerobic pulmonary infections (especially empyema), and, rarely, after penetrating head trauma, brain abscess.

Diagnosis

Clostridial food poisoning is diagnosed by recovery of *C. perfringens* organisms from suspected food, or stool from patients collected within 48 hours of symptom onset. Cultures with greater than 10^6 colony-forming units (CFUs)/g and demonstration of the CPE gene are considered diagnostic.

Diagnosis of *enteritis necroticans* is primarily clinical and should be suspected in a patient with the risk factors outlined above, in the setting of anorexia, vomiting, abdominal pain, and bloody diarrhea. Absence of colonic involvement and rapid progression of the illness to sepsis and shock favor the diagnosis.

Diagnosis of gas gangrene is primarily clinical but is supported by evidence of myonecrosis seen at surgery. Wound exudates may reveal Gram-positive or Gram-variable rods with a typical "box-car" appearance and few white blood cells. This may be the earliest laboratory confirmation of this disease. Spores are not seen on Gram stain. Only 15% of gas gangrene cases have bacteremia.

Treatment

For *C. perfringens* food poisoning, rehydration, and other supportive measures are required in exceptional cases of severe diarrhea. Otherwise, the illness resolves without treatment. Treatment for *enteritis necroticans* includes chloramphenicol or penicillin G, supportive care, and bowel decompression. Small bowel resection may be required for persistent paralytic ileus, septicemia, peritonitis, persistent pain, or a palpable mass lesion.

The cornerstone of treatment of gas gangrene is early and extensive surgical debridement, with wide excision for abdominal wall involvement, and usually amputation if an extremity is involved. Penicillin remains the antibiotic of choice, although clindamycin, metronidazole, and the carbapenems are acceptable alternatives.

The use of hyperbaric oxygen in the treatment of necrotic tissue infections generally is controversial[8] but may be considered as an additional therapy. Surgery should not be delayed. The mortality rate of gas gangrene treated with surgery and antibiotics still approaches 25%.

Medical surveillance

There are no recommended medical screening activities for *C. perfringens*. Outbreaks of *Clostridia* food poisoning should be reported promptly to the local health authorities.

Prevention

Foodborne clostridial disease can be prevented by adequate cooking temperatures and rapid cooling and by adequate reheating of foods. Division of large batches of food into smaller units facilitates rapid cooling, which prevents germination of clostridial spores. A vaccine against *C. perfringens* beta toxin is in use in previously endemic countries, where it has been quite effective in preventing *enteritis necroticans*.[9] Cleaning grossly contaminated wounds may help prevent gas gangrene, but most cases are not easily prevented. Early detection and treatment may mitigate some of the more severe manifestations and reduce mortality.

References

1. Grass JE, Gould LH, and Mahon BE. Epidemiology of foodborne disease outbreaks caused by *Clostridium perfringens*, United States, 1998–2010. *Foodborne Pathog Dis* 2013; 10(2):131–6.
2. Centers for Disease Control and Prevention. Fatal foodborne *Clostridium perfringens* illness at a state psychiatric hospital—Louisiana, 2010. *MMWR Morb Mortal Wkly Rep* 2012; 61(32): 605–8.
3. Wu YE, Baras A, Cornish T, et al. Fatal spontaneous *Clostridium septicum* gas gangrene: a possible association with iatrogenic gastric acid suppression. *Arch Pathol Lab Med* 2014; 138(6):837–41.
4. Palmateer NE, Hope VD, Roy K, et al. Infections with spore-forming bacteria in persons who inject drugs, 2000–2009. *Emerg Infect Dis* 2013; 19(1):29–34.
5. Demianchuk AV and Vanat IM. Gas gangrene caused by agricultural trauma. *Klin Khir* 1974 (10):71–2.
6. Rudge FW. The role of hyperbaric oxygenation in the treatment of clostridial myonecrosis. *Mil Med* 1993; 158(2):80–3.
7. Gui FL, Subramony C, Fratkin J, et al. Fatal *enteritis necroticans* (pigbel) in a diabetic adult. *Mod Pathol* 2002; 15(1): 66–70.
8. Willy C, Rieger H, and Vogt D. Hyperbaric oxygen therapy for necrotizing soft tissue infections: contra. *Chirurg*. 2012; 83(11):960–72.
9. Lawrence GW, Lehmann D, Anian G, et al. Impact of active immunization against *enteritis necroticans* in Papua New Guinea. *Lancet* 1990; 336:1165–7.

CLOSTRIDIUM TETANI

Common names for disease: Tetanus, lockjaw

Occupational setting

Tetanus is caused by a toxin released by *Clostridium tetani*, a ubiquitous bacterium found in greatest numbers in soil, especially soil rich in fecal matter such as manure. It is a potential problem in any outdoor job, especially work in which minor skin trauma is frequent. Most occupational cases occur following minor injuries of the hands or fingers in the agriculture and forestry sectors.[1] Epidemics have occurred following natural disasters.[2] In the United States, there has been a progressive and dramatic decline in tetanus with 19 reported cases and two deaths in 2009 (Figure 22.4).[3]

Exposure (route)

Spores of *C. tetani* reproduce after inoculation into traumatized skin. Recent surveillance data from the United States indicate that only 71.7% of cases had antecedent acute trauma, and, of these, only 36.5% sought medical attention.[3] For 13% of cases, the entry was a chronic wound.[3] Therefore, minor or even unnoticed trauma can be a portal of entry.

Pathobiology

C. tetani is an anaerobic, Gram-positive rod that produces hardy spores found in large numbers in soil. When spores are inoculated into a wound, they germinate and produce

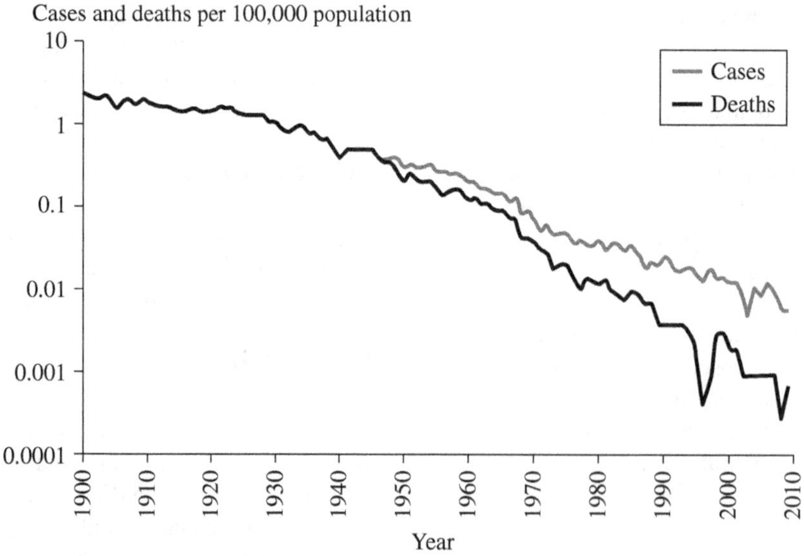

FIGURE 22.4 Tetanus – number of reported cases and deaths, United States, 1900–2010. *Source*: Centers for Disease Control and Prevention. Chapter 16: Tetanus. *Manual for the surveillance of vaccine-preventable diseases*. Centers for Disease Control and Prevention, Atlanta, GA, 2011. Available at http://www.cdc.gov/vaccines/pubs/surv-manual/chpt16-tetanus.html.

exotoxins, including tetanospasmin, responsible for the signs and symptoms of tetanus. Following hematogenous spread, the toxin travels via motor nerve axons and binds to receptors in muscle and the central nervous system, replicating the actions of neurotransmitters. The effects of the toxin last several weeks and recovery probably requires the growth of new synapses.

The nervous system effects vary, likely as a result of both the size of the inoculum and the degree of existing immunity. Effects can be classified into one of the following two categories: central motor control effects in which poisoning of inhibitory neuronal cells causes unopposed motor activity leading to rigidity and spasms, and autonomic instability, the leading cause of death in tetanus, which is manifested by increased sympathetic tone with massive catecholamine release from disinhibition of the sympathetic system neurons of the autonomic system. Occasionally, parasympathetic disinhibition is seen causing hypertension and cardiac dysrhythmias.

The incubation period varies from 1 day to several months, but the average is approximately 8 days. The shorter the period of time between spore inoculation and symptom onset, the poorer the prognosis. Other predictors of poor prognosis are autonomic dysfunction at presentation and burn or surgical site as the portal of entry.

Tetanus is divided clinically into a variety of presentations including localized, generalized, cephalic, and neonatal tetanus, which will not be discussed here. Localized tetanus is the least severe form and occurs when muscle at the site of the injury is fixed in spasm. Progression to generalized tetanus may occur. Generalized tetanus is the most common form and usually starts with nonspecific symptoms, such as malaise, restlessness, headache, insomnia, irritability, and profuse sweating, followed by stiffness, twitching and pain at the wound site, and fever. These progress to the classic signs of tetanus: trismus or lockjaw, *risus sardonicus*, the straightening of the upper lip which has reminded some of a sardonic smile, and opisthotonic posturing, in which spasm of the back muscles occur such that a patient placed supine would rest on his heels and the back of his head only. There is no loss of consciousness and the condition is extremely painful. In generalized tetanus, there may be airway compromise, diaphragmatic dysfunction, and autonomic dysfunction. Clinical progression can occur for up to 2 weeks despite administration of antitoxin, and full recovery may take months. The cephalic form of tetanus occurs following head trauma or ear infections and affects cranial nerves.

Diagnosis

Diagnosis is on clinical grounds only, based on the history and characteristic physical examination findings. Vaccination status should be established to increase the level of suspicion, since 92% of cases of tetanus among those with known vaccination status occurred in un- or under-vaccinated persons in the United States.[3]

Cultures lack both sensitivity and specificity. Laboratory measurement of tetanus antibodies can be performed prior to tetanus immune globulin (TIG) administration. However, there are numerous problems with this test, including disagreement on what levels are considered protective, limited availability, and numerous case reports of tetanus in those considered to have protective levels of antibodies.[4] Early diagnosis is essential to ensure prompt treatment.

Treatment

Passive immunotherapy with intramuscular human TIG is the standard therapy. TIG binds only free toxin and not that already bound, which explains the progression of symptoms once therapy has begun and the long recovery period. If human TIG is not available, equine antitoxin may be given intravenously after testing has ruled out hypersensitivity to horse serum. Tetanus toxoid should also be administered, as clinical tetanus does not confer subsequent immunity. Wounds should be cleaned and debrided. Full recovery may take a month or longer.

Patients with tetanus should be carefully monitored for signs of airway compromise from laryngospasm and may require extensive supportive therapy. Contractions can be severe enough to cause long-bone or spinal fractures. Because of prolonged debilitation, secondary complications are common and include pulmonary emboli and secondary healthcare-associated infections.

Medical surveillance

There are no recommended medical screening activities for tetanus. Tetanus is a nationally notifiable disease in the United States and cases must be reported to local public health authorities. Aggressive case finding has been recommended to ensure up-to-date boosters and reduce the size of the unvaccinated population.

Prevention

Active immunization with tetanus toxoid prevents disease. A three-dose schedule is part of the primary series for all infants and children with booster doses every 10 years. All adults with an unknown vaccination status should receive the three-dose series of vaccines. In cases of tetanus-prone injuries, passive immunization with human TIG should be given to patients with incomplete or unknown vaccine history. The current CDC recommendations for wound management are shown in Table 22.3.[3]

TABLE 22.3 CDC recommendations for tetanus wound management.

History of adsorbed tetanus toxoid (doses)	Clean minor wounds Tdap or Td†	Clean minor wounds TIG§	All other wounds* Tdap or Td†	All other wounds* TIG§
<3 or unknown	Yes	No	Yes	Yes
≥3¶	No**	No	No††	No

Source: MMWR 2011 Apr 1;60(12):365–9.
*Such as (but not limited to) wounds contaminated with dirt, feces, soil, and saliva; puncture wounds; avulsions; and wounds resulting from missiles, crushing, burns, and frostbite.
†For persons >10 years, Tetanus and diphtheria toxoids and acellular pertussis (Tdap) is preferred to tetanus and diphtheria toxoids (Td) if the patient has never received Tdap and has no contraindication to pertussis vaccine. For persons 7 years of age or older, if Tdap is not available or not indicated because of age, Td is preferred to TT.
§TIG is human tetanus immune globulin. Equine tetanus antitoxin should be used when TIG is not available.
¶ If only three doses of fluid toxoid have been received, a fourth dose of toxoid, preferably an adsorbed toxoid, should be given. Although licensed, fluid tetanus toxoid is rarely used.
**Yes, if it has been 10 years or longer since the last dose.
††Yes, if it has been 5 years or longer since the last dose. More frequent boosters are not needed and can accentuate side effects.

References

1. Luisto M and Seppäläinen AM. Tetanus caused by occupational accidents. *Scand J Work Environ Health* 1992;18(5):323–6.
2. Jeremijenko A, McLaws ML, and Kosasih H. A tsunami related tetanus epidemic in Aceh, Indonesia. *Asia Pac J Public Health* 2007;19:Spec No:40-4.
3. Centers for Disease Control and Prevention. Tetanus surveillance—United States, 2001–2008. *MMWR Morb Mortal Wkly Rep* 2011; 60(12):365–9.
4. Vollman KE, Acquisto NM, and Bodkin RP. A case of tetanus infection in an adult with a protective tetanus antibody level. *Am J Emerg Med* 2014; 32(4):392.e3–e4.

CORYNEBACTERIUM SPECIES

Common name for disease: Diphtheria

Occupational setting

Corynebacterium diphtheriae, the most important of the *Corynebacteria*, causes diphtheria. Diphtheria has been virtually eliminated in the working populations of many countries because of immunization. From 1980 to 2010, 55 cases of diphtheria were reported in the United States.[1] Most of these cases were in adults.

Nevertheless, a large population of susceptible adults due to declining immunity without booster doses, together with importation of cases from travelers to endemic countries has heightened concern about a possible resurgence of this disease in the United States.[2]

C. ulcerans, frequently present in farm animals, may also cause diphtheria in humans.[3] A variety of other species, several of which are zoonoses, have been described as sources of infection in occupational settings, but are very rare. These include *C. striatum* causing septic arthritis following a scalpel injury in a surgeon,[4] *C. aquaticum* infection of a high-pressure injection injury,[5] and *C. pseudotuberculosis* (also known as *C. ovis*) among those occupationally exposed to large animals.[6]

Corynebacteria have also been cultured from the smoke plume of an operating room laser.[7]

Exposure (route)

Person-to-person transmission occurs through inhalation of airborne respiratory droplets. Transmission from contact with infected skin lesions or fomites is unusual.

Pathobiology

Corynebacteria are pleomorphic, Gram-positive, aerobic bacilli. By far, the most pathogenic species is *C. diphtheriae*, for which humans are the only known reservoir. Most infections are asymptomatic.

Signs and symptoms of infection develop after an incubation period of 2–5 days, locally at either mucous membranes (respiratory, ocular, or genital diphtheria) or the superficial layers of the skin through pre-existing skin breaks (cutaneous diphtheria). *C. diphtheria* may or may not produce an exotoxin, depending on whether or not the bacterium has itself been infected by a bacteriophage containing the gene mediating toxin production. The toxin causes both local tissue necrosis and systemic effects with absorption. In addition to nonspecific signs of shock (tachycardia, stupor), the most common systemic effects include myocarditis and neuritis.

Local infection with toxigenic diphtheria is followed by hyperemia, edema, and development of the characteristic gray exudative pseudomembrane. Although virtually any mucous membrane can be infected, the most frequent and well-known sites are the pharynx and tonsils. Fever is low-grade and a classic finding is of a "bull neck" appearance from a combination of submandibular edema and lymphadenopathy. Extensive membrane formation may lead to respiratory obstruction.

Cutaneous diphtheria is less severe than infection of the respiratory tract. Lesions usually occur in the setting of primary infection with other organisms (typically, *Staphylococcus aureus* and group A streptococci). The characteristic lesion is a non-healing ulcer with a gray membrane.

Infections caused by non-toxigenic diphtheria are milder and confined to local effects. The usual picture is pharyngitis and tonsillitis, although endocarditis has also been described.[8]

Diagnosis

In endemic areas, toxigenic diphtheria may be diagnosed on clinical grounds alone based on the relatively specific clinical picture. Definitive diagnosis requires selective culture from nasal and throat swabs. Because this procedure may not be routinely performed in some laboratories, communication with laboratory personnel regarding a suspicion of diphtheria is advisable to ensure that appropriate isolation and identification techniques are applied. An immunodiffusion test for the toxin (known as the Elek test) as well as a PCR-based assay for the toxin gene can also be used. Direct examination through microscopy of stained samples may be unreliable because of the presence of commensals with a similar appearance.

Treatment

Treatment of active diphtheria consists of diphtheria antitoxin, antibiotics directed against the organism, and supportive care. Diphtheria antitoxin is no longer licensed for use in the United States and must be obtained through an Investigational New Drug protocol by contacting the CDC Emergency Operations Center (770-488-7100). Penicillin and erythromycin are the drugs of choice, although antibiotic resistance is an increasing concern. Supportive therapy should pay particular attention to maintenance of airway.

Medical surveillance

Although cases of respiratory diphtheria only are nationally notifiable in the United States, prompt involvement of public health authorities is advisable in any suspected case of infection with toxigenic diphtheria. Respiratory cases should be reported promptly by telephone to the CDC Emergency Operations Center (770-488-7100).

Some have advocated the routine culturing of all throat swabs for *C. diphtheria*, since non-toxigenic strains have been isolated with increasing frequency and the potential exists for conversion of these organisms to toxigenic forms.[8] Such a program identified three cases of non-toxigenic *C. diphtheria* tonsillitis in British military personnel.[9]

Prevention

Active immunization prevents disease. A three-dose schedule is part of the primary series for all infants and children with booster doses every 10 years. All adults with an unknown vaccination status should receive the three-dose series of vaccines. Since infection may not confer immunity, vaccination is also recommended for those with a history of the illness. Travelers should ensure that vaccinations are up-to-date prior to departure.

Patients with respiratory diphtheria should be strictly isolated. After 48 hours of antibiotic therapy, the disease is no longer contagious. Close contacts should be traced and receive a booster (or full series of the vaccine if unimmunized or vaccine status is unknown) and antibiotic prophylaxis. Antitoxin is reserved for use at early signs of illness. Two consecutive negative cultures following therapy of cases and carriers should be obtained to document elimination of the organism.

References

1. Tiwari TSP. Diphtheria. In Roush SW, Baldy LM, Centers for Disease Control and Prevention (eds). *Manual for the Surveillance of Vaccine-Preventable Diseases*, 5th ed. Centers for Disease Control and Prevention, Atlanta, GA, 2011. Available at: http://www.cdc.gov/vaccines/pubs/surv-manual/chpt01-dip.html (accessed on June 30, 2016).
2. Centers for Disease Control and Prevention. Fatal respiratory diphtheria in a U.S. traveler to Haiti—2003. *MMWR* 2003; 52:1285–6.
3. Sangal V, Nieminen L, Weinhardt B, et al. Diphtheria-like disease caused by toxigenic *Corynebacterium ulcerans* strain. *Emerg Infect Dis* 2014; 20(7):1257–8.
4. Cone LA, Curry N, Wuestoff MA, et al. Septic synovitis and arthritis due to *Corynebacterium striatum* following an accidental scalpel injury. *Clin Infect Dis* 1998; 27(6):1532–3.
5. Larsson P, Lundin O, and Falsen E. "*Corynebacterium aquaticum*" wound infection after high-pressure water injection into the foot. *Scand J Infect Dis* 1996; 28(6):635–6.
6. Peel MM, Palmer GG, Stacpoole AM, et al. Human lymphadenitis due to *Corynebacterium pseudotuberculosis*: report of ten cases from Australia and review. *Clin Infect Dis* 1997; 24(2):185–91.
7. Capizzi PJ, Clay RP, and Battey MJ. Microbiologic activity in laser resurfacing plume and debris. *Lasers Surg Med* 1998; 23:172–4.
8. Wilson AP. The return of *Corynebacterium diphtheriae*: the rise of non-toxigenic strains. *J Hosp Infect* 1995; 30(suppl):306–12.
9. Sloss JM and Faithfull-Davies DN. Non-toxigenic *Corynebacterium diphtheriae* in military personnel. *Lancet* 1993; 341(8851):1021.

ERYSIPELOTHRIX RHUSIOPATHIAE

Common names for diseases: Erysipeloid, Erysipelothricosis, fish poisoning, fish-handler's disease, seal finger, crab dermatitis, Baker–Rosenbach syndrome, Klauder disease.

Occupational setting

Erysipelothrix rhusiopathiae has been termed an "occupational pathogen" since the majority of infections occur through work with a wide variety of animals, including mammals, birds, fish, and crustaceans.[1] Those at increased risk include fishermen,[2] farmers,[3] and meat processors.[4] Human infection has also occurred in a laboratory setting.[5] An outbreak occurred among workers at a shoe factory, with *E. rhusiopathiae* isolated from washings of leather and casein glue.[6] The organism is found in animal waste and can remain viable in the environment for several months, thus providing an additional source of exposure in farm workers.[7]

Exposure (route)

Cutaneous inoculation of bacteria from a contaminated source is the usual mode of transmission when scales, shell, or bone fragments puncture the skin surface. Exposure following an animal bite has also been documented.[8]

Pathobiology

E. rhusiopathiae is a nonmotile, non-sporulating, Gram-positive rod found as a commensal or pathogen in many animal species. Swine are the primary reservoir and are also particularly susceptible to disease. The organism is harbored in the pharynx and excreted in feces. Other potential reservoirs include turkeys, chickens, ducks, deer, emus, sheep, crab, and fish.[1]

Infection can result in three clinical entities in humans.[1] The most common is erysipeloid, a localized skin infection at the site of contact (generally the back of the hand or fingers). The incubation period is 2–7 days. The involved area is a clearly demarcated, edematous, violaceous lesion that fades centrally as it spreads peripherally. Swelling can be extensive. Lesions can be asymptomatic, painful, or mildly pruritic. Fever and arthralgia occur in ~10% of cases, and lymphangitis and lymphadenopathy can be associated as well.[1] Cellulitis from *E. rhusiopathiae* can be distinguished from that commonly caused by streptococci or *Staphylococcus aureus* by the absence of pitting on pressure and lack of suppuration in the lesion.

The diffuse cutaneous form is an unusual presentation in which multiple skin lesions occur in a generalized pattern. The individual lesions are similar in appearance to those seen in localized presentations. Although patients may have constitutional symptoms, blood cultures are negative.

The third form is a rare, severe systemic illness consisting of sepsis with seeding of infection in a variety of sites, most commonly acute or subacute endocarditis. In a review of 45 *E. rhusiopathiae* endocarditis cases, 36% were accompanied by the characteristic skin lesion, 89% had an identifiable occupational association, and the aortic valve was preferentially affected in 61%.[9]

Diagnosis

The diagnosis is largely clinical, based on a history of trauma involving the typical sources of infection such as fish or animal bones together with the characteristic skin lesions.

The organism can be cultured from skin biopsies of the lesions or from blood cultures in patients with endocarditis. Identification can be difficult, since the organism resides deep in skin lesions, is difficult to culture, and may be incorrectly identified.[10] A PCR-based assay offers a more rapid means of identification, but experience on the performance of this test is limited.[10]

Treatment

Penicillin is the recommended therapy for both localized and systemic infections.[10] The bacterium is not sensitive to vancomycin, which is commonly used to treat endocarditis due to Gram-positive organisms. Therefore, in patients with endocarditis and a history of compatible occupational exposure, empirical antibiotic therapy should include coverage of this organism until the bacterial etiology can be established.

Medical surveillance

There are no specific medical screening or surveillance activities for this pathogen.

Prevention

E. rhusiopathiae is readily killed by commonly available disinfectants that can be used to clean contaminated surfaces.[11]

Containment and control measures should be in place wherever potentially infected animals are kept, slaughtered, or processed, or where animal waste is used.[1] Prevention efforts should also focus on avoidance of skin inoculation and worker education about the bacterium and its clinical presentations. Guards for cutting instruments and gloves with metal mesh or other reinforcement are useful in preventing skin abrasions and lacerations. Work practices designed to reduce contact with bone fragments, fish scales, and knife tips should be encouraged. Handwashing after contact with infected animals and patients or their bacteriologic specimens is essential.

References

1. Brooke CJ and Riley TV. *Erysipelothrix rhusiopathiae*: bacteriology, epidemiology and clinical manifestations of an occupational pathogen. *J Med Microbiol* 1999; 48(9):789–99.
2. Rocha MP, Fontoura PR, Azevedo SN, et al. Erysipelothrix endocarditis with previous cutaneous lesion: report of a case and review of the literature. *Rev Inst Med Trop Sao Paulo* 1989; 31(4):286–9.
3. Andrychowski J, Jasielski P, Netczuk T, et al. Empyema in spinal canal in thoracic region, abscesses in paravertebral space, spondylitis: in clinical course of zoonosis *Erysipelothrix rhusiopathiae*. *Eur Spine J* 2012; 21(Suppl 4):S557–63. Available at: http://www.ncbi.nlm.nih.gov/pmc/articles/PMC3369048/pdf/586_2012_Article_2289.pdf (accessed on June 30, 2016).
4. Hill DC and Ghassemian JN. *Erysipelothrix rhusiopathiae* endocarditis: clinical features of an occupational disease. *South Med J* 1997; 90(11):1147–8.
5. Ajmal M. A laboratory infection with *Erysipelothrix rhusiopathiae*. *Vet Rec* 1969; 85(24):688.
6. Popugaĭlo VM, Podkin IuA, Gurvich VB, et al. Erysipeloid as an occupational disease of workers in shoe enterprises. *Zh Mikrobiol Epidemiol Immunobiol* 1983 (10):46–9 [Russian].
7. Chandler DS and Craven JA. Persistence and distribution of *Erysipelothrix rhusiopathiae* and bacterial indicator organisms on land used for disposal of piggery effluent. *J Appl Bacteriol* 1980; 48(3):367–75.
8. Abedini S and Lester A. *Erysipelothrix rhusiopathiae* bacteremia after dog bite. *Ugeskr Laeger* 1997; 159(28):4400–1 [Danish].
9. Gorby GL and Peacock JE. *Erysipelothrix rhuriopathiae* endocarditis: microbiologic, epidemiologic, and clinical features of an occupational disease. *Rev Infect Dis* 1988; 10:317–25.
10. Veraldi S, Girgenti V, Dassoni F, et al. Erysipeloid: a review. *Clin Exp Dermatol* 2009; 34(8):859–62.
11. Fidalgo SG, Longbottom CJ, and Rjley TV. Susceptibility of *Erysipelothrix rhusiopathiae* to antimicrobial agents and home disinfectants. *Pathology* 2002; 34(5):462–5.

ESCHERICHIA COLI

Common names for disease: Traveler's diarrhea, turista, food poisoning

Occupational setting

Escherichia coli is a normal commensal of the human intestinal tract. Disease is caused by novel or particularly virulent strains transmitted primarily by the fecal–oral route. Occupations with particular risk for symptomatic *E. coli* infection include workers in long-term care facilities, childcare centers, hospitals, and schools.[1] Jobs requiring travel to other countries, especially low income countries, put workers (including military personnel) at risk for traveler's diarrhea due to *E. coli*.[2] The gastrointestinal tract of animal handlers will be colonized with the *E. coli* strains in the feces of the animals they handle, with evidence of an acquired protective immunity.[3] Large outbreaks have occurred among visitors at farms from animal contact.[4]

Numerous outbreaks of serious diarrheal disease have been caused by *E. coli* transmitted by consumption of undercooked meats, especially improperly handled and prepared ground beef.[5] This may be a risk for workers eating in institutional cafeterias.[6] Like agricultural workers, food handlers have been found to be colonized with potentially pathogenic *E. coli*,[7] although the risk of disease in this group is unclear. Finally, laboratory[8] and healthcare-associated[9] *E. coli* infections, presumably due to failure to follow standard precautions, have been documented.

Exposure (route)

The primary route of exposure is fecal–oral. A review of outbreaks from *E. coli* O157:H7 over a 20-year period in the United States reported that 52% of cases were foodborne, 14% person-to-person, 9% waterborne, 3% from animal contact, and 0.3% laboratory-related.[5] Approximately 20% of cases are thought to arise from secondary transmission.[10]

Pathobiology

E. coli is an aerobic, non-spore-forming Gram-negative rod that has hundreds of serotypes classified by various antigens. Although many of these serotypes can cause a wide variety of diseases, certain types have been implicated more frequently in specific diarrheal syndromes. There are six major categories of diarrheagenic *E. coli*, each producing disease through a different mechanism as follows:

1. Enterotoxigenic *E. coli* (ETEC) is implicated in most cases of traveler's diarrhea. It is also an important cause of childhood diarrhea in low-income countries. Infection results from consumption of contaminated food or water. Bacteria adhere to the intestinal mucosa and produce an enterotoxin that causes massive fluid secretion into the gut. Disease can range from an asymptomatic carrier state to severe, watery diarrhea with cramping abdominal pain following an incubation period of hours to a few days. Symptoms last for 3–5 days and usually resolve without specific treatment. The adult resident population is not affected, because regular exposure leads to the development of immunity to the bacterial adhesive factor.
2. Another group variously referred to as enterohemorrhagic *E. coli* (EHEC), Shiga toxin-producing *E. coli* (STEC), or verocytotoxic *E. coli* (VTEC) is typified by the strain O157:H7, which causes regular foodborne outbreaks in the United States (Figure 22.5).[5]

   Following ingestion of contaminated food, usually undercooked ground beef, there is an incubation

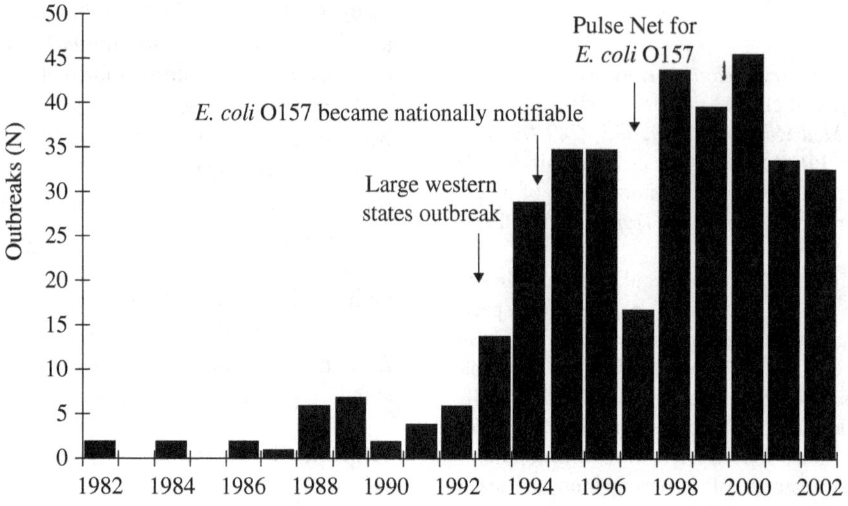

FIGURE 22.5 *E. coli* O157 outbreaks in the United States, 1982–2002 (N = 350). *Source*: Emerg Infect Dis 2005 Apr;11(4):603–9.

period of 1–10 days, followed by abdominal cramping and frequently bloody diarrhea without fever or fecal leukocytes lasting less than 7 days. The course may be complicated by hemolytic-uremic syndrome with subsequent renal failure, especially in children, or by thrombotic thrombocytopenic purpura.

3. Enteroaggregative *E. coli* (EAEC) is becoming increasingly implicated in cases of traveler's diarrhea, acute diarrhea among children in both low- to high-income countries as well as chronic diarrhea, especially among the immune-compromised. The precise mode of transmission is not known. The mechanism of disease is not well understood, although it involves damage to colonic mucosa with loss of microvilli. Diarrhea is mucoid and may be protracted.
4. Enteropathogenic *E. coli* (EPEC) is the major cause of infantile diarrhea in low-income countries. Transmission is from person-to-person. These bacteria disrupt the protective mucous gel coating the intestinal cells, bind to the cells, and cause characteristic mucosal lesions. After a 2–6 day incubation period, the clinical response is acute watery diarrhea with vomiting lasting 1–3 weeks. Dehydration can be severe enough to be fatal.
5. Enteroinvasive *E. coli* (EIEC) is similar to ETEC with invasion of intestinal cells, evoking an inflammatory response that destroys the intestinal mucosa. Transmission is from contaminated food and this class is implicated in large outbreaks in low-income countries. The clinical syndrome includes watery diarrhea or a more severe form resembling bacterial dysentery and is rare in the United States. The incubation period is usually 2–3 days, followed by fever and diarrhea, which can be bloody with numerous fecal leukocytes lasting 1–2 weeks.
6. Diffusely-adhering *E. coli* (DAEC) has been identified in cases of diarrhea among infants and children in both low- and high-income countries. The mode of transmission, clinical features, and mechanism of disease are not well-known.

E. coli is also the most frequent cause of urinary tract infections and healthcare-associated bacteremia. It is a common cause of healthcare-associated pneumonia in severely ill patients.

Diagnosis

In cases of diarrhea, stool samples should be obtained for culture and sensitivity, toxin identification, and smear for fecal leukocytes.

Various immunoassays, bioassays, and DNA probes are used to differentiate among the serotypes. Many of these specialized tests are only available through local health department laboratories. When a specific *E. coli* serotype is suspected, the laboratory should be informed so that appropriate tests may be run. Diagnosis of *E. coli* infection in other sites, such as blood or urinary tract, is made by culture and Gram stain of the appropriate samples.

Treatment

The first consideration in treating all forms of *E. coli* diarrhea should be supportive care consisting of electrolyte and fluid replacement to prevent dehydration. Severe cases of traveler's diarrhea can be presumed to be due to ETEC and may be treated with a short course of antibiotics such as trimethoprim–sulfamethoxazole, doxycycline, rifaximin, or a fluoroquinolone, which shorten the duration of illness by 1–2 days. Anti-motility agents are useful for symptomatic relief but are contraindicated when diarrhea is bloody due to an increased risk of hemolytic-uremic syndrome. Similar choices of antibiotics can be used, as appropriate, in protracted cases of EPEC and EAEC diarrhea. Antibiotics are currently considered contraindicated for EHEC diarrhea as they may also increase the risk of hemolytic-uremic syndrome and enhance toxin production.[11]

Medical surveillance

There are no recommended medical screening activities for infections due to *E. coli*. STEC is a nationally notifiable disease in the United States. Foodborne outbreaks of diarrhea due to *E. coli* should be reported to the local public health authorities.

Prevention

Meticulous handwashing with strict avoidance of high-risk foods and untreated water is the mainstay of prevention of traveler's diarrhea. Unfortunately, because of transmission from food handlers serving travelers, cases will still occur despite these measures.

Bismuth subsalicylate has been shown to be effective in preventing traveler's diarrhea, although the four times a day dosing with blackening of the tongue and stool make this option unappealing to many. Prophylaxis should not be extended beyond 3 weeks due to the risk of salicylate toxicity.

Various antibiotics have also been shown to be effective. However, they are generally not recommended prophylactically, due to the cost, the brief, self-limited nature of most cases, the increasing emergence of resistant bacteria, and the potential for serious complications such as pseudomembranous colitis. An antibiotic with poor oral bioavailability called rifaximin has been shown to be effective in preventing traveler's diarrhea but is currently only FDA-approved for treatment.[12]

Proper preparation and storage of food at the workplace will prevent this and other causes of infectious gastroenteritis. Work practices should emphasize good hygiene with appropriate glove use, strict handwashing after contact with potentially infected material, and disinfecting potentially contaminated surfaces. Unpasteurized dairy products and juices as well as undercooked meats should always be avoided. There is evidence that diarrheal disease can be reduced among both children and staff in childcare centers through use of handwashing, proper diaper changing, and food preparation equipment specifically designed to reduce the transmission of enteric infection.[13]

References

1. Wikswo ME, Hall AJ, and Centers for Disease Control and Prevention. Outbreaks of acute gastroenteritis transmitted by person-to-person contact—United States, 2009–2010. *MMWR Surveill Summ* 2012; 61(9):1–12.
2. Nada RA, Armstrong A, Shaheen HI, et al. Phenotypic and genotypic characterization of enterotoxigenic *Escherichia coli* isolated from U.S. military personnel participating in Operation Bright Star, Egypt, from 2005 to 2009. *Diagn Microbiol Infect Dis* 2013; 76(3):272–7.
3. Quilliam RS, Chalmers RM, Williams AP, et al. Seroprevalence and risk factors associated with *Escherichia coli* O157 in a farming population. *Zoonoses Public Health* 2012; 59(2):83–8.
4. Wise J. Outbreak of *E. coli* O157 is linked to Surrey open farm. *BMJ* 2009; 339:b3795.
5. Rangel JM, Sparling PH, Crowe C, et al. Epidemiology of *Escherichia coli* O157:H7 outbreaks, United States, 1982–2002. *Emerg Infect Dis* 2005; 11(4):603–9.
6. Welinder-Olsson C, Stenqvist K, Badenfors M, et al. EHEC outbreak among staff at a children's hospital—use of PCR for verocytotoxin detection and PFGE for epidemiological investigation. *Epidemiol Infect* 2004; 132(1):43–9.
7. Oundo JO, Kariuki SM, Boga HI, et al. High incidence of enteroaggregative *Escherichia coli* among food handlers in three areas of Kenya: a possible transmission route of travelers' diarrhea. *J Travel Med* 2008; 15(1):31–8.
8. Spina N, Zansky S, Dumas N, et al. Four laboratory-associated cases of infection with *Escherichia coli* O157:H7. *J Clin Microbiol* 2005; 43(6):2938–9.
9. Burke L, Humphreys H, and Fitzgerald-Hughes D. The revolving door between hospital and community: extended-spectrum beta-lactamase-producing *Escherichia coli* in Dublin. *J Hosp Infect* 2012; 81(3):192–8.
10. Snedeker KG, Shaw DJ, Locking ME, et al. Primary and secondary cases in *Escherichia coli* O157 outbreaks: a statistical analysis. *BMC Infect Dis* 2009; 9:144.
11. Wong CS, Jelacic S, Habeeb RL, et al. The risk of the hemolytic uremic syndrome after antibiotic treatment of *Escherichia coli* O157:H7 infections. *N Engl J Med* 2000; 342:1930–6.
12. Alajbegovic S, Sanders JW, Atherly DE, et al. Effectiveness of rifaximin and fluoroquinolones in preventing travelers' diarrhea (TD): a systematic review and meta-analysis. *Syst Rev* 2012; 1:39.
13. Kotch JB, Isbell P, Weber DJ, et al. Hand-washing and diapering equipment reduces disease among children in out-of-home child care centers. *Pediatrics* 2007; 120(1):e29–36.

FRANCISELLA TULARENSIS (INCLUDING *F. NOVOCIDA*)

Common names for disease: Tularemia, rabbit fever, deer fly fever

Occupational setting

Tularemia was first described as a zoonotic disease, resembling plague, in ground squirrels in Tulare County, California, from which the disease and species names are derived.

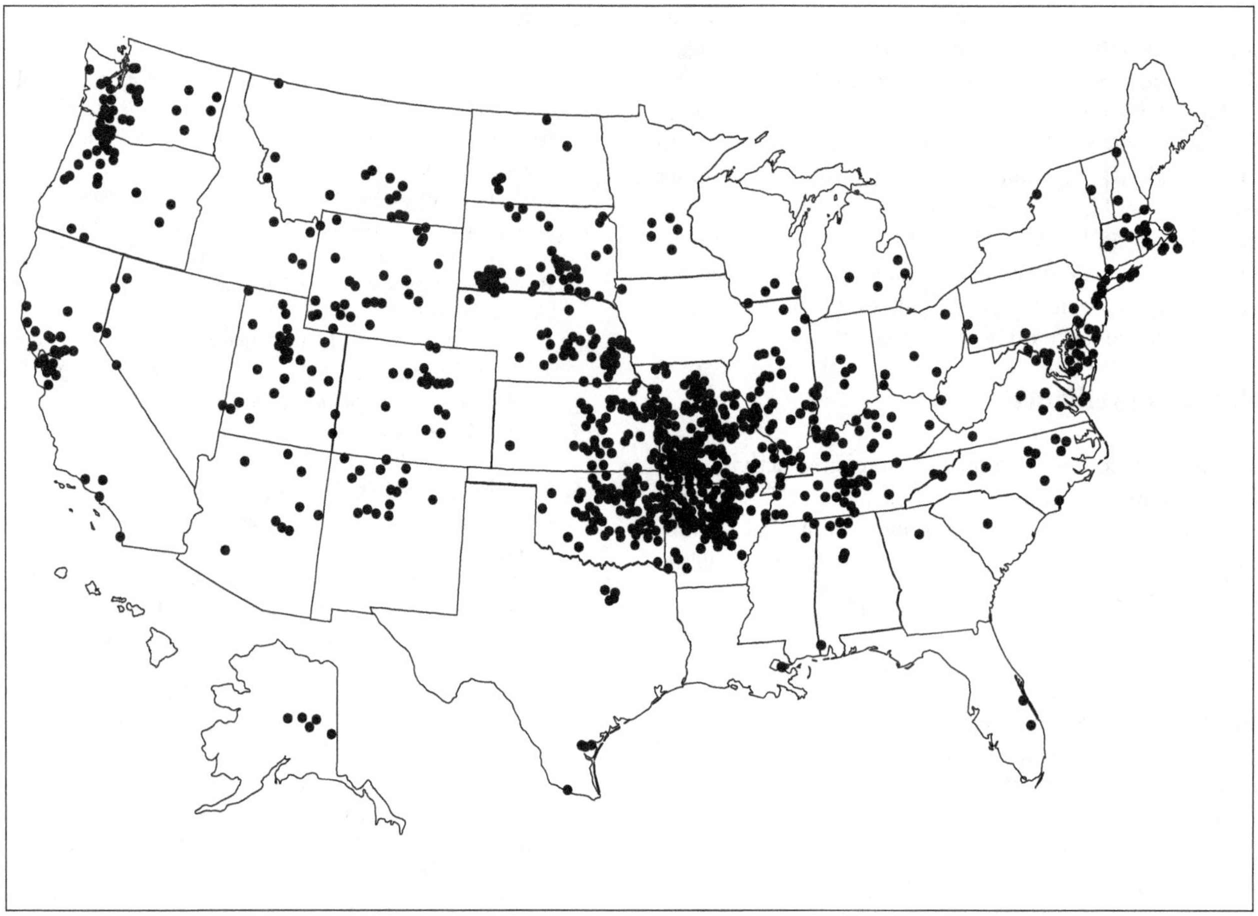

FIGURE 22.6 Reported cases of tularemia — United States, 2001–2010. *Source*: MMWR 2013 Nov 29; 62(47):963-6.

Classically, tularemia is described as a disease of small-game hunters.[1] Cases have also been described among trappers,[2] farm workers,[3] professional landscapers,[4] and laboratory workers.[5]

Approximately 125 cases are reported in the United States per year.[6] Most cases are in males, children, and the elderly, and reported in summer, corresponding to increased human activity outdoors.[6] Cases have been reported from almost every state, with a large proportion from Missouri, Arkansas, Oklahoma, Massachusetts, South Dakota, and Kansas (Figure 22.6).[6]

Exposure (route)

The organism is both highly infectious and transmissible to humans through a wide variety of routes. Direct transmission occurs through handling of infected animal carcasses, hides, or fur, through non-intact skin or after bites by infected animals.[6] Inhalation is an important route of exposure in outbreaks from cleaning and handling dead animals[1] and for laboratory technicians. Although a large number of arthropods transmit disease between various animal hosts, hard ticks, deer flies, horse flies, and mosquitoes are thought to be the principle vectors for humans.[7] Disease can also arise from ingestion of contaminated food or water. Person-to-person transmission has not been reported.

Pathobiology

Francisella tularensis is a Gram-negative, pleomorphic, nonmotile, non-spore-forming bacterium that has been found in more than 100 species of animals.[8] Two tularemia strains (types A and B) have been identified on the basis of virulence. Type A is highly virulent, is found only in North America, and is most often associated with lagomorphs (hares or rabbits).[7] Type B tularemia is less infectious, found throughout Europe, Asia, and North America, is associated with rodents as well as aquatic environments, and causes mild or even subclinical human diseases.[7]

A common source outbreak of disease from the closely related species, *F. novicida*, thought to be due to contaminated ice in a prison, has recently been reported.[9]

The incubation period is usually 3–7 days. As many as six different clinical syndromes with overlapping features are recognized, all of which include a high fever, each arising from different portals of entry for the bacteria. *Ulceroglandular tularemia* is the most common form, accounting for approximately 75% of cases with initial symptoms consisting of headaches, myalgias, and rigors.[10] Skin inoculation of the organism from an arthropod bite or infected wound produces a cutaneous ulcer with a depressed, blackened center, and well-demarcated, elevated margins. Proximal to this lesion, painful lymphadenopathy may ensue. If there is no cutaneous ulcer, the condition is simply referred to as *glandular tularemia*. *Oculoglandular tularemia* consists of conjunctivitis with pre-auricular lymphadenopathy and arises from splashes or direct inoculation with infected material into the eyes. *Oropharyngeal tularemia* involves a non-exudative pharyngitis, oral ulcers, and tonsillitis with cervical lymphadenopathy following ingestion of contaminated food or water. The mortality rate for these untreated forms of tularemia is less than 5%, but fever can last for weeks, the ulcer heals slowly over weeks to months, and lymphadenopathy can persist for months.

Typhoidal tularemia (also known as *septicemic tularemia*) is the second most common form with hepatosplenomegaly in the absence of cutaneous or lymph node involvement and a heart rate lower than expected given the high fever (pulse temperature disassociation).[10] The mode of transmission is uncertain but is thought to be from ingestion.

Pneumonic tularemia is a severe, atypical pneumonia that has a high case fatality ratio when untreated, approaching 60% in the pre-antibiotic era. Appropriate therapy has decreased the case fatality ratio to less than 1%. This form arises either from primary inoculation of the lungs by inhalation of infected aerosols or after hematogenous dissemination.

Diagnosis

Tularemia should be suspected in patients with a compatible exposure history, especially children and men over age 55 years, with acute fever and regional lymphadenopathy.[6]

Because of the risk of transmission in the laboratory, notification of all personnel handling specimens sent for testing is essential. Cultures from appropriately collected clinical specimens are considered definitive.[11] However, blood cultures are usually negative. A variety of rapid testing methods using PCR, direct fluorescent antibody, or immunohistochemical-based assays can be used but are not widely available.

Treatment

Fluoroquinolones are the antibiotics of choice for mild to moderate cases, while severe cases should receive intravenous streptomycin or gentamicin.[12] Tetracyclines such as doxycycline are an alternative therapy, but are less desirable due to the common contraindications, need for more prolonged therapy, and increased frequency of relapses. Treatment for up to 3 weeks may be indicated depending upon the stage of disease and choice of antibiotic.

Relapses are common and are more likely to occur following treatment that is delayed or of insufficient duration.[12]

Medical surveillance

There are no recommended medical screening activities. Tularemia is a nationally notifiable disease in the United States and cases must be reported to public health authorities. Because tularemia is a CDC category A bioterrorism agent, immediate notification is recommended for cases caused by suspected intentional release.

Prevention

An attenuated live vaccine was withdrawn in the United States due to concerns about variable efficacy and reversion to wild-type virulence. The use of protective clothing and gloves is recommended during skinning or handling of potentially infected animals. Particular attention should be given to avoid inhalation of aerosols when carcasses are rinsed.[1] Prevention of arthropod-borne disease includes strategies to decrease bites with protective clothing, repellent use, and careful inspection of the skin. Laboratories should be alerted about samples from suspected cases and apply biosafety level 2 conditions.

CDC has provided detailed guidance on managing suspected laboratory exposures.[13] Antibiotic prophylaxis with either doxycycline or ciprofloxacin can be given to exposed workers.

References

1. Hauri AM, Hofstetter I, Seibold E, et al. Investigating an airborne tularemia outbreak, Germany. *Emerg Infect Dis* 2010; 16(2):238–43.
2. Lévesque B, De Serres G, Higgins R, et al. Seroepidemiologic study of three zoonoses (leptospirosis, Q fever, and tularemia) among trappers in Québec, Canada. *Clin Diagn Lab Immunol* 1995; 2(4):496–8.
3. No authors listed. Case records of the Massachusetts General Hospital. Weekly clinicopathological exercises. Case 14-2000. A 60-year-old farm worker with bilateral pneumonia. *N Engl J Med* 2000; 342(19):1430–8.
4. Hofinger DM, Cardona L, Mertz GJ, et al. Tularemic meningitis in the United States. *Arch Neurol* 2009; 66(4):523–7.
5. Lawler A. Biodefense labs. Boston University Under Fire for Pathogen Mishap. *Science* 2005; 307(5709):501.
6. Centers for Disease Control and Prevention. Tularemia – United States, 2001–2010. *MMWR Morb Mortal Wkly Rep* 2013; 62(47):963–6.

7. Petersen JM, Mead PS, and Schriefer ME. *Francisella tularensis*: an arthropod-borne pathogen. *Vet Res* 2009; 40(2):7.
8. Hopla CE. The ecology of tularemia. *Adv Vet Sci Comp Med* 1974; 18:25–53.
9. Brett ME, Respicio-Kingry LB, Yendell S, et al. Outbreak of *Francisella novicida* bacteremia among inmates at a Louisiana correctional facility. *Clin Infect Dis* 2014; 59(6):826–33.
10. Nigrovic LE and Wingerter SL. Tularemia. *Infect Dis Clin North Am* 2008; 22(3):489–504.
11. Centers for Disease Control and Prevention. Tularemia: for Clinicians: Diagnostic Testing. Available at: http://www.cdc.gov/tularemia/clinicians/index.html (accessed on June 6, 2016).
12. Boisset S, Caspar Y, Sutera V, et al. New therapeutic approaches for treatment of tularaemia: a review. *Front Cell Infect Microbiol* 2014; 4:40.
13. Centers for Disease Control and Prevention. Tularemia Fact Sheet. Available at: http://www.cdc.gov/tularemia/resources/lab/TularemiaLabExposureFactSheet.pdf (accessed on June 6, 2016).

HAEMOPHILUS DUCREYI

Common names for disease: Chancroid, soft chancre

Occupational setting

In the United States, there was a steady decline in cases of chancroid, a sexually transmitted infection, between 1987 and 2001, with 15 cases reported in 2012.[1] However, due to the difficulty in culturing *Haemophilus ducreyi*, cases may be underreported.

Outbreaks occur in association with high-risk behavior, often involving sex trade workers, and in conjunction with other sexually transmitted infections, most notably HIV.[2]

Exposure (route)

Chancroid is sexually transmitted.

Pathobiology

H. ducreyi is a small, pleomorphic coccobacillus that causes genital and perianal ulcerative lesions. After sexual exposure to an infected person, there is a variable incubation period of a day to several weeks. The chancroidal lesion begins as a tender erythematous papule, which becomes pustular and then ulcerates, and lesions are often multiple. The ulcers are usually painful and ragged in appearance, with easy bleeding upon manipulation. Tender inguinal adenopathy is common and may progress to abscess (bubo), which often spontaneously drains.

Diagnosis

Presumptive diagnosis on clinical grounds without laboratory confirmation is often inaccurate, since other ulcerating sexually transmitted diseases, including syphilis and genital herpes simplex, may mimic chancroid.[2] Material taken from the base of an ulcer, or aspirated from a bubo, may be cultured and identified using special techniques to isolate *H. ducreyi*. However, the appropriate culture media is not widely available and the sensitivity of this method is less than 80%.[3] There is no FDA-approved PCR-based test in the United States, although laboratories may develop their own assay. A presumptive diagnosis can be made on clinical grounds in a patient with the characteristic ragged, purulent, painful genital ulcers and tender inguinal lymphadenopathy after ruling out (or concomitantly treating for) syphilis and *Herpes simplex*.

Treatment

Recommended antibiotics include oral azithromycin or IM ceftriaxone, both of which offer the advantage of being effective in a single dose. HIV coinfected patients should be carefully monitored due to an increased risk of treatment failure with single dose regimens. Oral ciprofloxacin and erythromycin are alternatives.

Fluctuant nodes should be drained by needle aspiration to prevent rupture and fistula formation.

Medical surveillance

There are no recommended medical screening activities. Chancroid is a nationally notifiable disease in the United States and cases must be reported to the local health department. Partners should be identified and treated, even if asymptomatic.

Prevention

Practicing safe sexual practices can prevent chancroid. Male circumcision has been shown to reduce the risk of transmission.[4] Identification and treatment of partners, who are often asymptomatic carriers, may help curtail outbreaks.

References

1. Centers for Disease Control and Prevention. 2012 Sexually Transmitted Diseases Surveillance: Other Sexually Transmitted Diseases. Available at: http://www.cdc.gov/std/stats12/other.htm (accessed on June 30, 2016).
2. Mertz KJ, Weiss JB, Webb RM, et al. An investigation of genital ulcers in Jackson, Mississippi, with use of a multiplex PCR assay: high prevalence of chancroid and human immunodeficiency virus infection. *J Infect Dis* 1998; 178(4):1060–6.
3. Lockett AE, Dance DA, Mabey DC, et al. Serum free media for the isolation of *Haemophilus ducreyi*. *Lancet* 1991; 338:326.
4. Weiss HA, Thomas SL, Munabi SK, et al. Male circumcision and risk of syphilis, chancroid, and genital herpes: a systematic review and meta-analysis. *Sex Transm Infect* 2006; 82(2):101–9.

HAEMOPHILUS INFLUENZA

Common name for disease: None

Occupational setting

Invasive disease due to *Haemophilus influenzae* is more common in young children, among whom nasopharyngeal carriage rates are high. When disease occurs in adults, it is usually among those with impaired immune function or other chronic medical conditions. Outbreaks of disease have occurred in childcare centers (even among fully vaccinated populations),[1] long-term care facilities,[2] and hospitals.[3]

Exposure (route)

Person-to-person transmission occurs through airborne droplets or by direct contact with infectious secretions.

Pathobiology

H. influenzae is a pleomorphic, Gram-negative coccobacillus found only in humans, and there are no other known hosts. *H. influenzae* is dichotomized into encapsulated and nonencapsulated (also known as nontypeable) strains. Encapsulated bacteria are further divided into the following types: a–f. While all six types are capable of causing human disease, the most clinically relevant is type b, which produces a polysaccharide capsule, a polymer of polyribitol ribose phosphate (PRP) allowing bacteria to evade opsonization and spread systemically.

Nontypeable strains are commonly part of the normal flora in the upper respiratory tract, with colonization occurring shortly after birth. The prevalence of colonization declines into adulthood. Nontypeable *H. influenzae* is responsible for many mild illnesses in adults, including sinusitis, conjunctivitis, otitis media, and exacerbation of COPD.

Encapsulated *H. influenzae* is more likely to cause invasive disease, through hematogenous spread to distant sites. The most common form of invasive disease is meningitis, which is associated with antecedent head trauma, sinusitis, otitis, or cerebrospinal fluid leak. The clinical course resembles other forms of purulent meningitis. Epiglottitis is also unusual in adults, but it is the only invasive *H. influenzae* disease to affect healthy adults without underlying or preceding illness.[4] It presents with sore throat, fever, and dyspnea, progressing to dysphagia, drooling, and an upright posture with the neck extended and chin protruding to maintain airflow. Death may result due to airway obstruction. Pneumonia due to *H. influenzae*, usually nontypeable, may occur in adults with lung disease or alcoholism. The radiographic picture varies, but a pleural effusion, usually sterile, is common. Septic arthritis occurs on occasion, usually in adults with impaired immunity or a pre-existing arthritic condition. Bacteremia may accompany invasive disease in adults.

Diagnosis

A diagnosis of *H. influenzae* invasive disease is made by confirmatory Gram stain and culture of appropriately collected clinical samples. All isolates should be serotyped, in view of the important distinction between type b and other serotypes. A variety of rapid testing methods are available to detect the PRP capsular polysaccharide found on the type b strain.

Identifying nontypeable *H. influenza* as the cause of noninvasive disease is challenging, as it is often present as a colonizer in the setting of polymicrobial growth. Therefore, these diagnoses are usually made on clinical grounds. Blood cultures in cases of pneumonia are usually negative.

Treatment

Invasive infections should be treated with a cephalosporin such as ceftriaxone or cefotaxime. Steroids are indicated to reduce neurological sequelae in cases of meningitis. Epiglottitis is a medical emergency requiring airway maintenance.

Medical surveillance

There are no recommended medical screening activities for diseases due to *H. influenzae*. Cases of invasive *H. influenzae* disease should be reported to the local public health authorities.

Prevention

Invasive *H. influenzae* type b disease has almost disappeared in those countries incorporating the vaccine into the primary childhood schedule.[5] Vaccination of adults is not routinely indicated because of high rates of natural immunity but should be considered in special populations with underlying conditions at high risk for infection.[6] The vaccine for *H. influenzae* type b does not confer protection against other encapsulated or nontypeable strains.

In cases of *H. influenzae* type b outbreaks, rifampin chemoprophylaxis is indicated for all household contacts of persons less than 4 years who are not fully vaccinated or those less than 18 years who are immunocompromised, irrespective of vaccination status.[6] In childcare centers with two or more cases of invasive disease within 60 days and unimmunized/underimmunized children, prophylaxis is indicated for all attendees, regardless of age or vaccine status, and all childcare providers.[6]

N95 respirators have been shown to be protective against bacterial colonization among healthcare workers.[7]

References

1. McVernon J, Morgan P, Mallaghan C, et al. Outbreak of *Haemophilus influenzae* type b disease among fully vaccinated children in a day-care center. *Pediatr Infect Dis J* 2004; 23(1):38–41.
2. Van Dort M, Walden C, Walker ES, et al. An outbreak of infections caused by non-typeable *Haemophilus influenzae* in an extended care facility. *J Hosp Infect* 2007; 66(1):59–64.
3. Yang CJ, Chen TC, Wang CS, et al. Nosocomial outbreak of biotype I, multidrug-resistant, serologically non-typeable *Haemophilus influenzae* in a respiratory care ward in Taiwan. *J Hosp Infect* 2010; 74(4):406–9.
4. Takala AK, Eskola J, and Van Alphen L. Spectrum of invasive *Haemophilus influenzae* type b disease in adults. *Arch Intern Med* 1990; 150:2573–6.
5. Bisgard KM, Kao A, Leake J, et al. *Haemophilus influenzae* invasive disease in the United States, 1994–1995: near disappearance of a vaccine-preventable childhood disease. *Emerg Infect Dis* 1998; 4(2):229–37.
6. Briere EC, Rubin L, Moro PL, et al. Prevention and control of *Haemophilus influenzae* type b disease: recommendations of the Advisory Committee on Immunization Practices (ACIP). *MMWR Recomm Rep* 2014; 63(RR-01):1–14.
7. MacIntyre CR, Wang Q, Rahman B, et al. Efficacy of face masks and respirators in preventing upper respiratory tract bacterial colonization and co-infection in hospital healthcare workers. *Prev Med* 2014; 62:1–7.

HELICOBACTER PYLORI

Common name for disease: None

Occupational setting

Numerous studies are available that have examined occupational risk factors for *Helicobacter pylori* infection. However, since the vast majority of the studies are cross-sectional and the risk of infection also varies markedly with numerous other factors, such as geographic location, socioeconomic status, age, and gender, occupational studies must be interpreted with caution due to the possibility of confounding.

The evidence for infection among those in gastroenterology departments has recently been reviewed and provides a mixed picture.[1] While a statistically significant increased overall risk was reported, this was only observed when compared to nonmedical controls rather than hospital staff from other departments. Further, a consistent increase risk was only seen in studies from Asia, not Europe or America/Australia. Therefore, there may be a modest increased risk of infection from those in the healthcare setting generally, rather than from exposure to specific activities such as endoscopy. Healthcare workers in long-term care facilities[2] and laboratory workers handling specimens[3] may also be at increased risk. There is concern that infection can occur from contaminated water, and one study has reported an increased prevalence of seropositivity among fish handlers.[4] No association has been found for dentists[5] or sewage workers.[6]

Exposure (route)

Humans are the only important host for *H. pylori* and are estimated to have been infected for approximately 100 000 years or earlier.[7] It is considered to be the most common human chronic infection, with an overall global prevalence of approximately 50%.

While the reservoir is the stomach, person-to-person transmission occurs by routes that are incompletely understood. The fecal–oral route is considered to be more important in low-income countries than the oral–oral route in high-income countries.[8] Most individuals acquire infection in childhood. Although the prevalence of infection increases with age, this is not because of new infections in adults (which is rare) but a birth cohort effect reflecting declines in infection thought to be due to improved sanitation and hygiene. Much research has focused on contaminated water as a vehicle for infection, but studies have been hampered by difficulties in culturing the organism from this source.

Pathobiology

H. pylori is a Gram-negative, spiral-shaped rod found within the gastric mucosal layer or adhering to the epithelium of the stomach. Although chronic superficial gastritis occurs in essentially all those infected, less than 15% go on to develop an associated disease due to a variety of environmental, host, and bacterial strain-related factors.

H. pylori infection has been causally associated with peptic ulcer disease, gastric adenocarcinoma, and mucosa-associated lymphoid tissue (MALT) lymphoma. Infection appears to reduce the risk of gastroesophageal reflux disease. It is now understood that *H. pylori* infection also has profound systemic effects. Accordingly, the list of associated conditions continues to grow and includes immune thrombocytopenic purpura, idiopathic sideropenic anemia, vitamin B12 deficiency, diabetes mellitus, cardiovascular disease, hepatobiliary disease, and neurologic disease.[9]

Diagnosis

The gold standard approach to diagnosis includes histologic examination and culture of gastric biopsy specimens for *H. pylori*. Additional non-endoscopic tests include the urea breath test, serology for IgG, and the fecal antigen test.

Treatment

For many years, triple therapy consisting of a proton pump inhibitor, clarithromycin, and amoxicillin or metronidazole was standard. However, increasing antibiotic resistance, especially to clarithromycin, has resulted in unacceptably high failure rates. As a result, therapy based on susceptibility testing is preferred. Empiric therapy should be guided by local resistance patterns.[10] There are now concomitant therapies, sequential therapies, and hybrid therapies with four drug combinations.

Medical surveillance

There are no recommended medical screening or surveillance activities.

Prevention

In view of the many unresolved issues concerning transmission of *H. pylori*, specific preventive recommendations cannot be provided. General good hygiene precautions, applicable to prevent the spread of many different organisms, should be in place. Efforts to develop a vaccine have been unsuccessful to date.[11]

References

1. Peters C, Schablon A, Harling M, et al. The occupational risk of *Helicobacter pylori* infection among gastroenterologists and their assistants. *BMC Infect Dis* 2011; 11:154.
2. De Schryver A, Cornelis K, Van Winckel M, et al. The occupational risk of *Helicobacter pylori* infection among workers in institutions for people with intellectual disability. *Occup Environ Med* 2008; 65(9):587–91.
3. Matysiak-Budnik T, Briet F, Heyman M, et al. Laboratory-acquired *Helicobacter pylori* infection. *Lancet* 1995; 346(8988):1489–90.
4. Ullah SS, Shamsuzzaman SM, Ara MN, et al. Seropositivity of *Helicobacter pylori* among the fish handlers. *Mymensingh Med J* 2010; 19(2):219–24.
5. Lin SK, Lambert JR, Schembri MA, et al. The prevalence of *Helicobacter pylori* in practising dental staff and dental students. *Aust Dent J* 1998; 43(1):35–9.
6. Tschopp A, Joller H, Jeggli S, et al. Hepatitis E, *Helicobacter pylori* and peptic ulcers in workers exposed to sewage: a prospective cohort study. *Occup Environ Med* 2009; 66(1):45–50.
7. Moodley Y, Linz B, Bond RP, et al. Age of the association between *Helicobacter pylori* and man. *PLoS Pathog* 2012; 8(5):e1002693.
8. Bruce MG and Maaroos HI. Epidemiology of *Helicobacter pylori* infection. *Helicobacter* 2008; 13(Suppl 1):1–6.
9. Roubaud Baudron C, Franceschi F, Salles N, et al. Extragastric diseases and *Helicobacter pylori*. *Helicobacter* 2013; 18(Suppl 1):44–51.
10. Graham DY and Shiotani A. Which therapy for *Helicobacter pylori* infection? *Gastroenterology* 2012; 143(1):10–2.
11. Koch M, Meyer TF, and Moss SF. Inflammation, immunity, vaccines for *Helicobacter pylori* infection. *Helicobacter* 2013; 18(Suppl 1):18–23.

LEGIONELLA SPECIES (*LEGIONELLA PNEUMOPHILA, LEGIONELLA LONGBEACHAE*)

Common names for disease: Legionellosis, Legionnaires' disease, Pontiac fever

Occupational setting

Legionella organisms are ubiquitous in natural aquatic sources and proliferate easily in water supply systems, cooling towers, evaporative condensers, and distribution lines. Disease caused by *Legionella pneumophila* has been reported in many diverse settings including workers cleaning steam turbine condensers[1] and cooling towers,[2] from aerosols generated by a leaking coolant system,[3] a water tank cooling system for welding,[4] in sewage treatment workers,[5] crews[6] or those repairing ships[7], from showers in long-distance truck drivers,[8] in well excavators,[9] in workers at an automobile engine manufacturing plant,[10] and in workers at an offshore drilling facility.[11]

In the healthcare setting, outbreaks in hospitals[12] and long-term care facilities[13] are well known. Although the vast majority of cases are reported among patients, especially the immunocompromised including neonates, employees have also been affected.[12] Contamination of dental water supplies has been extensively documented but case reports of disease among staff are limited.[14]

More than 20% of cases of Legionnaire's disease reported to CDC are travel associated, usually from cruise ships or hotels.[15]

Pontiac fever and Legionnaire's disease can also be caused by *L. longbeachae*. Unlike other *Legionella* species, this organism is found in natural soil, commercial potting soil, and compost, especially in New Zealand and Australia, although disease occurs globally. Outbreaks have occurred among gardeners[16] and nursery workers.[17]

In a study of Legionnaires' disease in New York City, there was a 230% increase in reported cases from 2002 to 2009. Cases followed a socioeconomic gradient, with the highest incidence occurring in the highest poverty areas.[18] Work in transportation, repair, protective services, cleaning, or construction was associated with a significantly higher risk for Legionnaires' disease compared with the general working population.[18]

Exposure (route)

Transmission of *L. pneumophila* occurs through inhalation of aerosols or aspiration of water droplets, while transmission of *L. longbeachae* is through inhalation of soil dust. Person-to-person transmission does not occur.

Pathobiology

Organisms in the *Legionellaceae* family are aerobic, Gram-negative rods, which do not grow in routine culture media. *L. pneumophila*, responsible for approximately 90% of infections, exists in 16 different serogroups, with serogroups 1, 4, and 6 most frequently causing infection. Almost 20 other *Legionella* species, most notably *L. longbeachae*, have also been identified as human pathogens.

Following aspiration or inhalation, bacteria use pili to adhere to the epithelial lining of the respiratory tract. Those reaching the lungs are phagocytized by alveolar macrophages. However, they are able to block mechanisms that cause intracellular killing, multiply in these cells resulting in lysis, and release of more bacteria.

Legionella species cause two distinct clinical syndromes. The first, Pontiac fever, is a flu-like illness that occurs after a short incubation period of 1–2 days. The attack rate is over 90%, but the illness is usually mild and self-limited with a rapid recovery within about a week. Symptoms consist of fever, myalgias, headache, chills, and fatigue.

The other syndrome is Legionnaires' disease, a pneumonia with high morbidity and mortality rates. The incubation period is longer, varying from 2 to 10 days, and the attack rate is less than 5%. There is a prodrome similar to Pontiac fever, but nonproductive cough is more prominent and lung examination and chest radiographs may reveal evidence of consolidation. Myalgias and a fever greater than 40°C are almost universally present. Chest pain and hemoptysis, which can be confused with a pulmonary embolus, may occur. Watery diarrhea is common. A number of neurologic abnormalities may be present but altered mental status is the most common. Hyponatremia occurs more frequently in pneumonia due to *Legionella* species than in pneumonia due to other pathogens. Although there is no classic radiographic presentation, the chest x-ray may worsen during the initial treatment and take several months to resolve.

Diagnosis

For Legionnaire's disease, culture from appropriately collected respiratory specimens, ideally from bronchoalveolar lavage, is the gold standard diagnostic test. However, *Legionella* organisms are particularly difficult to grow in the laboratory and require selective media and specialized techniques. Results are not available for 3–5 days. Bacteria cannot be isolated in cases of Pontiac fever.

A urinary antigen assay is inexpensive, has high sensitivity and specificity with the additional advantage of same day results. However, it is only available for *L. pneumophila* serogroup 1, which nevertheless causes the majority of cases. CDC recommends both culture and urinary antigen testing as the preferred diagnostic approach.

Acute and convalescent antibody titers can be drawn, which require paired blood samples to be taken at 3 and 6 weeks postexposure. A fourfold or greater rise in titers is diagnostic. Single titers must be interpreted with caution due to the lack of specificity. Positive serologies indicate exposure to the organism only, which may or may not be associated with the disease. Clinical correlation is therefore essential.

Additional tests that are less widely available include PCR and direct fluorescent antibody testing of biological specimens.

Treatment

Pontiac fever is self-limited and requires no specific treatment. Azithromycin is the treatment of choice for Legionnaires' disease. Other macrolides, quinolones, or erythromycins are also effective.

Medical surveillance

Legionellosis is a nationally notifiable disease in the United States and other countries.

Prevention

For the prevention of *L. pneumophila*, proper maintenance of water storage, distribution, and coolant systems is essential. The American Society of Heating, Refrigerating, and Air-Conditioning Engineers (ASHRAE) provides detailed guidance for water systems available at ashrae.org. Routine testing of hospital water supplies is recommended, and facilities with known contamination should have the specialized testing needed for *Legionella* available for cases of healthcare-associated pneumonia.

Respiratory protection during manual cleaning operations of water systems may be helpful. The prevention of *L. longbeachae*, especially in the higher risk countries of Oceania, includes use of gloves, a disposable N95 respirator, keeping potting soil damp while in use, and handwashing.

References

1. Fraser DW, Deubner DC, Hill DL, et al. Nonpneumonic, short-incubation-period legionellosis (Pontiac fever) in men who cleaned a steam turbine condenser. *Science* 1979; 205:690–1.
2. Girod JC, Reichman RC, Winn WC Jr, et al. Pneumonic and nonpneumonic forms of legionellosis. The result of a common-source exposure to *Legionella pneumophila*. *Arch Intern Med* 1982; 142(3):545–7.
3. Allen KW, Prempeh H, and Osman MS. Legionella pneumonia from a novel industrial aerosol. *Commun Dis Public Health* 1999; 2(4):294–6.

4. O'Keefe NS, Heinrich-Morrison KA, and McLaren B. Two linked cases of legionellosis with an unusual industrial source. *Med J Aust* 2005; 183(9):491–2.
5. Gregersen P, Grunnet K, Uldum SA, et al. Pontiac fever at a sewage treatment plant in the food industry. *Scand J Work Environ Health* 1999; 25(3):291–5.
6. Rowbotham TJ. Legionellosis associated with ships: 1977 to 1997. *Commun Dis Public Health* 1998; 1(3):146–51.
7. Caylà JA, Maldonado R, González J, et al. A small outbreak of Legionnaires' disease in a cargo ship under repair. *Eur Respir J* 2001; 17(6):1322–7.
8. Public Health Laboratory Service (UK). Legionnaires' disease in long distance lorry drivers. *Commun Dis Rep* 1998; 10:13–4.
9. Miragliotta G, Del Prete R, Sabato R, et al. Legionellosis associated with artesian well excavation. *Eur J Epidemiol* 1992; 8(5):748–9.
10. Fry AM, Rutman M, Allan T, et al. Legionnaires' disease outbreak in an automobile engine manufacturing plant. *J Infect Dis* 2003; 187(6):1015–8.
11. Lapiński TW and Kruminis-Lozowski J. Infection with *Legionella pneumophila* among workers of Polish sea drilling platforms. *Wiad Lek*1997; 50(1–3):11–5.
12. Ozerol IH, Bayraktar M, Cizmeci Z, et al. Legionnaire's disease: a nosocomial outbreak in Turkey. *J Hosp Infect* 2006; 62(1):50–7.
13. Trop Skaza A, Beskovnik L, Storman A, et al. Epidemiological investigation of a legionellosis outbreak in a Slovenian nursing home, August 2010. *Scand J Infect Dis* 2012; 44(4):263–9.
14. Chikte UM, Khondowe O, and Gildenhuys I. A case study of a dental receptionist diagnosed with Legionnaires' disease. *SADJ* 2011; 66(6):284–7.
15. de Jong B, Payne Hallström L, Robesyn E, et al. Travel-associated Legionnaires' disease in Europe, 2010. *Euro Surveill* 2013; 18(23):pii: 20498.
16. Potts A, Donaghy M, Marley M, et al. Cluster of Legionnaires disease cases caused by *Legionella longbeachae* serogroup 1, Scotland, August to September 2013. *Euro Surveill* 2013; 18(50):20656.
17. Cramp GJ, Harte D, Douglas NM, et al. An outbreak of Pontiac fever due to *Legionella longbeachae* serogroup 2 found in potting mix in a horticultural nursery in New Zealand. *Epidemiol Infect* 2010; 138(1):15–20.
18. Farnham A, Alleyne L, Cimini D, et al. Legionnaires' disease incidence and risk factors, New York, New York, USA, 2002–2011. *Emerg Infect Dis* 2014; 20(11):1795–1802.

LEPTOSPIRA INTERROGANS

Common names for disease: Leptospirosis, Weil's disease or syndrome (name applied to severe, icteric disease), milker's fever

Occupational setting

Leptospirosis is an enzootic infection that is ubiquitous in nature. Important reservoirs include cattle, swine, dogs, rodents, and fish. Many infections in animals are not clinically apparent, and prolonged urinary shedding of the organism can occur. In an Australian study of 208 laboratory-confirmed cases of leptospirosis, 56% had a clear association with occupational exposure.[1] High-risk groups include farmers,[1] sanitation and sewage workers,[2] rodent control workers,[3] laboratory animal handlers,[4] forestry workers,[5] trappers,[5] zoo workers,[6] veterinarians,[7] slaughterhouse and other meat workers,[1] and fish farmers.[8]

Dairy farmers are at high risk during milking. Aerosols from bovine urination may contain leptospires that can infect humans through inhalation or entry through the eyes, nose, or throat.[9]

More recent case series suggest that multiuse land development, with water from farmlands draining into recreational bodies of water, may be contributing to an increasing proportion of cases from nonoccupational exposures. Waterborne disease is the single most important source of infection.[10] Widespread epidemics of leptospirosis have been described following floods.[11]

Exposure (route)

Humans contract the infection from contaminated fluids, tissues, or waters through direct contact with breaks in skin or mucous membranes. Urinary shedding of organisms from infected animals is the most common source of these pathogens, but meat handling is also an important route of exposure. Although person-to-person transmission is generally not thought to occur, a recent case series raise the possibility of disease transmission to healthcare workers.[12]

Pathobiology

The organism, a spirochete that is an obligate aerobe, is easily visualized by phase contrast and dark-field microscopy but grows slowly in culture. *Leptospira* consists of several species, only one of which, *L. interrogans*, is pathogenic in humans. *L. interrogans* consists of almost 300 serovars, arranged into 25 major related serogroups. This classification is important, since the animal reservoirs, clinical picture of infection, and geographic distribution vary between serogroups. Active and passive immunity is also serovar-specific. The icterohaemorrhagiae serogroup is usually carried by rats and is associated with the more severe, classic form of leptospirosis known as Weil's disease. The Australis serogroup is a common cause of infection in many parts of the world such as Australia, New Zealand, and Asia. Serogroup Pomona is frequently found in pigs and cattle. The Canicola serogroup causes canine leptospirosis.

Untreated leptospirosis is most often a self-limited illness. Because of the nonspecific and often mild presentation, the disease is underdiagnosed and underreported. The incubation period is usually about 10 days but can vary from

2 days to 4 weeks. The clinical presentation is variable, but common symptoms include the abrupt onset of fever, headache, muscular pain, nausea, vomiting, and diarrhea. Conjunctival suffusion is more specific than other findings and should raise concern for leptospirosis.

Leptospirosis is a biphasic illness. The first phase is a flu-like illness as described above. A minority of patients will go on to develop a more serious illness. The second phase of the disease is characterized by both liver and kidney failure (Weil's disease). Pulmonary hemorrhage and cardiac arrhythmia (most commonly AV nodal block) are also seen in more severe disease. The overall case fatality rate as reported by the CDC is 1–5%. However, in patients that go on to develop end organ damage, case fatality rate may be as high as 15%.[13]

Diagnosis

Leptospirosis should be considered in any patient with fever, myalgias, headache, and nausea or vomiting. Culture of leptospires from blood and cerebrospinal fluid during the first 10 days of the illness, and of urine beyond the first week can aid diagnosis, but the organism grows slowly and cultures may take up to 8 weeks to become positive. The only screening test approved for use in the United States is the indirect hemagglutination assay (IHA), which utilizes pooled antigens from all serogroups of leptospirosis, and is broadly available. However, since the prevalence of different serovars varies geographically, this test may not be sufficiently sensitive in some regions.[14] The microscopic agglutination test (MAT) requires paired sera and is generally only available in reference laboratories. ELISA-based screening tests are also available. They have shown satisfactory sensitivity but results should be confirmed with MAT.[15] A dipstick assay for serum has been developed which is suitable for widespread field use.[16] PCR and multiplex PCR tests have also been developed.[17,18] Other laboratory results will show signs of systemic infection with possible renal and liver dysfunction.

Treatment

When given within the first 4 days of infection, penicillin and doxycycline are both effective in shortening the duration of illness and decreasing symptoms of fever, headache, and myalgias. Supportive therapy and careful management of renal, hepatic, hematologic, and central nervous system complications are also important. Therapy has been reported to be infective when administered after day 4 of infection, and the efficacy of other antimicrobial agents has not been rigorously studied in randomized trials. The Jarisch–Herxheimer reaction, an inflammatory reaction induced during antibiotic treatment as a result of rapid release of antigen, is commonly observed during treatment.

Medical surveillance

Screening is not recommended. In the United States, leptospirosis has been reinstated as a nationally notifiable disease as of January 2013.

Prevention

Primary prevention strategies should focus on both animal reservoirs and humans. Animal preventive activities mainly consist of vaccines that are available for cattle and pigs. Vaccines are serovar-specific and are only useful where a small number of serovars are prevalent. Rodent control is important.

Strategies to prevent leptospirosis in humans have included environmental control measures, protective clothing, and antibiotic prophylaxis. A dramatic decline in *L. icterohaemorrhagiae* infections in Great Britain between 1978 and 1985 was attributed to vigorous rodent control programs, protective clothing use, attention to personal hygiene, and worker education in coal workers, sewer workers, and fish workers.[19] Personal protective equipment consisting of gloves and boots, together with careful work practices around domestic animals to avoid contact with potentially contaminated tissues and fluids (particularly urine), are recommended.

A randomized trial of chemoprophylaxis in military personnel at high risk for leptospirosis in Panama revealed that 200 mg of doxycycline administered once weekly was 95% effective in preventing the disease.[20] This strategy would seem useful only for populations at high risk for relatively brief periods, such as travelers. There are no currently available human vaccines.

References

1. Swart KS, Wilks CR, Jackson KB, et al. Human leptospirosis in Victoria. *Med J Aust* 1983; 14:460–3.
2. De Serres G, Levesque B, Higgins R, et al. Need for vaccination of sewer workers against leptospirosis and hepatitis A. *Occup Environ Med* 1995; 52(8):505–7.
3. Demers RY, Frank R, Demers P, et al. Leptospiral exposure in Detroit rodent control workers. *Am J Public Health* 1985; 75(9):1090–1.
4. Natrajaseenivasan K and Ratnam S. An investigation of leptospirosis in a laboratory animal house. *J Communicable Dis* 1996; 28(3):153–7.
5. Moll van Charante AW, Groen J, Mulder PG, et al. Occupational risks of zoonotic infections in Dutch forestry workers and muskrat catchers. *Eur J Epidemiol* 1998; 14(2):109–16.
6. Anderson DC, Geistfeld JG, Maetz HM, et al. Leptospirosis in zoo workers associated with bears. *Am J Trop Med Hyg* 1978; 27(1 Pt 1):210–1.
7. Kingscote BF. Leptospirosis in two veterinarians. *CMAJ* 1985; 133(9):879–80.

8. Gill ON, Coghlan JD, and Calder IM. The risk of leptospirosis in United Kingdom fish farm workers. Results from a 1981 serological survey. *J Hyg (Lond)* 1985; 94(1):81–6.
9. Skilbeck NW and Miller GT. A serological survey of leptospirosis in Gippsland dairy farmers. *Med J Aust* 1986; 144:565–7.
10. Ciceroni L, Stepan E, Pinto A, et al. Epidemiological trend of human leptospirosis in Italy between 1994 and 1996. *Eur J Epidemiol* 2000; 16(1):79–86.
11. Trevejo RT, Rigau-Perez JG, Ashford DA, et al. Epidemic leptospirosis associated with pulmonary hemorrhage—Nicaragua, 1995. *J Infect Dis* 1998; 178(5):1457–63.
12. Ratnan S and Seenivasan N. Possible hospital transmission of leptospiral infection. *J Commun Dis* 1998; 30(1):54–6.
13. Bharti, AR, Nally JE, Ricaldi JN, et al. Leptospirosis: a zoonotic disease of global importance. *Lancet Infect Dis* 2003; 3(12):757–71.
14. Effler PV, Domen HY, Bragg SL, et al. Evaluation of the indirect hemagglutination assay for diagnosis of acute leptospirosis in Hawaii. *J Clin Microbiol* 2000; 38(3):1081–4.
15. Winslow WE, Merry DJ, Pirc ML, et al. Evaluation of a commercial enzyme-linked immunosorbent assay for detection of immunoglobulin M antibody in diagnosis of human leptospiral infection. *J Clin Microbiol* 1997; 35(8):1938–42.
16. Smits HL, Hartskeerl RA, and Terpstra WJ. International multi-centre evaluation of a dipstick assay for human leptospirosis. *Trop Med Int Health* 2000; 5(2):124–8.
17. Ahmed SA, Sandai DA, Musa S, et al. Rapid diagnosis of leptospirosis by multiplex PCR. *Malays J Med Sci* 2012; 19(3):9–16.
18. Backstedt BT, Buyuktanir O, Lindow J, et al. Efficient detection of pathogenic leptospires using 16S ribosomal RNA. *PLoS One* 2015; 10(6):e0128913. doi:10.1371/journal.pone.0128913.
19. Waitkins SA. Leptospirosis as an occupational disease. *Br J Ind Med* 1986; 43:721–5.
20. Takafuji ET, Kirkpatrick JW, Miller RN, et al. An efficacy trial of doxycycline chemoprophylaxis against leptospirosis. *N Engl J Med* 1984; 310:497–500.

LISTERIA MONOCYTOGENES

Common names for disease: Listeriosis

Occupational setting

Listeria monocytogenes is a common environmental bacterium, and has been recovered from soil, dust, and water, and from mammals, birds, fish, ticks, and crustaceans. Most cases of listeriosis in the United States affect urban dwellers without specific occupational exposures to the bacterium; and many are linked to the ingestion of contaminated food, particularly in pregnant women, newborns, and those with compromised immunity[1]. Veterinarians handling infected calves have developed skin infections, as have laboratory personnel following accidental direct skin inoculation.[2,3] Mild cases of listeria conjunctivitis occur occasionally in laboratory and poultry workers.[4] Although slaughterhouse workers have been found to have five times the normal fecal carriage rate of *L. monocytogenes* (5 versus 1%), an increased risk of disease among animal handlers other than veterinarians has not been identified.

Exposure (route)

Exposure occurs by ingestion of contaminated food, by direct skin or eye inoculation, and by transplacental transmission from an infected mother to her fetus. Foods that have been implicated are unpasteurized dairy products, undercooked meats, and vegetables grown in fields fertilized with manure from infected animals.[5]

Pathobiology

L. monocytogenes is a Gram-positive non-spore-forming aerobic rod, which can cause a variety of clinical syndromes. Transient asymptomatic carriage in the stool is common. Serious symptomatic infection occurs almost exclusively in neonates and immunocompromised adults.[1] Infection during pregnancy (a state of relative immunodeficiency) is often unrecognized and may lead to preterm labor, intrauterine fetal demise, or a critically ill baby.

In adults, *L. monocytogenes* infection can present as sepsis of unknown origin. Symptomatic illness in adults usually occurs in those who are immunosuppressed, including those with acquired immune deficiency syndrome (AIDS) and malignancies. Bacteremia may lead to seeding of the meninges or brain. *L. monocytogenes* is the leading cause of meningitis in immunosuppressed adults and should be considered in the differential diagnosis of any adult with meningitis. The onset is usually subacute, with low-grade fever and personality changes. Infrequently, there are focal neurologic findings; typical meningeal signs are usually absent. Cerebritis may present with headache and fever or as a paresis resembling a cerebrovascular accident.

In cases of direct inoculation, ulcerating skin lesions have occurred, as well as purulent conjunctivitis, and, rarely, acute anterior uveitis.[2] Focal internal infections, which may arise from dissemination, include lymphadenitis, subacute bacterial endocarditis, osteomyelitis, spinal abscess, peritonitis, cholecystitis, and arthritis. Disseminated listeriosis may be accompanied by hepatitis.

Diagnosis

Diagnosis is made by isolation of the organism from cultures of blood, cerebrospinal fluid, skin ulcer, conjunctival pus, or other specimens from an infected site. Presumptive diagnosis may be made pending culture results if a Gram-stained specimen reveals Gram-positive rods resembling diphtheroids (or sometimes diplococci). Large samples of infected fluid are required (at least 10 mL of cerebrospinal fluid) because the bacteria are often sparse and are difficult to isolate. A direct fluorescent antigen test is available but is difficult to interpret and so has little practical use in most laboratories. Cerebritis is diagnosed with CT or MRI scans showing focal areas of increased uptake, without ring enhancement, and a positive blood culture; cerebrospinal fluid culture is usually negative.

Treatment

There have been no controlled studies of the efficacy of various treatment regimens, but clinical experience with penicillin and ampicillin has shown these to be usually effective, although there have been rare cases of resistance to each. Because of the refractory nature of Listeria to the action of most antibiotics, an aminoglycoside, usually gentamicin, is added for synergy.[6] For patients with penicillin allergy, the best alternative is probably trimethoprim–sulfamethoxazole, although erythromycin, tetracycline, and chloramphenicol have all been used successfully. If gentamicin is used for central nervous system infection, it should be administered both intravenously and intrathecally. There may be progression of disease despite appropriate antibiotic therapy, and the optimum duration of therapy is unknown. Although 2 weeks of therapy is usually effective, there have been relapses in immunosuppressed patients, who may require 3–6 weeks of treatment. Effective treatment remains difficult in the immunocompromised patient, and mortality remains high (approximately 30%) despite appropriate choice of therapy.

Medical surveillance

There are no recommended medical screening activities for this disease. Cases are required to be reported to local health authorities in many parts of the United States and in some other countries. Prompt reporting of outbreaks is also required.

Prevention

Animal handlers, including veterinarians, should wear gloves and splash goggles, and should wash their hands frequently. Avoiding unpasteurized dairy foods and undercooked meats can prevent foodborne listeriosis. Vegetables and fruits grown near the ground should be washed thoroughly before consumption. Uncooked meats should not be stored near vegetables or ready-to-eat foods. Hands, knives, and cutting boards should be washed after handling uncooked foods.[5]

References

1. Centers for Disease Control and Prevention. Update: multistate outbreak of listeriosis—United States, 1998–1999. *MMWR* 1999; 47:1117–8.
2. McLauchlin J and Low JC. Primary cutaneous listeriosis in adults: an occupational disease of veterinarians and farmers. *Vet Rec* 1994; 135:615–7.
3. Zelenik K, Avberšek J, Pate M, et al. Cutaneous listeriosis in a veterinarian with the evidence of zoonotic transmission—a case report. *Zoonoses Public Health* 2014; 61(4):238–41. doi:10.1111/zph.12075.
4. Jones D. Foodborne illness: foodborne listeriosis. *Lancet* 1990; 336:1171–74.
5. Centers for Disease Control. Update: Foodborne listeriosis—United States, 1988–1990. *MMWR* 1992; 41:251–8.
6. Jones EM and MacGowan AP. Antimicrobial chemotherapy of human infection due to *Listeria monocytogenes*. *Eur J Clin Microbiol Infect Dis* 1995; 14:165–75.

MYCOPLASMA PNEUMONIAE

Common names for disease: Mycoplasma pneumonia, atypical pneumonia, walking pneumonia

Occupational setting

Transmission of *Mycoplasma pneumoniae* is thought to require close, prolonged contact. Therefore, closed populations such as those in military barracks[1] or on college campuses are subject to *M. pneumoniae* infections.[2]

Although epidemiologic data are limited and somewhat inconsistent, outbreaks in hospitals have been described.[3] Since many cases in such instances are quite mild, it is possible that many outbreaks go unrecognized.[3] A cluster of cases among workers in a prosthodontics laboratory was suspected to be caused by an aerosol generated during abrasive drilling on the false teeth of a patient who was diagnosed with *M. pneumoniae* 11 days later.[4] Since the organism also causes respiratory disease in animals, veterinarians, animal handlers, and farmers may be exposed if contact is prolonged.[5]

Exposure (route)

Transmission occurs through contact with infected respiratory secretions.

Pathobiology

Mycoplasma are the smallest free-living organisms. They are neither true bacteria nor viruses, because, unlike the former, they lack a cell wall, and, unlike the latter, they do not require other cells to grow. *M. pneumoniae* is the most important species in this group and is a frequent cause of respiratory disease. Other pathologic mycoplasmas such as *M. hominis* and the closely related *Ureaplasma urealyticum* are common etiologic agents in infections of the urogenital tract. Infections at other sites by *M. hominis* are rare but have been described in the immunocompromised.[6] Occult infection with *M. fermentans* has been proposed as a cause of illness in Persian Gulf War veterans, but this association has not been confirmed in serologic studies.[7]

M. pneumoniae is a common cause of respiratory infections. Illness develops gradually over a period of several days following an incubation period of 1–4 weeks. Symptoms are flu-like and include fever, malaise, headache, sore throat, and cough. Children and young adults are the most frequently affected age groups. Most infections result in relatively benign illness that cannot be distinguished from viral etiologies and may include tracheobronchitis, pharyngitis, and otitis. However, 3–10% of *M. pneumoniae* infections result in pneumonia.[8] Because infection with this organism is so common, its contribution to community-acquired pneumonia is significant, causing an estimated 500 000 cases per year.[8]

M. pneumoniae pneumonia classically presents with a non-productive cough, fever, and upper respiratory tract symptoms such as sore throat and rhinitis. Pleuritic chest pain, dyspnea, and rigors are less common than in other bacterial pneumonias, and the white blood cell count is often normal. Chest radiographic findings are variable, but lower lobe involvement, patchy infiltrates, and pleural effusions are common. Approximately 5–10% of patients require hospitalization.[8]

These cases can be severe, resulting in respiratory insufficiency and a number of extrapulmonary complications ranging from otitis media and bullous myringitis to significant neurologic and cardiac disease. Central nervous system involvement occurs in up to 7% of hospitalized patients and includes meningitis, meningoencephalitis, and neuritis.[9] Cardiac manifestations due to myocarditis or pericarditis may result in arrhythmias and heart failure. Nausea, vomiting, or diarrhea is reported in 14–44% of patients.[9] Skin rashes are common. This infection is commonly believed to be associated with erythema multiforme; however, a systematic review of the case literature found that the association is with Stevens–Johnson syndrome and not erythema multiforme.[10]

Autoantibodies that agglutinate red blood cells at 40°C (cold agglutinins) can result in hemolytic anemia. Mycoplasma infections have also been associated with a variety of other autoimmune disorders, including rheumatoid arthritis and Guillain–Barré syndrome. A link between infection with *M. pneumoniae* and asthma has recently been established.[11]

Diagnosis

Three traditional methods have been used to diagnose *M. pneumoniae* infections. The organism can be cultured from biological specimens, but since growth is slow, 14–21 days may be needed before Mycoplasma species can be detected. The cold hemagglutinins test detects the IgM autoantibody responsible for the agglutination of red blood cells, but it is neither sensitive nor specific. The complement fixation test measures antibody production to the mycoplasma lipid membrane but is only positive late in the course of illness. Rapid PCR-based assays can be performed on respiratory tract samples and are reported to be both sensitive and specific.[12,13]

Treatment

Erythromycin and tetracycline are traditionally used in treating *M. pneumoniae* respiratory infections. However, tetracycline resistance has been described with increasing frequency. The newer macrolides and quinolones have been shown to be effective.[14] Supportive care may be needed in the setting of complications.

Medical surveillance

Outbreaks of Mycoplasma infection should be promptly reported to local public health authorities. There are no recommended medical screening activities.

Prevention

Transmission of the organism generally involves prolonged close contact. Therefore, the risk is greatest in households, barracks, and dormitories. Healthcare workers and those exposed to infected animals should practice good hygiene with strict handwashing after contact. Gloves and respiratory protection may be useful if contact is prolonged. In a study of health care workers in Beijing, China, N95 respirators showed a significant reduction in relative risk of mycoplasma pneumonia over medical masks.[15]

One study has concluded that antibiotic prophylaxis of contacts with azithromycin during outbreaks may significantly reduce secondary attack rates.[16] An infected person is generally not regarded as contagious beyond 3 weeks.

References

1. Gray GC, Callahan JD, Hawksworth AW, et al. Respiratory diseases among US military personnel: countering emerging threats. *Emerg Infect Dis* 1999; 5(3):379–85.
2. Feikin DR, Moroney JF, Talkington DF, et al An outbreak of acute respiratory disease caused by *Mycoplasma pneumoniae* and adenovirus at a federal service training academy: new implications from an old scenario. *Clin Infect Dis* 1999; 29(6):1545–50.
3. Kleemola M and Jokinen C. Outbreak of *Mycoplasma pneumoniae* infection among hospital personnel studied by a nucleic acid hybridization test. *J Hosp Infect* 1992; 21(3):213–21.
4. Sande MA, Gadot F, and Wenzel RP. Point source epidemic of *Mycoplasma pneumoniae* infection in a prosthodontics laboratory. *Am Rev Respir Dis* 1975; 112:213–7.
5. Jordan FT. Gordon Memorial Lecture: People, poultry and pathogenic mycoplasmas. *Br Poult Sci* 1985; 26(1):1–15.

6. Mattila PS, Carlson P, Sivonen A, et al. Life-threatening *Mycoplasma hominis* mediastinitis. *Clin Infect Dis* 1999; 29(6):1529–37.
7. Gray GC, Kaiser KS, Hawksworth AW, et al. No serologic evidence of an association found between Gulf War service and *Mycoplasma fermentans* infection. *Am J Trop Med Hyg* 1999; 60(5):752–7.
8. Mansel JK, Rosenow EC, Smith TF, et al. *Mycoplasma pneumoniae* pneumonia. *Chest* 1989; 95:639–46.
9. Cassell GH and Cole BC. Mycoplasmas as agents of human disease. *N Engl J Med* 1981; 304:80–9.
10. Tay YK, Huff JC, and Weston WL. *Mycoplasma pneumoniae* infection is associated with Stevens–Johnson syndrome, not erythema multiforme. *J Am Acad Dermatol* 1996; 35(5 Pt 1):757–60.
11. Daian CM, Wolff AH, and Bielory L. The role of atypical organisms in asthma. *Allergy Asthma Proc* 2000; 21(2):107–11.
12. Abele-Horn M, Busch U, Nitschko H, et al. Molecular approaches to diagnosis of pulmonary diseases due to *Mycoplasma pneumoniae*. *J Clin Microbiol* 1998; 36(2):548–51.
13. Nilsson AC, Björkman P, and Persson K. Polymerase chain reaction is superior to serology for the diagnosis of acute *Mycoplasma pneumoniae* infection and reveals a high rate of persistent infection. *BMC Microbiol* 2008; 8:93.
14. Taylor-Robinson D and Bebear C. Antibiotic susceptibilities of mycoplasmas and treatment of mycoplasmal infections. *J Antimicrob Chemother* 1997; 40(5):622–30.
15. MacIntyre CR, Wang Q, Rahman B, et al. Efficacy of face masks and respirators in preventing upper respiratory tract bacterial colonization and co-infection in hospital healthcare workers. *Prev Med* 2014; 62:1–15.
16. Klausner JD, Passaro D, Rosenberg J, et al. Enhanced control of an outbreak of *Mycoplasma pneumoniae* pneumonia with azithromycin prophylaxis. *J Infect Dis* 1998; 177(1):161–6.

NEISSERIA GONORRHOEAE

Common name for disease: Gonorrhea, clap

Occupational setting

Gonorrhea is primarily a sexually transmitted disease. There are few occupations outside of prostitution where a true occupational risk has been demonstrated, although seafarers are known to be at increased risk of several types of sexually transmitted diseases.[1] A theoretical risk exists among dentists caring for patients with oral gonorrhea. Cutaneous infection has been reported in a laboratory worker who cut his finger on a test tube containing *Neisseria gonorrhoeae* prepared for lyophilization.[2]

Exposure (route)

Transmission is by direct contact of mucous membranes with the organism, usually during sexual intercourse. Oral or conjunctival splashing, as well as direct inoculation, can transmit the bacteria.

Pathobiology

N. gonorrhoeae is a nonmotile, aerobic, Gram-negative diplococcus. The urethra, endocervix, anal canal, pharynx, and conjunctiva are infected directly by contact with *N. gonorrhoeae*. The organism penetrates the mucosal epithelium and causes a local inflammatory response within 72 hours (although appearance of clinical symptoms may be delayed). Infection in men results in urethritis, with local extension in the urogenital tract if untreated. Infection in women results in vaginitis, cervicitis, and urethritis, can extend locally to the ducts of Skene, and Bartholin's glands, and if untreated can cause endometritis, salpingitis, and pelvic inflammatory disease. In both sexes, *N. gonorrhoeae* can directly infect the anorectal mucosa, pharynx, and conjunctiva, causing symptomatic disease. If untreated, infection can become systemic, resulting in bacteremia, arthritis, tenosynovitis, endocarditis, meningitis, or a disseminated rash.[3]

Diagnosis

Diagnosis is made by nucleic acid amplification tests from infected body fluid. Gram stain is also used, usually in conjunction with culture. Bacterial culture allows determination of antibiotic resistance patterns, which are increasingly important.

Treatment

Because of widespread resistance, penicillin and oral cephalosporins are no longer the recommended drug of choice for gonorrhea. The current CDC recommendations are for a single dose of ceftriaxone 250 mg intramuscularly and either azithromycin 1 g orally as a single dose or doxycycline 100 mg orally twice daily for 7 days.[4]

Medical surveillance

There are no recommended medical screening activities for gonorrhea. It is a reportable disease in the United States.

Prevention

Prevention of gonorrhea in the general population consists of treatment of infected individuals and their partners, and the use of condoms or avoidance of sexual contact with infected individuals.[4] Individuals who have direct contact with potentially infectious individuals or material, including healthcare and laboratory workers, should use appropriate personal protective equipment such as gloves.

References

1. International Labor Organization—World Health Organization. Joint ILO–WHO committee on the hygiene of seafarers. *WHO Tech Rep Ser* 1961; 224:1–14.

2. Collins CH and Kennedy DA. Microbiological hazards of occupational needlestick and 'sharps' injuries. *J Appl Bacteriol* 1987; 62:385–402.
3. Cheng DSF. Gonorrhea. In: Parish LC and Gschnait F (eds.), *Sexually Transmitted Diseases: A Guide for Clinicians*. New York: Springer-Verlag, 1988:59–77.
4. Centers for Disease Control and Prevention. Update to CDC's sexually transmitted diseases treatment guidelines, 2010: oral cephalosporins no longer a recommended treatment for gonococcal infections. *Morb Mortal Wkly Rep* 2012; 61(31):590–4.

NEISSERIA MENINGITIDIS

Common names for diseases: Meningococcal meningitis, meningococcemia, cerebrospinal fever

Occupational setting

Serious disease due to Neisseria meningitidis, including meningitis and septicemia, usually occurs sporadically. Epidemics usually occur in institutional settings or such places as dormitories, schools, and military barracks. Occupational groups at risk during epidemics would include those who work at close quarters or with institutionalized individuals: military recruits, day care workers, prison personnel, employees of chronic care facilities, and dormitory supervisors. There have also been infections, some fatal, among laboratory personnel working with *N. meningitidis*.[1]

Exposure (route)

Spread is by direct contact with the nose, throat, and upper respiratory secretions of persons infected with (or asymptomatically carrying) the bacteria or via inhalation of respiratory droplets from coughing or sneezing carriers. Transmission may be more efficient to persons already suffering with a viral upper respiratory infection.

Pathobiology

N. meningitidis is a Gram-negative diplococcus with a polysaccharide capsule. Capsular antigens form the basis for serogroup typing of the bacteria into different strains. Four serogroups are responsible for the majority of diseases. Serogroups B, C, and Y cause most disease in the United States. However, among those older than 11 years of age, serogroups C, Y, and W cause 73% of disease. These are the serogroups included in available vaccines. Rates of meningococcal disease have been declining in the United States since the late 1990s (Figure 22.7).

Asymptomatic carriage of meningococcus in the nose and throat is common and may be chronic, intermittent, or transient. Approximately 10% of the human population harbor meningococci in the nose or throat.[2] Nasopharyngeal carriage immunizes the host, with antibody production occurring within 2 weeks. Antibodies are serogroup-specific but confer some cross-immunity against other serogroups. There is also some cross-immunity between Neisseria and

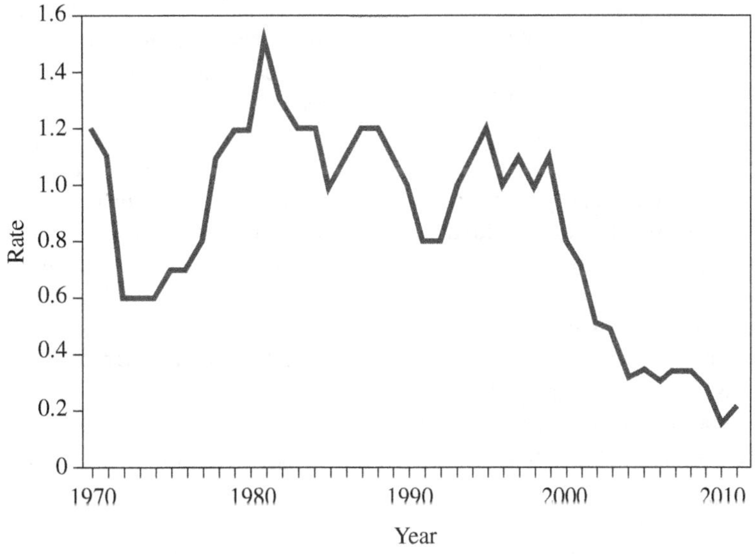

FIGURE 22.7 Meningococcal disease – reported cases per 100 000 population, by year, United States, 1970–2011. *Source*: MMWR 2013; 62(RR02):1–22.

other species of bacteria, which probably promotes natural immunity against meningococcus in a population. Serogroup B is poorly immunogenic, possibly because antigens in the polysaccharide capsule resemble neonatal host antigens.

Cases of serious meningococcal disease occur more often in the winter and spring and are more common in very young children. Males are affected more than females.

When transmission of meningococcus occurs, it most commonly results in asymptomatic nasopharyngeal carriage. When illness results from transmission, it usually takes the form of a mild pharyngitis. Pharyngitis may precede serious illness but not reliably. Serious meningococcal infection can progress to death within a few hours of symptom onset, making a high index of suspicion for the disease, and a low threshold to treat essential.

The most common meningococcal syndromes are the following:

1. Bacteremia without sepsis. The patient presents with symptoms similar to those of a viral upper respiratory infection or rash and recovers without specific therapy. Diagnosis is sometimes made when blood cultures are later found to be positive.
2. Meningococcemia without meningitis. The patient presents with signs and symptoms of bacteremia and sepsis, including fever, chills, malaise, and petechial rash. Leukocytosis and hypotension may be found on examination.
3. Meningitis with or without meningococcemia. The patient presents with headache, fever, and meningeal signs; a petechial rash may be present in meningococcemia. Mental status may range from normal to comatose. Reflexes are usually normal. Lumbar puncture will disclose cloudy cerebrospinal fluid.
4. Meningoencephalitis. The patient is obtunded, with meningeal signs, abnormal reflexes, and septic cerebrospinal fluid; pathologic reflexes may be present.

Petechial rash is common in meningococcal sepsis and meningitis. The rash is usually predominantly on the trunk and lower body, but petechiae are also often found on mucous membranes, including the palpebral conjunctiva. The lesions sometimes coalesce, especially at pressure points, to form ecchymotic-appearing lesions. A pink papular rash is an infrequent variant, and a vesicular rash is rare.

Endotoxemia may occur as a result of release of bacterial components of the Gram-negative cell wall. Endotoxemia may give rise to disseminated intravascular coagulation, septic shock with heart failure, myocarditis, pericarditis, peripheral hypoperfusion, adult respiratory distress syndrome, and adrenal hemorrhage (Waterhouse–Friderichsen syndrome). Less common manifestations of meningococcal disease include chronic meningococcemia with low-grade fever, rash, and arthritis, recurrent meningococcemia in persons with various complement deficiencies, meningococcal pneumonia, and meningococcal urethritis.

Ten to fifteen percent of cases of meningitis caused by the bacteria are fatal according to CDC. Of those that survive permanent hearing loss, mental retardation, and loss of limbs is not uncommon.[3]

Diagnosis

Treatment should not be withheld pending definitive diagnosis if meningococcal meningitis or meningiococcemia is suspected as time to antibiotic therapy is lifesaving. Presumptive diagnosis is made based on clinical presentation and by finding typical organisms (Gram-negative diplococci) on Gram-stained smear of cerebrospinal fluid or by recovering meningococcus from cerebrospinal fluid, or blood culture. Occasionally, stained smears from petechiae reveal the organisms. Various antigen detection techniques can be used to identify group-specific polysaccharides. Although these techniques are rapid and specific, they are not sensitive, so they cannot be used to exclude the diagnosis of meningococcal infection. Other findings on cerebrospinal fluid examination vary, although neutrophils predominate in untreated cases. Cerebrospinal fluid chemistry usually reveals low glucose and high protein, findings seen in many infections and not specific to meningococcal meningitis.

Treatment

If the diagnosis is seriously considered, antibiotic therapy should be administered within 30 minutes. In adults, empiric treatment is with a third-generation cephalosporin, ceftriaxone, or cefotaxime administered intravenously. If there is proven susceptibility, penicillin can be used. Dexamethasone is indicated in cases of haemophilus influenza type b and should be administered until that diagnosis is disproved.[4] Patients need close monitoring early in the course of meningococcal disease for endotoxin-related complications, including disseminated intravascular coagulation (DIC) and septic shock. Because penicillin does not eradicate nasopharyngeal carriage in patients with meningococcal illness, patients should be given rifampin prior to discharge from the hospital.

Medical surveillance

Household and intimate contacts should be closely watched for signs of meningococcal infection, so that treatment can be started promptly. Surveillance of nasopharyngeal cultures is generally not indicated, since carriage is common, often transient, and not a consistent risk factor for infection. Cases must be reported immediately to local health authorities so that contact tracing can begin.

Prevention

Antibiotic prophylaxis is indicated for household and intimate contacts of cases. Household contact includes contacts in crowded quarters such as day care centers, dormitories, military barracks, prisons, and chronic care facilities. Hospital personnel are not considered to be at increased risk unless there has been intimate contact such as mouth-to-mouth resuscitation.

Rifampin, given as a 600-milligrams dose twice daily for 2 days, is the drug of choice for chemoprophylaxis. If the isolate is appropriately sensitive to sulfadiazine, this may be used as an alternative. Other alternatives for chemoprophylaxis in adults are single-dose ceftriaxone or ciprofloxacin. Improved building ventilation may help prevent or control epidemics in close quarters.

Conjugate and polysaccharide quadrivalent vaccines are currently available against serogroups A, C, Y, and W-135. Vaccination with the conjugate vaccine is recommended for all children aged 11–12 years with a booster at 16 years of age. Patients older than 56 years of age should receive the unconjugated polysaccharide vaccine when they are traveling to an endemic area or if they work with *N. meningitides* in a laboratory. Travelers to endemic areas, particularly the "meningitis belt" of sub-Saharan Africa extending from Mauritania to Ethiopia, should be vaccinated if their work or journey brings them into prolonged contact with the local populace.[5] In 2014, an unvaccinated 25-year-old laboratory worker died after working with *N. meningitidis*. It was later found that there were several breaches of recommended laboratory safety practices. Work with preparations of meningococcus should only be performed in a BSL 3 laboratory. Laboratory personnel expected to be exposed should be protected by use of standard microbiology laboratory procedures, including double gloves, N-95 fit tested respirator use and closed front laboratory coat, and work under a Class II biological safety cabinet when performing mechanical manipulations with the potential for aerosolization. All laboratory workers should be vaccinated.[6] According to CDC, laboratory workers who were infected at work in the United States had a 50% fatality rate.[7] Any incident or exposure involving meningococcus should receive prompt medical attention. In cases of percutaneous exposure, penicillin should be used for prophylaxis; for mucosal exposure, rifampin should be used.[1]

References

1. Centers for Disease Control. Laboratory-acquired meningococcemia—California and Massachusetts. *MMWR* 1991; 40(3):46–47, 55.
2. van Deuren M, Brandtzaeg P, and van der Meer JW. Update on meningococcal disease with emphasis on pathogenesis and clinical management. *Clin Microbiol Rev* 2000; 13:144–66.
3. Rosenstein NE, Perkins BA, Stephens DS, et al. Meningococcal disease. *N Engl J Med* 2001; 344:1378–88.
4. Brouwer MC, McIntyre P, Prasad K, et al. Corticosteroids for acute bacterial meningitis. *Cochrane Database Syst Rev* 2013; 6:CD004405.
5. Cohn AC, MacNeil JR, Clark TA, et al. Prevention and control of meningococcal disease: Recommendations of the Advisory Committee on Immunization Practices (ACIP). *MMWR Recomm Rep* 2013; 62(RR-2):1–28.
6. Centers for Disease Control (CDC). Recommendation of the Immunization Practices Advisory Committee: meningococcal vaccines. *MMWR Morb Mortal Wkly Rep* 1985; 34(18):255–259. Available at: http://www.cdc.gov/mmwr/preview/mmwrhtml/00020273.htm (accessed on June 30, 2016).
7. Centers for Disease Control and Prevention. *Biosafety in Microbiological and Biomedical Laboratories (BMBL)*, 5th ed. Washington, DC: US Government Printing Office, 2009.

PASTEURELLA MULTOCIDA

Common name for disease: Pasteurellosis

Occupational setting

Increased risk of exposure is associated with work involving contact with animals. Occupations with exposure include animal laboratory personnel, veterinarians, pig, cattle and chicken breeders, zoo personnel, and abattoir workers.[1]

Exposure (route)

The bacteria are usually transmitted through animal bites or scratches, although exposure to animal secretions has been implicated in cases without a history of skin trauma. Pulmonary pasteurellosis is thought to occur via inhalation of aerosolized contaminated secretions; these may be from animals or via nosocomial spread from animals.[2,3]

Pasteurella multocida infection results in three clinical presentations in humans, all of which are more common in older or immunocompromised individuals. The most common is local infection following a bite or scratch. Evidence of infection develops rapidly and generally can be noted within 24 hours. Local erythema, warmth, and tenderness are accompanied by purulent drainage in about 40% of cases.[4] Lymphangitis and lymphadenopathy may be present. Bites or scratches from cats are the most common cause of these infections. Wound cellulitis can be complicated by osteomyelitis, from either local extension to bone or direct inoculation, and septic arthritis. The latter occurs proximal to the wound, often in a joint previously damaged by osteoarthritis or rheumatoid arthritis.

Respiratory tract infections represent the second clinical syndrome caused by *P. multocida*. The spectrum of disease in respiratory Pasteurella infection ranges from sinusitis to tracheobronchitis to pneumonia with lung abscess or empyema. Pulmonary pasteurellosis is primarily a problem in individuals with underlying lung disease including chronic obstructive pulmonary disease and malignancy. Although unusual, colonization of the respiratory tract has been reported in occupationally exposed individuals without apparent infection or underlying disease.

Third, systemic infections may develop and, as with the other types of *P. multocida* illnesses, are primarily diagnosed in patients with underlying medical disease. *P. multocida* meningitis affects the very young and the elderly. Although endocarditis is rare, bacteremia can occur associated with localized sites of infection. Both respiratory and systemic infections can result without a clear history of animal-associated trauma or even documented animal exposure.

Pathobiology

The organism is a small, nonmotile, Gram-negative coccobacillus that frequently exhibits bipolar staining. It is present in the gastrointestinal or respiratory tracts of a number of animal species, including rodents, cats, dogs, and larger domestic and wild animals ranging from cattle to lions. Although it can cause significant disease in animals, it is commonly found in apparently healthy carriers. Approximately 70–90% of oral and nasal secretions in cats have been shown to carry the bacteria.[4] Virulence of the organism is thought to be related to a microbial capsule.

Diagnosis

Initial Gram stain of purulent material may suggest the diagnosis; culture of blood, wound, respiratory secretions, or other body fluid is confirmatory. Serologic tests to detect antibody against *P. multocida* have been used in animals and as research tools in humans.

Treatment

The preferred treatment for *P. multocida* infection is penicillin; doxycycline is also effective. Other oral therapeutic agents that may be used include amoxicillin, amoxicillin–clavulanate, cefuroxime, tetracycline, and ciprofloxacin.[5] Penicillin G and its derivatives or second- and third-generation cephalosporins are effective intravenous therapy.

Medical surveillance

There are no recommended medical screening or surveillance activities.

Prevention

Those working with small animals should wear gloves, use animal-handling techniques designed to avoid bites, and practice careful handwashing after contact. Exposed workers should be educated about the hazards of *Pasteurella* infection. Bites, if they occur, should be thoroughly cleansed, and attention sought if the worker becomes ill or if the injured site becomes infected. Suturing of bite wounds is controversial, and in general not recommended, especially if the hand is involved, as it may make early detection and management of infection difficult. Prophylactic antibiotics may be beneficial in high-risk wounds where infection is suspected, in hand wounds, in deep puncture wounds, and in immunocompromised patients. Recent microbiological analysis of animal bite wounds found *P. multocida* to be the most common constituent isolated from cat bites.[6] Prophylaxis, if initiated, should usually be with a β-lactam antibiotic plus a β-lactamase inhibitor, such as amoxicillin–clavulinate. Particular care should be taken with the pregnant worker. A 2003 case report described a pregnant veterinarian who presented with fever and vaginal bleeding and was infected with *Pasteurella*.[7]

References

1. Choudat D, Le Goff C, Delemotte B, et al. Occupational exposure to animals and antibodies against *Pasteurella multocida*. *Br J Ind Med* 1987; 44:829–33.
2. Beyt BE, Sondag J, Roosevelt TS, et al. Human pulmonary pasteurellosis. *JAMA* 1979; 242:1647–48.
3. Itoh M, Tierno PM, Milstoc M, et al. A unique outbreak of *Pasteurella multocida* in a chronic disease hospital. *Am J Public Health* 1980; 70:1170–3.
4. Weber DJ, Wolfson JS, Swartz MN, et al. *Pasteurella multocida* infections: report of 34 cases and review of the literature. *Medicine* 1984; 63:133–54.
5. Weber DJ and Hansen AR. Infections resulting from animal bites. *Infect Dis Clin North Am* 1991; 5:663–80.
6. Talan DA, Citron DM, Abrahamian FM, et al. Bacteriologic analysis of infected dog and cat bites. *N Engl J Med* 1999; 340:85–92.
7. Waghorn DJ and Robson M. Occupational risk of *Pasteurella multocida* septicaemia and premature labour in a pregnant vet. *BJOG* 2003; 110(8):780–1.

PSEUDOMONAS AND BURKHOLDERIA SPECIES

Common name for disease: Glanders (*B. mallei*), melioidosis, Whitmore's disease, pseudoglanders (*B. pseudomallei*)

Occupational setting

Pseudomonas and *Burkholderia* species are free-living, ubiquitous bacteria that are a particularly common contaminant of moist environments and microenvironments. While some species are very important causes of nosocomial infections in immunocompromised patients, these organisms are opportunistic pathogens that rarely cause infection in healthy persons. Therefore, occupationally acquired infection is uncommon.

Nevertheless, *Pseudomonas* can be a concern either through direct cutaneous infection or indirectly through reactions from exposure to pseudomonas-contaminated media. *P. aeruginosa* has been associated with skin infections in commercial divers[1] and nosocomial keratitis in a nurse.[2] *P. fluorescens* has been implicated in an outbreak of hypersensitivity pneumonitis from exposure to contaminated metalworking fluids termed "machine operator's lung."[3]

Although *B. mallei* and *B. pseudomallei* are no longer found in North America and Europe, they may be transmitted through laboratory work.[4,5] A US Army microbiologist was reported to have acquired *B. pseudomallei* infection from working in a biological weapons' defense facility.[6] *B. pseudomallei* is well known to occur in immigrants and travelers from endemic areas, including Vietnam veterans, and this infection may be a hazard to workers returning from these regions.[7] Unlike many other species in these two genera, *B. pseudomallei* has been associated with infection in otherwise healthy groups, such as military personnel.[8]

Exposure (route)

The route of exposure depends on the occupational setting. Transmission through direct contact, often because of improper handwashing, is a well-known means of propagation for nosocomial pathogens in this group.[9] Inhalational exposure applies to potential respiratory effects from media contaminated with pseudomonas.

Direct contact of soil onto broken skin is the principal route of exposure for *B. mallei* and *B. pseudomallei*. *B. pseudomallei* may additionally be acquired by inhalation and a recent outbreak in Australia has been attributed to contaminated water.[10]

Pathobiology

Pseudomonas and *Burkholderia* are Gram-negative, aerobic, slightly curved, or straight rods. Some species previously grouped under the genus *Pseudomonas* have been transferred to the genus *Burkholderia*. These include *B. cepacia*, *B. mallei*, and *B. pseudomallei*.

Community-acquired *P. aeruginosa* infection is usually localized and occurs as the result of exposure to large number of organisms from a contaminated water source. Examples include folliculitis from swimming pools or hot tubs, otitis externa ("swimmer's ear"), and eye infections associated with contact lens use. As noted, *P. aeruginosa* is a common and dreaded nosocomial pathogen infecting multiple sites in immunocompromised hosts.

Exposure to aerosols in contaminated environments has been associated with occupational asthma in several settings. These microenvironments usually have a mixed flora, including Pseudomonas species, and exposure to endotoxin, produced by Gram-negative bacteria, has been implicated as the agent of bronchospasm, as well as symptoms of fever, diarrhea, fatigue, headache, nausea, and eye and nasal irritation.

In 2014, a case of corneal ulcer from *P. aeruginosa* was described in a 26-year-old nurse anesthetist after work place exposure to body fluids.[11] Workplace exposures in the healthcare setting are often multidrug resistant.

B. mallei produces a disease known as glanders. This is primarily an infection of horses, mules, and donkeys. Sporadic cases occur in humans in parts of South America, Africa, and Asia. Three acute clinical pictures are possible: a rapidly fatal sepsis, a pulmonary form, or an ulcerative infection of the mucosa of the nose, mouth, and conjunctiva. A chronic cutaneous form is also recognized. There is concern that B. mallei could be used as an agent of agroterrorism.[12]

B. pseudomallei causes melioidosis or pseudoglanders, a disease endemic to Southeast Asia and northern Australia. Its range is expanding. The presentation of melioidosis is protean with systemic and localized forms involving virtually any organ system. Most commonly, it presents as pneumonia, mimicking tuberculosis infection, or as an ulcerative skin lesion. The diagnosis is made even more challenging by a variable time course for disease progression. There are acute, subacute, and chronic forms as well as the potential for a latency period lasting several years prior to disease manifestation. It is a category B bioterrorism agent as determined by CDC.

Diagnosis

Standard isolation and identification techniques can be used to diagnose infections with *Pseudomonas* and *Burkholderia* species.

Treatment

Therapy may consist of topical antibiotics with systemic antibiotics reserved for use in refractory cases or cases with complicated courses. Because of high rates of drug resistance, nosocomial infections require multiple broad-spectrum antibiotics.

Infections with *B. mallei* and *B. pseudomallei* are highly antibiotic resistant. Currently, a 14-day course of ceftazadime combined with either meropenem or imipenem followed by a course of trimethoprim–sulfamethoxazole is recommended.[13] Systemic infection has a mortality rate of up to 40%.

Medical surveillance

There are no recommended medical screening or surveillance activities.

Prevention

Engineering controls and work practices should be aimed at reducing microbial contamination of water and other media. Outdoor work practices in endemic regions of *B. pseudomallei* should avoid soil contact with non-intact skin.

References

1. Ahlen C, Mandal LH, Johannessen LN, et al. Survival of infectious *Pseudomonas aeruginosa* genotypes in occupational saturation diving environments and the significance of these genotypes for recurrent skin infections. *Am J Ind Med* 2000; 37(5):493–500.
2. Bowden JJ and Sutphin JE. Nosocomial pseudomonas keratitis in a critical-care nurse. *Am J Ophthalmol* 1986; 101: 612–3.
3. Bernstein DI, Lummus ZL, Santilli G, et al. Machine operator's lung. A hypersensitivity pneumonitis disorder associated with exposure to metalworking fluid aerosols. *Chest* 1995; 108(3):636–41.
4. Howe C and Miller WR. Human glanders: report of six cases. *Ann Intern Med* 1947; 26:93.
5. Schlech WF 3rd, Turchik JB, Westlake RE Jr, et al. Laboratory-acquired infection with *Pseudomonas pseudomallei* (melioidosis). *N Engl J Med* 1981; 305(19):1133–5.
6. Centers for Disease Control. Laboratory-acquired human glanders – Maryland. *MMWR* 2000; 49(24):532–5.
7. Koponen MA, Zlock D, Palmer DL, et al. Melioidosis. Forgotten, but not gone! *Arch Intern Med* 1991; 151(3): 605–8.
8. Lim MK, Tan EH, Soh CS, et al. *Burkholderia pseudomallei* infection in the Singapore Armed Forces from 1987 to 1994—an epidemiological review. *Ann Acad Med Singapore* 1997; 26(1):13–7.
9. Doring G, Jansen S, Noll H, et al. Distribution and transmission of *Pseudomonas aeruginosa* and *Burkholderia cepacia* in a hospital ward. *Pediatr Pulmonol* 1996; 21(2):90–100.
10. Inglis TJ, Garrow SC, Henderson M, et al. *Burkholderia pseudomallei* traced to water treatment plant in Australia. *Emerg Infect Dis* 2000; 6(1):56–9.
11. Darouiche MH, Baccari T, Hammami KJ, et al. Keratitis after corneal projection of biological fluids: a possible occupational prejudice? *Workplace Health Saf* 2014; 62:400–2.
12. Gill KM. Agroterrorism: the risks to the United States food supply and national security. *US Army Med Dep J* 2015 (January–March):9–15.
13. Wiersinga WJ, Currie BJ, Peacock SJ. Melioidosis. *N Engl J Med* 2012; 367:1035–44.

RAT-BITE FEVER: *STREPTOBACILLUS MONILIFORMIS* AND SPIRILLUM MINOR

Common names for diseases: Streptobacillary fever, Haverhill fever, sodoku

Occupational setting

Rat-bite fever results from infection with *Streptobacillus moniliformis*, most frequently transmitted by the bite of a rat. Although it is rare in North America, cases continue to be reported, mostly in children.[1]

Work involving exposure to rodents, particularly rats and mice, or small animals that prey on them, such as cats and dogs, confers increased risk. Animal laboratory personnel,[2] veterinarians, animal breeders,[3] and agricultural workers[4] are included in this group. Work in heavily rat-infested areas has also been implicated, either through non-bite trauma or without recognized trauma.

Exposure (route)

Bites from wild or laboratory rats,[2] whose oral cavities and upper respiratory tracts are commonly colonized with these organisms, can lead to infection, as can bites or scratches of other animal carriers. In 2004, a 24-year-old pet shop employee died of *S. moniliformis* endocarditis thought due to a rat scratch.[4] Haverhill fever occurs from ingesting food products presumably contaminated with rat excreta containing the organism. Recent outbreaks have implicated contaminated milk.[5]

Pathobiology

S. moniliformis is a pleomorphic, Gram-negative bacillus that may exhibit branching filaments. Rats are the most common reservoir, shedding the bacteria in saliva and urine.

Rat-bite fever follows an incubation period that can span from 1 to 22 days but is usually less than 10 days.[1] The illness consists of a prodrome of relapsing fever, chills, headache, vomiting, myalgias, and polyarthralgia.[6] The initial wound often appears to be healed at this point. Within 2–4 days, a rash, usually maculopapular in nature, develops over the extremities. This is followed by polyarthritis. The most frequently described complication in case reports is septic arthritis.[7] There are also reports, usually in children, of more serious complications, including septicemia,[8] endocarditis,[9] localized abscess,[10] and death.[11]

Spirillum minus is a Gram-negative spirochete responsible for rat-bite fever (sodoku) primarily in Asia, although cases are reported in the Americas.[12] It has a longer incubation period of 1–3 weeks. The symptoms are similar to those caused by *S. moniliformis*, except that ulceration at the site of the bite with associated lymphadenopathy and lymphangitis is common, while arthritis is unusual.

Diagnosis

The differential diagnosis for a patient presenting with the signs and symptoms of rat-bite fever is broad. Eliciting a history of bite or rodent exposure is essential in narrowing the possibilities but is not always present.[13] Atypical presentations occur, which may delay the diagnosis.[14]

S. moniliformis is confirmed by culture of biological specimens. The organism has fastidious growth requirements, making culture and isolation difficult. Specific agglutinins appear ~10 days after the onset of illness; a fourfold rise in titer during the following 2 weeks, or an initial titer of 1 : 80 is diagnostic.

Spirillum minus cannot be cultured *in vitro* and requires inoculation of body fluids intraperitoneally into laboratory animals with subsequent identification of the organism in peritoneal fluid by dark-field microscopy. No serologic test is available for *S. minus*.

Treatment

Penicillin is the antibiotic of choice.[4] Therapy should be IV for at least 7 days depending on response.

Medical surveillance

There are no recommended medical screening or surveillance activities.

Prevention

Control of rat populations at work sites is important. Persons working with small animals should wear gloves and use handling techniques designed to avoid bites. Routine handwashing after contact is essential.

Laboratory rats should be separated from other rodents to reduce the risk of transmission between animals. Bites, if they occur, should be thoroughly cleaned. The utility of prophylactic antibiotics after a bite has not been investigated. Studies of rat bites show that only a small minority becomes infected,[13] but some authors recommend a short course of penicillin to reduce the potential morbidity of infection.[14]

References

1. Centers for Disease Control. Rat-bite fever—New Mexico, 1996. *MMWR* 1998; 47(5):89.
2. Anderson LC, Leary SL, and Manning PJ. Rat-bite fever in animal research laboratory personnel. *Lab Anim Sci* 1983; 33(3):292–4.
3. Wilkins EG, Millar JG, Cockcroft PM, et al. Rat-bite fever in a gerbil breeder. *J Infect* 1988; 16(2):177–80.
4. Hagelskjaer L, Sorensen I, and Randers E. *Streptobacillus moniliformis* infection: 2 cases and a literature review. *Scand J Infect Dis* 1998; 30(3):309–11.
5. Shvartsblat S, Kochie M, Harber P, et al. Fatal rat bite fever in a pet shop employee. *Am J Ind Med* 2004; 45(4):357–60.
6. McEvoy MB, Noah ND, and Pilsworth R. Outbreak of fever caused by *Streptobacillus moniliformis*. *Lancet* 1987; 2:1361–3.
7. Rumley RL, Patrone NA, and White L. Rat-bite fever as a cause of septic arthritis: a diagnostic dilemma. *Ann Rheum Dis* 1986; 46(10):793–5.
8. Rygg M and Bruun CF. Rat bite fever (*Streptobacillus moniliformis*) with septicemia in a child. *Scand J Infect Dis* 1992; 24(4):535–40.
9. McCormack RC, Kaye D, and Hook EW. Endocarditis due to *Streptobacillus moniliformis*. *JAMA* 1967; 200(1):77–9.
10. Vasseur E, Joly P, Nouvellon M, et al. Cutaneous abscess: a rare complication of *Streptobacillus moniliformis* infection. *Br J Dermatol* 1993; 129(1):95–6.
11. Sens MA, Brown EW, Wilson LR, et al. Fatal *Streptobacillus moniliformis* infection in a two-month-old infant. *Am J Clin Pathol* 1989; 91(5):612–6.
12. Hinrichsen SL, Ferraz S, Romeiro M, et al. Sodoku—a case report. *Rev Soc Bras Med Trop* 1992; 25(2):135–8.
13. Fordham JN, McKay-Ferguson E, Davies A, et al. Rat bite fever without the bite. *Ann Rheum Dis* 1992; 51(3): 411–72.
14. Weber DJ and Hansen AR. Infections resulting from animal bites. *Infect Dis Clin North Am* 1991; 5:663–80.

RELAPSING FEVER: BORRELIA SPECIES (other than *B. burgdorferi*)

Common names for disease: Relapsing fever, endemic (or sporadic) and epidemic forms

Occupational setting

Relapsing fever, with few exceptions, occurs in countries throughout the world. Louse-borne, or epidemic, disease is transmitted from person to person by the human body louse (*Pediculus humanus*). Epidemics are traditionally associated with war, famine, and other catastrophic events; migrant workers and soldiers have been particularly prone to this infection. At present, the epidemic disease is found only in Ethiopia and neighboring countries and occurred in association with the wars there in the 1980s and 1990s.[1] Tick-borne, or endemic, disease occurs worldwide, with the largest outbreak in the western hemisphere occurring in 62 campers in Arizona in 1973.[2] As in many outbreaks, the disease appeared to be acquired through vacationing in cabins were rodents have nested. Many species of rodents and small mammals including chipmunks, squirrels, rabbits, mice, and rats serve as reservoirs of the infection. The vectors of the disease, ticks of the genus *Ornithodoros*, are found preferentially in forested mountain habitats, frequently above 3000 ft, especially caves, decaying wood, rodent burrows, and animal shelters.[3] Outdoor workers in selected environments, particularly those in remote natural settings, would seem to be at risk for the disease.

B. miyamotoi may be an under-recognized causative agent of a relapsing fever in the United States. The sera of healthy New Englanders were examined between 1991 and

2012, 3.9% were found to be positive for *B. miyamotoi* antibody. The ixodes tick that carries Lyme disease also carries *B. miyamotoi.*[4]

Exposure (route)

The disease is vector-borne, transmitted to humans by the human body louse and ticks of the genus *Ornithodoros* or *Ixodes* in the case of *B. miyamotoi.*

Pathobiology

The relapsing fevers (epidemic and endemic) are caused by several species of *Borrelia* spirochetes. *B. recurrentis* is the sole cause of epidemic relapsing fever and is transmitted by the human body louse. At least 15 other species of *Borrelia* are transmitted by ticks and cause the endemic, or sporadic, variety of the disease; examples include *B. turicatae*, *B. hermsii*, *B. parkeri*, and *B. duttonii*. The clinical manifestations of louse-borne and tick-borne disease are similar, characterized by acute onset of high fever with rigors, severe headache, myalgias, arthralgias, lethargy, photophobia, and cough, after an incubation period of about 7 days.[3] Fever is intermittent, with the initial episode typically lasting 3–6 days, followed by an asymptomatic period of 7–10 days. The patient is often unaware of a tick bite. In the absence of treatment, three to five relapses occur, with the duration and intensity of symptoms decreasing with each relapse of tick-borne disease. A single relapse is characteristic of louse-borne disease.

Diagnosis

Definitive diagnosis requires identification of spirochetes in peripheral blood smears by dark-field microscopy or appropriately stained specimens. Serologic tests are not standardized and are of limited diagnostic value other than to demonstrate rising titers in convalescent sera.

Treatment

As is the case in other *Borrelia* infections, relapsing fever is best treated successfully with tetracyclines, such as doxycycline. The disease may also respond to erythromycin and chloramphenicol. Penicillin has been associated with an increased rate of relapse. Treatment of tick-borne disease requires 7–10 days. Antibiotic treatment may produce a Jarisch–Herxheimer reaction characterized by fever, chills, tachycardia, and hypotension. This serious complication most likely represents an inflammatory reaction induced during treatment, believed to be due to a rapid release of antigen with associated cytokine or other mediator response. Death is rare in tick-borne relapsing fever and is limited to infants and older individuals. The case fatality ratio of untreated epidemic disease can approach 40%.

Medical surveillance

There are no recommended medical screening or surveillance activities.

Prevention

Prevention of the epidemic form of the disease requires appropriate response to natural and artificial disasters, good personal hygiene, and delousing procedures. In epidemic situations, short-term use of prophylactic antibiotics can contain the spread of infection to persons at high risk.[2] Endemic disease can be prevented by activities that limit tick exposure, including vector control, rodent control, and personal protective measures. Environmental application of insecticides and use of tick repellents can also decrease the potential for tick exposure.

References

1. Raoult D and Roux V. The body louse as a vector of reemerging human diseases. *Clin Infect Dis* 1999; 29:888–911.
2. Centers for Disease Control. Relapsing fever. *MMWR* 1973; 22:242–6.
3. Spach DH, Liles WC, Campbell GL, et al. Tick-borne diseases in the United States. *N Engl J Med* 1993; 329:936–45.
4. Krause PJ, Narasimhan S, Wormser GP, et al. Human *Borrelia miyamotoi* infection in the United States. *N Engl J Med* 2013; 17:291–3.

SALMONELLA SPECIES

Common names for disease: Salmonellosis, typhoid fever, paratyphoid fever

Occupational setting

Non-typhoidal *Salmonella* species can cause infections in most animal species, including poultry, cattle, swine, cats, dogs, and turtles, and individuals in occupations with animal contact are therefore at increased risk. Furthermore, large community outbreaks have occurred in which dairy products or meat from farms has been traced as the primary source (Figure 22.8).[1] Such incidents are of particular concern because of the frequency of multiply resistant pathogens due to heavy antibiotic use in animal feeds.[1]

Food handlers are at risk through contact with contaminated animal products. Organisms causing both non-typhoidal and typhoidal illnesses represent an occupational hazard to healthcare workers. Employees in patient care and laboratory settings experience increased exposure from infected patients, resulting in documented clinical infections of both typhoid fever and enterocolitis.[2, 3]

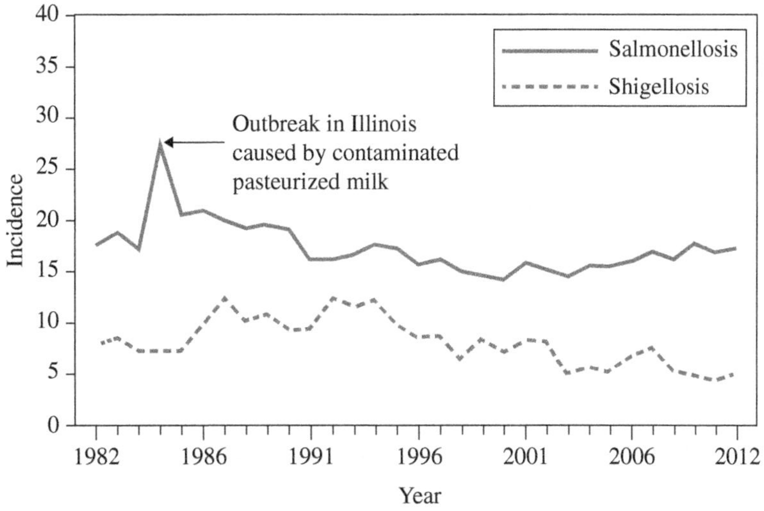

FIGURE 22.8 Salmonellosis and Shigellosis – reported cases per 100 000 population by year, United States, 1982–2012. *Source*: MMWR 2014; 61(53):87.

In 2006, 21 employees of a facility that made poultry vaccine against *Salmonella enteritidis* contracted the disease from their workplace.[4] Contact with *S. typhi* in proficiency tests of clinical laboratories and training exercises for medical laboratory technician students has also resulted in typhoid fever.[5] International travelers may be exposed to *Salmonella* species due to poor sanitation.

Exposure (route)

Ingestion of bacteria is the usual route of exposure. This occurs through contamination of food or water with infected fecal material or inadequate handwashing by personnel after contact with infected humans or animals. In 2013, a phlebotomist in Minnesota was infected with salmonella from a patient.[6]

Transmission has also been described through fomites in a report of laundry workers infected by contact with soiled linen in a nursing home.[7]

Pathobiology

Salmonellae are Gram-negative, flagellated rods. The nomenclature for this genus has changed as a result of new information gained from DNA studies. Previously distinct species are now all classified under the species *S. choleraesuis*. There are seven subgroups in this species, with subgroup I containing nearly all human pathogens. The serotypes are generally referred to in shortened form as if they were species instead of the longer, but strictly correct, designation of genus–species–serotype (i.e., *S. typhi* versus *S. choleraesuis* (group I) serotype *typhi*).

A variety of illnesses are caused by *Salmonella* species; however, they can be divided into two general groups: typhoidal and nontyphoidal. Typhoid fever is caused by *S. typhi*; the less severe illness, paratyphoid fever, is due to *S. paratyphi* A, *S. paratyphi* B (*S. schottmuelleri*), and *S. paratyphi* B (*S. hirschfeldii*). Other Salmonella species are occasionally responsible for this clinical presentation. Humans are the only reservoir for *S. typhi* and *S. paratyphi*.

Once ingested, *S. typhi* penetrates the intestinal wall, causing necrosis and ulceration, and eventually gains access to the bloodstream. After an incubation period of 1–3 weeks, fever, headache, abdominal pain, and constipation or diarrhea may develop. Respiratory symptoms may also be present. Physical examination may reveal intestinal ileus with abdominal tenderness and palpable bowel loops. Rose spots, caused by leakage from capillary endothelial cells due to bacterial infiltration, may be noted on the anterior chest and abdomen. Hepatosplenomegaly is common, and some patients may have decreased levels of consciousness. Laboratory abnormalities include anemia, leukopenia, liver function test alterations, and subclinical clotting abnormalities.

The illness is prolonged in the absence of antibiotic treatment and may be fatal. Regardless of treatment, an extended convalescence is frequently necessary. Complications are numerous including gastrointestinal perforation, hemorrhage, and localized infections, such as pneumonia and meningitis. The chronic carrier state (excretion of bacteria in feces for >1 year) develops in 1–3% of those infected and may not be preceded by a serious initial illness.[8] Although typhoid fever has become quite rare in the United States, (Figure 22.9), it should be suspected in recent travelers to undeveloped countries or in exposed healthcare workers.

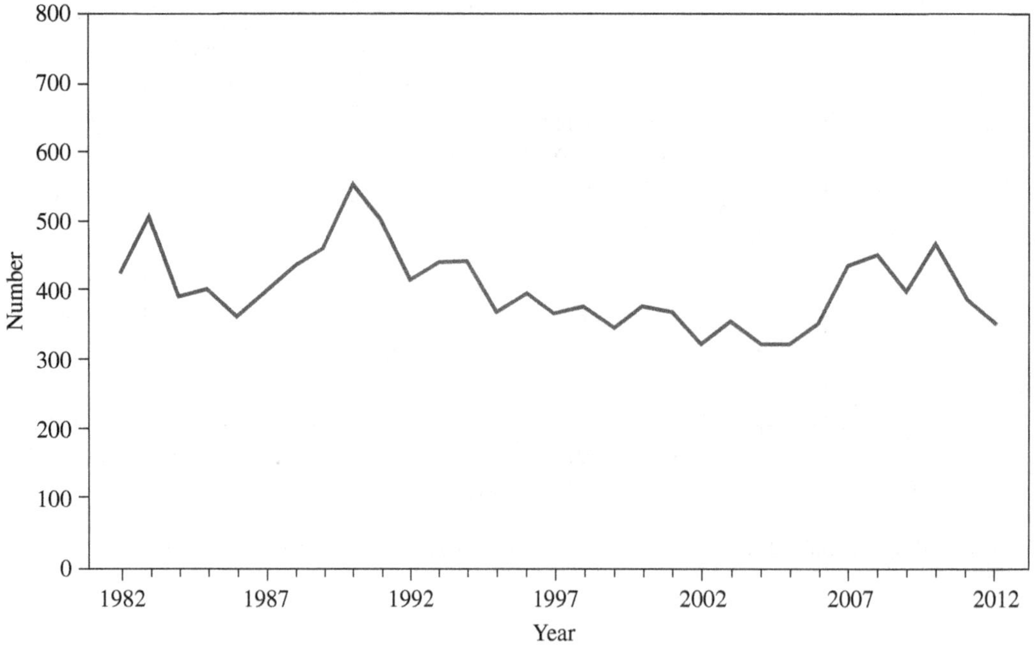

FIGURE 22.9 Typhoid fever – reported cases by year, United States, 1982–2012. *Source*: MMWR 2014; 61(53):97.

Non-typhoidal illness, in contrast, is an increasing problem in the United States. From a public health perspective, it is one of the most important zoonotic infections. Most infections are due to contaminated food, especially poultry and their products. The most common presentation is enterocolitis, and *S. typhimurium* and *S. enteritidis* are the most frequently responsible serotypes. The incubation period is 6–48 hours after bacterial ingestion. The organisms reach the lower intestinal tract and multiply there. After mucosal invasion, the bacteria may be ingested by macrophages and multiply within these cells. Other factors contributing to bacterial virulence include elicitation of a secretory response, which may be mediated by enterotoxin; tissue destruction, which may be cytotoxin-induced; and antimicrobial resistance, which is increasingly plasmid-related. Host defenses include gastric acidity, which kills bacteria; presence of normal intestinal flora, which decreases bacterial multiplication; and cellular immunocompetence.

Symptoms are fever, nausea, vomiting, and headache, followed by abdominal pain and diarrhea. The diarrhea is generally of moderate volume and nonbloody; it lasts 3–7 days, although bacteria are shed in the stool for an average of 5 weeks. The abdominal pain may localize to the right lower quadrant, mimicking appendicitis. Bacteremia may accompany the illness. Approximately 1% of the infected persons continue to shed bacteria in the stool for over 1 year. Bacteremia without enterocolitis or typhoid fever may occur. Localized abscesses, meningitis, pneumonia, endocarditis, arteritis, and osteomyelitis can result.

Diagnosis

Diagnosis is made by cultures of blood or stool or, in the case of localized infections, specimens from the affected area. Serologic studies are neither sensitive nor specific, although the enzyme-linked immunosorbent assay shows promise. Plasmid and phage typing of the causative organism are useful to determine the source of infection in non-typhoidal illnesses.

Treatment

Multidrug-resistant *S. typhi* is now widespread. A recent cross-sectional study showed that there is wide spread antibiotic resistance including floroquinolone resistance. Antibiotic choice should be driven by resistance pattern where the patient became infected with *S. typhi*. Ciprofloxacin is still a reasonable choice for empiric antibiotic therapy but confirmed resistance pattern or treatment failure should prompt reconsideration of antibiotic choice.[9]

Antibiotic therapy for healthy patients with non-typhoidal salmonella enteroclitis is not indicated[10] and, in fact, may prolong fecal excretion of bacteria and increase adverse effects. However, antibiotic therapy can be considered in patients with underlying medical conditions, pediatric patients, elderly, and those requiring hospitalization. Of note was an outbreak with 634 cases of multidrug resistant *S. heidelberg* in the United States in 2013 and 2014.[11] Likely there is emerging widespread antibiotic resistance in non-typhoidal salmonella.

Medical surveillance

Immediate case reporting to the proper health authorities is required in the United States and most other countries.

Prevention

Work practices should emphasize good hygiene with careful handwashing after contact with infected patients, animals, or their feces. Equipment used in patient procedures is potentially infectious as well. Food handlers should wear gloves. Laboratory personnel should utilize good work practices including, in addition to handwashing, the use of gloves, laboratory coats, and mechanical pipetting devices. Eating and drinking in the workplace should be prohibited. Travelers should avoid fresh, peeled fruits and vegetables, and drink only bottled water. Individuals in sensitive jobs who develop infection, such as those in patient care and food handlers, should not return to work until their stool cultures are negative.

Worker education is essential in the management of this occupational hazard. Typhoid vaccines are available for travelers to high-risk areas, individuals persistently exposed to chronic carriers, and laboratory workers having frequent contact with *S. typhi*. There is an oral live vaccine (Ty21A) and two injectable killed vaccines available (Vi polysaccharide and Vi-rEPA). A Cochrane review showed better protection overall from the two injectable vaccines, with the oral vaccine only protecting against half of the cases.[12] New vaccines with fewer side-effects have been developed but do not appear to provide the same level of long-term protection as the older whole-cell vaccine.

The control of foodborne Salmonella requires a number of interventions including the control of bacteria in animal feed, a decrease in the use of antibiotic supplements in animal feed, and proper food-preparation practices. Careful evisceration practices and physical separation of this area from the rest of the slaughterhouse is beneficial.

References

1. Molbak K, Baggesen DL, Aarestrup FM, et al. An outbreak of multidrug-resistant, quinolone-resistant *Salmonella enterica* serotype typhimurium DT104. *N Engl J Med* 1999; 341(19):1420–5.
2. Grist NR and Emslie JAN. Infections in British clinical laboratories, 1984–5. *J Clin Pathol* 1987; 40:826–9.
3. Pike RM. Laboratory-associated infections: incidence, fatalities, causes, and prevention. *Annu Rev Microbiol* 1979; 33:41–66.
4. Centers for Disease Control and Prevention. Salmonella serotype enteritidis infections among workers producing poultry vaccine—Maine, November–December 2006. *MMWR Morb Mortal Wkly Rep* 2007; 56 (34):877.
5. Hoerl D, Rostkowski C, Ross SL, et al. Typhoid fever acquired in a medical technology teaching laboratory. *Lab Med* 1988; 19:166–8.
6. Centers for Disease Control and Prevention. Occupationally acquired Salmonella I 4,12:i:1,2 infection in a phlebotomist—Minnesota, January 2013. *MMWR Morb Mortal Wkly Rep* 2013; 62(25):525.
7. Standaert SM, Hutcheson RH, and Schaffner W. Nosocomial transmission of Salmonella gastro-enteritis to laundry workers in a nursing home. *Infect Control Hosp Epidemiol* 1994; 15(1): 22–6.
8. Ackers ML, Puhr ND, Tauxe RV, et al. Laboratory-based surveillance of Salmonella serotype Typhi infections in the United States: antimicrobial resistance on the rise. *JAMA* 2000; 283(20):2668–73.
9. Lynch MF, Blanton EM, Bulens S, et al. Typhoid fever in the United States, 1999–2006. *JAMA* 2009; 302:859–65.
10. Onwuezobe IA, Oshun PO, and Odigwe CC. Antimicrobials for treating symptomatic non-typhoidal Salmonella infection. *Cochrane Database Syst Rev* 2012; 11:CD001167.
11. Centers for Disease Control and Prevention. Multistate Outbreak of Multidrug-Resistant *Salmonella* Heidelberg Infections Linked to Foster Farms Brand Chicken (Final Update). Available at: http://www.cdc.gov/salmonella/heidelberg-10-13/ (accessed on June 6, 2016).
12. Anwar E, Goldberg E, Fraser A, et al. Vaccines for preventing typhoid fever. *Cochrane Database Syst Rev* 2014; 1:CD001261.

SHIGELLA SPECIES

Common names for disease: Shigellosis, dysentery, bacillary dysentery

Occupational setting

Shigella outbreaks have been described in laboratories,[1] child day care centers,[2] cruise ships,[3] primate animal handlers,[4] and hospitals.[5] Military personnel[6] and travelers from endemic areas[7] may also be at increased risk. The overall incidence of shigellosis has declined in the United States over the past 5 years, although significant outbreaks continue to occur (Figure 22.8).

Exposure (route)

Transmission is through the fecal–oral route. This occurs most often through direct person-to-person contact. Less commonly, the organism is water- or foodborne.

Pathobiology

Shigellae are nonmotile, nonencapsulated Gram-negative rods of the Enterobacteriaceae family. There are four species that can produce diarrhea. *S. sonnei* accounts for the overwhelming majority of cases in the United States.[8] *S. dysenteriae* produces a toxin (the shiga toxin) and generally results in more severe disease. The shiga toxin is also produced by some serotypes of *E. coli*, such *E. coli* O157, and is associated with hemolytic-uremic syndrome in infections from either species of bacteria.[9] *S. flexneri* and *S. boydii* may also cause shigellosis.

Symptoms develop 24–48 hours after infection and include fever, abdominal cramping, tenesmus, and watery diarrhea that may contain blood, pus, and mucus. Mild cases are self-limited and resolve within a few days. Complications are rare and generally occur in children or immunocompromised adults. They include intestinal perforation,[10] sepsis,[11] and hemolytic-uremic syndrome.[9] Delayed sequelae include reactive arthritis and Reiter's syndrome.[12] *Shigella* demonstrates a high degree of pathogenicity; disease may result from an inoculation with as few as 10–100 bacteria. After ingestion and passage through the stomach, the organism invades the mucosa of the colon. Infection is thought to involve primarily local cell invasion and destruction, although the bacilli also produce several toxins. The superficial mucosal layer of the colon then ulcerates and sloughs. Toxins may also contribute to a moderate secretory diarrhea.

Diagnosis

Since bleeding may not be clinically apparent, *Shigella* should be included within the differential diagnosis of any compatible acute diarrhea. Fecal leukocytes are a frequent finding. A definitive diagnosis is made by culturing the organism from stool samples.

Treatment

Treatment is primarily supportive and consists of oral or, if necessary, intravenous rehydration. Antibiotic therapy may shorten the course of disease and prevent transmission of infection. However, there is growing antibiotic resistance. In 2011, 6% of isolates in the United States were resistant to ciprofloxacin or azithromycin.[13] From September 2014 through April 2015, five cases of extremely drug resistant shigellosis were identified in Illinois and Montana residents.[14] The isolates were resistant to ampicillin, ciprofloxacin, nalidixic acid, streptomycin, sulfisoxazole, tetracycline, and trimethoprim/sulphamethoxazole; had azithromycin minimum inhibitory concentrations >16 μg/mL; and harbored macrolide resistance genes mphA and ermB.[14]

Quinolones or cefixime are reasonable choices when empiric antibiotic therapy is indicated.[15]

Medical surveillance

There are no recommended medical screening activities for infections due to *Shigella*. Reporting of cases to local public health authorities is required in most jurisdictions.

Prevention

Work practices common to prevent transmission of any enteric pathogen should be in place. These measures include handwashing, glove use, and appropriate preparation and storage of food. Chlorination of water supplies and provision of sanitary facilities are also important.

Persons with active infections can easily transmit the disease, so adequate precautions are essential. These include enteric (or universal) precautions for healthcare workers caring for the infected patient and verification of noninfectious status (with two successive negative stool cultures) before allowing the individual to prepare food or resume child or patient care. In Belgium, an asymptomatic cafeteria worker was thought to be the index case in an outbreak of 52 cases of *Shigella*.[16]

Current research offers the hope of a vaccine for *Shigella* in the future.[17]

References

1. Mermel LA, Josephson SL, Dempsey J, et al. Outbreak of *Shigella sonnei* in a clinical microbiology laboratory. *J Clin Microbiol* 1997; 35(12):3163–5.
2. Mohle-Boetani JC, Stapleton M, Finger R, et al. Community-wide shigellosis: control of an outbreak and risk factors in child day-care centers. *Am J Public Health* 1995; 85(6):812–6.
3. Centers for Disease Control. Outbreak of Shigella flexneri 2a infections on a cruise ship. *MMWR* 1994; 43(35):657.
4. Kennedy FM, Astbury J, Needham JR, et al. Shigellosis due to occupational contact with non-human primates. *Epidemiol Infect* 1993; 110(2):247–51.
5. Hunter PR and Hutchings PG. Outbreak of *Shigella sonnei* dysentery on a long stay psychogeriatric ward. *J Hosp Infect* 1987; 10(1):73–6.
6. Mikhail MM, Mansour MM, Oyofo BA, et al. Immune response to *Shigella sonnei* in US marines. *Infect Immun* 1996; 64(9):3942–5.
7. Aleksic S, Bockemuhl J, and Degner I. Imported shigellosis: aerogenic Shigella boydii 74 (Sachs A 12) in a traveller followed by two cases of laboratory-associated infections. *Tropenmed Parasitol* 1981; 32(1):61–4.
8. Centers for Disease Control. Outbreaks of *Shigella sonnei* infection associated with eating fresh parsley—United States and Canada, July–August 1998. MMWR 1999; 48(14):285–9.
9. Bhimma R, Rollins NC, Coovadia HM, et al. Post-dysenteric hemolytic uremic syndrome in children during an epidemic of Shigella dysentery in Kwazulu/Natal. *Pediatr Nephrol* 1997; 11(5):560–4.
10. Upadhyay AK and Neely JA. Toxic megacolon and perforation caused by Shigella. *Br J Surg* 1989; 76(11):1217.
11. Trevett AJ, Ogunbanjo BO, Naraqi S, et al. Shigella bacteraemia in adults. *Postgrad Med J* 1993; 69(812):466–8.
12. Finch M, Rodey G, Lawrence D, et al. Epidemic Reiter's syndrome following an outbreak of shigellosis. *Eur J Epidemiol* 1986; 2(1):26–30.

13. Centers for Disease Control and Prevention. Antibiotic Resistance Threats 2013. Available at: http://www.cdc.gov/drugresistance/pdf/ar-threats-2013-508.pdf (accessed on June 6, 2016).
14. Replogle ML, Fleming DW, and Cieslak PR. Emergence of antimicrobial-resistant shigellosis in Oregon. *Clin Infect Dis* 2000; 30(3):515–9.
15. Centers for Disease Control. Ciprofloxacin- and Azithromycin-Nonsusceptible Shigellosis in the United States. CDCHAN-00379. Distributed via the CDC Health Alert Network, June 4, 2015. Available at: http://emergency.cdc.gov/han/han00379.asp (accessed on June 6, 2016).
16. Gutiérrez Garitano I, Naranjo M, Forier A, et al. Shigellosis outbreak linked to canteen-food consumption in a public institution: a matched case–control study. *Epidemiol Infect* 2011; 139:1956–64.
17. Shata MT, Stevceva L, Agwale S, et al. Recent advances with recombinant bacterial vaccine vectors. *Mol Med Today* 2000; 6(2):66–71.

STAPHYLOCOCCUS SPECIES

Common name for diseases: Common names for some staphylococcal illnesses include impetigo, toxic shock syndrome (TSS), scalded skin syndrome, and staphylococcal food poisoning

Occupational setting

Staphylococcal skin infections can occur in any setting where trauma to the skin occurs, including agricultural workers,[1] construction workers, poultry process workers, and meat-packers.[2,3] Outbreaks of skin infection outside of healthcare settings usually occur as a result of direct physical trauma, but have occurred in some unusual outdoor occupations such as river rafting guides.[4] Staphylococcal infections and asymptomatic carriage of *S. aureus* in healthcare employees, either of which can cause outbreaks of staphylococcal infection in patients, are of particular concern in hospitals.[5] Food handlers or preparers with staphylococcal skin infections have caused food-poisoning outbreaks,[6] and much current food-handling regulation is directed at reducing this risk to the general public.

Exposure (route)

Humans are the reservoir for Staphylococcus species. The coagulase-negative staphylococci (*S. epidermidis* and *S. saprophyticus*) are ubiquitous inhabitants of the skin; *S. aureus* may colonize the skin in individuals who come into contact with it through work or other activities. Carriage of the bacteria may be either chronic or transient, and it can be transmitted from carriers to other individuals or objects. *S. aureus* may be cultured from 30 to 50% of healthy individuals from the nasopharynx, skin, gastrointestinal and urogenital tracts, and perineum; persistent colonization occurs in 10–20% of cases.[7] Infection occurs by direct invasion of a tissue, usually through traumatic breaks in the skin or through conditions such as eczema that disrupt the skin's integrity as a barrier.

Pathobiology

Staphylococcus is a Gram-positive, nonmotile coccus that grows in clusters. It is a facultative anaerobe that colonizes human skin. The most important human pathogen, *S. aureus*, is the only coagulase-positive staphylococcus species. *S. epidermidis* and *S. saprophyticus* are clinically the most important of the coagulase-negative staphylococci. They are often contaminants in culture from skin flora.

S. aureus is the most common staphylococcal pathogen in the occupational setting. Factors that predispose to infection with *S. aureus* include injury to normal skin, prior viral infections, immunologic compromise, indwelling foreign bodies including sutures and catheters, prior antibiotic treatment, and pre-existing illness.[8] *S. aureus* infection usually occurs through a break in the skin. The bacteria produce a wide variety of enzymes and toxins that enhance their virulence and pathogenicity. Nonspecific toxins, including hemolysins and leukocidins, and those specific to some strains, including epidermolytic toxins (which cause bullous impetigo and scalded skin syndrome), increase the invasive properties of staphylococci. Additional strains produce very specific exotoxins, which include toxic shock toxin and the enterotoxin responsible for food poisoning.[9]

The spectrum of disease arising from *S. aureus* depends on the specific organism, the location of invasion, and host characteristics. Superficial infections include folliculitis, furunculosis, skin abscesses, impetigo, mastitis (in nursing mothers), wound infections, and spreading pyodermas. Systemic infections include superficial scalded skin syndrome, a serious condition that can result in desquamation of the entire skin, and staph toxic shock syndrome, a severe illness characterized by fever, hypotension, rash with subsequent desquamation, and involvement of several organ systems.[8] Toxic shock syndrome is caused by strains of *S. aureus* that produce a unique toxin, TSST-1 or toxic shock toxin. The disease may arise from local staphylococcal infections such as blisters[10] and may be introduced from a site of colonization rather than infection. Historically toxic shock syndrome resulting from staph was associated with tampon use. However, about half of infections seen today are not related to tampon use and can occur in either sex. Organ infections caused by *S. aureus* include endocarditis, pericarditis, pneumonia, osteomyelitis, septic arthritis, septic bursitis, and pyomyositis. Staphylococcal bacteremia and endocarditis have occurred as a result of accidental needle sticks.

Food poisoning is caused by an *S. aureus* enterotoxin that produces vomiting, diarrhea, fever, and abdominal pain. The source of the organism can be direct contact of food with the skin or infectious discharge of someone harboring the organism, with subsequent incubation of the organisms and production of enterotoxin while in storage, or from an animal source, such as meat or milk that has been inadequately processed or stored. The onset of symptoms is typically only a few hours after ingestion; the enterotoxin is heat-stable and is not inactivated by subsequent cooking of food after contamination.

S. epidermidis and *S. saprophyticus* are not of major concern in occupational settings. *S. epidermidis* is typically associated with infections in patients with prosthetic or intravenous access devices. It is often a contaminant from skin flora when found in blood culture. *S. saprophyticus* is usually associated with urinary tract infections.

Diagnosis

Staphylococcal disease can be diagnosed by demonstrating the organism on Gram stain and culture. Specific diagnostic criteria have been established for the diagnosis of toxic shock syndrome, which must be distinguished from other diseases with similar clinical presentations, including Rocky Mountain spotted fever, leptospirosis, scarlet fever, and measles.[7]

Treatment

Treatment of all staphylococcal infections includes appropriate antibiotic therapy. Selection of antibiotics depends on the type of infection, the host, and the likely resistance patterns of the infecting organism, but choices include penicillins with β-lactamase inhibitors in combination (i.e., amoxicillin–clavulanate), certain cephalosporins, erythromycin and macrolide antibiotics, (clarithromycin or azithromycin) or quinolones, such as ciprofloxacin. Methicillin-resistant *S. aureus* (MRSA) is of concern primarily in healthcare settings because of selection pressures for bacterial antibiotic resistance in these locations. Vancomycin is indicated for the treatment of MRSA or suspected MRSA infections requiring hospitalization. Resistance to quinolone antibiotics has rapidly developed within the past decade, and their use in staphylococcal infections may be limited. Doxycycline or clindamycin administered orally is a reasonable choice for outpatient care of suspected MRSA infection.

Medical surveillance

Reporting to local health authorities is required for cases or outbreaks of staphylococcal food poisoning, for community outbreaks (especially in schools or camps) of other staphylococcal infections, for epidemics in hospitals, and for cases of toxic shock syndrome in most of the Unites States and in other countries. Patient screening, with nasal swab for staphylococcal carriage, may be performed in medical facilities where nosocomial transmission of *S. aureus* is suspected. Screening of medical personal is controversial and there is no proven reduction in MRSA transmission in a hospital setting.[11]

Prevention

Food handlers should follow strict hygienic practices, including proper handwashing technique and use of gloves. Additional precautions, including temporary work removal, should be taken if they have purulent lesions of the hands, nose, or face. Food itself should be appropriately cooked, processed, and rapidly refrigerated when not used. Meatpackers should be provided with appropriate tools and personal protective equipment to minimize the risk of exposure and skin trauma, such as cut-resistant gloves. Equipment should also be cleaned regularly.

Healthcare workers are more likely to be asymptomatic carriers of *S. aureus* than the general population[12,13] and must be scrupulous in the use of handwashing techniques and personal protective equipment to prevent transmission to susceptible individuals. The optimal management, including surveillance guidelines decolonization therapy and work restrictions, has not been clearly determined.[14] It is important to reduce the opportunity for needlesticks and other trauma through the use of work practices, equipment redesign, and personal protective equipment, and to sterilize equipment and other fomites such as microscopes and ocular eyepieces that may be contaminated to eliminate potential reservoirs for infection.

References

1. Pardo-Castello V. Common dermatoses in agricultural workers in the Caribbean area. *Indust Med Surg* 1962; 31:305–7.
2. Barnham M and Kerby J. A profile of skin sepsis in meat handlers. *J Infect* 1984; 9:43–50.
3. Fehrs LJ, Flanagan K, Kline S, et al. Group A beta-hemolytic streptococcal skin infections in a US meat-packing plant. *JAMA* 1987; 258:3131–4.
4. Decker MD, Lybarger JA, Vaughn WK, et al. An outbreak of staphylococcal skin infections among river rafting guides. *Am J Epidemiol* 1986; 124:969–76.
5. Patterson WB, Craven DE, Schwartz DA, et al. Occupational hazards to hospital personnel. *Ann Intern Med* 1985; 102:658–80.
6. Eisenberg MS, Gaarslev K, Brown W, et al. Staphylococcal food poisoning aboard a commercial aircraft. *Lancet* 1975; ii:595–9.
7. Lowy FD. *Staphylococcus aureus* infections. *N Engl J Med* 1998; 339:520–32.

8. Noble WC. Skin bacteriology and the role of *Staphylococcus aureus* in infection. *Br J Dermatol* 1998; 139(Suppl 53):9–12.
9. Berkeley SF, McNeil JG, Hightower AW, et al. A cluster of blister-associated toxic shock syndrome in male military trainees and a study of staphylococcal carriage patterns. *Military Med* 1989; 154:496–9.
10. Godfrey ME and Smith IM. Hospital hazards of staphylococcal sepsis. *JAMA* 1958; 166:1197–201.
11. Hawkins G, Stewart S, Blatchford O, et al. Should healthcare workers be screened routinely for methicillin-resistant *Staphylococcus aureus*? A review of the evidence. *J Hosp Infect* 2011; 77:285–9.
12. Rongpharpi SR, Hazarika NK, and Kalita H. The prevalence of nasal carriage of *Staphylococcus aureus* among healthcare workers at a tertiary care hospital in Assam with special reference to MRSA. *J Clin Diagn Res* 2013; 7:257–60.
13. Dulon M, Peters C, Schablon A, et al. MRSA carriage among healthcare workers in non-outbreak settings in Europe and the United States: a systematic review. *BMC Infect Dis* 2014; 14:363.
14. Calfee DP, Salgado CD, Milstone AM, et al. Strategies to prevent methicillin-resistant *Staphylococcus aureus* transmission and infection in acute care hospitals: 2014 update. *Infect Control Hosp Epidemiol* 2014; 35 Suppl 2:S108–32.

STREPTOCOCCUS SPECIES

Common names for diseases: Impetigo, erysipelas, strep throat, scarlet fever, rheumatic fever

Occupational setting

Like staphylococcal infections, the most common occupational streptococcal infections are skin infections, which are prominent among workers with frequently traumatized or abraded skin, including construction workers, foresters, and farmers. Group A β-hemolytic streptococcal and *S. pyogenes* infections have been reported in slaughterhouse workers after pig bites.

Streptococcal septicemia from *S. pyogenes* has also been reported in a mortuary technician who punctured himself while conducting a post-mortem examination.[1] Healthcare workers can be asymptomatic carriers of streptococci (group A or B) and transmit nosocomial infection to patients.[2] There are several other species of *Streptococcus*, especially those found in several animal species, that rarely cause skin infections or systemic disease in humans. *S. agalactiae* (group B), *S. milleri* (α-hemolytic), and *S. equisimilis* (group C) have been associated with local and systemic infections in persons with exposure to pigs and meat-packers and poultry handlers.[3-6]

An excess of pneumococcal pneumonia has been noted in welders, although these workers were considered to have an increased susceptibility to infection from unspecified welding fumes rather than direct exposure to *S. pneumoniae*.[7]

Streptococcus suis Type II (group R β-hemolytic streptococci) was first noted in 1968 as the cause of a syndrome of meningitis and sepsis in both pigs and humans.[8] Groups at risk include pig farmers and handlers of raw pork, such as meat-packers and butchers.[9] A 2014 meta-analysis reported that *S. suis* was most prevalent in Asia and that occupation along with risky food consumption was the main risk factor. They found a case fatality rate of 12%.[10] Meningitis, endocarditis, and arthritis are seen in *S. suis* infection.

Exposure (route)

Streptococci are ubiquitous human pathogens. Skin infection probably occurs through breaks in the skin arising from laceration, trauma, surgery, or skin breakdown, although in some cases the breaks may be unnoticed. Respiratory exposure occurs through inhalation of droplets. *S. suis* colonizes the snout and pharynx of healthy pigs; diseased pigs may exhibit a bacteremia. Transmission to humans occurs from exposure to work with pigs or raw pork products, most likely through minor skin breaks, although transmission may also take place through respiratory exposure.

Pathobiology

Streptococci are Gram-positive, non-spore-forming bacteria that typically grow in pairs or chains of spherical cells. They are classified as β-hemolytic if a clear zone of hemolysis surrounding bacterial colonies is seen on sheep blood agar medium. If the zone is only partly clear (usually noted as a greenish tint, giving rise to the term "viridans" streptococci), they are considered α-hemolytic. Another classification scheme, developed by Lancefield, designates groups (A–D, G, and R) based on antigenic differences.

Group A streptococcus (*Streptococcus pyogenes*) is the most important human pathogen, causing both streptococcal skin infections and pharyngitis. Streptococcal skin infections include erysipelas, a rapidly progressive skin infection accompanied by fever and, in some cases, bacteremia; pyoderma or impetigo, a localized purulent skin infection; and cellulitis.[11] Lymphangitis and lymphadenitis may also occur. The most severe form of group A streptococcal infection is necrotizing fasciitis, an infection of subcutaneous tissue with relative sparing of overlying skin and underlying muscle. Symptoms include severe local pain and tenderness, fever, and systemic toxicity. There is rapid progression to tissue gangrene and death unless the infection is quickly treated.[11,12] Elaboration of pyrogenic exotoxins in strains of *S. pyogenes* is responsible for the streptococcal toxic shock syndrome, characterized by tachycardia, tachypnea, fever, chills, and diarrhea, progressing to septic shock and organ failure. Approximately half the cases are seen in association with necrotizing fasciitis. Skin or vaginal mucosa appears to be the portal of entry in 60% of cases, with the rest probably arising from bacteremia originating in the pharynx.[11]

Streptococcal pharyngitis, along with skin infection, is one of the most common streptococcal illnesses; its importance lies in its potential sequelae. Although nasopharyngeal carriage of group A streptococci declines somewhat from childhood to adulthood, it is not uncommon to see pharyngitis in adults, and outbreaks in crowded conditions (i.e., military barracks) can occur. Outbreaks may also arise from foodborne or waterborne transmission. Clinically, an incubation period of several days is followed by fever and sore throat. The posterior pharynx is red and edematous, the tonsils are enlarged and frequently have a patchy white exudate, and the cervical lymph nodes are swollen and tender. Although very early treatment may shorten the duration of symptoms and the period of communicability, the primary purpose and importance of treatment is to prevent complications of the infection.

Complications and sequelae of streptococcal pharyngitis include local head and neck infections, including otitis media, sinusitis, peritonsillar cellulitis or abscess, suppurative lymphadenitis, and bacteremia. Scarlet fever, which is caused by a toxin (erythrogenic toxin) produced by some strains, is characterized by a distinctive rash and, in more severe cases, systemic toxicity. Acute rheumatic fever and acute post-streptococcal glomerulonephritis are two delayed complications of streptococcal infection. Both are inflammatory diseases that develop after the streptococcal infection itself has resolved. Acute rheumatic fever is a systemic illness that involves connective tissue, primarily in the joints, manifesting as acute arthritis. Other organs may be affected, including the heart, skin, and blood vessels, leading to carditis, erythema marginatum and subcutaneous nodules, and chorea. Acute post-streptococcal glomerulonephritis can follow either pharyngitis or pyoderma and consists of a proliferative glomerular disease that is manifested clinically by edema, proteinuria, hematuria, and hypertension.

S. suis infection is usually manifested by a flu-like prodromal illness, followed by fever and meningismus. Hearing loss and vestibular dysfunction with ataxia occur in approximately 50% of cases. Other manifestations of infection may include endocarditis, arthritis, septicemia with shock and disseminated intravascular coagulation, and rhabdomyolysis.[13]

Diagnosis

Diagnosis in most cases of infection is made by Gram stain and culture of the organism. Kits are available to detect group A antigen for rapid diagnosis of streptococcal pharyngitis. Throat culture is the preferred method of diagnosis; however, to distinguish streptococcal pharyngitis from other causes of exudative pharyngitis such as *C. diphtheriae*, *C. haemolyticum*, *N. gonorrhoea*, *M. pneumoniae*, *Yersinia enterocolitica*, and several species of viruses. Mononucleosis should be suspected in cases of exudative pharyngitis, particularly in adolescents and younger adults. Detection of specific streptococcal antibodies such as antistreptolysin O is not useful in the diagnosis of acute infections.

S. suis infection should be considered in cases of systemic illness where a history of exposure to pigs or raw pork products is obtained or suspected. Signs of meningitis or septicemia with characteristic findings of hearing loss and vestibular dysfunction are nonspecific but may lead to a higher index of suspicion for the infection. The bacteria can be cultured from both blood and cerebrospinal fluid.

Treatment

Treatment of streptococcal infections requires antibiotic therapy. Penicillin is the antibiotic of choice, macrolides like erythromycin and β-lactam antibiotics as well as clindamycin are also effective; addition of the latter may be advisable in cases of streptococcal toxic shock or toxin-related illness. Steptococci are usually resistant to sulfonamides, flouroquinolones, and tetracyclines. Pharyngitis should be treated for 10 days to prevent post-streptococcal complications, unless a single dose of a long-acting penicillin is used intramuscularly. Most strains of *S. suis* are penicillin-sensitive, though consideration should be given to addition of a second antibiotic until culture and sensitivity results are obtained. Aggressive treatment and surgical debridement is essential in necrotizing fasciitis to reduce the mortality from this condition. Additional supportive measures, including fluid resuscitation and mechanical ventilation, may be necessary for serious infections and streptococcal toxic shock.

Medical surveillance

There are no recommended medical screening activities for streptococcal diseases. Community or school screening programs for identifying group A streptococcal carriers have not been shown to be effective. Epidemics must be reported to local health authorities in the United States.

Prevention

Prevention of streptococcal disease requires good hygiene practices. Food handlers should use strict personal hygiene, and food handlers with active streptococcal respiratory infections should be considered for job reassignment. Food itself should be appropriately cooked, processed, and refrigerated. Meat-packers and meat-handlers should be provided with appropriate tools and personal protective equipment to minimize the risk of exposure and skin trauma, such as cut-resistant gloves. Equipment should also be cleaned regularly. Antibiotic prophylaxis of asymptomatic swine herds and changes in breeding conditions have not been well studied.[9]

Pneumococcal vaccine (the polyvalent vaccine for *S. pneumoniae*) is indicated for those at risk of pneumococcal disease on the basis of underlying health status and age, not occupational exposure.[14]

References

1. Hawky PM, Pedler SJ, and Southall PJ. *Streptococcus pyogenes*: a forgotten occupational hazard in the mortuary. *Br Med J* 1980; 281:1058.
2. Patterson WB, Craven DE, Schwartz DA, et al. Occupational hazards to hospital personnel. *Ann Intern Med* 1985; 102:658–80.
3. Fehrs LJ, Flanagan K, Kline S, et al. Group A beta-hemolytic streptococcal skin infections in a US meat-packing plant. *JAMA* 1987; 258:3131–4.
4. Phillips G, Efstratiou A, Tanna A, et al. An outbreak of skin sepsis in abattoir workers caused by an "unusual" strain of Streptococcus pyogenes. *J Med Microbiol* 2000; 49:371–4.
5. Barnham M, Kerby J, Skillin J. An outbreak of streptococcal infection in a chicken factory. *J Hyg* 1980; 84:71–5.
6. Barnham M. Pig bite injuries and infection: report of seven human cases. *Epidemiol Infect* 1988; 101:641–5.
7. Coggon D, Inskip H, Winter P, et al. Lobar pneumonia: an occupational disease in welders. *Lancet* 1994; 344:41–3.
8. Zanen HC and Engel HW. Porcine streptococci causing meningitis and septicaemia in man. *Lancet* 1975; 1(7919):1286–8.
9. Dupas D, Vignon M, and Geraut C. *Streptococcus suis* meningitis: a severe noncompensated occupational disease. *J Occup Med* 1992; 34:1102–5.
10. Huong VT, Ha N, Huy NT, et al. Epidemiology, clinical manifestations, and outcomes of *Streptococcus suis* infection in humans. *Emerg Infect Dis* 2014; 20:1105–14.
11. Bison AL and Stevens DL. Streptococcal infections of skin and soft tissue. *N Engl J Med* 1996; 334:240–5.
12. Green RJ, Dafoe DC, and Raffin TA. Necrotizing fasciitis. *Chest* 1996; 110:219–29.
13. Tambyah PA, Kumarasinghe G, Chan HL, et al. *Streptococcus suis* infection complicated by purpura fulminans and rhabdomyolysis: case report and review. *Clin Infect Dis* 1997; 24:710–2.
14. Centers for Disease Control. Update on adult immunization: recommendations of the Immunization Practices Advisory Committee (ACIP). *MMWR* 1991; 40(RR-12):43–4.

TREPONEMA PALLIDUM

Common name for disease: Syphilis

Occupational setting

Treponema pallidum is the causative agent of syphilis, a sexually transmitted disease. The incidence of primary and secondary syphilis is increasing in the United States. According to CDC, reported cases increased by 10% between 2012 and 2013.[1] Occupational groups at increased risk of syphilis are those who have direct, primarily genital, contact with infectious lesions, such as sex workers.

In the early part of the century, infection was reported among some laboratory workers who were handling animals infected with strains of *T. pallidum*. Transmission may have occurred through scratches, bites, or self-inoculation.[2]

Exposure (Route)

Transmission of the organism among adults takes place through direct contact of mucous membrane or skin with an infectious lesion and almost always involves sexual contact. Infectious lesions include chancres, skin rashes, mucous patches, or condylomata lata. Syphilis can also be transmitted vertically, resulting in congenital syphilis, and by contact with infected human blood through transfusion or autoinoculation.

Pathobiology

After an incubation period of 9–90 days (median of 21 days), a primary lesion develops at the site of infection. The chancre of primary syphilis is typically painless and indurated with a well-defined border. It usually heals within 2 months. Secondary syphilis, a systemic disease, typically occurs 6–8 weeks after the primary lesion is healed. A wide variety of clinical manifestations may be present, including involvement of the skin, mucous membranes, and lymphatic, renal, gastrointestinal, and skeletal systems. The most characteristic lesion at this stage is the punctate or pox-like rash, which appears most often on the palms and soles.[3] Late or tertiary syphilis occurs years after primary infection. Consequences of syphilis at this stage include neurosyphilis, with numerous clinical manifestations including ataxia and ocular lesions, cardiovascular syphilis, which may be seen as aortitis and valvular disease, and a "benign" form in which the characteristic syphilitic gumma may be found in organs including liver and bone systems.

Diagnosis

As *T. pallidum* is extremely difficult to grow in culture, diagnosis is made either by direct examination or serologic testing. Identification of the organism is made by dark-field microscopic examination of material (scrapings or exudate) from an active lesion or lymph node. Serologic tests fall into two categories. Those that detect antigenic indicators of host tissue damage (the Venereal Disease Research Laboratory (VDRL), rapid plasma reagin or RPR tests) are nonspecific but inexpensive and can be used for initial testing of individuals suspected of having disease or for population screening. Specific treponemal antigen testing is used to confirm the diagnosis in individual patients. These tests include microhemagglutinin assays for *T. pallidum* antibody (MHA-TP) and the fluorescent treponemal antibody absorption (FTA-ABS) test.[3,4]

Treatment

All stages of syphilis can be treated with one 2.4-million-unit dose of intramuscular benzathine penicillin.[5] It is notable however that there are no clear guidelines about duration of treatment and so RPR or VDRL titers must be followed to ensure they are decreasing. Penicillin desensitization is recommended for those that are allergic to penicillin. Those with suspected neurosyphilis should undergo a lumbar puncture to assess *T. pallidum* infection and titer, and will need an extended course of intravenous penicillin. As with treatment of other spirochetes, there is a risk of Jarisch–Herxheimer reaction during therapy.

Medical surveillance

There are no recommended occupational screening activities for syphilis. Syphilis is a reportable disease in the United States and many other countries.

Prevention

Infection with *T. pallidum* is prevented by treatment of infected individuals and the use of condoms and antiseptic prophylactic agents. In addition, healthcare workers should use appropriate personal protective equipment in cases where the organism is being handled when examining patients in whom clinical syphilis is suspected.

References

1. Centers for Disease Control and Prevention, Division of STD Prevention. Sexually Transmitted Disease Surveillance 2013. Available at: http://www.cdc.gov/std/stats13/surv2013-print.pdf (accessed on June 6, 2016).
2. Collins CH and Kennedy DA. Microbiological hazards of occupational needlestick and "sharps" injuries. *J Appl Bacteriol* 1987; 62:385–402.
3. Brown TJ, Yen-Moore A, and Tyring SK. An overview of sexually transmitted diseases. *J Am Acad Dermatol* 1999; 41:511–32.
4. Larsen SA, Steiner BM, and Rudolph AH. Laboratory diagnosis and interpretation of tests for syphilis. *Clin Microbiol Rev* 1995; 8:1–21.
5. Clement ME, Okeke NL, and Hicks CB. Treatment of syphilis: a systematic review. *JAMA* 2014; 312:1905–17.

VIBRIO CHOLERAE

Common name for disease: Cholera

Occupational setting

Persons at risk of cholera include travelers to areas where cholera is endemic[1] and those in occupations where exposure to contaminated seawater or food is possible.[2] Those who handle or consume undercooked shellfish may also be at risk.[3] Healthy commercial divers have been found to become colonized following dives at contaminated sites.[4] Outbreaks have occurred in hospitals and affected laboratory workers.[5,6] In 2012, six states reported cases of cholera serogroup O1 to CDC; most were international travelers returning from endemic areas, while one was laboratory-acquired.[7]

Exposure (route)

Exposure is through ingestion of food or water containing live organisms, or through direct contact with water bearing the organisms.

Pathobiology

Vibrios are curved, flagellated, Gram-negative rods. The organism is found in surface waters (both fresh and salt water) all over the world. *V. cholerae* is a diverse species, consisting of numerous different strains. Strains are primarily grouped according to the type of cell wall O antigen present. Most cases of epidemic cholera are due to serogroup O1. Within this serogroup, there are also subdivisions into three serotypes: Ogawa, Inaba, and Hikojima, as well as two biotypes: classical and El Tor. However, not all O1 strains are pathogenic. While some members of serogroups O2–O138 have the potential to cause isolated cases of cholera,[8] they are not thought to result in epidemics. *V. cholerae* O139 (Bengal strain) has been isolated from a large proportion of cases in the recent cholera epidemic in Asia.[9] New serotypes are likely to be implicated in future epidemics.[10]

When ingested with water or food, *V. cholerae* passes through the stomach into the small bowel, where it adheres to the mucosal lining. There, it secretes an enterotoxin that acts on intestinal cell receptors, mediated by cyclic adenosine monophosphate, to cause the active secretion of sodium chloride into the gut lumen. This, in turn, causes a voluminous watery diarrhea, which may occur at a rate of 1 L/h.

The incubation period after exposure ranges from 6 hours to 5 days. Prodromal symptoms consist of abdominal discomfort and anorexia. This is followed by diarrhea that progresses from brown to a "rice water" appearance (due to mucus secretion). The complications of cholera all derive from the rapid loss of fluids and electrolytes and include hypotension, hypoglycemia, electrolyte imbalance, renal failure, and acidosis. Rarely, an ileus may occur. Death is most often due to the loss of glucose, electrolytes, and fluid. It should be noted that in most studies of institutional outbreaks, over 75% of those infected are asymptomatic.[5]

Diagnosis

Diagnosis of cholera can be made by dark-field examination of stool or by culture. A history of travel to areas with endemic cholera is also helpful.

Treatment

The cornerstone of cholera treatment is fluid and electrolyte replacement, either orally or by intravenous rehydration. With adequate rehydration, the case fatality rate is low, 2% in a case series from the United States.[10] Oral rehydration solutions can be made from salt and sugar (5 g sodium chloride and either 20 g glucose or 40 g sucrose per liter of water) or according to the World Health Organization formulation (3.5 g sodium chloride, 2.5 g sodium bicarbonate, 1.5 g potassium chloride, and 20 g glucose per liter of water). Since this formulation does not decrease (and may even increase) the duration and volume of diarrhea, formulations that add starch from a variety of sources have been advocated.[11] Most cases can be successfully treated with oral rehydration alone; however, more severe diarrhea may require intravenous rehydration. Antibiotics are useful in shortening the duration of infection in moderate to severe cases. Tetracycline, doxycyline, erythromycin, and trimethoprim–sulfamethoxazole have all been used successfully. However, multiple antibiotic resistance has emerged among strains of *V. cholerae*. Currently, the quinolones, such as ciprofloxacin and norfloxacin, have shown excellent results and offer the advantages of efficacy with a short course or even a single dose.[12]

Medical surveillance

Reporting of cholera cases and epidemics to local health authorities is mandated in virtually all countries. There are no recommended medical screening activities for this disease.

Prevention

Prevention of cholera requires avoiding contaminated food and water. The absence of municipal water chlorination has been identified as a major contributor to the reemergence of cholera in South America.[13]

An inactivated bacteria cholera vaccine is available, but it is not normally recommended. The vaccine is about 50% effective in preventing illness for 3–6 months, does not prevent transmission of infection, and may not offer cross-protection against diarrhea caused by *V. cholerae* O139.[14] Newer oral live attenuated vaccines have been developed which appear to offer protection for up to 2 years.[15]

Persons with cholera infections should not return to work until clearance of the organism is complete as documented by negative stool cultures. In outbreaks, asymptomatic carriers should be identified and treated to eradicate the organism. Healthcare workers caring for patients with cholera should apply appropriate enteric precautions.

References

1. Cooper G, Hadler JL, Barth S, et al. Cholera associated with international travel, 1992. *MMWR* 1992; 41:664–7.
2. Hunt MD, Woodward WE, Keswick BH, et al. Seroepidemiology of cholera in Gulf coastal Texas. *Appl Environ Microbiol* 1988; 54(7):1673–7.
3. Weber JT, Mintz ED, Canizares R, et al. Epidemic cholera in Ecuador: multidrug-resistance and transmission by water and seafood. *Epidemiol Infect* 1994; 112(1):1–11.
4. Huq A, Hasan JA, Losonsky G, et al. Colonization of professional divers by toxigenic *Vibrio cholerae* O1 and *V. cholerae* non-O1 at dive sites in the United States, Ukraine and Russia. *FEMS Microbiol Lett* 1994; 120(1–2):137–42.
5. Goh KT, Teo SH, Lam S, et al. Person-to-person transmission of cholera in a psychiatric hospital. *J Infect* 1990; 20(3):193–200.
6. Huhulescu S, Leitner E, Feierl G, et al. Laboratory-acquired *Vibrio cholerae* O1 infection in Austria, 2008. *Clin Microbiol Infect* 2010; 16:1303–4.
7. Newton A, Kendall M, Vugia DJ, et al. Increasing rates of vibriosis in the United States, 1996–2010: review of surveillance data from 2 systems. *Clin Infect Dis* 2012; 54 Suppl 5:S391–5.
8. Dalsgaard A, Forslund A, Bodhidatta L, et al. A high proportion of *Vibrio cholerae* strains isolated from children with diarrhoea in Bangkok, Thailand are multiple antibiotic resistant and belong to heterogenous non-O1, non-O139 serotypes. *Epidemiol Infect* 1999; 122(2):217–26.
9. Bhattacharya SK, Bhattacharya MK, Nair GB, et al. Clinical profile of acute diarrhoea cases infected with the new epidemic strain of *Vibrio cholerae* O139: designation of the disease as cholera. *J Infect* 1993; 27:11–15.
10. Weber JT, Levine WC, Hopkins DP, et al. Cholera in the United States, 1965–1991. Risks at home and abroad. *Arch Intern Med* 1994; 14:551–6.
11. Rabbani GH. The search for a better oral rehydration solution for cholera. *N Engl J Med* 2000; 342(5):345–7.
12. Usubutun S, Agalar C, Diri C, et al. Single dose ciprofloxacin in cholera. *Eur J Emerg Med* 1997; 4(3):145–9.
13. Ries AA, Vugia DJ, Beingolea L, et al. Cholera in Piura, Peru: a modern urban epidemic. *J Infect Dis* 1992; 166(6):1429–33.
14. Albert MJ, Alam K, Ansaruzzaman M, et al. Lack of cross-protection against diarrhea due to *Vibrio cholerae* O139 (Bengal strain) after oral immunization of rabbits with *V. cholerae* O1 vaccine strain CVD103-HgR. *J Infect Dis* 1994; 169:230–1.
15. Graves P, Deeks J, Demicheli V, et al. Vaccines for preventing cholera. *Cochrane Database Syst Rev* 2000; 2:CD000974.

VIBRIO SPECIES OTHER THAN *V. CHOLERAE* (*V. PARAHEMOLYTICUS*, *V. VULNIFICUS*)

Common name for disease: None

Occupational setting

Individuals at risk for non-cholera *Vibrio* infection are those in close contact with both aquatic environments such as commercial divers,[1] fishermen,[2] fish farmers,[3] and workers who handle seafood or shellfish.[4] Outbreaks have also occurred aboard cruise ships.[5] Bacteria in this group have been described as "occupational pathogens," since asymptomatic carriage rates approaching 4% have been documented in high-risk worker groups, which result in sporadic outbreaks.[6] In some coastal areas of the United States, non-cholera *Vibrio* infections are increasing, possibly due to warming ocean temperatures.[7]

Exposure (route)

Exposure is through ingestion or direct contact with marine organisms, or with water containing live organisms.

Pathobiology

Vibrios are curved, flagellated, Gram-negative rods. These organisms are found in surface waters (fresh and salt water) around the world. A number of *Vibrio* species have been identified which may cause disease in humans. Most, like *V. cholerae*, cause toxic gastrointestinal disease. Several species, however, also have varying degrees of predilection for causing soft tissue infections, usually at the site of pre-existing skin breaks or wounds. Sepsis and death are well-known complications of either presentation, but typically only in those with chronic disease.[8] In cases complicated by sepsis, there may be lower extremity edema and bullae.[9]

The most important species to consider are *V. parahaemolyticus* and *V. vulnificus*. *V. parahaemolyticus* is a common cause of diarrheal disease throughout the world. The most frequent route of exposure is consumption of raw shellfish, usually oysters.[9] Four such outbreaks have occurred in the United States since 1997.[9] Other Vibrio species primarily associated with gastroenteritis include *V. mimicus*,[10] *V. hollisae*,[11] and *V. fluvialis*.[12]

V. vulnificus causes local wound infections that may follow an aggressive course with rapid spread and necrosis of surrounding tissue.[13] Other conditions associated with this organism include corneal ulcers[4] and a fulminant systemic illness characterized by a hemorrhagic rash, fever, gastroenteritis, and hypotension.[14] Other vibrios that cause local wound infections are *V. alginolyticus*,[15] *V. damsela*,[16] and *V. metschnikovii*.[17] The infections occur frequently in pre-existing wounds or skin breaks, or they develop as an acute otitis.[1] The most common route of exposure is direct contact with open seawater.[13]

Diagnosis

All *Vibrio* infections are diagnosed by culturing the organism from stool, blood, or wound samples. A history of exposure to seawater or raw shellfish, or ingestion of raw or under-cooked shellfish, is helpful in the diagnosis. Development of a severe cellulitis of the extremities after exposure to seawater, especially if the cellulitis does not respond to aminoglycosides, should raise the suspicion of infection with *Vibrio* species.

Treatment

Most of the gastrointestinal diseases caused by non-cholera vibrios are self-limited and do not require specific therapy other than rehydration.

Local wound infections generally respond to antibiotics such as tetracycline, with chloramphenicol or penicillin as a second choice. Surgical debridement is frequently required when soft-tissue necrosis is present.[13] Systemic infection requires parenteral antibiotic therapy.

Medical surveillance

There are no recommended medical screening or surveillance activities for non-cholera *Vibrio* infections.

Prevention

There is no effective vaccine for any of the non-cholera *Vibrio* infections. Prevention of foodborne disease consists primarily of proper handling, cooking, and refrigeration. Immunocompromised individuals, particularly those with chronic liver disease, should avoid ingestion of uncooked shellfish.[13] The concentration of *Vibrio* species in seawater increases with increasing water temperature, increasing the risk of contamination for seafood harvested during the summer.[18]

Individuals with pre-existing wounds should take precautions to avoid exposure to sea water or other potentially contaminated material.

References

1. Tsakris A, Psifidis A, and Douboyas J. Complicated suppurative otitis media in a Greek diver due to a marine halophilic Vibrio sp. *J Laryngol Otol* 1995; 109(11):1082–4.
2. Hoi L, Dalsgaard A, Larsen JL, et al. Comparison of ribotyping and randomly amplified polymorphic DNA PCR for characterization of *Vibrio vulnificus*. *Appl Environ Microbiol* 1997; 63(5):1674–8.
3. Bisharat N, Agmon V, Finkelstein R, et al. Clinical, epidemiological, and microbiological features of *Vibrio vulnificus* biogroup 3 causing outbreaks of wound infection and bacteraemia in Israel. Israel Vibrio Study Group. *Lancet* 1999; 354(9188): 1421–4.
4. Massey EL and Weston BC. *Vibrio vulnificus* corneal ulcer: rapid resolution of a virulent pathogen. *Cornea* 2000; 19(1): 108–9.
5. Centers for Disease Control. Gastroenteritis caused by *Vibrio parahaemolyticus* aboard a cruise ship. *MMWR* 1978; 27:65–6.

6. Morris JG Jr. Non-O group 1 *Vibrio cholerae*: a look at the epidemiology of an occasional pathogen. *Epidemiol Rev* 1990; 12:179–91.
7. Jones EH, Feldman KA, Palmer A, et al. Vibrio infections and surveillance in Maryland, 2002–2008. *Public Health Rep* 2013; 128:537–45.
8. Klontz KC. Fatalities associated with *Vibrio parahaemolyticus* and *Vibrio cholerae* non-O1 infections in Florida (1981 to 1988). *South Med J* 1990; 83(5):500–2.
9. Centers for Disease Control. Outbreak of *Vibrio parahaemolyticus* infection associated with eating raw oysters and clams harvested from Long Island Sound—Connecticut, New Jersey, and New York, 1998. *MMWR* 1999; 48(3):48–51.
10. Campos E, Bolanos H, Acuna MT, et al. *Vibrio mimicus* diarrhea following ingestion of raw turtle eggs. *Appl Environ Microbiol* 1996; 62(4):1141–4.
11. Carnahan AM, Harding J, Watsky D, et al. Identification of *Vibrio hollisae* associated with severe gastroenteritis after consumption of raw oysters. *J Clin Microbiol* 1994; 32(7): 1805–6.
12. Klontz KC, Cover DE, Hyman FN, et al. Fatal gastroenteritis due to *Vibrio fluvialis* and nonfatal bacteremia due to *Vibrio mimicus*: unusual vibrio infections in two patients. *Clin Infect Dis* 1994; 19(3):541–2.
13. Howard RJ and Bennett NT. Infections caused by halophilic marine Vibrio bacteria. *Ann Surg* 1993; 217(5):525–30.
14. Serrano-Jaen L and Vega-Lopez F. Fulminating septicaemia caused by *Vibrio vulnificus*. *Br J Dermatol* 2000; 142(2): 386–7.
15. Mukherji A, Schroeder S, Deyling C, et al. An unusual source of *Vibrio alginolyticus*-associated otitis: prolonged colonization or freshwater exposure? *Arch Otolaryngol Head Neck Surg* 2000; 126(6):790–1.
16. Tang WM and Wong JW. Necrotizing fasciitis caused by *Vibrio damsela*. *Orthopedics* 1999; 22:443–4.
17. Hansen W, Freney J, Benyagoub H, et al. Severe human infections caused by *Vibrio metschnikovii*. *J Clin Microbiol* 1993; 31(9):2529–30.
18. Shapiro RL, Altekruse S, Hutwagner L, et al. The role of Gulf Coast oysters harvested in warmer months in *Vibrio vulnificus* infections in the United States, 1988–1996. Vibrio Working Group. *J Infect Dis* 1998; 178(3):752–9.

YERSINIA PESTIS

Common names for disease: Plague

Occupational setting

Human plague arises from infection with *Yersinia pestis*, a bacterium that is maintained in a natural reservoir of small rodents and the fleas that infest them. Since 1925, plague in the United States has been associated with exposures to wild rodents only in the southwestern states, primarily around the Four Corners region where New Mexico, Arizona, Utah, and Colorado join, and in California.[1] The disease is also endemic in many areas of South America, Africa, and Southeast Asia. Historically, throughout the world, urban and domestic rats have been the most important reservoirs for epidemic plague. As control measures have reduced the proximity of rats to humans in urban areas, foci of the sporadic illness have shifted to rural areas where burrowing mammals make their habitat. The oriental rat flea (*Xenopsylla cheopis*) is the most important vector. In the United States, plague is maintained in well-established enzootic foci among wild rodents, including rock squirrels, the California ground squirrel, prairie dogs, chipmunks, and woodrats. Although they are not part of the enzootic cycle, are only incidentally infected, and rarely develop overt illness, rabbits and hares, deer, antelope, gray fox, badger, bobcat, and coyote have been occasionally associated with human plague in hunters and trappers. Plague associated with rock squirrels is the most important cause of human disease in North America, as housing developments have been introduced into habitats where plague was enzootic. There were 13 cases of human plague in four states in the United States in 2006, the highest number in recent years (Figure 22.10).[2] Hunters, trappers, foresters, rangers, and others working in remote locations where contact with rodent habitats might be expected are at risk for contracting plague, as are veterinarians in enzootic areas. In 2009, a wildlife biologist in the Grand Canyon died of plague after conducting a necropsy. Military personnel and individuals who might be stationed in areas where rats or other animal reservoirs are present would also be at increased risk.

Exposure (route)

Plague is transmitted from infected animals to humans by several species of rodent fleas, including in the United States the ground squirrel fleas *Diamana montana* and *Thrassis bacchi*. Domestic cats can also transmit the infection after consuming plague-infected animals, or after bites from infected rodent fleas, to humans by bites or scratches.[3] Human-to-human transmission can occur in pneumonic plague, and such transmission constitutes a public health emergency. Human-to-human pneumonic plague transmission has not been reported in the United States since a 1925 epidemic in Los Angeles. Human pneumonic plague can be acquired from domestic cats with secondary pneumonic plague; this mode of transmission has been associated with disease in veterinarians. *Y. pestis* can also be acquired through infectious fluids or tissues entering through cuts or abrasions in the skin, a mode of transmission most commonly seen in hunters and trappers who skin infected animals.

Pathobiology

Y. pestis is a Gram-negative, nonmotile coccobacillus. Human infection is associated with four common clinical presentations: bubonic plague, septicemic plague, pneumonic plague, and meningitis. From 2 to 6 days after the bite of an infected

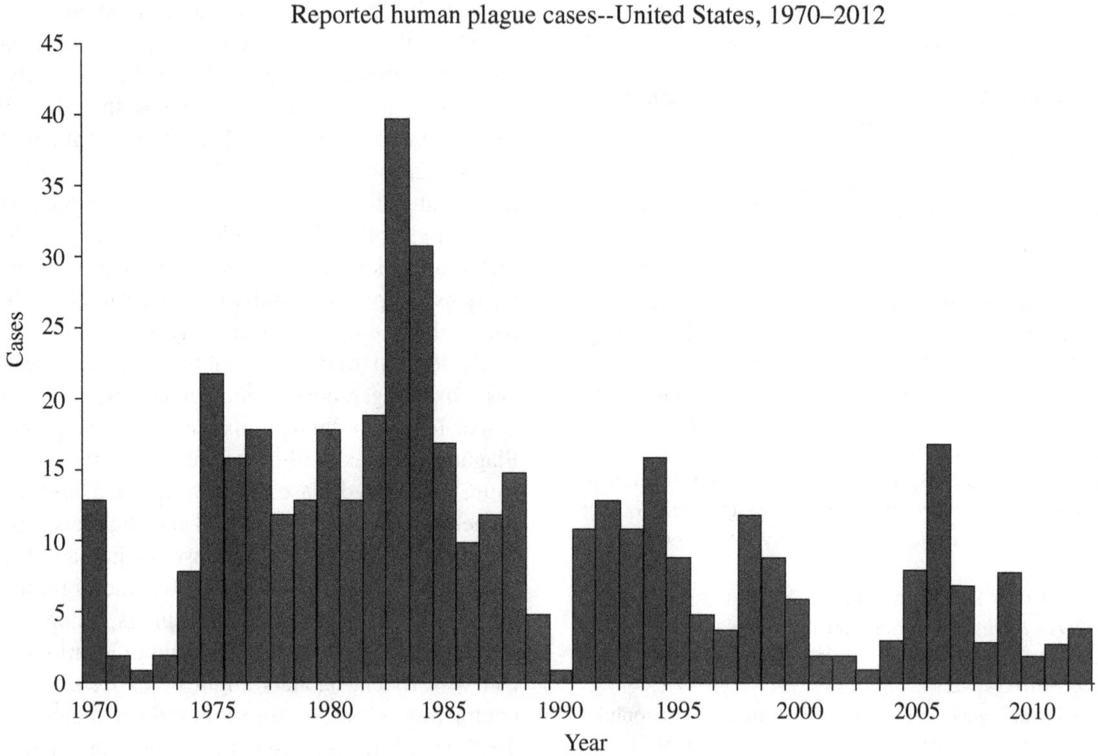

FIGURE 22.10 Plague – reported cases among humans, United States, 1970–2012. *Source*: http://www.cdc.gov/plague/maps.

flea, the patient develops a febrile illness, followed by the development of very painful suppurating lymphadenopathy (buboes) proximal to the bite. These are seen most commonly in the groin, axillae, or cervical region.[1] Septicemic plague is usually secondary to untreated bubonic plague and is usually rapidly fatal without treatment. Hematogenous dissemination can affect any organ system, but most commonly involves lungs, eyes, meninges, joints, and skin. Occlusion of small cutaneous blood vessels can result in necrosis and gangrene of the fingers and skin (which gave rise to the term "Black Death" in the fourteenth century). Meningitis is a rarer complication that follows inadequately treated bubonic plague and is more commonly associated with axillary buboes. Case fatality rates for untreated bubonic and septicemic plague exceed 50% in most reports.[4]

Pneumonic plague can be secondary to septicemia or arise as a primary infection after the inhalation of infectious droplet nuclei. Primary pneumonic plague has a short incubation period and can spread rapidly in close contacts; its presence is therefore a public health emergency, as the infected case will generate infective aerosols. In the United States, the few reported human cases of primary pneumonic plague have been acquired from domestic cats that developed secondary pneumonic infection. Survival is unlikely unless treatment is initiated within 18 hours after onset of respiratory symptoms.[4]

Diagnosis

Infectious clinical material can be examined with light microscopy after appropriate staining. Buboes and blood can also provide specimens for culture. Large quantities of bacteria in lesions and blood typically make bacteriologic diagnosis relatively easy when the etiology is suspected. However, the use of automated systems for laboratory detection has resulted in diagnostic errors and delayed diagnosis.[5] In patients with symptoms associated with plague, identification of unusual organisms (e.g., *P. luteola*, *Y. pseudotuberculosis*, or *A. lwoffii*) in the blood using automated systems should trigger further clinical and laboratory evaluation.[5] If plague is suspected, treatment should be started and samples should be sent to public health laboratories for confirmation.[5] A rapid fluorescent antibody test is available at certain reference laboratories.

Treatment

Streptomycin has been the traditional treatment of choice, but due to its toxicity and limited availability, gentamicin has become a more accepted initial treatment. Tetracycline can be used if gentamicin is contraindicated. Chloramphenicol is the preferred treatment for meningitis and endophthalmitis.

Medical surveillance

There are no recommended human medical screening activities for plague. Suspected and confirmed cases of plague are required to be immediately reported to local health authorities worldwide. Monitoring of rodent populations for increased die-offs and evidence of epizootics should be performed in endemic areas.

Prevention

Discussion of the prevention of human plague can be grouped into three categories: (i) very high-risk populations such as laboratory workers or medically underserved endemic or epidemic areas; (ii) exposure to human plague cases or epizootic foci; and (iii) environmental management of hyperendemic residential or recreational foci.[1] A 2013 survey of wildlife workers' use of personal protective equipment, found there was significant room for improvement and that equipment was rarely used and seldom accessible to workers.[6] An inactivated plague vaccine is available in the United States, consisting of a primary immunization series of two injections and boosters every 6 months, and is recommended for laboratory workers with frequent exposure to *Y. pestis* and persons such as mammalogists, ecologists, and other field workers who have regular contact with wild rodents or their fleas in areas in which plague is enzootic or epizootic.[4]

Management of contacts of human plague cases and of exposures to epizootic plague are potential public health emergencies. Contacts should be carefully identified, and appropriate follow-up, including consideration of prophylactic antibiotic therapy, should be ensured. Available environmental control measures for management of epizootic plague include application of pesticides to control fleas; closure of recreational areas to humans; and judicious use of rodenticides. Control may also be exercised through management of the environment near the interface of human and rodent activities, such as appropriate handling of trash, and removal of rock piles, and dilapidated buildings that provide habitats for rodents. Catastrophic events, such as war or natural disasters, which disrupt normal sanitary activities, may lead to spread of plague from rural foci into urban centers; control measures should be reinstituted as rapidly as possible to prevent this scenario.

References

1. Craven RB and Barnes AM. Plague and tularemia. *Infect Dis Clin North Am* 1991; 5:165–75.
2. Centers for Disease Control and Prevention. Human plague—four states, 2006. *MMWR Morb Mortal Wkly Rep* 2006; 55:940–3.
3. Doll JM, Zeitz PS, Ettestad P, et al. Cat-transmitted fatal pneumonic plague in a person who traveled from Colorado to Arizona. *Am J Trop Med Hyg* 1994; 51:109–14.
4. Centers for Disease Control and Prevention. Prevention of plague: recommendations of the Advisory Committee on Immunization Practices (ACIP). MMWR 1996; 45(RR-14):1–15.
5. Tourdjman M, Ibraheem M, Brett M, et al. Misidentification of *Yersinia pestis* by automated systems, resulting in delayed diagnoses of human plague infections—Oregon and New Mexico, 2010–2011. *Clin Infect Dis* 2012; 55(7):e58–60.
6. Bosch SA, Musgrave K, and Wong D. Zoonotic disease risk and prevention practices among biologists and other wildlife workers—results from a national survey, US National Park Service, 2009. *J Wildl Dis* 2013; 49:475–585.

YERSINIA PSEUDOTUBERCULOSIS AND *ENTEROCOLITICA*

Common names for disease: None

Occupational setting

Yersinia enterocolitica and *Y. pseudotuberculosis* are widespread in natural settings and have been isolated from wild and domestic animals, foods, water, and soil. *Y. enterocolitica* is a frequent cause of diarrhea and gastroenteritis in European countries and North America. Swine and pigs are common asymptomatic carriers of the bacteria, with Yersiniae frequently isolated from pigs' tongues and tonsils, and abattoir workers have been found to have an elevated risk of infection.[1] Other animal reservoirs of the infection include rodents, rabbits, sheep, cattle, horses, dogs, and cats. Workers who come in contact with refrigerated meat after slaughter including butchers, meat-handlers, and meat-packers, are also at risk, as Yersiniae can propagate at low temperatures. *Y. pseudotuberculosis* is a zoonosis of the aforementioned animals and also several species of birds (turkeys, ducks, geese, pigeons). It is an uncommon disease, more frequent in children and during the winter months and is associated with exposure to animals or common-source outbreaks from contaminated food or water.

Exposure (route)

Transmission occurs via ingestion of contaminated food or water and by direct contact with infected animals, possibly by a fecal–oral route. Fecal–oral transmission between humans has not been documented.

Pathobiology

These Yersiniae are facultatively anaerobic Gram-negative bacilli that are motile when grown at 25°C. Asymptomatic infection with either *Yersinia* species is common. The prevalence of antibodies to Yersiniae is up to 10% in the general population and up to 40% in slaughterhouse workers. The incubation period for *Y. enterocolitica* enterocolitis ranges from 1 to 11 days.[2] Symptomatic infection most commonly is a diarrheal illness with fever and severe abdominal pain that can mimic acute appendicitis. Septicemia is less common, has an untreated case fatality ratio of 50%, and is usually associated with moderate to severe underlying medical problems such as diabetes or cancer. Interestingly, in one study, abattoir workers were reported to have almost a fourfold increased risk, and pig farmers a twofold increased risk of appendectomy compared to grain or berry farmers. The authors hypothesized that the severe abdominal pain associated with Yersinia infections in these workers could have accounted for the increased risk.[3] Postinfection complications include reactive polyarthritis, erythema nodosum, and eye inflammation (e.g., iridocyclitis).[1] Person-to-person transmission of *Y. enterocolitica* has also been reported as a cause of septicemia arising from blood transfusions.[4] Patients developed the abrupt onset of fever and hypotension within 50 minutes after transfusion had begun; four of six patients reported between 1989 and 1991 died from the infection. Transient bacteremia in donors with proliferation of the bacterium under cold storage conditions was considered responsible for the contamination of blood products. *Y. pseudotuberculosis* most commonly causes mesenteric adenitis in adults, mimicking acute appendicitis, which is often self-limiting.

Diagnosis

Culture of appropriate clinical specimens (usually stool samples) often yields Yersiniae. Cold enrichment increases the yield of cultures by selectively favoring the growth of Eurasian, although, because of asymptomatic colonization, care must be taken with culture results that become positive only after prolonged culture. Serologic tests using adsorption methods (to remove cross-reacting antibodies) are also useful in diagnosis. If transfusion-associated bacteremia is suspected, the residual blood in the bag should be examined by Wright–Giemsa or other hematologic stain, and cultured.[4]

Treatment

Enterocolitis and mesenteric adenitis are usually self-limited, and the need for antibiotic therapy in these conditions is unclear. Doxycycline and trimethoprim–sulfamethoxazole are effective in complicated gastrointestinal infection or focal extraintestinal infection.[2] There is resistance to quinolones.[5] Septicemia should be treated with a combination of doxycycline and an aminoglycoside. Laparotomy for suspected appendicitis should be avoided if *Yersinia* infection is a likely diagnosis.

Medical surveillance

There are no recommended medical screening activities. Reporting of cases is mandatory in many areas in the United States and in many other countries.

Prevention

Prevention should focus on the animal reservoirs of the infection. Institution of work practices to minimize contamination of meat, such as altered methods of slaughter of pigs and avoidance of prolonged refrigeration of meat before consumption, is advised. Careful handwashing and cleaning of surfaces after food preparation is essential to prevent bacterial spread to other foods. Personal protective equipment use may also afford some protection from infection, but its effectiveness has not been evaluated.

References

1. Merilahti-Palo R, Lahesmaa R, Granfors K, et al. Risk of Yersinia infection among butchers. *Scand J Infect Dis* 1991; 23:55–61.
2. Cover TL, Aber RC. *Yersinia enterocolitica. N Engl J Med* 1989; 321:16–22.
3. Seuri M. Risk of appendicectomy in occupations entailing contact with pigs. *Br Med J* 1991; 301:345–6.
4. Centers for Disease Control and Prevention. Epidemiologic Notes and Reports Update: *Yersinia enterocolitica* bacteremia and endotoxin shock associated with red blood cell transfusions—United States, 1991. *MMWR* 1991; 40:176–8.
5. Capilla S, Ruiz J, Goñi P, et al. Characterization of the molecular mechanisms of quinolone resistance in *Yersinia enterocolitica* O:3 clinical isolates. *J Antimicrob Chemother* 2004; 53:1068–71.

Chapter 23

MYCOBACTERIA

GREGG M. STAVE*

MYCOBACTERIUM TUBERCULOSIS (M. tb.)

Common names for disease: Tuberculosis, consumption

Occupational setting

Tuberculosis (TB) exposure may occur in healthcare facilities, including hospitals, dental clinics and nursing homes, and in clinical or research laboratories processing tuberculosis cultures or infected specimens. Exposure may occur in funeral homes and is a significant risk in drug treatment centers, in correctional institutions, and in facilities for the homeless, alcoholics, or persons with AIDS. Animal caretakers can contract tuberculosis from primates even though the animal may not appear ill. Maintenance and construction workers may be exposed while manipulating ventilation systems for patient care isolation rooms or for biologic safety cabinets in which infectious samples are handled.

Exposure (route)

Tuberculosis is contracted after inhalational exposure via droplet nuclei that are 1–5 μm in size and thus remain airborne for long periods of time. TB is not contracted by skin contact with surfaces such as hospital room furniture, equipment, or walls. Infection by gastrointestinal exposure is not a significant risk.

Pathobiology

There are more than 30 members of the genus *Mycobacterium*, many of which are saprophytes that cause no human disease. *Mycobacterium tuberculosis* is the organism that causes tuberculosis. The surface lipids of mycobacteria cause them to be resistant to decolorization by acid alcohol during staining procedures. This property gives rise to the name *acid-fast bacilli*. Mycobacteria will not grow in common culture media but require techniques and reagents found in specialized laboratories.

TB may be insidious in onset, causing symptoms that the affected individual may ignore. These nonspecific symptoms include anorexia, weight loss, low-grade fever, fatigue, and cough. Pulmonary and pleural TB is the most common acute manifestation, presenting with pleuritic chest pain, cough productive of bloody sputum, high fever, and profuse sweating.

Multidrug-resistant tuberculosis (MDR-TB) and extensively drug-resistant tuberculosis (XDR-TB, also referred to as extremely drug-resistant TB) have high case-fatality rates.

Tuberculosis can affect organs other than the lungs. Tuberculous infection in the genitourinary system can cause ureteral obstruction or irregular menses. Lymphatic infection can cause the swelling of the lymph nodes known as

*The author wishes to thank Linda M. Frazier for her contribution to an earlier version of this chapter.

Physical and Biological Hazards of the Workplace, Third Edition. Edited by Gregg M. Stave and Peter H. Wald.

scrofula. If vertebral bodies are affected, pain and compression fractures may occur. Meningeal TB is associated with abnormal behavior, headaches, or seizures. Tuberculous peritonitis causes abdominal pain and ascites. Tuberculosis can affect the pericardium, causing heart failure, and the larynx, causing persistent hoarseness. Adrenal involvement may cause Addison's disease, and tuberculus skin infiltration has been reported. If infection is overwhelming and disseminated (miliary tuberculosis), the presentation can mimic acute leukemia.[1,2]

Eliminating occupational exposure to tuberculosis has become crucial because large numbers of strains have become resistant to several of the commonly used antituberculous agents (MDR-TB, XDR-TB). Intermittent compliance in taking antituberculous drugs increases the risk that initially susceptible strains will develop resistance. After a century of decline, the incidence of tuberculosis in the United States began rising in the 1980s, especially in inner cities and certain states.[3] Although the incidence has declined, several states continue to have above-average rates of infection (Figure 23.1). The majority of cases occur in foreign-born persons (Figure 23.2). The resurgence of tuberculosis has been attributed to economic deprivation, homelessness, alcoholism, drug use, and the rising incidence of AIDS. Public health workers and clinicians subsequently increased their efforts to trace contacts and ensure completion of therapy among individuals with tuberculosis. After reaching a peak of >26 000 confirmed cases of tuberculosis reported per year in the United States, incidence rates fell by 2012 to under 10 000 cases per year. In 2014, there were 9 412 reported cases (Figure 23.2).[3,4]

Nosocomial spread of tuberculosis has been documented; some outbreaks have been associated with tuberculin skin test conversions in >50% of healthcare workers in a single year. These significant occupational exposures have been associated with failure to comply with basic personal protective practices, delayed diagnosis of infected patients, and inadequate hospital room ventilation.[5–7] Healthcare facilities have made improvements in the availability of isolation rooms for potentially infectious patients, in use of proper respiratory protective devices, and in training healthcare workers; even so, optimal practices are not always followed.[3,8–11] Along with the patient care setting, the risk for occupational infection is also present in the autopsy room and the clinical laboratory.[12–14]

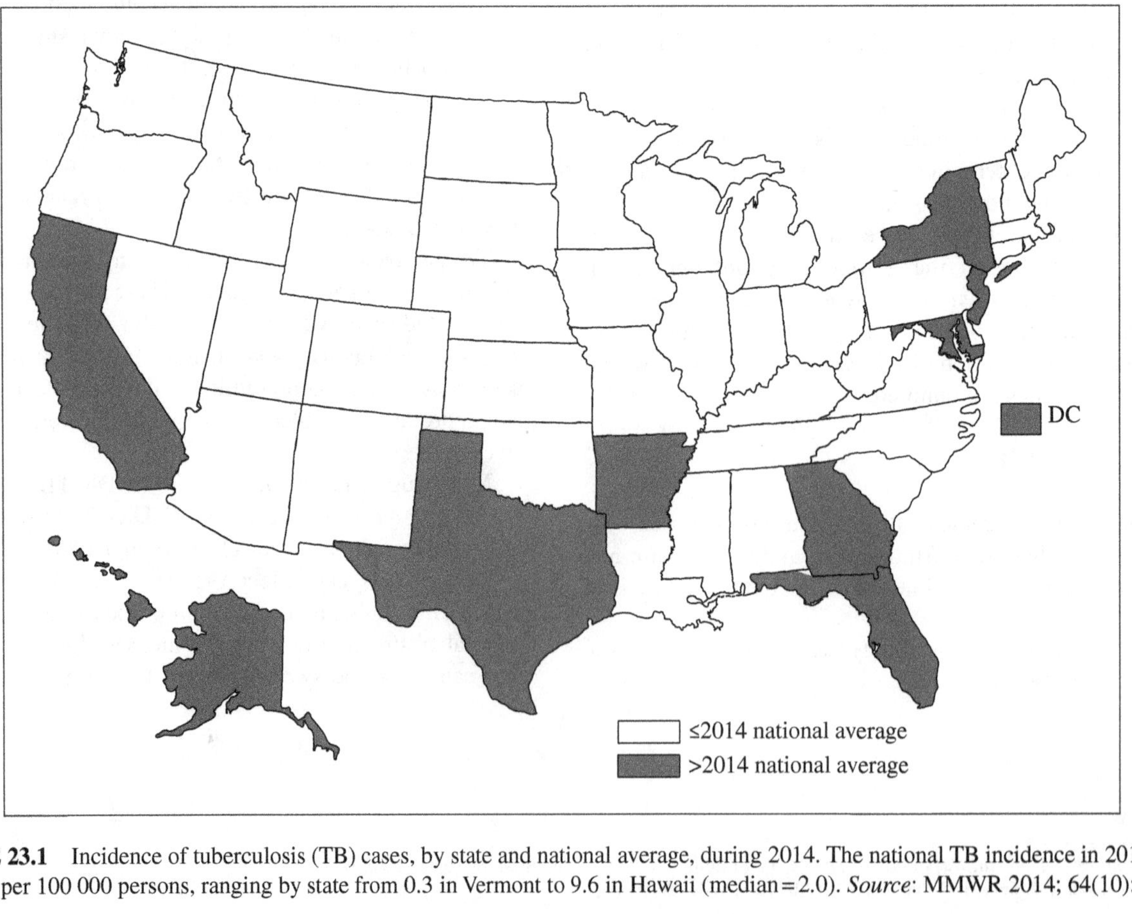

FIGURE 23.1 Incidence of tuberculosis (TB) cases, by state and national average, during 2014. The national TB incidence in 2014 was 3.0 cases per 100 000 persons, ranging by state from 0.3 in Vermont to 9.6 in Hawaii (median = 2.0). *Source*: MMWR 2014; 64(10):267.

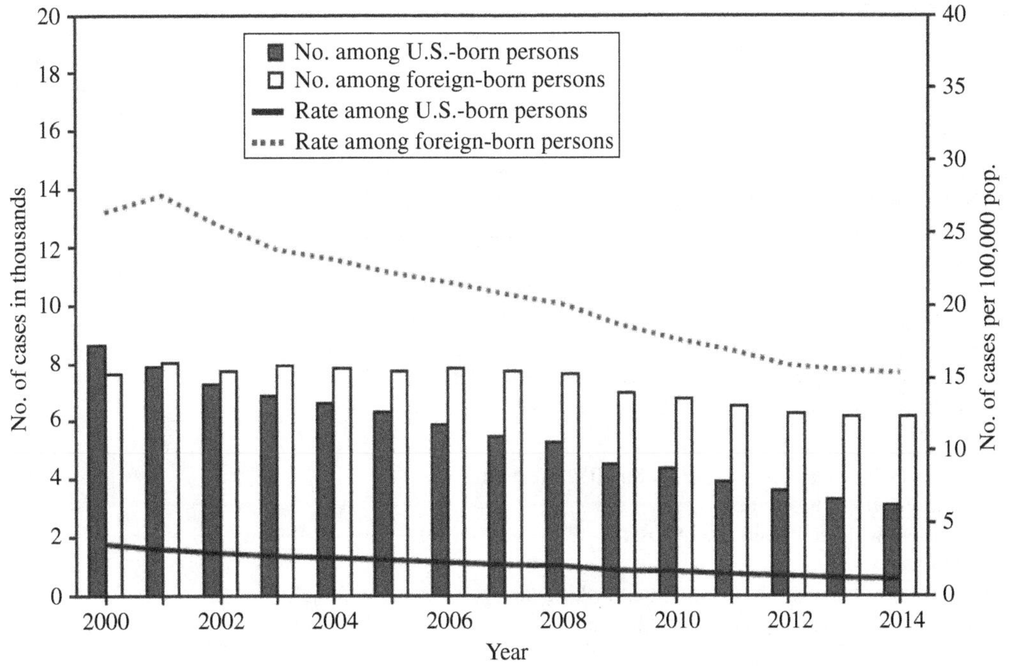

FIGURE 23.2 Number and rate per 100 000 of tuberculosis (TB) cases among U.S.-born and foreign-born persons, by year reported — United States, 2000–2014. *Source*: MMWR 2014; 64(10):267.

Diagnosis

Skin testing coupled with clinical evaluation serves to distinguish between infection with the tubercle bacillus and disease caused by the infection. When first exposure results in the entry of tubercle bacilli into the body, bacilli that are not immediately cleared can reside in lymph nodes and other tissues for long periods. A cell-mediated immune response develops that will cause a tuberculin skin test to register positive in ~2–10 weeks. Most individuals never develop disease manifestations. An estimated 5% of newly infected individuals develop clinical illness within 1 year. Later in life, particularly during advanced age, another 5% of infected people develop "reactivation" tuberculosis. In these cases, dormant tubercle bacilli begin multiplying and clinical TB develops, often in the lungs.[1–3]

When symptoms are present, organ system evaluations coupled with cultures for acid-fast bacilli are needed to confirm the diagnosis. Tuberculous lung infection typically causes lesions in the upper lung fields, such as granulomatous infiltrates. Radiographs may also show a diffuse miliary pattern or sterile pleural effusions. Urinary system tuberculosis can cause white blood cells in the urine with negative cultures for routine urinary pathogens. Skin tests can be used to help diagnose tuberculosis, but individuals with extensive TB or other immunosuppressing conditions may have false-negative skin tests.[1–3]

Sputum, urine, or other tissues can be evaluated rapidly by Ziehl-Nielsen (acid-fast) staining for tubercle bacilli. Biopsy of infected tissue may show caseating granulomas as well as acid-fast organisms. Acid-fast organisms can be detected and further classified by culture. A proficient and experienced microbiology laboratory should be used. Specimens must be prepared correctly and cultures may require 1 or more months to grow. New techniques, such as DNA polymerase reactions and immunoassays, are under development to allow more rapid identification of mycobacterium species and to determine drug sensitivity.[1,15] DNA fingerprinting of tuberculosis isolates using IS6110 RFLP and spoligotyping has been used to study how tuberculosis is transmitted through a population.[16] The technique was able to determine that contact with a patient who had tuberculosis was not the source of a healthcare worker's infection in many cases, and to confirm in other cases that unsuspected transmission in occupational, nosocomial, and school settings had occurred.

Active TB is a reportable condition. Local health departments can provide information about this procedure and will assist with tracing contacts that might also be infected. The Centers for Disease Control and Prevention (CDC) has published Guidelines for the Investigation of Contacts of Persons with Infectious Tuberculosis to assist with these investigations.[17]

Screening for latent tuberculosis infection

CDC guidelines and the Joint Commission require screening of healthcare workers and those at risk of exposure in healthcare facilities. Two approaches are available for screening: tuberculosis skin testing (TST) and blood testing.

TABLE 23.1 Criteria for positive Mantoux test.

Clinical situation	Minimum size of positive Mantoux test
No sign of illness, no risk factors for tuberculosis, no contact with tuberculosis	15 mm
Employees in facilities where a person with disease would pose a hazard to large numbers of susceptible persons (i.e., healthcare facilities, schools, childcare facilities, certain food services) Individuals with potential occupational exposure to tuberculosis including Mycobacteriology laboratory personnel Employees with silicosis People with other nonoccupational risk factors for tuberculosis (foreign-born persons from endemic area; medically underserved low-income populations; residents of long-term care facilities; persons with certain medical conditions, such as gastrectomy, severe underweight, renal failure, diabetes mellitus, immunosuppressive diseases other than HIV infection, corticosteroid, or other immunosuppressive therapy, etc.)	10 mm
HIV infection or unknown HIV status with risk factors for HIV infection Close recent contact with infectious tuberculosis case Chest radiography consistent with old healed tuberculosis Patients with organ transplants Persons who are immunosuppressed for other reasons (e.g., taking the equivalent of >15 mg/day of prednisone for 1 month or longer, taking TNF-a antagonists)	≥5 mm

Skin testing by the Mantoux method using purified protein derivative (PPD) is preferred over other skin test methods because of its greater sensitivity and reproducibility. Contact with nonpathogenic mycobacteria can cause small cross reactions to tuberculin skin tests, so minimum skin induration diameters have been established to define a tuberculin skin test as positive (Table 23.1). Using the Mantoux method, 5 tuberculin units (5 TU) are injected intradermally by a syringe according to the manufacturer's instructions. The 1 TU and the 250 TU products have limited clinical applications. Improper administration, such as failure to administer the skin injection soon after drawing the solution from the vial, can lead to false-negative results. Other causes of false-negative PPD results are immunosuppression, including AIDS, hematogenous malignancy, concurrent serious infection, recent live virus vaccination (measles, mumps, polio), and improper storage or denaturation of the reagent. One can differentiate between general anergy and a truly negative PPD by simultaneously skin testing for reaction to other common exposures, such as *Candida albicans*.[1,2] Diagnostic methods and other information about treatment and prevention of tuberculosis can be found on the CDC website (www.cdc.gov/tb).

Prior vaccination with bacillus Calmette–Guérin (BCG) does not completely protect against TB infection and is not a likely cause of a positive skin test. Among individuals who have had infections with the tubercle bacillus, the PPD skin response may wane after age 55. To assess for this phenomenon, PPD boosting can be performed by administering two 5 TU tests 1–2 weeks apart. Immunocompetent individuals who carry the tubercle bacillus but who have experienced a wane in PPD response will react to the second PPD. This is known as "the booster phenomenon." Repeated skin testing will not cause a positive PPD among individuals who are uninfected.[1,4] Newly employed healthcare workers who will be screened with PPD testing should receive a baseline two-step TST upon hire, unless they have documentation of either a positive TST result or treatment for latent TB infection or TB disease.[18] This requires a four-visit process, unless the worker screens positive after the first test.

Since screening using PPD requires at least two visits to place and read the result, as well as the availability of a trained reader, this has led to a rapid adoption of use of interferon-gamma release assays (IGRAs) for screening for infection. These blood tests use a mixture of synthetic peptides representing TB antigens that cause immune cells to release interferon-gamma when someone is infected with *M. tuberculosis*. IGRAs are unlikely to produce false-positive results for persons who have been vaccinated with BCG. Two tests are available: QuantiFERON®-TB Gold and T-SPOT.TB. While widely used, concerns have been raised about false positives and a lack of reproducibility with IGRAs. In a study of over 9000 healthcare workers at Stanford University Medical Center, investigators found that the conversion rate using QuantiFERON®-TB Gold was higher than the historical TST conversion rate.[19] More than half of workers with a positive result who had a repeat within 60 days had a subsequent negative result, suggesting that these were false positives. The investigators determined that if they increased the cut-off level for the tests (from 0.35 to 5.3 IU/mL), this eliminated most of the false positives. While this reduced the conversion rate to historic norms, it also resulted in missing more than 50% of those with persistent positives on repeat testing.[19] The use of IGRAs is an appropriate option in all circumstances in which TST can be used.[20] Decisions about whether to use TST or IGRA should be based on consideration of cost, time lost from work for employees needing repeated visits, availability of trained personnel to apply and read the TST, and issues relating to false-positive and false-negative results.

Treatment

A physician familiar with the most up-to-date treatment recommendations should always be consulted when an employee needs antituberculous drug therapy. When therapy is initiated, it is crucial that the individual comply with therapy, both to combat his or her disease and to reduce the likelihood that multidrug resistance will develop. Initial therapy should follow established guidelines; pregnant women may also be safely treated.[21,22]

If an individual who has had a skin test conversion does not have evidence of active tuberculosis, prophylactic treatment should be considered. The purpose of prophylaxis is to reduce the likelihood of tuberculous infection progressing to active disease. The usual prophylaxis regimen is isoniazid 300 mg per day for 6 months; however, certain groups, such as immunosuppressed or pregnant individuals, should be treated for 9 months.[6,21,22] Pyridoxine (vitamin B_6) should be taken with isoniazid to prevent peripheral neuropathy, especially in individuals with poor nutrition or other diseases that may cause peripheral neuropathy.[6,21–23]

Isoniazid can cause hepatic toxicity, particularly among individuals over age 35 or those exposed to other hepatotoxicants. Isoniazid can rarely cause liver failure and death, so patients should receive clear instructions to report promptly malaise, nausea, abdominal pain, or jaundice and to discontinue isoniazid until cleared by a physician. Baseline liver function tests and periodic surveillance for symptoms should be performed. Surveillance should include a contact within the first month of beginning isoniazid. At baseline, providers should take a complete medication history to assess concurrent use of hepatotoxic drugs, including acetaminophen. Because alcohol abuse can increase the risk of hepatic toxicity, abstention should be recommended. Close follow-up of alcoholics is warranted.[23]

Active TB is treated with a combination of drugs. The most commonly used drugs are isoniazid, rifampin, pyrazinamide, ethambutol, and streptomycin. An accepted protocol must be used. Microbiologic tests for drug resistance are crucial.[2,3,24] An effective duration of therapy can be as short as 6 months for low-risk, compliant individuals with no evidence of drug resistance. Each drug has potential side effects that should be monitored. To prevent peripheral neuropathy, concomitant use of pyridoxine (vitamin B_6) should be considered when administering isoniazid. Persons at greatest risk include individuals with poor nutrition or other diseases, such as diabetes and alcoholism, that may cause peripheral neuropathy.[6,21–23] Patients with active tuberculosis should be evaluated clinically at least monthly. They should also have sputum cultures and sensitivities performed monthly for the first 3 months. At the conclusion of therapy for pulmonary tuberculosis, sputum cultures should be repeated if sputum can be obtained and a chest radiograph performed. Monthly chest radiographs are not necessary if the patient is progressing well.

Individuals with high likelihood of poor compliance should be placed on observed therapy, that is, they should report to an outpatient healthcare facility for each dose of medication.

Any individual may be noncompliant with therapy but most at risk are homeless people, migrants, alcoholics, and drug abusers. Local health departments can assist with providing observed therapy.

TB is common among people with AIDS. A smaller skin test response is positive for people with HIV infection (Table 23.1) than for individuals without immunosuppression. Even if a skin test is negative, the index of suspicion should be high for any AIDS patient with symptoms. Treatment, which should be managed by a physician expert in AIDS, may require more aggressive drug therapy than is needed for individuals without HIV infection.

Employees with active pulmonary tuberculosis should be restricted from work until they are documented to be noninfectious. Coworkers of employees with active tuberculosis should be assessed for skin test conversion if there was a likelihood of exposure to aerosolized bacilli. Employees with skin test conversion in the absence of active pulmonary tuberculosis do not need to be restricted from work or from patient care activities (Table 23.2).

Medical surveillance

Medical surveillance should be conducted among selected employee groups at risk of exposure. Skin testing using the Mantoux technique and blood testing using interferon-gamma release assays can be used as a surveillance test. Chest radiographs should not be used for routine surveillance, but they should be performed if skin test conversions are documented or if symptoms suspicious of tuberculosis develop.

Workers who have direct contact with patients, clients, animals, or tissues known to have a high risk of infectiousness require surveillance. These workers should have skin testing annually. The frequency of skin testing can be increased based on the worksite's recent past history of skin test conversions. For workers at very high risk, skin testing every 6 months has been recommended. Some individuals who work in healthcare facilities may not require tuberculosis surveillance because they do not have patient contact or risk of contracting TB. Guidelines from CDC can help determine which worker groups should be included in surveillance.[18]

The size of skin reaction considered to be a positive response among populations that have various pretest probabilities of

TABLE 23.2 Courses of action for typical scenarios when occupational tuberculosis surveillance is conducted.

Surveillance scenario	Course of action
Previously had negative skin or blood test; now test is also negative	No action unless false-negative reaction is suspected.
False-negative skin test is suspected	If aged 55 or over, repeat 5 TU PPD in 2 weeks. If immunosuppressive condition is suspected, refer to experienced personal physician for evaluation. Restrict from work pending evaluation only if active pulmonary tuberculosis) is suspected regardless of skin test results.
Previously had negative skin test; now has positive skin test	Consider booster phenomenon (false-positive result). Refer to experienced personal physician to evaluate for active tuberculosis Consider prescribing prophylactic therapy with isoniazid if active tuberculosis is ruled out (consider need for pyridoxine with isoniazid). Restrict from work pending evaluation only if active pulmonary tuberculosis is suspected. If skin test conversion is likely to have been caused by work, check infection control practices and improve them where needed.
Untoward reaction to PPD (large, ulcerated reaction, lymphangitis, regional adenopathy, fever)	Refer to experienced personal physician to evaluate for active tuberculosis; restrict from work pending this evaluation if index of suspicion for active pulmonary tuberculosis is high. Exclude from further routine PPD skin testing, can use blood test.
History of vaccination with bacillus Calmette–Guérin (BCG)	OK to skin test or blood test. If positive test, evaluate for active tuberculosis If active tuberculosis excluded with positive skin test, consider prophylaxis with isoniazid.
Signs of possible tuberculosis	Refer to experienced personal physician. Exclude from work pending this evaluation if index of suspicion is high.
Active pulmonary tuberculosis	Refer to experienced physician for treatment and follow-up. May not return to work until acid-fast bacteria have been cleared from the sputum; this usually requires 2 or more weeks of antituberculous therapy and follow-up sputum examination by Ziehl-Nielsen staining for confirmation. If high likelihood of noncompliance with drug therapy, observed treatment may be warranted to prevent relapse and multidrug-resistant tuberculosis Evaluate coworkers for skin test conversion if history of exposure to aerosolized bacilli from the index case is likely. If tuberculosis is likely to have been caused by work, check infection control practices and improve them where needed
Active extrapulmonary tuberculosis	Experienced personal physician should treat and follow up. If well enough to tolerate working, may work unless there a draining skin lesion. If high likelihood of noncompliance with drug therapy, observed treatment may be warranted to prevent relapse and multidrug-resistant tuberculosis.
Recent significant exposure to infectious tubercle bacilli from a person, specimen, culture, primate, or other source	Skin test immediately to establish baseline reactivity status. Retest in 6–12 weeks to determine if PPD converts to positive. In the interim, if any significant new symptoms develop, refer to personal physician for evaluation. Check infection control practices and improve them where needed.

tuberculous infection is shown in Table 23.1. Many clinicians, however, feel uncomfortable with reading as negative an 11–14 mm PPD, even if the patient is not at increased risk for tuberculosis. When skin test surveillance is indicated in occupational settings, employees by definition have some increased risk of tuberculosis exposure or transmission, so a 10-mm reaction is the correct definition of positive. In addition, employees with skin test reactions of 5–9 mm should be considered positive if they are immunosuppressed or have had close recent contact with an infectious tuberculosis case. Courses of action for typical scenarios when tuberculosis surveillance is conducted are listed in Table 23.2.

Prevention

Methods to prevent tuberculosis transmission in occupational settings focus on using a combination of tactics involving exposure control. Reducing individual susceptibility by means of vaccination with BCG has not been recommended in the United States as a general public health measure. Nor has BCG vaccination been recommended as a general form of worker protection, although its use could be considered in settings where the likelihood is high that transmission and subsequent infection with *M. tuberculosis* strains resistant to isoniazid and rifampin may occur.[25,26] BCG vaccination should not be used, however, in lieu of infection control precautions.

The Occupational Safety and Health Administration (OSHA) is considering a new infectious disease standard that would include coverage of tuberculosis. OSHA developed and subsequently withdrew a tuberculosis standard in 2003 (29 CFR 1910.1035).[27] A potential OSHA Infectious Disease standard could incorporate coverage of tuberculosis. In 2008, OSHA issued a memorandum that it will enforce respiratory protection requirements for tuberculosis under the Respiratory Protection standard, including the requirement for medical evaluation, fit testing, a written program, training, and record keeping.[28] Additionally, OSHA will look to the CDC's Guidelines for Preventing the Transmission of *Mycobacterium tuberculosis* in Health-Care Settings for enforcement under the General Duty Clause.

The updated CDC guidelines include an expanded scope by incorporating healthcare-associated settings.[18] These include:

- Inpatient settings (patient rooms, emergency departments (EDs), intensive care units (ICUs), surgical suites, laboratories, laboratory procedure areas, bronchoscopy suites, sputum induction or inhalation therapy rooms, autopsy suites, and embalming rooms)
- Outpatient settings (medical offices, ambulatory-care settings, dialysis units, and dental-care settings)
- TB clinics
- Settings in correctional facilities in which healthcare is delivered
- Settings in which home-based healthcare is delivered
- Long-term care settings (hospices, skilled nursing facilities)
- Settings in which emergency medical services are provided
- Laboratories handling clinical specimens that might contain *M. tuberculosis*
- Other settings in which suspected and confirmed TB patients might be encountered (cafeterias, kitchens, laundry areas, maintenance shops, pharmacies, and law enforcement settings)

The CDC guidelines recommend that every healthcare setting should have a written TB infection-control plan that is part of an overall infection-control program. The plan should ensure prompt detection, use of environmental and respiratory precautions, and prompt treatment. The first steps in developing the plan are assigning responsibility and performing a risk assessment. The results of the TB risk assessment determine which administrative, environmental, and respiratory protection controls are needed. The risk assessment should begin with a review of past experience with TB, past occupational exposure to tuberculosis, and the frequency with which active tuberculosis occurs in the surrounding community.[18]

The following preventive practices are important for worker protection and regulatory compliance: a written Exposure Control Plan, training for employees, a medical surveillance program, procedures for prompt identification of individuals with suspected or confirmed infectious tuberculosis, use of isolation rooms employing engineering controls for such individuals, placement of warning signs at the entrance to high-risk areas, use of specific work practices including Respiratory Protection, procedures to evaluate employee exposures, medical removal protection for workers who contract tuberculosis occupationally, and a record-keeping system that protects confidential medical information appropriately.

Employee training should be provided on hire and annually for those workers with occupational exposure. The training should include the signs and symptoms of tuberculosis, hazards of transmission, the purpose and nature of medical surveillance, and site-specific controls. The primary method of transmission of tuberculosis through aerosols should be reviewed. Methods to ensure early identification of suspected cases of tuberculosis should be in place at the facility and should be emphasized during training. Medical surveillance should include pre-placement evaluation, periodic Mantoux or IGRA testing, and management of persons with positive test results. The CDC guidelines recommend the use of risk classification to determine the need for and frequency of screening[18] (Table 23.3).

Along with efforts to control exposure at the source, isolation rooms with a negative pressure gradient and preferably a single-pass air system should be available. This will prevent tubercle bacilli from being blown into halls, neighboring patient care rooms, and other work areas. Contaminated air from isolation rooms should not be exhausted near air intake vents or public walkways. An appropriate air exchange rate should be maintained and monitored. Ultraviolet light may be used to supplement the engineering controls in a facility, but ultraviolet light is not a substitute for negative pressure, exhaust ventilation, or HEPA filtration. Isolation rooms that are engineered correctly need regular maintenance to ensure that clogged filters do not prevent negative pressure from being achieved.

TABLE 23.3 Risk classifications for various health-care settings and recommended frequency of screening for *Mycobacterium tuberculosis* infection among health-care workers (HCWs).*

Setting	Risk classification†		
	Low risk	**Medium risk**	**Potential ongoing transmission§**
Inpatient <200 beds	<3 TB patients/year	≥3 TB patients/year	Evidence of ongoing *M. tuberculosis* infection, regardless of setting
Inpatient ≥200 beds	<6 TB patients/year	≥6 TB patients/year	
Outpatient; and nontraditional facility-based	<3 TB patients/year	≥3 TB patients/year	
TB treatment facilities	Settings in which • persons who will be treated have been demonstrated to have latent TB infection (LTBI) and not TB disease • a system is in place to promptly detect and triage persons who have signs or symptoms of TB disease to a setting in which persons with TB disease are treated • no cough-inducing or aerosol-generating procedures are performed	Settings in which • persons with TB disease are encountered • criteria for low risk are not otherwise met	
Laboratories	Laboratories in which clinical specimens that might contain *M. tuberculosis* are not manipulated	Laboratories in which clinical specimens that might contain *M. tuberculosis* might be manipulated	
Recommendations for Screening Frequency			
Baseline two-step TST or one BAMT¶	Yes, for all HCWs upon hire	Yes, for all HCWs upon hire	Yes, for all HCWs upon hire
Serial TST or BAMT screening of HCWs	No**	At least every 12 months††	As needed in the investigation of potential ongoing transmission§§
TST or BAMT for HCWs upon unprotected exposure to *M. tuberculosis*	Perform a contact investigation (i.e., administer one TST or BAMT as soon as possible at the time of exposure, and, if the result is negative, give a second test [TST or BAMT, whichever was used for the first test] 8–10 weeks after the end of exposure to *M. tuberculosis*)¶¶		

*The term Health-care workers (HCWs) refers to all paid and unpaid persons working in health-care settings who have the potential for exposure to *M. tuberculosis* through air space shared with persons with TB disease.

†Settings that serve communities with a high incidence of TB disease or that treat populations at high risk (e.g., those with human immunodeficiency virus infection or other immunocompromising conditions) or that treat patients with drug-resistant TB disease might need to be classified as medium risk, even if they meet the low-risk criteria.

§A classification of potential ongoing transmission should be applied to a specific group of HCWs or to a specific area of the health-care setting in which evidence of ongoing transmission is apparent, if such a group or area can be identified. Otherwise, a classification of potential ongoing transmission should be applied to the entire setting. This classification should be temporary and warrants immediate investigation and corrective steps after a determination has been made that ongoing transmission has ceased. The setting should be reclassified as medium risk, and the recommended timeframe for this medium risk classification is at least 1 year.

¶All HCWs upon hire should have a documented baseline two-step tuberculin skin test (TST) or one blood assay for *M. tuberculosis* (BAMT) result at each new health-care setting, even if the setting is determined to be low risk. In certain settings, a choice might be made to not perform baseline TB screening or serial TB screening for HCWs who 1) will never be in contact with or have shared air space with patients who have TB disease (e.g., telephone operators who work in a separate building from patients) or 2) will never be in contact with clinical specimens that might contain *M. tuberculosis*. Establishment of a reliable baseline result can be beneficial if subsequent screening is needed after an unexpected exposure to *M. tuberculosis*.

**HCWs in settings classified as low risk do not need to be included in the serial TB screening program.

††The frequency of screening for infection with *M. tuberculosis* will be determined by the risk assessment for the setting and determined by the Infection Control team.

§§During an investigation of potential ongoing transmission of *M. tuberculosis*, testing for *M. tuberculosis* infection should be performed every 8–10 weeks until a determination has been made that ongoing transmission has ceased. Then the setting should be reclassified as medium risk for at least 1 year.

¶¶Procedures for contact investigations should not be confused with two-step TSTs, which are used for baseline TST results for newly hired HCWs.

Source: Centers for Disease Control and Prevention. Guidelines for Preventing the Transmission of Mycobacterium tuberculosis in Health-Care Settings, MMWR 2005; 54:RR-17 Appendix C.

Maintenance and construction employees who are potentially exposed to tuberculosis require training and need to use personal protective equipment. The CDC guidelines provide detailed guidance on environmental controls.[18]

Respirator use must comply with the provisions of the OSHA Respiratory Protection standard as well as specific requirements for prevention of occupational tuberculosis. Surgical masks are not respirators, as they are designed to prevent the healthcare provider from contaminating the patient, but they do not prevent the wearer from being exposed to ambient organisms.[6,7,28] An N95 respirator approved by the National Institute for Occupational Safety and Health (NIOSH) is acceptable in most exposure settings. Combination products incorporating both a surgical mask and an N95 disposable respirator (respirator portion certified by CDC/NIOSH and surgical mask portion listed by FDA) are available that provide both respiratory protection and bloodborne pathogen protection.[18]

Workers must receive medical clearance prior to using the N95 respirator or other respirators. Training should include instructions on the correct way to wear the respirator, how to self-fit test each time a disposable respirator is worn, and how to store a respirator properly. A worker with a full beard cannot achieve an adequate fact-to-respirator seal and will need to use an alternative respirator, such as powered air-purifying respirators (PAPRs), to achieve protection.

Individuals with active pulmonary tuberculosis whose sputum contains acid-fast bacilli should wear a fitted mask (a surgical mask is acceptable) when they leave the isolation room to be transported. They should be instructed to cover their mouth and nose with a tissue when coughing. When these measures are not possible to achieve, for instance, when a patient is combative, the worker should wear a respirator. Attendants should also wear a respirator when transporting a person with active pulmonary tuberculosis in an enclosed vehicle, or when working in the home of such a person, even if the infected individual is wearing a mask.

Sputum induction, administration of aerosolized medications, bronchoscopy, and other procedures likely to produce airborne bacilli should be performed in a properly ventilated setting, such as a sputum induction booth, and attendants must wear proper respiratory protection. In the proposed standard, posting a specific sign at the entrance to high-risk areas will be required. The proposed sign will be red, in the shape of a stop sign, and will contain this text: "No admittance without wearing a Type N95 or more protective respirator."

In work settings where tissues or cultures are the source of infectious organisms, the use of engineering controls such as biologic safety cabinets, instead of respirators, is the preferred method to prevent exposure. To ensure that the correct air balance occurs in a biologic safety cabinet, the HEPA-filtered exhaust air must be discharged either directly to the outside, or by means of proper connections with the building exhaust system. Equipment such as continuous-flow centrifuges used to process specimens containing tubercle bacilli may produce aerosols, and so these should be exhausted in such a way as to prevent contamination of the laboratory.

In all work settings where there is potential exposure to tuberculosis, workers need to follow standard infection control practices, especially since infectious agents other than tuberculosis may be present. These practices include handwashing and use of protective clothing. Specimen containers should be transported in a plastic bag using gloves. Biohazardous waste should be stored in labeled containers and disposed of properly. Work areas should be cleaned with a disinfectant. Although extensive activities to prevent exposure to fomites are not necessary, the manipulation of biological materials containing high concentrations of infectious tubercle bacilli can create hazardous aerosols. Therefore, trained individuals who use a respirator and other appropriate personal protective equipment should clean up sputum, body fluids, or tissues containing tubercle bacilli.

Engineering controls, administrative controls, patient care procedures, and other work practices should be evaluated periodically for areas that need improvement, with attention to results from medical surveillance such as skin test conversion rates. In the event that an employee contracts tuberculosis at work, procedures should be in place to address medical removal. The now-withdrawn OSHA standard stipulated that medical removal procedures should include protection of the person's job with full medical earnings and all other rights and benefits until the worker becomes noninfectious and can return to work, or for 18 months, whichever comes first. It is not known whether a future OSHA infectious disease standard will have similar worker protections. However, a workplace-acquired infection would likely be covered under workers' compensation laws.

References

1. American Thoracic Society and CDC. Diagnostic standards and classification of tuberculosis in adults and children. *Am J Respir Crit Care Med* 2000; 161(4 Pt 1):1376–95.
2. Raviglione MC, O'Brien RJ. Tuberculosis. In: Kasper DL, Fauci AS, Longo DL, et al., eds. Harrison's Principles of Internal Medicine, 19th ed. New York: McGraw-Hill, 2015.
3. Centers for Disease Control, National Center for HIV/AIDS, Viral Hepatitis, STD, and TB Prevention Division of Tuberculosis Elimination. *Core Curriculum on Tuberculosis: What the Clinician Should Know*, 6th ed., 2013. Available at: http://www.cdc.gov/tb/education/corecurr/pdf/corecurr_all.pdf (accessed on June 16, 2016).
4. Centers for Disease Control and Prevention. Tuberculosis Trends—United States, 2014. *MMWR* 2014; 64(101):265–9.
5. Centers for Disease Control. Outbreak of multidrug-resistant tuberculosis at a hospital—New York City, 1991. *MMWR* 1993; 42:427, 433–4.
6. Mahmoudi A, Iseman MD. Pitfalls in the care of patients with tuberculosis: common errors and their association with the acquisition of drug resistance. *JAMA* 1993;270:65–8.

7. Dooley SW, Villarino ME, Lawrence M, et al. Nosocomial transmission of tuberculosis in a hospital unit for HIV-infected patients. *JAMA* 1992; 267:2632–5.
8. Manangan LP, Simonds DN, Pugliese G, et al. Are US hospitals making progress in implementing guidelines for prevention of *Mycobacterium tuberculosis* transmission? *Arch Intern Med* 1998; 158:1440–4.
9. Sutton PM, Nicas M, and Harrison RJ. Tuberculosis isolation: comparison of written procedures and actual practices in three California hospitals. *Infect Control Hosp Epidemiol* 2000; 21:28–32.
10. Porteous NB and Brown JP. Tuberculin skin test conversion rate in dental healthcare workers—results of a prospective study. *Am J Infect Control* 1999; 27:385–7.
11. Steenland K, Levine J, Sieber K, et al. Incidence of tuberculosis infection among New York State prison employees. *Am J Public Health* 1997; 87:2012–7.
12. Templeton GL, Illing LA, Young L, et al. The risk for transmission of *Mycobacterium tuberculosis* at the bedside and during autopsy. *Ann Intern Med* 1995; 122:922–5.
13. Flavin RJ, Gibbons N, and O'Briain DS. *Mycobacterium tuberculosis* at autopsy—exposure and protection: an old adversary revisited. *J Clin Pathol* 2007; 60:487–91.
14. Collins CH and Grange JM. Tuberculosis acquired in laboratories and necropsy rooms. *Commun Dis Public Health* 1999; 2:161–7.
15. McNerney R, Maeurer M, Abubakar I, et al. Tuberculosis diagnostics and biomarkers: needs, challenges, recent advances, and opportunities. *J Infect Dis* 2012; 205(Suppl 2):S147–58.
16. Bauer J, Kok-Jensen A, Faurschou P, et al. A prospective evaluation of the clinical value of nation-wide DNA fingerprinting of tuberculosis isolates in Denmark. *Int J Tuberc Lung Dis* 2000; 4:295–9.
17. Centers for Disease Control and Prevention. Guidelines for the investigation of contacts of persons with infectious tuberculosis: recommendations from the National Tuberculosis Controllers Association and CDC. *MMWR* 2005; 54(RR-15): 1–47. Available at: http://www.cdc.gov/mmwr/pdf/rr/rr5415.pdf (accessed July 1, 2016).
18. Centers for Disease Control and Prevention. Guidelines for preventing the transmission of *Mycobacterium tuberculosis* in healthcare settings, *MMWR* 2005; 54(RR-17): 1–141. Available at: http://www.cdc.gov/mmwr/pdf/rr/rr5417.pdf (accessed July 1, 2016).
19. Slater ML, Welland G, Pai M, et al. Challenges with QuantiFERON-TB gold assay for large-scale, routine screening of U.S. healthcare workers. *Am J Respir Crit Care Med* 2013; 188:1005–10.
20. Centers for Disease Control and Prevention. Guidelines for using the QuantiFERON®-TB gold test for detecting *Mycobacterium tuberculosis* infection, United States. *MMWR* 2005; 54(RR-15): 49–55. Available at: http://www.cdc.gov/mmwr/pdf/rr/rr5415.pdf (accessed July 1, 2016).
21. Centers for Disease Control. Initial therapy for tuberculosis in the era of multidrug resistance: recommendations of the Advisory Council for the Elimination of Tuberculosis. *MMWR* 1993; 42(No. RR-7):l–8.
22. Centers for Disease Control. Tuberculosis among pregnant women–New York City, 1985–1992. *MMWR* 1993; 605:611–2.
23. Centers for Disease Control. Severe isoniazid-associated hepatitis—New York, 1991–1993. *MMWR* 1993; 42:545–7.
24. Bloch NB, Cauthen GM, Onorato IM, et al. Nationwide survey of drug-resistant tuberculosis in the United States. *JAMA* 1994; 271:665–71.
25. Centers for Disease and Prevention. The role of BCG vaccine in the prevention and control of tuberculosis in the United States: a joint statement by the Advisory Council for the Elimination of Tuberculosis and the Advisory Committee on Immunization Practices. *MMWR* 1996; 45(RR4):1–18.
26. Marcus AM, Rose DN, Sacks HS, et al. BCG vaccination to prevent tuberculosis in healthcare workers: a decision analysis. *Prev Med* 1997; 26:201–7.
27. Occupational Safety and Health Administration. *Occupational Exposure to Tuberculosis; Proposed Rule – 62-54159-54309.* Available at: http://www.osha-slc.gov/FedReg_osha_data/FED19971017.html (accessed on June 16, 2016).
28. Occupational Safety and Health Administration. Tuberculosis and Respiratory Protection Enforcement. 29 CFR 1910.134; 1910.134(f)(2). Washington, DC: Department of Labor, 2008. Available at: https://www.osha.gov/pls/oshaweb/owadisp.show_document?p_table=Interpretations&p_id=26013 (accessed on June 16, 2016).

MYCOBACTERIA OTHER THAN *MYCOBACTERIUM TUBERCULOSIS*

Occupational setting

Several mycobacteria other than *Mycobacterium tuberculosis* can cause disease in humans. Significant occupational exposure is much less common for these bacilli than for *M. tuberculosis*.

Potential contact with *M. kansasii* and *M. avium intracellulare* could occur in patient care settings, especially during care of immunosuppressed patients with pulmonary infections from these organisms. However, person-to-person transmission may not be the only route of exposure.[1] The organisms are also found in the environment and in animal reservoirs. *M. kansasii* outbreaks have occurred among miners, particularly when the miners are in poor health, water used during mining is contaminated with the organism and the miners have evidence of dust-related pulmonary disease or silicosis.[2,3] *M. avium intracellulare* infections and hypersensitivity pneumonitis have been described in spa workers, referred to as "hot tub lung," as well as in a swim instructor and lifeguards.[4–9] Water aerosolization during filter cleaning and inadequate ventilation may have contributed to these cases.[7,10]

Occupational contact with *M. bovis* may occur from working with dairy cattle. Infected beef cattle and game animals such as deer can also transmit the disease.[11,12] Tuberculin testing of cattle and pasteurization of milk have greatly reduced the incidence of *M. bovis* infection, but in some developed and developing countries, infection rates remain high in cattle herds, so that farm and slaughterhouse workers may be at risk.[13–17] A case of animal-to-human transmission has also occurred in a veterinarian.[18]

M. marinum and *M. abscessus* can be acquired via skin exposure to infected fish or to freshwater or saltwater aquatic environments. Individuals who are occupationally infected often give a history of minor skin trauma to the hands or other skin immersed in water.[19–21]

M. immunogen and other *Mycobacterium* species have been identified as a contaminant in metal-working fluid and a possible cause or contributor to hypersensitivity pneumonitis.[10,22]

Occupational contact with *M. leprae* is rare in the United States. In the 1980s, more than 900 immigrants and refugees from Southeast Asia had leprosy, but there was no evidence that these individuals transmitted the disease to others after their arrival.[23] The incidence among immigrants has declined significantly and the incidence in U.S. borns has remained low (Figure 23.3).[24] The organism is difficult to transmit, although nasal colonization can occur occupationally. In one study, a polymerase chain reaction test revealed *M. leprae* genetic material in nasal swab specimens among 55% of untreated patients, 19% of occupational contacts, and 12% of controls from an endemic region.[25] Less than 10% of close family members of affected patients contract the disease.[1,26]

If infected specimens are handled improperly, laboratory workers can be exposed. Contact during travel to underdeveloped nations, where the incidence of mycobacterial disease may be high, is a theoretical possibility, but it is unlikely unless the traveler has prolonged close physical contact with infected, untreated individuals.

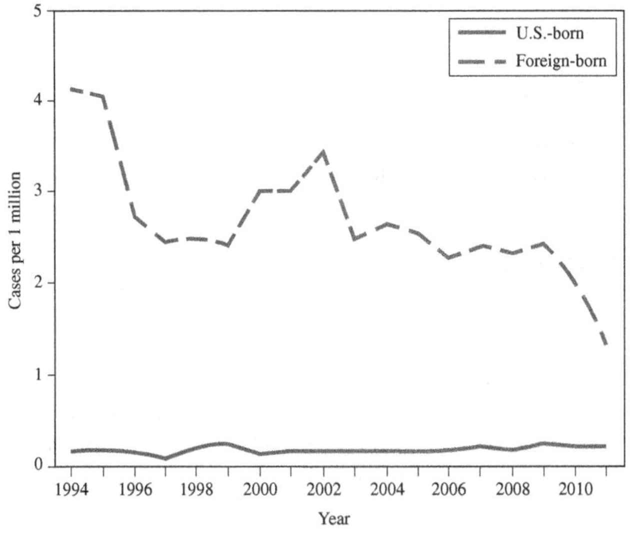

FIGURE 23.3 Rate of new diagnoses of Hansen's disease, by U.S. birth status — United States, 1994–2011. *Source*: MMWR 2014:63(43):970.

Exposure (route)

Routes of exposure have not been entirely elucidated; however, person-to-person contact is not believed to be as high a risk as with *M. tuberculosis*.[1] Most pathogenic *Mycobacterium* species are thought to be absorbed from the upper respiratory tract. Infection with some species can occur after skin contact (*M. marinum*, *M. abscessus*, *M. ulcerans*, *M. fortuitum*, and *M. leprae*). The gastroenterologic route of exposure is important for *M. bovis* (contaminated milk), as is respiratory exposure to aerosols produced during the slaughtering process.[1,4,5,26] Some armadillos in the southern United States are naturally infected with Hansen's disease.[27] While the risk of transmission is considered low, infections can occur following contact.[27] There is also evidence that *M. leprae* can also survive in amebal cysts representing another potential source for exposure.[28]

Pathobiology

M. avium intracellulare, *M. kansasii*, and other species can cause a potentially serious pulmonary infection similar to that caused by *M. tuberculosis*. Disseminated disease is more common among immunosuppressed individuals, especially those with AIDS or organ transplants. *M. bovis* can cause pulmonary disease. Localized skin infection or lymphadenitis can occur after a percutaneous exposure from *M. fortuitum*, *M. marinum*, or *M. ulcerans*. Lymphadenopathy, primarily cervical, appears after oral exposure to *M. scrofulaceum*.[1,3]

M. leprae affects the skin, soft tissues, and peripheral nerves. The average incubation period is 2–4 years. It is common in some underdeveloped nations but rare in the United States. Anesthetic or paresthetic skin lesions with a typical appearance are usually the first clinical sign of infection with *M. leprae*. In advanced cases, destruction of the soft tissues of the nose and face, fingers, toes, and other structures leads to cosmetic deformities. The granulomatous reaction of the host, rather than a neurotoxic effect of the bacilli themselves, is thought to cause nerve damage.[25]

Diagnosis

Pulmonary infection is diagnosed in the same manner as *M. tuberculosis*-using microscopic examination of sputum (Ziehl-Nielsen staining), sputum cultures, and chest radiographs. Sputum cultures should be sent to a qualified laboratory. Biopsy material from skin or lymph node lesions suspected of harboring *Mycobacterium* species should be evaluated similarly with staining and culture. Skin tests with PPD can be positive, but tissue diagnosis is more definitive, especially since many infected patients are immunosuppressed.[1,3]

Hansen's disease (leprosy) is diagnosed among individuals with suggestive symptoms by demonstrating *M. leprae* organisms on biopsy of skin or peripheral nerves. Skin testing is of no use, but a serodiagnostic test to a specific phenolic glycolipid from the surface of the bacillus has a high diagnostic specificity. The organism can be cultured in footpads of mice and in the armadillo. In the United States, the National Hansen's Disease Program is responsible for Hansen's disease (leprosy) care and research. Care is provided for patients at its facility at the Ochsner Medical Center in Baton Rouge, LA, and the program also oversees an ambulatory care network with 11 clinics in seven states and Puerto Rico. It also conducts biomedical research.[29]

Treatment

With the exception of Hansen's disease, infections with the most common species are treated with the many of the same drugs used with *M. tuberculosis*. *M. avium intracellulare* is commonly multidrug-resistant; therefore, two to six drugs are usually chosen from among isoniazid, ethambutol, rifampin, ethionamide, pyrazinamide, cycloserine, amikacin, and clofazimine. Surgical excision is sometimes used for lymphadenitis and some other localized infections. A physician who is experienced with infectious diseases should direct the patient's care.[1,30]

Three drugs have been traditionally used to treat the more severe forms of Hansen's disease—dapsone, rifampin, and clofazimine. At least 2 years of drug therapy are required, and indefinite treatment may be necessary. Cosmetic deformities can often be surgically corrected. For less-severe cases, dapsone and rifampin are given for at least 6 months.[26] In 1998, the US Food and Drug Administration approved the use of thalidomide to treat skin lesions in leprosy known as erythema nodosum leprosum.

Medical surveillance

Many workers at risk for exposure to these *Mycobacterium* species will already be in a medical surveillance program for *M. tuberculosis*, which should be adequate. CDC can be used as a resource for program planning. Recommended medical surveillance for the rare worker who may have significant occupational contact with *M. leprae* should consist at least of a clinical examination focusing on detection of typical skin lesions. Serologic testing may be a useful adjunct.[26]

Prevention

A prudent strategy for prevention of possible exposure to bioaerosols of *Mycobacterium* species that cause pulmonary infection follows the same principles as the program for *M. tuberculosis*. Because of concomitant risk of exposure to *M. tuberculosis*, control measures should already be in place in most targeted workplaces. Control measures include rapid diagnosis of infectious patients, respiratory isolation, negative pressure room ventilation, and maintenance of ventilation systems. Worker training should focus on the use of gloves, protective clothing, and an approved respirator. Biological wastes should be disposed of properly. Patient care procedures likely to produce bioaerosols should be conducted in a properly ventilated area, such as a sputum induction booth. Laboratory workers should use biologic safety cabinets. Workers in the fishing industry and in agriculture should use protective clothing and gloves.

To prevent exposure to *M. leprae*, standard infection control measures should be followed. Prophylactic postexposure chemotherapy under the direction of a physician experienced in caring for patients with Hansen's disease should be considered for individuals who have had a significant occupational exposure.[26]

References

1. Holland SM. Nontuberculous mycobacterial infections. In: Kasper DL, Fauci AS, Longo DL, et al., eds. Harrison's principles of internal medicine, 19th ed. New York: McGraw-Hill, 2015.
2. Corbett EL, Blumberg L, Churchyard GJ, et al. Nontuberculous mycobacteria: defining disease in a prospective cohort of South African miners. *Am J Respir Crit Care Med* 1999; 160:15–21.
3. Chobot S, Malis J, Sebakova H, et al. Endemic incidence of infections caused by *Mycobacterium kansasii* in the Karvina district in 1968–1995 (analysis of epidemiological data & review). *Cent Eur J Public Health* 1997; 5:164–73.
4. Moraga-McHaley SA, Landen M, Krapfl H, et al. Hypersensitivity pneumonitis with *Mycobacterium avium* complex among spa workers. *Int J Occup Environ Health* 2013; 19(1):55–61.
5. Rose CS, Martyny J, Huitt J, et al. Hot tub associated granulomatous lung disease from mycobacterial aerosols. *Am J Respir Crit Care Med* 2000; 161:A730.
6. Agarwal R and Nath A. Hot-tub lung: hypersensitivity to *Mycobacterium avium* but not hypersensitivity pneumonitis. *Respir Med* 2006; 100:1478.
7. Fjallbrant H, Akerstrom M, Svensson E, et al. Hot tub lung: an occupational hazard. *Eur Respir Rev* 2013; 22(127):88–95.
8. Yu TC, Ahmed R, Yap E, et al. Dyspnoea in a 17-year-old swim instructor: a diagnosis of hot tub lung. *NZ Med J* 2008; 121:78–80.
9. Rose CS, Martyny JW, Newman LS, et al. "Lifeguard lung": endemic granulomatous pneumonitis in an indoor swimming pool. *Am J Public Health* 1998; 88:1795–1800.
10. Falkinham III JO. Mycobacterial aerosols and respiratory disease. *Emerg Infect Dis* 2003; 9(7):763–7.
11. Cousins DV, Williams SN, and Dawson DJ. Tuberculosis due to *Mycobacterium bovis* in the Australian population: DNA typing of isolates, 1970–1994. *Int J Tuberc Lung Dis* 1999; 3:722–31.

12. Liss, GM, Wong L, Kittle DC, et al. Occupational exposure to *Mycobacterium bovis* infection in deer and elk in Ontario. *Can J Public Health* 1994; 85:326–9.
13. de Kantor IN and Ritacco V. Bovine tuberculosis in Latin American and the Caribbean: current status, control and eradication programs. *Vet Microbiol* 1994; 40:5–14.
14. de la Rua-Domenech, R. Human *Mycobacterium bovis* infection in the United Kingdom: incidence, risks, control measures and review of the zoonotic aspects of bovine tuberculosis. *Tuberculosis* 2006; 86:77–109.
15. Torres-Gonzalez P, Soberanis-Ramos O, Martinez-Gamboa A, et al. Prevalence of latent and active tuberculosis in dairy farm workers exposed to cattle infected by *Mycobacterium bovis*. *PLoS Negl Trop Dis* 2013; 7(4):e2177.
16. Evans JT, Smith EG, Banarjee A, et al. Cluster of human tuberculosis caused by *Mycobacterium bovis*: evidence for person-to-person transmission in the UK. *Lancet* 2007; 369: 1270–76.
17. Thoen C, LoBue P, and de Kantor I. The importance of *Mycobacterium bovis* as a zoonosis. *Vet Microbiol* 2006;112: 339–45.
18. Corcoran JP, Hallifax RJ, Bettinson HV, et al. Tuberculous pleuritis secondary to *Mycobacterium bovis* in a veterinarian. *Clin Respir J* 2014; Oct 22. 10.1111/crj.12231. [Epub ahead of print]
19. Iredell J, Whitby M, and Blacklock Z. *Mycobacterium marinum* infection: epidemiology and presentation in Queensland 1971–1990. *Med J Aust* 1992; 157:596–8.
20. Cheung JP, Fung B, Wong SS, et al. Review article: *Mycobacterium marinum* infection of the hand and wrist. *J Orthopaedic Surg* 2010; 18(1):98–103.
21. Kang GC, Gan AW, Yam A, et al. Mycobacterium abscessus hand infections in immunocompetent fish handlers: case report. *J Hand Surg Am* 2010; 35(7):1142–5.
22. Shelton BG, Flanders WD, and Morris GK *Mycobacterium* sp. as a possible cause of hypersensitivity pneumonitis in machine workers. *Emerg Infect Dis* 1999; 5:270–3.
23. Mastro TD, Redd SC, and Breiman RF. Imported leprosy in the United States, 1978 through 1988: an epidemic without secondary transmission. *Am J Public Health* 1992; 82:1127–80.
24. Nolen L, Haberling D, Scollard D, et al. Incidence of Hansen's disease—United States, 1994–2011. *MMWR* 2014; 63(43): 969–72.
25. de Wit MY, Douglas JT, McFadden J, et al. Polymerase chain reaction for detection of *Mycobacterium leprae* in nasal swab specimens. *J Clin Microbiol* 1993; 31:502–6.
26. Gelber RH. Leprosy. In: Kasper DL, Fauci AS, Longo DL, et al, eds. Harrison's Principles of Internal Medicine, 19th ed. New York: McGraw-Hill, 2015.
27. Centers for Disease Control and Prevention. Hansens Disease (Leprosy): Risk of Exposure. Available at: http://www.cdc.gov/leprosy/exposure/armadillos.html (accessed on June 16, 2016).
28. Wheat WH, Casali AL, Thomas V, et al. (2014) Long-term survival and virulence of *Mycobacterium leprae* in amoebal cysts. *PLoS Negl Trop Dis*; 8(12):e3405. 10.1371/journal.pntd.0003405.
29. Health Resources and Services Administration. National Hansen's Disease (Leprosy) Program. Available at: http://www.hrsa.gov/hansensdisease/ (accessed on July 1, 2016).
30. Griffith DE, Aksamit T, Brown-Elliott BA, et al. (2007). An official ATS/IDSA statement: diagnosis, treatment, and prevention of nontuberculous mycobacterial diseases. *Am J Resp Critic Care Med*, 175(4):367–416.

Chapter 24

FUNGI

Craig S. Glazer and Cecile S. Rose

A variety of fungi are associated with occupational illnesses (see Tables 24.1, 24.2, and 24.3 on page 426.)

ALTERNARIA SPECIES

Common name for disease: Wood pulp workers' disease

Occupational setting

Woodworkers exposed via inhalation to wood dusts contaminated with the mold *Alternaria* may develop hypersensitivity pneumonitis, asthma, and allergic rhinitis.

Exposure (route)

Inhalation of the pyriform-shaped spores with an average length of 18 µm and a diameter of 5 µm (Figure 24.1) can cause hypersensitivity lung disease. Infection of the nails and cornea in an immunocompromised wood pulp worker has been reported,[1] as have corneal infections in farmers working in humid environments.[2]

Pathobiology

In 1930, Hopkins described a 37-year-old man whose asthma was associated with exposure to damp musty areas contaminated with *Alternaria spp.* and confirmed by bronchial challenge.[3] Bronchial challenge with both fungal extracts and whole spores of *Alternaria* can trigger both immediate and late asthmatic reactions in sensitized asthmatics.[4] Immunologic sensitization to *Alternaria* has been linked to increased bronchial hyperreactivity and to life-threatening asthma.[5–7] Alternaria is also a rare cause of allergic bronchopulmonary mycosis.[8] Progressive hypersensitivity pneumonitis leading to chronic interstitial fibrosis has been documented in two workers with prolonged exposure to *Alternaria* during the manufacture of wood pulp.[9] Eosinophilic pneumonia related to *Alternaria* exposure in a water-damaged home has been reported.[10]

Diagnosis

Diagnosis of sensitizing occupational asthma relies on symptom and exposure histories, findings on pulmonary function testing (typically airflow limitation with a significant bronchodilator response or positive methacholine challenge), and the results of peak flow monitoring at and away from work. Allergy skin prick testing to *Alternaria* is often but not always positive.[7] Diagnosis of hypersensitivity pneumonitis relies on a constellation of clinical findings, including a careful symptom and occupational history.[11] Physical examination may be normal or show basilar crackles; the chest radiography may be normal or show diffuse alveolar or interstitial opacities; pulmonary function tests show restriction, obstruction, or a mixed picture. Fiber-optic bronchoscopy with bronchoalveolar lavage and transbronchial biopsies may be helpful when the clinical suspicion is strong but routine tests are nondiagnostic. Exercise physiology testing may be helpful in patients who have dyspnea but normal resting pulmonary function. Serum precipitating antibodies to *Alternaria* are often found in asymptomatic exposed workers and may be negative if the wrong antigen preparation is used. Infections are diagnosed by culture.

Physical and Biological Hazards of the Workplace, Third Edition. Edited by Gregg M. Stave and Peter H. Wald.

TABLE 24.1 Common fungi associated with hypersensitivity diseases.

Fungus	Exposure	Syndrome
Alternaria spp.	Wood pulp	Wood pulp worker's lung, mold allergy, asthma
Aspergillus spp.	Moldy hay	
	Water	Farmer's lung
Aspergillus fumigatus	General environment	Ventilation pneumonitis
Aspergillus clavatus	Barley	Allergic bronchopulmonary aspergillosis
Aureobasidium pullulans	Water	
Cladosporium spp.	Hot tub mists	Malt worker's lung
	General environment	Humidifier lung
Cryptostroma corticale	Wood bark	Hot tub HP
Graphium, Aureobasidium pullulans	Wood dust	Mold allergy, asthma
Merulius lacrymans	Rotten wood	Maple bark stripper's lung
Penicillium spp.	Fuel chips, sawmills, tree cutting	Sequoiosis
		Dry rot lung
	Shiitake mushroom manufacturing, humidifier water	Hypersensitivity pneumonitis, asthma
Penicillium frequentans	Cork dust	Suberosis
Penicillium casei, Penicillium roqueforti	Cheese	Cheese washer's lung
Rhizopus, Mucor	Damp basements	Asthma, mold allergy
Trichosporon cutaneum	Damp wood and mats	Japanese summer-type HP

TABLE 24.2 Common toxigenic fungi.

Toxin	Fungal Source	Occupational Exposure
Aflatoxin	*Aspergillus flavus*	Farmers
	Aspergillus parasiticus	Peanut handlers
Fumitoxin	*Aspergillus fumigatus*	Compost workers
Satratoxin	*Stachybotrys chartarum*	Maintenance workers (from insulated pipes)
Sterigmatocystin	*Avicularia versicolor*	Housekeepers
T-2 toxin	*Fusarium*	Machinists
		Farmers

TABLE 24.3 Occupational fungal infections.

Fungus	Source	Exposure
Blastomyces dermatitidis	Acid soil near rivers, streams, and swamps	Hunters
		Campers
Coccidioides immitis	Semiarid or desert soils	Construction workers
		Farmers
		Archeologists
		Laboratory workers
		Textile workers
Cryptococcus neoformans	Pigeon/avian droppings	Pigeon breeders
Histoplasma capsulatum	Bat-infested caves, staring/chicken roosts	Spelunkers
		Construction workers
		Laboratory workers
Sporothrix schenckii	Contaminated soil and vegetation in warm or tropical areas	Gardeners
		Farmers
		Florists
		Hunters
		Gold miners
		Laboratory workers

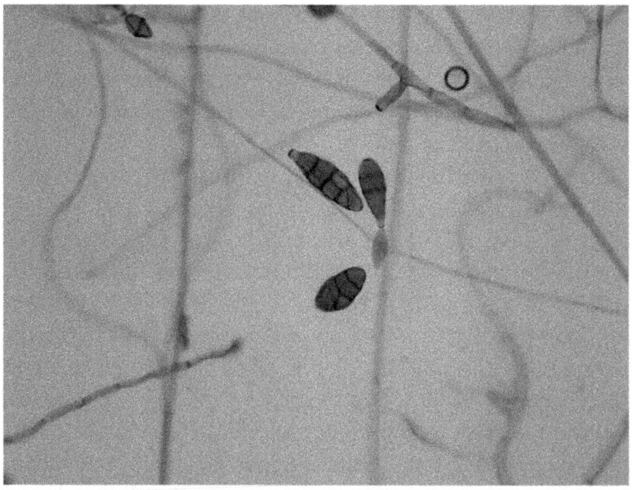

FIGURE 24.1 Photomicrograph showing Alternaria spores. Courtesy of Dr. Dominick Cavuoti and Dr. Francesca Lee.

Treatment

The treatment of hypersensitivity lung diseases begins with removal from antigen exposure. Inhaled corticosteroids are useful as first-line pharmacotherapy for asthma, often in combination with inhaled bronchodilators. Oral corticosteroids may be indicated in patients with hypersensitivity pneumonitis or eosinophilic pneumonia manifested by severe symptoms and radiographic or functional abnormalities. Amphotericin B and itraconazole are the most frequently used agents to treat infections, but recent data suggest posaconazole may have better *in vitro* activity.[12]

Prevention

Adequate ventilation and process enclosure, appropriate respiratory protection, and work practices that reduce airborne dust levels are recommended. Workers in at-risk industries should be educated regarding exposure risks and encouraged to seek early medical attention for persistent respiratory and/or systemic symptoms.[11]

References

1. Arrese JE, Pierard-Franchimont C, Pierard GE. Onychomycosis and keratomycosis caused by *Alternaria spp.* A bipolar opportunistic infection in a wood-pulp worker on chronic steroid therapy. *Am J Dermatopathol* 1996; 18:611–3.
2. Jiang K, Brownstein S, Baig K, et al. Clinicopathologic case reports of *Alternaria* and *Fusarium keratitis* in Canada. *Can J Ophthalmol* 2013; 48:e151–4.
3. Hopkins J, Benham R, Kesten B. Asthma due to a fungus – *Alternaria. JAMA* 1930; 94:6–11.
4. Licorish K, Novey H, Kozak P. Role of *Alternaria* and *Penicillium* spores in the pathogenesis of asthma. *J Allergy Clin Immunol* 1985; 76:819–25.
5. Nelson HS, Szefler SJ, Jacobs J, et al. The relationships among environmental allergen sensitization, allergen exposure, pulmonary function, and bronchial hyperresponsiveness in the Childhood Asthma Management Program. *J Allergy Clin Immunol* 1999; 104:775–85.
6. Black PN, Udy AA, Brodie SM. Sensitivity to fungal allergens is a risk factor for life-threatening asthma. *Allergy* 2000; 55:501–4.
7. Fernandez C, Bevilacqua E, Fernandez N, et al. Asthma related to Alternaria sensitization: an analysis of skin-test and serum-specific IgE efficiency based on the bronchial provocation test. *Clin Exp Allergy* 2011; 41:649–56.
8. Chowdhary A, Agarwal K, Randhawa HS, et al. A rare case of allergic bronchopulmonary mycosis caused by *Alternaria alternata. Med Mycol* 2012; 50:890–6.
9. Schlueter D, Fionk J, Hensley G. Wood pulp workers' disease: a hypersensitivity pneumonitis caused by *Alternaria. Ann Intern Med* 1972; 77:907–14.
10. Ogawa H, Fujimura M, Amaike S, et al. Eosinophilic pneumonia caused by *Alternaria alternata. Allergy* 1997; 52:1005–8.
11. Rose C, Lara A. Hypersensitivity pneumonitis. In: Murray JF, Nadel JF (eds.), Textbook of Respiratory Medicine, 5th edn. Philadelphia: Saunders, 2010:1587–600.
12. Alastruey-Iazquierdo A, Cuesta I, Ros L, et al. Antifungal susceptibility profile of clinical *Alternaria* spp. identified by molecular methods. *J Antimicrob Chemother* 2011; 66:2585–7.

ASPERGILLUS SPECIES

Common names for diseases: Hypersensitivity diseases: Extrinsic allergic alveolitis, farmer's lung, malt worker's lung (*Aspergillus clavatus*), allergic bronchopulmonary aspergillosis (ABPA), allergic *Aspergillus* sinusitis, allergic fungal sinusitis (AFS), baker's asthma, mushroom grower's asthma; Infections: Invasive aspergillosis, aspergilloma. Mycotoxins: Mycotoxicosis, aflatoxin-induced liver cancer.

Occupational setting

There are over 600 species in the genus *Aspergillus.* Most *Aspergillus* species are found in soil. Many species are found on a wide variety of substrates, including forage products, food products, cotton, and other organic debris. *Aspergillus fumigatus,* the most common species, accounts for most disease, both allergic and infectious. Farmers, sawmill workers, mushroom workers, greenhouse workers, tobacco workers, bakers, garbage workers, and bird hobbyists are among the many groups at risk from this fungal exposure.[1–12] Workers who deal with compost piles, decomposing haystacks, or moldy grains may develop hypersensitivity responses and are at increased risk of developing liver cancer.[13–16] Power plant workers are an emerging at-risk group in plants that are transitioning to biomass fuels as an alternative to traditional carbon-based fuels.[17] Allergy to *Aspergillus*-derived enzymes has been associated with baker's asthma.[18]

Exposure (route)

Aspergillus species produce conidia in chains measuring 2–5 µm in diameter that are readily airborne and easily respirable (Figure 24.2). *Aspergillus* aflatoxin may become airborne as well.[19] Most *Aspergillus* diseases are acquired from inhalation of spores, but airborne spores probably also infect tissues exposed during surgery. Hospital renovations may increase the risk of aspergillosis in immunocompromised hosts due to release of spore-bearing dust. This risk can be attenuated by the addition of either high-efficiency particulate air filtration or laminar airflow.[20] Contaminated ventilation systems and compost sites have been associated with case clusters.[21] Other portals of entry (including the eye, paranasal sinuses, skin, GI tract, and via intravenous drug use) have been associated with invasive aspergillosis.

Pathobiology

There are four main categories of disease involving *Aspergillus* spp.—allergic aspergillosis, colonizing aspergillosis, invasive infections (both pulmonary and nonpulmonary), and aflatoxin-induced malignancy.

ALLERGIC ASPERGILLOSIS: The four major allergic manifestations of *Aspergillus* sensitization are asthma, hypersensitivity pneumonitis, allergic bronchopulmonary aspergillosis, and allergic sinusitis.

The clinical manifestations of *Aspergillus*-related asthma are no different from other forms of extrinsic asthma: symptoms of cough, wheezing, chest tightness, and dyspnea; wheezing on examination; and obstructive changes on pulmonary function testing during acute exacerbations. Peripheral eosinophilia is common, and *Aspergillus*-specific IgE antibodies are detectable with serum RAST or ELISA.

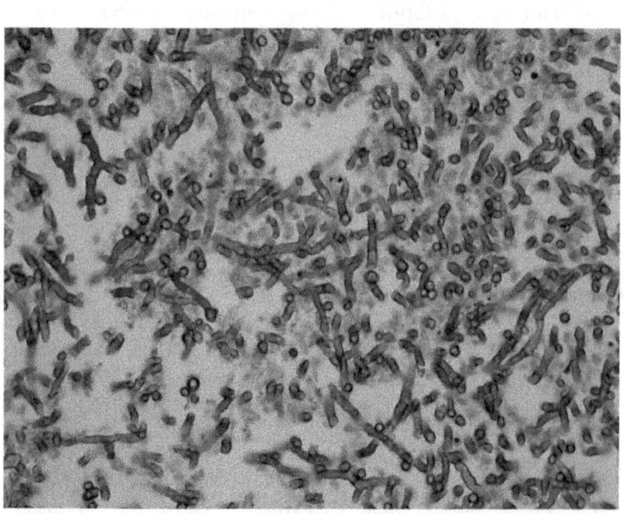

FIGURE 24.2 Tissue sample hematoxylin and eosin stain revealing Aspergillus hyphae in tissue. Courtesy of Dr. Dominick Cavuoti and Dr. Francesca Lee.

Hypersensitivity pneumonitis (extrinsic allergic alveolitis) can occur in individuals with repeated exposure to organic dusts containing *Aspergillus* species. Symptoms may resemble an acute, self-limited, flu-like illness occurring 6–12 hours after exposure. Signs include crackles on lung exam, peripheral leukocytosis, infiltrates on chest radiographs, and normal, restrictive, or obstructive changes on lung function testing. Symptoms of hypersensitivity pneumonitis may also be subacute or chronic. They include myalgias, weight loss, fatigue, chest tightness, cough, and subtle, progressive dyspnea on exertion. The chest radiograph and pulmonary function tests are often normal in early disease. More sensitive diagnostic testing, such as fiber-optic bronchoscopy, may be needed to detect the characteristic lymphocytic alveolitis and granulomatous lung disease. Continued antigen exposure typically leads to worsening symptoms, restrictive physiologic changes, and permanent interstitial fibrosis.

Allergic bronchopulmonary aspergillosis (ABPA) is an inflammatory disease caused by an immunologic response to *Aspergillus fumigatus* (Af) and other *Aspergillus* species growing in the bronchi of patients with asthma and cystic fibrosis.[22] Classically, patients present with chest radiographic infiltrates and peripheral blood eosinophilia. ABPA can range from mild asthma to end-stage fibrotic lung disease. The latest ABPA diagnostic criteria include (i) the presence of a predisposing condition such as asthma or cystic fibrosis; (ii) obligatory criteria including both elevated total IgE (>1000) and immunologic sensitization as shown by immediate cutaneous reactivity to *Aspergillus* or elevated serum IgE antibodies to *Aspergillus*; *and* (iii) two out of the following three other criteria: (a) precipitating (IgG) antibodies to *Aspergillus*, (b) radiographic abnormalities consistent with ABPA, or (c) total eosinophil count greater than 500. [22]

Allergic fungal sinusitis (AFS) due to *Aspergillus* species typically occurs in immunocompetent atopic patients. Most patients have asthma; 85% have nasal polyposis.[23] Patients present with chronic recurrent sinusitis are unresponsive to conventional therapy. Many describe blowing dark, rubbery plugs out of their nose. Fungal hyphae are identified with Gomori methenamine silver stain of the allergic mucin. Diagnostic criteria are similar to those for ABPA and include peripheral eosinophilia, increased serum IgE levels, and precipitating antibodies to fungal extracts.

COLONIZING ASPERGILLOSIS: Saprophytic colonization of air spaces by *Aspergillus* species can result in a number of outcomes, including otomycosis, fungus ball of the paranasal sinuses,[24] endobronchial colonization, and fungus ball of the lung (aspergilloma). Fungus balls of the lung typically occur in patients with preexisting cavitary lung diseases or chronic allergic aspergillosis, with an aspergilloma forming in an ectatic bronchus. The condition is associated with eosinophilic pneumonia and bronchiectasis. Primary aspergillomas may also occur in patients without allergic disease; they present as fuzzy infiltrates that progress to rounded cavities on the chest radiograph. The major symptom of pulmonary aspergilloma is recurrent hemoptysis. Masses of fungal elements are found on surgical resection. Resolution with residual fibrosis is uncommon. Aspergillomas may occur in association with other cavitary diseases, including tuberculosis and sarcoidosis. Fungus balls occur rarely in the urinary bladder, gallbladder, and bile ducts.

INVASIVE ASPERGILLOSIS: This is an uncommon form of *Aspergillus*-related disease. It is also the most serious. It typically occurs in immunocompromised patients, most notably those with leukemia and lymphoma. Fatal invasive microgranulomatous aspergillosis occurred after shoveling moldy cedar wood chips in a patient with chronic granulomatous disease.[25] An unusual case of fatal invasive pulmonary aspergillosis occurred in a nonimmunocompromised gardener exposed to heavy environmental concentrations of *A. fumigatus*.[3] The disease presents as pneumonia with fever, cough, leukocytosis, and respiratory distress. Radiographically, there may be diffuse, patchy, alveolar infiltrates, or consolidation with mass effect (lung abscess). The necrotizing pneumonia that develops is usually fulminant, and death typically occurs in 1–2 weeks. Diagnosis is difficult because *Aspergillus* antibodies are often not detectable in the immunocompromised patient. Depending on the patient's clinical status, invasive diagnostic procedures, such as bronchoscopy or surgical lung biopsy, early in the course of disease may enable early diagnosis and treatment with intravenous amphotericin B. Recent advances in the diagnosis of invasive disease include serum and respiratory sample testing for aspergillus galactomannan and aspergillus PCR.[26] The combination of both has the best positive predictive values.[27]

Extrapulmonary forms of invasive aspergillosis may involve the eye (keratitis), ear, paranasal sinuses, skin, CNS, heart, and GI tract. Intravenous drug injection and iatrogenic procedures (i.e., dialysis, cardiac surgery, and intrathecal medication) can create portals of entry for fungal dissemination.[28]

MYCOTOXIN DISEASE: Aflatoxin, a mycotoxin produced by many aspergillus species, is an IARC Class I carcinogen.[29] Specifically, aflatoxin increases the risk for developing liver cancer in exposed subjects. Increased standardized mortality ratios (SMRs) for liver cancer have been described in Swedish grain millers, animal feed production plant workers, and other agricultural occupations in both Denmark and the United States where aflatoxin exposures likely occur.[14–16]

Treatment

Management of *Aspergillus*-related asthma includes treatment with inhaled steroids and bronchodilators and avoidance of exposure to high airborne concentrations of conidia. Similarly, the mainstay of management of hypersensitivity pneumonitis is the removal from further antigen exposure. Treatment with oral corticosteroids may be indicated in persistently or severely symptomatic individuals with abnormal physiology. For patients with ABPA, treatment with systemic corticosteroids is indicated to prevent or minimize bronchiectasis that occurs with each episode of pneumonia. Antifungal therapy with itraconazole may allow a reduction in corticosteroid dose.[22,30] Therapy for AFS usually includes surgical debridement followed by systemic corticosteroids; however, evidence of efficacy is largely anecdotal. Early diagnosis and treatment are important in the management of invasive aspergillosis. Voriconazole is now the preferred agent for treatment of invasive aspergillosis based on head to head trials showing superiority to amphotericin B.[30] The voriconazole dose is 6 mg/kg IV every 12 hours on the first day, followed by 4 mg/kg IV every 12 hours. Alternative agents include caspofungin, posaconazole, and liposomal amphotericin.[30] Combination therapy is only recommended for salvage after initial treatment failure.[30]

Prevention

Although *Aspergillus* spp. are ubiquitous, immunocompromised patients should be protected from exposure to high concentrations of fungal conidia and should avoid occupational settings where high spore concentrations are likely. High-efficiency filters remove *Aspergillus* from the air of operating rooms and laminar flow rooms. Isolating immunosuppressed patients from dusty hospital renovation and construction appears useful, as does keeping potted plants out of their rooms. High-efficiency particulate air filtration has been used during renovations to minimize exposure risk.[20] Occupational groups such as agricultural workers, bird breeders, sawmill workers, and brewery workers should be provided with adequate ventilation or respiratory protection in circumstances where exposure concentrations are likely to be high. These occupational groups should be educated regarding the exposure risks and instructed to seek early medical attention for recurrent or persistent respiratory and systemic symptoms. Aggressive diagnostic evaluation of suspect hypersensitivity pneumonitis (including early bronchoscopy) is essential, as is the removal from exposure if disease is confirmed.

References

1. Yoshida K, Ueda A, Yamasaki H, et al. Hypersensitivity pneumonitis resulting from *Aspergillus fumigatus* in a greenhouse. *Arch Environ Health* 1993; 48:260–262.

2. Yoshida K, Ando M, Ito K, et al. Hypersensitivity pneumonitis of a mushroom worker due to *Aspergillus glaucus*. *Arch Environ Health* 1990; 45:245–247.
3. Zuk J, King D, Zakhour HD, et al. Locally invasive pulmonary aspergillosis occurring in a gardener; an occupational hazard? *Thorax* 1989; 44:678–679.
4. Land C, Hult K, Fuchs R, et al. Tremorigenic mycotoxins from *Aspergillus fumigatus* as a possible occupational health problem in sawmills. *Appl Environ Microbiol* 1987; 58:787–790.
5. Reijula K, Sutinen S. Immunohistochemical identification of *Aspergillus fumigatus* in farmer's lung. *Acta Histochem* 1984; 75:211–213.
6. Mehta S, Sandhu R. Immunological significance of *Aspergillus fumigatus* in cane-sugar mills. *Arch Environ Health* 1983; 38:41–46.
7. Palmas F, Cosentino S, Cardia P. Fungal airborne spores as health risk factors among workers in alimentary industries. *Eur J Epidemiol* 1989; 5:239–243.
8. Anonymous. Fatal pulmonary aspergillosis following a farm accident (Letter). *Chest* 1984; 85:448–449.
9. Riddle H, Channell S, Blyth W, et al. Allergic alveolitis in a maltworker. *Thorax* 1968; 23:271–280.
10. Huuskonen M, Husman K, Jarvisalo J, et al. Extrinsic allergic alveolitis in the tobacco industry. *Br J Ind Med* 1984; 41:77–83.
11. Wiszniewska M, Tymoszuk D, Nowakowska-Swirta E, et al. Mould sensitization among bakers and farmers with work-related respiratory symptoms. *Ind Health* 2013; 51:275–284.
12. Hagemeyer O, Bunger J, van Kampen V, et al. Occupational allergic respiratory diseases in garbage workers: relevance of molds and actinomycetes. *Adv Exp Med Biol* 2013; 788:313–320.
13. Vincken W, Roels P. Hypersensitivity pneumonitis due to *Aspergillus fumigatus* in compost. *Thorax* 1984; 39:74–75.
14. Alvanja MC, Malker H, Hayes RB. Occupational cancer risk associated with the storage and bulk handling of agricultural foodstuff. *J Toxicol Environ Health* 1987; 22:247–254.
15. Autrup JL, Schmidt J, Autrup H. Exposure to aflatoxin B1 in animal-feed production plant workers. *Environ Health Perspect* 1993; 99:195–197.
16. Peraica M, Radic B, Lucic A, et al. Toxic effects of mycotoxins in humans. *Bull World Health Organ* 1999; 77:756–766.
17. Lawniczek-Watczyk A, Golofit-Szymczak M, Cyprowski M, et al. Exposure to harmful microbiological agents during the handling of biomass for power production purposes. *Med Pr* 2012; 63:395–407.
18. Quirce S, Cuevas M, Diez-Gomez M, et al. Respiratory allergy to *Aspergillus*-derived enzymes in bakers' asthma. *J Allergy Clin Immunol* 1992; 90:970–978.
19. Selim MI, Juchems AM, Popendorf W. Assessing airborne aflatoxin B1 during on-farm grain handling activities. *Am Ind Hyg Assoc J* 1998; 59:252–256.
20. Cornet M, Levy V, Fleury L, et al. Efficacy of prevention by high-efficiency particulate air filtration or laminar airflow against *Aspergillus* airborne contamination during hospital renovation. *Infect Control Hosp Epidemiol* 1999; 20:508–513.
21. Topping MD, Scarisbrick D, Luczynska C, et al. Clinical and immunological reactions to *Aspergillus niger* among workers at a biotechnology plant. *Br J Ind Med* 1985; 42:312–318.
22. Agarwal R, Chakrabarti A, Shah A, et al. Allergic bronchopulmonary aspergillosis: review of literature and proposal of new diagnostic and classification criteria. *Clin Exp Allergy* 2013; 43:850–873.
23. Schwietz L, Gourley D. Allergic fungal sinusitis. *Allergy Proc* 1992; 13:2–6.
24. Robb P. Aspergillosis of the paranasal sinuses: a case report and historical perspective. *J Laryngol Otol* 1986; 100:1071–1077.
25. Conrad DJ, Warnock M, Blanc P, et al. Microgranulomatous aspergillosis after shoveling wood chips: report of a fatal outcome in a patient with chronic granulomatous disease. *Am J Ind Med* 1992; 22:411–418.
26. Hayes GE, Denning DW. Frequency, diagnosis and management of fungal respiratory infections. *Curr Opin Pulm Med* 2013; 19:259–265.
27. Reinwald M, Spiess B, Heinz WJ, et al. Diagnosing pulmonary aspergillosis in patients with hematological malignancies: a multicenter prospective evaluation of an Aspergillus PCR assay and a galactomannan ELISA in bronchoalveolar lavage samples. *Eur J Haematol* 2012; 89:120–127.
28. Leon EE, Craig TJ. Antifungals in the treatment of allergic bronchopulmonary aspergillosis. *Ann Allergy Asthma Immunol* 1999; 82:511–517.
29. Anonymous. Aflatoxins. *IARC Monogr Eval Carcinog Risks Hum* 1993; 56:245–395.
30. Walsh TJ, Anaissie EJ, Denning DW, et al. Treatment of aspergillosis: clinical practice guidelines of the Infectious Diseases Society of America. *Clin Infect Dis* 2008; 46:327–360.

BASIDIOMYCETES (INCLUDING MERULIUS LACRYMANS, LYCOPERDON, AND MUSHROOMS)

Common names for disease: Mushroom spore asthma, hypersensitivity pneumonitis, extrinsic allergic alveolitis, mushroom worker's lung, mushroom compost worker's lung, lycoperdonosis.

Occupational setting

Basidiomycetes are common in nature but they are rarely associated with human disease. Allergic reactions (asthma, rhinoconjunctivitis, and hypersensitivity pneumonitis) to inhaled spores have been described for a number of species.[1] Basidiomycetes are also rare causes of allergic bronchopulmonary mycosis.[2,3] Contact dermatitis has also been described, but infection is very rare.[4] Certain basidiospores (e.g., *Merulius lacrymans*) contaminate wood with 20–25% water content, and mycelia typically extend in sheets over timber and adjacent brickwork.[5,6] Mushroom workers may develop asthma, allergic rhinitis, or hypersensitivity pneumonitis from inhalation of several species of mushroom spores.[7–11] Mushroom soup workers have been known to develop asthma and allergic rhinoconjunctivitis from inhaled mushroom dusts.[12] Removal of stored spent mushroom compost may lead to the release of hazardous concentrations of hydrogen sulfide gas, a respiratory irritant.[13,14]

Exposure (route)

Inhalation of basidiospores causes hypersensitivity lung disease. Contact dermatitis after skin exposure to *Hericium erinaceum* has been reported.[4]

Pathobiology

The *Basidiomycetes* are known as club fungi because, after mycelial growth, a fruiting body is formed and club-shaped structures called *basidia* develop where basidiospores are produced. Many diverse forms are included in this class, including puffballs, common rusts and smuts, mushrooms, and certain yeasts. Spores may be discharged in bursts during times of high humidity.

Atopic asthmatics may demonstrate IgE-mediated reactivity to *Basidiomycete* aeroallergens. *Merulius lacrymans,* a basidiomycete found in buildings in cool temperate climates, has been associated with both asthma and hypersensitivity pneumonitis.[5,6,15] In one report, a teacher developed insidious onset of symptoms of dyspnea, cough, malaise, fever, and weight loss. He also had rales on physical exam. His chest radiograph showed diffuse micronodular infiltrates. Pulmonary function tests showed low diffusing capacity for carbon monoxide. Serum precipitins to *M. lacrymans* present in his home were positive, as was inhalation challenge with the fungus. Clinical recovery progressed slowly following cessation of antigen exposure.

Inhalation of dried powder from the fleshy basidiomycete *lycoperdon* (puffball) has been associated with lycoperdonosis, a form of hypersensitivity pneumonitis that developed following treatment of epistaxis.[16]

In another report, eight workers exposed to dried mushroom soup developed symptoms of asthma and rhinoconjunctivitis.[9] A Type I hypersensitivity reaction was suggested by clinical history, positive immediate skin prick test reactivity to mushroom extracts, and immediate response to inhalation challenge.

Mushroom worker's lung typically occurs during the first few months of employment, though sensitization after many years of exposure is also described.[11] Seven workers at a mushroom farm in Florida developed episodic dyspnea, cough, fever, chills, and myalgias. Pulmonary function tests and chest radiographs showed evidence of hypersensitivity pneumonitis. The workers were subsequently removed from exposure, but a specific causative antigen was not identified.[7] In a similar case series from Japan, specific precipitating antibodies were found to mushroom spore extracts.[11]

Diagnosis

Diagnosis of sensitizing occupational asthma relies on the symptom and exposure histories, findings on pulmonary function testing (including positive methacholine challenge), and results of peak flow monitoring at and away from work. Allergy skin prick testing to mushroom antigen is often (but not always) positive. Diagnosis of hypersensitivity pneumonitis relies on a constellation of clinical findings, including a careful symptom and occupational history.[17] Physical examination may be normal or show basilar lung crackles; the chest radiograph may be normal or show diffuse alveolar or interstitial opacities; pulmonary function tests show restriction, obstruction, or a mixed picture. Fiber-optic bronchoscopy with bronchoalveolar lavage and transbronchial biopsies may be helpful when the clinical suspicion is strong but routine tests are normal. Exercise physiology testing may be helpful in patients who have dyspnea but normal resting pulmonary function. Serum precipitating antibodies to mushroom extracts may be found in asymptomatic exposed workers and may be negative if the wrong antigen preparation is used, limiting their clinical utility.

Treatment

The treatment of hypersensitivity lung diseases begins with early recognition and prompt removal from exposure to the immunogen. Following removal, inhaled corticosteroids are useful as first-line pharmacotherapy for asthma. The efficacy of oral corticosteroids in the treatment of hypersensitivity pneumonitis is unclear, but treatment is probably indicated in patients with severe symptoms, radiographic, or functional abnormalities.

Prevention

Once sensitization has occurred, the prognosis for recovery from occupational asthma is directly related to the duration of exposure. If removal from exposure is delayed, individuals with hypersensitivity pneumonitis are at risk for disease progression. A high index of clinical suspicion for workers in at-risk environments is therefore crucial. Reduction of workplace exposure to high fungal concentrations through engineering and process controls reduces disease risk.

References

1. Rivera-Mariani FE, Nazario-Jimenez S, Lopez-Malpica F, et al. Sensitizationto airborne ascospores, basidiospores, and fungal fragments in allergic rhinitis and asthmatic subjects in San Juan, Puerto Rico. *Int Arch Allergy Immunol* 2011; 155:322–334.
2. Singh PK, Kathuria S, Agarwal K, et al. Clinical significance and molecular characterization of nonsporulating molds isolated from

the respiratory tracts of bronchopulmonary mycosis patients with special reference to basidiomycetes. *J Clin Microbiol* 2013; 51:3331–3337.
3. Ogawa H, Fujimura M, Takeuchi Y, et al. A case of sinobronchial allergic mycosis; possibility of basidiomycetous fungi as a causative antigen. *Intern Med* 2011; 50:59–62.
4. Maes MF, van Baar HM, van Ginkel CJ. Occupational allergic contact dermatitis from the mushroom White Pom Pom (*Hericium erinaceum*). *Contact Dermatitis* 1999; 40:285–290.
5. Herxheimer H, Hyde H, Williams D. Allergic asthma caused by basidiospores. *Lancet* 1969; 2:131–133.
6. O'Brien I, Bull J, Creamer B, et al. Asthma and extrinsic allergic alveolitis due to *Merulius lacrymans*. *Clin Allergy* 1978; 8:535–542.
7. Sanderson W, Kullman G, Sastre J, et al. Outbreak of hypersensitivity pneumonitis among mushroom farm workers. *Am J Ind Med* 1992; 22:859–872.
8. Michils A, DeVuyst P, Nolard J, et al. Occupational asthma to spores of *Pleurotus cornucopiae*. *Eur Resp J* 1991; 4:143–147.
9. Mori S, Nakagawa-Yoshida K, Tsuchihashi H, et al. Mushroom worker's lung resulting from indoor cultivation of *Pleurotus osteatus*. *Occup Med* 1998; 48:465–468.
10. Helbling A, Gayer F, Brander KA. Respiratory allergy to mushroom spores: not well recognized, but relevant. *Ann Allergy Asthma Immunol* 1999; 83:17–19.
11. Akizuki N, Inase N, Ishiwata N, et al. Hypersensitivity pneumonitis among workers cultivating *Tricholoma conglobatum* (Shimeji). *Respiration* 1999; 66:273–278.
12. Symington I, Kerr J, McLean D. Type I allergy in mushroom soup processors. *Clin Allergy* 1981; 11:43–47.
13. Velusami B, Curran TP, Grogan HM. Hydrogen Sulfide gas emissions during disturbance and removal of stored spend mushroom compost. *J Agric Saf Health* 2013; 19:261–275.
14. Velusami B, Curran TP, Grogan HM. Hydrogen sulfide gas emissions in the human-occupied zone during disturbance and removal of stored spent mushroom compost. *J Agric Saf Health* 2013; 19:277–291.
15. Horner WE, Helbling A, Lehrer SB. Basidiomycete allergens. *Allergy* 1998; 53:1114–1121.
16. Strand R, Neuhauser E. Lycoperdonosis. *N Engl J Med* 1967; 277:89–91.
17. Rose C, Lara A. Hypersensitivity pneumonitis. In: Murray JF, Nadel JF (eds.), Textbook of Respiratory Medicine, 5th edn. Philadelphia: Saunders, 2010:1587–1600.

BLASTOMYCES DERMATITIDIS

Common names for disease: Blastomycosis. North American blastomycosis, Gilchrist's disease, Chicago disease, Namekagon fever

Occupational setting

Exposure to the fungus *Blastomyces dermatitidis* can cause the infection blastomycosis. Blastomycosis is most prevalent in the southeastern United States and Ohio-Mississippi River Valley area. However, the geographic range may be broader than previously believed as blastomycosis has been reported in Colorado following prairie dog relocation.[1,2] An African form of blastomycosis has been reported. Disease is much more common in men than in women (9 : 1).

Epidemiologic studies suggest that patients often work outdoors and have intimate contact with soil. A horticulturist developed progressive blastomycosis from exposure to contaminated fertilizer.[3] A tobacco worker in Switzerland and a packing material handler in England developed the illness after handling fungal fomites.[4] There have been occasional reports of small clusters or disease outbreaks in many areas of the United States and Canada.[5] In Virginia, four hunters were infected while raccoon hunting at night in swampy, wooded areas. In a Minnesota outbreak, four cases developed in three families constructing a cabin in a wooded area near a lake.[6] In a 1979 Wisconsin outbreak, seven of eight individuals camping near a river developed acute pneumonia.[7] In a larger outbreak in Wisconsin in 1984, numerous elementary schoolchildren and several adults who visited a beaver pond at a campground developed symptomatic blastomycosis with an incubation period of 21–106 days after exposure to the presumed point source.[8,9] A technician working for several years in a small, dusty, wooden petroleum filtering shed in southwest Ontario developed systemic blastomycosis with meningeal involvement; *B. dermatitidis* was isolated from the earthen floor of the shed.[10]

These data suggest that *B. dermatitidis* survives in wet soil of acid pH with a high organic content and probably exists in point sources close to rivers, streams, or swamps. Environmental conditions during cool months may be more favorable for the saprophytic growth and survival of the fungus; disturbance of these sites either through occupational or avocational activities may lead to airborne dispersal of the spores.

Several cases of laboratory-acquired disease have been reported; the majority resulted from finger inoculation with the yeast form during autopsy by pathologists who developed primary cutaneous blastomycosis.[11–13] Primary pulmonary blastomycosis also can be a laboratory-acquired infection.[14]

Exposure (route)

The most important route of exposure leading to infection from *B. dermatitidis* is disturbance of contaminated point sources leading to airborne dispersal of spores. Accidental skin inoculation of the yeast form has been reported, as has transmission via dog bites. Venereal transmission of genitourinary infection and *in utero* transmission are rare.

Pathobiology

Blastomyces dermatitidis is a dimorphic fungus that grows as a mycelial form at room temperature and as a yeast form at 37°C (Figure 24.3).

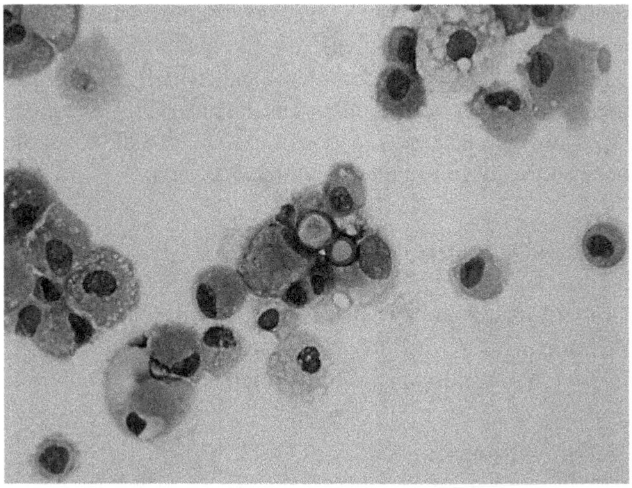

FIGURE 24.3 Bronchoalveolar lavage sample demonstrating typical yeast forms of Blastomycosis. Courtesy of Dr. Dominick Cavuoti and Dr. Francesca Lee.

The lung is the organ most commonly infected by *B. dermatitidis;* the resulting illness is usually indolent in onset and course. Symptoms may be present for weeks, months, or even years before diagnosis. Symptoms typically include cough, weight loss, chest pain, skin lesions, fever, hemoptysis, and localized swelling. In almost half the patients with pulmonary infection, respiratory symptoms are mild or absent. It is usually the systemic symptoms or extrapulmonary lesions that lead to medical attention.

A number of chest radiographic patterns have been described, including a patchy alveolar airspace process with air bronchograms, fibronodular densities, miliary nodules, linear interstitial infiltrates, and cavitation. Pleural effusions and hilar adenopathy are uncommon. Laryngeal, tracheal, or endobronchial lesions are seen occasionally. The rate of progression of indolent disease may be gradual or sudden and rapid. Occasionally, patients present with acute symptoms of fever, productive cough, and pleuritic chest pain. The chest radiograph typically shows single or multiple nodular or patchy infiltrates. Spontaneous improvement of acute blastomycotic pneumonia may occur after 2–12 weeks of symptoms.

Blastomycotic skin lesions involving the face, extremities, neck, and scalp are common and provide ready access for biopsy and culture. Most lesions arise from hematogenous seeding from the lung, although local inoculation may occur in researchers, pathologists, and morticians who handle infected tissue.

Osteomyelitis involving vertebrae, skull, ribs, and distal extremities are found in 14–60% of cases. Osseous lesions may produce symptoms from abscess development in adjacent soft tissue, by spread to contiguous joints, or by vertebral collapse. Radiography shows a sharply defined area of osteolysis.

Blastomycotic arthritis, typically monoarticular, is not rare and first appears as swelling, pain, and limited range of motion in an elbow, knee, or ankle. Infection of the prostate, epididymis, or kidney can be documented in cultured urine in one-fourth of cases. Hematogenous spread to the brain occurs in 3–10% of cases and may present as meningitis, brain abscess, spinal epidural lesions, or cranial lesions. Blastomycotic lymphadenitis and intraocular infection have been reported.

Diagnosis

Diagnosis of blastomycosis requires isolation of the fungus in culture or the demonstration of characteristic yeast-like cells in pus, sputum, or tissue. Mycelial growth is usually evident within 3–14 weeks of incubation at 25–30°C, but cultures should be kept for at least 4 weeks before recording them as negative.

Treatment

Oral itraconazole is the treatment of choice for patients with indolent extracranial blastomycosis. Amphotericin B remains the treatment of choice for patients with severe, rapidly progressive, or CNS infection.[15]

Prevention

Since environmental point sources in soils close to rivers and swamps are difficult to identify and since occupational inhalational exposures are rare, few preventive methods have been identified. Careful handling of laboratory specimens using BSL2 practices and procedures[16] and wearing impermeable gloves will minimize the risk of aerosol exposure and hand inoculation.

References

1. De Groote MA, Bjerke R, Smith H, et al. Expanding epidemiology of blastomycosis: clinical features and investigation in Colorado. *Clin Infect Dis* 2000; 30:582–584.
2. Anonymous. From the Centers for Disease Control and Prevention. Blastomycosis acquired occupationally during prairie dog relocation—Colorado, 1998. *JAMA* 1999; 282:21–22.
3. Sarosi G, Serstock D. Isolation of *Blastomyces dermatitidis* from pigeon manure. *Am Rev Respir Dis* 1976; 114:1179–1193.
4. Anonymous. Blastomycosis. In: Rippon JW (ed.), Medical Mycology: The Pathogenic Fungi and the Pathogenic Actinomycetes. Philadelphia: WB Saunders, 1982:428–458.
5. Dwight PJ, Naus M, Sarsfield P, et al. An outbreak of human blastomycosis: the epidemiology of blastomycosis in the Kenora catchment region of Ontario, Canada. *Can Commun Dis Rep* 2000; 26:82–91.

6. Vaaler A, Bradsher R, Davies S. Evidence of subclincal blastomycosis in forestry workers in northern Minnesota and northern Wisconsin. *Am J Med* 1990; 89:470.
7. Cockerill F, Roberts G, Rosenblatt J, et al. Epidemic of pulmonary blastomycosis (Namekagon fever) in Wisconsin canoeists. *Chest* 1984; 86:688–692.
8. Klein B, Vergeront J, Weeks R, et al. Isolation of *Blastomyces dermatitidis* in soil associated with a large outbreak of blastomycosis in Wisconsin. *N Eng J Med* 1986; 314:529–534.
9. Klein B, Vergeront J, DiSalvo A, et al. Two outbreaks of blastomycosis along rivers in Wisconsin: isolation of *Blastomyces dermatitidis* from riverbank soil and evidence of its transmission along waterways. *Am Rev Respir Dis* 1987; 136:1333–1338.
10. Bakerspigel A, Kane J, Schaus D. Isolation of *Blastomyces dermatitidis* from an earthen floor in southwestern Ontario, Canada. *J Clin Microbiol* 1986; 24:890–891.
11. Larson C, Eckman M, Alber R, et al. Primary cutaneous (inoculation) blastomycosis: an occupational hazard to pathologists. *Am J Clin Pathol* 1983; 79:523–525.
12. Larsh H, Scharz J. Accidental inoculation blastomycosis. *Cutis* 1977; 19:334–337.
13. Kantor G, Roenigk R, Mailin P, et al. Cutaneous blastomycosis. Report of a case presumably acquired by direct inoculation with carbon dioxide laser vaporization. *Cleve Clin J Med* 1987; 54:121–124.
14. Baum G, Lerner P. Primary pulmonary blastomycosis: a laboratory-acquired infection. *Ann Intern Med* 1970; 73:263–269.
15. Champman SW, Dismukes WE, Proia LA, et al. Clinical practice guidelines for the management of blastomycosisi: 2008 update by the infectious diseases society of America. *Clin Infect Dis* 2008; 46:1801–1812.
16. Centers for Disease Control and Prevention, National Institutes of Health. Biosafety in Microbiological and Biomedical laboratories, 5th edn. HHS Publication no. (CDC) 21-1112. New York: U.S. Department of Health and Human Services, 2009.

CANDIDA SPECIES

Common names for disease: Candidiasis, candidosis, thrush, moniliasis

Occupational setting

Candida species are found in soils, especially those with heavy organic debris, and have been recovered from hospital environments and inanimate objects. The intact integument is the most important defense against cutaneous candidiasis. Environmental factors that lead to increased moisture, such as prolonged immersion of hands in water or tight clothing worn in hot climates, increase the risk for cutaneous candidiasis by compromising the tissue.[1] *Candida* paronychia often arises after continued immersion and mechanical irritation of the hands. Homemakers, dishwashers, bartenders, cannery workers, and nurses are at risk for cutaneous candidiasis.[2] Nonoccupational iatrogenic factors (antibiotics, immunosuppressants, barrier breaks, prostheses) and chronic disease such as diabetes are the most common causes of systemic candidiasis. Sensitization to candida antigens has been demonstrated in both farmers and bakers with respiratory symptoms, indicating the potential for candida to cause occupational rhinitis and occupational asthma.[3]

Exposure (route)

Organisms are normal commensals and the vast majority of human infections are of endogenous origin. Person-to-person transmission has been described.

Pathobiology

Candida albicans and the other *Candida* species are budding yeasts that produce mycelia with continued growth (Figure 24.4). *Candida* invasion of the moist areas of the skin produces a red, "scalded skin" lesion with a scalloped border. Satellite pustular lesions surround the primary lesion, and dry scaly lesions may also occur.

Diagnosis

Skin scrapings examined microscopically in potassium hydroxide (KOH) show budding yeast and mycelial hyphae.

Treatment

Nystatin ointment, topical amphotericin, gentian violet, and a number of other topical treatments are effective in the treatment of *Candida* paronychia and intertriginous candidiasis.

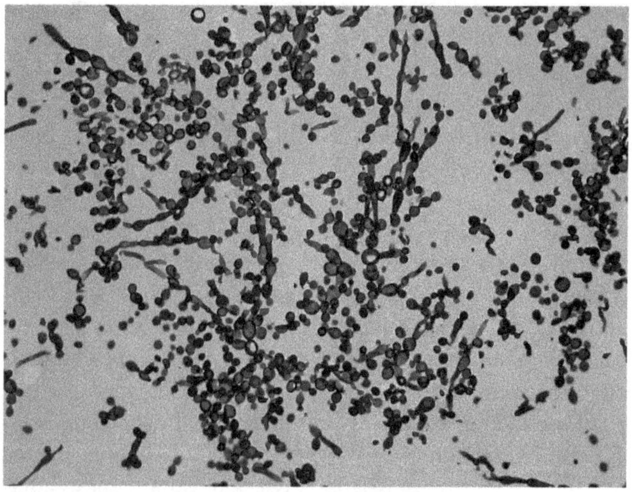

FIGURE 24.4 Grocott's methenamine silver stain showing Candida yeast forms. Courtesy of Dr. Dominick Cavuoti and Dr. Francesca Lee.

Prevention

Avoidance of tight clothing in tropical climates, tight boots that macerate the skin, and prolonged immersion of hands will prevent tissue compromise and diminish the risk of cutaneous candidiasis.

References

1. Campbell M, Stewart J. The Medical Mycology Handbook. New York: John Wiley & Sons, Inc., 1980:244–252.
2. Hunter P, Harrison G, Fraser C. Cross-infection and diversity of *Candida albicans* strain, carriage in patients and nursing staff on an intensive care unit. *J Med Vet Mycol* 1990; 28:317–325.
3. Wiszniewska M, Tymoszuk D, Nowakowska-Swirta E, et al. Mould sensitization among bakers and farmers with work-related respiratory symptoms. *Ind Health* 2013; 51:275–284.

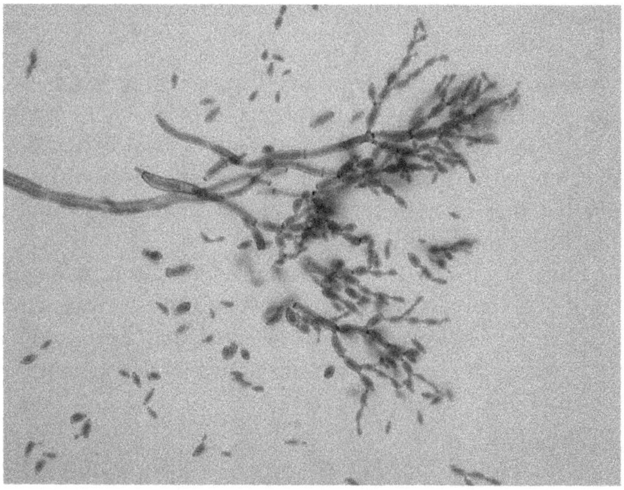

FIGURE 24.5 Photomicrograph of Cladosporium conidia. Courtesy of Dr. Dominick Cavuoti and Dr. Francesca Lee.

CLADOSPORIUM SPECIES

Common names for disease: asthma, allergic rhinoconjunctivitis, and hypersensitivity pneumonitis (HP)

Occupational setting

Cladosporium spp. are ubiquitous in nature. Peak ambient air levels generally occur in summer and early fall.[1,2] Farmers, agricultural workers, and occupants of water-damaged buildings are at risk for exposure and associated hypersensitivity lung diseases.[3] Reports in sawmills have also described allergic disease due to cladosporium exposure.[4] Workers cleaning mold off of salami during production developed hypersensitivity pneumonitis attributed to cladosporium.[5] Cladosporium was one of the molds associated with hypersensitivity pneumonitis in a saxophone player from contamination of the instrument.[6] Rarely, skin, corneal, CNS, and pulmonary infections may occur.

Exposure (route)

Exposure occurs primarily through inhalation of airborne conidia or spores (Figure 24.5). Skin infection may occur with direct inoculation.

Pathobiology

Cladosporium sensitization occurs with a prevalence of about 3%[7] and is associated with allergic rhinitis and eczema.[8] *Cladosporium* sensitization has also been associated with increased bronchial hyperresponsiveness and is a risk factor for both the development of asthma and fatal asthma attacks.[7,9,10] Occupational asthma related to *Cladosporium* sensitization is rarely reported. Other hypersensitivity reactions to *Cladosporium* include allergic bronchopulmonary mycosis and hypersensitivity pneumonitis (HP).[11,12] HP developed in a 48-year-old woman after exposure to a contaminated indoor hot tub, and *Cladosporium* was confirmed as the cause by positive specific challenge.[12]

There are approximately 30 reports of brain abscess caused by *Cladosporium trichoides* in the literature, primarily in immunocompromised hosts.[13] Skin and pulmonary infections also occur but are uncommon.

Diagnosis

Diagnosis of asthma relies on symptom and exposure histories, findings on pulmonary function testing (including positive nonspecific bronchial challenge), and the results of peak flow monitoring. Allergy skin prick testing and specific IgE to *Cladosporium* are often but not always positive. Diagnosis of hypersensitivity pneumonitis relies on a constellation of clinical findings, including a careful symptom and exposure history.[14] Physical examination may be normal or show basilar lung crackles; the chest radiography may be normal or show diffuse alveolar or interstitial opacities; and pulmonary function tests show restriction, obstruction, or a mixed picture. Fiber-optic bronchoscopy with bronchoalveolar lavage and transbronchial biopsies may be helpful when the clinical suspicion is strong but routine tests are normal. Exercise physiology testing may be helpful in patients who have dyspnea but normal resting pulmonary function.

Diagnosis of infectious disease related to *Cladosporium* requires positive culture or the demonstration fungal forms in histological specimens.

Treatment

As with all hypersensitivity lung diseases, prompt removal from exposure is essential. Inhaled corticosteroids are useful as first-line pharmacotherapy for asthma. Oral corticosteroids may be indicated in patients with hypersensitivity pneumonitis manifested by severe symptoms and radiographic or functional abnormalities.

Fluconazole at a dose of 400 mg/day in combination with surgery has successfully treated CNS infection.[15] Progressive disease is generally treated with intravenous amphotericin B; however, this has not been shown to alter the outcome.[13]

Prevention

Rapid remediation of water damage in homes and office buildings will prevent fungal growth and minimize exposure. Proper attention to building practices during new construction will help prevent subsequent leaks and water damage. In agricultural settings, engineering and process controls can reduce exposure to high fungal concentrations. Workers in at-risk industries should be educated regarding exposure risks and encouraged to seek early medical attention for persistent respiratory and systemic symptoms.[14]

References

1. Mediavilla MA, Angulo RJ, Dominguez VE, et al. Annual and diurnal incidence of *Cladosporium* conidia in the atmosphere of Cordoba, Spain. *J Invest Allerg Clin Immunol* 1997; 7:179–182.
2. Ren P, Nakun TM, Leaderer BP. Comparisons of seasonal fungal prevalence in indoor and outdoor air and in house dusts of dwellings in one Northeast American county. *J Expos Analysis Environ Epidemiol* 1999; 9:560–568.
3. Kotimaa MH, Terho EO, Husman K. Airborne moulds and actinomycetes in work environment of farmers. *Eur J Respir Dis Suppl* 1987; 152:91–100.
4. Klaric MS, Varnai VM, Calusic AL, et al. Occupational exposure to airborne fungi in two Croatian sawmills and atopy in exposed workers. *Ann Agric Environ Med* 2012; 19:213–219.
5. Marvisi M, Balzarini L, Mancini C, et al. A new type of Hypersensitivity Pneumonitis: salami brusher's disease. *Monaldi Arch Chest Dis* 2012; 77:35–37.
6. Metzger F, Haccuria A, Reboux G, et al. Hypersensitivity pneumonitis due to molds in a saxophone player. *Chest* 2010; 138:724–726.
7. Chinn S, Jarvis D, Luczynska C, et al. Individual allergens as risk factors for bronchial responsiveness in young adults. *Thorax* 1998; 53:662–667.
8. Bundgaard A, Boudet L. Reproducibility of early asthmatic response to *Cladosporium herbarum*. *Eur J Resp Dis Suppl* 1986; 143:37–40.
9. Abramson M, Kutin JJ, Raven J, et al. Risk factors for asthma among young adults in Melbourne, Australia. *Respirology* 1996; 1:291–297.
10. Black PN, Udy AA, Brodie SM. Sensitivity to fungal allergens is a risk factor for life-threatening asthma. *Allergy* 2000; 55:501–504.
11. Moreno-Ancillo A, Diaz-Pena JM, Ferrer A, et al. Allergic bronchopulmonary cladosporiosis in a child. *J Allergy Clin Immunol* 1996; 97:714–715.
12. Jacobs RL, Thorner RE, Holcomb JR, et al. Hypersensitivity pneumonitis caused by *Cladosporium* in an enclosed hot-tub area. *Ann Int Med* 1986; 105:204–206.
13. Dixon DM, Walsh TJ, Merz WG, et al. Infections due to *Xylohypha bantiana (Cladosporium trichoides)*. *Rev Infect Dis* 1989; 11:515–525.
14. Rose C, Lara A. Hypersensitivity pneumonitis. In: Murray JF, Nadel JF (eds.), Textbook of Respiratory Medicine, 5th edn. Philadelphia: Saunders, 2010:1587–1600.
15. Turker A, Altinors N, Aciduman A, et al. MRI findings and encouraging fluconazole treatment results of intracranial *Cladosporium trichoides* infection. *Infection* 1995; 23:60–62.

COCCIDIOIDES IMMITIS

Common names for disease: Coccidioidomycosis, valley fever, desert rheumatism, valley bumps, California disease, Posada's mycosis

Occupational setting

Coccidioides immitis is a soil fungus that is endemic in the semiarid or desert-like regions of the United States, Mexico, Guatemala, Honduras, Colombia, Venezuela, Bolivia, Paraguay, and Argentina.[1] *Coccidioides immitis may* be dispersed by wind or by disruptions from construction work, farming, or archeological digs. Outbreaks after natural disasters (e.g., earthquakes) have been reported.[2] Occupations at risk for developing coccidioidomycosis include agricultural workers, construction crews, telephone post diggers, archeologists, and military personnel traveling to endemic areas.[3,4] Other outdoor activities in arid endemic regions also carry risk. A recent outbreak occurred in the cast and crew of a television program filming outdoors.[5] Outbreaks in armadillo hunters in Brazil have also been reported.[6] Laboratory workers are at risk for infection from inhalation of the arthroconidia.[7] Although there are no racial, gender, or age differences in susceptibility to primary coccidioidomycosis, dark-skinned races and pregnant women are more prone to severe primary illness and disseminated disease.[1]

Cases of occupational person-to-person transmission are very rare. Six medical staff members were infected after inhaling arthrospores that had grown on the plaster cast of a patient with coccidioidal osteomyelitis. An embalmer developed disease after accidentally piercing his skin with a needle during preparation of the body of a victim of disseminated coccidioidomycosis.[8] There have been occasional reports of coccidioidomycosis in Georgia, Virginia, and North Carolina among textile workers who inhaled dust particles from wool or cotton shipped from the San Joaquin Valley.

Exposure (route)

The primary route of exposure is inhalation. Skin inoculation and transplacental infection from mothers with disseminated disease have rarely been reported.[9]

Pathobiology

Coccidioides immitis exists in the mycelial phase in soil, where it matures to form arthroconidia that can be inhaled. In the host, these spores swell to form thick-walled, nonbudding, round cells that contain endospores (Figure 24.6).

In most cases, coccidioidomycosis is a mild respiratory infection or is completely asymptomatic. Primary coccidioidomycosis occurs in ~40% of patients with positive coccidioidin skin tests. Symptoms typically begin 7–21 days after exposure; they include fever, pleuritic or dull chest pain, cough, white or blood-streaked sputum, and constitutional symptoms of malaise, headache, anorexia, myalgia, and fever. A fine, diffuse, erythematous skin rash often occurs within the first few days of illness. Erythema nodosum and erythema multiforme are more common in Caucasian women with primary coccidioidomycosis; they are accompanied by arthralgias of the knee or ankle in a third of cases. The rash, arthralgias, and mild conjunctivitis that often occur are probably all manifestations of exuberant delayed-type hypersensitivity reactions to *C. immitis* antigens. The chest radiograph typically shows alveolar infiltrates, with or without hilar adenopathy. Paratracheal or mediastinal adenopathy suggests that infection may be spreading beyond the lung. Small pleural effusions may occur, but large effusions are uncommon. Laboratory studies often show a mild leukocytosis, elevated sedimentation rate, and eosinophilia. Conversion of the coccidioidin skin test to positive occurs 2–21 days after onset of symptoms. The appearance of complement fixing antibodies in serum is often delayed.

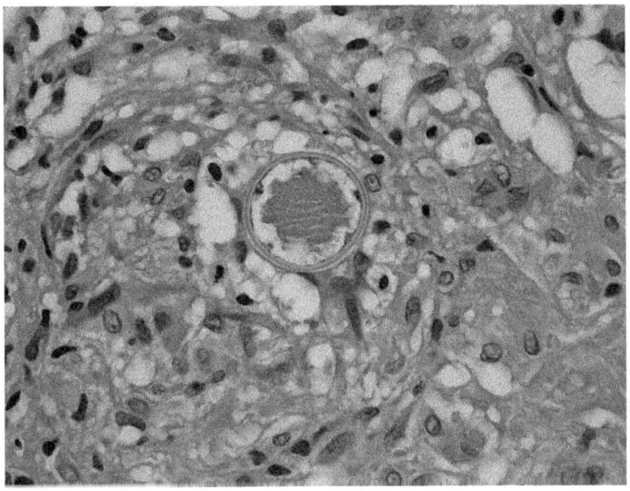

FIGURE 24.6 Tissue sample hematoxylin and eosin stain revealing a typical spherule of Coccidioidomycosis. Courtesy of Dr. Dominick Cavuoti and Dr. Francesca Lee.

Coccidioidal pneumonia may resolve by forming dense spherical nodules (coccidioidomas) in the area of infiltrate. The mass may cavitate, leaving a single thin-walled cavity; approximately half of these cavities close spontaneously within 2 years. Though hemoptysis can occur, most are asymptomatic. Extension of the cavity to the pleura can cause bronchopleural fistula, pneumothorax, and coccidioidal empyema.[10] Acute coccidioidal pneumonia may disseminate rapidly; it is potentially fatal. Some individuals develop a chronic progressive form of pulmonary coccidioidomycosis that mimics tuberculosis, with apical fibronodular lesions and cough, weight loss, fever, and chest pain of many months duration.

Extrapulmonary dissemination occurs in <5% of cases, most often in dark-skinned men. Pregnancy also appears to increase the risk of dissemination. Infection in later stages of pregnancy results in increasing morbidity and mortality for the mother. In disseminated infection, skin and subcutaneous lesions are the most common manifestations. Bone lesions occur in 20% of patients with disseminated disease. Meningitis occurs in one-third to one-half of cases of disseminated coccidioidomycosis, usually with a subacute or chronic presentation including headache, lethargy, confusion, or decreased memory. Anorexia, nausea, weight loss, and ataxia may occur. The most valuable diagnostic test for meningitis is the complement fixation test for cerebrospinal fluid (CSF) antibody to *C. immitis,* which is positive in 75–95% of cases. Multiple organ systems may be involved in disseminated coccidioidomycosis. Patients are at risk for anterior and posterior uveitis, lymphadenitis, cystitis, renal abscess, orchitis, epididymitis, peritonitis, urethroscrotal fistula, laryngitis, and otitis.

Diagnosis

Together with the typical clinical manifestations, a positive coccidioidin skin test is suggestive for disease in someone who has recently traveled to an endemic area. Coccidioidal serology on acute and convalescent sera is a reliable means of diagnosis. Current tests that use enzyme-linked immunoassay test kits with proprietary antigens are more sensitive than traditional serologies and can convert to positive at an earlier disease phase.[1] Recovery of *C. immitis* from sputum, urine, or bronchial washing is definitive, but a negative culture does not exclude the diagnosis. Microscopic identification of *C. immitis* spherules on wet smear can aid in diagnosis. In disseminated coccidioidomycosis with meningitis, CNS serologic tests are helpful. Demonstration of

C. immitis by culture, smear, or biopsy is the most definitive diagnostic test in disseminated disease. Fungemia is detected in ~12% of disseminated cases; it is an extremely grave prognostic sign.

Treatment

Because spontaneous cure is common, treatment is not usually necessary for acute pulmonary coccidioidomycosis.[11] Intravenous amphotericin B may be indicated in some very ill patients with primary illness without proof of dissemination in an effort to prevent extrapulmonary foci. The only effective treatment for cavitary disease is surgical resection, with intravenous amphotericin B serving an adjunctive role. Repeated bacterial superinfection and the presence of an expanding cavity near the pleural surface are indications for resection of the cavities. Treatment of patients with chronic, indolent apical coccidioidal pneumonia is difficult and requires at least 1 year of therapy. Initial treatment with azoles is generally preferred. For those patients who fail initial therapy, options include switching to an alternative azole or to intravenous amphotericin B. When infection is confined to one lobe, combination treatment with intravenous amphotericin B and resection may be useful. Adjunctive interferon gamma may also be useful in refractory disease.[12]

In the treatment of disseminated coccidioidomycosis, therapy with an oral azole is begun unless the patient is severely ill, in which case intravenous amphotericin B is used. When amphotericin B is used initially, an oral azole is usually substituted once the patient has stabilized.[11] Oral ketoconazole, itraconazole, or fluconazole following a course of treatment with intravenous amphotericin B can help in management of skin lesions, subcutaneous abscesses, and joint effusions. Complete cure is elusive, and improvement often takes many months. Oral fluconazole is the preferred treatment for coccidioidal meningitis, often beginning at high doses of 800 mg/day. Intrathecal amphotericin B may be added initially.[11] For those patients who respond, lifelong therapy with oral azoles is required.[11,13]

Prevention

Given the considerable danger of laboratory infection by *C. immitis,* precautionary measures in handling cultures should be emphasized. The organism should be handled using BSL2 procedures and practices in clinical laboratories.[14] Petri dishes should not be used for isolation of the organism from clinical specimens. Subculturing and harvesting of the arthrospores should be carried out under a laminar flow hood or other isolation hood using BSL3 practices and facilities. Viable plate cultures should never be discarded or sent through the mail. For outdoor work in endemic areas, dust control procedures should be followed; the California Department of Public Health has published guidelines.[15]

References

1. Nguyen C, Barker BM, Hoover S, et al. Recent advances in our understanding of the environmental, epidemiological, immunological, and clinical dimensions of coccidioidomycosis. *Clin Microbiol Rev* 2103; 26:505–525.
2. Schneider E, Hajjeh RA, Spiegel RA, et al. A coccidioidomycosis outbreak following the Northridge, Calif, earthquake. *JAMA* 1997; 277:904–908.
3. (a)Johnson W. Occupational factors in coccidioidomycosis. *J Occup Med* 1981; 23:367–374. (b)El-Ani A, Elwood C. A case of coccidioidomycosis with unique clinical features. *Arch Intern Med* 1978; 138:1421–1422.
4. Stander SM, Schooner W, Galgiani JN, et al. Coccidioidomycosis among visitors to a *Coccidioides immitis*-endemic area: an outbreak in a military reserve unit. *J Infect Dis* 1995; 171:1672–1675.
5. Wilken JA, Marquez P, Terashita D, et al. Coccidioidomycosis among cast and crew memebers at an outdorr television filming event – California 2012. *MMWR Morb Mortal Weekly Rep* 2014; 63:321–324.
6. Brillhante RS, Moreira Filho RE, Rocha MF, et al. Coccidioiomycosis in armadillo hunters from the state of Ceara, Brazil. *Mem Inst Oswaldo Cruz* 2012; 107:813–815.
7. Kohn G, Linne S, Smith C. Acquisition of coccidioidomycosis at necropsy by inhalation of coccidioidal endospores. *Diagn Microbiol Infect Din* 1992; 15:527–530.
8. Canoil F, Haley K, Brown J. Primary cutaneous coccidioidomycosis: a review of the literature and a report of a new case. *Arch Dermatol* 1977; 113:933–936.
9. Charlton V, Ramsdell K, Sehring S. Intrauterine transmission of coccidioidomycosis. *Pediatr Infect Dis J* 1999; 18:561–563.
10. Shekhel TA, Ricciotti RW, Blair JE, et al. Surgical pathology of pleural coccidioidomycosis: a clinicopathological study of 36 cases. *Hum Pathol* 2014; 45:961–969.
11. Galgiani JN, Ampel NM, Blair JE, et al. Coccidioidomycosis: IDSA guidelines. *Clin Infect Dis* 2005; 41:1217–1223.
12. Duplessis CA, Tilley D, Bavaro M, et al. Two cases illustrating successful adjunctive interferon-gamma immunotherapy in refractory disseminated coccidioidomycosis. *J Infect* 2011; 63:223–228.
13. Dewsnup DH, Galgiani JN, Graybill JR, et al. Is it ever safe to stop azole therapy for *Coccidioides immitis* meningitis? *Ann Intern Med* 1996; 124:305–310.
14. Centers for Disease Control and National Institutes of Health. Biosafety in Microbiological and Biomedical Laboratories, 3rd edn. Washington, DC: U.S. Government Printing Office, 1993:79.
15. Das R, McNary J, Fitzsimmons K, et al. Occupational coccidioidomycosis in California: outbreak investigation, respirator recommendations, and surveillance findings. *J Occup Environ Med* 2012; 54:564–571.

CRYPTOCOCCUS NEOFORMANS AND CRYPTOCOCCUS GATTII

Common names for disease: Cryptococcosis, torulosis, European blastomycosis

Occupational setting

The most important natural source of *Cryptococcus* is weathered droppings from pigeons and soil contaminated with avian droppings. The organism is most likely to be found in old pigeon droppings that have accumulated over years in roosting sites such as towers, window ledges, hay mows of barns, and upper floors of old buildings. *Cryptococcus neoformans* has also been isolated from the droppings of other birds, from dairy products, soil, wood, rotting vegetables and fruits, and swallows' nests.[1] Cryptococcal antibodies are detected more commonly in pigeon breeders than in other occupational groups,[2] but the infection rate is no greater because the disease mainly affects immunocompromised hosts (including patients with AIDS, sarcoidosis, lymphoma, and those requiring chronic steroids). For example, a recent report describes a case of cryptococcosis secondary to exposure to contaminated chicken manure in a patient with Crohn's disease on immunosuppression.[3] The organism was cultured from bagpipes played by a patient with leukemia who developed pulmonary cryptococcosis.[4] Cases of cryptococcosis rarely occur in clusters, and there is no clear occupational predisposition. Histories of exposure to pigeons or dust are usually unhelpful. *Cryptococcus gattii* is responsible for an unexplained outbreak of disease among immunocompetent individuals in the Pacific Northwest.[5]

Exposure (route)

Inhalation of fungal spores is the major route of entry.

Pathobiology

Cryptococcus is an encapsulated yeast-like fungus that reproduces by budding into 4–6 μm diameter cells (Figure 24.7) that can cause disease when they are aerosolized and inhaled.

The most common clinical manifestation of cryptococcosis is infection of the cerebral cortex, brain stem, cerebellum, or meninges.[6] Symptom onset may be insidious (with headache, dizziness, irritability, subtle altered mental status, personality change, and visual symptoms) or explosive (with rapid deterioration and death within 2 weeks of onset).

Pulmonary cryptococcosis has a variety of clinical manifestations and an unpredictable course. A self-limited pneumonia with indolent onset and symptoms of dry cough, chest pain, and little or no fever can occur. The chest radiograph typically shows one or more well-circumscribed areas of pneumonitis, occasionally with central cavitation. Pleural effusions, hilar adenopathy, and calcification are rare. Resolution often requires months of treatment and occasionally progresses to chronic pneumonia. The most serious outcome in cryptococcal pneumonia is silent dissemination to the central nervous system. Patients with underlying lung disease may develop asymptomatic colonization of the bronchial tree with *C. neoformans*.

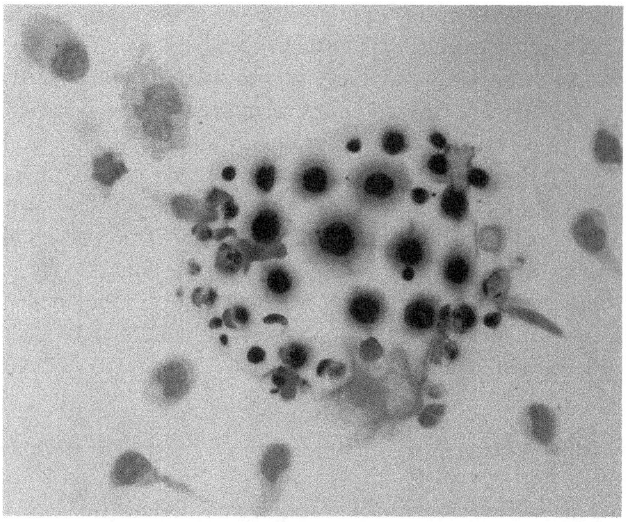

FIGURE 24.7 Gram stain of fluid obtained from lumbar puncture demonstrating encapsulated *Cryptococcus* yeast forms. Courtesy of Dr. Dominick Cavuoti and Dr. Francesca Lee.

Papular skin lesions from hematogenous dissemination occur in ~10% of patients with cryptococcosis; such lesions are more common in immunocompromised patients. Local cutaneous cryptococcosis as a result of direct inoculation may occur in immunocompetent individuals.[7] Bone and joint involvement may also occur following hematogenous dissemination; vertebral lesions are the most common site. Ocular lesions of cryptococcosis include chorioretinitis, papilledema, optic atrophy, scotomata, and ocular motor palsies. Rarely, cryptococcosis may involve the genitourinary tract, heart valves, liver, sinuses, and GI tract.

Diagnosis

The most important procedure in the diagnosis of cryptococcal meningitis is lumbar puncture, which characteristically shows pleocytosis, elevated protein, and hypoglycorrhachia. India ink smear of CSF shows the encapsulated yeast in 50% of cases, and cryptococcal antigen is detected in 94% of cases. Diagnosis of cryptococcal pneumonia is often challenging, since cryptococci are scanty in sputum except in cases of cavitary lung disease or widely disseminated infection. Cultures of sputum and bronchoalveolar lavage (BAL) are occasionally helpful. Cryptococcal antigen can be measured in BAL fluid. Early studies showed high diagnostic sensitivity of BAL cryptococcal antigen at 98%; however, follow-up investigations indicate a sensitivity closer to 70%.[8,9] Positive serum cryptococcal antigen is suggestive, but it occurs only in cases with extensive pulmonary infiltrates or extrapulmonary dissemination. Transbronchial biopsy is occasionally helpful, but surgical lung biopsy is often necessary to confirm the diagnosis. Central punch biopsy of cutaneous cryptococcosis with culture and histology has a high diagnostic yield.

Treatment

An aggressive search for disseminated disease (including cerebrospinal fluid culture, multiple urine cultures, and blood cultures) is necessary in patients with pulmonary cryptococcosis. If results are negative, chemotherapy can be withheld except in patients at risk for dissemination, such as those with steroid therapy, diabetes, HIV infection, or underlying malignancy. However, careful follow-up of immunocompetent patients with suspected illness is required. Therapy for pulmonary cryptococcosis is recommended if the serum level of cryptococcal antigen is greater than 1 : 8.[5] For immunosuppressed patients or patients with severe involvement, pulmonary cryptococcosis should be treated the same as meningeal disease.[5] Fluconazole at a dose of 400 mg/day for 6–12 months is the recommended therapy for mild to moderate pulmonary disease in normal hosts.[5] Itraconazole, voriconazole, and posaconazole are acceptable alternatives. For CNS or severe disseminated disease, guidelines recommend induction therapy with 2 weeks of amphotericin B plus flucytosine followed by at least 8 weeks of fluconazole at 400 mg/day and then 6–12 months of 200 mg/day of fluconazole for maintenance.[5] For nonimmunosuppressed patients, a longer induction of 4 weeks is recommended if no neurologic complications are present, with an additional 2 weeks if neurologic complications are present.[5] Intrathecal amphotericin B has been used for refractory CNS disease. All patients with meningitis should be evaluated for elevated intracranial pressure. Daily large volume lumbar puncture to reduce intracranial pressure until a normal opening pressure is achieved on several consecutive days is recommended. Symptomatic hydrocephalus should be treated by ventriculoperitoneal shunt even if viable cryptococci are still present in CSF. Treatment for cryptococcosis in AIDS patients is beyond the scope of this discussion.

Prevention

Since cryptococcosis is typically a disease of the immunocompromised host and since occupational cases are rare, few preventive strategies have been identified. In the laboratory, BSL2 procedures and practices should be followed.[10] In endemic areas, pigeon dropping controls should be utilized.

References

1. Gordon M. Cryptococcosis: a ubiquitous hazard. *Occup Health Saf* 1980; 49:61–63.
2. Tanphaichitra D, Sahaphongs S, Srirnuang S. Cryptococcal antigen survey among racing pigeon workers and patients with cryptococcosis, pythiosis, histoplasmosis and penicilliosis. *Int J Clin Pharmacol Res* 1988; 8:433–439.
3. Fraison JB, Guilpain P, Schiffmann, A, et al. Pulmonary cryptococcosis in a patient with Crohn's disease treated with prednisone, azathioprine and adalimumab: exposure to chicken manure as a source of contamination. *J Crohns Colitis* 2013; 7:e11–e14.
4. Cobcroft R, Kronenberg H, Wilkinson T. *Cryptococcus* in bagpipes [Letter]. *Lancet* 1978; 1:1368–1369.
5. Perfect JR, Dismukes WE, Dromer F, et al. Clinical practice guidelines for the management of cryptococcal disease: 2010 update by the Infectious Diseases Society of America. *Clin Infect Dis* 2010; 50:291–322.
6. White P, Kaufman L, Weeks R, et al. Cryptococcal meningitis: a case report and epidemic- logic study. *J Med Assoc Ga* 1982; 71:539–542.
7. Micalizzi C, Persi A, Parodi A. Primary cutaneous cryptococcosis in an immunocompetent pigeon keeper. *Clin Exp Dermatol* 1997; 22:195–197.
8. Baughman RP, Rhodes JC, Dohn MN, et al. Detection of cryptococcal antigen in bronchoalveolar lavage fluid: a prospective study of diagnostic utility. *Am Rev Respir Dis* 1992; 145:1226–1229.
9. Kralovic SM, Rhodes JC. Utility of routine testing of bronchoalveolar lavage fluid for cryptococcal antigen. *J Clin Microbiol* 1998; 36:3088–3089.
10. Centers for Disease Control and Prevention, National Institutes of Health. Biosafety in Microbiological and Biomedical Laboratories, 5th edn. HHS Publication no. (CDC) 21–1112. New York: U.S. Department of Health and Human Services, 2009.

CRYPTOSTROMA CORTICALE

Common names for disease: Maple bark disease, maple bark stripper's disease

Occupational setting

Wood and sawmill workers engaged in the debarking of logs prior to cutting are at risk for developing hypersensitivity pneumonitis or asthma from a variety of fungi that contaminate wood, including *Cryptostroma corticale*.[1,2]

Exposure (route)

Inhalation of respirable spores can cause sensitization and subsequent occupational lung disease.

Pathobiology

Maple bark disease is a rare disorder that can affect both trees and humans. Both sensitizing asthma and hypersensitivity pneumonitis from the fungus contaminating maple bark have been described.[1,2] In one report, five workers in the wood room of a paper mill, where logs were debarked and cut, developed cough, dyspnea, chest tightness, fever, and weight loss during the winter months, when workplace ventilation was minimal.[3] Physical examination showed pulmonary crackles; chest radiographs showed a diffuse reticulonodular infiltrate, occasionally with patchy alveolar infiltrates; and arterial oxygen saturations were reduced in three of the five patients. Lung histology demonstrated granulomas and scattered fibrosis; fungal spores were detected by methenamine silver staining of lung tissue in four individuals.

Diagnosis

Diagnosis of sensitizing occupational asthma relies on symptom and exposure histories, findings on pulmonary function testing (including positive nonspecific bronchial challenge), and results of peak flow monitoring at and away from work. Diagnosis of hypersensitivity pneumonitis relies on a constellation of clinical findings, including a careful symptom history.[4] Physical examination may be normal or show basilar crackles; the chest radiograph may be normal or show diffuse alveolar or interstitial opacities; and pulmonary function testing may be normal or show restriction, obstruction, or a mixed picture. Fiber-optic bronchoscopy with bronchoalveolar lavage and transbronchial biopsies may be helpful when the clinical suspicion is strong but routine tests are normal. Exercise physiology testing may be helpful in patients with dyspnea but normal resting pulmonary function. Serum precipitins are often found in asymptomatic exposed workers and may be negative if the wrong antigen preparation is used.

Treatment

A strong index of clinical suspicion and prompt removal of a sensitized worker from the antigen-containing environment are the mainstays of therapy. Treatment with inhaled steroids for asthma and with oral corticosteroids for severe hypersensitivity pneumonitis may be indicated in some patients.

Prevention

Removal of symptomatic individuals from the spore-laden environment and changes in the manufacturing process to reduce spore concentrations should lead to eradication of the disease.

References

1. Towey J, Sweany H, Huron W. Severe bronchial asthma apparently due to fungus spores found in maple bark. *JAMA* 1932; 99:453–459.
2. Emanuel D, Lawton B, Wenzel F. Maple-bark disease; pneumonitis due to *Cryptostroma corticale*. *N Engl J Med* 1962; 266:333–337.
3. Emanuel D, Wenzel J, Lawton B. Pneumonitis due to *Cryptostroma corticale* (maple-bark disease). *N Engl J Med* 1966; 274:1413–1418.
4. Rose C, Lara A. Hypersensitivity pneumonitis. In: Murray JF, Nadel JF (eds.), Textbook of Respiratory Medicine, 5th edn. Philadelphia: Saunders, 2010:1587–1600.

FONSECAEA AND OTHER AGENTS OF CHROMOMYCOSIS

Common names for disease: Chromomycosis, chromoblastomycosis

Occupational setting

Chromomycosis is a chronic cutaneous and subcutaneous fungal infection that occurs most commonly in the tropics and subtropics among barefoot agricultural workers.[1] Corneal infections may also occur.[2] These opportunistic fungi are common in soil, decayed vegetation, and rotting wood. Handling lumber and sitting on wooden planks in Finnish saunas are additional documented sources of infection.[3]

Exposure (route)

Traumatic inoculation of fungi into the skin is the main mode of infection. Person-to-person transmission has not been documented.

Pathobiology

Fonsecaea pedrosi is the most commonly isolated agent of chromomycosis, accounting for the majority of cases in Brazil and 61% of the cases in Madagascar[1,4]; *Cladosporium carrionii* is the major pathogen in South Africa, Venezuela, and Australia.[5] All species produce slow-growing, 4–12 μm

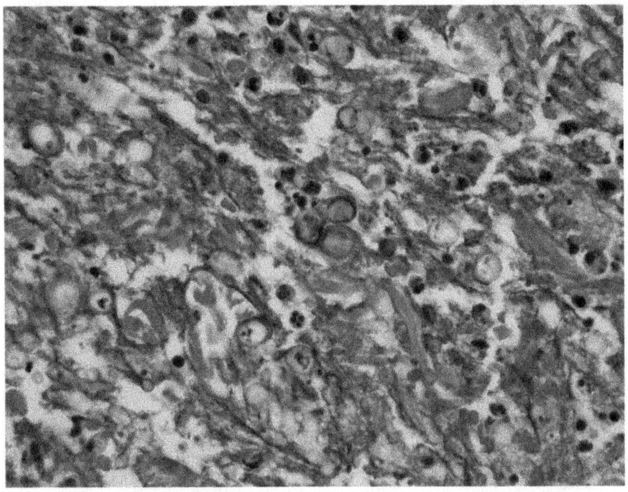

FIGURE 24.8 Tissue sample hematoxylin and eosin stain revealing a cluster of brown thick walled cells of *Fonsecaea*. Courtesy of Dr. Dominick Cavuoti and Dr. Francesca Lee.

round, brown, thick-walled cells, often in clumps; hyphae may be seen in crusts from lesions (Figure 24.8). The infections are slow growing, with an average duration of illness prior to therapy of 11 years.[6]

The most common manifestation is verrucous cutaneous infection. Over 90% of patients are male, and the lower limb is the site of infection in approximately 80%. Verrucous lesions are secondary to suppurative granulomas with large numbers of fungal cells.[7] Early ulcerated nodules develop into cauliflower-like masses. Small ulcerations ("black dots") are seen on the warty surface; they may be pruritic but are rarely painful. The second most common lesion is a well-delimited erythematous plaque or cicatricial lesion that on histology features tuberculoid-type granulomas with fewer fungal organisms.[7] Scarring can cause lymphostasis and lymphedema of the involved extremity. Hematogenous spread to brain, lymph nodes, and other organs is rare.

Diagnosis

Characteristic pigmented sclerotic bodies ("copper pennies" when seen microscopically) are present in tissue and exudate in all types of chromomycosis. Fungi appear as long, septate, and branched hyphal forms in crusts and exudate. Serologic testing is unhelpful, and culture is only positive in a minority of cases.[4]

Treatment

In early stages, when lesions are small, wide, and deep, surgical excision is the treatment of choice. Medical therapy for chromomycosis has been disappointing. Topical antifungals, potassium iodide, amphotericin B, 5-fluorocytosine, ketoconazole, and local thermotherapy, alone or in combination, have all had varying degrees of success. Treatment with terbinafine has shown promise, with cure rates of 82% at 1 year in open trials; however, randomized controlled trials are still needed.[8] Itraconazole, voriconazole, or combination therapy consisting of itraconazole with terbinafine or amphotericin plus terbinafine have reported success. The newer azoles have the best *in vitro* activity.[9] Photodynamic therapy has also shown good clinical responses in refractory cases.[10]

Prevention

Since close contact with soil is the most prevalent predisposing condition, appropriate protective clothing is recommended to prevent cutaneous inoculation.

References

1. Silva JP, de Souza W, Rozental S. Chromoblastomycosis: a retrospective study of 325 cases on Amazonic region. *Mycopathologia* 1998–1999; 143:171–175.
2. Barton K, Miller D, Pflugfelder SC. Corneal chromoblastomycosis. *Cornea* 1997; 16:235–239.
3. Sonck CE. Chromomycosis in Finland. *Dermatologia* 1975; 19:189–193.
4. Esterre P, Andriantsimahavandy A, Ramarcel ER, et al. Forty years of chromoblastomycosis in Madagascar: a review. *Am J Trop Med Hyg* 1996; 55:45–47.
5. McGinnis M. Chromoblastomycosis and phaeophyphomycosis: new concepts, diagnosis and mycology. *J Am Acad Dermatol* 1983; 8:1–16.
6. Pires CA, Xavier MB, Quaresma JA, et al. Clinical, epidemiological and mycological report on 65 patients from the Eastern Amazon region with chromoblastomycosis. *An Bras Dermatol* 2012; 87:555–560.
7. Avelar-Pires C, Simoes-Quaresma JA, Moraes-de Macedo GM, et al. Revisiting the clinical and histopathological aspects of patients with chromoblastomycosis from the Brazilian Amazon region. *Are Med Res* 2013:44:302–306.
8. Esterre P, Inzan CK, Ramarcel ER, et al. Treatment of chromomycosis with terbinafine: preliminary results of an open pilot study. *Br J Dermatol* 1996; 134(suppl 46):33–36.
9. Najafzadeh MJ, Badali H, Illnait-Zaragozi MT, et al. In vitro activities of eight antifungal drugs against 55 clinical isolates of *Fonsecaea* spp. *Antimicrob Agents Chemother* 2010; 54:1636–1638.
10. Lu S, Lu C, Zhang J, et al. Chromoblastomycosis in Mainland China: a systematic review on clinical characteristics. *Mycopathologia* 2013; 175:489–495.

HISTOPLASMA CAPSULATUM

Common names for disease: Histoplasmosis, Ohio Valley disease, cave disease, Tingo Maria fever, Darling's disease, reticuloendotheliosis

Occupational setting

Working on or under structures that have been habitats for birds or bats can lead to histoplasmosis.[1,2] Epidemics of acute pneumonia due to *Histoplasma capsulatum* result from group exposures to inhaled particulates containing high concentrations of the fungus.[3] For example, an outbreak occurred among 13 of 24 students on a biology field trip to an endemic area. The exposure source was a hollow bat-infested tree.[4] Common sites of outbreaks are bat-infested caves, starling roosts, and old chicken houses with dirt floors.[5] Outbreaks related to disposal of bird droppings also occur.[6] The endemic areas with the highest concentrations of disease are located in the eastern United States (Ohio River Valley) and Latin America. Activities such as exploring bat-infested caves, clearing bird roosts, or cleaning chicken houses are associated with the disease.[7] Cleanup, construction, or demolition activities in urban areas may be associated with inhalation of airborne conidia.[2,8] Cases of laboratory-acquired pulmonary histoplasmosis have been reported. Accidental inoculation in a hospital worker assisting on an autopsy and in a laboratory worker who accidentally pricked his thumb with a contaminated needle led to primary cutaneous histoplasmosis.

Infection caused by *H. capsulatum* variant *duboisii* occurs in tropical Africa. Human cases of *duboisii* infection have been associated with bat-infested caves and chicken roosts, suggesting that the var. *duboisii* shares the same ecological niche as the var. *capsulatum*.

Exposure (route)

The major route of exposure is inhalation of airborne conidia. Cutaneous inoculation has been reported.

Pathobiology

Histoplasma capsulatum is a yeast-like fungus with oval, budding, uninucleate cells measuring 1.5–2.0 μm × 3.0–3.5 μm that are often found within macrophages in viable tissue (Figure 24.9). The mycelial form is found in soil and bears infectious spores called microconidia (2–6 μm) and macroconidia (8–14 μm).

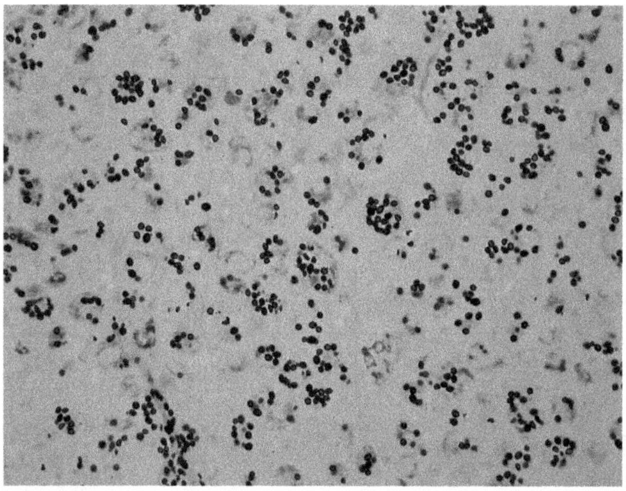

FIGURE 24.9 Grocott's methenamine silver stain showing Histoplasma yeast forms. Courtesy of Dr. Dominick Cavuoti and Dr. Francesca Lee.

Most infections are mild or subclinical. They are diagnosed in retrospect by a positive skin test or small, scattered pulmonary calcifications. Acute pulmonary histoplasmosis typically presents with symptoms of cough, pleuritic chest pain, fever, chills, myalgias, malaise, nausea, anorexia, and weight loss. The chest radiograph shows pulmonary infiltrates, usually patchy, finely nodular, and involving both lungs, often with hilar adenopathy. There are often slight elevations in peripheral blood leukocyte count and erythrocyte sedimentation rate. Illness may be mild or severe, accompanied by hypoxemia. Pleural effusions resolve over several weeks. Scattered calcification throughout the lung fields is a hallmark of healed acute pulmonary histoplasmosis. Microcalcifications of the spleen may be seen on the chest radiograph. Healing of a localized pulmonary infiltrate may lead to formation of a pulmonary nodule called a histoplasmoma.

Acute pulmonary histoplasmosis may result in lymphatic spread, leading to hilar or mediastinal lymphadenitis. In severe cases, granulomatous inflammation with central areas of caseation may replace entire mediastinal structures, causing fibrosing mediastinitis. Massive adenopathy typically appears in the hilar or right paratracheal area on chest radiograph. Obstruction of pulmonary veins causes hemoptysis and dyspnea. Heart failure and tracheobronchial hemorrhage are lethal complications of pulmonary venous obstruction. Histoplasmosis is the most common nonmalignant cause of superior vena cava syndrome, another potential sequela. Compression of the recurrent laryngeal nerve can cause hoarseness. Pericarditis may also occur, leading to potentially lethal complications such as tamponade and constrictive pericarditis. Esophageal complications of fibrosing mediastinitis may include ulceration, tracheoesophageal fistula, or traction diverticulum.

Chronic pulmonary histoplasmosis occurs most commonly in middle-aged men with underlying emphysema or chronic bronchitis. Symptoms include cough and sputum production, chest pain, dyspnea, malaise, weakness, fever, weight loss, and easy fatigability. Hemoptysis may also occur. The chest radiograph shows interstitial infiltrates, predominantly in the upper lobes. Nodular areas may slowly shrink or cavitate and expand. The course is marked by progressive hypoxemia, dyspnea, hemoptysis, bacterial pneumonia, and cor pulmonale. Laboratory abnormalities may include anemia of chronic disease, mild leukocytosis, and elevated alkaline phosphatase.

Hematogenously disseminated histoplasmosis is a rare but often lethal complication that occurs most commonly in immunosuppressed patients (e.g., those with hematologic malignancies, AIDS, or on high-dose corticosteroids). The clinical presentation may vary from acute to indolent. A variety of organ systems may be affected, causing meningitis, endocarditis, ulcerated lesions of the intestinal tract, hepatitis, and mucous membrane lesions of the oropharynx, face, and external genitalia. Ocular histoplasmosis occurs rarely. Laryngeal involvement can also occur.[9,10]

Diagnosis

A point source exposure to *H. capsulatum* should be suspected when several individuals develop respiratory illness two weeks following a common outdoor exposure. Serodiagnosis is presumptive but should be sought by immunodiffusion testing early in the disease course and by fourfold elevation of complement fixation titer between acute and convalescent sera. Cultures of blood, bone marrow, and urine should be obtained in hospitalized patients to rule out disseminated infection. Urinary antigen testing has a sensitivity approaching 90% for disseminated disease. Unfortunately, sensitivity is less than 50% for localized pulmonary infection. Specificity of urinary antigen testing is excellent although false positive reactions in patients with paracoccidioides can occur.[11] Serum and bronchoalveolar lavage (BAL) fluid testing for antigen are promising additions. In one study, the sensitivity of serum antigen testing was 83% for acute and 88% for chronic pulmonary histoplasma.[12] BAL antigen testing also has a higher yield than urine.[13]

In mediastinitis and pericarditis, Gomori methenamine silver (GMS) staining of biopsied lymph node sections has the best chance of demonstrating organisms. Sputum culture is recommended for diagnosis of chronic pulmonary histoplasmosis. If sputum is inadequate, bronchoalveolar lavage may be indicated. In disseminated histoplasmosis, taking a smear or biopsy or urinary antigen testing often leads to a faster diagnosis than awaiting culture result. Transbronchial biopsies may be useful in patients with diffuse radiographic infiltrates. In AIDS patients, GMS staining of bronchoalveolar lavage fluid has a high yield. Liver biopsy is useful in the setting of hepatic enlargement or abnormal liver function tests. Cultures of blood, bone marrow, and focal lesions have the highest yield for positive results.

Treatment

In most patients with acute pulmonary histoplasmosis, spontaneous improvement has begun at or before diagnosis, so no therapy is necessary. However, symptom persistence for more than 1 month is an indication for therapy.[14] Some immunocompetent patients will develop severe diffuse pneumonia and respiratory failure after high spore exposures, requiring therapy with liposomal amphotericin B for 2 weeks, followed by 3 months of itraconazole. Corticosteroids are recommended in the first 2 weeks of acute fulminant histoplasmosis in those with significant respiratory complications to attenuate the inflammatory response.[14] Early surgery to relieve pericardial tamponade and to confirm the diagnosis may be useful in pericarditis; however, antifungal therapy is often not required in this setting.[14] The role of surgery in early mediastinal infection has not been adequately evaluated, but late in the course of fibrosing mediastinitis, complications including massive hemoptysis from venous obstruction may require surgical management. Neither high-dose steroid therapy nor antifungal chemotherapy appears to be beneficial in fibrosing mediastinitis.

For chronic pulmonary histoplasmosis, itraconazole is the drug of choice at a dose of 200 mg 3 times daily for 3 days, followed by twice daily. The duration of therapy is typically 1 year but some experts treat for up to 2 years to reduce the relapse rate. Blood itraconazole levels should be obtained after 2 weeks. Ketoconazole is an effective alternative, but fluconazole has a higher rate of treatment failure. In patients where compliance is a problem or in those who have contraindications to azole therapy, intravenous amphotericin B may be useful.

In patients with disseminated histoplasmosis, liposomal amphotericin B is recommended for moderate to severe disease for 1–2 weeks, followed by itraconazole at the above dosing. Itraconazole is the treatment of choice for mild disease. The total duration of therapy is at least 1 year, but immunosuppressed patients may require lifelong suppressive therapy.[14]

Prevention

Avoidance of circumstances in which *H. capsulatum* is likely to be found in high concentrations is the best approach to prevention. Although there are no data on efficacy, fit-tested negative pressure respirators with HEPA filters or powered air purifying respirators probably decrease the risk for inhalation exposure when cleaning or bulldozing bird roosts or chicken houses, activities that tend to increase the number of airborne spores. BSL2 practices and facilities are recommended for handling and processing clinical specimens and for identifying cultures and isolates in diagnostic laboratories. BSL3 practices and procedures should be used for manipulating identified cultures and for processing soil or other environmental materials that contain infectious spores.[15]

References

1. Sorley DL, Levin ML, Warren JW, et al. Bat-associated histoplasmosis in Maryland bridgeworkers. *Am J Med* 1979; 67:623–6.
2. Anonymous. Case records of the Massachusetts General Hospital. *N Engl J Med* 1991;325:949–56.
3. Goodwin R, Loyd J, Des Prez RM. Histoplasmosis in normal hosts. *Medicine (Baltimore)* 1981; 60:231–66.
4. Cottle LE, Gkrania-Klotsas E, Williams HJ. A multinational outbreak of histoplasmosis following a biology field trip in the Ugandan rainforest. *J Travel Med* 2013;20:83–7.
5. Stobierski MG, Hospedales CJ, Hall WN, et al. Outbreak of histoplasmosis among employees in a paper factory – Michigan, 1993. *J Clin Microbiol* 1996;34:1220–3.
6. Tosh F, Doto I, Beecher S, et al. Relationship of starling-blackbird roosts and endemic histoplasmosis. *Am Rev Respir Dis* 1970;101:283–6.
7. Furcolow M. Environmental aspects of histoplasmosis. *Arch Environ Health* 1975;10:4–8.
8. Dean A, Bates J, Sorrels C, et al. An outbreak of histoplasmosis at an Arkansas courthouse with five cases of probable reinfection. *Am J Epidemiol* 1978;108:36–46.
9. Teoh JW, Hassan F, Yunus M. Laryngeal histoplasmosis: an occupational hazard. *Singapore Med J* 2013;54:e208–10.
10. Durkin MM, Connolly PA, Wheat LJ. Comparison of radioimmunoassay and enzyme-linked immunoassay methods for detection of Histoplasma capsulatum var. capsulatum antigen. *J Clin Microbiol* 1997;35:2252–5.
11. Taylor ML, Perez-Mejia A, Yamamoto-Furusho JK, et al. Immunologic, genetic and social human risk factors associated to histoplasmosis: studies in the State of Guerrero, Mexico. *Mycopathologia* 1997; 138:137–42.
12. Hage CA, Ribes, JA, Wengenack NL et al. A multicenter evaluation of tests for diagnosis of histoplasmosis. *Clin Infect Dis* 2011;53:448–54.
13. Hage CA, Knox KS, Davis TE, et al. Antigen detection in bronchoalveolar lavage fluid for diagnosis of fungal pneumonia. *Curr Opin Pulm Med* 2011;17:167–71.
14. Wheat JL, Freifeld AG, Kleiman MB, et al. Clinical practice guidelines for the management of patients with histoplasmosis: 2007 update by the Infectious Diseases Society of America. *Clin Infect Dis* 2007;45:807–25.
15. Centers for Disease Control and Prevention, National Institutes of Health. *Biosafety in microbiological and biomedical laboratories*, 5th edn. HHS Publication no. (CDC) 21-1112, 2009.

MADURELLA SPECIES AND OTHER AGENTS OF MYCETOMA

Common names of disease: Mycetoma, Madura foot, maduromycetoma, maduromycosis

Occupational setting

Cases of this chronic indolent infection are most often seen in tropical and subtropical countries, such as India, Mexico, sub-Saharan Africa, and Venezuela. In some cases, saprophytic soil fungi enter the hands or feet after local trauma such as a thorn prick, or they enter the chest wall and back from soil-contaminated sacks carried on the shoulders; in other cases, mycetomas form on the head and neck as a result of carrying bundles of wood.[1] Mycetomas (Figure 24.10) occur most frequently in male farmers and other rural laborers exposed to penetrating wounds from thorns and splinters. Inadequate nutrition and hygiene are probably contributory factors.

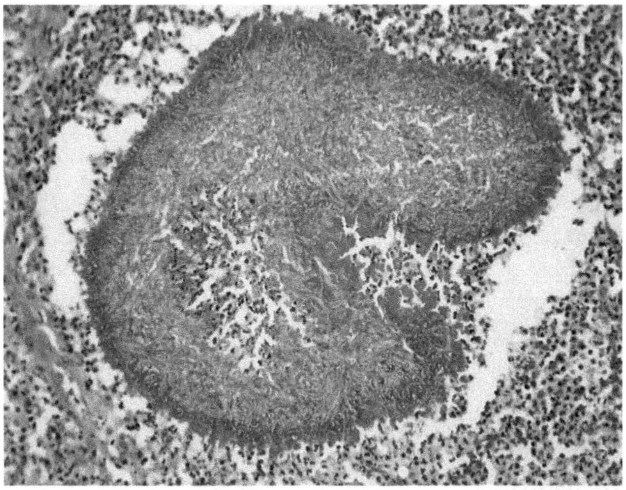

FIGURE 24.10 Tissue sample hematoxylin and eosin stain revealing a fungal mycetoma from Madurella. Courtesy of Dr. Dominick Cavuoti and Dr. Francesca Lee.

Exposure (route)

The route of infection is through skin inoculation. Hematogenous spread has been rarely reported. Recent phylogenetic research indicates *Madurella* lives nested with other fungi in animal dung and enriched soil.[2]

Pathobiology

Causal fungi include *Pseudallescheria (Petriellidium) boydii, Madurella mycetomatis, Madurella grisea, Acremonium (Cephalosporium)* species, *Fusarium* species, *Exophiala (Phialophora) jeanselmei*, and a number of others.

A triad of signs—indurated swelling, multiple sinus tracts draining grain-filled pus, and localization to the foot (the most common site of infection)—characterize mycetomas.[3] The primary lesion is a locally invasive, indolent, and painless subcutaneous swelling that slowly enlarges, causing subsequent distortion, pain, and disability. The radiographic findings include necrosis, osteolysis, and bone fusion.

Diagnosis

In addition to the classic clinical triad, characteristic grains in draining sinuses can be seen on hematoxylin–eosin staining. Gomori methenamine silver or periodic acid–Schiff staining will detect hyphae in tissue. Speciation requires culture of the grain and isolation of the organism.

Treatment

Surgical resection of a localized mycetoma may be necessary. Medical therapy is often unsuccessful, but newer triazoles may be of benefit. *In vitro* studies show isavuconazole has strong activity against *Madurella*.[4] Likewise, posaconazole has been used with reported success.[5] In a larger series, itraconazole at high dose improved the lesions and appeared to enhance encapsulation of the organisms making surgical resection easier.[6] *Madurella* is not susceptible to the echinocandins.[7]

Prevention

Since the major predisposing factor is inoculation through bare feet in contact with soil, adequate shoes and clothing are recommended. Improvements in nutrition and hygiene would undoubtedly diminish the risk for infection as well. Reducing contact with cow dung or cow dung enriched soil may also be beneficial.[2]

References

1. Sugar AM. Agents of mucormycosis and related species. In: Mandell GL, Douglas RG, Bennett JE (eds.), Principles and Practice of Infectious Disease, 3rd edn. New York: Chruchill Livingstone, 1990:1962–1972.
2. De Hoog GS, Ahmed SA, Najafzadeh MJ, et al. Phylogenetic findings suggest possible new habitat and routes of infection of human eumycetoma. *PloS Negl Trop Dis* 2013; 7:e2229.
3. Butz W, Ajello L. Black grain mycetoma. *Arch Dermatol* 1971; 104:197–201.
4. Kloezen W, Meis JF, Curfs-Greuker I, et al. In vitro antifungal activity of isavuconazole against *Madurella mycetomatis*. *Antimicrob Agents Chemother* 2012:56:6054–6056.
5. Difonzo EM, Massi D, Vanzi L, et al. *Madurella mycetomatis* mycetoma treated successfully with oral posaconazole. *J Chemother* 2011; 23:243–244.
6. Fahal AH, Rahman IA, El-Hassan AM, et al. The safety and efficacy of itraconazole for the treatment of patients with eumycetoma due to *Madurella mycetomatis*. *Trans R Soc Trop Med Hyg* 2011; 105:127–132.
7. Van de Sande WW, Fahal AH, Bakker-Woudenberg IA, et al. *Madurella mycetomatis* is not susceptible to the echinocandin class of antifungal agents. *Antimicrob Agents Chemother* 2010; 54:2738–2740.

PARACOCCIDIOIDES BRASILIENSIS

Common names for disease: Paracoccidioidomycosis, South American blastomycosis, Brazilian blastomycosis, paracoccidioidal granuloma, Lutz's disease

Occupational setting

Paracoccidioidomycosis is a chronic granulomatous disease that is geographically restricted to areas of Central and South America. The etiologic agent, *Paracoccidioides brasiliensis*, is found in soil in humid mountain forests. Over 5000 cases have been reported, the majority from Brazil.[1] Disease is much more common in men than women (7–70:1) and typically occurs between the ages of 20 and 50. Women appear to acquire the disease at a younger age than men.[2] Most cases occur in rural workers such as tree cutters (46% of all cases in one region of Brazil occurred in rural occupations[2]), and most patients are at least moderately malnourished.

Exposure (route)

The primary route of entry is by fungal inhalation into the lungs. Local trauma leading to inoculation with the organism is less common.

Pathobiology

Paracoccidioides brasiliensis is a dimorphic fungus that forms 2–30 μm round budding cells that are released when small.

The pulmonary infection is usually subclinical. It then disseminates to form ulcerative granulomata of the buccal, nasal, and occasionally gastrointestinal mucosa. In clinically evident lung disease, the alveolitis is manifested as patchy bilateral radiographic infiltrates with hilar adenopathy. Occasionally, chronic progressive pulmonary disease can occur, leading to diffuse cavitary and alveolar involvement. Symptoms and signs include dyspnea, productive cough, chest pain, fever, and rales. Hematogenous and lymphatic dissemination without lung involvement is more common, especially to mucous membranes and mucocutaneous junctions. Papules first become vesicles, then granulomatous ulcers. The spleen, liver, CNS, bones, lymph nodes, and intestine may be involved.

Diagnosis

Sputum, crusts, material from the granulomatous bases of ulcers, biopsies of lesions, and pus from draining lymph nodes contain fungal elements, typically budding yeast forms. Serologic studies (complement fixation, immunodiffusion) are usually positive,[3] although there is some concern about interlaboratory variability.[4] Antigen tests for gp43 and gp70 are available and can be used for diagnosis and to monitor treatment.[5,6]

Treatment

Most cases are self-limiting. Itraconazole is the drug of choice for cases requiring treatment.[4] Amphotericin B may be required for extensive pulmonary and severe disseminated forms of infection.

Prevention

Disease is limited to endemic areas in Central and South America, where prevention of malnutrition and other diseases (such as Chagas' disease and tuberculosis) may decrease the risk for paracoccidioidomycosis in rural workers.

References

1. Franco M, Montenegro M, Mendes R, et al. Paracoccidioidomycosis: a recently proposed classification of its clinical forms. *Rev Soc Bras Med Trop* 1987; 20:129–132.
2. Blotta MH, Mamoni RL, Oliveira SJ, et al. Endemic regions of paracoccidioidomycosis in Brazil: a clinical and epidemiologic study of 584 cases in the southeast region. *Am J Trop Med Hyg* 1999; 61:390–394.
3. Restrepo A, Robledo M, Giraldo R, et al. The gamut of paracoccidioidomycosis. *Am J Med* 1976; 61:33–42.
4. Vidal MS, Del Negro GM, Vicentini AP, et al. Serological diagnosis of paracoccidioidomycosis: high rate of inter-laboratorial variability among medical mycology reference centers. *PLoS Negl Trop Dis* 2014; 8(9):e3174.
5. de Camargo ZP. Serology of paracoccidioidomycosis. *Mycopathologia* 2008; 165:289–302.
6. Dos Santos PO, Rodrigues AM, Fernandes GF, et al. Immunodiagnosis of paracoccidioidomycosis due to paracoccidioides brasiliensis using a latex test: detection of specific antibody anti-gp43 and specific antigen gp43. *PLoS Negl Trop Dis* 2015; 9(2):e0003516.

PENICILLIUM SPECIES

Common names for disease: Humidifier lung, suberosis (*Penicillium frequentans*), cheese washer's disease (*Penicillium casei*, *Penicillium roqueforti*), cheese worker's lung, woodman's disease, allergic bronchopulmonary mycosis (ABPM), Penicilliosis (*Penicillium marneffei*), Peat moss worker's lung (*Penicillium citreonigrum*).

Occupational setting

Because the blue-green *Penicillium* molds (Figure 24.11) are ubiquitous in nature, they are common contaminants of indoor environments. Exposure to *Penicillium* spp. has been associated with hypersensitivity lung disease in cork workers,[1] cheese workers,[2,3] laboratory workers, farmers,[4] tree cutters,[5] sawmill workers, other handlers of mold-contaminated wood,[6] peat moss workers,[7] salami factory workers,[8] and sausage production workers.[9] Contaminated humidifier water and moldy HVAC (heating, ventilation, and air conditioning) systems have been associated with *Penicillium*-induced hypersensitivity pneumonitis.[10,11] *Penicillium* spores are also encountered in poultry farming and coconut production.[12,13]

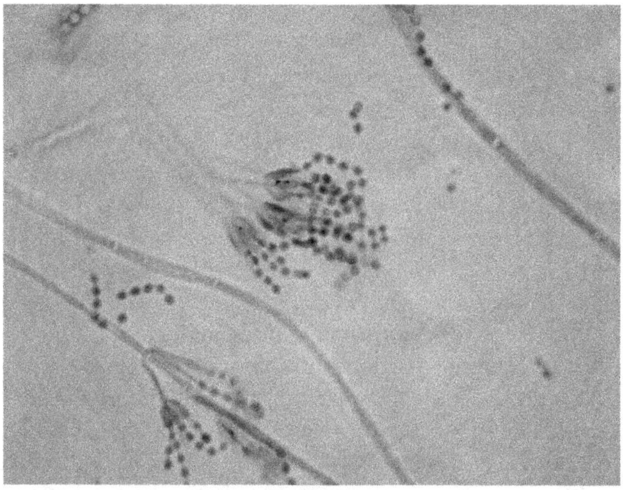

FIGURE 24.11 Photomicrograph of Penicillium conidia. Courtesy of Dr. Dominick Cavuoti and Dr. Francesca Lee.

Exposure (route)

Inhalation of airborne spores is the major route of entry.

Pathobiology

A variety of *Penicillium* species have been associated with hypersensitivity lung diseases; the most common is hypersensitivity pneumonitis, although asthma has also been described. *Penicillium frequentans* spores in the air of factories where cork bark is processed can cause suberosis, a form of hypersensitivity pneumonitis. *Penicillium casei* and *P. roqueforti* have been associated with hypersensitivity pneumonitis in cheese workers exposed to *Penicillium*-contaminated cheese. Several *Penicillium* species isolated from contaminated humidifier water were shown to induce a precipitating antibody response in an entomologist exposed to mists generated by a reservoir type of humidifier. Woodworkers, including those exposed to mold-contaminated fuel chips, debarking of live trees, and sawmill particulates, are at risk for *Penicillium*-induced hypersensitivity pneumonitis. Workers exposed to *Penicillium*-contaminated peat moss have also developed hypersensitivity pneumonitis.

The clinical presentation of hypersensitivity pneumonitis is variable, ranging from severe, acute respiratory, and systemic symptoms to subtle, chronic symptoms.[13,14] Acute illness is manifested by fevers, chills, cough, dyspnea, abnormal chest radiograph, and leukocytosis; improvement is seen within a few days following removal from exposure. The more subacute or chronic illness is manifested by insidious onset of cough and progressive dyspnea on exertion. Pulmonary physiology may be normal, show isolated obstruction, or show the more classic restrictive or mixed restrictive and obstructive pattern. Exercise physiology often demonstrates gas exchange abnormalities. The chest radiograph may be normal or show inhomogeneous, patchy alveolar infiltrates, or interstitial opacities High-resolution CT scans typically show diffuse, fine, and poorly defined centrilobular micronodules. Serum precipitating antibodies to *Penicillium* species are often positive.

Rare cases have been reported of allergic bronchopulmonary penicilliosis causing intermittent airways obstruction, transient pulmonary infiltrates, blood and sputum eosinophilia, a positive dual skin test (Types I and III), and precipitating antibodies in serum.[15] Proximal saccular bronchiectasis is often found in the segment of lung containing the infiltrate. Bronchial hygiene alone or in combination with inhaled or oral corticosteroids is usually effective treatment.

Penicillium infections of clinical importance are very rare. There have been case reports of *Penicillium* infection of the ear, foot, urinary bladder, and lung. *Penicillium* presenting as a solitary pulmonary nodule in a nonimmunocompromised host has also been reported. *Penicillium marneffei* is endemic in Southeast Asia and is also found in Africa. It has been associated with recurrent episodes of hemoptysis attributed to bronchitis and bronchiectasis. Histopathologically, lung tissue shows granulomata with central areas of necrosis and neutrophilic infiltration with many yeast-like tissue-forming cells of *P. marneffei.* In addition, *P. marneffei* has caused disseminated infection manifested by fever, weight loss, anemia, and skin lesions (Penicilliosis) in immunocompromised hosts.[16,17] *Penicillium marneffei* can cause peritonitis in peritoneal dialysis patients.[18]

Treatment

Treatment of *Penicillium*-induced hypersensitivity pneumonitis relies on removal of the affected individual from exposure to the contaminated environment. Systemic steroids have been used in severely ill patients with interstitial pneumonitis, but controlled clinical trials are lacking. Treatment of *Penicillium*-induced asthma involves elimination of antigen exposure and use of inhaled steroids and bronchodilators. Allergic bronchopulmonary penicilliosis is rare, but treatment should be similar to that for ABPA, with inhaled or oral corticosteroids to prevent recurrent pneumonitis and subsequent bronchiectasis. Treatment of *P. marneffei* generally includes induction with amphotericin B followed by itraconazole therapy.[19]

Prevention

Attention should be paid to safe handling of all types of solid fuel (wood, chips, and peat) and other materials in which mold may grow. Dry storage, prevention of mold growth, and use of appropriate respiratory protection are important.

Avoidance or elimination of water damage to HVAC systems is important in preventing significant indoor mold contamination. Regular HVAC maintenance and inspection, appropriate filtration of outside air, and provision of indoor environments free from water intrusion are crucial to prevent fungal amplification and dissemination. Hard surfaces supporting fungal growth in indoor environment should be cleaned with dilute bleach (1 : 10–1 : 50 dilution); then, the surface should be rinsed with clean water and dried. Mold-contaminated materials such as furniture, draperies, and insulation material should be discarded.

A variety of measures have been introduced in the cheese production industry to reduce contamination with airborne molds. These measures include wrapping cheese in foil or plastic film before entering the aging room, thus preventing surface mold formation; careful temperature and humidity control in aging rooms; and removal of surface mold contamination before it becomes an aerosolized dust.

References

1. Avila R. Lacey T. The role of *Penicillium frequentans* in suberosis (respiratory disease in workers in the cork industry). *Clin Allergy* 1974; 4:109–117.
2. Campbell J, Kryda M, Treuhaft M, et al. Cheese worker's hypersensitivity pneumonitis. *Am Rev Respir Dis* 1983; 127:495–496.
3. Schlueter D. "Cheesewasher's disease": a new occupational hazard? *Ann Int Med* 1973; 78:606.
4. Nakagawa-Yoshida K, Ando M, Etches RI, et al. Fatal cases of farmer's lung in a Canadian family. Probable new antigens, *Penicillium brevicompactum* and *P. olivicolor. Chest* 1997; 111:245–248.
5. Dykewicz M. Laufer P, Patterson R, et al. Woodman's disease: hypersensitivity pneumonitis from cutting live trees. *J Allergy Clin Immunol* 1988; 81:455–460.
6. Van Assendelft A, Raitio M, Turkia V. Fuel chip-induced hypersensitivity pneumonitis caused by *Penicillium* species. *Chest* 1985; 87:394–396.
7. Cormier Y, Israel-Assayag I, Bedard G, et al. Hypersensitivity pneumonitis in peat moss workers. *Am J Respir Crit Care Med* 1998; 158:412–417.
8. Marvisi M, Balzarini L, Mancini C, et al. A new type of hypersensitivity pneumonitis: salami brusher's disease. *Monaldi Arch Chest Dis* 2012; 77:35–37.
9. Morell F, Cruz MJ, Gomez FP, et al. Chacinero's lung – hypersensitivity pneumonitis due to dry sausage dust. *Scand J Work Environ Health* 2011; 37:349–356.

10. Baur X, Behr J, Dewair M, et al. Humidifier lung and humidifier fever. *Lung* 1988; 166:113–124.
11. Bernstein K, Sorenson W, Garabrant D, et al. Exposures to respirable, airborne *Penicillium* from a contaminated ventilation system: clinical, environmental and epidemiological aspects. *Am Ind Hyg Assoc J* 1983; 44:161–169.
12. Richerson H, Bernstein I, Fink J, et al. Guidelines for the clinical evaluation of hypersensitivity pneumonitis. *J Allergy Clin Immunol* 1989; 84:839–844.
13. Rimac D, Macan J, Varnai VVM, et al. Exposure to poultry dust and health effects in poultry workers: impact of mould and mite allergens. *Int Arch Occup Environ Health* 2010; 83:9–19.
14. Nascimento Mdo D, Leitao VM, Neto Silva MA, et al. Eco-epidemiologic study of emerging fungi related to the work of babacu coconut breakers in the State of Maranhao, Brazil. *Rev Soc Bras Med Trop* 2014; 47:74–78.
15. Knutsen AP, Bush RK, Demain JG, et al. Fungi and allergic lower respiratory tract diseases. *J Allergy Clin Immunol* 2012; 129:280–291.
16. Rose C, Lara A. Hypersensitivity pneumonitis. In: Murray JF, Nadel JF (eds.), Textbook of Respiratory Medicine, 5th edn. Philadelphia: Saunders, 2010:1587–1600.
17. Chang HR, Shu KH, Cheng CH, et al. Peritoneal-dialysis-associated *Penicillium* peritonitis. *Am J Nephrol* 2000; 20:250–252.
18. Lo Y, Tintelnot K, Lippert U, et al. Disseminated *Penicillium marneffei* infection in an African AIDS patient. *Trans R Soc Trop Med Hyg* 2000; 94:187.
19. Kurup A, Leo YS, Tan Al, et al. Disseminated *Penicillium marneffei* infection: a report of five cases in Singapore. *Ann Acad Med Singapore* 1999; 28:605–609.

SPOROTHRIX SCHENCKII

Common name for disease: Sporotrichosis

Occupational setting

Although sporotrichosis occurs worldwide, it is found mainly in temperate, warm, and tropical areas. *Sporothrix schenckii* is isolated most often from soil and living plants or plant debris. Cutaneous infection develops where the organism is introduced to sites of skin injury. Subsequent nodular lymphangitic spread is common. Pulmonary sporotrichosis is a rare condition caused by inhalation of fungal spores. Infection of bones and joints also rarely occurs in immunocompetent hosts.[1]

Occupations that predispose to infection include gardening, farming, masonry, outdoor work, floral work, and other activities with exposure to contaminated soil or vegetation such as sphagnum moss, prairie hay, salt marsh hay, or roses.[2–4] Outbreaks have occurred among miners, nursery workers, and forestry workers who handle contaminated timbers, seedlings, mulch, hay, or other plant materials.[5–7] In one study, risk for infection was related to exposure to moss and to seedlings from a particular nursery.[8] Arm and finger infections have been reported in laboratory workers through contact with experimentally infected animals or contaminated material.[9] Two laboratory workers developed conjunctival and eyelid infections after mycelial elements were accidentally splattered into the eyes. Transmission to humans from infected cats has also been reported.[10] Armadillo hunting in Uruguay has been associated with sporotrichosis, presumably from exposure to the fungus isolated from the dry grass used by armadillos and rodents to prepare their nests. An outbreak of the illness in South African gold miners was traced to contaminated mine timbers.[11]

Exposure (route)

Sporothrix schenckii usually enters the body through traumatic implantation,[12] but inhalation of fungal conidia is occasionally associated with pulmonary infection.

Pathobiology

Sporothrix schenckii is a dimorphic fungus with 4–6 μm round, oval, or cigar-shaped cells.

Cutaneous disease arising at sites of minor trauma begins as a small, erythematous papule that enlarges over days or weeks. It usually remains painless, but it may secrete a clear discharge. Typically, discrete nodular lesions spread along lymphatic channels. Skin lesions may also result from hematogenous dissemination. Such lesions may herald the onset of osteoarticular sporotrichosis, which is manifested by stiffness and pain in a large joint, particularly the knee, elbow, ankle, or wrist. Radiologic evidence of osteomyelitis develops slowly, and additional joints may become involved in untreated patients. An indolent tenosynovitis of the wrist or ankle with pain, limitation of motion, and nerve entrapment can occur. Endophthalmitis, brain abscess, chronic meningitis, and other manifestations of disseminated disease are rare. Patients with a history of alcoholism or immunosuppression are at increased risk for dissemination.[13]

Pulmonary sporotrichosis typically presents with cough, low-grade fever, weight loss, hemoptysis, and an upper lobe single cavitary lesion with or without surrounding parenchymal infiltrate. Pleural effusion, hilar adenopathy, and calcification are rare. In the absence of treatment, the lung lesion gradually progresses to death. Chronic obstructive pulmonary disease (COPD) is a risk factor for pulmonary infection.[13]

Diagnosis

Accurate diagnosis requires detection of the fungus in clinical specimens including skin biopsy, joint aspirate, or sputum, either through culture or by fluorescent antibody staining. Recent data suggest PCR techniques may be valuable for more rapid diagnosis.[14,15]

Treatment

The key to appropriate diagnosis and treatment of sporotrichosis is a high index of clinical suspicion combined with culture to confirm results. Itraconazole is the drug of choice for lymphocutaneous infection. 200 mg/day for 3–6 months results in a >90% cure rate.[16] Fluconazole at a dose of 400 mg/day is second-line treatment; ketoconazole should not be used. Terbinafine may be as effective as itraconazole.[17] Intravenous amphotericin B is reserved for treatment failures. Itraconazole at a dose of 200 mg twice daily can be used for mild pulmonary disease. However, surgery combined with antifungal therapy is more effective than medical therapy alone in cavitary disease.[18]

In osteoarticular sporotrichosis, itraconazole at a dose of 200 mg twice daily is recommended, as accompanying systemic illness is unusual. Cure rates of 60–80% have been achieved with this regimen. Amphotericin B is indicated for severely ill patients or for those in whom itraconazole fails. Amphotericin B is also the drug of choice for disseminated or meningeal sporotrichosis.[13]

Prevention

Use of heavy, impermeable gloves and long-sleeved shirts in at-risk occupations has been shown to limit traumatic fungal implantation.[8] The use of alternative packing materials for plant products, such as shredded paper or cedar wood chips, has been recommended. Laboratory workers who handle contaminated material should work under appropriate hoods and utilize BSL2 work practices and facilities that minimize inhalation of conidia.[19] Early recognition and prompt treatment of disease limit morbidity and mortality.

References

1. Bariteau JT, Warysz GR, Mcdonnell M, et al. Fungal osteomyelitis and septic arthritis. *J Am Acad Orthop Surg* 2014; 22:390–401.
2. Centers for Disease Control. Sporotrichosis among hay-mulching workers—Oklahoma, New Mexico. *MMWR* 1984; 33:682–683.
3. Dixon D, Salkin I, Duncan R, et al. Isolation and characterization of *Sporothrix schenckii* from clinical and environmental sources associated with the largest U.S. epidemic of sporotrichosis. *J Clin Microbiol* 1991; 29:1106–1113.
4. Cox R, Reller L. Auricular sporotrichosis in a brickmason. *Arch Dermatol* 1979; 115:1229–1230.
5. Grotte M, Younger B. Sporotrichosis associated with sphagnum moss exposure. *Arch Pathol Lab Med* 1981; 105:50–51.
6. Powell K, Taylor A, Phillips B, et al. Cutaneous sporotrichosis in forestry workers. Epidemic due to contaminated sphagnum moss. *JAMA* 1978; 240:232–235.
7. Dooley DP, Bostic PS, Beckius ML. Spook house sporotrichosis. A point-source outbreak of sporotrichosis associated with hay bale props in a Halloween haunted-house. *Arch Intern Med* 1997; 157:1885–1887.
8. Hajjeh R, McDonnell S, Reef S, et al. Outbreak of sporotrichosis among tree nursery workers. *J Infect Dis* 1997; 176:499–504.
9. Cooper C, Dixon D, Salkin I. Laboratory- acquired sporotrichosis. *J Med Vet Mycol* 1992; 30:169–171.
10. Reed K, Moore F, Geiger G, et al. Zoonotic transmission of sporotrichosis: case report and review. *Clin Infect Dis* 1993; 16:384–387.
11. Einstein H. ACCP Committee Report: occupational aspects of deep mycoses. *Chest* 1978; 73:115.
12. Tan T, Field C, Faust B. Cutaneous sporotrichosis. *J La State Med Soc* 1988; 140:41–45.
13. Kauffman CA, Hajjeh R, Chapman SW. Practice guidelines for the management of patients with sporotrichosis. *Clin Infect Dis* 2000; 30:684–687.
14. Liu X, Zhang Z, Hou B, et al. Rapid identification of *Sporothrix schenckii* in biopsy tissue by PCR. *J Eur Acad Dermatol Venereol* 2013; 27:1491–1497.
15. De Oliveira MM, Sampaio P, Almeida-Paes R, et al. Rapid identification of Sporothrix Specis by T3B fingerprinting. *J Clin Microbiol* 2012; 50:2159–2162.
16. De Lima Barros MB, Schubach AO, de Vasconcellos Carvalhaes de Oliveira R, et al. Treatment of cutaneous sporotrichosis with itraconazole – study of 645 patients. *Clin Infect Dis* 2011; 52:e200–e206.
17. Francesconi G, Francesconi do Valle AC, Passos SL, et al. Comparitive study of 250 mg/day terbinafine and 100 mg/day itraconazole for the treatment of cutaneous sporotrichosis. *Mycopathologia* 2011; 171:349–354.
18. Aung AK, Teh BM, McGrath C, et al. Pulmonary sporotrichosis: case series and systematic analysis of literature on clinic-radiological patterns and management outcomes. *Med Mycol* 2013; 51:534–544.
19. Centers for Disease Control and Prevention, National Institutes of Health. Biosafety in Microbiological and Biomedical Laboratories, 5th edn. HHS Publication no. (CDC) 21–1112. New York: U.S. Department of Health and Human Services, 2009.

STACHYBOTRYS CHARTARUM

Common name for disease: None

Occupational setting

Stachybotrys is a rare fungal contaminant of water-damaged cellulose materials. High exposure levels have been reported among farmers, woodworkers, and composting workers exposed to moldy plant products.[1–4] High concentrations have also been described in water-damaged office buildings, courthouses, and homes.[5–8]

Exposure (route)

Exposure occurs through inhalation of mycotoxin-containing spores. Disease in animals has been reported after ingestion of contaminated plant feeds.

Pathobiology

Stachybotrys chartarum is a rare contaminant of nitrogen-poor straw and cellulose-based water-damaged materials.[6] Pathogenic effects were first described in the veterinary literature of the 1920s when cattle and horses developed anemia and gastrointestinal hemorrhage after ingesting contaminated grain.[9]

Stachybotrys is believed to cause disease through the production of powerful trichothecene mycotoxins. The mechanism of trichothecene toxicity is potent inhibition of DNA, RNA, and protein synthesis. These mycotoxins act as powerful immunosuppressants and can induce hemolysis and bleeding in animal models.[10,11]

Many possible human health effects have been described but none definitively proven. Occupational exposure to *Stachybotrys* in farmers and agricultural workers handling contaminated straw has been associated with epistaxis, hemoptysis, skin irritation, and alterations in white blood cell counts.[12] *Stachybotrys* has been implicated (along with many other toxigenic fungi) in sick building syndrome,[5,6,8,13,14] although causation has not been definitively proven.[15] A possible association has been described between *Stachybotrys* contamination of homes and idiopathic pulmonary hemorrhage/hemosiderosis in infants.[7,16] A cluster of 10 infants suffering from idiopathic pulmonary hemorrhage/hemosiderosis occurred in Cleveland between January 1993 and December 1994. Only three cases had been reported in the prior 10 years. Symptoms began with a prodrome featuring abrupt cessation of crying, limpness, and pallor. The prodrome was followed by hemoptysis, grunting, and respiratory failure. Chest radiographs revealed diffuse, bilateral alveolar infiltrates, and laboratory examination revealed decreased hematocrits and hemolysis on blood smears. Fifty percent of the cases recurred after returning home, and one infant died. Initial epidemiologic investigations implicated water-damaged homes as a significant risk factor.[17] Follow-up industrial hygiene studies revealed increased levels of *Stachybotrys* in case versus control homes. However, reexamination by the Centers for Disease Control and Prevention (CDC) concluded the data were not strong enough to definitively support an association between pulmonary hemorrhage in infants and exposure to *S. chartarum.*[18]

Diagnosis

Clinical evaluation requires careful occupational and environmental histories to elicit circumstances of water damage and environmental sampling to confirm the exposure.

Treatment

Removal of contaminated materials and prevention of circumstances leading to moisture intrusion in indoor moisture environments are keys to management of *Stachybotrys* exposure.

Prevention

Rapid remediation of water damage in homes and office buildings will diminish exposure risks. Proper attention to building practices during new construction will help prevent subsequent leaks and water damage. In agricultural settings, engineering and process controls can reduce exposure to high fungal concentrations.

References

1. Croft WA, Jarvis BB, Yatawara CS. Airborne outbreak of trichothecene toxicosis. *Atmos Environ* 1986; 20:549–552.
2. Mainville C, Auger PL, Smoagiewica W, et al. Mycotoxins and chronic fatigue syndrome in a hospital. In: Andersson K (ed.), Healthy Buildings Conference. Stockholm: Swedish council of Building Research, 1988:1–10.
3. Auger PL, Gourdeau P, Miller JD. Clinical experience with patients suffering from a chronic fatigue-like syndrome and repeated upper respiratory infections in relation to airborne molds. *Am J Ind Med* 1994; 25:41–42.
4. Johanning E, Auger PL, Reijula K. Building-related illnesses. *N Eng J Med* 1998; 338:1070.
5. Johanning E, Landsbergis P, Gareis M, et al. Clinical experience and results of a sentinel health investigation related to indoor fungal exposure. *Environ Health Perspect* 1999; 107(suppl 3):489–494.
6. Hodgson MJ, Morey P, Leung WY, et al. Building-associated pulmonary disease from exposure to *Stachybotrys chartarum* and *Aspergillus versicolor. J Occup Environ Med* 1998; 40:241–249.
7. Etzel RA, Montana E, Sorenson WG, et al. Acute pulmonary hemorrhage in infants associated with exposure to *Stachybotrys atra* and other fungi. *Arch Pediatr Adolesc Med* 1998; 152:757–762.
8. Sudakin DL. Toxigenic fungi in a water-damaged building: an intervention study. *Am J Ind Med* 1998; 34:183–190.
9. Hinitikka EL. Stachybotryotoxicosis as a veterinary problem. In: Rodricks JV, Hesseltine CW, Mehlman MA (eds.), Mycotoxins in Human and Animal Health. Park Forest South: Pathotox Publishers, 1977:277–284.
10. Pang VF, Lambert RJ, Felsburg PJ, et al. Experimental T-2 toxicosis in swine following inhalation exposure: clinical signs and effects on hematology, serum biochemistry, and immune response. *Fundam Appl Toxicol* 1988; 11:100–109.
11. Ueno Y. Trichothecene mycotoxins – mycology, chemistry and toxicology. *Adv Nutr Res* 1980; 3:301–353.
12. Hintikka EL. Human stachybotrytoxicosis. In: Wyllie TD, Morehouse LG (eds.), Mycotoxigenic Fungi, Mycotoxins, Mycotoxicoses. New York: Marcel Dekker, 1987:87–89.

13. Johanning E, Biagini R, Hull D, et al. Health and immunology study following exposure to toxigenic fungi (*Stachybotrys chartarum*) in a water-damaged office environment. *Int Arch Occup Environ Health* 1996; 68:207–218.
14. Mahmoudi M, Gershwin ME. Sick building syndrome. III. *Stachbotyrs chartarum*. *J Asthma* 2000; 37:191–198.
15. Hardin, BD, Kelman BJ, Saxon A. Adverse human health effects associated with molds in the indoor environment. *J Occup Environ Med* 2003; 45(5):470–478.
16. CDC. Update: pulmonary hemorrhage/hemosiderosis among infants – Cleveland, Ohio, 1993–1996. *MMWR* 1997; 46:33–35.
17. Montana E, Etzel RA, Allan T, et al. Environmental risk factors associated with pediatric idiopathic pulmonary hemorrhage and hemosiderosis in a Cleveland community. *Pediatrics* 1997; 99:117–124.
18. CDC. Update: pulmonary hemorrhage/hemosiderosis among infants – Cleveland, Ohio, 1993–1996. *MMWR* 2000; 49:180–184.

TRICHOPHYTON AND OTHER DERMATOPHYTES

Common names for diseases: Dermatophytosis, tinea corporis (ringworm), tinea glabrosa, Majocchi's granuloma, tinea cruris (jock itch), tinea pedis (athlete's foot).

Occupational setting

Cutaneous infections occur from exposure to the fungal dermatophytes, most commonly species of *Trichophyton*. The primary sources of dermatophytes are animals (zoophilic), humans (anthropophilic), and soil (geophilic). Dermatophytes originating from soil are occasionally responsible for outbreaks of human disease in exposed occupational groups such as gardeners and farm workers. Zoophilic outbreaks among cattle workers and rabbit farmers have also been reported.[1,2] Coaches, trainers, and professional athletes are another high-risk occupational group.[3,4] Spread of the anthropophilic organisms that infect glabrous skin is typically through contact with infected desquamated skin scales, such as in bathing or shower facilities in military barracks or factories.[5] As many as 30–35% of British coal miners were found in one study to have dermatophyte infections of the feet.[6] *Trichophyton tonsurans* was responsible for an outbreak of dermatophytosis in hospital personnel exposed to an infected patient.[7] *Trichophyton verrucosum*, the cause of cattle ringworm, and *Microsporum canis*, in cats and dogs, are the most common zoophilic dermatophytes that cause human infection in temperate countries.[8]

Podiatrists exposed to toenail dust generated when drills and burrs are used to reduce the thickness of hyperkeratotic nails can develop hypersensitivity reactions to *T. rubrum* (including nasal and eye symptoms, restrictive changes on pulmonary function tests, and specific IgG-precipitating antibodies).[9,10]

Exposure (route)

Dermatophytes invade the stratum corneum of the skin, most commonly of the feet, groin, scalp, and nails. Hypersensitivity reactions of the nasal mucus membranes and lungs can occur from exposure to *Trichophyton* dusts.

Pathobiology

The three genera of pathogenic dermatophyte fungi are *Trichophyton* (Figure 24.12), *Microsporum*, and *Epidermophyton*. The classic lesion of dermatophytosis is an annular scaling patch with a raised edge and a less inflamed central area.

Tinea refers to dermatophyte infection and is followed by the Latin word for the affected site. *Trichophyton rubrum* is the most common cause of tinea cruris, the dermatophyte infection of the groin. Scaling and irritation are the usual presenting findings. The disease is most common in young adult males. The leading edge extending onto the thighs is prominent and may contain follicular papules and pustules. Tinea pedis usually begins in the lateral interdigital spaces of the foot. The main symptom is itching; the skin usually cracks and may macerate. Tinea corporis usually involves the trunk or legs. Tinea capitis involves the scalp and can cause alopecia.[11]

Diagnosis

Fungal hyphae are easily observed as chains of arthrospores in wet mount preparations from skin scrapings. It is important to sample the edge of skin lesions and to allow the material to soften in potassium hydroxide before microscopic examination.

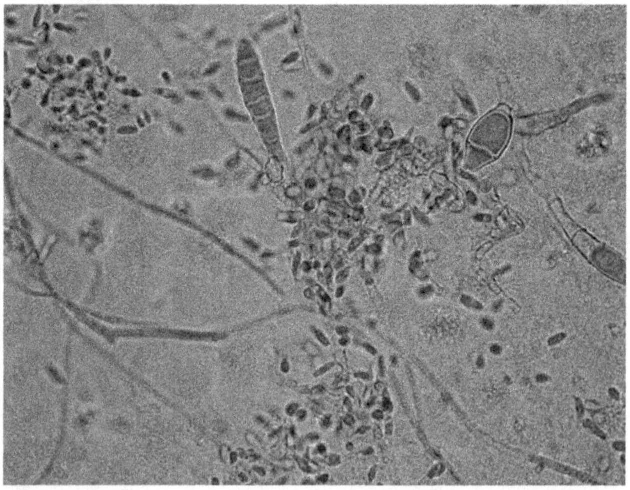

FIGURE 24.12 Lactophenol cotton blue wet mount revealing Trichophyton tonsurans. Courtesy of Dr. Dominick Cavuoti and Dr. Francesca Lee.

Treatment

Topical treatment with keratolytics and compounds with specific antifungal activity is usually successful. Nail, hair, and widespread skin infections may require oral agents such as griseofulvin. *Trichophyton tonsurans* infection may respond to oral ketoconazole. Griseofulvin and terbinafine are effective for tinea capitis.[12,13]

Prevention

To prevent tinea cruris, cool and loose-fitting clothing should be worn in hot and humid environments, where perspiration and irritation of skin are contributing factors. Avoidance of contact with contaminated clothing and towels is helpful. The floors of locker rooms or showers contaminated with dermatophytes should also be avoided. Sanitization of contaminated footwear and contaminated equipment can also help prevent the spread of infection.[4,14]

References

1. Lehenkari E, Silvennoinen-Kassinen S. Dermatophytes in northern Finland in 1982–90. *Mycoses* 1995; 38:411–414.
2. Weigl S, Figueredo LA, Otranto D. Molecular identification and phylogenesis of dermatophytes isolated from rabbit farms and rabbit farm workers. *Vet Microbiol* 2012; 154:395–402.
3. Collins CJ, O'Connell B. Infectious disease outbreaks in competitive sports, 2005–2010. *J Athl Train* 2012; 47:516–518.
4. Bassiri-Jahromi S, Sadeghi G, Paskiaee FA. Evaluation of the association of superficial dermatophytosis and athletic activities with special reference to its prevention and control. *Int J Dermatol* 2010; 49:1159–1164.
5. Korting H, Zienicke H. Dermatophytoses as occupational dermatoses in industrialized countries. Report on two cases from Munich. *Mycoses* 1990; 33:8609.
6. Gugnani H, Oyeka C. Foot infections due to *Hendersonula toruloidea* and *Scytalidium hyaline* in coal miners. *J Med Vet Mycol* 1989; 27:167–179.
7. Arnow P, Houchins S, Pugliese G. An outbreak of tinea corporis in hospital personnel caused by a patient with *Trichophyton tonsurans* infection. *Pediatr Infect Dis J* 1991; 10:355–359.
8. Chmel L, Buchvald J, Valentova M. Ringworm infection among agricultural workers. *Int J Epidemiol* 1976; 5:291–295.
9. Davies R, Ganderton M, Savage M. Human nail dust and precipitating antibodies to *Trichophyton rubrum* in chiropodists. *Clin Allergy* 1983; 13:309–315.
10. Abramson C, Wilton T. Nail dust aerosols from onychomycotic toenails. Part II. Clinical and serologic aspects. *J Am Podiatr Med Assoc* 1992; 82:116–123.
11. El-Khalawany M, Shaaban D, Hassan H, et al. A multicenter clinicomycological study evaluating the spectrum of adult tinea capitis in Egypt. *Acta Dermatovenerol Alp Pannonica Adriat* 2013; 22:77–82.
12. Grover C, Arora P, Manchanda V. Comparative evaluation of griseofulvin, terbinafine and fluconazole in the treatment of tinea capitis. *Int J Dermatol* 2012; 51:455–458.
13. Tey HL, Tan AS, Chan YC. Meta-analysis of randomized controlled trials comparing griseofulvin and terbinafine in the treatment of tinea capitis. *J Am Acad Dermatol* 2011; 64:663–670.
14. Gupta AK, Brintnell WC. Sanitization of contaminated footwear from onychomycosis patients using ozone gas: a novel adjunct therapy for treating onychomycosis and tinea pedis. *J Cutan Med Surg* 2013; 17:243–249.

ZYGOMYCETES

Including the order Mucorales, and *Hyphomycetes* (*Verticillium, Fusarium,* and *Neurospora*)

Common names for diseases: Hypersensitivity diseases: Paprika splitter's disease (*M. stolonifer*), Wood trimmer's disease (Rhizopus, Mucor), Tomato grower's asthma (*Verticillium albo-atrum*), Sinus fusariosis (*Fusarium*) Infections: Mucormycosis, phycomycosis, hyphomycosis, zygomycosis

Occupational setting

Hypersensitivity pneumonitis can occur in workers exposed to respirable *Mucorales* in paprika and from contaminated wood bark in sawmills.[1,2] In addition to occupational hypersensitivity lung disease, various genera and species of the class Zygomycete, order Mucorales, can cause infectious mucormycosis. The Zygomycetes grow in the environment and in tissue as hyphae. They are thermotolerant fungi that are commonly found in decaying organic debris. Despite their ubiquity, human infection is infrequent. Typically, it is associated with severe immunocompromise, malnutrition, iron chelation therapy, diabetes mellitus, or trauma.[3,4] The hyaline hyphomycete *V. albo-atrum* has been associated with cases of occupational asthma in tomato and tobacco growers exposed to crop outbreaks.[5–7] *Fusarium* species are very common soil organisms. Maxillary sinus fusariosis has been described in agricultural workers exposed to *Fusarium*.[8] Occupational asthma from immune sensitization to *Neurospora* spp. occurred in a plywood factory worker exposed to moldy wood products.[9] Occupational sensitization was demonstrated by allergy prick skin testing, the presence of specific serum IgE antibodies, and inhalation challenge with the *Neurospora* mold growing on plywood. Suggested preventive strategies included sealing of the wood drying machine to minimize dust concentrations and shorter outdoor storage times for the plywood to prevent fungal growth.

Exposure (route)

For the hypersensitivity diseases (asthma and hypersensitivity pneumonitis) and for respiratory tract infection, the route of exposure is inhalation of respirable airborne hyphae.

Cutaneous and subcutaneous mucormycosis can occur by direct implantation from "barrier breaks" or by hematogenous dissemination. Several cases have been associated with contaminated bandages and surgical dressings. Rarely, gastrointestinal transmission may occur.[4]

Pathobiology

All of the Zygomycetes grow as 4–15 μm wide hyphae in the environment and tissue and are identified microscopically by their morphology (Figure 24.13). As with other forms of hypersensitivity pneumonitis (HP), the symptoms of hypersensitivity diseases caused by these organisms may be acute and flu-like or subtle, chronic, and predominantly respiratory in nature. Acute illness is manifested by fevers, chills, cough, dyspnea, abnormal chest radiograph, and leukocytosis; improvement follows within a few days after removal from exposure. The more subacute or chronic forms of HP are typically manifested by insidious onset of cough, progressive dyspnea on exertion, and weight loss. Pulmonary physiology may be normal, show isolated obstruction, or show the more classic restrictive or mixed restrictive and obstructive pattern. Exercise physiology often shows gas exchange abnormalities. The chest radiograph may be normal or show inhomogeneous, patchy alveolar infiltrates, or interstitial opacities. Serum precipitating antibodies to Zygomycete antigens species are often positive.

Many different species in the order Mucorales have been implicated in infectious disease, including *Rhizopus*, *Mucor*, *Mortierella, and Absidia* species. Mucormycosis is the most acute, fulminant fungal infection known. Organisms invade arterial vessels and may infect the face and cranium, lungs, GI tract, or skin. Rhinocerebral disease typically occurs in the acidemic patient with uncontrolled diabetes; it begins in the nasal turbinates, paranasal sinuses, palate, pharynx, or ears and spreads to the central nervous system. Renal dialysis patients treated with deferoxamine may also be at increased risk as are immunocompromised patients (those with chronic steroid use, posttransplant, and hematologic malignancies). Presenting signs usually include a thick, dark, and blood-tinged discharge that often reveals hyphal strands on KOH mount. Rapid invasion of surrounding tissue with sloughing and cranial nerve dysfunction follows, and death usually occurs in a few days. Pulmonary mucormycosis occurs after inhalation of spores or from aspiration of nasopharyngeal secretions, leading to bronchitis and pneumonia with subsequent arterial invasion, often massive cavitation, and rapid death. Primary gastrointestinal and pelvic mucormycosis are uncommon, and malnourished patients or those with hematologic malignancies are most at risk. Cerebral zygomycosis has been described in intravenous drug abusers.[10] This infection most likely spreads through the bloodstream following the intravenous injection of infectious organisms.

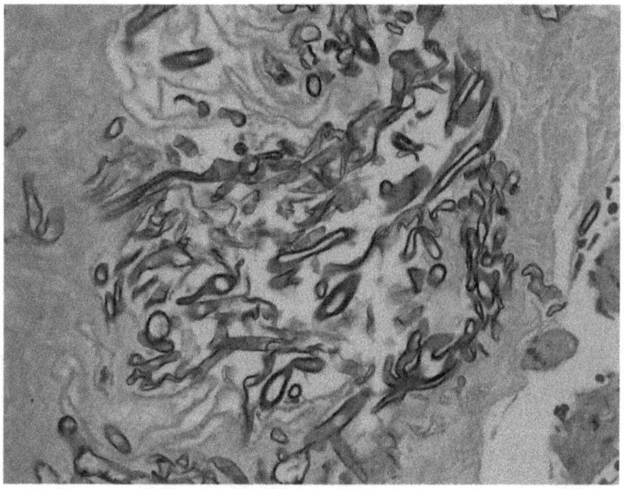

FIGURE 24.13 Tissue sample hematoxylin and eosin stain showing Zygomycetes hyphae. Courtesy of Dr. Dominick Cavuoti and Dr. Francesca Lee.

Treatment

Treatment of hypersensitivity lung disease focuses on early disease recognition and elimination of further exposure to the offending antigen. Oral corticosteroids may be useful for patients with severe symptoms or physiologic abnormalities.

Treatment of the underlying disease (e.g., diabetes) is the most effective method to control infectious mucormycosis. The prognosis is poor (80–90% mortality rate). Treatment requires a combination of surgical debridement of necrotic tissue and antifungal therapy with intravenous amphotericin B.[4] Recently posaconazole has emerged as a potential salvage therapy.[11]

Prevention

Prevention of occupational sensitization to these organisms involves the provision of adequate ventilation, process controls, and respiratory protection to limit exposure to high fungal bioaerosol levels.

References

1. Eduard W, Sandven P, Levy F. Relationships between exposure to spores from *Rhizopus*, *Microsporus* and *Paecilomyces variotii* and serum lgG antibodies in wood trimmers. *Int Arch Allergy Appl Immunol* 1992; 97:274–282.
2. Hedenstierna G, Alexandrsson R, Belin L, et al. Lung function and *Rhizopus* antibodies in wood trimmers. A cross-sectional and longitudinal study. *Int Arch Occup Environ Health* 1986; 58:167–177.
3. Gordon G, Indeck M, Bross J, et al. Injury from a silage wagon accident complicated by mucormycosis. *J Trauma* 1988; 28:866–867.

4. Ribes JA, Vanover-Sams CL, Baker DJ. Zygomycetes in human disease. *Clin Microbiol Rev* 2000; 13:236–301.
5. Eaton K, Hannessy T, Snodin D, et al. *Verticillium lecanii.* Allergological and toxicological studies on work-exposed personnel. *Ann Occup Hyg* 1986; 30:209–217.
6. Anonymous. Occupational asthma in tomato growers. *Occup Health (Lond)* 1989; 41:70–71.
7. Davies P, Jacobs R, Mullins J, et al. Occupational asthma in tomato growers following an outbreak of the fungus *Verticillium albo-atram* in the crop. *J Soc Occup Med* 1988; 38:13–17.
8. Kurien M, Anandi V, Raman R, et al. Maxillary sinus fusariosis in immunocompetent hosts. *J Laryngol Otol* 1992; 106:733–736.
9. Cote J, Chan H, Brochu G, et al. Occupational asthma caused by exposure to neurospora in a plywood factory worker. *Br J Ind Med* 1991; 48:279–282.
10. Stave GM, Heimberger T, Kerkering T. Cerebral zygomycosis in intravenous drug abuse. *Am J Med* 1989; 86:115–117.
11. Cronely OA, Vehreschild JJ, Ruping MJ. Current experience in treating invasive zygmycosis with posaconazole. *Clin Microbiol Infect* 2009; 15(suppl 5):77–81.

Chapter 25

ANAPLASMA, CHLAMYDOPHILA, COXIELLA, EHRLICHIA, and RICKETTSIA

DENNIS J. DARCEY*

Anaplasma, *Coxiella*, *Ehrlichia*, and *Rickettsia* spp. were previously considered as members of the same family but are now considered distinct entities based on genetic analysis. These groups of organisms are bacteria but only grow inside living cells.

ANAPLASMA

Common names for disease: Anaplasmosis, human granulocytic anaplasmosis (HGA), previously described as human granulocytic ehrlichiosis (HGE)

Occupational setting

Ranchers, farmers, foresters, rangers, hunters, lumberjacks, landscapers, outdoor workers, veterinarians, and research and laboratory technicians are at increased risk of exposure.

Exposure (route)

Anaplasma is transmitted to humans by the bite of an infected tick. The blacklegged tick (*Ixodes scapularis*) is the vector of *Anaplasma phagocytophilum* in the northeast and upper midwestern United States, and the western blacklegged tick (*Ixodes pacificus*) is the primary vector in Northern California (Figure 25.1).[1] Cases have also been confirmed in several European and Asian countries.

*The author wishes to thank Ricky L. Langley for his contribution to earlier versions of this chapter.

Pathobiology

In the early 1990s, patients from Michigan and Wisconsin with a history of tick bites were described as having an illness similar to human monocytotropic ehrlichiosis but were notable for the presence of inclusion bodies in granulocytes rather than monocytes.[2] The syndrome was initially called human granulocytic ehrlichiosis. The disease is now called human granulocytic anaplasmosis (HGA) after phylogenetic analysis reclassified *Ehrlichia phagocytophilum* as a member of the genus *Anaplasma*.[2]

Anaplasma phagocytophilum selectively inhabit and replicate within intracellular vacuoles of their host cells.[2] They have an ultrastructural similarity to gram-negative bacteria but lack a lipopolysaccharide endotoxin.[3]

The symptoms caused by infection with *A. phagocytophilum* usually develop 1–2 weeks after being bitten by an infected tick. The tick bite is usually painless, and about half of the people who develop anaplasmosis may not recall being bitten by a tick. Symptoms vary from mild to severe, but common presentations include fever, headache, confusion, cough, chills, malaise, muscle pain, nausea, and abdominal pain.[1] Rash is seen in less than 10% of patients with anaplasmosis[3] and may indicate coinfection with *Borrelia burgdorferi* and/or *Rickettsia rickettsii*.

Most patients develop transient reductions in white cell and platelet counts. Relative granulocytosis with a left shift and lymphopenia is common during the first week of illness.[4] During the acute phase, serum hepatic transaminase concentrations usually increase two- to fourfold, and inflammatory markers, such as C-reactive protein and the erythrocyte sedimentation rate, are increased.[4] Deaths do occur but

Physical and Biological Hazards of the Workplace, Third Edition. Edited by Gregg M. Stave and Peter H. Wald.

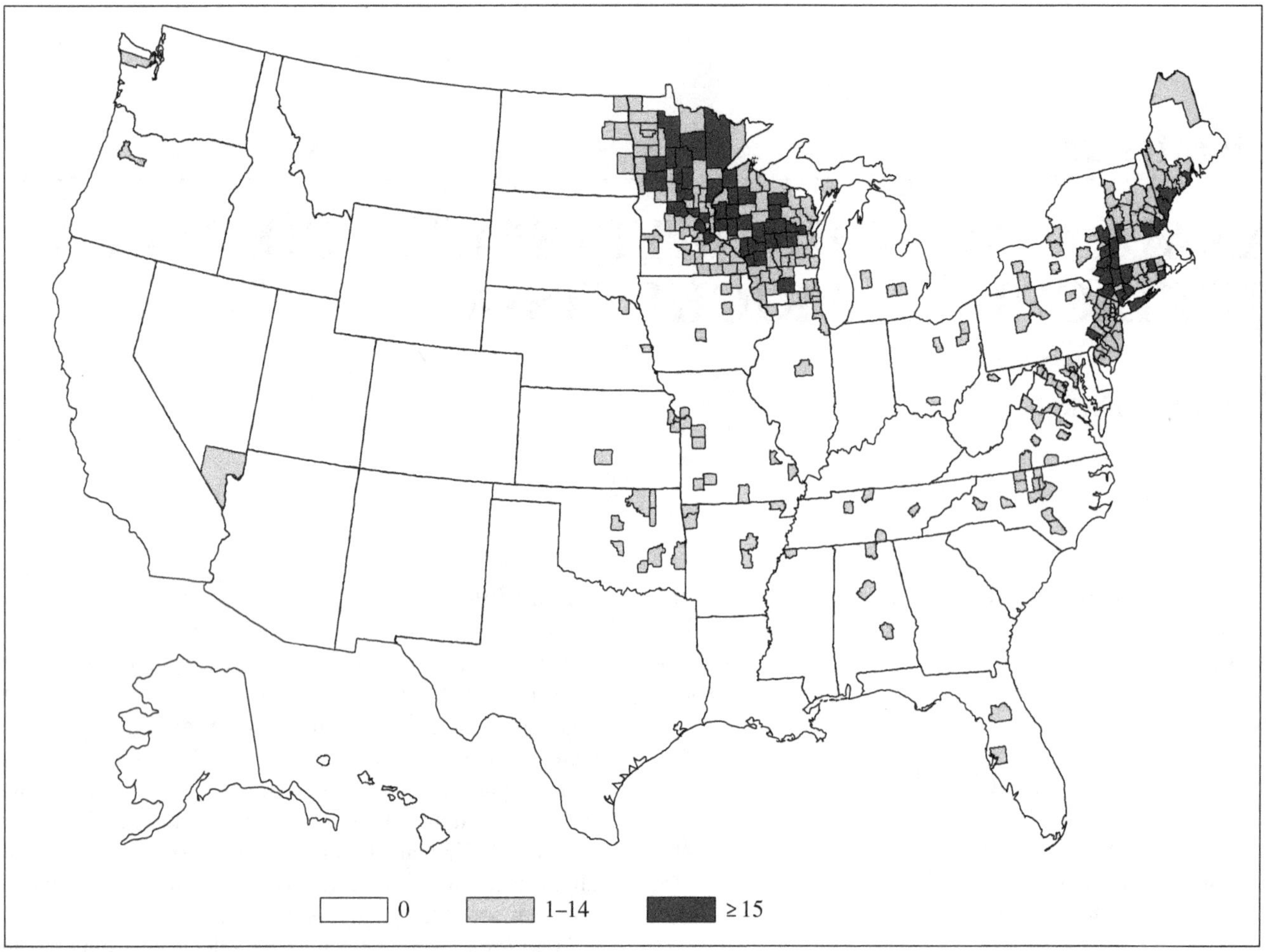

FIGURE 25.1 Map of the United States showing the number of *Anaplasma phagocytophilum* cases by county in 2012. Cases are reported primarily from the upper Midwest and coastal New England, reflecting both the range of the primary tick vector species, Ixodes scapularis—also known to transmit Lyme disease and babesiosis—and the range of preferred animal hosts for tick feeding. *Source*: MMWR 2015; 62(53):67.

are rare. Persons with compromised immunity caused by immunosuppressive therapies (such as corticosteroids, chemotherapy, or immunosuppressive therapy following organ transplant), HIV infection, or splenectomy appear to develop more severe disease and may also have higher case fatality rates.[1]

Diagnosis

The incidence of infection is highest in the United States during the summer months when human tick encounters are at their peak with increased outdoor activities. The diagnosis of anaplasmosis is made based on clinical signs and symptoms consistent with the disease and history of tick bite, illness in an area where ticks are known to be present, or recent travel to an area where anaplasmosis is endemic. Early treatment is more likely to be effective and should be initiated if anaplasmosis is suspected.[1] Treatment should not be delayed while awaiting laboratory test results, nor should it be withheld or withdrawn on the basis of an initial negative laboratory test result.[1]

Cytoplasmic inclusion bodies (morulae) are seen on a peripheral blood smear in 25–75% of cases and can aid in early diagnosis (Figure 25.2).[2]

During the acute phase of illness, a sample of whole blood can be tested by polymerase chain reaction (PCR) assay to determine if a patient has anaplasmosis. This method is most sensitive in the first week of illness and quickly decreases in sensitivity following the administration of appropriate antibiotics. Although a positive PCR result is helpful, a negative result does not completely rule out the diagnosis.[1] The gold standard serologic test for diagnosis of anaplasmosis is the indirect immunofluorescence assay (IFA) using *A. phagocytophilum antigen*, performed on paired serum samples to demonstrate a significant (fourfold) rise in antibody titers.[1] The first sample should be taken as early in the disease as possible, preferably in the first week of symptoms, and the second sample should be taken 2–4 weeks later. In most cases of anaplasmosis, the first IgG IFA titer is typically low, or negative, and the second typically shows

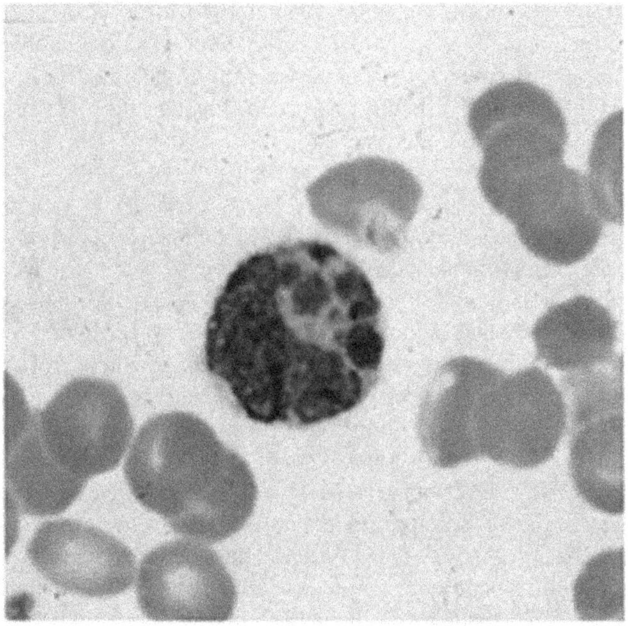

FIGURE 25.2 Morulae detected in a granulocyte on a peripheral blood smear, associated with *A. phagocytophilum* infection. *Source*: Centers for Disease Control and Prevention: Anaplasmosis, http://www.cdc.gov/anaplasmosis/symptoms/index.html.

a significant (fourfold) increase in IgG antibody levels. IgM antibodies usually rise at the same time as IgG near the end of the first week of illness and remain elevated for months or longer. IgM antibodies are less specific than IgG antibodies and more likely to result in a false positive result. For these reasons, physicians requesting IgM serologic titers should also request a concurrent IgG titer.[1]

Serologic tests based on enzyme immunoassay (EIA) technology are available from some commercial laboratories. However, these tests are qualitative and less useful to measure changes in antibody titers between paired specimens. Some EIA assays rely on the evaluation of IgM antibody alone, which may have a higher frequency of false-positive results.[1]

Treatment

Doxycycline is the treatment of choice, and patients should be treated for 7–14 days or at least 3 days after defervescence.[1] This treatment course will also treat the possibility of coinfection with *B. burgdorferi* (patients with coinfection should be treated for 14 days).[4] Fever typically subsides within 24–48 hours after treatment when the patient receives doxycycline or another tetracycline during the first 4–5 days of illness.[5] If a patient fails to respond to early treatment with a tetracycline antibiotic (i.e., within 48 hours), an alternate diagnosis should be considered.[5] Recovery may be prolonged in severely ill patients.[5]

Medical surveillance

Routine medical surveillance is not recommended.

Prevention

To prevent infection from tick bites, avoid wooded and bushy areas with high grass. The use of tick repellants such as DEET (*N,N*-diethyl-3-methylbenzamide) on exposed skin and permethrin on clothing coupled with timely skin examination and removal of ticks after exposure to tick-infested environments can prevent infection. Antibiotic treatment is not recommended following a tick bite to prevent anaplasmosis.[1] There are no licensed vaccines for anaplasmosis. In the laboratory, *Anaplasma* should be handled at biosafety level 2 (BSL2).[6]

References

1. Centers for Disease Control and Prevention. Anaplasmosis. http://www.cdc.gov/anaplasmosis/symptoms/index.html (accessed June 17, 2016).
2. Ismail N, Bloch KC, McBride JW. Human ehrlichiosis and anaplasmosis. *Clin Lab Med* 2010; 30(1):261–92.
3. Borjrsson DL, Barthold SW. The mouse as a model for investigation of human granulocytic ehrlichiosis: current knowledge and future directions. *Comp Med* 2002; 52(5):403–13.
4. Bakken JS, Dumler JS. Clinical diagnosis and treatment of human granulocytotropic anaplasmosis. *Ann NY Acad Sci* 2006; 1078:236–47.
5. Chapman AS, Backend JS, Folk SM, et al. Diagnosis and management of tickborne rickettsial diseases: rocky mountain spotted fever, ehrlichioses, and anaplasmosis—United States: a practical guide for physicians and other health-care and public health professionals. *MMWR Recomm Rep* 2006; 55(RR-4):1–27. Available at: http://www.cdc.gov/mmwr/preview/mmwrhtml/rr5504a1.htm (accessed on July 2, 2016).
6. Chosewood LC, Wilson DE, Centers for Disease Control and Prevention, National Institutes of Health. *Biosafety in microbiological and biomedical laboratories*, 5th edn. HHS publication no. (CDC) 21-1112, 2009. Washington, DC: U.S. Department of Health and Human Services, Public Health Service, Centers for Disease Control and Prevention, National Institutes of Health. Available at: http://www.cdc.gov/biosafety/publications/bmbl5/ (accessed on July 2, 2016).

CHLAMYDOPHILA PSITTACI

Common names for disease: Psittacosis, ornithosis, parrot fever, previously known as *Chlamydia psittaci*

Occupational setting

Bird handlers, pet shop workers, zoo attendants, poultry workers, laboratory personnel, veterinarians, and veterinary technicians are at risk from exposure.

Exposure (route)

Inhalation of secretions or excrement of infected domestic birds (parakeets, parrots, macaws, cockatiels, pigeons, turkeys, chickens, ducks, etc.) is the most common route of exposure. Person-to-person transmission has been reported, including infections in healthcare workers.[1–3]

Pathobiology

Chlamydophila psittaci (previously known as *Chlamydia psittaci*) is a gram-negative obligate intracellular bacterium. Once inhaled, *C. psittaci* rapidly enters the bloodstream and is transported to the reticuloendothelial cells of the liver and spleen. It replicates in these sites and then invades the lungs and other organs by hematogenous spread. The incubation period ranges from 5 to 15 days. Disease often starts abruptly and causes influenza type symptoms with chills, headache, malaise, and anorexia. Myalgias and arthralgias are common. A persistent dry hacking cough is prominent and complications can result in severe pneumonia. Respiratory symptoms are often mild compared to the extensive changes present on chest X-ray. Occasionally, changes in mentation may be noted. Hepatosplenomegaly may also occur in a significant number of patients. Myocarditis and encephalitis complications are rare but may occur.[4] There are no characteristic laboratory or chest X-ray changes. Since 1996, fewer than 50 confirmed cases per year have been reported in the Unites States. Many more cases may occur that are not correctly diagnosed or reported.[5]

Diagnosis

The diagnosis can be confirmed either by isolation of the organism in culture or by serologic studies. Most diagnoses are established by clinical presentation and positive antibodies against *C. psittaci* in paired sera using microimmunofluorescence (MIF) methods. MIF is more sensitive and specific than the previously used complement fixation (CF) tests; however, there is still some cross-reactivity with other chlamydiae (*Chlamydophila pneumoniae*, *Chlamydia trachomatis*, and *Chlamydophila felis*) so a titer result less than 1:128 should be interpreted with caution. Acute-phase serum specimens should be obtained as soon as possible after the onset of symptoms, and convalescent-phase serum specimens should be obtained at least 2 weeks after the first specimen. Because antimicrobial treatment can delay or diminish the antibody response, a third serum sample 4–6 weeks after the acute sample might help confirm the diagnosis. Real-time polymerase chain reaction (rt-PCR) assays have been developed for use in the detection of *C. psittaci* in respiratory specimens. These assays can distinguish *C. psittaci* from other chlamydial species and identify different genotypes. *Chlamydia psittaci* can also be isolated from the patient's sputum, pleural fluid, or clotted blood during acute illness and before treatment with antimicrobial agents; however, few laboratories perform cultures because of the technical difficulty and occupational health concerns.[4]

Treatment

Tetracycline antibiotics are the drugs of choice for *C. psittaci* infection in humans. Mild to moderate cases can be treated with oral doxycycline (100 milligrams every 12 hours) or tetracycline hydrochloride (500 milligrams every 6 hours) for a minimum of 10 days. Severely ill patients should be treated with intravenous (IV) doxycycline hyclate (4.4 mg/kg/day divided into two infusions, maximum 100 mg/dose). Antibiotic therapy should be continued for at least 10–14 days after fever abates. Most *C. psittaci* infections are responsive to antibiotics within 1–2 days; however, relapses can occur. Although *in vivo* efficacy has not been determined, macrolide antibiotics are considered the best alternative agents in patients for whom tetracyclines are contraindicated (e.g., children <8 years of age, pregnant women, and persons allergic to tetracyclines). Prophylactic antibiotics are not routinely administered after a suspected exposure to *C. psittaci*, but may be considered in some circumstances.[4]

Medical surveillance

Routine medical surveillance is not recommended. If cases can be linked to a source of exposure, then surveillance of the pet shop, aviary, farm, or healthcare setting should be undertaken. Infected birds should be treated or destroyed. The area where they are housed should be cleaned and disinfected.

Prevention

To prevent transmission of *C. psittaci* to humans and among bird populations, aviary and pet shop owners should follow specific control measures recommended by the National Association of State Public Health Veterinarians.[4] When cleaning cages or handling potentially infected birds, caretakers should wear protective clothing, which includes gloves, eyewear, a disposable surgical cap, and an appropriately fitted respirator with N95 or higher rating. In addition, necropsies of potentially infected birds should be performed in a biological safety cabinet. The carcass should be moistened with detergent and water to prevent aerosolization of infectious particles during the procedure. Bird caretakers with respiratory or influenza-like symptoms should seek prompt medical attention and inform their healthcare provider about bird contact.

For healthcare workers treating patients with psittacosis, enhanced Respiratory Protection and isolation may be needed when caring for severely ill patients.[3]

In the laboratory, *C. psittaci* is considered as biosafety level 2 (BSL2) and animal biosafety level 3 (ABSL3).[6] For procedures that have a high potential for aerosolization, it should be handled at biosafety level 3 (BSL3).[6]

There is currently no vaccine available to prevent psittacosis.

References

1. Hughes C, Maharg P, Rosario P, et al. Possible nosocomial transmission of psittacosis. *Infect Control Hosp Epidemiol* 1997; 18:165–168.
2. McGuigan CC, McIntyre PG, Templeton K. Psittacosis outbreak in Tayside, Scotland, December 2011 to February 2012. *Euro Surveill* 2012; 17(22):pii: 20186.
3. Wallensten A, Fredlund H, Runehagen A. Multiple human-to-human transmission from a severe case of psittacosis, Sweden, January–February 2013. *Euro Surveill* 2014; 19(42):pii: 20937. http://www.eurosurveillance.org/ViewArticle.aspx?ArticleId=20937 (accessed June 17, 2016).
4. National Association of State Public Health Veterinarians (NASPHV). Compendium of Measures to Control *Chlamydophila psittaci* Infection among Humans (Psittacosis) and Pet Birds (Avian Chlamydiosis), 2010. http://www.nasphv.org/Documents/Psittacosis.pdf (accessed June 17, 2016).
5. Centers for Disease Prevention and Control. Psittacosis. http://www.cdc.gov/pneumonia/atypical/psittacosis.html (accessed June 17, 2016).
6. Chosewood LC, Wilson DE, Centers for Disease Control and Prevention, National Institutes of Health. *Biosafety in microbiological and biomedical laboratories*, 5th edn. HHS publication no. (CDC) 21-1112, 2009. Washington, DC: U.S. Department of Health and Human Services, Public Health Service, Centers for Disease Control and Prevention, National Institutes of Health. Available at: http://www.cdc.gov/biosafety/publications/bmbl5/ (accessed on July 2, 2016).

COXIELLA BURNETII

Common name for disease: Q fever

Occupational setting

Q fever is a nationally notifiable disease caused by the bacterium *Coxiella burnetii*. The Centers for Disease Control and Prevention (CDC) receives approximately 150 reports of this illness in the United States each year (Figure 25.3). Occupations at risk include abattoir and livestock workers, dairy workers, veterinarians, veterinary and laboratory technicians, laboratory animal handlers, farmers, ranchers, and hide and wool handlers. Farmers who are in contact with cattle, sheep, and goats, assisting in the birthing of lambs, or exposed to birth by-products of other animals such as dogs and cats are at particularly high risk. Although Q fever has been historically considered an occupational disease, most cases reported to CDC each year involve individuals who do not report livestock contact and do not work with animals. These cases are likely due to airborne transmission of the bacteria after environmental contamination by infected animals.

Exposure (route)

Exposure occurs through inhalation of aerosolized particles or direct contact with infected animals, primarily cattle, sheep, or goats. Placentas from infected sheep are extremely infectious. In addition to domesticated livestock, a broad range of domestic and wild animals are natural hosts for *C. burnetii*, including horses, dogs, swine, pigeons, ducks, geese, turkeys, squirrels, deer, mice, cats, and rabbits. Exposure to *C. burnetii* in research laboratories and veterinary hospitals has resulted in large outbreaks of Q fever.[1, 2] The largest known reported Q fever outbreak involved approximately 4000 human cases and occurred during 2007–2010 in the Netherlands. This outbreak was linked to dairy goat farms near densely populated areas and presumably involved human exposure via a wind-borne route.[3]

Approximately 200 cases of acute Q fever were reported in US military personnel who had been deployed to Iraq since 2003. Investigations of these cases linked illness to tick bites, sleeping in barns, and living near helicopter zones with environmental exposure resulting from helicopter-generated aerosols.[4, 5]

Pathobiology

Coxiella burnetii is an obligate intracellular bacterium that appears as a short pleomorphic gram-negative rod. *Coxiella burnetii* grows in the phagosomes of the cell. It is very resistant to inactivation and can survive in the environment for long periods of time in the spore stage. *Coxiella burnetii* is extremely infectious.

Frequently underreported and misdiagnosed, acute Q fever often resembles a nonspecific viral illness. Symptomatic acute Q fever, which occurs in approximately half of infected persons, is characterized by a wide variety of clinical signs and symptoms. Q fever usually presents as a mild respiratory illness and is often described as one of the atypical pneumonias. The incubation period typically ranges from 2 to 3 weeks. Clinical signs and symptoms include fever, malaise, headache, weakness, and transient pneumonitis with cough, chest pain, myalgias, and arthralgias. Physical examination is often unremarkable, and the most common physical finding is inspiratory crackles. Fever lasts a median of 10 days in untreated patients but when treated fever defervesces within 72 hours of antibiotic administration. The majority of infections are mild and self-limiting. Although mortality is <2% in patients with acute Q fever, in the Netherlands outbreak, which included approximately 4000 reported cases, up to

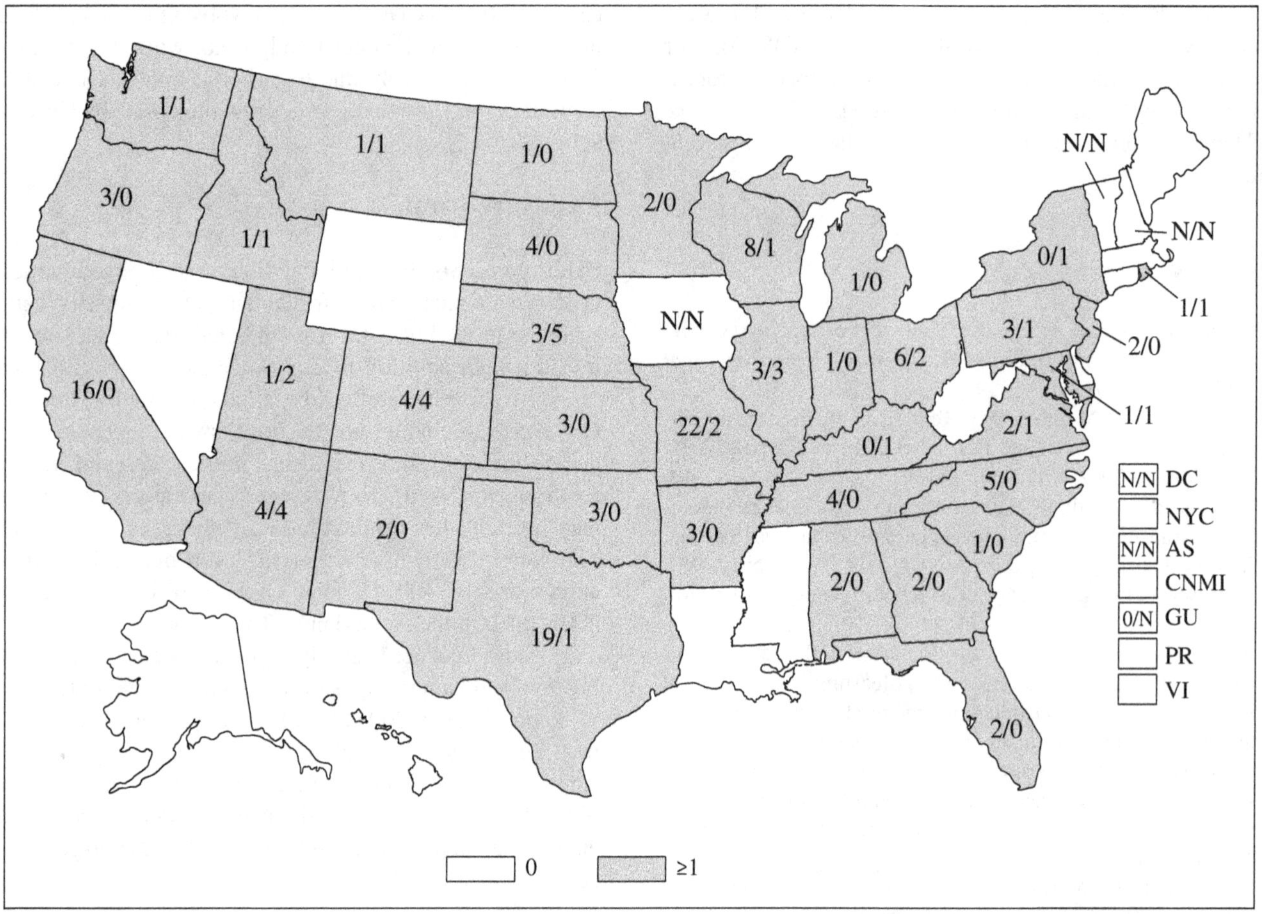

FIGURE 25.3 Map of the United States and U.S. territories showing the number of acute and chronic Q fever cases in each state and territory in 2013. Q fever, caused by *Coxiella burnetii*, is reported throughout the United States. Human cases of Q fever most often result from contact with infected livestock, especially sheep, goats, and cattle. *Source*: MMWR 2015; 62(53):88.*Number of Q fever acute cases/number of Q fever chronic cases.

50% of acute Q fever patients were hospitalized.[6] Q fever infections in women that occur shortly before conception or during pregnancy increase the risk for miscarriage, stillbirth, premature birth, intrauterine growth retardation, and low birth weight. Less frequently described clinical symptoms include pericarditis, myocarditis, aseptic meningitis, encephalitis, and cholecystitis.

Chronic Q fever is rare, occurring in <5% of persons with acute infection with onset as soon as months, or as late as decades following an acute illness. The patients at highest risk for chronic Q fever are those with valvular heart disease, a vascular graft, or an arterial aneurysm.

Diagnosis

The diagnosis of Q fever can be made by isolating the organism in the laboratory or by serologic demonstration of infection. The most common serologic tests include complement fixation (CF) and indirect fluorescent (IFA) antibody procedures.

Seroconversion typically occurs 7–15 days after symptoms appear, and 90% of patients seroconvert by the third week of illness.[7] In acute infection, a phase II IgG antibody is higher than the antibody response to phase I. The reverse is true in chronic infection. The most commonly used means of confirming the diagnosis of acute Q fever is demonstration of a fourfold rise in phase II IgG by IFA between serum samples from the acute and convalescent phases taken 3–6 weeks apart. However, IFA antibody results are not usually helpful in guiding immediate treatment decisions.

Treatment

Treatment for acute illness should begin immediately and not be delayed while awaiting laboratory results. The treatment of choice for acute Q fever is a 2-week course of doxycycline. Alternative regimens for specific groups of patients (e.g., pregnant women) and for chronic infection are discussed in detail within the complete Recommendations from

CDC and the Q Fever Working Group.[7] Management of chronic Q fever requires long-term treatment with multiple antibiotics and intensive patient monitoring for years for possible relapse. Surgical intervention may be required, and consultation with an infectious disease specialist is recommended.

Medical surveillance

Routine medical surveillance is not recommended.

Prevention

Coxiella burnetii organisms are widespread in the environment and resistant to inactivation. Control of major animal reservoirs of the organism is impractical. Personal protective equipment when handling affected animals, bedding, and their by-products and respiratory protection when working in dusty environments contaminated with organisms are recommended. Other preventive measures include pasteurization of milk to reduce the potential risk for transmission through milk and cheese products. Workers with prosthetic heart valves and liver disease are at particularly high risk for the sequelae of infection and are best restricted from high-risk environments. Biosafety level 2 (BSL2) practices, containment equipment, and facilities are recommended for nonpropagative laboratory procedures, including serological examinations and staining of impression smears. Biosafety level 3 (BSL3) practices and facilities are recommended for activities involving the inoculation, incubation, and harvesting of embryonated eggs or tissue cultures, the necropsy of infected animals, and the manipulation of infected tissues. Experimentally infected rodents should also be maintained under animal biosafety level 3 (ABSL3).[8] Laboratory safety and containment recommendations for *C. burnetii* should be followed as described in the CDC *Biosafety in Microbiological and Biomedical Laboratories* Manual 5th edition[8] and following the guidelines in Diagnosis and Management of Q Fever—United States: Recommendations from CDC and the Q Fever Working Group.[7] Q fever is a nationally notifiable disease in the United States. Healthcare providers should report suspected or confirmed cases through local or state reporting mechanisms in place for notifiable disease conditions.

References

1. Johnson JE II, Kadull PJ. Laboratory-acquired Q fever. *Am J Med* 1966; 41:391–403.
2. Hall CJ, Richmond SJ, Caul EO, et al. Laboratory outbreak of Q fever acquired from sheep. *Lancet* 1982; 1:1004–6.
3. Schimmer B, Dijkstra F, Velllema P, et al. Sustained intensive transmission of Q fever in the south of the Netherlands, 2009. *Euro Surveill* 2009; 14:3.
4. Anderson AD, Baker TR, Littrell AC, et al. Seroepidemiologic survey for *Coxiella burnetii* among hospitalized U.S. troops deployed to Iraq. *Zoonoses Public Health* 2011; 58:276–83.
5. Faix DJ, Harrison DJ, Riddle MS, et al. Outbreak of Q fever among U.S. military in western Iraq, June–July 2005. *Clin Infect Dis* 2008; 46:e65–8.
6. Van der Hoek W, Dijkstra F, Schimmer B, et al. Q fever in the Netherlands: an update on the epidemiology and control measures. *Euro Surveill* 2010; 15:pii: 19520. Available at: http://www.eurosurveillance.org/ViewArticle.aspx?ArticleId=19520 (accessed on July 2, 2016)
7. Anderson A, Bijlmer H, Fournier PE, et al. Diagnosis and management of Q fever—United States: recommendations from CDC and the Q Fever Working Group. *MMWR Recomm Rep* 2013; 62(RR-3):1–30. Available at: https://www.cdc.gov/mmwr/preview/mmwrhtml/rr6203a1.htm (accessed on July 3, 2016).
8. Chosewood LC, Wilson DE, Centers for Disease Control and Prevention, National Institutes of Health. *Biosafety in microbiological and biomedical laboratories*, 5th edn. HHS publication no. (CDC) 21-1112, 2009. Washington, DC: U.S. Department of Health and Human Services, Public Health Service, Centers for Disease Control and Prevention, National Institutes of Health. Available at: http://www.cdc.gov/biosafety/publications/bmbl5/ (accessed on July 2, 2016).

EHRLICHIA SPECIES

Common name for disease: Ehrlichiosis, when caused by *Ehrlichia chaffeensis*, is also described as human monocytic ehrlichiosis (HME); infection with *Ehrlichia ewingii* is sometimes referred to as human ehrlichiosis ewingii (HEE).

Occupational setting

Ranchers, farmers, foresters, rangers, hunters, lumberjacks, landscapers, outdoor workers, veterinarians, and research and laboratory technicians are at increased risk of exposure.

Exposure (route)

Ehrlichia chaffeensis and *Ehrlichia ewingii* are transmitted by the lone star tick in the southeastern and south central United States (Figures 25.4 and 25.5). In addition, a third *Ehrlichia* species provisionally called *Ehrlichia muris*-like (EML) has been identified in a small number of patients residing in or traveling to Minnesota and Wisconsin; a tick vector for the EML organism has not yet been established.[1]

Pathobiology

Ehrlichiae are obligate intracellular bacteria that grow within membrane-bound endosomes in human and animal phagocytic or endothelial cells and establish persistent infection in their vertebrate reservoir hosts.[2] Previously, *Ehrlichiae* were divided into human monocytic ehrlichiosis (HME) and

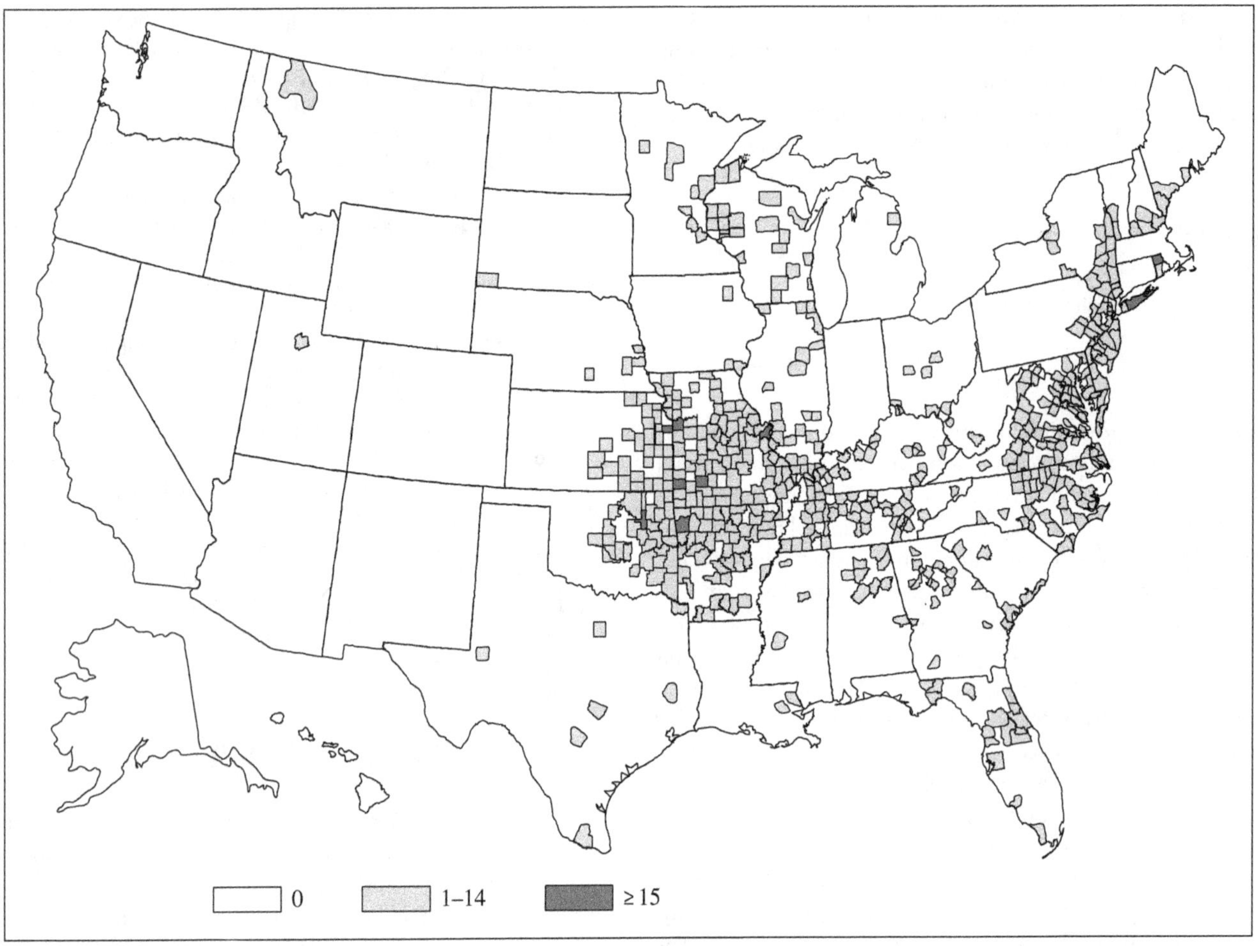

FIGURE 25.4 Map of the United States showing the number of Ehrlichiosis (*Ehrlichia chaffeensis*) cases by county in 2013. *E. chaffeensis* is the most common type of ehrlichiosis infection in the United States. This tick-borne pathogen is transmitted by *Amblyomma americanum*, the lonestar tick, whose geographic range extends from the Southeast into parts of the Northeast and Midwest. The majority of cases of *E. chaffeensis* are reported from the Midwest, South, and Northeast regions. *Source*: MMWR 2015; 62(53):64.

granulocytic ehrlichiosis (HGE). However, the causative organism for infection of granulocytes has been reclassified and is now called *A. phagocytophilum*, and the disease is termed anaplasmosis or human granulocytic anaplasmosis (HGA) and is discussed in the section on "Anaplasma" at the beginning of this chapter.

Ehrlichia ewingii was considered to be exclusively a canine pathogen until the first human cases of infection were described in 1999. Most infections reported to date have occurred in patients with HIV or who were immunosuppressed following organ transplantation.[3] Although the infection occurs in immunocompromised patients, symptoms have generally been milder with *E. ewingii*.[3]

The symptoms caused by infection with these *Ehrlichia* species usually develop 1–2 weeks after being bitten by an infected tick. The tick bite is usually painless, and about half of the people who develop ehrlichiosis may not remember being bitten by a tick. Symptoms vary from mild to severe, but common presentations include fever, meningitis, meningoencephalitis, headache, chills, malaise, muscle pain, nausea, vomiting, diarrhea, confusion, conjunctival injection, and rash (in up to 60% of children, less than 30% of adults).[1]

About 40% of those with *E. chaffeensis* infection require hospitalization.[3] Severe infections may resemble toxic shock syndrome or Rocky Mountain spotted fever.[3]

Laboratory findings may include hyponatremia, thrombocytopenia, leucopenia, anemia, and elevated aminotransferase levels. For patients that undergo lumbar puncture, the CSF WBC count is typically <100 cells/mm^3 and protein may be mildly elevated.[3]

Deaths do occur but are rare. Persons with compromised immunity caused by immunosuppressive therapies (such as corticosteroids, chemotherapy, or immunosuppressive therapy following organ transplant), HIV infection, or splenectomy appear to develop more severe disease with *E. chaffeensis* and may have higher case fatality rates.

FIGURE 25.5 Map of the United States that presents the number of Ehrlichiosis (*Ehrlichia ewingii*) cases in by county in 2013. *Ehrlichiosis ewingii* is the least common cause of ehrlichiosis. *E. ewingii* is carried by *Amblyomma americanum*, the lonestar tick, which is the same vector that transmits *E. chaffeensis*. Currently, no serologic tests are used to distinguish between the two species, and differentiation can only be made by molecular genotyping. *Source*: MMWR 2015; 62(53):53.

Diagnosis

The incidence of infection is highest in the United States during the summer months when human tick encounters are at their peak with increased outdoor activities. The diagnosis of ehrlichiosis is made based on clinical signs and symptoms consistent with the disease and history of tick bite, illness in an area where ticks are known to be present, or recent travel to an area where ehrlichiosis is endemic. Treatment should never be delayed pending the receipt of laboratory test results or be withheld on the basis of an initial negative laboratory result.

Cytoplasmic inclusion bodies (morulae) are seen on a peripheral blood smear in up to 20% of cases and can aid in early diagnosis (Figure 25.6).[1] *Ehrlichia chaffeensis* usually infects monocytes, while *E. ewingii* more commonly infects granulocytes.[1] However, morulae in a particular cell type cannot conclusively identify the infecting species.[1] Routine hospital blood cultures cannot detect *Ehrlichia*.[1]

During the acute phase of illness, a sample of whole blood can be tested by polymerase chain reaction (PCR) assay to determine if a patient has ehrlichiosis. This method is most sensitive in the first week of illness and quickly decreases in sensitivity following the administration of appropriate antibiotics. Although a positive PCR result is helpful, a negative result does not completely rule out the diagnosis.[1] The gold standard serologic test for diagnosis of ehrlichiosis is the indirect immunofluorescence assay (IFA) using *E. chaffeensis* antigen, performed on paired serum samples to demonstrate a significant (fourfold) rise in antibody titers. The first sample should be taken as early in the disease as possible, preferably in the first week of symptoms, and the second sample should be taken 2–4 weeks later. In most cases of ehrlichiosis, the first IgG IFA titer is typically low, or "negative," and the second typically shows a significant (fourfold) increase in IgG antibody levels. IgM antibodies usually rise at the same time as IgG near the end of the first week of illness and remain elevated

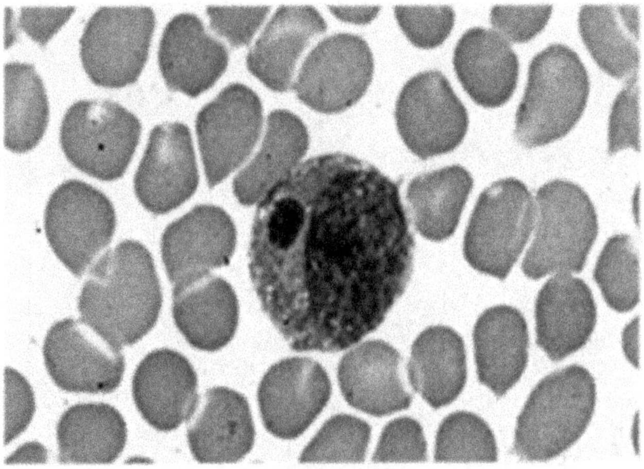

FIGURE 25.6 Morulae detected in a monocyte on a peripheral blood smear, associated with *E. chaffeensis* infection. *Source*: Centers for Disease Control and Prevention: Ehrlichiosis. http://www.cdc.gov/ehrlichiosis/symptoms/index.html

for months or longer. IgM antibodies are less specific than IgG antibodies and more likely to result in a false positive. For these reasons, physicians requesting IgM serologic titers should also request a concurrent IgG titer.[1]

Serologic tests based on enzyme immunoassay (EIA) technology are available from some commercial laboratories. However, these tests are qualitative and less useful to measure changes in antibody titers between paired specimens. Some EIA assays rely on the evaluation of only IgM antibodies, which may have a higher frequency of false-positive results.[1]

Treatment

Doxycycline is the treatment of choice and patients should be treated for at least 10 days or for 3–5 days after defervescence.[3] This treatment course will also treat the possibility of coinfection with *B. burgdorferi* and/or *R. rickettsii*. Fever typically subsides within 24–48 hours after treatment when the patient receives doxycycline or another tetracycline during the first 4–5 days of illness.[3] If a patient fails to respond to early treatment with a tetracycline antibiotic (i.e., within 48 hours), an alternate diagnosis should be considered.[4] Recovery may be prolonged in severely ill patients.[4]

Prevention

To prevent infection from tick bites, avoid wooded and bushy areas with high grass. The use of tick repellants such as DEET (*N,N*-diethyl-3-methylbenzamide) on exposed skin and permethrin on clothing coupled with timely skin examination and removal of ticks after exposure to tick-infested environments can prevent infection. Antibiotic treatment is not recommended following a tick bite to prevent ehrlichiosis.[1]

There are no licensed vaccines for ehrlichiosis or anaplasmosis. In the laboratory, *Ehrlichia* should be handled at biosafety level 2.[5]

References

1. Centers for Disease Control and Prevention. Ehrlichiosis. http://www.cdc.gov/ehrlichiosis/symptoms/index.html (accessed June 17, 2016).
2. Walker DH, Ismail N, Olano JP, et al. *Ehrlichia chaffeensis*: a prevalent, life-threatening, emerging pathogen. *Trans Am Clin Climatol Assoc* 2004; 115:375–382.
3. Ismail N, Bloch KC, McBride JW. Human Ehrlichiosis and Anaplasmosis. *Clin Lab Med* 2010; 30(1):261–292.
4. Chapman AS, Backend JS, Folk SM, et al. Diagnosis and management of tickborne rickettsial diseases: rocky mountain spotted fever, ehrlichioses, and anaplasmosis—United States: a practical guide for physicians and other health-care and public health professionals. *MMWR Recomm Rep* 2006; 55(RR-4):1–27. Available at: http://www.cdc.gov/mmwr/preview/mmwrhtml/rr5504a1.htm (accessed on July 2, 2016).
5. Chosewood LC, Wilson DE, Centers for Disease Control and Prevention, National Institutes of Health. *Biosafety in microbiological and biomedical laboratories*, 5th edn. HHS publication no. (CDC) 21-1112, 2009. Washington, DC: U.S. Department of Health and Human Services, Public Health Service, Centers for Disease Control and Prevention, National Institutes of Health. Available at: http://www.cdc.gov/biosafety/publications/bmbl5/ (accessed on July 2, 2016).

RICKETTSIA RICKETTSII

Common name for disease: Rocky Mountain spotted fever (RMSF)

In addition to *Rickettsia rickettsii*, other tick-borne species of Rickettsia, broadly grouped under the heading "Spotted Fever Group Rickettsia" (SFGR), have been shown to cause human infections. Tick-borne SFGR are transmitted to humans by the bite of an infected tick and may cause similar signs and symptoms to those observed for Rocky Mountain spotted fever (RMSF). These pathogens include several species of Rickettsia found in the United States, including *R. parkeri* and *Rickettsia* species 364D.

Occupational setting

Persons in outdoor occupations, including farming, forestry, landscaping, logging, construction, telephone lineman, environmental technicians, and some laboratory workers, are at risk from exposure.

Exposure (route)

Transmission of infection occurs primarily following bites from infected ticks or from skin contamination with tick tissue or feces when removing ticks from humans or animals, especially when the tick is crushed between the fingers. In the United States, these include the American dog tick (*Dermacentor variabilis*), Rocky Mountain wood tick (*Dermacentor andersoni*), and brown dog tick (*Rhipicephalus sanguineus*). The American dog tick, *D. variabilis*, is the most prevalent vector in the Eastern United States. The Rocky Mountain wood tick, *D. andersoni*, is the prevalent vector in the Western United States. Transmission of infection occurs after about 6–10 hours of feeding. There have also infections associated with needle-stick injuries, blood transfusions, and among laboratory workers handling rickettsiae. Laboratory-acquired infection has also occurred following exposure to infectious aerosols.[1]

Pathobiology

Rickettsia rickettsii belongs to the spotted fever group of rickettsiae that are genetically related but different in their surface antigenic proteins. The *Rickettsia* is small obligate intracellular bacteria measuring approximately 0.3 to 1.0 microns in size. The cell wall has an ultrastructural appearance of a gram-negative bacterium and contains lipopolysaccharide (LPS).[2] The LPS contains immunogenic antigens that are shared among the *Rickettsia* and cross-react with Proteus and Legionella. Cross-reactivity with Proteus is the basis for the Weil–Felix diagnostic test.

First recognized in the Rocky Mountain region, cases have been reported in almost every state. The incidence of disease varies, but around 2000 cases are reported each year to CDC. Although RMSF cases have been reported throughout most of the contiguous United States, five states (North Carolina, Oklahoma, Arkansas, Tennessee, and Missouri) account for over 60% of RMSF cases. The disease is most prevalent in the South and Midwest with the highest rates reported in southeastern North Carolina and southwestern Oklahoma (Figure 25.7). The disease is more prevalent in the spring and summer months but has been reported throughout the year in the United States.

The incubation period ranges from 2 to 14 days. Onset of symptoms is abrupt with fever, headache, chills, myalgia, and malaise. Frequently this is accompanied by gastrointestinal symptoms of nausea, vomiting, abdominal pain, and diarrhea, which can lead to confusion and delay in the diagnosis. The characteristic rash usually appears 3–5 days after the onset of fever. In many cases it appears first on the ankles and wrist and then becomes generalized. Involvement of the palms and soles is considered characteristic but often appears late in the course of the acute illness. The rash is initially maculopapular and becomes petechial and hemorrhagic as the illness progresses. Vasculitis involving the brain, heart, liver, and kidneys may cause complications from seizures to congestive heart failure and acute renal failure.

Fatality rates have decreased over the years with improvements in early detection and treatment. Risk factors associated with a higher mortality rate include increasing age, delay in initiation of chemotherapy, male gender, and glucose-6-phosphate dehydrogenase deficiency.

Diagnosis

The initial diagnosis of Rocky Mountain spotted fever is largely based upon clinical signs and symptoms including fever, headache, myalgia, rash, and epidemiology (geographic and seasonal variability). Treatment should be initiated before laboratory confirmation of the diagnosis. In endemic regions, an appropriate constellation of symptoms and signs is sufficient justification to begin treatment, even in those with no discernible history of tick bite.

Rickettsia rickettsii infects the endothelial cells that line blood vessels and does not circulate in large numbers in the blood unless the patient has progressed to a very severe phase of infection. For this reason, blood specimens (whole blood, serum) are not always useful for detection of the organism through polymerase chain reaction (PCR) or culture. If the patient has a rash, PCR or immunohistochemical (IHC) staining can be performed on a skin biopsy taken from the rash site. This test can often deliver a rapid result. These tests have good sensitivity (70%) when applied to tissue specimens collected during the acute phase of illness and before antibiotic treatment has been started, but a negative result should not be used to guide treatment decisions.[3]

Biopsy of a skin lesion obtained with a 3 mm punch biopsy can establish the diagnosis of RMSF. Direct immunofluorescence testing or immunoperoxidase staining can be performed for *R. rickettsii* on a skin biopsy specimen and may offer a more timely diagnosis if the laboratory is equipped with the appropriate methodology. A skin biopsy has been shown to be approximately 70–90% sensitive.[3]

The gold standard serologic test for diagnosis of RMSF is the indirect immunofluorescence assay (IFA) with *R. rickettsii* antigen, performed on two paired serum samples to demonstrate a significant (fourfold) rise in antibody titers. The first sample should be taken as early in the disease as possible, preferably in the first week of symptoms, and the second sample should be taken 2–4 weeks later. In most RMSF cases, the first IgG IFA titer is typically low or negative, and the second typically shows a significant (fourfold) increase in IgG antibody levels. IgM antibodies usually rise at the same time as IgG near the end of the first week of illness and remain elevated for months or even years. IgM antibodies are less specific than IgG antibodies and more likely to result in a false positive. Physicians requesting IgM serologic titers should also request a concurrent IgG titer.[3]

Treatment

Rocky Mountain spotted fever responds to treatment with doxycycline. Starting treatment early in the course of the disease before the rash is apparent is critical in reducing

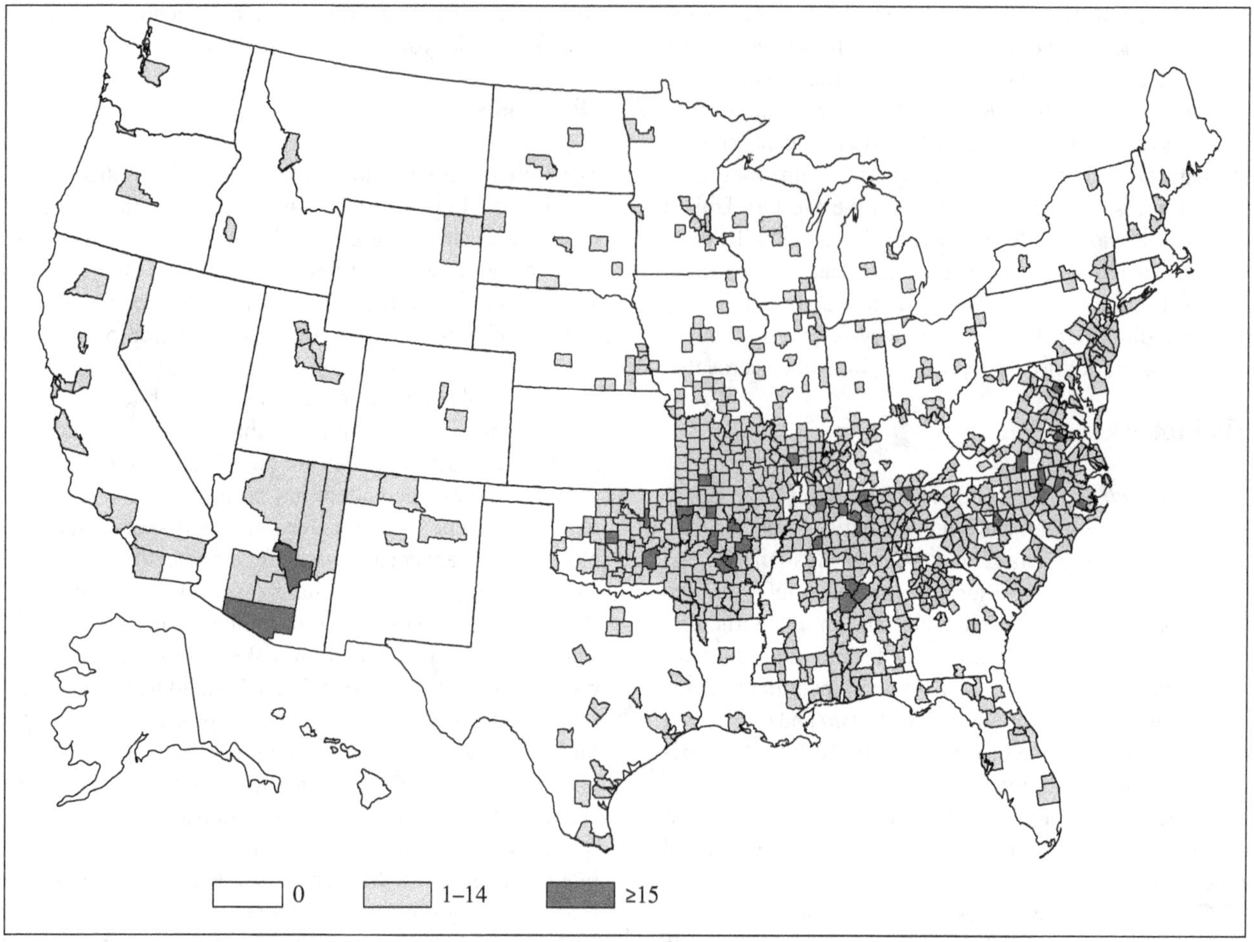

FIGURE 25.7 Map showing the number of spotted fever rickettsiosis cases by county in the United States in 2012. In the United States, the majority of cases of spotted fever rickettsiosis are attributed to infection with *Rickettsia rickettsii*, the causative agent of Rocky Mountain spotted fever (RMSF), but might also be from other agents such as *Rickettsia parkeri* and *Rickettsia* species 364D. RMSF is ubiquitous across the United States, which represents the widespread nature of the three tick vectors known to transmit RMSF: *Dermacentor variabilis* in the East, *Dermacentor andersoni* in the West, and *Rhipicephalus sanguineus*, recently recognized as a new tick vector in parts of Arizona. Historically, much of the incidence of RMSF has been in the Central Atlantic region and parts of the Midwest; however, endemic transmission of RMSF in Arizona communities has led to a substantial reported incidence rate. *Source*: MMWR 2015; 62(53):92.

mortality and morbidity. Patients should be treated for at least 3 days after the fever subsides and until there is evidence of clinical improvement. The standard duration of treatment is 7–14 days. Doxycycline or another tetracycline is considered the drug of choice, but chloramphenicol is preferred during pregnancy because of tetracycline's effects on fetal bones and teeth.[4]

Medical surveillance

It is essential that laboratories working with *R. rickettsii* have an effective system for reporting febrile illness in laboratory personnel. A medical surveillance program should include evaluation of potential cases and, when indicated, institution of appropriate antibiotic therapy.

Prevention

At present there is no commercially available vaccine for Rocky Mountain spotted fever. The best means of prevention remains avoidance of contact with ticks, wearing protective clothing, and using insect repellents. Body checks with particular attention to the scalp, pubic, and axillary hair should be conducted daily for at-risk workers. Ticks can be removed using forceps and gentle traction, being careful to remove all of the mouth parts from the skin. Tick bites should be cleansed and care taken during removal to prevent crushing the tick and contaminating fingers with tick tissue and feces that is potentially infectious. In endemic areas, physicians caring for injured and ill workers should be periodically reminded of the importance of early diagnosis and treatment to prevent serious sequelae. In the laboratory, biosafety level 2 (BSL2) practices, containment equipment, and facilities are recommended for all nonpropagative laboratory procedures including serological and fluorescent antibody tests and staining of impression smears. Biosafety level 3 (BSL3) practices and facilities are recommended for all other manipulations of known or potentially infectious materials, including necropsy of experimentally infected animals and trituration of their tissues, and inoculation, incubation, and harvesting of embryonated eggs or tissue cultures.[5]

References

1. Johnson JE, Kadull PJ. Rocky Mountain spotted fever acquired in a laboratory. *N Engl J Med* 1967; 227:842–847.
2. Walker DH. *Rickettsia rickettsii* and other spotted fever group Rickettsiae (Rocky Mountain spotted fever and other spotted fevers). In: Mandell GL, Bennett JE, Dolin R, (eds). *Mandell, Douglas, and Bennett's principles and practice of infectious diseases*, 7th edn., 2010:2499–2508. Philadelphia: Churchill Livingstone.
3. Centers for Disease Control and Prevention. Rocky Mountain spotted fever. http://www.cdc.gov/rmsf/symptoms/index.html (accessed June 17, 2016).
4. Chapman AS, Backend JS, Folk SM, et al. Diagnosis and management of tickborne rickettsial diseases: rocky mountain spotted fever, ehrlichioses, and anaplasmosis—United States: a practical guide for physicians and other health-care and public health professionals. *MMWR Recomm Rep* 2006; 55(RR-4):1–27. Available at: http://www.cdc.gov/mmwr/preview/mmwrhtml/rr5504a1.htm (accessed on July 2, 2016).
5. Chosewood LC, Wilson DE, Centers for Disease Control and Prevention, National Institutes of Health. *Biosafety in microbiological and biomedical laboratories*, 5th edn. HHS publication no. (CDC) 21-1112, 2009. Washington, DC: U.S. Department of Health and Human Services, Public Health Service, Centers for Disease Control and Prevention, National Institutes of Health. Available at: http://www.cdc.gov/biosafety/publications/bmbl5/ (accessed on July 2, 2016).

Chapter 26

PARASITES

William N. Yang

CRYPTOSPORIDIUM PARVUM

Common names for disease: Cryptosporidiosis

Many species of *Cryptosporidium* exist that infect humans and a wide range of animals. Although *Cryptosporidium parvum*, found in young cattle and other herbivores, and *Cryptosporidium hominis*, the human-adapted species, (formerly known as *C. parvum* anthroponotic genotype or genotype 1) are the most prevalent species causing disease in humans, infections by *Cryptosporidium felis*, *Cryptosporidium meleagridis*, *Cryptosporidium canis*, and *Cryptosporidium muris* have also been reported.[1]

Occupational setting

Farmers, animal handlers, veterinarians, laboratory personnel, healthcare workers, daycare workers, and divers are at risk from exposure.[1–4]

Exposure (route)

Human cryptosporidiosis was first reported in 1976. Its characteristics are having a very low infectious dose (10–1000 oocysts), being severe in immunocompromised populations, being very resistant to antiparasitic therapy, and having human and nonhuman reservoirs. Cryptosporidiosis is most common in developing countries, with waterborne, foodborne, person-to-person, and zoonotic transmission pathways. It still remains a significant pathogen of children and the elderly in industrialized nations and can cause major outbreaks when public health measures break down. It is a common cause of wasting and lethal diarrhea in HIV-infected persons, and acute and persistent disease among farmers, animal handlers, and veterinarians results most commonly from exposure to infected calves and other farm animals. The prevalence of infection is higher in young animals, such as calves and lambs.[5] An outbreak of cryptosporidiosis occurred among firefighters responding to a fire involving a barn that contained preweaned calves.[6] Other sources of infection are contaminated water supplies, canals, lakes or rivers, and swimming and wading pools, which have been contaminated by animal or human sewage. Occupational and recreational divers are at risk. Person-to-person (fecal–oral) infection may also occur in healthcare and daycare settings.[4,7,8] The 1993 cryptosporidiosis waterborne outbreak in Milwaukee, Wisconsin, was the largest waterborne outbreak in US history, affecting over 400 000 people. The outbreak was estimated to cost over 96 million dollars in lost productivity and medical costs.[9]

Pathobiology

Cryptosporidium parvum is an intracellular but extracytoplasmic protozoan parasite. When a human ingests the oocyst that has been passed in the feces of an infected host, the oocyst wall dissolves, and sporozoite forms invade the host gastrointestinal (GI) epithelial cells. They pass through a trophozoite stage and an asexual multiplication stage that results in schizonts. These schizonts can reinvade the host or continue to a sexual multiplication stage that leads to new oocysts that are passed outside the body. The sporulated oocyst is the only developmental stage that occurs extracellularly. Since this part of the cycle is completed before the oocyst is excreted, the oocysts are immediately infectious when passed in feces.[5]

Physical and Biological Hazards of the Workplace, Third Edition. Edited by Gregg M. Stave and Peter H. Wald.

In immunocompromised persons, the ability to reinfect the host can lead to continuing infection (Figure 26.1).

Cryptosporidium hominis and *C. parvum* are the cause of about 75 and 20% of human disease.

Cryptosporidium hominis is more virulent in humans than other species and only infects humans.

Cryptosporidium parvum infects cattle and sheep. A third common human pathogen is *C. meleagridis*, an avian species.[1] In developing countries ~20% of children with diarrhea have cryptosporidiosis compared with 1–5% in North America and Europe. Cryptosporidiosis prevalence is higher in rural areas. Malnutrition, HIV/AIDS, and other immunocompromising conditions significantly increase the risk, severity, and persistence of cryptosporidiosis. It is an AIDS opportunistic infection and often afflicts malnourished children with persistent diarrhea. Cryptosporidiosis-affected children may have impaired growth and cognitive function and have multiple symptomatic episodes before they acquire partial protective immunity. While *Cryptosporidium* parasites infect the colonic mucosa and, less commonly, the small intestine and stomach, replicating parasites can infect the entire gut from the oropharynx to the anus.

The incubation period ranges from 4 to 10 days after oocyst ingestion. Excretion of infectious oocysts in feces begins when symptoms begin and can persist for months. Oocysts of *Cryptosporidium* spp. are immediately infectious when they are passed in stool. Internal autoinfection is common and may be the reason for the persistent nature of the disease in immunocompromised people. The immune status of the person determines the severity and length of the symptoms. In immunocompetent persons, the symptoms begin quickly, after a 2–12-day incubation period, and then last from 7 to 14 days with a mean of 9–11 days.[1,5]

Symptoms of disease include watery diarrhea (the most frequent symptom), cramping abdominal pain, weight loss, and flatulence. Less common symptoms are nausea, vomiting, myalgias, and fever. While the small intestine is the site most commonly affected, symptomatic *Cryptosporidium* infections have also been found in other organs including other digestive tract organs, the lungs, and possibly conjunctiva. [1,5]

In immunocompetent persons, symptoms are usually self-limited (1–2 weeks). In immunocompromised patients, especially those with CD4 counts <200/μL, symptoms can be chronic and more severe. The diarrhea can last for months and lead to dehydration, electrolyte imbalance, and malnutrition and can be life threatening.[1] While cryptosporidial cysts can be found in the stool, blood and white blood cells are usually absent. Oocysts may remain in the stool for 8–50 days (mean 12–14 days) after the clinical symptoms have ended.

AIDS patients with *Cryptosporidium* sclerosing cholangitis and biliary involvement usually have elevated serum alkaline phosphatase levels and right upper quadrant pain. Symptoms in patients with reversible causes of immunodeficiencies usually resolve quickly when the cause of the immunosuppression is eliminated.[1,5]

Diagnosis

Cryptosporidiosis is usually mild and underdiagnosed since often no diagnostic testing is done. Suspicion for *Cryptosporidium* as the cause of persistent or chronic diarrhea is increased if the individual is malnourished, has an immunosuppressive condition, or has another risk factor. The incubation period is longer (~1 week) than it is for viral or bacterial diseases (1–3 days) yet shorter than for giardiasis (~2 weeks). *Giardia* and *Isospora* infections may also mimic cryptosporidiosis.[1]

The diagnosis of cryptosporidiosis can be made by microscopic examination of stool specimens or by examination of biopsied tissue from infected intestine, biliary system, or respiratory tract. With a modified acid-fast technique, the oocysts are red-stained and can be differentiated from similar-appearing yeast forms that do not take up the acid-fast stain. Strict morphologic criteria must be applied to the diagnosis to avoid confusion with other oocysts, such as *Cyclospora* oocysts, which are much larger (10 mm).[10] Antigen detection by ELISA, PCR, or a direct fluorescent antibody (DFA) assay are also available and have the advantage of high sensitivity.[1,5]

Treatment

In the immunocompetent host, the infection is self-limited. Oral or intravenous rehydration, antimotility agents, and nutritional support are used for the symptomatic treatment of diarrhea. Nitazoxanide (NTZ) is the only Food and Drug Administration (FDA)-approved drug for cryptosporidiosis. NTZ can shorten clinical disease by a day or so on average and reduce parasite loads in immunocompetent patients, but it is not consistently effective for immunodeficient persons. Those on corticosteroids or cytotoxic drugs will recover if the agents are stopped.[1,5] Since highly active antiretroviral therapy (HAART) restores cell-mediated immunity and has been shown to decrease the prevalence of cryptosporidiosis in HIV-infected patients, it is considered the most effective treatment and prophylaxis for AIDS patients.[1,5]

Medical surveillance

Occupationally exposed or at-risk workers should be monitored for signs or symptoms of diarrhea. In conjunction with medical surveillance, workers should receive education about the sources of exposure and preventive measures.

Prevention

Prevention is based on practicing good hygiene and educating staff about the way cryptosporidiosis is spread in childcare or animal settings. Good hygiene includes good hand

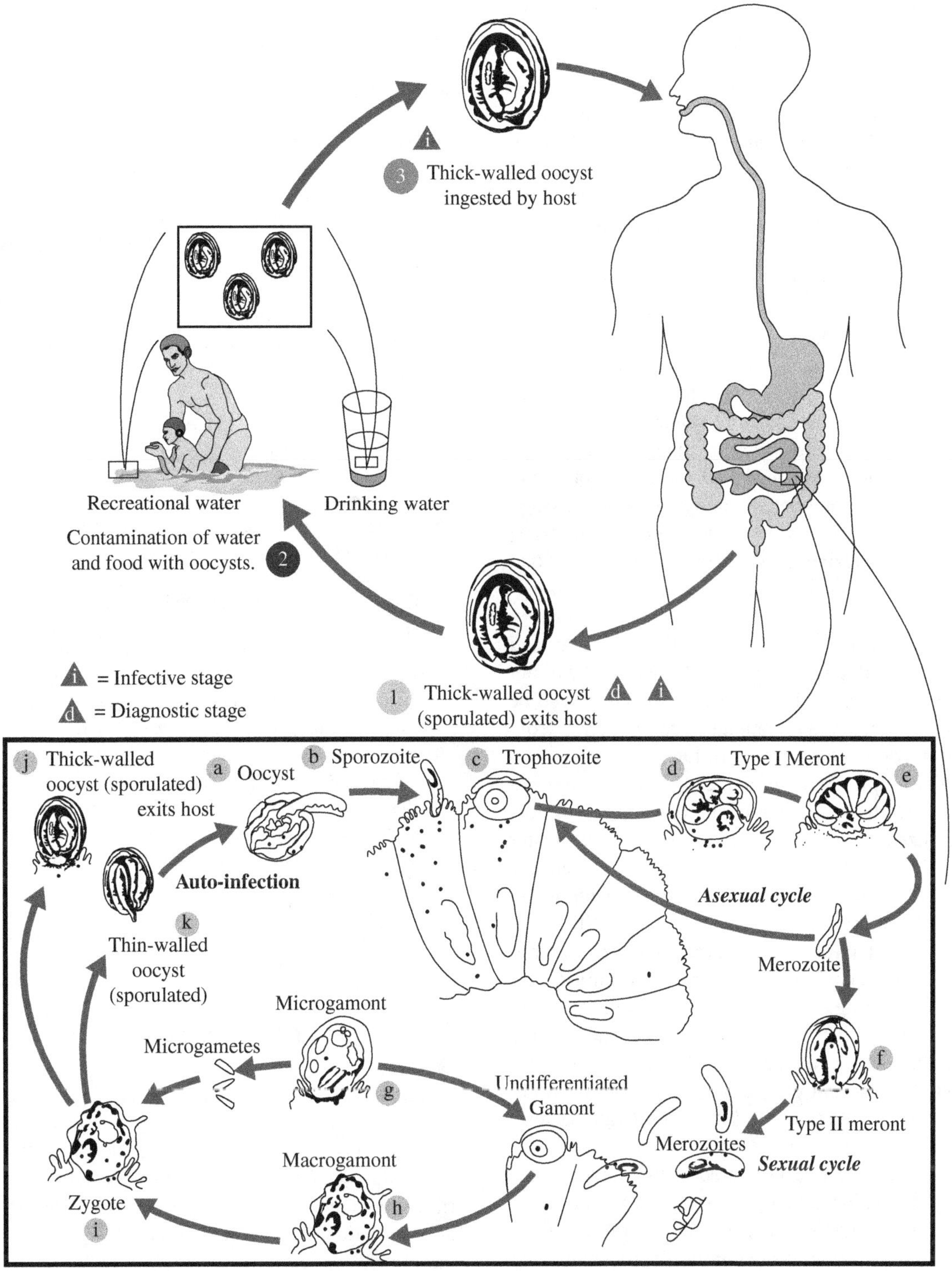

FIGURE 26.1 *Cryptosporidium* Life Cycle. Sporulated oocysts, containing 4 sporozoites, are excreted by the infected host through feces and possibly other routes such as respiratory secretions (1). Transmission of *Cryptosporidium parvum* and *C. hominis* occurs mainly through contact with contaminated water (e.g., drinking or recreational water). Occasionally food sources, such as chicken salad, may serve as vehicles for transmission. Many outbreaks in the United States have occurred in waterparks, community swimming pools, and day care centers. Zoonotic and anthroponotic transmission of *C. parvum* and anthroponotic transmission of *C. hominis* occur through exposure to infected animals or exposure to water contaminated by feces of infected animals (2). Following ingestion (and possibly inhalation) by a suitable host (3), excystation (a) occurs. The sporozoites are released and parasitize epithelial cells (b, c) of the gastrointestinal tract or other tissues such as the respiratory tract. In these cells, the parasites undergo asexual multiplication (schizogony or merogony) (d, e, f) and then sexual multiplication (gametogony) producing microgamonts (male) (g) and macrogamonts (female) (h). Upon fertilization of the macrogamonts by the microgametes (i), oocysts (j, k) develop that sporulate in the infected host. Two different types of oocysts are produced, the thick-walled, which is commonly excreted from the host (j), and the thin-walled oocyst (k), which is primarily involved in autoinfection. Oocysts are infective upon excretion, thus permitting direct and immediate fecal-oral transmission. Centers for Disease Control and Prevention. Parasites – *Cryptosporidium*. *Source*: http://www.cdc.gov/parasites/crypto/

washing before preparing or eating food, after toilet use, in daycare settings after changing diapers or cleaning up a child who has used the toilet, and in animal settings after handling animals or animal waste. Good hand washing is defined as washing hands with soap and water for at least 20 seconds, rubbing hands together vigorously, and scrubbing all surfaces.[5]

In farm and institutional settings, personal hygiene practices, appropriate use of disposable gloves, and environmental sanitation should be emphasized. Strict enteric precautions should be practiced around infected animals or humans. Eliminating the disease in farm animals decreases the risk for farmers and veterinary personnel.[1]

Cooking kills *Cryptosporidium* in food, but vegetables and fruits to be consumed uncooked must be thoroughly washed with potable water before eating. Oocysts are killed by pasteurization (72°C for 5 seconds), heating to 60°C for 30 minutes, or freezing at −7°C for 1 hour. No safe and effective disinfectant has been identified for decontaminating produce.[1]

Cryptosporidium is extremely tolerant to chlorine, and chlorine-based disinfectants (e.g., bleach solutions) will not kill *Cryptosporidium.*[1] Drinking water sources must be protected from human and animal fecal material from surface runoff. Maintaining proper filtration procedures for drinking water supplies will minimize infection but may not prevent it due to the small size of the oocysts.[10] Travelers should minimize their exposure by avoiding untreated water and uncooked fruits and vegetables. The oocyst will survive in 3% hypochlorite solution, iodophors, benzalkonium chloride, and 5% formaldehyde. Reverse osmosis filters can remove all oocysts but require low-turbidity source water to prevent plugging. Contaminated water can be purified for personal use by heating the water to a rolling boil for 1 minute or using a filter that has been tested and rated by the National Sanitation Foundation (NSF) Standard 53 or NSF Standard 58 for cyst and oocyst reduction; filtered water will need additional treatment to kill or inactivate bacteria and viruses.[5]

Cryptosporidiosis can be prevented in the immunocompromised by education and by drinking only filtered or boiled water, ingesting only cooked food, thorough hand washing, avoiding bathing or swimming in water used by other people or animals, and avoiding contact with people or animals (especially calves and lambs) with diarrhea. Companion dogs and cats should be examined and cleared of infection by a veterinarian.[1]

Immunocompromised persons should not have contact with any animal that has diarrhea. They should also realize that there is risk of infection from the accidental ingestion of recreational lake, river, or swimming pool water due to intermittent contamination with cryptosporidia from human or animal waste.[1,5]

References

1. Xiao L, Griffiths JK. Cryptosporidiosis. In: Magill AJ, Hill DR, Solomon T, et al., eds. Hunter's Tropical Medicine and Emerging Infectious Diseases, 9th edn. London/New York: Saunders/Elsevier, 2013:673–80.
2. Current VVL, Reese NC, Ernst JY, et al. Human cryptosporidiosis in immunocompetent and immunodeficient persons. Studies of an outbreak and experimental transmission. *N Engl J Med* 1983; 308:1252–7.
3. Pohjola S, Oksanen H, Jokipii L, et al. Outbreak of cryptosporidiosis among veterinary students. *Scand J Infect Dis* 1986; 18:173–8.
4. Artieda J, Basterrechea M, Arriola L, et al. Outbreak of cryptosporidiosis in a child day-care centre in Gipuzkoa, Spain, October to December 2011. *Euro Surveill* 2012; 17(5):pii: 20070.
5. Centers for Disease Control and Prevention. Parasites – *Cryptosporidium.* http://www.cdc.gov/parasites/crypto/ (accessed October 15, 2014).
6. Centers for Disease Control and Prevention. Outbreak of cryptosporidiosis associated with a firefighting response — Indiana and Michigan, June 2011. *MMWR* 2012; 61(9):153–6.
7. Koch KL, Phillips DJ, Aber RC, et al. Cryptosporidiosis in hospital personnel. Evidence for person-to-person transmission. *Ann Intern Med* 1985; 102(5):593–6.
8. Xiao G, Wang Z, Chen J, et al. Occurrence and infection risk of waterborne pathogens in Wanzhou watershed of the Three Gorges Reservoir, China. *J Environ Sci (China)* 2013; 25(9):1913–24.
9. Corso PS, Kramer MH, Blair KA, et al. Cost of illness in the 1993 waterborne *Cryptosporidium* outbreak, Milwaukee, Wisconsin. *Emerg Infect Dis* 2003; 9(4):426–31.
10. MacKenzie WR, Hoxie NJ, Proctor ME, et al. A massive outbreak in Milwaukee of cryptosporidium infection transmitted through the public water supply. *N Engl J Med* 1994; 331:161–7.

CYCLOSPORIASIS

Common names for disease: None

Cyclosporiasis is an intestinal infection with the coccidian organism *Cyclospora cayetanensis*. Symptoms of cyclosporiasis begin an average of 7 days (range 2 days to ≥2 weeks) after ingestion of sporulated cysts—the infective form of the parasite.

Symptoms can include watery diarrhea, fatigue, anorexia, nausea, cramping, increased gas, and bloating. If a person ill with cyclosporiasis is not treated, symptoms can persist for several weeks to a month or more. Some symptoms, such as diarrhea, can return, and other symptoms, such as muscle aches and fatigue, may continue after the gastrointestinal symptoms have gone away. The infection usually is not life threatening.

The organism was definitively identified in 1993, after being independently discovered by four different researchers dating back to 1979. Before 1996, cyclosporiasis in North

America was associated with overseas travelers. In 1996, a large outbreak in North America occurred in the spring and summer and was associated with eating Guatemalan raspberries.[1–3] Cyclosporiasis is most common in tropical and subtropical regions. While the risk for infection is seasonal, there is no consistent pattern regarding the time of year or the environmental conditions, such as temperature or rainfall. In the United States foodborne outbreaks of cyclosporiasis since the mid-1990s have been linked to various types of imported fresh produce including raspberries, basil, snow peas, and mesclun lettuce. No commercially frozen or canned produce has been implicated.[1]

Even though here have not been any large outbreaks of cyclosporiasis recently, in 2013 the Centers for Disease Control and Prevention (CDC) reported a total of 631 persons infected with *C. cayetanensis* from 25 states and New York City. Most illness occurred between mid-June and mid-July with eight percent of ill persons hospitalized and no deaths reported. Ill persons ranged from less than 1 year to 94 years, with a median age of 52 years. [1]

Occupational setting

Cyclosporiasis can be an important cause of diarrhea domestically and internationally as a result of contaminated food or water. *Cyclospora* has been found throughout the world in both immunocompetent and immunosuppressed persons. *Cyclospora* infected 7% of the Nepal American Embassy community during the 1992 *Cyclospora* season and was the cause of 11% of the cases of diarrhea seen at the Canadian International Water and Energy Consultants Clinic in Kathmandu. *Cyclospora cayetanensis* should be considered as a possible cause of diarrhea in an individual with prolonged watery diarrhea with a history of travel to a developing country.[4]

Cyclospora has also been linked to an outbreak of foodborne illness on a passenger ship.[5] Epidemiology studies done in Guatemala have found *Cyclospora* in farm workers in one study but not in another. With the globalization of food systems, a large part of the US food supply now comes from areas where *Cyclospora* is endemic and where methods for inactivation and complete removal of *Cyclospora* oocysts from contaminated produce have not been developed. Countries where crops are grown for local consumption and exportation need to practice good agricultural methods and use filtered or otherwise decontaminated irrigation water. Worldwide food safety education and training is important to insure safe food and water for all.[6]

Exposure (route)

While the transmission of *Cyclospora* appears to be fecal–oral, there is no definite evidence of person-to-person spread. Waterborne transmission and foodborne transmission evidence does exist. A water supply from a hospital building at a Chicago hospital was implicated as the source of a 1990 summer outbreak in the hospital staff, and water was associated with a case–control study of cyclosporiasis in Kathmandu.[5] The large outbreak that occurred in 1996 was related to the consumption of raspberries imported from Guatemala.[3] In 1997, several other outbreaks in Canada and the United States were also linked to raspberries from Guatemala and to basil and mesclun lettuce.[1]

The mode of contamination of the raspberries is unclear. However, following the 1996 outbreak, berry growers and exporters in Guatemala, with assistance from the Food and Drug Administration (FDA) and CDC, began voluntary water quality and sanitary control measures.[1,4] After another outbreak in spring 1997, Guatemala began classifying the farms, and only "low-risk" farms were allowed to export to North America.[4] In 1998, the FDA banned the importation of fresh raspberries into the United States, but Canada continued to import them until June 1998. Two surveys of indigenous Peruvian children under the age of 30 months found 6 and 18% with *Cyclospora* cysts in their feces.[1,2] Of those infected, 11 and 28%, respectively, presented with diarrhea.

Of interest is the seasonality of *Cyclospora*. In Nepal, the organism is present from May, just before the rainy season, until October, just after the rainy season, with the peak risk in June and July. In Peru, the season extends from January to July, with a peak from April to June. In Guatemala, the main season of risk is May to September.[2]

Since a large part of the US food supply comes from areas where *Cyclospora* is endemic and methods for inactivation and complete removal of *Cyclospora* oocysts from contaminated produce have not been developed, countries where crops are grown for local consumption and exportation need to practice good agricultural methods and use filtered or otherwise decontaminated irrigation water.[6]

Pathobiology

Cyclospora cayetanensis completes its life cycle in humans. However, the oocysts shed in the feces of infected persons must mature (sporulate) outside the host, in the environment, to become infective for someone else. Therefore, direct person-to-person (fecal–oral) transmission of *Cyclospora* is unlikely. The process of maturation (sporulation) is thought to require days to weeks (Figure 26.2).

Persons of all ages living or traveling in the tropics and subtropics may be at increased risk because cyclosporiasis is endemic in some countries in these zones. In some regions, infection appears to be seasonal, but the seasonality varies in different settings and is not well understood.

Some infected persons are asymptomatic, particularly in settings where cyclosporiasis is endemic. Among symptomatic persons, the incubation period averages ~1 week (range

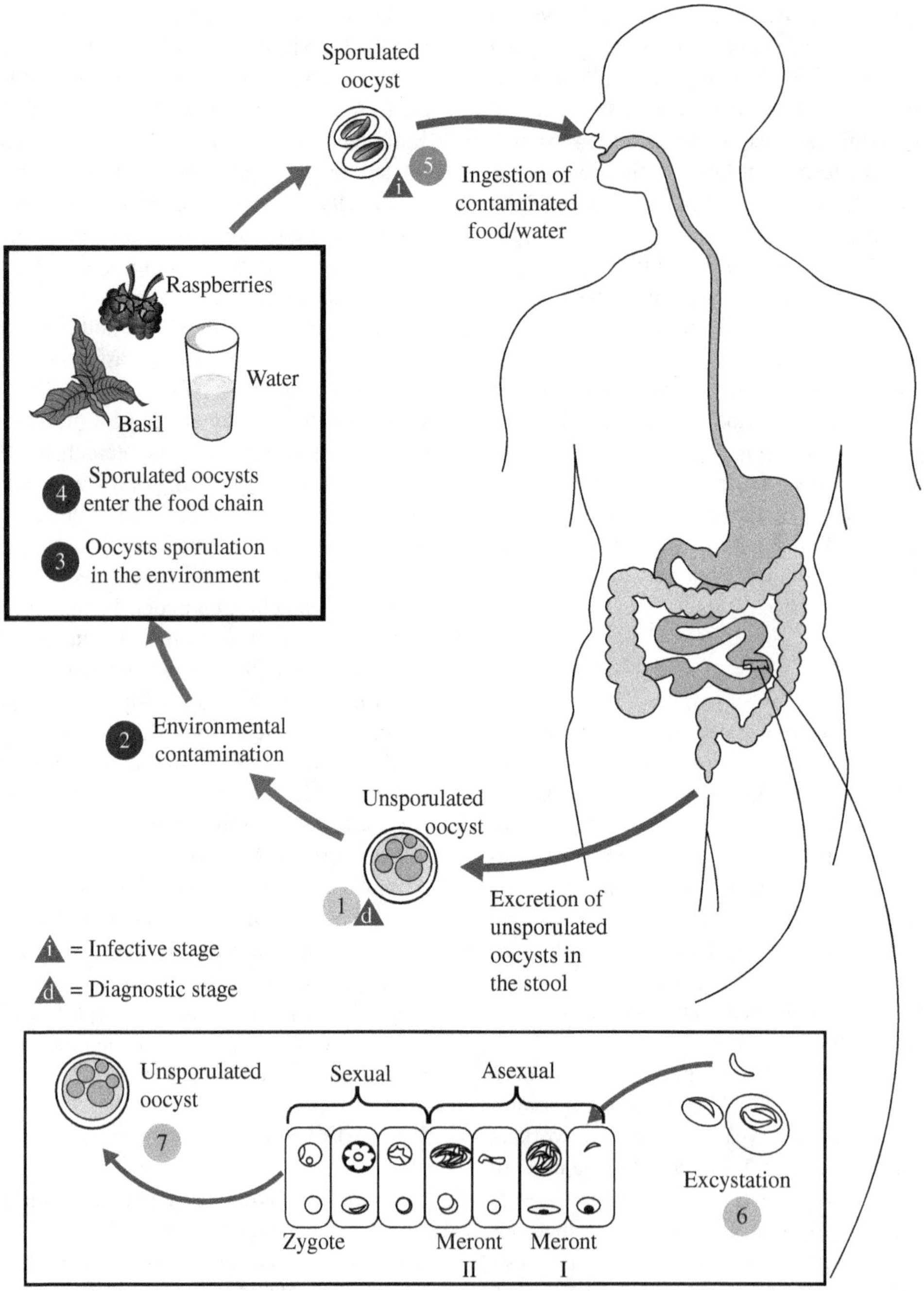

FIGURE 26.2 *Cyclospora* Life Cycle. Some of elements of this figure were created based on an illustration by Ortega et al. *Cyclospora cayetanensis*. In: Advances in Parasitology: opportunistic protozoa in humans. San Diego: Academic Press; 1998. p. 399–418. When freshly passed in stools, the oocyst is not infective 1 (thus, direct fecal-oral transmission cannot occur; this differentiates *Cyclospora* from another important coccidian parasite, *Cryptosporidium*). In the environment 2, sporulation occurs after days or weeks at temperatures between 22°C to 32°C, resulting in division of the sporont into two sporocysts, each containing two elongate sporozoites 3. Fresh produce and water can serve as vehicles for transmission 4 and the sporulated oocysts are ingested (in contaminated food or water) 5. The oocysts excyst in the gastrointestinal tract, freeing the sporozoites which invade the epithelial cells of the small intestine 6. Inside the cells they undergo asexual multiplication and sexual development to mature into oocysts, which will be shed in stools 7. The potential mechanisms of contamination of food and water are still under investigation. *Life cycle image and information courtesy of* ***DPDx***. *Source*: http://www.cdc.gov/parasites/cyclosporiasis/biology.html

~2–14 or more days). *Cyclospora* infects the small intestine and typically causes watery diarrhea, with frequent, sometimes explosive, stools. Other common symptoms include loss of appetite, weight loss, abdominal cramping and bloating, increased flatus, nausea, and prolonged fatigue. Vomiting, body aches, low-grade fever, and other flu-like symptoms may be also be noted. If untreated, the illness may last for a few days to a month or longer and may follow a remitting–relapsing course. Although cyclosporiasis usually is not life threatening, reported complications have included malabsorption, cholecystitis, and Reiter's syndrome—a reactive arthritis.[1]

If untreated, clinical manifestations can last for an average of 6–7 weeks in immunocompetent hosts, and in HIV-positive individuals, the diarrhea can last for months. The organism can be detected in stools up to 8 weeks after the onset of symptoms.[2] Asymptomatic individuals who are excreting cysts have been documented, particularly in areas where it is endemic.[1]

Diagnosis

The main method is identifying the typical oocysts in stools after using concentration methods and then staining stool smears with a modified acid-fast stain. Concentration techniques can increase the diagnostic yield by over 40%, which is very helpful in individuals with acute and severe disease who are shedding low number of organisms.[2] Fluorescence microscopy can assist in the diagnosis, since the oocysts autofluoresce. The diagnosis can also be made by observing the oocysts in wet preparation from jejunal aspirates.[5] Current CDC guidelines for the confirmation of *C. cayetanensis* foodborne disease outbreaks require the demonstration of the organism in stools of two or more ill persons.[7] Molecular diagnostic methods, such as polymerase chain reaction (PCR) analysis, can be used to look for the parasite's DNA in the stool. The lack of DNA information of the various strains of the parasite hinders surveillance and outbreak response.

Treatment

Most people with healthy immune systems will recover without treatment. Without treatment, the illness may last for a few days to a month or longer. Symptoms may seem to go away and then return one or more times (relapse). Antidiarrheal medicine may help reduce diarrhea, but a healthcare provider should be consulted before such medicine is taken. People who are in poor health or who have weakened immune systems may be at higher risk for severe or prolonged illness.

The medical treatment for immunocompetent adults is TMP 160 mg plus SMX 800 mg (one double-strength tablet), orally, twice a day, for 7–10 days. HIV-infected patients and others with immune compromise may need longer courses of therapy. No other alternate medications have been identified for persons who are allergic to or cannot take TMP/SMX.[1]

Medical surveillance

CDC conducts surveillance for cyclosporiasis. Cyclosporiasis is a nationally notifiable disease and is reportable in over 35 states. Regardless of whether cyclosporiasis is reportable in a particular state, the public and clinicians need to inform their local health department about potential cases and clusters of the disease so that appropriate investigative actions can be taken to help prevent additional cases of illness.

CDC, in collaboration with public health authorities, analyzes each reported case for epidemiological evidence of linkage to other cases to facilitate rapid identification and investigation of outbreaks. There are no molecular tools available for linking *C. cayetanensis* cases.[1]

Future sequencing of the DNA of samples of the parasite that circulate in the United States and different parts of the world and analyzing the DNA of parasites collected from individual outbreak-related cases—as well as cases not known to be linked to an outbreak—will identify potential genotyping markers and develop a new DNA-based surveillance system for cyclosporiasis.[1]

Prevention

On the basis of the currently available information, avoiding food or water that may have been contaminated with feces is the best way to prevent cyclosporiasis. *Cyclospora* oocysts are killed by boiling water but not by treatment with chlorine or iodine. No vaccine for cyclosporiasis is available. Fresh fruits, especially raspberries and salads, should be thoroughly washed before eating.[1]

References

1. Centers for Disease Control and Prevention. Cyclosporiasis. http://www.cdc.gov/parasites/cyclosporiasis/ (accessed October 21, 2014).
2. Shlim DR, Connor BA. Cyclosporoiasis. In: Magill AJ, Hill DR, Solomon T, et al., eds. Hunter's Tropical Medicine and Emerging Infectious Disease, 9th edn. New York: Elsevier, 2013:680–3.
3. Herwaldt BL, Ackers M. The cyclospora working group. An outbreak in 1996 of cyclosporiasis associated with imported raspberries. *N Engl J Med* 1997; 336:1548–56.
4. Hoge CW, Shlim DR, Rajah R, et al. Epidemiology of diarrhoeal illness associated with coccidian-like organism among travelers and foreign residents in Nepal. *Lancet* 1993; 341:1175–9.
5. Roisin M, Cramer E, Mantha S, et al. A review of outbreaks of foodborne disease associated with passenger ships: evidence for risk management. *Public Health Rep* 2004; 119:427–34.

6. Ortega YR, Sanchez R. Update on cyclospora cayetanensis, a food-borne and waterborne parasite. *Clin Microbiol Rev* 2010; 23:218–34.
7. Centers for Disease Control and Prevention. Guide to Confirming an Etiology in Foodborne Disease Outbreak 2015. Table B-3. Guidelines for Confirmation of Foodborne-Disease Outbreaks (Parasitic). http://www.cdc.gov/foodsafety/outbreaks/investigating-outbreaks/confirming_diagnosis.html (accessed July 3, 2016).

CUTANEOUS, MUCOCUTANEOUS, AND VISCERAL LEISHMANIASIS

Common names for cutaneous disease (CL) are Baghdad or Delhi boil, oriental sore, or espundia in the Old World and uta, chiclero ulcer, or forest yaws in the New World.[1] Mucocutaneous (MCL) disease is primarily a disease of South America. Visceral leishmaniasis (VL) is also commonly known as kala-azar. The distribution of leishmaniasis is divided into Old World or the Eastern Hemisphere and New World or the Western Hemisphere (Figures 26.3, 26.4, and 26.5).

Cutaneous leishmaniasis (CL) is the most common form of leishmaniasis with about 95% of CL cases occurring in the Americas, the Mediterranean basin, and the Middle East and Central Asia. Over two thirds of CL new cases occur in six countries: Afghanistan, Algeria, Brazil, Colombia, the Islamic Republic of Iran, and the Syrian Arab Republic.

Visceral leishmaniasis (VL) occurs in rural tropical and subtropical areas, with over 90% of (VL) cases occurring in Bangladesh, Brazil, Ethiopia, India, South Sudan, and Sudan. VL has been reported in US veterans deployed to Iraq, Kuwait, and Afghanistan.

Almost 90% of MCL cases occur in the Plurinational State of Bolivia, Brazil, and Peru.[2]

Occupational setting

Leishmaniasis is endemic in 88 countries and found in all continents except Australia and Antarctica. Two million new cases occur yearly—1.5 million cases of CL and MCL and 500 000 cases of VL. The odds of encountering cases of leishmaniasis in nonendemic countries is rising because of an increase in travel by the public sector, occupational exposure through military deployments to disease-endemic areas, and factors that change the pathogenicity of the disease, such as immunodeficiency states (e.g., HIV).[3,4] In the Western Hemisphere CL is found in parts of Mexico, Central America, and South America. It is not found in Chile, Uruguay, or Canada. In the United States occasional cases have been reported in Texas and Oklahoma.[5] In the Western Hemisphere, workers in forested areas or workers living next to or working next to forested areas—loggers, road builders, agricultural workers, hunters, explorers, scientists, missionaries, and military personnel undergoing jungle training—are at risk. Individuals living and working in farming communities in newly cleared forest areas and workers in road construction and mining are also at risk.[1] Workers who spend months in the forests of southern Mexico collecting chewing gum latex, "chicleros," have a high incidence of infection—30% during the first year of employment.[1]

For those who are born in geographic areas of Old World CL, the disease is usually a childhood disease, and adults have acquired immunity and are not affected.[1] American Regular, Reserve, and National Guard troops and civilian support personnel who have been deployed to Iraq, Kuwait, and Afghanistan have been exposed to leishmaniasis. In 2004 over 600 cases of CL and 4 cases of VL were diagnosed in American soldiers. As these soldiers return to their communities and civilian life, the presentation, diagnosis, and treatment of leishmaniasis have become a challenge for civilian physicians.[6]

Ecotourists, adventure travelers, bird watchers, and Peace Corps volunteers to endemic areas are also at risk.[5]

There are other factors that are linked to the risk of occupational exposure.

Poor socioeconomic conditions are a risk factor because poor housing and sanitary conditions (e.g., lack of waste management, open sewerage) may increase sandfly breeding and resting sites. A diet lacking protein energy, iron, vitamin A, and zinc increases the risk that an infection will progress to kala-azar. Epidemics of both CL and VL are often associated with population migration and the movement of nonimmune people into areas with existing transmission cycles. Occupational exposure of people who settle or work in areas that used to be forests and sandfly habitats can lead to an increase in cases. Similarly environmental changes that affect the incidence of leishmaniasis include urbanization, domestication of the transmission cycle, and the incursion of agricultural farms and settlements into forested areas. Leishmaniasis is very climate sensitive and can be affected by changes in rainfall, temperature, and humidity. Small fluctuations in temperature can affect the developmental cycle of *Leishmania* promastigotes in sandflies, allowing transmission of the parasite in areas not previously endemic for the disease.

Drought, famine, and flood resulting from climate change can lead to massive displacement and migration of people to areas with transmission of leishmaniasis, and poor nutrition could compromise their immunity.[2]

Exposure (route)

Humans acquire cutaneous leishmaniasis (CL), VL, and MCL through the bite of an infected female phlebotomine sandfly. Sandflies typically feed (bite) at night and during twilight hours and are less active during the hottest time

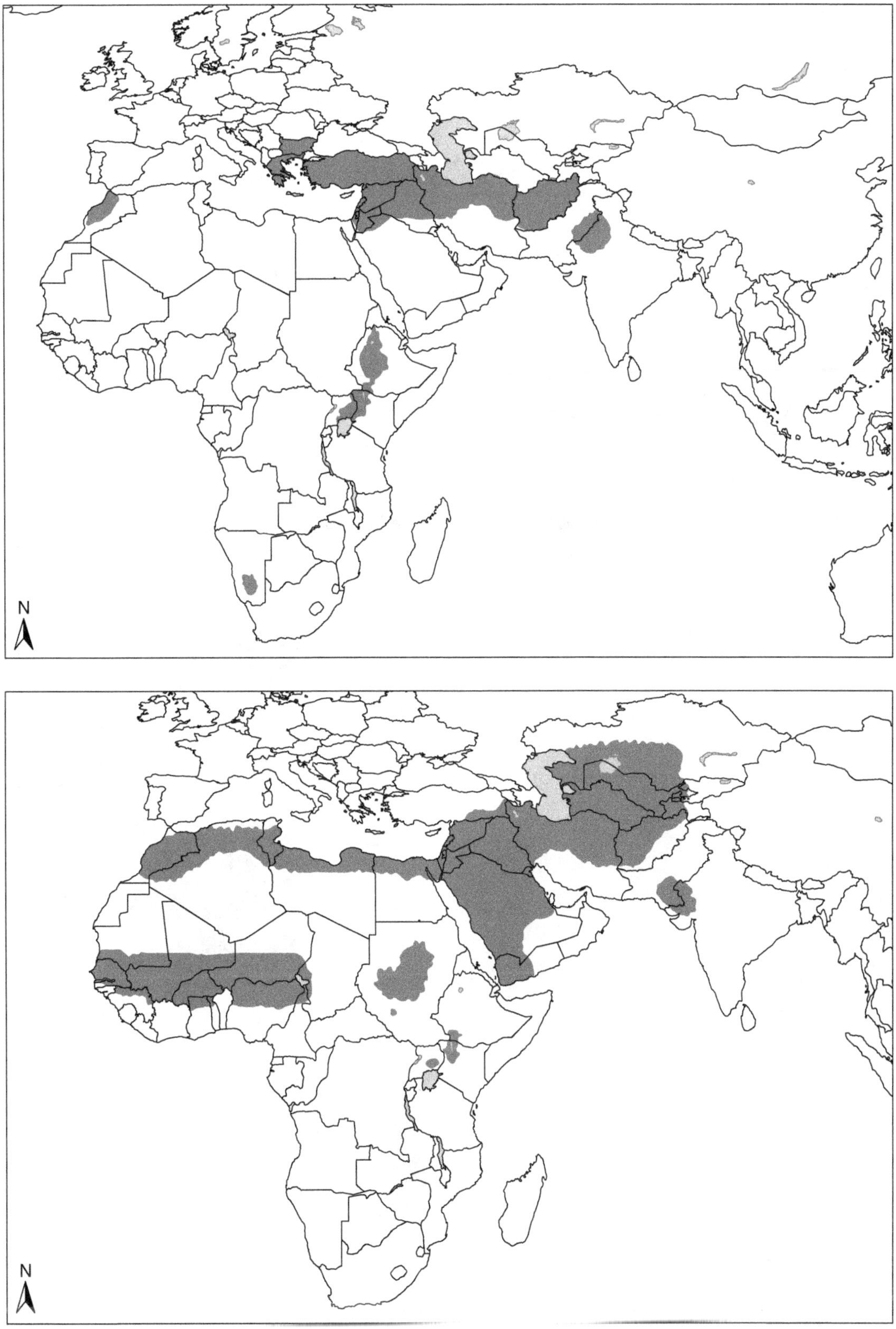

FIGURE 26.3 Cutaneous leishmaniasis. Geographical distribution of Old World cutaneous leishmaniasis due to *L. tropica* and related species and *L. aethiopica*. Geographical distribution of Old World cutaneous leishmaniasis due to *L. major*. *Source*: World Health Organization http://www.who.int/leishmaniasis/leishmaniasis_maps/en/ Reprinted with permission.

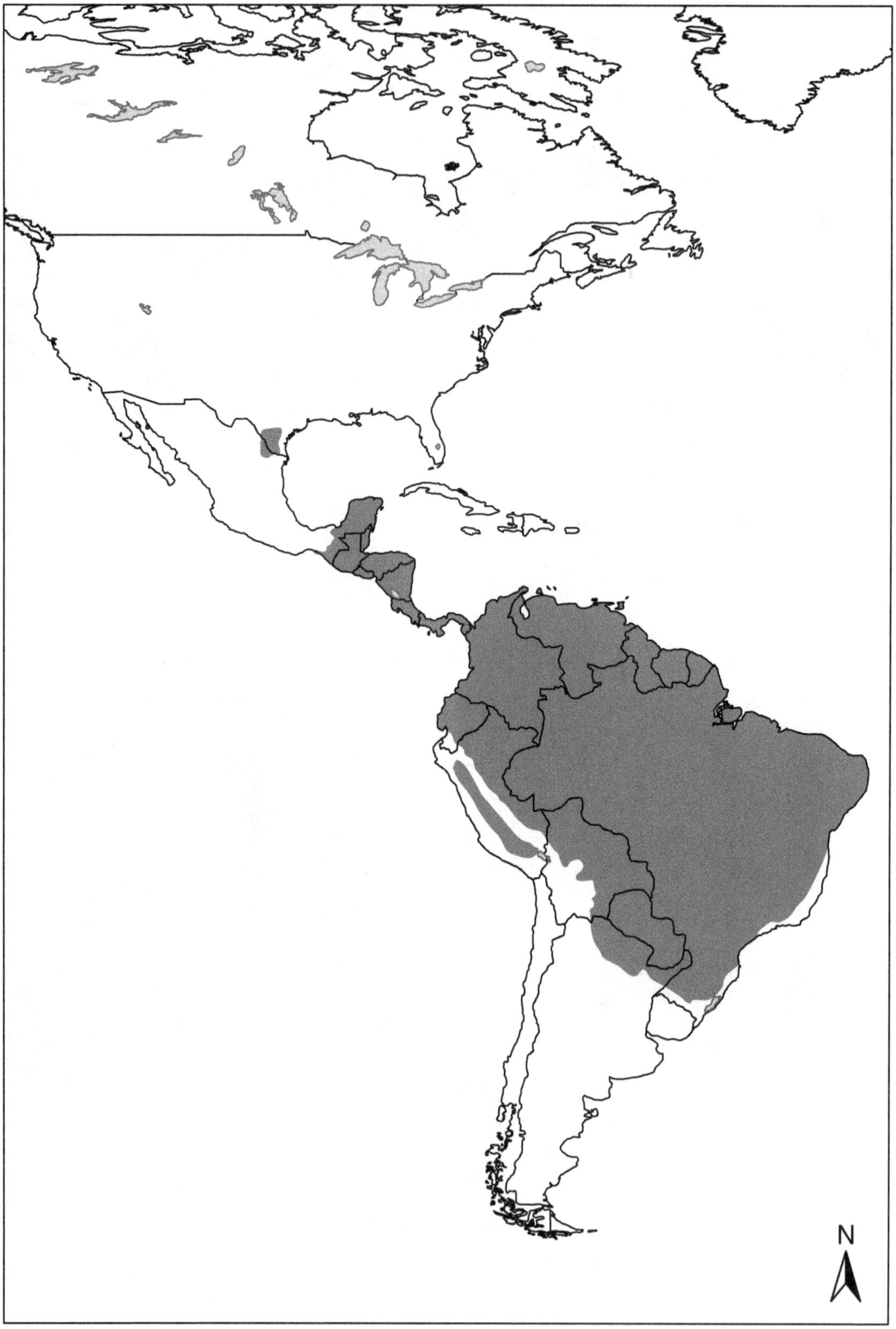

FIGURE 26.4 Geographical distribution of cutaneous and mucocutaneous leishmaniasis in the New World. *Source*: World Health Organization http://www.who.int/leishmaniasis/leishmaniasis_maps/en/ Reprinted with permission.

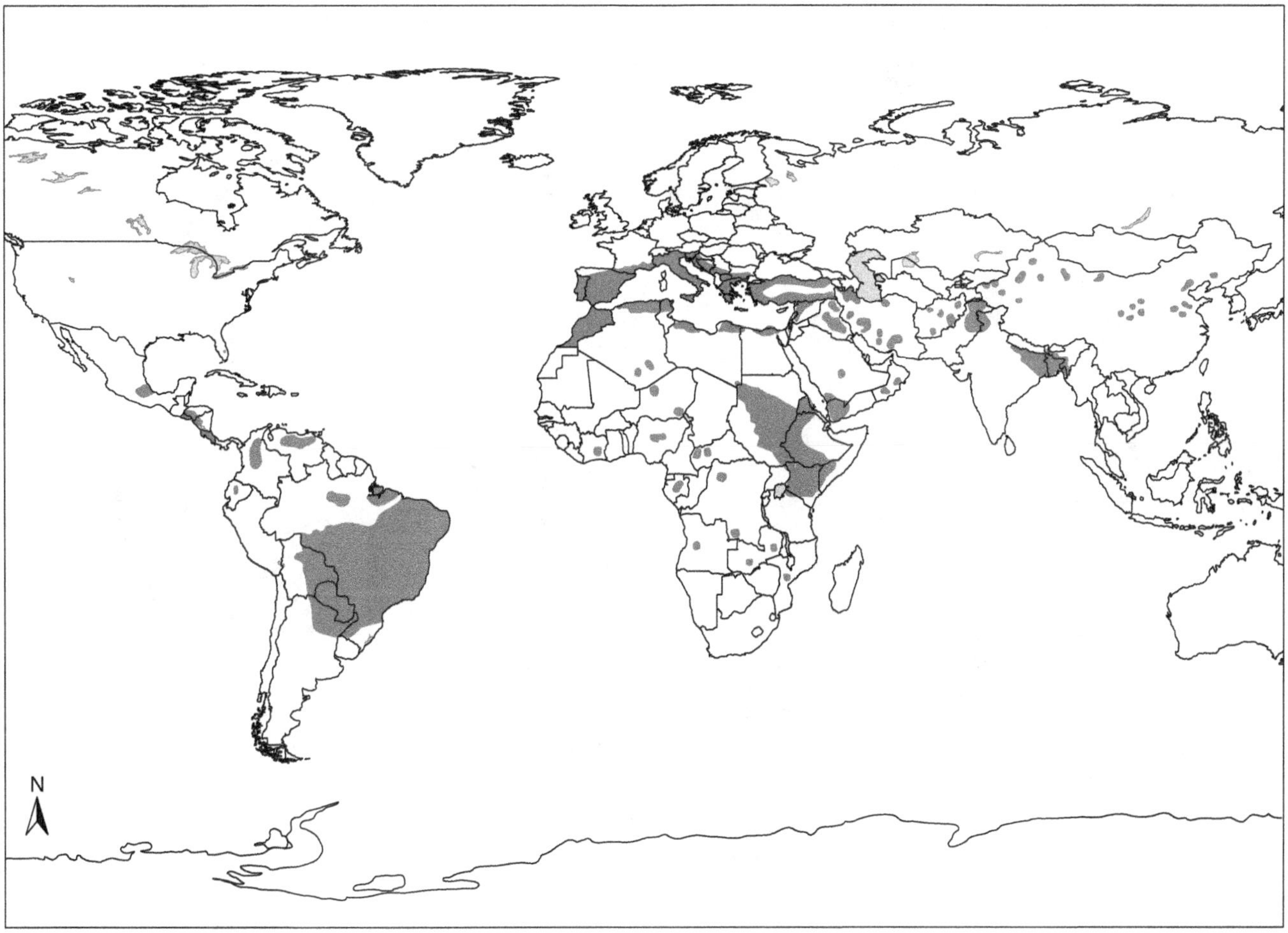

FIGURE 26.5 Geographical distribution of visceral leishmaniasis in the Old and New world. *Source*: World Health Organization http://www.who.int/leishmaniasis/leishmaniasis_maps/en/ Reprinted with permission.

of the day, though they may bite if they are disturbed, such as when individuals brush against tree trunks or other sites where sandflies are resting. The risk of exposure by sandfly bites is often overlooked because sandflies do not make noise, they are small (approximately one third the size of mosquitoes), and their bites might not be noticed.[5]

Humans, dogs (wild and domesticated), rodents (gerbils), cattle, bats, birds, lizards, edentates (sloths), and marsupials serve as reservoirs. Human reservoir hosts may remain infectious even after clinical symptoms have resolved. In India humans are the only known reservoir. Besides occupational exposure, transmission by blood transfusions, contaminated needles, congenital transmission, and sexual contact have been reported.[1]

Current American Red Cross recommendations are that any travelers to Iraq wait 12 months before donating blood and that anyone who has been diagnosed with leishmaniasis cannot ever donate blood.[7]

Pathobiology

CL and MCL are infections that affect the skin and mucous membranes, respectively; they are caused by the vector-borne intracellular protozoan *Leishmania*. In Asia, Africa, and southern Europe, the agents are *Leishmania tropica*, *Leishmania major*, and *Leishmania aethiopica*. In the Western Hemisphere, the *Leishmania braziliensis* complex and *Leishmania mexicana* cause cutaneous and mucocutaneous lesions. The *Leishmania donovani* complex can cause single cutaneous lesions in both hemispheres; it also can cause visceral disease.

After the sandfly feeds on an infected host, flagellated forms develop and multiply in the sandfly gut. After 8–20 days, infective parasites are present and can be transmitted to another host during a blood meal. After the parasites are injected, they are taken up by macrophages, where they can become amastigotes. The amastigotes multiply, causing macrophage rupture and leading to further spread to other

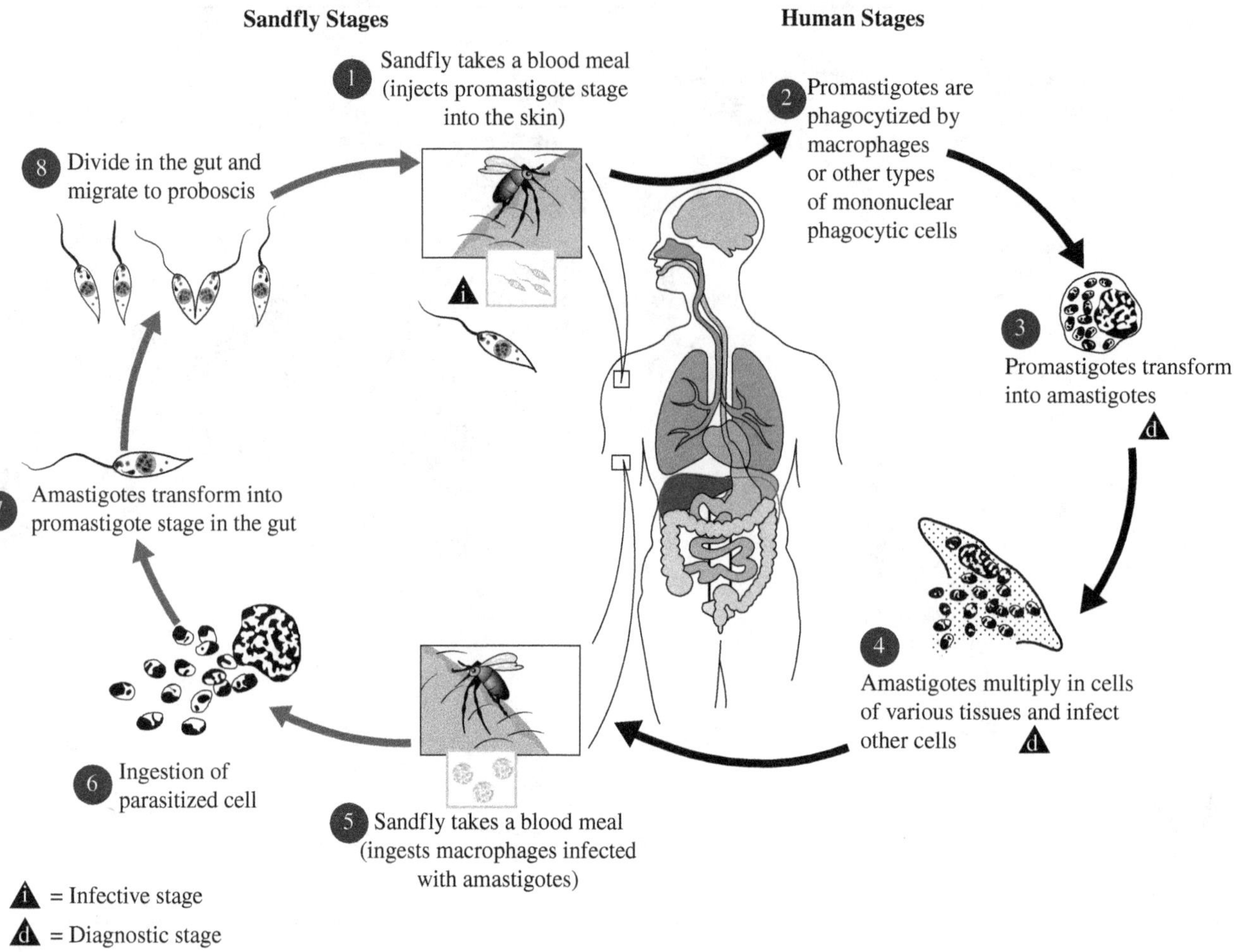

FIGURE 26.6 Leishmaniasis Life Cycle: Leishmaniasis is transmitted by the bite of infected female phlebotomine sandflies. The sandflies inject the infective stage (i.e., promastigotes) from their proboscis during blood meals 1. Promastigotes that reach the puncture wound are phagocytized by macrophages 2 and other types of mononuclear phagocytic cells. Progmastigotes transform in these cells into the tissue stage of the parasite (i.e., amastigotes) 3, which multiply by simple division and proceed to infect other mononuclear phagocytic cells 4. Parasite, host, and other factors affect whether the infection becomes symptomatic and whether cutaneous or visceral leishmaniasis results. Sandflies become infected by ingesting infected cells during blood meals (5, 6). In sandflies, amastigotes transform into promastigotes, develop in the gut 7 (in the hindgut for leishmanial organisms in the *Viannia* subgenus; in the midgut for organisms in the *Leishmania* subgenus), and migrate to the proboscis 8. *Life cycle image and information courtesy of* ***DPDx***. *Source*: http://www.cdc.gov/parasites/leishmaniasis/biology.html

macrophages (Figure 26.6). When an infected sandfly bites and feeds on exposed skin, a small erythematous papule appears after an incubation period of 1–2 weeks to 1–2 months.

All forms of leishmaniasis have a range of clinical disease. In cutaneous leishmaniasis, the scope varies from a progressive nonhealing lesions associated with anergy (diffuse cutaneous leishmaniasis) to the exaggerated hypersensitivity seen in mucosal leishmaniasis and leishmaniasis recidivans, where severe tissue damage is a result of the immune response. In visceral leishmaniasis, many infections are subclinical and self-healing, although malnutrition or immunosuppressive process can reactivate latent infection. In those who develop the syndrome of visceral leishmaniasis, there is suppression of delayed hypersensitivity specifically to leishmanial antigens and nonspecifically to tuberculin and other unrelated antigens. There is also an increase of reticuloendothelial cells and an amplified humoral immune response with the production of nonprotective polyclonal immunoglobulins.[1]

In cutaneous leishmaniasis, eventually, the papule becomes a nodule, and then an ulcer with characteristic firm, raised, and reddened edges. The ulcer can be dry with a central crust or it may weep seropurulent fluid. It is not usually painful. Subcutaneous nodules may develop along lymphatics, but they represent collections of infected macrophages, not lymph nodes.

Most Old World cutaneous leishmaniasis is caused by infections with *L. tropica* and *L. major*. It is characterized by chronic slow-to-heal ulcerative and nodular skin lesions that heal spontaneously, with scarring, after several weeks to several months to as long as 1 year later.[1]

CL of the New World is caused by parasites of the species complexes *L. mexicana* and *L. braziliensis*. The site and appearance of the lesion may be specific for an occupational work group and a geographic region. Chiclero ulcer occurs in Central American forest workers who harvest chicle gum from plants and characteristically develop CL from *L. mexicana* on the pinna of the ear.[1]

Mucocutaneous leishmaniasis (MCL), also known as espundia, occurs when parasites from cutaneous lesions metastasize and cause destructive lesions in the oronasopharynx. MCL is found in Brazil, Bolivia, Ecuador, Peru, and other countries of northern and central South America. The most common cause is the *L. braziliensis* complex. MCL usually occurs after the initial cutaneous lesion has healed, often as long as several years later. The nose is commonly involved, and initial symptoms include stuffiness and intermittent nosebleed. Tissue destruction can involve just the nasal septum, or it can destroy the nose. The upper lip, soft and hard palate, and larynx can also be involved. The parasites in MCL may be difficult to find; in spite of the extensive tissue involvement, VL can have many manifestations. The onset can be acute, subacute, or chronic. The incubation period ranges from weeks to months, but an asymptomatic infection can develop clinical signs and symptoms years to decades after the exposure in people who become immunocompromised for other medical reasons (such as HIV/AIDS). Visceral leishmaniasis usually is caused by the species *L. donovani* and *Leishmania infantum* (*L. chagasi* is considered synonymous with *L. infantum*) and affects internal organs—particularly the spleen, liver, and bone marrow.

Clinical signs and symptoms of visceral infection include fever, weight loss (cachexia, wasting), hepatosplenomegaly (the spleen is often more prominent than the liver), pancytopenia, and a high total protein level and a low albumin level with hypergammaglobulinemia.

Lymphadenopathy may be seen, particularly in some geographic regions, such as Sudan. HIV-coinfected patients can have atypical manifestations, such as involvement of the gastrointestinal tract and other organ systems. The Hindu term kala-azar—which means black (kala) fever (azar) in Hindi—refers to severe or advanced cases of visceral leishmaniasis. Untreated severe cases of visceral leishmaniasis are often fatal directly from the disease or indirectly from complications, such as secondary mycobacterial infection or hemorrhage.[8]

The syndrome post-kala-azar dermal leishmaniasis (PKDL) refers to a condition characterized by erythematous or hypopigmented macules, papules, nodules, and patches of the skin. The skin lesions appear first on the face and develop at varying intervals during or after therapy for VL. PKDL is best described in cases of *L. donovani* infection in South Asia and East Africa. PKDL is more common, develops earlier, and is less chronic in patients in East Africa. In Sudan, PKDL is noted in up to 60% of patients, typically from 0 to 6 months after therapy for visceral leishmaniasis, and often heals spontaneously, while in South Asia, PKDL is noted in ~5–15% of patients several years after initial therapy and usually requires additional treatment. Persons with chronic PKDL can serve as important reservoir hosts of infection.[8]

Diagnosis

No single diagnostic test for leishmaniasis will give a definitive answer in all clinical settings. The diagnosis is made by finding *Leishmania* parasites (or DNA) in tissue specimens—such as from scrapings or biopsies of skin lesions for CL or from bone marrow for visceral leishmaniasis. Techniques include light microscopic examination of stained slides, specialized culture techniques, or molecular methods.

Identification of the *Leishmania* species also can be important, particularly if more than one species is found where the patient lived or traveled since they can have different clinical and prognostic implications. The species can be identified by molecular methods and biochemical techniques (isoenzyme analysis of cultured parasites).

For VL, serologic testing can provide supportive evidence for the diagnosis. The performance of various serologic assays may vary by geographic region and by host factors (e.g., the sensitivity of serologic testing generally is lower in HIV-coinfected patients, particularly if the HIV infection predated the *Leishmania* infection). Most serologic assays do not reliably distinguish between active and quiescent infection.[8]

No leishmanin skin-test preparation has been approved for use in the United States.

In the United States, CDC provides reference diagnostic services for leishmaniasis.[1,8] Clinicians seeing current or past US military members can find instructions for obtaining help at the following Walter Reed Army Institute of Research (WRAIR) website.[9]

Treatment

Treatment can produce clinical outcomes (absence of clinical signs or findings) or a parasitologic endpoint cure (the eradication of parasites). The initial clinical response may be followed by relapse requiring retreatment. Treatment choices are often based on the particular parasite strain, the area where the infection occurred, host factors such as a normal or abnormal immune system, and special treatment groups such as young children, elderly persons, and pregnant/lactating women. Specific treatment of leishmaniasis should be individualized and discussed with experienced experts.

Most clinical cases of visceral leishmaniasis and mucosal leishmaniasis should be treated. Even though not all cases of cutaneous leishmaniasis require treatment since many clinically heal on their own over months to years, many CL cases are treated to avoid scarring or the risk of becoming MCL for cases acquired in the New World. Systemic chemotherapy in CL is usually indicated for New World and Old World patients with large or multiple lesions. [1,8]

Currently treatment with sodium stibogluconate (Pentostam), a pentavalent antimonial, speeds the time to healing. The entavalent antimonial (SbV) compound sodium stibogluconate (Pentostam®) for IV use is available to US-licensed physicians through the CDC Drug Service (404-639-3670), under an Investigational New Drug (IND) protocol approved by the Food and Drug Administration (FDA) and by CDC's institutional review board.

Liposomal amphotericin B (AmBisome®), which is administered by IV infusion, is FDA approved for the treatment of visceral leishmaniasis. In 2014, FDA approved the oral agent miltefosine for the treatment of cutaneous, mucosal, and visceral leishmaniasis caused by particular *Leishmania* species in adults and adolescents at least 12 years of age who weigh at least 30 kg (66 lb). Some medications that might have merit for treating selected cases of leishmaniasis are available in the United States, but the FDA-approved indications do not include leishmaniasis. Examples of such medications include the parenteral agents amphotericin B deoxycholate and pentamidine isethionate, as well as the orally administered "azoles" (ketoconazole, itraconazole, and fluconazole).

Some cases of cutaneous leishmaniasis without risk for mucosal dissemination/disease might be candidates for local therapy, in part depending on the number, location, and characteristics of the skin lesions. Examples of local therapies that might have utility in some settings include cryotherapy (with liquid nitrogen), thermotherapy (use of localized current field radio-frequency heat), intralesional administration of pentavalent antimonial (SbV) compounds (not covered by CDC's IND protocol for Pentostam®), and topical application of paromomycin (such as an ointment containing 15% paromomycin/12% methylbenzethonium chloride in soft white paraffin, not commercially available in the United States).[8]

Prevention

Successful prevention and control of leishmaniasis require a combination of strategies because transmission occurs in a complex biological environment involving the human host, parasite, sandfly vector, and in some causes an animal reservoir. Key elements of prevention and control strategies include early diagnosis and effective case management, vector control, effective disease surveillance, control of reservoir hosts, and mobilization and education of the at-risk communities with effective behavioral change interventions and locally tailored communication strategies. Partnership and collaboration with various stakeholders and other vector-borne disease control programs are critical at levels.[2]

No vaccine or prophylactic medications to prevent leishmaniasis are available at this time. Preventive measures in endemic areas are based on minimizing contact with sandflies. Personal protective methods include avoiding outdoor activities when sandflies are active (dusk to dawn), covering exposed skin, sleeping under fine mesh mosquito netting (18 holes to the linear inch), using insect repellents containing DEET, and saturating mosquito nets and clothing with permethrin. Sleeping in air-conditioned areas is recommended if possible since the fine mesh mosquito netting can be uncomfortable. The use of fans and ventilators is also helpful because sandflies are weak fliers.

Workers and travelers with the potential for exposure should be educated about the transmission and clinical manifestations of leishmaniasis, as well as control methods for the vector phlebotomines (sandflies). Insecticides with residual activity can be used to control sandfly populations. Sandfly breeding sites should be eliminated, as well as the control of principal animal reservoir-infected dogs and rodents.[4]

For some endemic geographic areas where leishmaniasis is found in people, infected people are not needed to maintain the transmission cycle of the parasite in nature. The animal reservoir hosts (such as rodents or dogs), along with sandflies, maintain the cycle.

In other parts of the world, infected people are needed to maintain the cycle; this type of transmission (human–sandfly–human) is called anthroponotic. In the Indian subcontinent (South Asia), the transmission of *L. donovani* is anthroponotic. Here early detection and effective treatment of patients can serve as a control measure, while suboptimal treatment can lead to the development and spread of drug resistance. Because the transmission is between and within homes, spraying dwellings with residual-action insecticides and using bed nets treated with long-lasting insecticides may be protective.[8]

References

1. Magill AJ. Leishmaniasis. In: Magill AJ, Hill DR, Solomon T, et al., eds. Hunter's Tropical Medicine and Emerging Infectious Disease, 9th edn. New York: Elsevier, 2013:739–60.
2. WHO. http://www.who.int/mediacentre/factsheets/fs375/en/ (accessed December 1, 2014).
3. Crum N, Aronson N, Lederman E, et al. History of U.S. military contributions to the study of parasitic diseases. *Mil Med* 2005; 170(4 suppl):7–29.
4. Pesho E, Wortmann G, Neafie R, et al. Cutaneous Leishmaniasis: Battling the Baghdad Boil. Federal Practitioner website, October 2004. http://www.thoracicsurgerynews.com/

fileadmin/qhi_archive/ArticlePDF/FP/021100058.pdf (accessed November 3, 2014).
5. Centers for Disease Control and Prevention. Infectious Diseases Related to Travel: Leishmaniasis, Cutaneous. http://wwwnc.cdc.gov/travel/yellowbook/2014/chapter-3-infectious-diseases-related-to-travel/leishmaniasis-cutaneous (accessed January 4, 2015).
6. Weina P, Neafie R, Wortmann G, et al. Old world leishmaniasis: an emerging infection among deployed US military and civilian workers. *Clin Infect Dis* 2004; 39:1674–80.
7. American Red Cross. Donating Blood: Eligibility Requirements. http://www.redcrossblood.org/donating-blood/eligibility-requirements/eligibility-criteria-alphabetical-listing (accessed January 4, 2015).
8. CDC. Parasites – Leishmaniasis. http://www.cdc.gov/parasites/leishmaniasis/ (accessed November 1, 2014).
9. WRAIR. Leishmaniasis. http://www.pdhealth.mil/downloads/Leishmaniasis_DS_04272004.pdf (accessed June 20, 2016).

NANOPHYETUS

Common name for disease: Human nanophyetiasis

Occupational setting

Fish handlers working with salmon and trout are at risk from exposure.[1]

Exposure (route)

Eating raw or incompletely cooked, smoked, or salted salmon or steelhead trout causes most cases of human intestinal infection with *Nanophyetus salmincola salmincola*. Infections have also been reported from ingestion of raw steelhead trout eggs, as well as from handling infected salmonid fish. Infection has been reported in the Pacific Northwest from *Nanophyetus salmincola salmincola* and in Siberia from *Nanophyetus salmincola schihobalowi*.[2,3]

Pathobiology

This zoonotic disease is caused by the trematode *N. salmincola salmincola*. Disease usually results from ingestion of raw, undercooked, or undersmoked salmonid fishes. Recently, a case was reported in a biological technician due to hand contamination from handling freshly killed, infected juvenile Coho salmon.[4]

Nanophyetus salmincola salmincola can also infect dogs through the ingestion of infected raw fish. Although it does not cause clinical disease in dogs, it can be the vector of a rickettsial organism, *Neorickettsia helminthoeca*, which causes a systemic infection called salmon poisoning of dogs. This infection can be fatal in dogs; however, it does not cause disease in humans.[2]

When *N. salmincola* eggs from the adult worm are shed in the feces of fish-eating animals such as raccoons, otters, spotted skunks, coyotes, foxes, herons, and diving ducks, miracidia hatch which then penetrate an intermediate snail host. In the snail, the parasite grows; it is shed from the snail as xiphidiocercaria that can penetrate and encyst in 34 species of fish.[4] Salmonid fishes seem to be more susceptible.[4] The cycle is completed when fish containing the encysted metacercaria are ingested by another animal, allowing the fluke to mature in the intestine. If humans ingest infected fish, they become definitive hosts.[2,4]

In humans, the clinical findings can range from no symptoms to abdominal pain, bloating, diarrhea, nausea and vomiting, and fatigue. Symptoms seem to be related to worm burdens.[3] Eosinophilia can be present and may be significant.[2] Eggs appear in the stool 1 week after eating infected fish. The number of eggs in the stool is related to the number of worms causing the infection.[2]

In the case of the biological technician who was handling infected Coho salmon and removing the posterior one third of the kidney of each fish, infection occurred by accidental hand-to-mouth ingestion of infectious metacercariae.[4]

Diagnosis

The diagnosis is made in patients with GI symptoms or unexplained eosinophilia by examining the stool for eggs or mature flukes.

Treatment

Bithionol (50 mg/kg orally on alternate days for a total of two doses), niclosamide (2 g orally on alternate days for a total of three doses), and praziquantel (20 mg/kg three times daily for 1 day) are effective treatments for *Nanophyetus* infection.[3,4] In the series of patients treated with praziquantel, stool examinations done 2–12 weeks after treatment were negative for eggs.[3]

Medical surveillance

Fish handlers who clean and eviscerate infected salmonid fishes should be monitored for symptoms of diarrhea and eosinophilia.[4]

Prevention

Workers at risk for exposure should wear gloves and practice regular hand washing and good personal hygiene. Fish viscera should be disposed of safely. Thorough cooking, or freezing at −20°C for 24 hours, inactivates metacercarial cysts. Individuals should be advised to avoid eating incompletely cooked, salted or smoked, or raw salmon or steelhead trout.

References

1. Dieckhaus KD, Garibaldi RA. Occupational infections. In: Rom WN, ed. Environmental Occupational Medicine, 3rd edn. Philadelphia: Lippincott-Raven, 1998:768.
2. Eastburn RL, Fritsche TR, Terhune CA Jr. Human intestinal infection with *Nanophyetus salmincola* from salmonid fishes. *Am J Trop Med Hyg* 1987; 36:586–91.
3. Fritsche TR, Eastbum RL, Wiggins LH, et al. Praziquantel for treatment of human *Nanophyetus salmincola* (*Troglotrerna salmincola*) infection. *J Infect Dis* 1989; 160:896–9.
4. Harrel LW, Deardorff TL. Human nanophyetiasis: transmission by handling naturally infected coho salmon (*Oncorhynchus kisutch*). *J Infect Dis* 1990; 161:146–8.

PFIESTERIA PISCICIDA

Common names for disease: Possible estuary-associated syndrome (PEAS)

In the autumn of 1996, watermen (commercial fisherman) reported seeing fish with "punched-out" skin ulcers and erratic swimming behavior in the Pocomoke and neighboring estuaries on the eastern shore of the Chesapeake Bay, Maryland. Sightings continued and increased in the spring and summer of 1997.[1,2]

Watermen in Maryland who had environmental exposure to water from the affected waterways began reporting learning and memory difficulties. Other complaints included headache, skin lesions, and skin burning on contact with water. A study of 24 exposed individuals showed a dose-related reversible clinical syndrome consisting of difficulties with learning and memory. The symptoms resolved in 3–6 months after stopping exposure.[2] The affected waterman had had high-level occupational exposure to waterways where the one-celled dinoflagellated (alga) *Pfiesteria piscicida* had been identified in association with several fish kill events.

No other reports of clusters of disease attributed to *P. piscicida* or other *Pfiesteria*-like organisms (PLOs) have been reported.[3] A team of North Carolina medical specialists investigated 67 persons exposed to fish kill waters in North Carolina and found no evidence of adverse health effects from their exposure.[3]

Since there is no evidence for a diffusible *Pfiesteria* spp. toxin causing fish death in nature, there is no biomarker of exposure, and the cause and effect relationship of *P. piscicida* and *P. piscicida*-related illness in humans remains speculative. *Pfiesteria piscicida* has been found in waters where there were no reports of symptoms or findings in fish in the waters or in persons exposed to the waters.

In January 1998, a CDC-sponsored work group suggested using the term "possible estuary-associated syndrome" (PEAS) for the possible human health problems that may occur after exposure to estuarine waters. The work group developed specific surveillance criteria (see Diagnosis).[3–5]

Research after the 1997 event has led to an alternate cause for the fish kill. The dinoflagellate *Karlodinium veneficum* was also present in Pocomoke in 1997. *Karlodinium veneficum* has a worldwide distribution and has been implicated in numerous fish kill events around the world since 1950. Late summer fish kills associated with *K. veneficum* blooms on the Corsica River, an eastern shore subestuary of the Chesapeake Bay, occurred in 2005 and 2006. A *K. veneficum* toxin has been isolated, its structure has determined, and it has been quantified at specific fish kill events. An alternate explanation for the fish kill events of 1997 and subsequent similar events attributed to *Pfiesteria* spp. is the co-occurrence of *K. veneficum* and *Pfiesteria* spp. The second explanation does not explain the cause(s) of the human health effects related to these events.[5]

Occupational setting

Fisherman and crabbers (watermen), environmental workers, and laboratory workers who come in contact with *P. piscicida* toxins in water from river or estuaries during periods of "fish kills" or when fish with *Pfiesteria*-like lesions are present may be at risk.

Exposure (route)

Skin contact with water in affected waterways and exposure to aerosolized spray from affected waters are the routes of exposure.[6]

Pathobiology

The relationship of PEAS to the dinoflagellate species and the toxins has not been fully characterized. It is not clear that the "clinical neurotoxic" symptoms are directly due to the toxins produced by *P. piscicida* or *Pfiesteria*-like dinoflagellates.[2] It is unknown whether individuals exposed to *P. piscicida* while swimming, boating, or engaging in other types of recreational activities in coastal waters are at risk for developing PEAS. PEAS does not appear to be infectious, since there is no association with the consumption of fish or shellfish caught in waters containing *P. piscicida*.[4] The evidence that suggests a relationship between PEAS and toxins produced by *P. piscicida* is based on the reports that individuals exposed to estuarine water in Maryland prior to and during fish kills associated with *P. piscicida* toxin developed symptomatic neurocognitive deficits. All deficits resolved by 3–6 months after stopping exposure.[2,3]

While there is a report of learning difficulties in laboratory rats associated with exposure to water from aquaria containing *P. piscicida* toxins (2,3), a study designed and conducted under the guidance of an independent expert Task Force on Health Risk of Exposure to Fish Kill Waters by a team of medical specialists from North Carolina found no evidence of severe, chronic, or widespread adverse health effects from exposures to fish kill waters in North Carolina.[3]

The lack of a definitive link between *Pfiesteria* species and their toxins with criteria for PEAS in humans has been the result of both the difficulties in the field identification of these organisms and the species complexity of local algal blooms. [5] While *Pfiesteria* spp. can now be detected and identified in water and sediment samples and also at fish kill events, there has been no indication of involvement of *Pfiesteria* spp. in any fish kill event since 1998.

Diagnosis

The presence of *P. piscicida* does not indicate risk to fish or humans. Fish lesions can also result from a variety of biological physical and environmental factors. CDC, along with other federal, state, and local government agencies and academic institutions, has set up organized surveillance for PEAS.[4] The surveillance system tracks PEAS rather than *P. piscicida*-related illness, since *P. piscicida* toxin has not been linked to the criteria for PEAS.[5,7]

The CDC PEAS criteria are:

- Developing symptoms within 2 weeks after exposure to estuarine water.
- Reporting memory loss or confusion of any duration and/or three or more selected symptoms (i.e., headache, skin rash at the site of water contact, sensation of burning skin, eye irritation, upper respiratory irritation, muscle cramps, and GI symptoms) that—with the exception of skin rash at the site of water contact and sensation of burning skin—persist for 2 or more weeks.
- A healthcare provider cannot identify another cause for the symptoms.[4]

Treatment

Due to the lack of evidence of the specific cause of symptoms, current treatment is avoiding further exposure. During the initial description of PEAS, there are case reports of treatment with cholestyramine, a toxin-binding polymer.[8]

Medical surveillance

Since 1998 there has been no indication of the involvement of *P. piscicida* in any fish kill event. Individuals who experience health problems or providers who see patients with symptoms after exposure to estuarine water, a fish disease event, or a fish kill site should contact their healthcare provider and their state or local public health agency.

Prevention

Even though no definitive cause or toxin has been identified, from a commonsense point of view, individuals who are occupationally exposed to fish kill waters should avoid areas with diseased, dying, or dead fish or shellfish. Persons should not go in or near water in areas that have been closed by governmental agencies and should not harvest or eat fish or shellfish from such areas.[3,4] People who come in contact with water from fish kill areas should wash well with plain soap and water and wash any clothes that come in contact with such water. Use waterproof gloves when handling items or objects contaminated with fish kill waters.

References

1. Centers for Disease Control and Prevention. Results of the public health response to Pfiesteria workshop—Atlanta, Georgia, September 29–30, 1997. *MMWR* 1997; 46:951–2.
2. Grattan LM, Oldach D, Tracy JK, et al. Learning and memory difficulties after environmental exposure to waterways containing toxic-producing Pfiesteria or Pfiesteria-like dinoflagellates. *Lancet* 1998; 352:532–9.
3. Smith CG, Music SI. Pfiesteria in North Carolina: the medical inquiry continues. *N C Med J* 1998; 59:216–9.
4. CDC. Possible estuary-associated syndrome. *MMWR* 1999; 48:381–2.
5. Robledo JAF, Deeds JR, Place JR, et al. Unresolved questions and an alternative hypothesis. In: Botana LM, ed. Seafood and Freshwater Toxins: Pharmacology, Physiology, and Detection, 2nd edn. Boca Raton: CRC Press/Taylor & Francis Group, 2008:717–51.
6. Warrell DA. Fish poisoning: gastrointestinal and neurotoxic syndromes. In: Magill AJ, Hill DR, Solomon T, et al., eds. Hunter's Tropical Medicine and Emerging Infectious Diseases, 9th edn. New York: Elsevier, 2013:925–7.
7. CDC. Surveillance for possible estuary-associated syndrome (Pfiesteria), 6 states. *MMWR* 2000; 49:372–3.
8. Shoemaker RC, Hudnell HK. Possible estuary-associated syndrome: symptoms, vision and treatment. *Environ Health Perspect* 2001; 109(5):539–45.

PLASMODIUM SPECIES

Common name for disease: Malaria

Occupational setting

Malaria is a serious worldwide disease. 3.4 billion people live in areas at risk of malaria transmission in 106 countries and territories. The majority (56%) of malaria infections are in sub-Saharan Africa, followed by Southeast Asia (27%), the Eastern Mediterranean (12%), and South America (3%).[1] The World Health Organization estimates that in 2012 malaria caused 207 million clinical episodes and 627 000 deaths. An estimated 91% of deaths in 2010 were in the African region.[2]

Workers or travelers in endemic areas are at risk from exposure to malaria through bites from mosquitoes. Occupational factors that increase the risk of malaria include working in rural areas where mosquito control methods are not effective or nonexistent, working during mosquito biting hours (dusk to dawn for *Anopheles* sp.), and working without protective clothing or insect repellents.

One interesting, although uncommon, occupational exposure is airport malaria. Airport malaria refers to malaria case reports of individuals who never traveled to malarious areas and who also lacked other risk factors for malaria, such as a history of blood transfusions or intravenous drug abuse. In several reported cases, the victims worked near or at an international airport and were thought to have been infected by the bite of an infected tropical *Anopheles* sp. mosquito when cabin or cargo hold doors were opened. During the summer of 1994, six cases of airport malaria were found around Roissy Charles de Gaulle Airport.[3] Four of the cases were in airport workers and two other cases lived in Villeparisis about 7.5 km away. The nonairport worker cases were thought to be from *Anopheles* mosquitoes that were carried in the cars of airport workers who lived next door. Hot, humid summer weather is thought to be a factor that allows the survival of infected *Anopheles* mosquitoes brought by airplanes.[2]

A 2002 review found 89 cases of airport malaria between 1969 and 1989.[4] Airport malaria does not include cases in persons who became infected during brief stops at airports in malaria-endemic areas, nor does it include those who may have acquired the disease from an infected *Anopheles* sp. mosquito during a flight.[5]

Human population movement is a significant factor in the reemergence of malaria worldwide. Besides the intercontinental movement of infected people from endemic areas to nonmalarious areas through modern transportation, population movement may lead to environmental changes that increase the risk of malaria by creating better habitats for *Anopheles* mosquitoes. Examples are deforestation, the creation of irrigation systems, and rapid, unregulated urbanization accompanied by poor housing and sanitation.[2]

Malaria was eradicated from the United States in the 1940s. However, 1400–1900 cases are reported to CDC each year—with 1925 cases in 2011.[6] The largest number of cases is typically seen in New York City and other ports of entry (Figure 26.7). The majority of cases are acquired during international travel. This group includes business travelers, pleasure travelers, active duty military members serving in endemic areas, and first- and second-generation immigrants (including their spouses) who travel back to their country of origin to visit friends and relatives (VFR travelers). While 75% of the cases are associated with failure to use recommended prophylaxis, approximately half of all cases of malaria in US travelers are among VFR travelers.[2]

VFR travelers as a group have many factors that place them at greater risk for getting malaria. Their duration of travel tends to be longer than for other types of travel such as business trips or tourist travel tours. They also are more likely to stay at the houses of these friends and relatives versus hotels, and based on the destination or socioeconomic status, the private homes may be less likely to be air conditioned or have screened windows. VFR travelers are less likely to use the recommended malaria prevention measures such as insect repellent and chemoprophylactic medicines. Reasons for this behavior include socioeconomic factors such as access to healthcare or health insurance. The lack of access can limit appropriate preventive medical interventions including malaria prophylaxis. VFR travelers often consider themselves to be at low or no risk for infection because they grew up in a malaria-endemic country and consider themselves to be immune. Part of this mistaken perception is the belief that even if they are infected, the infection will be mild and can be easily treatable with medicines they can acquire while abroad.

Any partial immunity that VFR travelers may have developed while growing up in a malaria-endemic country is lost very quickly after moving away making them as vulnerable to infection as people who grew up in nonendemic countries. Likewise their children and spouses who may be accompanying them on the trip will also have no immunity. The medicines that are available overseas to treat malaria may not be appropriate or effective and may not meet the same quality standards as those found in the United States. VFR travelers need to follow the same preventive measures that are recommended for all travelers.[2]

Rarely, cases in the United States occur through exposure to infected blood products, by congenital transmission or by local mosquito-borne transmission.[6]

Exposure (route)

Infected female *Anopheles* sp. mosquitoes transmit malaria. The malaria sporozoite is introduced into humans when the mosquito punctures the skin to feed. Malaria transmission is definitely influenced by climate. Sporogony does not occur at temperatures below 16°C or higher than 33°C. The optimal conditions are between 20 and 30°C, and the mean relative humidity is at least 60%.[7]

Pathobiology

Historically, four species of *Plasmodium* parasites were considered capable of infecting humans:

Plasmodium falciparum, *Plasmodium vivax*, *Plasmodium ovale*, and *Plasmodium malariae*. Recently, a fifth species, *Plasmodium knowlesi*, that causes malaria among monkeys and occurs in certain forested areas of Southeast Asia has been recognized as a significant human pathogen.[7]

The life cycle of the *Plasmodium* sp. that cause human malaria involves humans and the *Anopheles* sp. mosquito. The mosquito becomes infected after feeding on an infected human. If *Plasmodium* gametocytes of both sexes are present, the sexual cycle leads to the creation of an ookinete

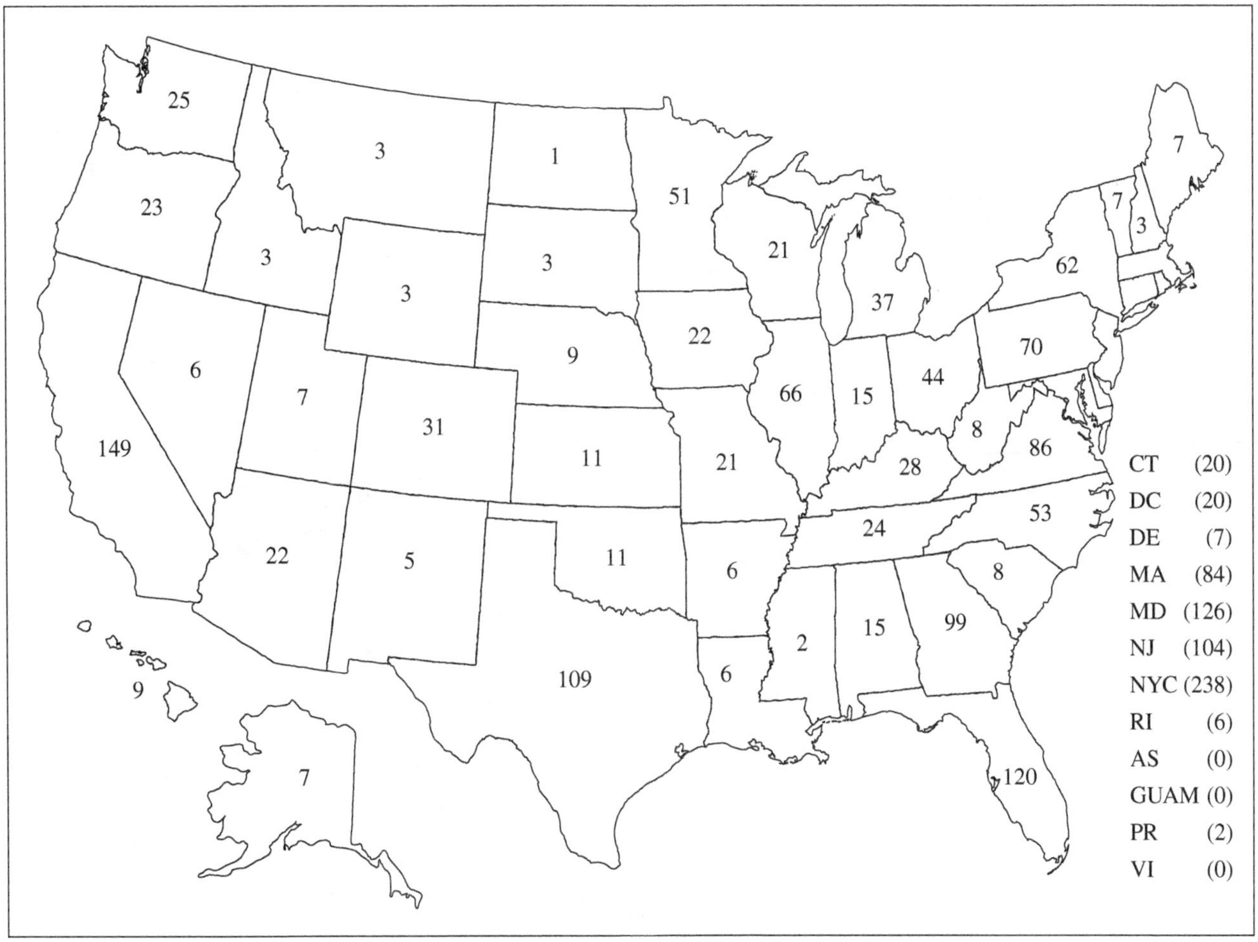

FIGURE 26.7 Number of malaria cases by state or territory in which case was diagnosed—United States, 2011. Number of malaria cases ($N = 1925$) by state or territory in which case was diagnosed—United States, 2011. Nearly all of these cases are imported, though there are rare instances of transmission due to blood transfusion or organ transplantation, congenital transmission, or mosquito-borne cases. Approximately 3–5 malaria deaths occur annually in the U.S. Most of these deaths are due to delayed diagnosis and treatment. *Source*: http://www.cdc.gov/mmwr/preview/mmwrhtml/ss6205a1.htm.

(a mobile, fertilized egg). The ookinete penetrates through the gut, where an oocyst forms. After 2 weeks, this oocyst ruptures, releasing sporozoites. Many of these make their way into the mosquito's salivary glands. An infected female *Anopheles* sp. mosquito is then capable of inoculating malaria sporozoite forms into humans while she feeds. The sporozoites rapidly enter the bloodstream and within hours enter liver cells (hepatocytes) and develop into liver-stage schizonts. After an asexual multiplication stage, each shizont ruptures, releasing 10 000–40 000 uninucleate merozoites, which can invade red blood cells.

Inside the red blood cell, each merozoite develops asexually before rupturing the red blood cell, releasing more merozoites and continuing the cycle. The red blood cell cycle takes 48–72 hours, depending on the species. *Plasmodium falciparum, P. vivax*, and *P. ovale* take 48 hours, and *P. malariae* takes 72 hours. The one exception is *P. knowlesi* whose cycle is 24 hours. The red blood cell stages of the malaria parasite are primarily asexual. Occasional merozoites become male or female gametocytes, permitting the life cycle to continue.[6] Both *P. vivax* and *P. ovale* can have dormant forms of the parasites, hypnozoites, which can remain in the liver and become liver schizonts months or years after the initial inoculation. The liver schizonts can start new cycles (relapses) of red blood cell infections—an important fact when treating malaria caused by these two species [7] (Figure 26.8).

Plasmodium falciparum is the species most commonly associated with severe disease. *Plasmodium vivax* is the dominant species found outside of Africa (the Middle East, Asia, the Western Pacific, and Central and South America). *Plasmodium ovale* is found in West Africa. While *P. malariae* has a worldwide distribution, it is usually only found in isolated pockets. *Plasmodium knowlesi* so far is restricted to South and Southeast Asia, usually in areas harboring macaque monkeys. Statistically *P. falciparum* and *P. vivax* are the most common, and *P. falciparum* causes the most deaths.[1,7]

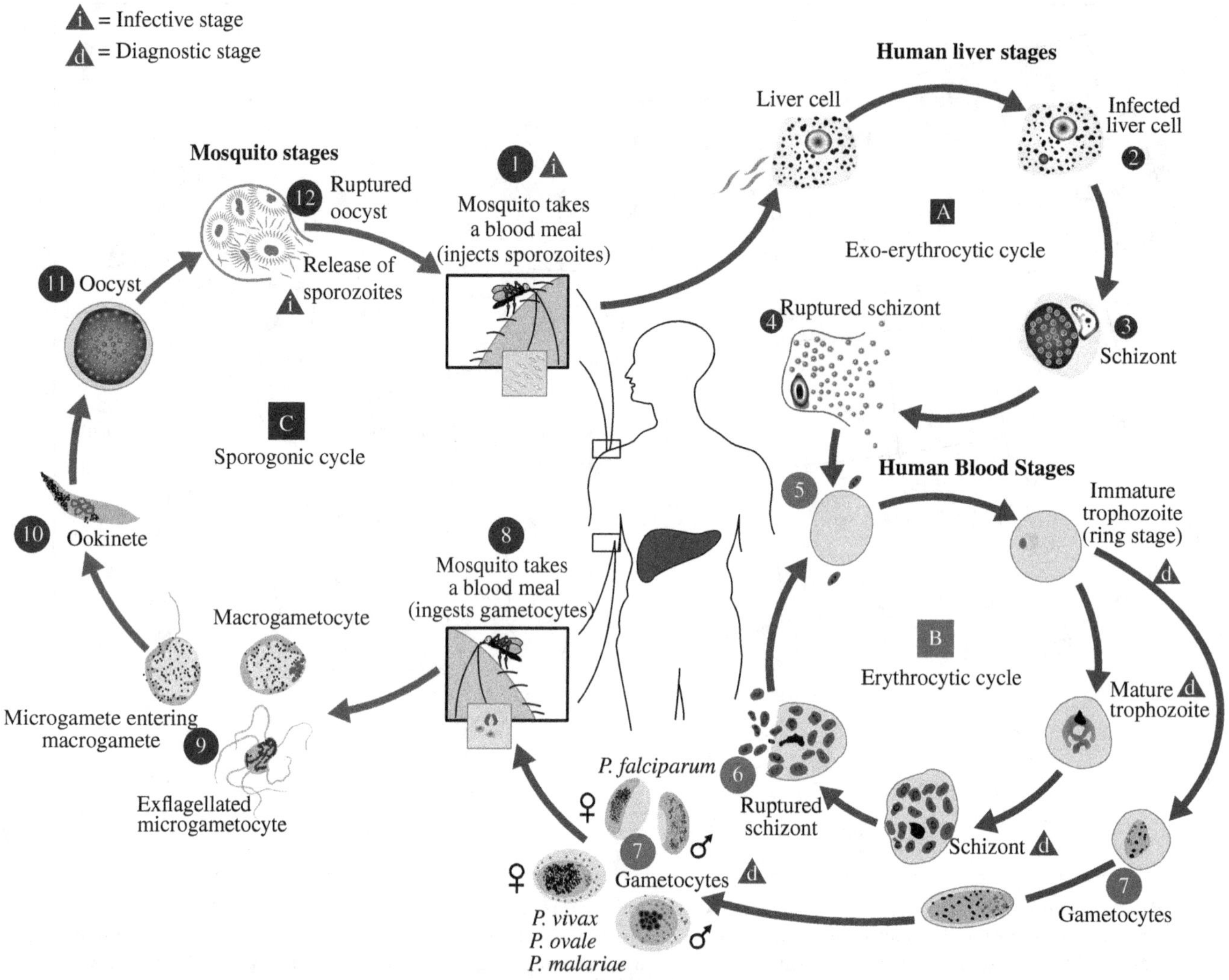

FIGURE 26.8 Malaria parasite life cycle. The malaria parasite life cycle involves two hosts. During a blood meal, a malaria-infected female *Anopheles* mosquito inoculates sporozoites into the human host ❶. Sporozoites infect liver cells ❷ and mature into schizonts ❸, which rupture and release merozoites ❹. (Of note, in *P. vivax* and *P. ovale* a dormant stage [hypnozoites] can persist in the liver and cause relapses by invading the bloodstream weeks, or even years later.) After this initial replication in the liver (exo-erythrocytic schizogony Ⓐ), the parasites undergo asexual multiplication in the erythrocytes (erythrocytic schizogony Ⓑ). Merozoites infect red blood cells ❺. The ring stage trophozoites mature into schizonts, which rupture releasing merozoites ❻. Some parasites differentiate into sexual erythrocytic stages (gametocytes) ❼. Blood stage parasites are responsible for the clinical manifestations of the disease. The gametocytes, male (microgametocytes) and female (macrogametocytes), are ingested by an *Anopheles* mosquito during a blood meal ❽. The parasites' multiplication in the mosquito is known as the sporogonic cycle Ⓒ. While in the mosquito's stomach, the microgametes penetrate the macrogametes generating zygotes ❾. The zygotes in turn become motile and elongated (ookinetes) ❿ which invade the midgut wall of the mosquito where they develop into oocysts ⓫. The oocysts grow, rupture, and release sporozoites ⓬, which make their way to the mosquito's salivary glands. Inoculation of the sporozoites ❶ into a new human host perpetuates the malaria life cycle. *Source*: http://www.cdc.gov/malaria/about/biology/

Even though most symptoms occur within 2 weeks of exposure and 95% of cases occur within 6 weeks of inoculation, an occasional case of *P. vivax* may not develop symptoms for years after exposure.[7] In a person with no prior exposure to malaria, the presenting symptoms go through four successive stages of chills, fever, sweats, and remission of fever. Nonspecific symptoms associated with malaria include headache, back pain, muscle pain, and malaise. When partial immunity has developed due to repeated exposure, the symptoms of infection may be much milder or, on occasion, nonexistent.[7]

Malaria is classically described with cyclical episodes of fever every 3 days (tertian) for *P. falciparum, P. vivax*, and *P. ovale* and every 4 days (quartan) for *P. malariae*. Fever follows the release of blood merozoites from rupture of erythrocytes after schizogony. The cyclical fevers are only evident when a population of parasites predominates and the majority of the schizogony ends cyclically and in synchronization.

Plasmodium falciparum causes the most severe complications. It can cause neurologic symptoms, from headaches to seizures and cerebral edema. *Plasmodium falciparum* can also cause pulmonary edema, glomerulonephritis, renal failure, liver dysfunction, anemia, and hypoglycemia. *Plasmodium vivax*, *P. ovale*, and *P. malariae* can lead to anemia, splenomegaly, and headache; however, they generally do not cause the severe symptoms and mortality associated with *P. falciparum*.[7]

Patients diagnosed with malaria are generally categorized as having either uncomplicated or severe malaria. Patients diagnosed with uncomplicated malaria can be effectively treated with oral antimalarials. The diagnosis of severe malaria is made when the patient has one or more of the following clinical criteria: impaired consciousness/coma, severe normocytic anemia [hemoglobin <7], renal failure, acute respiratory distress syndrome, hypotension, disseminated intravascular coagulation, spontaneous bleeding, acidosis, hemoglobinuria, jaundice, repeated generalized convulsions, and/or parasitemia of $\geq 5\%$. Patients with severe disease should be treated aggressively with parenteral antimalarial therapy.[1]

While individuals of all ages remain susceptible to infection, immunity to severe disease can develop over time, from repeated exposure to bites from infected anopheline mosquitoes. While antidisease immunity is not fully understood, transmission intensity seems to be a major factor. Young children in sub-Saharan Africa, tourists, and soldiers in other malaria-endemic areas are considered to be nonimmune and are at risk of developing severe and complicated malaria. Semi-immune people can be infected (i.e., parasitemic) but can be asymptomatic. If semi-immune people develop a malaria illness, their only symptoms may be fever and malaise. They rarely develop symptoms of severe life-threatening malaria.[7]

Pregnant women have an increased susceptibility to *P. falciparum* malaria. Pregnant women who have partial immunity due to repeated infections may lose some of their immunity because of immune system changes during pregnancy and the presence of the placenta and additional place where malaria parasites can bind.[3]

In malaria-endemic countries, *P. falciparum* contributes to 8–14% of low birth weight, which decreases the chance of a baby's survival.[3]

Diagnosis

Patients suspected of having malaria infection should be urgently evaluated. Suspicion should lead to a search for parasites in Giemsa-stained blood smears. Thick blood smears contain 20–40 times more blood than thin smears and are more sensitive. Thin blood smears are better for species identification once malaria parasite forms have been seen. A negative blood smear makes the diagnosis of malaria unlikely. However, because nonimmune individuals may be symptomatic at very low parasite densities that may be undetectable by blood smear, blood smears should be repeated every 12–24 hours for a total of three sets. If all three are negative, the diagnosis of malaria has been essentially ruled out.[2]

After malaria parasites are detected on a blood smear, the parasite density should be estimated. The thin smear can be used to identify the species, since it is important to determine if the infection is due to *P. falciparum*, since this species causes the greatest morbidity and mortality and is often resistant to chloroquine. Identifying *P. vivax* and *P. ovale* is also important, because they require treatment of the hypnozoites in the liver to prevent relapses, as well as treatment of the blood-stage infection.[2] Mixed cases of malaria can also occur.

Besides microscopy, other laboratory diagnostic tests that are available are rapid diagnostic tests (RDT), polymerase chain reaction (PCR) tests, and serologic tests.[8] RDTs can more rapidly determine that the patient is infected with malaria, but they cannot confirm the species or the parasitemia. PCR is more sensitive and specific than microscopy, but it can only be performed in reference laboratories—a disadvantage since it may take time to get the results from a reference lab. An advantage of PCR is that it is a very useful tool for confirmation of species and detecting of drug resistance mutations. Serologic tests, which are also only performed in reference laboratories, can assess past malaria experience but not current infection. CDC offers malaria drug testing for all malaria diagnosed in the United States at no charge.[2]

Treatment

While malaria should be considered in anyone who has recently traveled in a malaria-endemic area who presents with a fever, whether they took prophylaxis or not, the treatment for malaria should not be started until the diagnosis of malaria has been definitely established. Treatment should be based on three main factors: the infecting *Plasmodium* species, the clinical status of the patient, and the drug susceptibility of the infecting parasites based on the geographic area where the infection was acquired and the previous use of antimalarial medicines.[2]

Identifying the *Plasmodium* species for treatment purposes is important because *P. falciparum* and *P. knowlesi* infections can rapidly lead to progressive severe illness or death, while the other species, *P. vivax, P. ovale*, or *P. malariae*, are less likely to cause severe manifestations. *Plasmodium vivax* and *P. ovale* infections also require treatment for the hypnozoite forms that remain dormant in the liver and can cause a relapsing infection. Finally, *P. falciparum* and *P. vivax* species have different drug resistance patterns in differing geographic regions. For *P. falciparum* and *P. knowlesi* infections, the urgent initiation of appropriate therapy is especially critical.

The clinical status of the patient—whether they have uncomplicated or severe malaria—determines whether they can be treated with oral antimalarials or in the case of severe malaria should be treated aggressively with parenteral antimalarial therapy.

Knowledge of the geographic area where the infection was likely acquired provides information on the chances of drug resistance of the infecting parasite. Considering this information the treating clinician can choose an appropriate drug or drug combination. If a malaria infection occurred despite use of a specific medicine for chemoprophylaxis, that medicine should not be a part of the treatment regimen. If faced with the clinical situation where the diagnosis of malaria is suspected and cannot be confirmed or if the diagnosis of malaria is confirmed but species determination is not possible, antimalarial treatment effective against chloroquine-resistant *P. falciparum* should be started immediately.[2]

The treatment choice should rapidly reduce or eliminate the parasitemia, and in the case of severe malaria it should prevent the complications of severe malaria. Drug therapy must reduce or eliminate the asexual blood stage and in patients with *P. vivax* or *P. ovale* must also include a drug such as primaquine, which eliminates the liver stage, to prevent relapses.[6] Treatment decisions should include input from an infectious disease or tropical disease expert or discussion with CDC for specific treatment regimens. (CDC clinicians are on call 24 hours to provide advice to clinicians on the diagnosis and treatment of malaria and can be reached through the Malaria Hotline 770-488-7788 (or toll free 855-856-4713) Monday–Friday, 9:00 a.m. to 5:00 p.m. On off-hours, weekends, and federal holidays, call 770-488-7100 and ask to have the malaria clinician on call to be paged.)

The current medications for the treatment of malaria are active against the parasite forms in the blood and include:

- Chloroquine
- Atovaquone/proguanil (Malarone®)
- Artemether/lumefantrine (Coartem®)
- Mefloquine (Lariam®)
- Quinine
- Quinidine
- Doxycycline (used in combination with quinine)
- Clindamycin (used in combination with quinine)
- Artesunate (not licensed for use in the United States, but available through the CDC malaria hotline)

Patients who have severe *P. falciparum* malaria or who cannot take oral medications should be given the treatment by intravenous infusion. The current intravenous drugs regimens are quinidine gluconate plus one of the following: doxycycline, tetracycline, clindamycin, or artesunate (see above), followed by one of the following: atovaquone/proguanil (Malarone®), doxycycline (clindamycin in pregnant women), or mefloquine. Consultation with an infectious disease or tropical disease expert or discussion with CDC for specific treatment regimens is recommended.[2] While primaquine is active against the dormant parasite liver forms (hypnozoites) and prevents relapses, it should not be taken by pregnant women or by people who are deficient in glucose-6-phosphate dehydrogenase (G6PD). Primaquine can cause severe hemolysis in persons with low activity of G6PD, and patients should undergo G6PD testing before receiving the medication. Primaquine should not be given to pregnant women, because the drug may cross the placenta to a G6PD-deficient fetus and lead to hemolysis in utero.[2,7]

Prevention

Prevention strategies include avoidance and control of the mosquito vector, as well as chemoprophylaxis for travel to endemic areas. Chemoprophylaxis should always be used in conjunction with personal protective measures to avoid mosquito bites. Individuals can minimize their chance of contact with the *Anopheles* sp. mosquito by using barrier methods that include wearing long-sleeve shirts and full-length pants, sleeping under insecticide-treated bed nets, staying in accommodations with screened windows or air conditioning, and avoiding exposure by staying indoors, especially at dusk and dawn when mosquitoes are most active (Figure 26.9).

Personal protective measures include the use of mosquito repellents containing the following ingredients: DEET up to 50% (products containing DEET include Off!, Cutter, Sawyer, and Ultrathon); picaridin, also known as KBR 3023, Bayrepel, and icaridin (products containing picaridin include Cutter Advanced, Skin So Soft Bug Guard Plus, and Autan [outside the United States]); oil of lemon eucalyptus (OLE) or PMD (products containing OLE include Repel and Off! Botanicals); and IR3535 (products containing IR3535 include Skin So Soft Bug Guard Plus Expedition and SkinSmart). Spraying rooms with pyrethrum-containing spray and spraying clothing with permethrin are other methods to minimize mosquito contact.[2]

Vector control methods include reducing the breeding areas of mosquitoes by draining or eliminating standing water and by killing the adult and (larval) stages of the mosquito.

Individuals traveling to malaria-endemic areas should see a healthcare provider with expertise in travel medicine for a travel evaluation and risk assessment. The risk assessment,

FIGURE 26.9 How travelers can protect themselves against malaria. This picture shows some things that travelers can use to protect themselves against malaria: malaria pills, insect repellent, long-sleeved clothing, bed net, and flying insect spray. (Not shown, but also protective: air conditioned or screened quarters.) Based on the risk assessment, specific malaria prevention interventions should be used by the traveler. Often this includes avoiding mosquito bites through the use of repellents or insecticide treated bed nets, and specific medicines to prevent malaria. *Source*: http://www.cdc.gov/malaria/travelers/

including the need for antimalarial prophylaxis, should take into account the destination country; detailed itinerary, including specific cities, types of accommodation, season, the style of travel, and length of stay; age; pregnancy and breastfeeding status; current medications; and general medical history. Additional considerations when choosing an antimalarial include the following: Does the traveler have a medical condition (including, but not limited to, G6PD deficiency for chloroquine and primaquine, renal impairment for atovaquone/proguanil, or seizure disorders for mefloquine) or allergy that would be a contraindication for certain antimalarials? Does the traveler currently take any medication that might interact with certain antimalarials? Does the traveler have preferences that may affect adherence (a daily vs. a weekly medication)? Is cost of medication a consideration for the patient? Are there logistical issues? Prophylactic drugs need to be started prior to leaving for the trip—anywhere from 1 day to 2 weeks prior depending on the medication.[2] Commonly used drugs for chemoprophylaxis include atovaquone/proguanil (Malarone®), chloroquine, doxycycline, mefloquine (Lariam®), and primaquine.[2] CDC maintains country-specific tables that include resistance patterns and current recommendations.[2] These should be consulted prior to providing a prescription.

A final aspect of malaria prevention is the necessity for prompt diagnosis and treatment of malarial disease. Ninety percent of all travelers who develop malaria do not show symptoms until they return home. The number of cases of *P. falciparum* infections from travel to sub-Saharan Africa continues to increase, even though it represents less than 2% of all travel. The increase in cases is due to the high risk of malaria to travelers in both urban and rural areas of sub-Saharan Africa. Malaria can still occur in persons who use personal protective measures and take chemoprophylaxis correctly.

In the United States malaria transmitted through blood transfusion is very rare with a rate of less than 1 per 1 million units of blood transfused. Prevention of transfusion cases of malaria depends on collecting a travel history from presenting donors and following FDA screening guidelines:

- Most travelers to an area with malaria are deferred from donating blood for 1 year after their return.
- Former residents of areas where malaria is present will be deferred for 3 years.
- People diagnosed with malaria cannot donate blood for 3 years after treatment, during which time they must have remained free of symptoms of malaria.[2]

References

1. Taylor T, Agbenyega T. Malaria. In: Magill AJ, Hill DR, Solomon T, et al., eds. Hunter's Tropical Medicine and Emerging Infectious Disease, 9th edn. New York: Elsevier, 2013:695–717.
2. Centers for Disease Control and Prevention. Travelers Health: Malaria. http://wwwnc.cdc.gov/travel/diseases/malaria (accessed January 5, 2015).
3. Martens P, Hall L. Malaria on the move: human population movement and malaria transmission. *Emerg Infect Dis* 2000; 6:103–9.
4. Guillet P, Germain MC, Chandre F, et al. Origin and prevention of airport malaria in France. *Trop Med Int Health* 1998; 3:700–5.
5. Thang HD, Elsas RM, Veenstra J. Airport malaria: report of a case and a brief review of the literature. *Neth J Med* 2002; 60(11):441–3.
6. Isaacson M. Airport malaria: a review. *Bull WHO* 1989; 67:737–43.
7. Centers for Disease Control and Prevention. Malaria surveillance, 2011. *MMWR* 2013; 62(SS-5):1–18.
8. World Health Organization. Malaria: Diagnostic Testing. http://www.who.int/malaria/areas/diagnosis/en/ (accessed January 5, 2015).

TOXOPLASMA GONDII

Common name for disease: Toxoplasmosis

Occupational setting

Toxoplasmosis is caused by a single-celled parasite called *Toxoplasma gondii*. It was considered to be benign in immunocompetent persons and potentially severe in immunocompromised patients and for the fetus. More recent research shows that the clinical spectrum and severity of disease depend on the host and the pathogen's genetic background, the immune status of the host, and the host–parasite interaction. Toxoplasmosis is now recognized as a possible etiology for a severe infectious syndrome in travelers living in, or returning from, tropical countries.[1,2]

The parasite is found throughout the world, but the incidence depends on local eating habits, environmental conditions, and the presence of definitive hosts. Prevalence increases with age. In Western Europe the prevalence is 30–50%. In France, the prevalence declined from more than 80% in the 1960s to 54.3% in 1995 and to 43.8% in 2003.[1] The past high prevalence of infection in France has been related to a preference for eating raw or undercooked meat, particularly lamb; and the high prevalence in Central America has been related to the frequency of stray cats in a climate favoring the survival of oocysts and soil exposure. Reasons for the declines are believed to be improved hygiene, more widespread freezing of meat—which kills cysts—and better awareness in the general population regarding the risks of consuming undercooked meat. In the United States the seroprevalence has declined from 14.1% in the 1988–1994 National Health and Nutrition Examination Survey (NHANES) to 9% in the 1999–2004 NHANES.[1] Even with the decline in the United States, toxoplasmosis is estimated to be the second leading cause of death from foodborne illness and the fourth leading cause of foodborne illness hospitalization.[3,4]

When a person with a healthy immune system becomes infected, their immune system usually keeps the parasite from causing illness. In pregnant women and individuals who have compromised immune systems, *Toxoplasma* infection can cause serious health problems.

Occupations such as butchers, slaughterhouse workers, laboratory workers (occupational exposure and transmission by needlestick can occur when working with live *Toxoplasma*), landscapers, gardeners, farmers, veterinarians and their associates, pet store owners, and cat breeders are at risk from exposure.[1,3,5] In the United States a case–control study of adults recently infected with *T. gondii*, working with meat, was associated with recent *T. gondii* infection. The authors postulated that some persons who work with meat could inadvertently ingest undercooked meat (e.g., if gloves were not worn or were removed incorrectly). They also suggested that cooks who work with meat could taste dishes before the meat was fully cooked or could inadvertently ingest undercooked meat from their hands.[6]

Workers in stables or grain elevators where feral or house cats are present also can be exposed.[7] Seronegative missionaries, volunteers, and workers for agencies in endemic areas are also at risk for toxoplasmosis.

Exposure (route)

Human infection is usually through contact with and ingestion of parasite oocysts from:

- Eating undercooked, contaminated meat especially pork, lamb, and venison
- Accidental ingestion of undercooked, contaminated meat after handling it and not washing hands thoroughly (*Toxoplasma* cannot be absorbed through intact skin.)
- Eating food that was contaminated by knives, utensils, cutting boards, and other foods that have had contact with raw, contaminated meat
- Drinking water contaminated with *T. gondii* or drinking unpasteurized milk (goat) contaminated with *T. gondii*
- Accidentally ingesting the parasite through contact with cat feces that contain *Toxoplasma.* This can happen by:
  1. Cleaning a cat's litter box when the cat has shed *Toxoplasma* in its feces
  2. Touching or ingesting anything that has come into contact with cat feces that contain *Toxoplasma*
  3. Accidentally ingesting contaminated soil (e.g., not washing hands after gardening or eating unwashed fruits or vegetables from a garden)

Other mechanisms are mother-to-child (congenital) transmission and rarely by receiving an infected organ transplant or infected blood by transfusion.[3]

The only known definitive hosts for *T. gondii* are members of family Felidae (domestic cats and their relatives). A cat that has eaten a single infected bird or mouse can greatly increase the risk to other animals and humans by shedding millions of oocysts in its feces during the 2-week period of the intestinal phase of its infection. The oocysts can survive for more than a year in moist soil. Most people who become infected do not have symptoms, and those who do become may only have mild illness often with only swollen lymph nodes.[3,5]

However, recently illness with more severe symptoms has been seen in immunocompetent travelers to South America (French Guiana) and Africa.[2,3,8]

Pathobiology

Members of the cat family (Felidae) serve as the definitive host and do not require an intermediate host to become infected. Cats become infected after consuming intermediate hosts harboring tissue cysts or become directly infected by ingestion of sporulated oocysts. One unique characteristic of the infection is the transmission of infection from one intermediate host to another by carnivorism, without need for passage through a definitive cat host. Domestic animals bred for human consumption and wild game may also become infected with tissue cysts after ingestion of sporulated oocysts in the environment. Common intermediate hosts are pigs, sheep, deer, rabbits, rats, mice birds, cattle, and humans.[1]

Toxoplasma gondii can exist in three forms: the tachyzoite and the trophozoite forms of acute infections; the bradyzoite, found in the tissue cyst; and the oocyst, shed in cat feces. Oocysts transform into tachyzoites shortly after ingestion and multiply rapidly during the acute infection. Tachyzoites localize in the neural and muscle tissue and are the cause of acute illness. They destroy their host cells, enter adjacent cells, and result in focal necrosis and a vigorous local inflammatory reaction. Tachyzoites divide every 5–12 hours and can cause significant cell destruction until immunity is acquired. After 3–6 weeks, the infection persists with only slowly dividing bradyzoites inside tissue cysts. Neither chemotherapy nor host immunity affects the parasite once this transformation occurs. The long-lived bradyzoites persist inside cysts during the remainder of the hosts' life and are responsible for latency and reactivation. *Toxoplasma* infection is the presence of the trophozoite form or the cyst form in tissues whether there is clinical disease or not. The tachyzoite is present in the host tissues during the acute stage of the infection. Tissue cysts containing the bradyzoites develop in host tissues and are the main mode of transmission of the organism to carnivores, including humans.[1] The oocyst is produced only through an enteroepithelial cycle in the intestines of the cat family. Cats are important in the spread of toxoplasmosis, because they can deposit massive numbers of oocysts that remain infective for months. Cats develop a primary infection during their first year of life, when they begin hunting and ingest tissue cysts in rodents, birds, or oocysts from other cat feces. Kittens and cats can shed millions of oocysts in their feces for as long as 3 weeks after infection, and the oocysts can remain infective for over a year. Mature cats are less likely to shed *Toxoplasma* if they have been previously infected.[3] The oocysts sporulate 1–5 days after excretion. Oocysts are easily spread by wind and rain or on fomites. Oocysts tend to survive longer in moist, shaded soil. Their survival is decreased under temperature extremes, low humidity, and exposure to direct sunlight. The common chemical disinfectants are not effective in inactivating oocysts. Once sporulated oocysts or tissue cysts are ingested, the extraintestinal cycle can occur. Trophozoite forms are released from the oocysts. They invade the GI mucosa and reach all organs by means of the bloodstream. Within the organs, the organisms invade and kill cells (Figure 26.10).

Differentiation of tachyzoites into bradyzoites correlates with the onset of protective immunity. When the immune system gains control of the infection, the organism encysts and remains as a tissue cyst in the host. In the setting of T-cell immune deficiency, bradyzoites can reactivate into tachyzoites. The genetic background of the host may influence the risk of reactivation. In HIV-infected patients the risk of reactivated toxoplasmosis is higher or lower depending on the host's human leukocyte antigen (HLA) haplotype.[1] About 80% of immunocompetent persons who become infected by sporulated oocysts are asymptomatic. Among the 20% who develop symptoms, the incubation period is 1–2 weeks, and symptoms can resemble a flu-like illness.

When symptoms develop, the most common are fever and transient head and neck lymphadenopathy, most commonly of the posterior cervical nodes. Other systemic nonspecific symptoms can include fatigue, sore throat, myalgia, and headache. The usual time course for symptoms is 2–4 weeks. The disease is usually self-limited and resolves spontaneously within a few weeks or months. The differential diagnosis includes infectious mononucleosis and lymphoma. Occasionally, patients develop myocarditis, splenomegaly, encephalitis, hepatomegaly, and retinochoroiditis. When antibody appears, it results in lifelong immunity against reinfection. Tissue cysts are not eradicated by antibody, and maintenance of an antibody titer can be an indicator of the continuing presence of live organisms in the cysts.[1] Cases of severe primary toxoplasmosis caused by an atypical or recombinant strain of *T. gondii* in immunocompetent patients have been reported. The majority of patients were infected in French Guiana, but cases have also been reported from other areas. Some of the European cases are linked to undercooked horsemeat imported from South America. The clinical presentation is prolonged fever higher than 39°C, weight loss of >5% body weight, generalized lymphadenopathy, elevated liver enzyme levels, and lung involvement. Fifty percent of the patients have severe headache, and a third have diarrhea, hepatosplenomegaly, or chorioretinitis. Without treatment, this severe presentation is often fatal.[1]

Differentiation of tachyzoites into bradyzoites correlates with the onset of protective immunity and begins after a few days. In the setting of T-cell immune deficiency, bradyzoites can reactivate into tachyzoites. The genetic background of the host may have influence the risk of reactivation. This is well known in HIV-infected patients for whom the risk of reactivated toxoplasmosis is higher or lower depending on the host's human leukocyte antigen (HLA) haplotype. The absence of effective T-cell immunity in the fetus explains the gravity of congenital toxoplasmosis.

In immunocompromised patients, toxoplasmosis is a severe disease, and most cases are reactivation of chronic infection resulting from impaired of T-cell immunity. High-risk groups include HIV-infected patients with a CD4 cell count <100/mm, patients treated with immunosuppressive agents, and bone marrow transplant patients.

The most common presentation is encephalitis and another common clinical presentation is pneumonia (which can resemble pneumocystosis), but any organ can be involved because *T. gondii* infects all nucleated cell types. Disseminated toxoplasmosis usually presents with fever, altered mental status, arthralgia, rash, and focal findings, such as CNS abscess, pneumonia, or elevated liver enzymes.[1]

Severe generalized toxoplasmosis with high fatality rates occurs when recipient seronegative persons receive donor

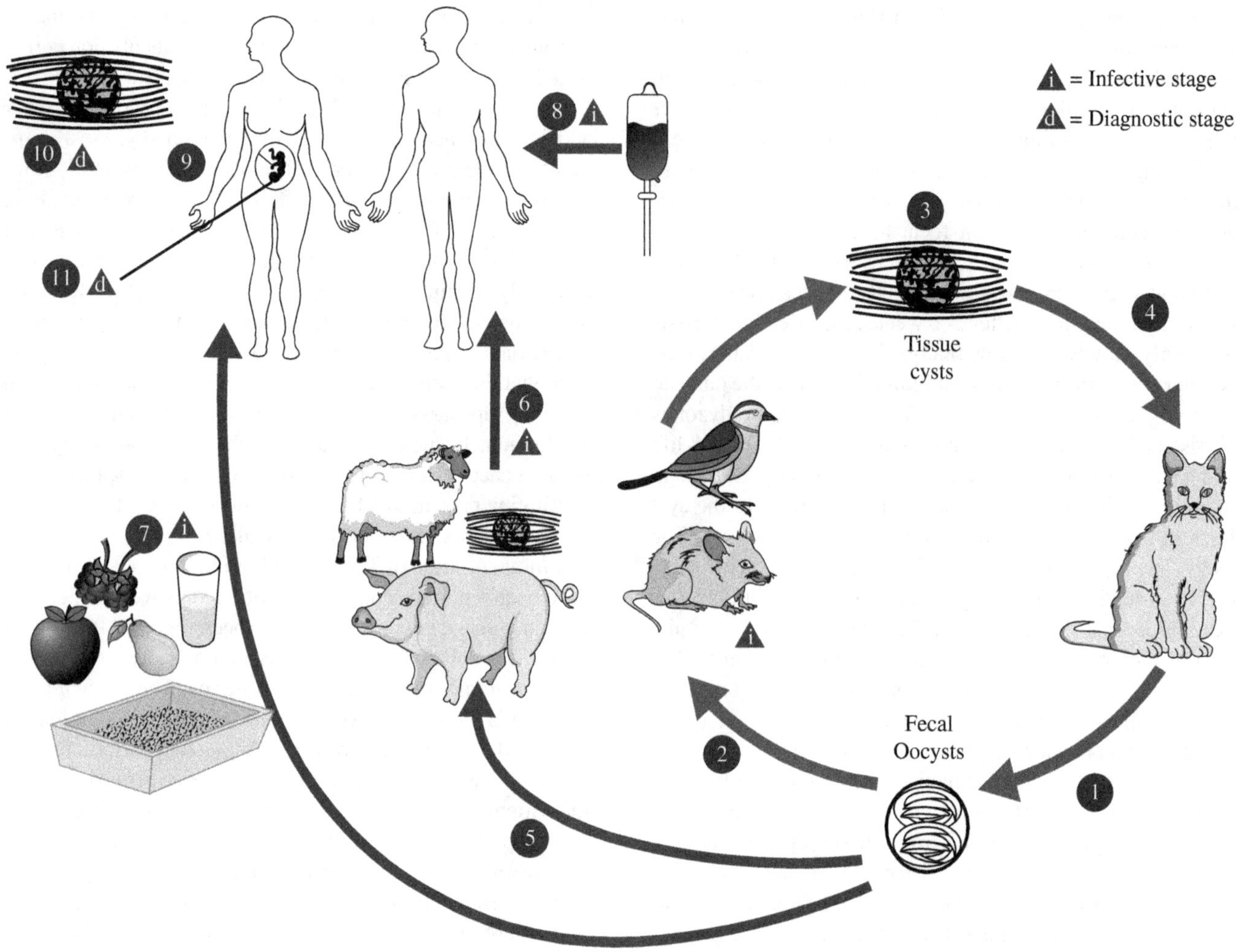

FIGURE 26.10 *Toxoplasma gondii* Life Cycle: The only known definitive hosts for *Toxoplasma gondii* are members of family Felidae (domestic cats and their relatives). Unsporulated oocysts are shed in the cat's feces ❶. Although oocysts are usually only shed for 1–2 weeks, large numbers may be shed. Oocysts take 1–5 days to sporulate in the environment and become infective. Intermediate hosts in nature (including birds and rodents) become infected after ingesting soil, water or plant material contaminated with oocysts ❷. Oocysts transform into tachyzoites shortly after ingestion. These tachyzoites localize in neural and muscle tissue and develop into tissue cyst bradyzoites ❸. Cats become infected after consuming intermediate hosts harboring tissue cysts ❹. Cats may also become infected directly by ingestion of sporulated oocysts. Animals bred for human consumption and wild game may also become infected with tissue cysts after ingestion of sporulated oocysts in the environment ❺. Humans can become infected by any of several routes:

- eating undercooked meat of animals harboring tissue cysts ❻.
- consuming food or water contaminated with cat feces or by contaminated environmental samples (such as fecal-contaminated soil or changing the litter box of a pet cat) ❼.
- blood transfusion or organ transplantation ❽.
- transplacentally from mother to fetus ❾.

In the human host, the parasites form tissue cysts, most commonly in skeletal muscle, myocardium, brain, and eyes; these cysts may remain throughout the life of the host. Diagnosis is usually achieved by serology, although tissue cysts may be observed in stained biopsy specimens ❿. Diagnosis of congenital infections can be achieved by detecting *T. gondii* DNA in amniotic fluid using molecular methods such as PCR ⓫.
Source: http://www.cdc.gov/parasites/toxoplasmosis/biology.html

positive organs. Universal prophylaxis with trimethoprim/sulfamethoxazole (TMP/SMX) is effective in preventing posttransplant toxoplasmosis.[1]

Ocular toxoplasmosis is the leading cause of posterior uveitis worldwide. In the past it was believed to be a late consequence of congenital toxoplasmosis; the current thought is that the majority of ocular disease in immunocompetent patients is from postnatally acquired infection. Ocular involvement can be linked with recent primary infection and is more frequently reported in South America, particularly Brazil.[1] It can also be delayed several months or even years after primary infection.[1]

Pregnant women

A woman who is newly infected with *Toxoplasma* during pregnancy can pass the infection to her unborn child (congenital infection). The woman may not have symptoms, but there can be severe findings in the unborn child, such as diseases of the nervous system and eyes.[3] The rate of transplacental infection and the severity of disease in the fetus are related to the time of infection during gestation. First trimester maternal infection carries a risk of fetal infection that is only about 10%, but the resulting disease is severe. Third trimester maternal infection increases the risk of fetal infection to 65% and approaches 100% at term, but the neonatal infection is usually asymptomatic.[1]

Diagnosis

The diagnosis of acute toxoplasmosis can be made by the serologic pattern, a finding of trophozoites in tissue, isolation of trophozoites from blood or body fluids, lymph node histology, or a finding of cysts in tissue.[1]

The Sabin–Feldman dye test is very specific and sensitive and is considered the gold standard for detecting *Toxoplasma* antibodies in humans, but it creates a potential occupational hazard because it requires the use of live parasites. Since so many cases are asymptomatic, the dye test is not helpful in distinguishing recent from past infection, since the titer peaks early and remains elevated. It is currently done only in select reference laboratories.[9]

IgG and IgM antibody (Ab) tests by IFA or by ELISA are available. The IgG Ab test is specific and sensitive, with levels peaking in 1–2 months and persisting for several months or years. The IgM antibodies are the first to be detected and are useful in diagnosing acute infection (1,3). IgG Ab levels are helpful in determining the prevalence of infection, the risk of reactivation in the immunocompromised, and the risk of acquiring *Toxoplasma* infection. Diagnosis of acute acquired infection can be based on seroconversion, a fourfold rise, two-tube rise in IgG antibody titer, or the presence of a high level of IgM antibody. Because they do not show a rise in titer, diagnosis of reactivated infection in immunocompromised patients is more difficult.[1]

Molecular techniques that can detect the parasite's DNA in the amniotic fluid can be useful in cases of possible mother-to-child (congenital) transmission.

Ocular disease is diagnosed based on the appearance of the lesions in the eye, symptoms, course of disease, and often serologic testing.[3]

Treatment

Most of the medications used to treat toxoplasmosis are active only against the tachyzoite forms of the parasite and do not eradicate the infection. Atovaquone does have limited activity against the cyst forms.[1]

The most common side effect is dose-related suppression of the bone marrow, which may be decreased by concomitant administration of folinic acid (calcium leucovorin).[10]

Lymphadenopathic toxoplasmosis symptoms in immunocompetent adults are usually self-limited. With visceral disease that is clinically evident or if symptoms are severe or persistent, treatment may be indicated for 2–4 weeks. Pyrimethamine is considered to be the most effective agent which should be included in drug regimens used against the parasite. Pyrimethamine is a folic acid antagonist. Leucovorin protects the bone marrow from the toxic effects of pyrimethamine. A second drug, such as sulfadiazine or clindamycin, should be added.[10] Other medications include a fixed combination of trimethoprim with sulfamethoxazole as well as less extensively studied drugs such as atovaquone and pyrimethamine plus azithromycin. Corticosteroids are sometimes prescribed along with antiparasitic agents.[3]

For ocular diseases treatment should be based on a complete ophthalmologic evaluation, and then the decision to treat ocular disease is dependent on factors such as acuteness of the lesion, degree of inflammation, visual acuity, lesion size, location, and persistence.[3]

Management of maternal and fetal infection varies and should be discussed with infectious disease specialists. Spiramycin is often recommended for the first and early second trimesters or pyrimethamine/sulfadiazine and leucovorin for late second and third trimesters for women with acute *T. gondii* infection diagnosed at a reference laboratory during gestation. PCR is often performed on the amniotic fluid at 18 gestation weeks to determine if the infant is infected. If the infant is likely to be infected, then treatment with pyrimethamine, sulfadiazine, and leucovorin is typical. Congenitally infected newborns are usually treated with pyrimethamine, sulfonamide, and leucovorin.

Persons with AIDS who develop active toxoplasmosis (usually toxoplasmic encephalitis) need treatment along with antiretroviral therapy to significantly improve their immunologic status.[3]

Medical surveillance

Surveillance in areas or countries with a high prevalence of *Toxoplasma* may be appropriate for workers in specific high-risk settings listed above and particularly for women of childbearing age. Initial screening may include IgG and IgM antibodies. Periodic screening can be done annually. Seronegative women at high risk of acquiring toxoplasmosis who become pregnant should be screened early in pregnancy, again at 20–22 weeks, and near or at term to find those who may have acquired primary infections during pregnancy.[1] Another alternative would be to offer removal from potential exposure to all seronegative pregnant women for the duration of their pregnancy.

In China *T. gondii* serological surveys done from the late 1990s to the early to mid-2000s in various provinces show a range of 25–45% positivity in slaughterhouse workers, 10–24% positivity in dairy workers, 12.5–26% positivity in veterinarians, 14–21% positivity in meat-processing workers, 10–30% positivity in cooks, and 12–20% positivity in animal breeders.[11]

Prevention

Currently there is no vaccine against toxoplasmosis and there is no justification for chemoprophylaxis. Prevention for people at risk is education about the organism, the routes and symptoms of infection, and the oocysts in the environment. For those working outdoors, gloves should be worn in environments where fecal contamination may be found. Workers should avoid touching facial mucous membranes with gloves.

Prevention is most important in seronegative pregnant women and immunodeficient patients. Again it is most readily accomplished through education.

Specific educational points include:

- Avoid contact with materials potentially contaminated with cat feces, especially handling of cat litter and gardening. Gloves are advised when these activities are necessary. Because oocysts require 1–2 days to mature, dispose of all cat feces daily.
- Keep outdoor sandboxes covered.
- Feed cats only canned or dried commercial food or well-cooked table food, not raw or undercooked meats.
- Change the litter box daily if you own a cat. The *Toxoplasma* parasite does not become infectious until 1–5 days after it is shed in a cat's feces.
- Since epidemiology studies have identified water as a potential source for *T. gondii* infection both in humans and animals, river and lake water should be boiled prior to drinking.
- Cook food to safe temperatures, and since tachyzoites cannot survive desiccation, freezing and thawing freeze meat for several days at subzero (0°F) temperatures before cooking greatly reduce chance of infection.
- Meat that is smoked, cured in brine, or dried may still be infectious.
- Peel or wash fruits and vegetables thoroughly before eating.
- Wash countertops, cutting boards, dishes, counters, utensils, and hands with hot soapy water after contact with raw meat, poultry, seafood, or unwashed fruits or vegetables.
- Avoid drinking untreated drinking water.
- Wear gloves when gardening and during any contact with soil or sand because it might be contaminated with cat feces that contain *Toxoplasma*. Wash hands with soap and warm water after gardening or contact with soil or sand.
- Teach children the importance of washing hands to prevent infection.[1,10]

If you are pregnant or immunocompromised:

- Avoid changing cat litter if possible. If no one else can perform the task, wear disposable gloves and wash your hands with soap and warm water afterward.
- Keep cats indoors.
- Do not adopt or handle stray cats, especially kittens. Do not get a new cat while you are pregnant.[3]

On farms, seronegative pregnant women should avoid in lambing activities. Cats should not be used to control rodents; traps or rodenticides should be used instead. Cats should not be allowed inside feed-storage facilities, and all cat feces should be removed from buildings and stalls. Farm cats should be fed palatable dry or canned food, not forced to forage and hunt. Pigs or cats should never be permitted to eat uncooked scraps of pork or sheep. Dead animals should be removed promptly to prevent cannibalism.[1]

References

1. Paris L. Toxoplasmosis. In: Magill AJ, Hill DR, Solomon T, et al., eds. Hunter's Tropical Medicine and Emerging Infectious Disease, 9th edn. New York: Elsevier, 2013:765–75.
2. Anand R, Jones CW, Ricks JH, et al. Acute primary toxoplasmosis in travelers returning from endemic countries. *J Travel Med* 2012; 19:57–60.
3. Centers for Disease Control and Prevention. Parasites: Toxoplasmosis. http://www.cdc.gov/parasites/toxoplasmosis/ (accessed January 5, 2015).
4. Jones LJ, Praise ME, Fiore AE. Neglected parasitic infections in the United States: toxoplasmosis. *Am J Trop Med Hyg* 2014; 90:794–9.

5. The Canadian Centre for Occupational Health and Safety (CCOHS). Toxoplasmosis. http://www.ccohs.ca/oshanswers/diseases/toxoplasmosis.html (accessed January 5, 2015).
6. Teutsch SM, Juranek DD, Sulzer A, et al. Epidemic toxoplasmosis associated with infected cats. *N Engl J Med* 1979; 300:695–9.
7. Jones JL, Dargelas V, Roberts J, et al. Risk factors for *Toxoplasma gondii* infection in the United States. *Clin Infect Dis* 2009; 49(6):878–84.
8. Khan A, Ajzenberg D, Mercier A, et al. Geographic separation of domestic and wild strains of *Toxoplasma gondii* in French Guiana correlates with a monomorphic version of chromosome1a. *PLoS Negl Trop Dis* 2014; 8(9):e3182. doi:10.1371/journal.pntd.0003182. http://www.ncbi.nlm.nih.gov/pmc/articles/PMC4169257/pdf/pntd.0003182.pdf (accessed December 14, 2014).
9. Udonsom R, Buddhirongawatr R, Sukthana Y. Is Sabin-Feldman dye test using *T. gondii* tachyzoites from animal inoculation still the best method for detecting *Toxoplasma gondii* antibodies? *Southeast Asian J Trop Med Public Health* 2010; 41(5):1059–64.
10. Montoya JG, Boothroyd J, Civics JA. Toxoplasma gondii. In: Bennett JE, Dolin R, Blaser MJ, et al., eds. Mandell, Douglas, and Bennett's Principles and Practice of Infectious Diseases, 8th edn., Philadelphia: Saunders/Elsevier, 2015:3122–53.
11. Zhou P, Chen Z, Li H, et al. *Toxoplasma gondii* infection in humans in China. *Parasit Vectors* 2011; 4:165. doi:10.1186/1756-3305-4-165. http://www.parasitesandvectors.com/content/4/1/165 (accessed December 14, 2014).

Chapter 27

ENVENOMATIONS

James A. Palmier*

ARTHROPOD ENVENOMATIONS

HYMENOPTERA

Common names: Apoidea (bumblebee, honeybee), Vespoidea (wasp, hornet, yellow jacket), Formicoidea (fire ants)

Occupational setting

Gardeners, orchard workers, farmers and agricultural workers, florists and workers in the flower industry, forestry workers, and sanitation workers are potentially at risk.

Exposure (route)

Exposure occurs through physical trauma to the skin and occasionally the oral mucosa as a result of a bite or sting.

Pathobiology

Bees have a barbed stinger that becomes attached to the human skin after a sting, when it deposits its venom. In honeybee stings, the stinger apparatus becomes disengaged from the bee with its venom sac intact. Wasps also have a stinging apparatus but it does not contain barbs; therefore, they are able to withdraw the stinger and sting again. Ants have powerful jaws that grasp the skin and cause the release of venom locally; they are capable of inflicting multiple bites.

*The author wishes to thank Catherine E. Palmier for her contribution to previous versions of this chapter.

The Hymenoptera venom of the bee (Apoidea) consists of histamine, dopamine, enzymes (phospholipase A_1 and hyaluronidase), and peptides, including neurotoxins and hemolysins.[1] The Vespoidea venom is very similar to the bee venom, except that both wasp venom and hornet venom contain serotonin, and the venom of the hornet also contains acetylcholine.[1] The Formicoidea venom contains alkaloids and a small amount of proteins. Histamine is released when these alkaloids come in contact with the mast cells, causing a local reaction. The hypersensitivity-type reaction appears to be caused by the low-molecular-weight proteins found in the venom.[2,3] Hymenoptera stings most often cause pain characterized as sharp stabbing or burning; they are accompanied by a local reaction consisting of mild erythema, edema, and pruritus.

Stings can cause severe local reactions, systemic or anaphylactic reactions, as well as delayed and unusual reactions. Severe local reactions consist of increased swelling (edema) around the sting site. The severity depends on the sting location. Stings in the mouth, upper airway, or esophagus may cause obstruction of the airway or dysphagia.[4] Stings that occur on the eye structures or around the eye may cause local complications, including cataract formation and glaucoma. Severe local reactions on an extremity may include swelling of one or more joints without systemic toxicity.

Systemic or anaphylactic reactions may consist of mild symptoms and signs, including malaise, urticaria, pruritus, and anxiety. A moderate response may present with tightness in the throat or chest, generalized angioedema, nausea, vomiting, and abdominal discomfort. A severe reaction consists of cyanosis, hypotension, loss of consciousness, and other signs of shock, as well as dyspnea, hoarseness, increased malaise, and incontinence. A systemic or

Physical and Biological Hazards of the Workplace, Third Edition. Edited by Gregg M. Stave and Peter H. Wald.

anaphylactic reaction may occur from single or multiple stings; its severity may increase rapidly within minutes to cardiovascular collapse and death.

Delayed hypersensitivity reactions may occur within the first 2 weeks after an envenomation. Symptoms often include fever, urticaria, arthralgias, malaise, headache, lymphadenopathy, and myalgias. Unusual reactions include neurologic complications of encephalopathy, neuritis, Guillain–Barré-type reaction, myasthenia gravis, multiple sclerosis exacerbations, extrapyramidal disorders,[5–7] hemolytic anemia,[8] and the nephrotic syndrome.[9]

Diagnosis

The specific identification of the organism causing the sting is difficult and is usually dependent on history. The history should include the geographic location where the sting took place, and the location of any nests should be carefully noted. Wasps usually make nests under eaves, hornets choose bushes or tree limbs, and yellow jackets usually make nests in the ground. Pictures or photographs of the various Hymenoptera species often help the patient to identify the insect.

The site of the sting is usually not helpful, except in the stings of the honeybee and the fire ant. The honeybee almost always leaves its stinger at the site of the sting with its venom sac attached. The fire ant usually inflicts a distinctive pattern of multiple bites, which subsequently produce pustules. Laboratory tests are not usually helpful in the diagnosis of Hymenoptera envenomations, except in the few unusual cases of hemolytic anemia subsequent to a wasp bite.

Treatment

Most local reactions can be treated with local wound care. If a stinger is in place at the envenomation site, it should be removed immediately by gently scraping or teasing the affected area, thereby minimizing the spread of venom into the wound. All sites should be washed thoroughly with an antiseptic, and the tetanus status of the patient should be assessed. All potentially constrictive jewelry should be removed. Ice packs at the sting site decrease venom absorption and edema. Oral analgesics and antihistamines (e.g., diphenhydramine) usually relieve the pain and subsequent pruritus associated with a mild reaction. Elevation and rest are indicated if there is increased edema of a limb. Antibiotics are indicated only if a secondary infection occurs.

Anaphylactic reactions are medical emergencies. Initial mild systemic symptoms may progress very rapidly to respiratory obstruction, cardiovascular compromise, and death. Any patient who displays signs of a systemic reaction should immediately be placed on a cardiac monitor, an intravenous line should be started, and epinephrine should be administered. In patients who are not in shock, 0.3–0.5 mL for an adult, or 0.01 mL/kg (~0.3 mL) for a child, of 1:1000 epinephrine hydrochloride should be administered subcutaneously and the site of administration massaged. A second injection of epinephrine may be needed in 10–15 minutes if symptoms do not improve or have become more severe. For severe anaphylactic reactions, epinephrine should be given slowly intravenously in a 1:10000 solution and the patient observed and monitored closely. A repeat dose may be needed in 20–30 minutes.

Antihistamines such as diphenhydramine may be given intramuscularly or intravenously, depending on the severity of the anaphylactic reaction. Corticosteroids may also be helpful; but, as with antihistamines, their effect may be delayed and they should be used only as adjuncts to epinephrine therapy. Bronchospasm usually responds well to intravenous aminophylline after a loading dose (5–6 mg/kg over 20 minutes). The use of nebulized β-agonists may also be considered for patients with significant bronchospasm. Vasopressors are indicated in situations where hypotension is unresponsive to intravenous fluids and epinephrine.

Individuals who have moderate to severe systemic reactions to Hymenoptera venom with positive skin reactions should undergo desensitization with venom immunotherapy.[10,11] Immunotherapy usually consists of extracts containing purified insect venom rather than whole-body extracts, which appears to limit substantially the number of systemic reactions to the therapy.[12,13]

Prevention

Preventive measures, especially in allergic individuals, include the avoidance of bright or flowered clothing and the avoidance of scented deodorants, perfumes, and shampoos. Individuals should avoid going outdoors barefoot. Nests around the home or frequented dwellings should be destroyed, preferably with professional assistance. Sensitized individuals should carry commercial epinephrine kits for self-administration; they should be available for use in their home and car and carried along with other personal belongings, such as golf bags). Sensitized patients should wear a medical alert bracelet or tag. Potential victims should stand still or retreat slowly when a Hymenoptera species is encountered. They should be instructed to brush the insect off their skin but avoid swatting or moving quickly. Sensitized individuals should avoid gardening. They should use caution when visiting places with open garbage cans, such as picnic areas, or other locations where insects are likely to be found.

References

1. Hoffmann D. Hymenoptera venom allergens. *Clin Rev Allergy Immunol* 2006; 30(2):109–28.
2. Dongol Y, Dhananjaya BL, Shrestha RK, et al. Pharmacological and Immunological Properties of Wasp Venom. In: Gowder SJT, ed. *Pharmacology and Therapeutics*. Rijeka: Intech, 2014.

3. Kemp SF. Expanding habitat of the imported fire ant (*Solenopsis invicta*): a public health concern. *J Allergy Clin Immunol* 2000; 105(4):683–91.
4. Farivar M. Bee sting of the esophagus. *N Engl J Med* 1981; 305:1020.
5. Sun Z, Yang X, Ye H, et al. Delayed encephalopathy with movement disorder and catatonia: a rare combination after wasp stings. *Clin Neurol Neurosurg* 2013; 115:1506–9.
6. Incorvaia C. Clinical aspects of Hymenoptera venom allergy. *Allergy* 1999; 54(Suppl 58):50–2.
7. Dionne AJ. Exacerbation of multiple sclerosis following wasp stings. *Mayo Clin Proc* 2000; 75(3):317–8.
8. Lee, Y, Wang JS, Shiang JC, et al. Haemolytic uremic syndrome following fire ant bites. *BMC Nephrol* 2014; 15:5.
9. Tareyeva JE, Nikolaev AJ, Janushkevitch TN. Nephrotic syndrome induced by insect sting. *Lancet* 1982; 2:825.
10. Antolin-Amerigo D, Moreno Aguilar C, Vega A, et al. Venom immunotherapy: an updated review. *Curr Allergy Asthma Rep* 2014; 14:449.
11. Rieger-Ziegler V. Hymenoptera venom allergy: time course of specific IgE concentrations during the first weeks after a sting. *Int Arch Allergy Immunol* 1999; 120(2):166–8.
12. Patella V. Hymenoptera venom immunotherapy: tolerance and efficacy of an ultrarush protocol versus a rush and a slow conventional protocol. *J Allergy* 2012; 192.
13. Ebner C. Immunological mechanisms operative in allergen-specific immunotherapy. *Int Arch Allergy Immunol* 1999; 119(1):1–5.

LATRODECTUS SPECIES

Common names: Black widow spider, brown widow spider, redlegged spider, hourglass spider, poison lady spider, deadly spider, red-bottom spider, T-spider, gray lady spider

Occupational setting

Farmers, gardeners, and forestry, construction, and sanitation workers are at risk for exposure.

Exposure (route)

Exposure occurs through physical trauma (bite) to the human skin with injection of venom.

Pathobiology

The female black widow spider is almost twice the size of its male counterpart. Although both are considered venomous, only the female spider is able to bite and envenomate humans. During the summer months, the female black widow spider is the most venomous. The body of the female can range from 8 to 15 mm long, with a leg span of 5 cm. The spider undergoes multiple moltings throughout the year and often changes color. The female is most often shiny black in color and has a rounded abdomen with a red distinctive hourglass on its ventral surface. Occasionally, two red spots may be seen instead of the hourglass configuration.

The five species in the United States are *Latrodectus mactans*, *Latrodectus hesperus*, *Latrodectus variolus*, *Latrodectus bishopi*, and *Latrodectus geometricus*. Although variations occur between species, most *Latrodectus* venoms contain multiple proteins and a potent neurotoxin. The proteins include hyaluronidase, phosphodiesterase, GABA, and 5-hydroxytryptamine. This neurotoxin is one of the most potent human neurotoxins. Neurotransmitters (GABA, acetylcholine, and norepinephrine) appear to be released secondary to the *Latrodectus* venom.[1,2] The clinical presentation usually begins with a pinprick sensation, followed by the appearance of mild swelling and erythema around the bite wound. It is not unusual for the patient to be unaware of the bite until a local reaction has occurred. Close evaluation of the site may reveal two erythematous fang marks. In the first hour after the bite, pain often increases around the area of the bite and spreads to the entire body. Symptoms peak at 3–4 hours. Upper extremity bites often lead to spasm of the upper trunk muscles; bites of the lower extremity often lead to abdominal spasms. These muscle spasms can become very intense and painful and have been misdiagnosed as an acute surgical abdomen. However, cases of abdominal spasms secondary to the black widow bite do not present with associated fever or a leukocytosis.

Other common symptoms include paresthesias and hyperesthesia, especially of the lower extremities and feet. Increased deep tendon reflexes, headache, anxiety, nausea, vomiting, diaphoresis, tremor, restlessness, and seizures may also be seen. Symptoms usually resolve within 24–48 hours. Pregnancy,[3] infancy, increased age, and chronic debilitating illness appear to increase the risk of complications.

Diagnosis

Laboratory studies are usually not helpful in diagnosis. A description of the spider's appearance, identification of the geographic area where the bite took place, and the clinical presentation and evaluation are the most helpful factors in diagnosing a black widow spider bite. The bite area usually consists of two closely approximated erythematous fang marks. In *L. hesperus* bites, there may be a target or halo lesion.[4] Severe chest and back spasms must be carefully distinguished from myocardial infarction or an acute surgical abdomen.

Treatment

In general, treatment is supportive and includes local wound care, tetanus prophylaxis, and pain control (i.e., salicylates, opioids). All victims should be monitored for 6–8 hours for progression of symptoms. Airways, breathing, and circulation should be closely monitored and assisted, if necessary. The pain of severe generalized spasm may be treated with intravenous morphine. Intravenous calcium gluconate (10 mL of 10% solution) can be used to decrease cramping and spasm in select individuals. Methocarbamol and diazepam have also been used as muscle relaxants and for their anxiolytic effect. The use of intravenous dantrolene sodium for muscle relaxation is controversial and needs further study.[5,6]

Most cases can be appropriately managed without the equine-derived antivenom. However, in cases where there is severe hypertension, or in the case of a pregnant woman at risk for premature labor or spontaneous abortion because of severe muscle spasms, the lyophilized antivenom may be indicated. Other indications for antivenom include respiratory distress, protracted symptoms despite treatment, cardiovascular symptoms, and symptomatic patients younger than 16 years and older than 65 years.[2,7–9] The *Latrodectus* antivenom is effective for all species. Unfortunately, it has the potential to cause immediate hypersensitivity, including anaphylaxis, as well as delayed onset serum sickness within 2 weeks. One vial of antivenom is usually necessary for envenomations. All patients should be skin tested for horse serum sensitivity prior to administration.

Prevention

Prevention measures should focus on the use of personal protective equipment such as gloves, heavy garments that are fully buttoned, and protective footwear when working in endemic areas. These spiders generally inhabit places that are dark and protected, such as woodpiles, barns, stables, garages, homes, outdoor sanitation facilities, and walls made of rock. Areas suspected of harboring black widows should be subjected to professional pest control.

References

1 King GF, Hardy MC. Spider-venom peptides: structure, pharmacology and potential control of insect pests. *Annu Rev Entomol* 2013; 58:475–96.
2 Saez NJ, Senff S, Jensen JE, et al. Spider-venom peptides as therapeutics. *Toxins* 2010; 2:2851–71.
3 Russell FE, Marcus P, Streng JA. Black widow spider envenomation during pregnancy: report of a case. *Toxicon* 1979; 17:620–3.
4 Vetter RS, Isbister GK. Medical aspects of spider bites. *Annu Rev Entomol* 2008; 53:409–29.
5 Ahmed N, Pinkham M, Warrell DA. Symptom in search of a toxin: muscle spasms following bites by Old World tarantula spiders with review. *Q J Med* 2009; 102:851–67.
6 Ryan PJ. Preliminary report: experience with the use of dantrolene sodium in the treatment of bites by the black widow spider *Latrodectus hesperus*. *J Toxicol Clin Toxicol* 1984; 21:487.
7 Dart R, Bogdan G, Heard K, et al. A randomized double-blind, placebo-controlled trial of a highly purified equine F(ab)2 antibody black widow spider antivenom. *Ann Emerg Med* 2013; 61: 458–67.
8 Rahmani F, Banan Khojasteh SM, Ebrahimi Bakhtavar H, et al. Poisonous spiders: bites, symptoms, and treatment: an educational review. *Emergency* 2014; 2:54–8.
9 Heard K. Antivenom therapy in the Americas. *Drugs* 1999; 58(1):5–15.

LOXOSCELES SPECIES

Common names: Brown recluse spider, violin spider, Arizona brown spider, fiddle spider, necrotizing spider

Occupational setting

Gardening, housekeeping, forestry, farming, and agricultural workers are at risk of exposure.

Exposure (route)

Exposure occurs through physical trauma (bite) to the human skin with injection of venom.

Pathobiology

The brown recluse spider is ~1 cm in body length, with a leg span of up to 2.5 cm. The color of these spiders is usually tan to brown. In the United States, there are 13 species of *Loxosceles*, but *Loxosceles reclusa* is the most commonly found. The female is more dangerous than the male. These spiders contain a very distinctive violin-shaped, dark brown to yellow marking on the dorsal cephalothorax.

The *Loxosceles* venom contains a number of enzymes, including sphingomyelinase D, the agent believed to be responsible for skin necrosis and hemolysis by its direct effect on the cell plasma membrane as well as sulfated nucleosides.[1–3] The venom also contains hyaluronidase, which does not appear to cause skin necrosis. *Loxosceles* envenomation is initially painless for most victims. Within the first few hours, pain and erythema occur at the bite site. A central bleb or vesicle forms. It is surrounded by a blanched, whitish ring of ecchymosis and ischemia. The appearance resembles a bull's-eye and it is most often 1–5 cm in diameter. Over the next few days, the central vesicle ulcerates, necroses, spreads in diameter, and involves the skin and subcutaneous fat. In ~1 week, an eschar develops that eventually sloughs and can require many months to heal.

Systemic reactions to *Loxosceles* envenomation, while uncommon, can occur. Systemic symptoms and signs usually occur within the first 2 days after envenomation; they include fever, chills, generalized rash, nausea, vomiting, and hemolysis with possible renal failure. Disseminated intravascular coagulation and thrombocytopenia are rare but have been reported.[4] The size of the necrotic ulcer does not seem to correlate with the frequency or severity of the systemic symptoms.[5]

Diagnosis

Laboratory studies are not usually helpful in diagnosing a *Loxosceles* envenomation. A description of the spider's appearance, identification of the geographic area where the bite took place, and the clinical presentation and evaluation are the most helpful factors in diagnosing a brown recluse spider envenomation. Moreover, it should be noted that there are other species that may reproduce the symptoms of the brown recluse spider.[6]

Treatment

In general, treatment should begin with local wound care, tetanus prophylaxis, immobilization, elevation, observation, and antipruritic agents, as needed. Local wound care has been controversial. Some authorities advocate early surgical excision of the wound; others recommend allowing complete delineation of the necrotic area before surgical intervention.[7] The use of dapsone in *Loxosceles* envenomations is controversial and may cause methemoglobinemia in some cases.[8] The use of hyperbaric oxygen to decrease the extent of the necrosis has been successful in a number of cases.[9]

All patients with systemic symptoms must be hospitalized and followed closely for hemolysis, coagulopathy, and renal failure. Treatment for systemic symptoms is supportive. There is no commercially available antivenom.

Prevention

Prevention measures should focus on the use of personal protective equipment like gloves, fully buttoned heavy garments, and protective footwear when working in endemic areas. The spider is usually not aggressive. Its habitat includes woodpiles, barns, the underside of rocks, closets, attics, clothing, carpets, linen, and other dry, dark places. Areas suspected of harboring these spiders should undergo professional pest control.

References

1 Vetter R, Isbister GK. Medical aspects of spider bites. *Annu Rev Entomol* 2008; 53:409–29.
2 Gomez HF, Miller MJ, Desai A, et al. *Loxosceles* spider venom induces the production of alpha and beta chemokines: implications for the pathogenesis of dermonecrotic arachnidism. *Inflammation* 1999; 23(3):207–15.
3 Dos Santos L, Dias NB, Roberto J, et al. Brown recluse spider venom: proteomic analysis and proposal of a putative mechanism of action. *Protein Pept Lett* 2009; 16:933–43.
4 Levin C, Bonstein L, Lauterbach R, et al. Immune-mediated mechanism for thrombocytopenia after *Loxosceles* spider bite. *Pediatr Blood Cancer* 2014; 61:1466–8.
5 Wasserman CS, Siegel C. Loxoscelism (brown recluse spider bite): a review of the literature. *Clin Toxicol* 1979; 14:353–8.
6 Isbister G, Fan H. Spider bite. *Lancet* 2011; 378:2039–47.
7 Anderson R, Campoli J, Johar SK, et al. Suspected brown recluse envenomation: a case report and review of different treatment modalities. *J Emerg Med* 2011; 41:31–7.
8 Hubbard J, James LP. Complications and outcomes of brown recluse spider bites in children. *Clin Pediatr* 2011; 50:252–8.
9 Tambourgi D, Gonçalves-de-Andrade RM, van den Berg CW. Loxoscelism: from basic research to the proposal of new therapies. *Toxicon* 2010; 56:1113–9.

SCORPIONIDA (SCORPIONS)

Common names: Scorpion, *Centruroides*

Occupational setting

Persons employed in farming, agriculture, construction, forestry, and other outdoor occupations in the southwestern part of the United States are at risk of exposure.

Exposure (route)

Exposure occurs through physical trauma (sting) to the human skin with injection of the venom.

Pathobiology

The scorpion consists of anterior pincers, a pseudoabdomen, and a tail that ends in a bulbous structure called the telson. Within the telson is the stinger and venom. The scorpion usually grasps the victim using its pincers, then arches its body and strikes the victim with the stinger located in its tail, injecting the venom. It ranges in size from 1 to 7 cm. A nocturnal creature, it prefers to take shelter during the day under rocks, debris, or objects outside the house. It generally climbs rather than burrows and therefore is often found among trees.

The venom consists of enzymes (hyaluronidase, acetylcholinesterase, and phospholipase), proteins, serotonin, and neurotoxins, which appear to affect the sodium channel.[1] It results in increased stimulation at the neuromuscular junction and increased axonal discharges.[2]

The clinical presentation usually consists of immediate, severe burning pain at the site of the sting. Paresthesias around the area of the sting may occur. There may be some local inflammation around the site, but most often there is no erythema, swelling, or evidence of ecchymosis.

Severe systemic symptoms may occur in some victims, with a greater incidence in victims under the age of 10 years. Systemic symptoms may occur within minutes or hours after the envenomation. Symptoms result from excitation of the sympathetic and parasympathetic nervous systems, as well as the neuromuscular junction. Symptoms include hyperexcitability, diplopia, nystagmus, restlessness, diaphoresis, muscle fasciculations, tachycardia, hypertension, opisthotonus, and convulsions. Cardiovascular collapse, respiratory arrest, disseminated intravascular coagulation, renal failure, and death are rare.

A clinical grading system for *Centruroides sculpturatus* envenomation uses the following 4-point scale: I (local pain), II (pain or paresthesia remote from sting site), III (either cranial nerve or somatic skeletal neuromuscular dysfunction), and IV (both cranial nerve and somatic skeletal neuromuscular dysfunction).[3] There is a case report of a scorpion sting that appeared to relieve the symptoms of multiple sclerosis.[4]

Diagnosis

Laboratory studies are usually not helpful in diagnosis. Description of the scorpion's appearance, identification of the geographic area where the sting took place, and the clinical presentation are the most helpful factors in making the diagnosis. In *Centruroides* stings, tapping or pressure on the site of the wound ("tap test") usually produces severe discomfort.

Treatment

Most *Centruroides* stings can be managed with local wound care, tetanus prophylaxis, and the application of ice to the affected area. Systemic reactions should be treated as necessary, with advanced life support and supportive measures, and should be monitored very closely in the hospital. Antivenom has been used successfully in many severe cases.[5,6] Complications of the use of antivenom include immediate hypersensitivity (anaphylaxis) and delayed hypersensitivity (serum sickness) reactions.

Prevention

Prevention measures should focus on the use of personal protective equipment like gloves, heavy garments that are fully buttoned, and protective footwear when working in areas that the scorpion inhabits. Areas suspected of harboring scorpions should undergo professional pest control.

References

1 Nencioni AL, Carvalho FF, Lebrun I, et al. Neurotoxic effects of three fractions isolated from *Tityus serrulatus* scorpion venom. *Pharmacol Toxicol* 2000; 86(4):149–55.

2 Petricevich V. Scorpion venom and the inflammatory response. *Mediat Inflamm* 2010; 3:1–16.

3 Khattabi A, Soulaymani-Bencheikh R, Achour S, et al. Classification of clinical consequences of scorpion stings: consensus development. *Trans R Soc Trop Med Hyg* 2011; 105:364–9.

4 Breland AE, Currier RD. Scorpion venom and multiple sclerosis. *Lancet* 1983; 2:1021.

5 Kamerkar N, Kamerkar SB, Geeta K, et al. Efficacy of anti-scorpion venom serum over prazosin in the management of severe scorpion envenomation. *J Postgrad Med* 2010; 56:275–80.

6 Krifi MN, Amri F, Kharrat H, et al. Evaluation of antivenom therapy in children severely envenomed by *Androctonus australis* garzonii (Aag) and *Buthus occitanus* tunetanus (Bot) scorpions. *Toxicon* 1999; 37(11):1627–34.

MARINE ENVENOMATIONS

CATFISH

Common names: Catfish (channel, blue, bullhead, brown) and Carolina madtom catfish[1]

Occupational setting

Occupations with potential exposure include commercial fishing, diving, seafaring, and fish handling.

Exposure (route)

Exposure occurs through physical trauma (sting) to the skin with release of venom.

Pathobiology

The catfish has dorsal and pectoral fins with venom-containing spines. The most common affected sites are the hands of fishermen and fish handlers.[2–5] The venom contains agents that can cause skin necrosis and vasoconstriction, as well as other heat-labile agents.

The clinical presentation includes burning and throbbing at the envenomation site that usually resolves within a few hours but can last for days.[4] The wound site may contain fragmented spines and is often edematous or may appear ischemic. Gangrene is possible secondary to severe envenomation.

Diagnosis

There are no known laboratory tests that are helpful in diagnosing these envenomations. Positive identification of the offending organism is helpful. Barring positive identification, the clinical presentation and geographic location of the envenomation are used to determine the appropriate treatment modality. Since most catfish spines are radiopaque, a radiograph of the affected area may be useful in identifying retained spines.

Treatment

Local wound care, including careful removal of the spines and tetanus prophylaxis, is extremely important. Immediate immersion in hot water (110–115°F) gives adequate pain relief.[2–5] Analgesics are a helpful adjunct in pain relief. Close observation of the wound site for infection is indicated.

Prevention

Preventive measures include wearing gloves and long-sleeved garments when handling organisms or specimens, avoiding any contact with reefs, and never contacting or touching an unknown organism. Persons at risk should always dive with a skilled diver (never alone) in unknown waters. Before working in unknown waters, they should become familiar with the potential marine envenomations and the local medical facilities in case treatment is needed.

References

1 Eastaugh J, Shepherd S. Infectious and toxic syndromes from fish and shellfish consumption: a review. *Arch Intern Med* 1989; 149:1735–40.
2 Kizer KW. Marine envenomations. *J Toxicol Clin Toxicol* 1983–1984; 21:527–55.
3 Blomkalns AL, Otten EJ. Catfish spine envenomation: a case report and literature review. *Wilderness Environ Med* 1999; 10(4):242–6.
4 Auerbach PS. Stings of the deep. *Emerg Med* 1989; 21:27–31.
5 Dorooshi G. Catfish stings: a report of two cases. *J Res Med Sci* 2012; 17:578–81.

COELENTERATE: ANTHOZOA

Common names: Soft coral, true coral, sea anemones

Occupational setting

Workers with potential exposure include sailors, divers, fishermen, and seafood producers.

Exposure (route)

Exposure occurs through physical trauma (sting) to the skin caused by nematocysts that release venom.

Pathobiology

Sea anemones contain stinging structures, called nematocysts, which contain venom. They are abundant, flower-like structures that are usually sessile and have the ability to envenomate with their tentacles but rarely do so.[1] The soft and true corals do not contain nematocysts and therefore cannot envenomate. The venom of coelenterates is not well characterized, but serotonin (5-hydroxytryptamine (5-HT)) appears to be a major component.[2, 3] It produces pain, vasoconstriction, and histamine release.

The clinical presentation of sea anemone envenomation is stinging, pain, and a burning rash. Systemic reactions are rare and the envenomations are infrequent. The coral usually causes injury secondary to trauma sustained when the victim brushes against the extremely sharp, calcified exoskeleton, suffering lacerations.

Diagnosis

There are no known laboratory tests that are helpful in diagnosing these envenomations. Positive identification of the offending organism is helpful. Barring positive identification, the clinical presentation and geographic location of the envenomation are very useful in determining the appropriate treatment modality.

Treatment

Treatment for the sea anemone envenomation includes inactivating the nematocysts by applying alcohol (i.e., isopropyl) or salt water. Alternatively, dilute vinegar, papain, or baking soda can be used. Fresh water should be avoided because it can cause increased venom release from the nematocysts. Immobilization of the affected area is also useful. Baking soda and abrasion of the area with a sharp knife can help to remove any remaining tentacles. Soaking the affected area in hot water (110–115°F) will alleviate severe pain.[4] Narcotics, such as codeine, may also be given for pain control. Tetanus prophylaxis and local wound care should be administered. Oral antihistamines are often given routinely and can be especially effective with symptoms of pruritus. Tissue trauma caused by the corals should be treated with appropriate wound care and tetanus prophylaxis.

Prevention

Preventive measures include wearing gloves and long-sleeved garments when handling organisms or specimens, avoiding contact with reefs, and never contacting or touching an unknown organism. At-risk workers should always dive with a skilled diver (never alone) in unknown waters. Before working in unknown waters, they should become familiar with the potential marine envenomations and the local medical facilities, in case treatment is needed.

References

1 Weinstein S, Dart R, Staples A, et al. Envenomations: an overview of clinical toxinology for the primary care physician. *Am Fam Physician* 2009; 80:793–802.

2 Witzany G, Madl P. Biocommunication of corals. *Int J Integr Biol* 2009; 5:152–63.
3 Grotendorst GR, Hessinger DA. Enzymatic characterization of the major phospholipase A2 component of sea anemone (*Aiptasia pallida*) nematocyst venom. *Toxicon* 2000; 38(7): 931–43.
4 Warrell D. Venomous bites, stings and poisoning. *Infect Dis Clin N Am* 2012; 26:207–23.

COELENTERATE: HYDROZOA

Common names: Portuguese man-of-war (*Physalia*), fire coral, hydroids

Occupational setting

Sailors, divers, fishermen, and seafood producers are at risk for exposure.

Exposure (route)

Exposure results from physical trauma (sting) to the skin inflicted by a stinging structure (nematocyst) containing the venom.

Pathobiology

All members capable of inflicting stings possess stinging structures, called nematocysts, which contain venom. The Portuguese man-of-war consists of a body and tentacles. On these tentacles, there are millions of nematocysts that can release venom upon contact. The tentacles of the *Physalia* can be up to 10 ft long. This creature can be found in all oceans. The hydroids and fire coral, also members of the Coelenterata, possess calcareous growths and inhabit coral reefs in tropical oceans. They also have venom-containing nematocysts.

The venom of coelenterates is not well characterized, but serotonin (5-hydroxytryptamine) appears to be a major component.[1,2] It produces pain, vasoconstriction, and histamine release. In particular, the venom of the Portuguese man-of-war contains a neurotoxin that produces severe pain and possible systemic symptoms.[3]

The clinical presentation of the Portuguese man-of-war envenomation is severe pain with multiple, linear, erythematous macules and papules, which may progress to vesicular lesions. Systemic reaction to envenomation may cause nausea, vomiting, chills, myalgias, and respiratory and cardiovascular depression. The clinical presentation of the envenomation caused by the fire coral and hydroids consists of erythema, pruritus, and edema of the affected skin area. Vesicular lesions and ulcer formation can occur. Systemic reactions of fever, chills, fatigue, myalgias, and abdominal discomfort occur infrequently.[4]

Diagnosis

There are no known laboratory tests that are helpful in diagnosing these envenomations. Positive identification of the offending organism is useful. Barring positive identification, the clinical presentation and geographic location of the envenomation are very helpful in determining the appropriate treatment modality.

Treatment

Treatment includes inactivating the nematocysts by applying alcohol (i.e., isopropyl) or salt water. Alternatively, dilute vinegar, papain, or baking soda can be used. Vinegar appears to be especially effective in Portuguese man-of-war envenomations.[5,6] Fresh water should be avoided because it can cause increased venom release from the nematocysts. Immobilization of the affected area is also helpful. Baking soda and abrasion of the area with a sharp knife can help to remove any remaining tentacles. Soaking the affected area in hot water (110–115°F) will alleviate severe pain. Narcotics, such as codeine, can be given for pain control. Tetanus prophylaxis and local wound care should be administered. Oral antihistamines are often given routinely and can be especially helpful with symptoms of pruritus.

Prevention

Preventive measures include wearing gloves and long-sleeved garments when handling organisms or specimens, avoiding contact with reefs, and never contacting or touching an unknown organism. Persons at risk should always dive with a skilled diver (never alone) in unknown waters. Before working in unknown waters, they should become familiar with the potential marine envenomations and the local medical facilities, in case treatment is needed.

References

1 Edwards L, Hessinger DA. Portuguese man-of-war (*Physalia physalis*) venom induces calcium influx into cells by permeabilizing plasma membranes. *Toxicon* 2000; 38(8):1015–28.
2 Burnett JXV, Carlton GJ. The chemistry and toxicology of some venomous pelagic colenterates. *Toxicon* 1977; 15:177–96.
3 Haddad V, Lupi O, Lonza JP, et al. Tropical dermatology: marine and aquatic dermatology. *J Am Acad Dermatol* 2009; 61:733–50.
4 Kizer KW. Marine envenomations. *J Toxicol Clin Toxicol* 1983–1984; 21:527–55.
5 Cegolon L, Heymann WC, Lange JH, et al. Jellyfish stings and their management: a review. *Mar Drugs* 2013; 11:523–59.
6 Fenner PJ. Dangers in the ocean: the traveler and marine envenomation. I. Jellyfish. *J Travel Med* 1998; 5(3):135–41.

COELENTERATA: SCYPHOZOA

Common names: True jellyfish, sea nettles, sea wasp (box jellyfish)

Occupational setting

Sailors, divers, fishermen, and seafood producers are at risk for exposure.

Exposure (route)

Exposure occurs through physical trauma (sting) to the skin inflicted by a stinging structure (nematocyst) containing venom.

Pathobiology

All members contain stinging structures called nematocysts that contain venom. Jellyfish are found in many shapes, sizes, and colors. They can ride the waves or swim freely on their own. Their tentacles contain numerous nematocysts. Sea wasps are found in Australian waters; the sea nettle is a type of jellyfish found in waters off the Middle Atlantic States.

The venom of coelenterates is not well characterized, but serotonin (5-hydroxytryptamine) appears to be a major component.[1] It produces pain, vasoconstriction, and histamine release.

Jellyfish envenomations usually cause erythematous papular lesions followed by vesicles within the first day. The victim experiences pain, regional lymphadenopathy, and burning. Anaphylactic reactions can occur in some envenomations. The sea wasp, which is found only in Australian waters, causes the most poisonous of all marine envenomations. It can result in death within minutes.[2] Envenomation by this creature can cause pain, nausea, vomiting, erythematous lesions, headache, chills, and occasionally immediate cardiovascular collapse.

Diagnosis

There are no laboratory tests that are helpful in diagnosing these envenomations. Positive identification of the offending organism is helpful. Without positive identification, the clinical presentation and geographic location of the envenomation are very helpful in determining the appropriate treatment.

Treatment

Treatment includes inactivating the nematocysts by applying alcohol (i.e., isopropyl) or salt water. Alternatively, dilute vinegar, papain, or baking soda can be used. Fresh water should be avoided because it can cause increased venom release from the nematocysts. Immobilization of the affected area is also helpful. Baking soda and abrasion of the affected area can be effective in removing any remaining tentacles. Soaking the affected area in hot water (110–115°F) will alleviate severe pain. Narcotics, such as codeine, may be given for pain control. Tetanus prophylaxis and local wound care should be administered. Oral antihistamines are given routinely and can be especially helpful with symptoms of pruritus.

An antivenom is available for sea wasp envenomations. It should be given immediately after skin testing in all severe envenomations. Cardiovascular support (e.g., fluids, vasopressors) may also be needed.[3, 4]

Prevention

Preventive measures include wearing gloves and long-sleeved garments when handling organisms or specimens, avoiding contact with reefs, and never contacting or touching an unknown organism. Persons at risk should always dive with a skilled diver (never alone) in unknown waters. Before working in unknown waters, they should become familiar with the potential marine envenomations and the local medical facilities, in case treatment is needed.

References

1 Balhara K, Stolbach A. Marine envenomations. *Emerg Med Clin North Am* 2014; 32:223–43.

2 Cegolon L, Heymann WC, Lange JH, et al. Jellyfish stings and their management: a review. *Mar Drugs* 2013; 11:523–59.

3 Tibballs J. Australian venomous jellyfish, envenomation syndromes, toxins and therapy. *Toxicon* 2006; 48:830–59.

4 Tibballs J. The effects of antivenom and verapamil on the haemodynamic actions of *Chironex fleckeri* (box jellyfish) venom. *Anaesth Intensive Care* 1998; 26(1):40–5.

DASYATIS (STINGRAY)

Common name: Stingray

Occupational setting

Commercial fishermen, divers, and sailors are at risk of exposure.

Exposure (route)

Exposure occurs through physical trauma to the skin with the release of venom.

Pathobiology

The stingray is a bottom dweller that usually envenomates by raising its barbed, stinger-like tail in a defensive posture after it has been touched or stepped on. The tail apparatus penetrates the skin, usually fracturing the creature's spine in the wound, and envenomates the victim. The laceration and wound are usually deep and irregular.[1] The venom appears to contain cardiotoxins, convulsants, and respiratory depressants.[1]

The stingray envenomation is followed by almost immediate severe pain. The pain's intensity is grossly out of proportion to the degree of trauma visualized; thus, it is helpful in making the diagnosis when the organism was not seen. Edema, erythema, and cyanosis often appear around the wound site. The fragmented spine is also often seen. Systemic symptoms include nausea, vomiting, muscle cramps, syncope, tachycardia, abdominal pain, cardiac arrhythmias, convulsions, hypotension, and (rarely) death.[2]

Diagnosis

There are no known laboratory tests that are helpful in diagnosing these envenomations. Positive identification of the offending organism is useful. Barring positive identification, the clinical presentation and geographic location of the envenomation are very helpful in determining the appropriate treatment.

Treatment

Initial treatment includes local wound care, including removal of any of the fragmented spines, tetanus prophylaxis, and immersion in hot water (110–115°F) for 1 hour.[3,4] Analgesics may be an important adjunct for pain control. Prophylactic antibiotics are not indicated, but these wounds must be observed closely for evidence of infection. Systemic symptoms should receive prompt supportive care and close observation in a hospital setting.

Prevention

Preventive measures include wearing gloves and long-sleeved garments when handling organisms or specimens, avoiding contact with reefs, and never contacting or touching an unknown organism. Persons at risk should always dive with a skilled diver (never alone) in unknown waters. Before working in unknown waters, they should become familiar with the potential marine envenomations and the local medical facilities, in case treatment is needed.

References

1 Kizer KW. Marine envenomations. *J Toxicol Clin Toxicol* 1983–1984; 21:527–55.
2 Forrester M. Pattern of stingray injuries reported to Texas poison centers from 1998 to 2004. *Hum Exp Toxicol* 2005; 24:639–42.
3 Clark R, Girard RH, Rao D, et al. Stingray envenomation: a retrospective review of clinical presentation and treatment in 119 cases. *J Emerg Med* 2007; 33:33–7.
4 Baldinger PJ. Treatment of stingray injury with topical becaplermin gel. *J Am Podiatr Med Assoc* 1999; 89(10):531–3.

ECHINODERMATA

Common names: Starfishes, sea urchins, sea cucumbers

Occupational setting

Sailors, divers, fishermen, and seafood producers are at risk from exposure.

Exposure (route)

Exposure occurs through physical trauma to the skin with the release of venom.

Pathobiology

Sea urchins are bottom-dwelling organisms that have calcareous exoskeletons. They are covered with spines that become embedded in the victim's skin upon contact. Some species contain venom in their spines. The venom contains steroid glycosides, acetylcholine-like substances, and serotonin.[1]

The clinical manifestation of sea urchin envenomation includes both an immediate and a delayed reaction. The immediate reaction consists of severe pain and localized edema and erythema. Dye within the sea urchin spine (usually black or purple) may stain the surrounding skin. Paresthesias around the untreated wound site are common. Occasionally, systemic symptoms of nausea, myalgias, fatigue, syncope, and respiratory difficulties occur.[2,3] The delayed reaction consists of diffuse inflammation or local granuloma formation a few months after the initial injury.[2] Bone destruction is a rare occurrence. Sea urchins can also cause direct injury to joints, nerves, and soft tissue secondary to their sharp spines.[2]

Starfish have sharp spines and secrete a slimy venom. Envenomation can cause nausea, vomiting, paresthesias, and muscle paralysis.[4] Sea cucumbers are sausage-shaped organisms that produce a venom called holothurin, which is a cardiac glycoside. The venom causes a macular and papular rash. Contact with the eyes can cause severe conjunctivitis, opacities, and blindness. The sea cucumber also ingests nematocysts from other species and can secrete them later in intact form as a self-defense measure, causing symptoms of coelenterate envenomation.

Diagnosis

There are no laboratory tests that are helpful in diagnosing these envenomations. Positive identification of the offending organism is useful. Barring positive identification, the clinical presentation and geographic location of the envenomation are very helpful in determining the appropriate treatment.

Treatment

Treatment should begin with local wound care and tetanus prophylaxis. Immersion in hot water (110–115°F) should be done and analgesics should be administered. Spines should be removed carefully in order to avoid fragmentation. Granulomas secondary to delayed reaction may need specific surgical or dermatologic procedures. Steroids, ammonia, antibiotics, and acetone have all been used with limited success.[5]

Prevention

Preventive measures include wearing gloves and long-sleeved garments when handling organisms or specimens, avoiding contact with reefs, and never contacting or touching an unknown organism. Persons at risk should always dive with a skilled diver (never alone) in unknown waters. Before working in unknown waters, they should become familiar with the potential marine envenomations and the local medical facilities, in case treatment is needed.

References

1 Ritchie KB. A tetrodotoxin-producing marine pathogen. *Nature* 2000; 404(6776):354.
2 Schwartz S. Venomous marine animals of Florida: morphology, behavior, health hazards. *J Fla Med Assoc* 1997; 84(7):433–40.
3 Reese E, Depenbrock P. Water envenomations and stings. *Curr Sports Med Rep* 2014; 13:126–31.
4 Liram N, Gomori M, Perouansky M. Sea urchin puncture resulting in PIP joint synovial arthritis: case report and MRI study. *J Travel Med* 2000; 7(1):43–5.
5 Warrell D. Venomous bites, stings and poisoning. *Infect Dis Clin North Am* 2012; 26:207–23.

MOLLUSCA

Common names: Cone shell (Conidae), blue-ringed octopus

Occupational setting

Sailors, divers, fishermen, and seafood producers are at risk from exposure.

Exposure (route)

Exposure occurs through physical trauma to the skin with injection of venom.

Pathobiology

Conidae, which are found mainly in tropical waters, are cone-shaped, univalve organisms surrounded by a shell. They contain a radula tooth apparatus that envenomates its victims. The blue-ringed octopus is found mainly in Australian waters. It injects its venom through a beak-like apparatus. The venom appears to contain a number of neurotoxins.

The initial symptoms of envenomation include pain, stinging, burning, numbness, and paresthesias around the wound site. In severe envenomations, local ischemia and cyanosis can occur. Systemic reactions include fatigue, dysphagia, diplopia, paresthesias, dyspnea, coma, and cardiovascular collapse.[1] Systemic reactions to severe envenomations include paresthesias, slurred speech, weakness, dysphagia, and respiratory depression. Allergic reactions to the envenomation have been reported.[2,3]

Diagnosis

There are no known laboratory tests that are helpful in diagnosing these envenomations. Positive identification of the offending organism is useful. Barring positive identification, the clinical presentation and geographic location of the envenomation are very helpful in determining the appropriate treatment.

Treatment

Conidae envenomations respond to hot water immersion (110–115°F) as well as analgesics. All patients should be observed for 6 hours for systemic reactions. Blue-ringed octopus envenomations do not respond to hot water immersion. These victims also need to be observed for systemic symptoms.

Prevention

Preventive measures include wearing gloves and long-sleeved garments when handling organisms or specimens, avoiding contact with reefs, and never contacting or touching an unknown organism. At-risk persons should always dive with a skilled diver (never alone) in unknown waters. Before working in unknown waters, they should become familiar with the potential marine envenomations and the local medical facilities, in case treatment is needed.

References

1 Anderson P, Bokor G. Conotoxins: potential weapons from the sea. *J Bioterr Biodef* 2012; 3:120.
2 Edmonds C. A nonfatal case of blue-ringed octopus bite. *Med J Aust* 1969; 2:601.
3 Bonnet MS. The toxicology of *Octopus maculosa*: the blue-ringed octopus. *Br Homeopath J* 1999; 88(4):166–71.

PORIFERA

Common names: Caribbean and Hawaiian fire sponge (*Tedania ignis*), poison burn sponge (*Fibula nolitangere*), red sponge (*Microciona prolifera*)

Occupational setting

Sailors, divers, fishermen, and seafood producers are at risk from exposure.

Exposure (route)

Exposure occurs through physical and chemical irritation of the skin.

Pathobiology

Sponges contain calcareous spines that can cause direct trauma to the skin. The components of the sponge's venom are unknown, but it does cause symptoms similar to those of *Rhus* dermatitis.[1,2] Initially, there is pruritus and erythema in the area of contact, followed by progressive swelling and vesicle formation within a few hours. Lymphadenopathy can also occur. Anaphylactoid reactions and erythema multiforme have been reported.[3]

Diagnosis

There are no laboratory tests that are helpful in diagnosing these envenomations. Positive identification of the offending organism is useful. Barring positive identification, the clinical presentation and geographic location of the envenomation are very helpful in determining the appropriate treatment.

Treatment

Local wound care and tetanus prophylaxis should be administered. Topical steroids and antihistamines may be helpful. Tape can be used to remove spines. Dilute vinegar (30 mL in 1 L) or 5% acetic acid applications may be helpful with the acute dermatitis.

Prevention

Preventive measures include wearing gloves and long-sleeved garments when handling organisms or specimens, avoiding contact with reefs, and never contacting or touching an unknown organism. Persons at risk should always dive with a skilled diver (never alone) in unknown waters. Before working in unknown waters, they should become familiar with the potential marine envenomations and the local medical facilities, in case treatment is needed.

References

1 Balhara K, Stolbach A. Marine envenomations. *Emerg Med Clin North Am* 2014; 32:223–43.
2 Kobayashi M, Kitagawa I. Marine spongean cytotoxins. *J Nat Toxins* 1999; 8(2):249–58.
3 Yaffee S, Stargardter F. Erythema multiforme from *Tedania ignis*: report of a case and an experimental study of the mechanism of cutaneous irritation from the fire sponge. *Arch Dermatol* 1963; 87:601–4.

SCORPAENIDAE

Common names: Lionfish, turkeyfish (*S. pterois*), stonefish (*S. synanceja*), scorpionfish (*S. scorpaena*)

Occupational setting

Workers with potential exposure include commercial fishermen, divers, sailors, aquarists, and fish handlers.

Exposure (route)

Exposure occurs through physical trauma (sting) to the skin, with the potential release of venom.

Pathobiology

The Scorpaenidae are usually bottom dwellers with fins that are supported by venomous spines. They live in shallow waters around coral reefs and rocky formations. Envenomation can occur when the fish is stepped on or touched by the hand. A fragmented spine is often left in the wound site. The venom is a heat-labile mixture that can cause hypotension, weakness, vasodilation, and respiratory depression.[1,2]

The clinical presentation of envenomation includes local effects such as edema, erythema, and ecchymosis at the wound site with intense, severe pain that can last up to 2 days if not treated. Paresthesias of the affected extremity are common. Neuropathy is rarely seen. Severe envenomation can cause hypotension, cardiac arrhythmias, syncope, nausea, vomiting, diaphoresis, convulsions, and paralysis. These systemic symptoms can develop very quickly.

Diagnosis

There are no known laboratory tests that are helpful in diagnosing these envenomations. Positive identification of the offending organism is useful. Barring positive identification, the clinical presentation and geographic location of the envenomation are very helpful in determining the appropriate treatment. Laboratory studies such as complete blood count (CBC), electrolytes, glucose, blood urea nitrogen (BUN), creatine phosphokinase, creatinine, ECG, chest radiograph, and urinalysis will not be diagnostic, but they are helpful adjuncts to care of the patient.

Treatment

The local wound site should be examined, and local wound care and tetanus prophylaxis should be given. The extremity should be immediately immersed in hot water at 110–115°F for at least 1 hours. This treatment usually decreases the severe pain of the sting.[3,4] Analgesics can also be used as an adjunct. The wound site should be examined carefully for any fragmented spines, which should be removed with care. Radiograph of the affected site may be helpful in revealing these fragmented spines; however, a negative radiograph does not rule out the presence of spines, because not all spines are radiopaque.

In severe envenomations, Scorpaenidae antivenom may be indicated. Assistance can be obtained from the local/regional poison control center. Immediate supportive care is needed for patients with severe envenomations who present with hypotension, arrhythmias, convulsions, or paralysis.

Prevention

Preventive measures include wearing gloves and long-sleeved garments when handling organisms or specimens, avoiding contact with reefs, and never contacting or touching an unknown organism. At-risk persons should always dive with a skilled diver (never alone) in unknown waters. Before working in unknown waters, they should become familiar with the potential marine envenomations and the local medical facilities, in case treatment is needed.

References

1 Kizer KW, McKinney HE, Auerbach PS. Scorpaenidae envenomation. A five-year poison center experience. *JAMA* 1985; 253:807–10.
2 Gomes H, Andrich F, Mauad H, et al. Cardiovascular effects of scorpionfish venom. *Toxicon* 2010; 55:580–9.
3 Patel MR, Wells S. Lionfish envenomation of the hand. *J Hand Surg (Am)* 1993; 18(3):523–5.
4 Balhara K, Stolbach A. Marine envenomations. *Emerg Med Clin North Am* 2014; 32:223–43.

SNAKE ENVENOMATIONS

COLUBRIDAE

Common names: Garter snake (*Thamnophis elegans vagrans*), king snake (*Lampropeltis*), hognose snake (*Heterodon*), boomslang snake (*Dispholidus*), racer snake (*Coluber*), and red-necked keelback snake (*Rhabdophis*)

Occupational setting

Farming and agricultural workers, gardeners and orchard workers, forestry workers, construction workers, veterinary office workers, pet store workers, and zoo personnel are at risk of exposure.

Exposure (route)

Exposure occurs through physical trauma (bite) to the human skin, with injection of venom.

Pathobiology

There are 1400 species in the Colubridae family. The majority contains teeth and no venom-injecting apparatus and is therefore nonpoisonous. However, a small percentage of colubrids, like the hognose snake, have rear fangs with a venom-injecting apparatus.[1] Because of its small size and the posterior location of its fangs, these snakes do not usually envenomate their victims, but they are capable of doing so, given the right circumstances. All colubrids, whether rear-fanged or not, contain saliva that has been shown to be toxic.[2]

The colubrid bite usually presents as a minor abrasion or an imprint of teeth on the skin surface. Occasionally, there is evidence of a puncture wound inflicted by rear-fanged colubrids, but this would take prolonged skin contact time and can usually be elicited with the history. Occasionally, erythema, pruritus, and swelling are seen around the bite site; however, most bites cause only minor discomfort for the victim secondary to the abrasion.

Rare envenomations usually produce low-grade, mild symptoms. Fibrinolysis was reported secondary to a bite of a red-necked keelback snake that was purchased in a pet store.[3] Fibrinolysis has also been reported with other colubrid bites.[1,4,5] In a number of reports, a garter snake produced local hemorrhage, pain, and swelling of an upper extremity as well as neurologic complications without systemic symptoms.[6,7]

Diagnosis

The identification of the snake that inflicted the bite is extremely helpful in the management of all snakebites. The actual snake should be brought to the treating physician for identification, if possible, or the snake should be identified using well-defined photographs of different varieties of snakes. A herpetologist or regional poison control center may need to be consulted. Envenomation can occur hours after a snake's death; therefore, these snakes should be handled carefully.

The victim should be evaluated carefully for fang marks, swelling, erythema, pain, local reaction (i.e., blebs, vesicles, petechiae), and any systemic reactions. Most colubrid bites present with minor abrasions or the imprint of teeth on the skin surface. Any deviation from this minor abrasion, without positive identification of the offending snake, should make the treating physician suspicious of a more venomous species. Laboratory data are usually not helpful in making the diagnosis. Coagulation abnormalities suggest a severe envenomation.

Treatment

Treatment usually consists of local wound care and tetanus prophylaxis. Antibiotics may be indicated if there is any suspicion of infection. After an uncomplicated colubrid bite, observation for at least 4 hours is indicated to rule out systemic symptoms. If fang marks are present and the snake has been identified as a colubrid, then at least 6 hours of observation is prudent to rule out systemic symptoms and a significant envenomation. Fibrinolysis and hemorrhage should be treated with conventional therapy. There is no approved antivenom currently available. The vast majority of bites need only local wound care. Any systemic symptoms respond well to supportive measures.

Prevention

Whether dead or alive, snakes should always be handled carefully, especially when the treating personnel are attempting to identify the snake. Snakes can still envenomate hours after their death. Snakes are ubiquitous in all outdoor areas. Areas like rock crevices, caves, and unusual terrain should be avoided. Trousers, gloves, boots, and long-sleeved shirts should be worn in areas where snakes may be present. Pest control professionals should be consulted when work must be performed in high-risk areas. One should never awaken or try to molest a snake. Venomous snakes should not be sold or kept as pets.

References

1 Mandell F, Bates J, Mittleman M, et al. Major coagulopathy and "non-poisonous" snake bites. *Pediatrics* 1980; 65:314–17.
2 McKinstry DM. Evidence of toxic saliva in some colubrid snakes of the United States. *Toxicon* 1978; 16:523–34.
3 Cable D, MeGehee XV, Wingert WA, et al. Prolonged defibination after a bite from a "non-venomous" snake. *JAMA* 1984; 25(1):925–6.
4 Hill RE, Mackessy SP. Venom yields from several species of colubrid snakes and differential effects of ketamine. *Toxicon* 1997; 35(5):671–8.
5 Vank F, Jackson K, Doley R, et al. Snake venom: from fieldwork to the clinic. *Bioessays* 2011; 33:269–79.
6 Vest DK. Envenomation following the bite of a wandering garter snake (*Thamnophis elegans vagrans*). *Clin Toxicol* 1981; 18:573–9.
7 Del Brutto O, Del Brutto VJ. Neurological complications of venomous snake bites: a review. *Acta Neurol Scand* 2012; 125:363–72.

CROTALIDAE

Common names: Rattlesnake (*Crotalus*), copperhead, cottonmouth, water moccasin (*Agkistrodon*), fer-de-lance (*Bothrops*)

Occupational setting

Farming and agricultural workers, gardeners and orchard workers, construction workers, veterinary and ancillary staff workers, pet store personnel, and zoo workers are at risk of exposure.

Exposure (route)

Exposure occurs through physical trauma (bite) to the human skin, with injection of venom.

Pathobiology

All pit vipers (Crotalidae) have a vertical, elliptical pupil; facial pits located between the nostril and the eye, which are used to sense heat and vibration; a triangular head; and a single row of subcaudal plates (in contrast to the double row found in nonvenomous snakes). Most true rattlesnakes also contain a terminal segment called a rattle, but not all species contain this rattle. In some cases, trauma or congenital anomaly can also contribute to the lack of the rattle apparatus. Cottonmouths (*Agkistrodon piscivorus*) have a white buccal mucosa and usually a dark green to black appearance; they are often found in swampy marshes, lakes, or other aquatic environments. Copperheads have a head that is copper to brown in color and a distinctive band-like appearance on their bodies. Copperheads can be found in any outdoor area. Neither copperheads nor cottonmouths have rattles.

Pit vipers have anterior fangs on the roof of their mouths that can be up to 20 mm in length, allowing for easy envenomation and injection of venom. The venom contains numerous components, with a predominance of enzymes and neurotoxins that are used to immobilize and kill the animal prey.[1] The potency and amount of venom injected varies within the individual species. In general, the potency of the venom of the true rattlesnake is greater than that of the cottonmouth, which is, in turn, greater than that of the copperhead.

After a Crotalidae bite, there are typically two or more fang or puncture marks on the victim's skin. In minutes, the victim experiences pain, edema, and erythema around the site of envenomation. Over the next few hours, the edema progresses proximally, the erythema increases, and there is evidence of hemorrhagic blebs or vesicles and ecchymosis. An affected extremity can become quite large, secondary to edema. Hypovolemic shock may occur in some cases. If progressive swelling has not occurred in the first few hours, then envenomation probably has not occurred.[2] The exception is the Mojave rattlesnake (*Crotalus scutulatus scutulatus*), whose envenomation produces little or no local symptoms but may cause the delayed systemic symptoms of muscle fatigue, ptosis, and respiratory depression.

Systemic effects of Crotalidae envenomation include fatigue, muscle fasciculations, light-headedness, minty or metallic taste, nausea, vomiting, and paresthesias (both perioral and peripheral). Thrombocytopenia and coagulopathy can occur.[3,4] Hypotension, pulmonary edema, and renal failure may also be seen.[5]

Diagnosis

The identification of the snake that inflicted the bite is critical in the management of all snakebites. The snake should be brought to the treating physician for identification, if possible, or it should be identified using well-defined photographs of different varieties of snakes. A herpetologist located at a local zoo, park, or aquarium may be consulted. The regional poison control center or the Antivenom Index of the Arizona Poison Control Center may be helpful. Snakes should be handled carefully even when dead, since envenomation can occur hours after a snake's death.

The victim should be evaluated for fang marks, swelling, erythema, pain, local reactions (i.e., blebs, vesicles, petechiae), and any systemic reactions. The classic symptoms and signs of Crotalidae envenomation include puncture or fang wounds, pain, erythema, and edema around the bite site. The evaluation of the site may reveal one or more puncture wounds, which could represent the number of fangs the snake had or the number of strikes that occurred.[6]

There is no specific assay or laboratory test that can reliably identify and thereby assist in the diagnosis of Crotalidae envenomation. However, cell blood count, coagulation studies, disseminated intravascular coagulopathy (DIC) panel, blood type and screen, platelet count, creatine phosphokinase, and urine for myoglobin can be helpful.

Treatment

In the field, prior to definitive care by medical personnel, the affected bite area should be immobilized in a dependent position. Ice should not be applied. Tourniquets as well as incision and suction are controversial and can be deleterious.[7]

Medical personnel should monitor all Crotalidae envenomation for at least 4 hours for local and systemic effects. In the case of the Mojave rattlesnake, observation should be for 12–24 hours because of the delayed effects of the venom in this species. The affected limb should be closely monitored for local swelling, ecchymosis, and circulatory compromise. Tetanus prophylaxis and local wound care should be administered. Any complication regarding the wound, especially a compartment syndrome, should be referred to an experienced surgeon.

Systemic effects should be monitored closely and treated immediately. Respiratory depression requires immediate airway maintenance and assisted ventilation. Coagulopathy, hypotension, and rhabdomyolysis also require immediate treatment.

If documented envenomation has occurred, antivenom should be administered. Most local and systemic symptoms improve with the antivenom. The Crotalidae antivenom is equine based and is therefore contraindicated in individuals with known hypersensitivity to the antivenom or horse serum. Antivenom may need to be given cautiously in severe envenomations despite a positive allergic history. All patients should be initially skin tested. Both immediate hypersensitivity (anaphylaxis) and delayed hypersensitivity (serum sickness) can occur.

The amount of antivenom to administer is determined by the degree of envenomation.[8,9] It is wise to consult the regional poison control center prior to its use. For bites that produce very minor local symptoms and no systemic symptoms after 6 hours, no antivenom is recommended. In moderate envenomations where there is progression of local symptoms beyond the bite site and minor systemic symptoms such as nausea, anxiety, or paresthesias, fewer vials of Crotalidae antivenom can be administered, depending on the severity of the symptoms. In severe envenomations, where there is progressive local pathology and severe systemic symptoms, more vials may be needed. All patients receiving antivenom should be closely observed and monitored in a hospital setting.

Prevention

Whether dead or alive, snakes should be handled carefully, especially when the treating personnel are attempting to identify the snake. Dead snakes can still envenomate hours after their death. Snakes are ubiquitous in all outdoor areas. Areas like rock crevices, caves, and unusual terrain should be avoided. Trousers, gloves, boots, and long-sleeved shirts should be worn in areas where snakes may be present. Pest control professionals should be consulted when work must be performed in high-risk areas. One should never awaken or try to molest a snake. Venomous snakes should not be sold or kept as pets.

References

1 Segura A, Herrera M, Villalta M, et al. Assessment of snake antivenom purity by comparing physicochemical and immunochemical methods. *Biologicals* 2013; 41:93–7.

2 Russell FE, Carlson RW, Wainschel J, et al. Snake venom poisoning in the United States: experience with 550 cases. *JAMA* 1975; 233:341–4.
3 Lavonas E. Coagulopathy: the most important thing we still don't know about snakebite. *West J Emerg Med* 2012; 13:75–6.
4 Gibly RL, Walter FG, Nowlin SW, et al. Intravascular hemolysis associated with North American crotalid envenomation. *J Toxicol Clin Toxicol* 1998; 36(4):337–43.
5 Pinho F, Zanetta DM, Burdmann EA. Acute renal failure after *Crotalus durissus* snakebite: a prospective survey on 100 patients. *Kidney Int* 2005; 67:659–67.
6 Snyder CC, Knowles RP. Snakebites: guidelines for practical management. *Postgrad Med* 1988; 83:53–75.
7 Wasserman GS. Wound care of spider and snake envenomation. *Ann Emerg Med* 1988; 17:12.
8 Spiller H, Bosse GM, Ryan ML. Use of antivenom for snakebites reported to United States poison centers. *Am J Emerg Med* 2010; 28:780–5.
9 Weant K, Bowers RC, Reed J, et al. Safety and cost-effectiveness of a clinical protocol implemented to standardize the use of Crotalidae polyvalent immune Fab antivenom at an academic medical center. *Pharmacotherapy* 2012; 32:433–40.

ELAPIDAE

Common names: Coral snake (*Micrurus*), cobra (*Naja*), krait (*Bungarus*), mamba (*Dendroaspis*)

Occupational setting

Farming and agricultural workers, gardeners and orchard workers, forestry workers, construction workers, veterinary and ancillary staff workers, pet store personnel, zoo workers, and research personnel are at risk from exposure.

Exposure (route)

Exposure occurs through physical trauma (bite) to the human skin with injection of venom.

Pathobiology

Coral snakes, the only members of the Elapidae family native to the United States, are found in the southeastern and south central states. They include the Sonoran or Arizona coral snake (*Micruroides euryxanthus*), the Eastern coral snake (*Micrurus fulvius fulvius*), and the Texas coral snake (*Micrurus fulvius tenere*). The eyes are round and the mouth is small, with a row of teeth and rear fangs. This anatomy forces the coral snake to hold on and "chew" its victim in order to get its rear fangs into the skin and cause envenomation. Because of its anatomy, the coral snake is unable to envenomate its victim in most circumstances. The venom is a complex mixture of many components, including a neurotoxin.[1] Coral snake venom causes very little local destruction, but it can cause systemic reactions that are primarily neurologic.

Coral snakes have a distinct pattern of red and black bands that are interspaced with narrower yellow bands. This is often remembered as: "Red on yellow, kill a fellow"—coral snake; "Red on black, venom lack"—nonvenomous snake.

There are usually very mild (if any) swelling, erythema, and pain around the bite site. Occasionally, paresthesias occur around the site. A few hours after envenomation, the victim experiences nausea, vomiting, confusion, euphoria, or drowsiness. The systemic effects, which are predominantly neurologic, can progress to diplopia, fasciculations, slurred speech, dysphagia, paralysis, and respiratory depression. It is important to realize that the systemic symptoms of a coral snake envenomation may be delayed for up to 10 hours.[2] The Arizona coral snake appears to cause greater neurologic symptomatology than the Texas or Eastern coral snakes.[3]

Diagnosis

The identification of the snake that inflicted the bite is critical in the management of all snakebites. The actual snake should be brought to the treating physician for identification, or the snake should be identified using well-defined photographs of different varieties of snakes. A herpetologist located at a local zoo, park, or aquarium may need to be consulted. The regional poison control center or the Antivenom Index of the Arizona Poison Control Center may be helpful. Because envenomation can occur even hours after a snake's death, snakes should be carefully handled even when dead.

The victim should be evaluated for fang marks, swelling, erythema, pain, local reaction (i.e., blebs, vesicles, petechiae), and any systemic reaction. Most victims of coral snake bites present with minor abrasion or the imprint of teeth on the skin surface. Fang or puncture marks reveal possible severe envenomation. The victim should be observed for up to 12 hours for systemic symptoms, which are predominantly neurologic and would be consistent with a coral snake envenomation.

Treatment

Most coral snake envenomations can be treated with local wound care and tetanus prophylaxis. The presence of fang or puncture marks is a sign of a potentially serious envenomation and necessitates observation for at least 12–24 hours. The affected area should be immobilized in a dependent position to decrease the spread of the venom. Cryotherapy should not be used, and incision and suction are usually not warranted.

In severe envenomations, consultation with a regional poison center and local herpetologist is recommended before using the *Micrurus fulvius* antivenom. In general, any victim with severe envenomation from an Eastern coral snake or Texas coral snake, which causes systemic neurologic symptomatology or coagulopathy, should receive the antivenom.[4] It is not effective for envenomation by the Arizona or Sonoran coral snake. Hypersensitivity to horse serum or the antivenom is a relative contraindication, but it may be given with extreme caution in severe envenomations. All victims should be skin tested with the antivenom prior to administration. Adverse effects include immediate hypersensitivity and possible anaphylaxis as well as delayed hypersensitivity (serum sickness).

Prevention

Whether dead or alive, snakes should be handled carefully, especially when the treating personnel are attempting to identify the snake. Dead snakes can still envenomate hours after their death.

Snakes are ubiquitous in all outdoor areas. Areas like rock crevices, caves, and unusual terrain should be avoided. Trousers, gloves, boots, and long-sleeved shirts should be worn in areas where snakes may be present. Pest control professionals should be consulted when work must be performed in high-risk areas. One should never awaken or try to molest a snake. Any snake that has the potential of envenomation should not be sold or kept as a pet.

References

1 Silveira de Oliveira J, Rossan de Brandão Prieto da Silva A, Soares MB, et al. Cloning and characterization of an alpha-neurotoxin-type protein specific for the coral snake *Micrurus corallinus*. *Biochem Biophys Res Commun* 2000; 267(3):887–91.

2 Deer PJ. Elapid envenomation: a medical emergency. *J Emerg Nurs* 1997; 23(6):574–7.

3 Russell FE. Bites by the *Sonoran coral* snake *Micruroides euryxanthus*. *Toxicon* 1967; 5:39–42.

4 Tanaka G, Furtado Mde F, Portaro FC, et al. Diversity of Micruris snake species related to their venom toxic effects and the prospective of antivenom neutralization. *PLoS Negl Trop Dis* 2010; 4:e622.

HYDROPHIDAE

Common name: Sea snake

Occupational setting

Fishermen and diving personnel, especially those in the Pacific and Indian oceans, are at risk from exposure. There do not appear to be any sea snakes in the Atlantic Ocean or the Caribbean.

Exposure (route)

Exposure occurs through physical trauma (bite) to the human skin, with potential envenomation.

Pathobiology

Sea snakes have a compressed, fin-like tail that allows them to propel themselves in the water. They have two to four fangs, which are attached to venomous glands, but the fangs are short; therefore, most bites result in no envenomation. There are at least 52 species and all are venomous. *Enhydrina schistosa* is the most dangerous. The venom is a potent neurotoxin[1,2] that contains myotoxins, such as phospholipase A, which can cause striated muscle necrosis.[3] Many other substances are present in this complex venom.

Most often, the initial bite is painless and the victim asymptomatic. If envenomation did occur, symptoms can develop within 10 minutes and will usually be present within an 8-hour period. In actual envenomations, there are puncture wounds on the victim's skin. The classic envenomation results in muscle pain, difficulty with muscle movement, and myoglobinuria.[1] Ophthalmoplegia, trismus, ptosis, paralysis (flaccid or spastic) of the lower extremities, and respiratory failure can occur.[2]

Diagnosis

The identification of the snake that inflicted the bite is critical in the management of all snakebites. The snake should be brought to the treating physician for identification, if possible, or it should be identified using well-defined photographs of different varieties of snakes. Specifically, with sea snakes, the bite must have been inflicted in the Pacific or Indian Ocean. The evaluation of the bite site should reveal two or more puncture wounds secondary to the fang marks. The clinical presentation of a painless bite followed by trismus, ptosis, myalgias, difficulty with muscle movement, or lower extremity paralysis should make the treating personnel suspicious of a sea snake envenomation.

Treatment

Treatment consists of immobilizing the victim with the affected site in a dependent position to decrease the spread of the venom. Cryotherapy is not indicated nor is incision and drainage of the affected site in most cases. Most victims can be treated with supportive care.

For severe envenomations, there is antivenom available from Australia, as well as a polyvalent tiger snake antivenom.[3,4] As with all antivenom, sensitivity testing should be performed prior to administration and the victim observed for signs of anaphylaxis and serum sickness. In all systemic reactions to the sea snake bite, the victim should be closely monitored in a hospital setting, since respiratory failure is possible.

Prevention

Trousers, gloves, boots, and long-sleeved shirts should be worn in areas where the snakes may be present. Sea snakes are often found in fishing nets. Fishermen should be extremely careful when removing entangled sea snakes from these nets. Whether dead or alive, sea snakes should be handled carefully. Dead snakes can still envenomate hours after their death.

References

1 Hodgson W, Wickramaratna JC. Snake venoms and their toxins: an Australian perspective. *Toxicon* 2006; 48:931–40.
2 Tu AT. Biotoxicology of sea snake venoms. *Ann Emerg Med* 1987; 16:1023–8.
3 Warrell D. Snake bite. *Lancet* 2010; 375:77–88.
4 Amarasekera N. Bite of a sea snake (*Hydrophis spiralis*): a case report from Sri Lanka. *J Trop Med Hyg* 1994; 97(4):195–8.

Chapter 28

ALLERGENS

David C. Caretto*

ENZYMES

Common names: Detergents, digestive aids, dough improvers, papain, others

Occupational setting

Enzymes are used in the chemical, pharmaceutical, cosmetic, textile, medical, detergent, and food and beverage industries. Paper and pulp industries use enzymes to break down wastes. Cellulase is used as a digestive aid. Aspergillus-derived alpha-amylase and cellulase are added to flour as dough improvers. Enzymes derived from *Bacillus subtilis*, such as alcalase (subtilisin A), are used in the detergent industry as an aid in removing stains from clothing. Papain and other enzymes are used in pharmaceutical products where the manufacturing process may involve sieving, blending, and compressing powders. Papain also has uses as a meat tenderizer (which use has resulted in illness in industrial kitchens), in the treatment of wool and silk for textiles, and in clarifying beer. Trypsin, a pancreatic enzyme, is used in the rubber industry.

Exposure (route)

Enzyme powders tend to be fine and easily airborne. Exposure to enzymes may occur through inhalation of enzyme dusts or through skin contact with liquid enzyme preparations or airborne dust. Dust may impact the upper airways or be carried directly into the lungs.

* The author wishes to thank Gwendolyn S. Powell for her contribution to an earlier version of this chapter.

Pathobiology

Enzymes are proteins that catalyze chemical reactions. They are usually high-molecular-weight proteins and are effective in very small quantities. Cellulase catalyzes the cellulose to glucose reaction, xylanase catalyzes the breakdown of xylan, and alpha-amylase also has carbohydrate-cleaving activity. Papain and bromelain are proteolytic enzymes derived from the latex of ripe fruit of the pawpaw tree and the pineapple, respectively. Alcalase is a proteolytic enzyme derived from *B. subtilis* bacteria. Enzyme preparations may be contaminated with by-products of their production, such as growth media material, microorganisms, and preservatives.

Enzymes are potent respiratory sensitizers that may cause immediate, delayed, or dual allergy consisting of rhinoconjunctivitis and several patterns of asthma. In addition to respiratory allergy, immediate and delayed types of skin allergy have been described from enzyme exposure. Respiratory allergy may occur from exposure to cellulase, xylanase, papain, flaviastase, trypsin, bromelain, pepsin, alpha-amylase, and other enzymes. Asthmatic illnesses occur in a portion of sensitized workers who all typically have rhinitis and conjunctivitis along with nasal congestion. The pattern of asthma symptoms may be immediate (Type I, IgE mediated), biphasic with both early and late reactions, or delayed. Hypersensitivity pneumonitis may also occur. The occurrence of systemic symptoms along with delayed resolution of symptoms after removal from exposure and the presence of precipitins provide evidence for Type III hypersensitivity. In papain-induced asthma, the pathology seems to involve small airways, alveolar, and interstitial lung

Physical and Biological Hazards of the Workplace, Third Edition. Edited by Gregg M. Stave and Peter H. Wald.

tissue in an inflammatory manner. Higher doses of enzyme seem to cause more intense pulmonary symptoms. Atopic individuals are thought to be at higher risk of developing enzyme-related allergies, such as that to papain, and to develop antibody and symptoms sooner than nonatopics.

Skin disorders from enzyme exposure may be irritant, digestive (commonly), or allergic. Skin allergy may be of the immediate (Type I, IgE mediated) variety, manifested by urticaria following skin contact with enzymes in liquid or powder form. Papain, cellulase, and xylanase are among the enzymes that cause contact urticaria.

Contact dermatitis, a delayed (Type IV) allergy that can be verified by positive patch tests, may also occur following exposure to enzymes. Contact dermatitis consists of an itchy, raised, red skin rash that may be slow to resolve. Although alcalase skin disorders may be irritant or digestive in nature, the combination of detergent with alcalase may be more allergenic than alcalase alone. Some enzymes, such as cellulase and xylanase, have been found to cause both urticaria and contact dermatitis. Patients with enzyme-related contact dermatitis may or may not be able to eat related products without problem.

Allergy to enzymes may develop after months to years of exposure. Allergic individuals exhibit rhinorrhea, conjunctivitis, and often shortness of breath, cough, chest tightness, and wheezing that occurs minutes to hours following exposure to enzymes. Oculonasal complaints typically precede the development of chest symptoms.

Anaphylaxis has followed the ingestion of papain in sensitized individuals. Cellulase and xylanase exhibit some cross-reactivity, as do chymopapain, papain, and bromelain. There is conflicting evidence regarding cross-reactivity between alcalase, sarinase, and esperase.[1]

Diagnosis

Medical and occupational history can be used to diagnose this disorder when the temporal relationship to exposure is elucidated in association with allergic symptoms. Symptoms that appear following exposure may resolve on weekends and holidays. Physical examination may reveal mucous membrane inflammation and respiratory wheeze. In patients with skin allergy, an itchy, red skin rash may be observed on exposed surfaces.

Pre- and postshift spirometry can be used to document the association of bronchoconstriction with exposure. One case report documented a 48–50% decrement in peak expiratory flow rate (PEFR) in a patient following entry into rooms where xylanase was handled.[2] Confirmation of the diagnosis of enzyme allergy may be obtained through specific immunologic testing. RAST, skin prick, and skin patch testing may demonstrate specific IgE and hyperreactivity, respectively. Skin prick tests have been used to document reactivity of allergic patients to enzymes including cellulase, xylanase, papain, bromelain, and trypsin. In some cases, skin prick testing has resulted in a systemic reaction.[1] Specific IgE measurements using RAST have been done for antibodies to cellulase, xylanase, alpha-amylase, alcalase, and papain.

ELISA has been reported to measure specific IgE antibody to alcalase in exposed detergent workers. In one study, ELISA was more sensitive than RAST, detecting 85% versus 68.4% of skin test positive workers.[1] IgG antibodies to papain have been documented. Bronchial provocation has been used to diagnose alpha-amylase, cellulase,[3] trypsin, and papain asthma.

Treatment

The preferred treatment for enzyme allergies is removal from exposure until corrective measures in work practice or engineering controls are implemented. In sensitized workers with symptoms, job change should be the ultimate goal. In severe cases, there is no alternative to a job change. In workers who remain in exposure situations, conventional allergy treatments may be tried. Cromolyn sodium reportedly prevented immediate papain-related symptoms in allergic workers, although steroids were needed to control the late response.[4] Long-acting bronchodilators may abate late symptoms.

Medical surveillance

In the detergent industry, the goals of the medical surveillance program is to identify employees who have become sensitized to enzymes prior to the development of symptoms and to identify problems with the enzyme industrial hygiene program.[5] All employees undergo annual pulmonary function testing, respiratory questionnaires, and monitoring by prick testing for specific IgE antibodies to enzymes. Total IgE and enzyme-specific IgE antibodies are also measured, typically by RAST.[1] Although persons with IgE antibodies to enzymes are at higher risk of developing symptoms during prolonged, repeated, or high-dose exposures, skin prick test positive workers can continue to work with enzymes as long as they remain symptom-free.[6] If an employee is found to be sensitized and symptomatic, an investigation of work practices and engineering controls should be conducted. The employee should be removed from the workplace until symptoms resolve, and the exposure is identified and corrected.

Prevention

The problem of respiratory sensitization in the detergent industry has led to the implementation of engineering controls that reduced exposures and the incidence of disease. Enzyme encapsulation, fume hoods, and other containment devices can prevent powder from becoming airborne and inhaled. When possible, enzymes should be handled in liquid preparations rather than powders to prevent airborne exposures to the respiratory tract and skin. Personal protective equipment may be used to further reduce the risk of exposure.

References

1. Sarlo K, Clark ED, Ryan CA, et al. ELISA for human IgE antibody to subtilisin A (alcalase): correlation with RAST and skin test results with occupationally exposed individuals. *J Allergy Clin Immunol* 1990; 86:393–9.
2. Tarvoinen K, Kanerva L, Tupasela O, et al. Allergy from cellulase and xylanase enzymes. *Clin Exp Allergy* 1991; 21:609–15.
3. Quirce S, Caevas M, Diez-Gome ML, et al. Respiratory allergy to *Aspergillus*-derived enzymes in bakers asthma. *J Allergy Clin Immunol* 1992; 90:970–8.
4. Novey HS, Keenan WJ, Fairshter RD, et al. Pulmonary disease in workers exposed to papain: clinico-physiological and immunological studies. *Clin Allergy* 1980; 10:721–31.
5. Sarlo K, Kirchner DB. Occupational asthma and allergy in the detergent industry: new developments. *Curr Opin Allergy Clin Immunol* 2002; 2(2):97–101.
6. Vanhanen M, Tuomi T, Tupasela O, et al. Cellulase allergy and challenge tests with cellulase using immunologic assessment. *Scand J Work Environ Health* 2000; 26(3):250–6.

FARM ANIMALS

Common names: Cows, horses, pigs, reindeer, sheep, chickens

Occupational setting

The raising of livestock occurs in various settings ranging from indoor to outdoor and small, family-run enterprises to large, commercial facilities. Reindeer herding is an important and prevalent occupation in Scandinavia. Horse exposure occurs in mounted law enforcement personnel and racetrack workers as well as agricultural workers. Livestock exposure is also a risk for veterinarians.

Exposure (route)

Exposure to farm animal antigens may be through inhalation of airborne particles or through skin contact with animals. Housekeeping tasks cause increased airborne dust concentrations, which may result in inhalation exposure. Fomite transmission of livestock allergens can occur when work clothes are brought into household living spaces.

Pathobiology

Animal proteins from farm animals may cause occupational allergy. Farm animals include various species of domesticated animals. Animals whose antigenic dander is associated with occupational allergy include horses, cows, sheep, pigs, and reindeer.

Skin contact with farm animal antigens may cause immediate or delayed skin allergy that produces dermatitis. Delayed dermatitis can be predicted to be an IgG-mediated process following skin exposure. One case report described delayed hypersensitivity in a piggery worker to pig epithelium that resulted in hand and body eczema.[1] This contact dermatitis was substantiated by positive patch test to pig epithelium.[1]

Allergic respiratory disease has been associated with exposure to farm animals including cows, horses, and reindeer. Dander from these animals is allergenic. Respiratory disorders in hog farmers are not thought to be IgE or IgG allergic illnesses[2] but rather inflammatory responses consisting of cytokine-mediated lymphocytes directed against endotoxin and dust associated with swine containment facilities.[3] Pork allergens are associated with occupational asthma in hog farmers and pork processing plant employees. In one case study, a pork processing plant employee presented with occupational asthma from Type I hypersensitivity and contact dermatitis from Type IV hypersensitivity.[4]

Poultry workers have been found to have rhinitis and asthma in association with poultry-related antigens.[5] This IgE antibody mediated allergy may result from chicken antigens such as serum, dander, and droppings. Other related antigens include the northern fowl mite antigen (see section on "Mites"). Bird fanciers and pigeon breeders have been shown to have bird-related immediate respiratory symptoms associated with immediate wheal and flare reactions. Although 19% of allergic patients reportedly react to horse allergens by intracutaneous testing, clinical disorders from horse allergy seem to be relatively infrequent.[6]

Diagnosis

Contact dermatitis from farm animal exposure may be diagnosed through a thorough medical and occupational history and physical examination along with patch skin testing. Workers with occupational dermatitis develop red, raised, itchy, often lichenified patches of skin inflammation in areas where direct skin contact with an allergen has occurred. Hand dermatitis is more common. This condition can also occur on other exposed areas, such as limbs, face, and thorax. The dermatitis can be chronic, with clearing associated with time away from the work site.

In immediate hypersensitivity disorders, rhinitis and asthma may develop following months to years of exposure. Workplace challenge can be used to demonstrate signs of allergy, such as bronchoconstriction as measured by pre- and postshift peak flow measurements. Skin prick test and specific IgE antibody measurements with reindeer epithelium have been used to document reindeer allergy in herders.[7] Skin prick test antigens from cow, sheep, goat, and horse are also available. Nasal provocation has been used to confirm cow dander allergy.[6] Skin prick test and

RAST have also been used to demonstrate antibody to poultry-related antigens in poultry workers and cow dander in farmers.

Treatment

Avoidance of antigen exposure should be the ultimate goal in the treatment of occupational animal allergies. Since this solution is normally not practical, conventional treatments for dermatitis and asthma may be tried.

Medical surveillance

Farm workers tend to be self-employed or employed in relatively small numbers per facility. Medical surveillance using allergy symptom questionnaires would be appropriate for larger employers.

Prevention

Farmers need to be educated about the risks of allergy in association with farm work. Informed workers are more likely to avoid antigen exposure. Gloves, protective clothing, and respirators may decrease exposure. Work practice controls that minimize the generation of airborne dust and the need for direct skin contact should be instituted. Additional controls include showering and changing of clothes prior to entry into the home to prevent fomite transmission. An increased risk of symptoms in winter associated with closed quarters may necessitate increased vigilance in protective measures.

References

1. Bovenschen HJ, Peters B, Koetsier MI, et al. Occupational contact dermatitis due to multiple sensitizations in a pig farmer. *Contact Dermatitis* 2009; 61(2):127–8.
2. Matson SC, Swanson MC, Reed CE, et al. IgE and IgG-immune mechanisms do not mediate occupation-related respiratory or systemic symptoms in hog farmers. *Allergy Clin Immunol* 1983; 72:299.
3. Dosman JA, Fukushima Y, Senthilselvan A, et al. Respiratory response to endotoxin and dust predicts evidence of inflammatory response in volunteers in a swine barn. *Am J Ind Med* 2006; 49(9):761–6.
4. Labrecque M, Coté J, Cartier A, et al. Occupational asthma due to pork antigens. *Allergy* 2004; 59(8):893–4.
5. Viegas S, Faísca VM, Dias H, et al. Occupational exposure to poultry dust and effects on the respiratory system in workers. *J Toxicol Environ Health A* 2013; 76:230–9.
6. Bardana EJ Jr. Occupational asthma and related conditions in animal workers. In: Bardana EJ Jr., Montanaro A, O'Hollaren MT, eds., *Occupational Asthma*. Philadelphia: Hanley & Belfus, 1992:225–35.
7. Reijula K, Halmepuro L, Hannaksela M, et al. Specific IgE to reindeer epithelium in Finnish reindeer herders. *Allergy* 1991; 46:577–81.

GRAIN DUST

Common names: Rye grass, soybeans, buckwheat, oat grass, barley

Occupational setting

Grain dust allergy occurs in any occupation where grains are handled, stored, or used. Farming, cereal manufacturing, and grain loading and unloading operations are examples of settings associated with occupational grain dust allergy. Rye flour is used in agglomerate board manufacturing glue and therefore may be an allergen in wood dust. Buckwheat hulls are sometimes used as fillers in pillow and cushion manufacturing.[1] Aspergillus-derived enzymes are used to enhance baked products.[2] Brewery workers have developed occupational asthma after exposure to grain dust in the workplace.[3]

Exposure (route)

Inhalation exposure to respirable dust is the route of exposure in the development of occupational respiratory sensitization to grain dust. Grain dust results from the abrasion of kernels during handling; it forms at an estimated rate of 3–4 lb per ton of grain handled.[4] Airborne dust measurements made during wheat and oat loading in Canada showed a mean particle size of 1.7–3.1 μm, which is respirable.[5] Some grain-associated allergens, such as cellulase and buckwheat, may also cause skin reactions from skin contact.

Pathobiology

Grain dust is a product of various grass species used primarily in food manufacturing. It contains fractured grain kernels as well as various contaminants and additives, including molds, fungi, Aspergillus-derived enzymes (high-molecular-weight proteins with catalytic activity), bacteria, insects such as grain weevil, storage mites, pollens, fractured weed seeds, and mineral particles.[2,6] Wheat dust and storage mites are covered separately in other sections of this chapter.

Because there are many allergenic components of grain dust, the etiology of grain-related allergy varies. Potential allergens range from contaminants such as insects, bacteria, and fungi to components of the grains themselves, such as soybean flour (which contains a number of antigens) and buckwheat flour. Wheat, rye, and triticale grasses are closely related species that have significant cross-antigenicity. Rye, barley, and oat are less closely related but still exhibit some cross-reactivity.[6] Allergic individuals may be sensitized to more than one grain-related antigen. Most grain-associated antigens cause an immediate-type IgE-mediated hypersensitivity that may result in grain-associated rhinoconjunctivitis and asthma. Symptoms typically occur within minutes of exposure. A late-phase aggravation of symptoms after the workday may occur. Buckwheat and cellulase have also been associated with contact urticaria. One patient with prior symptoms related to occupational buckwheat and kapok exposure experienced an anaphylactic episode after buckwheat ingestion. Community asthma epidemics have been associated with soybean loading and unloading in Barcelona, Spain.[7,8] A case of occupational allergy to oilseed rape dust has been documented.[9]

Among grain workers, grain dust exposure is a well-documented cause of nonallergic respiratory symptoms, including cough, shortness of breath, and decrease in lung function. Grain elevator workers have been found to have lower forced expiratory volume in 1 second (FEV_1) and lower forced vital capacity (FVC) than nonexposed controls on a chronic basis.[10] Severity of symptoms tends to be related to duration of exposure. Nonallergic acute grain-related respiratory disorders also occur, including an asthma-like acute syndrome and grain fever. These conditions of the bronchial airways should be considered when assessing a case of possible allergic disease from grain dust.

Diagnosis

The presence of rhinitis, conjunctivitis, sneezing, coughing, wheezing, and shortness of breath upon exposure to grain dust provides evidence for grain dust allergy. However, a medical history should be taken to exclude the existence of an underlying respiratory illness. Grain-exposed workers are known to have a high prevalence of respiratory symptoms, including productive cough, wheezing, and shortness of breath from nonallergic grain-related pathology (e.g., exposure to toxic gases, pesticides, fertilizers)

Conjunctivitis, rhinitis, edematous mucosa, and wheezing may be present on physical examination. Serial peak flow measurements may be used to establish reversibility of airflow obstruction and the temporality of symptoms in relation to exposure. Also, a trial of removal from the workplace to watch for resolution of symptoms may be useful. Demonstration of atopic status by history or prick skin test to common allergens is useful, since atopic individuals are at increased risk of some types of grain dust allergy.

Demonstrating positive skin prick test or RAST or positive bronchial challenge to grain-related antigens can support the diagnosis of grain dust hypersensitivity. Prick skin tests and reverse enzyme immunoassay have been used to demonstrate hypersensitivity to fungal alpha-amylase and cellulase in bakers who were also sensitized to wheat flour.[2] Specific IgE antibodies to some cereal grains, including wheat and rye, can be measured through commercially available tests. Prick skin test methods and specific IgE tests have been developed for a number of research applications in grain dust allergy. Prick skin tests and RAST have been used for barley and rye flour. Bronchial challenge has been used to demonstrate reactions in individuals sensitized to alpha-amylase, cellulase, and wheat flour. Because of coincident exposures in the grain industry, nonallergenic causes of respiratory pathology should be ruled out.

Treatment

Treatment for grain dust allergy focuses on the avoidance of exposure. If workers are unable or unwilling to change occupations, various standard preventive and symptomatic therapies for allergies and asthma may be instituted. However, when the allergic disorder is severe, avoidance of exposure is prudent. If specific antigens are identified to which the patient is allergic, immunotherapy could be considered; however, this is not standard treatment.

Medical surveillance

An occupational history, concentrating on recent tasks, exposures, and respiratory symptoms, may be taken as part of routine medical surveillance. Periodic spirometry is indicated in workers exposed to grain dust. Although other screening tests for surveillance purposes have not been well studied, prick skin testing or RAST should be considered.

Prevention

Occupational allergy develops following exposure to antigenic materials. Therefore, avoidance of inhalation of and skin contact with these substances is the best way to prevent the development of hypersensitivity. The American Conference of Governmental Industrial Hygienists (ACGIH) has recommended 4 mg/m^3 8-hour TWA (time-weighted average) as an exposure limit for oat, wheat, and barley dust. For buckwheat, however, allergic illness has been documented among workers exposed to as little as 1–2 mg/m^3 of airborne dust.[11] Work practice, engineering, housekeeping, and personal protective equipment controls can help to minimize dust exposure in the food and other grain handling industries. Recent studies have identified the N95 elastomeric respirator as preferable for protecting agricultural workers from aerosolized particulates of all particle size ranges.[12,13]

References

1. Heffner E, Nebiolo F, Asero R, et al. Clinical manifestations, co-sensitizations, and immunoblotting profiles of buckwheat-allergic patients. *Allergy* 2011; 66(2):264–70.
2. Quirce S, Cuevas M, Díez-Gómez M, et al. Respiratory allergy to Aspergillus-derived enzymes in bakers' asthma. *J Allergy Clin Immunol* 1992; 90:970–8.

3. Godnic-Cvar J, Zuskin E, Mustajbegovic J, et al. Respiratory and immunological findings in brewery workers. *Am J Ind Med* 1999; 35(1):68–75.
4. Chan Yeung M, Enarson D, Kennedy S. The impact of grain dust on respiratory health. *Am Rev Respir Dis* 1992; 145:476–87.
5. Williams N, Skoulas A, Merriman JE. Exposure to grain dust. I. A survey of the effects. *J Occup Med* 1964; 6:319–29.
6. Quirce S, Diaz-Perales A. Diagnosis and management of grain-induced asthma. *Allergy Asthma Immunol Res* 2013; 5(6):348–56.
7. Anto JM, Sunyer J, Rodrigues-Roisin R, et al. Community outbreaks of asthma associated with inhalation of soybean dust. *N Engl J Med* 1989; 320:1097–102.
8. Codina R, Ardusso L, Lockey RF, et al. Sensitization to soybean hull allergens in subjects exposed to different levels of soybean dust inhalation in Argentina. *J Allergy Clin Immunol* 2000; 105(3):570–6.
9. Suh CH, Park HS, Nahm, DH, et al. Oilseed rape allergy presented as occupational asthma in the grain industry. *Clin Exp Allergy* 1997; 28:1159–63.
10. Wild P, Dorribo V, Pralong J, et al. Respiratory effects of an exposure to wheat dust among grain workers and farmers: a longitudinal study. *Occup Environ Med* 2014; 71(Suppl 1):A18–9.
11. Goehte CJ, Wieslander G, Ancker K, et al. Buckwheat allergy: health food, an inhalation health risk. *Allergy* 1983; 38:155–9.
12. Lee SA, Adhikari A, Grinshpun SA, et al. Respiratory protection provided by N95 filtering facepiece respirators against airborne dust and microorganisms in agricultural farms. *J Occup Environ Hyg* 2005; 2(11): 577–85.
13. Cho KJ, Jones S, Jones G, et al. Effect of particle size on respiratory protection provided by two types of N95 respirators used in agricultural settings. *J Occup Environ Hyg* 2010; 7(11):622–7.

INSECTS

Common names: Moths, bees, beetles, cockroaches, others

Occupational setting

Exposure to insects occurs in a variety of occupational settings. Workers at risk of exposure include those who work directly with insects, those who work with materials that may be contaminated by insects, and those who work in outdoor environments where insects coincidentally live. Occupations involving direct work with insects include entomologists, lepidopterists, ecologists, aquarists, toxicologists, beekeepers, spice or dye factory workers, organic farmers, and pest control researchers. Occupations involving work with potentially contaminated materials include bakers, process and warehouse workers (e.g., honey processors, grain mill workers), silk weavers, dock loaders, and sewage treatment workers. Outdoor occupations where insects coincidentally occur include farmers, fishermen, gardeners, fire fighters, forestry workers, environmental researchers, hydroelectric plant workers, fruit pickers, and those involving waterside work.

Exposure (route)

Insect-related organic materials readily become airborne. Shedded insect exoskeletons and scales are thin, readily dried, and very friable. In environments where these and related materials are found, dust may be raised through nearly any activity. Particulates from insects may be inhaled, resulting in respiratory illnesses.

Skin contact with certain insect parts may result in irritant or allergic disorders. Such contact can arise through direct handling of insects or by contact with contaminated objects or surfaces. Bites, such as from beetles, can cause urticaria. Insect stings may also cause allergy.

Pathobiology

Insects are small, invertebrate animals that have an adult stage characterized by three pairs of legs, a segmented body with three major divisions, and usually two pairs of wings. Insect parts and by-products are the source of allergenic proteins that cause insect-related allergies and asthma.

Allergenic particles can arise from shed skeleton, scales, excretions, and secretions of insects. Caterpillar hairs and moth scales are known skin irritants that, upon repeated contact, can cause dermal sensitization. Insect hairs that are not irritants may also give rise to allergic disorders following repeated exposure. Larvae may contain the same antigenic material found in adult insects.[1]

Inhalation of insect fragments may result in respiratory sensitization, which may occur after weeks to years of exposure. Intensity and duration of exposure are important determinants of allergy development. The mechanism for most insect allergies is immediate-type IgE-related hypersensitivity. In one survey of entomologists, 25% of workers directly exposed to insects had allergic conditions.[2] In evaluating other documents that estimate incidence, 30% is a representative figure. In moth and butterfly-rearing laboratories, an incidence of allergy of 53–75% has been documented.[3]

Many insect species have been implicated in occupational allergic disorders. Some examples include moths of various species, grasshoppers, locusts, screwworms, blowflies, beetles, parasitic wasps, *Drosophila*, yellow jackets, honeybees, bumblebees, houseflies, caddis flies, cochineal insect (source of carmine dye), red midges and larvae, and bloodworms, as well as cockroaches.[4–7] Reactions that occur upon development of insect-related occupational allergy include eye symptoms, rhinitis, nasal congestion, urticaria, and often cough, wheezing, and shortness of breath. Once sensitized, workers develop illness within minutes of exposure to insect-related allergens.

The exoskeleton of many insect species is allergenic. The scales of butterflies and moths produce allergy. Hemoglobins are major allergens of red midge larvae and adults.[1] In species where feces cause sensitization, the allergen arises from gut-derived cellular material, possibly the peritrophic membrane.[8,9] Other insect by-products such as "bee dust" and cocoons are also allergenic.

Diagnosis

Diagnosis of insect allergy can be made presumptively by a thorough medical and occupational history along with physical examination. Immunological testing can be used to confirm that the suspected allergenic material will cause the signs and symptoms of the illness.

For example, prick skin tests, RAST, and bronchial provocation with honeybee whole-body extract have been used to document IgE-mediated occupational asthma in a honey processor.[10] Intracutaneous tests produce reactions in bumblebee-allergic patients.[11] ELISA has been used for the measurement of specific IgE and IgG antibodies to insect extracts from locust, mealworms, cockroaches, spring stick insects, and mulberry moon moths.[8] Specific IgE can be documented by immunoblot. Skin prick tests and bronchial challenge have been used to document occupational asthma to cockroaches in international shipping deckhands, crickets in greenhouse staff, and blowflies in researchers.[5,12,13] Nasal provocation in cockroach-sensitized workers demonstrated decreased nasal flow rates.[3] Caution must be used in interpretation of skin prick test results in chironomid-exposed workers since cross-sensitization with other insect or crustacean species may occur.[14]

Treatment

Definitive treatment of occupational insect allergies is removal from exposure. Treatment by conventional therapies such as antihistamines, bronchodilators, and inhaled steroids may be used. In one survey, 28% of individuals reporting allergy indicated that job transfer or discontinuation was necessary.[3] Desensitization injections have been successfully used to treat occupational insect allergies. Patients with anaphylactic reactions to bumblebees have been desensitized using honeybee venom.[15] Injectable adrenaline kits are indicated for workers with history of anaphylaxis.

Medical surveillance

No literature is available on the application of surveillance methods to insect-handling workers. In theory, a program including a respiratory questionnaire and examination along with pulmonary function testing would be useful. Research programs could be conducted utilizing prick skin testing or RAST methods.

Prevention

Prevention of inhalation of or skin contact with insects and insect by-products will prevent insect-related allergies. Engineering and work practice controls may reduce the amount of airborne dust in occupational settings where insects are used. This includes institution of a formal integrated pest management program at the work site, which has been shown to decrease incidence of sensitization and occupational asthma in employee populations.[16] Laboratories should be designed with appropriate air circulation, segregation of insects, and ease of maintenance and housekeeping in mind. Good hygiene is imperative.

Personal protective equipment may be useful in avoiding exposure and preventing symptoms. However, one survey found that individuals at institutions that reported no insect allergies were less likely to use protective equipment on either routine or as-needed bases than individuals at institutions that reported allergies.[8] Protective equipment may include gloves, respirators, face masks, head nets, and lab coats.

Educating workers about the risk of insect allergy is important in preventing these disorders, particularly since the illnesses are not well known. Workers may be at risk of nonoccupational exposure to insects and may have illness because of antigenic cross-reactivity. In one case report, a butcher allergic to carmine developed food allergy from exposure to the dye while manufacturing sausages, which had initially produced rhinitis and respiratory symptoms from inhalation.[17] Workers who are aware of the possibility of developing respiratory allergy are more likely to protect themselves from exposure.

References

1. Galindo PA, Feo F, Gomez E, et al. Hypersensitivity to chironomid larvae. *Invest Allergol Clin Immunol* 1998; 8(4):219–25.
2. Bauer M, Patnode R. NIOSH HHE Report No. HETA-81-121-1421, Insect Rearing Facilities, Agricultural Research Service, U.S. Department of Agriculture, Cincinnati, OH, 1984. Available at: http://www.cdc.gov/niosh/nioshtic-2/00149683.html (accessed on July 13 2016).
3. Wirtz RA. Occupational allergies to arthropods – documentation and prevention. *Bull Entomol Soc Am* 1980; 26:356–60.
4. Arruda LK, Vailes LD, Ferriani VP, et al. Cockroach allergens and asthma. *J Allergy Clin Immunol* 2001; 107(3):419–28.
5. Oldenburg M, Latza U, Baur X. Occupational health risks due to shipboard cockroaches. *Int Arch Occup Environ Health* 2008; 81(6):727–34.
6. Focke M, Hemmer W, Wöhrl S, et al. Specific sensitization to the common housefly (*Musca domestica*) not related to insect panallergy. *Allergy* 2003; 58(5):448–51.
7. Linares T, Hernandez D, Bartolome B. Occupational rhinitis and asthma due to crickets. *Ann Allergy Asthma Immunol* 2008; 100(6):566–9.
8. Edge G, Burge PS. Immunological aspects of allergy to locusts and other insects. *Clin Allergy* 1980; 10:347.
9. Tee RD, Gordon DJ, Hawkins ER, et al. Occupational allergy to locusts: an investigation of the sources of the allergen. *J Allergy Clin Immunol* 1988; 81(3):517–25.

10. Ostrom NK, Swanson MC, Agarwal MK, et al. Occupational allergy to honey bee-body dust in a honey-processing plant. *J Allergy Clin Immunol* 1986; 77:736–40.
11. de Groot H. Allergy to bumblebees. *Curr Opin Allergy Clin Immunol* 2006; 6(4):294–7.
12. Lopata AL, Fenemore B, Jeebhay MF, et al. Occupational allergy in laboratory workers caused by the African migratory grasshopper *Locusta migratoria*. *Allergy* 2005; 60(2):200–5.
13. Kaufman GL, Baldo BA, Tovey ER, et al. Inhalant allergy following occupational exposure to blowflies. *Clin Allergy* 1986; 16:65–71.
14. Galindo PA, Lombardero M, Mur P, et al. Patterns of immunoglobulin E sensitization to chironomids in exposed and unexposed subjects. *Invest Allergol Clin Immunolog* 1999; 9(2):117–22.
15. Kochuyt AM, Van Hoeyveld E, Stevens EAM. Occupational allergy to bumble bee venom. *Clin Exp Allergy* 1993; 23:190–5.
16. Portnoy J, Chew GL, Phipatanakul W, et al. Environmental assessment and exposure reduction of cockroaches: a practice parameter. *J Allergy Clin Immunol* 2013; 132(4):802–8.
17. Ferrer A, Marco FM, Andreu C, et al. Occupational asthma to carmine in a butcher. *Int Arch Allergy Immunol* 2005; 138(3):243–50.

LABORATORY ANIMALS

Common names: Rats, mice, guinea pigs, rabbits, hamsters, monkeys

Occupational setting

Laboratory animal allergy (LAA) is an important work-related illness that occurs in 11–44% of exposed workers. It occurs in workers who have laboratory animal contact ranging from casual, indirect exposure, through infrequent animal handling, to full-time daily care of animals and their housing facilities. Laboratory animals are housed and handled at many types of research institutions, including academic centers, medical schools, and private sector research facilities such as pharmaceutical companies. LAA may also result from work in other settings where rodents are present such as pet shops, vivariums, and veterinary offices.

Exposure (route)

Laboratory animal allergy may result from skin exposure or respiratory exposure. Percutaneous exposures may result from animal bites or allergen contamination of wounds, which may result in anaphylaxis. Different routes of exposure result in different disorders with different mechanisms.

Exposure to allergens may result from direct contact through handling contaminated animal bedding and laboratory animals as well as from indirect contact through airborne dust.

Inhalation exposure may result from general contamination of air within the facilities or from airborne dust created through animal handling or care. Allergens, such as those from rodent urine, may contaminate animal cage bedding and other materials, and these allergens may become airborne through any type of disturbance. Additionally, allergens are carried on animal fur and dander and easily become airborne when the animal is handled.

Any factors that increase the concentration of allergens in the workplace also increase potential exposure. These factors include increased numbers of animals, decreased frequency of cage changing, increased manipulation of animals, and accumulation of dust. Cage washing and cleaning are associated with significantly increased environmental allergen concentration. Using a vacuum without a HEPA filter can result in allergens being expelled into the environment.

Pathobiology

Laboratory animals reared for use in research consist mainly of rodents. Those laboratory animals best studied regarding occupational allergy include mice, rats, guinea pigs, hamsters, and rabbits. In addition, frogs and monkeys that are laboratory reared may also cause occupational allergy.

Exposure to laboratory animals and associated materials may result in development of immediate-type IgE-mediated respiratory hypersensitivity. Immediate skin reactions including wheal and flare from skin contact (contact urticaria) may also occur.

The prevalence of allergy among lab animal handling workers is ~11–44%. About 4–22% of these workers develop asthma.[1] Associations exist between the level of allergen exposure and the development of laboratory animal allergy.[1,2] Evidence that exposure–response relationships exist had been reported with respect to rat allergy.[2] Although many cases have been reported that developed more than 20 years after initial exposure, most laboratory animal allergies develop within the first few years of exposure. It has been suggested that of those who develop laboratory animal allergies, 30% do so within 1 year of employment.[1] This percentage increases to 70% within 3 years of employment.

The intensity of allergen exposure may be a far greater determinant in the development of laboratory animal allergies compared to duration of exposure.[3] Sensitized workers are more likely than nonsensitized workers to develop pulmonary symptoms[3] and secondary hypersensitization to additional laboratory animals.[4] These findings suggest that decreasing the level of exposure will limit sensitization. Due to a greater emphasis on workplace controls, the incidence of progression to occupational asthma secondary to laboratory animal allergy has decreased from a peak in the early 1980s.[5]

Allergy typically presents as rhinitis and conjunctivitis, which progresses into asthma in a minority of patients. Symptoms typically occur within 5–30 minutes after exposure. Asthma may occur immediately or may exhibit dual or delayed patterns. Skin reactions include contact urticaria from direct contact of the animal with exposed skin and a maculopapular pruritic rash on exposed skin in association with airborne exposure and respiratory symptoms.

The antigens responsible for laboratory animal allergy come from urine, dander, saliva, and serum. These antigens are typically small acidic glycoproteins with molecular weights of 15–30000 kDa.[6,7] For mice, rats, rabbits, and guinea pigs, urinary proteins and saliva are thought to cause the majority of hypersensitivity problems. The main allergen in mice is a urinary protein, possibly prealbumin,[1] whereas the main allergens in rats are serum albumin, α2-urinary globulins, and prealbumin.[3] Monkey dander has caused occupational asthma in researchers.[8] The allergenic proteins share sequence homology with proteins from the *Schistosoma* parasite, which may explain why they can trigger an immune response.[1]

The presence of several general allergic symptoms and several historical indicators of atopy is moderately predictive of the new onset of laboratory animal allergy.[1] IgE antibody is produced in response to allergen exposure. IgE antibodies bind allergen and this complex binds to mast cells. Absorption of the corresponding allergen triggers the release of histamine and other mediators from the sensitized mast cell. Other antigens, such as those derived from molds or mites, cause laboratory animal-related allergies and are found in the same settings. When evaluating a suspected case of LAA, these other allergens should also be considered as they may be the cause or symptoms, or the worker could be allergic to more than one source.

Several human leukocyte antigen (HLA) and the major histocompatibility complex (MHC) genes are associated with the laboratory animal allergy and sensitization.[9] HLA class II molecules are involved in the presentation of allergen to the T cell. HLA-DR7 was found to be strongly associated with sensitization, respiratory symptoms at work, and sensitization with symptoms, while HLA-DR3 was found to be protective from sensitization.[9]

High levels of allergen-specific IgG have been associated with clinical efficacy in immunotherapy studies. Among workers with detectable mouse IgE, higher mouse IgG, and mouse IgG4 levels are associated with a decreased risk of mouse-related symptoms.[10] IgG4 is postulated to block IgE-antigen binding complexes. Laboratory animal workers with high levels of IgG4 were found to have less circulating IgE-antigen binding complexes and the absence of pulmonary symptoms in the setting of the highest laboratory animal allergen exposures.[11]

Diagnosis

A clinical diagnosis of LAA can be made based on a careful medical and occupational history along with a physical examination. A detailed respiratory and dermatologic history will reveal symptoms of allergic disorders such as wheal and flare reaction to contact such as a rat's tail wrapping around the hand. Respiratory symptoms may include rhinitis, conjunctivitis, sneezing spells, nasal congestion, cough, shortness of breath, and wheezing.

It is critical to document the fact that the symptoms are temporally associated with exposure to laboratory animal antigens and that they resolve when the individual is away from work. Objective evidence of temporality may be obtained by doing a physical examination or peak flow measurements prior to exposure and again at the end of the work shift. The finding of objective evidence of bronchoconstriction, such as wheezes or a decreased FEV_1 and FVC, supports the diagnosis of occupational asthma.

Appropriate immunological testing for IgE antibodies to laboratory animal allergens are used to confirm a diagnosis. Skin prick testing with relevant allergens or RAST may be used to demonstrate IgE antibody. RAST to detect mouse- and rat-specific IgE are commercially available and have been developed for other animal allergens. Nonlaboratory animal antigens, such as house dust mite or molds, may be used to rule out disease from these animal-associated allergens or to establish coincidental allergies. Bronchial challenge testing with animal antigens can be used to confirm reversible airway constriction related to exposure.

Treatment

Avoidance of exposure to animal antigens should be the goal of treatment. Sensitized workers who have rhinitis should be counseled on their risk of developing secondary allergy, asthma, and the risk of eventual intolerance of exposure.

Symptomatic care may include antihistamines or corticosteroids for upper tract symptoms and bronchodilators, corticosteroids, and theophylline for asthma. Cromolyn may prevent asthma from allergen exposure. Long-acting bronchodilators may be tried as preventive therapy. Immunotherapy is not likely to be helpful for most sensitized individuals who remain in an environment where regular exposure occurs.[12,13]

Medical surveillance

Recent literature supports the notion that atopy predisposes to laboratory animal allergy and is an important risk factor to assess in medical surveillance.[1,14] One review reports a statistically significant odd ratio of 3 : 2 for the development of laboratory animal allergies in atopic individuals compared to nonatopic individuals.[1] Conversely, it has been suggested that HLA-B16 may confer a protective effect against the development of LAA.[15] Other theoretical protective factors include early childhood exposure to dogs, which is reported in mouse to modulate the immune system against allergic disease development.[16]

Skin prick testing and RAST have been suggested as surveillance methods that may prove useful in detecting preclinical allergy to laboratory animals; however, these would only be useful in the setting of significant ongoing exposure and cannot be recommended for routine use. Respiratory history may be used to document early symptoms of laboratory animal allergy. If nasal symptoms are detected early and further exposure is eliminated, subsequent progression to asthma or development of secondary allergies may be prevented.

Prevention

Containment of the source protein is the cornerstone of animal allergy prevention measures. A comprehensive program including education for engineering controls and administrative controls, use of personal protective equipment, and medical surveillance can prevent the development of laboratory animal allergy.[17] Efforts should be made to limit airborne dust, such as by housing and transporting animals in filter-top cages or separately ventilated cages. For shaving animals, a wet prep can be used if it does not lower their body temperature too far or a shaver with an attached HEPA-filtered vacuum can be used. Room vacuums should have HEPA-filtered exhaust. Engineering controls such as negative pressure rooms, construction of clean corridors, provision of high ventilation rates with limited recirculation of air, use of downdraft (ceiling to floor) ventilation, use of robots or mechanical systems for cage cleaning and washing, and installation of HEPA filters are helpful.

Airborne antigen levels have been found to be related to litter type and stock density. In one study, significant reduction in rat allergen concentrations were achieved by replacing wood-based (sawdust) contact litter with noncontact absorbent pads.[5] The concentration of airborne animal allergen is dependent on the activities in the contaminated areas. For example, rate allergen concentrations in air have been found to be 10–100 times higher during animal handling or disturbance of bedding than at quiet times.[2]

Because allergens from laboratory animals adhere to all types of particulate matter, contamination of facilities occurs easily, as evidenced by the presence of antigen in all types of dust. Therefore, a reservoir of allergens exists outside the immediate animal housing areas.[12] Removal of animals from work areas will not eliminate exposure. In homes where pets are removed, it takes 4–6 months before allergen levels are at clinically insignificant levels.

Personal protective equipment can be used to supplement engineering and hygiene controls. Gloves and laboratory coats should be worn routinely for animal handling. Respiratory protection should be required in situations when engineering controls are not adequate to prevent exposure because of the high incidence of LAA. Requiring the use of hairnets reduces the transfer of laboratory animal allergens from an employee's workplace to home.[18] Protective equipment should be confined to the animal facility. Workers should remove protective equipment and wash their hands before leaving to perform other activities or go to common areas such as the cafeteria.

Many animal-allergic workers also have household pets, which may complicate treatment. Whenever possible, pets in the home should be removed. It is possible that cromolyn or combined treatment with inhaled corticosteroids and long-acting bronchodilators may prevent asthma exacerbations.

References

1. Bush RK, Stave GM. Laboratory animal allergy: an update. *ILAR J* 2003; 44(1):28–51.
2. Hollander A, Heederick D, Doekes G. Respiratory allergy to rats: exposure–response relationships in laboratory animal workers. *Am J Resp Crit Care Med* 1997; 155:562–7.
3. Nieuwenhuijsen MJ, Putcha V, Gordon S, et al. Exposure–response relations among laboratory animal workers exposed to rats. *Occup Environ Med* 2003; 60(2):104–8.
4. Goodno LE, Stave GM. Primary and secondary allergies to laboratory animals. *J Occup Environ Med* 2002; 44(12):1143–52.
5. Folletti I, Forcina A, Marabini A, et al. Have the prevalence and incidence of occupational asthma and rhinitis because of laboratory animals declined in the last 25 years? *Allergy* 2008; 63(7):834–41.
6. Gordon S, Tee RD, Lowson D, et al. Reduction of airborne allergenic urinary proteins from laboratory rats. *Br J Ind Med* 1992; 49:416–22.
7. Heederik D, Doekes G, Nieuwenhuijsen MJ. Exposure assessment of high molecular weight sensitisers: contribution to occupational epidemiology and disease prevention. *Occup Environ Med* 1999; 56(11):735–41.
8. Petry RW, Voss MJ, Kroutil LA, et al. Monkey dander asthma. *Allergy Clin Immunol* 1985; 75:268–71.
9. Jeal H, Draper A, Jones M, et al. HLA associations with occupational sensitization to rat lipocalin allergens: a model for other animal allergies? *J Allergy Clin Immunol* 2003; 111(4):795–9.
10. Matsui EC, Diette GB, Krop EJ, et al. Mouse allergen-specific immunoglobulin G4 and risk of mouse skin test sensitivity. *Clin Exp Allergy* 2006; 36(8):1097–103.
11. Jones M, Jeal H, Schofield S, et al. Rat-specific IgG and IgG4 antibodies associated with inhibition of IgE-allergen complex binding in laboratory animal workers. *Occup Environ Med* 2014; 71(9):619–23.
12. Eggleston PA, Wood KA. Management of allergies to animals. *Allergy Proc* 1992; 13:289–92.
13. Bush RK. Assessment and treatment of laboratory animal allergy. *ILAR J* 2001; 42(1):55–64.
14. Hollander A, Doekes G, Heederik D. Cat and dog allergy and total IgE as risk factors of laboratory animal allergy. *J Allergy Clin Immunol* 1996; 98:545–54.
15. Sjostedt L, Willers S, Orbaek P. Human leukocyte antigens in occupational allergy: a possible protective effect of HLA-B16 in laboratory animal allergy. *Am J Indust Med* 1996; 30:415–20.

16. Fujimura KE, Demoor T, Rauch M, et al. House dust exposure mediates gut microbiome Lactobacillus enrichment and airway immune defense against allergens and virus infection. *Proc Natl Acad Sci USA* 2014; 111(2):805–10.
17. Stave GM, Darcey DJ. Prevention of laboratory animal allergy in the United States: a national survey. *J Occup Environ Med* 2012; 54(5):558–63.
18. Krop EJ, Doekes G, Stone MJ, et al. Spreading of occupational allergens: laboratory animal allergens on hair-covering caps and in mattress dust of laboratory animal workers. *Occup Environ Med* 2007; 64(4):267–72.

MITES

Common names: Dust mites, storage mites, red spider mites, citrus red mites

Occupational setting

Occupational illness from exposure to mites may occur in a variety of settings. Workplaces associated with mite-related illnesses include warehouses, barns, poultry houses, greenhouses, flower farms, fruit orchards, livestock farms, grain storage facilities, bakeries, and animal housing facilities.

Exposure (route)

Occupational allergy to mites results from inhalation exposure or skin contact. Because mites are tiny creatures, entire animals can be airborne and readily inhaled. In addition, skin exposure to mite-infested materials may result in contact allergy, occupational dermatitis, or contact urticaria.

Pathobiology

Mites are a class of arthropods. Storage mites are wingless, translucent, microscopic invertebrates. Dust mites, or pyroglyphid mites, are more abundant than nonpyroglyphid or storage mites. In optimal conditions, mites can self-multiply at rates of four- to 10-fold weekly.[1]

Various mite species are ubiquitous in our environment. Dust mites infest organic particles in dusts such as mattress, bedding, pillow, carpet, or floor dust. Storage mites infest stored food and vegetable products as well as hay and straw. Red spider mites, *Tetranychus* species, parasitize flowers such as carnations in greenhouse cultivation as well as in open fields. Mites, especially storage types, thrive in damp environments. The major species in infestations vary by season.

Inhalation of mites or their by-products may result in immediate-type IgE-mediated respiratory hypersensitivity. Mites as well as their feces are allergenic.[2] Mites that have been implicated in occupational allergy include storage mites, grain mites, dust mites, red spider mites, citrus red mites, fowl mites, and hay itch mites. Spider silks have also been implicated as workplace allergens. Exposure to antigens from these arachnids may cause rhinoconjunctivitis and asthma upon exposure in sensitized individuals. With *Dermatophagoides pteronyssinus*, late-phase asthma (occurring hours after exposure) is associated with a Th17 inflammatory response mediated by IL-17.[3] Acute hypersensitivity reactions to house dust mites and storage mites are orchestrated by a Th2-mediated inflammatory response that stimulates B-cells to produce IgE antibodies.[4] Contact allergy has been reported with two-spotted spider mites and red spider mites.[5] Case reports have described occupational allergy to cheese mites (*Blomia kulagini*), chorizo mites (*Euroglyphus maynei*), and salty ham mites (*Tyrophagus putrescens*).[6]

Storage mites contain the most important antigens that have been related to asthma and allergic rhinoconjunctivitis among grain farmers.[7] A strong association between storage mite allergy and house dust allergy has been noted in areas or occupations with high exposure to both.[8] In one survey, >80% of patients thought to be at risk of occupational exposure had storage mite allergy. Bakers appear to be excluded from this risk as recent research indicates that cosensitization to storage mites is less likely to exist among bakers who are allergic to wheat flour.[9]

In one study, *D. pteronyssinus* was the most allergically potent of four mite species evaluated. The *Dermatophagoides* genus provides the major allergens of house dust. Up to three-quarters of serum IgE to mites are directed against antigen P_1, which is associated with fecal particles.[2] Cross-reactivity may occur between several species of house dust mite. There may also be shared antigens among storage mite species.

Diagnosis

Medical history may reveal symptoms of nasal and ocular itching, rhinorrhea, sneezing, shortness of breath, wheezing, and dry or productive cough.[7] Mite-induced skin disorders include contact urticaria and dermatitis. These symptoms in association with exposure to mite-infested material suggest the diagnosis of allergy. Immunological testing may provide supportive evidence for IgE-mediated hypersensitivity, or, in the case of contact dermatitis, delayed-type allergy.

Skin prick test materials are available for house dust mite (*Dermatophagoides farinae* and *D. pteronyssinus*), storage mite (*Acarus siro*, *Tyrophagus putrescentiae*, *Glycyphagus domesticus*, and *Lepidoglyphus destructor*), and northern fowl mite. Skin prick testing may be the most sensitive immunologic test in confirming a diagnosis of occupational mite allergy. Skin patch testing (Finn Chamber, urticaria-inducing open test, and acute eczema-inducing open test) is used to detect allergic contact dermatitis or overlapping contact cutaneous syndrome to red spider mite.[10] Specific

IgE can be measured by RAST for house dust mites, storage mites, and northern fowl mites. ELISA and EAST have been used to document citrus mite and red spider mite allergy, respectively. Correlation between skin prick testing with RAST and skin prick/RAST testing with symptoms is variable. Conjunctival provocation can be used to substantiate rhinoconjunctivitis from specific allergens. Bronchial challenge with mite antigen may cause immediate, dual type, or delayed reactions.[7] Methacholine challenge, documenting bronchial hyperreactivity, may be useful but must be conducted by a clinician with expertise in the procedure.

Treatment

Avoidance of further exposure to allergen will prevent symptoms of mite allergy. Since changing vocations is often difficult, medical management of symptoms may be tried, such as conventional therapies including antihistamines and inhaled beta agonists. Immunotherapy has been used with some success. Pretreatment with cromolyn sodium can prevent the occurrence of asthma from exposure to mite allergens.[3] Because of the abundance of various mites in the environment, attempts to avoid exposure must extend to nonoccupational settings.

Medical surveillance

Respiratory symptoms can be assessed periodically in mite-exposed workers. Skin prick testing and RAST have no proven utility in surveillance for occupational arachnid allergy but may be considered on an experimental basis.

Prevention

Approximately 30% of allergic farm workers have symptoms consistent of allergy, hay fever, or respiratory problems to spider mites, and the intensity of symptoms is associated with high levels of total and mite allergen-specific IgE.[11,12] This finding suggests that atopic individuals may be at increased risk of mite allergy. Also, sensitization to house dust mites may indicate increased susceptibility to spider mite allergy.[11,12] Preventive measures should be aimed at reducing the mite population in the workplace. Intensification in such fields as poultry husbandry may have actually increased the concentration of mites in some workplaces.[13] Open field cultivation may be associated with a lower prevalence of clinical spider mite allergy than greenhouse environments because of ventilation.[14,15] To decrease mite populations, decreasing reservoirs, decreasing humidity, and using miticides may be helpful. HEPA filters on air ducts and vacuum cleaners can be used. Once the population has been minimized through housekeeping measures and other types of dust control, secondary means of preventing exposure can be undertaken. Secondary exposure control should aim at minimizing the opportunities for mites and related dusts to become airborne.

Educating workers about the risk of allergy to mites is important in preventing the illness. Workers who are aware of the risks are better prepared to prevent exposure. Home-related exposures should be assessed in workers with occupational allergic symptoms to mites.

References

1. Wraith DG, Cunningham AM, Seymour WM. The role and allergenic importance of storage mites in house dust and other environments. *Clin Allergy* 1979; 9:545–62.
2. Tovey ER, Chapman MD, Platt Mills TXE. Mite faeces are a major source of house dust allergens. *Nature* 1981; 289:592–3.
3. Bajoriuniene I, Malakauskas K, Lavinskiene S, et al. Th17 response to *Dermatophagoides pteronyssinus* is related to late-phase airway and systemic inflammation in allergic asthma. *Int Immunopharmacol* 2013; 17(4):1020–7.
4. Yu SJ, Liao EC, Tsai JJ. House dust mite allergy: environment evaluation and disease prevention. *Asia Pac Allergy* 2014; 4(4):241–52.
5. Wirtz RA. Occupational allergies to arthropods – documentation and prevention. *Bull Entomol Soc Am* 1980; 26:356–60.
6. Armentia A, Fernandez A, Perez-Santos C, et al. Occupational allergy to mites in salty ham, chorizo, and cheese. *Allergol et Immunopathol* 1994; 22:152–4.
7. Armentia A, Tapias J, Bowber D, et al. Sensitization to the storage mite *Lepidoglyphus destructor* in wheat flour respiratory allergy. *Ann Allergy* 1992; 68:398–406.
8. Morales M, Iraola V, Leonor JR, et al. Different sensitization to storage mites depending on the co-exposure to house dust mites. *Ann Allergy Asthma Immunol* 2015; 114(1):36–42.
9. Droste J, Myny K, Van Sprundel M, et al. Allergic sensitization, symptoms, and lung function among bakery workers as compared with a nonexposed work population. *Occup Environ Med* 2003; 45(6):648–55.
10. Astarita C, Di Martino P, Scala G, et al. Contact allergy: another occupational risk to *Tetranychus urticae*. *J Allergy Clin Immunol* 1996; 98:732–8.
11. Gargano D, Romano C, Manguso F, et al. Relationship between total and allergen-specific IgE serum levels and presence of symptoms in farm workers sensitized to *Tetranychus urticae*. *Allergy* 2002; 57(11):1044–7.
12. Jeebhay MF, Baatjies R, Chang YS, et al. Risk factors for allergy due to the two-spotted spider mite (*Tetranychus urticae*) among table grape farm workers. *Int Arch Allergy Immunol* 2007; 144(2):143–9.
13. Rimac D, Macan J, Varnai VM, et al. Exposure to poultry dust and health effects in poultry workers: impact of mould and mite allergens. *Int Arch Occup Environ Health* 2010; 83(1):9–19.
14. Burches E, Pelaez A, Morales C, et al. Occupational allergy due to spider mites: *Tetranychus urticae* (Koch) and *Panonychus citri* (Koch). *Clin Exp Allergy* 1996; 26:1262–7.
15. Kronqvist M, Johansson E, Kolmodin-Hedman B, et al. IgE-sensitization to predatory mites and respiratory symptoms in Swedish greenhouse workers. *Allergy* 2005; 60(4):521–6.

PLANTS

Common names: Poison ivy (*Toxicodendron radicans*), eastern poison oak (*Toxicodendron quercifolium*), western poison oak (*Toxicodendron diversilobum*), poison sumac (*Toxicodendron vernix*), coffee beans, tobacco, psyllium, tea leaves, ipecac, colophony, others (natural rubber latex is covered in Chapter 29)

Occupational setting

Plant allergies may result from exposure in agricultural settings such as outdoor farming areas. In addition, facilities that process plant materials may provide a setting for the development of occupational allergy. Tobacco and garlic farmers, coffee bean processors, tea blenders, plant-leasing farm workers, plant wholesalers, woodwork teachers, spice processors, paper and rubber workers, florists, gardeners, and beekeepers may have exposure to plant-derived antigens. Exposure to plant products occurs in the pharmaceutical industry and in healthcare settings. For example, ispaghula husks (psyllium) and senna pods are used as bulk laxatives. In addition to these settings, occupations where workers may be exposed to natural vegetation also provide an opportunity for plant-derived antigen exposure. Examples of outdoor occupations where workers may be exposed to poison ivy or poison oak include forest rangers (and other forest workers), surveyors, and utility company field workers.

Exposure (route)

Plant allergy results from dermal or respiratory exposure. For those substances that cause skin allergy, exposure occurs via direct contact of the plant material with exposed skin. Alternatively, contamination of objects such as clothing and tools with antigen such as that from rhus plants (of the family *Anacardiaceae*) may result in exposure when these objects come into contact with skin. Inhalation of plant-derived allergens is necessary for the development of respiratory sensitization to plants.

Pathobiology

The plant kingdom contains organisms with a vast array of characteristics, and many plants have sensitizing capabilities. The plants considered here are ones that have been documented to be capable of inducing occupational allergy. Included are tobacco, tea, poison oak and ivy, colophony from *Pinus* species, ispaghula, senna, *ipecacuanha*, natural rubber latex, and others. Many other plants have been implicated in occupational allergy including spathe flower, saffron, compositae species (sesquiterpene lactone allergen: e.g., chicory, camomile), weeping fig, Christmas cactus, carnation, umbrella tree, mugwort, alstroemeria, and narcissus.[1–4] Cross-sensitization to other flowering plants is common.[5,6]

Exposure to plant materials may cause skin or respiratory allergy. There may be some overlap of respiratory allergy and skin symptoms; for example, contact or generalized urticaria can occur along with respiratory symptoms in IgE-mediated disorders.

Contact dermatitis is caused by a delayed IgG-mediated hypersensitivity. It is manifested by an itchy, red, raised skin rash that results from repeated contact with an allergen. Once allergy has developed, contact dermatitis occurs one to several days after exposure. Repeated allergen exposure may cause a chronic dermatitis characterized by thickening and lichenification of the skin. Occupational contact dermatitis typically involves the hands and other exposed skin such as arms and face.

Toxicodendron (*Rhus*) dermatitis from poison ivy, poison oak, and poison sumac may occur in outdoor workers. *Rhus* rash occurs within 48 hours after exposure in sensitized individuals. It is characterized by intense itching followed by inflammation and grouped vesicles resulting from contact with plant oil. No spread of the dermatitis occurs after the oil has been removed; however, new patches may occur because of the delayed nature of this disorder. Attacks usually last 2–3 weeks, with patches becoming crusted and dry. The antigen involved in *Rhus* dermatitis is pentadecylcatechol, a component of the oleoresin urushiol.[7]

Colophony is a resin that is derived from *Pinus* and other species of trees. Rosin may be obtained from living trees or may be a by-product of paper pulp manufacturing.[8] It is used as a resin in solder. Colophony is an important cause of contact allergy, especially hand eczema. Positive patch test results are frequent.

Propolis, a sticky, resinous material collected by bees from the bud scales of plants and trees, is a well-known cause of allergic eczematous contact dermatitis in beekeepers.[9] The major allergens in propolis seem to come from poplar species, including 1,1-dimethylallyl caffeic acid ester from poplar bud extracts.

In one case report, a saxophonist developed cheilitis due to a musical reed. In this patient, a skin prick test to cane reed scraping was positive, indicating an IgE-mediated dermatitis. Contact skin allergy to reed has been noted in other types of musicians, as well as workers who handle the reed *Arundo donax*.[10]

Although reports of cutaneous reactions are rare, olive oil has been reported to cause contact allergy, including cases of hand eczema in an aromatherapy masseuse,[11] and occupational asthma in an olive oil mill worker.[12] The sensitizers in olive oil are largely unknown, but thaumatin-like proteins, which are found in plants and in pollen, are thought to play a role in sensitization.[12] Coffee bean dust may cause occupational contact dermatitis. Dandelion, a member of the daisy family, has been reported to cause allergic contact dermatitis in gardeners.[13] Cross-reactivity between dandelions and other allergenic *Compositae* may occur.

Occupational asthma has been found to occur from tea leaves, ficus, ispaghula, senna, green tobacco, green coffee beans, baby's breath (*Gypsophila*), and many other plants. These hypersensitivities are in general of the Type I IgE-mediated type.

Asthma from inhalation exposure to green tobacco leaf has been documented in tobacco workers. In one study, green tobacco leaf asthma was found to result from an immediate, IgE-mediated hypersensitivity, as evidenced by positive RAST, nasal and bronchial provocation, and histamine release assay.[14] Because this illness is related to green tobacco exposure, it is postulated that the antigen degrades in the curing process. Other allergens found in association with tobacco plants—for example, microfungi—may also cause asthma. Asthma associated with green tobacco is a distinct entity from green tobacco sickness. Green tobacco sickness is a form of nicotine poisoning resulting from dermal absorption during the handling of wet plant leaves.

Respiratory sensitization (IgE mediated) from green coffee beans may result in upper respiratory symptoms as well as asthma. As with tobacco, the processing of coffee beans destroys some antigen, although roasted coffee allergy also occurs. An estimated 10% of exposed workers develop allergy, a small proportion of who also develop asthma. Immediate- and delayed-type allergy may coexist in coffee bean-sensitized workers.[15]

Inhalation allergy to pharmaceutical products such as psyllium (from ispaghula husks) and senna pods seems to cause asthma less frequently than other respiratory allergies. One cross-sectional study found a prevalence of asthma of 3.2% in a population of whom 7.6% were allergic to ispaghula.[16] Anaphylaxis and eczema have also been reported following ispaghula husk exposure.

Descriptions in the literature of respiratory allergy and asthma from plants are varied and may be documented in only an article or two for each plant. For example, a few studies have described mushroom workers' lung, a hypersensitivity pneumonitis that may result from mushroom spores or microorganisms.[17,18] Farm workers who harvest garlic bulbs and spice factory workers have been shown to have garlic allergy. Pectin has been found to cause IgG4-mediated occupational asthma in a candy maker.[19] Pectin is a large-molecular-weight product of fruits and fruit rinds that contains methyl-esterified galacturonan, galactan, and araban. Tea leaf allergy typically occurs at facilities where tea leaves are mixed together or blended, generating a fine dust.

Diagnosis

Medical and occupational history coupled with physical examination may be used to make a presumptive diagnosis of contact dermatitis from plants. The presence of an itchy, red skin rash occurring hours or days after contact with allergenic material should lead to a suspicion of allergic dermatitis. Acute eruptions are usually vesicular and edematous. *Rhus* contact occurs in a linear fashion, such as from brushing against twigs or leaves. Rashes associated with these plants are usually linear.

Patch skin testing should usually be done to confirm a diagnosis of allergic contact dermatitis when the causative agent is in question or if the consequences of the diagnosis may affect decisions about the risk of further exposure. Usually, the allergen is applied to the skin for 24–48 hours, after which a delayed rash appears in positive cases. Care must be taken to distinguish between irritant and allergic causes. Other allergenic exposures such as insects and mites should be considered in appropriate settings. Castor bean allergy may complicate the diagnosis of green coffee bean allergy; onion or pollen allergy may be comorbid in garlic allergic patients.[20]

Respiratory sensitization to plants results in similar symptoms to those found with other inhalant allergens. These symptoms, which must be temporally associated with exposure, include rhinitis, conjunctivitis, nasal congestion, cough, wheezing, and shortness of breath.

Skin prick tests and RAST, among other immunological studies, may be used to confirm plant allergies of the immediate, respiratory type. Pulmonary function testing is also helpful in diagnosing inhalant allergy. Bronchial provocation has been used to demonstrate coffee bean, garlic, and tea leaf allergy, among others. Significant differences have been documented in peak expiratory flow rates between tobacco workers and unexposed controls.[21] Coffee bean-exposed workers may have a higher than average incidence of chronic respiratory problems.

Treatment

For allergic contact dermatitis, treatment with local topical corticosteroid creams or ointments is recommended. For more extensive and severe cases, systemic steroids should be used. For *Rhus* dermatitis, standard steroid protocols usually begin with 40–60 mg of prednisone orally in a single daily dose with a 3-week taper.[7] Calamine lotion may ease the symptoms. Local antihistamine and anesthetic ointments should be avoided because of the possibility of contact sensitization.

Allergic respiratory disorders from plant exposure may be treated with conventional allergy therapies. These include antihistamines, bronchodilators, and steroids. Avoidance of exposure is the only definitive treatment.

Prevention

Atopy may be a risk factor in green coffee bean allergy.[22] Protecting workers from skin and respiratory exposure can prevent occupational plant allergies. Engineering controls and personal protective equipment as well as sound work practices may be helpful. Persons with latex allergy of the immediate type should be removed from exposure because of the risk of serious reactions. Substitute gloves are available.

Use of mechanical blending processes with extraction ventilation rather than hand mixing reduces personnel exposure to tea leaf dust. Wearing cotton glove liners reduces contact with skin sensitizers and therefore prevents allergy. Personal protective equipment to prevent inhalation and skin exposure to airborne materials, such as respirators and barrier clothing, should also be considered.

References

1. Pirson F, Detry B, Pilette C. Occupational rhinoconjunctivitis and asthma caused by chicory and oral allergy syndrome associated with bet v 1-related protein. *J Investig Allergol Clin Immunol* 2009; 19(4):306–10.
2. Rudzki E, Rapiejko P, Rebandel P. Occupational contact dermatitis, with asthma and rhinitis, from camomile in a cosmetician also with contact urticaria from both camomile and lime flowers. *Contact Dermatitis* 2003; 49(3):162.
3. Sánchez-Fernández C, González-Gutiérrez ML, Esteban-López MI, et al. Occupational asthma caused by carnation (*Dianthus caryophyllus*) with simultaneous IgE-mediated sensitization to *Tetranychus urticae*. *Allergy* 2004; 59(1):114–5.
4. Grob M, Wuthrich B. Occupational allergy to the umbrella tree (*Schefflera*). *Allergy* 1998; 53:1008–9.
5. deJong NW, Vermeulen AM, Gerth van Wijk R, et al. Occupational allergy caused by flowers. *Allergy* 1998; 53:204–9.
6. Akpinar-Elci M, Elci OC, Odabasi A. Work-related asthma-like symptoms among florists. *Chest* 2004; 125(6):2336–9.
7. Ellenhorn MJ, Barceloux DG. *Medical Toxicology*. New York: Elsevier, 1988:1299–304.
8. Downs AM, Sansom JE. Colophony allergy: a review. *Contact Dermatitis* 1999; 41(6):305–10.
9. de Groot AC. Propolis: a review of properties, applications, chemical composition, contact allergy, and other adverse effects. *Dermatitis* 2013; 24(6):263–82.
10. Ruiz-Hornillos FJ, Alonso E, Zapatero L, et al. Clarinetist's cheilitis caused by immediate-type allergy to cane reed. *Contact Dermatitis* 2007; 56(4):243–5.
11. Williams JD, Tate BJ. Occupational allergic contact dermatitis from olive oil. *Contact Dermatitis* 2006; 55(4):251–2.
12. Palomares O, Alcántara M, Quiralte J, et al. Airway disease and thaumatin-like protein in an olive-oil mill worker. *N Engl J Med* 2008; 358(12):1306–8.
13. Lovell CR, Rowan M. Dandelion dermatitis. *Contact Dermatitis* 1991; 25:185–8.
14. Gleich GJ, Welsh PW, Yunginger JW, et al. Allergy to tobacco: an occupational hazard. *N Engl J Med* 1980; 302(11):617–9.
15. Treudler R, Tebbe B, Orfanos CE. Coexistence of type I and type IV sensitization in occupational coffee allergy. *Contact Dermatitis* 1997; 36:109.
16. Marks GB, Salome SM, Woodcock AJ. Asthma and allergy associated with occupational exposure to ispaghula and senna products in a pharmaceutical work force. *Am Rev Resp Dis* 1991; R4:1065–9.
17. Vereda A, Quirce S, Fernández-Nieto M, et al. Occupational asthma due to spores of *Pleurotus ostreatus*. *Allergy* 2007; 62(2):211–2.
18. Foti C, Nettis E, Damiani E, et al. Occupational respiratory allergy due to *Boletus edulis* powder. *Ann Allergy Asthma Immunol* 2008; 101(5):552–3.
19. Kraut A, Peng Z, Becker NB, et al. Christmas candy maker's asthma. IgG5-mediated pectin allergy. *Chest* 1992; 102:1605–7.
20. Anibarro B, Fontela JL, De La Hoz F. Occupational asthma induced by garlic dust. *J Allergy Clin Immunol* 1997; 100:734–8.
21. O'Holleran MT. Byssinoses and tobacco related asthma. In: Bardana EJ Jr., Montanaro A, O'Holleran MT, eds., *Occupational Asthma*. Philadelphia: Hanley & Belfus, 1992:77–85.
22. Larese F, Fiorito A, Casasola F, et al. Sensitization to green coffee beans and work-related allergic symptoms in coffee workers. *Am J Ind Med* 1998; 34:623–7.

SHELLFISH AND OTHER MARINE INVERTEBRATES

Common names: Crustacean, crab, lobster, shellfish, prawn

Occupational setting

Food processing facilities that handle prawns, lobster, crabs, and other shellfish are a potential source of exposure to crustacean-derived allergens. Oyster farming may also cause exposure to invertebrate allergens. Because of the diversity of sea animals associated with occupational and nonoccupational allergy, any setting where marine animals are handled may pose risks from exposure. Exposure to horseshoe crab-derived antigen may occur in laboratory settings where assays for bacterial endotoxins are performed. The snow crab industry has high prevalence of sensitization (18.4%) and occupational asthma (15.8%) among workers in Canadian processing plants.[1] In one study of 107 employees of one snow crab processing plant, 26% reported experiencing asthma-like symptoms over the course of one processing season.[2] The rates of sensitization are associated with cumulative and dose exposure.

Exposure (route)

Inhalation of allergenic components of shellfish may result in respiratory sensitization. Allergens become airborne in food processing facilities by aerosolization or by being carried along with steam and water vapor. Dust from the processing of dried products may also cause inhalant allergy.

Pathobiology

Shellfish, such as mollusks and crustaceans, are aquatic invertebrate animals with a shell or exoskeleton. Crustaceans such as crabs and lobsters are arthropods. A variety of other marine invertebrates may cause illnesses similar to those seen with shellfish.

Allergy to marine invertebrates typically results from an IgE-mediated immediate hypersensitivity. As with other IgE-mediated respiratory allergies, these disorders are manifested by watery, itchy eyes and nose and sneezing; less frequently, they produce cough and wheezing, with shortness of breath.

Many marine species have allergenic components. Occupational allergy has been reported from exposure to marine sponge,[3] clam,[4] abalone,[5] brine shrimp, and daphnia. Sea squirt-induced asthma has been associated with oyster farming.[6] Limulus amoebocyte lysate (LAL), a horseshoe crab-derived product, used in a laboratory assay, has been reported to cause occupational allergy.[7] Cross-reactivity between shrimp, crab, lobster, scallops, and crayfish antigens has been documented.[8,9]

Diagnosis

The presence of respiratory allergic symptoms in temporal association with exposure to marine animal products suggests the diagnosis of occupational allergy. A thorough history, including qualitative assessment of exposure and the relationship of symptoms to exposure, is critical. Physical examination may support the diagnosis of occupational allergy if mucosal change or wheezing is found. Resolution of symptoms upon removal from the workplace can provide sufficient evidence for a working diagnosis.

Immunological studies such as skin prick testing may be employed. RAST is available for crab, shrimp, chironomids, lobster,[4] and sea squirt. Skin prick testing has been used with horseshoe crab[7] and others. Bronchial provocation may be used to demonstrate asthma in the presence of the suspected allergen. One case study demonstrated this principal by confirming sensitization to octopus in a chef working in a seafood restaurant.[10]

Treatment

There is no specific treatment for marine animal allergy. Definitive treatment involves removal of the patient from exposure to the allergen. Conventional allergy therapies such as antihistamines and bronchodilators may be used. Immunotherapy has been used in treating sea squirt asthma with reported success.[6]

Medical surveillance

Surveillance methods should be directed toward detecting respiratory symptoms. History and pulmonary function testing may be useful.

Prevention

Inhalation of dusts and vapors containing marine animal allergen should be avoided. Engineering and administrative controls can help prevent personnel exposure. Personal protective equipment including respirators may be useful.

References

1. Gautrin D, Cartier A, Howse D, et al. Occupational asthma and allergy in snow crab processing in Newfoundland and Labrador. *Occup Environ Med* 2010; 67(1):17–23. doi:10.1136/oem.2008.039578. Epub 2009 Sep 6.
2. Ortega HG, Daroowalla F, Petsonk EL, et al. Respiratory symptoms among crab processing workers in Alaska: epidemiological and environmental assessment. *Am J Ind Med* 2001; 39(6):598–607.
3. Baldo BA, Krils S, Taylor KM. IgE mediated acute asthma following inhalation of a powdered marine sponge. *Clin Allergy* 1982; 12:171–86.
4. Desjardins A, Malo JL, L'Archevêque J, et al. Occupational IgE-mediated sensitization and asthma caused by clam and shrimp. *J Allergy Clin Immunol* 1995; 96(5 Pt 1):608–17.
5. Masuda K, Tashima S, Katoh N, et al. Anaphylaxis to abalone that was diagnosed by prick test of abalone extracts and immunoblotting for serum immunoglobulin E. *Int J Dermatol* 2012; 51(3):359–60.
6. Montanaro A. Asthma in the food industry. In: Bardana EJ Jr., Montanaro A, O'Hollaren MT, eds., *Occupational Asthma*. Philadelphia: Hanley & Belfus, 1992:125–30.
7. Ebner C, Kraft D, Prasch F, et al. Type I allergy induced by limulus amoebocyte lysate (LAL). *Clin Exp Allergy* 1992; 22:417–9.
8. Lopata AL, Jeebhay MF. Airborne seafood allergens as a cause of occupational allergy and asthma. *Curr Allergy Asthma Rep* 2013; 13(3):288–97.
9. Rosado A, Tejedor MA, Benito C, et al. Occupational asthma caused by octopus particles. *Allergy* 2009; 64(7):1101–2.
10. Goetz DW, Whisman BA. Occupational asthma in a seafood restaurant worker: cross-reactivity of shrimp and scallops. *Ann Allergy Asthma Immunol* 2000; 85(6 Pt 1):461–6.

WHEAT FLOUR AND EGG

Common names for disease: Bakers' asthma, wheat flour asthma, egg allergy

Occupational setting

Baker's asthma has been associated with occupational wheat flour exposure in bakers since the 1700s. Other settings in which occupational grain dust allergy occurs include flour mill work, grain handling (including loading, unloading, and storage operations), pastry factories, cereal factories, and animal feed facilities. Confectionary and baking industry workers also have exposure to egg products.

Exposure (route)

Flour and grain dust and powdered egg readily become airborne. Inhalation exposure may result in occupational allergic disorders, including asthma. Extrinsic allergic alveolitis may result from egg allergy.[1]

Pathobiology

Wheat flour is a grass product that contains albumin, globulin, gliadin, and glutenin proteins. Contaminants of and additives to flour are considered separately in the section on "Grain Dust." Egg proteins such as ovalbumin are allergenic.[1]

Repeated inhalation of wheat flour may result in IgE-mediated, immediate-type allergic respiratory disorders in susceptible individuals. These allergic disorders, which may become manifest after months or years of exposure, occur in 10–30% of workers and include rhinoconjunctivitis and asthma.[2] Asthma is typically immediate but may also be delayed. Although >40 wheat antigens have been documented, many of which produce disease; albumin in wheat flour is most closely linked to bakers' asthma.[2] Many wheat antigens are insoluble and therefore of questionable significance in bakers' asthma, but the strongest IgE response is associated with water-soluble albumins and globulins with in the 12–17 kD range.[3]

In baker's asthma, individual strongly react to wheat thioredoxin-hB (*Triticum aestivum* allergen 25 (Tri a 25)), a class of wheat allergens. Other allergens associated with flour dust, including egg, molds, fungi, insects, storage mites, pollens, and bacteria, may also cause asthma in flour dust workers. Workers who are skin tested with combinations of flour and related allergens rather than wheat flour alone have been shown to have a high prevalence of positivity.[4] In addition, cross-reactivity has been reported among wheat, rye, and barley.[5]

The risk of development of wheat flour allergy is related to the intensity and duration of exposure. Working conditions important to the development of allergy include dust concentration, ventilation, lack of engineering controls on machinery, and varying levels of allergen.[6,7] Rhinitis and conjunctivitis typically precede the development of asthma. The incidence of bakers' asthma increases with longer duration of wheat flour exposure.[8] Exposure to wheat flour at levels of 1–2 mg/m^3 results in a significant risk of allergy development.[9] Individuals with atopic characteristics are at increased risk.

Diagnosis

Bakers' asthma may be diagnosed through a thorough history and physical examination, along with demonstration of reversible airway obstruction in association with flour dust exposure. Rhinitis, conjunctivitis, cough, wheezing, and shortness of breath are typical. Symptoms develop after varying periods of exposure; rhinitis usually precedes asthma. Once sensitized, workers become ill within minutes to hours after starting the workday and are generally well during weekends and holidays. Late responses often occur, typically 4–8 hours after exposure; antecedent rhinorrhea, congestion, and sneezing are typical. Atopic individuals are at higher risk of developing disease.

Workplace physical examination and serial peak flow testing may be useful in documenting reactive airways. Diagnostic tests include prick skin testing with commercially available wheat flour standardized extract, specific IgE RAST, leukocyte histamine release, and inhalation challenge with wheat flour extract. In a study of 71 symptomatic bakers with high IgE concentrations and positive skin prick test, 37 subjects were found to have a positive bronchial challenge test.[5] Those with positive prick tests are more likely to have reactive airways. Bakers may have hypersensitivity to other flour-associated antigens; therefore, care must be taken not to miss multiple hypersensitivities. In one study, five patients with wheat flour hypersensitivity also had allergy to alpha-amylase or cellulose.[10] Nasal provocation has been used to reproduce symptoms from egg allergy.[1] Nasal challenge test has been validated for use in flour allergic subjects through the measurement of increases in eosinophil and basophil numbers, albumin/total protein ratio, eosinophil cationic protein, and tryptase levels.[5, 11]

Treatment

Definitive treatment of wheat flour-related allergic disorders and asthma requires prevention of further exposure. Since changing occupations is often not possible, affected workers often rely on palliative therapy with careful monitoring. Cromolyn sodium or combined long-acting beta agonist and corticosteroid pretreatment, along with short-acting beta agonist treatment of symptoms may allow patients to continue with work. Respirator use may reduce work-related symptoms. Allergen-specific immunotherapy (SIT) with wheat flour extract injections has been shown to produce a decrease in hyperresponsiveness to methacholine, skin sensitivity, and specific IgE to wheat flour as well as subjective improvement in symptoms.[12] When considering SIT, all potential hypersensitivities such as enzymes, mites, and multiple grains should be investigated and ruled out prior to therapy.

Medical surveillance

Surveillance for bakers' asthma should include a periodic detailed respiratory history that may be supplemented by pulmonary function testing. Although the relationship of positive skin prick tests to the development of asthma is not known, consideration should be given to the use of skin prick tests as a marker of exposure. An increase in the prevalence of positive skin tests among bakers from 9% initially to >30% by the fifth year has been documented.[13] Some of these workers reverted to negative within 12 months.

Prevention

The demonstration of exposure–response relationships in bakers' asthma indicates that preventive efforts are worthwhile. Measurements of inhalable flour dust particulate from personal sampling have a strong correlation to levels of wheat and rye exposure.[7]

Avoiding inhalation exposure to dusts can prevent occupational allergy in flour workers. Implementation of engineering controls can help to ensure proper dust containment and exhaust ventilation and is associated with a twofold reduction decrease in alpha-amylase exposures.[7] Administrative controls include starting mixers at slow speeds until the addition of wet ingredients is completed, the elimination of dry sweeping and flour dusting, the application of mixer lids, formal training, and supervision.[6,7,14] An 80% reduction of flour dust was observed with a combination of these controls.[14]

As with any inhalation exposure, respirator use will decrease the dose of antigen received. The use of dust/mist respirators, half face masks for short-term activities, and PAPR for individuals with facial hair for long-term activities is recommended.[6]

References

1. Valero A, Lluch M, Amat P, et al. Occupational egg allergy in confectionary workers. *Allergy* 1996; 51:588–92.
2. O'Holleran MT. Baker's asthma and reactions secondary to soybean and grain dust. In: Bardana EJ Jr., Montanaro A, O'Holleran MT, eds., *Occupational Asthma*. Philadelphia: Hanley & Belfus, 1992:107–16.
3. Brant A. Baker's asthma. *Curr Opin Allergy Clin Immunol* 2007; 7(2):152–5.
4. Quirce S, Fernández-Nieto M, Escudero C, et al. Bronchial responsiveness to bakery-derived allergens is strongly dependent on specific skin sensitivity. *Allergy* 2006; 61(10):1202–8.
5. van Kampen V, Rabstein S, Sander I, et al. Prediction of challenge test results by flour-specific IgE and skin prick test in symptomatic bakers. *Allergy* 2008; 63(7):897–902.
6. Patouchas D, Sampsonas F, Papantrinopoulou D, et al. Determinants of specific sensitization in flour allergens in workers in bakeries with use of skin prick tests. *Eur Rev Med Pharmacol Sci* 2009; 13(6):407–11.
7. Baatjies R, Jeebhay MF. Sensitisation to cereal flour allergens is a major determinant of elevated exhaled nitric oxide in bakers. *Occup Environ Med* 2013; 70(5):310–6.
8. Cullinan P, Cook A, Nieuwenhuijsen MJ, et al. Allergen and dust exposure as determinants of work-related symptoms and sensitization in a cohort of flour-exposed workers: a case–control analysis. *Ann Occup Hyg* 2001; 45(2):97–103.
9. Baur X, Chen Z, Liebers V. Exposure–response relationships of occupational inhalative allergens. *Clin Exp Allergy* 1998; 28(5):537–44.
10. Quirce S, Cuevas M, Díez-Gómez M, et al. Respiratory allergy to Aspergillus-derived enzymes in bakers' asthma. *J Allergy Clin Immunol* 1992; 90(6 Pt 1):970–8.
11. Gorski P, Krakowiak A, Pazdrak K, et al. Nasal challenge test in the diagnosis of allergic respiratory diseases in subjects occupationally exposed to a high molecular allergen (flour). *Occup Med* 1998; 48:91–7.
12. Quirce S, Diaz-Perales A. Diagnosis and management of grain-induced asthma. *Allergy Asthma Immunol Res* 2013; 5(6):348–56.
13. Herxheimer H. The skin sensitivity to flour of bakers' apprentices. *Acta Allergol* 1973; 28:42.
14. Baatjies R, Meijster T, Heederik D, et al. Effectiveness of interventions to reduce flour dust exposures in supermarket bakeries in South Africa. *Occup Environ Med* 2014; 71(12):811–8.

Chapter 29

LATEX

CAROL A. EPLING*

OCCUPATIONAL SETTING

Contemporary awareness of the problem of occupational latex hypersensitivity (LH) began in 1979 with Nutter's report of a case of contact urticaria produced by latex gloves.[1] Over the next 10 years there were a few literature reports of occupational latex allergies, but the problem remained a minor curiosity to occupational physicians. About 1990, however, the trickle of reports expanded into a steady stream, and LH became recognized as a significant occupational problem, particularly for healthcare workers.

Various theories have been advanced to explain this phenomenon. Perhaps the most plausible is that occupational exposure to latex antigens was low until the implementation of universal precautions in the late 1980s prompted an increase of more than 10-fold in the use of powdered natural rubber latex gloves. As the usage increased, manufacturers reduced quality control in an attempt to keep pace with demand, and new manufacturers with little experience entered the market. This led to a marked rise in the antigen content of latex gloves. The juxtaposition of elevated antigen content and increased wear time probably increased the latex antigen exposure of the typical healthcare worker by more than an order of magnitude. Meanwhile, the prevalence of latex-related problems rose markedly throughout the 1990s. While case reporting may have increased somewhat due to greater physician awareness, there can be little doubt that there was a dramatic increase in both the prevalence and severity of latex-related occupational illness.

* The author wishes to thank Charles C. Goodno, for his significant contribution to a previous version of this chapter.

The occupational population at risk for latex allergy comprises primarily end users of latex gloves, predominantly healthcare workers. As a group, healthcare workers account for almost all the occupational epidemiology studies of latex allergy. However, numerous other groups have significant occupational exposure to latex. These include morticians, hairdressers, greenhouse workers, food handlers, and a variety of manufacturing workers. Interestingly, the literature contains few reports of latex allergy associated with the manufacture of latex products. This may be related to the movement of much of latex product manufacturing to overseas locations and to relatively low exposure to latex aerosols during production. As a result, this chapter will focus on latex allergy and latex glove dermatitis in healthcare workers. It should also be noted that persons with spina bifida also have a significant risk of latex allergy, attributed to repeated mucous membrane exposures to latex from operations early in life.

EXPOSURE (ROUTE)

Exposure to latex antigens typically occurs through skin contact and inhalation. Although ingestion is a theoretical possibility, there have been no reports of occupational exposure by this route. Skin exposure occurs primarily by direct contact with latex products such as gloves. Latex gloves contain highly variable amounts of extractable protein, which acts as the antigenic material. The wearer's skin moisture provides an effective mechanism for extraction of these antigens. Glove powder also provides an effective vehicle for the transfer of latex antigen from the glove matrix to the wearer's skin. Although the proteins that comprise the latex antigen are nonvolatile, they bind to glove powder. When the glove

Physical and Biological Hazards of the Workplace, Third Edition. Edited by Gregg M. Stave and Peter H. Wald.

TABLE 29.1 Prevalence of latex hypersensitivity according to exposure group.

Population	Prevalence (%)
General population	0.5–1.0
Healthcare workers	3–15
Clinical laboratory workers	10–20
Spina bifida patients	35–65

Note that reported prevalence varies with measurement technique as well as population.

is donned or removed, a puff of glove powder with bound antigen is released into the air as an aerosol that can be inhaled by the wearer or nearby coworkers. When this process is extrapolated to an area such as an operating room or a laboratory, where dozens of workers may don and remove dozens of pairs of gloves each day, the potential for inhalation exposure can be magnified. Thus, glove powder can act as a classical vector. Latex antigen exposure can be estimated by the latex content of gloves, the duration of wear, the route of exposure (with mucosal and parenteral routes thought to be most significant), and the average daily number of glove pairs worn by the worker and coworkers. Considering these variables, the prevalence of LH in subsets of the population correlates roughly with the intensity of latex antigen exposure (Table 29.1).

A meta-analysis of the evidence of LH among healthcare workers using powdered latex gloves found the prevalence of clinical allergy to latex confirmed by IgE diagnostic testing ranged between 4.01 and 4.63%.[2] The study by Kelly et al.[3] highlighted the importance of healthcare worker exposure to airborne latex antigen leading to the development of LH. The investigators measured airborne latex antigen levels in the workplace before and after replacing powdered latex gloves with nonpowdered latex gloves. Not only was there a quantitative relationship between the prevalence of LH and the concentration of latex antigen in air ducts, but also there was a significant reduction of new onset latex sensitization following glove substitution.

PATHOBIOLOGY

The raw material for natural rubber latex is the milky sap of the rubber tree *Hevea brasiliensis* (Hev b) grown in Africa and Southeast Asia. This sap is an aqueous emulsion of isoprene droplets (which polymerize to form natural rubber) and a variety of other substances, including over 250 distinct proteins, some 60 of which have been shown to be antigenic. Researchers have identified at least 14 Hev b latex allergens most commonly associated with clinical allergy. Hev b 5 and 6.02, found in dipped rubber products including gloves, are of major importance inducing allergy in latex glove users.[4,5] As the sap is processed, various stabilizers and vulcanizing agents are added. These include thiurams, mercaptobenzothiazoles, carbamates, and phenol derivatives. When the emulsion is deposited on hand-shaped ceramic forms and polymerized to form a glove, some of the proteins are adsorbed and trapped in the matrix. Depending on the manufacturing process and the subsequent washing steps, the protein content of gloves can range from undetectable to over 1000 μg/g of finished glove.[6]

Latex gloves can produce three specific types of problems: irritant contact dermatitis, allergic contact dermatitis, and LH. In addition, all types of gloves can produce nonspecific irritation in the form of hyperhidrosis and maceration of the stratum corneum from accumulation of heat and moisture inside the glove. The vast majority of latex glove problems fall into the category of irritant contact dermatitis (ICD), which is produced by direct local tissue irritation without activation of the immune system. ICD may be immediate or slowly progressive and is characterized by erythema limited to skin areas in direct contact with gloves. It does not spread to noncontact areas or produce more serious sequelae.

Allergic contact dermatitis (ACD) is a Type IV (cell mediated) immune reaction and is typically characterized by a delay of one to several days between exposure and development of local erythema. With chronic glove wear, however, the delay may be less apparent. Although latex proteins may occasionally incite ACD, the antigens for this reaction are usually the low-molecular-weight additives such as thiurams and carbamates. ACD is usually limited to areas in direct contact with gloves, but the rash can occasionally generalize. Frequently, ACD and ICD are difficult to distinguish clinically.

LH is a Type I (IgE mediated) immune reaction and typically occurs as an immediate reaction (within minutes of exposure). LH may occur in a spectrum of presentations (listed in rough order of increasing severity):

- Positive skin prick test in an asymptomatic patient
- Contact urticaria
- Rhinoconjunctivitis
- Angioedema
- Asthma
- Anaphylaxis

Since it is somewhat unusual to see serious reactions such as asthma and anaphylaxis as the initial presentations of LH, some investigators believe that susceptible individuals who experience ongoing exposure to latex will tend to show a progression toward serious manifestations with time. In this respect, LH appears to act like other IgE-mediated responses to protein antigens.

While LH is the least common of latex glove problems, it is potentially the most serious and expensive. When LH is defined by positive skin prick testing, which is the current gold standard, the prevalence is between 3% and 15% for healthcare workers exposed to powdered latex gloves, in general, and possibly higher for laboratory workers. Unfortunately, skin prick testing simply identifies workers with sufficient latex-specific IgE to produce a skin reaction but does not provide a clear estimate of the prevalence of symptomatic disease. In a recent prospective cohort study of healthcare workers exposed to nonpowdered latex gloves, Filon et al.[7] reported the incidence of latex sensitization, urticaria, rhinitis, and asthma as 1.0, 0.72, 0.12, and 0.21 per 1000 person-years, respectively. While prevalence of LH was 5.2% among a group using powdered latex gloves, it was 3.2% among newly hired workers exposed to nonpowdered latex gloves.

Several studies have detected a correlation between airborne latex antigen levels and measures of LH.[8–10] The corollary of this is that the indices of LH appear to decrease when airborne latex antigen is reduced. These indices include severity of rhinoconjunctivitis, latex-specific IgE levels, and frequency of asthma attacks. Equally importantly, the study of Allmers et al. showed that removal of powdered latex gloves led to dramatic reduction of airborne latex antigen levels within a few days.[10]

An association between LH and allergies to certain foods has been observed by a number of investigators. The most prominent of these associations are listed in Table 29.2.[11] The studies of Mikkola et al. and Posch et al. provide a biochemical model for such cross-reactions.[12,13] Using sera from banana- and avocado-sensitized latex-hypersensitive patients, respectively, they identified 31–33 kDa proteins from banana and avocado that cross-react with hevein (one of the major latex antigens).

Amino acid sequencing revealed that these proteins (class I endochitinases) contain a domain of sequence homology with hevein, which explains the IgE cross-reactivity. Since endochitinases are common plant proteins, this type of cross-reactivity may offer a general explanation for associations between LH and allergies to fruits and vegetables. Other associations with LH include an atopic history, multiple operations, mucosal exposure to latex, and spina bifida.[14]

TABLE 29.2 Major food allergies associated with latex hypersensitivity and their prevalence among latex-hypersensitive patients. Adapted from Kim and Hussain.[11]

Food	Prevalence %
Banana	18
Avocado	16
Kiwi fruit	12
Shellfish	12
Fish	8
Tomato	6
Watermelon, peach, carrot (each)	4
Apple, chestnut, cherry, coconut, apricot, strawberry, loquat (each)	2

DIAGNOSIS

Nonspecific glove irritation is generally characterized by mild local itching or burning in the distribution of glove contact that resolves rapidly with cessation of glove wear. The skin appears normal or pale with possible mild edema and mild fissuring. Excoriation and lichenification are minimal. Usually this can be distinguished clinically from ICD or ACD.

Both ICD and ACD present with local itching, burning, and erythema in the glove distribution. Marked erythema, vesicles, and weeping may be present in severe cases. In chronic cases, fissuring and lichenification may be prominent. ICD and ACD can sometimes be distinguished on the basis of the history of an early reaction in the case of ICD and a delayed reaction for ACD. However, most workers are unable to give a sufficiently detailed history to allow the distinction to be made with confidence. Rashes that begin in the glove distribution but then generalize are not ICD. Both ACD and LH should be considered in such cases. Patch testing with a kit of typical latex additives may clarify the diagnosis by confirming ACD.

LH, like all allergic reactions, is primarily a clinical diagnosis. This is illustrated by analysis of the results of the Multi-Center Latex Skin Testing Study Task Force.[15] These results show that a clinical evaluation provides about 90% sensitivity and specificity for diagnosis of LH (defined by subsequent positive skin prick test). In the evaluation, historical features of exposure, proximate symptoms, nature of symptoms, and family history are most important. Reported symptoms should fit into a recognizable pattern of Type I reactions (rhinoconjunctivitis, urticaria, angioedema, asthma, or anaphylaxis). A useful early sign of LH is development of a papular rash consistent with contact urticaria. Extension of the papular rash beyond the area of the glove is further evidence of LH. Physical examination should be used to provide objective confirmation of signs and symptoms but may be unremarkable in the asymptomatic patient. In the occupational setting, consistent provocation of typical symptoms by known work exposure to latex is frequently sufficient to make the diagnosis. In the case of latex-induced asthma, serial peak expiratory flow measurements are sometimes useful.

Laboratory testing is useful but primarily serves to confirm the clinical evaluation. Probably the ultimate diagnostic test is the wear and puff test.[16] This involves starting with a finger cut from a high-antigen-content latex glove and wearing progressively larger portions until the entire glove is worn. Development of symptoms is assessed at each stage. If the glove can be worn without symptoms, a glove is inflated like a balloon, and the air inside (with suspended glove powder) is exhausted into the patient's breathing zone. While highly sensitive, this test can provoke severe asthma or anaphylaxis in susceptible individuals and is not recommended except at specialized centers where occupational aerosol challenge testing is routinely performed.

Skin prick testing is a more practical confirmatory test. A standardized latex skin testing antigen manufactured by Greer Laboratories was evaluated by the Multi-Center Latex Skin Testing Study Task Force and found to be sensitive (sensitivity 95%), specific (specificity about 99%), and safe (no epinephrine-requiring reactions).[15] Because the FDA has not licensed this antigen preparation, skin testing for LH in the United States is presently limited to either experimental or foreign (e.g., Canadian) antigen preparations.

In vitro testing for latex-specific IgE has also been used in evaluation of LH. Currently, three FDA-licensed test kits are commercially available in the United States. The characteristics of two of these kits (Diagnostic Products Corp. AlaSTAT and Pharmacia–Upjohn CAP) are comparable, with sensitivity in the range of 75% and specificity in the range of 97% in a population with more than 40% prevalence of LH.[17] However, in a population with low prevalence of LH (5% by skin prick testing), the CAP assay demonstrated only 35% sensitivity and 98% specificity.[18] Thus, serologic testing is recommended only for confirmation of a clinical suspicion of LH and not for screening in a population at low risk for LH.

TREATMENT

Nonspecific glove irritation usually responds to a glove holiday and/or use of glove liners made from materials such as white cotton or Dermapore.[1] Liners should be changed whenever they become damp. ICD and ACD will sometimes respond to the same regimen, but the ultimate treatment is substituting a different glove material for the one that provokes the problem. In the acute phase, a glove holiday combined with treatment using a low- to medium-potency topical corticosteroid is beneficial. This may be combined with astringent soaks in the case of weeping lesions or emollient preparations in the case of dryness. The use of barrier creams is not a suitable alternative to removal from latex gloves, and some authors advise that creams may act as a vehicle for extraction of antigens from latex gloves.[14]

Symptomatic LH poses a special problem, since the definitive treatment is a complete removal from exposure to latex antigens. Sending a latex-hypersensitive worker back to work in a medical institution usually involves a degree of risk. As mentioned in the 1997 National Institute for Occupational Safety and Health (NIOSH) Alert, one of the most critical issues is whether the workplace is free of powdered latex gloves.[19] Institution-wide substitution of powder-free latex gloves for powdered ones has been reported to reduce the level of airborne latex antigen to undetectable levels. Thus, it seems reasonably safe to send workers with LH back to work in institutions that have made this change. Care must still be exercised with activities that tend to release latex particles from reservoirs such as HVAC filters and plenums. The remaining issue is one of helping the worker avoid direct contact with latex. Given the variety of alternative glove materials, selecting alternative gloves need not be a problem. Avoiding contact with latex-containing supplies and equipment (Table 29.3), however, may require detailed examination of specifications for these items. In addition to these measures, LH workers should be educated to follow the recommendations in Table 29.4.

TABLE 29.3 Common items containing latex.

Hospital supplies:
Gloves, barium enema catheters, adhesive tape, EKG electrodes, blood pressure cuffs, stethoscopes, anesthesia masks, surgical masks, goggles, catheters, injection ports
Office supplies:
Rubber bands, erasers, adhesives
Household objects:
Automobile tires, elastic waistbands, dishwashing gloves, condoms, diaphragms, balloons, pacifiers, baby bottle nipples

TABLE 29.4 Recommendations for healthcare workers with latex hypersensitivity.

1. Inform employer, healthcare providers, and family members of latex allergy
2. Identify latex-containing products and avoid contact
3. Wear medical alert bracelet with latex allergy notation
4. Carry and understand the use of an epinephrine injector (e.g., EpiPen)
5. Notify employer's occupational health service of changes in latex-related symptoms

MEDICAL SURVEILLANCE

Given the prevalence of LH, and that early diagnosis is almost certainly beneficial, LH is suitable for a medical surveillance program.[20] Screening can be implemented with a simple symptom inventory for exposed workers, since history is the key feature in LH diagnosis. Workers classified as high risk by screening might undergo subsequent clinical history and physical examination, serial peak expiratory flow measurements, or nonspecific inhalation challenge testing for confirmation. Assuming the existence of an appropriate index of suspicion, clinical evaluation alone is probably sufficient to make the diagnosis in many cases. When a standardized skin testing antigen becomes available, skin prick testing will be the confirmatory test of choice in view of its sensitivity, specificity, and safety. For latex-hypersensitive workers who return to work with some degree of ongoing latex exposure, periodic follow-up in a medical surveillance program is indicated to detect development of new symptoms. If new symptoms occur, further intervention to eliminate latex exposure is vital.

PREVENTION

Primary prevention is aimed at avoidance of high-level and long-term latex exposure in all workers. Careful selection of barrier protection according to risk for contact with infectious materials may reduce the use of latex gloves among workers who are unlikely to encounter infectious substances (i.e., food services, maintenance, and housekeeping). Strict avoidance would mean eventually converting from latex to nonlatex gloves and perhaps making substitutions for latex articles such as Foley catheters and tourniquets that are routinely handled by workers. The cost of substituting nonlatex gloves can be substantial for medical institutions. In addition, wearer satisfaction and barrier properties of nonlatex gloves are significant issues for workers who wear gloves for prolonged periods during direct contact with blood.[21] A compromise solution is to convert to low-antigen powder-free latex gloves as suggested by the NIOSH Alert.[19] This provides a significant measure of protection from inhalational exposure for latex-hypersensitive individuals and decreases risk for new cases of sensitization.[3,22,23]

A powder-free latex workplace also affords a measure of secondary prevention for those who already have LH. In this situation, workers with LH, once identified, can return to work wearing nonlatex gloves while exercising caution in tasks that involve handling latex-containing supplies and equipment. One can be reasonably confident that these affected workers will have minimal latex aerosol exposure in this situation. Close medical follow-up, however, is indicated to monitor for possible progression of symptoms.

References

1. Nutter AF. Contact urticaria to rubber. *Br J Dermatol* 1979; 101:597–8.
2. Bousquet J, Flahault A, Vandenplas O, et al. Natural rubber latex allergy among health care workers: a systematic review of the evidence. *J Allergy Clin Immunol* 2006; 118:447–54.
3. Kelly KJ, Wang ML, Klancnik M, et al. Prevention of IgE sensitization to latex in health care workers after reduction of antigen exposures. *J Occ Environ Med* 2011; 53:934–40.
4. Rolland JM, O'Hehir RE. Latex allergy: a model for therapy. *Clin Exp Allergy* 2008; 38:898–912.
5. Yagami A, Suzuki K, Saito H, et al. Hev b 6.02 is the most important allergen in health care workers sensitized occupationally by natural rubber latex gloves. *Allergol Int* 2009; 58: 347–55.
6. Williams PB, Halsey JF. Endotoxin as a factor in adverse reactions to latex gloves. *Ann Allergy Asthma Immunol* 1997; 79:303–10.
7. Filon FL, Bochdanovits L, Capuzzo C, et al. Ten years incidence of natural rubber sensitization and symptoms in a prospective cohort of health care workers using non-powdered latex gloves 2000–2009. *Int Arch Environ Health* 2014; 87:463–9.
8. Swanson MC, Bubak ME, Hunt LW, et al. Quantification of occupational latex aeroallergens in a medical center. *J Allergy Clin Immunol* 1994; 94:445–51.
9. Tarlo SM, Sussman G, Contala A, et al. Control of airborne latex by use of powder-free latex gloves. *J Allergy Clin Immunol* 1994; 93:985–9.
10. Allmers H, Brehler R, Chen Z, et al. Reduction of latex aeroallergens and latex-specific IgE antibodies in sensitized workers after removal of powdered natural rubber latex gloves in a hospital. *J Allergy Clin Immunol* 1998; 102: 841–5.
11. Kim KT, Hussain H. Prevalence of food allergy in 137 latex-allergic patients. *Allergy Asthma Proc* 1999; 20:95–7.
12. Mikkola JH, Alenius H, Kalkkinen N, et al. Hevein-like protein domains as a possible cause for allergen cross-reactivity between latex and banana. *J Allergy Clin Immunol* 1998; 102:1005–12.
13. Posch A, Wheeler CH, Chen Z, et al. Class 1 endochitinase containing a hevein domain is the causative allergen in latex-associated avocado allergy. *Clin Exp Allergy* 1999; 29:667–72.
14. Taylor JS, Erkek E. Latex allergy: diagnosis and management. *Derm Therapy* 2004; 17:289–301.
15. Hamilton RG, Adkinson NF Jr. Diagnosis of natural rubber latex allergy: multi-center latex skin testing efficacy study. The Multi-Center Latex Skin Testing Study Task Force. *J Allergy Clin Immunol* 1998; 102:482–90.
16. Hamilton RG, Adkinson NF Jr. Validation of the latex glove provocation procedure in latex-allergic subjects. *Ann Allergy Asthma Immunol* 1997; 79:266–72.
17. Hamilton RG, Biagini RE, Krieg EF. Diagnostic performance of food and drug administration-cleared serologic assays for natural rubber latex-specific IgE antibody. The multi-center latex skin testing study task force. *J Allergy Clin Immunol* 1999; 103:925–30.
18. Accetta Pedersen DJ, Klancnik M, Elms N, et al. Analysis of available diagnostic tests for latex sensitization in an at-risk population. *Ann Allergy Asthma Immunol* 2012; 108:94–7.
19. National Institute for Occupational Safety and Health. *Preventing allergic reactions to natural rubber latex in the workplace.* NIOSH publication no. 97-135. Cincinnati: Government Printing Office, 1997.
20. Epling C, Duncan J, Archibong E, et al. Latex allergy symptoms among health care workers: results from a university health and safety surveillance system. *Int J Occup Environ Health* 2011; 17:17–23.
21. Aldlyami E, Kulkarni A, Reed MR, et al. Latex-free gloves. Safer for whom? *J Arthroplasty* 2010; 25(1):27–30.
22. Allmers H, Schmengler J, John SM. Decreasing incidence of occupational contact urticaria caused by natural rubber latex allergy in German health care workers. *J Allergy Clin Immunol* 2004; 114:347–51.
23. LaMontagne AD, Radi S, Elder DS, et al. Primary prevention of latex related sensitization and occupational asthma: a systematic review. *Occup Environ Med* 2006;63:359–64.

Chapter 30

MALIGNANT CELLS

AUBREY K. MILLER

OCCUPATIONAL SETTING

Workers at risk include (i) laboratory workers who handle malignant cells during the performance of in vitro and in vivo research, (ii) histology and pathology workers involved in the preparation and processing of neoplastic tissues, (iii) medical and nursing staff involved in surgical procedures (i.e., aspirates, biopsies, and resections) on cancer patients, (iv) surgical scrub personnel handling sharp instruments contaminated with malignant cells, and (v) housekeeping workers (especially those in cancer research areas) exposed to sharp objects contaminated with malignant cells.

EXPOSURE (ROUTE)

The occupational risk of cancer occurring in humans from exposure to malignant cells is not well recognized, owing to the few cases reported in the medical literature. Based on these reports, the most likely route of occupational transmission involves needlestick or sharp object injuries whereby malignant cells are cutaneously injected or possibly implanted into an open wound. This risk is best understood by the well-described occurrence of occupational transmission of infectious diseases such as HIV, hepatitis B, and hepatitis C via needlesticks or other sharp objects (i.e., surgical instruments, histologic tissue cutters, broken capillary tubes, and pipettes).

There are currently more than 8 million healthcare workers in the United States in hospitals and other healthcare settings. While precise data are not available with respect to the actual number of annual needlestick injuries or other percutaneous injuries among healthcare workers, it is estimated that 600 000–800 000 occur annually and that half of these go unreported.[1] It is estimated that as many as 2 800 injuries may occur each year from handling glass capillary tubes.[2]

A review of studies reporting needlestick injuries found that 34–50% of healthcare workers were injured and that 10–70% of those injuries were due to recapping of needles.[3] Studies of hospital workers have shown that the highest incidence of needlestick injuries occurs in housekeeping personnel (during trash disposal) and laboratory and nursing personnel (during needle disposal or recapping).[4,5] Pathologists and surgeons have also been shown to be at increased risk for cutaneous injuries from sharp instruments and needlesticks (especially involving the distal fingers of the nondominant hand) during operative procedures.[6,7] Although the incidence of cutaneous injuries resulting from sharps contaminated with viable cancer cells is unknown, it probably represents only a small fraction of the cutaneous injuries incurred by potentially exposed workers.

PATHOBIOLOGY

Transplantation of foreign human tissue to a healthy recipient normally leads to an immune response resulting in destruction of the transplanted tissue (rejection).[8] Southam et al. [9,10] showed that normal recipients given subcutaneous injections of human cancer cells responded with a marked local inflammatory reaction and a rapid complete regression of the cancer implants within 3–4 weeks. In contrast, cancer cell injections given to advanced cancer patients showed little or no acute inflammatory reaction; the cancer cells typically grew for 3 weeks or longer before regression, and in some recipients growth continued beyond 6 weeks.[10] One

Physical and Biological Hazards of the Workplace, Third Edition. Edited by Gregg M. Stave and Peter H. Wald.

recipient exhibited local recurrence of tumor growth even after three excisional biopsies, and another recipient had lymph node metastasis.[9] In another study, local cancer growth occurred in two patients who received small allogeneic tumor implants as part of an immunotherapy protocol for advanced cancer.[11]

Scanlon et al.[12] reported that some patients with advanced cancer have even tolerated tissue grafts from other animal species. Growth of transplanted cancer cells has also been reported to occur in healthy immunocompetent individuals. In one case, death from metastatic disease was reported in a woman who received a small melanoma graft taken from her daughter as part of an immunotherapy protocol.[12] In another case, a healthy 19-year-old laboratory worker developed an actively growing adenocarcinoma of colonic origin on her hand following a needlestick injury. At the time of the injury, only a small superficial wound was noted, with no apparent injection of the cancer cell suspension. The tumor, which was widely excised after 19 days, showed no evidence of an inflammatory response or necrosis. The worker was noted to be free from recurrence 4 years after the injury.[13]

The occurrence of transplanted cancer cell growth and metastasis in some individuals appears to be related to alterations of immune functioning. Rejection of foreign tissue depends upon recognition of major cell surface histocompatibility (HLA) antigens and involves both cell-mediated and humoral immunities.[8] The most important cell-mediated reactions involve both $CD4^+$ T- helper cells and $CD8^+$ cytotoxic T-cells, which play a crucial role in the recognition of foreign tissue cells and regulation of the immune response.[8] Humoral immune reactions to transplantation antigens appear to be mediated by antibodies formed against foreign class I and class II HLA antigens.[8] Therefore, HIV/AIDS patients, other immunocompromised individuals, and those on immunosuppressive medications (i.e., steroids, cyclosporine, and azathioprine) may be at increased risk for developing viable neoplasms when exposed to malignant cells. The occurrence of cancerous growth in two apparently healthy immunocompetent adults is not well understood. Immune tolerance, lack of an immune response to specific antigens, under these conditions may be due to certain mechanisms that allow the tumor cells to escape immunosurveillance, such as loss or reduced expression of histocompatibility antigens, shedding or modulation of tumor antigens, and production of immunosuppressive factors.[8]

TREATMENT

Injuries should be medically treated as with other needlestick or cutaneous injuries. In addition, the injury site should be periodically evaluated for any tumor growth for at least the ensuing 3–4 weeks (immunocompromised individuals may require longer follow-up). If tumor growth occurs, wide excision of the tumor with close follow-up should be considered. This treatment was apparently effective in at least one of the reported cases.[13]

MEDICAL SURVEILLANCE

All percutaneous injuries should be handled in accordance with the Occupational Safety and Health Administration (OSHA) Bloodborne Pathogens Standard, which covers workers occupationally exposed to unfixed human tissues or blood products.[14]

PREVENTION

All nonessential sharps should be eliminated where possible, especially in laboratory situations. Eliminating all needle recapping can reduce the risk of needlestick injury and nonessential unprotected needle use; needleless or protected needle devices should be used where possible.[15] Where the use of needles or sharp instruments is indicated, workers should be trained in the safe techniques for handling and disposal (i.e., using puncture-resistant containers) of these objects. Additionally, workers should be encouraged to report all needlesticks and contaminated cutaneous injuries so that appropriate postexposure treatment can be given and so that the incident can be studied to prevent similar accidents in the future. Further, worker education and training, needle handling, and sharps disposal should be conducted in accordance with the OSHA Bloodborne Pathogens Standard and the National Institute for Occupational Safety and Health (NIOSH) recommendations to reduce the likelihood of worker injuries.[1,14–16]

References

1. NIOSH. *Preventing needlestick injuries in health care settings.* DHHS (NIOSH) Publication No. 2000–108. Cincinnati, OH: Department of Health and Human Services, Public Health Service, Centers for Disease Control and Prevention, National Institute for Occupational Safety and Health, 2000.
2. Jagger J, Bentley M, Perry J. Glass capillary tubes: eliminating an unnecessary risk to healthcare workers. *Adv Exp Prev* 1998; 3(5):49–55.
3. Martin LS, Hudson CA, Strine PW. Continued need for strategies to prevent needlestick injuries and occupational exposures to bloodborne pathogens. *Scand J Work Environ Health* 1992; 18:94–6.
4. McCormick JD, Maki DG. Epidemiology of needles-stick injuries in hospital personnel. *Am J Med* 1981; 70:928–32.
5. Neuberger JS, Harris JA, Kundin WD, et al. Incidence of needlestick injuries in hospital personnel: implications for prevention. *Am J Infect Control* 1984; 12:171–6.
6. O'Brian DS. Patterns of occupational hand injury in pathology. *Arch Pathol Lab Med* 1991; 115:610–3.
7. Tokars JI, Bell DM, Culver DH, et al. Percutaneous injuries during surgical procedures. *JAMA* 1992; 267:2899–904.

8. Cotran RS, Kumar V, Robbins SL. *Robbins pathologic basis of disease*, 5th edn. Philadelphia, PA: WB Saunders Company, 1994, 175–7, 190–7.
9. Southam CM. Homotransplantation of human cell lines. *Bull N Y Acad Med* 1958; 34:416–23.
10. Southam CM, Moore AE. Induced immunity to cancer cell homografts in man. *Ann N Y Acad Sci* 1958; 73:635–53.
11. Nadler SH, Moore GE. Immunotherapy of malignant disease. *Arch Surg* 1969; 99:376–81.
12. Scanlon EF, Hawkins RA, Fox WW, et al. Fatal homotransplantated melanoma: a case report. *Cancer* 1965; 18:782–9.
13. Gugal EA, Sanders ME. Needle-stick transmission of human colonic adenocarcinoma. *N Engl J Med* 1986; 315:1487.
14. OSHA Bloodborne Pathogens Standard. 29 CFR 1910.30; 56 Federal Register 64004:1991.
15. NIOSH. *What every worker should know: how to protect yourself from needlestick injuries*. DHHS (NIOSH) Publication No. 2000-135. Department of Health and Human Services, Public Health Service, Centers for Disease Control and Prevention, National Institute for Occupational Safety and Health, Cincinnati, OH, 2000. Available at: https://www.cdc.gov/niosh/docs/2000-135/pdfs/2000-135.pdf (accessed on July 3, 2016).
16. NIOSH. *Guidelines for protecting the safety and health of healthcare workers*. DHHS (NIOSH) Publication No. 88-119. Department of Health and Human Services, Public Health Service, Centers for Disease Control and Prevention, National Institute for Occupational Safety and Health, Cincinnati, OH, 1988. Available at: https://www.cdc.gov/niosh/docs/88-119/pdfs/88-119.pdf (accessed on July 3, 2016).

Chapter 31

RECOMBINANT ORGANISMS

JESSICA HERZSTEIN AND GREGG M. STAVE*

OCCUPATIONAL SETTING

Recombinant organisms are used routinely in the biotechnology and pharmaceutical industries and in academic laboratories. They are the source of many of the most innovative biopharmaceuticals that have contributed to medical science in the past 40 years. Because of the relative effectiveness of production techniques, certain processes may move from laboratory to pilot plant (often still a laboratory) to production with very little apparent change. Occasionally, relatively small facilities can produce large quantities of complex and previously rare or unattainable products. Increasingly, though, the scale of commercial manufacturing has grown significantly in order to support these valuable products.

The workforce at these facilities may be small but is very highly trained. Most biotechnology research and pilot production personnel have advanced degrees. Commercial facilities tend to be staffed with a mixture of individuals with high school or technical education and managers with advanced degrees. As a group, biotechnology workers are highly invested, both emotionally and economically, in the success of their new enterprises. Long work hours, secrecy concerning processes and product development, emphasis on rapid progress toward production, and very high economic stakes increase the potential for worker hazards.

*The authors wish to thank Ed Fritsch, and John L. Ryan, for their contributions to an earlier version of this chapter.

EXPOSURE (ROUTE)

Exposure to three types of hazards must be considered in biotechnology research and production: recombinant organisms, biological (human or animal derived) reagents used for recombinant organism growth, and nonbiological hazards such as chemicals, radiation, and recombinant products. Of these, biological reagents and chemicals/radiation pose the greatest concern. Exposure may theoretically occur through ingestion, inhalation, skin, and mucous membrane contact or skin penetration by a contaminated needle or other sharp object.

PATHOBIOLOGY

Recombinant biology is the ability to insert specific pieces of DNA into selected organisms for the purpose of creating desired products. A variety of techniques are used to edit or delete genes, including endonucleases, zinc finger nucleases, transcription activator-like effector nucleases (TALENs), and clustered regularly interspaced short palindromic repeats (CRISPR/Cas9). Recombinant organisms are the resulting genetically modified bacteria, fungi, and cells. Humans have selectively bred animal and plant species for desired traits for thousands of years. Recombinant biology extends that activity to the molecular level and can produce very precise outcomes.

More than 80 biopharmaceutical products have now been approved in the United States, Europe, or both.[1] These fall into multiple broad categories including blood factors, hormones,

Physical and Biological Hazards of the Workplace, Third Edition. Edited by Gregg M. Stave and Peter H. Wald.

TABLE 31.1 Selected products from recombinant organisms.

Biopharmaceutical class	Examples	Date first approved (EU or USA)
Blood factors	Factor VIII	1993
	Factor IX	1997
	Hirudin	1997
Hormones	Insulin	1982
	Growth hormone	1985
	Glucagon	1998
	Glucagon-like-peptide-1 (GLP-1)	2014
	Follicle stimulating hormone	1995
Hematopoietic growth factors	Erythropoietin	1989
	Granulocyte colony stimulating factor	1991
	Alpha and beta interferon	1992
	Interleukin-2	1994
	Granulocyte–macrophage colony stimulating factor	1995
	Megakaryopoiesis-stimulating protein	2005
Vaccines	Hepatitis B surface antigen	1986
	Cholera toxin B subunit/vibrio cholerae 01	2004
	Human papillomavirus	2006
Monoclonal antibodies	Anti-CD-3	1986
	Anti-GII_bIII_a	1994
	Anti-CD20	1997
	Anti-HER 2	1998
Enzymes	DNase	1993
	b-Glucocerebrosidase	1994
	Tissue plasminogen activator	1996
Chimeric molecules	IL-2 Diphtheria toxin	1999
	TNF Receptor-IgG	1998

Data are from the following sources: http://www.fda.gov, http://www.ema.europa.eu

hematopoietic growth factors, vaccines, monoclonal antibodies, enzymes, and chimeric molecules (Table 31.1). Many more products are currently in development, and the human genome and bioinformatics efforts are likely to result in an explosion of new possibilities. In addition, new approaches to the production of biopharmaceutical products such as transgenic animals and gene therapy are aggressively being developed, extending the scope of technologies, facility design, and environmental impact. Nonmedical products include pesticide-resistant and pest-resistant plants, industrial or food-processing enzymes, chemicals, fuels, and foods.

In the future, specialty products may appear in clothing, construction, or transportation materials. Even before the public became aware of commercial recombinant technology in the mid-1980s, hundreds of start-up enterprises existed in the United States (and elsewhere). Continued progression from laboratory to pilot plant to commercial products is inevitable for a wide new array of products. As in every new industry, the potential also exists for health and safety hazards.

One of the fastest-growing areas in the biotechnology industry is gene therapy. Gene therapy employs viral and bacterial vectors to deliver specific human genes to patients with hereditary and acquired diseases. Recombinant vectors are often replication-deficient viruses, for example, retrovirus, adenovirus, or herpes virus. These are used in gene therapy both for protein or enzyme replacement and for oncology. There are serious concerns that these viruses may become replication competent in the host. Recognizing that large numbers (10^{10} or more) of viruses are often delivered emphasizes the need for caution, because there may be risks to the caregivers in preparation of the therapy or to the patient from the vector carrying the intended therapeutic gene.

Simple infection is the most significant biological hazard for workers. At the research stage, hazardous organisms such as drug-resistant microorganisms, hepatitis viruses, and oncogenic viruses are frequently used. Potentially infectious organisms such as adenovirus have specialty uses in the field of gene therapy. Careful attention to biological containment and good microbiological practice is critical. However, many organisms used in recombinant work have been attenuated by the loss of specific gene functions or modified so that they cannot synthesize an organic compound required for their growth. Commonly, tissue culture work is carried out with nonpathogenic *Escherichia coli*, *Bacillus subtilis*, *Saccharomyces cerevisiae*, *Aspergillus niger*, Baculovirus, or mammalian cells. Exposure to the relatively nonpathogenic

E. coli used for much recombinant technology work has not been associated with diarrheal or other gastrointestinal illness. Several studies have demonstrated the lack of colonization of workers with recombinant organisms.[2]

For production of recombinant proteins, the most common infection risk comes from exposure to reagents used to support growth and productivity of the recombinant organism itself. Mammalian cells typically used for production are not dangerous themselves, as nonpathogenic recombinant organisms are used exclusively and they have been extensively tested for a variety of known and possible infectious agents. However, material of animal or human origin, such as raw serum, partially purified blood components (human or bovine serum albumin), or even highly purified reagents from animals (bovine or porcine insulin) are frequently required. The risk of known or previously unknown infectious agents in the source animals or humans must be carefully assessed. The devastating impact of unknown HIV infection in the blood supply on the hemophilia population in the 1980s is ample demonstration of the concern becoming reality. The emergence of a new variant of a human transmissible spongiform encephalopathy agent, new variant Creutzfeldt–Jakob Disease (nvCJD), following from the epidemic of bovine spongiform encephalopathy in the United Kingdom in the late 1990s, has also highlighted the possibility of infectious risk. Although the exposure of workers to such infectious agents is probably lower than for those receiving the products, the need for appropriate precautions is clear, especially when the volume of materials handled or stored may be significant.

Another potential biological hazard involves special applications of hybridoma technology. Typically, antibodies are produced from two fused murine cells to produce specific antibodies, creating an immortal cell line. Atypically, a theoretical hazard of this activity exists when one of the two cell lines is human, which carries the risk of latent or unapparent viral infection. In vitro human cells, such as lymphocytes, are susceptible to infection with tumorigenic virus. In addition, the murine fusion partner can carry mouse type C leukemia and sarcoma. Although these viruses are typically not infectious of human cells, it is not known whether residence in hybrid cell lines could alter the host range. If so, this may create a new hazard whereby pathogenic animal viruses extend their host range to humans (and vice versa). Studies to date have shown no evidence of clusters of cancer or other diseases or elevated rates of disease in workers.

The pharmaceutical industry creates final recombinant products as well as intermediate biological products that are highly concentrated and physiologically active. Normal physiologic responses to inhaled or absorbed substances may include hormonal and allergic responses. This is the same problem encountered in the conventional pharmaceutical industry.[3] Workers exposed to estrogens, growth hormone, or other physiologically active chemicals can be expected to have unintended but predictable responses.

Allergy to the concentrated product has been seen in biotechnologists. Products and intermediate products may elicit immune responses. In addition, product development involves laboratory animal handling, which has its own set of allergic issues. Engineering controls, with use of personal protective devices when engineering controls are insufficient, represent the best means of preventing sensitization or other unintended physiologic responses.

In addition to these concerns surrounding recombinant organisms and their products, other important nonbiological hazards exist in the biotechnology industry. These may be of a substantially greater magnitude than in traditional research laboratories. For example, many laboratory research operations use radioisotopes, including tritium (^{3}H), carbon-14 (^{14}C), sulfur-35 (^{35}S), calcium-45 (^{45}Ca), phosphorus-32 (^{32}P), chromium-51 (^{51}Cr), and iodine-125 (^{125}I). Several of these are volatile and therefore quite dangerous. Safe handling and dose reduction techniques should be collaborative efforts between a laboratory radiation safety liaison and a consulting certified health physicist. Given the relatively young age of the biotechnology workforce and the reproductive implications of hazards such as ^{125}I, additional emphasis should be placed on workplace safety. Most monitoring is done by radiation badge, a process comparable to industrial hygiene monitoring. Health physicists should perform thyroid and whole-body scanning when needed.

A surprising number of physical hazards are associated with both research and product preparation in this highly technical industry. Biotechnologists must deal with compressed gas, cryogenics, humidified atmospheres (with associated building biological hazards), high-voltage electricity, as well as some flammable and corrosive materials. Heat and cryogenic burns are a common hazard of laboratory and start-up production facilities, especially if laboratory personnel transport laboratory-scale hoses for production use. Ultraviolet light and associated skin hazard may be encountered in manufacturing suites, research biological safety cabinets, and electrophoresis operations. Laboratory and even manufacturing operations can be hands on, with lots of lifting and frequent exposure to glassware. In industry operations, glassware handling (and cleaning) present opportunities for sharps exposure. For one biotechnology company, lacerations comprised 41% of injuries, and back injuries accounted for 93% of lost time.

Biotechnology may require shift work. Relatively small labor forces are engaged in both research and production, yet the cell cycle is continuous in both settings. Round-the-clock care of cultures is dictated by biological realities. Shift work may be informal in the laboratory or formal in production facilities.

Biotechnology processes require significant chemical use, with the potential for exposure.[4] Classes of chemicals common in biotechnology include culture additives and antibiotics, solvents and extractants (including those used in chromatography and sequencing), and the corrosive additives needed to maintain a working chilled-water facility. Ethidium bromide is a genotoxic compound used for fluorescent staining in sequencing operations. Cell culture additives are frequently mutagenic. The most common acute chemical overexposures have been to acetonitrile, which is used as an extractant. Patients with mild cyanide-like toxicity of acetonitrile have been reported anecdotally in several emergency situations following spills and inappropriate handling during cleaning operations. One potentially hazardous repetitive process involves the use of acrylamide in gel preparations. The purchase of preformed gels can reduce the hazard of handling the neurotoxic bis-acrylamide powder. Finally, many production processes, especially those using *E. coli*, result in a fusion protein as the primary product. The fusion protein requires peptide bond cleavage with a strongly reactive and specific chemical in order to prepare the final product. This cleavage is carried out at large volume with highly reactive chemicals, including hydrofluoric acid. Precautions to prevent both exposure of workers and inadvertent release into the environment must be in place.

MEDICAL SURVEILLANCE

Reduction of health risks in biotechnology is attained through control of work hazard and exposure, which requires in-depth focus on work processes and the methods of containment. Medical surveillance is a strategy for disease prevention that is based on a risk assessment of the jobs performed by the worker.[5] Medical surveillance is usually targeted to specific chronic risks.

Medical surveillance is not routinely indicated for workers simply because they work with recombinant organisms. There is no prescribed medical surveillance related directly to work with recombinant organisms. Studies of workers in the biotechnology industry have concluded that there is no evidence of adverse health effects related to the unique aspects of recombinant DNA technology.[6,7] Medical surveillance data have documented neither the occurrence of clusters of disease nor an increased incidence of diseases in workers who are frequently exposed to recombinant organisms.[5] Medical surveillance for health effects related to exposure is best conceived when certain criteria are fulfilled:

1. The exposure can potentially cause an identifiable health effect.
2. It is reasonably likely that the disease or effect may occur (and it is related to work).
3. There is an acceptable and scientifically sound methodology for diagnosing the condition or disease.
4. Early diagnosis has the potential to reduce morbidity or mortality.

When data collection and understanding the prevalence of a work-related condition are the goals, surveillance may target specific conditions even if therapeutic approaches are not yet available.

Some common exposures that may warrant medical surveillance in biotechnology laboratories and industrial settings include infectious agents, sensitizing chemicals, radioisotopes, and animal handling (a potential for zoonotic infections and allergic responses). The goal is detection of early health effects and prevention of long-term morbidity related to agents such as animal dander, enzymes, and endotoxin.

Work with recombinant organisms that present no identifiable risk to human health (group I in the European hazard classification scheme) would not entail routine medical surveillance. For work with highly pathogenic organisms or biologically active substances such as enzymes that are expressed by genetically modified microorganisms, periodic medical evaluation may be desirable. Medical surveillance for infectious disease end points emphasizes the identification of workers at risk and the early diagnosis of an infectious process.[8]

The goals of periodic surveillance are primarily to (i) detect early signs and symptoms of disease and (ii) detect changes in the health of employees indicating a need for changes in job functions and/or work process. The genetically modified organism does not typically have human health effects different from those associated with the unmodified organism. However, the genetically modified organism may cause allergenic responses or the expressed products may have toxic effects, and these may warrant medical surveillance. The periodic evaluation usually includes a health questionnaire. Other components of the evaluation depend on the type of exposure and health effect and should be targeted to specific exposures and health risks.[9] For example, a lung function assessment is appropriate if there is potential exposure to asthmagen(s) or endotoxin.

In general, implementation of the hazard control plan at work protects the workers from a significant risk of health effects related to work. However, removal from work with certain pathogenic microorganisms or biologically active products may be necessary in the case of an allergic or a susceptible individual.

Is medical surveillance also indicated for a special subpopulation of workers who are more susceptible to health effects from recombinant DNA technology? Who is potentially susceptible to health effects related to work with genetically modified microorganisms? Similar to any work with pathogenic microorganisms, individuals with reduced immunocompetence (including steroid treatment) and individuals with less effective barriers to infection (usually related to disease of the respiratory tract or gastrointestinal tract or illness or injury of the skin) represent a potentially susceptible population. Preplacement evaluations are therefore recommended to identify persons with medical conditions that may increase risk of adverse health effects in work with infectious organisms, including recombinant microorganisms. The focus of the preplacement evaluation is the potential for altered host defenses.

PREVENTION

Guidelines for laboratory animal use were formulated in the 1960s, and with the expanding use of recombinant organisms in research laboratories, the NIH has developed guidelines for research involving recombinant DNA, which have been revised annually.[10,11] The Centers for Disease Control and Prevention (CDC)/NIH revised biosafety guidelines were issued in 2009.[12] There are four levels of biosafety containment (BSL).[12,13] Level 1 is for well-characterized agents not known to cause disease in healthy adults and of minimal hazard to personnel. Level 2 is for agents with moderate potential hazard and requires special training for personnel. Level 3 is for potentially lethal infectious agents and requires specific containment, protective clothing, and special safeguards, such that many laboratories are not able to handle this work. Level 4 is for exotic highly pathogenic, poorly understood pathogens and is extremely rare in either the academic or industrial setting. Very recently, a new European Council Directive on contained use of genetically modified agents was approved.[14] Member states have the option of requiring even more stringent safety measures.[15] The European Directive focuses on appropriate training and supervision in the workplace where genetically modified organisms are used. In the United States, the NIH Recombinant DNA Advisory Committee (RAC) has set up strict guidelines for most recombinant DNA technology including gene therapy. Thus it is clear that biological safety has become a paramount issue on a global scale and that recombinant organisms have provided the impetus for these developments.

References

1. Walsh G. Biopharmaceutical benchmarks. *Nature Biotechnol* 2000; 18:831–3.
2. Cohen R, Hoerner CL. Recombinant DNA technology: a 20-year occupational health retrospective. *Rev Environ Health* 1996; 11(3):149–65.
3. Klees JE, Joines R. Occupational health issues in the pharmaceutical research and development process. *Occup Med: State of the Art Rev* 1997; 12(2):5–27.
4. Ducatman Alan M, Coumbis John J. Chemical hazards in the biotechnology industry. *Occup Med: State of the Art Rev*, 1991; 6(2):193–208.
5. Liberman DF, Israeli E, Fink R. Risk assessment of biological hazards in the biotechnology industry. *Occup Med: State of the Art Rev* 1991; 6:2.
6. Vidal DR, Paucod JC, Thibault F, et al. Biological safety in the laboratory. Biological risk, standardization and practice. *Ann Pharmaceut Francaises* 1993; 51(3):154–66.
7. Finn AM, Scott AJ, Stave GM. Genetic modification and biotechnology. In: Baxter PJ, Adams PH, Aw T-C, et al., eds. *Hunter's diseases of occupations*, 9th edn. Arnold, London, 2000, 521–35.
8. Rosenberg J, Clever LH. Medical surveillance of infectious disease endpoints. *Occup Med: State of the Art Rev* 1990; 5(3):583–605.
9. Goldman RH. Medical surveillance in the biotechnology industry. *Occup Med: State of the Art Rev*, 1991; 6(2):209–25.
10. Department of Health and Human Services, National Institutes of Health. NIH Guidelines for research involving recombinant DNA molecules; November 2013. 78 Federal Register 66751. Available at: http://osp.od.nih.gov/office-biotechnology-activities/biosafety/nih-guidelines (accessed on June 17, 2016).
11. McGarrity GJ, Hoerner C. Biological safety in the biotechnology industry. In: Fleming DO, Richardson JH, Tulis JT, et al., eds. *Laboratory Safety: Principles and Practices*, 2nd edn. 1995. American Society of Microbiology, Washington, DC, 119–31.
12. Chosewood LC, Wilson DE, Centers for Disease Control and Prevention, National Institutes of Health. *Biosafety in Microbiological and Biomedical Laboratories*, 5th edn. HHS Publication no. (CDC) 21-1112. U.S. Department of Health and Human Services, Public Health Service, Centers for Disease Control and Prevention, National Institutes of Health, Washington, DC, 2009. Available at: http://www.cdc.gov/biosafety/publications/bmbl5/bmbl.pdf (accessed on July 2, 2016).
13. Gilpin RW. Research activities including pathogens, recombinant DNA, and animal handling. In: McCunney RJ, Barbanel CS, eds. *Medical Center Occupational Health and Safety* 1999. Lippincott Williams & Wilkins, Philadelphia, 115–36.
14. Vranch S. New directive on biosafety: the contained use of genetically modified microorganisms. *Pharmaceut Technol Eur* 2000; 12(5):42.
15. Health and Safety Executive. The SACGM Compendium of guidance: Guidance from the Scientific Advisory Committee on Genetic Modification 2014. HSE Books, London. Available at: http://www.hse.gov.uk/biosafety/gmo/acgm/acgmcomp/ (accessed on July 4, 2016).

Chapter 32

PRIONS: CREUTZFELDT–JAKOB DISEASE (CJD) and RELATED TRANSMISSIBLE SPONGIFORM ENCEPHALOPATHIES (TSEs)

Dennis J. Darcey

Common names for disease: Sporadic Creutzfeldt–Jakob disease, new variant Creutzfeldt–Jakob disease, fatal familial insomnia, kuru, Gerstmann–Sträussler–Scheinker disease

Classification: Prion diseases

OCCUPATIONAL SETTING

There are theoretical but unproven risks to healthcare workers including physicians, surgeons, pathologists, nurses, dentists, laboratory technicians, veterinarians, veterinary technicians, agriculture workers, farmers, meat processors, butchers, abattoir workers, and cooks exposed to blood and uncooked animal products.

EXPOSURE (ROUTE)

Transmission of Creutzfeldt–Jakob disease (CJD) from human to human has been reported for patients receiving corneal transplants, dural grafts, and human growth hormone and gonadotropins derived from pooled human cadaver pituitary glands. Transmission has also been linked to inadequately sterilized instruments and stereotactic electrodes. In addition, transmission of kuru by ritual cannibalism has been documented. Although there have been case reports of healthcare personnel, veterinarians, and farmers developing CJD, there is no firm epidemiologic evidence to date to suggest that work in these occupations increases the risk for developing CJD.

A new variant Creutzfeldt–Jakob disease (nvCJD) has been associated with consumption of brain and spinal cord in sausages, hamburger, and other processed meats from cattle with bovine spongiform encephalopathy (BSE), or "mad cow disease," in the United Kingdom. Transmission of CJD through human blood or blood products has not been documented but is of some concern.

In August 1999 the FDA issued guidelines to reduce the possible risk of transmission of CJD and nvCJD by blood and blood products.[1] As a result, the Red Cross has implemented restrictions on blood donors who have been diagnosed with CJD, have relatives with CJD, received a dura mater (brain covering) transplant, or received human pituitary growth hormone. Potential blood donors who spent a total time of 3 months or more during 1980–1996 in the United Kingdom during the mad cow disease epidemic or who received a blood transfusion in the United Kingdom or France during this time frame are also restricted from blood donation. In addition members of the of the US military, civilian military employees, and dependents of a member of the US military who spent a total time of six months on or associated with a certain military bases during the specified time frames are also restricted from blood donation.

PATHOBIOLOGY

Prion diseases or transmissible spongiform encephalopathies (TSEs) are a family of rare progressive neurodegenerative disorders that affect both humans and animals. They are distinguished by long incubation periods, characteristic spongiform changes associated with neuronal loss, and a failure to induce an inflammatory response.

A number of transmissible spongiform encephalopathies or prion diseases have been described in animals and humans. Animal diseases include scrapie in sheep and goats, transmissible mink encephalopathy, wasting disease of deer and elk,

Physical and Biological Hazards of the Workplace, Third Edition. Edited by Gregg M. Stave and Peter H. Wald.

bovine spongiform encephalopathy, and feline spongiform encephalopathy. Human diseases include sporadic CJD, nvCJD, Kuru, and genetically transmitted familial Creutzfeldt–Jakob disease, Gerstmann–Sträussler–Scheinker disease, and fatal familial insomnia.

The causative agents of TSEs are believed to be prions, a small proteinaceous infectious particle. The term "prion protein" (PrP) was coined by Prusiner in 1982 to describe a novel host membrane sialoglycoprotein that caused scrapie in sheep.[2] This PrP was apparently able to "replicate" without DNA or RNA. The PrP was resistant to treatments that inactivate nucleic acids and viruses including ionizing radiation, alcohol, formalin, proteases, and nucleases. It was inactivated by treatments that disrupt proteins including phenol, detergents autoclaving, and extremes of pH. The protease-resistant protein associated with disease proved to be an isoform of a protease-sensitive normal host cellular protein. The functions of normal prion proteins are still not completely understood. The abnormal folding of the prion proteins leads to brain damage and the characteristic signs and symptoms of the disease. Prion diseases are usually rapidly progressive and always fatal.[3]

DIAGNOSIS

CJD was first described in 1920 and occurs worldwide with an incidence of 0.5–1.5 cases per million population per year. CJD generally occurs between the ages of 50 and 70. The incidence of the disease has been constant over many decades, and there is no apparent geographic or seasonal clustering of the sporadic form. A genetically determined familial variant has also been reported. To date no environmental risk factors for the disease have been identified although dietary factors, exposure to animals, and occupational exposures have been evaluated in epidemiologic studies. Surgeons and pathologists handling infected human brain tissue, as well as abattoir workers and cooks exposed to blood and uncooked animal products do not appear to have an increased risk for disease.[4,5]

Patients with CJD complain of fatigue, disordered sleep, decreased appetite, memory loss, and confusion. Focal neurological signs, such as ataxia, aphasia, visual loss, hemiparesis, and amyotrophy are reported. A diagnosis of CJD is suggested by a progressive loss of cognitive abilities, the development of mild chronic jerking, and pyramidal tract, cerebellar, and extrapyramidal signs. During the latter stages of disease, the patient becomes mute and akinetic. The mean survival time is 5 months, and 80% of patients with sporadic disease die within 1 year.[6]

There appear to be no diagnostic peripheral blood abnormalities or abnormalities in the cerebral spinal fluid that are helpful in the diagnosis. The electroencephalogram can be normal early in the disease but has been shown to be abnormal in 90% of patients if repeated during the course of the illness. As the disease progresses, computed tomography may show progressive generalized atrophy, and magnetic resonance imaging may show hyperintense signals in the basal ganglion on T2-weighted images.

Abnormal protein patterns in the cerebral spinal fluid of patients with CJD were found with two-dimensional electrophoresis, but the method was not practical for routine use.[7] Histological examination of the brain and immunostaining for the abnormal PrP are the gold standards for diagnosis. Spongiform changes accompanied by neuronal loss and gliosis are common features. Amyloid plaques are found in 10% of the brains in the sporadic form of CJD. In contrast, plaques are common in kuru, some of the familial spongiform encephalopathies, and new variant Creutzfeldt–Jakob disease. Tonsillar biopsy has been suggested as a possible pathological diagnostic technique in the living patient, particularly for nvCJD.[8] A confirmatory diagnosis of CJD requires neuropathological and/or immunodiagnostic testing of brain tissue obtained either at biopsy or autopsy.

In the United States, the National Prion Disease Pathology Surveillance Center (NPDPSC) was established in 1997 at Case Western Reserve University in response to concerns about the transmission to humans of prion diseases, or spongiform encephalopathies, acquired from animals, as well as transmission between humans during medical procedures.

TREATMENT

CJD and the related transmissible spongiform encephalopathies are universally fatal. There is no effective treatment or vaccine.

PREVENTION

Although iatrogenic transmission of CJD has been reported, and possible transmission of bovine spongiform encephalopathy to humans has been proposed, the cause of sporadic CJD that comprises 90% of the cases remains unknown. To date, there does not appear to be an increased risk of transmission from CJD patients or tissues to surgeons, pathologists, dentists, or other healthcare professionals. Nor does there appear to be an increased risk to veterinarians, meat packers, abattoir, or agricultural workers handling infected TSE animals or tissues.

Unlike most other pathogens, prions resist ordinary disinfection and sterilization techniques and can remain infectious for years even in a dried state. Thus, infected spinal fluid and tissue from the brain, spine, and eyes are theoretically of concern. Healthcare personnel caring for patients with CJD should follow universal precautions for blood-borne pathogens. Gloves should be worn when in contact with central nervous system tissue, contaminated instruments, and surfaces and when handling blood and body fluids, particularly CSF. Gowns and plastic aprons are indicated if splattering of blood or tissue is anticipated. Masks or protective goggles should be worn if aerosol generation or splattering is likely to occur, such as in dental or surgical procedures, wound irrigations, postmortem examinations, bronchoscopy, or endoscopy. Needles and other sharp objects should never be clipped, recapped, or removed from disposable syringes. Needles should be disposed of in a puncture-proof, leak-proof, and rigid plastic biohazard container.

Extensive precautions should be taken when handling cerebral spinal fluid and brain tissue. Where practicable disposable instruments are preferable. The Centers for Disease Control and Prevention and the World Health Organization have published guidelines for sterilizing instruments, tissues, or other materials known to contain prions.[9–11]

References

1. Food and Drug Administration. Guidance for Industry: Revised Preventive Measures to Reduce the Possible Risk of Transmission of Creutzfeldt-Jakob Disease (CJD) and Variant Creutzfeldt-Jakob Disease (vCJD) by Blood and Blood Products, 2016. Available at: http://www.fda.gov/downloads/BiologicsBloodVaccines/GuidanceComplianceRegulatoryInformation/Guidances/Blood/UCM307137.pdf (accessed on July 4, 2016).
2. Prusiner SB. Novel proteinaceous infectious particles cause scrapie. *Science* 1982; 216:136–44.
3. Centers for Disease Control and Prevention. Prions. http://www.cdc.gov/prions/index.html (accessed August 24, 2016).
4. Harris-Jones R, Knight R, Will RG, et al. Creutzfeldt-Jakob disease in England and Wales, 1980–1984: a case–control study of potential risk factors. *J Neurol Neurosurg Psychiatry* 1988; 51:1113–9.
5. Wientjens DP, Davanipour Z, Hofman A, et al. Risk factors for Creutzfeldt-Jacob disease: a reanalysis of the case control studies. *Neurobiology* 1996; 46:1287–91.
6. Brown P, Gibbs CJ Jr., Rodgers-Johnson P, et al. Human spongiform encephalopathy; the National Institutes of Health series of Health series of 300 cases of experimentally transmitted disease. *Ann Neurol* 1994; 35:513–29.
7. Harrington MG, Merril CR, Asher DM, et al. Abnormal proteins in the cerebrospinal fluid of patients with Creutzfeldt-Jakob disease. *New Engl J Med* 1986; 315:279–83.
8. Hill AF, Zeidler M, Ironside J, et al. Diagnosis of new variant Creutzfeldt-Jakob disease by tonsil biopsy. *Lancet* 1997; 349:99–100.
9. Centers for Disease Control and Prevention. Creutzfeldt-Jakob Disease Infection-Control Practices. http://www.cdc.gov/prions/cjd/infection-control.html (accessed August 24, 2016).
10. WHO. WHO Infection Control Guidelines for Transmissible Spongiform Encephalopathies. Report of a WHO Consultation, Geneva, Switzerland, March 23–26, 1999.
11. Chosewood LC, Wilson DE, Centers for Disease Control and Prevention, National Institutes of Health. *Biosafety in microbiological and biomedical laboratories*, 5th edn. HHS Publication no. (CDC) 21-1112, 2009. Washington, DC: U.S. Department of Health and Human Services, Public Health Service, Centers for Disease Control and Prevention, National Institutes of Health. Section VIII—H: Prion Diseases. Available at: http://www.cdc.gov/biosafety/publications/bmbl5/bmbl.pdf (accessed July 13, 2016).

Chapter 33

ENDOTOXINS

Robert R. Jacobs*

Common names for disease: Cotton dust—byssinosis and brown lung. Additionally, an active febrile response to endotoxin has been given many names depending on the source of exposure such as mattress maker's fever, mill fever, and card room fever.

OCCUPATIONAL SETTING

Agricultural workers and processors of vegetable fibers are most likely to be at risk. Workplaces with potentially high airborne concentrations of endotoxins include cotton and flax mills, grain storage and handling operations, poultry houses and processing plants, saw and paper mills, sewage treatment plants, and swine confinement buildings. Workers may also be exposed during animal handling in various facilities or during composting operations. Contamination of cutting fluids used in machining operations, workplace humidifying systems, or biotechnology processes using Gram-negative bacteria can also result in significant exposure.

EXPOSURE (ROUTE)

Occupational exposures are predominantly by inhalation. Significant exposure can occur wherever aerosols of materials contaminated with Gram-negative bacteria are generated, including office buildings and residential structures. There is no evidence to support a dermal route of exposure for endotoxins.

*The author wishes to thank Brian A. Boehlecke for his contribution to a previous version of this chapter.

PATHOBIOLOGY

Endotoxin refers to the lipopolysaccharide (LPS) complex and associated proteins found in the outer layer of the cell wall of Gram-negative bacteria. Although the lipid component (lipid A) is responsible for most of the toxic effects, variability of the biological responses to endotoxin exposures in natural settings may be due in part to influences from associated cell wall components, which differ among Gram-negative bacterial species.

Most clinical attention has focused on the toxicity of endotoxins reaching the bloodstream from endogenous Gram-negative bacterial infections of the host or via contaminated parenteral products.[1,2] However, little endotoxin is detectable in the blood after inhalational exposure. Local uptake in the respiratory system appears to account for the manifestations observed from exposures in the workplace.

ACUTE EXPOSURES

An acute febrile response may follow inhalation of aerosols containing high concentrations of endotoxin.[3] This is especially true for individuals with no prior occupational exposure or after a hiatus in exposure for workers with chronic low-level exposures. Historically, process or material-specific descriptors such as mattress maker's fever, card room fever, mill fever, and grain fever were used to describe acute responses in environments with endotoxin. Monday fever is named for its occurrence on the first workday after a weekend break. In 1986, the generic term organic toxic dust syndrome (OTDS) was proposed to emphasize the

Physical and Biological Hazards of the Workplace, Third Edition. Edited by Gregg M. Stave and Peter H. Wald.

commonality of this response to many agents and the nonimmunologic nature of its pathobiology.[4]

The fever usually begins several hours after exposure. Other symptoms include chills, myalgias, malaise, anorexia, headache, cough, and chest tightness. The condition is self-limited and usually lasts only a day or two, but it may be more prolonged if the exposure is massive. Tolerance develops with repeated exposures, and symptoms diminish despite similar doses of endotoxin.

A syndrome commonly described in association with exposure to cotton dust mimics many of the symptoms associated with acute inhalation exposure to endotoxins. Symptoms associated with exposure to cotton dust consist of recurrent chest symptoms that begin several hours into the workshift, often on the first workday after a break. Chest tightness with or without cough or shortness of breath develops gradually and may persist for several hours after the exposure ceases. Symptoms generally remit by the following day and do not recur or are markedly diminished on subsequent days, despite continued exposure. Some, but not all, workers with these symptoms also show a significant decline in ventilatory function over the workshift, with a decrease in the forced expiratory volume in the first second (FEV_1) on spirometry. This value generally returns to pre-exposure levels by the following day. The term "byssinosis" has been used when this syndrome is associated with exposure to cotton dust.

The exact mechanism for the acute symptoms and ventilatory changes has not been established. Animal and human exposure studies have shown that inhaled endotoxins interact with alveolar macrophages, initiating a cascade of events that result in the expression and release of inflammatory mediators such as IL-1 and TNF-alpha.[5,6] Thus, endotoxin inhalation can produce airway inflammation and bronchoconstriction, which could account for the symptoms associated with cotton dust.

Although not extensively studied, inhalation of organic dusts or purified endotoxin has been shown to transiently increase nonspecific airway reactivity, as measured by methacholine or histamine bronchoprovocation. Both healthy people and persons with mild atopy or frank asthma have shown bronchial hyperresponsiveness lasting from a few hours to 24 hours after exposure. The clinical significance of these findings is uncertain. Acute airway inflammation may be a contributing factor to these changes.[7]

Some epidemiologic studies of textile workers have found better correlation between chest symptoms and estimated doses of inhaled endotoxin than with crude measures of dust exposure. Controlled exposures of subjects preselected for acute ventilatory responses to cotton dust showed a strong correlation between drop in FEV_1 during exposure and airborne endotoxin concentration, but no association with respirable dust concentration uncorrected for endotoxin content.[8] These findings suggest that the endotoxin content of cotton dusts and other organic dust aerosols is likely to be an important factor in producing the acute recurrent chest symptoms in exposed persons. This also suggests that measures of airborne endotoxin may be a better predictor of the acute responsiveness to aerosols of cotton dust than gravimetric dust measures. However, other studies have shown biological activity of endotoxin-free extracts of organic dusts. Therefore, the mechanism for responses to natural organic dust exposures may be complex and dependent on interactions of multiple agents.

CHRONIC EXPOSURE

Numerous epidemiologic studies have shown that workers exposed to cotton dust are at increased risk for symptoms of chronic bronchitis and for airway obstruction detectable by spirometry. Retrospective autopsy studies have confirmed the increased prevalence of pathologic changes of chronic bronchitis, even in nonsmoking textile workers. Prospective studies have now confirmed that decrements in ventilatory function in textile workers that exceed the expected rate are correlated with dust exposure. Cigarette smoking is clearly an important factor; however, accelerated loss in FEV_1 also occurs in nonsmokers. Although this condition has been referred to as byssinosis, or brown lung, when associated with cotton dust exposure, workers exposed to other organic dusts are also at risk. A recent review concluded that chronic exposure to organic dust in the textile industry has characteristics similar to both asthma and COPD and that cessation of exposure may lead to improved respiratory function.[9]

The role of endotoxin in the pathogenesis of this chronic condition is not clear.[10] In animals, the inflammatory response seen acutely after endotoxin inhalation appears to diminish with repeated exposures. Also, chronic high-level exposures to organic dusts containing endotoxin do not consistently reproduce the pathologic changes of chronic bronchitis in animal models. Tolerance to the effects of endotoxin given intravenously is well documented in humans as well as animals. In healthy young volunteers, the Th1 type of lymphocyte cytokine response is decreased after intravenous LPS challenge.[11] A recent study analyzed the inflammatory response associated with persistent airflow obstruction resulting from chronic exposure to low doses of endotoxin that would be similar to exposure that might occur in occupational environments. After 8 weeks of chronic low-dose exposure, there was an increase in airway hyperresponsiveness and an increase in lung neutrophils was observed that correlated with an increase in proinflammatory cytokines. Evaluation of inflammatory cell subsets showed an increase in proinflammatory dendritic cells (DCs) with a reduced percentage of macrophages. Gene expression profiling showed up-regulation of genes consistent with DC recruitment and lung histology revealed an accumulation of inflammatory aggregates of DCs around the airways. Thus, repeated, low-dose endotoxin resulted in airway hyperresponsiveness, associated with a failure to resolve the proinflammatory response, an inverted macrophage to dendritic cell ratio, and an increase in the inflammatory dendritic cell population. These results demonstrate a possible underlying mechanism of airway obstruction for chronic endotoxin exposure that is consistent with symptoms observed for cotton and other organic dust.[12]

Respiratory effects similar to those seen in textile workers have also been reported in workers exposed to non-cotton organic dust environments with endotoxin. Poultry house workers had drops in FEV_1 over the workshift that correlated with endotoxin exposure.[13] However, the correlation of lung function decrement with total dust exposure was stronger than that with the endotoxin component. Sewage treatment plant workers had increased nonspecific airway responsiveness measured by methacholine challenge compared to controls, but the difference was small and no difference in lung function was found.[14]

DIAGNOSIS

The nonspecific nature of these syndromes makes definitive diagnosis difficult. A careful history is critical to establish a pattern consistent with an association between symptoms and the putative causal exposure. For clinical evaluations, self-measurement and recording of peak flows using portable devices several times per day for 2 weeks may be helpful. Consistent acute worsening of ventilatory function during exposure or a progressive decline during the workweek with improvement on weekends suggests an association with exposure. However, lack of significant findings does not exclude an association between the symptoms and the exposure. Auscultation of the chest will generally not reveal wheezing or rales unless another underlying condition is present, which accounts for the finding. Persons with underlying hyperreactive airways may develop symptomatic bronchospasm with wheezing in response to organic aerosol exposures without detectable sensitization to any components of the aerosol.

No specific diagnostic tests are available for the syndrome of recurrent chest symptoms. No evidence for a direct immunologic mechanism has been found in textile workers with either underlying recurrent chest symptoms (byssinosis) or acute declines in FEV_1 over the workshift. RAST for specific IgE is therefore of no value. Chest radiographs will be normal unless there is other pulmonary disease. Clinical judgment is, of course, necessary to determine if other causes for chest symptoms, such as coronary artery disease, should be pursued. Longer term trends in ventilatory function may show losses in FEV_1 greater than expected from aging alone. Normal variability and the influence of non-work-related exposures, including cigarette smoking, must be considered when interpreting these findings. However, an accelerated decline in function associated with recurrent work-related chest symptoms or acute decrements in function during exposure merits attention and consideration of reduction in exposure.

Demonstration of environmental airborne concentrations of endotoxin of the magnitude that has been associated with these findings in other settings would support a causal relationship. Threshold concentrations of airborne endotoxin for the syndromes described have been postulated for cotton dust environments.[3] They vary from 0.5–1 milligram per cubic meter for the febrile reaction to 0.1–0.5 milligram per cubic meter for chest tightness and acute declines in ventilatory function in chronically exposed individuals. Previously nonexposed persons may require higher concentrations (2–3 milligram per cubic meter) to develop acute symptoms of chest tightness. Endotoxin concentrations associated with increased airway reactivity and bronchitis are postulated to be lower, but they are more speculative, given the interaction of the many potential agents to which a worker may be exposed over the course of many years.

Established chronic bronchitis and airway obstruction associated with organic dust exposure has no inherent pathophysiologic features that allow differentiation from that caused by cigarette smoking. A history of prior recurrent chest symptoms or acute ventilatory declines associated with the exposure is helpful, but its absence does not eliminate the possibility that exposure is a causal factor. The magnitude and duration of exposure must always be considered; the greater the cumulative dose, the more likely it is the exposure contributed to the development of the condition. When a history of significant cigarette consumption is also present, the contribution of exposure to the pathogenesis of the condition is difficult to assess.

The threshold endotoxin levels described above are based on the measurement of endotoxin by the Limulus amebocyte lysate (LAL) method. However, the bioavailability of endotoxin in aerosols may vary depending on the nature of the materials. Therefore, concentrations measured in extracts by LAL may not be accurate predictors of the toxic potential in the natural setting. Based on experimental exposures, cell-bound endotoxin has been estimated to be three times as biologically active as isolated endotoxin in equal amounts, as measured by LAL. Also, inter-laboratory variation in measurement of endotoxin is large.

The problem of assessing exposure to airborne endotoxins has been studied for years, but sampling and analysis procedures are still not adequately standardized. The different protocols mean there is broad inter-laboratory variability in the results of endotoxin analyses. Standardized methods for sample collection, storage, extraction, and analysis have been proposed and round robin testing, using standardized protocols, samples, and reagents, has been done. For example, a comparative study showed a sixfold range in reported endotoxin content of the same cotton dust by 11 laboratories experienced with this measurement, despite the use of a common extraction protocol and commercial lots of analytical reagents.[15] A review of the methods for measuring airborne endotoxin and their need for standardization has been published.[16]

Until standardization is achieved, endotoxin analyses should be done by a laboratory familiar with measurement of endotoxin in the type of material sampled. Caution should be exercised when comparisons are made between observed concentrations and proposed effect threshold levels.

TREATMENT

Reduction in exposure to the organic aerosol is the treatment of choice. The acute febrile reaction is usually self-limited and can be treated symptomatically if necessary. Recurrent chest symptoms may improve with pre-exposure use of an inhaled bronchodilator, or regular use of inhaled steroid medication, but these methods should not be used in lieu of reduction in exposure.

Established chronic bronchitis and airway obstruction may be treated with inhaled bronchodilators and steroids. Courses of antibiotics may be given when bacterial superinfections occur. Cessation of smoking and avoidance of other respiratory irritants are also indicated. Vaccination against pneumococcal pneumonia and yearly influenza vaccination are recommended for persons with moderate or severe airway obstruction.

MEDICAL SURVEILLANCE

Periodic medical questionnaires focusing on respiratory symptoms are helpful. The association of symptoms with exposure should be explored carefully with questions referring to the intensity of symptoms at home compared to at work, improvement during days off or vacations, and changes in severity related to specific job activities or use of personal protective equipment.

Periodic measurement of pre- and post-shift spirometry may show significant reductions in ventilatory capacity after exposure. Normal variability in spirometry must be considered when interpreting these results. The acute response to inhalation of organic dust may vary significantly on different occasions. Therefore, repeated observations are most helpful to determine if a meaningful pattern is present.

Traditionally, most surveillance examinations have been done on a yearly basis. However, modification of the interval based on the intensity of exposure and the presence or absence of symptoms is justifiable. If work-related symptoms are absent and exposure is relatively low, ventilatory function testing every 2 years is probably adequate to detect important long-term trends. Changes in intensity of exposure or the onset of symptoms are indications for consideration of more frequent monitoring.

PREVENTION

Currently, there are no enforceable standards that limit exposure to endotoxins. However, there are legal limits for exposure to selected types of organic dust. In Denmark, the exposure limit for "organic dust" is 3 mg/m^3 of "total" dust; in Norway and Sweden, it is 5 milligram per cubic meter. In the United States, the permissible exposure limit for grain dust is 10 mg/m^3 for total grain dust (established by the Occupational Safety and Health Administration (OSHA) in 1989) and the OSHA cotton dust standard limits exposure to 0.2 mg/m^3 PEL 8 hours time weighted average. Suggestions for other threshold limit values for different types of organic dust have been made; however, these were outside a formal standard making process.

Prior to 2010, the only recommendation for a health-based occupational exposure limit for endotoxin was 50 EU/m^3 as an 8 hours TWA by the Health Council of the Netherlands. [17] (Note: There are approximately 10 endotoxin units (EUs) per nanogram of endotoxin.) This recommendation was modified in 2010 to 90 EU/m^3 (8 hours TWA) and was based on the lung function results of a series of experimental cross-sectional studies exposing healthy individuals to cotton-derived endotoxins[8] and of an epidemiological cohort study among grain processing and animal-feed industry workers.[18,19] In addition to the exposure limit, the Health Council also recommended a specific endotoxin analysis method.[20]

However, until a reliable and reproducible method for measuring endotoxin is developed, the most effective primary preventive measure is to reduce exposure by decreasing workplace concentrations of endotoxin-containing materials. Traditional methods of reducing airborne contaminants include enclosure of aerosol-generating processes and increasing general ventilation. Decreasing the endotoxin content of organic dusts by selective removal has not yet reached commercial feasibility. Water washing of cotton prior to processing has been shown to reduce both the endotoxin content of the dust generated during processing and the acute dilatory functional decline associated with exposure. Other methods of detoxification, such as heat treatment, have not been successful due to the relative stability of endotoxin and the toxicity of cell wall components of nonviable bacteria.

The use of personal protective equipment to reduce the dose of inhaled aerosol can be effective. However, because the higher concentration of endotoxin appears to be associated with the smallest particulates in an organic dust-contaminated environment and workers dislike using respirators, due to their discomfort and interference with job activities, reliance on this technique is questionable. Nevertheless, a respirator program with adequate training and monitoring of use can be a valuable adjunct to other measures and may be necessary for adequate protection of especially sensitive workers. There is both *in vitro* and *in vivo* evidence that atopic persons have an increased risk for adverse effects. Atopic persons showed an increase in nonspecific airway reactivity when exposed experimentally to cotton dust.[21] Atopic individuals also had increased nasal airway eosinophils when challenged with LPS compared to after saline challenge.[22] Lymphocytes from persons allergic to birch pollen had increased expression of CD154, a component of the signaling pathway for allergic inflammatory responses, after incubation with pollen and LPS compared to incubation with pollen alone.[23] However, despite these findings, at present there is not sufficient justification for excluding atopic persons from jobs with potential exposure for endotoxin. If work-related symptoms and acute decrements in ventilatory function are found, reduction in exposure is clearly indicated.

References

1. Brigham KL, Meyrick B. Endotoxin and lung injury. *Am Rev Respir Dis* 1986; 133:913–27.
2. Ghosh S, Latimer RD, Gray BM, et al. Endotoxin-induced organ injury. *Crit Care Med* 1993; 21:519–24.
3. Rylander R. Health effects of cotton dust exposures. *Am J Ind Med* 1990; 17:30–45.
4. doPico GA. Health effects of organic dusts in the farm environment. Report on diseases. *Am J Ind Med* 1986; 10:261–5.
5. Imrich A, Yu Ning Y, Koziel H, et al. Lipopolysaccharide priming amplifies lung macrophage tumor necrosis factor production in response to air particles. *Toxicol Appl Pharmacol* 1999; 159:117–24.
6. Liebers V, Raulf-Heimsoth M, and Bruning T; Health effects due to endotoxin inhalation: *Arch Toxicol* 2008; 82:203–10.
7. Michel O, Ginanni R, LeBon B, et al. Inflammatory response to acute inhalation of endotoxin in asthmatic patients. *Am Rev Respir Dis* 1992; 146:352–7.
8. Castellan RM, Olenchock SA, Kinsley KB, et al. Inhaled endotoxin and decreased spirometric values. *N Engl J Med* 1987; 317:605–10.
9. Lai PS, Christiani DC. Long term respiratory health effects in textile workers. *Curr Opin Pulm Med* 2013; 19(2): 152–7.
10. Rylander R. The role of endotoxin for reactions after exposure to cotton dust. *Am J Ind Med* 1987; 12:687–97.
11. Lauw FN, Ten Hove T, Dekkers P, et al. Reduced Th1 but not Th2 cytokine production by lymphocytes after in vivo exposure of healthy subjects to endotoxin. *Infect Immun* 2000; 68:1014–8.
12. Lai PS, Fresco JM, Pinilla MA, et al. Chronic endotoxin exposure produces airflow obstruction and lung dendritic cell expansion, *Am J Respir Cell Mol Biol* 2012; 47(2):209–17.
13. Donham KJ, Cumro D, Reynolds SJ, et al. Dose–response relationships between occupational aerosol exposures and cross-shift declines of lung function in poultry workers: recommendations for exposure limits. *J Occup Environ Med* 2000; 42:260–9.
14. Rylander R. Health effects among workers in sewage treatment plants. *Occup Environ Med* 1999; 56:354–7.
15. Chun DTW, Chew V, Bartlett K, et al. Preliminary report on the results of the second phase of a round-robin endotoxin assay study using cotton dust. *Appl Occup Environ Hyg* 2000; 15(1):152–7.
16. Duquenne P, Marchand G, Duchaine C. Measurement of endotoxins in bioaerosols at workplace: a critical review of literature and a standardization issue. *Ann Occup Hyg* 2012; 57: 137–72.
17. Health Council of the Netherlands, Dutch Expert Committee on Occupational Standards (DECOM). *Endotoxins*. Rijswijk: Health Council of the Netherlands, 1998; publication no. 1998/O3WDG.
18. Post W, Heederik D, Houba R. Decline in lung function related to exposure and selection processes among workers in the grain processing and animal feed industry. *Occup Environ Med* 1998; 55: 349–55.
19. Smid T, Heederik D, Houba R, et al. Dust- and endotoxin-related respiratory effects in the animal feed industry. *Am Rev Respir Dis* 1992; 146: 1474–9.
20. Spaan S, Heederik DJ, Thorne PS, et al. Optimization of airborne endotoxin exposure assessment: effects of filter type, transport conditions, extraction solutions, and storage of samples and extracts. *Appl Environ Microbiol* 2007; 73: 6134–43.
21. Jacobs R, Boehlecke B, Van Hage-Hamsten M, et al. Bronchial reactivity, atopy and air-way response to cotton dust. *Am Rev Respir Dis* 1993; 148:19–24.
22. Peden DB, Tucker K, Murphy P, et al. Eosinophil influx to the nasal airway after local, low level LPS challenge in humans. *J Allergy Clin Immunol* 1999; 104(2 Pt 1):388–94.
23. Nakstad B, Kahler H, Lyberg T. Allergen-stimulated expression of CD154 (CD40 ligand) on CD3+ lymphocytes in atopic but not in nonatopic individuals. Modulation by bacterial lipopolysaccharide. *Allergy* 1999; 54:722–9.

Chapter 34

WOOD DUST

Harold R. Imbus and Gregg M. Stave

OCCUPATIONAL SETTING

Exposure to wood dust occurs in many industries, including logging and sawmill operations, furniture manufacturing, paper manufacturing, construction of residential and commercial buildings, and especially carpentry and cabinet making. Workers are exposed when wood is sawed, chipped, routed, or sanded.

Wood may also contain biological or chemical contaminants. Biological contaminants include molds and fungi, which often grow on the bark of wood. Exogenous chemicals include those used in treating the wood. Common wood preservatives are arsenic, chromium, copper, creosote, and pentachlorophenol. Wood also contains many endogenous chemicals that are responsible for its biological actions.

EXPOSURE (ROUTE)

Wood dust exposure occurs through inhalation. In general, the finer particles of wood dust are more biologically active due to their greater surface area and their ability to penetrate and adhere to the respiratory mucosa. Furniture manufacturing and cabinet making are operations that produce the finer particles. Contact with skin or mucous membranes may also have health consequences.

PATHOBIOLOGY

Wood dust is composed of wood particles generated by the processing or handling of wood. Hardwoods, such as maple, oak, and cherry, come from deciduous trees with broad leaves. Softwoods come from evergreen trees such as pine, spruce, and fir. The terms are somewhat misleading in that some of the hardwoods may be soft and some of the softwoods may be relatively hard.

Health effects of wood dust may be classified primarily as irritation, sensitization, and cancer. In the case of irritation, these effects can involve the skin, eyes, or respiratory tract. Allergic manifestations can involve skin or respiratory tract, and cancer associated with wood dust exposure involves the sinonasal tract.

Skin irritation caused by wood dust is often mechanical. Splinters or tiny particles of wood can get under the skin and cause irritation or infection. Soaps and chemicals can add to the irritation. Particles may lodge between the folds of skin, and sweat and rubbing can result in inflammation. Good personal hygiene practices and protection of exposed areas can obviate this type of problem.

Some woods, mostly foreign and exotic species, contain chemicals that are irritants. These can cause dermatitis, resulting in redness and blistering. Eyes may also become irritated. Teak, mansonia, and radiata pine have been reported to cause such reactions.

Allergic contact dermatitis can result from some species, again mostly from foreign woods such as teak and African mahogany. However, some of the North American woods, such as Douglas fir, western red cedar, poplar, airborne pine dust due to colophony,[1] and rosewood, may cause allergic contact dermatitis.[2] Specific chemicals used in glues and resin binders, such as urea or phenol–formaldehyde, potassium dichromate, ethylene glycol, and propylene glycol, can also be responsible for allergic contact dermatitis.

Wood dust is a particulate that causes irritation of the eyes and the upper and lower respiratory tracts. As with all

Physical and Biological Hazards of the Workplace, Third Edition. Edited by Gregg M. Stave and Peter H. Wald.

particulates, the magnitude of the effect depends on the size of the particulate and the amount of exposure. Some woods with strong chemicals are more irritating, and most wood is more irritating than inert dust. For example, wood dust has been found to be almost four times as irritating as plastic dust in the same concentration.[3]

Clinical epidemiologic studies of woodworkers have found frequent symptoms and physical findings of nasal irritation. Symptoms often consist of continued colds, nosebleeds, sneezing, and sinus inflammation. Unfortunately, most of the studies lack adequate controls and dose–response information. In one study, the employees in a dusty furniture plant, including office workers, had decreases in 1-second forced expiratory volume (FEV1) and forced vital capacity (FVC) during the workshift.[4] A large Vermont study that also measured pulmonary function found an inverse association between pulmonary flow (FEV1/FVC ratio) and indexes of wood dust exposure.[5] Similarly, a study of sawmill workers in Canada exposed to pine and spruce showed lower average values for FEV1 and FEV1/FVC(%).[6] A recent study in India showed that carpenters had lower peak expiratory flow rates compared with controls.[7] Mucociliary clearance in the nose has also been measured in woodworkers. Andersen et al. found a higher percentage of individuals with mucostasis among workers exposed to higher wood dust levels.[8] Mucociliary clearance is important in cleansing the nasal passages. Individuals with impaired mucous flow may be more susceptible to infection and other problems.

Western red cedar is well known for its potential to cause asthma. Workers in British Columbia, Canada, have been studied extensively.[9–20] Plicatic acid, an extract of western red cedar, was shown to be the cause of bronchial asthma.[20] Other domestic species capable of causing asthma include oak, ash, redwood, mahogany, eastern white cedar, and spruce.[21–24] A number of foreign exotic species are capable of causing allergic asthma. In a case report, Kespohl et al. determined that the allergens in spruce wood were peroxidases.[25]

Microorganisms on wood bark can cause Type III allergic reactions, or extrinsic allergic alveolitis.[26–28] Maple bark disease, sequoiosis, and suberosis are caused by the inhalation of fungal spores associated with maple, redwood, and cork, respectively.

Wood dust may also cause other lung diseases. High levels of small dust particles that penetrate into the bronchial tubes or smaller airways may produce irritation and bronchitis. Whether they produce irreversible changes is not known. Most reports of serious or permanent lung disease associated with wood dust exposure, until recently, have been anecdotal. However, later studies showed lower pulmonary function in wood mill workers,[29] higher prevalence of respiratory impairment,[30] and significant cross-shift declines of pulmonary function in wood dust-exposed workers.[31]

An English otolaryngologist was the first to note an unusually high incidence of nasal cancer in the chair making and furniture industry of High Wycombe, Buckinghamshire.[32] This finding was confirmed by Acheson et al.,[33,34] who found an incidence of nasal adenocarcinoma from 1956 to 1965 of 0.7 cases per 1000 per year, an incidence approximately 1000 times greater than that seen in the general population. Dust depositions were noted on the nasal mucosa of furniture workers at the anterior part of the nasal septum and at the anterior ends of the middle turbinates.[35] A biopsy of these areas revealed squamous metaplasia. Excesses of nasal adenocarcinoma have also been described in France, Australia, Finland, Italy, Holland, Denmark, and Belgium.[36–44] An early study in Canada showed no excess of nasal cancer in woodworkers; however, few furniture workers were involved in the study.[45] A 1967 Canadian study showed an odds ratio of 2.5 for occupations involving exposure to wood dust when compared to the general population.[46] A more recent study in five Nordic countries that followed 2.8 million incident cancer cases in 15 million people through 2005 observed a standardized incidence ratio (SIR) for nasal cancer of 1.84 (95% CI 1.66–2.04) in male and 1.88 (0.90–3.46) in female woodworkers. For nasal adenocarcinoma, the SIR in males was 5.50 (4.60–6.56).[47]

In the United States, a case-control study published in 1977 showed an approximate fourfold increase of nasal cancer in furniture workers in North Carolina.[48] Also, a Connecticut study showed an odds ratio of 4.0 for nasal cancer among persons exposed to wood dust.[49] In a follow-up study, Brinton et al. found an overall relative risk of nasal cancer among furniture workers of 0.74 in men and 0.91 in women; however, the relative risk of the rarer adenocarcinoma in male furniture workers was 5.68.[50] These excesses of adenocarcinoma, a very rare disease, are far lower than those noted in the United Kingdom and other countries. A review of nasal cancer in furniture manufacturing and woodworking in North Carolina suggested that there was some excess of nasal cancer in the industry prior to World War II, but it found little or no evidence of continuing excess risk,[51] while another review of US woodworkers confirmed this disparity between the United States and the United Kingdom.[52] Another pooled reanalysis observed excesses of nasal cancer only among workers in British furniture manufacturers as compared with those in the United States.[53] Likewise, in the United Kingdom, Acheson postulated that the factor in furniture manufacturing that gave rise to the nasal adenocarcinomas was present only between 1920 and 1940, since cases did not occur among workers who entered the industry after World War II.[54]

It is unlikely that exogenous chemicals were responsible for the excess nasal cancer, since they were not widely used prior to World War II. It may well be that differences in processing, including the use of tools that create more heat, or dust, or different particle sizes, account for the increased disease of the earlier years.[51]

It is less clear whether there is an association between wood dust exposure and lung cancer as studies have produced conflicting results.[55]

Beach and teak (hardwoods) and pine and spruce (softwoods) dusts have been shown to effect expression of cytokines and chemokines in an *in vitro* murine macrophage cell line.[56] They were all shown to induce TNF-α and inhibit IL-1β expression. Similarly, they induced the expression of CCL2, CCL3, CCL4, and CXCL2/3 chemokines and inhibited CCL24 expression. In a human epithelial cell line, wood dusts from beech, oak, teak, birch, pine, spruce, and medium-density fiberboard (MDF) dusts all induced a cytokine response with increased expression of cellular *IL-6* and *IL-8* mRNA.[57] There was also evidence of genotoxicity with an increase in DNA strand breaks after incubation with beech, teak, pine, and MDF dusts. No difference in genotoxicity was seen between hardwoods and softwoods in this study. Genotoxicity of beech and other woods has also been demonstrated in other *in vitro* models.[58–60] In respiratory epithelial cells, dusts from birch, oak, and pine were cytotoxic and stimulated the production of reactive oxygen species, and also caspace-3, one of the components of the apoptotic cascade.[61]

In 1995, the International Agency for Research on Cancer (IARC) classified wood dust: "Group I, carcinogenic to humans".[62] A 2009 review reaffirmed the classification (IARC 2011).[63]

DIAGNOSIS

A clinical diagnosis of allergy or asthma can be made based on a careful medical and occupational history along with a physical examination. A detailed respiratory and dermatologic history will reveal symptoms of allergic disorders. Respiratory symptoms may include rhinitis, conjunctivitis, sneezing spells, nasal congestion, cough, shortness of breath, and wheezing.

Documenting the occurrence of these symptoms as temporally associated with exposure to wood dust and the resolution of symptoms while away from work is critical to the diagnosis. Objective evidence of temporality may be obtained by doing a physical examination or peak flow measurements prior to exposure and again at the end of the workshift. The finding of objective evidence of bronchoconstriction, such as wheezes or a decrement in FEV1 and FVC, supports the diagnosis of occupational asthma. However, asthma due to allergenic wood dust may be delayed, occurring during evening or night after the work shift.

To evaluate suspected contact dermatitis, a patch test can be done with the sawdust of the wood itself or, in some cases, with an extract of woods such as teak. Patch testing can also be performed with exogenous chemicals if they are suspected of being the causative agent.

TREATMENT

Avoidance of exposure to wood dust should be the goal of treatment of allergy. Sensitized workers who have rhinitis should be counseled on their risk of developing asthma and the risk of eventual intolerance of exposure.

Symptomatic care may include antihistamines and/or nasal topical corticosteroids for upper tract symptoms. Treatment of wood dust asthma is the same as for other types of asthma, which, in addition to avoidance of exposure, includes short-term inhaled bronchodilators for acute attacks. For chronic wood dust asthma, standard treatments for asthma include inhaled corticosteroids alone or combined with long-acting bronchodilators. Oral medications include leukotriene modifiers, such as montelukast, and theophylline.

Skin disease can be treated with topical or, in severe cases, systemic corticosteroids. Work practices and controls should be modified to prevent further exposure.

MEDICAL SURVEILLANCE

Respiratory history may be used to document early symptoms of allergy. If nasal symptoms are detected early and further exposure is eliminated, subsequent progression to asthma may be prevented. Pre-placement, periodic, and possibly cross-shift pulmonary function testing may be useful, depending upon the type and quantity of wood dust, and any indicators of employee health problems related to their exposure.

PREVENTION

Containment of wood dust should be the cornerstone of allergy and irritation prevention measures. Efforts should be made to limit airborne dust through engineering controls such as ventilation systems. Personal protective equipment, including skin coverings and respirators, should be considered where significant exposure cannot be engineered out. The concentration of airborne dust is dependent on the activities in the contaminated areas.

Dust exposure evaluation should be made in accordance with good industrial hygiene practices, which will include personal inspection of the work site and use of a dust monitor for screening purposes. Personal sampling for both total and respirable dust should be done when the result indicates potential overexposure or if employee(s) complaints have been received. The American Conference of Governmental Industrial Hygienists (ACGIH) standard for western red cedar is 0.5 mg/m^3 (which is labeled as a sensitizer).[64] For all other wood dusts, the occupational exposure level (OEL) is 1 mg/m^3. The National Institute for Occupational Safety and Health (NIOSH) has set the recommended exposure limit for wood dust at 1 mg/m^3.

Until January 19, 1989, the Occupational Safety and Health Administration (OSHA) regulated wood dust as a nuisance dust with a permissible exposure level (PEL) of 15 mg/m^3. After that date, OSHA adopted the PEL standard, which at that time regulated wood dust at 5 mg/m^3 for hardwood and softwood; for western red cedar, the PEL was 2.5 mg/m^3. On July 7, 1992, the US Court of Appeals vacated the PEL standard. As a result, OSHA regulates wood dust as a nuisance dust for total dust at 15 mg/m^3 and respirable dust at 5 mg/m^3.[65] However, the use of lower exposure levels is prudent to reduce the risk of illness. The ACGIH recommendations are followed in many locations outside the United States.

Though this chapter has dealt with health issues pertaining to wood dust exposure, it should be noted that excess accumulation of wood dust in a workplace is a risk of fire and explosion. Therefore, work surfaces, machinery, floors, and rafters should be regularly cleaned of accumulations, and ambient air levels kept low. *A Guide to Combustible Dust* can be obtained from the North Carolina Department of Labor.[66]

References

1. Watsky KL. Airborne allergic contact dermatitis from pine dust. *Am J Contact Dermat* 1997; 8:118–20.
2. Weber LE. Dermatitis veneata due to native woods. *AMA Arch Dermatol Syphilol* 1953; 67:388–94.
3. Andersen I. Effects of airborne substances on nasal function in human volunteers. In: Carrow CS, ed. *Toxicology of the nasal passages*. Washington, DC: Hemisphere, 1986, 143–54.
4. Zuhair YS, Whitaker CJ, Cinkotai EF. Ventilatory function in workers exposed to tea and wood dust. *Br J Ind Med* 1981; 38:339–45.
5. Whitehead LW, Asbikaga T, Vacek P. Pulmonary function status of workers exposed to hardwood or pine dust. *Ann Ind Hyg Assoc* 1981; 42:1780–6.
6. Hessel, PA, Herbert FA, Melenka, LS, et al. Lung health in sawmill workers exposed to pine and spruce. *Chest* 1995; 108(3), 642–6.
7. Mohan M, Aprajita, Panwar NK. Effect of wood dust on respiratory health status of carpenters. *J Clin Diagn Res* 2013; 7(8): 1589–91.
8. Andersen HC, Solgaard J, Andersen I. Nasal cancer and nasal mucus transport rates in woodworkers. *Acta Otolayngol* 1976; 82:263–5.
9. Brooks SM, Edwards JJ, Edwards FE. An epidemiologic study of workers exposed to western red cedar and other wood dusts. *Chest* 1981; 80(suppl 1):30–2.
10. Chan-Yeung M. Maximal expiratory flow and airway resistance during induced bronchoconstriction in patients with asthma due to western red cedar (*Thuja plicata*). *Am Rev Respir Dis* 1973; 108:1103–10.
11. Chan-Yeung M. Fate of occupational asthma. A follow-up study of patients with occupational asthma due to western red cedar (*Thuja plicata*). *Am Rev Respir Dis* 1977; 116:1023–9.
12. Chan-Yeung M. Immunologic and nonimmunologic mechanisms in asthma due to western red cedar (*Thuja plicata*). *J Allergy Clin Immunol* 1982; 70:32–7.
13. Chan-Yeung M, Abboud R. Occupational asthma due to California redwood (*Sequoia sempervirens*) dusts. *Am Rev Respir Dis* 1976; 114:1027–31.
14. Chan-Yeung J, Ashley MJ, Corey P, et al. A respiratory survey of cedar mill workers. I. Prevalence of symptoms and pulmonary function abnormalities. *J Occup Med* 1978; 20:323–7.
15. Chan-Yeung M, Barton GM, MacLean L, et al. Bronchial reactions to western red cedar (*Thuja plicata*). *Can Med Assoc J* 1961; 105:56–8.
16. Chan-Yeung M, Barton GM, MacLean L, et al. Occupational asthma and rhinitis due to western red cedar (*Thuja plicata*). *Am Rev Respir Dis* 1973; 108:1094–102.
17. Chan-Yeung M, Lam S, Koener S. Clinical features and natural history of occupational asthma due to western red cedar (*Thuja plicata*). *Am J Med* 1982; 72:411–5.
18. Chan-Yeung M, Veda S, Kus J, et al. Symptoms, pulmonary function, and bronchial hyperreactivity in western red cedar workers compared with those in office workers. *Am Rev Respir Dis* 1984; 130:1038–41.
19. Vedal S, Chan-Yeung M, Enarson D, et al. Symptoms and pulmonary function in western red cedar workers related to duration of employment and dust exposure. *Arch Environ Health* 1986; 41:179–83.
20. Chan-Yeung M, Gicias PC, Henson PM. Activation of complement by plicatic acid, the chemical responsible for asthma due to western red cedar (*Thuja plicata*). *J Allergy Clin Immunol* 1980; 65:333–7.
21. Malo JL, Cartier A, Desjardins A, et al. Occupational asthma caused by oak wood dust. *Chest* 1995; 108:856–8.
22. Fernandez-Rivas M, Perez-Carral C, Senent CJ. Occupational asthma and rhinitis caused by ash (Fraxinus excelsior) wood dust. *Allergy* 1997; 52:196–9.
23. Malo JL, Cartier A, L'Archeveque J, et al. Prevalence of occupational asthma among workers exposed to eastern white cedar. *Am J Respir Crit Care Med* 1994; 150:1697–701.
24. Wittczak T, Dudek W, Walusiak-Skorupa J, et al. Occupational asthma due to spruce wood. *Occup Med* 2012; 62:301–4.
25. Kespohl, S., Kotschy-Lang, N., Tomm, J. M., et al. Occupational IgE-mediated softwood allergy: characterization of the causative allergen. *Int Arch Allergy Immunol* 2012; 157:202–8.
26. Rask-Andersen A, Land CJ, Enlund K, et al. Inhalation fever and respiratory symptoms in the trimming department of Swedish sawmills. *Am J Ind Med* 1994; 25:65–7.
27. Dahlqvist M, Johard U, Alexandersson R, et al. Lung function and precipitating antibodies in low exposed wood trimmers in Sweden. *Am J Ind Med* 1992; 21:549–9.
28. Dahlqvist M, Ulfvarson U. Acute effects on forced expiratory volume in one second and longitudinal change in pulmonary function among wood trimmers. *Am J Ind Med* 1994; 25:551–8.
29. Rastogi SK, Gupta BN, Husain T, et al. Respiratory health effects from occupational exposure to wood dust in sawmills. *Am Ind Hyg Assoc J* 1989; 50:574–8.
30. Liou SH, Cheng SY, Lai FM, et al. Respiratory symptoms and pulmonary function in mill workers exposed to wood dust. *Am J Ind Med* 1996; 30:293–9.

31. Mandryk J, Alwis KU, Hocking AD. Work related symptoms and dose–response relationships for personal exposures and pulmonary function among woodworkers. *Am J Ind Med* 1999; 35:481–90.
32. Macbeth R. Malignant disease of the paranasal sinuses. *J Laryngol Otol* 1965; 79:592–612.
33. Acheson ED, Hadfleld EH, Macbeth RG. Carcinoma of the nasal cavity and accessory sinuses in wood workers. *Lancet* 1967; 1:311–2.
34. Acheson ED, Cowdell RH, Hadfield F, et al. Nasal cancer in woodworkers in the furniture industry. *Br Med J* 1968; 2:587–96.
35. Hadfield EH. A study of adenocarcinoma of the paranasal sinuses in woodworkers in the furniture industry. *Ann R Coll Surg Engl* 1970; 46:301–19.
36. Cignoux M, Bernard P. Malignant ethmoid bone tumors in woodworkers. *J Med Lyon* 1969; 25:92–3.
37. Ironside P, Matthews J. Adenocarcinoma of the nose and paranasal sinuses in woodworkers in the state of Victoria, Australia. *Cancer* 1975; 36:1115–21.
38. Klintenberg C, Olofsson J, Hellquist H, et al. Adenocarcinoma of the ethmoid sinuses: a review of 28 cases with special reference to wood dust exposure. *Cancer* 1984; 54:482–8.
39. Cecchi F, Buiatti E, Kriebel D, et al. Adenocarcinoma of the nose and paranasal sinuses in shoemakers and woodworkers in the province of Florence, Italy (1963–77). *Br J Ind Med* 1980; 37:222–5.
40. Battista G, Cavallucci F, Comba P, et al. A case-referent study on nasal cancer and exposure to wood dust in the province of Sienna Italy. *Scand J Work Environ Health* 1983; 9(1):25–9.
41. Debois JM. Tumors of the nasal cavity in woodworkers. *Tjidschr Diergeneeskd* 1969; 25:92–3.
42. Mosbech J, Acheson ED. Nasal cancer in furniture makers in Denmark. *Dan Med Bull* 1970; 18:34–5.
43. Van den Oever R. Occupational exposure to dust and sinonasal cancer. An analysis of 386 cases reported to the NCCSF Cancer Registry. *Acta Otorhinolaryngol Belg* 1996; 50:19–24.
44. Demers PA, Kogevinas M, Boffetta P, et al. Wood dust and sinonasal cancer: pooled reanalysis of twelve case–control studies. *Am J Ind Med* 1995; 28:151–66.
45. Ball MJ. Nasal cancer and occupation. *Lancet* 1967; 2:1089–90.
46. Elwood JM. Wood exposure and smoking: association with cancer of the nasal cavity and paranasal sinuses in British Columbia. *Can Med Assoc J* 1981; 124:1573–7.
47. Pukkala E, Martinsen, JI, Lynge E, et al. Occupation and cancer – follow-up of 15 million people in five Nordic countries. *Acta Oncol*, 2009; 48: 646–790.
48. Brinton LA, Blot WJ, Stone BJ. A death certificate analysis of nasal cancer among furniture workers in North Carolina. *Cancer Res* 1977; 37:3473–4.
49. Roush GC, Meigs JW, Kelly J. Sinonasal cancer and occupation: a case control study. *Am J Epidemiol* 1980; 111:183–93.
50. Brinton LA, Blot WJ, Becker JA. A case–control study of cancers of the nasal cavity and paranasal sinuses. *Am J Epidemiol* 1984; 119:896–906.
51. Imbus HR, Dyson WE. A review of nasal cancer in furniture manufacturing and woodworking in North Carolina, the United States, and other countries. *J Occup Med* 1987; 29:734–40.
52. Blot WJ, Chow WH, McLaughlin JK. Wood dust and nasal cancer risk. A review of the evidence from North America. *J Occup Environ Med* 1997; 39:148–56.
53. Demers PA, Boffetta P, Kogevinas M, et al. Pooled reanalysis of cancer mortality among five cohorts of workers in wood-related industries. *Scand J Work Environ Health* 1995; 21:179–90.
54. Acheson ED. Nasal cancer in the furniture and boot and shoe manufacturing industries. *Prez'ı Med* 1976; 5:295–315.
55. Vallieres E, Pintos J, Parent M-E, Siematycki J. Occupational exposure to wood dust and risk of lung cancer in two population-based case–control studies in Montreal, Canada. *Environ Health* 2015; 14:1.
56. Määttä J, Luukkonen R, Husgafvel-Pursiainen K, et al. Comparison of hardwood and softwood dust-induced expression of cytokines and chemokines in mouse macrophage RAW 264.7 cells. *Toxicology* 2006; 218:13–21.
57. Bornholdt J, Saber AT, Sharma AK, et al. Inflammatory response and genotoxicity of seven wood dusts in the human epithelial cell line A549. *Mutat Res/Genet Toxicol Environ Mutagen* 2007; 632:78–88.
58. Nelson E, Zhou Z, Carmichael PL, et al. Genotoxic effects of subacute treatments with wood dust extracts on the nasal epithelium of rats: assessment by the micronucleus and 32P-postlabelling. *Arch Toxicol* 1993; 67:586–9.
59. Mohtashamipur E, Norpoth K, Ernst H, et al. The mouse-skin carcinogenicity of a mutagenic fraction from beech wood dusts. *Carcinogenisis* 1989; 10:483–7.
60. Mohtashamipur E, Norpoth K, Hallerberg B. A fraction of beech wood mutagenic in the Salmonella/mammalian microsome assay. *Int Arch Occup Environ Health* 1986; 58:227–34.
61. Pylkkänen L, Stockmann-Juvala H, Alenius H, et al. Wood dusts induce the production of reactive oxygen species and caspase-3 activity in human bronchial epithelial cells. *Toxicology* 2009; 262:265–70.
62. IARC Working Group. Wood dust and formaldehyde. In: *IARC Monographs on the evaluation of carcinogenic risks to humans*, Vol. 62. Lyon: World Health Organization, 1995.
63. IARC. *Monograph Volume 100: A review of human carcinogens—Part C: metals, arsenic, dusts, and fibres*. Lyon: World Health Organization, 2012. Available at: http://monographs.iarc.fr/ENG/Monographs/vol100C/mono100C.pdf (accessed on June 21, 2016).
64. American Conference of Governmental Industrial Hygienists. *2014 TLVs and BEIs*. Cincinnati, OH, 2014.
65. Compliance and Enforcement Activities Affected by the PELs Decision. Available at: https://www.osha.gov/pls/oshaweb/owadisp.show_document?p_table=INTERPRETATIONS&p_id=21220 (accessed on June 21, 2016).
66. Occupational Safety and Health Division, 2012. *A Guide to Combustible Dust*, NC Department of Labor, Raleigh, NC.

INDEX

Physical and Biological Hazards of the Workplace, Third Edition. Edited by Gregg M. Stave and Peter H. Wald.